NSF FACTBOOK

NSF FACTBOOK

ALVIN RENETZKY, Ph.D.
Editor-in-Chief

BARBARA J. FLYNN, M.L.S.
Editor

and

**THE STAFF
OF
ACADEMIC MEDIA**

Guide to National Science Foundation Programs and Activities

ACADEMIC MEDIA · ORANGE, NEW JERSEY

Academic Media, Division of Computing and Software, Inc.
32 Lincoln Avenue
Orange, New Jersey 07050

Library of Congress Card Number: 77-165273
ISBN 0-87876-011-3

Printed in the United States of America

INTRODUCTION

CREATION AND AUTHORITY The National Science Foundation was established by the National Science Foundation Act of 1950 (64 Stat. 149; 42 U.S.C. 1861-1875), and was given additional authority by the National Defense Education Act of 1958 (72 Stat. 1601; 42 U.S.C. 1876-1879) and by the National Sea Grant College and Program Act of 1966 (80 Stat. 998; 33 U.S.C. 1121-1124). The Foundation consists of the National Science Board of 24 members, a Director, Deputy Director, and four Assistant Directors, each appointed by the President with the advice and consent of the Senate. The Director is the Chief Executive Officer of the Foundation and serves ex officio as a member of the Board and as Chairman of its Executive Committee.

PURPOSE The fundamental purpose of the National Science Foundation is to strengthen research and education in the sciences in the United States.

ACTIVITIES Among the activities of the Foundation are:

1. The development and dissemination of information relating to scientific resources, including manpower, aimed at facilitating national decisions relating to strengthening the scientific effort of the Nation.

2. The award of grants and contracts primarily to universities and other non-profit institutions in support of scientific research. Awards include those made for small and large research projects, for the construction of laboratories or specialized facilities, and for generally strengthening an institution's scientific endeavors. This activity also includes support of concerted research efforts that are planned, coordinated, and funded on a national program basis because of the scope of the research being performed and its relationship to national goals.

3. The support, through contracts, of national centers where large facilities are made available for the use of qualified scientists. At the present time, the Foundation is supporting the Kitt Peak National Observatory, the Cerro Tololo Inter-American Observatory, the National Radio Astronomy Observatory, and a National Center for Atmospheric Research.

4. Maintenance of a current register of scientific and technical personnel and provision of a central clearinghouse for data on the supply and needs for scientific and technical resources.

5. The award of graduate fellowships in the mathematical, physical, medical, biological, engineering, and social sciences, and the provision of support for graduate student traineeship programs at educational institutions.

6. Programs aimed at improving scientific education in the United States through providing support for: special institutes to improve the competence of teachers of science, mathematics, and engineering; projects to modernize materials of instruction and courses of study; and projects to afford opportunities for high-ability secondary school and college students to secure added scientific experiences.

7. A program aimed at strengthening research, education, and training in oceanography and exploitation of the marine environment. This area includes the activities of the International Decade of Ocean Exploration.

8. Programs supporting the development and use of computer and other scientific methods and technologies.

9. A program aimed at improving the coordination of the various scientific information activities within the Federal Government; developing new or improved methods of making scientific information available; fostering the interchange of scientific information among scientists of the United States and foreign countries; and providing support for the translation of foreign scientific information.

CONTENTS

Part 1

NSF IN REVIEW

Research Support Activities

The National Science Foundation supports scientific research primarily through grants to colleges and universities for projects proposed by the scientists who will conduct the research. Activities covered in this chapter are divided into four [1] major categories:

- Research grants to institutions for scientific investigations by individual scientists or groups of scientists.
- Grants to academic institutions for acquisition of specialized research equipment and facilities.
- Support of cooperative National (and International) Research Programs.
- Support of National Research Centers funded by the Foundation and operated under university management.

Scientific Research Support

In fiscal year 1970, the Foundation awarded 3,817 grants for research projects amounting to a total of $161.7 million. Comparable figures for fiscal 1969 were 4,053 grants for a total of $176.0 million. Table 1 gives the distribution, number, and amount of grants according to fields of science for fiscal years 1968, 1969, and 1970. Of all the actions taken by the Foundation on research project proposals in fiscal year 1970, 47 percent were awards as compared to 51 percent in 1969. Grants were awarded to 410 institutions, including 310 colleges and universities, in all 50 States, the District of Columbia and Puerto Rico. Ninety-five percent of the funds went to academic institutions. Of these, 211 received two or more research grants and 117 received at least $200,000.

[1] Research undertaken in connection with the National Sea Grant Program and computing activities in research are discussed in separate chapters.

Specialized Research Facilities and Equipment

The purpose of the Specialized Research Facilities and Equipment Program is to help institutions obtain the scientific equipment and facilities required for the conduct of very advanced research projects. These range from facilities such as nuclear accelerators and oceanographic ships to equipment of a specialized nature, such as electron microscopes, mass spectrometers and cryogenic equipment, required for common use by several investigators at an institution. The availability of these facilities and equipment enables research scientists to be more productive as well as to make possible the conduct of some scientific investigations which would not be possible otherwise.

National and Special Research Programs

National Research Programs are specifically identified as major research efforts undertaken to accomplish a designated objective related to one or more fields of science. Some of these programs include aspects of applied science, and some may be interdisciplinary in nature. In some cases, the work may be performed in a specific geographical area, and some activities involve international cooperation and coordination. In some instances the Foundation has been assigned responsibility for these programs by the President, or by the Congress, or by agreement within the Executive Branch of Government.

Table 3 which summarizes grants and contracts for National Research Programs over the past 3 years, includes the first awards for Interdisciplinary Research Relevant to Problems of Our Society (IRRPOS), a new program initiated by the Foundation in fiscal year 1970.

Table 1
Scientific Research Projects
Fiscal Years 1968, 1969, and 1970

[Dollars in millions]

	Fiscal year 1968		Fiscal year 1969		Fiscal year 1970	
	Number	Amount	Number	Amount	Number	Amount
Astronomy:						
Optical		$4.06		$3.85		$3.71
Radio		2.14		2.96		2.09
Subtotal	119	6.19	125	6.82	108	5.80
Atmospheric Sciences:						
Aeronomy		1.89		1.65		1.69
Meteorology		3.94		4.32		3.95
Solar-Terrestrial		1.74		2.25		2.28
Subtotal	103	7.57	116	8.21	118	7.92
Biology:						
Cellular Biology		10.02		9.28		8.68
Ecology and Systematic Biology		8.65		7.96		8.60
Molecular Biology		10.34		9.88		9.76
Physiological Processes		11.18		10.04		9.53
Psychobiology		4.27		4.02		4.30
Subtotal	1,130	44.46	1,173	41.18	1,072	40.87
Chemistry:						
Chemical Analysis		1.07		1.48		1.71
Chemical Dynamics		3.73		4.16		3.58
Chemical Thermodynamics		1.60		1.56		1.86
Quantum Chemistry		3.38		3.54		3.39
Structural Chemistry		3.73		3.22		2.80
Synthetic Chemistry		4.26		3.90		4.05
Subtotal	454	17.77	484	17.85	449	17.40
Earth Sciences:						
Geology		1.56		1.31		1.42
Geochemistry		2.97		3.31		3.07
Geophysics		3.28		3.30		3.36
Subtotal	214	7.81	200	7.92	196	7.85
Engineering:						
Engineering Chemistry		2.84		2.73		2.82
Engineering Energetics		3.11		2.94		2.86
Engineering Materials		3.44		3.23		3.29
Engineering Mechanics		5.93		6.39		6.55
Engineering Systems		3.44		3.00		[1]
Special Engineering Programs		.63		.98		1.17
Subtotal	506	19.40	491	19.27	463	16.70
Mathematics:						
Algebra and Topology		4.44		4.39		4.49
Analysis, Foundations, and Geometry		4.32		4.37		4.34
Applied Mathematics and Statistics		3.94		3.94		3.83
Subtotal	405	12.70	462	12.70	489	12.66
Oceanography: [2]						
Biological Oceanography		2.41		3.13		3.66
Physical Oceanography		1.94		2.16		2.07
Geological Oceanography		2.91		2.55		3.18
Support, Ship Operations		6.88		8.64		[3]
Subtotal	239	14.14	280	16.48	218	8.91
Physics:						
Atomic, Molecular, and Plasma Physics		2.33		2.46		2.72
Elementary Particle Physics		6.48		11.53		11.24
Nuclear Physics		9.07		8.01		6.45
Solid State and Low Temperature Physics		4.40		4.61		4.42
Theoretical Physics		3.63		3.73		3.34
Subtotal	236	25.90	283	30.35	245	28.18
Social Sciences:						
Anthropology		3.50		3.42		3.48
Economics		3.58		4.29		4.34
Geography		.59		.19		.48
Sociology and Social Psychology		3.73		3.29		3.35
Political Science		.76		1.28		1.19
History and Philosophy of Science		.73		.87		.83
Special Projects		1.76		1.90		1.74
Subtotal	426	14.67	474	15.24	459	15.42
Total	3,832	170.61	4,053	176.02	3,817	161.71

[1] Included in National and Special Research Programs for FY 1970.
[2] Includes marine biology.
[3] Included in National and Special Research Programs for FY 1970.

Table 2
Specialized Research Facilities and Equipment Fiscal Years 1968, 1969, and 1970

[Dollars in millions]

	Fiscal year 1968		Fiscal year 1969		Fiscal year 1970	
	Number	Amount	Number	Amount	Number	Amount
Astronomy	4	$0.662	5	$0.324	5	$0.190
Atmospheric Sciences	11	.788	8	.298	4	.199
Biological and Medical Sciences	30	1.709	34	.880	11	.918
Chemistry	113	4.296	57	1.700	63	1.697
Earth Sciences	0	0	0	0	3	.103
Engineering	43	1.073	26	.880	28	.600
Oceanography	9	4.711	1	1.397	[1]	[1]
Physics	19	4.697	25	1.300	12	2.499
Social Sciences	3	1.006	2	.438	1	.298
Total	232	18.942	158	7.216	127	6.504

[1] Included in National and Special Research Programs for FY 1970.

Table 3
National and Special Research Programs Fiscal Years 1968, 1969, and 1970

[Dollars in millions]

	Fiscal year 1968		Fiscal year 1969		Fiscal year 1970	
	Number	Amount	Number	Amount	Number	Amount
Arctic Research program	0	0	0	0	2	$0.13
Ocean Sediment Coring program	4	$4.17	5	$2.43	25	6.56
Global Atmospheric Research program	1	.20	9	.54	19	1.49
Interdisciplinary Research Relevant to Problems of Our Society	0	0	0	0	21	5.98
International Biological program	1	.70	16	1.22	24	4.00
U.S. Antarctic Research program	149	7.64	145	6.88	128	7.28
Weather Modification program	32	2.77	24	2.43	27	2.63
Engineering systems					92	3.30
Oceanographic Ship Operations and Facilities					31	7.60
Total	187	15.48	199	13.48	369	38.97

Foundation support for National and Special Research Programs more than doubled in fiscal year 1970 over the previous year. Most research supported by these programs is associated with the environment, an area of expanded emphasis in many other Foundation programs as well in fiscal year 1970. In addition to the programs listed, the Foundation in fiscal year 1970 was designated lead agency for the International Decade of Ocean Exploration (IDOE), an international program proposed by the United States and endorsed by the General Assembly of the United Nations. Initial planning for IDOE places emphasis on environmental quality, environmental forecasting, and seabed assessment. It is expected that the first awards for IDOE will be announced early in calendar year 1971.

National Research Centers

The National Science Foundation provides support for the development and operation of National Research Centers established to meet national needs for research in specific areas of science requiring facilities, equipment, staffing, and operational support which are beyond the financial capabilities of private or State institutions and which would not appropriately be provided to a single institution to the exclusion of others. Unlike many federally sponsored research laboratories, the NSF-supported National Research Centers do not perform specific research tasks assigned by or for the direct benefit of the Government. They are established and supported for the purpose of making available, to all qualified scientists, the facilities, equipment, skilled personnel support, and other resources required for the performance of independent research of their own choosing, in the applicable areas of science.

In recent years, the Foundation has supported three astronomy centers (Cerro Tololo Inter-American Observatory, Kitt Peak National Observatory, and National Radio Astronomy Observatory) and one atmospheric research center (National Center for Atmospheric Research). In fiscal year 1970, the Foundation assumed principal funding responsibility for the Arecibo Observatory in Puerto Rico, and established it as a fifth NSF-sponsored National Research Center. This observatory was built with funds provided by the Department of Defense and, prior to fiscal year 1970, principal funding support was furnished by the Department of Defense.

Funding levels for the National Research Centers during fiscal years 1968, 1969, and 1970 are given in the table below.

Table 4
National Research Centers
Fiscal Years 1968, 1969, and 1970

	Fiscal year 1968			Fiscal year 1969			Fiscal year 1970		
	Capital obligations	Research operations and support services	Total	Capital obligations	Research operations and support services	Total	Capital obligations	Research operations and support services	Total
Cerro Tololo Inter-American Observatory	$1,502,000	$823,000	$2,325,000	$3,449,000	$1,101,000	$4,500,000	$365,000	$1,535,000	$1,900,000
Kitt Peak National Observatory	8,331,176	4,144,192	12,475,368	1,137,700	4,561,809	5,699,510	46,000	6,379,000	6,425,000
National Radio Astronomy Observatory	874,300	3,989,700	4,864,000	483,212	6,795,001	7,278,214	675,000	5,125,000	5,800,000
Arecibo Observatory	----	----	----	----	----	----	150,000	1,400,000	1,550,000
National Center for Atmospheric Research	2,041,100	9,758,612	11,799,712	425,000	10,611,736	11,036,737	212,840	11,367,000	11,536,800
Total	12,748,576	18,715,504	31,464,080	5,494,912	23,069,547	28,564,461	1,448,840	25,857,960	27,211,800

MATHEMATICAL AND PHYSICAL SCIENCES

The pressure from the scientific research community for support of research projects in all of the mathematical and physical sciences—mathematics, physics, astronomy, and chemistry—rose sharply during fiscal year 1970. This was attributed largely to reorientation in the priorities of other Federal agencies supporting similar research. Another element adding to the increased pressure on the mathematical and physical sciences from the scientific community was the increased number of proposals from scientists trained in other disciplines. For example, sugar and protein chemistry projects were proposed by biologists, and a number of projects in mathematics was proposed by engineering faculty. Insofar as the total amount of funding available has remained nearly constant, a larger number of the proposals received was necessarily declined.

While the last few years of basic research in the physical sciences have not been easy ones, nonetheless progress continues to be made. While it is never possible for the Foundation to predict what area of research will produce the answer to a particular significant problem, some approaches and techniques appear to have greater possibilities of fruitfulness within a broadly defined problem area. Without receding from its policy of supporting high quality fundamental research across the disciplinary spectrum, certain areas are receiving special emphasis.

A major factor in a whole class of important chemical reactions is catalysts, a group of substances which influence the reaction rate without being permanently changed themselves. Perhaps the most important of these catalytic materials are the enzymes which regulate many of the chemical reactions of living systems. It is believed that most catalysts perform their chemical magic by providing a physical surface which temporarily holds and positions the reacting molecules so as to facilitate their combination. In enzymes, which are themselves large molecules, the catalytic surface is thought to be a particular segment or "active site" on the molecular chain. The physical structure and properties of these active sites can now be studied by a promising new technique known as spin labeling. In this technique, a molecule is synthesized to have properties similar to those of the molecules on which the enzyme operates and in addition to have magnetic properties. Detailed study of the magnetic properties of the artificial molecule when bound to the enzyme gives detailed information about the properties of the active site on the enzyme.

In physics, concomitant advances in astrophysics, general relativity, and gravitational radiation are bringing improved understanding about the universe and the nature of its evolution.

The technological areas associated with physics—long a source of advanced technology and instrumentation for other sciences, pure and applied, and for industry—have advanced steadily. In cryophysics, the study of physical behavior near temperatures of absolute zero, studies made for the improvement of particle acceleration and the detection of gravitational radiation have markedly increased our ability to deal with low temperature phenomena such as superconductivity on a large scale. The latter phenomenon will eventually find major applications in the generation and transmission of heavy-load electricity and in higher speed computers.

Technology relating to the confinement of plasmas—hot ionized gases—has produced several new developments raising hopes for eventual production of electric power by nuclear fusion, a pollution-free technique which also will draw upon hydrogen rather than our diminishing supply of fossil fuel as a source of energy.

In astronomy, new techniques of working in the far infrared—light of very long wavelength merging into the microwave segment of the radio spectrum—have enlarged the scientist's view of the universe considerably. The successful flight of Stratoscope II, a balloon-borne, high-resolution telescope, has markedly increased our ability to study the planets of our solar system—knowledge which will not be improved upon until planetary probes "fly by" these planets in the late 1970's. For the observing astronomer, a new technique of computer-guidance will cut drastically the time now spent in aiming telescopes at desired sectors of the sky and increase efficiency and actual observing time—in some cases by as much as 50 percent. Discovery of the water molecule and several new polyatomic organic molecules in interstellar space has broadened our view about the conditions under which life itself can evolve.

In applied mathematics, which uses the highly abstract tools of pure mathematics for applied ends, new techniques of nonlinear analysis are being applied to problems such as multiple inputs of municipal effluent into the Hudson River and the flow mechanisms of blood. The theory of differential games is being used in economics and in the study of transportation and traffic problems. Statistical theory is being applied to quality control testing in industry, experimental design in the laboratory, and to studies of genetics.

Eventually, many of the abstract relationships discovered in pure mathematics facilitate man's conceptual mastery of real phenomena in all of science, which in turn finds concrete form in physical instrumentation and experiments. Finally, these relationships emerge, with regular and lasting impact, as new processes, new instruments and mechanisms, and new materials. The availability and relative cheapness of paint and plastic, of solid state circuitry, of long distance telephone and television communication, of improved traffic and transportation control all find their roots in basic research in the mathematical and physical sciences.

CHEMISTRY

The Association Reactions of Borane

The relation between the nature of a chemical species and the reactions it undergoes has always been a major research thrust in chemistry. There are two levels at which one can understand a chemical reaction. The first level of understanding involves knowing what stable products are produced from stable reactants under the specified conditions of the reaction. A second and deeper level of understanding is achieved when one can answer the question: What is the detailed motion of the atoms in the chemical species during the course of an overall reaction? An approach to answering this question is to break down a given chemical reaction into smaller steps that can be scrutinized individually.

There are many ways to break down a chemical reaction into smaller steps. One simple and direct way is by detecting intermediate chemical species. These are species which are formed and subsequently destroyed during the course of the reaction. They are neither reactants nor products. Knowing the structure and the speed with which these intermediates react is equivalent to having "snap shots" of the atomic motions which control the total chemical reaction. The identification and characterization of important types of intermediates has historically had a large impact on chemistry.

Thomas P. Fehlner of the University of Notre Dame is currently involved with the characterization of a new kind of intermediate, borane (BH_3), and the comparison of its unique qualities with those of related, yet different species. In most stable boron compounds, the boron atom is surrounded by eight electrons, usually shared in chemical bonds with other atoms. Borane has only six electrons, so it seeks other molecules which can share two electrons with it. Because of this characteristic, it is so reactive that it does not normally exist in a free state for a lifetime of more than a fractional part of a second.

Two things were prerequisites to the study of borane: a pure and intense source of borane, and a system suitable for observing its reactions. The first requirement was fulfilled after much effort in qualitatively examining the production and destruction of borane in a variety of systems. By heating complexes where borane was weakly bound to compounds which would readily give up a pair of electrons, Dr. Fehlner produced quite pure borane at relatively high concentrations for long times (0.001 seconds) compared to the measurement time (0.00005 seconds).

Dr. Fehlner was able to observe the reactions of borane by constructing a mass spectrometer system and coupling it directly to a borane production device. The mass spectrometer acts by converting all of the chemical species entering it into ions, which are then separated according to their mass, and counted. The relative numbers of ions of various masses yield information both as to the identity and the number of the species ionized. Consequently, as borane is reacted with various species, this instrument can be used (1) to identify unambiguously the products and (2) to measure the amounts. Finally, the time of reaction can be varied so as to measure the speed of the reaction.

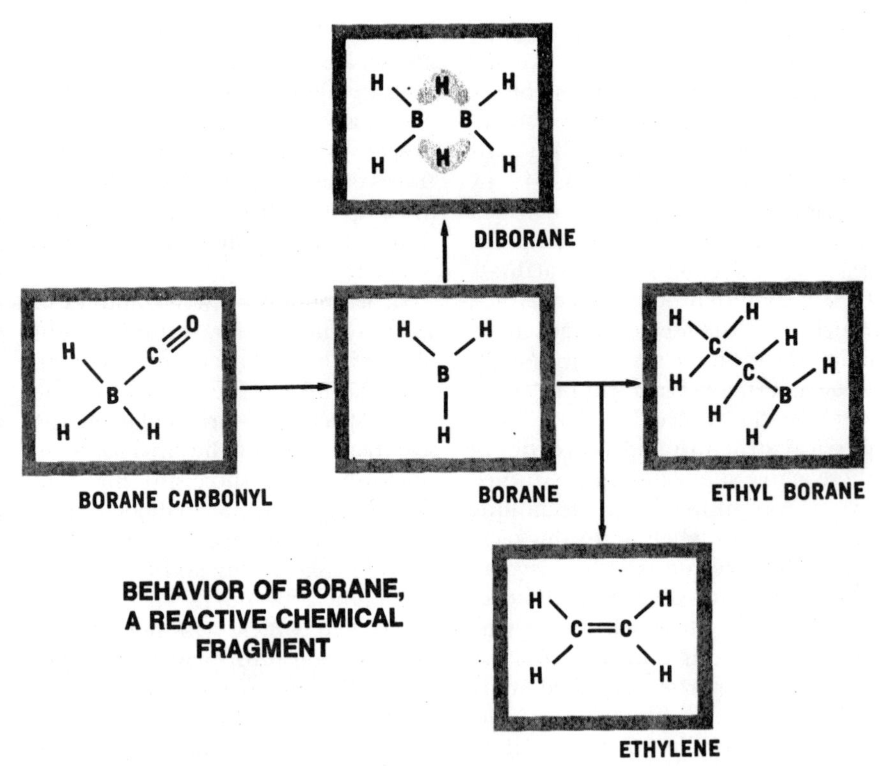

"Snapshots" of the behavior of borane, a reactive chemical fragment. In the reaction on the left, borane is released, along with carbon monoxide, by heating borane carbonyl. It reacts rapidly with another borane fragment to form diborane, with its unusual hydrogen-bridged structure (top). Borane also reacts with ethylene, which has an extra pair of electrons in its central bond.

During the past year the Notre Dame group succeeded in observing some of the characteristic association reactions of borane. The first reaction examined was the combination of two borane fragments to form the stable compound, diborane (B_2H_6). As far as chemical reactions go, this reaction is very efficient in that about one out of every ten borane-borane collisions yields the product. The second reaction Dr. Fehlner has examined is the association of borane with electron pair donors to form complexes. The products are already known in most cases, but measurements of the relative reaction speeds will characterize those aspects of borane reactivity involving the acceptance of electrons. The product of the reaction of borane with ethylene, a compound containing a double bond, has recently been observed. The product, ethyl borane ($C_2H_5BH_2$), was itself a previously unknown compound. This type of reaction is important in the synthesis of many new chemical compounds. Dr. Fehlner's work will significantly broaden our understanding of fast reactions in general and the relationship between chemical reactivity and electronic structure.

Nitrogen Fixation

Life on earth would not be possible without the fundamental biological process by which certain plants convert atmospheric nitrogen, which is chemically inert, into ammonia (NH_3). From this ammonia, proteins are constructed which are essential components of all living cells. In nature, the conversion of nitrogen to ammonia is accomplished by nitrogen-fixing bacteria through the agency of a complicated enzyme catalyst called nitrogenase. Industrially, the fixation of nitrogen to ammonia is done by a high temperature (300°–600° C.) high pressure reaction (several hundred atmospheres) known as the Haber process.

Chemists have long sought a way to reduce molecular nitrogen to ammonia under moderate conditions without requiring the high temperature or pressures of the Haber process. Research to date indicates that iron and molybdenum, which are present in many natural systems, seem to play an important transition role in the process of fixation and reduction of nitrogen to ammonia. Apparently, these transition metals form a metal-nitrogen complex which weakens the bond which binds the inert nitrogen molecule together.

Although a number of well-characterized metal-nitrogen complexes have been made, it has not been possible to reduce the nitrogen in these systems down to the ammonia stage. The difficulty undoubtedly lies in the thermodynamic stability of molecular nitrogen and the relative thermodynamic instability of the intermediate chemical species necessary in the successive stages of reduction.

Recently, Fred Basolo and Ralph G. Pearson at Northwestern University made a discovery which bears on the reduction problem and promises to throw light on the requirements for a metal ion catalyst which can reduce molecular nitrogen to ammonia. Drs. Basolo and Pearson prepared two different metal salts, one containing the element ruthenium and the other containing the element iridium each of which contains a nitrogen complex corresponding to one of the apparently required intermediate chemical species. This intermediate product can be reduced to ammonia complexes by mild reducing agents. The secret of success apparently was to start with metal in the reduced (having a surplus of electrons) rather than oxidized (deficient in electrons) form thereby enabling the reaction to proceed towards ammonia rather than reverting back to the stable molecular nitrogen stage.

The new metal-nitrogen complexes have some analogy to the molybdenum-iron system which is presumed to be the catalyst in the enzyme nitrogenase. It will be of interest to see if iron can be used to replace ruthenium. While a system capable of fixing nitrogen in this manner has not been developed, systems are now available in which the several stages of reduction of nitrogen can be studied.

The work by Drs. Basolo and Pearson appears to be a significant step forward in the elucidation of the complete mechanism of biological nitrogen fixation which, once achieved, may ultimately make possible breeding of nitrogen-fixing powers into crops such as wheat and corn, minimizing the problems of run-off of excess fertilizer, a possible cause of river and lake pollution and an uneconomic use of fertilizer.

Chemical Instrumentation

The acquisition of modern instruments is recognized both by the Foundation and chemistry departments at institutions of higher learning to be of crucial importance. During the past year the Foundation received 202 requests from colleges and universities for purchase of chemical instrumentation. Highest priority requests totaled $10.4 million, with lower priority items bringing the total request to over $30 million. The Foundation was able to support 63 of these requests by contributing $1.7 million toward the purchase of $3.39 million worth of instrumentation. The difference of $1.69 million was provided by institutional contributions, which this year were 25 percent higher than the average contribution over the past 6 years. These very significant contributions provide substantive evidence that colleges and universities recognize the importance of complex instrumentation in chemical education and the pursuit of basic chemical research.

PHYSICS

The Solid State Physics of Pulsars

Pulsars, the recently discovered, rapidly pulsating radio sources observed in the sky, have attracted interest outside the domain of the astronomers who first observed them. These phenomena appear to represent such an extreme form of behavior that what they do and are may be of fundamental importance to our understanding of the basic physical laws of the universe.

Less than 2 years ago, Malvin Ruderman of Columbia University pointed out that pulsars would most likely exhibit phenomena such as superconductivity and superfluidity which are more normally associated with materials that are manipulated by solid state physicists on laboratory bench tops.

Recently, physicists and astronomers have leaned strongly toward the theory that these objects are the long predicted neutron stars (first postulated by the Russian physicist, L. D. Landau, and the American physicist, J. Robert Oppenheimer in the late 1930's). The reason for this is the short time-scale of the "pulsations." Just as a pendulum has a characteristic period which is determined by its length and the gravitational field strength at the surface of the earth, any astronomical object will have pulsation periods which are uniquely deter-

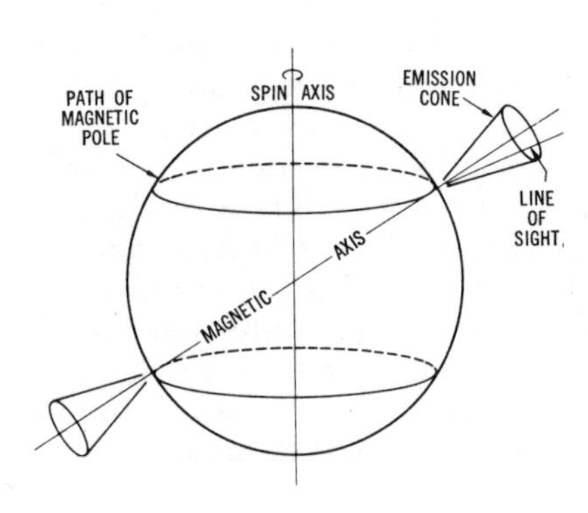

A theoretical model of a pulsar, pictured as a rotating neutron star with magnetic and spin axes out of alignment. It is thought that intense surface magnetic fields in such an object would be responsible for beaming electromagnetic radiation along the direction of the magnetic axis. If the star is properly oriented, this beam would sweep through the observer's line of sight once (or twice) for every rotation of the star, thus giving rise to the "pulsar" effect. Other models which have been suggested differ in detail and involve emission regions further out from the surface of the star and "beaming" perpendicular to the field axis, but all current theories emphasize the rotating searchlight effect as the basic source of the pulses.

mined by its density. In order to pulsate once a second, it must have a mean density of 10^8 to 10^{10} grams per cubic centimeter. The densities of the heaviest naturally occurring elements are of the order of 20 grams per cubic centimeter.

Because of the unique way in which atoms are being "squeezed" at these densities, such an object would not be stable, and it would continue to collapse to a density of 10^{13} to 10^{14} grams per cubic centimeter, at which point the individual nuclei would be touching each other and would cease to exist as separate entities. At this higher density, pulsation periods would be only 10^3 seconds, but such a collapsed star with these properties could easily be in rotation once a second without being disrupted by centrifugal forces, and it is now commonly accepted that it is the magnetic field (which the neutron star is likely to have associated with it if it has evolved from a normal star) rotating with the star which somehow produces the periodic radio bursts.

At these high densities, the individual nuclei are expected to break up, and the most stable form of matter will be a neutron "liquid" with perhaps a few percent of protons and electrons, also in a "liquid" state. The solid state physicist commonly describes the liquid and solid states of normal matter in terms of interacting atoms, but he can also discuss the interactions of large numbers of particles independent of their precise internal nature. Through the use of such techniques, Dr. Ruderman was able to conclude that a typical neutron star should have a core-mantle structure much like that of the earth. In the outermost layer, where the density is not yet as high as the density of the nucleus, one would expect the individual nuclei to form a crystal lattice structure, just as atoms commonly form such structures at normal densities. As one goes further toward the center, the density continues to increase and the number of subnuclear particles or nucleons in each individual lattice nucleus decreases, until finally the lattice is made up solely of unbound nucleons. Deeper into the star, the still higher density of nucleons forces them to have sufficiently large kinetic energy so that they will no longer be fixed on individual lattice sites. The proton, neutron, and electron components

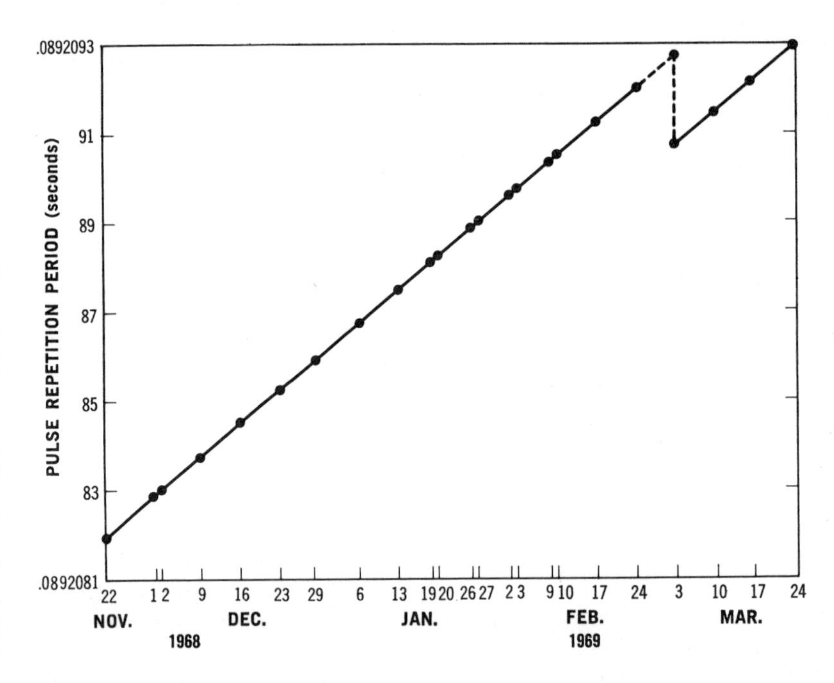

The pulse period of a pulsar in the constellation Vela is plotted as a function of time. Up until February 24, 1969, the period is steadily increasing. One week later it has decreased drastically, and in following weeks again shows a gradual increase at a slightly different rate. If the model of a pulsar is correct, the pulsar had to undergo a sudden increase in rotational velocity during the week of February 29—March 3. This could have been triggered by a "starquake." (Courtesy of P. E. Reichley and G. S. Downs at the California Institute of Technology Jet Propulsion Laboratory)

will then all be in the liquid state, and it is highly likely that the neutron component will be superfluid and the proton and electron components will be superconducting.

Two new observations made during the past year and a half have given increasing weight to this theoretical picture. The first of these was the report of the sudden increase in rotational velocity or "spin up" of the pulsar located in the constellation Vela. Observations made 1 week apart indicated that its period had decreased by one part in 10^6, or equivalently, that its rotational velocity had increased by that amount. After this "spin up" occurred, the object was again observed to follow a gradual slowing-down law, but at a different rate than it had prior to spin-up. A sudden change in the moment of inertia of the pulsar appears to be the most likely explanation; a 1-centimeter decrease in the mean radius would be sufficient to account for the observations, and Ruderman and collaborators suggest that the speed-up is due to a "starquake" which occurs in the solid crust of the pulsar when the centrifugal forces which would cause it to bulge after its initial solidification have relaxed enough to make sufficiently large stress. They find that the observed changes in period and rate of slow-down are consistent with this picture, provided they make one additional hypothesis about the viscous forces which tend to damp out motions of the pulsar's crust relative to the fluid core—they must be so small that the core can only be interpreted as existing in a superfluid state. A number of months ago, another pulsar "spin-up," this time of the pulsar in the Crab Nebula, was observed by an NSF-sponsored research team at Princeton University. The observed data were again in agreement with the superfluid model.

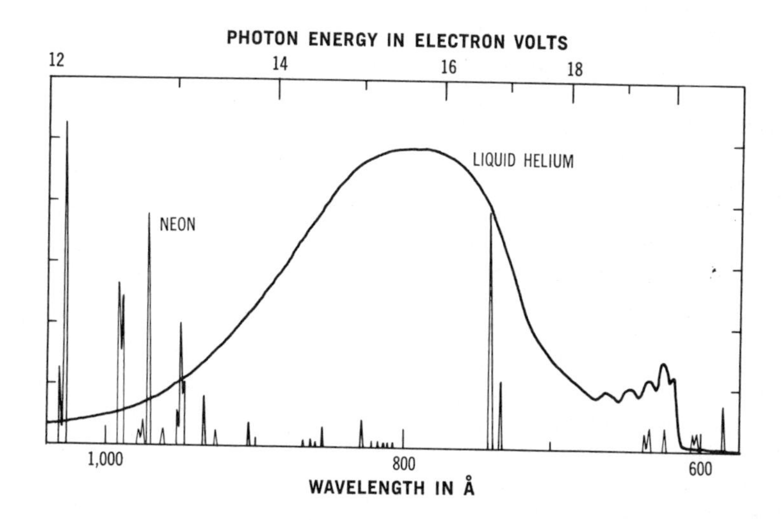

Spectrum of ultraviolet (u.v.) radiation observed from liquid helium bombarded by 160 keV electrons, showing the broad range of photon energies emitted. For comparison, the narrow line spectrum of neon (from a calibration lamp) is also shown, to illustrate typical atomic u.v. spectra. For many industrial and scientific applications, a broad emission spectrum and high efficiency are required of a u.v. source.

Ultraviolet from Liquid Helium

Liquid helium is of interest to scientists in widely separated fields because it is the coldest fluid known to man. Liquid helium is also of interest because it is a superfluid; that is, it has no viscosity and no surface tension—which means that it can flow through holes no other liquid can permeate, and can climb the sides of containers in which it is placed. Further, liquid helium is practically a perfect conductor of heat, which makes it an ideal refrigerant for super-cold applications since it quickly removes heat from the object being cooled. Finally, it is of interest to scientists because it shows forms of behavior (superfluidity, superconductivity, etc.) on a gross or "macroscopic" scale that are usually only associated with behavior of matter on an atomic scale.

Interested in learning more about the physical properties of this liquid and what it could reveal about the structure of the atoms of which it is made, a young scientist W. A. Fitzsimmons of the University of Wisconsin has been probing the effects of shooting an electron beam into a container of liquid helium. Dr. Fitzsimmons is interested in what the helium does after absorbing energy from the electron beam, since this reveals to him something of the nature of the helium immediately surrounding the atom which has absorbed energy and then released it.

In his apparatus the beam enters the liquid through a thin foil (which separates the liquid from the vacuum of the electron source), and the light then emitted is observed at a 90° angle with a specially designed monochromator, a device which measures the intensity of light at a given wavelength.

The light emitted from the liquid along the path of the beam was found to be characteristic of various excited states of the helium (He) atoms and He_2 molecules formed by the beam. The latter is in itself of interest since helium is usually thought of as so totally inert as to never form molecular combinations. The subsequent interaction of these excited atoms and molecules with the body of the liquid serves as a probe of the microstructure of the liquid helium.

A surprising result was an efficient and intense production of ultraviolet light. Dr. Fitzsimmons' first measurements showed that about 10 percent of the electron beam energy appears as such light. He estimates that his first experiments have produced an ultraviolet source on the order of 100 times the intensity of previous ones operating in this portion of the ultraviolet spectrum. He observes that the helium can tolerate an intense local absorption of energy without boiling because as a superfluid and an unusually good conductor of heat, the heat deposited by the beam is carried away before local boiling can occur.

An ultraviolet source of this novel type would be inexpensive, compact, and efficient. It could be operated steadily, or pulsed on quickly. Its intensity can be easily and continuously changed over a wide range. In addition to a number of scientific uses, this source could be considered for application in industrial processes where it can serve to initiate bulk chemical reactions or to heat the body of a liquid from within or to sterilize liquids.

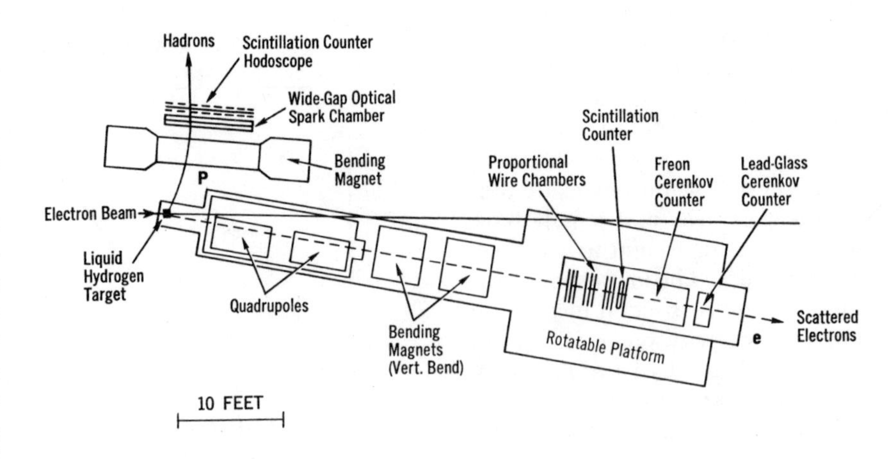

Experimental arrangement of the 10 BeV spectrometer and wide aperture magnet and spark chamber. This arrangement is used for the detection and identification of deep inelastic scattering electron interactions with production of hadrons.

Cornell Electron Accelerator

Since the Cornell Electron Synchrotron reached its design energy of 10 billion electron volts (BeV) in 1968, a vigorous experimental program has been pursued. This machine is the latest in a group of pioneering electron accelerators built and used at Cornell under the direction of Robert R. Wilson, now Director of the National Accelerator Laboratory in Batavia, Ill.

The Cornell synchrotron, presently under the direction of Boyce D. McDaniel, has a capability which is unique in the world—it produces photon and electron beams of up to 10 BeV energy which are sufficiently uniform in time (compared to the one-thousandth of a second beam bunches presently available in linear accelerators) to permit the detailed study of interactions in which several particles are detected simultaneously.

This property of the accelerator gives Cornell a unique capability to study many interesting fundamental processes. Among the most interesting of these is the phenomenon called "deep inelastic scattering." This is a process in which the scattering of a high energy electron by a target nucleon produces energetic particles known as hadrons. Hadrons are the particles which are involved in the "strong" interactions (one of the four basic interactions of nature, which include also the "weak," the "electromagnetic," and the "gravitational" interactions). This deep inelastic scattering is now the subject of much theoretical investigation because the probability for such scattering occurring is inexplicably substantially higher than for particle collisions without hadron production ("elastic scattering"). One theory to explain this high probability proposes that the target nucleon (proton or neutron) is made up of parts rather than being a single elementary particle. Such a composite structure for the proton would require modifications to our present interpretation of experimental data and major revision of our theoretical ideas, which presently view the proton as a fundamental building block of matter and truly elementary in nature. Hadron states of considerable interest are now being observed at Cornell, providing data on what is now one of the most important unsolved problems in particle physics.

Major Physics Research Facilities

Research in the technology of extremely low temperatures is an essential part of the work of William M. Fairbank, H. Allan Schwettman, and collaborators at the High Energy Physics Laboratory of Stanford University. This has led to major advances in the areas of high efficiency power transfer systems for particle acceleration and large-scale refrigeration near absolute zero with superfluid helium.

Superconductivity is that phenomenon which manifests itself at extremely low temperatures by the disappearance of resistance to the passage of electric currents. Extreme efficiency in power transfer and large voltage gradients, as much as 8 million volts per foot, have already been achieved. The Stanford group has also pioneered in the use of superfluid helium as a refrigerant on a large scale. The high heat capacity and remarkable mass transport properties of superfluid helium lead to applications on a scale impractical with ordinary liquid helium. A closed-cycle superfluid helium refrigerator capable of liquefying 450 liters of helium per hour at 1.8° K. is now in operation, a major advance in low-temperature technology. One possible application of these developments is in the use of superconducting systems for confinement and acceleration of plasmas, with potential consequences of the highest significance for the development of new power sources.

A novel type of beam collection and focussing channel is also being developed at this laboratory to provide high-intensity beams of pi-mesons, particularly suited for cancer radiation therapy. This is being designed in conjunction with the construction of a 500-foot-long superconducting linear accelerator, which represents the fruition of many years of research in low temperature technology, and is a development that will have considerable impact on future particle accelerators. The anticipated performance characteristics are a continuous electron beam with a current of at least 100 microamperes, energy spread and stability of one part in 10,000, and final energy above 2 billion electron volts. The initial stages of this system are now being tested. Beams at intermediate energies will be available for research purposes in 1971.

Work is also progressing on the design of superconducting systems for the acceleration of protons and heavy ions, both at Stanford University and the California Institute of Technology. Preliminary results indicate that these techniques show great potential for a practical solution to the problem of obtaining intense beams of heavy ions.

ASTRONOMY

Kitt Peak National Observatory

Kitt Peak National Observatory (KPNO), with headquarters located in Tucson, Ariz., and telescope installations on Kitt Peak on the Papago Indian Reservation, is operated under contract with the National Science Foundation by the Association of Universities for Research in Astronomy (AURA), Inc.

150-inch Telescopes

Construction of the two 150-inch telescopes—the world's second largest—one for Kitt Peak and one for the Cerro Tololo Inter-American Observatory in Chile has progressed during fiscal year 1970 according to plan. On Kitt Peak, the building and dome are nearing completion and are scheduled for acceptance from the contractor by the end of September 1970. Fabrication of the mechanical mountings for these telescopes is well underway and optical "figuring" on the GE fused quartz mirror blank for the Kitt Peak telescope is almost completed in the KPNO Optical Shop in Tucson. After completion, grinding and polishing of the blank for Cerro Tololo will begin on the same grinding machine.

KPNO Research Projects

Stellar Astronomy

Much of the research done using the smaller telescopes is in photoelectric photometry. KPNO staff significantly increased the capabilities in this field by the development of pulse-counting techniques, and by the introduction of equipment suitable for work in the near infrared region of the spectrum. A major part of the bright-moon time schedule of the 50-inch reflector was used for infrared spectrophotometry, with the auxiliary equipment of visiting scientists. These studies obtained new evidence of radiation from circumstellar dust shells.

The major part of the bright-moon time use of the 84-inch reflector was devoted to spectroscopic studies of the chemical composition of the stars, of their rotational velocities, radial and orbital velocities, and dynamics of their atmospheres. Quasars, galaxies, and faint stars were subjects for study during dark-of-the-moon use of this instrument. During a collaborative study of optical counterparts of radio sources located with the 210-foot Parkes telescope in Australia, an object having a red shift of record size was observed. This object is receding at more than 80 percent of the velocity of light, and the radiation now being observed left the object when the universe was one-tenth its present age.

Solar Astronomy

Continued improvements and new additions to the McMath Solar Telescope make it one of the world's most powerful and versatile instruments for the study of the sun's surface and atmosphere. With the addition this year of the auxiliary mirrors the installation is now really three telescopes in one, and has the capability of providing simultaneous studies of the same solar phenomenon with different tools and techniques. For example, the structures of the solar atmosphere can be photographed with one instrument, their spectral characteristics recorded with a second instrument, and a third can obtain the magnetic fields in the same region.

After 3 years' work, the 40-channel magnetograph has been brought into operation. It worked from the first flick of the switch, and gives detailed maps of solar magnetic fields. Simultaneously the brightness and velocity fields of each area are plotted.

A Harvard University graduate student from Australia detected fluorine in the sun. This is an important observation since it represents the first halogen detected. A search will now be made for chlorine in the farther infrared portion of the spectrum.

The eclipse of March 7, 1970, was successfully observed by Solar Division staff at instrument sites in southern Mexico and from Kitt Peak.

Planetary Sciences

Although the major emphasis of staff work is on other planetary atmospheres, additional contributions come from a program of terrestrial aeronomy. Investigations included studies of the production and transport of atomic nitrogen and nitric oxide in the upper atmosphere, and the maintenance of the earth's nighttime ionosphere. Related rocket flights have observed the day airglow, and twilight observations from Kitt Peak contribute similar information for certain emissions.

Three Aerobee sounding rockets were launched at White Sands Missile Range, N. Mex. Two of these were successful in returning scientific data on galactic x-rays, spectrometry and photometry of dayglow emissions.

Exterior view (March 17, 1970) and interior cut-away drawing of the 150-inch telescope installation on Kitt Peak. (KPNO photo)

Previous solar rocket spectra of this wavelength region gave questionable photometry, since they were measured piecewise, photographically with different instruments on different rocket flights. Thus, good photometry of this region of the solar spectrum is of particular importance in determining the albedos, the ratio of reflected to incident radiation, of the planets. Although several good UV planetary spectra are available, there is still skepticism that their analysis is satisfactory because of the available solar comparison spectra.

In planetary aeronomy, much of the KPNO work used information from the Mariner and Venera space probes, and from observatory rocket flights by visitors and staff. M. B. McElroy's studies of the thermal structure of CO_2 atmospheres have been fundamental to nearly all recent work on Mars and Venus.

Rocket work, partly by visitors, has produced calibrated ultraviolet spectra of Venus, Mars, and Jupiter. Scattering and absorption by CO_2 dominate the Mars spectrum; for the other two planets it has been shown that scattering by aerosols suspended high in their atmospheres is clearly important.

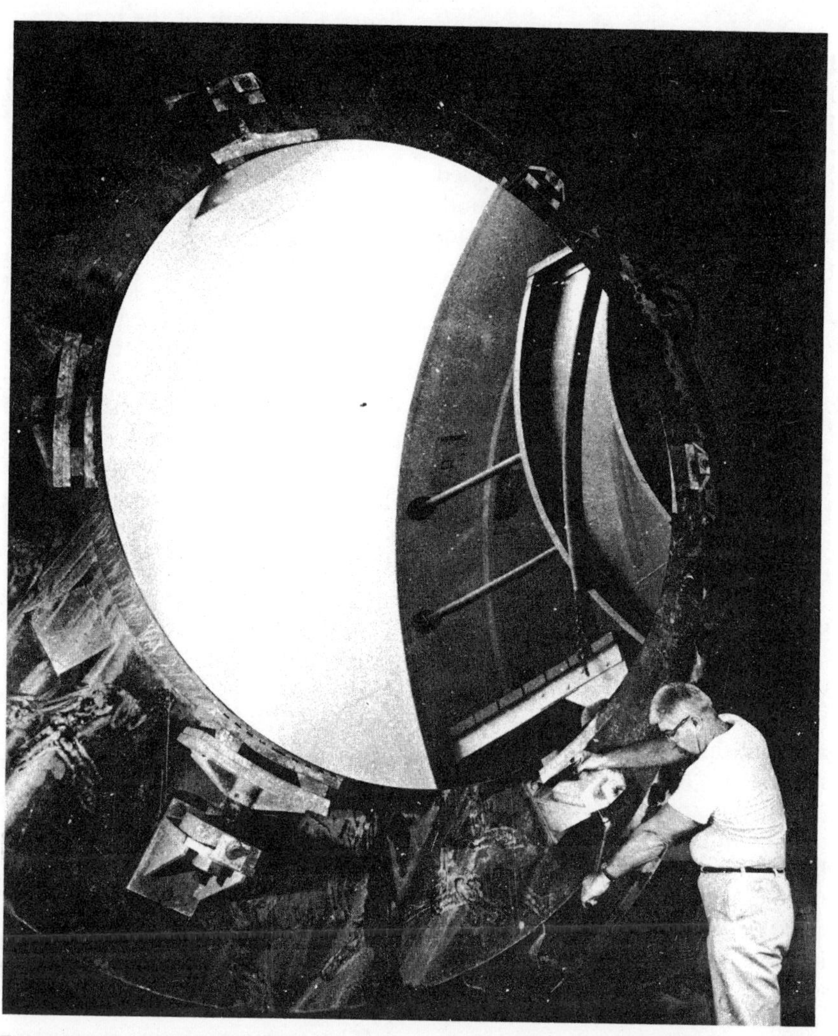

This 100-inch-aluminum-cast mirror will be used for testing the optical features of secondary mirrors on the new 150-inch telescope at KPNO. The 158-inch quartz mirror for the new 150-inch telescope will be finished on this same grinding and polishing machine at the observatory optical shop. (KNPO photo)

Cerro Tololo Inter-American Observatory

The Cerro Tololo Inter-American Observatory (CTIO) was established and is operated in the Republic of Chile by the Association of Universities for Research in Astronomy (AURA), Inc., under contract with the National Science Foundation. At a southern latitude of 30°, CTIO provides astronomers with opportunities to observe scientifically relevant southern sky objects.

The observing facilities are located at an elevation of 7,200 feet on Cerro Tololo in the foothills of the Andean Cordillera. Astronomical observing conditions at Cerro Tololo are superb, as at other nearby sites where other organizations have followed CTIO's example.

CTIO Research Projects

During fiscal year 1970 visitors and staff carried out photometric, spectroscopic, and photographic researches on the moon, planets, asteroids, stars, pulsars, gaseous nebulae, clusters, quasars, and galaxies. For example, T. McCord of the Massachusetts Institute of Technology investigated the optical properties of lunar and planetary surfaces. His observations of the Sea of Tranquility, *Mare Serenitatis,* yield reflectivity properties identical to those found in the Apollo 11 soil samples. M. F. Walker of the University of California at Santa Cruz did a photometric study of faint stars in globular clusters of the Magellanic Clouds. With image tube techniques, he was able to observe stars to visual magnitude 23.7, the faintest so far reached in these nearby galaxies. Finally, an example of the cooperation between CTIO and KPNO is Malcolm

Smith's observation of gaseous nebulae in the Carina region with Kitt Peak's pressure scanning Fabry-Perot interferometer. Dr. Smith's preliminary findings indicate that the great Carina nebula is formed by a single expanding gas cloud, rather than by two or more gaseous bodies in the line of sight.

Facilities

The principal construction effort during fiscal year 1970 was aimed at completion of the building to house the 150-inch telescope. All basic structural work on the building was completed. Fabrication of the telescope mounting is progressing satisfactorily in the United States, and first shipment of mounting parts is expected during fiscal year 1971. The 150-inch mirror blank has been cast, and grinding and polishing will begin as a next step at the KPNO Optical Shop.

Six telescopes were operational throughout the fiscal year. The largest of these is a reflector with a 60-inch aperture, specially designed to be used in a wide variety of investigations. Other telescopes have apertures of 36, 24, and 16 inches (two of the latter) and classical Cassegrain optics. The sixth telescope has Schmidt-type optics with a 24-inch aperture and is on loan from the University of Michigan.

A computer-controlled data acquisition system was developed during fiscal year 1970. The more rapid data collection possible with this system will result in a marked increase in the productivity of the telescopes.

During fiscal year 1970, a total of 69 astronomers and graduate students observed at Cerro Tololo, including 55 visitors from the United States, seven of whom were graduate students. Thirteen of the visitors were from Latin America, principally from Chile and Argentina; these included four graduate students. Altogether, visitors were assigned 68 percent of the total available observing time; the rest was used by the resident staff, KPNO staff, and for maintenance work.

National Radio Astronomy Observatory

The National Radio Astronomy Observatory (NRAO) is a national research facility funded by the National Science Foundation under contract with Associated Universities, Inc. Now in its 14th year of operation as a national center for basic research in radio astronomy, the observatory has two observing sites—Green Bank, W. Va., and Tucson, Ariz.

NRAO Research Projects

Very Long Baseline (VLB) Interferometry

In October 1969, the 140-foot radio telescope in Green Bank was linked together in an intercontinental experiment with a 72-foot radio telescope in the Crimea, U.S.S.R., over 6,000 miles away. For this experiment three NRAO staff members took the VLB equipment into the Soviet Union.

The purpose of the experiment was to observe two dozen quasi-stellar radio sources to determine the angular diameters of the sources and to investigate their fine structure. These radio sources of very small angular size are of great interest because radio sources are apparently born small and expand with increasing age. In order to understand the physical processes associated with the births of strong radio sources, many of which may be located at the edge of the observable universe, it is necessary to know more about their physical properties. From the variations in intensity of many of these sources, it may be inferred that their linear diameters often do not exceed a few light-months and once their angular diameters are measured, their distances can be determined. Then, from observations of their radio spectra, their magnetic field strengths can be calculated.

Discovery of More Interstellar Molecules at NRAO

Following their discovery of the carbon-12 isotope of formaldehyde ($H_2C^{12}O$) in ionized gaseous regions of the Milky Way and in many dark, cool nebulae, David Buhl (NRAO), Lewis E. Snyder (Virginia), Patrick Palmer (Chicago), and Benjamin Zuckerman (Maryland) discovered another concentrated source of formaldehyde containing the carbon-13 isotope, $H_2C^{13}O$, toward the center of the Milky Way and in many of the ionized regions in which the carbon-12 isotope had been found. The normal amount of carbon-13 in our solar system is one atom for every 90 carbon-12 atoms. In our galactic center, however, apparently the ratio is 1 to 10. The investigators believe that the over-abundance of carbon-13 in the center of our galaxy indicates that formation and, later, explosion of massive stars is taking place in our galactic nucleus since large quantities of carbon-13 can be produced in the cores of very massive stars as a by-product of the cycle which converts hydrogen to helium in a stellar interior.

The 36-foot millimeter wave telescope was first used for radio spectrum line work during the spring of 1970. Keith B. Jefferts, Arno A. Penzias, and Robert W. Wilson of the Bell Telephone Laboratories, discovered a radio emission line from carbon monoxide (CO), the sixth atom or molecule to be detected in inter-stellar space, using a receiver built jointly by Bell Laboratories and NRAO. The CO line was seen in the galactic center as well as in a number of other Milky Way sources where other molecules had previously been

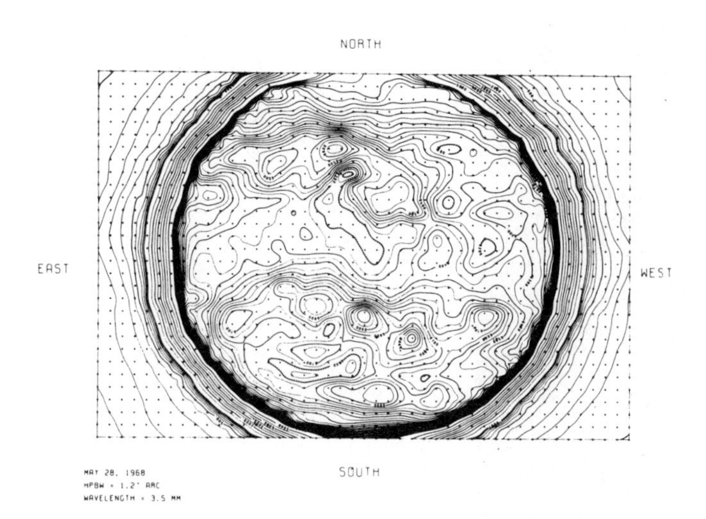

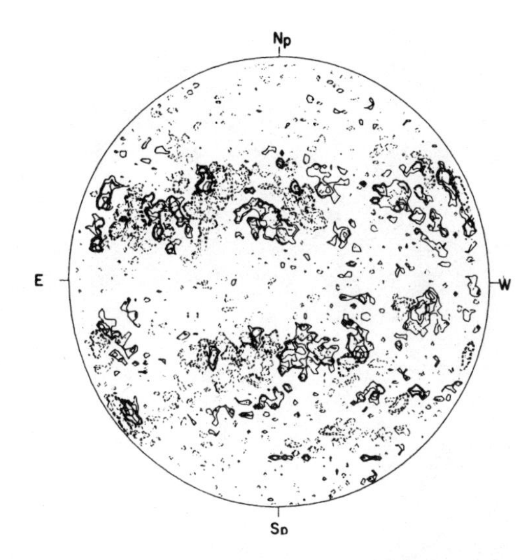

The NRAO 36-foot telescope at Kitt Peak was used to record 3.5 mm. contours of the sun. A solar magnetogram is shown at the right for comparison. (NRAO photo)

found. The CO line in the Orion Nebula is exceptionally strong and extends over an angular extent much larger than the nebula itself. The great strength of the CO line shows that the molecule appears in great abundance among many of the stars of our galaxy. The CO line at 115 GHz (2.6-millimeter wavelength) is observable only with a radio telescope that has an extremely precise paraboloidal surface. The 36-foot telescope is the largest such telescope in the world and affords scientists a unique opportunity to pursue the discovery still further. On another telescope run, the same group of scientists discovered a seventh molecule, CN, as well as two other isotopes of carbon monoxide, $C^{12}O^{18}$ and $C^{13}O^{16}$.

Finally, Buhl and Snyder detected an eighth molecule, hydrogen cyanide (HCN) in a number of galactic sources, including the Orion Nebula and the galactic center.

Facilities

The major observatory telescopes include a 300-foot meridian transit telescope that after November 1970 will have a new surface capable of operating down to wavelengths of 10 centimeters; an interferometer which operates at 3- and 11-centimeter wavelengths consisting of three 85-foot telescopes with a portable 42-foot telescope for remote operation; a 140-foot fully steerable telescope that will operate at 1 centimeter wavelength; and a 36-foot millimeter wave telescope that will operate down to wavelengths of 1 millimeter. The 36-foot telescope is located at Kitt Peak, Ariz., while the other systems are located in Green Bank, W. Va.

Each telescope is equipped with an on-line computer for limited analysis of data as they are received. Data are usually recorded on magnetic tape and are later processed in a general-purpose computer.

During the past fiscal year a new traveling feed was installed at the 300-foot transit telescope for use at low frequencies. This new feed moves at the telescope focus in such a way that a radio source can be tracked at transit, trebling the time during which a source can be studied at each meridian passage.

A new dual channel, low noise spectral line receiver was placed in operation at the wavelength of the 18-centimeter OH line which permits any two of the OH lines to be observed simultaneously.

The interferometer was converted during the spring of 1970 to a two-frequency system and is now operated at a 3-centimeter wavelength as well as 11 centimeters. Maps of radio sources may be made at 3 centimeters with a resolution of 2 seconds of arc by aperture synthesis techniques for the study of the detailed structure of astronomical objects.

Computer programs have been completed that enable observers to use the 413-channel autocorrelation receiver to improve baselines and increase the overall efficiency of the receiver. A new antenna measuring instrument has been built and tested that will monitor the shape of an antenna surface using a continuous-wave radar, and promises to be of value in studies of thermal deformations and other surface calibrations of antennas.

Subsequent to the completion of the report "A 300-Foot High Precision Radio Telescope" in June 1969, it was decided to incorporate the principles of homologous design of radio telescopes into a smaller but more precise instrument. In this approach, the defor-

Aerial view of the Arecibo Observatory showing the feed platform (upper center) suspended by cables between three concrete towers 435 feet above the reflector surface. Also shown are the operations, office, and visitors' quarters complex (lower right), and warehouse and maintenance buildings (lower left). (Photo Cornell University)

mation of the telescope due to the pull of gravity, different for different pointings of the instrument, is made use of to bring the reflecting surface to the desired shape to focus the incoming radio beam. A design is being made of a 65-meter (213-foot) diameter, fully steerable paraboloid designed to operate at wavelengths as short as 3 millimeters under stable atmospheric conditions.

Arecibo Observatory

The Arecibo Observatory (AO) located approximately 12 miles south of the city of Arecibo, Puerto Rico, has the largest radio reflector in the world. It was initially constructed with funds supplied to Cornell University by the Advanced Research Projects Agency, Department of Defense. Commencing in October 1969, the National Science Foundation assumed sponsorship of the AO and it was designated a National Research Center with Cornell University continuing to operate and manage the AO. The AO itself provides on-site scientific management and administration, with additional administration and planning conducted by the Arecibo Project Office at the Ithaca campus of Cornell University.

Research Projects

Planetary Radar

The program of detailed mapping of the surface of Venus has begun and the first map is now available. Since the atmosphere of Venus is optically quite impenetrable, radar holds out the best promise of obtaining information about the structure of its surface. The resolution obtainable is only limited by the precision with which the echo power can be analyzed. With an improvement of the AO facilities, one can expect such maps in the future to have a resolution much higher than those now being obtained and therefore to show the nature of the topography of Venus.

Pulsars

In radio astronomy, work on pulsars has been a major occupation at the AO. After the discovery of the pulsar in the Crab Nebula and the further discovery that the frequency of the pulses decreased with time, much more attention has been paid to this object. In addition to giving the hint that the energy source for the high energy particles in that nebula is to be found in the rotational energy of a neutron star, the detailed observations of the Crab Nebula have now given much in-

Photo by National Radio Astronomy Observatory.

formation about these high density and high energy regions in the universe. Details of the variations in rotation speed of this and other pulsars are carefully recorded, and slight changes in the number of free electrons in the line of sight from us to the pulsars are also measured.

Facilities

The AO makes available to atmospheric physicists and to astronomers a major research instrument which can function either actively as a radar telescope, or passively as a radio telescope.

As a radar telescope, the instrument transmits a pulsed signal, and receives that portion of the signal which is reflected back by electrons in the ionosphere, or from the moon, or the planets Mercury, Venus, and Mars. Planetary radar studies at AO have revealed the previously unknown rotations of Mercury and of Venus, as well as giving most precise distances accurate to about 1 kilometer for these planets. Surface properties and topography of the moon and planets can also be investigated, and a beginning is being made to obtain a detailed map of the surface of Venus.

The beam of the antenna—located at a position of 18° north latitude—can be swung over an angle of 20° in any direction from the zenith. About 40 percent of the sky can be surveyed and the sun and the planets can be observed on approximately half the number of days in each year.

The AO is the only operational spherical antenna system in the world. It has shown itself to be relatively cheap to construct, and convenient and versatile in operation. The general development of this type of antenna system will be influenced by the progress made at the AO.

In a spherical antenna many feeds can be used at the same time; thus, AO has a large number of different frequency feeds permanently mounted, and it also has a multibeam system where ten beams can be used simultaneously to speed up a detailed survey of the sky.

During fiscal year 1970, a total of 21 scientific visitors from 12 different organizations utilized the AO facilities. At present, a majority of the telescope operating time is used by graduate students in residence and observatory staff.

University Research

Stratoscope II—The Airborne Observatory

Perhaps the single greatest problem encountered by astronomers using telescopes on the ground is the earth's atmosphere. Turbulence, pollution, and absorption of light at certain wavelengths severely limit the resolution of even the largest and most carefully constructed telescope mirrors.

To a very great extent this difficulty has been overcome for at least one telescope which observes astronomical phenomena while floating at an altitude of 80,000 feet suspended from a giant plastic balloon. The telescope, Stratoscope II, has a 36-inch mirror and, with the use of photographs with guidance and radio command from the ground, is able to obtain a resolution of 0.1 second of arc. This resolution, roughly equivalent to the ability to distinguish between two golf balls 30 inches apart at a distance of 1,000 miles, is close to the theoretical limit for a telescope of this size.

After several flights during the past few years in which technical difficulties hampered astronomical observations, Stratoscope II was successfully launched from the National Scientific Balloon Flight Station in Palestine, Tex., the night of March 26, 1970, under the direction of Martin Schwarzschild and

Airborne 36-inch optical telescope is completely remote-controlled from the ground. (Photo Princeton University)

Robert Danielson of Princeton University. Dr. Schwarzschild has been director of the Stratoscope II project and its predecessor, Stratoscope I, since inception of the latter in 1956. The project is co-sponsored by the National Science Foundation and the National Aeronautics and Space Administration, following initial support by the Office of Naval Research.

During the flight of the night of March 26–27, Stratoscope II obtained photographs of unprecedented sharpness of both the planet Uranus and of the nucleus of a rare Seyfert galaxy. Both sets of photographs will markedly increase the state of our knowledge of the two celestial objects.

A Seyfert galaxy—in this case the one known as NGC 4151, about 30 million light years from the earth—has a very small but extremely bright nucleus characterized by variability in light intensity, emission of radio waves, and a spectrum of broad bright lines produced by extremely hot gases in rapid motion. The sharpness of the Stratoscope II photographs establishes the upper limit of the diameter of the nucleus at about 12 light years. This volume of space in the immediate vicinity of earth contains five stars, including our own sun. In contrast, the same volume in the Seyfert galaxy contains around ten billion stars. One implication of this high density of stars is that there must be collisions between stars on the average of every 4 months, which may account both for the very great brightness and the observed variability of the light from the nucleus, since such high-speed collisions between stars can be expected to produce intense heating and large amounts of radiation.

The photographs of the planet Uranus are the sharpest yet obtained. They reveal none of the surface features of the planet occasionally reported by visual observers, but do reveal the planet as a slightly flattened disk, somewhat less bright around the rim. Measurement of the latter phenomenon, called limb darkening by astronomers, will provide a test of a theory that Uranus —unlike Jupiter and Saturn—has no clouds in its atmosphere. The photographs will be further enhanced by combining several of them by means of an electronic computer, which will compensate for the known optical properties of the telescope. This analysis should conclusively establish whether or not Uranus has surface features.

University Astronomy Research Instruments Program

The erection of the new 120-foot radio telescope at the Vermillion River radio observatory of the University of Illinois was begun on July 9, 1970. The 120-foot steerable reflector is near completion, with the mechanical and electrical components purchased and in a final state of assembly. The radio telescope will be used for sky surveys, continuum mapping, and spectroscopy of the OH molecule. Polarization studies of galactic sources will be possible, and observing time will be available to guest investigators.

A 60-inch optical telescope for the Hale Observatories (Carnegie Institution of Washington) is set to be completed in 1970, and there is auxiliary equipment for modern operation of this telescope on Palomar Mountain in California.

The fiscal year 1970 funds for astronomy facilities and equipment included support of spectrographs for the New Mexico State University and the University of Oregon, a spectrum scanner and a three-element aperture synthesis radio interferometer for the Massachusetts Institute of Technology, modernization of the 150-foot solar telescope of the Carnegie Institution on Mount Wilson, and facilities for photographic plate storage at Swarthmore College.

MATHEMATICS

Technique for Statistical Analysis

In almost all real life situations, there is an element of randomness to the behavior of parts of the system. That is to say that even if all other factors are relatively constant, as for example, a group of people of the same age, sex, ethnic, socioeconomic, and geographical background, some element of chance will enter into their choices, actions, and answers to questions. If this behavior is truly random, and if the sample is large enough, the mathematical laws of statistics describe accurately certain things about the distribution of their actions, i.e., a certain number of the group will behave in a given manner, another number in a different manner. These laws of statistical analysis have proved valuable in making decisions where there is an element of randomness in the selection, evaluation, or compilation of data. Recently, a new method of testing has been developed which should considerably improve the decision process.

Suppose we wish to make a decision (e.g., drug B should be adopted in place of drug A; or a machine should be stopped and adjusted). First we establish a criterion in the form of a hypothesis that drug B is more effective than drug A; or that the items produced by the machine are out of tolerance. An experiment is then designed to select and evaluate samples, and a method of computation of the results established to accept or reject the hypothesis. In practice, the hypothesis and computations are designed so that if the computed average of the samples is positive, the hypothesis is rejected; if it is zero or negative, the hypothesis is accepted. In the design of the experiment, a certain confidence level—a probability that the decision is correct—is prescribed. In all tests in current use, the probability of rejection has a small value when the computed sample average is close to zero. The probability of rejection approaches a certainty when the computed sample average gets very large. This means if the average is positive but small, the probability of rejection will be close to the prescribed confidence level. However, there is no way of testing the hypothesis at a given confidence level which can take into account the possibility of future observations. Thus, for example, if we repeatedly test a hypothesis at a given confidence level with fresh sets of data from the same process, we are sure to reject it due to the cumulative effects of errors in the data and the continuity of the computation method for the probability of rejection.

Herbert Robbins, at Columbia University, has developed a theory of testing which is of a radically different nature. His procedure starts in the traditional manner, but he has a new method of computing the probability of rejection in such a way that rejection will be certain for every computed sample average greater than zero, while the probability of rejection for any average not greater than zero will be less than the (small) probability for a zero average. The heart of this new technique lies in finding a sequence of numbers with certain mathematical properties and such that the probability that a zero average will be rejected is less than the given confidence level. Such sequences have been known to exist, but it has not previously been possible to evaluate the probability that a zero average would be rejected, and thereby permit testing at a given level of confidence.

To illustrate the advantages of this new method, consider the following example. A machine produces an item which must be within certain tolerances. Samples are drawn to test the hypothesis that the machine is working properly. Assume that the machine works perfectly forever. Under any tests currently in use, the machine will be stopped once every hundred tests, say, because the hypothesis is rejected. Under the tests developed by Robbins, there is only a probability of one in one hundred that the machine would *ever* be stopped.

The new method has a wide range of applications in such fields as quality control, drug testing, etc., and will possess considerable advantages over current methods in other important practical problems.

BIOLOGICAL AND MEDICAL SCIENCES

During fiscal year 1970, the effects of reorientation and cutbacks in the support of research in the life sciences by other agencies of the Executive Branch have become apparent in proposals to the NSF. At the same time, the number of proposals which were successful in obtaining NSF grants in biology dropped from 1,173 in fiscal year 1969 to 1,072 in fiscal year 1970.

More proposals have been received from investigators who were previously supported by the National Institutes of Health, the Office of Naval Research, the Air Force, Army, and the Atomic Energy Commission. Since the number of grants which NSF will be able to make in 1971 will not be substantially larger than in 1970, it is anticipated that the fraction of proposals which it will be possible to fund will drop substantially below the figure of 50 percent which has prevailed in the past few years. The next few years will clearly be a time when the National Science

Foundation should maintain the maximum flexibility in deploying its funds in order to be able to respond to a changing pattern of support by other agencies.

In spite of the current difficulties and the uncertainty about the long-term future, biologists are excited about the strides their science has made in the last few years, and a variety of new fields and approaches which are clearly ripe for further exploration. Within the biological programs of the Foundation, it is planned to place further emphasis on environmental research. Major increases in ecological research will occur within the International Biological Program (IBP), bringing the desert, deciduous forest, and coniferous forest biome studies to a fully operational level, and initiating an integrated research project on the evolution of ecosystems. Both within the IBP and in other programs, increased attention will be given to the biological control of populations. Some of this research will be directed toward improved understanding of the factors which operate generally to influence the balance between different species of plants and animals, while other research will be designed to promote particularly promising approaches to the biological control of pests of particular economic importance.

Other planned programmatic efforts include an increased emphasis on psychobiology and neurobiology, reproductive biology, the molecular biology of the human cell; and the development of an improved base of support for resource centers such as museum collections, genetic stock centers, and controlled environment laboratories.

These initiatives on the part of the Foundation are matched by counterpart trends within the scientific community. The "invisible college" centered on a given problem area is not new to biology, but an increased willingness to formalize such arrangements is appearing. One example is provided by the organization of the Integrated Research Projects of the International Biological Program. Another example is provided by the formal organization of systematic biologists with the objective of producing a Flora of North America and at the same time cooperating in the development of a computer-assisted system for handling taxonomic data of this type. A more recent grouping has emerged among molecular biologists who are proposing to coordinate their efforts in order to make an effective attack upon the problems of the molecular biology of the human cell.

The prospects in psychobiology and neurobiology seem particularly exciting because it now appears to be possible to approach problems of learning, memory, behavior, and perception at the level of mechanisms. The techniques for measuring parameters of behavior have been greatly refined, recording of electrical events can be made from highly localized regions in the central nervous system, the mapping of functional regions and pathways has progressed to a very substantial extent, and tools are available for examining the chemical basis of structure and function. The identification of sensory pathways and events has opened the exciting possibility of direct stimulation of the central nervous system with the electrical output of a sensory prosthesis—an artificial sense organ. Although recognized as possible in a speculative sense for many years, such an undertaking is now clearly possible with predictable improvements in our understanding of the functional anatomy of the central nervous system and the normal electrical output of sense organs. Thus, by substituting an artificial sensor, coupled appropriately with the central nervous system, it will be possible to restore some degree of sight to the blind or hearing to the deaf.

At the molecular level, we can now anticipate a developing understanding of the chemical basis of learning and memory, and the basis of chemical effects on behavior. Many such chemical effects have been identified as a part of the normal regulatory processes of behavior and as desired or undesired effects of drugs, but we do not yet understand the mechanisms by which these chemical effects are mediated. There is, for example, no understanding of the mechanisms responsible for drug addiction or dependence, and until these mechanisms are understood, there is no hope of dealing in an effective way with this frightfully expensive social problem.

Remote sensing techniques have been applied by ecologists for studies of the distribution of animals and plants as rapidly as they have had access to this technology. There is great interest in the expansion, improvement, and increased access to these techniques because it is clear that they will be essential to effective ecological studies of any substantial magnitude, as well as to improved wildlife management, forestry, and agriculture — when coupled with the required fundamental research.

Finally, although the National Science Foundation cannot propose to greatly expand support of tropical biology in the near future, we wish to recognize the necessity of continuing with the modest investment in this area. Aside from the inherent interest of the rich flora and fauna of the tropics to biologists, the tropics also represent the greatest undeveloped potential for food production. The use of extensive monocultures, which has been so effective in the temperate zones, has been less effective or actually disasterous in tropical agriculture.

Insights currently being developed into the differences between tropical and temperate ecology suggest the possibility of using multicultural farming methods in the tropics—the technique of raising several food crops simultaneously on the same plot of ground. Perhaps in this way, food production may be greatly increased without inviting ecological catastrophe in tropical areas.

INTERNATIONAL BIOLOGICAL PROGRAM

The International Biological Program (IBP) has as its worldwide theme "The Biological Basis of Productivity and Human Welfare." U.S. participation in this international program has taken the form of multi-investigator integrated research projects dealing with two of the important topics facing mankind in a world of burgeoning population—scientific management of biological resources and human adaptation to the stresses of the physical environment.

Rational use of the environment requires a better understanding of how ecosystems operate. Such an understanding has long been sought by individual investigators probing important aspects of plant and animal ecology. The new dimension added by IBP is the integrated attack on complex ecological systems by teams of investigators representing a variety of disciplines and, often, many institutions. Each investigator pursues his own specialty, but shares his data with scientists in neighboring fields. The objective is to achieve a fuller understanding of the processes and rates of nutrient cycling, water movement, energy flow, and population dynamics in natural and man-dominated ecosystems than can be obtained by individual investigators working alone. This additional knowledge is essential if man is to cope adequately with the twin challenges of producing enough food and fiber to feed a hungry world and of maintaining and enhancing the quality of the environment.

The problem is being attacked through intensive studies of ecosystems in four distinct life zones, or biomes: deciduous forest, coniferous forest, grassland, desert. The intensive study of the Grassland Biome moved from the planning and preparatory stage into full operation during the year; expanded field research in the Desert Biome began in May 1970. Planning was completed for the other biome studies, and they are expected to begin operations during fiscal year 1971.

In addition to the biome studies, IBP includes a wide range of other environmental research. An integrated research program in biological control of insect pests began during the year under the direction of Carl Huffaker of the University of California at Berkeley. Emphasis will be put on natural factors regulating key groups of insects, with the aim of learning how to maintain pest populations at noneconomic densities in such a manner as to optimize cost-benefit relations and to minimize environmental degradation. Collaborative research is being undertaken in the ecology of upwelling areas, which comprise only about 1 percent of the sea surface but are responsible for the productivity of perhaps half of the world's fisheries. Comparative studies of upwelling in the Pacific off Peru and in the eastern Mediterranean are in progress.

The human adaptability component of IBP has developed more slowly than the ecosystem research. A collaborative research effort involving anthropologists, archeologists, geologists, and marine and terrestrial ecologists was begun to study the adaptations of Aleut Indians to the changing conditions of the Bering Sea land bridge during and since the Pleistocene. Research in the genetics of South American Indian populations little influenced by Europeans continued under the direction of James Neel of the University of Michigan. As Dr. Neel has put it, "This is the first generation of scientists to have the tools to do this kind of sophisticated research in the genetics of primitive populations, and the last to have the opportunity to do so." Dr. Neel has pointed out that since man's genetic diversity arose while he was living as a hunter and gatherer with simple agriculture, many insights into the population genetics of civilized man can be gained only from primitive tribes. It is important to study them before their genetic constitution and social structure have been altered by extensive contact with other populations.

Molecular Biology

Two of the most exciting achievements of the year arose from studies on the molecular biology of the gene. A chemist, starting with simple molecules, and a geneticist starting with living cells accomplished in principle the same result. The end product of their respective experiments was, in tangible form, a single gene.

In the fall of the year, Jonathan Beckwith and his co-workers at Harvard Medical School reported the successful isolation of a gene from the bacterium *Escherichia coli*. More recently, H. Gobind Khorana (recipient of the 1968 Nobel prize for his contributions to unraveling the genetic code) and his colleagues at the University of Wisconsin synthesized a gene from elementary chemical units. In essence, the two groups exploited the chemical properties of DNA, the genetic material to achieve their goals (but from opposite starting points).

The Watson-Crick model of DNA is a long molecule consisting of two intertwined "strands" held toegther

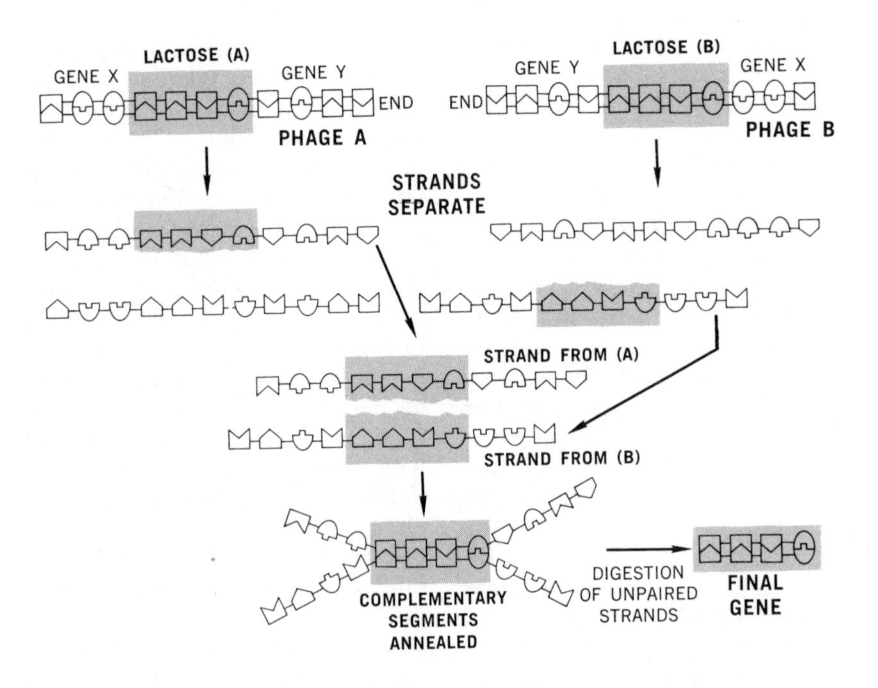

In the Beckwith isolation of a single gene, DNA from two different bacteriophages is uncoiled, and single molecular strands from the two sources mixed. The complementary sections of the two strands which together make up the lactose region are annealed and the unpaired strands digested enzymatically to leave only this region. The diagram is highly schematic.

throughout their length by specific pairs of chemical units, the nucleotide bases. The sequence of these units in one strand defines the chemical code for a given gene. Because of chemical considerations, the sequence of bases on one strand dictates the sequence of bases in the opposite strand. With careful manipulation, the two strands of intact molecules of DNA can be separated. If separated strands are mixed and placed under conditions which allow accurate reassociation, a strand finds its complementary partner and reforms a normal two-stranded DNA molecule.

In essence, Dr. Beckwith and coworkers performed surgery on the genetic material of the bacteria using a bacteriophage (virus which infects bacteria) to dissect out and carry the bacterial gene to the experimenters' test tube. Upon infection of bacterial cells, certain types of phage are capable of exising small pieces of the host cell's DNA, incorporating it into their own genetic material. As the phage multiplies, this DNA is replicated together with the phage DNA.

Specifically, the Harvard investigators used two different phages, both of which are known to carry in their chromosomes different bacterial genes and a single region —common to both—that participates in the metabolism of the sugar lactose. The principal difference between the two types of phage—and this is the key to the success of the experiments—is that in one type the bacterial DNA is incorporated into the phage DNA in a left to right direction while in the second type the bacterial DNA is inverted by being incorporated into the phage DNA in a right to left direction.

The experiments performed by the Harvard investigators consisted of extracting the DNA of the two types of phage separately and treating each DNA preparation so that each of the double-stranded molecules uncoiled, permitting the strands to separate. They then brought one strand of DNA from each type of phage together under conditions in which reassociation could occur thereby restoring the double-stranded state of complementary DNA chains. Since the region of the DNA strands representing the bacterial lactose gene was the only opposite mirror image or complementary region, only this part of the mixed chains came together immediately to form double-stranded DNA. The neighboring DNA segments remained dangling as single strands and were digested by an enzyme that degrades single-stranded DNA. This resulted in a preparation of purified bacterial DNA segments that represented only the lactose gene.

Dr. Khorana's synthetic assembly of a gene also depended on the complementarity of DNA strands, but his experimental approach was quite different and involved an additional property of DNA. The information encoded in DNA is used as a template and transcribed by living cells into a chemically related class of molecues, RNA. The RNA molecules are also complementary to the strand of DNA transcribed, so if the sequence of bases in the RNA is known, one can deduce the sequence of bases in the template DNA strand. Given the sequence of one DNA strand, it is possible to predict the composition of the complementary partner strand by the base-pairing rule.

Several years ago, Khorana began experiments starting with alanine transfer-RNA from yeast. The sequence of the 77 nucleotides in the RNA was known from the work of Robert Holley and it was a simple enough matter to visualize the expected sequence of the 154 paired nucleotides in the corresponding segment of DNA which constituted the gene that codes for the transfer-RNA.

To achieve the synthesis of this segment of DNA, however, was far from simple. Starting with the simple chemical building blocks, Khorana synthesized short lengths of strands of correct sequence, and for each such sequence, produced a strand of partial complementarity. This partial complementarity was crucial to the success of the experiment. Pairing of the short strands produced a DNA segment with a two-stranded middle portion where the base sequences were complementary and at each end an unpaired strand remained. Enough of these partially two-stranded pieces of DNA were synthesized to mimic the entire sequence of the nucleotides in the gene. Thus, when pieces that occurred in consecutive order were mixed, a single strand of one complemented a single strand of the other. By mixing the pieces in successive order, pairing between the overlapping ends produced progressively longer two-stranded segments. After base-pairing had ordered the short segments of DNA, an enzyme was used to form the chemical linkage between adjacent ends, resulting, finally, in the intact synthetic gene for alanine transfer-RNA.

It remains to be shown that these isolated genes can be reinserted into a cell and express their chemical information. However, these successes provide biologists with techniques to allow the test tube study of gene action to proceed at a highly sophisticated level. It becomes possible to imagine that we shall soon have considerable new biochemical information on how genes are regulated in living cells.

Microtubules

In the last 20 years, the view of the cell as an undifferentiated bit of protoplasm containing a few specialized microscopic organelles has given way to a picture of a highly structured system whose parts are intricately interdependent. Improved techniques such as microsurgery and better observational instrumentation such as the electron microscope have enabled biologists to see and work directly with the components.

The study of one such type of structural component has bloomed so rapidly that "microtubule biology" might now be called a subdiscipline of biology. This area of work deals with a variety of filament-like structures which have a seemingly ubiquitous distribution in cells. These structures are assemblies of macromolecules and have been named on the basis of their diameters: the largest, greater than 200 Ångstrom (Å) units, are called microtubules; the next sizes, simply "100 Å filaments" and "50 Å filaments." (An Ångstrom unit is equal to about four one-billionths of an inch.)

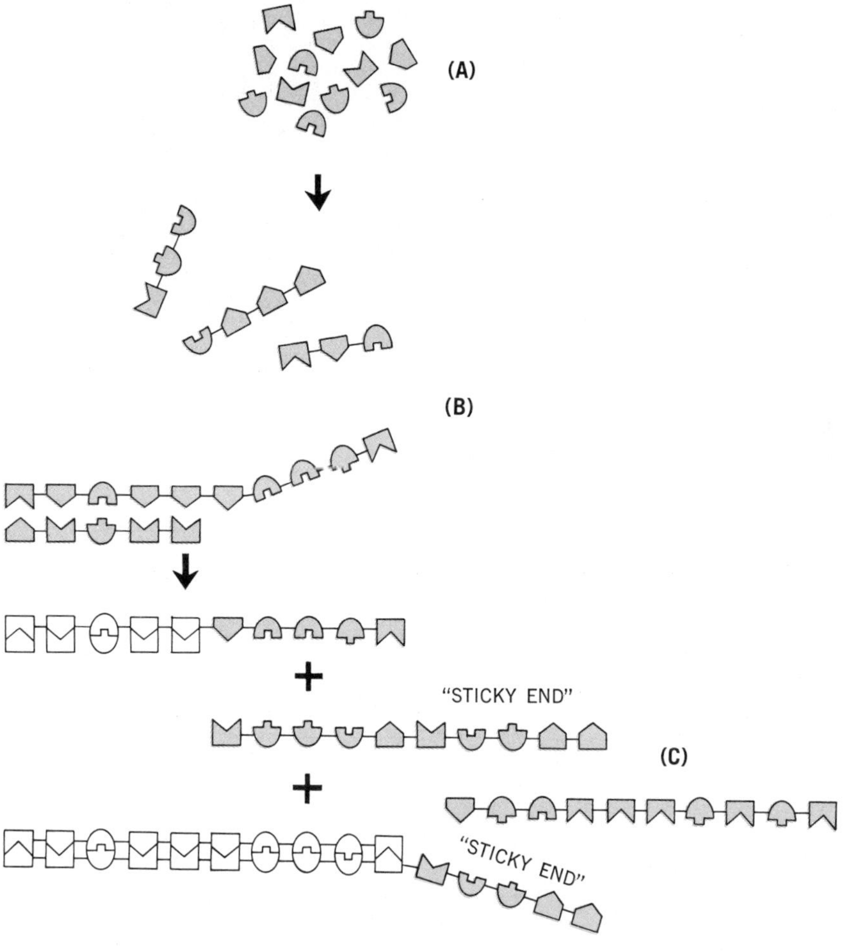

(A)

(B)

"STICKY END"

(C)

"STICKY END"

Starting with individual nucleotides (A) Dr. Khorana synthesized short lengths of nucleotide chains (B). Using complementary base pairing, he annealed part of a longer chain to a short chain, leaving a "sticky end" to which to anneal a third short length, again leaving a "sticky end" and so on (C). The overlapping produced progressively longer two-stranded segments which were enzymatically linked to form the complete gene of 77 nucleotide pairs. The diagram in this case is intended to show regions of double-stranded DNA as open symbols and those fragments not yet base-paired as darker symbols.

The filaments are of special interest because of their probable role in influencing the shape and movement of living cells either in migration or displacement of cells from one location to another or in movement of materials within cells.

The problem of cell shape has tantalized biologists for many years. How a definitive cell shape is acquired is a central question in differentiation, the event in embryonic development which, generally, is the time when a cell attains its characteristic adult function. The cell shape problem also bears on a popular biological generalization; namely, that there is a correlation between the architecture of a cell and its physiological function.

Norman K. Wessells at Stanford University has provided important new information on the way in which microtubules affect cell shape. Dr. Wessells has found that the application of a chemical, cytochalasin, causes the selective disappearance of the 50 Å class of filaments. Studies of the salivary gland during the differentiating phase of development reveal that in the presence of cytochalasin, the gland does not assume its characteristic shape. The effect is reversible; when the chemical is removed the 50 Å filaments reappear and the tissue undergoes cell rearrangements which are characteristic of the differentiated tissue. In a second series of investigations Dr. Wessells has found that the 50 Å filaments associated with the growing tip of a nerve cell are also disorganized by cytochalasin. In this case the tip of the cell becomes rounded and further extension of the nerve cell axon—the long extension of a nerve cell that conducts nervous impulses away from the cell body—is inhibited. Interestingly enough, the microtubules in the axon of the nerve cell remain intact and except for the tip, the cell retains its characteristic shape. If a second chemical, colchicine, a drug used in the treatment of gout, is applied simultaneously with cytochalasin to the nerve culture, the microtubules also become disorganized and the entire cell becomes rounded.

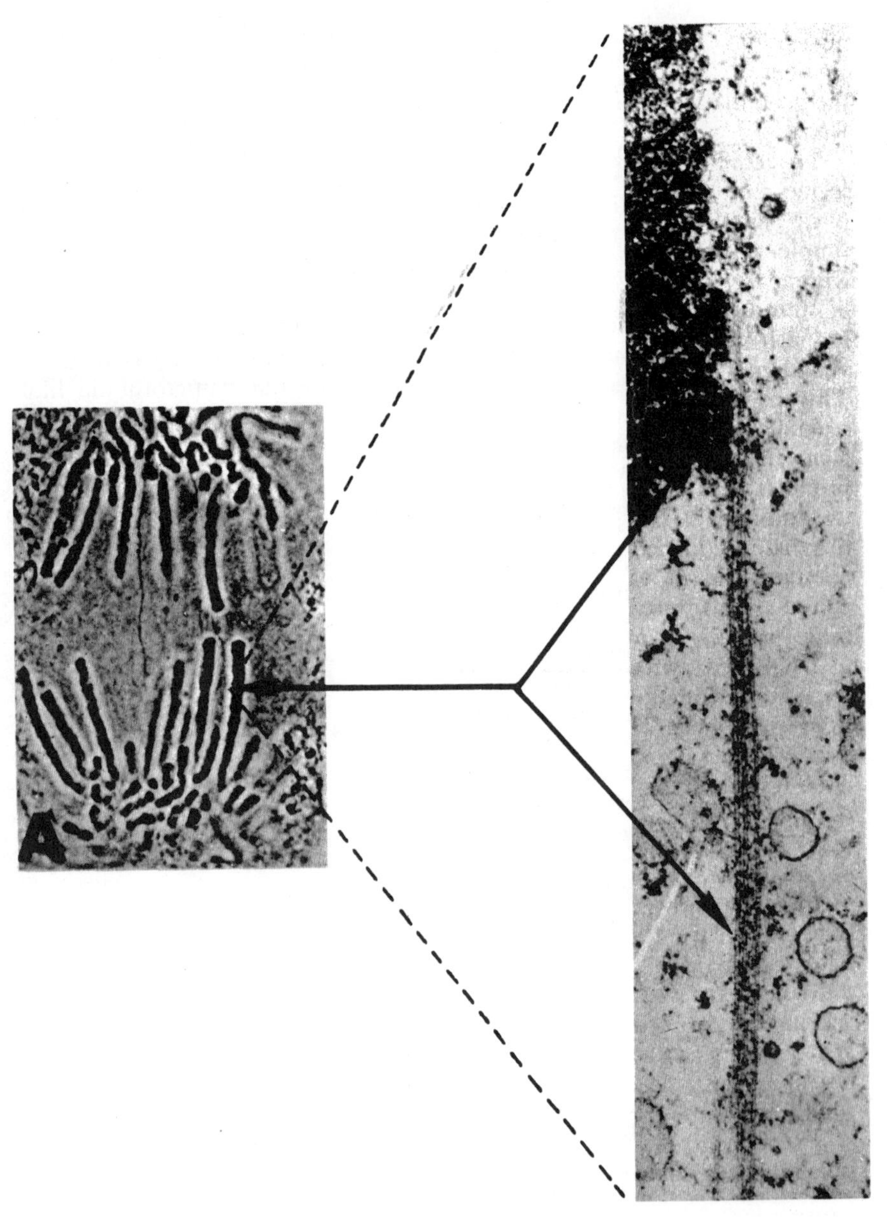

Using refined microscopic techniques, Dr. Andrew Bajer at the University of Oregon is able to trace microtubules from the light microscope level to the fine structure level of the electron microscope. (The cell shown is *Haemanthus*, the African blood lily; the arrow indicates the microtubule and associated (darker) chromosomes.)

Howard Holtzer at the University of Pennsylvania has found that application of cytochalasin also inhibits division of the cytoplasm of cells. In each of the foregoing examples, the 50 Å filaments appear to have a functional association with the cell membrane, yet in each case the biological event affected is unique to the given cell type.

The understanding and control

over the development and movement of cells gives the biologist a profound and important tool for further advances in embryology and perhaps, in the far future, presents possibilities for directing the regeneration of damaged tissues and even organs.

Ecological Gradients

In disentangling the complex net of relationships in natural ecological systems, the techniques of controlled manipulation, so common in laboratory science, have so far been extensively used only in lakes and ponds where manipulation of the fish stocks can modify the composition and functioning of the entire ecosystem.

A generally more useful strategy for the ecologist pursuing terrestrial studies has been to analyze the properties of the system whenever it is disposed along a measurable gradient, a change in some value per unit of distance in a specified direction, in the physical environment. The addition or removal of a single species population wherever its limit of tolerance is reached can cause a modification in the entire system. Should this happen, the nature and extent of the modification suggests the ecological role which that species plays within the system.

The marked gradient in water temperatures within hot springs such as those of Yellowstone Park provides opportunity for employing this strategy in the understanding not only the ecophysiology—organic processes specifically related to adaptation to a particular environment—of the thermophilic or "heat loving" organisms inhabiting the hot springs but also the community dynamics of the limited life forms of this environment in which most freshwater organisms would perish. The water is hottest where it issues from its underground source. In some springs the hottest water may be above 90° C.; only certain filamentous and unicellular bacteria can exist in such hot water, just a few degrees below the boiling point. But photosynthetic blue-green algae mixed with bacteria flourish, forming thick mats, where the temperatures range between 50° and 75° C. These mats are colorful — the browns, yellows, rich greens, and blue-greens where blue-green algae predominate contrast with the oranges, pinks, and reds of the bacteria.

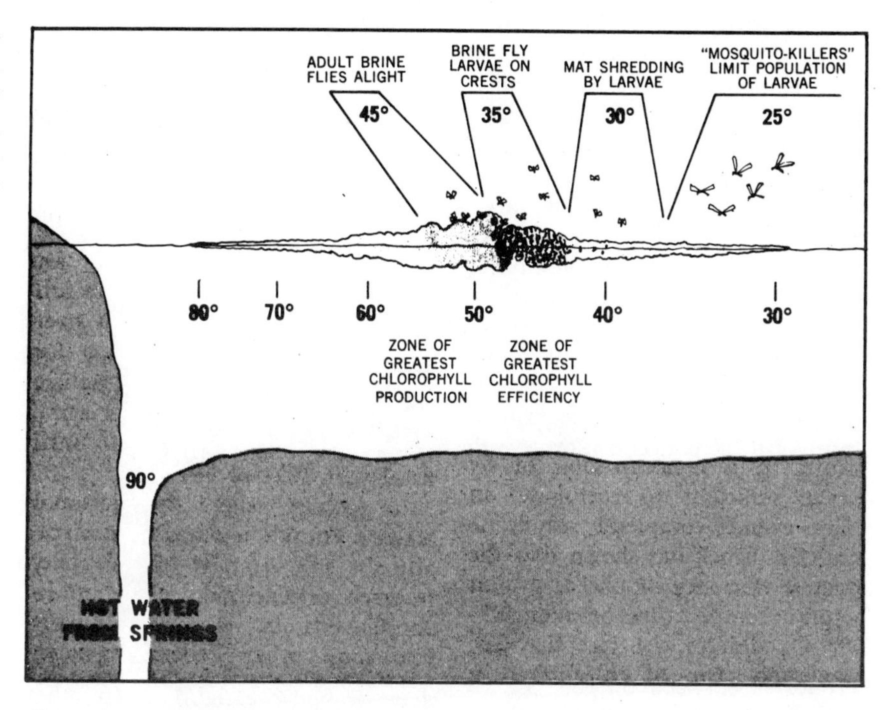

Temperature gradient extends both downstream from hot spring and upwards from water surface as algae-bacterial mat cools. Perforation of mat by grazing larvae increases efficiency of photosynthesis, so maximum growth occurs at lower temperature than that of maximum production of chlorophyll. At still lower temperature, population of brine fly larvae is limited by predators. Vertical scale in the diagram has been exaggerated to show detail.

Richard Castenholz of the University of Oregon has been extending his field studies of the blue-green algae by examining the growth of these simple plants under controlled laboratory conditions simulating those of the field. The research of Thomas Brock of Indiana University on the ecology and physiology of the thermophilic bacteria of Yellowstone springs, and that of Castenholz on the algae, have been planned to complement each other. While each of these investigators has also been concerned with the effects of grazers on the microbial mats, the community ecology has been the major concern of Richard Wiegert of the University of Georgia, who has worked closely with the other two.

None of the small arthropods that live in and on the algal-bacterial mat—ostracods (small freshwater crustaceans), mites, and flies—occur where the temperature is over 50° C. In the Yellowstone springs, the most common animal seen on the mats is a brine fly, the adults of which can be seen alighting here and there on bits of the mat that prove to have cooled to at least 45° C. by being raised above the surface. Radioactive carbon incorporated into the microbial mat has been recovered from both the adults of these brine flies and their larva, demonstrating that they do consume the mat. The adults commonly lay eggs and feed in those patches where the mat has cooled to 30°–40° C. Eggs laid where the mat is hotter do not survive. But the concentration of the eggs in the cooler patches results in a concentration of fly larvae sufficient to corrugate, in places even to completely shred, the mat. Dr. Brock has shown that the greatest efficiency of photosynthesis occurs at temperatures between 40°–50° C., despite the fact that the maximum crop of chlorophyll is produced between 50°–60° C. He believes that the greater efficiency at the lower temperature range is due to the grazers' perforation of the still cooler (30°–40° C.) part of the mat above the water surface. These perforations promote circulation of the nutrient-laden water and, in part, help the mat to maintain its integrity in spite of patchy grazing.

At a lower temperature level on the thermal gradient where the microbial mat grows less well, it is often saved from excessive consumption by brine fly larvae because these grazers are subject to predation by another species of fly, one that becomes a conspicuous member of the community where the mat becomes cooler and the brine fly larvae more numerous. This dolichopodid (literally, "long-legged" fly), locally known as a mosquito killer, by eating the eggs and larvae of the brine fly, stabilizes the community by controlling the excessive development of the population of these insect grazers which might otherwise destroy the energy-fixing basis of the system.

ENVIRONMENTAL SCIENCES

In the environmental sciences field, as elsewhere in the Foundation, fiscal year 1970 saw an upswing in the number of research scientists applying to the NSF for support. Also, as elsewhere, the full impact is expected to be felt over the next 2 years. The reason for this delay is the continuing nature of existing grants from other agencies, which will not run out until fiscal year 1971 or 1972.

In geology, where the National Science Foundation has been virtually the sole support of university research, retrenchment elsewhere in the Government was not so deeply felt—most other Federal agencies with an interest in field geology have in-house capability. In other areas, however, requests for support were up 50 percent over the previous year. It should be noted that new requests in all fields were of the same or higher caliber as the previous year.

In spite of handicaps, the fields dealing with the environment have made substantial gains during the past year, and clear lines of possible advance for the future were identified.

In the atmospheric sciences, there has been a definite trend towards working on physical rather than statistical concepts of weather phenomena, particularly through the process of comparing predictions from theoretically derived models to actual results, and using the results of this comparison to improve the model. Meteorologists are investigating methods of monitoring effluents on a national and global basis. The subfield of atmospheric chemistry has been very active in devoting its attention to pollution-related research.

In weather modification, a major program is being launched to learn how to mitigate the destructive hailstorms of the Great Plains. Scientists are also interested in learning more about warm fogs and warm clouds in general. Methods of producing precipitation from cold clouds and dissipating cold fogs are well known, and many techniques are operational and in commercial use. This is not true of those warm clouds where the water content is above the dew point.

The earth sciences are moving ahead rapidly in organic geochemistry, a relatively new field, and in seismology. Many aspects of the earth sciences are rapidly being pulled together by the unifying concepts of global tectonics—the theory of continental drift and seafloor spreading. The resounding successes to date of the Ocean Sediment Coring Program have added to the mounting evidence in support of this theory and also to our knowledge of the geology of dry land.

There is a growing awareness on the part of urban planners of the importance of input from the earth sciences in planning for the most beneficial and efficient use of land for our new cities. It is estimated that by the year 2000, tens of billions of dollars worth of new engineering structures will be built in areas of known earthquake activity. Knowledge of the local geology applied to the siting and engineering of these structures could result in reducing earthquake losses by up to 50 percent.

New experimental techniques are becoming available which permit subjecting materials to high temperatures at pressures—accurate to within 1 percent—of up to 150 kilobars (150 times atmospheric pressure—about 1,000 tons per square inch). These temperatures and pressures are equivalent to those deep in the earth's crust which cause the metamorphosis of minerals to various crystalline rock formations. The technique has been developed for geological experimentation but should have profound impact on research and possible synthesis of new "mineral-like" materials of exceptional properties.

In oceanography, new techniques are coming to the fore such as a systems approach using theoretical models, particularly through the use of on-board computers which simultaneously record and operate on multiple sources of data. Oceanographers see the results of their experience with their traditional integrated team approaches to research producing even better results and affecting the methodology of scientists in a variety of other fields. New techniques for measuring the physical parameters of the ocean are continuously being developed and tested.

POLAR PROGRAMS

Fundamental to the National Science Foundation role in polar activities was the assignment to it in fiscal year 1970 of the responsibility for the development of a national Arctic Research Program and for the coordination of the research activities among the several Federal agencies having an interest in Arctic research. The Office of Antarctic Programs was redesignated the Office of Polar Programs to handle the new responsibility. The second half of the year was devoted to an assessment of current arctic work and the development of a long-range plan. The Arctic Research Program will focus on seven areas: polar pack ice, the delicately balanced tundra ecosystem, perennial ground ice, glaciology, the active polar geomagnetic field, the geological structure underlying the area, and the complex interrelations between man and his activities and the arctic environment.

The unpredictability of antarctic weather was forcefully demonstrated at the opening of the last austral season when unusually heavy deposits of snow were dumped at McMurdo Station, completely deranging plans for the orderly movement of the parties into the field.

The highlight of the season was the discovery in the Transantarctic Mountains of a large number of tetrapod fossils confirming the evidence for the theory of drifting continents and the existence of Gondwanaland, the hypothetical ancestral supercontinent that broke up to form the land masses we know today. Coalsack Bluff, the site of the discovery, is by far the most productive fossil locality in Antarctica discovered so far, and undoubtedly will be the focus of further paleontological investigations.

The oceanographic program suffered two setbacks during the season: the first was the inability, as a result of very heavy sea ice, to recover the current buoys emplaced in the Weddell Sea in 1967–68, the second was the damage suffered by the Argentine icebreaker, *San Martin,* which prevented her from joining the icebreaker, *Glacier,* in the International Weddell Sea Oceanographic Expedition 1970.

At Byrd Station the cable suspending the drill in the deep borehole had to be cut when recovery of the drill bit which stuck in the 1968–69 season proved to be impossible. The entrapped atmosphere of past ages was taken from the upper section of the hole as well as from other drill holes at the station for radio-carbon dating and for assessing changes in atmospheric composition since the Great Ice Age.

In the 1969–70 season, 65 individual field projects were carried out by 193 scientists and technicians representing 47 institutions and Government bureaus. The geographic range of the projects was widespread over West Antarctica. Three U.S. exchange scientists accompanied foreign expenditions while 14 foreign scientists joined the U.S. Antarctic Research Program.

In November 1969 at 75°55′S. 83°55′W. a new U.S. station, Siple, was established for upper atmospheric research particularly on the plasmapause. The station was named after the late Dr. Paul Siple, who first gained fame as the Boy Scout on Admiral Byrd's 1928–30 expedition to Little America, and who devoted most of his scientific career to the Antarctic.

Ionospheric rockets were launched for the first time by the United States at Byrd Station to obtain information on particle bombardment, geomagnetic effects and ionospheric structure. Balloon launchings were also made for the same purpose.

The R/V *Hero* completed its second year of activity in the Antarctic Peninsula and southern South America areas. Logistic support was provided by *Hero* for cooperative projects in biology and geology with Chilean and Argentine scientists during the course of the year. The USNS *Eltanin* continued her circumnavigation of Antarctica with multidisciplinary cruises in the Pacific and Indian Ocean areas.

U.S. Antarctic Research Projects

Antarctic Vertebrate Fossils

One of the most significant scientific events during 1970 was the discovery of fossils of land-dwelling amphibians and reptiles in central Antarctica, in rocks of Triassic age. This fossil deposit was found by David Elliott of the Ohio State University and Edwin Colbert of the Museum of Northern Arizona. The fossil locality is in cross-bedded sandstones of the Beacon Formation at Coalsack Bluff in the central Transantarctic Mountains. Several of the fossil types, particularly the reptilian genus *Lystrosaurus,* are especially characteristic of the lowest Triassic period in South Africa. *Lystrosaurus* and several of the other fossil types, including genera of the amphibian group, Labyrinthodont, are key fossils for this particular period throughout the ancient supercontinent of Gondwanaland. The presence of the same genera of land-dwelling and freshwater-dwelling amphibians and reptiles in various parts of Gondwanaland, and in particular, in Antarctica, which is so widely separated from any other continent by deep ocean basins, would seem to definitely indicate the former existence of Gondwanaland as a single continent, made up of all or major parts of present Antarctica, Africa, South America, India, and Australia. Gondwanaland broke up and the fragments drifted apart subsequent to Triassic time, and some of this continental drift seems still to be underway.

Productivity of Antarctic Waters

The study of the productivity of the Antarctic waters could have profound significance with regard to the world's food supply. It is well known that the food chain in the Antarctic waters is simple and direct. Baleen whales thrive mainly on a shrimp-like organism named krill *(Euphausia superba)*, which also furnishes food for a vast host of animals, including winged marine birds as well as penguins, crabeater seals, squid, and fish. These krill are in turn supported by phytoplankton, free-floating microscopic marine plants. Thus, the productivity of these waters resides primarily in the food-building activities of these minute plants.

In studying food chain relationships in the Antarctic ecosystem, it is imperative to know the amount of carbon fixed annually by the marine phytoplankton. Based on the extensive observations made in the Atlantic and Pacific sectors of the Antarctic during the past 5 years, Sayed El-Sayed calculated the annual production of the Antarctic waters as 3.03×10^9 tons of carbon. This estimate does not take into account the amount of organic production in the pack ice region—a region which fluctuates between 10 million square miles (in late winter) and 1 to 2 million square miles (in late summer). Recent investigations on the productivity of the water in the pack ice regions, using icebreakers, suggest that it is much higher than hitherto suspected. The enormous bloom of phytoplankton off the Filchner Ice Shelf (in the southeast Weddell Sea) encountered during the International Weddell Sea Oceanographic Expedition (February 1968), seems to bear this out. Dr. El-Sayed's primary productivity studies in the Antarctic underscore the striking differences between the productivity of the oceanic (offshore) and neritic (inshore) regions. This had led to the conclusion that the proverbial richness of the Antarctic waters is true only with regard to coastal and inshore regions, but not with regard to the oceanic regions.

ATMOSPHERIC SCIENCES

National Center for Atmospheric Research

The breadth of activity at the National Center for Atmospheric Research (NCAR) allows scientists from diverse disciplines to join in attacking complex atmospheric problems. Interdisciplinary study and combined observational techniques are necessary to deal with atmospheric processes whose dimensions vary widely and escape definition from a single viewpoint. During fiscal year 1970 NCAR concentrated much effort on theoretical research and on the development of measuring systems to deal with atmospheric problems on scales appropriate to their complexity.

NCAR continues to be involved in activities to attract students into the atmospheric sciences and to include visiting scientists in its research activities. Predoctoral and postdoctoral fellowships are offered to scientists from the United States and abroad, and nine NCAR scientists serve as affiliate or adjoint professors in university research and teaching. The Research Aviation and Computing Facilities hold work study programs to teach students practical research skills. Each summer the Advanced Study Program sponsors a colloquium to explore some topic related to atmospheric science—in 1970 the topic was microphysics and dynamics of convective clouds.

NCAR is sponsored by the NSF and operated by the University Corporation for Atmospheric Research (UCAR), a nonprofit consortium of 27 universities which have graduate programs in the atmospheric sciences. The principal laboratory is at Boulder, Colo.

Research on the Atmospheres of the Earth and Sun

Global Modeling. — Numerical simulation of atmospheric motions and weather behavior has progressed steadily at NCAR to include an increasing number of interacting processes. A new method for treating the effects of mountain ranges on large-scale flow patterns has added realism to the diagnosis of high and low pressure areas. Expansion of the model to six vertical layers has allowed closer study of motions in the lower stratosphere, and has improved the simulation of tropospheric motions by depicting, for example, the separation of the westerly and polar jets. The model is used with observational data for studying short-term weather processes, and for exploration of special problems such as the influence of southern hemisphere meteorological data on weather predictability in the northern hemisphere.

Progress in modeling oceanic circulation continued in preparation for developing a combined atmosphere-ocean model applicable to studies of climate change. NCAR ocean models now include the main features of the North and South Pacific ocean flow, the Antarctic circumpolar current, and the effects of ocean bottom topography. During the summer of 1969, NCAR held a symposium on physical oceanography to gain a unified view of recent research trends in a field that has grown increasingly specialized.

Tropical Convection.—The tropics often develop massive downdraft systems which bring cool, dry air to lower levels and cause sharp contrasts in temperature and dewpoint; warm updrafts in turn accelerate the rate of heat and water vapor transfer from sea to air. Studies of tropical disturbances, using satellite and conventional data, showed that the water and energy budgets of updraft and downdraft cycles closely resemble those of a Midwest squall line about 300 kilometers in length; further exploration of these parallels can thus reveal important details about the transfer of energy through deep layers of the tropical atmosphere.

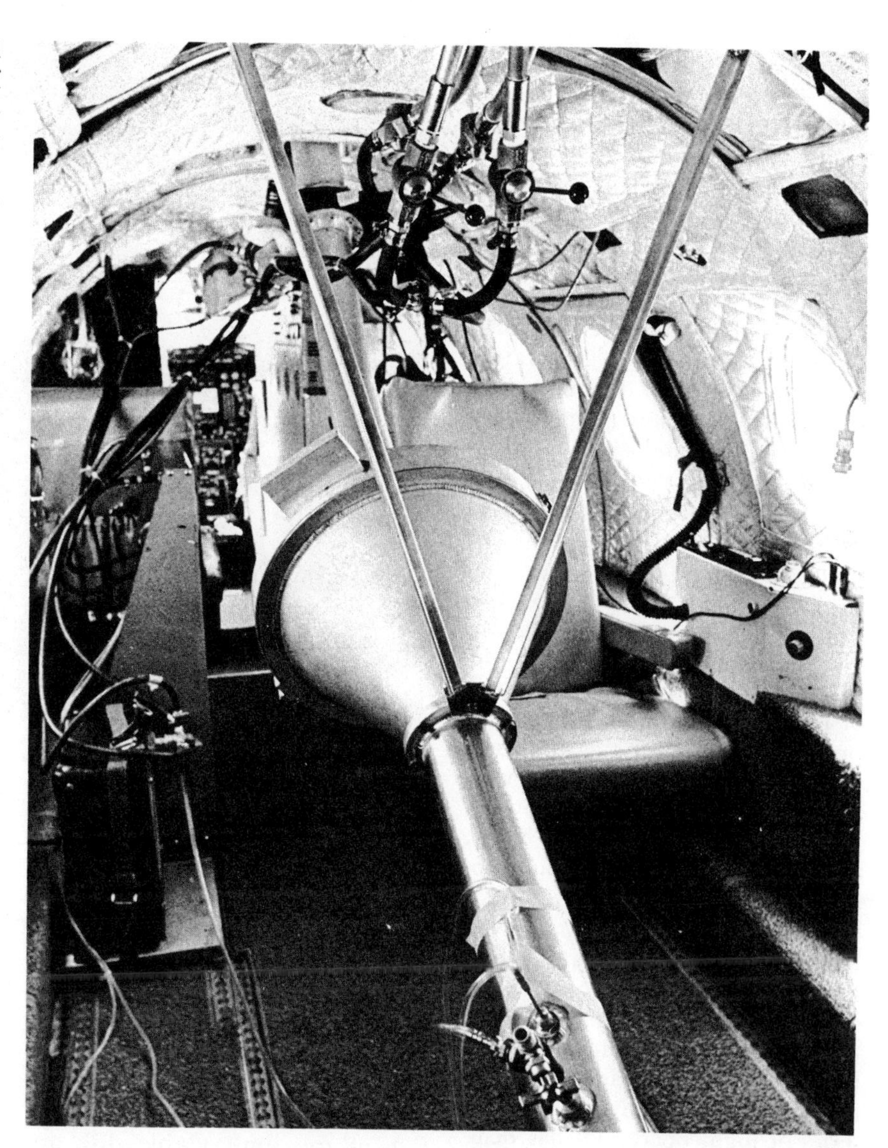

External ducts on NCAR Sabreliner jet aircraft bring air samples to the filter and impaction collectors located inside the cabin. Measurement and analysis of particulate matter in the atmosphere is necessary to characterize the global distribution of natural trace constituents and manmade pollutants. (NCAR photo)

Turbulence.—Accurate measurement of turbulent air motions whose wavelengths are larger than 1 or 2 kilometers was for the first time made possible by an airborne system developed jointly by NCAR and the University of Nevada Desert Research Institute. Installed on NCAR's Buffalo aircraft, the system can measure air velocity to an accuracy of 10 centimeters per second for the duration of a flight. External sensors measure angles between the airstream and the aircraft, true airspeed, and temperature; an inertially stabilized reference platform is coupled to the sensors and continuously measures the velocity and orientation of the aircraft. The system has already been used in two research programs.

Theoretical studies of turbulence have shown that in some important respects the large-scale atmosphere behaves like a two-dimensional rather than a three-dimensional fluid. Two-dimensional turbulence theory therefore allows simpler but no less rigorous exploration of many of the atmosphere's characteristics. One of its important applications has been the investigation of how errors grow in predictions of the flow field, a topic crucial to assessing the general limits on long-term atmospheric predictability.

Atmospheric Chemistry.—Ground-based laser equipment

and aircraft sampling devices were used to study the stratospheric sulfate layer which lies over most of the earth at an altitude of about 18 kilometers. Some investigators have questioned whether the layer is predominantly sulfate, and various sources have been suggested to account for particulate accumulation at such a high altitude. Airborne sampling, carried out in cooperation with the USAF Air Weather Service, verified that most particles consist of sulfate, and that most are formed in the stratosphere by oxidation of sulfur dioxide gas from manmade sources and volcanoes.

In addition to collecting and analyzing air samples over several continental and maritime regions, NCAR chemists completed a detailed study of the trace chemistry of moist tropical air in Panama. A primary objective of the study was to determine the amounts and variability of constituents in the atmospheric nitrogen and sulfur cycles, and their correlation with meteorological conditions.

Hole in this supercooled stratus deck is the result of the application of dry ice, followed by precipitation from the seeded area in the form of snow. (NSF photo)

Wave Cloud Experiment. — NCAR studies of the processes that lead to precipitation from convective clouds require broad application of theoretical, field, and laboratory research, including investigation of nonconvective clouds. A series of flight operations has demonstrated that mountain wave (lenticular) clouds can serve as steady-state cloud "laboratories" for a variety of aerosol and cloud physics experiments. Lenticular clouds form at the peaks of large waves that frequently develop when strong winds blow across mountain barriers. These clouds remain at relatively stable positions but lose moisture on their downwind sides and are replenished on their upwind sides. Air operations have shown that prolonged flights in and around these isolated clouds are feasible. Release of chemical vapors from aircraft on the upwind side was found to retard the rate of droplet evaporation on the downward side. This technique provides a tracer for investigating cloud droplet migration; it has also given support to the theory that surface impurities may affect droplet lifetimes.

Granules.—Granules on the surface of the sun's visible disk are frequently thought to represent convection cells flowing upward at their bright centers and horizontally outward away from their centers. To determine the average flow pattern within a granule, NCAR investigators devised a new technique of "velocity-grams" averaged for 1,100 observed granules which yielded a picture of the flow pattern accurate within 20 meters per second, and allowed separation of the vertical and horizontal motions of material. Upflow had been established observationally before, but the existence of outflow has now been clearly established for the first time. Maximum upflow velocity was found to be 0.5 kilometer per second, and maximum outflow velocity 0.3 kilometer per second.

Facilities Operations

The Scientific Balloon Facility conducted theoretical and experimental studies of balloon design, flight dynamics, and inflation and deployment systems. Eighty large superpressure balloons were launched for university and Government scientists in support of astronomy and physics programs.

The Computing Facility published the first of four atlas volumes on the climatology of the southern hemisphere, and added to its computer-assembled set of microfilm analyses and grids. A 26-minute computer-generated film derived from the atlas was made available for purchase or loan.

The Field Observing Facility supported numerous field programs during fiscal year 1970; the largest of these were the High Altitude Observatory eclipse expedition to Mexico in March 1970, and Colorado State University's VIMHEX field operation in Venezuela from May to October 1969. Both involved logistics support and field management.

Extensive testing by staff of the Global Atmospheric Measurement Program (GAMP) at its Christchurch, New Zealand, flight station showed that constant-level balloons are capable of flying for longer than 6 months in the stratosphere above 100 millibars (53,000 feet) to measure large-scale circulation patterns. Nine constant-level balloons were flown from Ascension Island in the equatorial Atlantic in preparation for testing a balloon-satellite location system.

This mobile pulsed-Doppler radar laboratory was developed with National Science Foundation support at the University of Arizona for use in cloud-physics studies. Development and construction of the equipment was directed by John B. Theiss, in doorway, senior engineer of the university engineering experiment station. (NSF photo)

WEATHER MODIFICATION

Unintentional Modification of the Weather

The effect of man's activities upon the weather is receiving increasing attention following the observations that rainfall patterns appear to have changed in the wake of large urban and industrial development. The University of Illinois has started a search of climatological records of eight urban and industrial areas to determine whether changes similar to those discovered at La Porte, Ind., and St. Louis, Mo., have occurred. The University of Washington has conducted a series of aircraft measurements using nuclei counters, and has determined that paper mills and other industrial plants are prolific sources of cloud condensation nuclei. Clouds are often observed to form downwind of these industrial sources and particles large enough to fall as rain appear to form readily in them. A comparison of precipitation and stream flow records in the State of Washington for the period 1929–46 with those for 1947–66 has shown that areas in the vicinity of these large industrial sources of cloud condensation nuclei have experienced a mean annual precipitation during the second period which is 33 percent greater than that during the first period. However, the inference that the precipitation increase is a direct consequence of industrial development requires further study.

Hail

Although hail can occur anywhere, it achieves its most dramatic—and most destructive—form over the great farming plateau of the Great Plains. During the summer months, hot air formation over the mountains in the late afternoon frequently combines with abundant moisture at high altitudes to produce the towering white convective clouds commonly called thunderheads. From these, tons of hailstones can come slashing down to rip leaves off growing corn in Nebraska, and to beat the heads off harvest-ready grain in the Dakotas. One hailstorm in Rapid City, S. Dak., on July 10, 1969, caused over $2 million in property damage alone.

During 1970, the Foundation established the National Hail Research Experiment in northeastern Colorado under the field management of the National Center for Atmospheric Research (NCAR). This experiment is designed to provide a coordinated and intensive study of hailstorms which occur over the Great Plains of Colorado in order to determine how the hail-forming mechanisms of severe storms may be modified to reduce hail damage on the ground. The field phase of the experiment will begin in the summer of 1972 with the cooperative participation of the Departments of Commerce, Agriculture, Interior, Defense, and Transportation, the National Aeronautics and Space Administration, and the Atomic Energy Commission. Support has been provided to the South Dakota School of Mines and Technology, the University of Wyoming, and Colorado State University to provide specialized aircraft measurements of the convective cloud systems, and a unique dual wave radar—which can differentiate between water and ice—is being developed jointly by the University of Illinois for detecting the presence of hail and estimating the liquid water content of the storm. Measurements on hailstorms in the vicinity of Rapid City, S. Dak., by the South Dakota School of Mines have resulted in a mathematical model of a typical Great Plains hailstorm 15 miles wide, and 10 miles high, which ingests approximately 5 million tons of water vapor per hour from the surrounding atmosphere. The updrafts at the center of these large hail-bearing clouds have been estimated to reach velocities between 30 to 100 miles per hour. The National Hail Research Experiment will consult such mathematical models to estimate the proper time and place for the injection of silver iodide into the storm to limit the growth to harmless size hailstones. The critical areas of the storm will be seeded by rockets fired by jet aircraft which will be accurately positioned by ground and airborne radar. Rocket delivery systems are presently being developed and tested by the Colorado State University and NCAR.

Public Acceptance

Regardless of the state of readiness of technology for modifying the weather, its eventual employment for social and economic benefit must consider public opinion. The Foundation has requested the University of Colorado to find out what rural Americans think about plans to conduct weather modification experiments to produce more rain or snow in the areas where they live. Sociological researchers from the University of Colorado have carried out studies of this and related questions in such areas of the United States as western New York, Montana, and Utah. Most citizens believed that, in general, scientific experimentation was beneficial to mankind. This view seemed to carry over to the consideration of weather modification experiments. By the end of the experiment, only 9 percent of these rural residents were opposed to local weather modification experiments.

Global Atmospheric Research Program

The Global Atmospheric Research Program (GARP) is an international program designed to study the transient behavior of the atmosphere and the factors that determine the statistical properties of the atmosphere's general circulation. Oceanographic and Meteorological Experiment (BOMEX) was carried out from May 1-July 31, 1969, in the Atlantic Ocean east of Barbados. BOMEX was a field experiment designed to explore the interactions at the air-sea interface and above, which govern the transfer of momentum, heat, and water vapor between the tropical ocean and the atmosphere. In 1970 the total Foundation support for GARP from research funds was approximately $1.5 million. These funds supported studies on several aspects of GARP, including analysis and interpretation of data observed during BOMEX.

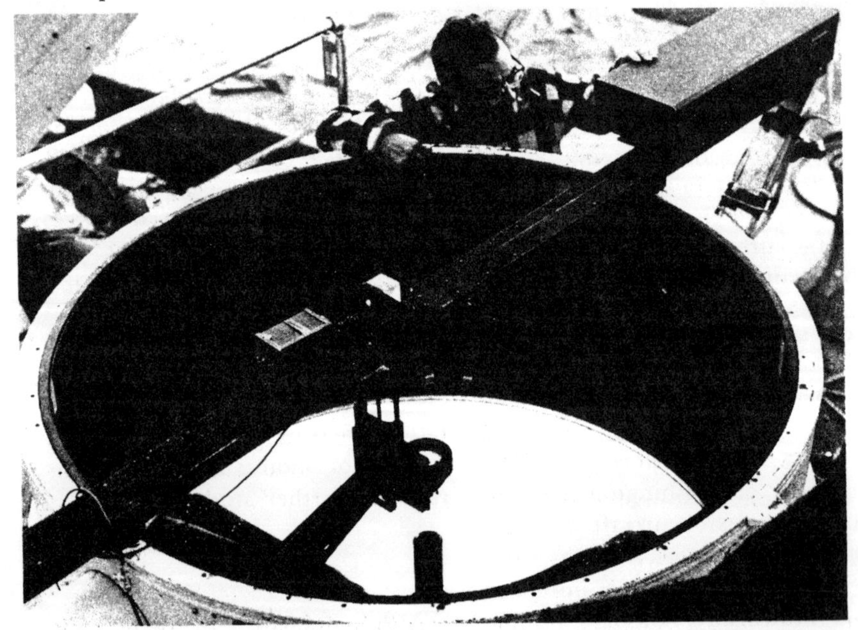

This laser device is used by the National Center for Atmospheric Research in the study of atmospheric aerosols. The laser head (source of light) and a prism rest on a bar over a large searchlight mirror. Laser impulses, directed skyward by the prism, are reflected by layers of atmospheric aerosols, and returning impulses are collected by the mirror and measured electronically. Time in transit indicates the height of the aerosol layers. (NCAR photo)

Analysis of BOMEX data has proceeded throughout the past year culminating in scientific reports that have been given by a number of investigators at two symposia devoted to BOMEX results, one at the University of Washington, Seattle, one at the American Geophysical Union meeting in Washington, D.C. Several other meetings of specialists have also been held.

GARP has as one of its operational goals the improvement of long-range weather forecasting through numerical prediction methods that use inputs and data from field experiments such as BOMEX. One of the less complex models of the general circulation of the atmosphere was that developed by Yale Mintz and Akio Arakawa of the University of California at Los Angeles. Their model has been used by other scientists such as Robert Jastrow of the Goddard Institute for Space Sciences in New York City and Jules Charney of the Massachusetts Institute of Technology to investigate effects on the dynamics when simple parameters are changed.

In March 1970 a planning conference on GARP was held in Brussels. It was the consensus of that conference that an experiment to investigate the atmospheric energy cycle in the tropical atmosphere and, in particular the convection in cloud clusters, should be planned in the eastern Atlantic for the fall of 1974.

At least eight nations will take part using up to 24 ships, aircraft, balloons, buoys, and satellites as measurement platforms.

Meteorology

Weather Predictions

A major goal of meteorological research is to improve short-term prediction of local or regional weather. This involves not only the gathering and presentation of data from which forecasts can be made but also the analysis and interpretation of the data.

Generally nowadays, the structure of the atmosphere is represented in pressure coordinates and the analysis of its motions are conducted in these terms. At the University of Wisconsin, Donald Johnson and Frank Sechrist are reviving the idea of representing the atmosphere in terms of its entropy, a thermodynamic variable related to the energy content of a weather system, so as to understand better how storm systems form, develop, and decay.

From an analytic point of view this results in a more vivid representation, in three dimensions, of the evolution of the atmospheric structure. From a diagnostic point of view the result is to reveal simpler relationships between the variables which govern the dynamical processes. This technique is well suited to the type of data provided by satellites, as well as the formulation of equations and relationships which describe the transport of mass, momentum, angular momentum, and energy in the atmosphere.

The University of Wisconsin team has demonstrated the advantages of this technique by a case study of the role of the strong upper-level wind known as the polar jet stream in triggering a line of severe storms—a squall line —and in the development of a cyclone in April 1968. While satellite pictures indicated the proximity of the polar jet to the severe weather, it was through the analysis of equations cast in terms of entropy that the conditions for squall line formation were clarified and the role of the polar jet was established. Such a role was never noted using the conventional pressure coordinate system for analysis because the interactions of the dynamic components of the weather system were then not so clearly delineated.

How much data does a meteorologist need to make a forecast? The economically important answer to this question depends in part on the physical conditions of the atmosphere, the skill of the meteorologist, and possibly by the way in which the data are handled. The latter variable was the object of a study at the University of Michigan undertaken by Edward Epstein, Allan Murphy, and Glenn Trapp in collaboration with staff members of the Department of Psychology.

Two alternative forecasting systems were tested on experienced weather forecasters who were presented with sequences of weather information taken from historical records. In one system known as POP (Posterior Odds Processing) which represents the forecasting system in use at present, a forecaster interprets the data directly, in the customary manner, and forms a subjective judgment, for example, about the probability of precipitation. In the other system called PIP (Probabilistic Information Processing) the forecaster interprets the data by assigning levels of significance or diagnostic impact to the information he has, and then the computer is used to make an objective forecast using mathematical decision theory. The hypothesis being tested by comparison of the results of applying both systems is that forecasters working intuitively without the aid of computer products tend to be conservative and require more data before making judgments.

The preliminary results of a pilot experiment indicate that forecasters using the PIP system may indeed require fewer data, but may not necessarily produce more accurate forecasts. A more definitive statement about the PIP and POP systems must await the completion of the analyses of the results of the experiment as well as the results of experiments conducted at the Detroit Metropolitan Airport in a truly operational setting. Studies such as that being conducted at the University of Michigan emphasize the broad role meteorological research plays in increasing our understanding, measurement, and prediction capability of atmospheric processes which affect our whole environment.

Aeronomy

Earthquakes and the Ionosphere

Seismic waves produced by earthquakes on the ground can produce measurable motions of the ionosphere, and this fact is being exploited for practical purposes by Paul Yuen and his colleagues at the University of Hawaii. This unusual blend of normally independent geophysical phenomena in aeronomy and seismology promises to be useful in the early identification of and warning on ocean-borne tsunami waves, which can bring catastrophe to shoreline victims thousands of miles away from the earthquake source.

Using a very sensitive radio-sounding technique for continuously monitoring changes of height of the E and F layers (the two principal layers of the ionosphere containing free electrons) by the Doppler shift of reflected frequencies, Dr. Yuen has a constantly available measure of motions at heights as great as 300 kilometers. A Doppler shift is the change in the frequency of received waves caused by a changing path length from the source of the observer; an example is the apparent change in pitch of a railway horn as it passes an observer at a crossing. Although the ionospheric measurements primarily give detailed information on atmospheric motions relating to charged particles, they also reveal the rise and fall caused by the pumping action of an earthquake displacement. Vertical air motion at ground level is translated upward into much larger motion at the lower pressures at high altitude, giving characteristic Doppler shifts at certain frequencies.

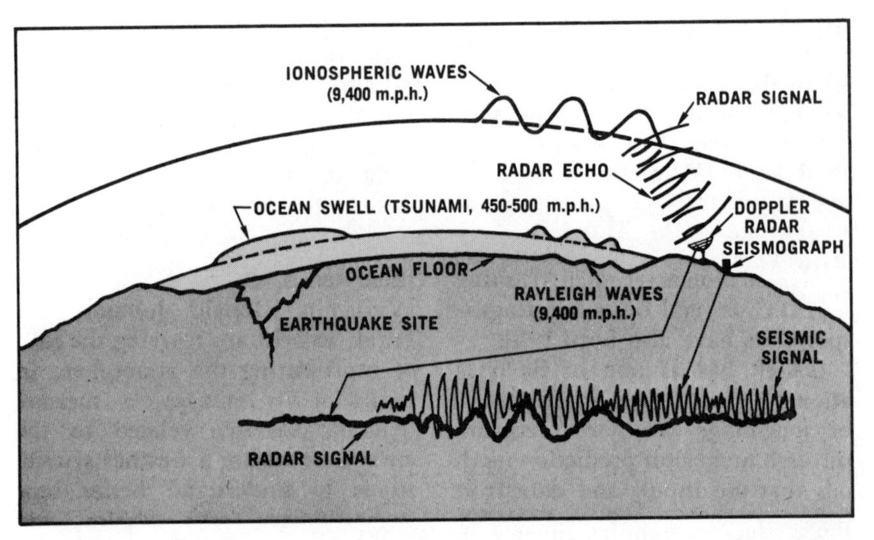

Rayleigh waves generated by submarine earthquake propagate along sea floor and are transmitted acoustically to sea surface and ionosphere. Rate of travel of these waves is around 9,400 m.p.h., while tsunami travels at about 450–500 m.p.h. Doppler radar signals, damped by atmosphere, show only major motions, in comparison to more complex seismic signal. (Tsunami damage occurs when wave builds up on reaching land.)

As the seismic Rayleigh wave—one of the two principal types of earthquake waves which travel on the surface of the earth—travels outward along the earth's surface at speeds between 3 and 4 kilometers per second, the resulting ionospheric displacement travels similarly at ionospheric heights. Because of the acoustical filtering of the atmosphere, the recorded Doppler wave loses most of the confusing short period components present in seismograms, and the identification is not only quicker, but simpler.

Both waves travel many times faster than the destructive tsunamis and can therefore provide early warning. Ionospheric waves have been observed in Hawaii from the Kurile Island earthquake of August 1969, the Japanese earthquake of May 1968, and the Alaska earthquake of March 1964. These aeronomic research techniques are being developed as a multifaceted tool for obtaining basic information on the coupling of atmospheric motions, learning the nature of deep-ocean Rayleigh waves, and providing a working new method for early warning of tsunamis.

Solar-Terrestrial Research

The Plasmapause

The space between the earth and the sun is not a vacuum but a highly tenuous plasma. A plasma, sometimes called the "fourth state of matter," is a gas consisting of ionized particles and electrons. The importance of plasma has received recent attention because of its role in future controlled thermonuclear fusion reactions as a source of electric power.

In the early 1960's, Donald Carpenter at Stanford University, using ground-based measurements, and K. I. Gringauz of the U.S.S.R., using rocket-borne probes, discovered that the density of plasma in the atmosphere which immediately surrounds the earth decreases abruptly beyond a border region called the plasmapause, which surrounds the earth as a shell several times the earth's diameter. This boundary fluctuates because it is sensitive to the solar wind—a stream of energized particles, itself a plasma, flowing from the sun—and it transmits energy to the earth in the form of electric currents, particles, and fields.

The plasma within this boundary has been investigated through study of whistler waves—radio signals produced by lightning near the earth's surface which arch out into space along the lines of the earth's magnetic field and are modified by the medium through which they pass. The results of these studies and satellite measurements sampling this environment show that the plasmasphere has a relatively large bulge in the late afternoon hours. The earth's magnetosphere—that portion of space influenced by the earth's magnetic field has an elongated tail which always points away from the sun and is therefore over the "night" side of the earth. The afternoon bulge in the plasmapause location is believed to arise from large-scale motion of the plasma within the boundary of the magnetosphere which is referred to as "magnetospheric convection."

The plasmapause and magnetospheric convection.—In the outermost regions of the earth's magnetosphere the solar wind flowing away from the sun is the dominant influence. It drags the plasma very near the magnetosphere boundary away from the sun, and this in turn establishes a return flow toward the sun within the magnetosphere. On the afternoon side of the earth, this return flow opposes the clockwise flow of plasma which is produced by the rotation of the earth. The interaction between these opposing flows produces a large "backwater" or eddy which is observed as the

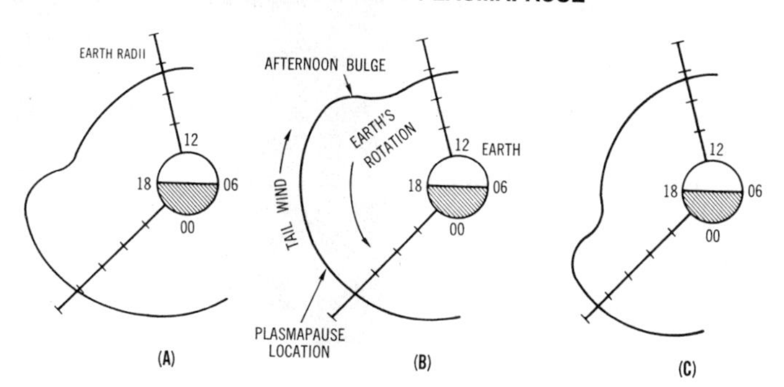

The afternoon bulge in plasmapause location arises from the opposing counterclockwise rotation of the earth and the "tail wind" blowing from the night side of the earth. During steady conditions, the bulge is seen at about 6 p.m. (18 hours) (a). If the tail wind increases in intensity, the bulge is blown forward into the early afternoon hours (b). A decrease in tail wind allows the earth's rotation to sweep the bulge into the night hours (c).

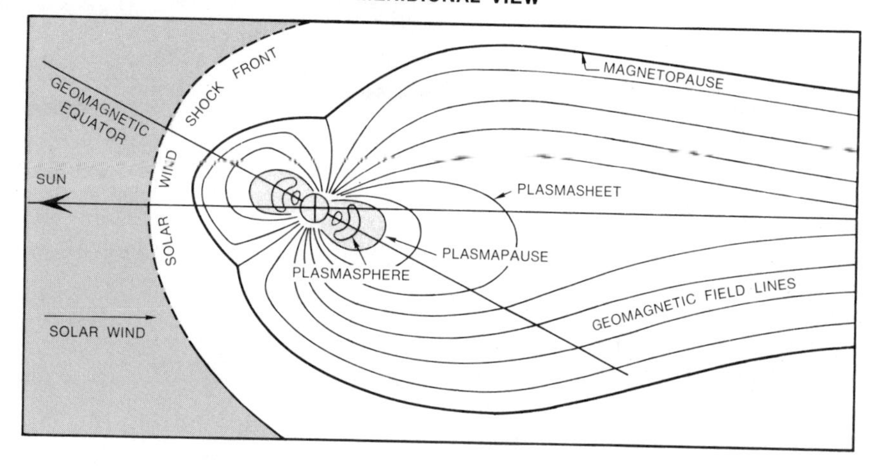

relatively large bulge in the location of the plasmapause during the late afternoon and early evening hours. The above model, proposed by Neil Brice at Cornell University in 1967 has now been substantially verified. However, recent measurements indicate that the convective plasma flow, like the solar wind, is far from steady, but tends to occur in gusts. The whistler measurements referred to above have shown that when the "tail wind" from the night side increases in intensity, it blows the afternoon bulge in the plasmapause forward into the early afternoon hours, while a sudden decrease in the "tail wind" allows the earth's rotation to sweep the bulge into the night time hours.

Sudden increases in the "tail wind" are associated with "magnetospheric substorms." This term was introduced by Dr. Brice and

Kinsey Anderson and co-workers at the University of California at Berkeley to emphasize their conclusion that increases in geophysical disturbance activity occurred simultaneously throughout the entire magnetosphere. Magnetospheric substorms occur when energy that has accumulated in the magnetospheric tail is released abruptly, producing widespread aurora over the polar regions and causing disruption of some radio communication circuits.

Some of the substorm energy goes to increasing the intensity of the Van Allen radiation belts and to producing a "ring current" of lower energy charged particles moving around the earth.

New results on the dynamic behavior of the plasmapause.—When the earth's environment is subjected to a magnetic disturbance, the outer part of the plasmasphere is initially convected away into space, leaving a smaller core during times of magnetic storms. During the much longer recovery period, the region outside the reduced storm-time plasmapause radius is replenished from the ionosphere below. Many days may be required to refill the tenuous outer region back to its normal extent, and thus in intermediate periods the plasmapause can be difficult to detect. Processes which depend sensitively on local electron density may begin to occur in the latter phases of magnetic storms while the electron density rises and the storm-time boundary is gradual and irregular.

These and other insights into the structural and dynamic behavior of the plasma surrounding earth increase the value of this "laboratory" to those interested in the generalized behavior of plasmas for eventual application, and provide deeper knowledge of the manner in which energy is transferred from the sun to the earth.

X-ray emitting areas on the sun photographed outside the March 7, 1970, eclipse shadow from a NASA Aerobee rocket are shown superimposed on a picture of the white light corona. The relation of the out-flowing coronal streamers to the active X-ray regions on the solar surface is evident. (Photo courtesy of American Science and Engineering Inc., and the High Altitude Laboratory)

Coordination of the 1970 Solar Eclipse

The total eclipse of the sun on March 7, 1970, was an unprecedented success in regard to wide scientific participation and the near-perfect weather conditions prevailing over most of the path which crossed Mexico and most of the eastern seaboard including Canada. The Federal Council for Science and Technology, in setting plans for these studies, asked the National Science Foundation to coordinate Federal activities relating to this eclipse.

The coordination was accomplished through conferences, negotiations, publications, and other exchanges of information. The cooperation which developed allowed the Foundation to serve not only Federal activities but also to a considerable extent the serious participation by academic, amateur, and foreign groups. The final *Eclipse Bulletin* reports over 200 projects representing 17 nations.

Research techniques used by U.S. scientists included ground-based optical, electronic, and acoustic equipment of many kinds, two instrumented jet planes, 12 gun-launched probes, nearly 70 rockets, and the ESSA Applications Technology Satellite ATS–3. As soon as 1 day afterwards, reports from Mexico to Canada made it clear that the coordinated exploitation of the 1970 eclipse was a success. The quick and total removal of solar energy input not only had revealed new features of the solar corona reaching out to extreme distances from the sun, but had also shown solar influences on marine organisms, on constituent gases of the atmosphere, on temperatures, winds, and clouds, on airglow emission, and on the electrically charged layers of the atmosphere which allow radio communication.

Five teams of investigators from NCAR's High Altitude Observatory (HAO) obtained excellent observations from a mountain field site in southern Mexico. HAO scientists spent over a year preparing the special instruments for experiments which examined the fine structure of solar prominences and spicules and the mechanisms by which they condense from the solar corona; the magnitude and direction of coronal magnetic fields and their role in determining coronal structure, temperature distributions, transient wave phenomena, polarization of major emission lines, and excitation of ions in the corona; and the infrared spectrum of the T-corona, composed of small particles of interplanetary dust.

Scientific publications and international conferences over the following year will present detailed results, and planning will commence for exploiting the few minutes of eclipse totality over central Africa on June 30, 1973.

OCEANOGRAPHY

Ocean Sediment Coring Program

Activities under the Ocean Sediment Coring Program during the past year consisted of a Deep Sea Drilling Project being conducted by Scripps Institution of Oceanography from the Drilling Vessel *Glomar Challenger*. The purpose of the project is to explore the floors of the deep ocean basins by means of coring through the sedimentary layer. Following publication of the first volumes of the *Initial Reports of the Deep Sea Drilling Project*, distribution of samples of the core material was started, making them widely available to scientists for pursuit of individual research projects.

The initial 18-month term of the drilling project was completed in February 1970. It had generally been acclaimed as an exemplary scientific and technological success. The initial term comprised several traverses across the Atlantic and Pacific Oceans and adjacent seas. The term of drilling was extended to provide an additional 30 months' work, and the activities continued then without interruption. Plans for the extended project include a broader geographic range (including the Indian Ocean and Mediterranean Sea) and a closer investigation of continental margins in the Atlantic and Pacific.

The re-entry cone is hoisted over the side of D/V *Glomar Challenger* into the Atlantic Ocean during Deep Sea Drilling Project re-entry trials. The cone is 16 feet in diameter and 14 feet tall. The three sonar reflectors are visible. After the cone was submerged to a depth of 22 feet, it was keel-hauled to the opening in the middle of the vessel, directly beneath the drilling derrick, and sent to the bottom on the drill string. (Photo Scripps Institution of Oceanography)

Diagram shows how re-entry sytsem was successfully tested in 10,000 feet of water on 14 June, 1970. A transducer is lowered on a conductor cable through the core-hole in the bit, and the transducer scans 360 degrees in search of the cone. Reflections are displayed on a scope mounted on the ship's bridge. The bit can then be maneuvered over the cone by moving the vessel and by "jetting" or pumping water down the pipe and expelling it through a small hole just above the bit. (Photo Scripps Institution of Oceanography)

At the completion of the initial 18-month program, the vessel had drilled 149 holes at 84 sites, drilled a total of 87,919 feet below the sea floor and recovered 21,983 feet of sediment cores. She had drilled in 20,146 feet of water, suspended a drill string of 20,760 feet, and had penetrated to 3,231 feet below the ocean floor. Subsequently, a new penetration record of 3,320 feet has been established.

The latter part of the 18-month program was concerned with drilling in the Pacific Ocean where 140-million-year-old sediments were recovered. Further evidence for continental drift or seafloor spreading was acquired in the Pacific by dating of ocean floor sediments.

During the first 4 months of the program extension, the Gulf of Mexico and western North Atlantic basins were drilled. In the Gulf of Mexico it was established that the water has been deep for the past 100 million years, suggesting that if subsidence from a shallow saline basin occurred it was prior to the late Cretaceous, the period during which the Rocky Mountains were formed. In the western North Atlantic, the oldest sediment recovered from the deep ocean, middle Jurassic limestones (160 million years old), overlies basalt, which is probably basement. The limestones appear to have been deposited in increasing deep water through time, suggesting that the early Atlantic was a shallow-water sea. Minerals such as native copper, zinc sulfide, and iron carbonate were recovered.

Coring in the young, soft sediment has proven to be exceptionally rewarding, but within the older sediments, penetration and recovery have been thwarted by the presence of widespread hard chert (flint) layers. Consequently, our knowledge of the early history of the ocean basins still remains fragmentary. A system to replace worn-out drill bits was needed. A review indicated

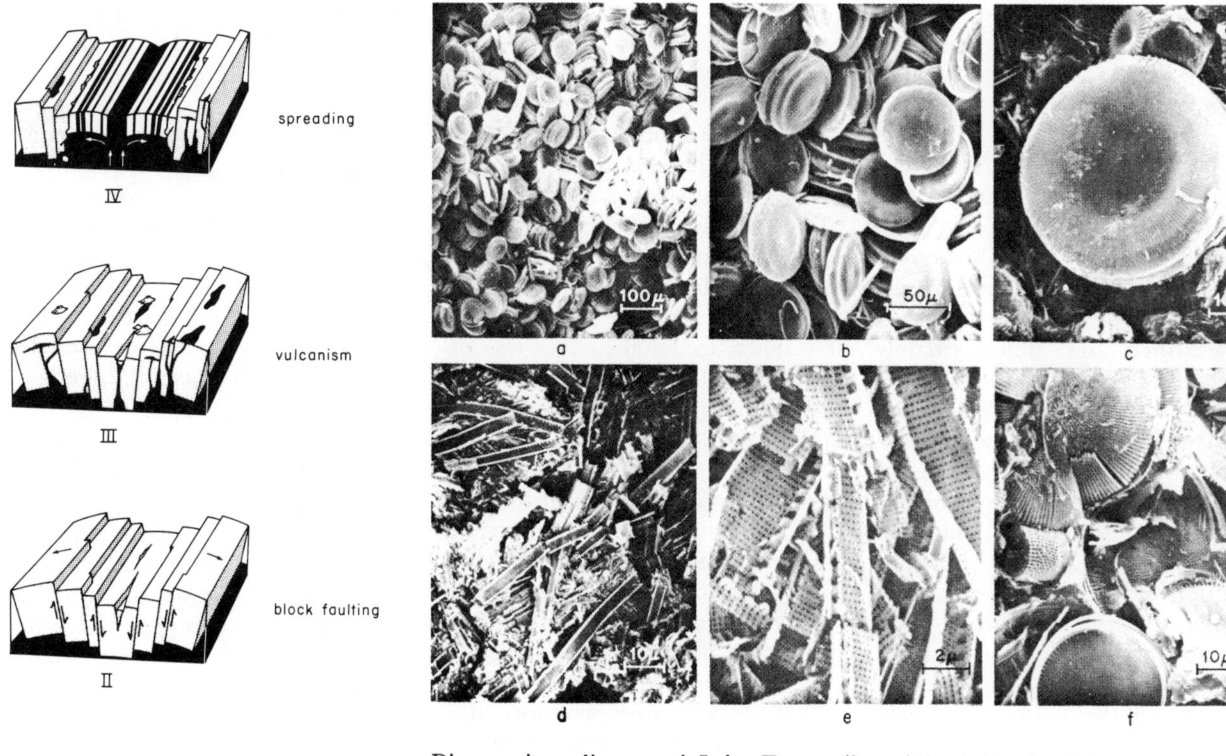

Stages in evolution of an oceanic rift. (Photo Woods Hole Oceanographic Institution)

Diatoms in sediments of Lake Tanganyika. (Photo Woods Hole Oceanographic Institution)

that a hole re-entry capability would best meet this need. The necessary technological development was accomplished, and resulted in a successful re-entry test in 10,000 feet of water, 180 miles southeast of New York on June 14, 1970.

Oceanography on Lake Tanganyika

The mid-ocean ridges are long scars of global proportions that run down the center of the sea basins and mark the zone where the seafloor is spreading apart, widening the oceans, and separating the continents. The Red Sea and the Gulf of Aden basins are a rift that started to open about 25 million years ago and a new ocean basin may be in the making. Some American geophysicists believe that this great split continues into Africa, forming the deep East African lake systems of which Lake Tanganyika is the most outstanding one.

Building on the experience and results of the 1966 and 1969 expeditions to the Red and Black Seas by Woods Hole Oceanographic Institution, a joint cruise on Lake Tanganyika was undertaken in April 1970 by the U.S. Navy Underwater Sound Laboratory at New London, and three scientists from Woods Hole, E. T. Degens, R. von Herzen, and H. K. Wong. The objectives were twofold. The first objective was to determine the structure of the lakes. Are they simple grabens or are they like an oceanic rift which typically goes through three developmental stages—uplift, block faulting, and volcanic and hydrothermal activity? Since the lakes of East Africa are likely a southern extension of the Red Sea and the Gulf of Aden, which show spreading of the seafloor, the geologic structure of the lakes has broad regional implications.

The second objective of the cruise was to ascertain the sedimentation history of Lake Tanganyika as it relates to the paleoclimatology and ecology of the lake.

The work was accomplished successfully despite the acute logistics problems attendant to working in remote areas with antiquated ships. Seismic profiles gave clear evidence that the topography of the lake bottom is strictly controlled with graben-type basins at the north and south ends of the lake—separated by an uplifted block in the middle which is also structurally produced. Magnetic surveys revealed no magnetic lineations which are typical of rifting. The observed magnetic pattern is structurally controlled and gives no evidence for active seafloor spreading in the past or the present.

Sediment fill in the lake is very massive and may reach thicknesses of several kilometers. The sediments are stratified and almost entirely composed of organic matter and the skeletal remains of diatoms, which are a type of algae. Because the sediments are almost entirely biological in origin, changes in the fossil inventory are undoubtedly linked to the chemical and, in turn, biological evolution of the lake. By means of scanning electron microscopy, it is hoped to relate changes in the type of fossil organisms to changes in the chemistry of the sediments which, in turn, will provide evidence on the paleoclimatology of the area.

Deep-Water Formation in the Mediterranean

Improved instrumentation and research from submersible vehicles have increased our knowledge of life on the very deep sea floors. Since these organisms need oxygen to maintain their life, from where, how, and when the dissolved oxygen reaches the deep ocean is an important question for oceanographers. Since future undersea human habitats may need to utilize the dissolved oxygen in sea water for life support, the dynamics of dissolved oxygen may become of direct concern to human welfare.

The source of all oxygen in the sea including the deep-sea oxygen is the surface region of the ocean where oxygen produced by phytoplankton—free floating microscopic marine plants—and atmospheric oxygen are dissolved in the water. This oxygen-laden surface water then either sinks down as a body or mixes with deep water, a process often triggered by the onset of winter. The cooling of seawater increases its density, causing it to sink.

One of the most exciting oceanographic observations recently carried out is the direct observation of the beginning of the formation of oxygenated deep water in the Mediterranean during later winter. A sinking rate of about 200 meters per day down to 1,400 meters was observed by Henry Stommel of Massachusetts Institute of Technology and his coworkers.

Aboard R/V *Atlantis II* of Woods Hole Oceanographic Institution, the U.S. scientists made a hydrographic survey south of the Gulf of Lyons late in January 1969. During the survey period, the weather was quite calm and the surface mixing zone, where atmospheric oxygen penetrates with ease, extended to a depth of only 200 meters. On February 3, 1969, the *Mistral,* the cold, dry, winter northerly wind of southern France, began to blow. After 7 days of strong wind, the oceanographers observed that the surface mixing layer extended to a depth of 1,400 meters whereas 1 week earlier it was less than 200 meters deep.

Oceanographic data from British R/V *Discovery* and French R/V *Charcot* also showed a deep mixed layer at about the same latitude but at different longitudes. Therefore, it appears that intense vertical mixing occurs in a narrow band 10 to 20 miles wide in north-south extent, and somewhat larger in the east-west direction. By the end of February, this deep mixed layer had been driven down to within 100 meters of the bottom, but with moderating winds in March it was quickly sealed off at the surface by a thin layer of fresher water which overlays the surface everywhere except at the small region of deep mixing. It will be an interesting theoretical problem to explain the smallness of the region in which vertical mixing is allowed, since the winds blew strongly over a much larger region.

The fact that the mixing of oxygen-rich surface waters with deep water can occur as dramatically as it does in this case provides oceanographers with valuable new knowledge of ocean dynamics. Furthermore, areas in which this mixing occurs regularly could prove rich in ocean life, providing a valuable resource for commercial fisheries.

Specialized Oceanographic Research Facilities

The National Science Foundation continues to be the major funding agency for the operation of the U.S. academic fleet of 32 ships operated by 18 academic institutions. In fiscal year 1970, NSF funds committed to the fleet's operation totaled $7.4 million. In the same period, the oceanographic facilities allotment was $0.2 million, one-half of which was used for the construction of the shore facilities for the University of the Pacific Marine Laboratory and the remaining sum for the purchase of several minor shipboard research facilities for other institutions.

EARTH SCIENCES

Earthquake Hazards

The recent flurry of predictions of a major earthquake and landsliding in California into the Pacific Ocean has generated widespread interest, and even alarm, at the possibility of geologic catastrophies in this country. Although these predictions were without direct scientific basis and major disasters failed to materialize, the dangers of earthquake hazards are attracting increasing public attention. That earthquakes of relatively low magnitude may produce major damage, or even collapse of modern structures, points up the urgent need for a further detailed study of earthquakes, their cause and effects.

In order to understand why and how earthquakes occur it is necessary to study the mechanics of what happens at the original focus, or source, of the earthquake. Previous theories on the origins of earthquakes have relied heavily on the "elastic rebound" theory, which considered that the shock at the source of an earthquake came from the two sides of a fault-line rebounding as the strain between them was abruptly released. Recent studies have shown, however, that the theory does not account for all the energy released by an earthquake.

James Brune, Clarence Allen, and associates at the California Institute of Technology have been conducting research on movements along active fault zones in the California-Nevada region. They noted, for example, that along the San Andreas fault system, rupture of the earth's surface is associated with earthquakes too weak to produce vibration damage and too small to be felt except over very small areas. These earthquakes have magnitudes as low as 3.6 on the Richter scale, where 8.5 represents the strongest shocks so far recorded. The fault displacements included both slippage accompanied by earthquakes, some of which were predicted in advance, and others by slower creeping movement without accompanying earthquakes. Either sudden or slow displacements may cause structural damage.

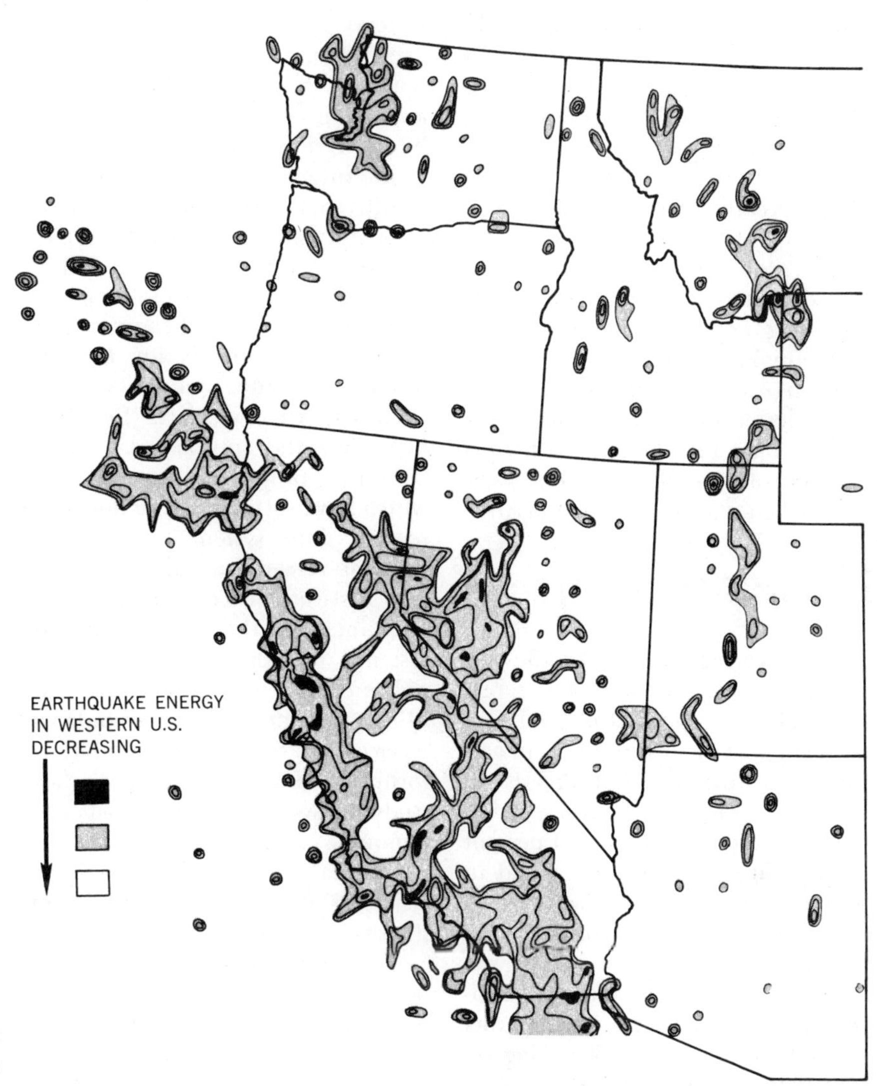

In the laboratory, Drs. Brune and Allen studied the effects of certain types of physical shock that produce rock fracture on a microscopic scale. They have discovered a variety of initial shock conditions where, as the microfractures in the brittle rock propagate themselves, they produce a seismic event sequence similar to that of a natural earthquake, i.e., foreshocks, a main event, and a decaying sequence. This has opened the door to meaningful study of earthquake mechanisms by carefully controlled laboratory experiments.

Earthquake prediction with a scientific basis and the triggering of earthquakes are relatively new fields of study. Jack Oliver and associates at Lamont-Doherty Geological Observatory are exploring this important problem in a series of microearthquake studies. By studying the source mechanisms of the frequent minor earthquakes which occur in seismically active areas such as Iceland, the rift valleys of eastern Africa, and the islands of the South Pacific, they are able to compress into short periods studies that would ordinarily take many years, or tens of years of observation. The Lamont studies along with those of Alan Ryall at the Mackay School of Mines, University of Nevada, have shown that even minor changes in forces such as the earth tides, caused by the sun and moon, affect the frequency of earthquakes. Ryall and his colleagues have also discovered that nuclear detonations in southern Nevada can trigger natural earthquakes in nearby areas. Such observations have led to the exciting thought that man may someday be able to prevent some catastrophic earthquakes by triggering the release of seismic energy by many smaller earthquakes.

These several independent studies have greatly increased our understanding of faulting and its relation to earthquakes. It suggests that we are drawing closer to a basic understanding of source mechanisms of earthquakes and their magnitude and frequency. This in turn also provides encouragement that in the future man may be able to predict or control on a scientific basis where and when earthquakes will occur.

ENGINEERING

Engineering support by the National Science Foundation represents about 10 percent of total Federal support of such research in academic institutions, but this is an overall average and the NSF portion has much greater than a 10 percent impact in some institutions and in some areas of engineering.

Because of the Foundation's established program of making research initiation grants in engineering to younger investigators (those within 3 years of having received the Ph.D.), the general tightening of funds is not selectively affecting younger investigators in terms of research support. Younger faculty members are making innovative contributions to research in areas of current concern.

Many engineers see their profession as being one by which the fruits of scientific research are put to human use, and this philosophy is reflected not only in the schools of engineering, but in the basic trends of engineering research in the country today. In the engineering schools, much thought is being put to the orientation of future engineers in the social sciences to increase their understanding of the societal interactions between engineering and social sciences. This is not easy because of the immense amount of engineering material to be covered as well.

An interest in relating research results to human and social needs is emerging more strongly in the research proposed to the Foundation. Not surprisingly, the most marked occurrence of this phenomenon is among the proposals for research initiation grants, whose younger proposers are the source for many new ideas and approaches.

Many proposals for research which have societal interaction tend to fall into systems engineering. Current research at the Massachusetts Institute of Technology in this area includes the development of systems for the optimization of police cruiser utilization and scheduling, and analysis of ambulance services. At the University of California at Berkeley, engineers are studying systems for more efficient removal of automobile wreckage from roadways after throughway accidents and of traffic diversion for the duration of these emergencies.

The field of biomedical engineering has drawn on electronics and mechanical engineering for many years. The bypass equipment, heart-lung machines, and artificial kidneys, without which organ transplants and the long surgical procedures they entail would be impossible, are the product of skilled engineers working with medical personnel in pursuit of solutions to specific problems. Materials engineering as well has contributed to medicine through the development of materials for implantation in the body—artificial heart valves, pacemakers, bone pins, and other prosthetics—which have both the engineering strength and durability required for lifelong service and are physiologically acceptable to the host body.

Other trends in engineering involve design of better and stronger buildings. Recent discoveries, stemming from scientific analyses of damage done by the 1970 tornado which hit Lubbock, Tex., indicate that the destructive forces of high winds have effects not unlike ground shaking during earthquakes. Earthquake engineering studies may be expanded to include effects of wind forces to permit better designed structures that can successfully withstand both sorts of damaging forces.

In the field of engineering chemistry, the future may well see the development of the subdiscipline of enzyme engineering. Enzymes, the catalysts of chemical reactions within living systems, are now well enough understood that synthesizing them lies within the realm of possibility. When this becomes a reality, the industrial use of enzymes to catalyze a series of reactions for the synthesis of edible protein food for human use may become economically feasible, and a major new source of food for a hungry world would become available.

Effects of Forest Clearcutting on Slope Stability

Clearcutting is a timber harvesting procedure in which all the vegetation is felled in a selected area. This is the usual logging practice in the redwoods of the North Coast ranges of California and in the vast tracts of Douglas fir in the Cascade Range of Oregon and Washington. Denudation is made more awesome and complete by burning the slash remaining after a logging or cutover operation. Controlled slash burning is justified by various arguments, the foremost being that it eliminates a potentially serious fire hazard later on.

What impact do clearcutting,

road building, and other forest practices have on slope stability? What is the role of a forest cover and other types of slope vegetation in

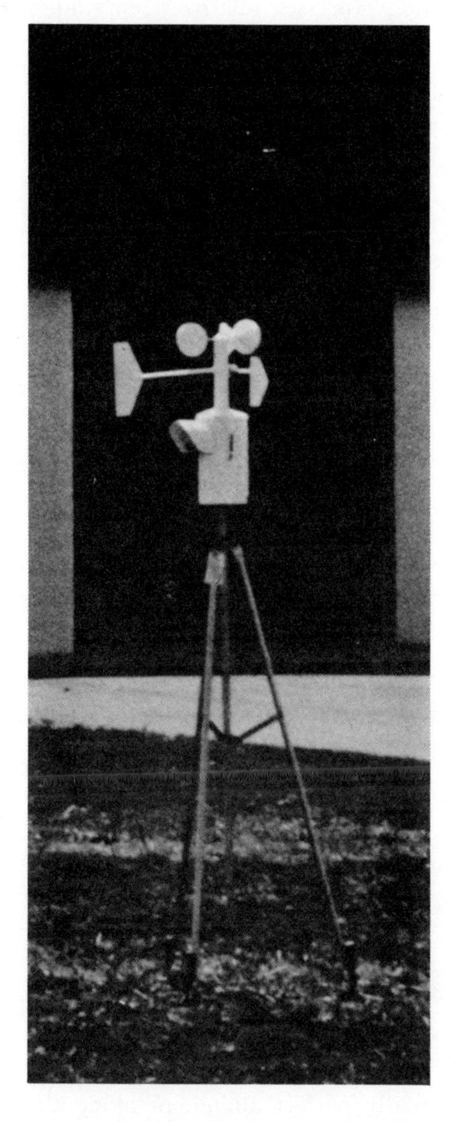

Micrometerological station in newly urbanized area at Columbia, Md., will help to determine effects of urbanization on local weather conditions. (University of Maryland photo)

preventing soil erosion and mass soil movement? These are timely and important questions because pressures are mounting from many sides to increase allowable timber cuts and to accelerate construction of access roads in our national forests.

A forest cover appears to affect deep-seated stability in two principal ways: by modifying the hydrologic regime in the soil mantle and by mechanical reinforcement from its root system. This is a difficult problem, but it is partly amenable to slope stability analyses based on principles of soil mechanics and on knowledge of soil-water-plant interactions. Current research under the direction of Donald H. Gray of the Department of Civil Engineering at the University of Michigan is developing a theoretical analysis which should make it possible to predict stability of a forested slope and assess the probable effects of denudation.

The conventional slope stability analysis in current engineering practice relies upon information of *in situ* (in place) soil information such as moisture content and distribution, moisture stress, strength parameters, physiochemical properties and structure, both microscopic and macroscopic. The real problem lies in determining how much and how quickly these change after clearcutting. To this end, Dr. Gray is instrumenting and sampling actual logging sites in the Cascade Range of central Oregon which have histories of slope instability. On these sites, he has installed inclinometers — instruments which measure and record miniscule slippages in the soil—to indicate incipient instability in the slope. Other field measurements include the installation of recording tensiometer-piezometers, which give a record of stresses produced by soil moisture during and after major rainstorms —and soil core sampling and analysis. This allows the physiographic data as well as the usual field and laboratory determination of engineering soil properties to be correlated as input into the mathematical model.

The data which Dr. Gray records should add significantly to the store of available quantitative knowledge on slope stability and allow advance prediction of the long-term effects of clearcutting on mountain slopes.

Neural Mapping with the Scanning Electron Microscope

In order for neurophysiologists—scientists concerned with the growth and development of nerves—to be able to work with and extend their knowledge of nervous systems, they need to know what the physical layout of the system is. The lack of good maps for nervous systems—or of methods for obtaining them efficiently—has been one of the major obstacles to progress in this field.

Edwin R. Lewis at the University of California at Berkeley has been applying a scanning electron microscope in search of a solution to this problem and has obtained a wealth of micrographs showing the beautiful and mysterious world found in a spot of biological tissue no bigger than the point of a pin.

The research team, which is in the Department of Electrical Engineering, is using the scanning electron microscope to examine specimens of nerve tissue taken from the abdomen of a marine snail, *Aplysia californica,* chosen because of the simplicity of its nervous system.

They have obtained the first photographs of what are identified as synaptic knobs—the crucial point where the nerve impulse is passed along from one cell to another.

These photographs, taken at magnifications of about 20,000 times life size, show with remarkable three-dimensional clarity a number of such knobs at the ends of fibers

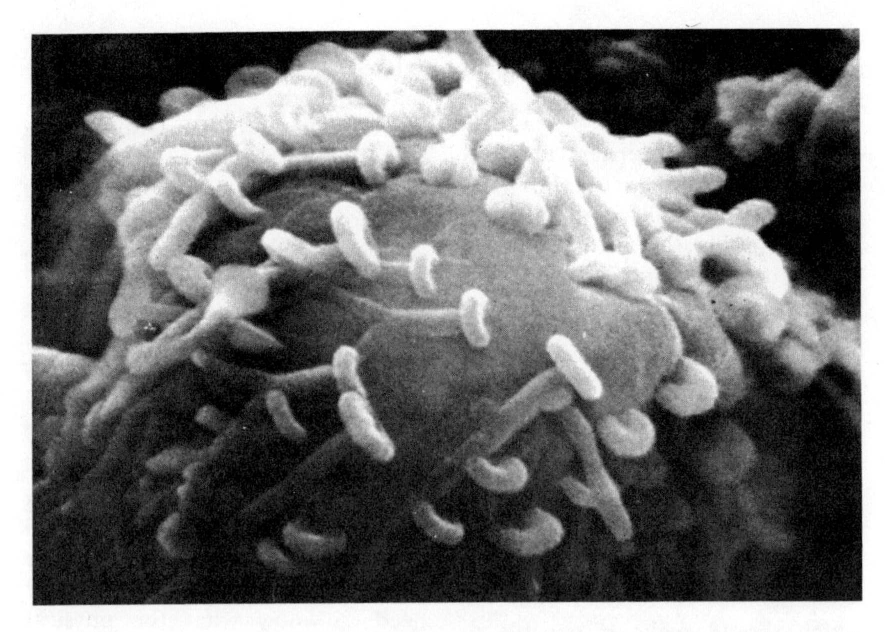

Synaptic knobs in nerve tissue taken from the abdomen of a marine snail, *Aplysia californica.* (Photo University of California, Berkeley)

which seem to lie across each other like a random pile of logs. Other photographs at lower magnifications show complex bundles of such fibers and knobs lying together in clusters at the point where a large "trunkline" fiber from one cell meets a similar fiber from another cell.

The engineers noticed that the knobs seemed to have five or six spots which were firmly attached to other knobs or nerve tissue. A montage of photographs taken as the microscope moved along the specimen traced the complete linkage from cell to cell. The conclusion that the knobs serve a synaptic function (that is, that they form the communications junction between nerve cells) must remain tentative until additional evidence is obtained.

Use of the scanning electron microscope for biological studies is less than 10 years old. Such studies in the past have been limited almost entirely to the conventional light microscope, which has useful magnifications of surface features to only about 100 times life size, barely reaching down to the level of the cell, and to the transmission electron microscope, which offers magnifications of several hundred-thousand times life size but must use extremely thin slices of tissue, and produces a two-dimensional shadow image of the specimen, much like an x-ray.

The scanning electron microscope, on the other hand, sees only the surface of specimens, and produces images and photographs with three-dimensional qualities. The techniques developed by this group point to the possibility of developing neural maps on a cellular basis. From this knowledge of neural anatomy, a whole new branch of medicine and surgery, microneurosurgery could spring.

Partially Crystalline Polymers

In a number of practical applications, polymers, such as polyethylene, are extruded through dies to form desired shapes. A polymer is a large molecule synthesized by linking together many smaller, identical subunits and is the basic kind of material of which modern plastic is made. Because of the scientific complexities involved in the extrusion processes, relatively few studies have been on the flow properties of partially crystalline polymers. And yet study was needed because of the fundamental importance and practical importance of flow in partially crystalline systems.

Roger Porter at the University of Massachusetts has been investigating what unusual structures and properties might be achieved by shearing polymers near their melting points. His initial studies provide guidelines for producing very strong and clear polyethylene. Capillary extrusion was conducted over a sensitive temperature range near 138° C. Crystalline filaments could be continuously extruded which had both unusual clarity—complete visual transparency—and conventional tensile strength over six times that of higher temperature extrusion of the same polymer. The unusual properties are the result of pronounced molecular extension, orientation, and crystallization in the entrance region of the capillary extruder. X-ray, calorimetry, and a variety of other techniques have shown that the perfection (84 percent crystalline) and orientation of the polyethylene crystals are as high as has ever been documented. The axes of the molecules and of the crystals are in virtually perfect alignment along the filament length, and the unusual properties result from this feature. This research has provided the first documented example of a transparent and extended chain crystal structure in polyethylene.

Dr. Porter also found that the transparent material was relatively tough, and resisted fracture at extremely low temperatures, while the opaque material was brittle and broke easily. These improved engi-

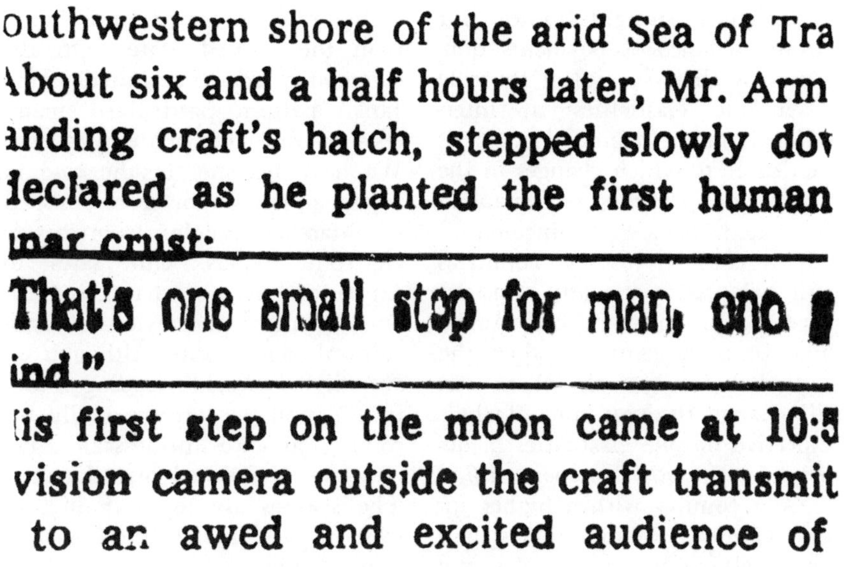

outhwestern shore of the arid Sea of Tra
About six and a half hours later, Mr. Arm
anding craft's hatch, stepped slowly dov
leclared as he planted the first human
unar crust:
That's one small step for man, one
ind"
His first step on the moon came at 10:5
vision camera outside the craft transmit
to an awed and excited audience of

A strand of the special morphology polyethylene is shown over a section of conventional newsprint. The high clarity and lens effect are apparent. The same polyethylene prepared under conventional conditions is entirely opaque. (Photo University of Massachusetts)

neering properties can be achieved at virtually no increase in processing costs over that for producing conventional polyethylene. Polyethylene is already used in a variety of applications as a packaging film and as a structural plastic, and these improvements will increase the number of uses to which this useful material can be applied.

SOCIAL SCIENCES

The number of research proposals from the disciplines which make up the social sciences rose markedly from 823 in fiscal year 1969 to 1,087 during fiscal year 1970. This increase reflects not only changes in the ways by which social science research is conducted but also an increased awareness of the possibility of Federal support for the social sciences through the National Science Foundation, in large part stemming from the Foundation's new mandate to strengthen its efforts to improve the social sciences.

Despite the fact that the current trend in actual Federal support for the social sciences has not matched the interest evidenced in what the social sciences can do, the social sciences themselves are surging forward in substance, partly as a result of a methodological revolution. Prior to a very few years ago, social scientists designed their experiments so as to minimize the number and complexity of calculations they would have to perform—always a massive and time-consuming job where large data bases are involved. However, with the advent of the computer and the availability of relatively cheap computational ability, social scientists now no longer need avoid the massive computational loads and much of the field is rapidly becoming more quantified and more scientific in its outlook and methodology.

It is difficult to generalize about the social sciences as a group or even about some of the disciplines within that group. But it is certainly true of all these disciplines that, insofar as basic research in the social sciences bears directly on human social welfare, the advances and breakthroughs in knowledge in the field, while they may be modest with respect to the total problem, constitute a real and lasting benefit. (For example, improvement in public and private policies resulting from research in economics which would increase the gross national product by only one-tenth of 1 percent would add $1 billion yearly to our nation's economy.)

While law as a discipline has not traditionally been considered as one of the social sciences, it is nonetheless closely related to them and is essential to social change and social regulation. Law and the traditional social sciences have had many points of contact in sociology (for example, criminology), social psychology, and economics. This trend toward cooperative research is continuing, and the Foundation has helped to foster interaction between the two fields and is supporting both lawyers and social scientists who are doing research on social problems involving both legal and social scientific operations.

Recent public interest has drawn attention to the study of social indicators — strategic and identifiable measures which indicate significant social changes. "Replication studies" in which sociological research is repeated after 10 or 15 years on a subject, such as the relative prestige of certain job categories, or the influence of education on income make it possible to determine if there has been any basic change. In fact, many such surveys now being conducted for the first time are being designed so as to permit exact replication in the future. A larger number and variety of such measures would supplement the base of our current understanding of change in America which we have from data such as the Census and economic indicators and enhance our knowledge of social phenomena.

Increasingly, problem-focused research draws its personnel from a variety of disciplines and traditional disciplinary work has broadened to accommodate new problems. This is particularly true of urban problems and of foreign area studies. The bulk of NSF research grants in geography, for example, is no longer for the exploration and mapping of exotic lands; geographers are intimately involved with urban planning, with the spatial distribution of income, with community locational decision, and with the perception of neighborhood by individuals. Grantees are affiliated not only with departments of geography but also with schools of design and departments of regional science and of urban planning.

As part of its effort to strengthen scientific research in the social sciences, the Foundation supports work in or related to foreign areas. Some of this research is inherent in the nature of the disciplines—notably anthropology, geography, and linguistics—but much reflects the strong interest in techniques of comparative studies and in problems of social development which is burgeoning in contemporary economics, sociology, social psychology, and political science. While research on foreign areas is of value to those who are interested in the subject areas themselves, it is also a useful means of acquiring a better general scientific understanding of human beings and social behavior.

International Trade and the Balance of Payments

The trade problems which the U.S. economy has encountered in the postwar period are well known, but the causes and cures have been inadequately understood. Hendrik S. Houthakker at Harvard University and his associate, Stephen Magee, have developed a number of valuable insights into our chronic balance of payments difficulties by their studies of the role of income elasticities in international trade. Income elasticity is the degree to which changes in the quantity of a commodity demanded are related to changes in income.

If we consider just two countries with balanced trade and constant prices, and if income growth is the same in both countries, then the trade balance between them can still change through time if their respective income elasticities of demand for the *other's exports differ*. Thus, a country with a higher income elasticity for its imports than the corresponding income elasticity for its exports will sustain more rapid import growth than export growth. Ultimately, this may be followed by a deterioration in its trade balance and eventual pressure on its exchange rate.

Based on this model, Drs. Houthakker and Magee have investigated the demand elasticities for both imports and exports with respect to income and price for selected countries over the period 1951–66. In addition to the analysis of total imports and exports by country, more detailed studies were made of U.S. trade by country of origin or destination and by commodity class.

Drs. Houthakker and Magee's studies indicate that the U.S. income elasticity of demand for total imports is about the same as that of the other developed countries, but that the income elasticity of other countries' demand for U.S. exports is unusually low. Therefore, the U.S. trade balance — other things equal—will tend to worsen over time.

The prospective deterioration in the U.S. trade balance will probably be especially marked with respect to Japan and Canada, according to Drs. Houthakker and Magee, unless these countries develop much higher rates of growth or inflation than the United States. On the commodity side, the United States-Japan pattern particularly manifests overall U.S. trade problems. We have become, in the case of Japan, a net importer of finished manufactures ranging from cars to electronic goods. Our sales to Japan, on the other hand, contain an increasing proportion of agricultural commodities. Although we are still the world's leading industrial nation, we are gradually becoming on a worldwide scale a net importer of finished manufactures. The reasons are to be found, at least in part, in the differing long-term elasticities which Drs. Houthakker and Magee have discovered to exist among the major classes of commodities.

Better understanding of the nature and quality of pressures on the U.S. balance of payments is, of course, not the end of the story. Rather, it offers a framework of fact and analysis for other research in progress on alternative monetary arrangements. It is reasonable to expect that the latter will lay the groundwork for a more stable means of international adjustment and, ultimately, an alleviation of our balance of payments difficulties.

Early Irrigation Patterns

Since the dawn of civilization, man has had to struggle with and control his environment. As time has gone by, many of the techniques he has used to manage the world around him have become more sophisticated, and older methods have been discarded and forgotten with the passage of time.

The science of archeology can help to determine what these discarded patterns were and—how well they succeeded, as compared to the patterns which supplanted them. This perspective over a very long time scale can often provide valuable knowledge of the limitations as well as the unrealized potential of environments for constructive human uses that may not be easily seen from recent and current experiences.

As an example, in his archaeological investigation of Hay Hollow Valley in eastern Arizona, Fred Plog of the University of California at Los Angeles has found evidence of prehistoric irrigation in three localities. These irrigation systems include ditches, canals, a check dam, and basalt walls which probably were a device to slow slope wash as it approached a series of sand dunes. On the basis of radiocarbon dating, these features range in time from about the 10th to the 18th century.

The topography of the valley varies considerably from one point to another, and the old irrigation systems appear to have been designed in response to these differences. For example, the southwestern exposures of what is called Point of the Mountain have trapped great quantities of windblown sand. These dunes, which catch surface water from Point of the Mountain, were used for dune farming. Structures—like the basalt walls—which slowed water down as it came off the mountain or which protected plants from the wind were far more adaptive to farming in these circumstances than canals would have been. In Hay Hollow Wash there is characteristically a difference of several meters between the flow of the channel and its banks, except where such channels flow over bedrock, and most of the points where water was taken out of Hay Hollow Wash are ones where the wash has a bedrock bottom.

Forms of surface water available in the area depend at any time on weather patterns. Slope wash occurs when rain falls in the valley; the main wash runs when water is falling to the south of the valley, and tributary arroyos (or gulches) run when rain is falling to the west of the valley. Given the classic southwestern thunderstorm pattern, a single thunderstorm rarely covers all the possibilities, and the variation in locations and structures of the old irrigation systems seem to have been geared to the diversity of meteorological as well as topographical influences.

Plog's findings have significant implications for modern farmers in the area where communities generally occur sufficiently close together for them to share water resources, reservoirs, and some main canals. Because the streams tapped are characterized by flash flooding, dams are frequently washed out. None of the contemporary irrigation systems are differentiated as the old systems seem to have been. Arroyo runoff is largely unexploited since it is rarely sufficient to provide water for a whole community. However, it is apparent that for a single farm or a few farms, arroyos could be tapped with minimal technological and capital requirements. Based on the conclusions of this archaeological project, it would appear that further large-scale irrigation projects may be less efficient than concentration on small-scale water resources, modeled on some of the ancient patterns.

INTERDISCIPLINARY RESEARCH RELEVANT TO PROBLEMS OF OUR SOCIETY

Fiscal year 1970 marked the initiation of a new program of Interdisciplinary Research Relevant to the Problems of Our Society (IRRPOS). The IRRPOS program was explicitly designed to mobilize the intellectual skills of the nation's scientists to conduct research on major societal problems. Through the IRRPOS program, the Foundation seeks to support interdisciplinary research needed to provide a fuller understanding of major societal problems and to develop new and improved ways to deal with them.

The IRRPOS program does not replace nor merely supplement existing Foundation programs for the support of problem-oriented research; nor does it in any way represent a change in the Foundation's objective to support fundamental scientific research. The IRRPOS program is intended instead to supply a focus within the Foundation for the encouragement and support of scientific research on complex societal issues that require the contributions of diverse scientific disciplines.

Between December 11, 1969, when the program was announced and the end of the fiscal year, over 200 preliminary proposals were submitted to the Foundation. A total of 42 formal proposals requesting over $18.5 million were reviewed, and 21 awards for $5,984,099 were made. Among these are included such projects as:

- The Oak Ridge National Laboratory is conducting research into the potential for genetic mutations in man resulting from the introduction of manmade chemicals into the environment. Oak Ridge is developing several approaches for mass screening of mutations in man. Other studies will develop techniques for performing overall ecological evaluations of the environment, increasing public awareness of environmental quality, projecting the costs and consequences of alternate environmental policy actions, and development of a computer simulation model to predict the influences of alternate environmental policies in the Tennessee River Valley. Additionally, Oak Ridge is investigating possible methods of moderating energy demand and beneficial use of waste heat in a regional energy system.

• Drawing on the disciplines of engineering, applied mathematics, economics, political science, and city and regional planning, the Environmental Systems Program at Harvard University is conducting research on the technical, economic, social, and political aspects of problems of environmental quality. Major aims are to develop an improved framework for analysis of urban environmental problems and to train graduate students and postdoctoral fellows in the conduct of interdisciplinary research.

The research program involves the collaboration of the School of Public Health, the Graduate School of Design, and the John F. Kennedy School of Government. It includes analyses of environmental management institutions; economic analysis of the household as an individual decision unit; studies of municipal financing of environmental control system; expansion of current studies of solid waste disposal; and studies of the epidemiology of urban fires. The program involves graduate students and postdoctoral fellows from a variety of specialist backgrounds in order to develop the talent required for intelligent management of environmental quality.

• In collaboration with the Rand Institute of New York City and with the cooperation of public agencies of Nassau and Suffolk Counties, the Urban Systems Engineering group of the State University of New York at Stony Brook is initiating a program of interdisciplinary research on the flow of solid wastes and on fire protection.

Data obtained from existing literature and a case study of the sources and disposal of solid wastes are being used to construct a benefit-cost model of the waste disposal system.

Building on the experience of the Rand Institute with research on fire protection, the Urban Systems Engineering group is using operations analysis techniques to deal with problems of fire protection in an area characterized by rapid urbanization. Computer simulations will be used to develop estimates of benefits and costs associated with alternative systems.

The aims are to develop the analytical investigations that are likely to be of immediate benefit to public agencies, to develop a close working relationship with the Rand Institute, to strengthen the interdisciplinary research resources of the Urban Systems Engineering group, and to construct a series of models describing the essential features of selected urban systems. The research is closely tied to a graduate program involving students in engineering, economics, and physical sciences.

• The University of Illinois and Colorado State University are investigating the sources and magnitude of lead pollution in the environment. The Illinois project emphasizes the study of lead from gasoline but will also include study of the movement of all forms of lead through the environment. The Colorado State project will stress a systems approach for studying environmental contamination, and the techniques developed should have broad application to studies of other contaminants such as mercury, combustion-produced carcinogens, and pesticides.

Science Education Support

Since the start of its activities in science education, the Foundation has been concerned with the training of enough scientific and technical manpower of high quality to meet the nation's needs. Science and technology have become, however, such integral parts of our society that an understanding of their processes is now recognized as essential even for those who are not and do not expect to become professional scientists or technologists. Clearly, the improvement of the nation's scientific research potential, through the education of scientists, continues as a priority goal for the Foundation. But fiscal year 1970 has seen new emphasis on efforts to educate all citizens in both the uses and the limitations of science and technology, particularly as these bear on the analysis of societal problems, finding of alternative solutions, and rational decision making. Thus, the Foundation's science education programs are evolving toward meeting the two goals of educating the nation's scientists and technologists and improving the quality of education in the sciences for all students.

Science education is a cumulative process which begins at the first-grade level and may extend beyond the earned doctorate. Since each successive level rests upon earlier levels, one cannot hope to improve materially the quality of education in the sciences, let alone the production of highly trained professionals, by concentrating efforts at any one academic level. Even massive effort at the graduate level produces limited results in the absence of improved preparation of students prior to that point. On the other hand, efforts at the early levels of education are ineffectual unless momentum can be sustained thereafter. The education activities of the Foundation are designed at each major level to address those components of the system which exert the most leverage in terms of improving science education, but with varying emphasis. At the graduate level, a major portion of support funds is invested in professional training of future scientists; at the undergraduate level, in the improvement of instructional programs and institutional capability; and at the pre-college level, in

Table 5
Education in Science
Fiscal Year 1970

[Dollars in thousands]

	Number of proposals received	Dollar amount requested	Number of awards made	Funds obligated
Graduate Education in Science:				
Fellowships	10,914[1]	$73,063	3,082[1]	$15,877
Traineeships	279	100,126	368[2]	27,269
Advanced science education program	166	8,476	92	2,393
Undergraduate Education in Science:				
College teacher program	474	13,833	394	4,161
Science curriculum improvement	1,795	36,517	685	9,806
College science improvement program	104	18,610	46	6,829
Undergraduate student program	836	9,300	432	3,817
Pre-College Education in Science:				
Institutes	1,472	60,500	1,049	36,936
Cooperative college school science program	325	14,075	136	4,654
Course content improvement	101	11,504	77	6,507
Student science training program	305	4,873	130	1,931

[1] Individuals involved.
[2] Includes 89 same-year supplementary amendments not included in proposal count.

teacher training—but all these elements are represented at each level. Table 5 reflects NSF support for educational activities in fiscal year 1970.

The Quality of Science Education

Early Foundation efforts to improve the quality of science education in schools and colleges emphasized resources and personnel from outside the normal educational system or institution—particularly the improvement of teacher competency and the creation of high-quality instructional programs. Several years ago, it became clear, not unexpectedly, that the use of these external resources would be neither widespread nor—in many instances—effective without financial and professional assistance. Accordingly, some five years ago the Foundation began programs to help with actual implementation, e.g., the College Science Improvement Program at the undergraduate level and the Cooperative College-School Science Program at the pre-college level. Simultaneously, the existing teacher institute programs expanded their efforts to include increased participation by training specialists who could assist their schools in the adoption of new courses and methods. Looking ahead, the Foundation plans to support the generation and growth of capability *within* educational systems and institutions for locally initiated improvement of educational programs, relying on broadly constituted curriculum and advisory groups and some of the resources developed over the last decade.

Assessments and Priorities

While the Foundation's education programs have evolved over the past several years from a limited manpower goal to include a concern with improved science educa-

In a unique project combining several program elements, the Foundation in 1970 supported a research project which included superior high school and undergraduate students working with a college teacher of science to explore the ecology of lands above the timberline. (Photos NSF)

tion for all, and from the development of external resources to increasing emphasis on implementation and development of capability for locally generated change, the past year was one of particularly intensive review. A number of factors sparked this evaluation and planning: the recurring need to examine the efficacy of established programs, changes in the demands for and supply of different types of scientific and technological manpower, fiscal retrenchment both at government and local institutional levels, and—overarching all—ever more insistent pressures for reform at all levels of education.

NSF staff and consultant review and planning for educational programs was augmented by a study commissioned by the National Science Board and conducted by the Advisory Committee for Science Education under the leadership of Dr. Joseph B. Platt (President, Harvey Mudd College) and Dr. Herbert S. Greenberg (University of Denver). The study report develops a theme for the next decade: "To educate scientists who will be at home in society and to educate a society that will be at home with science." Recommendations to achieve this aim include:

> Grants to support faculty and graduate students for research in science education equivalent to those for research in science;
>
> Experimentation and innovation in the application of technology to education to develop not only better hardware systems specially designed for educational needs but also an adequate range of high-quality materials suitable for exploration of television and computers as instructional tools; and
>
> Increasing support for science education outside the formal classroom such as museum activities, travel exhibits, films, and television programs.

The report also makes specific recommendations at each level of education. At the graduate level, support is urged for advanced practitioners degree programs; revitalization of the master's degree for those who will be engaged in the teaching and application of existing knowledge rather than creating new knowledge through research; and innovation in interdisciplinary course content. At the undergraduate level, the report recommends improvement of science courses for the non-science major; development for science majors of interdisciplinary problem-oriented science programs; increase and improvement of the science component in the education of prospective teachers; greater emphasis on two-year programs for the training of technical personnel; and some experiments in restructuring the total undergraduate experience. Recommendations for the pre-college level include second-generation course development focused on interdisciplinary, vertically integrated course sequences; production of outlines and suggestions to be adapted by teachers to fit local conditions; increased support for social and behavioral science curricula and teacher training; and support of experimental schools exploring major changes in the school environment.

New Directions

The Committee's recommendations and the staff planning have found expression in specific program activities during fiscal year 1970 designed to meet the current challenges in education. A number of projects supported in the Advanced Science Education Program are experimenting with alternatives to the traditional research Ph.D. programs. These experiments are given impetus by the changes both in manpower needs and in the interests of students oriented more and more toward interdisciplinary problem-solving rather than research in a narrow specialty. A newly established program at the undergraduate level — Student-Originated Studies — will support projects by creative and able students who wish to take a hand in their own education while gaining an understanding of how to analyze science components of a problem and formulate possible approaches toward solutions. An increasing number of projects are concerned with developing new instructional patterns for the undergraduate preparation of prospective technical and teaching professionals. At the pre-college level, fiscal year 1970 was marked by growing coordination between curriculum development and the training of teachers who can implement new curricula. Also, an increasing number of principals, science supervisors, and science education faculty from colleges and universities have received special information and training instruction so that a cadre of supervisory and resource personnel will be able to aid individual classroom teachers.

In addition to its investment in individuals to recognize and develop scientific talent, and in instructional programs and the training of instructional personnel to improve the quality of science education, the Foundation is also concerned with exploring and defining the role of educational technology (computers, television, film, programmed instruction) and with support of basic research that has educational significance. Projects in support of these two aims are funded through several of the education programs, often in conjunction with the Office of Computing Activities or the Division of Social Sciences. The following reports for each education division illustrate in greater detail the direction that the Foundation's science education activities are taking.

GRADUATE EDUCATION IN SCIENCE

The Foundation's concern with graduate education in science is apparent not only in its graduate education programs, but also in the emphasis given it in research and institutional programs. Support is extended to talented individuals for graduate and postdoctoral study through a variety of fellowship and traineeship programs. From 1952 through 1970, the Foundation will have funded about 65,000 awards for 9 or 12 months' study at the graduate level. Research grants to institutions of higher education usually carry a component of support for graduate students—some 6,000 being so supported in fiscal year 1970; special grant programs to support in-depth field work directly pertinent to dissertation projects contribute to the quality of the individual's training; large development grants to institutions—or to departments within institutions—are designed to improve their graduate education programs. Graduate students are also supported through the National Research Centers, the Sea Grant Program, and in various National Research Programs. Through these mechanisms, NSF provides support for nearly one-fourth of all science students holding federally funded awards.

In direct support of graduate education, fellowship programs designed as open national competitions for graduate students identify scientific talent and provide support to develop that talent to its fullest potential. Similar programs serve to further advanced training for postdoctoral and senior postdoctoral scientists and members of the science faculties of colleges. Particularly in the senior postdoctoral and science faculty programs, a major effect is on higher education itself, since fellows from those programs return with new training and insight to their teaching responsibilities. In addition to those programs, traineeship grants permit institutions to make their own selection of recipients for support of graduate studies and develop graduate departments. A program that brings outstanding foreign scientists in leadership roles to U.S. faculty positions for periods of up to one year adds new perspectives and experience to U.S. faculty members and graduate students, and strengthens scientific cooperation and understanding on an international basis. Table 6 shows the number of individuals supported with fiscal year 1970 funds under each of these programs.

Through the Advanced Science Education Program (ASEP), grants are made to institutions for the development of innovative graduate-level course offerings, to experiment with new kinds of educational techniques, and to examine the needs and problems in various disciplines so that graduate education in science can evolve to keep step with the changing needs of individuals and society as a whole. A particular responsibility of ASEP is the funding of Advanced Training Projects which provide educational opportunities for graduate and postdoctoral students and for graduate-level faculty where no training is available through regular university offerings.

Scientific Manpower and Graduate Education

The Selective Service Act had its most severe impact on the NSF Graduate Fellowship Program during 1969–70. Of the 2,500 persons offered fellowships for the year, a total of 101 requested that the awards be deferred because of military obligations. It appears that the Congressional action in December 1969, which made possible the order of call on a lottery basis, may result in a reduction of this impact, since the rate of deferment requests to date has been half that of last year.

Fiscal year 1970 was a period of intense re-examination of the whole manpower question as it relates to graduate education in science. In some areas of science, particularly physics and mathematics, it appeared that the number of Ph.D.'s produced during the last decade had satisfied or even surpassed demand for their services in the usual professions associated with academic careers or industrial research. Conflicting statements as to the seriousness of the situation have appeared, but the supply of traditional Ph.D. manpower seems to be coming into balance with the need. Recent studies have also shown that, since the mid-1960's, there has been a trend toward disinclination on the part of graduate students to enter the science fields, with a resulting decrease in enrollments. The decrease in Federal support for research and graduate study is likely to result in an even sharper decrease in the overall graduate school enrollment beginning in the fall of 1970. The Foundation is planning readjustments in its support of graduate students to take account of these projections and trends while still meeting anticipated needs for the 1970's. With less Federal support for graduate students, an increase in the number of part-time graduate students is anticipated; accordingly, the graduate traineeship program has been adjusted to permit institutions to award part-time traineeships beginning in the fall of 1970.

Another likely result of decreased Federal support is an increase in the number of students who will pursue the master's degree as a terminal degree. Some graduate training will probably be necessary for many scientific and technological professions immediately below the

Table 6
NSF Fellowship and Traineeship Programs
Fiscal Year 1970

	Awards requested by institutions	Individuals involved in applications	Fellowships awarded	Net amount
Graduate traineeships	17,510 ([1]224)	----	5,301 ([1]224)	$26,240,000
Summer traineeships for graduate teaching assistants	8,737 ([1]211)	----	938 (207)	1,029,290
Graduate fellowships	----	8,201	2,212	10,374,817
Postdoctoral fellowships	----	1,295	109	1,000,000
Senior postdoctoral fellowships	----	338	50	686,000
Science faculty fellowships	----	994	209	3,083,889
Senior foreign scientist fellowships	----	86	61	779,518
Total	26,247	10,914	8,880	43,146,514

[1] Number of institutions involved.

top levels, so that the master's rather than the doctorate degree may well become the target of a large segment of the student population. The Advanced Science Education Program is supporting several projects designed to explore the role of the master's degree. For instance, Georgia State University is developing a master's degree physics program to serve both part-time students from industry and students preparing to become secondary school and junior college teachers.

Improving the Quality of Graduate Education

An important resource to institutions trying to strengthen their graduate programs is the Senior Foreign Scientist Fellowship Program. It brings to institutions in this country foreign scientists whose training, experience, and formal accomplishments enable them to make significant contributions to the education and research programs of the host institutions. The increasing importance of this program for U.S. institutions is indicated by the extent of their participation this year, the greatest since inception of the program in fiscal year 1963: nearly 85 percent of the eligible institutions nominated candidates. The selection process is such that individuals nominated for the fellowships are able to exert in-depth influence within the departments they join, usually for a year, where they are considered members of the senior science faculty. They teach, lead faculty seminars, collaborate with and guide research activities of faculty and graduate students, contribute to professional society meetings in the United States, lecture at nearby institutions, and participate fully in the departmental development programs of the U.S. universities. They also bring to the institutions different views in their fields of science, particularly in smaller departments, and afford educational opportunity and scientific expertise often not available in this country to faculty personnel as well as to students. In addition to being an important resource in upgrading graduate science education in this country, the association between the outstanding foreign scientists and their colleagues in the United States adds a significant increment to the improvement in international understanding and cooperation.

The Senior NATO Fellowships in Science Program (administered but not funded through the Foundation) also provides resources for strengthening graduate education by allowing a limited number of senior staff members of U.S. institutions to spend short terms (usually 1 to 3 months) in other NATO countries to learn new developments in their fields of specialization. In fiscal year 1970, 36 individuals received support for senior NATO fellowships.

In attempting to improve and introduce new directions into graduate education, the Foundation is particularly concerned with the societal problems facing the nation in the next decade. Solutions of these problems will require broadly trained, creative individuals capable of working in interdisciplinary teams. In its fellowship programs at the postdoctoral level, the Foundation has invited applicants whose training and experience are in one field of science to propose plans of study or research in different but related fields. For instance, a person trained in chemistry might wish to tackle problems associated with air or water pollution and, to be effective, might need additional training in atmospheric or oceanic sciences. In another area, a person whose training is in business-cycle economics might now wish to undertake research in urban planning or some other sociological aspects of the inner city.

The training provided through Advanced Training Projects is frequently interdisciplinary in nature and aimed at current problems. Some of the fields of science in which courses, seminars, symposia, or field work were supported this year are: molecular techniques in developmental biology, geographical analysis of U.S. metropolitan areas, behavioral and social science in legal education, earthquake engineering, pest management, planetary atmospheres, and marine paleontology.

In addition to training scientists who can work in multidisciplinary areas, the Foundation is concerned with new interdisciplinary fields presently emerging as scientific disciplines. For example, a grant to the Greater Los Angeles Consortium, made up of liberal arts colleges in California, supported a conference in May 1970 on urban studies. Academicians from approximately 25 colleges and universities in all regions of the United States who had been involved in urban and regional studies met with professional urban specialists for the purpose of evaluating the present state of urban studies as an area of science. Considered were such topics as: Should this be an undergraduate or graduate field of study? What preparation should its instructors receive? What kind of graduates does it seek to produce? Is interest in the field transitory, or will it become a distinctive discipline? A report of the conference is expected to have major impact in developing a degree-granting curriculum.

Alternatives in Graduate Education

A number of experimental projects supported through the Advanced Science Education Program are exploring new approaches to curricula or seeking alternatives to the traditional Ph.D. research degree so as to meet a variety of needs for highly educated science and technology professionals. An example of the kind of experimentation now underway is the recently established "SESAME" (Search for Excellence in Science and Mathematics Education) Ph.D. program at the University of California at Berkley. In this program, a student takes all the course work required for a Ph.D. in a specific discipline and a few courses in education. After passing a qualifying examination in his scientific field, the student undertakes a thesis research project dealing with educational improvement and innovation. The program seeks to encourage work in the area of science education by science faculty members and to train students who will be qualified to teach at the college level. In fiscal year 1970, the Foundation made a modest grant for this program to the university to help with faculty-released time, educational research assistantships, and related expenses.

Several projects are trying to introduce greater relevance into graduate training for students who are both scientifically sophisticated and intensely concerned with societal problems. A student-initiated research project at Stanford University will permit graduate students to conduct a study of mass media coverage of environmental problems. The students will first construct an environmental picture of the San Francisco Bay area from available technical information and their own study of specific aspects of pollution. They will then analyze coverage of environmental problems in the newspapers, radio, and television. The two sets of information will be compared to determine the accuracy and scope of media coverage. The researchers plan to conclude the study with a seminar at Stanford University for editorial writers, environmental reporters, station managers, news directors, ecologists, and environmental specialists.

New graduate curricula in technology have received particular emphasis, especially those seeking to increase opportunities for continuing education on a part-time basis for employed individuals—generally industrial scientists and engineers. For instance, a grant to Creighton University will support a joint educational-industrial-governmental symposium to explore how universities can cooperate with industries and other organizations employing research scientists in a specific geographic area. A project at Colorado State University started out in 1967 developing videotapes for graduate instruction in several engineering areas to be taught at in-plant locations. Demand for the courses increased rapidly, so that by 1969–70, some 40 courses in a wide variety of fields were being made available to nearly 800 students at 14 off-campus locations. Moreover, the expanding clientele began to demand new graduate curriculum development. In the area of remote sensing of material resources, off-campus response was so great in 1969–70 that the enrollment exceeded the total interest in the subject at the university in all prior years combined. Hence, the university, with partial support through a 1970 grant, is now developing a major new graduate curriculum on the technology and application of remote sensing through new techniques using sound, light, radio, radar, heat, X-rays, and magnetism to monitor the environment.

UNDERGRADUATE EDUCATION IN SCIENCE

Foundation activity in support of undergraduate education in science must be sensitive to the critical nature of undergraduate education within the educational chain, as well as the diversity of the institutions in which it operates. In its position between the two other major educational "establishments" —pre-college education and graduate education—undergraduate education is the crucial connecting link between the generally uncommitted and the practitioner, for it is at the undergraduate level that serious preparation for a career begins. At the pre-college level the student may decide for or against "science;" at the college level he decides *which science,* if any, and in the 4-year college period must either acquire the knowledge and training that will enable him to pursue a career for which an undergraduate degree constitutes the necessary formal requirement, or build a foundation on which his specialized graduate training will rest.

Undergraduate colleges face a formidable task. They must take students from secondary schools with very good to very poor preparation, and they must recognize the wide variety of career choices possible and offer the necessary preparation. Further, they cannot overlook the fact that modern society requires that there be within the citizenry at least a core of educated nonscientists who understand science and its interactions with society.

The 2,550 undergraduate colleges are far more diverse in nature than either the secondary schools or the graduate schools: 12 percent are parts of major institutions offering graduate education to the Ph.D. level; another 21 percent offer graduate work through the master's degree; for 30 percent of them the baccalaureate is the highest degree granted; and another 37 percent offer something less than the baccalaureate, usually the first two years of undergraduate education. In each of these categories, institutions range in size from enrollments of less than 100 to thousands or tens of thousands. The quality of science education provided varies from totally inadequate to excellent.

Even the best of the undergraduate colleges and universities find it difficult to maintain their positions as high quality institutions—not only because of increasing enrollments and the rate at which new knowledge is being developed but also because of the changing demands imposed by increasing expectations. The orientation of many science students toward courses of study relevant to societal problems is likely to force major changes in science curricula. Advances in science and a rapidly changing technology demand curricula that will permit far greater flexibility in career choices. Scientists are also increasingly recognizing that they can no longer concentrate only on reproducing their own kind—that they must take a hand in giving the great body of students, many of them heading toward nonscience careers, an understanding of what science is, of its impact on society, and of its importance to the nation's future.

During fiscal year 1970, the Foundation assisted with the maintenance and improvement of the quality of undergraduate science education through several programs, some of them of long standing. Over 3,000 faculty members from 4-year and junior colleges participated in programs to enhance or update their knowledge and capabilities through Summer Institutes, Short Courses, or the Research Participation programs operated through College Teacher Programs. Support for the improvement of undergraduate science curriculum was provided through the Science Curriculum Improvement Program, supporting the development of teaching materials adaptable to use by a variety of undergraduate colleges; the Instructional Scientific Equipment Program, providing matching funds for the purchase of instructional equipment for use in science laboratories and demonstration lectures (suspended for fiscal year 1971); and the Pre-Service Teacher Education Program, assisting in the development and modernization of programs to produce adequately prepared teachers of science for the secondary and elementary schools.

The College Science Improvement Program is aimed at improving the overall instructional programs of institutions. Through its three sections, it provides (A) comprehensive support for the improvement of a wide range of instructional activities in individual colleges; (B) support for associations of 4-year colleges to carry on cooperative projects beyond the capabilities of an institution working independently; and (C) support for associations of junior colleges working with a major college or university on problems of curriculum and curriculum articulation. Table 7 records fiscal year 1970 funding of the respective components.

Through the Undergraduate Research Participation Program, over 3,100 students shared in the research and study activities of university scientists, thus gaining intensive exposure to both the satisfactions and frustrations that scientific work can bring. A small but significant num-

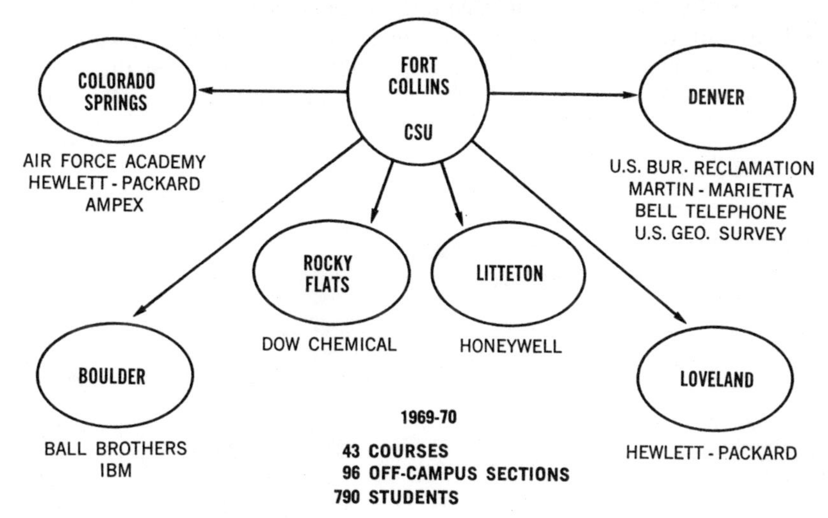

ber of special projects were funded to encourage development and testing of new ideas, concepts, and techniques in the teaching of undergraduate science.

In its programs to support undergraduate science education, the Foundation, while recognizing the need for maintaining and indeed improving the preparation of those who will be the nation's future scientists, has also turned its attention to new problems and ways to help undergraduate institutions meet their changing responsibilities. In this endeavor, several areas have begun to receive particular emphasis.

THE NEED FOR RELEVANCE

One of the major issues in undergraduate education is the need to face the question of what science has to offer in the solution of problems of intense concern to many undergraduate students, particularly the contamination and pollution of the physical, biological, and social environments. Several of the NSF undergraduate programs are encouraging the introduction into curricula of a consideration of the applications of science to a wide variety of these emerging problems.

An increasing number of college teacher programs will respond to the desire of undergraduate teachers for better understanding of the interactions of science and technology with society. For example, a Short Course on "Models of Urban Spatial Structure and Ecology" at Ohio State University is being given in response to the recent convergence of research themes in economics, geography, sociology, and social psychology which focus on the structure and organization of urban space. Problems of metropolitan government, transportation, land use, housing, employment, social conditions, and health all have important spatial dimensions which at present are only partially understood. It is the intent of the Short Course to review critically ongoing research into the spatial structure of cities and to promote the dissemination of urban research findings among a wider group of social scientists and policymakers. Another Short Course on "Engineering and the Technological Society" at Ohio University assists social science teachers in assessing the social consequences of technological development—such phenomena as machines, automation, energy resources, the computer—and encourages them to develop curricula on the role of technology in our society. A Summer Institute on "History of Technology" at the Smithsonian Institution surveys the evolution of technology from antiquity to the early 20th century, with emphasis on the interactions between technology and the physical sciences and on the unsolved problems. Discussion of these problems—requiring no costly equipment or personnel—is particularly well suited to exploitation by teachers at smaller colleges. The institute also includes opportunity for original investigations using the notable resources and personnel of the Smithsonian Institution.

At several institutions, support under the College Science Improvement Program has facilitated introduction of more "relevance" into courses and curricula. For example, the Departments of Physics and Biology at Albion College (Michigan) are concerned with revising curricula so that they recognize and attack significant current problems. Former science students and other visiting scientists, including at least one Nobelist, have served as advisors and lecturers while the college faculty are carrying out the curriculum changes. At Sweet Briar College (Virginia), the focal point of the project is the establishment of a center for ecological studies utilizing the college's exceptional campus of 3,400 acres and the surrounding rural and urban communities. Participating in this multidisciplinary activity are the Departments of Biology, Psychology, Economics, Government, Sociology, and, to a lesser extent, Chemistry and Physics.

Several proposals reaching the Undergraduate Special Projects Program indicate that, in faculty-student interactions, students are often the prime moving force. During fiscal year 1970, five grants were made to support student-originated and -managed research or study projects directed toward problems of the physical, biological, and/or social environment. For example, at Heidelberg College, five undergraduate students will conduct a study of pollution of the Sandusky River. The students plan to investigate the extent of agricultural fertilizer runoff and its contribution to the nutrient pollution of the river, the rate at which biodegradable detergents in the river break down as the water

Table 7
College Science Improvement Program

	Requests		Grants	
	Number	Amount	Number	Amount
Individual Institutions	58	$12,812,984	25	$5,235,900
College Associations	20	3,869,344	11	887,500
Junior College Cooperative Projects	26	1,927,400	10	680,800
Evaluation Contract				25,000
Total	104	18,609,728	46	6,829,200

temperature changes, and oxygen depletion below the towns of Bucyrus and Upper Sandusky. Through the project, the students hope to obtain results useful in local pollution controls, but with applications to more widespread problems; and to arouse interest and gain experience in environmental research. In recognition of the strong student interest in such problems, and to encourage students to express productively their concern for the environment, the Foundation announced (on Earth Day, April 22, 1970) its intention to initiate a new program, Student-Originated Studies (SOS), to provide support for interdisciplinary groups of college and university students prepared to undertake a search for solutions. It is expected that the first grants under SOS will be made in fiscal year 1971, with project activities beginning in the summer of 1971.

The Junior Colleges and Technician Education

Since 1965, when NSF institutes and short courses for college teachers were opened to applications from junior college teachers, the number of participants from junior colleges has risen steadily. In the spring of 1968, at a combined meeting of project directors of all college teacher programs, a full session was devoted to a discussion of the college-parallel programs offered in junior colleges, the project directors being joined by a large number of junior college representatives. That meeting culminated in suggestions for an entirely new cooperative program for junior colleges, directed toward developing better articulation between the programs of the 2-year colleges and the upper-level programs of 4-year colleges and universities. The proposed program, now identified as one component of the College Science Improvement Program, was initiated in the following year and has, since its beginning, provided funds amounting to $2.7 million for support of 41 associations of junior colleges (a total of 467 colleges) in 23 States, each association working in cooperation with a major nearby college or university.

The Foundation recognized, even then, that in concentrating its attention on college-parallel science courses, it was essentially ignoring one of the important fields of activity of many of the junior colleges—that of providing training for students in technical fields. The extent to which the Foundation should become involved in the education of technicians and technologists was at one time a question of major concern. The question as to "whether" was resolved in 1969 with the decision to support several institutes dealing with technician-training subjects. Fiscal year 1970 saw expanding support for technical education, most of it directed toward development of curricula and teaching materials for training of the kinds of technicians now needed to provide adequate backup for scientists and engineers. For example, the Chemical Technician Curriculum Project (ChemTec) has begun development of curriculum materials for a 28- to 30-semester-hour chemistry core for a 2-year college level program in chemical technology. Located at the Lawrence Hall of Science in Berkeley, Calif., the project developed in summer 1970 texts, laboratory experiments, film-loops, and other teaching materials, which are being field tested in 12 pilot schools during academic year 1970–71. The experience accumulated will be used to revise the materials in the summer of 1971 and prepare them for release through conventional commercial channels. So that the training will be consistent with the abilities and interests of the target group of students, the curriculum will emphasize laboratory work and direct "hands-on" experience. The aim is comprehensive coverage of basic chemical subject matter which, through modular organization of the content units, will encourage the inclusion of locally important topics and modifications.

A major study of engineering technology education is being conducted by the American Society for Engineering Education as a guide to later developments and to institutions engaged in the training of technicians and technologists. Included is a broad survey of all technology education — 2-year, 4-year, and graduate programs—together with an assessment of the industrial demand for the personnel output at the various levels of training, and suggested curriculum accreditation standards.

Based on experience with projects thus far supported, and in response to a number of recent studies indicating a continuing shortage of adequately trained technicians and technologists, plans have been developed for introduction, in fiscal year 1971, of a new program specifically oriented toward Technical Education Development.

Pre-Service Teacher Education

The Pre-Service Teacher Education Program has, up to fiscal year 1970, been concerned with the education of prospective elementary and secondary school teachers. During the past year, however, concern about the preparation of teachers at the college level increased considerably—not about their knowledge of the subject matter of science, but about their acquaintance with the teaching-learning process. There is questioning, too, of the preparation for teaching offered in the course of earning the Ph.D., even though many who take this course will eventually become teachers in undergraduate colleges. Fellows and research assistants may have little chance to learn about the rewards of a career as a teacher;

teaching assistants often have a distressing or disappointing experience, as too many are expected to function without benefit of guidance, to the detriment of their undergraduate students. Several conferences held during the year have addressed the problems in current patterns of using teaching assistants. Consideration is being given to providing support in some form that, while meeting the needs of the graduate students acting as assistants, will focus on providing better instruction for the undergraduates.

Problems of Communication with the Academic Community

Because adequate communication between the National Science Foundation and the college and university community is an integral part of developing effective support programs for undergraduate science education, strong efforts have been undertaken to further exchange of information. Mass mailings of lists of projects conducted under the Undergraduate Research Participation Program and the various Programs for College Teachers seem to be effective in apprising prospective participants of available opportunities, judged by the number of applications for participation. On the other hand, preliminary proposals for special projects and curriculum improvement indicate that the academic community is not sufficiently informed about projects and developments already under way. It is in this area—the dissemination of information about ongoing activities or about materials already developed and in many cases available for distribution—that avenues of communication seem ineffective.

Attempts were made during fiscal year 1970 to broaden these avenues in two ways. In February 1970, abandoning the project directors' meetings usually conducted program by program, arrangements were made for a more comprehensive meeting, bringing together for the first time project directors of all NSF undergraduate programs. This group, almost 1,000 in number, spent 3 days in Washington hearing about and discussing topics of current importance in undergraduate science education. Some of the topics covered were: instructional technology, new course patterns, science for nonscience students, student-originated research, graduate teaching assistants, preservice teacher education, and technology education.

The other attempt took NSF staff members out of Washington, closer to the scene of activity in the colleges and universities. For periods of 2 weeks, NSF staff were present in each of three major cities—Atlanta, Boston, and Minneapolis. Colleges in each of the areas were notified well in advance, and appointments were scheduled enabling individual faculty members or, more often, groups of faculty members, from area schools to discuss Foundation-related matters with NSF staff members on duty during the period. The response was greater than expected; during each 2-week period the staff were visited by, on the average, some 200 faculty members representing 40 colleges. Plans are now being developed to extend the operation to seven other areas during academic year 1970–71.

PRE-COLLEGE EDUCATION IN SCIENCE

Through the Division of Pre-College Education in Science, the Foundation administers programs for the development of scientific talent, the supplementary training of teachers, and the improvement of school science programs for all students. By far the largest portion of funds available for this level, nearly 75 percent in fiscal year 1970, is devoted to in-service training for secondary school teachers and supervisors through the various institute programs. These programs cover all the scientific disciplines and include a broad range of activities from short briefing conferences on new teaching materials to intensive studies during the academic year and/or adjacent summers. Nearly 50,000 individual teachers and science supervisors will participate in institute programs during the summers and academic year of 1970–71.

In addition, about 6,300 elementary and secondary school teachers will receive in-service training through Cooperative College-School Science (CCSS) projects. The primary objective of CCSS is the effective introduction of new teaching materials and methods in a school system or related group of schools through a plan developed jointly with a university whose staff then helps with implementation.

Course Content Improvement projects are concerned with developing better instructional materials for science education from kindergarten through the 12th grade, ranging from single-topic pamphlets to multimedia courses, from equipment for students to resource materials for teachers. Included also are Resource Personnel Workshops to train leaders who will then initiate local in-service programs for the effective use of new curriculum materials.

The Student Science Training Program for developing scientific talent is a relatively small component of pre-college activities, accounting in fiscal year 1970 for 3.7 percent of the total budget to support training for some 5,500 students. These special study opportunities in science and mathematics for high school juniors of outstanding ability not only further their science interests, but help them in career decisions which usually begin to crystallize at this level.

The way in which the various pre-college programs interact is il-

Students investigate gravitational potential energy in Physical Science II class. This course is intended for the middle segment of the high school population but can serve as a preparation for more specialized courses in science. (Photo Educational Development Center)

lustrated by several examples of the kind of approach made to problem areas that represent priorities for fiscal year 1970 and the coming years.

Second Generation Curriculum Efforts

During the past decade, Foundation support has been predominantly focused on strengthening the quality of instruction in the schools as they now function. The new courses developed and the programs for teacher training have emphasized up-to-date content and student experimentation and inquiry, but they were designed to fit established curriculum guidelines. The disciplinary structure of subject matter characteristic of the senior high school was maintained, with the middle school (7th through 9th grades) representing a transition stage to the basic disciplines of chemistry, biology, and physics. Mathematics is unique in the sciences in having a separate disciplinary niche throughout all levels of instruction.

With Foundation support, curriculum materials have been developed in mathematics and the natural sciences which can now be sequenced to provide a variety of curricula from kindergarten through high school, and a substantial start has been made in assisting schools in their implementation. The present stage of development can be regarded as a first generation effort. Two basic problems must now be confronted:

—Should there be a next round of curriculum effort, and, if so, what should be its directions?

—How can the Foundation help improve the learning environment in the schoolroom? This includes not only problems of teaching competence and implementation of better curricula but also support of new ways to better utilize time and space.

In the consideration of these problems, new objectives have been identified and activities initiated or planned.

Scientific Literacy vs. Scientific Manpower

The first efforts supported in curriculum development in the early sixties were aimed at developing materials for the students who generally enroll in high school science courses, ranging from 80 percent in biology to 25 percent in physics. Some critics argue that these courses are too rigorous and sophisticated for nonscience students. For the junior high school and earlier, where all students take science courses, the materials have been specifically designed to fit the requirements of all students insofar as that can be done. It is now clear that a second generation of materials is required which provides more options and is geared to the needs of students not necessarily intending a scientific career. At the same time, efforts will continue to support development of some materials for scientifically gifted students when such needs are documented. For example, in fiscal year 1970 a grant was awarded to Columbia

University for an integrated mathematics sequence in grades 10–12. This is deliberately designed to serve high-ability students who are likely to become mathematicians. Over the last several years, however, support of materials for talented students has amounted to less than 10 percent of total pre-college course improvement, a proportion likely to be maintained for the present.

Interdisciplinary Courses

A substantial investment has been made in the development of interdisciplinary courses for the junior high school level. An example of one of these will illustrate how curriculum development proceeds, and what subsequent steps are necessary to implement the new courses.

The Earth Science Curriculum Project (ESCP) was initiated in the summer of 1964 through a grant to the American Geological Institute. At that time, spurred by the nation's space exploration program and by increasing concern with the physical environment, rapid growth in earth science instruction took place in junior high school, but it was severely hampered by a dearth of good instructional material. Hence, earth scientists and educators planned, developed, and tested curriculum materials for a modern, interdisciplinary course. The textbook, *Investigating the Earth,* published in 1967 by Houghton Mifflin, is intended for use in the 8th or 9th grade, and covers the disciplines of astronomy, meteorology, oceanography, geophysics, geology and geography. The materials also include a laboratory manual, teacher's guide, field study guides, laboratory equipment and experiments. Current concerns of the study group are teacher training, both in-service and pre-service, implementation, and evaluation related to the use of the materials.

The Science Curriculum Improvement Study is developing ungraded, sequential physical and life science programs for the elementary school—programs which in essence turn the classroom into a laboratory. A child's elementary school years are a period of transition as he continues the exploration of the world he began in infancy. Extensive laboratory experiences at this time will enable him to relate scientific concept to the real world in a meaningful way. (Photos University of California, Berkeley)

Assistance to schools in the effective use of the materials has been provided through teacher training institutes, cooperative college-school science projects, and resource personnel workshops which provide training for teachers who then become leaders for in-service training projects in their own areas.

An illustration of the implementation of ESCP materials is exemplified in the State of Missouri. A resource personnel workshop held at the University of Maryland in the summer of 1968 trained teams from all sections of the country in the use of ESCP materials. The teams consisted of a college or university scientist, a school administrator, and a classroom teacher-leader. The team from Missouri organized a "second generation" leadership conference supported by the Foundation in the summer of 1969 which was designed to prepare teams from eight Missouri colleges and universities for a State-wide in-service ESCP training program. This was followed by an In-Service Institute project, starting with a 1-week preliminary session involving the participants from the eight separate in-service institutes scheduled to begin locally in September 1969. The result of these activities is described by John Hooser, Missouri State Science Supervisor and a member of the initial Missouri team.

> "Prior to the earth science project, only a few schools offered earth science. The State Department of Education did not offer certification in earth science. Since this project, there are 126 school districts, 231 teachers, 761 sections, and 21,000 students involved in earth science in Missouri. The State Department of Education has certified requirements for earth science. The first year of funding has made it possible to involve 110 to 115 teachers with eight colleges in this State-wide earth science project. This is approximately half of the teachers presently teaching earth science in the State. The teachers enrolled in this program will fulfill about half of the required hours of certification."

Grants have been awarded in fiscal year 1970 for the continuation of the in-service program in the eight colleges. Grants for ESCP teacher training and implementation activities awarded in fiscal year 1970 throughout the country include 18 summer institutes and conferences, 54 in-service institutes, and 11 cooperative college-school science projects. These projects will provide a variety of training opportunities in the use of ESCP materials for over 2,000 junior high school teachers. It is estimated that over a third of the million students now taking earth science in the secondary schools of the country are using ESCP materials.

Finally, to round out the set of activities ultimately required for curriculum and teaching reform, the ESCP has this year received an NSF grant through the Division of Undergraduate Education in Science for a teacher education project in which the ESCP will assist and coordinate a consortium of colleges across the country in the establishment of undergraduate curricula for prospective earth science teachers geared to current materials and techniques. Included as participants in the project will be Boston College, Southern Illinois University, Minot (North Dakota) State College, California State College at Fullerton, the State University of New York College at Oswego, Western Connecticut State College, and Colorado State College.

Social Sciences Curriculum Materials

The social sciences represent a major curriculum area in which development of suitable materials has lagged rather far behind that for mathematics and the natural sciences, with large gaps over the whole kindergarten to 12th grade range. Because there is no discernible consensus as to what an integrated sequence might be, the Foundation has supported the development of specific high school courses in geography and sociology, and of supplementary materials in sociology and anthropology for use in social studies courses. An interdisciplinary effort was initiated this year, led by Dr. Irvin De Vore of Harvard, to create a behavioral science course for the intermediate level on the general theme, *Exploring Human Nature.*

At the elementary level, an innovative 5th grade course neared completion in fiscal year 1970. This course, *Man—A Course of Study,* is based on three questions framed by Jerome S. Bruner, the principal developer: "What is human about human beings? How did they get that way? How can they be made more so?" The first half of the course concentrates on the life cycles and behaviors of salmon, herring gulls, and baboons. These studies lead students to assess the significance of generational overlap and parental care, innate and learned behavior, group structure and communication, and their relevance to the varying life styles of animal species, including the human species. The second half of the course is an intensive study of man in society—as culture-building, ethical creatures, toolmakers and dreamers. The Netsilik Eskimos of the Canadian Artic are studied in depth, because their society is small and technologically simple, yet universal in the problems it faces. Course materials rely heavily on research sources and present subject matter through a variety of media, including films, filmstrips, records, posters, and booklets.

This project posed an unusual distribution problem. Although the teachers and children involved in the trial phase of the development were enthusiastic about the course, commercial book publishers and film distributors were unwilling to contract for publication because of the variety of materials to be handled and the unconventional subject matter. Support therefore was made available to the sponsor, on the basis of a revolving fund award, to conduct a quasi-commercial publication and distribution operation to demonstrate the general public acceptance and the commercial feasibility of distributing the materials. The success of this venture is attested to by the fact that a publication contract has recently been executed with Curriculum Development Associates. Further implementation is being aided through the development of resource teams in teacher training institutions and through cooperative college-school science projects.

Teacher Participation in Course Development

One criticism of curriculum materials developed by nationally constituted groups of scientists and educators, and distributed through commercial channels, is that this procedure deprives local teachers of the opportunity to contribute their own creative efforts and develop their own ideas. As one approach to the problem, a grant was made this year to Indiana University for a project directed by Robert A. Hanvey to develop materials for supplementary units in cultural anthropology for secondary school social studies courses. These units will treat the topics of *Biological and Social Differentiation of Man* and *Science, Technology and Change,* each occupying from two to four weeks of class time. The final materials, instead of commercial textbooks, will be "unfinished" outlines, syllabi, and resource materials that may be fleshed out and refined to suit the teacher's specific needs and taste. Opportunities can then be provided through institutes or other means for individual teachers to complete adaptations for their own classes. To serve this kind of approach, the Summer Institute Program is encouraging prospective directors to submit proposals for special institutes in all disciplines that will permit teachers, with leadership from university scholars, to develop their own curriculum ideas which they do not have the time or resources to pursue during the teaching year.

Supervisory and Resource Personnel

There are two major problems in the introduction of new courses and methods into school curricula. One of these is the dissemination of information to administrators and supervisors sufficient to enable them to reach decisions on curriculum adoption. To try to meet this problem, the Foundation has begun support of short courses for administrators and science supervisors to acquaint them with new materials that are available. During the summer of 1970 the Foundation is supporting nine conferences for secondary school principals designed to provide information on curriculum developments. Also, the Association of Secondary School Principals cooperated this year in arranging informational workshops which were conducted during the annual meeting of the association. In addition, conferences for State science supervisors and State mathematics supervisors have been supported for the past 3 years. These conferences are concerned with current science education activities and problems; for example, this year's theme of the science conference was environmental education in the secondary school curriculum.

A second aspect of the implementation problem is the training of supervisors, subject-matter specialists, and resource personnel in the content and methods of new courses so that they can serve effectively as leaders in teacher training and implementation. Two avenues of support have been provided through pre-college programs. One of these is the Academic Year Institutes Program. Last year, the first academic year institute designed expressly for experienced, practicing supervisors was held at the University of Maryland, under the leadership of David Lockard. The applicant response to this innovation was so encouraging that a second such project is to be conducted by Ohio State University this year. Other academic year institutes stress intern training for science and mathematics supervisors-to-be, the placement of project graduates, and rigorous discipline orientation for subject-matter specialists in specific areas needing this kind of expertise.

The Resource Personnel Workshops approach the same problem in a different way by developing leaders in colleges and schools with sufficient in-depth understanding of one or more new curricula to initiate teacher training activities in their own schools and colleges reflecting the content and spirit of the new materials. This program, initiated as an experiment in 1967 with six grants, has expanded this year to 27 projects at a cost of over $1 million with provision for approximately 1,800 participants. The need for leadership training has been particularly acute at the elementary level since there is no existing cadre of experts for the interdisciplinary courses at this level analogous to college physicists, chemists, or biologists in the case of new high school courses in those disciplines. Demand for the elementary school materials is increasing as they become widely available, but familiarity with them has tended to be restricted to those science educators and teachers who had participated in their development.

Hence, the workshops are filling a real void in developing the resources necessary for implementing improved curricula.

One example of the "multiplier effect" of the workshop projects has been cited earlier in the Missouri implementation of ESCP materials. Another instance is the leadership development project in *Science—A Process Approach,* the American Association for the Advancement of Science elementary science curriculum, at Pennsylvania State University. The initial grant was made in 1968; follow-up studies in May 1969 (85 percent of those trained responded) revealed that the participants had trained 2,050 teachers, who in turn instructed almost 56,000 children in 1969–70. Moreover, those participants who are members of college faculties have already incorporated about 200 hours of instruction from the in-service course into their pre-service course for approximately 550 college students preparing to teach.

Experimental School Activities

To help each student learn at his own pace and to the extent of his own abilities, schools need to modify their rigid concepts of how to organize time and facilities, particularly at the elementary level. This is a difficult problem area for the Foundation which has historically been concerned with the support of science and mathematics education activities only. And yet, the problem is so critical and its scope so broad that joint funding with other agencies deserves consideration to support integrated educational efforts, with the Foundation responsible for the science and mathematics components. In this connection, Max Beberman at the University of Illinois received support this year for an intensive study of the suitability for U.S. elementary schools of the English "Integrated Day" approach. The main objective of the project is to establish at an experimental public school a working model of a total educational program which takes into account individual differences through using broad themes as vehicles for integrating the various traditional school subjects. The children work on projects, both individually and in groups. Emphasis is to be placed on interrelations between mathematics and science, the use of laboratory equipment and experimentation, and the invention of teaching procedures and student practices which develop the ability to reason.

Experiments of this kind appear to have great educational effectiveness in England, and offer promise for U.S. schools. It is therefore of prime importance to investigate the exportability of this kind of school experience to the American scene. The Foundation expects to extend this type of support to other experiments for reorganizing the structure of educational processes.

Program Evaluation

With rapid changes in the current climate of education, the present time is a critical one for reconsideration of NSF programs in precollege education. The educational structure and its modes of operation for which the established NSF educational programs were conceived seem to be undergoing irresistible pressures for reform. Earlier evaluations are now for the most part obsolete and often inadequate to provide guidance for planning purposes. Hence, the Foundation intends to place increased emphasis on evaluation. Initial steps were taken in fiscal year 1970 to evaluate the impact of the Academic Year Institutes Program with distribution of a questionnaire to all participants in this program since its inception in 1956. From the returns, the staff will derive information on how influential this program has been in effecting changes in secondary schools to date, and whether the investment in time and money for the teacher participants has paid off and will be likely to pay off in the near future.

Data are also being collected from a sample population of science and social science teachers. This study, undertaken by contract with Vitro Corporation, will establish a baseline of teacher characteristics of the present era, so that past and future comparisons will be possible. Student participants in the pre-college programs are also being followed up in a contract with the American Council on Education. Of particular interest is a comparison of the influence of the Student Science Training Program on the current generation of high school and college students as compared with those of a decade ago, since comparable data were collected in 1960.

Beyond the appraisal of the effects of individual programs, the Foundation is now exploring ways of studying the total impact of precollege programs through observation and data collection of change in science education in individual classrooms and school systems.

PUBLIC UNDERSTANDING OF SCIENCE

The overall objective of this program is public education with respect to science and technology, so that citizens may function more effectively in a technological society. This involves communicating not only the "facts" of science but also some appreciation of the relationship of science to other forms of scholarly investigation and some understanding of the scientific and technological aspects of societal problems. In fiscal year 1970 the Foundation made 15 awards, amounting to $212,488, for public understanding of science projects. The mechanisms employed included conferences, summer courses, an exhibit, design of a film series, curriculum development for an adult education program, and a State-wide information program on the scientific aspects of environmental pollution.

A recent issue of *Impact of Science on Society* discusses the problems of bringing about a public understanding of science. In one article Miguel Angel Asturias, 1967 winner of the Nobel prize for Literature, states: "In our day science and literature seem so far removed, so widely separated from one another that a poet or writer like myself looks with timid respect on everything relating to science, scarcely daring to inquire into, to glance at, the awesome discoveries of the scientists. There are those who speak, not unjustifiably, of a veritable schism in what is called Western culture, a schism which, at its most extreme, leads not a few men of letters and artists to ignore and despise the scientists and the technicians . . . ."

The problem is enormous and is compounded of a lack of knowledge of the humanistic origins of science, a confusion between science and technology, and a concern with the misuse of technology. What is worse—a misunderstanding of science and technology (and the difference between the two) is shared by the uneducated at all ages, by educated adults, and by many of our brightest youth.

The support of two seminars on the Impact of Science and Technology on Society for women community leaders and undergraduate women students represents an approach to the less scientifically oriented of our two sexes. These are to be carried out (one in September 1970 and one in January 1971) by the Oak Ridge Associated Universities, which in the past conducted similar programs for humanities professors and practicing clergymen.

Joint support of a Dialogue on the Identity and Dignity of Man with the National Endowment for the Humanities was an attempt to focus scientific knowledge and humanistic wisdom on such problems as Control of Population and Regulation of Behavior, Extension of Life Through Organ Replacement, and the Improvement of Life Through Genetic Manipulation. The dialogue was conducted at Boston University in conjunction with the annual meeting of the American Association for the Advancement of Science.

A Summer Session on the Quality of Life, supported by a grant to the Institute on Man and Science, helped a group of scientists, educators, doctors, lawyers, publishers, politicians, clergy, housewives, and students to explore the interactions of science, technology, and human values as they converge on environmental concerns. Predictions concerning the environmental state of the world in the near future were examined from the standpoint of solid experimental evidence, and participants sought solutions to real case problems such as the decision on where, if anywhere, in a given State to build nuclear reactors for increased electrical power.

Institutional Programs

Foundation programs for improving and sustaining science in institutions of higher education began to undergo substantial reorientation in fiscal year 1970. A major change in the basis for computing Institutional Grants for Science greatly increased the number of institutions eligible to receive these flexible funds. Two separate programs designed to develop science in doctoral-level universities were replaced by a single Science Development program. And the oldest of the Foundation's institutional programs —Graduate Science Facilities—was discontinued as a discrete grant-making activity and became contributory to the new emphases planned for institutional development. Most of the changes in organization and function came late in the year, however, and their results will provide the content of future annual reports.

NSF obligations under the institutional programs discussed below appear in table 8.

SCIENCE DEVELOPMENT

For several years the Foundation has been making a large and sustained effort to increase the number of universities capable of conducting distinguished programs of education and research in the sciences. The widely known University Science Development (USD) activity, initiated in March 1964, aimed to help very good universities to become excellent. USD has normally provided funds to improve several science departments in an institution. A related program, Departmental Science Development (DSD), begun in fiscal year 1967, has focused on a single department or area of science within a university. In both programs the Foundation's intention has been to assist universities in the achievement of their long-range science goals, and the grants have been predicated on substantial commitments of the institutions' own resources to the execution of their development plans. The 3-year grants under the DSD program have not been renewable for the same department, but the USD program has usually offered the prospect of 2 years of supplementary support if the initial grant resulted in the anticipated progress.

In fiscal year 1970 the Foundation obligated $15.9 million through the University Science Development program. Supplementary awards to the University of Colorado, the University of Georgia, Louisiana State University at Baton Rouge, the University of Oregon, and the University of Rochester accounted for $10.3 million; 3-year grants to Brandeis University, Northwestern

Table 8
Obligations for Fiscal Years 1968, 1969, and 1970
(Millions of dollars)

Program	Fiscal year 1968		Fiscal year 1969		Fiscal year 1970	
	Number of awards	Amount	Number of awards	Amount	Number of awards	Amount
Science Development:						
University Science Development	9	$29.6	9	$23.1	9	$15.9
Departmental Science Development	29	12.0	15	8.6	18	10.6
Graduate Science Facilities	50	17.8	14	6.0	15	4.0
Institutional Grants for Science	497	14.2	(1)	(1)	634	14.5
Total	585	73.6	38	37.7	676	45.0

[1] A change in the timing of awards from June 1969 to fall 1969 resulted in no obligations in fiscal year 1969.

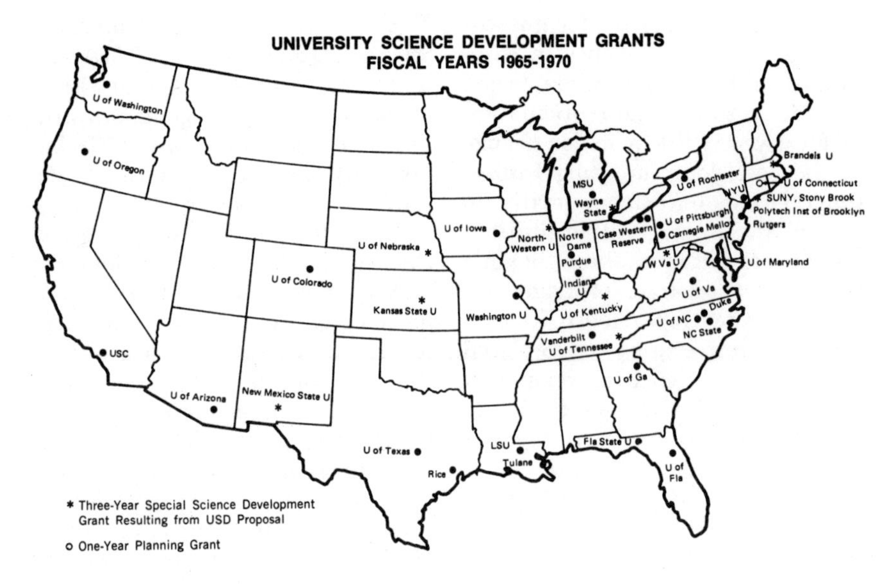

University, and the State University of New York at Stony Brook, and a 1-year grant for curriculum development and planning to the University of Connecticut, accounted for the remainder of the obligations. The Foundation has awarded $168.7 million through the USD program since its beginning. Thirty-one institutions have received large, multidisciplinary awards, and of this group 11 have thus far qualified for supplementary grants. Counting the supplements, the average NSF development support for the 31 institutions already amounts to $5 million. Eleven other institutions have received special awards based on parts of their science development proposals. These special grants are somewhat less comprehensive and are smaller in amount, although one amounted to $2 million and three others to more than $1 million.

Through the Departmental Science Development program the Foundation obligated $10.6 million in 18 awards in fiscal year 1970. Since the program began in fiscal year 1967, 59 grants amounting to $33.1 million have been made to 54 institutions in 32 States. (Five universities have received two awards.) The grants have averaged $560,000.

In both programs public universities have received more funds than private institutions—62 percent of the total in the USD program and 65 percent in DSD.

The DSD program has especially emphasized the improvement of the quality of faculty and graduate students as the principal means of institutional development. Nearly two-thirds (65 percent) of the DSD funds have been allocated for manpower. The comparable figure for the USD program is 40 percent. One important difference between the two programs is that only small amounts of DSD grants have been for renovation of facilities, whereas nearly one-fourth of the funds of the USD grants has been for construction or renewal of science buildings. Both programs have allocated a substantial share of their funds for the purchase of equipment and supplies and some funds for library resources, computer costs, and travel.

By field of science, slightly over half of the USD program funds was for physical sciences and about one-sixth for life sciences. Engineering accounted for 12 percent, mathematics for 11, the social sciences for 6, and environmental sciences for 3. A considerably larger share of DSD program funds has been awarded for the social sciences—13 percent in all years and 22 percent in fiscal year 1970.

The Foundation has expected institutions receiving development grants to make contributions of their own funds to their improvement. Thus, the institutional commitment under the DSD program amounts to nearly one-half of the estimated total development cost; the $33.1 million in DSD grants amounts to only 17 percent of the total; the remainder (about one-third) of the development costs is expected to come from other sources. Similarly, under the more expensive development activities aided by USD grants, institutions have committed themselves to make contributions greater than the Foundation. Thus far, most institutions in the USD program have contributed from their own resources at least as much as they had initially projected, and sometimes considerably more. Experience under this program, which is greater than under the newer DSD program, indicates that universities can construct realistic plans which set important and attainable goals; that the NSF grants have stimulated improvement not only in the departments supported but often in other parts of the institution; and that the achievement of an institution's primary goals justifies supplementary NSF investment in support of its further planned development.

One of the chief hopes for the USD program was that it would eventuate in the emergence of very high quality universities in areas of the country having none or too few such centers. A similar goal to strengthen resources for advanced scientific education and research in as many regions and population centers of the nation as possible animated the DSD program. The accompanying maps show how this purpose has thus far been attained through the development programs. Underlying this policy of geographic distribution has been the desire to further the national goal of equality of opportunity for higher education and to help achieve equitable distribution of the beneficial effects of strong educational and research centers.

The separate University and Departmental Development programs were incorporated in a new Science Development program near the close of fiscal year 1970. Besides making departmental grants, this reshaped institutional development program will provide continuing opportunities for supplementary grants, as well as other forms of developmental support, to the institutions that have already received awards through the USD program. The program will also expand institutional development activities of the Foundation in new directions by seeking to stimulate the development of institutional capabilities in the social sciences and interdisciplinary areas so that the recipient universities can effectively participate in the solution of important problems confronting society. The nation's foremost universities, which have formerly been discouraged from applying for NSF development funds, will be eligible for support under some of the new categories of institutional development. Also, in attempting to develop centers or institutes focusing on national problems, the Foundation will foster concerted efforts of a variety of institutions, nonacademic as well as educational.

GRADUATE SCIENCE FACILITIES

Graduate Science Facilities, the first of the Foundation's institutional programs, completed its 11th year in fiscal year 1970. Aimed at sustaining the strength of graduate-level science departments, the program provided funds for the renovation and construction of academic facilities for research and research training. In fiscal year 1970 the Foundation obligated through the program $4 million to 15 institutions. The average grant was $267,178, substantially below the average of $446,000 in fiscal year 1969. As noted above, the Foundation has decided to discontinue, for a year at least, Graduate Science Facilities as a separate program and to use its resources in the reoriented science development activity.

During the 11 years since its inception, the Graduate Science Facilities program provided $186 million to 179 different institutions of higher education. The recipient institutions were required at least to match the NSF funds. They usually overmatched. The actual facilities constructed with NSF support have cost about half a billion dollars and accommodate approximately 40,000 academic personnel in research and graduate education. The 10 million net square feet of space they use is comparable in size to the entire academic facilities of Michigan State University (East Lansing) and the University of Maryland (College Park) combined.

DEPARTMENTAL SCIENCE DEVELOPMENT GRANTS
FISCAL YEARS 1967-1970

INSTITUTIONAL GRANTS FOR SCIENCE

Through its program of Institutional Grants for Science the Foundation provides funds for the general support of science in U.S. colleges and universities. Campus officials determine how the grants will be used, and this discretionary nature of the funds makes them uniquely adaptable to local circumstances.

In fiscal year 1970 the Foundation made an important change in the program. In earlier years the grants had been computed by applying a graduated arithmetical formula to the amount of NSF research

Dreyfus Chemistry Building at Massachusetts Institute of Technology. (Photo MIT)

and research-training support received by an institution. The fiscal 1970 awards were computed on a much wider Federal base—the research obligations to colleges and universities of the Departments of Agriculture; Commerce; Defense; Health, Education, and Welfare (exclusive of the Public Health Service); Housing and Urban Development; Interior; Transportation; the Atomic Energy Commission; the National Aeronautics and Space Administration; the National Science Foundation; and the Office of Economic Opportunity. (The Public Health Service awards were excluded to prevent overlap with similar formula-grant programs of the National Institutes of Health.) In addition to these Federal research obligations (for fiscal 1968) the Foundation continued to include in the computation base awards made through the NSF programs of Undergraduate Research Participation and Research Participation for College Teachers.

The extension to a broad Federal base resulted in the eligibility of many institutions that had not participated in the program before. The largest number of Institutional Grants in any earlier year had been 517; in fiscal 1970 grants totaling $14.5 million were made to 634 institutions. The formula continued to provide 100 percent of the first $10,000 of an institution's base figure, but the subsequent percentages were much smaller than in earlier years. The largest grant ($138,967) amounted to less than one-fifth of one percent of the amount on which it was based. Some institutions that had participated in the program before benefited from the shift to a broader Federal base, but most did not. The average grant dropped from $28,410 to $22,894, and the median grant from $13,256 to $10,800. Eighty institutions that had received fiscal year 1968 grants suffered reductions of 30 percent.

Since the beginning of the pro-

Table 9
Uses of Institutional Grant Funds
Fiscal Years 1962–69
[Millions of dollars]

A. Type of use:	Amount spent [1]	Percent of total expenditures
Equipment and supplies	$34.8	50.6
General	32.0	46.5
Libraries	2.8	4.1
Facilities	11.2	16.3
General	8.1	11.7
Computers	3.2	4.6
Personnel	19.3	28.1
Faculty salaries	9.0	13.1
Graduate assistants	3.9	5.6
Other student stipends	1.9	2.7
Visiting lecturers	1.1	1.6
Technicians' salaries	1.8	2.6
Other	1.7	2.5
Travel	1.7	2.5
All other	1.7	2.5
Total	68.9	100.0

B. Field of science:	Amount spent [1]	Percent of total expenditures
Physical sciences	$24.6	35.7
Astronomy	1.0	1.5
Chemistry	12.5	18.2
Physics	10.1	14.7
Other	0.9	1.3
Mathematical sciences	3.4	5.0
Environmental sciences	5.5	8.0
Atmospheric sciences	0.8	1.1
Earth sciences	3.9	5.6
Oceanography	1.0	1.4
Engineering	8.6	12.5
Life sciences	14.7	21.4
Psychology	2.7	3.9
Social sciences	4.2	6.1
All other (inter- and multidisciplinary)	5.2	7.5
Total	68.9	100.0

[1] From awards made fiscal years 1961–68. Total amount of awards, $79.4 million; total expenditures fiscal years 1962–69, $68.9 million.

NOTE: Totals do not add because of rounding.

gram in fiscal year 1961, the Foundation has made Institutional Grants amounting to $94 million to 820 colleges and universities. Many of these institutions have participated in the program every year.

As table 9 shows, about half of the funds has been spent for equipment and supplies; about one-sixth for construction, renovation, and computer costs; over one-fourth for personnel; and small amounts for travel and other uses. By field of science, more than one-third of the funds has been allocated to the physical sciences and more than one-fifth to the life sciences. The social sciences and psychology if combined accounted for one-tenth of the total expenditures.

Although the grants are not large, their flexibility makes them unusually useful for such purposes as the following: ensuring a backup for commitments and freedom from normal budgetary constraints; making available small research grants for new faculty members; providing means of keeping graduate students on campus during the summer and of speeding up the earning of degrees; encouraging undergraduate research and interest in scientific careers; facilitating the employment of new faculty members; bolstering neglected departments or areas and maintaining balance; breaking down traditional barriers between departments and colleges; fostering the development of central services used by several departments; and exploring new means of instruction and new fields of research. Local control and ready availability of the funds have permitted institutional officials to respond quickly to unanticipated needs and opportunities. In such ways the grants have helped to maintain the strength of academic science during a period of growing financial constraints and have helped to uphold institutional autonomy.

Computing Activities in Education and Research

Computers and related methodologies exert a pervasive influence on research and educational efforts in many disciplines, and particularly in interdisciplinary projects. Hence, various programs throughout the Foundation are involved with computer-related activities, but the Office of Computing Activities is the primary focus for support and coordination of such projects. Since this office was established in 1967, it has developed a variety of programs to promote research, to foster educational innovations, to explore training techniques, to assist the improvement of academic resources, and to promote institutional cooperation in the area of computing.

The figure below shows the distribution of fiscal year 1970 funds by major program categories, and table 10 gives the history of awards. Areas of emphasis during the year are illustrated through a sample of projects supported.

EDUCATION, RESEARCH, AND TRAINING

The technologically advanced computing industry is built on a rather narrow research base, so academic work in computer science helps broaden this base while training future specialists. One of the

COMPUTING ACTIVITIES IN EDUCATION AND RESEARCH BY AREAS OF EMPHASIS, FISCAL YEAR 1970

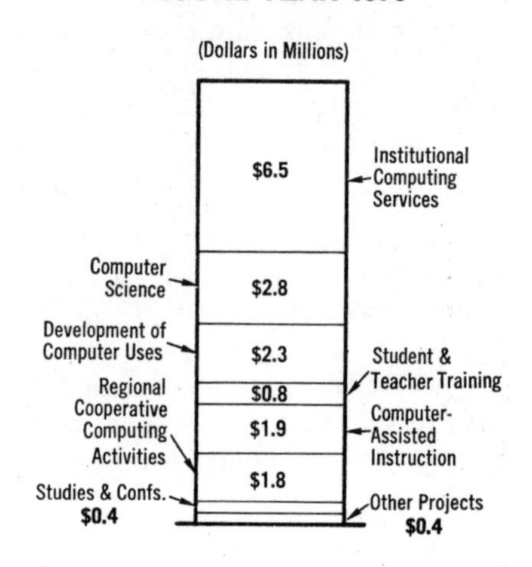

principal features of current research is the attempt to bring structure and definition to computer science through providing the experimental evidence and theoretical understanding which will permit guided development. Other thrusts involve studies of the implications of new technology for hardware and software and efforts to extend the utility of computers.

Investigations of theoretical foundations of computing are being supported at various universities throughout the country. A grant to the University of California at Los

Table 10

Computing Activities in Education and Research
Awards by Program Categories, Fiscal Years 1968, 1969, and 1970

[Millions of dollars]

Section	1968		1969		1970	
	Number	Obligations	Number	Obligations	Number	Obligations
Education, research and training	64	$6.1	116	$5.9	89	$6.0
Institutional computing services	42	10.6	23	6.5	23	6.5
Special projects	67	5.3	55	4.6	75	4.5
Total	173	22.0	194	17.0	187	17.0

Angeles will enable experiments with computer systems to determine parameters which can be measured as sensitive indicators of systems performance. Also, the development of very complex integrated circuits with components a few thousandths of an inch in size points to new hardware and software possibilities. A project at the University of Texas is concerned with micro-programming, in which the operations of the computer are built up in a flexible way from very simple, fundamental, logical instructions. Other grants, to Rice University and to the Universities of Iowa and Michigan, support research in the application of repetitive arrays of basic logical elements to theoretical aspects of system design.

The use of computers to restructure the teaching of subjects in many disciplines and to develop interdisciplinary, problem-oriented curricula is in the ascendancy. At Dartmouth, the Departments of Sociology and Political Science are developing data bases and an inquiry system which permit students to investigate, through a computer link, various relationships among data elements. In a short time, a student can develop good intuitive understanding of the relative significance of data elements, how to formulate questions to study relations, and how to pursue an evolving direction of investigation based on earlier results. This Project IMPRESS is jointly supported by NSF, the Alfred P. Sloan Foundation, and the Carnegie Corporation.

At the University of Michigan, computer simulations of living systems such as animal populations will serve as the basis for a course in natural resource management. A student will sample important parameters of the simulated populations, analyze data, formulate management programs, and evaluate the effects of his decisions. Through simulations of increasing sophistication, the student can be exposed to the complexities of real situations, challenged to make decisions based on the incomplete data available, and confronted with the long-term consequence of his actions —all at a computer terminal.

A project at Tulane University presents a unique opportunity to foster the utilization of surplus Minuteman I general-purpose digital computers. One hundred of these $234,000 systems have already been declared surplus, and it is expected that over a thousand will be available in the next few years. These computers have significant potential for educational use, but considerable hardware interfacing with external devices is required. This project will develop tested hardware interface designs which will be available to others, and will explore various computational and control applications of this machine.

INSTITUTIONAL COMPUTING SERVICES

The dynamic growth rate of academic computing and an accompanying increase in sophistication of computer applications in education and research is reflected in the number, substance, and quality of the proposals received in fiscal year 1970. Over 90 proposals were considered for improvement of computing facilities, a greater number than for any other year in the history of the program. Awards were made to 22 institutions for a total Foundation commitment of $6.5 million. Grants ranged in size from $5,000 awarded to Western Michigan University for improvement of its computer printing facilities to a $1 million grant made to the University of California at Los Angeles as partial support of a major new computer system to meet significant new

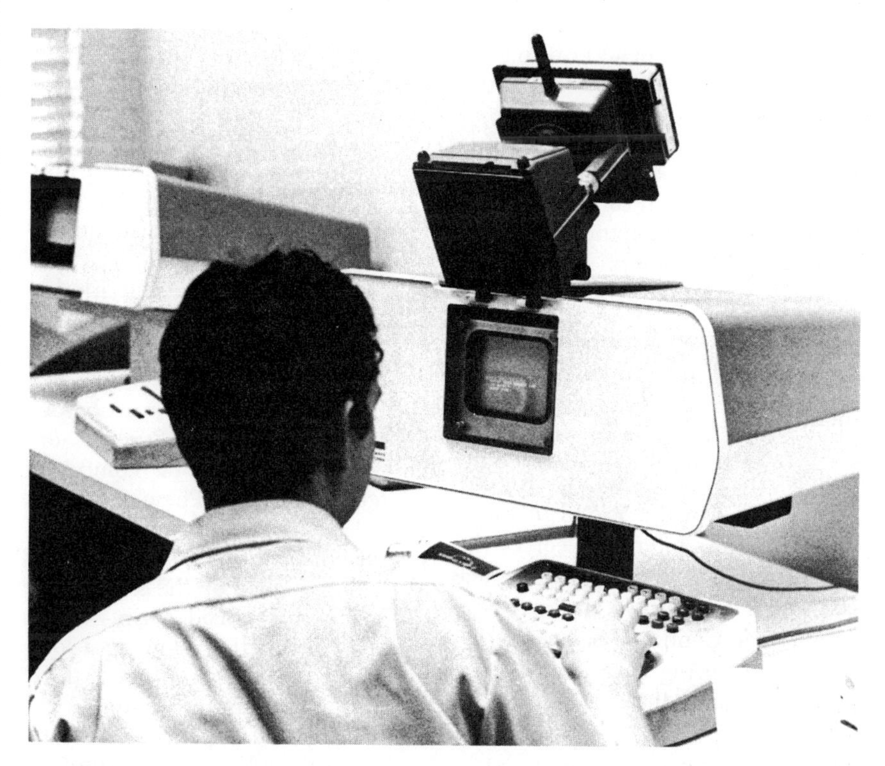

A camera is mounted on the graphics display terminal, connected to the UCLA computer, for permanent recording purposes. (Photo University of California, Los Angeles)

and innovative computing requirements of the institution. Three awards are described in detail to illustrate program activity and its role in institutional development.

The University of Tennessee at Knoxville, with more than 22,000 undergraduate and 4,500 graduate students, had a medium-size second generation computer as the primary facility to service exploding computing demands. In addition to growing research computing requirements, a new degree program in Computer Science emerged in late 1969 to add to the already heavy demands on an overtaxed system. A substantial upgrading with planned future expansion was needed. To help accomplish this, a Foundation grant of $500,000 was awarded to support a program with a 3-year budget of $3.5 million. A large third generation computer has been installed, with memory expansions scheduled at regular intervals, along with periodic additions of remote terminals and on-line peripheral devices. A new building is scheduled for completion in the third year to house the central computer and staff. The result of this program is a modern facility with a planned growth consistent with the developing computer demands of a major institution.

North Carolina A&T State University illustrates a situation where a change in academic curricula and research activity has caused the small but satisfactory computer facility of 5 years ago to be completely inadequate today. Recent accreditation of the School of Engineering, now offering degrees in architectural, electrical, and mechanical engineering, an increased emphasis in Computer Science activity in a growing mathematics department, and increased research activity in physics and social sciences made the establishment of a major computing facility a high priority objective of the university.

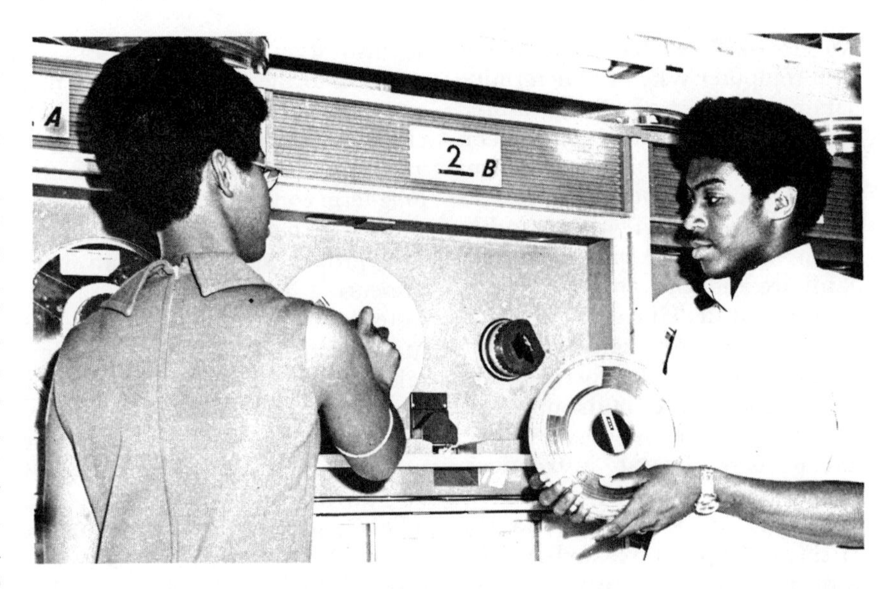

Students at North Carolina A&T State University load magnetic tapes on a Control Data tape drive. A Foundation grant will assist in significant expansion of computing facilities at this growing institution. (Photo North Carolina A&T State University)

A large-scale computer with batch and time-sharing capabilities will be installed in the fall of 1970, with new space to be available as permanent quarters for the new Computer Science Center in the spring of 1971. The staff size will be increased from two to 21 in a program with a 3-year operational budget exceeding $1 million. An NSF grant of $175,000 will assist this significant expansion, with two private foundations providing another $280,000.

Bucknell University acquired a small IBM 1620 computer in 1961, and by 1968 there was an obvious need for improved facilities to support a broad class of computer-related activities. These activities extend beyond the university to smaller institutions in the vicinity as a consequence of the keen sense of community leadership which exists at Bucknell. Following a long period of careful planning, a third generation computer system was selected capable of providing a variety of local and remote computing services. The equipment acquisition was closely coupled with a strong emphasis on the strengthening of faculty and senior professional staff to encourage further developments of educational and research computing applications. A Foundation grant of $395,000, representing approximately 25 percent of the estimated 3-year project costs, was made to assist Bucknell University in the program. The regional significance of a strong computing center extends beyond the educational institution itself, and this led the Appalachian Regional Commission to provide funds to improve the equipment configuration.

SPECIAL PROJECTS

Regional Cooperative Computing Activities

In fiscal years 1968 and 1969, the Foundation explored the merit of various computer-based cooperative arrangements, principally at the college level. Typically, each regional activity was centered about a major university which provided computer services and technical assistance to help a cluster of nearby institutions introduce computing to faculties and students, thereby developing a potential for further educational innovation. Altogether, 15 regional activities were established including 12 major universities, 116 participating colleges, 11 junior colleges, and 27 secondary schools located in 21 States.

In July 1969, a regional project directors' meeting was held at Oregon State University in Corvallis to study successes and failures and to assemble a reservoir of useful data for others. A *First Report on An Exploratory Program of Regional Cooperative Computing Activities,* available from the Office of Computing Activities, includes descriptions of the participating institutions, hardware and software systems utilized in the various projects, some cost figures, and indications of the educational impact of computer use.

In fiscal year 1970, 47 additional grants totaling approximately $1.8 million were awarded involving 15 major universities and 79 participating colleges in 24 States. Three new regional activities were established, two of which are unique in that they provide models for State-wide cooperative computing activities, one in North Carolina and one in Georgia. (See figure.)

In Georgia, 19 grants totaling $519,300 enabled the University System of Georgia to extend by telephone lines the computing resources of the Georgia Institute of Technology and the University of Georgia to other institutions throughout the State. One grant for $233,200 to the University System of Georgia provided partial support of its central staff of curricular experts and computer specialists. Two grants of $66,000 each were made to the University of Georgia and the Georgia Institute of Technology, and 16 grants were made to participating institutions ranging in size from $1,500 to $14,000. Twenty-eight institutions are currently participating in the project. State and institutional contributions to this project now exceed $1,675,000.

In June 1970, a conference on "Computers in the Undergraduate Curricula" was held at the University of Iowa, sponsored jointly by the University of Iowa and the National Science Foundation. Seventy-five papers were presented to 800 attendees representing 48 States. About one-third of these papers were from institutional participants of the regional program, thereby transmitting the experience of NSF grantees to those starting their own programs.

Computer-Assisted Instruction

The Foundation has been supporting research and development in computer-assisted instruction since fiscal year 1968. In fiscal year 1970, eight grants were awarded in

UNIVERSITY SYSTEM OF GEORGIA NETWORK

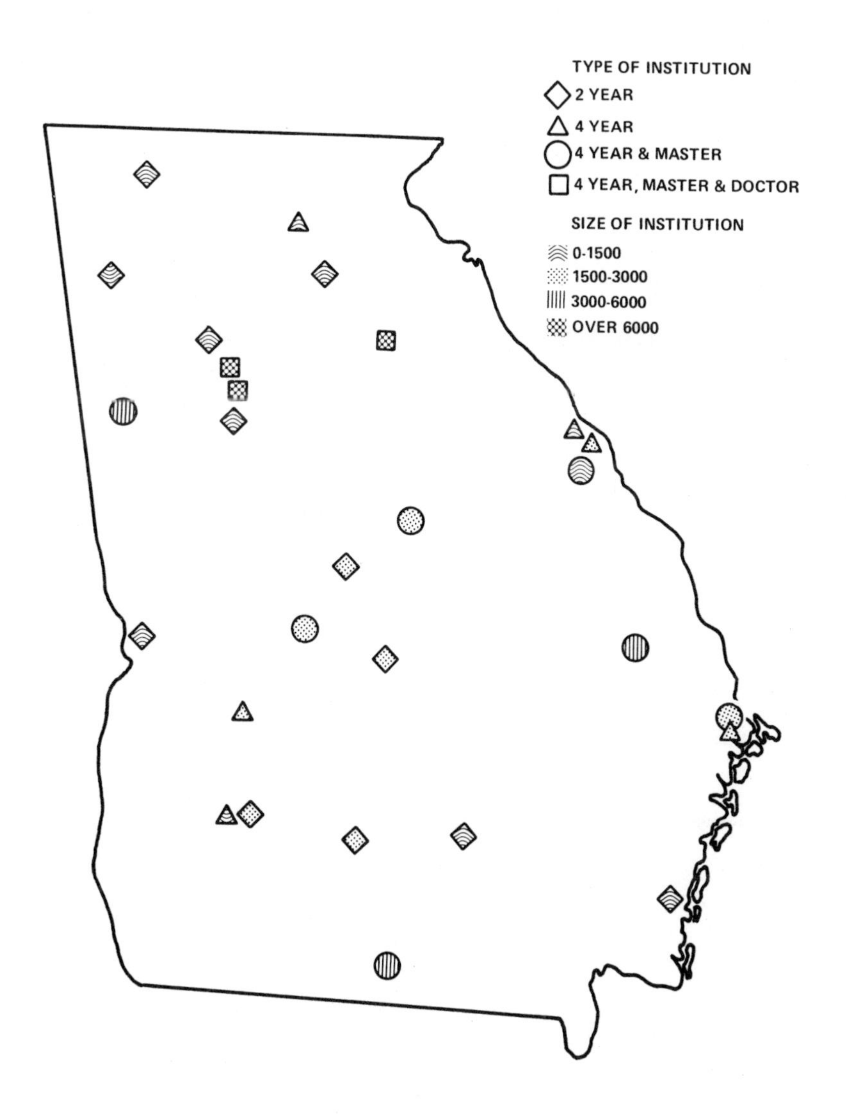

UNIVERSITY OF ILLINOIS TERMINAL DEVELOPED FOR PLATO SYSTEM

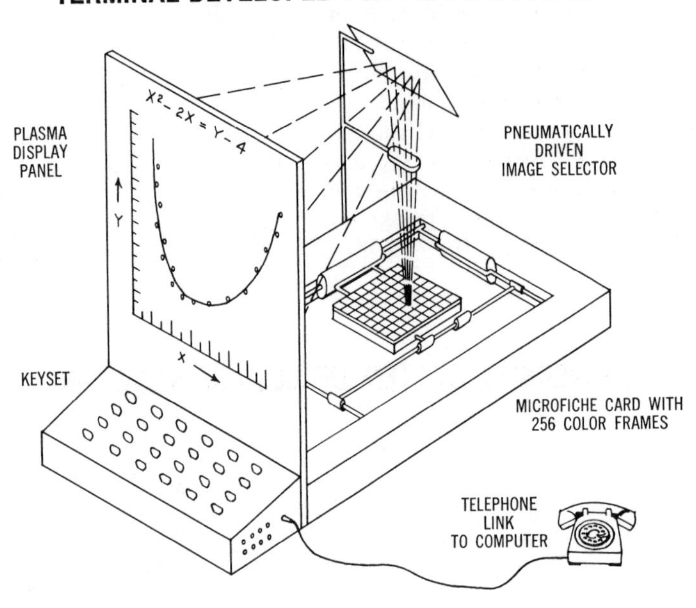

the total amount of $1.9 million.

Of particular interest among these awards is one in the amount of $430,000 to the University of Illinois to accelerate the development of a prototype educational system called PLATO IV. A full-scale system is designed to service simultaneously 4,000 student-terminals based on a novel plasma-display device invented at Illinois and being developed commercially. The prototype system will include up to 10 terminals.

Sea Grant Programs

The National Science Foundation Sea Grant Program supports activities in research, education and training, and advisory services for development of the nation's marine resources. In fiscal year 1970, the Foundation awarded approximately $9 million for these purposes as shown in the table below.

One of the principal objectives of the National Sea Grant College and Program Act of 1966 is the creation of a strong base of institutions dedicated to development of marine resources comparable to the Land Grant College efforts in the field of agriculture. In its third year of operation, the Foundation's National Sea Grant Program has continued active development of the National Sea Grant network of institutions with the addition of the University of Southern California, bringing the total of major universities now in the program to nine.

The other eight institutions which received initial support in fiscal years 1968 and 1969 are: Oregon State University, the University of Rhode Island, the University of Washington, Texas A&M University, the University of Hawaii, the University of Wisconsin, the University of Miami, and the University of Michigan.

The nine institutions are engaged in comprehensive Sea Grant programs involving research in all fields important to marine conservation, management, and development, including law, economics, and other social sciences as well as engineering and the natural sciences. All institutions have major educational programs at both the graduate and undergraduate level, and all engage in programs designed to communicate the results of research to such users as fishermen, seafood processors, ocean engineering firms, and State governments.

In addition to the major institutions, the Sea Grant Program has initiated comprehensive activities in several smaller institutions for the purpose of developing major marine competence in geographical areas where no broadly based marine research and education programs now exist. Grants are made to universities with a strong core of activities in limited marine fields for the purpose of applying the existing competence to local and regional marine problems while developing additional competence for the future. Such grants, known as "Coherent Project Grants," were made in 1968 to Louisiana State University and the University of Delaware. In 1969, the University of California at Santa Barbara, Humboldt State College, Calif., and the Virginia Institute of Marine Sciences were added. In fiscal year 1970, major coherent project grants were made to the University of Alaska and the Massachusetts Institute of Technology. These grants, in company with the institutional grants, form the base of the rapidly growing Sea Grant family of institutions.

Activities funded in previous years, under institutional, coherent project, and general project grants, began to show positive results of value to the national marine effort during fiscal year 1970.

Sea Grant efforts to develop techniques for the mass culture of high-value marine food organisms are continuing with a number of problems in aquaculture already solved.

Table 11
National Science Foundation
Sea Grant Program
Awards Fiscal Year 1970

Category of grant	Number	Amount
Institutional support	7	$5,675,400
Project support:		
Coherent area projects	3	797,400
Educational projects	7	561,512
Research projects	17	1,776,144
Study and planning projects	2	33,700
Advisory services projects	2	130,000
Total	38	8,974,156

Notable advances have been made in the development of techniques for mass culture of marine food organisms, including the successful raising of shrimp from egg to adult. (Photo University of Miami)

Tanks and artificial ponds are used by scientists at Oregon State University in a project designed to introduce exotic species of salmon into Oregon waters. (Photo Oregon State University)

One of these is the raising of shrimp from egg to adult, accomplished at the University of Miami. Substantial progress in the introduction of exotic species of salmon into Oregon waters was made during the year by fisheries scientists at Oregon State University. Louisiana State University scientists determined the salinity and temperature tolerances necessary for the culture of pompano.

Disease of fish and shellfish will be an increasing problem as aquaculture on a large scale is attempted. Scientists at Texas A&M University identified what could be a serious problem for commercial shrimp growers when it was noted that shrimp being used for nutritional experiments developed a high mortality. The Texas A&M Sea Grant team determined the cause to be a pathogenic bacterium, *Vibrio parahemolyticus*. This organism could be especially lethal in closed ponds where young shrimp are raised. While the bacterium also could cause food poisoning in America (as it does in Japan), this is unlikely because little seafood is eaten raw in this country; cooking destroys the disease organism.

Scientists at the Lamont-Doherty Geological Observatory of Columbia University successfully installed the first stage of a system that will not only provide nutrients for aquaculture, but should also produce fresh water and electric power through the raising of deep, cold seawater from near the ocean bottom. A mile-long pipeline was laid from the shore of St. Croix, in the Virgin Islands, into deep water where the temperature is only 41 degrees, and far more nutrient-laden than the warm surface waters. As the project continues, the cold water will be used in large condensors to remove fresh water from the warm trade winds by condensation. The cold water will also contribute to production of electric power by use of a steam generator powered by the temperature difference between the surface and bottom waters. Finally, the bottom water will be fed into a lagoon where its nutrients will support the start of a food chain that will end with commercially valuable marine organisms.

On the neighboring Virgin Island of St. John, Project Tektite II, a continuation of the man-in-the-sea program under Department of the Interior leadership, received major support from Sea Grantees. For the first time, student divers were given prime responsibility for logistic and diver safety support in a major operation. The divers were senior students from Highline Community College, Midway, Wash., trained and supported under the University of Washington institutional Sea Grant program. Cape Fear Technical Institute operated its own training vessel, *Advance II,* and the NSF vessel, *Undaunted,* in support of the operation. The Cape Fear Technical Institute at Wilmington, N.C., was responsible for coordinating and supporting a program of scientific oceanographic and meteorological research conducted in conjunction with the underwater scientific program, with over 400 persons and more than two dozen universities, industries, and Government agencies taking part. The University of Delaware Sea Grant program supplied the girl engineer, Margaret Lucas, to the first all-female aquanaut team. The Texas biomedical Institute at Galveston, under a Sea Grant project, conducted both medical support and a training program for diving doctors as part of Tektite. Southern Maine Vocational Technical Institute at South Portland, Me., provided both technician trainees and faculty members to the Tektite operations.

In advisory service operations, a significant forward step was the organization of the Pacific Coast Advisory Service. Centered at Oregon State University, the service involves institutions and agencies in California, Oregon, Washington, British Columbia, and Alaska. Cooperating with the Sea Grant program are the Bureau of Commercial Fisheries, the Environmental Science Services Administration, and a number of State agencies. The program is planned to improve information and extension services to marine activities throughout the Pacific coast.

During the year, other grants produced a broad spectrum of research and education activities, including the following:

- Scientists at Texas A&M University measured the effects of waves and currents on submerged pipelines and listed the data in a regular Sea Grant publication available to industry as a basis for improving pipeline operations.
- The cooperative work-study ocean engineering program at Florida Atlantic University resulted in 27 industry requests for the first 11 graduates. Fourteen graduates of the comparable bachelor-degree course at Mississippi State University/Gulf Coast Technical Institute were employed after graduation in June.
- Engineers at Stevens Institute of Technology developed a computer program for analysis of offshore floating platforms. By application of this technique, they expect to improve by 10 percent the design of offshore oil derricks for resistance to all elements of sea damage. This should eventually decrease insurance costs to industry and reduce platform loss.
- Sea Grantees at the University of Washington have successfully completed the first stage in de-

Undergraduate students from several Southern California colleges collect marine samples from the teaching-research vessel *Vantuna,* developed with the aid of a College Science Improvement Program grant. (Photo: Occidental College)

velopment of a system to locate biological targets in the sea more effectively through acoustic techniques—a project to help fishermen cut down the long periods spent hunting for fish.

- The University of Rhode Island has made extensive strides in developing a computerized lobster management model for optimizing the environment for lobster rearing.
- Sea Grantees at the University of Wisconsin have received a grant from oil companies to be used for unspecified research occurring as an outgrowth of the minerals research carried out under the Sea Grant program.
- The University of Maine Law School has identified and categorized the laws, regulations, and court decisions of Maine pertaining to recovery of living and mineral resources of State waters.
- Sea Grant food scientists and marine extension agents at Oregon State University brought shrimp and crab processors together for the first time, and conducted a program that has increased both the sanitation and efficiency of the entire Oregon shrimp and crab processing industry.
- The University of Hawaii uncovered deep shrimp resources about 1,500 feet under the sea. The shrimp may occur in commercially exploitable quantities. Tests indicate that, with heavy hauling gear, fishermen might catch as much as 1,000 pounds of shrimp per working day. Experiments at the University of Hawaii in growth rates of an octopus with commercial potential indicate that this is an extremely promising organism for aquaculture because of its rapid growth, good energy conversion, and high retail price.
- Scientists at the University of Rhode Island succeeded in raising Atlantic salmon from an average weight of 1½ ounces to a size of nearly 12 ounces in 6 months under relatively poor growing conditions. With improved environmental controls, it should be possible to effect an even greater growth rate. Progress is also being made in the raising of other so-called luxury fish, including chinook salmon from Alaska, rainbow trout, bluefish, and striped bass.

The Sea Grant Program is primarily geared to long-term results, and these "quick returns" represent a tiny fraction of the potential.

Science Information

Science information, as an integral part of the research and development process, must be easily accessible to scientists and engineers if science and technology are to make progress in improvement of the quality of man's physical and social environment. The Foundation's science information programs are directed toward ensuring that adequate information systems and services are available to the scientist and engineer. The long-term goal of the Foundation's Office of Science Information Service and its programs is to close the gap between the information needs of scientists now being served and those needs which must be met in the future as science and technology progress.

In pursuit of this goal the following major objectives have been set:

1. Investment in information system development for physics, chemistry, and other areas of science;

2. Aid to major universities to develop mechanisms which effectively serve research and education with present and new information products and services, including machine-readable tapes which are produced by professional societies, government agencies, and commercial organizations;

3. Short-term support to ongoing information activities, including translations, which are not yet self-sustaining;

4. Continued support of research and advanced development on science information problems; and

5. Fostering of cooperation, coordination and standardization among the various components of the present science communications complex which will lead to national and international networks of information services.

In fiscal year 1970, the Foundation awarded 104 grants and contracts and obligated $11.4 million for science information activities.

INFORMATION SYSTEMS DEVELOPMENT

The information systems development program was initiated in response to the needs of scientists and engineers for modernized information systems. The costs of developing modern computerized systems while simultaneously supporting existing services exceeds the financial resources of the scientific community. Therefore, to insure an adequate flow of information in the future, the Foundation has undertaken to provide support for the development of modernized systems.

Discipline-Oriented Science Information Activities

Support is being provided for the development of discipline-oriented science information systems in nine disciplines.

Chemical Information System.—Fiscal year 1970 marked the end of 5 years of intensive development of an information system for chemistry. By the end of June 1970, the American Chemical Society (ACS) had exceeded the 5-year objectives as stated in the Office of Science and Technology planning document issued in October 1965. Some of the major achievements during fiscal year 1970 were:

1. The American Chemical Society has concluded agreements with the West German Chemical Society and the Chemical Society of London for the processing of the primary publications of their respective countries for direct input into Chemical Abstracts Service (CAS) computer system. Similar agreements are being discussed with two other countries.

2. Agreements have been concluded with organizations in seven foreign countries for the utilization

of the computer tapes produced by the CAS system.

3. In the United States, the computer tapes are being used by commercial, industrial, and not-for-profit organizations as well as universities and Government agencies to provide scientists with a variety of services.

4. The Chemical Registry System now contains nearly 1.5 million substances with more than 1.75 million names and 3 million references.

5. The conversion of the file for CAS's Eighth Collective Index (1967 to 1971) to machine-readable form continued and the funding provided in fiscal year 1970 should be sufficient to complete this project.

The use of the CAS system has emphasized the need for better cooperation between the major abstracting and indexing services in order to avoid excessive duplication. Accordingly, CAS, Biological Abstracts, and Engineering Index, Inc. have undertaken a joint study to determine the areas of overlapping coverage and, if possible, to develop a plan to reduce the duplication of effort and effect operating economies.

National Information System for Physics.—The American Institute of Physics (AIP) continued its creation of a computerized file of the primary physics literature. The file contains the following items for each journal article: (1) bibliographic information—journal, volume, page, article title, author, and author's location; (2) abstract of article; (3) indexing information; and (4) citations of the article to other literature. About half of the world's primary physics literature is being entered into the file.

Four different services are either available or in the process of being made available. They consist of a magnetic tape service which covers the monthly additions to the file, a current awareness journal entitled *Current Physics Titles,* a series of bibliographies in special areas of physics, and the production of indexes to the various AIP journals.

The computer tapes produced by AIP are being used in a number of pilot operations which provide feedback information which will be used to improve the efficiency of the system.

Other Disciplines.—In the engineering sciences, indexes to electrical and electronics literature for both manual and automated usage have been developed. The American Psychological Association has defined a program of system development. The five remaining disciplines—linguistics, environmental sciences, life sciences, mathematics, and social sciences—are either in the process of defining their programs or in the preliminary study stage.

University-Centered Information Systems

The immediate objectives of support for university-centered information systems are threefold: (1) to meet the information requirements of academic scientists and the students they are training; (2) to establish "retail" campus-based terminals to accept the "wholesale" machine-readable tapes from the society-based, discipline-oriented systems, as well as the mission and problem-oriented products from Federal and private sources; and (3) to support the development of major nodes for the emerging national science information system.

During fiscal year 1970, the Foundation supported the development of discipline-oriented information service centers at six universities. Three of the centers—University of Georgia, University of Pittsburgh, and the Illinois Institute of Technology Research Institute—were originally established to develop systems to provide service for the tapes produced by CAS, but have now expanded their operations to cover tapes from commercial and mission-oriented systems. These centers together with other centers using tapes from Chemical Abtracts Service (CAS) and from other tape processors have formed the Association of Scientific Information Dissemination Centers (ASIDIC). A similar organization of distribution centers has been formed in Europe and is known as the European Association of Scientific Information Dissemination Centers (EUSIDIC). Both organizations include commercial and industrial organizations in addition to universities and other not-for-profit organizations.

Two other centers—University of Arizona and University of Washington—are concerned with the development of systems for the acquisition, processing, and distribution of interdisciplinary or subdisciplinary information. The University of Arizona is developing an Arid Lands Information System and is exploring the feasibility of establishing a worldwide arid lands information network with other institutions in the United States which are processing similar material and with institutions in Israel and Australia. The University of Washington continued to work on the development of a computerized data bank of the information in the U.N. Treaty Series and is investigating the extension of the system to cover maritime laws of interest to the Sea Grant project at the University of Washington.

OPERATIONAL SUPPORT FOR SERVICES AND PUBLICATIONS

The Foundation continued its support of existing information systems and services at an operational level, and extended its temporary support for the operation of developing systems in the major scientific disciplines. Altogether, support was provided for the operation of systems in six disciplines—psychology, engineering, geology, physics, mathematics, and atmospheric sciences. In addition support was provided for eight specialized bibliographies and indexing services.

The Foundation's support of publications was rigorously reduced. Only three monographs were supported as opposed to 22 in the previous year. Only one journal, one conference proceedings, and one critical review received support. Support was continued for the translation of 20 current primary journals by U.S. professional societies.

The science information activities conducted under the Agricultural Trade Development and Assistance Act of 1954 (Public Law–480) with eight foreign contractors resulted in the translation and republication in English of foreign primary journals, patents and monographs from Russian, East European, Japanese, and other languages; the preparation of abstracts; the compilation of annotated bibliographies, and the preparation of guides to foreign scientific institutions and information services. The combined activities of the Public Law–480 projects and the society-sponsored translation journals provided the scientific community with approximately 100,000 pages of foreign scientific and technological literature.

RESEARCH AND DEVELOPMENT

Support was provided for projects undertaken by individual research workers, research conducted by investigators associated with science information research centers, and the development of prototype experimental systems.

Cornell University has been doing research on procedures for the automation of indexing, classification, and construction of retrieval tools for indexers. These procedures and their effectiveness are being tested by comparison of manual and automatic processing of textual material.

Project INTREX at the Massachusetts Institute of Technology (MIT) has been studying the utilization of digital computers, communication systems and microphotography to enhance the effectiveness of the library as an information transfer center. During the past year, the remodeling of the Engineering Library at MIT provided INTREX with an opportunity to compare the conventional library services with the new information transfer techniques. INTREX terminals are being intermingled with bookstacks and study carrels in a variety of arrangements to determine the preferences of the users.

The Alfred P. Sloan School of Management at MIT has been studying how scientific and technical information passes from one person to another in industrial organizations. It was found that in any organization a few key people called "technological gatekeepers" are relied upon to provide information to other people. These key people read the professional literature and maintain close liaison with the experts in their fields. The extension of the "gatekeeper" concept to information transfer on an international scale is now being studied.

The Science Information Research Center at the Georgia Institute of Technology demonstrated its newly developed audiographic learning system at the 1970 International System Meeting, Las Vegas, Nev. The system provides access via telephone to a modular body of indexed, graphically supported, narrative presentations for a student controlled study. The existing facility is capable of supporting several telephone-connected student stations and providing each with random accessibility to learning ma-

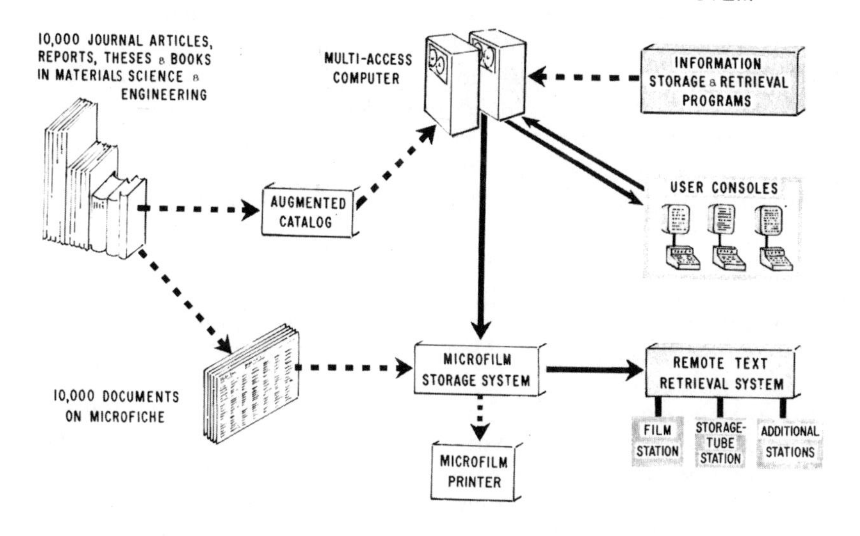

(Photo MIT)

terials or scientific information stored on computer-controlled tape recorders.

The Ohio State University Science Information Research Center reported that work on molecular cybernetics has led to the hypothesis that DNA stores programs or algorithms rather than "blueprint" or descriptive information. This suggests that the bridge between molecular and developmental biology is to be sought along lines similar to those developed for pattern recognition. Substantial progress has also been made on the theory of how people process information.

INTREX augmented catalog console at MIT. (Photo MIT)

PLANNING, COORDINATION, AND COOPERATION

The Office of Science Information Service continued to support studies and organizational activities and to provide assistance of a planning, coordinative, and cooperative nature to enhance science communication at the national and international levels.

During the year attention was given by the Committee on Scientific and Technical Communication (SATCOM) to the problem of determining the most appropriate planning and coordinating mechanism for the science communication complex. A SATCOM Task Group on the Economics of Primary Publication also prepared a study report on the present situation of primary journals, recent trends and problems, and a perspective for general national policies.

The Committee on Biological Sciences Information (COBSI), under sponsorship of the National Academy of Sciences-National Research Council, issued a report on *Information Handling in the Life Sciences*. The report concludes that the U.S. information system for the biological sciences, in the absence of any one monolithic information service, cannot be provided by a single Governmental or private organization but should inter-connect in a compatible manner the three existing major organizations in biological information—The National Agricultural Library, The National Library of Medicine, and the Biosciences Information Service of Biological Abstracts.

Continued recognition was given to the importance of standards to information system development and operation by support provided for the activities of the American National Standards Institute's Committee Z-39 on Library Work, Documentation and Related Publishing Practices. Among accomplishments of Z-39 during the year is a "Standard Identification Number for Serial Publications." This proposed national standard for serial numbering has been accepted by the International Standards Organization (ISO) as the basis for an international standard serial numbering scheme.

Foundation efforts continued to be directed in support of such international organizations as the Committee on Data for Science and Technology and the Abstracting Board of the International Council of Scientific Unions, and the U.S. National Committee for the International Federation for Documentation. The Foundation has continued also to participate in the International Council of Scientific Unions, UNESCO Joint Study Project UNISIST, on a worldwide science information system.

International Science Activities

Many of the research and science education activities supported by the Foundation have international significance. In addition to such major multinational projects as the International Biological Program, the Arctic and Antarctic research programs, among others, international aspects are also reflected in fellowship programs, support for attendance at scientific meetings, exchange of science information, and the translation into English of scientific literature published in foreign countries. These programs are discussed elsewhere in this report, and the activities summarized below represent only those programs administered by the Office of International Programs.

COOPERATIVE SCIENCE PROGRAMS

The Foundation's cooperative science programs include support for research projects, seminars, meetings, exchanges of scientists, and other scientific activities. The objective of these programs is to strengthen science in the United States. During fiscal year 1970, the Foundation acted as the lead agency for six bilateral cooperative science agreements (Australia, Republic of China, India, Italy, Japan and Romania). In addition, the Foundation supports the U.S.–U.S.S.R./Eastern European Exchange Program through the National Academy of Sciences. During the fiscal year, the Foundation and the National Center for Scientific Research (CNRS) of France made arrangements for a scientific exchange program. Highlights of these programs are presented below.

United States-Australia Agreement for Scientific and Technical Cooperation

No Foundation funds were awarded for projects under this agreement in fiscal year 1970. Most of the activity concerned plans for possible collaborative research projects in fiscal year 1971 and beyond. Some of the topics being discussed are scientific ballooning, drug abuse, biomedical research projects, photosynthesis, weather modification, and forest/brush fires.

United States-Republic of China (Taiwan) Cooperative Program

During fiscal year 1970, the Foundation provided travel support to U.S. scientists for 25 short-term visits to Taiwan for consultation, teaching, and research. A grant was awarded to the University of California to study the "Ecology of *Fusarium* Species in Taiwan with Special Reference to the *Gibberella* stage of *F. moniliforme* on Rice."

Foundation funding in this program amounted to $105,140; funds from other sources equalled $51,000. Total funds: $156,140.

United States-India Exchange of Scientists and Engineers

During fiscal year 1970 12 scientists from India visited the United States under this program. Fields of interest included engineering (soil mechanics and foundation engineering; integrated circuits; fire safety; desalination techniques; textile technology), paleontology; solid state physics, chemistry, and science information. Each of the visits consisted of a study tour of selected

NATIONAL SCIENCE FOUNDATION
OBLIGATIONS FOR INTERNATIONAL ACTIVITIES, FY 1963-71

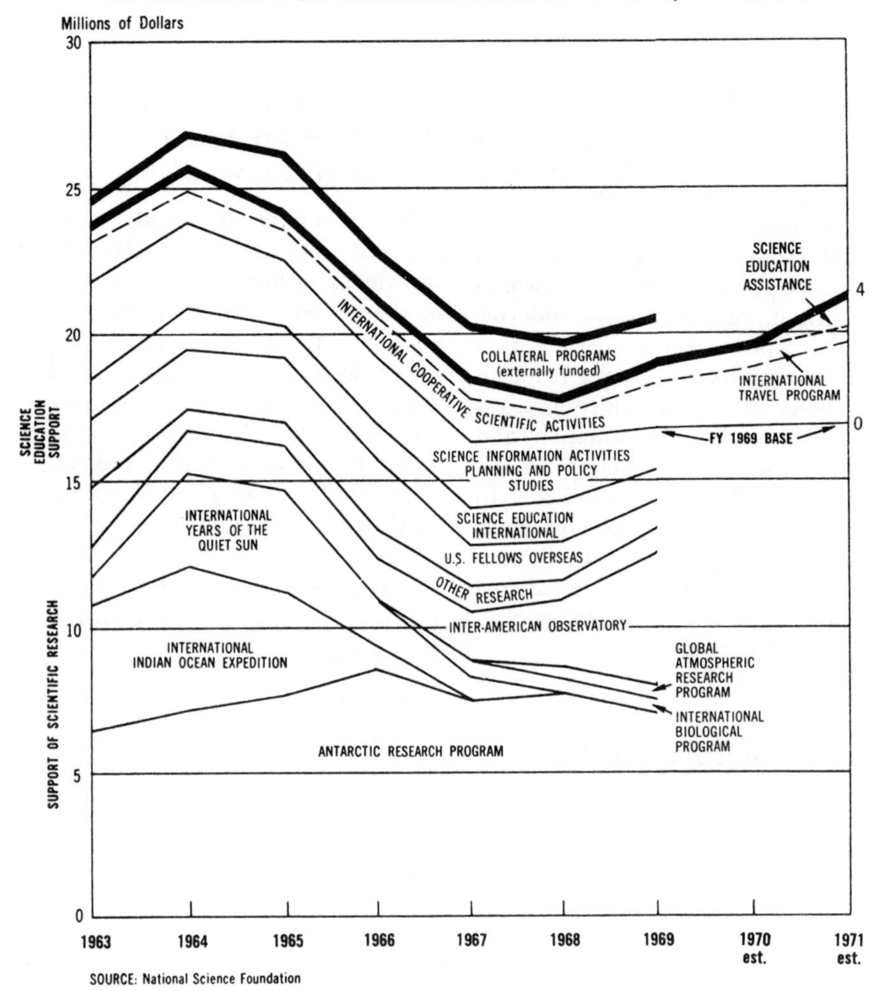

SOURCE: National Science Foundation

academic, industrial, and governmental laboratories and participation in conferences. Eleven American scientists participated in the program in 1970. Fields of interest included physics; engineering (aerodynamics, materials research, fuel research, foundation engineering); operations research; biomedicine (hematology, pharmacology); and mathematics.

United States-Italy Cooperative Program in Science

The Foundation provided $35,000 for the continuation of cooperative research under this program in fiscal year 1970. There are presently 27 active research projects as follows: physics, 10; biology, 5; agricultural sciences, 5; chemistry, 5; geological sciences, 1; and engineering, 1. An example of collaborative research is the study of proton channeling through gold crystals by American and Italian physicists. The Proton Channeling Spectrometer, located at the University of Bologna, is constructed partly with U.S. and partly with Italian funds.

United States-Japan Cooperative Science Program

The Foundation awarded grants to 13 U.S. scientists during fiscal year 1970 to visit and conduct research in Japan. Twenty-two seminars were held—12 in the United States and 10 in Japan—in which 224 American and 226 Japanese scientists participated. The Foundation funded seven new research grants in chemistry, engineering, biology, geology, and meteorology. Two ongoing projects were extended with additional funds.

An example of a cooperative research project is one being conducted by the Lamont-Doherty Geological Observatory and the Maisuru Marine Observatory. The objective of the project is to study the tectonic development of the Pacific Ocean floor by geothermal and geomagnetic investigations, with particular emphasis on the Philippine Sea basin. American scientists joined Japanese scientists on the Research Vessel *Seifu Maru* to make a geophysical survey in the seas around the Ryukyu Island Arc. The American scientists worked at the Earthquake Research Institute, University of Tokyo, on the results of the U.S.-Japan field work. A Japanese scientist joined American researchers on the Research Vessel *R. D. Conrad* for geothermal and geomagnetic studies in the seas around the Aleutian Islands. The Japanese scientists then visited the Lamont-Doherty Geological Observatory to work on the results of this trip and to plan future cooperative research.

Total NSF funds awarded under this program during fiscal year 1970 amounted to $421,315; other U.S. funding sources provided $164,937, for a total of $586,252 in U.S. funds.

United States-Romania Cooperative Science Program

During fiscal year 1970, the Foundation accepted the nomination of seven Romanian scientists for 48 man-months of study and travel in the United States. Two

U.S. scientists have visited Romania under the terms of this program. The Romanian National Council for Scientific Research has submitted applications for an additional 40 candidates for visits in fiscal year 1971. The Foundation obligated $35,000 under this program in fiscal year 1970.

United States-U.S.S.R./East European Exchange Program

These exchanges of scholars are conducted between the U.S. National Academy of Sciences (with Foundation funds) and the Academies of the Union of Soviet Socialist Republics, Poland, Yugoslavia, Romania, and Czechoslovakia. In fiscal year 1970, eight American scientists went to the U.S.S.R. for 1-month lecture and survey visits, and another 28 made research visits totaling 106 months. U.S. interest was divided evenly among biological, chemical, physical, and mathematical sciences; least interest was shown in engineering and the social sciences. Eleven Russian scientists made 1-month visits, and 25 spent a total of 92 months conducting research in the United States. Their emphasis was overwhelmingly on the physical, chemical, and engineering sciences.

In January 1970, a new Inter-Academy Exchange Agreement (National Academy of Sciences-Academy of Sciences of the U.S.S.R.) for 1970–71 was negotiated in Washington. The agreement continues the level of subsidized individual visits at 90 man-months per annum. A provision for joint research projects involving United States and Russian scientists was included for the first time. The U.S. National Academy of Sciences has submitted a proposal to the Academy of Sciences of the U.S.S.R. on behalf of an American zoologist who wishes to conduct a joint field trip to Siberia with a Russian colleague.

United States-France Exchange of Scientists Program

An additional bilateral agreement is expected to be implemented in the near future. During fiscal year 1970, representatives of the Foundation and the French Centre National de la Recherche Scientifique (CNRS) developed the terms for an exchange of scientists for study and research in the respective countries. Eligible individuals will be citizens or nationals of the United States and France who will have earned a doctoral degree or its equivalent normally not more than 5 years prior to the commencement of the exchange visit. The period of the visit will be normally between 5 and 15 months. The Exchange Agreement will be signed early in fiscal year 1971.

DEVELOPMENT ASSISTANCE PROGRAMS

In fiscal year 1970 the Foundation continued to manage two programs on behalf of the Agency for International Development (AID): (1) Science Education Improvement Program in India, and (2) Technical Cooperation and Evaluation Program—Worldwide Program.

Science Education Improvement Program in India

The collaborative program for the improvement of science education in India is defined by a contract between the United States and India. Funding is provided by the Agency for International Development and the Indian Ministry of Education. It is implemented jointly by the Foundation and the (Indian) National Council for Science Education. There are three main efforts: (1) the training of personnel through summer institutes, workshops, seminars, and short courses; (2) the development of new teaching materials including syllabi, textbooks, examinations, laboratory equipment, handbooks, and journals; and (3) the development of institutions which can sustain the improvement effort.

The jointly sponsored Summer Institute Program became a fully Indian institution in 1970, as this was the final year in which U.S. consultants are to be supplied to the Indian directors of the summer projects.

Technical Cooperation and Evaluation Program—Worldwide Program

With funds provided by AID's Technical Assistance Bureau, the Foundation in 1970 supported three continuing projects of worldwide scope and initiated a fourth:

1. Study of Low Cost Science Teaching Equipment, conducted at the Science Teaching Center of the University of Maryland, to gather information and materials from all parts of the world on science teaching equipment of low cost and easy manufacture from locally available materials;

2. Activities of the Biological Sciences Curriculum Study (BSCS) headquarters in support of adaptation of BSCS materials by local groups in AID countries;

3. Study by the International Education Committee of the Conference Board of Mathematical Sciences of the demand for and supply of U.S. mathematics educators for service in international projects, especially those in the developing nations;

4. Planning and monitoring an evaluation of BSCS adaptation activities in AID countries.

PLANNING AND DEVELOPING INTERNATIONAL PROGRAMS

The Foundation provides support for U.S. scientists and scientific organizations in their effort to organize, plan, and develop international scientific programs and activities.

U.S. National Committees for International Nongovernmental Scientific Organizations

Foundation funds supported the activities of the committees and staffs established by the National Academy of Sciences to represent the interests of the U.S. scientific community in the affairs and programs of the international scientific unions of which the Academy is the U.S. National Member.

Science Program and Policy Development

During fiscal year 1970, the Foundation awarded grants which provided partial funding to the American Academy of Arts and Sciences for the support of a U.S. Joint Committee for the International Center for Insect Physiology and Ecology (ICIPE). The joint committee participates in the planning and development of ICIPE (in Nairobi, Kenya) which will be administered by an international consortium of academies of sciences.

Table 12

U.S.S.R. and East European Exchange of Scientists Programs, Number and Duration of Individual Visits Initiated, Fiscal Year 1970

	Number	Length of time (in man-months)		Number	Length of time in the United States (in man-months)
U.S. Scientists to:			Foreign scientists from:		
Czechoslovakia	12	49	Czechoslovakia	7	38
Poland	12	33	Poland	8	52
Romania	16	26	Romania	10	45
Yugoslavia	4	12	Yugoslavia	5	13
U.S.S.R.	36	114	U.S.S.R.	36	103
Total	80	234	Total	66	251

Planning and Policy Studies

Through its science resources and policy studies, the Foundation identifies and analyzes important science policy issues; provides an adequate data and methodology base for sound decisionmaking; and develops science planning and policy study capabilities at various institutions in the United States. The information developed through these activities is used in assessing alternatives, in establishing priorities, and in arriving at recommendations concerning the national science effort. The results of these study activities serve not only the requirements of the National Science Foundation, in its concern with the scientific enterprise, but also serve other Federal agencies, Congressional groups, and non-Federal organizations.

SCIENCE POLICY ISSUES

The study of science policy issues is an essential component of any overview of the science picture in the United States. Insight is needed into the interrelationships between the scientific enterprise and the society it serves; the requirements of the various areas of science and engineering must be understood; and assessments of the impact of changes in the scientific resource base and the impact of current and potential changes due to scientific and technological advances must be available for guidance.

With limited resources to draw on, one of the critical and continuing problems facing science administrators is the question of establishing priorities for the competing areas of science. Through its support of the activities and special studies of the Committee on Science and Public Policy (COSPUP) of the National Academy of Sciences and the counterpart Committee on Public Engineering Policy (COPEP) of the National Academy of Engineering, the Foundation has been obtaining information concerning the current status and projected needs of the major areas of science as an aid to the determination of scientific priorities. Two COSPUP studies initiated at the end of fiscal year 1969 with Foundation support were fully underway in fiscal year 1970. One study concerned the status and needs of astronomy (both ground and space based), the other concerned the picture for physics. These studies represent the first efforts of COSPUP to update previous reviews of scientific disciplines. The new reviews are aimed at establishing not only the current and foreseeable needs and problems of these two disciplines but also at determining their relevance to other areas of science and technology and to society in general. The reviews will also concentrate on the establishment of priorities within these broad fields. Additional updating efforts, for other disciplines, are currently being planned in order to assure that information available keeps pace with rapidly changing scientific and technological developments.

Over the past year, support of the Committee on Public Engineering Policy continued. The committee addressed itself largely to the question of engineering as it relates to social utility. One activity culminated in the publication of a report, *Priorities in Applied Research,* which recommended that applied research be undertaken in the following major areas: the biosphere, techniques for applied social research, materials research, construction, and transportation.

Another important issue upon which attention was focused during the year concerned the effect of changes in Federal funding patterns on academic institutions in the United States. Federal obligations to universities and colleges totaled

$3.5 billion in academic year 1968–69, representing only a 2 percent rise over the previous year and the second consecutive year in which the growth rate was limited to this level. In contrast, the annual growth rate during the 1963–67 period was 24 percent.[1] Firm data have not been available on the effects of the changed funding pattern and the Foundation, with the encouragement of the Office of Science and Technology, initiated a survey to determine quantitatively the actual impact on academic institutions. The first phase of the study was conducted in fiscal year 1969 in about one hundred universities and some seven hundred graduate departments within these universities. Since the results of this survey indicated that 1968–69 constituted the first year of a major transitionary period, a second phase of this study was initiated in 1970 to provide a comparison over more than one time period. Results will be published in fiscal year 1971.

Also studied during this past year was the question of possible future imbalances between the pool of available Ph.D.'s in science and engineering and requirements for their utilization, a topic of particular importance when viewed within the context of a rapidly changing national situation. The National Science Foundation undertook to analyze and project the future relationship between the supply and utilization of science and engineering doctorates. The results of this analysis were made public in a report, *Science and Engineering Doctorate Supply and Utilization, 1968–80.* This and subsequent analyses seem to indicate that, by 1980, 320–350,000 science and engineering doctorates (compared with about 150,000 in 1968) might be available. However, present job markets for doctorates are not nearly as favorable as they have been in past years, though it is not clear in the current fluid situation whether the major present problem is a mismatch between opportunities and aspirations or actually an oversupply of Ph.D.'s. Thus, the 1980 pool of doctorates will depend on the extent to which the present situation will affect graduate school enrollments and thus the future rate of Ph.D. production. With regard to utilization, several projections were made on the basis of varying assumptions. The projected relationship between supply and utilization figures indicates that by 1980, the supply and utilization of science and engineering doctorates is likely to be in equilibrium. However, it would also appear that significant numbers of Ph.D.'s are likely to be engaged in activities which are markedly different from the primary ones practiced by most present doctorates. Examples of these different activities include non-R&D functions in industry and government as well as teaching in 2- and 4-year colleges. Recent evidence shows that the shift is already beginning. Implications of this analysis are that Ph.D. education should offer a variety of different programs including training most suitable for these new activities. It thus appears necessary for universities to examine their graduate programs and to develop new and different curricula for Ph.D.'s who do not intend to enter research careers.

[1] National Science Foundation, *Federal Support to Universities and Colleges and Selected Nonprofit Institutions, Fiscal Year 1969* (NSF 70–27) Washington, D.C. 20402: Supt. of Documents, U.S. Government Printing Office, 1970.

DEVELOPMENT OF BASIC TOOLS FOR SCIENCE PLANNING AND POLICYMAKING

The development of an adequate data and methodology base for the making of science policy decisions is carried out by the Foundation through a broad range of study activities. Included are studies for the collection and analysis of data concerning the nation's scientific resources; the development of concepts and projection and modeling techniques relating to these resources; and the correlation and synthesis of information from many sources on the flow of resources to scientific and technical activities. The more important results of these efforts are highlighted below.

The National Scene

A comprehensive review of total national expenditures for research and development over the period 1953 to 1970 was published in fiscal year 1970 in the NSF report, *National Patterns of R&D Resources.* The data in this publication were obtained through the periodic NSF surveys of all sectors of the economy. They revealed that in 1970 national R&D expenditures, from both Federal and non-Federal sources, reached an estimated record level of $27 billion. This amount is $1 billion higher than the estimated 1969 level and nearly $7 billion more than in 1965. However, the average annual rate of growth during the 1965–70 period was only 5.9 percent compared with a 9.4 percent average for 1958–65. Furthermore, this growth rate over the last 2-year period, 1968–70, has declined to a 3.6 percent level. The major reason for the decline has been a leveling off of R&D support by the Federal Government. Federal expenditures for R&D between 1958 and 1965 grew by 12.3 percent annually but increased by an annual average of only 3.4 percent between 1965 and 1970 and showed no increase at all in the period between 1968–70, remaining constant at about the $15 billion level. (See chart.)

The report on national R&D ex-

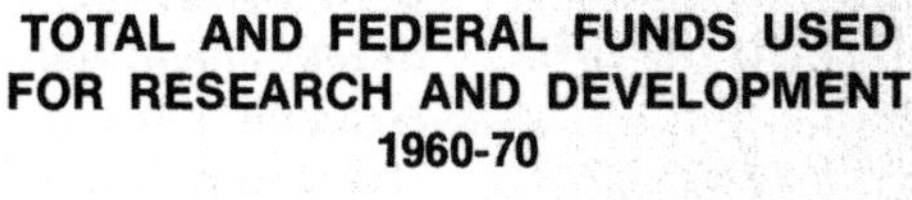

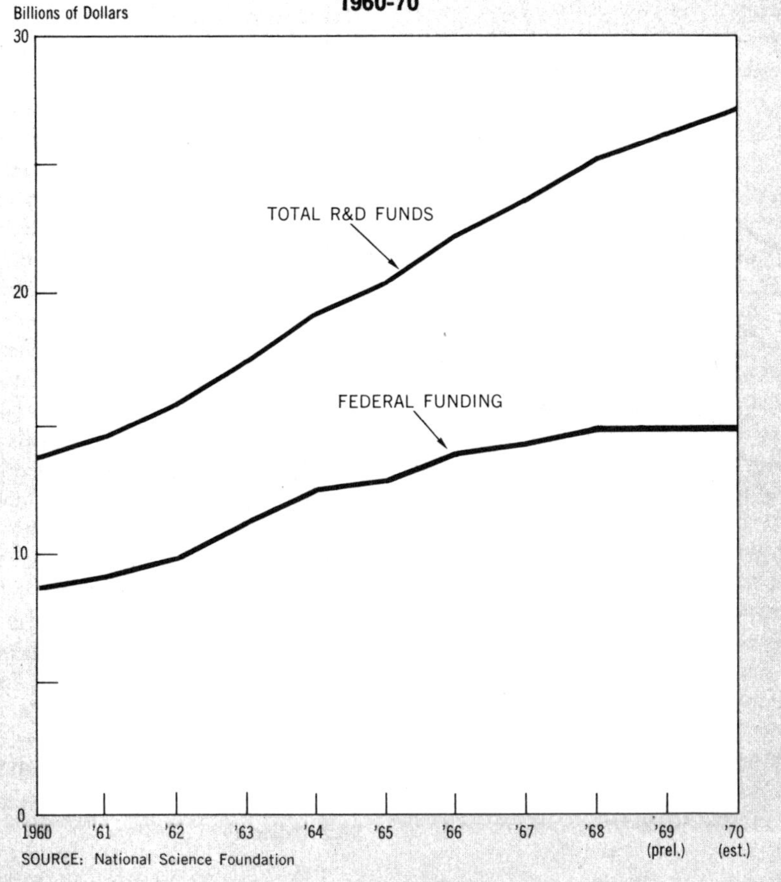

penditures also presented information on the full-time equivalent (FTE) number, and sectoral distribution, of scientists and engineers engaged in research and development. During 1968, an estimated 565,000 FTE scientists and engineers were engaged in research and development. Although this was nearly two and one-half times the number employed in R&D activities in 1954, the rate of increase of R&D scientists and engineers has been declining in recent years. Between 1954 and 1961, the annual average rate of growth of R&D scientists and engineers was 8.7 percent. This growth rate fell to 4.1 percent between 1961 and 1968.

Federal R&D Support

Federal obligations for research and development—as distinct from R&D expenditures discussed above—totaled $15.6 billion in fiscal year 1969 and were expected to total approximately the same amount in both fiscal years 1970 and 1971. (See volume XIX, *Federal Funds for Research, Development, and Other Scientific Activities.* NSF 69-31.) This represents a decline in support from the 1967 R&D obligation total of $16.5 billion—the year of the highest dollar funding of Federal R&D programs. The report on Federal funding also indicates that there has been a significant increase during the last decade in the proportion of federally sponsored R&D activities devoted to areas other than military, space, and atomic energy efforts. The share of total Federal funding in agencies other than DOD, NASA, and AEC has risen from 9 percent in 1960 to 18 percent in 1970.

In its second annual report to the President and Congress, entitled *Federal Support to Universities and Colleges and Selected Nonprofit Institutions, 1969,* the National Science Foundation reported that Federal support to universities and colleges for both academic science and nonscience activities totaled $3.5 billion in 1969, a gain of only 2 percent for the second consecutive year. (See chart.) This report also revealed that academic science programs in recent years have shown an even slower growth rate than nonscience programs, with Federal academic science obligations growing by only 1 percent in 1968 and one-half of 1 percent in 1969. The data for the universities and colleges appearing in this study are gathered under the auspices of the Committee on Academic Science and Engineering (CASE) of the Federal Council for Science and Technology.

In addition to the report to the President and Congress which presents detail for individual institutions, CASE is collecting data on individual federally sponsored university projects covering both funding and manpower associated with the projects. A publication covering this project reporting is expected during fiscal year 1971.

At the request of the Federal Council for Science and Technology, the National Science Foundation compiled a *Directory of Federal R&D Installations.* This directory, the first of its kind, was prepared and released for public use in 1970 and provides a comprehensive general reference to R&D

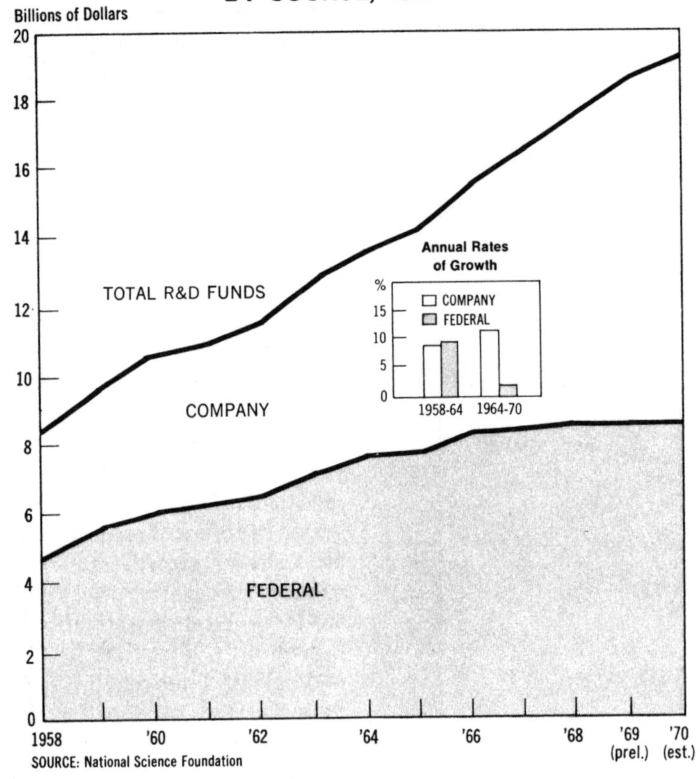

establishments owned and directly controlled by the Federal Government. More than 700 installations are listed in the directory with information provided concerning their location, size, functions, activities, and capabilities. It is anticipated that the directory will be an important mechanism for making more widely known the Federal installations capable of dealing with significant research and technological problems, and that it will also further interagency use of Federal R&D resources.

Non-Federal R&D Support

In contrast to the Federal R&D funding picture, industry has been increasing its financial contribution to R&D at an increasing rate of growth as reported in the NSF publication *Research and Development in Industry, 1968.* In 1968 industrial firms supported 51 percent of their R&D performance with their own funds compared with a decade earlier when companies funded only two-fifths of their R&D activities. In total, industry spent $8.9 billion in 1968 on company-financed research and development, and the Federal Government funded an additional $8.6 billion of industrial research and development. Indications are that the amount of Federal support for industrial research and development has been leveling off since 1968, while the amount of company support has been rising. Thus, it is likely that industrially financed research and development will continue to increase as a proportion of total R&D performance. (See chart.)

The report *Resources for Scientific Activities at Universities and Colleges, 1969,* the latest in a series providing information on scientific activities in the nation's academic institutions, will be published early in fiscal year 1971. This report reveals that in academic year 1968 a total of $7.0 billion was spent by universities and colleges (in current and capital expenditures) for science research and instruction. (This figure does not include funds expended by Federally Financed Research and Development Centers located at these institutions.) In 1968, R&D expenditures at academic institutions amounted to $2.6 billion, only 10 percent of the national total. In terms of basic research dollars spent, however, more than one-half of the nation's performance took place in colleges and universities (approximately $2.0 billion out of a national total of $3.7 billion).

Models and Methodology

In addition to having current, accurate data for planning and policy purposes, it is important to have available suitable models which can assist in putting the data to use. The development and testing of models and methodologies for use in science planning is still in its early stages and the Foundation has continued to encourage and support new exploratory efforts.

In 1969–70 the Institute for the Future completed a study designed to improve long-range forecasting techniques. New techniques were developed, tested, and applied prospectively and retrospectively to areas of economics, political science, and technology. Specific cases chosen for analysis included:

- the diffusion of hybrid corn in the midwestern United States in the 1930's;
- the dependence of population growth on family planning, health care, mortality, production, etc.;
- the Indian economic Five-Year Plan of 1956;
- future developments in low temperature physics and cryogenics and their interrelationship with certain social changes;
- the social, political, and economic impact of introducing certain technologies into developing countries.

Although the study primarily dealt with the development of methodological techniques, the advice of experts was used to develop the technical framework for consideration of the problem. For example, in the low temperature physics project, physicists were asked to predict the most likely scientific and technological developments occurring over various periods of time. Questions asked of a multidisciplinary team covered not only what was likely to happen, but also dealt with the effects of one development on others. This technique produced matrices which indicate whether one development is likely to increase or decrease another's likelihood. The results strongly indicate that this cross impact method has great potential in a comprehensive approach to forecasting.

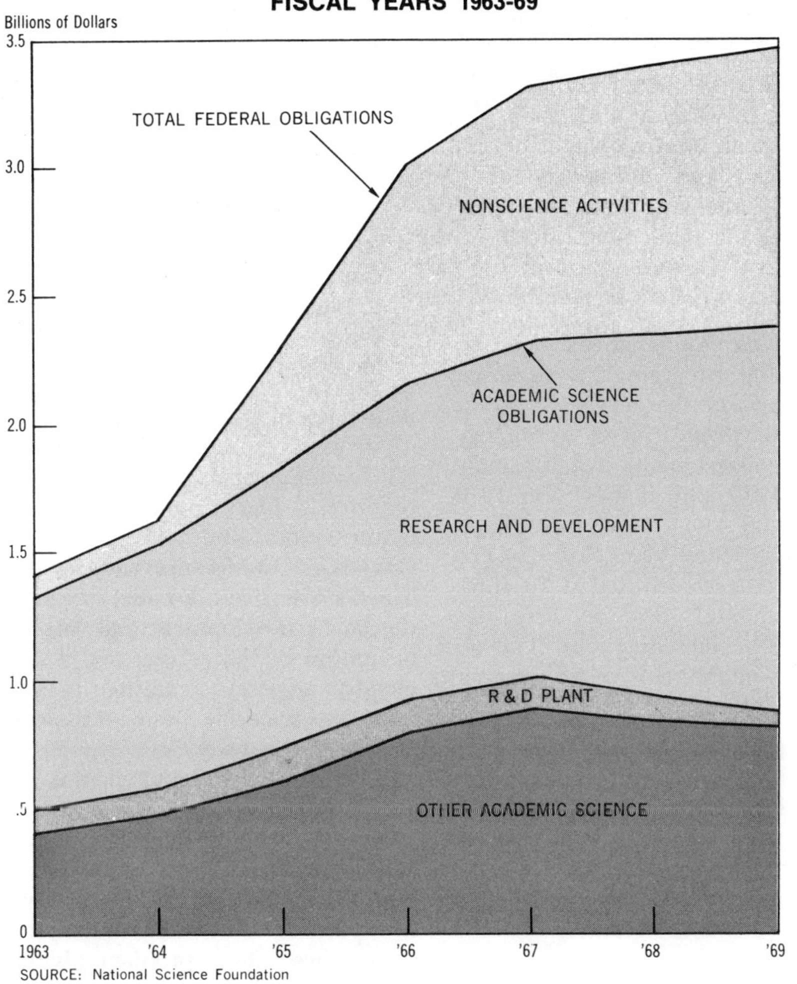

NSF has also been contributing to a nationwide effort among academic institutions to design, develop, and implement information systems for their individual use. Conducted by the Western Interstate Commission for Higher Education (WICHIE), the effort involves the development of resource allocation, cost, funding, student flow, and other planning models. In fiscal year 1969 a state-of-the-art seminar was held by WICHIE, supported by the Foundation and the U.S. Office of Education, regarding developments in management information systems and planning models. In fiscal year 1970, again joining with the Office of Education, NSF helped support a WICHIE seminar on the "outputs" of higher education. The conference brought together leaders in higher education research to discuss the identification and quantification of the products of higher education.

DEVELOPMENT OF SCIENCE PLANNING AND POLICY CAPABILITY

Intergovernmental Science Programs

Fiscal year 1970 saw a significant expansion and reorientation of an NSF activity formerly known as the State and Local Government Science Policy Planning Program. This

program was originated in fiscal year 1969 as an experimental effort to assist State governments in strengthening their science and technology planning programs. In its first year of operation, many universities requested Foundation support in working with State governments to improve the utilization of science and technology in their decisionmaking processes.

At the same time, efforts of the Federal Government over the past 8 years to share responsibility with State and local governments have tended to emphasize the need for a general program that examines relationships that exist, or should exist, between science, technology, and government. Accordingly, toward the end of fiscal year 1970, the State and Local Intergovernmental Science Policy Planning Program was reconstituted as the Intergovernmental Science Programs, with expanded functions. The objectives of the reorganized program are as follows:

1. To advance the understanding of public issues and problems having scientific and technological content at the State and local levels of government, and to assess needs and opportunities for more effective applications of science and technology;

2. To demonstrate innovative science and technology planning and decisionmaking processes related to State, local, and regional problems;

3. To stimulate selected State and local government experimentation, on a pilot basis, with science and technology systems in the context of their own needs and resources;

4. To encourage adoption of new systems which show promise for enhancing State and local ability to incorporate science and technology into public programs;

5. To improve communication between persons and groups concerned with science and technology at the Federal, State, and local levels of government.

During the past year, a substantial effort has been devoted to support of studies which examine appropriate roles of different levels of government in sponsoring and utilizing research and development. A $320,000 grant was given to the Council of State Governments (1) to analyze current and potential endeavors by Federal, State, and local governments to incorporate scientific considerations into governmental decisions and operations and (2) to assess the national machinery for development of problem-solving resources. The Council of State Governments and the National Governors' Conference have established a five governor panel to guide the policy development and implementation of the project and a 20-member advisory committee to advise on scientific and technical aspects of the study.

A regional science and technology award was given to the Southern Interstate Nuclear Board in association with the States of Georgia, North Carolina, and South Carolina to explore areas in which multistate approaches to problem-solving might be desirable either because of economics of scale or in areas where problems transcend State boundaries. This study, in addition to an earlier study supported at the University of Tennessee, will also examine how individual States might improve their program and policy relationships with the Federal Government.

Another grant was made to the California State Assembly to examine how scientific and technological considerations can be incorporated into the legislative process. In addition, several smaller grants were made to examine different areas of State science and technology policy relating to environmental quality (Louisiana), technological forecasting (Montana), development of new mechanisms for relating academic research outputs to government decisionmaking (Virginia), and Federal-State-local support and utilization of research and development in regulating air pollution (Pennsylvania). A grant to use a new technique of simulation to study Federal-State decisionmaking in regard to allocating governmental resources for science and technology was made to the Institute for the Future.

To improve communication between governmental leaders and the scientific community, a series of four regional (Southern, Eastern, Western, and Midwestern) and one national conference on Science, Technology and State Government were supported in conjunction with other Federal agencies and State organizations.

A survey of scientific and technological advice available to local governments was supported under a joint grant to the New York State Department of Education and the International City Managers Association. This will provide information at the local level to supplement information at the State level that will be obtained under nine State case studies of the science advisory mechanisms to State government supported under earlier NSF grants.

University Science Planning and Policy Program

The University Science Planning and Policy Program is designed to assist in the development of the resources and capabilities of academic institutions for training and research related to science planning and policy activities. The program was established in recognition of a critical need for a better understanding of the many complex science policy issues and the lack of adequately trained manpower to deal with these problems.

Institutions currently receiving grants under this program include Harvard University, Massachusetts Institute of Technology, University of Virginia, State University of New York at Albany, Cornell University, the University of Indiana, Stanford University, and the University of California at Berkeley.

These grants help to support teaching, research, and special seminars on such science policy problems as: the use of science in international affairs; scientific and technical manpower; environmental management; technology and the city; nuclear energy, the law and international affairs; the effects of new educational technology; legal and moral implications of modern biology and medicine; the effects of technology on economic growth; and the organization of large-scale technological projects.

The Cornell University Program on Science, Technology and Society has, during the first year of its grant, been successful in involving faculty and students from many disciplines, including the sciences and humanities, in new courses and seminars jointly sponsored with other units of the university, in such areas as Biology and Society, International Flows of Science and Technology, Social Implications of Technology, Law and Environmental Control, and Technology Assessment.

Under recent grants, Harvard will develop a series of case studies which demonstrate the application of analytical techniques to public policy problems; and Stanford will analyze the technical and policy alternatives involved in telecommunications and computer technology.

SOURCE: NSF, "National Science Foundation Annual Report 1970," 1971.

Part 2

NSF GRANTS AND AWARDS

RESEARCH PROJECT SUPPORT [1]

Biological and Medical Sciences Research Projects

Cellular Biology

ARIZONA

ARIZONA STATE UNIVERSITY
Justus, Jerry T.; *Study of the Hemoglobins, Erythropoiesis and Metamorphosis in Urodeles* (B014044); 24 months; $30,000

UNIVERSITY OF ARIZONA
Bartels, Paul G.; *Chloroplast Development—Inhibition by 3-Amino-1, 2,4-Triazole* (B016048); 24 months; $35,500
Mendelson, Neil H.; *DNA Replication and Cell Division in Bacillus Subtilis* (B017022); 24 months; $40,300

CALIFORNIA

CALIFORNIA INSTITUTE OF TECHNOLOGY
Mitchell, Herschel K.; *Regulation of Development in 'Drosophila'* (B023343); 24 months; $65,000

CALIFORNIA STATE COLLEGE, San Bernardino
Sokoloff, Alexander; *Maintenance of the Tribolium Stock Center* (B016310); 24 months; $35,000

SAN DIEGO STATE COLLEGE
Diehl, W. Paul; *Control of Chromosome Replication and Gene Expression in Phage lambda* (B020334); 24 months; $39,900

SCRIPPS CLINIC AND RESEARCH FOUNDATION
Crawford, Irving P.; *Organization and Genetic Control of the Enzymes of the Tryptophan Pathway in Bacteria* (B016388); 24 months; $64,100
Hoch, James A.; *Genetics and Regulation of the Citric Acid Cycle* (B015602); 24 months; $43,900

STANFORD UNIVERSITY
Kaiser, Armin D.; *Macromolecular Recognition Systems* (B017855); 24 months; $90,500
Lederberg, Joshua; *Exchange Program in Genetics and Molecular Biology Between the Universities of Stanford and Pavia* (GB7785 001); 8 months; $35,000
Stockdale, Frank E.; *Hormone Dependent Differentiation and Myogenesis* (GB6618X002); 12 months; $40,200
Wessells, Norman K.; *Mechanisms of Organ Morphogenesis and Cell Differentiation* (B015666); 12 months; $35,000
Yanofsky, Charles; *Genetic and Biochemical Studies of the Tryptophan Operon* (GB6790X001); 12 months; $70,500

UNIVERSITY OF CALIFORNIA, Berkeley
Collins, O'Neil R.; *Genetics of the Myxomycetes* (B015770); 24 months; $37,600
Stern, Curt; *Developmental Genetics of Drosophila* (B023024); 24 months; $55,000
Wilt, Fred H.; *Gene Expression in Sea Urchin Embryos* (B021412); 24 months; $28,000

UNIVERSITY OF CALIFORNIA, Davis
Allard, R. W.; *Genetics of Enzyme Polymorphisms in Plant Populations* (B013213); 24 months; $50,000
Edlin, Gordon J.; *Bacterial Nucleotide Pools and Nucleic Acid Synthesis* (B013528); 24 months; $75,000
Rick, Charles M.; *Cytogenetics of Tomato Species Hybrids* (B015765); 24 months; $41,800
Stebbins, G. Ledyard; *Developmental and Evolutionary Genetics of 'Hordeum Vulgare' and 'Plantago Insularis'* (GB5713X002); 12 months; $42,600

UNIVERSITY OF CALIFORNIA, Irvine
Schneiderman, Howard A.; *Control of Insect Development* (B016690); 24 months; $80,000
Tewari, Krishna K.; *Biogenesis of Cell Organelles—Studies on Chloroplasts* (B020674); 24 months; $41,500

UNIVERSITY OF CALIFORNIA, Los Angeles
James, Thomas W.; *Organelles and Morphology of Cells in Division—Synchronized Populations* (B018500); 24 months; $35,000
Ray, Dan S.; *Replication of Bacteriophage M13* (B018074); 24 months; $60,000
Siegel, Richard W.; *Genetics of Macroconidial Development in 'Neurospora Crassa'* (GB6642 001); 14 months; $6,000

UNIVERSITY OF CALIFORNIA, Riverside
Keen, Noel T.; *Molecular Basis of Gene-for-Gene Complementarity in Plant-Parasite Interactions* (B013559); 24 months; $51,600
Lewis, Lowell N.; *Hormonal Control of Cellulase in Cell Development and Senescence* (B017850); 24 months; $34,000

UNIVERSITY OF CALIFORNIA, San Diego
Abelson, John N.; *Bacteriophage Production of Mutations in its Host* (B020904); 24 months; $50,000
Brody, Stuart; *Morphological Mutants of Neurospora* (B021227); 24 months; $50,000
Sato, Gordon H.; *Hybridization of Clonal Strains of Functionally Differentiated Animal Cells* (B015788); 24 months; $70,000
Stern, Herbert; *Regulatory Mechanisms in Meiosis and Mitosis* (GB5173X003); 12 months; $100,000
———; *Advanced Science Seminar Summer Workshop on Molecular Techniques in Developmental Biology* (GZ1505*); 11 months; $26,200

UNIVERSITY OF CALIFORNIA, Santa Barbara
Kohl, David M.; *RNA Synthetic Patterns During Amphibian Development* (B020909); 24 months; $40,000

UNIVERSITY OF SOUTHERN CALIFORNIA
Lieb, Margaret; *Gene Interaction in Bacteria and Phage* (B016893); 24 months; $30,000
Zimmer, Russel L.; *Extra-Embryonic Nutrition in Bryozoa* (B018802); 24 months; $23,000

COLORADO

COLORADO STATE UNIVERSITY
Robertson, David W.; *Maintenance of a Barley Genetic Stock Center* (GB4482X003); 12 months; $19,700

UNIVERSITY OF COLORADO
Crumpacker, David W.; *Ecological Genetics of 'Drosophila Pseudoobscura' and its Relatives* (B018150); 24 months; $20,000
Ham, Richard G.; *Clonal Analysis of Differentiated Mammalian Cells* (B018502) 24 months; $59,700
Prescott, David M.; *Cell Growth and Reproduction* (GB7612X001); 13 months; $59,800

UNIVERSITY OF COLORADO, Denver Medical Center
Sauerbier, Walter; *Determinants in Selective Transcription of DNA Templates* (B021413); 24 months; $40,000

CONNECTICUT

TRINITY COLLEGE
Crawford, Richard B.; *Biochemistry of the Regulation of Embryogenesis* (B015141); 12 months; $20,000

UNIVERSITY OF CONNECTICUT
Berg, Claire M.; *Chromosome Replication in Escherichia Coli* (B018147); 24 months; $45,000
Laufer, Hans; *Control of Salivary Gland Function During Development* (B018606); 24 months; $63,500
Vasington, Frank D.; *Enymes Specified by the Histidine B Gene* (B015137); 24 months; $45,000

WESLEYAN UNIVERSITY
Kiefer, Barry I.; *Genetic Control of Sperm Development in 'Drosophila Melanogaster'* (B016254); 24 months; $50,000

YALE UNIVERSITY
Gehring, Walter J.; *Developmental Genetics of Imaginal Discs in 'Drosophila'* (B017267); 24 months; $68,000
Wilhelm, Robert C.; *Regulation of t-RNA Synthesis* (B018351); 24 months; $67,700

DELAWARE

UNIVERSITY OF DELAWARE
Sheppard, David E.; *Regulation in the L-Arabinose Operon* (B019676); 24 months; $37,600

DISTRICT OF COLUMBIA

HOWARD UNIVERSITY
Pipkin, Sarah B.; *Genetic and Biochemical Studies of Dehydrogenase Enzymes in the Genus Drosophila* (B008779); 24 months; $55,000

FLORIDA

UNIVERSITY OF MIAMI
Kelly, Douglas E.; *Developmental Ultrastructure of Amphibian Tissues—Adhesion Mechanisms and Intracellular Filament Systems* (B020277); 19 months; $41,000

GEORGIA

EMORY UNIVERSITY
Scott, June R.; *Lysogeny and Bacteriophage P1* (B016229); 24 months; $35,000
Shuster, Robert C.; *A New DNA Endonuclease Synthesized by Bacteriophage* (B015059); 24 months; $59,500

UNIVERSITY OF GEORGIA
Coward, Stuart J.; *Dynamics of Regeneration in Planaria* (B019630); 24 months; $40,000
Dure, Leon S. III; *Biochemistry of Growth and Development of Plant Embryos* (B016632); 24 months; $40,000

1. An asterisk following the NSF Grant Number indicates that funds for the projects described were provided by more than one program unit of the Foundation.

BIOLOGICAL AND MEDICAL SCIENCES RESEARCH PROJECTS

ILLINOIS

NORTHWESTERN UNIVERSITY
Burnett, Allison L.; *Control of Cellular Differentiation in Hydra* (B017917); 24 months; $50,000
Coe, Elmon L.; *Correlations Between Metabolic and Morphological Changes in Differentiating Cellular Slime Mold* (B017624); 24 months; $32,000
Whitten, Joan M.; *Cell Division, Growth, Differentiation, and Death* (B020901); 24 months; $55,000

NORTHWESTERN UNIVERSITY, Chicago Campus
Smith, David W. E.; *Regulation of Transfer Ribonucleic Acid Biosynthesis* (B018869); 18 months; $48,000

UNIVERSITY OF CHICAGO
Chiang, Kwen-Sheng; *Molecular Mechanism of Non-Mendelian Inheritance* (B019338); 24 months; $50,000
Haselkorn, Robert; *Blue-Green Algae—Mutants, Viruses, and Transduction* (B017514); 24 months; $49,800
Lewontin, Richard C.; *Genetic Heterozygosity in Natural Populations* (B7135X 001); 12 months; $47,100
Ravin, Arnold W.; *Molecular Genetics of Transformation in Pneumococci and Streptococci* (B018511); 24 months; $20,000

UNIVERSITY OF ILLINOIS, Urbana
Davenport, Richard; *Cellular Interactions During Oogenesis in 'Oncopeltus'* (B014894); 24 months; $29,300
Drake, John W.; *Studies on Mutation* (B015139); 24 months; $55,600
Hershberger, Charles L.; *Structure and Replication of Chloroplast NA in 'Euglena gracilis'* (B018933); 24 months; $50,000
Lambert, R. J.; *Maintenance of Genetic and Chromosomal Tester Stocks of Maize* (B014002); 24 months; $66,600

INDIANA

INDIANA UNIVERSITY, Bloomington
Starr, Richard C.; *Culture Collection of Algae at Indiana University* (B020680); 24 months; $40,000
White, David; *Morphogenesis in the Fruiting Myxobacteria* (B020516); 12 months; $18,000

PURDUE UNIVERSITY
Aronson, Arthur I.; *Macromolecule Synthesis During Early Sea Urchin Embryo Development* (B017947); 24 months; $42,400
Jaffe, Lionel F.; *Electrical Currents Through Developing Eggs* (B17571X); 24 months; $83,000

IOWA

UNIVERSITY OF IOWA
Kollros, Jerry J.; *Quantitative Control of Lateral Motor Column Cell Number* (B017626); 24 months; $40,000

KANSAS

KANSAS STATE UNIVERSITY
Bode, Vernon C.; *Arrangement of DNA in a Bacteriophage Head* (B025153); 24 months; $47,300

KENTUCKY

UNIVERSITY OF KENTUCKY
Arking, Robert; *Genetic and Hormonal Control of Specific Protein Synthesis During Insect Development* (B018803); 24 months; $27,000

LOUISIANA

LOYOLA UNIVERSITY
Lesseps, Roland J.; *Developmental Changes in the Cell Surface of Embryonic Chick Cells* (B017499); 24 months; $30,000

MARYLAND

JOHNS HOPKINS UNIVERSITY
Jacobson, Marcus; *Mechanism of Formation of Neuronal Connections* (GB8315 001); 3 months; $9,000
Rhoades, Marcus M. Jr.; *Role of Endonucleases in Genetic Recombination* (B020460); 24 months; $50,000

UNIVERSITY OF MARYLAND, Baltimore County Campus
Burchard, Robert P.; *Interaction of Bacteriophage with Myxococcus xanthus During Cellular Morphogenesis* (GB7808 001); 12 months; $10,900

MASSACHUSETTS

BOSTON UNIVERSITY
Margulis, Lynn; *Metabolism of Regenerating 'Stentor' Basal Bodies and Cilia* (B021408); 24 months; $36,600

BRANDEIS UNIVERSITY
Fulton, Chandler M.; *Cell Differentiation and Organelle Morphogenesis in the Amebo-Flagellate 'Naegleria'* (B016334); 24 months; $67,000
Klein, Attila O.; *Quantitative Biochemical Aspects of Light-Regulated Leaf Development* (B016047); 24 months; $40,000

HARVARD UNIVERSITY
Fraenkel, Dan G.; *Biochemical Genetics of Glycolysis and the Hexose Monophosphate Shunt in Bacteria* (B015958); 24 months; $79,800
Gorini, Luigi C. and Strigini, Paolo F. G.; *Ribosomal Role in Informational Suppression* (B016836); 24 months; $70,000
Levine, R. P.; *Genetic Control of Chloroplast Structure and Function in Chlamydomonas* (B018666); 24 months; $47,000

MASSACHUSETTS INSTITUTE OF TECHNOLOGY
Bell, Eugene; *Molecular Studies of Differentiating Cells* (B018142); 12 months; $70,000
Ingram, Vernon M.; *Control of the Biosynthesis of Specific Proteins During Development* (GB5181X003); 12 months; $75,300

RETINA FOUNDATION
Wright, Barbara E.; *Mechanism of Enhanced Enzyme Levels During Differentiation* (B015055); 24 months; $45,000

STATE COLLEGE AT BRIDGEWATER
Brennan, James R.; *Light and Electron Microscopic Studies of the Cytological Effects of Microtubule-Affecting Chemicals on 'Phleum' Root Tips* (B014503); 24 months; $27,200

TUFTS UNIVERSITY
Siegel, Eli G.; *Mutator Genes of Escherichia Coli* (B014279); 24 months; $45,800

UNIVERSITY OF MASSACHUSETTS
Dowell, Clifton E.; *Replication of Small DNA Bacteriophages* (B018639); 6 months; $15,000
Galinat, Walton C.; *Evolution of the American Maydeae* (B015767); 24 months; $28,300

MICHIGAN

MICHIGAN STATE UNIVERSITY
Kende, Hans J.; *Studies on the Site of Action of Gibberellins* (B013853*); 24 months; $15,000

UNIVERSITY OF MICHIGAN
Berns, Michael W.; *Laser Microbeam Studies on Chromosomes* (B024457); 24 months; $39,400
Nace, George W.; *Development of Defined Amphibian Strains* (GB8187X001); 12 months; $110,000

WAYNE STATE UNIVERSITY
Taylor, John D.; *Morphological and Physiological Aspects of Chromatophore Differentiation* (B016329); 24 months; $28,100

MINNESOTA

UNIVERSITY OF MINNESOTA
Enfield, Franklin D.; *Population and Quantitative Genetic Studies of Electrophoretic Variants in Tribolium Castaneum* (B012751); 24 months; $30,000

Hartl, D. L.; *Genetic and Population Studies of Segregation Distortion* (B018786); 24 months; $40,000
Zissler, James F.; *Genetic Recombination in Phage Lambda* (B020677); 24 months; $40,000

MISSOURI

ST. LOUIS UNIVERSITY
Eliceiri, George L.; *Studies on Mammalian RNA's* (B017546); 24 months; $60,800
Melechen, Norman E.; *Control of Bacteriophage Gene Function by Structural and Chemical Elements* (B022609); 24 months; $61,000

WASHINGTON UNIVERSITY
Apirion, David; *Genetic Approaches to Non-Ribosomal Factors in Escherichia coli Protein Synthesis* (B017888); 24 months; $60,000
Enders, Allen C.; *Mechanisms of Implantation in Mammals* (B020985); 24 months; $34,200
Kennell, David E.; *Gene Expression of Bacteria and Bacterial Viruses* (B019198); 24 months; $53,000
Levi-Monalcini, Rita; *'In vivo' and 'In vitro' Analysis of the Insect Nervous System* (B16330X); 24 months; $50,000

MONTANA

MONTANA STATE UNIVERSITY
Roemhild, George; *Equipment to Study Control of Development in Eggs Using Pulsed Laser Energy* (B019789); 12 months; $4,000

NEBRASKA

UNIVERSITY OF NEBRASKA, Lincoln
Brumbaugh, John A.; *Melanosome Fine Structure and Autoradiography—Genetic Variations* (B012429); 24 months; $25,000
Gupta, Naba K.; *Genetic Code, Protein Synthesis in Mammalian Systems* (B014160); 24 months; $80,000

NEW JERSEY

PRINCETON UNIVERSITY
Bonner, John T.; *Development in the Cellular Slime Molds* (B019930); 24 months; $78,000
Jacobs, William P.; *Control of Differentiation and Growth in Plants* (GB7783 001) 4 months; $5,000
Meins, Frederick Jr.; *Heritable Changes in the Tumor State of Plants* (B017916); 24 months; $35,000
Newton, William A.; *Mechanism and Control of Gene Expression* (B018510); 24 months; $55,000
Steinberg, Malcolm S.; *Cell Contact in Relation to Growth and Morphogenesis* (GB5759X-002); 12 months; $55,000

RIDER COLLEGE
Mayer, Thomas C.; *Developmental Control of Pigmentation in Mice* (B018271); 24 months; $21,000

NEW YORK

COLD SPRING HARBOR LABORATORY
Watson, James D.; *Symposium on RNA Transcription* (B019122); 12 months; $5,000
———; *Conference on Phage Lambda* (B021221); 12 months; $6,000
Zipser, David; *Polar Mutants and Lac Operon Function* (B020714); 24 months; $91,200

COLUMBIA UNIVERSITY
Marks, Paul A. and Rifkind, Richard A.; *Erythroid Cell Differentiation and Maturation* (GB4631X003); 12 months; $45,800
Sederoff, Ronald R.; *Properties of Y Chromosome Specific RNA and DNA* (B021225); 24 months; $40,100
Spiegelman, S.; *Relation of Morphogenesis to Genetic Transcription and Translation* (B019220); 12 months; $28,200
Zubay, Geoffrey L.; *Control of Bacterial Operons in Cell-Free Systems* (B018733); 24 months; $45,100

CELLULAR BIOLOGY

CORNELL UNIVERSITY
Calvo, Joseph M.; *Genetic and Metabolic Control of Leucine Biosynthesis in Bacteria* (B023180); 24 months; $50,000

CUNY, MT. SINAI SCHOOL OF MEDICINE
Capra, J. D.; *An Immunochemical and Immunogenetic Study of the FD Fragment of Human YG Globulins* (B017046); 24 months; $65,000

CUNY, LEHMAN COLLEGE
Jensen, Thomas E.; *Fine Structure and Physiology of the Blue-Green Algae* (B017700); 12 months; $4,000

PUBLIC HEALTH RESEARCH INSTITUTE
Dubnau, David; *Gene Conservation in Bacillus Species* (B018146); 24 months; $63,400
Oishi, Michio; *Molecular Events Associated with Gene Function and DNA Replication in Microorganisms* (B014313); 24 months; $74,900
Schaefler, Sam; *The B-Glucoside System in Bacteria* (B021234); 24 months; $20,000
Smith, Issar; *Regulation of Ribosome Synthesis in 'Bacillus Subtilis'* (B016782); 24 months; $58,100

ROCKEFELLER UNIVERSITY
Cerami, Anthony; *Characterization of Factors in Blood Cell Development* (B020900); 24 months; $42,200

SOCIETY FOR DEVELOPMENTAL BIOLOGY
Runner, Meredith N.; *Growth Symposium for 1970* (GB5794 003); $2,700

STATE UNIVERSITY AT ALBANY
Feldman, Jerry F.; *Circadian Clock Studies in Neurospora Crassa* (B021418); 24 months; $46,200
Mascarenhas, Joseph P.; *Synthesis of Wall Material During Pollen Tube Development* (B013115); 12 months; $18,000

STATE UNIVERSITY AT STONY BROOK
Arnheim, Norman; *Biochemical Analysis of Isoalleles in Natural Populations* (B017351); 24 months; $40,000
Leichtling, Ben H.; *Control of Transcription of Bacteriophage* (B016795); 24 months; $20,000
Lyman, Harvard; *Control Mechanisms in Chloroplast Biosynthesis and Replication* (B016922); 24 months; $60,000

SYRACUSE UNIVERSITY
Balbinder, Elias; *Regulation and Evolution of a Bacterial Operon* (B017609); 24 months; $71,000

UNIVERSITY OF ROCHESTER
Lockshin, Richard A.; *Breakdown of Tissues in Metamorphosing Moths* (B021237); 24 months; $40,000
Prakash, Satya; *Genic Variation in Natural Populations* (B016192); 24 months; $55,000

YESHIVA UNIVERSITY
Bloch, Eric; *Steroid Metabolism by the Reproductive Tract During Development* (B014034); 24 months; $35,000
Dorfman, Ben-Zion; *Genetic Regulation of Purine Biosynthesis in Saccharomyces Cerevisiae* (B018149); 24 months; $20,400
Levine, Elliott M.; *Molecular Biology of Cellular Interactions* (B017732); 24 months; $45,000

NORTH CAROLINA

DUKE UNIVERSITY
Gillham, Nicholas W. and Boynton, John E.; *Hereditary Control of Organelle Structure and Function* (B022769); 24 months; $75,000
Gross, Samson R.; *Regulatory Mechanisms of Enzyme Synthesis and Function in 'Neurospora'* (GB4489X003); 12 months; $49,200

UNIVERSITY OF NORTH CAROLINA AT CHAPEL HILL
Bleyman, Michael A.; *Transcriptional Mapping* (B021229); 24 months; $44,700
Stafford, Darrel W.; *Control of Genes for Ribosomal RNA Synthesis in Sea Urchin Development* (B020918); 24 months; $38,000
———; *Translation and Control in Developing 'Drosophila Melanogaster'* (GB8334 001); $6,600

OHIO

BOWLING GREEN STATE UNIVERSITY
Oster, Irwin I.; *Support of Drosophila Melanogaster Stock Center* (GB4700X003); 12 months; $34,700

KENYON COLLEGE
Jegla, Thomas C. and Costlow, John D. Jr.; *Control of Molting and Metamorphosis in Horseshoe Crab and Crustacean Larvae* (B016874); 24 months; $19,700

MEDICAL COLLEGE OF OHIO AT TOLEDO
Engelbrecht, Helen L.; *Sporulation—A Study of Cell Differentiation* (B019506); 24 months; $40,000

OREGON

OREGON STATE UNIVERSITY
Brookes, Victor J.; *Biochemistry of Insect Development* (B014093); 24 months; $37,000
Morris, John E.; *Cell Associations and Enzyme Induction in Development* (B019211); 24 months; $40,000

UNIVERSITY OF OREGON
Grant, Philip; *Role of Heterochromatin in Cytodifferentiation* (B016332); 24 months; $20,100

PENNSYLVANIA

BRYN MAWR COLLEGE
Oppenheimer, Jane M.; *Development Analysis of Fundulus* (B017944); 24 months; $18,200

INSTITUTE FOR CANCER RESEARCH
Loeb, Lawrence A.; *Control of DNA Synthesis in Synchronized Animal Cells* (B018419); 24 months; $28,100

TEMPLE UNIVERSITY
Brownstein, Barbara L.; *Ribosome Synthesis and Function* (B021414); 24 months, $47,000
Takats, Stephen T.; *Gametogenesis in Plants —Control of DNA Synthesis* (GB7822 001); 2 months; $4,900

UNIVERSITY OF PENNSYLVANIA
Frankel, Fred R.; *Nature of Two Enzymatic Activities Concerned with DNA Metabolism in Phage Infected Cells* (B015766); 24 months; $45,200
Green, Paul B.; *Growth and Cellular Morphogenesis in 'Nitella'* (B06055X002); 15 months; $41,500
Holtzer, Howard; *Mitotic Cycle and the Synthesis of Cell Specific Molecules* (GB5047X003); 12 months; $41,300

UNIVERSITY OF PITTSBURGH
Carell, Edgar F.; *Role of Vitamin B12 in Cell Development and Replication* (B020903); 24 months; $35,000

WISTAR INSTITUTE
Whittaker, J. Richard; *Regulation of Melanotic Expression in Pigment Cell Cultures—A Modulation Problem* (B015960); 24 months; $60,000

WOMAN'S MEDICAL COLLEGE OF PENNSYLVANIA
DiBerardino, Marie A.; *Analysis of Cellular Components During Development* (B019631); 24 months; $56,000

RHODE ISLAND

BROWN UNIVERSITY
Goss, Richard J.; *Mechanism and Control of Antler Regeneration* (B020166); 24 months; $40,000
Kimmel, Donald L.; *Expression of the Vermilion Locus During 'Drosophila melanogaster' Development* (B019513); 24 months; $30,000
Nei, Masatoshi; *Gene Substitution and Genetic Polymorphisms in Regulated Populations* (B021224); 24 months; $12,000
Shipp, William S.; *Respiratory Deficient Mutants of Escherichia Coli* (B015390); 24 months; $75,000

GORDON RESEARCH CONFERENCES
Cruickshank, Alexander M.; *Gordon Research Conference on Nucleic Acids* (B020075); 12 months; $3,000
———; *Gordon Research Conference on Lysosomes* (B020071); 12 months; $2,500
———; *Gordon Research Conference on Developmental Biology* (B023141); 12 months; $4,000

TENNESSEE

VANDERBILT UNIVERSITY
Bibring, Thomas; *Macromolecules of the Mitotic Apparatus* (B017741); 24 months; $55,000

TEXAS

M. D. ANDERSON HOSPITAL
Stubblefield, Elton T.; *Protein Synthesis in Synchronized Cells* (B016250); 24 months; $35,800

UNIVERSITY OF TEXAS MEDICAL BRANCH AT GALVESTON
Altmiller, Dale H.; *Structure and Function of Mitochondrial Proteins* (B018545); 24 months; $40,400

UNIVERSITY OF TEXAS AT AUSTIN
Hiraizumi, Yuichiro; *Genetic Studies on Prezygotic Selection* (B017986); 24 months; $45,000
Rode, Leonard J. Jr.; *Ultrastructure and Physiology of Bacterial Endospores* (B017677); 12 months; $15,000
Whaley, W. Gordon; *Developmental and Functional Studies of the Golgi Apparatus and Related Organelles in Secretion* (B017778); 24 months; $50,000

UNIVERSITY OF TEXAS AT DALLAS
Gutz, Herbert; *Molecular Genetics of Yeast* (B015148); 24 months; $50,000

UTAH

UNIVERSITY OF UTAH
Greenlee, Lorance L.; *Mechanism and Control of Replication of Small DNA Molecules* (B020908); 24 months; $50,000

VERMONT

MIDDLEBURY COLLEGE
Saul, George B. II; *Maintenance of a Mormoniella Stock Center* (B016132); 24 months; $21,900

VIRGINIA

UNIVERSITY OF VIRGINIA
Bodenstein, Dietrich; *Differentiation and Problems of Neuro-Development in Insects and Lower Vertebrates* (B020167); 24 months; $29,700
Konigsberg, Irwin R.; *Role of Collagen in the Development of Muscle Clones 'In Vitro'* (GB5963X002); 12 months; $48,200
Kutter, Elizabeth M.; *Regulation of Cytosine-Containing DNA in T4 Infection of E. Coli* (B019517); 24 months; $26,500
Wright, Theodore R. F.; *Genetic Control of Myogenesis in Drosophila* (B020910); 24 months; $40,000

VIRGINIA COMMONWEALTH UNIVERSITY
Collins, James M.; *Gene Amplification and Transfer RNA Restriction in the Regenerating Lens* (B018143); 24 months; $35,000

WASHINGTON

UNIVERSITY OF WASHINGTON
Fangman, Walton L.; *Cellular Synthesis of DNA* (B019792); 24 months; $45,000

BIOLOGICAL AND MEDICAL SCIENCES RESEARCH PROJECTS

Haskins, Edward F.; *Morphogenesis in the Myxomycete, Echinostelium* (B020517); 12 months; $15,000
Keller, John M.; *Changes in Animal Cell Membranes During Mitosis* (B019675); 24 months; $40,000
Kelly, Douglas E.; *Developmental Ultrastructure of Amphibian Tissues—Adhesion Mechanisms and Intracellular Filament Systems* (B014098); 24 months; $50,000
McCarthy, Brian J.; *Specificity and Application of DNA/DNA Duplex Formation* (B016751); 24 months; $70,000
Stadler, David R.; *Recombination and Gene Function in Fungi* (B018412); 24 months; $70,000

WISCONSIN

MARQUETTE UNIVERSITY
Krishnakumaran, Alapati; *Nucleic Acid Metabolism in Insect Embryogenesis* (B019629); 24 months; $34,300

UNIVERSITY OF WISCONSIN, Parkside
Goodman, Eugene; *Synchronization and Axenic Culture of Slime Mold Myxamoebae* (B020074); 24 months; $18,000

UNIVERSITY OF WISCONSIN, Madison
Dahlberg, James E.; *Nucleotide Sequence Analysis of Genetic Punctuation* (B015671); 24 months; $100,000
Fitch, Walter M.; *Special Approaches to the Study of Evolutionary Relationships* (B017841); 24 months; $65,000
Lilien, Jack E.; *Mechanism of Specific Cell Adhesion* (B017853); 24 months; $47,700
Mossman, Harland W.; *Vertebrate Fetal Membranes* (B019732); 24 months; $40,000
Nelson, Oliver E.; *The WX Locus in Maize and Genetic and Physiological Specialization* (B015104); 24 months; $68,800
Newcomb, Eldon H. and Becker, Wayne M.; *Studies in Plant Cell Biology* (B015246); 24 months; $95,000
Plaut, W. and Smith-Sonneborn, J.; *Pellicular DNA and the Mechanism of Aging in Paramecium* (B016333); 24 months; $55,600
Reznikoff, William S.; *Genetic Transcription and Regulation Signals in E. Coli* (B020462); 24 months; $45,000
Smithies, Oliver; *Genetic Control of Protein Structure* (GB4362 004); 12 months; $47,900
Wright, Sewall; *Evolution and the Genetics of Populations* (B019340); 12 months; $14,600

CANADA

UNIVERSITY OF CALGARY
Sanderson, Kenneth E.; *Maintenance of a Salmonella Genetic Stock Center* (B020464); 24 months; $10,000

Ecology and Systematic Biology

ALASKA

UNIVERSITY OF ALASKA
Peyton, Leonard J.; *Geographical Organization of Migratory Sparrow Nesting in Arctic and Subarctic Regions* (B020356); 24 months; $20,000
West, George C., Alexander, Vera, and Van Cleve, Keith; *Effects of Perturbations on Cold-Dominated Ecosystems in Alaska, Part 1* (B025024); 12 months; $162,500

ARIZONA

ARIZONA STATE UNIVERSITY
Gerking, Shelby D.; *Adaptation to Extremes of Temperature* (B013811); 24 months; $40,000
Minckley, W. L.; *Systematic and Biogeographic Studies of Aquatic and Semi-Aquatic Organisms in Northern Mexico* (GB6477X001); 12 months; $14,400

CALIFORNIA

CALIFORNIA ACADEMY OF SCIENCES
Brown, Walter C.; *Ecological Investigations of the Herpetofauna of the Philippine Islands* (B016972); 24 months; $18,300
Lindsay, George E.; *Care of Systematic Collections and Scientific Library* (B019367); 12 months; $38,000

CALIFORNIA STATE COLLEGE AT FULLERTON
McClanahan, Lon L.; *Ecology of Amphibians* (B019084); 12 months; $38,700

CALIFORNIA STATE COLLEGE AT LOS ANGELES
Allen, Richard K.; *Generic Revisions of the Ephemeroptera of Western North America* (B017570); 24 months; $23,000

SAN DIEGO STATE COLLEGE
Carpenter, Roger E.; *Energy and Water Relationships in Certain Chiropterans* (B015158); 24 months; $30,000
Coulombe, Harry N.; *Analysis of the Structure and Function of the Wet Tundra Ecosystem at Point Barrow, Alaska* (B017419); 12 months; $58,700
Coulombe, Harry N. and Brown, Jerry; *Analysis of the Structure and Function of the Wet Tundra Ecosystem* (GB17419001); $28,300

SAN FRANCISCO STATE COLLEGE
Thiers, Harry D.; *The Suillus Flora of the Southwestern United States* (B014937); 24 months; $28,000

STANFORD UNIVERSITY
Pattee, Howard H.; *Primeval Ecosystem—The Dynamics of Physical and Informational Networks at the Origin of Life* (B016563); 24 months; $45,000
Raven, Peter H.; *Systematics of the Onagraceae* (GB7949X001); 12 months; $26,700
Watt, Ward B.; *Evolutionary Adaptation in 'Colias'* (B018704); 24 months; $60,000

UNIVERSITY OF CALIFORNIA, Berkeley
Chaney, Ralph W.; *Extension of Geofloral Studies into Southern Asia* (B013522); 12 months; $11,350
Ornduff, Robert; *Evolution and Functional Significance of Heterostyly in Flowering Plants* (B017613); 24 months; $26,500
Pitelka, Frank A., Schultz, Arnold M. and Gersper, Paul L.; *Effects of Perturbations on Cold-Dominated Ecosystems in Alaska, Part 2* (B025025); 12 months; $35,000
Powell, Jerry A.; *Comparative Biology in Relation to Systematics of Microlepidoptera* (GB6813X001); 12 months; $16,200
Sarich, Vincent M.; *Immunological Studies of Protein Evolution in Mammals* (B020750); 24 months; $49,800
Wake, David B.; *Comparative Morphological and Evolutionary Studies on Amphibian Vertebrae* (B017112); 12 months; $17,000

UNIVERSITY OF CALIFORNIA, Davis
Axelrod, Daniel I.; *Miocene Floras from Western Nevada* (B019844); 24 months; $28,000
Barbour, Michael; *Factors Limiting the Distribution of a Species* (B014381); 24 months $24,000
Goldman, Charles R.; *Progressive Eutrophication of Lake Tahoe* (B019136); 15 months; $65,500
Hamilton, William J. III; *Behavior and Ecology of Tropical Mixed Avian Flocks* (B017180); 12 months; $4,000
——— ; *Ecological and Behavioral Adaptations in Lar and Pileated Gibbons* (B018651); 24 months; $30,000
Jain, S. K.; *Population Biology of Colonizing Plant Species* (GB8627 001); $2,000
Rudd. Robert L.; *Biological Characteristics of an Introgressed Population of Shrews at Junctions of Species Ranges* (B015916); 12 months; $18,600
Salt, George W.; *Interactions Between Predator and Prey Populations* (B017832); 24 months; $47,000

Williams, W. A.; *Modeling Foliage Canopies and Productivity of Plant Communities* (B018541); 24 months; $53,000

UNIVERSITY OF CALIFORNIA, Irvine
Arditti, Joseph; *Nature, Distribution and Taxonomic Significance of Anthocyanins in Some Orchids* (B013417); 24 months; $31,600
Atsatt, Peter R.; *Genetic and Evolutionary Significance of Natural Root Grafting* (B015574); 24 months; $28,000
Macmillen, Richard E.; *Water Regulation in Nocturnal California Rodents* (B017833); 24 months; $24,000

UNIVERSITY OF CALIFORNIA, Los Angeles
Ball, Gordon H.; *Life Histories of Sporozoan Parasites in the Blood of Reptiles* (B014587); 24 months; $39,000
Bartholomew, George A.; *Ecological Physiology and Behavior* (B018744); 24 months; $45,000
Brown, James H.; *Influences of Cholla Cactus on the Biology of Small Desert Mammals* (B008765); 24 months; $13,500
Cody, Martin L.; *Structure and Organization of Natural Communities* (B013651); 24 months; $22,000
Olson, Everett C.; *Studies of Permian Terrestrial Vertebrates* (B013910); 24 months; $31,400
Vaughn, Peter P.; *Late Pennsylvanian and Early Permian Vertebrates of the Four Corners States* (B019971); 24 months; $14,700

UNIVERSITY OF CALIFORNIA, Riverside
Debach, Paul; *Biosystematics and Phylogeny of Species of 'Aphytis'* (B017829); 24 months; $55,000
Newell, Irwin M.; *Cosystematics of the Genus Dinogamasus (Acari, Mesostigmata) and the Xylocopidae (Hymenoptera)* (B015840); 24 months; $22,300
Scora, Rainer W.; *Inheritance of Oils in the Rutaceae* (B014455); 12 months; $10,200
Van Gundy, Seymour D.; *Ecology of Terrestrial Nematodes* (B013248); 24 months; $32,000
Vasek, Frank C.; *Quantitative Estimates of Outcrossing in Clarkia* (GB5914X001); 12 months; $13,800

UNIVERSITY OF CALIFORNIA, San Diego
Hubbs, Carl L.; *Quaternary Environments and Faunas* (B013589); 24 months; $45,800
Soule, Michael E.; *West Coast Population Biologists' Workshop Conference Topical Problems in Population Biology* (B015102); 12 months; $2,300

UNIVERSITY OF CALIFORNIA, Santa Barbara
Muller, Cornelius H.; *Role of Natural Chemical Inhibitors in Plant Ecology* (B014891); 24 months; $45,000

COLORADO

COLORADO STATE UNIVERSITY
Moir, W. H.; *Nutrient Cycling in Pine Stands* (B020357); 15 months; $13,100
Reid, C. P. P.; *Role of Ectotrophic Mycorrhizal Structures in Tree Root Exudate Transport* (B019928); 24 months; $40,000

CONNECTICUT

CENTER FOR ENVIRONMENT AND MAN
Lord, Norman W.; *Effects of Perturbations on Cold-Dominated Ecosystems in Alaska, Part 6* (B025064); 12 months; $19,900

UNIVERSITY OF CONNECTICUT
Andrews, Henry N. Jr.; *Studies of Devonian Age Plants* (GB7102X001); 12 months; $17,500
Brush, Alan H.; *Evolutionary and Structural Studies on Feather Proteins* (B020086); 24 months; $33,000
Schultz, R. Jack; *Evolution of 'Poeciliopsis'* (B020686); 24 months; $55,500
Trainor, Francis R.; *Classification of Scenedesmus Species in Culture* (B018804); 24 months; $29,300

ECOLOGY AND SYSTEMATIC BIOLOGY

UNIVERSITY OF HARTFORD
Simpson, Tracy L.; *Ecology of Freshwater Sponges* (B018613); 18 months; $12,000
YALE UNIVERSITY
Bormann, F. H.; *Collaborative Research on Hydrologic-Mineral Cycle Interaction in Small Undisturbed and Man-Manipulated Watersheds* (B014289001); $4,000
———; *Collaborative Research on Hydrological-Mineral Cycle Interaction in Small Undisturbed and Man-Manipulated Watersheds* (B014289); 24 months; $156,000
Crompton, Alfred W.; *Transitional Thecodonts from Pre-Nordian Sediments of South America and Their Bearing on Middle Triassic Faunal History* (B015419); 12 months; $10,700
Delevoryas, Theodore; *Paleobotanical Research* (B20999X); 12 months; $36,900
Jordon, Peter A.; *Habitat Relationships and Distribution of Wild Ungulates in the Gir Forest, India* (B017676); 18 months; $3,000
Levin, Donald A.; *Potential Gene Flow in Phlox Pilosa L* (B017987); 24 months; $42,400
Ostrom, John H.; *Transitional Thecodonts from Pre-Nordian Sediments of South America and Their Bearing on Middle Triassic Faunal History* (B015419001); $5,100
———; *Osteology, Biomechanics and Aerodynamics of the Pterosauria* (B014033); 30 months; $25,000
Sibley, Charles G.; *Comparative Studies of Protein Structure* (GB6192X002); 12 months; $45,200
Thomson, Keith S.; *Lobe-Finned Fishes and the Fish-Tetrapod Transition* (GB7573X001); 12 months; $5,900

DISTRICT OF COLUMBIA

AMERICAN INSTITUTE OF BIOLOGICAL SCIENCES
Olive, John R.; *Flora North America Project* (GB8441 001); $943
———; *Status of Biological Field Stations in Teaching and Research* (B023195); 12 months; $6,250
GEORGETOWN UNIVERSITY
Colwell, Rita R.; *Systematic Study of the Genus Vibrio and Related Organisms* (B018274); 24 months; $66,500
SMITHSONIAN INSTITUTION
Cuatrecasas, Jose; *Taxonomic Study of Neotropical Phanerogams* (B017612); 24 months; $38,500
———; (GB5806X002); $10,116
McClure, F. A.; *Taxonomy of the Bambusoideae—Revision of the Bamboo Genera of the World* (GB8415 001); $3,240

FLORIDA

FAIRCHILD TROPICAL GARDEN
Tomlinson, P. B.; *Comparative Anatomy of Monocotyledons* (GB5762 002); 12 months; $31,600
FLORIDA STATE UNIVERSITY
Godfrey, Robert K.; *Systematics of Eastern North American Eupatorium (Compositae)* (GB7838X001); 12 months; $14,500
ORGANIZATION FOR TROPICAL STUDIES
Hooper, Emmet T.; *Small Mammal Faunas of Two Tropical Rain Forests* (B021439); 14 months; $33,200
UNIVERSITY OF FLORIDA
Gray, John; *Support of XVth Congress of International Union of Forestry Research Organizations* (B016410); 13 months; $10,000
McNab, Brian K.; *Ecological Basis of Marginal Temperature Regulation* (B020689); 24 months; $13,000
Walker, Thomas J.; *Biological Species in Gryllidae and Tettigoniidae (Orthoptera)* (B020749); 24 months; $32,000
UNIVERSITY OF MIAMI
Dodson, Calaway H.; *Systematic Studies in the Orchidaceae* (B017923); 24 months; $33,913
———; (GB17923001); $6,200
Taylor, Dennis L.; *Cellular Interaction—Algal—Invertebrate Symbiosis* (B019790); 24 months; $54,800

GEORGIA

GEORGIA SOUTHERN COLLEGE
Oliver, James H.; *Systematic Studies of 'Haemaphysalis Longicornis'* (B021008); 24 months; $45,000
UNIVERSITY OF GEORGIA
Atyeo, Warren T.; *The Supraspecific Taxa of the Analgoidea (Acarina-Sarcoptiformes)* (B015105); 24 months; $33,000
Coleman, James R.; *Systematics of Verbesina (Compositae)* (B013923); 24 months; $14,800
Golley, Frank B.; *Bioenergetics of a Food Chain* (B017113); 12 months; $1,200
Jones, Samuel B. Jr.; *Systematics of Vernonia (Compositae)* (B020687); 24 months; $12,200
Monk, Carl D.; *Species Diversity and Succession* (B019925); 12 months; $15,000
Odum, Eugene P.; *Acute Fire Stress on Old-Field Ecosystems* (B016507); 24 months; $22,700

HAWAII

BERNICE P. BISHOP MUSEUM
Gressitt, J. Lindsley; *World Monographs and Studies of Pupipara and Muscoid (Calyptrate) Flies* (B013731); 24 months; $45,000
Radovsky, Frank J.; *Systematics of Parasitic Mesostigmatic Mites* (B020087); 24 months; $35,400
St John, Harold *Support for Pandanus Research* (B17365X); 24 months; $15,900
Van Royen, Pieter; *Flora of the Alpine Regions of New Guinea* (B021438) 12 months; $8,200
UNIVERSITY OF HAWAII
Caperon, John; *Computer Simulation Studies of 'Littorina' Populations* (B019333); 18 months; $29,400

ILLINOIS

FIELD MUSEUM OF NATURAL HISTORY
Singer, Rolf; *Research on Agaricules with Special Reference to New Aspects of their Taxonomy* (B014492); 24 months; $21,200
Webber, E. Leland; *Center for Graduate Studies in Systematic Zoology and Paleontology* (GB8789); 12 months; $41,500
Williams, Louis O., Simpson, Donald R. and Izquierdo, Edwardo; *Neotropical Botanical Survey* (GB5588X003); 13 months; $28,200
NORTHWESTERN UNIVERSITY, Chicago Campus
Nadler, Charles F; *Chromosomal and Biochemical Studies of Evolution in Sciuridae and Gerbillinae (Mammalia-Rodentia)* (B016924); 24 months; $45,000
UNIVERSITY OF CHICAGO
Levins, Richard; *Genetic and Ecological Studies of Population Fitness* (B017737); 24 months; $51,000
Park, Thomas; *Population Ecology of 'Tribolium'* (B021246); 24 months; $16,200
Radinsky, Leonard B.; *Evolution of Three Orders of Mammals* (GB7071X001); 12 months; $19,700
Ziegler, Alfred M.; *Community Paleoecology of the Silurian Deposits of the Caledonian and Appalachian Geosynclines* (B017655); 24 months; $43,000
UNIVERSITY OF ILLINOIS, Chicago Circle
Mertz, David B.; *Interspecies Association* (B017028); 12 months; $10,800
UNIVERSITY OF ILLINOIS, Urbana
Levine, Norman D.; *Systematics of Sporozoan Protozoa* (B014096); 24 months; $44,000
Small, Eugene B.; *Differentiation and Morphogenesis on Scuticociliate Systematics* (B014828); 24 months; $44,000

INDIANA

INDIANA UNIVERSITY, Bloomington
Brock, Thomas D.; *Microbial Ecology of Thermal Environments* (B019138); 24 months; $65,000
Frey, David G.; *Limnology and Paleolimnology of a Tropical Lake* (B016054); 24 months; $64,900
Meinschein, Warren G.; *Biogeochemical Research* (B014867); 24 months; $50,000

IOWA

IOWA STATE UNIVERSITY
Pohl, Richard; *Gramineae of Central America* (GB7307X001); 12 months; $18,500

KANSAS

KANSAS STATE UNIVERSITY
Fretwell, Stephen D.; *Distribution and Abundance of Natural Populations* (B014293); 24 months; $22,000
Kramer, C. L.; *Revision of the Genus Taphrina* (B014972); 24 months; $15,100
Rettenmeyer, Carl W.; *Comparative Ecology of Neotropical Dorylines* (B014097); 24 months; $68,000
Robel, Robert J.; *Pesticide, Parasite, Energetic and Behavioral Interactions* (B016010); 24 months; $60,000
UNIVERSITY OF KANSAS
Byers, George W.; *A Systematic Study of Scorpion Flies* (GB7045X001); 12 months; $7,000
Jackson, Raymond C.; *Variation and Evolution of Races in Haplopappus Gracilis* (B014827); 24 months; $15,000
Johnston, Richard F.; *Evolution and Character Variation in Hybridizing Populations* (GB8781); 24 months; $34,000
Jones, J. Knox Jr.; *Coherent Area Research in Evolutionary Biology* (B008785); 12 months; $40,000
Lichtwardt, Robert W.; *Biological and Systematic Studies on Trichomycetes* (GB7072X001); 12 months; $13,750
Wells, Philip V.; *Biotic and Climatic Change in Western North America* (B014559); 24 months; $37,300

LOUISIANA

LOUISIANA UNIVERSITY MEDICAL CENTER
Dessauer, Herbert C.; *Molecular Data and Reptilian Systematics* (GB7294X001); 12 months; $16,300
TULANE UNIVERSITY
Welden, Arthur L.; *The Stereaceae—A Taxonomic Revision of Cymatoderma, Podoscypha, Stereum, and Other Stereum Segregates* (B014296); 24 months; $30,600
UNIVERSITY OF SOUTHWESTERN LOUISIANA
Thieret, John W.; *Floristic Studies in Louisiana* (GB4671X003); 12 months; $4,600

MARYLAND

AMERICAN TYPE CULTURE COLLECTION
Clark, William A.; *Curatorial Support for the American Type Culture Collection* (B019544); 12 months; $85,000
CATONSVILLE COMMUNITY COLLEGE
King, Robert M.; *Generic Analysis of the Tribe Eupatorieae (Compositae)* (B020502); 24 months; $6,500
UNIVERSITY OF MARYLAND
Highton, Richard; *Systematic Studies on the Salamanders of the Genus Plethodon* (B19566X); 24 months; $32,000
Stern, William L.; *Comparative Anatomy and Systematics of the Woody Saxifragaceae* (GB7431X001); 12 months; $15,300
Terborgh, John W.; *Ecological Relationships of Species in Neotropical Forests* (B020170); 24 months; $35,000

BIOLOGICAL AND MEDICAL SCIENCES RESEARCH PROJECTS

MASSACHUSETTS

INDIVIDUAL AWARD
Romer, Alfred S.; *Triassic Reptile Faunas of Chanares, Argentina* (B022658); 12 months; $19,200

HARVARD UNIVERSITY
Nevling, Loring I. Jr.; *Flora of Veracruz* (B020267); 24 months; $70,400
Rollins, Reed C.; *Research and Training in Evolutionary Biology* (B019922); 12 months; $54,900
Williams, Ernest E.; *Inter and Intra-Species Difference and Interaction in New World Saurians* (B019801); 24 months; $130,000
Wood, Carroll E. Jr.; *A Generic Flora of the Southeastern United States* (GB6459X001); 12 months; $43,000

UNIVERSITY OF MASSACHUSETTS
Mulcahy, David L.; *Ecological Significance of Flowering Time and the Influence of Reproductive Processes Upon Vegetative Growth* (B013806); 24 months; $27,000

WILLIAMS COLLEGE
Labine, Patricia A.; *Reproductive Strategies in Natural Populations* (B015210); 24 months; $33,700

MICHIGAN

ECOLOGICAL SOCIETY OF AMERICA
Cantlon, John E. and Cole, Lamont C.; *Feasibility Study of a National Institute of Ecology* (GB6890 001); $86,900

HOPE COLLEGE
Brady, Allen R.; *Ecology, Distribution and Behavior of the Lynx Spiders of Mexico and Central America (Araneae Oxyopidae)* (B013925); 24 months; $14,700

MICHIGAN STATE UNIVERSITY
Cooper, William E. and Hall, Donald J.; *Trophic Couplings of Aquatic Populations* (B020962); 24 months; $30,000
Cummins, Kenneth W.; *Trophic Relationships in Freshwater Ecosystems* (B015775); 24 months; $25,000
Johnson, John I. Jr.; *Determining Factors in Brain Evolution* (B013854); 24 months; $45,000
Lauff, George H.; *Coherent Area Research Project in Freshwater Ecosystems* (B015665); 24 months; $190,000
Scott, Harold W.; *Excavation of the Conodont Animal Beds* (B017826); 24 months; $14,000

UNIVERSITY OF MICHIGAN
Hairston, Nelson G.; *Graduate Training in Systematic and Evolutionary Biology* (B013104); 12 months; $52,300
Hibbard, Claude W. *Middle Pleistocene Fauna From Mid-Central Kansas and its Climatic Significance* (B020249); 24 months; $28,800
McVaugh, Rogers; *Revision of the International Code of Botanical Nomenclature* (B021007); 36 months; $12,400
Miller, Robert R.; *Systematics of Cenozoic Freshwater Fishes* (B014871); 24 months; $37,900
Smith, Alexander H.; *A Manual of the Higher Fleshy Fungi of the Western U.S.A.* (B016969); 24 months; $83,700
———; *Monograph of the Genus Corinarius (Agaricales)* (GB6876X001); 16 months; $9,600

WAYNE STATE UNIVERSITY
Goodman, Morris; *Effects of Evolution on Primate Macromolecules* (GB7426X001); 12 months; $24,000
———; *Molecular Systematics of the Primates* (B015060); 24 months; $56,400

MINNESOTA

UNIVERSITY OF MINNESOTA
Gorham, Eville, and Sanger, Jon; *History, Evolution and Eutrophication of Minnesota Lakes* (B018800); 24 months; $75,000
Hall, John W.; *Azolla and Other Heterosporous Ferns in the Cretaceous* (B014041); 24 months; $20,100
Marshall, William H.; *Biological Research, Lake Itasca Biology Sessions* (B019413); 12 months; $8,600
Shapiro, Joseph; *Phosphate Absorption and Utilization by Algae* (B015675); 24 months; $75,000

MISSOURI

MISSOURI BOTANICAL GARDEN
Gates, David M.; *Curatorial Support of the Missouri Botanical Garden* (B019217); 12 months; $30,000
Lewis, Walter H.; *Operating Support of Missouri Botanical Garden Systematic Facility* (B017006); 12 months; $10,680
Porter, Duncan M.; *Symposium on Systematics—Tropical Island Biogeography* (B014722); 12 months; $1,050

MONTANA

UNIVERSITY OF MONTANA
MacLean, Stephen F. Jr.; *Effects of Perturbations on Cold-Dominated Ecosystems in Alaska, Part 5* (B025028); 12 months; $10,200
Miller, Charles N. Jr.; *Conifer Cones from the Late Mesozoic and Early Tertiary* (B019737); 24 months; $23,600
Prescott, G. W.; *A Synopsis of North American Desmidieae* (B015903); 24 months; $10,800

NEW HAMPSHIRE

DARTMOUTH COLLEGE
Gilbert, John J.; *Organic Production in Lakes* (B019926); 24 months; $6,800

UNIVERSITY OF NEW HAMPSHIRE
Croker, Robert A.; *Population Studies of Marine Amphipod Crustaceans* (B018590*); 24 months; $10,000

NEW JERSEY

PRINCETON UNIVERSITY
MacArthur, Robert H.; *Ecological Coordinates for Avian Species* (B018743); 24 months; $25,000

RUTGERS UNIVERSITY
Lechevalier, Hubert A.; *Morphology and Chemistry of Actinomycetes* (B018705); 24 months; $34,500

NEW MEXICO

UNIVERSITY OF NEW MEXICO
Clark, George R. II; *Daily Growth Lines in the Bivalve Family Pectinidae* (B020692); 24 months; $31,400

NEW YORK

ADELPHI UNIVERSITY
West, Robert M.; *Paleofaunal Studies of the Middle Eocene Bridger Formation* (B020263); 24 months; $20,900

AMERICAN MUSEUM OF NATURAL HISTORY
Gertsch, Willis J.; *Systematics and Biology of Cave Spiders and the Genus Meioneta* (GB6524X001); 12 months; $27,000
Krishna, Kumar; *Revision of Genera of the Subfamily Termitinae* (B020684); 24 months; $25,100
Rindge, Frederick H.; *Systematic and Phylogenetic Studies of North American Ennominae (Lepidoptera)* (GB6478X001); 12 months; $20,600
Rozen, Jerome G. Jr.; *The Phylogeny and Higher Classification of the Apoidea (Hymenoptera)* (GB5407X002); 12 months; $23,500
Tedford, Richard H.; *Cenozoic Mammals* (B018273); 24 months; $51,000
Wygodzinsky, Pedro; *Taxonomy and Geographical Distribution of Black Flies (Simuliidae, Diptera, Insecta) of Western South America* (GB8783); 9 months; $16,800

BOYCE THOMPSON INSTITUTE FOR PLANT RESEARCH
Vite, J. P. and Renwick, J. A. A.; *Chemical Messengers in the Ecology of the Southern Pine Beetle* (B020822); 24 months; $45,000

COLLEGE OF FORESTRY AT SYRACUSE
Wang, C. J. K.; *Taxonomic Study of Hyphomycetes on Wood and Litter of New York* (B014050); 24 months; $15,000

COLUMBIA UNIVERSITY
Basile, Dominick V.; *Experimental Approach to the Systematics and Phylogeny of Leafy Liverworts* (B015498); 24 months; $24,000
Bock, Walter J.; *Evolution of the Feeding Apparatus in Songbirds* (GB6909X001); 12 months; $21,200

CORNELL UNIVERSITY
Clausen, Robert T.; *Sedum in North American* (GB4929X001); 12 months; $12,200
Likens, Gene E.; *Collaborative Research on Hydrologic-Mineral Cycle Interaction in Small Undisturbed and Man-Manipulated Watersheds* (B014325); 24 months; $200,000
Moore, Harold E.; *Studies Toward a Genera Palmarum* (B020348); 24 months; $86,700
Pimentel, David; *Population Ecology of the Genetic Feedback Mechanism* (B019239); 24 months; $66,000
Pough, F. Harvey; *Environmental Adaptations in Amphibian and Reptile Blood* (B018985); 24 months; $22,000
Raney, Edward C.; *Systematic Studies of Catostomidae and Percidae* (GB4865X001); 12 months; $15,600

CUNY, HUNTER COLLEGE
Szalay, Frederick S.; *Early Tertiary Insectivores and Primates* (B020085); 24 months; $20,000

CUNY, QUEENS COLLEGE
Aaronson, Sheldon; *Evaluation of Some Environmental Conditions Causing Algae Blooms* (B020825); 24 months; $28,400

NEW YORK BOTANICAL GARDEN
Koyama, Tetsuo; *A Taxonomic Monograph of the Cyperaceae Tribe Cypereae of Tropical America* (B017433); 24 months; $39,400
Liogier, Alain H.; *Floristic Studies of Hispaniola* (B017518); 24 months; $9,000
Maguire, Bassett and Irwin, Howard S.; *Herbarium Facility and Operation* (B019386); 12 months; $82,000
Maguire, Bassett; *Botany of the Guayana Highland* (GB5953X002); 12 months; $32,300
Prance, Ghillean T.; *Plant Survey of Amazonia* (B018655); 24 months; $50,200

RUSSELL SAGE COLLEGE
Sayre, Geneva; *Collections of Cryptogamic Plants Issued as Sets* (B017366); 9 months; $5,400

ST. LAWRENCE UNIVERSITY
Crowell, Kenneth L.; *Species Diversity of Insular Small Mammals* (B018232); 24 months; $6,500

STATE UNIVERSITY AT ALBANY
Rosenzweig, Michael L.; *Modes of Coexistence of Desert Rodent Species* (B021010); 24 months; $35,000

STATE UNIVERSITY AT BUFFALO
Gans, Carl; *Functional Morphology of Squamate Reptilia* (GB6521X002); 12 months; $23,000

STATE UNIVERSITY AT STONY BROOK
Gaudet, John J.; *Factors Controlling the Growth of the Aquatic Weed Salvinia* (B019051); 12 months; $5,600
Rohlf, F. James; *Analysis and Representation of Structure in Taxonomic Data* (B020496); 24 months; $35,000
Sokal, Robert R., Farris, James S. and Rohlf, F. J.; *Development and Testing of Methods of Numerical Taxonomy* (B012587); 24 months; $70,200

ECOLOGY AND SYSTEMATIC BIOLOGY

Sokal, Robert R.; *Experimental and Simulation Study of Ecogenetic Processes in Laboratory Insect Populations* (B015146); 24 months; $60,000

UNIVERSITY OF ROCHESTER
Muchmore, William B.; *Systematic Studies of Chelonethida* (B017964); 24 months; $17,000

NORTH CAROLINA

DUKE UNIVERSITY
Culberson, William L. and Kristinsson, Hordur; *Lichen Flora of Iceland* (GB6041X-002); 4 months; $4,800
Hellmers, Henry; *Operation of the Southeastern Plant Environment Laboratories* (B019634); 12 months; $77,000
Johnson, T. W. Jr.; *Systematic and Evolutionary Plant Sciences* (B023200); 12 months; $40,300
Strain, Boyd R.; *Physiological Ecology of Desert Flora* (B017357); 12 months; $9,400
Wilbur, Robert L.; *Gentianaceae of Mexico and Central America* (B013815); 24 months; $33,000

HIGHLANDS BIOLOGICAL STATION
Hardin, James W.; *Field Biology of the Southern Appalachians* (B019805); 12 months; $16,000
Wagner, Warren H.; *Conference on Pteridology of the Southern Appalachians* (B021275); 12 months; $2,500

NORTH CAROLINA STATE UNIVERSITY AT RALEIGH
Downs, Robert J. and Hellmers, Henry; *Operation of the Southeastern Plant Environment Laboratories* (B019650); 12 months; $33,000

OHIO

KENT STATE UNIVERSITY
Cooperrider, Tom S.; *Systematic Study of Chelone (Scrophulariaceae)* (GB7609X001); 12 months; $5,900
Foote, Benjamin A.; *Biology and Immature Stages of Otitidae* (B017502); 24 months; $12,000
———; *Biology and Immature Stages of Ephydridae* (B015483); 24 months; $19,000

MIAMI UNIVERSITY
Barrett, Gary W. and Wilson, Roger E.; *Pesticide Stress Effects on Experimental Ecosystems* (B020975); 24 months; $30,400

OBERLIN COLLEGE
Benzing, David H.; *Comparative Study of Nutritional Biology in the Bromeliaceae* (B008790); 24 months; $18,000

OHIO STATE UNIVERSITY
Colinvaux, Paul A.; *The Environmental History of the Galapagos Islands* (B017501); 24 months; $60,000
Delong, Dwight M.; *Monographic Study of the Gyponinae of the World* (GB5659X002); 12 months; $6,200

UNIVERSITY OF AKRON
Sheppe, Walter A.; *Use of Seasonal Floodplain by Mammalian Fauna* (B016508); 24 months; $32,000

OKLAHOMA

OKLAHOMA STATE UNIVERSITY
Dorris, Troy C.; *Plankton Productivity of Lake Atitlan in Guatemala* (B014823); 24 months; $57,500

UNIVERSITY OF OKLAHOMA
Riggs, Carl D. and Hill, Loren G.; *Grants-In-Aid for Research at the University of Oklahoma Biological Station* (B019633); 12 months; $2,800

OREGON

LINFIELD COLLEGE
Fender, Kenneth M.; *Revisional Studies in the Lampyroid Beetles* (GB6283X002); 12 months; $11,000

OREGON STATE UNIVERSITY
Bollen, W. B.; *Root Pathogen Regulation by Red Alder* (B018650); 24 months; $28,100
Dawson, Peter S.; *Behavior of Populations in Heterogeneous Environments* (B016131); 13 months; $16,500
Hansen, Henry P.; *Postglacial Palynology* (B015670); 24 months; $12,200
Ritcher, Paul O.; *Taxonomy of Scarabaeid Larvae and Related Adults* (GB6194X001); 12 months; $6,300

UNIVERSITY OF OREGON
Castenholz, Richard W.; *Ecology and Physiology of Thermophilic Blue-Green Algae* (B018799); 24 months; $62,000

PENNSYLVANIA

ACADEMY OF NATURAL SCIENCE OF PHILADELPHIA
Drouet, Francis; *Classification of Scytonema and Calothrix AG. (Blue-Green Algae)* (B018535); 24 months; $28,600
Patrick, Ruth and Goulden, Clyde E.; *Relationship of the Diversity of Organisms to the Stability and Pattern of the Ecosystem in a Stream* (B015144); 24 months; $75,100
Slifer, Eleanor H.; *The Structure of Arthopod Chemorceptors* (B017212); 24 months; $14,200

CARNEGIE INSTITUTE
Guilday, John E.; *Pleistocene Faunas of the Appalachian Karst Region* (B18706X); 24 months; $19,100

PENNSYLVANIA STATE UNIVERSITY
Casida, Lester E. Jr.; *Natural State of Microorganisms* (B014487); 24 months; $80,000
Grun, Paul; *Plasmon Analysis of Solanum Tuberosum* (B017838); 24 months; $20,500

SLIPPERY ROCK STATE COLLEGE
Mueller, Charles F.; *Bioenergetic Studies of an Amphibian, 'Plethodon Cinereus'* (B015894); 24 months; $13,192

SOUTH DAKOTA

AUGUSTANA COLLEGE
Tieszen, Larry L.; *Effects of Perturbations on Cold-Dominated Ecosystems in Alaska, Part 4* (B025027); 12 months; $24,600

SOUTH DAKOTA STATE UNIVERSITY
Greichus, Yvonne A.; *Physiological and Behavioral Effects of Insecticides on Cormorants* (B019121); 24 months; $60,000

TENNESSEE

ECOLOGICAL SOCIETY OF AMERICA
Deevey, Edward S.; *Feasibility Study of a National Institute of Ecology—Phase II* (GB6890 002); 6 months; $95,400

VANDERBILT UNIVERSITY
Kral, Robert; *Flora of Alabama and Central Tennessee* (GB6688X001); 12 months; $13,500

TEXAS

AMERICAN SOCIETY OF PARASITOLOGISTS
Otto, Gilbert F.; *Second International Congress of Parasitology* (B016959*); 12 months; $6,000

AMERICAN SOCIETY OF PARASITOLOGISTS
Otto, Gilbert F.; *Second International Congress of Parasitology* (GB16959001*); $3,000

M.D. ANDERSON HOSPITAL
Hsu, T. C.; *Cytogenetics of North American Cricetid Rodents* (B013661); 24 months; $47,000

SUL ROSS STATE UNIVERSITY
Powell, A. Michael; *Systematic Studies of Perityle and Related Genera Compositae* (B020361); 24 months; $17,200

UNIVERSITY OF TEXAS AT AUSTIN
Alexopoulos, Constantine; *Experimental Approach to the Taxonomy of the Slime Molds (Myxomycetes)* (GB6812X001); 12 months; $19,400

Blair, W. F.; *Amphibian Speciation and Evolutionary Relationships* (GB5406 002); 12 months; $29,100
———; *Program in Systematic and Environmental Biology* (B015479); 24 months; $150,000
Flake, Robert H.; *Computer-Aided Systematic Studies* (B020713); 12 months; $9,100
Selander, Robert K.; *Comparative Molecular Systematics of Vertebrates* (B015664); 24 months; $27,000

UTAH

BRIGHAM YOUNG UNIVERSITY
Wood, Stephen L.; *Taxonomy and Distribution of Scolytidae and Platypodidae—Coleoptera* (B012437); 24 months; $30,100

UNIVERSITY OF UTAH
Edmunds, George F. Jr.; *Classification of the Ephemeroptera* (B018656); 24 months; $54,000
Legler, John M.; *Taxonomy and Distribution of Chelonian Fauna in Central America* (B016249); 24 months; $20,000
Vickery, Robert K. Jr.; *Evolutionary Mechanisms and Biosystematics of Mimulus (Scrophulariaceae)* (B018139); 24 months; $21,200

UTAH STATE UNIVERSITY
Gessaman, James A.; *Evaluation of Heart Rate as an Indirect Monitor of Avian Metabolism* (B018158); 24 months; $32,000
West, Neil E. and Goodall, David W.; *Successional Status in Salt Desert Shrub Vegetation* (B019843); 24 months; $17,300

VERMONT

WINDHAM COLLEGE
Tseng, Charles C.; *Systematic Anatomy of the Schefflereae (Araliaceae)* (GB7988 001); $3,800

VIRGINIA

RADFORD COLLEGE
Hoffman, Richard L.; *Systematics and Zoogeography of Chelodesmoid Diplopoda* (GB7936 001); $900

UNIVERSITY OF VIRGINIA
Murray, J. J. Jr. and Riopel, James L.; *Summer Programs at Mountain Lake Biological Station* (B013902*); 12 months; $10,000

VIRGINIA POLYTECHNIC INSTITUTE
Benoit, Robert E.; *Effects of Perturbations on Cold-Dominated Ecosystems in Alaska, Part 3* (B025026); 12 months; $12,300
Kosztarab, Michael; *A Systematic Study on Scale Insects* (GB6885X001); 12 months; $15,900
Parker, Bruce C.; *Chemistry and Structure of Algal Cell Walls* (B017211); 12 months; $21,900

WASHINGTON

UNIVERSITY OF WASHINGTON
Del Moral, Roger; *Comparative Ecology of Closely Related Species* (B020492); 24 months; $19,000
Farner, Donald S.; *Control of Annual Cycles in 'Zonotrichia'* (GB5969X002); 12 months; $21,200
Henry, Dora P.; *Barnacles of the Eastern Pacific* (GB7074X001); 12 months; $18,200
Illg, Paul L.; *Systematics of Marine Symbiotic Crustacea from Invertebrates* (GB647X-001); 12 months; $11,200
Tsukada, Matsuo; *Past Vegetation and Dynamics of Lake Ecosystems* (B015300); 24 months; $55,000

WASHINGTON STATE UNIVERSITY
James, Maurice T.; *Systematic Study of Stratiomyidae* (B015774); 12 months; $21,600

WEST VIRGINIA

WEST VIRGINIA UNIVERSITY
Clarkson, Roy B.; *Chemotaxonomic Studies*

BIOLOGICAL AND MEDICAL SCIENCES RESEARCH PROJECTS

of Selected Taxa of Vascular Plants (B015386); 24 months; $20,000
Kowal, Norman E.; *The Identification of the Pine-Forest-Floor Temperature-Moisture System* (B017960); 24 months; $34,000

WISCONSIN

MARQUETTE UNIVERSITY
Houston, Arthur H.; *Physiological Responses to Thermal Pollution* (B014742); 24 months; $40,500

UNIVERSITY OF WISCONSIN, Madison
Emlen, John T.; *Ecology of the Andean Condor* (B019449); 18 months; $10,000
Keith, Lloyd B.; *Analysis of Natural Mortality in Cyclic Snowshoe Hare Populations Through Radio-Telemetry* (B012631); 24 months; $25,000
Magnuson, John J.; *Behavior and Survival of Freshwater Fishes Exposed to Severe Oxygen Depletion* (B018272); 24 months; $35,400
Neess, John C.; *Response of Soil Arthropod Populations to Disturbance* (B019564); 15 months; $8,400

WYOMING

UNIVERSITY OF WYOMING
Boyd, Donald W. and Newell, Norman D.; *Permian Species of Bivalved Mollusks from the Western United States* (GB6905X001); 12 months; $11,800
Parker, Michael; *Strategic Analysis and Phytoplankton Ecology* (B016847); 24 months; $15,000

Physiological Processes

ALABAMA

TUSKEGEE INSTITUTE
Siddique, Irtaza H.; *Pathogenesis of Listerial Infections* (B016968); 24 months; $20,200

UNIVERSITY OF ALABAMA, Birmingham
McKibbin, John M.; *Biosynthesis of the Glycolipids of Dog Intestine Mucosa* (B015605); 12 months; $15,000

UNIVERSITY OF ALABAMA, Tuscaloosa
Hiramoto, Raymond N.; *Cellular Antibody Response* (B013804); 24 months; $43,300

ARIZONA

ARIZONA STATE UNIVERSITY
Reeves, Henry C.; *Control and Function of Alternate Pathways of Propionate Metabolism* (B016265); 24 months; $53,000

UNIVERSITY OF ARIZONA
Calder, William A.; *Physiology of the Avian Respiratory System* (B013249); 24 months; $29,500

ARKANSAS

UNIVERSITY OF ARKANSAS, Little Rock Campus
Smith, W. Grady; *Amino Acid Metabolism in Bacteria* (B018015); 24 months; $44,100
Wadkins, Charles L.; *Oxidative Phosphorylation Studies* (B015383); 24 months; $50,000
Winter, Charles G.; *Mechanism of Cation Transport in 'Streptococcus faecalis'* (B015090); 24 months; $19,000

CALIFORNIA

CALIFORNIA INSTITUTE OF TECHNOLOGY
Van Harreveld, Anthonie; *Use of Freeze Substitution in Investigations of Physiological Processes with the Electron Microscope* (B016961); 24 months; $63,000
Wiersma, C. A. G.; *Nervous System of Crustaceans and Other Arthropods* (GB6931X-002); 12 months; $37,400

CALIFORNIA STATE COLLEGE AT LONG BEACH
Lincoln, Richard G., Carpenter, Bruce H. and Mayfield, Darwin L.; *Comparative Characterization of Florigenic Acid from Fungal and Higher Plant Sources* (GB7199 001); $1,800

CALIFORNIA STATE COLLEGE AT LOS ANGELES
Andreoli, Anthony J.; *Study of Enzyme Patterns in 'Bacillus megaterium' During Sporulation* (B015773); 24 months; $15,000
Sacher, Joseph A.; *Enzyme Synthesis During Senescence and Ripening* (B018071); 24 months; $30,000

CALIFORNIA STATE POLYTECHNIC COLLEGE
Shafia, Fred; *Autotrophy and Heterotrophy in 'Ferrobacillus ferrooxidans'* (B014457); 24 months; $19,500

CHICO STATE COLLEGE
Erpino, Michael J.; *Histophysiological Studies of the Adenohypophysis and Other Endocrines* (B017990); 24 months; $25,300
McNairn, Robert B.; *Callose and Phloem Translocation* (B020161); 24 months; $15,300
Pennington, Frank C.; *Reaction of Chlorophylls with Amines* (B019352); 10 months; $2,100

CHILDREN'S HOSPITAL MEDICAL CENTER OF NORTHERN CALIFORNIA
Abraham, Samuel; *Basic Aspects of Lipid and Carbohydrate Metabolism in Mammalian Tissues* (B015760); 24 months; $40,000

FRESNO STATE COLLEGE
Berg, Edward; *Ontogeny of Functional Antigens* (B020283); 24 months; $10,800

PROFESSIONAL STAFF ASSOCIATION, Lac-Usc
Russell, Findlay E.; *Electron Microscopy of the Black Widow Spider and Scorpion Venom Glands* (B017573); 24 months; $12,700

SALK INSTITUTE FOR BIOLOGICAL STUDIES
Gospodarowicz, Denis J.; *Interaction of Icsh with Luteal Cells* (B018029); 24 months; $40,000
Kaplan, Ann E.; *Cerebroside Biosynthesis in Vitro* (B016488); 24 months; $40,000

STANFORD UNIVERSITY
Ray, Peter M.; *Polysaccharide Synthetase-Bearing Organelle of Plant Cells* (B018499); 24 months; $89,600
Rosenquist, Grace L.; *Phylogeny of the Immune Response* (B015748); 24 months; $40,000

UNIVERSITY OF CALIFORNIA, Berkeley
Doudoroff, M.; *Biology of Pseudomonads and Blue-Green Algae* (B017517); 24 months; $81,400
Macey, Robert I.; *Studies in the Theory of Membrane Permeability* (B015810); 24 months; $25,000
McLaren, A. Douglas and Schroth, Milton N.; *Penetration of Pathogenic Enzymes, Proteins, and Other Macromolecules into Plant Tissue* (B016963); 12 months; $28,800
Steinbach, Alan B.; *Kinetics of Chemically Induced Ionic Conductance Change* (B017952); 24 months; $23,200
Valentine, Raymond C.; *Bacterial Ferredoxins as High Energy Electron Carriers* (B019328); 24 months; $52,200

UNIVERSITY OF CALIFORNIA, Davis
Kosuge, Tsune; *Regulation of Aromatic Amino Acid Biosynthesis in Higher Plants* (B015396); 24 months; $35,000
Stumpf, P. K.; *Enzymatic Mechanisms Involved in Lipid Metabolism In Higher Plants* (B019733); 12 months; $40,000
Thornton, Robert M.; *Photoresponse Mechanisms in the Mucorales* (B015110); 24 months; $29,400
Yang, Shang F.; *Biosynthesis and Biochemical Action of Ethylene* (B020336); 24 months; $33,000

UNIVERSITY OF CALIFORNIA, Irvine
Moyed, Harris S.; *Biochemical Basis for the Action of Indole-3-Acetic Acid* (B015898) 24 months; $69,300

UNIVERSITY OF CALIFORNIA, Los Angeles
Cooper, Edwin L.; *Immunogenesis in Invertebrates an Poikilotherms* (B017767); 24 months; $28,400
Engelmann, Franz; *Regulation of Reproduction and Protein Biosynthesis in an Insect* (B014965); 24 months; $60,700
Grossman, Morton I.; *Pancreatic and Biliary Physiology* (GB7285X001); 12 months; $13,100
Lascelles, June; *The Regulation of Tetrapyrrole Synthesis and Associated Complexes* (B014967); 24 months; $52,000
Rittenberg, S. C.; *Bacterial Metabolism and Physiology* (GB6223X002); 12 months; $46,900

UNIVERSITY OF CALIFORNIA, Riverside
Kaufmann, Merrill R.; *Water Relations of Growing Fruit* (B017215); 24 months; $25,000
Stolzy, Lewis H.; *Role of Oxygen in the Rooting Behavior of Plants* (B019916); 24 months; $22,000

UNIVERSITY OF CALIFORNIA, San Diego
Bullock, Theodore H.; *Studies in Comparative Neurology* (GB6969X002); 12 months; $60,000

UNIVERSITY OF CALIFORNIA, Santa Barbara
Holmes, William N.; *Endocrine Factors Associated with the Water and Electrolyte Regulation in Birds* (B020806); 24 months; $50,000

UNIVERSITY OF SOUTHERN CALIFORNIA
Brodie, Arnold F.; *Photo Catalyzed Reactions Involving Active Sites of Bioenergetic Enzymes* (GB6257X002); 12 months; $36,000
Hochstein, Paul; *Formation, Actions and Detoxication of Peroxides in Biological Systems* (B016777); 18 months; $16,300

COLORADO

COLORADO STATE UNIVERSITY
Dupont, Jacqueline and Mathias, Melvin M.; *Bio-Oxidation of Unsaturated Fatty Acids Via Gamma Cleavage and Methylmalonate Production* (B015482); 24 months; $30,000
Hendrix, John E.; *Determination of the First Free Sugar Produced Photosynthetically by Squash Plants* (B014773); 24 months; $19,000
Sobel, Harry; *Effect of Hyperbaric Oxygen on Connective Tissue in Vivo* (B016099); 12 months; $14,000

UNIVERSITY OF COLORADO
Winston, Paul W.; *Humidity Relations of Terrestrial Arthropods* (B014167); 24 months; $25,400

UNIVERSITY OF COLORADO, Denver Medical Center
Katsh, Seymour; *Mechanisms of Destruction of Testicular Cells Following Immunization with Specific Antigen* (B014830); 24 months; $65,000

CONNECTICUT

UNIVERSITY OF CONNECTICUT
Collins, Ralph P.; *Mechanisms of Host Plant Selection* (B012628); 12 months; $12,000

WESLEYAN UNIVERSITY
Lukens, Lewis N.; *Collagen Synthesis and Its Regulation* (B019794); 24 months; $73,000

YALE UNIVERSITY
Boulpaep, Emile L.; *Functional Structure of Proximal Renal Tubule* (B014173); 12 months; $15,000
Forbes, Thomas R.; *Action and Localization of Progestins* (B016749); 24 months; $33,000
Kantor, Fred S.; *Mechanism of Delayed Hypersensitivity—Effect of Antigen on Sensitized Leukocytes* (B013399); 24 months; $32,700
Schenkman, John B.; *Characterization of Beta-Diethylaminoethyl Diphenylpropylacetate* (B018660); 24 months; $9,800

PHYSIOLOGICAL PROCESSES

DISTRICT OF COLUMBIA

GEORGETOWN UNIVERSITY

Lewin Lawrence M.; *Biochemical Functions of Myo-Inositol in Yeast* (B015442); 24 months; $30,000

NATIONAL ACADEMY OF SCIENCES, NATIONAL RESEARCH COUNCIL

Yager, Robert H.; *Task Order for Support of the Institute of Animal Resources* (C310173); 12 months; $15,000

FLORIDA

PAPANICOLAOU CANCER RESEARCH INSTITUTE

Schultz, Julius; *2nd Annual Biochemistry Winter Symposia* (B018183); 12 months; $800

UNIVERSITY OF SOUTH FLORIDA

Linton, Joe R. and Robinson, Gerald G.; *Separation and Quantification of Prolactin in the Pituitary of the Black Mullet* (B015745); 12 months; $10,500

GEORGIA

UNIVERSITY OF GEORGIA

Fisher, Donald B.; *A Quantitative Cytological Investigation of Phloem Transport* (B014719); 24 months; $34,800

HAWAII

UNIVERSITY OF HAWAII

Mower, Howard F.; *Properties of Hydrogenase Enzymes* (B020479); 12 months; $15,000

ILLINOIS

ILLINOIS INSTITUTE OF TECHNOLOGY

Danforth, William F.; *Regulation of Carbon and Energy Metabolism in Heterotrophic 'Euglena'* (B015656); 12 months; $15,000

NORTHWESTERN UNIVERSITY

Hubbard, John I.; *Mobilization of Transmitter at Synapses* (B014294); 24 months; $44,000

Wolfson, Albert; *Environmental and Neuroendocrine Regulation of Reproductive Cycles and Migratory Behavior in Birds* (GB7100X-001); 12 months; $31,000

NORTHWESTERN UNIVERSITY, Chicago Campus

Corcoran, John W., *Biogenesis of Polyketides and Fatty Acids* (B018157); 24 months; $40,800

SOUTHERN ILLINOIS UNIVERSITY

Lefebvre, Eugene A.; *Collaborative Study of Albatross Flight and Bioenergetics with Richard C. Birkebak, University of Kentucky* (B015650); 24 months; $34,000

UNIVERSITY OF CHICAGO

Ruddat, Manfred; *Metabolism and Mode of Action of Tetracyclic Diterpenes, Steviol and Gibberellin* (B017304); 24 months; $48,200

Winfree, Arthur T.; *Dynamics of Biological Clocks* (B016513); 24 months; $30,000

UNIVERSITY OF ILLINOIS, Urbana

Beevers, Leonard; *Protein Metabolism in the Cotyledons of Pisum sativum During Development and Germination* (B015189); 24 months; $28,000

Gunsalus, I. C.; *Isoprenoid Metabolism—Regulation and Specificity* (B016312); 24 months; $100,000

Heath, James E.; *Comparative Study of Thermoregulation* (B013797); 24 months; $47,700

Larson, Bruce L.; *Specialized Synthesis and Dedifferentiation in the Mammary Secretory Cell* (B015758); 24 months; $45,000

Prosser, C. Ladd; *Comparative Physiology of Temperature Adaptation in Poikilotherms* (GB4005X003); 12 months; $30,000

INDIANA

INDIANA UNIVERSITY, Bloomington

Gest, Howard; *Mechanisms of Regulation of Biosynthetic Metabolism in Bacteria* (GB7333X 001); 12 months; $70,000

Hirs, C. H. W.; *Tetrapyrrole Chromoproteins From Photosynthetic Bacteria, Algae and Plants* (B020486); 24 months; $40,000

Ramaley, Judith A.; *Control of Gonadal Function in Prepuberal Rats* (B017438); 12 months; $16,000

Togasaki, Robert K.; *Regulation of Photosynthetic Processes in 'Chlamydomonas reinhardi'* (B018789); 24 months; $18,000

PURDUE UNIVERSITY

Cherry, Joe H.; *Control of Protein Synthesis at the Levels of Transcription and Translation* (B020483); 24 months; $40,000

Kuc, Joseph A. and Williams, E. V.; *Biochemical Expression of Genetic Determinants Controlling Disease Resistance in Plants* (B013994); 24 months; $40,000

IOWA

UNIVERSITY OF IOWA

Markovetz, Allen J.; *Bacterial Oxidation of Long-Chain Methyl Ketones* (B014052); 12 months; $18,600

KANSAS

UNIVERSITY OF KANSAS

Bovee, Eugene C.; *Chemical Effects on Motility and Locomotory Behavior in Protozoa* (B016616); 24 months; $45,000

Kitos, Paul A. and Hersh, Robert; *Hormone Effects on the Biosynthesis of Macromolecules* (B013924); 24 months; $30,000

KENTUCKY

UNIVERSITY OF KENTUCKY

Aleem, M. I. H.; *Mechanisms in Bioenergetics of Autotrophic Life* (B014661); 24 months; $40,000

Birkebak, Richard C.; *Collaborative Study of Albatross Flight and Bioenergetics with Eugene A. Lefebvre, Southern Illinois University* (B015579); 24 months; $39,000

Crawford, Eugene C. Jr.; *Physiology of Temperature Regulation in Lizards* (B019981); 12 months; $12,500

Rodriguez, J. G., *Nutrition of Plant-Feeding Mites* (B020339); 24 months; $20,000

Wekstein, David R. and Zolman, James F.; *Ontogeny of Thermoregulation* (B008775); 24 months; $45,000

LOUISIANA

LOUISIANA STATE UNIVERSITY, Baton Rouge

Lambremont, Edward N.; *Glyceryl Ethers and Their Metabolism in Insects* (B013315); 24 months; $17,000

Meier, Albert H.; *Temporal Synergisms in the Hormonal Control of Fat Storage and Reproduction in Vertebrates* (B020913); 12 months; $11,000

TULANE UNIVERSITY

Copeland, D. E.; *Fine Structure of Gas Secretion* (B014666); 24 months; $30,500

Johnson, Emmett J.; *Regulation of Autotrophic Carbon Dioxide Fixation* (B019042); 24 months; $39,000

Lumsden, Richard D.; *Environmental Physiological Aspects of Helminth Ultrastructure and Cytochemistry* (B017992); 24 months; $20,000

MAINE

MT. DESERT ISLAND BIOLOGICAL LABORATORY

Wilde, Charles E.; *Logistic Support and Maintenance of Facilities* (B019567); 12 months; $28,750

MARYLAND

JOHNS HOPKINS UNIVERSITY

Bang, Betsy G.; *Comparative Anatomy of the Vertebrate Nasal Organ* (B013233); 24 months; $19,100

Mayer, Manfred M.; *Immunological and Chemical Studies of the Complement System* (GB7406X001); 12 months; $40,200

Nason, Alvin; *Biochemistry of the Chemoautotrophic Nitrifying Bacteria* (B015577); 24 months; $60,000

Reissig, Magdalena; *Electron Microscopy of Schistosomes* (B017837); 24 months; $50,000

UNIVERSITY OF MARYLAND

Blake, William D.; *(NA+ — K+) Activated ATPase Kinetics and Possible Relationship of Enzyme Induction to Tissue NA in Salt Deficiency* (B013810); 12 months; $11,900

Fajer, Abram B.; *Effect of Prolactin on the Hormone Precursors in the Rat, Hamster and New World Primates* (BO13794); 12 months; $20,000

Grenell, Robert G.; *The Inhibition of Inhibition and Its Role in Regulation of Sensory Processing in the Cerebral Cortex* (B016042); 24 months; $16,000

MASSACHUSETTS

AMERICAN SOCIETY OF ZOOLOGISTS

Markert, Clement L.; *Symposium on the Physiology of Annelids* (B016745); 12 months; $2,450

Reiter, Russel J.; *Symposium on the Comparative Endocrinology of the Pineal* (B016304); 12 months; $1,000

BOSTON COLLEGE

Orlando, Joseph A.; *Energy-Linked Reactions of the Nonsulfur Purple Bacterium, 'Rhodopseudomonas spheroides'* (B014826); 24 months; $30,000

BOSTON UNIVERSITY

Curry, John J. III; *Olfactoy Influences—An Electrophysiological Study* (B017216); 24 months; $35,000

BRANDEIS UNIVERSITY

Abeles, Robert H.; *Mechanism of Enzyme Action* (B016413); 24 months; $56,600

CLARK UNIVERSITY

Ahmadjian Vernon; *Experimental Studies of the Lichen Symbiosis* (B015713); 21 months; $16,100

COLLEGE OF THE HOLY CROSS

Roffman, Berton; *In Vivo' Metabolism of Photosynthetic Compounds in an Algal-Invertebrate Symbiosis* (B017893); 24 months; $24,300

HARVARD UNIVERSITY

Baile, Clifton A. and Mayer, Jean; *Neural, Metabolic and Behavioral Aspects of the Control of Feed Intake of Ruminants* (B015812); 24 months; $60,000

Pappenheimer, Alwin M. Jr.; *I. Studies on Pneumococcal Anti-Polysaccharide Antibodies and II. Biology of Diphtheria and of the Diphtheria Bacillus* (B018919); 24 months; $52,000

MARINE BIOLOGICAL LABORATORY

Adelman, William J. Jr.; *Research-Training —Excitable Membrane Physiology and Biophysics* (B013805001); $1,400

Case, James F.; *Training Project in Experimental Invertebrate Zoology* (GZ1788*); 7 months; $20,000

MASSACHUSETTS GENERAL HOSPITAL

Martin, Donald B.; *Insulin-Stimulated Glucose Uptake by Particles from Adipose Tissue Cells* (B019636); 24 months; $25,000

MASSACHUSETTS INSTITUTE OF TECHNOLOGY

Brown, Gene M.; *Metabolism of B Vitamins and Related Compounds* (B12613X002); 12 months; $33,700

Buchanan, John M.; *Amino Acid Synthesis and Nitrogen Fixation* (B013586); 24 months; $100,000

Lax, Benjamin; *Measurement of the DC Magnetic Field Produced by the Human Heart* (B014040); 12 months; $10,000

BIOLOGICAL AND MEDICAL SCIENCES RESEARCH PROJECTS

NORTHEASTERN UNIVERSITY
Sturckow, Brunhild; *Site of Chemoreception in Gustatory Sensilla of Insects* (B013500); 24 months; $35,900

UNIVERSITY OF MASSACHUSETTS
Lockhart, James A.; *Study of Plant Cell Enlargement* (B013842); 24 months; $45,000

WORCESTER FOUNDATION FOR EXPERIMENTAL BIOLOGY
Bartosik, Delphine B.; *Luteoylsis-Physiological, Biochemical and Histological Studies* (B018068); 24 months; $50,000
Brunkhorst, Willa K.; *Effects of Glucocorticoids on Nucleohistones of Thymic Lymphocytes* (B015296); 12 months; $25,000
Cortes, Julio M.; *Effect of Ions on the Morphology and Function of the Adrenal Cortex* (B017993); 24 months; $42,000
McCracken, John A.; *Utero-Ovarian Function in the Sheep by Means of a Vascular Autotransplantation Technique* (B020473); 24 months; $88,000

MICHIGAN

KALAMAZOO COLLEGE
Evans, Michael L.; *Mode of Action of Auxin in the Promotion of Elongation* (B014969); 24 months; $22,300

MICHIGAN COLLEGE OF OSTEOPATHIC MEDICINE
Cohen, Leonard A.; *Cervical Proprioceptor Firing and Neck Position in Body Spatial Orientation* (B024049); 24 months; $39,100

MICHIGAN STATE UNIVERSITY
Bieber, Loran L.; *Studies on the Biosynthesis, Function, and Sites of Synthesis of Mitochondrial Phospholipids* (B015450); 24 months; $24,000
Tolbert, N. E.; *Leaf Peroxisomes* (B017543); 24 months; $78,000
Welsch, Clifford W.; *The Role of the Central Nervous System in the Etiology, Development and Growth of Tumors in the Rat* (B017034); 24 months; $25,000

UNIVERSITY OF MICHIGAN
Beyer, Robert E.; *Control of Heart Mitochondrial Energy Metabolism* (B013496); 24 months; $27,000

MINNESOTA

MOORHEAD STATE COLLEGE
Brummond, Dewey O.; *Enzymatic Cellulose Synthesis in Higher Plants* (B016055); 24 months; $34,900

UNIVERSITY OF MINNESOTA
Bernlohr, Robert W.; *Metabolism During Sporulation of Bacillus* (B019934); 24 months; $35,100
Gilbertson, Donald E.; *Metabolism of Larval Trematodes* (B014868); 24 months; $30,000
Halvorson, H. Orin; *Physiology of DPA-Less Mutant of 'Bacillus cereus'* (B013214); 24 months; $45,000
Holmes-Gray, Beulah; *Metabolism of Leukocytes During Phagocytosis* (B016835); 24 months; $20,000
Hooper, Alan B.; *Biochemistry and Physiology of Nitrification by Nitrosomonas* (B015101); 24 months; $40,000
Mirocha, Chester J.; *Structure and Action of an Activator of Beta Amylase* (B018497); 24 months; $20,000
Watson, Cecil James; *Chemistry and Biogenesis of the Urobilins* (B014833); 10 months; $12,000

MISSOURI

INSTITUTE OF MEDICAL EDUCATION AND RESEARCH
Nicholas, Harold J.; *Function of Sterols and Triterpenes in Plants* (B019113); 24 months; $30,000

ST. LOUIS UNIVERSITY
Coscia, Carmine J.; *Biosynthesis of Iridoid and Bicyclic Monoterpenes in Higher Plants* (B017957); 24 months; $35,000
Davis, J. Wendell; *Hormonal Control of Biosynthetic Pathways in Mammary Tissue* (B018735); 24 months; $36,600
Feir, Dorothy; *Hormonal Regulation of Nucleic Acid and Protein Synthesis in Insect Tissues* (B016688); 12 months; $15,300
Katzman, Philip A.; *Hormonal Control of Metabolism in Vaginal Tissue* (B008778); 24 months; $40,500

UNIVERSITY OF MISSOURI, Columbia
Goodman, Robert N.; *Ultrastructural Alterations in Plant Cells Undergoing the Hypersensitive Reaction Caused by Pathogenic Bacteria* (B017729); 24 months; $60,000
Levitt, Jacob; *Physiological Basis of Resistance of Plants to Frost and Drought* (B013222); 24 months; $51,600
McKenna, John M.; *Effects of Hibernation and Hypothermia on the Immune Response in Vivo and in Vitro* (B008776); 24 months; $31,200
Zatzman, Marvin L. and South, Frank E.; *Renal Function of the Marmot, 'Marmota' Flaviventris* (B017155); 24 months; $25,000

WASHINGTON UNIVERSITY
Chilson, Oscar P.; *Post-Hatching Development and Aging of Avian Erythrocytes* (B019397); 24 months; $40,000
Pickard, William F.; *Regulation of Electrogenic Pumps* (B020344); 24 months; $25,000
Suga, Nobuo; *Studies in Comparative Auditory Neurophysiology* (B013904); 24 months; $90,000
Vagelos, Pindaros R.; *Enzymatic Mechanisms in Lipid Metabolism* (GB5142X003); 12 months; $33,900

NEBRASKA

CREIGHTON UNIVERSITY
Magee, Donal F.; *Secretion of Gastrin and Duodenal Inhibitory Mechanisms* (B011985); 24 months; $25,600

UNIVERSITY OF NEBRASKA, Lincoln
Ferguson, Donald L. and Rhodes, Marvin B.; *Biochemical and Immunological Studies of Enzymes of Metazoan Parasites* (B019159); 24 months; $28,000
Stohs, Sidney J.; *Cholesterol Side Chain Cleavage Enzyme Systems in Cardenolide Producing Plants* (B013232); 24 months; $11,200

NEVADA

DESERT RESEARCH INSTITUTE
Went, Frits W.; *Plant Physiological and Biochemical Problems in Arid Regions* (B17731X); 24 months; $168,800

UNIVERSITY OF NEVADA, Reno Campus
Dill, David B.; *Adaptations in Mammals and Man to Desert and Mountain* (B017126); 24 months; $57,200

NEW HAMPSHIRE

DARTMOUTH COLLEGE
Dennison, David S.; *Photoresponses in Phycomyces* (GB6154X001); 12 months; $21,800

NEW JERSEY

RUTGERS UNIVERSITY
Chen, Tseh A.; *Nematode Vectors of Soil-Borne Viruses* (B016784); 24 months; $32,600
Myers, Ronald F.; *Culture and Nutrition of Plant Parasitic Nematodes* (B018012); 24 months; $24,000
Sturkie, Paul D.; *Role of Nerves on Heart Rate in Aves* (B019411); 12 months; $15,300

RUTGERS UNIVERSITY, Newark Campus
Wilhoft, Daniel C.; *Thermoregulation in a Heliotherm* (B014323); 12 months; $13,600
———; *Thermoregulation in Lizards — Multidisciplined Approach* (GB14323001); $1,900

NEW MEXICO

NEW MEXICO STATE UNIVERSITY
Wollum, A. G. II; *Host-Endophyte Interactions of Infection* (B020808); 24 months; $10,700

NEW YORK

BOYCE THOMPSON INSTITUTE FOR PLANT RESEARCH
App, Alva A.; *Regulation of Protein Synthesis During Seed Germination* (B019914); 24 months; $49,800
Staples, Richard C.; *Regulation of Protein Synthesis in Uredospores* (B017003); 24 months; $48,300
Streissle, Gert and Maramorosch, Karl; *Carcinogenesis by Wound Tumor Virus* (B017348); 24 months; $33,100
Turner, N. Joe; *Detoxication of Fungicides by Sulfhydryl Compounds in Relation to Specificity of Reaction* (B016837); 12 months; $10,000

COLLEGE AT BROCKPORT
Barr, Charles E.; *Relationship of the Resting Potential of Nitella to Light-Dependent H+ Fluxes* (B018069); 24 months; $25,000

COLLEGE OF FORESTRY AT SYRACUSE
Hartenstein, Roy C.; *Nitrogen Metabolism in Crustacea* (B015657); 24 months; $16,000
Silverstein, Robert M. and Borden, John H.; *Chemistry and Biology of Bark Beetle Pheromones* (B015959); 24 months; $53,900

COLGATE UNIVERSITY
Hoffman, Roger A.; *Influence and Interrelationships of Environmental Factors on Endocrine Systems* (B018350); 24 months; $19,600

COLUMBIA UNIVERSITY
Grundfest, Harry; *Fundamental Mechanisms of Bioelectric Activity* (GB6988X001); 12 months; $79,300
Jenkins; Farish A. Jr.; *Cineradiography of Locomotion in Primitive Mammals* (B013662); 24 months; $23,600
Loewenstein, Werner R.; *Study of Cellular Membranes* (B017768); 24 months; $75,000
Mancinelli, Alberto L.; *Phytochrome and Seed Germination* (B014749); 24 months; $53,000
Sawyer, Wilbur H.; *Comparative Physiology of the Neurohypophysis* (GB4923X003); 12 months; $17,200
Sprinson, David B.; *Biosynthesis of Aromatic Amino Acids* (B020996); 24 months; $60,000

CORNELL UNIVERSITY
Arion, William J.; *Regulation of Liver Microsomal Glucose 6-Phosphatase* (B014158); 24 months; $25,100
Hudson, Jack W.; *Thyroid Function and Temperature Regulation* (B017002); 24 months; $25,000
Musgrave, Robert B.; *Respiration of Various Genotypes of Maize Exhibiting Different Capacities for Apparent Photosynthesis* (B015904); 12 months; $20,000
Ross, A. F.; *Induced Resistance and Virus Localization in Hypersensitive Plant Hosts* (B013856); 24 months; $45,600

CUNY, HUNTER COLLEGE
Handler, Evelyn E.; *In Vitro Studies on Hematopoiesis in an Acute Myelogenous Leukemia of the Rat* (B017679); 24 months; $40,000

CUNY, MT. SINAI SCHOOL OF MEDICINE
Green, Jack P.; *The Influence of Pregnancy and Hormones on the Formation and Metabolism of Histamine by Human Beings* (GB7791 001); $14,300
Krulwich, Terry A.; *Gluconeogenesis and its Regulation in 'Arthrobacter'* (B020481); 24 months; $30,000

PHYSIOLOGICAL PROCESSES

DOMINICAN COLLEGE OF BLAUVELT
Klein, Sr. John M.; *Conservation of 'Ochromonas danica' and 'O. malhamensis'* (B015057); 24 months; $7,900

DOWNSTATE MEDICAL CENTER
Castells, Salvador; *Control PF Protein and RNA Synthesis in Fetal Adrenal Gland* (B016614); 24 months; $48,000
Nadler, Ronald D.; *Effect of Hormones on Brain Differentiation* (B014035); 24 months; $36,000

HEALTH RESEARCH INC.
Berns, Donald S.; *Isolation and Characterization of the Photoreceptor Pigment in Phycomyces* (B017991); 12 months; $9,000

INSTITUTE FOR MEDICAL RESEARCH AND STUDIES
Brodsky, William A.; *Ion Transport Mechanisms in the Reptilian Bladder* (B020912); 24 months; $54,000

MANHATTAN COLLEGE
Corbett, John J.; *Antigenic Comparison Between Classical Strains and Mating Varieties of 'Tetrahymena'* (GB7018X001); 12 months; $3,600

NEW YORK MEDICAL COLLEGE
Rappaport, Irving; *Immunocytes from Spleens of Non-Immunized Mice* (B016721); 24 months; $63,000

NEW YORK UNIVERSITY
Mitchell, Ormond G.; *Initiation and Control of Hair Growth in the Ventral Gland* (B017678); 24 months; $20,300
Uhr, Jonathan W.; *Regulation of Antibody Formation* (GB7473X001); 12 months; $59,300

PUBLIC HEALTH RESEARCH INSTITUTE
Osler, Abraham G.; *Mechanisms of Hypersensitivity Phenomena* (GB7404X003); 12 months; $35,800
———; (GB7404X002); $22,500

RENSSELAER POLYTECHNIC INSTITUTE
Diwan, Joyce J.; *Mechanism of Mitochondrial Ion Transport* (B017537); 24 months; $30,400

ROCKEFELLER UNIVERSITY
Deduve, Christian; *Biochemical Studies on Lysosomes and Peroxisomes* (GB5790X002), 12 months; $105,800
Rizack, Martin A.; *Regulation of Enzymatic Activity* (B015058); 12 months; $15,600

STATE UNIVERSTIY AT ALBANY
King, Tsoo E.; *Reconstitution of the Mitochondrial Respiratory Chain* (B013915); 24 months; $40,000

STATE UNIVERSITY AT STONY BROOK
Carlson, Albert D.; *Response of Luminescent Organs to Adrenergic Drugs* (B017347); 24 months; $46,300

SYRACUSE UNIVERSITY
Hainsworth, F. Reed; *Temperature Regulation and Water Metabolism of Birds and Mammals* (B012344); 24 months; $34,000

UNIVERSITY OF ROCHESTER
Adolph, E. F.; *Modification of the Onsets of Several Physiological Regulations* (B018546); 24 months; $40,500
Horowicz, Paul; *Electrochemical Properties of Muscle Membranes and Their Relation to Metabolism* (B015662); 24 months; $65,200

UPSTATE MEDICAL CENTER
Di Stefano, Henry S. and Dougherty, Robert M.; *Routes of Congenital Transmission of Avian Leukosis Viruses* (B014001); 24 months; $60,000

YESHIVA UNIVERSITY
Lev, Meir; *Vitamin K Metabolism in the K—Requiring Bacterium, 'Fusiformis nigrescens'* (B013590); 24 months; $58,000
Padawer, Jacques; *Cytological and Physiological Studies on Mast Cells* (GB8787); 24 months; $36,000

White, Abraham; *Biochemical Mechanisms of Control of Lymphoid Tissue Structure and Function* (GB6616X002); 12 months; $34,800

NORTH CAROLINA

NORTH CAROLINA STATE UNIVERSITY AT RALEIGH
Dobrogosz, Walter J.; *Cyclic Amp and Amino Sugar Metabolites in Catabolite Repression of Beta-Galactosidase Synthesis in Escherichia coli* (B018548); 24 months; $35,000

UNIVERSITY OF NORTH CAROLINA AT CHAPEL HILL
Penniall, Ralph; *Synthesis of Cytochrome Oxidase in Mammalian Cells* (B011589); 24 months; $35,000

NORTH DAKOTA

UNIVERSITY OF NORTH DAKOTA
Nordlie, Robert C.; *Inorganic Pyrophosphate-Glucose and Associated Phosphotransferase Activities of Mammalian Kidney and Small Intestine* (B014752); 24 months; $30,000

OHIO

CASE WESTERN RESERVE UNIVERSITY
Sable, Henry Z.; *Enzyme-Coenzyme Interactions* (B017094); 24 months; $53,500

CHILDREN'S HOSPITAL RESEARCH FOUNDATION
Addanki, Somasundaram; *Ion Transport and Intramitochondrial Ph* (B017634); 24 months; $37,800

MIAMI UNIVERSITY
McClure, Jerry W.; *Differentiation and Regulation of Secondary Phenolics in Barley Plumules* (B016636); 24 months; $30,100
Newman, David W.; *Developmental Control of Lipid Metabolism in Photosynthetic Tissues* (B017635); 24 months; $24,000

OHIO STATE UNIVERSITY
Brownell, Katharine A.; *Extra Adrenal Production of Adrenocortical Hormones in the Opossum* (B014462); 24 months; $30,000
Serif, George S.; *Biosynthesis of the Aglycones of Thioglucosides* (B015661); 24 months; $32,000

OHIO UNIVERSITY
Jaffe, Mordecai J.; *Mechanisms Responsible for Rapid Movement in Plants* (B020474); 24 months; $48,000

UNIVERSITY OF CINCINNATI
Gist, Daniel H.; *Control and Function of the Reptilian Adrenal* (B014987); 24 months; $22,300

OKLAHOMA

OKLAHOMA STATE UNIVERSITY
Mitchell, Earl D.; *Cell-Free Biosynthesis of the Methylcyclopentane Monoterpenoids from 'Nepeta cataria'* (B019561); 24 months; $30,000
Waller, George R.; *Mass Spectrometry in Biochemical Research* (B020926); 24 months; $50,000
———; *Metabolic and/or Physiologic Role of Diterpenoid Alkaloids in Plants* (B013126); 24 months; $25,000

UNIVERSITY OF OKLAHOMA
Bulmer, Glenn S.; *Light and Electron Microscopic Studies of the Cytological Effects of Microtubule-Affecting Chemicals on 'Phleum' Root Tips* (B015403); 12 months; $9,000
Wender, Simon H. and Smith, Eddie C.; *Studies of Enzymes Associated with Scopolin—Scopoletin Interconversions* (B014585); 24 months; $31,900

OREGON

OREGON STATE UNIVERSITY
Moore, Thomas C.; *Plant Hormone Metabolism and Mechanism of Auxin Action* (B018494); 24 months; $33,100

PORTLAND STATE UNIVERSITY
Simpson, Leonard and Calvin, Clyde L.; *Host Parasite Tissue Relations in Plants* (B016732); 24 months; $39,300

UNIVERSITY OF OREGON
Larrabee, Allan R.; *Mammalian Fatty Acid Synthetase* (B018998); 24 months; $49,900
Nabors, Murray W.; *Action Mechanism of Light in Lettuce Seed Germination* (B019046); 12 months; $4,000
Tunturi, Archie R.; *Anatomic Study of the Nerve Networks of the Auditory Cortex* (B016133); 24 months; $38,000

PENNSYLVANIA

ALBERT EINSTEIN MEDICAL CENTER
Goldberg, Harry and Sittel, Karl; *Theoretical-Experimental Development of Myocardial Model* (B017140); 12 months; $30,000
Suhadolnik, Robert J.; *Biosynthesis of the Nucleoside Antibiotics Sangivamycin, Nucleocidin and Nebularine* (B017441); 24 months; $50,600

PENNSYLVANIA HOSPITAL
Zacks, Sumner I.; *Properties of Muscle Surface Membrane and Basement Membrane in the Neuromuscular Junction* (B018709); 12 months; $25,400

PENNSYLVANIA STATE UNIVERSITY
Claus, George W.; *Nitrogen Metabolism in 'Acetobacter Suboxydans'* (B017493); 24 months; $33,500

PRESBETARIAN UNIVERSITY OF PENNSYLVANIA MEDICAL CENTER
Somlyo, Andrew P.; *Cellular Mechanisms of Vascular Smooth Muscle Contraction and Relaxation* (B020478); 12 months; $15,200

TEMPLE UNIVERSITY
Litwack, Gerald; *Molecular and Biochemical Fates of Cortisol in Liver* (B008784); 12 months; $25,000

THOMAS JEFFERSON UNIVERSITY
Toporek, Milton; *Plasma Protein Production in Tumor-Bearing Rats* (B012436); 12 months; $15,000

UNIVERSITY OF PENNSYLVANIA
Gasic, Gabriel J.; *Blood Platelets in Egg Implantation and Pregnancy* (B019325); 24 months; $40,000
Peachey, Lee D.; *Cellular Mechanisms of Muscle Contraction and of Antidiuretic and Steroid Hormone Action* (GB6975X002); 12 months; $50,900
Sterling, Peter; *Anatomy and Physiology of Visuo-Motor Systems* (B015970); 24 months; $39,300

WOMAN'S MEDICAL COLLEGE OF PENNSYLVANIA
Irving, Ronald E.; *Comparative Anatomical Investigation of the Auditory System* (B016485); 24 months; $35,000

SOUTH DAKOTA

SOUTH DAKOTA STATE UNIVERSITY
Schingoethe, David J.; *Growth Inhibitors in Soybeans* (B017779); 24 months; $30,000

TENNESSEE

UNIVERSITY OF TENNESSEE
Averill, Robert L. W.; *Hypothalamus and Lactation* (B019677); 12 months; $8,000
Hansard, Sam L.; *Placental Transfer of Minerals in Farm Animals* (B013116); 12 months; $20,100
Montie, Thomas C.; *Structure Synthesis and Regulation of the Protein Toxins of Pasteurella Pestis* (B012893); 24 months; $46,000

VANDERBILT UNIVERSITY
Hurwitz, Leon; *Drug Action on Isolated Smooth Muscle* (B013346); 24 months; $60,000

BIOLOGICAL AND MEDICAL SCIENCES RESEARCH PROJECTS

Touster, Oscar; *Particle-Bound Enzymes and the Glucuronate-Xylulose Pathway* (GB7425X-001); 12 months; $36,400

TEXAS

AMERICAN SOCIETY OF PARASITOLOGISTS
Otto, Gilbert F.; *Second International Congress of Parasitology* (B016959); 12 months; $6,000
Otto, Gilbert F.; *Second International Congress of Parasitology* (GB16959001); $3,000

GOUCHER COLLEGE
Habermann, Helen M.; *Pigment Deficient Mutants of 'Helianthus annuus'* (B014816); 24 months; $30,000

SOUTHWEST FOUNDATION FOR RESEARCH AND EDUCATION
Hagino, Nobuyoshi; *Base Line Study of Primate Neurophysiology* (GB8788); 12 months; $5,000

SOUTHWESTERN MEDICAL SCHOOL
Sulkin, S. Edward; *Latent Viral Infections and Temperature-Induced Alterations in Animal Viruses* (B012611); 24 months; $40,000
Vanatta, John C.; *Potassium Excretion and its Relationship to Sodium Reabsorption* (B017610); 24 months; $36,700

TEXAS A & M UNIVERSITY
Halliwell, Robert S.; *Biosynthesis of Tobacco Mosaic Virus Ribonucleic Acid* (B014174); 24 months; $20,000
Toler, Robert W. and Thomas, William B.; *Mycoplasma as a New Etiological Agent of Plant Diseases* (B014558); 24 months; $23,700

UNIVERSITY OF TEXAS, Medical Branch at Galveston
Nowinski, Wiktor W.; *Synthesis and Activity of Enzymes in Hypertrophy of the Kidney After Uninephrectomy* (B017830); 24 months; $46,300

UNIVERSITY OF TEXAS AT AUSTIN
Thompson, Guy A.; *Thiaminase Enzyme in Organisms* (B019933); 24 months; $36,000
———; *Biochemistry of Ether-Containing Lipids* (016363); 24 months; $25,300

UNIVERSITY OF TEXAS MEDICAL SCHOOL
Keenan, Roy W.; *The Biological Degradation of the Sphinolipid Bases* (B015586); 24 months; $30,000
Nishimura, Jonathan S.; *Biosynthesis of Actinomycins* (B017349); 12 months; $13,600

UTAH

BRIGHAM YOUNG UNIVERSITY
Hess, W. M.; *Fungus Ultrastructure and Biochemistry* (B014161); 12 months; $20,000

UNIVERSITY OF UTAH
Stevens, Walter; *Mechanism of Action of Cortisol on Lymphatic Tissue* (B019040); 24 months; $28,000
Strickland, Walter N.; *Regulation of Nitrogen Metabolism in Neurospora* (B013682); 24 months; $38,000

UTAH STATE UNIVERSITY
Salisbury, Frank B.; *Hormones and Time Measurement in Flowering of Xanthium* (B015291); 12 months; $25,100

VERMONT

PUTNAM MEMORIAL INSTITUTE FOR MEDICAL RESEARCH
Toolan, Helene W.; *Relationship of Virus Interplay to Cancer Induction* (B015751); 12 months; $50,000

VIRGINIA

UNIVERSITY OF VIRGINIA
Wagner, Robert R.; *The Interferons—Cellular Regulators of Viral-Infection* (GB6537X-002); 12 months; $25,900

VIRGINIA COMMONWEALTH UNIVERSITY
O'Neal, Charles H.; *Transfer Ribonucleic Acid and Protein Synthesis* (B014046); 24 months; $25,000

VIRGINIA POLYTECHNIC INSTITUTE
Schmidt, Robert R.; *Control of Enzyme Synthesis During the Cell Cycle of Eucaryotic Cells* (B017305); 24 months; $43,300

WASHINGTON

UNIVERSITY OF WASHINGTON
Meeuse, Bastiaan J. D.; *Cyclic (3',5')-AMP and the Metabolic Flare-up of Arum Lily Appendices* (B020158); 24 months; $39,000
Ordal, Erling J.; *Trace Inorganic Elements in the Metabolism of Bacteria* (B015107); 24 months; $50,000
Willows, A. O. Dennis; *Neuronal Mechanisms of Stereotyped Behavior* (B020351); 24 months; $31,000

WASHINGTON STATE UNIVERSITY
Higinbotham, N.; *Electrogenesis and Ion Transport in Higher Plant Cells* (B019201); 24 months; $35,000

WEST VIRGINIA

WEST VIRGINIA UNIVERSITY
Jung, Gerald A.; *Physiological and Biochemical Studies of Cold Hardiness in Plants* (B012618); 12 months; $20,000
Norman, Charles; *Regulatory Action of Exogenous Sterol and Nutrilites on Growth and Reproduction in the Fungus 'Phytophthora Cactorum'* (B011906); 33 months; $32,000

WISCONSIN

MARQUETTE SCHOOL OF MEDICINE
Ankel, Helmut K.; *Cell Wall Biosynthesis in Cryptococcus* (B018090); 24 months; $43,000

UNIVERSITY OF WISCONSIN, Madison
Bleier, Ruth H.; *Organization and Function of Certain Hypothalamic Cell Groups and Areas* (B017835); 24 months; $25,000
De Zoeten, Gustaaf A.; *Comparative Study of the Physiology and Cytology of Virus Infected Plant and Insect Cells* (B015381); 24 months; $32,500
Goldman, Dexter S.; *Manno-Lipids of 'Mycobacterium tuberculosis'* (B013598); 24 months; $25,000
Hanson, Richard S.; *Regulation of Respiration in Bacteria* (B019331); 24 months; $36,000
Porter, W. P., Myers, Glen E. and Mitchell, John W.; *Theoretical and Experimental Studies of Energy Budgets of Animal Appendages* (B015499); 24 months; $45,200

Molecular Biology

ALABAMA

UNIVERSITY OF ALABAMA, Tuscaloosa
Andreoli, Thomas E.; *Transport Processes in Lipid Membranes* (B022808); 7 months; $20,000

ARIZONA

UNIVERSITY OF ARIZONA
Hruby, Victor J.; *Structural Studies of Synthetic Hormones and Polypeptides* (B011781); 24 months; $23,000
Rupley, John A.; *Studies on Lysozyme* (B022632); 24 months; $80,000
Wyckoff, Ralph W. G.; *Physical Methods of Microanalysis of Biological Materials* (B017516); 24 months; $35,000

ARKANSAS

UNIVERSITY OF ARKANSAS, Little Rock Campus
York, J. L.; *Chemistry of the Active Site of Hemerythrin* (B015755); 24 months; $41,500

CALIFORNIA

CALIFORNIA INSTITUTE OF TECHNOLOGY
Dickerson, Richard E. and Stanford, Richard H.; *Structure and Function of Proteins and Related Macromolecules* (GB6617X001); 12 months; $108,800
Strauss, James H. Jr.; *Structure and Properties of Virus-Altered Membranes* (B013508); 24 months; $45,000

SALK INSTITUTE FOR BIOLOGICAL STUDIES
Holley, Robert W.; *Molecular Basis of the Control of Cell Division in Mammalian Cells* (B017912); 24 months; $100,000

SAN JOSE STATE COLLEGE
Weaver, Ellen C.; *Mechanism of Photosynthesis* (B018594); 24 months; $18,000

SCRIPPS CLINIC AND RESEARCH FOUNDATION
Dandliker, Walter B. and Levison, Stuart A.; *Fast Reaction Rate Studies of Antigen-Antibody Systems by Fluorescence Methods* (B015594); 24 months; $50,000

STANFORD UNIVERSITY
Jardetzky, Oleg; *Nuclear Magnetic Resonance Studies of Small-Molecule-Macromolecule Interactions* (B017812); 24 months; $163,000
Kornberg, Arthur; *Biochemistry of Sporulation and Germination in Bacteria* (GB7471X-001); 12 months; $45,000
Loew, Gilda H. and Lederberg, Joshua; *Electronic and Magnetic Properties of Home Proteins, and Ferredoxins* (B017980); 24 months; $65,000
McConnell, Harden M.; *Spin Label Study of Transport in Bacterial Membranes* (B019638); 24 months; $80,000
Shooter, Eric M.; *Structure and Mechanism of Action of the Nerve Growth Factor* (B014458); 24 months; $60,500

UNIVERSITY OF CALIFORNIA, Berkeley
Fraenkel-Conrat, Heinz; *Chemical and Biochemical Studies of Viral Rnas and Proteins* (GB6209X001); 12 months; $104,800
Koshland, Daniel E. Jr.; *Modification Studies on Lysozyme and Related Enzymes* (GB7057X-001); 13 months; $30,000
Packer, Lester; *Function of Subcellular Membranes* (B020951); 24 months; $80,000
Schachman, Howard K.; *The Functional Aspects of the Quaternary Structure of Proteins* (GB4810X002); 12 months; $32,000

UNIVERSITY OF CALIFORNIA, Davis
Deamer, David W.; *Substructure of Lipid Crystals by the Freeze-Etch Technique* (GB8600 001); $4,300

UNIVERSITY OF CALIFORNIA, Los Angeles
Eisenman, George; *Ion Specific Phenomena in Biology and Chemistry* (B016194); 24 months; $60,000
Martinez, Rafael J.; *Bacterial Flagella* (B015844); 24 months; $41,500

UNIVERSITY OF CALIFORNIA, San Diego
Butler, Warren L.; *Investigation of Phytochrome* (GB6557X001); 14 months; $50,000 $30,000
Kraut, Joseph; *Three-Dimensional Structure and Function of Biological Macromolecules* (B015684); 24 months; $73,000
Simon, Melvin I.; *Biosynthesis and Regulation of Flagella* (B015655); 24 months; $59,000

UNIVERSITY OF CALIFORNIA, Santa Barbara
Bruice, Thomas C.; *Mechanism of Pyriodoxal Catalysis* (GB5294X002); 12 months; $30,000

UNIVERSITY OF SOUTHERN CALIFORNIA
Kataja, Eva I.; *Recognition of Nucleotide Sequences by Proteins* (B016440); 24 months; $30,000

CONNECTICUT

YALE UNIVERSITY
Coleman, Joseph E.; *Electronic and Magnetic Properties of Macromolecules* (B013344); 24 months; $94,000

MOLECULAR BIOLOGY

Fruton, Joseph S.; *Enzymic Hydrolysis and Synthesis of Peptide Bonds* (B018268); 24 months; $90,000
Konigsberg, William; *Structure and Binding Specificity of Human and Mouse Myeloma Proteins* (B016061); 24 months; $93,000
Oeschger, Max P.; *Multiple Functions of Proteins During Virus Replication* (B018422); 12 months; $18,000
Soll, Dieter; *Recognition of t-RNA by Aminoacyl-t-RNA Synthetases* (B019085); 24 months; $46,000
Stryer, Lubert; *Optical Studies of Protein Structure and Function* (B019055); 12 months; $80,000
Sturtevant, Julian M.; *Calorimetric Measurements of the Heats of Biochemical Reactions* (GB6033 001); 12 months; $16,000
Wang, Jui H.; *Molecular Mechanisms of Biochemical Energy Conversion* (GB7459X-001); 12 months; $40,000

DISTRICT OF COLUMBIA

GEORGETOWN UNIVERSITY
Steinhardt, Jacinto; *Effects of Protein Interactions on Protein Stability* (B013391); 24 months; $59,000

FLORIDA

FLORIDA STATE UNIVERSITY
Fisher, James R.; *Enzyme Studies—Mechanism of Action and Evolution* (B016957); 24 months; $36,000
Homann, Peter H.; *Function and Organization of Photosystem II in Manganese Deficient, Fragmented and Mutant Chloroplasts* (B016301); 24 months; $43,000

PAPANICOLAOU CANCER RESEARCH INSTITUTE
Schultz, Julius; *2nd Annual Biochemistry Winter Symposia* (B018183*); 12 months; $800

UNIVERSITY OF FLORIDA
Chun, Paul W.; *Coat Protein Subunit Association and Self-Assembly of Rod-Shaped FD Virus* (B017847); 12 months; $13,000

GEORGIA

EMORY UNIVERSITY
Sophianopoulos, Alkis J.; *Conformation, Interactions and Function of Proteins* (B017842); 24 months; $57,000

UNIVERSITY OF GEORGIA
Brewer, John M.; *Comparison of Molecular Properties of Thermostable and Thermolabile Enzymes* (B013031); 24 months; $20,000
Dervartanian, Daniel V.; *Electron Transfer in Iron Deficient A. Vinelandii* (B013242); 24 months; $20,000
Lovins, Robert E.; *Quantitative Protein Sequencing by Mass Spectrometry* (B024542); 6 months; $6,000

HAWAII

UNIVERSITY OF HAWAII
Yasunobu, Kerry T.; *Mechanism of Action of Amine Oxidase* (B018739); 24 months; $40,000

ILLINOIS

NORTHWESTERN UNIVERSITY
Klotz, Irving M.; *Protein Interactions* (GB7122X001); 12 months; $35,000
Loach, Paul A.; *Photoreceptor Subunits and their Interaction with Membrane Structure* (B018420); 12 months; $17,000

UNIVERSITY OF CHICAGO
Kaiser, Emil T.; *The Catalytic Action of Pepsin* (B013208); 24 months; $25,000

UNIVERSITY OF ILLINOIS, Medical Center
Nisonoff, Alfred; *Structural Studies of Antibodies and Antigens* (GB5424X002); 12 months; $45,000

UNIVERSITY OF ILLINOIS, Urbana
Govindjee; *Mechanism of Photosynthesis in Corn (Zea Mays) and Its Mutants* (B014176); 24 months; $46,000
———; *Chlorophyll 'A' Fluorescence and Photosynthesis* (B019213); 14 months; $26,000
Hager, Lowell P.; *Biological Halogenation Mechanisms* (GB5442 002); 12 months; $42,000
Kaplan, Samuel; *Induction of Bacterial Chromatophore Formation* (B014966); 24 months; $43,000
Macleod, Roderick; *Immunochemical and Developmental Studies on a Plant Rhabdovirus* (B017629); 24 months; $22,000
Steffensen, Dale M.; *Genetic and Biochemical Studies with Drosophila Ribosomes and Related Problems* (B018922); 24 months; $30,000
Sybesma, Christiaan; *Primary Reactions in Bacterial Photosynthesis* (B015777); 24 months; $58,000
Wolin, Meyer J.; *Structure and Function of Bacterial Membranes* (B008777); 12 months; $16,000

INDIANA

INDIANA UNIVERSITY, Bloomington
Haurowitz, Felix; *Biosynthesis and Structure of Antibodies and Other Proteins* (B012797); 12 months; $7,500
Mahler, Henry R. and Moore, Walter J.; *Biochemistry and Biophysics of Neuronal Function* (B017157); 24 months; $50,000
Putnam, Frank W.; *Structural Study of the Type A Immunoglobulins of Man and Other Species* (B018483); 24 months; $50,000

INDIANA UNIVERSITY, Indianapolis
Roeskse, Roger W.; *Synthesis of Enzyme Models* (B015683); 24 months; $40,000
Steinrauf, Larry K.; *Structural Investigation of Biological Regulators* (B013422); 12 months; $22,000

PURDUE UNIVERSITY
Aronson, Arthur I.; *Biosynthesis of Bacterial Spores and the Regulation of Related Processes* (B020765); 18 months; $44,000
Lister, Richard M.; *Disassembly and Reassembly of Viruses* (B015573); 24 months; $26,000
Mortenson, Leonard E.; *Electron Transport and Energy Metabolism in Bacteria* (B022629); 24 months; $40,000
Rossmann, Michael G.; *X-Ray Determination of Proteins and Viruses* (GB5477 002); 12 months; $70,000
Weiner, Henry; *Mechanistic Studies with Dehydrogenases* (B018489); 24 months; $28,000

IOWA

UNIVERSITY OF IOWA
Swenson, Charles A.; *Interactions of Model Peptides and Related Molecules with Their Environment* (B018017); 24 months; $49,000

KANSAS

KANSAS STATE UNIVERSITY
Hedgcoth, Charlie Jr.; *Synthesis and Regulation of t-RNA* (B018024); 24 months; $20,000

UNIVERSITY OF KANSAS
Fisher, Harvey F.; *Structure and Mechanisms of Catalysis of Enzymes* (B020923); 24 months; $30,000

KENTUCKY

UNIVERSITY OF LOUISVILLE
Yankeelov, John A. Jr.; *Enzyme Structure and Function* (B017985); 24 months; $38,000

LOUISIANA

LOUISIANA STATE UNIVERSITY, Baton Rouge
Braymer, Hugh D.; *Genetic and Biochemical Studies of Multimeric Enzymes* (B015496); 12 months; $20,000

Chang, Simon H.; *Enzymatic Recognition Sites and Tertiary Structure of Transfer RNA* (B017124); 24 months; $32,000

TULANE UNIVERSITY
Li, Yu-Teh, *Isolation of Glycosidases Which Hydrolyze Carbohydrate Portion of Glycoproteins* (B018019); 24 months; $23,000
Steele, Richard H.; *Possible Role of Excited Singlet Molecular Oxygen Species in Biochemical Redox Energetics* (B015149); 24 months; $40,000

MARYLAND

JOHNS HOPKINS UNIVERSITY
Biltonen, Rodney L.; *Thermodynamic Studies of Protein Conformation* (B017680); 24 months; $50,000
Function Studies on the Control of Glyceralde-
Fenselau, Allan H.; *Comparative Structure-hyde-3-Phosphate Dehydrogenase* (B020672); 24 months; $30,000
Lehninger, Albert L.; *Biochemistry of Active Transport by Mitochondria* (GB7032X002); 12 months; $18,000
Lennarz, William J.; *Structure and Biosynthesis of Micrococcus Lysodeikticus* (B012768); (B012768); 24 months; $34,000

UNIVERSITY OF MARYLAND
Adelman, William J. Jr.; *Voltage Clamp Measurements of Membrane Conductance and Capacity by Oscillating Potential Control* (B015588); 24 months; $59,000
Aposhian, H. Vasken; *Sequential Cleavage of Dinucleotides From DNA* (B015673); 24 months; $67,000

UNIVERSITY OF MARYLAND, Baltimore County Campus
Schwartz, Martin; *Electron Transport, Proton Translocation and Phosphorylation in Photosynthesis* (B018076); 12 months; $18,000

MASSACHUSETTS

BETH ISRAEL HOSPITAL
Schlossman, Stuart F.; *Immunochemical Studies on Hapten Substituted Polylysines* (GB6675 002); 12 months; $29,600

BOSTON COLLEGE
Bade, Maria L.; *Biochemistry of Arthropod Chitin* (B017338); 24 months; $20,000

BRANDEIS UNIVERSITY
Grossman, Lawrence; *Effects of Ultraviolet Light and Other Mutagens on the Structure and Function of Nucleic Acids* (GB6208X002); 12 months; $37,200
Hollocher, Thomas C. Jr.; *Reactivities and Structure of Heart Muscle Succinic Dehydrogenase* (B019920); 12 months; $20,000
Jencks, William P.; *Chemistry and Biochemistry of Energy Transferring Reactions in Biological Systems* (GB5648 002); 12 months; $32,000
Levine, Lawrence; *Antigen-Antibody Reactions as Probes of Protein Macromolecular Structure* (B018538); 24 months; $30,000

HARVARD UNIVERSITY
Briggs, Winslow R.; *Biochemical and Physiological Studies of the Plant Pigment Phytochrome* (B015572); 24 months; $64,000
Doty, Paul; *Polypeptides and Proteins* (GB4563X001); 12 months; $95,000
Gill, David M.; *Diphtheria Toxin and Protein Synthesis* (B013217); 24 months; $34,000
Guidotti, Guido; *Interactions Between Macromolecules* (B017953); 24 months; $50,000
Holton, Gerald; *Thermodynamic Properties of Aqueous Solutions* (B018702); 24 months; $50,000
Strominger, Jack L.; *Structure and Biosynthesis of Bacterial Cell Walls* (B013795); 24 months; $190,000
Wald, George; *Basic Mechanisms of Photoreceptor Activities* (B013229); 24 months; $98,000

BIOLOGICAL AND MEDICAL SCIENCES RESEARCH PROJECTS

MASSACHUSETTS GENERAL HOSPITAL
Robinson, Dwight R.; *Mechanisms and Catalysis of Nucleophilic Reactions of Acyl Compounds* (B013428); 24 months; $50,000

MASSACHUSETTS INSTITUTE OF TECHNOLOGY
Baglioni, Corrado; *Hemoglobin Synthesis* (B014345); 24 months; $60,000
Khorana, H. Gobind; *Nucleic Acid Synthesis* (B021053); 12 months; $100,000
Luria, Salvador E.; *Molecular Aspects of Function and Organization of Viruses and Other Episomes* (GB5304X003); 12 months; $52,800
Magasanik, Boris; *Regulation of the Metabolic Processes of the Single Cell at the Molecular Level* (GB5322X002); 12 months; $42,000
Rich, Alexander; *Studies on the Mechanism of Protein Synthesis and on the Molecular Structure of the Nucleic Acids* (B015754); 24 months; $100,000

RETINA FOUNDATION
Gergely, John; *Biochemistry of Muscle Contraction* (B018417); 24 months; $46,000
Lam, Kwok-Wai; *Coupling Factors Concerned with Oxidative Phosphorylation* (B016419); 24 months; $22,000

TUFTS UNIVERSITY
Johnson, Brian J.; *Synthesis and Structure of Linear Antigenic Polypeptides* (B012114); 24 months; $45,000

WORCESTER FOUNDATION FOR EXPERIMENTAL BIOLOGY
Barber, Roger; *Binding of Charged Molecules to Polynucleotides* (B018022); 24 months; $42,000

MICHIGAN

EDSEL B. FORD INSTITUTE FOR MEDICAL RESEARCH
Shore, Joseph D.; *Mechanism of Action of Dehydrogenases* (B018027); 24 months; $34,000

MICHIGAN STATE UNIVERSITY
Kende, Hans J.; *Studies on the Site of Action of Gibberellins* (B013853); 24 months; $15,000
Kitchen, Hyram; *Sheep Polymorphic Hemoglobin Structure and Synthesis Within Individual Erythrocyte Precursors* (B017397); 12 months; $17,000
Rottman, Fritz M.; *Distribution of 2'-O-Methylnucleotides in RNA and Their Effect on Structure and Biological Function* (B020764); 24 months; $50,000
Tulinsky, Alexander; *X-Ray Structure Determination of Enzymes and Other Biological Molecules* (B015402); 24 months; $90,000
Varner, Joseph E.; *Hormonal Control of Enzyme Synthesis* (B008774); 24 months; $30,000
Wood, Willis A.; *Allosteric Activation of L-Threonine Dehydrase by 5'-Amp* (B020160); 24 months; $45,000

UNIVERSITY OF MICHIGAN
Krimm, Samuel; *Spectroscopy of Biological Macromolecules* (B015682); 24 months; $42,000

MINNESOTA

UNIVERSITY OF MINNESOTA
Anderson, Dwight L.; *Structure of Phage Phi 29 and Its Infectious DNA* (B015207); 24 months; $42,000
Liener, Irvin E.; *Structural Basis of Enzyme Action* (B015385); 24 months; $26,500
Lumry, Rufus, *Conference on Protein Calorimetry* (B018424); 12 months; $3,500

MISSOURI

WASHINGTON UNIVERSITY
Geller, David M.; *Mechanism of the Photophosphorylation Reaction in Biological Systems* (B016487); 24 months; $39,900

NEW JERSEY

NEW JERSEY COLLEGE OF MEDICINE AND DENTISTRY
Bartell, Pasquale F.; *Virus-Induced Modifications of Cell* (B020673); 12 months; $17,000

PRINCETON UNIVERSITY
Fresco, Jacques R.; *Physical Biochemistry of Polynucleotides and Ribonucleic Acids* (B018865); 24 months; $104,000
Sueoka, Noboru; *Molecular Aspects of Genetics* (B019560); 24 months; $60,000

RUTGERS UNIVERSITY
Crane, Robert K.; *Mechanism of Intestinal Absorption of Sugars* (B014464); 24 months; $64,000

NEW MEXICO

NEW MEXICO STATE UNIVERSITY
Birnbaum, Edward R. and Darnall, Dennis W.; *Interactions of Lanthanide Ions with Proteins* (B015192); 24 months; $39,000

NEW YORK

COLD SPRING HARBOR LABORATORY
Gesteland, Raymond F.; *Use of Cell Free Synthesis of Active T-4 Lysozyme to Investigate Protein Synthesis and its Control* (GB7209X001); 12 months; $40,000
Werner, Rudolf; *Investigation of the Mechanism of DNA Replication* (B018013); 24 months; $50,000

COLUMBIA UNIVERSITY
Benesch, Reinhold and Benesch, Ruth E.; *Role of Subunit Interactions in the Combination of Hemoglobin with Ligands* (GB4530 002); 12 months; $30,000
Beychok, Sherman; *Optical Activity of Biological Macromolecules in Relation to Conformation* (B022847); 12 months; $41,000
Erlanger, Bernard F.; *The Active Site of Pepsin* (B017915); 24 months; $48,000
Karlin, Arthur; *Chemical Characterization of the Acetylcholine Receptor* (B015906); 24 months; $43,000
Levinthal, Cyrus; *Protein Crystallography Using 'Partial Structures' for Phase Determination* (B018421); 24 months; $54,000
Spiegelman, S.; *The Mechanism of Enzyme Synthesis* (B017251); 15 months; $103,800
Tanenbaum, Stuart W.; *Structure-Function Biochemistry of Selected Antibodies and Antigens* (GB5316X002); 12 months; $35,000

CORNELL UNIVERSITY
Edelstein, Stuart; *Quaternary Structure of Hemoglobin* (B008773); 24 months; $43,000
Haschemeyer, Rudy H.; *Electron Microscopic Studies on the Macromolecular Arrangement of Subunits in Enzymes* (B015591); 24 months; $30,000
Heppel, Leon A.; *Mechanisms of RNA Breakdown in E. Coli with Special Reference to Rnase I.* (GB7093X001); 12 months; $40,000
Hess, George P.; *Structure and Function of Enzymes* (B016558); 24 months; $60,000
McCarty, Richard E.; *Roles of Ion Transport and Galactolipids in Photosynthetic Electron Transport and Phosphorylation* (B012960); 24 months; $38,500
Racker, Efraim; *The Role of Coupling Facots and the Chloroplast Membrane in Photophosphorylation* (B016198); 24 months; $28,700
Scheraga, Harold A.; *Calculations of Polypeptide Conformations* (B017388); 24 months; $40,000
———; *Thermodynamic and Statistical Mechanical Studies of Protein Reactions* (GB7571X001); 12 months; $63,500
Wu, Ray; *Deoxynucleotide Sequence Determination of DNA* (B013912); 24 months; $57,000

HEALTH RESEARCH INC., Roswell Park Division
Parsons, D. F.; *Cell Membranes—Electron Diffraction and X-ray Diffraction Studies* (B015389); 24 months; $45,000

NEW YORK UNIVERSITY
Brody, Seymour S.; *Model Systems for Primary Processes of Photosynthesis* (B018354); 12 months; $20,000
Heidelberger, Michael; *Chemical Constitution and Immunological Specificity of Poly-Saccharides and Natural Proteins* (B013592); 24 months; $60,000
Salton, Milton R. J.; *Biochemistry of Bacterial Membranes* (B017107); 24 months; $90,000

POLYTECHNIC INSTITUTE OF BROOKLYN
Goodman, Murray; *Synthesis, Properties and Reactions of Peptides and their Derivatives* (B022908); 24 months; $60,000

RENSSELAER POLYTECHNIC INSTITUTE
Birnboim, Meyer H.; *Viscoelasticity in Biophysics* (B008770); 24 months; $48,000
Landau, Joseph V.; *Hydrostatic Pressure Effects on the Biosynthesis of Proteins and Nucleic Acids* (B018567); 12 months; $10,000

ROCKEFELLER UNIVERSITY
Luck, David J. L.; *Origin and Development of Mitochondria in 'Neurospora Crassa'* (GB4878X002); 12 months; $14,500

STATE UNIVERSITY AT ALBANY
Myer, Yash P.; *Conformation of Cytochromes* (B013524); 24 months; $48,000
Rabinowitch, Eugene; *Mechanisms of Transfer and Photochemical Storage of Light Energy* (B015065); 24 months; $30,000
Walz, Frederick G. Jr.; *Relaxation Spectra of Myosin A* (B012750); 24 months; $42,000

STATE UNIVERSITY AT BUFFALO
McMenamy, Rapier H.; *Primary Sequence Determination of Human Serum Albumin* (B018707); 24 months; $40,000

STATE UNIVERSITY AT STONY BROOK
Lefevre, Paul G.; *Physicochemical Basis of Cell Membrane Sugar Transport* (B012685); 24 months; $63,900
Wishnia, Arnold; *Dynamic Aspects of Protein Structure* (B016060); 24 months; $74,000

SYRACUSE UNIVERSITY
Fondy, Thomas P.; *Structure and Evolution of NAD—and Flavin-Linked Glycerol-3-Phosphate Dehydrogenases* (B020943); 24 months; $22,000

YESHIVA UNIVERSITY
Eagle, Harry; *Investigations on Normal and Abnormal Mammalian Cells* (B024175); 24 months; $81,000
Ehrenfeld, Elvera; *Poliovirus RNA Polymerase* (B018026); 24 months; $34,000
England, Sasha; *Malic Dehydrogenases—Structure and Biological Control* (B012670); 24 months; $58,000
Horecker, B. L.; *Comparative Studies of Enzyme Structure and Enzyme Mechanisms in Carbohydrate Metabolism* (GB7140X001); 12 months; $75,000
Hurwitz, Jerard; *Enzymatic Reactions Involved in Nucleic Acid Metabolism* (B015596); 24 months; $100,000
Marmur, Julius; *Structure and Synthesis of Yeast Mitochondrial Nucleic Acids* (B012435); 24 months; $77,000
Scharff, Matthew D.; *Macromolecular Synthesis in Synchronized Mammalian Cells* (GB6364X002); 12 months; $40,000
Schildkraut, Carl L.; *DNA of Specialized Animal Cells* (B022589); 24 months; $80,000
Summers, Donald F.; *Translation of Polycistronic Messenger RNA of RNA Containing Animal Viruses* (B018025); 24 months; $58,000
Warner, Jonathan R.; *Ribosome Synthesis in Eukaryotes* (B017844); 24 months; $80,000

PSYCHOBIOLOGY

NORTH CAROLINA

DUKE UNIVERSITY
Hill, Robert L.; *Structure-Function Relationships of Enzymes* (B012676); 24 months; $69,000
Tanford, Charles; *Physico-Chemical Studies Related to the Structure and Function of Cell Membranes* (B014844); 24 months; $110,000

NORTH CAROLINA STATE UNIVERSITY AT RALEIGH
Leatherwood, James M.; *Investigations of Cellulase and a Disaccharide Epimerase* (B01751F); 24 months; $30,000
Swaisgood, Harold E.; *Characteristics of Modified Enzymes* (GB7855 001); 6 months; $2,700

UNIVERSITY OF NORTH CAROLINA AT CHAPEL HILL
Dearman, Henry H.; *Organic Free Radicals in Biological Systems* (B016423); 24 months; $28,000
Edgell, Marshall Hall; *Structure of the Bacteriophage Phi X 174* (B013397); 24 months; $12,000
Hermans, Jan; *Theoretical Aspects of Protein Structure* (B016420); 24 months; $44,000

NORTH DAKOTA

UNIVERSITY OF NORTH DAKOTA
Dyson, John E. D.; *Chemical and Physical Studies of 6-Phosphoguconate Dehydrogenases* (B012629); 24 months; $25,000

OHIO

OHIO STATE UNIVERSITY
Berliner, Lawrence J.; *Spin Label Investigation of Enzyme Conformation* (B016437); 24 months; $24,000
Scott, Roy Albert; *Conformational Studies of Biopolymers* (B008772); 24 months; $43,000

UNIVERSITY OF CINCINNATI
Prairie, Richard L.; *Camphorquinone Reductases* (B015584); 24 months; $44,000

UNIVERSITY OF TOLEDO
Diehn, Bodo; *Biochemistry and Biophysics of Phototaxis* (B018701); 24 months; $39,000

OKLAHOMA

UNIVERSITY OF OKLAHOMA
Chowdhury, Tushar K.; *Active Transport of Electrolytes in Living Systems* (B018726); 7 months; $7,100

OREGON

UNIVERSITY OF OREGON
Bernhard, Sidney A.; *Molecular Structure and Molecular Function of Catalytic Sites in Enzymes* (GB6173X001); 12 months; $50,000
Herbert, Edward; *Enzymatic Addition of the Pcpcpa Group to Transfer RNA* (GB6425X001); 12 months; $46,000
Litt, Michael; *Chemical Studies on the Structure and Function of Transfer RNA* (B016299); 24 months; $38,000
Peticolas, Warner L.; *Inelastic Two and Three Photon Scattering From Biological Polymers* (B013700); 24 months; $65,000
Schellman, John A.; *Optical Rotatory Properties of Biological Molecules* (GB6972X001); 12 months; $36,000

PENNSYLVANIA

ALBERT EINSTEIN MEDICAL CENTER
Boroff, Daniel A.; *Chemistry and Biological Activity of Botulinum Toxin* (B012376); 24 months; $35,000

INSTITUTE FOR CANCER RESEARCH
Cohen, Leonard H.; *Chemical Nature and Biological Significance of Fumarase Isozymes* (B013028); 24 months; $48,000
Kaji, Hideko; *Function and Synthesis of Ribosomal Protein* (B018418); 12 months; $15,000
Perry, Robert P.; *Synthesis, Organization, and Function of Ribonucleic Acid* (B015397); 24 months; $90,000

PENNSYLVANIA STATE UNIVERSITY
Benkovic, Stephen J.; *Folic Acid Mechanisms* (B016789); 24 months; $32,000

TEMPLE UNIVERSITY
Shockman, Gerald D.; *Bacterial Autolytic Enzymes and the Function and Structure of the Bacterial Cell Wall* (B020813); 24 months; $45,000

UNIVERSITY OF PENNSYLVANIA
Cohn, Mildred; *Phosphorylation and Phosphate Transfer Reactions* (B018487); 24 months; $51,000
George, Philip and Witonsky, Robert J.; *Solvation Effects in Ester and Anhydride Reactions* (B024409); 14 months; $22,000
Inoue, Shinya; *Analysis of Fine Structure in Living Cells* (B015658); 24 months; $96,000
Kaji, Akira; *Mechanism of Protein Synthesis* (B018355); 24 months; $60,000
Kallenbach, Neville R.; *Transition Studies of Nucleic Acid Conformation and Interactions* (B011782); 24 months; $42,000

UNIVERSITY OF PITTSBURGH
Hofmann, Klaus; *Binding Studies with the S-Peptide-S-Protein System* (B013647); 12 months; $15,900
Sydiskis, Robert J.; *Compartmentalization of Virus Replication* (GB7183 001); 4 months; $4,000

WOMAN'S MEDICAL COLLEGE OF PENNSYLVANIA
Rosenthal, Spencer; *Biosynthesis of Cell-Walls in Gram Positive Organisms* (B015589); 24 months; $34,000

RHODE ISLAND

GORDON RESEARCH CONFERENCES
Cruickshank, Alexander M.; *Gordon Research Conferences—'Membrane Transport' Meridan, New Hampshire During June, 1970, 'Biopoly-1970 and 'Proteins,' New Hampton, New Hampshire During June 1970* (B023862); 12 months; $4,500
Cohen, Paul S.; *RNA and Protein Synthesis During Unbalanced Growth* (B018077); 24 months; $22,000

TENNESSEE

MEMPHIS STATE UNIVERSITY
Marshall, David L.; *Polymer-Supported Peptide Synthesis* (B012689); 24 months; $29,000

VANDERBILT UNIVERSITY
Lerman, Leonard S.; *Structure Transition of DNA in Concentrated Polymer Solutions* (B015488); 24 months; $130,000
Park, Jane H.; *Relationship of Protein Structure to the Mechanism of Catalysis of 3-Phosphoglyceraldehyde Dehydrogenase* (B012236); 24 months; $34,000

TEXAS

M. D. ANDERSON HOSPITAL
Schlamowitz, Max; *Immunochemical and Immunological Studies of Glycoproteins* (B013650); 12 months; $12,000
Walter, Charles F.; *Kinetics of Open Biological Systems—Control and Oscillations* (B020612); 12 months; $23,800

TEXAS TECH UNIVERSITY
Song, Pill-Soon; *Chemistry of the Electronic Excited States of Photobiological Molecules* (B021266); 24 months; $20,000

UNIVERSITY OF TEXAS AT DALLAS
Lang, Dimitrij; *Quantitative Morphology of Nucleic Acids* (B014094); 24 months; $75,000
Werbin, Harold; *Photochemistry of Electron Transport Quinones* (B018023); 12 months; $15,000

VIRGINIA

UNIVERSITY OF VIRGINIA
Duckworth, Donna H.; *Effects of Bacteriophage on Membrane-Associated Functions of E. Coli* (B016296); 24 months; $30,000
Martin, R. Bruce; *Metal Ion Interactions With Compounds of Biological Interest* (B016365); 12 months; $29,000

WASHINGTON

UNIVERSITY OF WASHINGTON
Fischer, Edmond H.; *Pyridoxal Phosphate in the Structure and Function of Proteins* (B020482); 24 months; $60,000
Groman, Neal B.; *Dual Injection Studies With Specific Phages* (B015595); 24 months; $35,0000
Parson, William W.; *Laser Flash Studies of Photosynthesis and Pophyrin Spectroscopy* (B013495); 24 months; $44,900

WISCONSIN

UNIVERSITY OF WISCONSIN MADISON
Fulton, Robert W.; *Biological Properties of Heterogeneous Components of Plant Viruses* (B013911); 24 months; $37,000
Lardy, Henry A.; *Enzymes Involved in Energy Transfer Reactions* (GB6676 001); 12 months; $60,000
Nomura, Masayasu; *Structure, Function and Assembly of Ribosomes* (B015089); 24 months; $133,500
Tipper, Donald J.; *Polymer Synthesis During Microcyst Formation in 'Myxococcus Xanthus'* (B018597); 24 months; $48,000
Weisblum, Bernard; *Antibiotic Inhibitors of Ribosome Function* (B017108); 24 months; $65,600
Wells, Robert D.; *Defined Deoxyribonucleic Acids Studies* (B008786); 24 months; $65,000

Psychobiology

CALIFORNIA

CHICO STATE COLLEGE
Lathrop, Richard C.; *Modeling Human Decisions* (B019514); 12 months; $6,900

SAN DIEGO STATE COLLEGE
Johnson, Laverne C. and Lubin A.; *Effect of Total Sleep Deprivation and Selective Sleep Stage Deprivation on Autonomic and EEG Activity During Waking and Sleeping* (B014829); 24 months; $35,900

STANFORD RESEARCH INSTITUTE
Kelly, Donald H.; *Spatio-Temporal Modulation Sensitivity Studies in Vision* (B011571); 24 months; $70,600
———; (GB11571001); $3,700

STANFORD UNIVERSITY
Ganz, Leo; *The Topography of Attention at the Striate Cortex* (B15492X); 24 months; $94,100

UNIVERSITY OF CALIFORNIA, Irvine
Lynch, Gary S.; *Neurological and Pharmacological Studies of Recovery After Lesions to Brain Structure Regulating Behavioral Arousal* (B016973); 24 months; $26,000

UNIVERSITY OF CALIFORNIA, Los Angeles
Holman, Eric W.; *Mathematical Studies of Animal Learning* (B13588X); 24 months; $23,900

UNIVERSITY OF CALIFORNIA, Riverside
Sperling, Sally E.; *Reversal of a Discriminative Response* (B018851); 12 months; $19,500

UNIVERSITY OF CALIFORNIA, San Diego
Anderson, Norman H.; *Psychophysical Scaling* (B021028); 24 months; $49,600
Fantino, Edmund J.; *Factors Influencing Choice Behavior* (B013418); 24 months; $45,700

BIOLOGICAL AND MEDICAL SCIENCES RESEARCH PROJECTS

Reynolds, George S.; *Control of Inter-response Times and Latencies by Schedules of Reinforcement* (B014099); 24 months; $54,200

UNIVERSITY OF CALIFORNIA, Santa Barbara
Jacobs, Gerald H.; *Functional Organization of the Visual System* (B015969); 12 months; $16,500
Malecot, Andre; *Measurement of Selected Articulatory Events of Speech and Their Acoustic Correlatives* (B016251); 24 months; $31,600

UNIVERSITY OF CALIFORNIA, Santa Cruz
Berger, Ralph J.; *Rapid Eye Movement Sleep, Oculomotor Behavior, and Binocular Vision* (GB8782); 24 months; $39,600
Le Boeuf, Burney J.; *Ethological Studies of the Northern Elephant Seal* (GB16321X); 24 months; $35,100

UNIVERSITY OF SAN DIEGO
Green, David M.; *Detection with an Undefined Observation Interval* (B014049); 24 months; $73,000

UNIVERSITY OF SOUTHERN CALIFORNIA
Lewis, Donald J.; *Sources of Experimental Amnesia* (B012372); 24 months; $22,000

COLORADO

UNIVERSITY OF COLORADO, Denver Medical Center
Swett, John; *Third Annual Winter Conference on Brain Research* (B017892); 12 months; $3,564

CONNECTICUT

YALE UNIVERSITY
Chi, Carl C.; *Neural Pathways Mediating Aggressive Behavior* (B019687); 24 months; $16,000
Crowder, Robert G.; *Representation of Information in Short-Term Memory* (B015157); 24 months; $34,200
Davidovits, Paul and Egger, M. David; *Development of a Specialized Scanning Laser Microscope* (B018550); 24 months; $38,100
Wagner, Allan R.; *Acquisition-Limiting Factors in Associative Learning* (B014384); 24 months; $67,400

DISTRICT OF COLUMBIA

AMERICAN PSYCHOLOGICAL ASSOCIATION
Cook, Stuart W.; *Ethics of Research with Human Subjects* (B020066); 24 months; $11,500

FLORIDA

FLORIDA STATE UNIVERSITY
Beidler, Lloyd M.; *Physiological Properties of Taste Cells* (GB4068X003); 12 months; $41,000
Elfner, Lloyd F.; *Temporal and Intensive Cue-Function in the Auditory System* (B017769); 12 months; $15,000

UNIVERSITY OF FLORIDA
Dawson, William W.; *Behavioral Evaluation of Feline Vision as a Model for Human Strabismic Amblyopia* (B015382); 12 months; $32,800
———; (GB15382001); $4,600
Isaacson, Robert L.; *Expression of Infant Brain Damage in Behavior* (GB8060 001); 6 months; $17,100
———; *Behavioral and Anatomical Consequences of Early Brain Damage* (B017345); 24 months; $50,300

GEORGIA

EMORY UNIVERSITY
Rumbaugh, Duane M.; *Comparative Behavioral Studies of Primates* (B019146); 6 months; $6,800
———; (B018850); 6 months; $8,800
Winograd, Eugene; *Analysis of Mnemonic Systems Based on Imagery* (B018703); 24 months; $33,300

ILLINOIS

UNIVERSITY OF CHICAGO
McNeill, G. David; *Short Term Memory for Speech* (B015369); 24 months; $50,000
Zerlin, Stanley; *Acoustic Transients—A Study of Waveform Parameters* (B015154); 24 months; $34,700

UNIVERSITY OF ILLINOIS, Urbana
Hake, Harold W.; *Analysis of Recognition Performance in Perceptual Constancy* (B015163X); 24 months; $36,600

INDIANA

INDIANA UNIVERSITY, Bloomington
Guth, Sherman L.; *Psychophysical Studies of Color Vision* (GB8232X001); 12 months; $35,100
Jones, Arthur E. and Heath, Gordon G.; *Comparative Investigation of Primate Visual Characteristics and their Electrophysiological and Anatomical Correlates* (B017961); 12 months; $20,100
Neff, William D.; *Studies of Sensory Discrimination* (GB7134X001); 12 months; $38,800
Suthers, Roderick A.; *Sensory Physiology of Echolocating Animals* (B016731); 24 months; $39,000

PURDUE UNIVERSITY
Sorkin, Robert D.; *Studies of Auditory Signal Processing* (B014039); 24 months; $30,300

UNIVERSITY OF NOTRE DAME
Esch, Harald E.; *Neurophysiological Studies of Communication* (GB6528 001); 8 months; $546

IOWA

UNIVERSITY OF IOWA
Gormezano, Isidore; *Classical Defense and Classical Reward Conditioning* (GB7907X001); 12 months; $31,000

KANSAS

KANSAS STATE UNIVERSITY
Thompson, Charles P.; *Variables Affecting Inter-Trial Responding and Component Response Strength in Compound Conditioning* (B013347); 24 months; $30,000

LOUISIANA

TULANE UNIVERSITY
Mason, William A.; *Social Organization of Primates* (GB8202X001); 12 months; $44,100

MARYLAND

JOHNS HOPKINS UNIVERSITY
Green, Bert F. Jr.; *Computer Studies in Psychometrics: Sensitivity, Stability and Related Problems* (B015839); 22 months; $64,900

UNIVERSITY OF MARYLAND
Teitelbaum, Herman; *Interhemispheric Transfer and Neural Equivalence* (B012555); 24 months; $51,100

MASSACHUSETTS

BOSTON UNIVERSITY
Harrison, J. M.; *Behavioral and Anatomical Investigation of the Auditory System in Mammals* (B020103); 24 months; $40,000

CLARK UNIVERSITY
Stein, Donald G.; *Facilitation of Learning by Pharmacological Manipulation of CNS Development* (B018065); 12 months; $17,000

MASSACHUSETTS INSTITUTE OF TECHNOLOGY
Hein, Alan; *Acquisition of Visual-Motor Capacities* (B017045); 6 months; $22,600
Schiller, P. H.; *Neuropsychological Studies of Visuo-Motor Control* (B017047); 24 months; $47,000

TUFTS UNIVERSITY
McLaughlin, Samuel C. Jr.; *Sequence of Events in Adaptation to Prism* (B008477); 24 months; $52,000

UNIVERSITY OF MASSACHUSETTS
Appley, Mortimer H.; *Adaptation-Level Theory Conference* (B019643); 12 months; $3,000

MICHIGAN

MICHIGAN STATE UNIVERSITY
Denny, M. Ray; *Isolation of Critical Events in Avoidance Behavior* (B018415); 24 months; $39,100

UNIVERSITY OF MICHIGAN
Agranoff, Bernard W.; *Chemical Correlates of Memory* (GB8791); 24 months; $70,400
Birch, David; *Persisting Tendency Effects in Choice* (B015404); 24 months; $44,000
Coombs, Clyde H.; *Preference and Similarity Judgments* (B015653); 24 months; $90,000
Pollack, Irwin; *Perception of Temporal Microstructure of Auditory Signals* (B14036X); 24 months; $98,600

WAYNE STATE UNIVERSITY
Saltz, Eli; *Acquisition of Concept-Meaning* (B022665); 24 months; $35,300
———; *Role of Differentiation in Learning* (B016851); 12 months; $8,000

WESTERN MICHIGAN UNIVERSITY
Hutchinson, Ronald R. and Renfrew, John W.; *Environmental and Physiological Causes of Aggression* (B018413); 24 months; $51,200

MINNESOTA

UNIVERSITY OF MINNESOTA
Darley, John G.; *Coherent Area Support of the Center for Research in Human Learning* (B017590); 24 months; $162,000

MISSOURI

WASHINGTON UNIVERSITY
Finger, Stanley; *Skin Senses* (B012374); 24 months; $15,800

MONTANA

UNIVERSITY OF MONTANA
Jenni, Donald A.; *Patterns of Avian Organization* (B021279); 24 months; $35,000

NEVADA

UNIVERSITY OF NEVADA, Reno Campus
Gardner, R. Allen; *Acquisition of Sign Language by an Infrahauman Primate* (B019739); 12 months; $58,100
———; (GB19739001); $1,000

NEW JERSEY

EDUCATIONAL TESTING SERVICE
Kristof, Walter; *Rotational Problems in Factor Analysis* (B018230); 18 months; $32,000

PRINCETON UNIVERSITY
Gulliksen, Harold; *Mathematical Techniques in Psychology* (GB8023X001); 12 months; $13,100

NEW MEXICO

UNIVERSITY OF NEW MEXICO
Logan, Frank A.; *Conditions of Reinforcement* (B12954X); 12 months; $37,200

NEW YORK

ADELPHI UNIVERSITY
Paul, Coleman; *Transfer-Activated Response Sets in Discrimination Learning* (B021278); 24 months; $30,700

AMERICAN MUSEUM OF NATURAL HISTORY
Topoff, Howard R.; *Comparative Studies on the Biological Basis of Cyclic Colony Functions and Behavior in Two Genera of Doryline Ants* (B014724); 12 months; $20,900

ASTRONOMY

Swihart, Stewart L.; *An Electrophysiological Study of Sensory Discriminatory Mechanisms in Insects* (GB7097 002); 12 months; $16,800

COLUMBIA UNIVERSITY

Terrace, Herbert S.; *Acquisition of Stimulus Control* (GB8111X001); 12 months; $26,800

CORNELL UNIVERSITY

Capranica, Robert R.; *Neural Encoding of Species-Specific Signals* (B018836); 24 months; $59,000

Emlen, Stephen T. and Keeton, William T.; *Studies in Avian Orientation* (B013046); 24 months; $117,700

Morse, Roger A.; *Biology of Reproduction in Honey Bees* (B013840); 12 months; $20,600

CUNY, QUEENS COLLEGE

Bossom, Joseph; *Proprioceptive Mechanisms Within the Central Nervous System* (B023826); 6 months; $15,200

HOFSTRA UNIVERSITY

Buckley, Paul A.; *Adaptation to Environment and Population Dynamics* (GB7842 001); 2 months; $7,000

NEW YORK ZOOLOGICAL SOCIETY

Marler Peter R.; *Comparative Social Communciation* (B016606); 24 months; $64,500

Nottebohm, Fernando, *Natural Functions of Vocal Imitation* (B016609); 24 months; $31,600

Struhsaker, Thomas T.; *Field Study of Primate Behavior and Ecology* (GB15147X); 24 months; $37,700

NEW SCHOOL FOR SOCIAL RESEARCH

Festinger, Leon; *Effects of the Efferent System on Visual Perception* (GB8178 001); $2,700

NEW YORK UNIVERSITY

Schneider, Allen M.; *Effects of Electroconvulsive Shock and Spreading Depression on Memory Storage and Retrieval* (B019642); 24 months; $51,600

STATE UNIVERSITY AT BINGHAMTON

Richardson, Jack; *Abstraction in Paired-Associate Learning* (B017614); 24 months; $37,700

STATE UNIVERSITY AT STONY BROOK

Stamm, John S.; *Cortical Excitability Correlates of Learning and Memory* (B13859X); 24 months; $102,500

Wyers, Everett J.; *Retrograde Amnesia from Caudate Nucleus Stimulation* (B022921); 12 months; $22,400

SYRACUSE UNIVERSITY

Wayner, Matthew J.; *Neural Regulation of Ingestive Behavior* (B18414X); 24 months; $51,600

UNIVERSITY OF ROCHESTER

Dick, Alfred O.; *Visual Information Processing and Form Perception in Brief Visual Presentations* (B017138); 12 months; $10,400

Doty, Robert W.; *Multidimensional Analysis of Single Units in Area Striata of Primates* (GB7522X001); 12 months; $31,800

Ison, James R.; *Behavioral, Pharmacological and Physiological Approaches to Two-Process Learning Theory* (B014814); 24 months; $75,300

Kintz, Robert T.; *Symposium on Spatial and Temporal Interactions in Vision* (B024048); 12 months; $3,400

YESHIVA UNIVERSITY

Anisfeld, Moshe; *Lexical Organization and Memory* (B017856); 9 months; $6,100

——— ; *Coding of Lexical Material* (B021280); 24 months; $35,600

NORTH CAROLINA

DUKE UNIVERSITY

Erickson, Robert P.; *Neural Code for Taste* (B020767); 24 months; $47,400

Everett, John W.; *Neural Mechanisms Controlling the Pituitary Gland* (B021025); 24 months; $53,000

Staddon, J. E. R.; *Temporal Discrimination and Stimulus Control* (B013112); 24 months; $52,600

UNIVERSITY OF NORTH CAROLINA AT CHAPEL HILL

Glassman, Edward; *Molecular Approaches to Behavior* (B018551); 24 months; $70,000

Mueller, Helmut C.; *Study of Predatory Behavior* (B008771); 24 months; $36,900

OHIO

BOWLING GREEN STATE UNIVERSITY

Green, Phillip C.; *Role of Temporal Lobe Structures in Learning* (B017158); 12 months; $32,300

KENT STATE UNIVERSITY

Morin, Robert E.; *Information Processing and Memory* (GB4343X003); 12 months; $23,700

OREGON

PORTLAND STATE UNIVERSITY

Murch, Gerald M.; *The Nature of McCollough Afterimages* (B011817); 24 months; $20,100

UNIVERSITY OF OREGON

Fitzgerald, Robert D.; *Conditioned Emotional Factors in Self-Punitive Behavior* (B011594); 12 months; $15,900

PENNSYLVANIA

SUSQUEHANNA UNIVERSITY

Nagy, Z. Michael; *Thermal Adaptation* (GB8582 001); $3,900

UNIVERSITY CITY SCIENCE CENTER

Epple, Gisela; *Chemical Communication in Primates* (B012660); 24 months; $36,300

UNIVERSITY OF PENNSYLVANIA

Williams, David R.; *Interaction of Reinforcement Contingencies with Other Sources of Behavioral Control* (B014055); 24 months; $45,500

Winston, Harvey D.; *Genetic and Developmental Determinants of Behavior* (GB5284 001); $3,500

Zacks, James L.; *Visual Adaptation* (B016051); 24 months; $41,400

UNIVERSITY OF PITTSBURGH

Colavita, Francis B.; *Electrical Stimulation of Subcortical Auditory Centers* (B017418); 24 months; $27,500

RHODE ISLAND

BROWN UNIVERSITY

Schrier, Allan M.; *Discrimination Learning in Primates* (B018953); 24 months; $40,000

SOUTH DAKOTA

UNIVERSITY OF SOUTH DAKOTA

Kamback, Marvin C.; *Prefrontal Cortical Ablations and Attentional Deficits* (B13100X); 24 months; $50,100

TEXAS

UNIVERSITY OF TEXAS AT AUSTIN

Amsel, Abram; *Development and Transfer of Persistence Effects* (B014990X); 24 months; $93,500

UTAH

UNIVERSITY OF UTAH

Prokasy, William F.; *Models for Classical Conditioning* (B012688); 24 months; $44,200

UTAH STATE UNIVERSITY

Dixon, Keith L.; *Social Organization and Communication in Selected Species* (GB6535 001); $4,100

VIRGINIA

UNIVERSITY OF VIRGINIA

Cohen, David H.; *Neural Mechanisms of Cardiac Conditioning* (B013816); 24 months; $56,800

Homzie, Marvin J.; *Sequence Effects on Performance in Classical and Instrumental Conditioning* (B013341); 24 months; $41,900

Reid, L. Starling; *Information Processing and Human Memory* (B022664); 24 months; $48,900

WISCONSIN

LAWRENCE UNIVERSITY

Baker, Thomas W.; *Component Strength As a Function of the Type of Compound Conditioned Stimuli* (B015061); 24 months; $19,000

UNIVERSITY OF WISCONSIN, Madison

Brogden, W. J.; *Learning and Conditioning* (B017394); 24 months; $43,200

Emlen John T.; *Group-Specific Communication Patterns in Primates* (B015322); 21 months; $40,100

——— *Evolution of Vertebrate Social Systems* (B015304); 24 months; $45,000

Mathematical and Physical Sciences Research Projects

Astronomy

ARIZONA

UNIVERSITY OF ARIZONA

Cocke, W. J. and Disney, M. J.; *Optical Radiation from Pulsars* (P015069); 12 months; $29,100

Gehrels, A. M. J.; *Project Polariscope* (P019108); 12 months; $79,000

Johnson, Harold L.; *Nine-Color Astronomical Photometry* (GP7827X001); 12 months; $96,800

Lynds, Beverly T. and Weymann, Ray J.; *Symposium on Dark Nebulae* (P015266); 12 months; $11,600

Van Biesbroeck, George; *Astrometric Investigations* (P017784); 12 months; $13,800

CALIFORNIA

CALIFORNIA INSTITUTE OF TECHNOLOGY

Cohen, Marshall H.; *High Resolution Studies of Discrete Radio Sources* (P019400); 24 months; $87,300

Lauritsen, Thomas; *Nuclear Astrophysics* (P015911); 12 months; $172,000

Sargent, W. L. W. and Oke, J. B.; *Supernova Search* (P014756); 12 months; $21,100

Stanley, Gordon J.; *Operating Expenses for the 130-foot Telescope at the Owens Valley Radio Observatory* (P015271); 12 months; $190,000

STANFORD UNIVERSITY

Bracewell, Ronald N.; *Microwave Radio Telescope Design and Development* (P013696); 12 months; $34,600

UNIVERSITY OF CALIFORNIA, Berkeley

Anderson, Kinsey A.; *Solar X-Ray Spectra from NRL Spacecraft Data* (P020117); 12 months; $20,000

Field, George B.; *Interstellar Gas Dynamics* (P018476); 24 months; $103,200

Phillips, John G. and Davis, Sumner P.; *Analysis of Molecular Spectra* (P017928); 12 months; $39,800

Spinrad, Hyron; *Metal Abundance in the Galaxy and the Stellar Content of Nearby Extragalactic Systems* (P014588); 12 months; $16,900

UNIVERSITY OF CALIFORNIA, San Diego

REVELLE COLLEGE

Burbridge, E. Margaret and Burbridge, Geoffrey R.; *External Galaxies and Quasi-Stellar Objects* (P020280); 12 months; $61,800

MATHEMATICAL AND PHYSICAL SCIENCES RESEARCH PROJECTS

UNIVERSITY OF CALIFORNIA, Santa Cruz
Vasilevskis, Stanislaus; *Stellar Proper Motions and Parallaxes* (P015780); 12 months; $38,100

COLORADO

UNIVERSITY OF COLORADO
Garstang, Roy H.; *Atomic Oscillator Strengths and Cross Sections for Astronomical Applications* (P020696); 12 months; $44,800
Warwick, James W. and Dulk, George A.; *Radio Astronomy Studies* (P017602); 12 months; $121,500

UNIVERSITY OF DENVER
Riter, J. R., Jr.; *Estimate of the Ground State Lambda-Doubled Frequency in the CH Molecule* (P017756); 24 months; $26,300

CONNECTICUT

WESLEYAN UNIVERSITY
Page, Thornton and Rood, Herbert J.; *Structure and Content of Galaxies* (P013001); 12 months; $36,500
Upgren, Arthur R.; *Measurement of Parallaxes and Proper Motions* (P015779); 24 months; $48,500

DISTRICT OF COLUMBIA

CARNEGIE INSTITUTE OF WASHINGTON
Sandage, Allan R.; *Towards the Calibration of the Hubble Constant* (P014801); 12 months; $14,300

NATIONAL ACADEMY OF SCIENCES, NATIONAL RESEARCH COUNCIL
Clemence, Gerald M.; *Task Order for Astrometric Research in the Southern Hemisphere* (C310007009); 10 months; $32,500
Coleman, John S.; *Task Order for Committee on Radio Frequency Requirements for Scientific Research* (C310023008); 12 months; $26,500

FLORIDA

UNIVERSITY OF FLORIDA
Chen, Kwan-Yu and Gleim, J. K.; *Photoelectric Photometry of Selected Eclipsing Binary Stars* (P011176); 12 months; $16,100
Wood, Frank B.; *Variable Stars in the Southern Hemisphere* (P014755); 24 months; $41,300

UNIVERSITY OF SOUTH FLORIDA
Sofia, Sabatino; *Condensations in Planetary Nubulae, and the Nature of X-Ray Stars* (P017825); 24 months; $30,400
Wilson, Robert E. and Devinney, Edward J. Jr.; *Determination of Limb Darkening Coefficients for Selected Eclipsing Systems* (P018181); 24 months; $48,100

HAWAII

UNIVERSITY OF HAWAII
Jefferies, John T.; *Theoretical Studies in Formation and Analysis of Spectral Lines* (P016378); 12 months; $39,900
Wolff, Richard J.; *Transition Probabilities for Molecules of Astrophysical Interest* (P009672); 12 months; $30,000

ILLINOIS

UNIVERSITY OF CHICAGO
Chandrasekhar, S.; *Relativistic Astrophysics* (P015973); 12 months; $30,000
Morgan, William W.; *Morphological Investigations Based on Spectral Classification* (P19702X); 24 months; $93,500
Van Altena, W. F.; *Star Cluster Program at Yerkes Observatory* (P013771); 24 months; $40,000
Vandervoort, Peter O.; *Theoretical Investigations in Stellar Dynamics* (P017639); 24 months; $30,600

INDIANA

INDIANA UNIVERSITY, Bloomington
Atkinson, Robert D. E.; *Nutations in Obliquity and Longitude* (GP7860 001); 5 months; $8,000
Honeycutt, R. K.; *Observational Studies of Interstellar Extinction* (P011774); 12 months; $26,700
Mutschlecner, Joseph P.; *Solar Abundances for Rare Elements, and Solar Atmosphere Inhomogeneities* (P014963); 12 months; $12,500
Peery, Benjamin F., Jr.; *Spectroscopic Observations and Interpretations Relevant to Nucleosynthesis in Stars* (P017408); 12 months; $16,200

LOUISIANA

LOUISIANA STATE UNIVERSITY, Baton Rouge
Lee, P. and Perry, Clay L.; *Abundance Determinations in Late F-type Stars* (P014357); 24 months; $29,300

MARYLAND

UNIVERSITY OF MARYLAND
Bell, Roger A.; *Atmospheric Parameters of F, G and K Stars* (P013953); 12 months; $33,100
Erickson, William C.; *Astronomical Work at the Clark Lake Radio Observatory* (P019401); 12 months; $154,300
Kerr, Frank J.; *Large-scale Neutral Hydrogen Distribution in Our Galaxy* (P018626); 12 months; $101,500
Kundu, M. R.; *Physics of Supernova Remnants* (P016805); 12 months; $29,100
Moritz, Barry K.; *Dynamic Structure in the Solar Corona* (P021343); 12 months; $18,700
Prinz, Dianne; *Lyman-Alpha Radiation from the Sun* (P021346); 9 months; $14,100
Wentzel, Donat G.; *Small-Scale Structure of the Interstellar Medium* (P015218); 12 months; $47,300
Westerhout, Gart; *Maryland-Green Bank 21-CM Line Survey* (P019574); 12 months; $148,800

MASSACHUSETTS

BRANDEIS UNIVERSITY
Gilbert, Ira H.; *Statistical Stellar Dynamics* (GP9674); 24 months; $26,500

HARVARD UNIVERSITY
D'algarno, Alexander; *Theoretical Atomic and Molecular Physics* (P016681); 12 months; $70,000
Layzer, David; *Theoretical Studies in Cosmogony and Cosmology* (P019302); 12 months; $35,000
Lilley, A. Edward; *Harvard Radio Astronomy Research Programs, 1970–71* (P019717); 12 months; $350,900

MASSACHUSETTS INSTITUTE OF TECHNOLOGY
Iben, Icko, Jr.; *Stellar Physics and Stellar Evolution* (P019777); 12 months; $54,000
Morrison, Philip; *Theoretical Astrophysics and Cosmology* (P011453); 12 months; $15,000

NORTHEAST RADIO OBSERVATORY CORPORATION
Sebring, Paul B.; *Research Programs in Radio Astronomy Using the Haystack Facility* (P016918); 12 months; $198,500

UNIVERSITY OF MASSACHUSETTS
Arny, Thomas T.; *Star Formation in Interstellar Clouds* (P015857); 24 months; $20,800
Dent, William A.; *Measurements of the Time Variability of Extragalactic Radio Sources* (P014690); 24 months; $58,700
Harrison, Edward R.; *Cosmology and Galaxy Formation* (P016359); 24 months; $35,900
Huguenin, G. Richard and Taylor, Joseph H.; *Meter-Wave Radio Astronomy Program* (P017905); 12 months; $92,100

MICHIGAN

UNIVERSITY OF MICHIGAN
Cowley, Charles R.; *Atmospheric Structure and Chemical Abundances in Stars* (P017077); 12 months; $16,300
Miller, Freeman D.; *Operation of the University of Michigan Curtis Memorial Telescope at Cerro Tololo, Chile* (GP7259X001); 12 months; $37,500
Mohler, Orren C.; *Studies of Peculiar Stellar Spectra* (P020342); 12 months; $22,900

MONTANA

MONTANA STATE UNIVERSITY
Caughlan, Georgeanne R.; *Nuclear Reaction Rates and the Fast CNO Cycles in Stars* (GP9673); 24 months; $22,800

NEBRASKA

UNIVERSITY OF NEBRASKA, Lincoln
Refsdal, Sjur; *Theoretical Astrophysics* (P020698); 12 months; $27,600

UNIVERSITY OF VIRGINIA
Blaauw, Adriaan and Tolbert, Charles R.; *Kinematic Properties of Intermediate Age Galactic Population* (P018162); 24 months; $115,400

NEW JERSEY

PRINCETON UNIVERSITY
Schwarzschild, Martin; *Project Stratoscope II* (P006251007); 12 months; $370,000
Spitzer, Lyman, Jr.; *Observational Astronomy* (P014019); 12 months; $42,000

NEW MEXICO

NEW MEXICO INSTITUTE OF MINING and TECHNOLOGY
Colgate, Stirling A. and Moore, Elliott P.; *Digitized Astronomy and Supernova Search* (P018628); 12 months; $100,000

NEW MEXICO STATE UNIVERSITY
Cuffey, James; *Photometry of the Galactic Cluster NGC1893 and the Diffuse Nebulosity IC410* (P016698); 12 months; $9,800

NEW YORK

COLGATE UNIVERSITY
Aveni, Anthony F.; *Properties of Early Type Low Mass Condensations in the Interstellar Medium* (P016379); 18 months; $16,200

COLUMBIA UNIVERSITY
Prendergast, Kevin H. and Wolstar, J.; *Dynamics of Galaxies* (P017973); 12 months; $45,000

CORNELL UNIVERSITY
Gold, Thomas; *Design of a High Quality Reflector Surface for the Arecibo Radio-Radar Telescope* (P017934); 12 months; $6,500
Drake, Frank and Jauncey, David L.; *Very Long Baseline Interferometry* (P0154346); 12 months; $50,000

STATE UNIVERSITY AT ALBANY
Kaftan-Kassim, May A.; *High Resolution Mapping of Planetary Nebulae at Radio Frequencies* (P015910); 24 months; $36,400

STATE UNIVERSITY AT STONY BROOK
Shu, Frank H.; *Galactic Structure* (P013061); 24 months; $25,000

YESHIVA UNIVERSITY
Cameron, A. G. W.; *Stellar Physics* (P014527); 12 months; $41,000

OHIO

CASE WESTERN RESERVE UNIVERSITY
McCuskey, S. W.; *Low Dispersion Stellar Spectroscopy and Galactic Structure* (P021203); 12 months; $112,500

OHIO STATE UNIVERSITY
Czyzak, Stanley J.; *Spectrophotometry of Planetary Nebulae* (P014601); 24 months; $31,400

MATHEMATICS

Keenan, Philip C.; *Comparison of Spectra of Mira Variables of Type CE with Typical Carbon Stars* (P013466); 24 months; $14,800
Kraus, John D.; *Radio Astronomy* (P014904); 12 months; $230,000
Slettebak, Arne; *Standard Stars for Rotational Velocity Determinations* (P019357); 24 months; $107,700
——— *International Astronomical Union Colloquium on Stellar Rotation* (P013465); 12 months; $3,100

UNIVERSITY OF TOLEDO
Delsemme, A. H.; *Physico-Chemical Phenomena Near the Cometary Nucleus* (P017712); 24 months; $39,100

OREGON

UNIVERSITY OF OREGON
Donnelly, R. J.; *Research at Pine Mountain Observatory* (P019021); 12 months; $26,600

PENNSYLVANIA

PENNSYLVANIA STATE UNIVERSITY
Matsushima, Satoshi; *Theoretical Investigation of Stellar Atmospheres* (P018938); 24 months; $91,600
Usher, Peter D.; *Long-Term Optical Variability of Compact Galaxies* (P016852); 12 months; $8,000

SWARTHMORE COLLEGE
Van De Kamp, Peter; *Astrometric Study of Nearby Stars* (P017531); 12 months; $64,800

UNIVERSITY OF PENNSYLVANIA
Blitzstein, William and Koch, Robert H.; *Photoelectric and Photographic Observations of Southern Hemisphere Stars* (P013770); 24 months; $28,000
Koch, Robert H.; *Preparation of a Catalogue of Photometric Solutions of Eclipsing Binaries* (P014430); 12 months; $14,600

UNIVERSITY OF PITTSBURGH
Doschek, George A.; *Soft X-Ray Spectra of Solar Flares* (P018586); 12 months; $18,900
Wagman, Nicholas E.; *Spectroscopic and Astrometric Binaries* (P015265); 12 months; $54,700

TEXAS

RICE UNIVERSITY
Clayton, Donald D.; *Theoretical Astrophysics* (P018335); 24 months; $92,000
Haymes, Robert C.; *Nuclear Gamma Radiation from Solar Flares* (P011322); 12 months; $18,800
Low, Frank J. and Walters, C. King; *Cosmic Background Radiation at 1 MM* (P018678); 12 months; $35,700
Walters, G. K. and Goldwire, H. C. Jr.; *Measurement of the 3.46-CM Emissions from Ground State Hyperfine Transition* (P019079); 12 months; $27,700

UNIVERSITY OF TEXAS AT AUSTIN
De Vaucouleurs, Gerard; *Revision of Reference Catalogue of Bright Galaxies* (P013951); 24 months; $38,500
Douglas, James N.; *Solar System, Galactic and Extragalactic Radio Astronomy* (P018477); 12 months; $100,200
Nacozy, Paul E. and Smith, Harlan J.; *International Astronomical Union Symposium =40 on Planetary Atmospheres* (P017151); 12 months; $4,300
Szebehely, Victor G.; *Long-Term Motion of Pluto and Isoenergetic Numerical Integrations* (P017369); 24 months; $28,700
Warner, Brian; *Spectroscopic Studies of Late-Type Peculiar Stars* (P016406); 12 months; $16,400

UTAH

BRIGHAM YOUNG UNIVERSITY
McNamara, D. H.; *Intermediate-Band and Narrow-Band Photometry of Stars* (P014020); 24 months; $34,600

VIRGINIA

UNIVERSITY OF VIRGINIA
Fredrick, Laurence W.; *Parallaxes and Proper Motions of Selected Stars* (P014393); 12 months; $40,000
Tolbert, Charles R.; *Discrete Neutral Hydrogen Concentrations at High Galactic Latitudes* (P009663); 12 months; $13,400
Wood, H. John; *Magnetic and Related Stars* (GP13596001); 3 months; $7,500

WASHINGTON

BATTELLE MEMORIAL INSTITUTE, Battelle Northwest
Roach, Franklin E.; *Diffuse Galactic Light* (P017561); 12 months; $23,200

UNIVERSITY OF WASHINGTON
Bardeen, James M. and Peters, Philip C.; *General Relativistic Astrophysics* (P015267); 24 months; $70,600
Hodge, Paul W.; *Nearby Galaxies* (P016581); 24 months; $51,000
Wallerstein, George; *Stellar Atmospheres* (P019573); 12 months; $85,600

WISCONSIN

UNIVERSITY OF WISCONSIN, Madison
Forbes, Jack E.; *Stellar Structure and Evolution* (P009666); 12 months; $8,000

Mathematics

ALABAMA

AUBURN UNIVERSITY
Haynsworth, Emilie V.; *Conference on Matrix Theory* (P023220); 12 months; $4,000

ARIZONA

ARIZONA STATE UNIVERSITY
Hassett, Matthew J.; *Recursive Equivalence Types* (P019857); 12 months; $5,600

UNIVERSITY OF ARIZONA
Cohn, Harvey; *Modular Functions and Algebraic Numbers* (P013967); 12 months; $23,000

ARKANSAS

UNIVERSITY OF ARKANSAS
Summers, William H.; *Banach Algebras with Approximate Identities* (P011762); 24 months; $8,900

CALIFORNIA

CALIFORNIA INSTITUTE OF TECHNOLOGY
Dilworth, R. P.; *Combinatorial Algebra* (P013626); 12 months; $44,900
Lagerstrom, Paco A.; *Partial Differential Equations and Continuum Mechanics* (P018471); 12 months; $33,300
Luxemburg, W. A. J.; *Analysis and Nonstandard Analysis* (P023392); 12 months; $54,900
Todd, John; *Computational Problems in Algebra and Analysis* (P023306); 12 months; $42,900

CLAREMONT GRADUATE SCHOOL AND UNIVERSITY
James, Robert C.; *Functional Analysis and Related Algebraic and Topological Concepts* (P020838); 12 months; $25,000

STANFORD UNIVERSITY
Cohen, Paul J.; *Functional Analysis* (P022517); 12 months; $29,000
Deleeuw, Karel; *Harmonic and Abstract Analysis* (P018884); 12 months; $50,200
Gilbarg, David; *Classical and Applied Analysis* (P016115); 24 months; $134,000
Karlin, Samuel and McGregor, James; *Probability Theory and Applications in Analysis* (P017189); 12 months; $34,200
Moses, Lincoln E. and Stein, Charles M.; *Statistical Theory and Methodology* (P015909); 12 months; $54,400
Olkin, Ingram; *Multi-Dimensional Statistical Analysis* (P017172); 12 months; $40,100
Ornstein, Donald S.; *Probability and Ergodic Theory* (P021509); 12 months; $12,000
Samelson, Hans; *Topology and Complex Manifolds* (P022932); 12 months; $36,100

UNIVERSITY OF CALIFORNIA, Berkeley
Chern, Shiing-Shen; *Differential Geometry* (P020096); 12 months; $84,300
Feldman, Jacob; *Functional Analysis, Probability and Quantum Field Theory* (P015735;); 24 months; $70,500
Gale, David; *Mathematical Studies in Operations Research* (P015473); 12 months; $56,600
Kelley, John L. and Moore, Calvin C.; *Functional Analysis* (GP12997001); $6,100
Kobayashi, Shoshichi; *Differential Geometry* (GP8008 001); $7,000
Morrey, Charles B. Jr. and Lewy, Hans; *Analysis, Partial Differential Equations* (P020095); $68,000
Rhodes, John L.; *Applied and Pure Algebra* (P020837); 12 months; $7,400
Sachs, Rainer, K.; *General Relativistic Kinetic Theory* (P021495); 12 months; $17,400
Satake, Ichiro; *Algebraic Groups and Abelian Varieties* (P020436); 24 months; $19,400
Scott, Elizabeth L. and Lecam, Lucien; *Sixth Berkeley Symposium on Mathematical Statistics and Probability* (P022715); 17 months; $23,100
Seidenberg, Abraham and Rosenlicht, Maxwell A.; *Algebraic Geometry* (P020532); 12 months; $50,000
Smale, Stephen; *Global Analysis* (P014519); 24 months; $99,200
Tarski, Alfred; *Metamathematics and its Relation to Algebra, Set Theory, and Foundations of Geometry* (GP6232X005); 12 months; $64,700
Thomas, P. Emery; *Algebraic and Differential Topology* (P022723); 12 months; $40,000

UNIVERSITY OF CALIFORNIA, Davis
Barnette, David W.; *Combinatorial Structure of Convex Polytopes* (P019221); 12 months; $5,200
Chakerian, G. D. and Sallee, G. I.; *Covering and Intersection Properties of Convex Sets* (P019428); 12 months; $14,000

UNIVERSITY OF CALIFORNIA, Irvine
Cannonito, Frank B.; *Conference on Decision Problems in Group Theory* (P014593); 12 months; $6,200
Kalisch, Gerhard K., *Functional Analysis* (P021334); 12 months; $27,500
Kunze, Ray A.; *Infinite Dimensional Representations of Lie Groups* (P021081); 12 months; $12,500

UNIVERSITY OF CALIFORNIA, Los Angeles
Arens, Richard; *Functional Analysis and Applications* (P018127); 12 months; $60,000
Bade, William G. and Curtis, P. C. Jr.; *Embeddings of Banach Spaces in Spaces of Continuous Functions* (GP8383 001); 12 months; $9,100
Beckenbach, E. F.; *Analysis and Geometry* (P022518); 12 months; $31,600
Cantor, David and Gordon, Basil; *Combinatorics, Algebra and Number Theory* (P023113); 12 months; $32,000
Chang, Chen C.; *Foundations of Mathematics* (P022937); 12 months; $32,700
Coddington, Earl A.; *A Functional Analysis Year at UCLA* (P019067); 12 months; $58,500
Curtis, Philip C. Jr. and Bade, William G.; *Extension and Projection Problems for Spaces of Continuous Functions* (P022712); 12 months; $22,800
Ferguson, Thomas, Port, Sidney and Stone, Charles J.; *Dirichlet Processes, Infinite Particle Systems and I. D. Systems* (P017868); 12 months; $42,200
Montague, Richard; *Studies in Metamathe-*

MATHEMATICAL AND PHYSICAL SCIENCES RESEARCH PROJECTS

matics, Linguistics, and Philosophy of Science (S002785*); 24 months; $19,400

Motzkin, Theodore S.; *Convexity, Combinatorics and Inequalities* (P023482); 12 months; $20,000

Paige, Lowell J.; *An Algebra Year at UCLA* (P009661); 12 months; $45,000

Sanchez, David A., Sattinger, David H. and Fattorini, Hector O.; *Differential Equations and Applied Mathematics* (P021086); 12 months; $26,800

Straus, Ernst G.; *Albegra (Group Theory, Division Algebras), Number Theory and Combinatorics* (P023107); 12 months; $42,000

UNIVERSITY OF CALIFORNIA, Riverside

Block, R. E.; *Lie Algebras and Differential Ring Theory* (P023998); 24 months; $20,000

David, F. N.; *Estimation and Non-Parametric Discrimination* (P009671); 12 months; $14,300

Jones, F. Burton; *One-to-One Images of the Line and the Plane* (P018832); 12 months; $7,900

UNIVERSITY OF CALIFORNIA, San Diego

Shenk, Norman A.; *Scattering Theory for Wave Equations* (P023838); 12 months; $5,000

UNIVERSITY OF CALIFORNIA, San Diego, MUIR COLLEGE

Bishop, Errett; *Constructive Analysis, Programming Languages, and Combinatorial Theory* (P020047); 12 months; $15,300

Flanigan, Francis; *A Deformation-Theoretic Approach to Classical Problems in Algebra* (P023104); 12 months; $5,000

UNIVERSITY OF CALIFORNIA, Santa Barbara

Akemann, Charles A.; *C*-Algebras* (P019101); 12 months; $5,400

Berens, Hubert; *Pointwise Approximation of Fourier Series by Summation Methods* (P020125); 12 months; $5,500

Bruckner, Andrew M.; *Theory of Differentiation* (P018968); 12 months; $12,000

Ernest, John A.; *Von Neumann Algebras and Representation Theory* (P021196); 24 months; $13,000

Fan, Ky; *Functional Analysis (Topological Vector Spaces)* (P020632); 12 months; $28,100

UNIVERSITY OF SOUTHERN CALIFORNIA

Bellman, Richard; *New Analytic and Computational Techniques for Ordinary and Partial Differential Equations of Mathematical Physics* (P020423); 12 months; $37,100

Harris, Theodore E. and Pitcher, T. S.; *Stochastic Processes* (P014798); 12 months; $30,800

Martens, Phillip A.; *Representations and Imbeddings of Infinite Dimensional Spaces and Related Homology and Cohomology Functors* (P023103); 12 months; $4,100

Small, Lance W.; *Noncommutative Ring Theory* (P014787); 12 months; $5,200

Troesch, B. Andreas; *Variational Methods in Vibration Problems* (P022587); 12 months; $15,800

COLORADO

COLORADO STATE UNIVERSITY

Demeyer, Frank R.; *Groups with an Irreducible Character of Large Degree* (P020834); 12 months; $6,000

UNIVERSITY OF COLORADO

Gustafson, Karl; *Perturbations of Operators with Applications to Nonlinear Partial Differential Equations* (P015239); 24 months; $15,500

Jordan, Harry F.; *Collaborative Research in Computational Methods for Functional Integrals* (P015507); 12 months; $8,500

McKelvey, Robert W.; *Rocky Mountain Conference on Functional Analysis* (P019348); 12 months; $6,500

Monk, J. Donald; *Foundations of Mathematics* (P019655); 12 months; $21,300

Mycielski, Jan; *Foundation of Mathematics* (P019405); 12 months; $27,600

Sather, Duane P.; *Solutions of Nonlinear Equations in Banach Spaces* (P019712); 24 months; $22,100

Schmidt, Wolfgang; *Diophantine Approximation and Arithmetic of Polynjmials* (P022588); 24 months; $20,000

Ulam, Stanislaw; *Interdisciplinary Research in Mathematical Physics* (P025689); 12 months; $32,000

Williamson, John A.; *The Theory of Random Walks—Branching Processes* (P023111); 12 months; $4,900

CONNECTICUT

FAIRFIELD UNIVERSITY

Shaffer, Dorothy B.; *Distortion Theorems for Equipotential Curves and Surfaces* (P023504); 12 months; $4,000

WESLEYAN UNIVERSITY

Comfort, W. Wistar; *Rings of Continuous Functions* (P018825); 12 months; $26,600

Reid, James D.; *Abelian Groups and Group Rings of Arbitrary Groups* (P019351); 24 months; $12,000

YALE UNIVERSITY

Jacobson, Nathan; *Jordan and Lie Algebras, Finite Groups, and Related Topics* (P019960); 12 months; $66,400

Kakutani, Shizuo; *Mathematical Analysis* (P016392); 24 months; $99,000

Mostow, G. D.; *Theory of Lie Groups* (P012810); 12 months; $39,900

Robinson, Abraham; *Mathematical Logic* (P018728); 12 months; $56,500

DISTRICT OF COLUMBIA

CATHOLIC UNIVERSITY OF AMERICA

Saworotnow, Parfeny P.; *Abstract Analysis* (P011118); 12 months; $8,200

CONFERENCE BOARD OF THE MATHEMATICAL SCIENCES

Botts, Truman, A.; *Supporting Services Relating to the Promoting and Conducting of a Series of Ten Regional Conferences in the Field of Mathematics* (C543000001); $33,800

GEORGETOWN UNIVERSITY

Stokes, Arnold P.; *Non-Linear Oscillations and Control Theory* (P020143); 12 months; $13,000

FLORIDA

FLORIDA STATE UNIVERSITY

Harrold, O. G.; *Topology of Manifolds* (P019964); 12 months; $49,900

Heerema, Nickolas; *Ring Theory* (P019406); 12 months; $32,500

UNIVERSITY OF MIAMI

Connell, E. H.; *Generalized Cohomology Theories and Applications to Geometric Topology* (P018512); 24 months; $20,000

Curtiss, John H.; *Interpolation with Harmonic Polynomials* (P020049); 12 months; $13,700

UNIVERSITY OF SOUTH FLORIDA

Goodman, A. W.; *The Valence of Certain Means* (P018558); 12 months; $8,500

Saff, E. B.; *Best Approximation by Rational Functions* (P019275); 24 months; $12,100

GEORGIA

EMORY UNIVERSITY

Evans, Trevor; *Varieties of Algebras* (P020638); 24 months; $19,500

Neuberger, John W.; *Nonlinear Functional Analysis* (P015259); 24 months; $16,800

UNIVERSITY OF GEORGIA

Cantrell, J. C. and Edwards, C. H. Jr.; *Topology of Manifolds* (P019961); 12 months; $67,000

HAWAII

UNIVERSITY OF HAWAII

Bear, H. S.; *Function Spaces* (P019621); 12 months; $16,900

ILLINOIS

ILLINOIS INSTITUTE OF TECHNOLOGY

Bernstein, Barry and Edelstein, Warren S.; *Mathematical Investigations in the Mechanics of Continua* (P017506); 24 months; $39,500

Kass, Seymour; *Conference on Non-Associative Algebras* (P016881); 10 months; $4,950

NORTHERN ILLINOIS UNIVERSITY

Sons, Linda; *Value Distribution for Gap Power Series* (P016548); 24 months; $11,200

Wunderlich, M. C.; *Number Theoretic Properties of Sieve Generated Sequences* (P023299); 24 months; $16,700

NORTHWESTERN UNIVERSITY

Boas, R. P.; *Mathematical Analysis* (P019526); 12 months; $35,000

Mahowald, Mark; *Chicago Area Topology Year* (P011160); 12 months; $17,000

Sacks, Jerome and Austin, Donald G.; *Probability and Statistics* (P020043); 12 months; $25,000

Shinbrot, Marvin and Friedman, Avner; *Partial Differential Equations and Functional Analysis* (P019816); 12 months; $51,000

Simon, A. B. and Dwass, Meyer; *Analysis and Probability* (P018964); 12 months; $24,100

Williams, Robert F.; *Algebraic Topology* (P019815); 12 months; $43,000

Zelinsky, Daniel, Matlis, E. and Evens, L.; *Algebra and Algebraic Geometry* (P023861); 12 months; $66,500

UNIVERSITY OF CHICAGO

Baily, Walter L. Jr.; *Algebraic Groups and Analytical Methods in Number Theory* (P019350); 12 months; $35,000

Browder, Felix E.; *Partial Differential Equations and Nonlinear Functional Analysis* (P023564); 12 months; $60,000

Kaplansky, Irving; *Groups, Rings and Homological Algebra* (P020864); 12 months; $73,700

Kruskal, William; *Mathematical Statistics and Probability* (P016071); 12 months; $66,900

Lashof, Richard K.; *Differential Topology, Cobordism, Homotopy Theory and Algebraic K–Theory* (P023657); 12 months; $70,000

———; *Chicago Area Topology Year* (P011218); 12 months; $17,000

Maclane, Saunders; *Categorical Algebra and Analytical Dynamics* (P009670); 24 months; $44,700

Reid, William H.; *Applied Mathematics* (P020426); 12 months; $20,800

Zygmund, Antoni; *Mathematical Analysis* (P023563); 12 months; $75,000

UNIVERSITY OF ILLINOIS, Chicago Circle

Gugenheim, Victor; *Chicago Area Topology Year* (P011237); 12 months; $17,000

Gugenheim, Victor K. A. M.; *Topology and Geometry* (P021058); 12 months; $49,200

Soare, Robert I. and Hay, Louise; *Recursive Function Theory* (P019958); 24 months; $28,100

Twersky, Victor; *Applied Mathematics and Its Utilization in Biomedicine* (P021052); 12 months; $30,100

UNIVERSITY OF ILLINOIS, Urbana

Blyth, Colin R.; *Theory and Applications of Mathematical Statistics* (P023835); 12 months; $45,000

Boone, William W.; *Logic and Group Theory* (P023707); 12 months; $55,000

Cairns, Stewart S.; *Geometric and Differential Topology of Manifolds* (P017323); 12 months; $30,000

Day, Mahlon M. and Bartle, Robert G.; *Functional Analysis* (P020431); 12 months; $50,000

MATHEMATICS

Doob, J. L.; *Probability and Related Parts of Analysis* (P024573); 12 months; $33,000
Fosdick, Lloyd D.; *Collaborative Research in Computaitonal Methods for Functional Integrals* (GP9665); 12 months; $9,200
Jerrard, R. P.; *Differential Geometry* (P022929); 12 months; $50,000
Reiner, Irving and Bateman, Paul T.; *Algebra and Number Theory* (P021335); 12 months; $70,000
Suzuki, Michio and Walter, John H.; *Group Theory* (P023486); 12 months; $30,000
Ting, T. W. and Carroll, R. W.; *Differential Equations* (P019590); 12 months; $25,200

INDIANA

INDIANA UNIVERSITY, Bloomington
Azumaya, Goro; *Homological Investigations on Rings and Modules* (P020434); 24 months; $20,500
Brown, Arlen; *Operators in Hilbert Space and Banach Space* (P019955); 12 months; $40,200
Lenard, Andrew; *Mathematical Physics* (P022926); 12 months; $23,900
Lowengrub, Morton; *Mixed Boundary Value Problems in Elasticity* (P020149); 24 months; $22,300
Masani, P. R.; *Some Analytic and Stochastic Problems Suggested by Communication Theory* (P020435); 12 months; $12,300
Springer, George; *Geometric Analysis* (P019694); 12 months; $34,300
———; *Special Year in Functional Analysis dan Its Applications* (P012710); 12 months; $43,200
Thompson, Maynard; *Approximation Theory and Applications* (P020044); 12 months; $15,800

PURDUE UNIVERSITY
Flander, Harley; *Differential Geometry* (P020628); 12 months; $20,000
Hunt, Richard A.; *Convergence of Fourier Series and Related Problems* (P018831); 24 months; $26,700
Lin, Tsau-Young; *Stable Homotopy Theory* (P021062); 12 months; $4,900
Popp, Herbert; *Stratification of Quotient Varieties* (P019692); 24 months; $12,800
Rice, John R.; *Nonlinear Approximation* (P011695); 24 months; $45,000
Roberts, Joel; *Generic Projections of Algebraic Varieties* (P020550); 24 months; $9,600
Schultz, Reinhard; *Geometric Properties of Homotopy Spheres* (P019530); 24 months; $10,100
Silverman, Edward; *Variational Parametric Problems* (P016734); 24 months; $14,600
Studden, William J.; *Optimal Experimental Designs* (P020306); 12 months; $13,000

UNIVERSITY OF NOTRE DAME
Nagano, Tadashi and Matsushima, Yozo; *Lie Groups and Differential Geometry* (P020424); 12 months; $41,900
Stasheff, James D.; *Homotopy Theory of Classifying Spaces and Fibrings* (P023976); 12 months; $6,700
Stoll, Wilhelm and Siu, Yum-Tong; *Theory of Several Complex Variables* (P020139); 12 months; $49,600

IOWA

UNIVERSITY OF IOWA
Fuller, Kent R.; *QF-1 Rings and Double Centralizers of Modules* (P018828); 24 months; $11,100
Hogg, Robert V.; *Regional Conference on Weak Convergence* (P019871); 12 months; $7,920
Kirk, William A.; *Fixed Point Theory in Functional Analysis* (P018045); 24 months; $14,200
Kleinfeld, Erwin; *Algebra and Foundations of Projective Planes* (P023403); 12 months; $5,000
Price, T. M.; *Monotone Decompositions of Manifolds* (P019295); 12 months; $30,000
Waltman, Paul; *Two Point Value Problems* (P019071); 12 months; $7,800

KANSAS

KANSAS STATE UNIVERSITY
Greechie, Richard J.; *Orthomodular Lattices and Empirical Logic* (P011241); 24 months; $15,900
Grillet, Pierre A.; *Real or Rational Semigroups* (P020554); 24 months; $11,000

UNIVERSITY OF KANSAS
Aronszajn, N.; *Differential Problems and Functional Analysis* (P016292); 12 months; $81,300
Breuer, Manfred and Palmer, Theodore W.; *Operator Algebras and Their Applications in Analysis and Topology* (P009668); 12 months; $17,500
McClendon, James F.; *Obstruction Theory in Fiber Spaces* (P014364); 12 months; $5,000

KENTUCKY

UNIVERSITY OF KENTUCKY
Eberhart, C. A.; *Locally Compact Semigroups* (P021079); 12 months; $11,200
Enochs, E. E.; *Isomorphic Polynomial Rings* (P021186); 12 months; $20,000
Mack, John; *Representation of Rings by Sections* (P019549); 12 months; $7,200
Wells, James; *Complex Analysis* (P019533); 12 months; $52,900

LOUISIANA

LOUISIANA STATE UNIVERSITY, Baton Rouge
Anderson, R. D.; *Homeomorphisms on Infinite-Dimensional Spaces* (P014429); 24 months; $71,200
Heinzer, William J.; *Integral Closure and Ramification of Prime Ideals In Infinite Algebraic Field Extensions* (P019618); 12 months; $4,400
Koch, Robert J.; *Topological Semigroups* (P025014); 12 months; $24,900
Pall, Gordon; *Ideals, Modules and Forms* (P020220); 12 months; $10,000
Reid, K. B.; *Combinatorial Structures* (P019691); 24 months; $8,900
Retherford, James R. and Lazar, A. J.; *Functional Analysis* (P020844); 12 months; $11,400
Roselle, David P.; *Enumerative Combinatorial Analysis* (P019207); 24 months; $12,100

TULANE UNIVERSITY
Birtel, Frank T. and Quigley, Frank D.; *Uniform Algebras and Several Complex Variables* (P019825); 12 months; $26,600
Clifford, A. H. and Conrad, Paul F.; *Semigroups and Lattice Ordered Groups* (P019706); 12 months; $25,000
Fuchs, Laszlo; *Abelian Groups* (P019820); 12 months; $10,000
Hofmann, Karl and Mostert, Paul S.; *Topological Algebra* (P019846); 12 months; $26,000
Knill, Ronald J.; *Fixed Point Theory* (P019826); 12 months; $10,000
———; (P015521); 12 months; $15,300
Rosencrans, Steven, Goldstein, J. A. and Conway, Edward D.; *Nonlinear Partial Differential Equations* (P019824); 12 months; $17,500
Topping, David M.; *Lie Ideal Structure of Operator Algebras* (P019822); 12 months; $7,900

MARYLAND

JOHNS HOPKINS UNIVERSITY
Boardman, John M.; *Algebraic Topology* (P019481); 24 months; $14,000
Davis, Stephen H.; *Nonlinear Hydrodynamic Stability of Time-Dependent Flows* (P017562); 24 months; $16,700
Gastwirth, Joseph L.; *Methods of Statistical Inference* (P020527); 12 months; $15,100
Gibbs, John A.; *Conjugacy Classes of Unipotent Elements in Groups of Lie Type* (P019762); 12 months; $3,000
Igusa, Jun-Ichi; *Moduli and Automorphic Functions* (P016237); 24 months; $50,800
Kulkarni, R. S.; *Curvature and Topology of Riemannian Manifolds* (P020289); 24 months; $9,600
Ono, Takashi; *Arithmetic Properties of Algebraic Varieties* (P014397); 12 months; $18,000
Truesdell, Clifford A.; *Rational Thermomechanics of Materials* (P012184); 24 months; $88,800

UNIVERSITY OF MARYLAND
Adams, William W.; *Diophantine Approximations and Transcendental Numbers* (P020215); 24 months; $14,600
Auslander, Joseph; *Differential Equations and Dynamical Systems* (P019869); 12 months; $24,100
Berg, Kenneth R.; *Ergodic Theory and Dynamical Systems* (P020294); 24 months; $11,900
Brace, John W.; *Topological Vector Spaces* (P019862); 12 months; $10,000
Goldstein, L. J.; *Class Numbers of Algebraic Number Fields* (P020538); 24 months; $12,000
Hubbard, B. E.; *Fluid Dynamics and Applied Mathematics* (P018064); 12 months; $48,000
Hummel, James A. and Zedek, Michael; *Geometric Function Theory* (P017152); 12 months; $25,100
Jones, G. S.; *Topological Aspects of Generalized Dynamical Processes* (P017601); 12 months; $38,500
Kellogg, R. B.; *Symposium on the Numerical Solution of Partial Differential Equations* (P015238); 12 months; $13,200
Kellogg, R. Bruce; *Approximate Methods for Solving Functional Equations* (P020555); 12 months; $18,700
Olver, F. W. J.; *Asymptotic Approximations* (P020529); 24 months; $28,000
Schaefer, H. H.; *Functional Analysis* (P020543); 12 months; $20,800

UNIVERSITY OF MARYLAND, Baltimore County Campus
Gross, Fred; *Factorization of Meromorphic Functions and Functional Equations* (P021497); 12 months; $7,700

MASSACHUSETTS

BOSTON COLLEGE
Milnor, Tilla K.; *Differential Geomtry of Surfaces* (P016986); 24 months; $14,500

BRANDEIS UNIVERSITY
Brown, Edgar H. Jr.; *Topology of Manifolds* (P021510); 12 months; $45,700
Matsusaka, Teruhisa; *Global Algebraic Deformations of Polarized Algebraic Varieties and Related Topics in Algebra and Algebraic Geometry* (P023119); 12 months; $59,900
Rossi, Hugo; *Global Analysis* (P023117); 12 months; $55,400

CLARK UNIVERSITY
Kennison, John F.; *Universal Fibrations—A Categorical and Topological Approach* (P014962); 24 months; $14,300
Rubinstein, Zalman; *Theory of Polynomials* (P011881); 24 months; $21,300

HARVARD UNIVERSITY
Bott, Raoul; *Algebraic Geometry* (P019769); 12 months; $10,000
Carrier, George F.; *Mathematical Analysis of Non-Gray Gases* (P017383); 12 months; $30,000
Zariski, Oscar; *Algebraic Geometry* (P009667); 24 months; $32,300

MATHEMATICAL AND PHYSICAL SCIENCES RESEARCH PROJECTS

MASSACHUSETTS INSTITUTE OF TECHNOLOGY

Kleinman, Steven and Mattuck, A. P.; (P022931); 12 months; $26,400

Lin, Chia-Chiao; *Applied Mathematics and Mathematical Physics* (P022720); 12 months; $70,200

Milnor, John W. and Kostant, Bertram; *Topology and Number Theory* (P023305); 12 months; $38,700

Peterson, F. P.; *Algebraic Topology* (P023122); 12 months; $39,100

Rogers, Hartley Jr. and Schafer, Richard D.; *Logic and Algebra* (P021321); 12 months; $29,100

Segal, Irving E.; *Functional Analysis* (P022822); 12 months; $84,900

NORTHEASTERN UNIVERSITY

Bonic, Robert A.; *Global Analysis* (P019460); 12 months; $20,000

Cenkl, Bohumil; *Global Solutions of Linear Partial Differential Equations* (P016354); 24 months; $20,700

Kopell, Nancy; *Differential and Algebraic Topology* (P020851); 24 months; $11,300

Stolzenberg, Gabriel; *Constructive Mathematics* (P019767); 12 months; $18,400

TUFTS UNIVERSITY

Reynolds, William F.; *Modular Representations of Finite Groups* (P019661); 24 months; $13,800

UNIVERSITY OF MASSACHUSETTS

Becker, James C. and Sicks, Jon L.; *Differential Topology, Homotopy Theory* (P024498); 12 months; $10,800

Broshi, A. M.; *Finite Groups* (P021185); 12 months; $8,700

Chen, Yu Why; *Solutions of the Wave Equations with Data on the Characteristic Boundaries of Exterior Domains* (P015260); 12 months; $14,400

Fischer, Hans R. and Mann, Larry N.; *Topological Transformation Groups, Topological Dynamics, and Infinite Dimensional Differentiable Manifolds* (P024496); 12 months; $39,000

Hayes, David R.; *Adelic Analysis in Additive Number Theory* (P021085); 12 months; $7,000

Hurt, Norman E. and Stone, Marshall H.; *Mathematical Foundations of Quantum Field Theory* (P020856); 12 months; $12,800

Ku, Hsu-Tung and Hertz, Douglas N.; *Differentiable Transformation Groups and Differentiable Manifolds* (P019854); 12 months; $15,000

Schweizer, Berthold; *Multiplications on the Space of Probability Distribution Functions* (P022515); 24 months; $14,300

MICHIGAN

MICHIGAN STATE UNIVERSITY

Anderson, Glen D.; *Quasiconformal Mappings in Two- and Three-Dimensional Euclidean Space* (P013022); 24 months; $9,500

Blair, David E. and Ludden, Gerald D.; *Regional Conference on Differential Geometry (Kaehler Manifolds)* (P018827); 12 months; $6,000

Deskins, Wilbur E.; *Finite Groups* (P012404); 12 months; $15,900

Katz, Leo; *Mathematical Statistics and Probability Theory* (P023480); 12 months; $46,000

Kinney, John R.; *Fractional Dimensional Properties of Sets* (P020629); 12 months; $10,000

Kwun, Kyung W.; *Manifolds and Homotopy Theory* (P019462); 12 months; $33,700

Palmer, Edgar M.; *Graphical Enumeration and Its Applications* (P019292): 24 months; $11,200

Seiden, Esther; *Mathematical Theory of Design of Experiments and Their Applications* (P020537); 12 months; $14,900

Taylor, Gerald D.; *Approximation with Side Conditions* (P012088); 24 months; $11,300

OAKLAND UNIVERSITY

Bragg, Louis R. and Dettman, John W.; *Related Partial Differential Equations and Associated Topics in Analysis* (P017867); 12 months; $16,100

De Vore, Ronald A.; *Approximation by Positive Linear Operators* (P019620); 24 months; $11,900

UNIVERSITY OF MICHIGAN

Bell, Charles B., Woodroofe, Michael and Starr, Norman; *Statistical Inference and Control* (P018820); 12 months; $31,300

Brown, Morton; *Geometric Topology and Differential Geometry* (P021511); 12 months; $40,000

Coburn, Nathaniel; *Magnetic Non-Equilibrium Fluid Dynamics* (P019009); 12 months; $15,500

Dolph, Charles L.; *Complex Singularities of the Scattering Operator in Scalar and Electromagnetic Problems* (P019461); 12 months; $16,700

Gehring, Frederick W.; *Complex Analysis* (P019148); 12 months; $102,700

Hill, Bruce M. and Ericson, William A.; *Statistical Inference* (P018727); 12 months; $26,000

Lewis, Donald J.; *Number Theory* (P023435); 12 months; $57,400

Lyndon, Roger C.; *Algebra and Logic* (P020298); 12 months; $57,000

Pearcy, Carl M. Jr.; *Lattices of Invariant Subspaces and Non-Selfadjoint Operator Algebras* (P018903); 12 months; $18,500

Piranian, George; *Real and Complex Analysis* (P019259); 12 months; $40,800

Raymond, Frank; *Algebraic and Differential Topology* (P020038); 12 months; $29,000

Shields, Allen; *Conference in Function Algebra and Rational Approximation* (P013166); 12 months; $6,500

Ullman, Joseph L.; *Analytical Theory of Polynomials* (P019008); 12 months; $15,300

Wendel, James G.; *Stochastic Processes and Extremal Limit Theorems* (P019290); 12 months; $32,600

WESTERN MICHIGAN UNIVERSITY

Hsieh, Po-Fang and Stoddart, Arthur; *Differential Equations and Control Theory* (P014595); 24 months; $24,600

———; *Conference on the Analytic Theory of Differential Equations* (P016880); 12 months; $2,800

MINNESOTA

UNIVERSITY OF MINNESOTA

Chacon, Rafael V. and Orey, Steven; *Probability and Ergodic Theory* (P021188); $45,500

Das Gupta, Somesh; *Multivariate Analysis* (P021074); 12 months; $11,500

———; *Multivariate Analysis* (GP9593001); $1,400

Eagon, John A.; *Groups, Rings, and Algebras* (P020227); 12 months; $25,300

Fabes, E. B. and Riviere, N. M.; *Mathematical Analysis* (P015832); 24 months; $37,300

Green, Leon W.; *Dynamics and Topology* (P020871); 12 months; $101,500

Jeroslow, Robert G.; *Self-Reference in Logic* (P021067); 24 months; $9,900

Kallianpur, Gopinath and Striebel, Charlotte; *Stochastic Processes and the Theory of Stochastic Differential Equations in Estimation and Control* (P020429); 15 months; $38,500

McCarthy, C. A. and Gil De Lamadrid, J.; *Abstract Analysis* (P021330); 12 months; $56,000

Richter, Wayne H.; *Inductive Definitions in Recursion Theory* (P020846); 24 months; $13,200

Sagle, Arthur A.; *Homogeneous Spaces and Nonassociative Algebras* (P013328); 24 months; $20,500

MISSOURI

UNIVERSITY OF MISSOURI, Columbia

Beem, John K.; *Extremals in Indefinite Metric Spaces* (P011875); 24 months; $10,000

Taylor, Donald C.; *Banach Algebras* (P015736); 24 months; $10,500

UNIVERSITY OF MISSOURI, Rolla

Grimm, Louis J.; *Initial and Boundary Value Problems for Differential Equations with Deviating Arguments* (P020194); 12 months; $6,700

UNIVERSITY OF MISSOURI, St. Louis

Haimo, Deborah T.; *Generalized Heat Equations and Related Transforms* (P020536); 12 months; $10,800

Irwin, Ronald and Peterson, Gerald E.; *Statistical Methods in Tauberian Theory* (P017374); 12 months; $12,800

Irwin, Ronald L.; *Regional Conference on the Theory of Functions of Several Complex Variables* (P020133); 12 months; $8,420

WASHINGTON UNIVERSITY

Haimo, Franklin, Ford, Charles and Yohe, Cleon; *Groups and Rings* (P020291); 12 months; $20,000

Hirschman, Isidore I. Jr. and Nussbaum, A. Edward; *Mathematical Analysis* (P019588); 12 months; $30,000

Jenkins, James A.; *Geometric Function Theory* (GP9607 001); $4,800

NEW HAMPSHIRE

DARTMOUTH COLLEGE

Crowell, Richard H.; *Topology of Knots and H-Spaces* (P016820); 12 months; $32,000

Gross, Kenneth I.; *Regional Conference on Unitary Group Representations* (P016943); 12 months; $6,700

Snell, J. L.; *Theory of Stochastic Processes* (P015737); 24 months; $75,000

UNIVERSITY OF NEW HAMPSHIRE

Johnson, Richard E.; *Generalizations of the Faith-Utumi Theorem* (P021332); 12 months; $10,000

Nordgren, Eric A.; *Operator Theory* (P014784); 24 months; $20,000

NEW JERSEY

INSTITUTE FOR ADVANCED STUDY

Harish-Chandra, Montgomery, Deane and Whitney, Hassler; *Algebra, Analysis, and Topology* (GP7952X001); 12 months; $233,100

PRINCETON UNIVERSITY

Browder, William and Steenrod, Norman E.; *Manifolds and Algebraic Topology* (P021500); 12 months; $90,900

Gunning, Robert C.; *Topological Methods in Complex Analysis* (P020299); 12 months; $38,400

Kochen, Simon; *Mathematical Logic* (P022794); 12 months; $21,300

Nelson, Edward; *Functional Analysis* (P022592); 12 months; $36,000

Shimura, Goro and Washnitzer, Gerard; *Analytic and Algebraic Methods in Geometry and Number Theory* (P023855); 12 months; $70,000

RUTGERS UNIVERSITY

Cohen, Arthur; *Admissible and Minimax Estimation* (P013709); 24 months; $17,800

Dekker, Jacob C. E.; *Mathematical Logic* (P020134); 12 months; $40,000

Gorenstein, Daniel; *Finite Group Theory* (P016640); 24 months; $20,000

Gundy, Richard F. and Davis, Burgess J.; *Martingale Theory* (P019222); 24 months; $24,500

Kosinski, Antoni and Bredon, Glen; *Algebraic and Differential Topology* (P020870); 12 months; $35,000

MATHEMATICS

Landweber, Peter S.; *Cobordism Theory* (P021064); 12 months; $7,000
Levin, Frank and Sims, Charles C.; *Group Theory* (P020836); 12 months; $14,000
Mason, William K.; *Complex Planar Continua and Spaces of Homeomorphisms* (P020861); 12 months; $5,000
Muckenhoupt, Benjamin; *Orthogonal Series and Singular Integrals* (P020147); 24 months; $15,700
O'Nan, Michael; *Finite Group Theory* (P021071); 12 months; $5,000
Osofsky, Barbara L.; *Quotient Rings of Rings* (P019856); 12 months; $7,000
Petryshyn, Walter V.; *Nonlinear Functional Analysis* (P020228); 12 months; $15,000
Sackrowitz, Harold; *Decision Theoretic and Empirical Bayes Approaches to Inference for Monotone Parameter Sequences* (GP9664); 24 months; $9,700
Sibner, Robert J.; *Complex and Global Analysis* (P020835); 12 months; $6,400
Taft, Earl J.; *Invariant Subobjects of Groups and Algebras* (P019813); 12 months; $10,000
Tierney, Myles; *Applied Category Theory* (P020542); 12 months; $7,000
Vasconcelos, Wolmer V.; *Flat Modules Over Commutative Rings* (P019995); 12 months; $6,000
Wallach, Nolan; *Differential Geometry and Representation Theory* (P020631); 12 months; $5,900
Walsh, Bertram; *Axiomatic Theory Elliptic Differential Equations* (P021182); 12 months; $7,600

NEW MEXICO

New Mexico State University
Walker, Elbert A.; *Abelian Groups* (P021448); 12 months; $35,900
Williams, Francis D.; *Theory and Applications of Higher Homotopy Commutativity* (P017757); 24 months; $10,600

University of New Mexico
Koopmans, Lambert H.; *Theory of Probability* (P022825); 12 months; $13,300
Zacks, Shelemyahu; *Statistical Methods and the Design of Experiments* (P022721); 12 months; $14,500

NEW YORK

College at Buffalo
Barback, Joseph; *Regressive Isols* (P015309); 12 months; $5,800

Columbia University
Bers, Lipman and Lorch, Edgar R.; *Mathematical Analysis* (P021325); 12 months; $67,900
Gallagher, Patrick, Kolchin, E. R. and Bass, Hyman; *K Theory, Prime Nunmber Theory, and Differential Fields* (P021341); 12 months; $62,800
Levene, Howard; *Probability Theory and Mathematical Statistics* (P020867); 12 months; $24,200
Levene, Howard and Robbins, H. E.; *Probability Theory and Mathematical Statistics* (P020859); 12 months; $50,200
Littauer, Sebastian B.; *Statistical Theory of Extreme Values and its Applications* (P024439); 12 months; $11,900

Cornell University
Bramble, James H.; *Numerical Analysis and Differential Equations* (P022936); 12 months; $19,900
Dennis, J. E. Jr.; *Solutions of Nonlinear Equations by Newton-Like Methods* (J000844*); 12 months; $7,150
Fuchs, W. H. J. and Earle, C. J.; *Mathematical Analysis* (P022820); 12 months; $70,100
Hilton, Peter and Olum, Paul; *Topology* (P016862); 12 months; $71,800
Ito, K., Kesten, H. and Spitzer, F.; *Mathematical Probability* (P019658); 12 months; $50,200
Nerode, Anil; *Mathematical Logic* (P022719); 12 months; $39,100
Payne, Lawrence E.; *Differential Equations and Their Applications* (P022519); 12 months; $58,000
Weiss, Lionel I.; *Applied Stochastic Processes and Statistical Inference* (P021184); 12 months; $36,500

CUNY Central System Office
Auslander, Louis; *Analysis in the Large* (P021057); 12 months; $39,000
Dyer, Eldon and Heller, Alex; *Homological Algebra* (P021077); 12 months; $41,500

CUNY City College
Dressler, Robert; *Applied Partial Differential Equations* (P024619); 12 months; $20,000

CUNY, Lehman College
Lazorov, Connor; *Topics in K–Theory and Corbordism* (P012639); 24 months; $9,600

CUNY Queens College
Paulson, Edward; *Sequential Procedures for Interval Estimation and Multiple Decision Problems* (P019223); 24 months; $18,400

Fordham University
Dillemuth, Frederick J.; *Chemistry Research Instruments* (GP9593 001); $1,400

New York University
Blumenthal, Saul; *Estimation of the Largest Parameter and Sample Spacings* (P023171); 12 months; $10,000
Courant, Richard; *Research in Mathematical Physics* (P014764); 12 months; $40,600
Keller, Joseph B. and Ludwig, Donald A.; *Mathematical Problems of Electromagnetic Theory and Other Applied Fields* (P018682); 12 months; $98,600
Kervaire, M.; *Algebraic and Differential Topology* (P020307); 12 months; $20,100
Magnus, Wilhelm; *Group Theory* (P023565); 12 months; $16,000
Moser, J.; *Differential Equations and Continuum Mechanics* (P019617); 12 months; $130,300

Polytechnic Institute of Brooklyn
Guggenheimer, Heinrich; *Geometry & Differential Equations* (P019133); 12 months; $10,000

Rensselaer Polytechnic Institute
Habetler, George J.; *Numerical Analysis* (P016293); 12 months; $26,100
Handelman, George H.; *Differential Equations of Applied Mathematics* (P022576); 12 months; $55,800
Lemke, Carlton E. and Ecker, Joseph G.; *Mathematical Programming* (P015031); 24 months; $44,400

State University at Albany
Katz, Melvin; *Probability Theory* (P019225); 12 months; $15,100
Mac Gregor, Thomas H.; *Analytic Function Theory* (P019709); 12 months; $17,400
O'Neil, Richard; *Operators in Real and Functional Analysis* (P019185); 12 months; $13,800

State University at Binghamton
McAuley, Louis F.; *Topology* (P019589); 12 months; $24,900

State University at Buffalo
Balslev, Erik and Piech, M. Ann; *Differential Equations and Hilbert Space Theory* (P019651); 12 months; $12,600
Isbell, John R.; *Functioral Semantics* (P016951); 24 months; $20,000
Mitchell, Josephine M.; *Complex Analysis on Symmetric Domains* (P019656); 12 months; $15,000
Myhill, John and Vesley, Richard; *Mathematical Logic* (P021189); 12 months; $25,000
Rosenblatt-Roth, Millu; *Markov Chains and Fourier Transforms* (P019711); 12 months; $12,300
Schoenfeld, Lowell; *Number Theory* (P020533); 24 months; $22,500
Stroud, Arthur H.; *Topics in Approximate Integration* (P019613); 12 months; $8,100

State University at Stony Brook
Ax, James; *Number Theory and Logic* (P019569); 12 months; $55,000
Charlap, Leonard S.; *Topology of Flat Manifolds* (P019571); 12 months; $30,000
Kra, Irwin and Farkas, Hershel M.; *Riemann Surfaces—Kleinian Groups and Theta Functions* (P019572); 12 months; $21,200
Pincus, Joel D. and Douglas, Ronald G.; *Mathematical Analysis* (P019587); 12 months; $38,000
Rapaport, E. S.; *Group Theory* (P019752); 12 months; $9,000
Simons, James; *Differential Geometry* (P019585); 12 months; $85,000
Zemanian, Armen H.; *Generalized Integral Transformations, Hilbert Ports, and Linear Systems* (P018060); 12 months; $18,200

Syracuse University
Barth, Karl and Schneider, Walter J.; *Function Theory and Two-Dimensional Potential Theory* (P020703); 12 months; $11,900
Graver, Jack; *Combinatorial Mathematics* (P019404); 24 months; $13,100
Jurkat, W. B.; *Fourier Analysis, Combinatorial Analysis, and Prime Number Theory* (P019653); 12 months; $21,500
Richert, Hans-Egon; *Sieve Methods in Prime Number Theory* (P024066); 24 months; $20,000

University of Rochester
Alling, Norman L.; *Klein Surface and Real Algebraic Function Fields* (P019760); 12 months; $12,000
Blum, Peter; *Differential Varieties* (P019764); 24 months; $10,000
Harper, John R.; *Steenrod Algebras and Homotopy* (P017722); 12 months; $5,000
Kemperman, J. H. B.; *Probability and Analysis* (P021190); 12 months; $28,500
May, Warren L.; *Groups and Group Algebras* (P019756); 24 months; $10,000
Nachbin, Leopoldo; *Spaces of Holomorphic Mappings and Weighted Approximation* (P022713); 12 months; $41,600
Stone, Arthur H.; *General Topology* (P015262); 24 months; $19,900
Stone, Dorothy; *Abstract Ergodic Theory and Abstract Probability Theory* (P019483); 12 months; $12,000

Yeshiva University
Berger, Melvyn S.; *Nonlinear Analysis and Applied Mathematics* (P016578); 12 months; $18,800
Coburn, Lewis A.; *C*-Algebras, Topological Groups and Index Theorems* (P019963); 12 months; $9,700
Desapio, Rodolfo; *Differential Topology* (P016682); 24 months; $20,000
Koranyi, Adam; *Function Theory on Symmetric Spaces* (P022717); 12 months; $11,800
Schechter, Martin; *Mathematical Theory of Linear Operators Pertaining to Quantum Mechanical Systems of Particles* (P019757); 12 months; $23,800

NORTH CAROLINA

Duke University
Carlitz, Leonard; *Algebra and Number Theory* (P017031); 24 months; $21,300
Shoenfield, Joseph R.; *Foundations of Mathematics* (P016819); 12 months; $16,100
Smith, David A.; *Algebra and Combinatorial Theory* (P020092); 24 months; $12,000

University of North Carolina, Chapel Hill
Bose, R. C. and Wegman, E. J.; *Statistical and Mathematical Theory of Problems of Communications and Automata* (P023520); 12 months; $25,000

MATHEMATICAL AND PHYSICAL SCIENCES RESEARCH PROJECTS

Eisenman, Donald A.; *Holomorphic Self-Mappings of Analytic Spaces* (P019272); 12 months; $5,000

Petersen, Karl E.; *Mixing in Topological Dynamics* (P019758); 12 months; $4,900

OHIO

CASE WESTERN RESERVE UNIVERSITY

Hastings, Stuart P.; *Some Boundary Value Problems for Ordinary Differential Equations Which Arise in Fluid Mechanics* (P021181); 12 months; $5,800

Lazer, Alan C. and Hajek, Otomar; *Differential Equations and Dynamical Systems* (P022689); 12 months; $39,900

Wu, Ta-Sun and Lee, D. H.; *Topological Semigroups, Topological Dynamics and Topological Groups* (P021180); 24 months; $24,000

OHIO STATE UNIVERSITY

Brown, Harold; *Mathematical Crystallography of 4-Dimensional Space* (P016793); 24 months; $13,800

Janko, Z.; *Finite Group Theory* (P016949); 24 months; $20,000

Krengel, Ulrich; *Convergence Problems in Ergodic Theory and Probability Theory* (P014594); 12 months; $14,600

Miller, Leonhard; *The Hasse-Witt Matrix of an Algebraic Curve* (P020833); 24 months; $11,000

OKLAHOMA

UNIVERSITY OF OKLAHOMA

Levy, Gene; *Regional Conference on Recent Developments in the Theory of Rings* (P020135); 12 months; $7,800

OREGON

OREGON STATE UNIVERSITY

Pierce, Donald A.; *Regional Conference on Bayesian Statistical Inference* (P019864); 12 months; $8,600

Smith, J. Wolfgang; *Submersion Mappings on Manifolds* (P019861); 12 months; $8,500

Stalley, Robert D.; *Addition Theorems for Density Spaces* (P019349); 12 months; $13,400

UNIVERSITY OF OREGON

Andrews, Fred C. and Truax, Donald R.; *Probability and Statistics* (P023120); 12 months; $30,000

Civin, Paul, Ross, Kenneth A. and Yood, Bertram; *Normed Algebras and Abstract Harmonic Analysis* (P020226); 12 months; $23,900

Curtis, Charles W. and Harrison, David K.; *Problems on Groups and Rings* (P020308); 12 months; $24,700

Kuga, Michio and Leahy, John V.; *Problems in Geometry and Number Theory Arising from the Theory of Modular Functions* (P021178); 12 months; $15,000

Loeb, Henry L.; *Non-Linear Approximation Theory* (P018609); 12 months; $21,200

Wright, Charles R. B.; *Solvable Group Theory* (P019205); 12 months; $5,000

PENNSYLVANIA

CARNEGIE-MELLON UNIVERSITY

Degroot, Morris H.; *Statistical Analysis, Inference, and Programming* (P022595); 12 months; $17,000

MacCamy, Richard C.; *Diefferential and integral Equations in Continuum Mechanics* (P019208); 12 months; $9,100

Mizel, Victor J.; *Analysis and Continuum Mechanics* (P024339); 12 months; $16,000

Nehari, Zeev; *Differential Equations* (P023112); 12 months; $15,200

Pederson, Roger N.; *Schlicht Functions and Differential Equations* (P021512); 12 months; $22,300

Rao, Malempati M.; *Stochastic Equations, Generalized Random Fields, and Inference* (P015632); 24 months; $27,200

Schaffer, Juan J.; *Functional Analysis, Transition Systems and Differential Equations* (P019126); 12 months; $17,500

DREXEL UNIVERSITY

Duris, Charles S.; *Approximate Solutions of Overdetermined Linear Equations* (P015405); 12 months; $12,500

Trench, William F.; *Stability of Minimum Variance Polynomial Smoothing* (P023217); 12 months; $7,600

LEHIGH UNIVERSITY

Rivlin, R. S.; *Non-Linear Continuum Physics* (P015803); 24 months; $45,000

PENNSYLVANIA STATE UNIVERSITY

Chowla, Sarvadaman; *Number Theory and Allied Topics* (P009660); 12 months; $70,000

Deutsch, Frank; *Approximation in Normed Linear Spaces* (P013874); 24 months; $17,600

Mitchell, Josephine M.; *Classical Cartan Domains in the Space of N Complex Variables* (P011167); 12 months; $15,700

Petrich, Mario; *Dense Extensions of Reductive Completely O-Simple Semigroups and Applications* (P019069); 24 months; $14,100

SOCIETY FOR INDUSTRIAL AND APPLIED MATHEMATICS

Householder, A. S.; *Symposia on Numerical Linear Algebra and Bio-Mathematics* (P023652); 12 months; $7,600

SWARTHMORE COLLEGE

England, James W.; *Stable Manifold Theorem for Almost Periodic Minimal Sets* (P019103); 12 months; $5,300

TEMPLE UNIVERSITY

Grosswald, Emil; *Number Theory* (P023170); 12 months; $10,000

UNIVERSITY OF PENNSYLVANIA

Effros, Edward G.; *Functional Analysis and Boundary Value Problems* (P019860); 12 months; $18,600

Fell, J. M. G.; *Infinite-Dimensional Group Representations* (P020644); 12 months; $22,400

Gerstenhaber, Murray; *Modern Geometric Methods in Algebra* (P020138); 12 months; $42,000

Goldman, Oscar; *Ring Theory* (P025329); 12 months; $14,000

Kadison, Richard V.; *Functional Analysis and Mathematical Physics* (P020041); 12 months; $32,600

Nijenhuis, Albert; *Differential Structures* (P019693); 12 months; $51,600

Sakai, Shoichiro; *Operator Algebra* (P019845); 12 months; $21,000

Wallace, Andrew H.; *Singularities of Varieties* (P020094); 12 months; $19,800

Yang, Chung-Tao; *Transformation Groups* (P020042); 12 months; $10,000

UNIVERSITY OF PITTSBURGH

Thompson, Alan H.; *Regional Conference on Differential Topology in Relativity Theory* (P018904); 12 months; $8,000

RHODE ISLAND

AMERICAN MATHEMATICAL SOCIETY

Walker, Gordon L.; *Summer Institute on Algebraic Topology* (P019276); 18 months; $70,000

——— ; *Symposium on 'Computers in Algebra and Number Theory'* (P020418); 18 months; $2,600

——— ; *Symposium on Representation Theory of Finite Groups and Related Topics* (P017009); 12 months; $14,600

BROWN UNIVERSITY

Accola, Robert D. M.; *Open and Closed Riemann Surfaces* (P021191); 12 months; $16,000

Browder, Andrew and Wermer, John; *Function Algebras* (P021326); 12 months; $45,400

Clark, Allan H., Baum, Paul F. and Harris, Bruno; *Algebraic and Differential Topology* (P022580); 12 months; $37,300

Federer, Herbert; *Geometric Measure Theory* (P016807); 24 months; $40,000

Fleming, Wendell H.; *Optimal Stochastic Control* (P020868); 12 months; $30,000

Ito, Yuji; *Ergodic Theory* (P020309); 12 months; $8,000

Lasalle, Joseph P.; *Nonlinear Dynamical Systems* (P024338); 12 months; $41,300

Lubin, Jonathan D. and Rosen, Michael I.; *Arithmetic and Algebraic Geometry* (P023302); 12 months; $11,000

Nomizu, Katsumi; *Differential Geometry* (P019680); 12 months; $21,600

Strauss, Walter A.; *Nonlinear Partial Differential Equations* (P016919); 24 months; $48,000

SOUTH CAROLINA

CLEMSON UNIVERSITY

Cholewinski, Frank M.; *Generalized Weierstrass, Hankel and Ultraspherical Distributional, and Polynomial Transformations* (P023118); 12 months; $8,000

TENNESSEE

UNIVERSITY OF TENNESSEE

Bean, Ralph J.; *Mathematical Topology* (P019853); 24 months; $24,800

Daverman, Robert J. and Eaton, William T.; *Topology of Three Dimensional Manifolds* (P019966); 24 months; $16,600

Plemmons, Robert J. and Cline, Randall E.; *Generalized Matrix Inverses — an Algebraic Approach* (P015943); 24 months; $24,600

VANDERBILT UNIVERSITY

Jonsson, Bjarni; *Groups, Rings, Lattices and Logic* (P021059); 12 months; $35,000

TEXAS

RICE UNIVERSITY

Baumslag, Gilbert; *Classification Problems in Algebra* (P019586); 24 months; $45,000

Curtis, M. L.; *Manifold Theory and Homotopy* (P020552); 12 months; $19,900

Miele, Angelo and Huang, Ho-Yi; *Computer Algorithms for Optimization Theory* (P018522) 12 months; $22,600

Rachford, Henry R. Jr.; *Partial Differential Equations—Numerical Analysis, Functional Analysis—Weakly Holomorphic Functions on Varieties* (P023400); 12 months; $36,000

Veech, William A.; *Topological Dynamics and Ergodic Theory* (P018961); 12 months; $20,300

Wells, R. O. Jr.; *Complex Analysis* (P019011); 12 months; $59,600

TEXAS CHRISTIAN UNIVERSITY

Sanders, B. L.; *Regional Conference of Ten Problems in Hilbert Space* (P018939); 12 months; $7,100

UNIVERSITY OF TEXAS AT AUSTIN

Bernau, S. J.; *Lattice Groups and Operator Theory* (P020421); 12 months; $6,500

Cannon, John R.; *Numerical Continuation of Solutions of Partial Differential Equations* (P015724); 24 months; $27,600

Lorentz, George G.; *Approximation Theory and Functional Analysis* (P023566); 12 months; $11,000

UTAH

UNIVERSITY OF UTAH

Brooks, Robert M.; *Locally M-Convex Algebras* (P018729); 12 months; $7,300

Burgess, C. E.; *Mapping Manifolds into Manifolds* (P020292); 12 months; $15,000

Glaser, Leslie C.; *Geometric Piecewise Linear Topology* (P019812); 24 months; $20,000

Rushing, T. B.; *Topological Embedding Problems* (P019707); 24 months; $11,000

Scott, W. R.; *Group Theory* (P020850); 12 months; $16,800

PHYSICS

VERMONT

UNIVERSITY OF VERMONT
Cooke, Roger L.; *Spherical Summability of Trigonometric Series in Several Variables* (P012882); 24 months; $11,700
Wright, Robert K.; *Algebraic Differential Equations* (P013593); 24 months; $14,300

VIRGINIA

UNIVERSITY OF VIRGINIA
Simmonds, James G. and Danielson, Donald A.; *Some Mathematical Problems Arising in Thin Shell Theory* (P015333); 24 months; $39,600

VIRGINIA POLYTECHNIC INSTITUTE
Patty, C. Wayne; *Polyhedral and Piecewise Linear Topology* (P015357); 12 months; $13,600
——— ; *Regional Conference on Homological Algebra* (P019754); 12 months; $9,000

WASHINGTON

UNIVERSITY OF WASHINGTON
Arsove, Maynard G.; *Hardy Spaces and Potential Theory* (P022821); 12 months; $18,300
Beaumont, Ross A. and Pierce, Richard S.; *Structure of Modules, Rings and Groups* (P020195); 12 months; $68,000
Brownell, F. H.; *Applied Partial Differential Operators and Mathematical Quantum Mechanics* (P017526); 12 months; $23,200
Glicksberg, Irving; *Harmonic Analysis and Banach Algebras* (P019768); 12 months; $10,000
Hewitt, Edwin; *Functional Analysis* (P019274); 12 months; $15,000
Michael, Ernest A.; *Abstract Spaces* (P020849); 12 months; $34,500
Van Ness, John W. and Fisher, Lloyd D.; *Applications of Stochastic Processes* (P022591); 12 months; $11,700
Wulbert, Daniel E.; *Projections on Banach Spaces* (P019180); 12 months; $5,000

WASHINGTON STATE UNIVERSITY
Barnes, D. C.; *Isoperimetric and Related Inequalities* (P014428); 12 months; $5,100
Ostrom, T. G. and Kallaher, M. J.; *Translation Planes* (P017461); 12 months; $10,300

WISCONSIN

UNIVERSITY OF WISCONSIN, Madison
Martin, Joseph M.; *Geometric Topology* (P014396); 12 months; $16,200
Bleicher, Michael N.; *Studies in Geometry of Numbers, Number Theory, and Geometry* (P013970); 12 months; $17,800
Buck, R. C.; *Mathematical Analaysis* (P024182); 12 months; $59,600
Conley, Charles C.; *Global and Asymptotic Problems in Differential Equations* (P020858); 12 months; $12,900
Fadell, Edward R. and Husseini, Sufian Y.; *Algebraic Topology and Manifold Theory* (P020848); 12 months; $34,000
Gurland, John; *Distortion in Statistical Theory* (P013175); 24 months; $30,100
Hellerstein, Simon; *Meromorphic Function Theory and Related Topics* (P021340); 12 months; $23,300
Holland, W. Charles; *Lattice-Ordered Groups and Homological Ring Theory* (P019102); 12 months; $10,000
Hu, T. C.; *Theory and Computation of Integer Programming and Network Flows* (P020144); 12 months; $10,100
Keisler, H. Jerome; *Mathematical Logic and Foundations* (P023114); 12 months; $23,400
Knopp, M. I.; *Modular and Automorphic Functions* (P020219); 24 months; $21,000
Mcmillan, D. R. Jr.; *Geometric Topology* (P016093); 12 months; $17,900
Meyer, R. E.; *Asymptotic Methods for Partial Differential Equations and Applications* (P018641); 12 months; $45,600
Roussas, George G.; *Inference in Markov Processes* (P020036); 24 months; $24,000
Schneider, Hans; *Matrix and Combinatorial Theory* (P017815); 12 months; $5,200
Smith, Kennan T. and Turner, Robert E. L.; *Analysis and Differential Equations* (P021078); 12 months; $20,200
Solomon, Louis; *Groups and Geometry* (P018640); 12 months; $24,100
Young, Laurence C.; *Stochastic Integrals, Control Theory, Variational Methods and their Applications* (P018333); 12 months; $15,400

UNIVERSITY OF WISCONSIN, Milwaukee
Marden, Morris; *Geometry of Polynomials and Related Functions* (P019615); 13 months; $17,500

ISRAEL

WEIZMANN INSTITUTE OF SCIENCE
Pekeris, C. L.; *Stability of Laminar Flow to Periodic Axisymmetric Disturbances of Finite Amplitude* (P016073); 12 months; $20,200

Physics

ALABAMA

UNIVERSITY OF ALABAMA, Birmingham
Pearson, C. A.; *Nuclear Reactions* (P019015); 24 months; $19,600

UNIVERSITY OF ALABAMA, Tuscaloosa
Bartlett, James H.; *Periodic Motion and Stability of a Small Mass Under the Gravitational Attraction of Two Heavy Bodies* (P020122); 24 months; $15,600

ARKANSAS

UNIVERSITY OF ARKANSAS
Clayton, Glen T.; *X-Ray Diffraction Studies of Liquids* (P013148); 12 months; $18,900

CALIFORNIA

CALIFORNIA INSTITUTE OF TECHNOLOGY
Lauritsen, Thomas; *Nuclear Structure Physics and Nuclear Astrophysics* (P019887); 11 months; $979,600

CLAREMONT MEN'S COLLEGE
Klein, Stanley A.; *Theoretical Particle Physics* (P019553); 24 months; $11,500

STANFORD RESEARCH INSTITUTE
Smith, Felix T.; *Collision Processes Involving Metastable Atoms and Molecules* (P021314); 24 months; $46,200

STANFORD UNIVERSITY
Fairbank, W. M. and Hofstadter, Robert; *High Energy Physics and Cryogenic Research* (P022880*); 12 months; $1,750,000
Little, William A.; *Low Temperature Behavior of Bose and Fermi Systems* (P017278); 12 months; $10,000
Meyerhof, Walter E. and Hanna, Stanley S.; *Nuclear Structure Investigations with a 15-Mev Tandem Van de Graaff Accelerator* (GP8163 003); 12 months; $480,000
Schawlow, Arthur L.; *Spectroscopy and Quantum Electronics* (P024062); 12 months; $162,000

UNIVERSITY OF CALIFORNIA, Berkeley
Fretter, William B. and Bingham, Harry H.; *High-Energy Particle Physics* (P014507); 12 months; $125,000
Hahn, Erwin L.; *Coherent Optical Phenomena and Solid State Spin Resonance* (P016138); 24 months; $149,200

UNIVERSITY OF CALIFORNIA, Irvine
Parker, William H.; *Determination of h/m Using Macroscopic Quantum Phase Coherence in Superconductors* (P014240); 24 months; $39,800
Rynn, Nathan; *Alkali-Metal and Barium Plasmas, and Nonlinear Problems in Plasma Physics* (P024029); 12 months; $114,000
Shaw, Gordon L.; *Theoretical Studies in High Energy Physics* (P017170); 24 months; $74,600

UNIVERSITY OF CALIFORNIA, Los Angeles
Bommel, Hans E.; *Solid State Physics* (P023777); 24 months; $169,100
Orbach, Raymond L.; *Localized Moments in Metals* (P017014); 24 months; $76,000
Saxon, David S.; *Research in Theoretical Physics* (P015912); 12 months; $223,200
Wong, Eugene Y.; *Interaction of Lattice Vibrations and Pure Electronic States in Crystalline Solids* (P022818); 12 months; $10,000

UNIVERSITY OF CALIFORNIA, San Diego
Schultz, Sheldon and Feher, George; *Experimental Solid State Research* (P019809); 12 months; $119,300

UNIVERSITY OF CALIFORNIA, San Diego
REVELLE COLLEGE
Goodkind, John M.; *Quantum Fluids* (P017895); 24 months; $115,000

UNIVERSITY OF CALIFORNIA, Santa Barbara
Broida, H. P.; *Molecular and Solid State Spectroscopy Emphasizing the Use of Optical Lasers* (P014011); 24 months; $53,000
Sawyer, Raymond F.; Lewis, H. W. and Sugar, Robert; *Problems in Theoretical Physics* (P017668); 16 months; $140,000
Walker, W. C.; *Electronic Excitations in Condensed Matter* (P017382); 24 months; $55,000

UNIVERSITY OF SOUTHERN CALIFORNIA
Hu, Chia-Ren; *Microscopic Investigation of Non-Local, Non-Homogeneous Superconductors* (P020458); 24 months; $11,300
Porto, Sergio P. S.; *Laser Excited Spectroscopy of Solids* (P019109); 24 months; $135,200

COLORADO

COLORADO STATE UNIVERSITY
Raich, J. C.; *Solid Hydrogen* (P022553); 24 months; $25,200

UNIVERSITY OF COLORADO
Dreitlein, Joseph F. and Wyss, Walter; *A Program in Mathematical Physics* (P019479); 24 months; $22,100
Rogers, Robert N.; *Magnetic Interactions in Insulators* (P020284); 24 months; $54,500
Smith, Stephen J.; *Electron-Impact Excitation of Hydrogen Atoms* (P017174); 24 months; $69,300

CONNECTICUT

YALE UNIVERSITY
Hirshfield, J. L.; *Wave-Plasma Studies* (P014701); 12 months; $119,200
Hughes, Vernon W.; *Positronium, the Fine Structure of Hydrogen and Singly Ionized Helium* (P023722); 24 months; $60,000
Wheeler, Robert G.; *Far Infrared Spectroscopy of Spin Waves in Antiferromagnetic Crystals* (P016792); 24 months; $57,100

DELAWARE

UNIVERSITY OF DELAWARE
Cooper, Charles B.; *Low Energy Ion Bombardment of Solid Surfaces* (P014520); 24 months; $31,600
Kerner, Edward H. and Hill, Robert N.; *Relativistic Many-Particle Theory* (P014564); 24 months; $42,000
Murray, Richard B.; *Radiation-Induced Defects in Alkali Halides, and their Role in Recombination Processes* (P019130); 24 months; $45,600

MATHEMATICAL AND PHYSICAL SCIENCES RESEARCH PROJECTS

DISTRICT OF COLUMBIA

AMERICAN UNIVERSITY
Chertok, Benson T.; *Nuclear Excitation by Low Energy Electrons and Photons* (P016565); 12 months; $25,000
White, John A.; *Critical Point Phenomena in Fluids* (P014118); 24 months; $56,300

GEORGETOWN UNIVERSITY
Treado, Paul A. and Lambert, James M.; *Nuclear Structure and Interaction Studies with Low Energy Positive-Ion Accelerators* (P023233); 12 months; $42,400

NATIONAL ACADEMY OF SCIENCES, NATIONAL RESEARCH COUNCIL
Paul, Martin A.; *Task Order for Partial Support for Travel of Selected Participants in the International Conference on Precision Measurements and Fundamental Constants* (C310187); 6 months; $7,900
Reed, Charles K.; *Committee on Nuclear Science* (C310047007); 12 months; $19,000

FLORIDA

FLORIDA STATE UNIVERSITY
Edwards, Steve and Robson, Donald; *Theoretical Low Energy Nuclear Physics* (P015855); 24 months; $150,000

UNIVERSITY OF FLORIDA
Broyles, A. A.; *Theoretical Determination of Thermodynamic Quantities and X-Ray and Neutron Diffraction Intensities of Fluids* (P015976); 24 months; $50,000
Scott, Thomas A., Adams, E. D. and Rosenshein, J. S.; *High Pressure and Low Temperature Studies of Condensed Matter* (P015324); 24 months; $113,200
Slater, John C.; *Solid-State and Molecular Theory* (P016464); 24 months; $125,000

GEORGIA

MOREHOUSE COLLEGE
Chung, Ping L.; *Ultrasonic Wave in Semiconductors* (P018684); 12 months; $12,400

UNIVERSITY OF GEORGIA
Amos, Kenneth A.; *Theoretical Studies of Nuclear Reactions and of Gross Properties of the Nuclear Surface* (P022559); 24 months; $22,200

ILLINOIS

ILLINOIS INSTITUTE OF TECHNOLOGY
Burnstein, Ray A.; *High Energy Physics—Properties of Mesons and Hyperons* (P024128); 12 months; $67,100

NORTHERN ILLINOIS UNIVERSITY
Kimball, Clyde W.; *Intermediate Phases of Binary Rare Earth Alloys* (P23290); 12 months; $26,000

UNIVERSITY OF CHICAGO
Inghram, Mark G.; *Chemical Physics Studies by Mass Spectrometry* (P016587); 12 months; $83,000
Roothaan, Clemens C. J. and Mulliken, Robert S.; *Experimental and Theoretical Studies on Small Molecules* (P015216); 12 months; $70,000
Sachs, Robert G.; *High Energy Physics Research* (P024126); 12 months; $1,406,000
Thompson, R. W.; *Nature and Interactions of High Energy Elementary Particles* (P018468); 12 months; $188,700

UNIVERSITY OF ILLINOIS, Urbana
Allen, James S., Hanson, Alfred O. and Axel, Peter; *Nuclear Structure Research and Accelerator Development* (P016197); 12 months; $338,900
Baym, Gordon and Wortis, Michael; *Properties of Solids and Low Temperature Systems* (P016886); 24 months; $78,600
Frauenfelder, Hans and Debrunner, Peter; *Studies in Nuclear Physics and Related Subjects* (P017135); 12 months; $109,800
Ravenhall, David G.; *Electromagnetic Interactions* (P019433); 8 months; $40,000
Wyld, Henry W. Jr.; *Theoretical Elementary Particle Physics* (P013671); 12 months; $70,000

INDIANA

INDIANA UNIVERSITY, Bloomington
Bent, R. D., Miller, D. W. and Rickey, M. E.; *Nuclear Structure and Nuclear Processes* (P019721); 12 months; $479,100
Bron, Walter E.; *Electron-Lattice Coupling to Dynamical Phenomena in Crystals* (P020031); 24 months; $87,800
Chase, Lloyd L.; *Raman Scattering in Solids* (P019743); 24 months; $53,400
Emery, Guy T.; *Advisory Group Study of Projects and Priorities for Use of Indiana Cyclotron* (P024528); 12 months; $10,800

PURDUE UNIVERSITY
James, Hubert M.; *Molecular Orientations in Solid Hydrogen* (P019434); 24 months; $29,000
Mieher, Robert L.; *Endor and Optical Studies in Crystalline Solids* (P015799); 24 months; $65,100

UNIVERSITY OF NOTRE DAME
McGlinn, William D.; *1970 Midwest Conference on Theoretical Physics* (P020440); 12 months; $5,300

IOWA

UNIVERSITY OF IOWA
Carlson, Richard R.; *Basic Research in Nuclear Physics* (P019595); 12 months; $249,300
Savage, William R.; *Localized Moments Arising from Magnetic Impurities in Metallic Samples* (P020645); 24 months; $55,500

KANSAS

UNIVERSITY OF KANSAS
Ammar, Raymond G. and Stump, Robert; *High Energy k-d Interactions Using Bubble Chamber,* (P017600); 12 months; $92,900
Culvahouse, J. W.; *Origins of Spin-Spin Interactions in Insulating Solids* (P015256); 24 months; $62,000

KENTUCKY

UNIVERSITY OF KENTUCKY
Kern, Bernard D. and McEllistrem, Marcus T.; *Nuclear Structure and Nuclear Reaction Mechanisms* (P021088); 12 months; $120,000

LOUISIANA

LOUISIANA STATE UNIVERSITY, Baton Rouge
Hussey, R. G.; *Unsteady Flow in Viscous Fluids* (P015520); 24 months; $33,900

TULANE UNIVERSITY
Durham, Frank E., Buccino, Salvatore G. and Wilenzick, R. M.; *Nuclear Physics Research* (P017141); 12 months; $109,500

MARYLAND

JOHNS HOPKINS UNIVERSITY
Pevsner, A.; *High Energy Particle Physics* (P017148); 24 months; $575,000

UNIVERSITY OF MARYLAND
Davidson, Ronald C. and Krall, Nicholas A.; *Non-Neutral Plasmas* (P019695); 24 months; $25,000
Griem, Hans R. and DeSilva, Alan W.; *Plasma Shock Wave Structure, Light Scattering and Spectroscopy* (P016246); 24 months; $137,600
Krisher, Lawrence C.; *Microwave Spectroscopy* (P014306); 24 months; $44,000
Misner, Charles W.; *Theoretical Studies of Gravity* (P017673); 24 months; $63,800
Oneda, Sadao and Greenberg, O. W.; *Elementary Particle Theory and Quantum Field Theory* (P020709); 24 months; $166,100
Pechacek, Robert E.; *Diagnostics of a Relativistic Electron Beam* (P019890); 12 months; $31,000
Reiser, Martin P.; *Theoretical and Experimental Study of an Electron Ring Accelerator with Static Compression System* (P015798); 12 months; $37,900
Weber, Joseph; *Gravitational Radiation Experiments* (P022562); 12 months; $165,000
———; *Experimental and Theoretical Research on Gravitation* (GP8560 001); $19,100
Wilkerson, T. D., Coplan, M. A. and Benesch, W. M.; *Atomic and Molecular Processes Bearing on Atmospheric and Plasma Phenomena* (P023394); 12 months; $85,300

MASSACHUSETTS

AMHERST COLLEGE
Duffy, Richard J. and Romer, Robert H.; *Coupling of First and Second Sound in Liquid Helium* (P023127); 24 months; $35,500

BOSTON UNIVERSITY
Booth, Edward C.; *Nuclear Resonance Fluorescence Measurements of Electromagnetic Transition Probabilities* (P019061); 12 months; $20,000
Franzen, Wolfgang; *Electron Scattering and Polarization* (P022752); 24 months; $45,000

BRANDEIS UNIVERSITY
Amit, Daniel J., Gross, Eugene P. and Lange, Robert V.; *Theoretical Research in Many-Body Problems* (P017560); 24 months; $73,500
Schweber, Silvan S.; *Theoretical and Mathematical Physics* (P018721); 24 months; $59,900

HARVARD UNIVERSITY
Ehrenreich, Henry and Martin, Paul C.; *Theoretical Problems in Solid State and Statistical Physics* (P016504); 24 months; $141,200
Pipkin, Francis M.; *Studies of Nuclear Orientation* (P015072); 9 months; $25,600
———; *Atomic Physics Experiments Using Photon Coincidence Techniques* (P022787); 24 months; $54,200
Pound, Robert V.; *Resonance and Radiation Physics* (P024736); 6 months; $38,000
Ramsey, Norman F.; *Molecular Beam and Hydrogen Maser Research* (P13547X); 24 months; $180,000

MASSACHUSETTS INSTITUTE OF TECHNOLOGY
Sellmyer, David J. and Johnson, Keith H.; *Electrons in Complex Solids* (P021312); 24 months; $70,200
Shull, C. G.; *Magnetic Ordering and Neutron Physics Studies* (P022819); 24 months; $227,200
Stanley, H. Eugene; *Critical Phenomena and Statistical Mechanics* (P015428); 24 months; $13,600

TUFTS UNIVERSITY
Gunther, Leon; *Theory of Solids* (P016025); 24 months; $20,300
McCarthy, Kathryn A.; *Solids at Low Temperatures* (P015993); 24 months; $67,700

UNIVERSITY OF MASSACHUSETTS
Brehm, John J. and Cook, Leroy F. Jr.; *Strong Interaction Dynamics* (P015282); 24 months; $111,000
Engelsberg, Stanley and Guyer, Robert A.; *Equilibrium and Transport Properties in Many Body Systems* (P020896); 24 months; $36,800
Ford, Norman C. Jr.; *Liquid-Vapor Critical Points* (P020118); 12 months; $18,200
Jones, Phillips R.; *Inelastic Atomic Collisions at Energies Below 25 KeV* (P016024); 24 months; $39,900
Penchina, Claude M.; *Optical Properties of Semiconductors and Insulators in Electric Fields* (P021284); 24 months; $55,000
Peterson, Gerald A.; *Nuclear Structure Studies by the Scattering of High Energy Electrons* (P014130); 12 months; $13,700

PHYSICS

Sternheim, Morton M.; *Theory of Scattering of Pions and Nucleons by Nuclei—Tests and Applications of Quantum Electrodynamics* (P016377); 24 months; $25,400

UNIVERSITY OF MASSACHUSETTS, Boston Campus
Rao, D. V. G. L. N.; *Non-Linear Optical Properties of Liquid Crystals* (P012533); 24 months; $15,300

WILLIAMS COLLEGE
Crampton, Stuart J. B.; *Hydrogen Spin Exchange Parameters Using a Hydrogen Maser* (P014599); 24 months; $40,300

MICHIGAN

MICHIGAN STATE UNIVERSITY
Blatt, Frank J. and Schroeder, P. A.; *Electronic Properties of Metals and Alloys* (P016811); 12 months; $102,000
Blosser, Henry G.; *Research Program with 50 Mev Cyclotron* (GP6760 004); 12 months; $750,000
Edwards, T. H. and Hause, C. D.; *Molecular Spectra in the Near Infrared Region* (P023234); 24 months; $43,000

OAKLAND UNIVERSITY
McKinley, John M.; *Theoretical Physics* (P022566); 12 months; $9,400

UNIVERSITY OF MICHIGAN
Federbush, Paul and Williams, David; *Functional Analysis and Function Theoretic Methods in Elementary Particle Physics* (P017523); 24 months; $23,800
Jones, Lawrence W.; *Nucleon-Nucleon Total Cross Sections* (P025196*); 12 months; $235,500
Jones, Lawrence W., Longo, Michael J. and Overseth, Oliver E.; *Elementary Particle Physics* (P019257); 12 months; $311,200

WAYNE STATE UNIVERSITY
Beard, George B. and Kenealy, P. R.; *Properties of Low-Lying Nuclear Energy Levels* (P019270); 12 months; $22,500
Fradkin, David M.; *Dynamical Groups and Symmetries* (P015234); 24 months; $17,700
Gustafson, Daniel R.; *Angular Correlation of Positron Annihilation Radiation* (P022745); 24 months; $45,200
Piccirelli, Robert A.; *Molecular Theory of Fluids for Rapid Variation in Space and Time and for Strong Forces* (P016914); 24 months; $16,700

MINNESOTA

CARLETON COLLEGE
Butler, William A. and Reitz, Robert A.; *Thermoluminescence in Cesium Iodide* (P023718); 12 months; $14,800

ST. OLAF COLLEGE
Rossing, Thomas D.; *Ferromagnetic Resonance in Thin Films* (P020895); 24 months; $25,600

UNIVERSITY OF MINNESOTA
Moldover, Michael R.; *Critical Point of Phase Transitions* (P014431); 24 months; $66,100

MISSISSIPPI

UNIVERSITY OF MISSISSIPPI
Arnold, Roy T.; *Study of Phonon Self-Interactions in Ferroelectrics* (P015637); 24 months; $39,100

MISSOURI

UNIVERSITY OF MISSOURI, Columbia
Danner, Horace R.; *Molecular Vibrations in Polymers and Related Organic Compounds* (P017970); 24 months; $70,200
Hensley, Eugene B.; *Color Centers in Alkaline Earth Chalcogenides* (P022815); 24 months; $45,800

UNIVERSITY OF MISSOURI, Rolla
Park, John T.; *Heavy Ion Energy Loss Spectrometry* (P017425); 24 months; $51,700

WASHINGTON UNIVERSITY
Bolef, Dan I.; *Ultrasonic Studies in Solids* (P016713); 24 months; $75,500
Burgess, James H.; *Lattice Interactions in Free Radicals* (P023132); 24 months; $53,600
Feenberg, Eugene and Clark, John W.; *Microscopic Theory of Quantum Fluids and Extended Nuclear Systems* (P022564); 24 months; $70,600
Phillips, Peter R.; *Modern Analog to the Michelson-Morley Experiment to Look for a Preferred Frame of Reference in Space* (P015285); 12 months; $24,400
Shrauner, Ely; *Elementary Particles* (P020282); 12 months; $20,000

MONTANA

MONTANA STATE UNIVERSITY
Drumheller, John E.; *Forbidden Hyperfine Transitions in Electron Paramagnetic Resonance Spectra* (P022750); 12 months; $21,700
Schmidt, V. Hugo; *Proton Mobility in H-Bonded Crystals* (P022949); 24 months; $52,600

NEBRASKA

UNIVERSITY OF NEBRASKA, Lincoln
Katz, Robert; *Particle Tracks and Related Phenomena* (P023129); 12 months; $15,000
Rudd, M. E.; *Inelastic Processes in Atomic Collisions* (P015124); 24 months; $158,200
Smith, Kenneth; *Calculation of Atomic Collision Cross Sections* (P020559); 12 months; $50,000

NEW HAMPSHIRE

UNIVERSITY OF NEW HAMPSHIRE
Balling, L. C.; *RF Spectroscopy of Atoms and Ions* (P016814); 24 months; $40,100
Lambert, R. H.; *Optical Pumping at High Temperatures* (P014125); 14 months; $20,000

NEW JERSEY

FAIRLEIGH DICKINSON UNIVERSITY, Teaneck Campus
Moeller, Karl D.; *Torsional Vibrations of Large Molecules in the Far Infrared* (P012762); 12 months; $7,500

INSTITUTE FOR ADVANCED STUDY
Regge, Tullio; *Theoretical Physics* (P016147); 24 months; $130,000

PRINCETON UNIVERSITY
Carver, Thomas R.; *Magnetic and Optical Properties of Solids and Gases* (P019891); 24 months; $175,000
Daniels, W. B.; *Inelastic Scattering of Neutrons* (P018573); 24 months; $107,300
O'Neill, Gerard K.; *Colliding Beams* (P018624); 12 months; $127,700
Royce, Barrie S. H.; *Radiation Induced Reactions in Ionic Crystals* (P015945); 12 months; $19,000

RUTGERS UNIVERSITY
Luthi, Bruno; *Magnetic Materials Using Ultrasonic Techniques* (P016107); 24 months; $64,500
Temmer, Georges M.; *Nuclear Structure Research with a Tandem Accelerator* (P019742); 12 months; $425,500

STEVENS INSTITUTE OF TECHNOLOGY
Bernstein, Jeremy; *Elementary Particle Theory* (P013147); 24 months; $38,600
Daunt, John G.; *Two and Three Dimensional Solid Mixture of He_3 and He_4* (P017599); 24 months; $91,000
Meissner, Hans; *Time Dependent Phenomena in Superconductors* (P019696); 24 months; $50,000
Rothberg, Gerald M.; *The Mossbauer Effect in Studies of Electronic and Vibrational Structure* (P023125); 24 months; $81,200
Taylor, Snowden and Koller, Earl L.; *Properties of Elementary Particles* (P016714); 9 months; $37,000

NEW MEXICO

UNIVERSITY OF NEW MEXICO
Green, John R.; *Dielectric Properties and Phase Transformations in Plastic Solids* (P020230); 24 months; $48,100

NEW YORK

CLARKSON COLLEGE OF TECHNOLOGY
Helbig, Herbert; *Measurements of the Differential Scattering Crossection for Sub-KeV Protons from Hydrogen Atoms* (P023713); 12 months; $18,100

COLGATE UNIVERSITY
Holbrow, Charles H.; *Nuclear Reactions at Tandem Energies* (P009675); 24 months; $22,900

COLUMBIA UNIVERSITY
Devons, Samuel, Sachs, Allan M. and Wu, Chien-Shung; *Research Program and Synchrocyclotron Operation* (P022786*); 12 months; $1,050,000
Ruderman, Malvin H. and Spiegel, Edward A.; *Some Topics in Theoretical Astrophysics* (P018062); 24 months; $101,300

CORNELL UNIVERSITY
Bethe, Hans A.; *Fundamental Nuclear Physics and Astrophysics* (P014338); 24 months; $81,300
McDaniel, B. D.; *Contract for Operational Support of the 10 Bev and 2 Bev Electron Synchrotron Facilities at Cornell University* (C537000006); $2,600,000
———; (C537000007); $100,000
Orear, Jay and Peoples, John; *Large Angle Elastic Scattering at High Energies* (P015214); 24 months; $288,600

CUNY, BROOKLYN COLLEGE
Franco, Victor; *Scattering of Medium- and High-Energy Projectiles by Nuclei* (P016913); 24 months; $12,800

CUNY, CITY COLLEGE
Lea, Robert M.; *Particle Interactions* (GP9361 001); 8 months; $15,200
Rubin, Kenneth; *Low Energy Electron-Atom Elastic and Inelastic Differential Cross Section Measurements* (P023996); 24 months; $60,000

FORDHAM UNIVERSITY
Budnick, J. I.; *Nuclear Magnetic Resonance in Ferromagnetic Metals and Alloys* (P023283); 24 months; $53,400

NEW YORK UNIVERSITY
Bederson, Benjamin and Brown, Howard H.; *Atom Beam-Plasma Interaction Studies and Plasmas in Thermodynamic Equilibrium* (P017193); 24 months; $82,400
Richardson, Robert W.; *Many-Body Problem and Nuclear Models* (P016702); 24 months; $24,400
Rosenberg, Leonard; *Multiparticle Scattering Problems* (P016427); 24 months; $24,100
Stroke, H. Henry; *Atomic Hyperfine Structure and Isotope Shift* (P015258); 24 months; $93,000

POLYTECHNIC INSTITUTE OF BROOKLYN
Juretschke, Hellmut J.; *Magneto-Optical Properties of Ferromagnetic Metals* (P015257); 24 months; $48,300
Scarl, Donald B.; *Two-Photon Processes* (P023478); 24 months; $31,900

STATE UNIVERSITY AT BINGHAMTON
Raboy, Sol; *Nuclear Structure by Spectroscopy of X-Rays from Muonic Atoms and Gamma-Rays* (P023864); 12 months; $21,600

STATE UNIVERSITY AT BUFFALO
Fujita, Shigeji; *Statistical Physics of Various Materials—Gases, Plasmas, Metals, Carbons, Polymers, etc.* (P018972); 12 months; $11,000

STATE UNIVERSITY AT STONY BROOK
Lee, Linwood L. Jr.; *Operation of the Stony Brook Nuclear Structure Laboratory* (P017286); 12 months; $200,000

MATHEMATICAL AND PHYSICAL SCIENCES RESEARCH PROJECTS

SYRACUSE UNIVERSITY
Goldberg, Marvin and Moneti, Giancarlo; *High Energy Experimental Physics* (P016863); 12 months; $170,600
Goldberg, Joshua N. and Bergmann, Peter G.; *General Relativity* (P019378); 24 months; $70,000
Horwitz, Nahmin and Kalogeropoulos, Theodore; *Elementary Particle Physics with Track Chamber* (P018259); 9 months; $74,000
Rohrlich, Fritz; *Eighth Annual Eastern Theoretical Physics Conference* (P016109); 12 months; $5,100
———; *Quantum Field Theory and Related Problems* (P012536); 24 months; $73,300

UNIVERSITY OF ROCHESTER
Gove, Harry E.; *Research at the Nuclear Structure Research Laboratory* (GP6972 003); 12 months; $725,000
Kaplon, M. F. and Badhwar, G.; *Cosmic Ray Physics* (P024869); 12 months; $89,300
Thorndike, Edward H. and Lobkowicz, Frederick; *High Energy Photoproduction Studies* (P016737); 12 months; $163,800

YESHIVA UNIVERSITY
Finkelstein, David; *Quantum Theory of Elementary Processes* (P022952); 24 months; $23,900
Weinstein, Marvin; *Chiral Symmetry and its Breaking* (P017032); 24 months; $13,200

NORTH CAROLINA

DUKE UNIVERSITY
Biedenharn, L. C. and Evans, L. E.; *Theoretical Nuclear Structure and Elementary Particle Physics* (P014116); 24 months; $80,000
Meyer, Horst; *Thermal and Magnetic Properties of Some Compounds at Low Temperatures* (P017038); 24 months; $73,600

NORTH CAROLINA UNIVERSITY AT RALEIGH
Bennett, Willard H.; *Instabilities of Thin Streams in Plasmas* (P021282); 12 months; $38,200

UNIVERSITY OF NORTH CAROLINA AT CHAPEL HILL
Dewitt, Bryce S.; *The Role of Gravitation in Physics* (P015184); 24 months; $46,000
Hubbard, Paul S.; *Spin Resonance and Relaxation in Gases and Liquids* (P017930); 24 months; $56,000
Jarnagin, Richard C. and Silver, Marvin; *Generation of Free Electrons in Insulating Liquids and Films* (P017128); 24 months; $45,000

OHIO

CASE WESTERN RESERVE UNIVERSITY
Eck, Thomas G.; *Fine and Hyperfine Structure of Excited States of Atoms* (P019612); 24 months; $39,600

KENT STATE UNIVERSITY
Franklin, Wilbur M.; *Theoretical Studies of Diffusion in Solids* (P020152); 12 months; $12,000

OBERLIN COLLEGE
Warner, Robert E.; *Medium Energy Nuclear Physics;* (P019269); 12 months; $12,700

OHIO STATE UNIVERSITY
Hausman, H. J., Arns, R. G. and Blatt, S. L.; *Nuclear Physics Research with a 5.5 MeV Van de Graaff Accelerator* (GP10763002); 12 months; $189,900
Wigen, Philip E.; *Spin Resonance in Metals* (P020101); 24 months; $55,000

UNIVERSITY OF CINCINNATI
Russell, James E.; *De-Excitation of Mesonic Atoms* (P020889); 24 months; $13,700

OREGON

UNIVERSITY OF OREGON
Higgins, Richard J.; *Electronic Structure of Imperfect Metallic Systems* (P023399); 24 months; $56,000

Park, Kwangjai; *Two Photon Spectroscopy* (P023130); 24 months; $56,700

PENNSYLVANIA

CARNEGIE MELLON UNIVERSITY
DeBenedetti, Sergio; *Investigation of Solid State Properties with Radioactive Techniques* (P016268); 24 months; $67,400
Friedberg, Simeon A.; *Magnetic Interactions in Salts of the Transition Metals* (P008292001); 6 months; $8,000
Kisslinger, Leonard S. and Sorensen, Raymond A.; *Theoretical Nuclear Physics* (P013957); 12 months; $50,000
Nagle, John F.; *Materials Science and Statistical Mechanics* (P021093); 24 months; $12,000
Schumacher, Robert T. and Vander Ven, Ned S.; *Magnetic Resonance and Phase Transitions in Solids* (P017559); 24 months; $90,100

DREXEL UNIVERSITY
Rosen, Gerald; *Functional Calculus Solutional Methods for Classical Statistical and Quantum Field Theories* (P018885); 24 months, $56,000

FRANKLIN INSTITUTE
Van Patter, D. M.; *Nuclear Structure Physics* (P020556); 12 months; $40,000

PENNSYLVANIA STATE UNIVERSITY
McCammon, Robert D.; *Dielectric and Thermal Properties of Disordered Solids at Low Temperatures* (P017821); 24 months; $35,000
Mueller, Erwin W.; *Solid State Research with the Atom-Probe Field Ion Microscope* (P016773); 24 months; $60,000
Rank, D. H.; *Infrared Spectroscopy* (P013054); 24 months; $66,000

TEMPLE UNIVERSITY
Green, Melville S. and Swenson, Robert J.; *Studies in Statistical Mechanics* (P016336); 24 months; $72,700
Havas, Peter; *Studies in the Relativistic Theory of Interacting Particles* (P014123); 24 months; $44,000

UNIVERSITY OF PENNSYLVANIA
Ajzenberg-Selove, Fay; *Energy Levels of Light Nuclei* (P016451); 8 months; $31,900
Langenberg, D. N.; *Microwave Phenomena in Solids* (P024159); 24 months; $65,300
Middleton, Roy and Stephens, William E.; *Nuclear Research with Tandem Accelerator* (GP9281 002); 12 months; $387,700
Primakoff, Henry and Amado, Ralph D.; *Theoretical Physics* (P019346); 24 months; $176,700

UNIVERSITY OF PITTSBURGH
Cohen, Bernard L., Daehnick, Wilfried W. and McGruer, James N.; *Nuclear Structure and Nuclear Reactions* (P009349002); 12 months; $426,000
Fite, Wade L.; *Atomic Hydrogen Collision Studies* (P017637); 24 months; $103,000
Garfunkel, Myron P.; *Superconductors and Critical Point Phenomena* (P019131); 24 months; $128,800

SOUTH CAROLINA

UNIVERSITY OF SOUTH CAROLINA
Edge, Ronald D.; *Multiphase Nuclear Physics Research Program* (P014913); 12 months; $23,000

TENNESSEE

VANDERBILT UNIVERSITY
Hamilton, J. H. and Albridge, R. G.; *Experimental Nuclear Spectroscopy* (P017558); 12 months; $61,600
Pinkston, W. T.; *Nuclear Structure Aspects of Direct Reactions* (P015071); 24 months; $41,000
Salant, E. O. and Webster, M. S.; *Bubble Chamber Analysis* (P016984); 12 months; $176,500
———; (P024451); 12 months; $176,500

TEXAS

BAYLOR UNIVERSITY
Powers, Darden; *de/dx Studies of Heavy Ions in Solids and Gases* (P018722); 24 months; $41,200

RICE UNIVERSITY
Donoho, Paul L.; *Generation and Propagation of Elastic Waves in Solids* (P016137); 24 months; $43,800

TEXAS TECH UNIVERSITY
Quade, C. Richard; *Internal Rotation in Asymmetric-Asymmetric Molecules* (P023294); 24 months; $36,200

UNIVERSITY OF TEXAS AT AUSTIN
Davids, Cary N.; *Experimental Nuclear Astrophysics—A Study of Stellar Neutron Sources* (P023282); 24 months; $43,000
Frommhold, L. W.; *Atomic Processes in Gas Discharges* (P023519); 12 months; $24,100
Oakes, Melvin E.; *Low Temperature Plasmas* (P023738); 24 months; $63,200
Schild, Alfred; *Relativity and Gravitational Theories* (P020033); 24 months; $96,700

UTAH

UNIVERSITY OF UTAH
Ailion, David C.; *Magnetic Resonance Study of Ultraslow Atomic and Molecular Motions* (P017412); 24 months; $75,000
Das, Tara P.; *Theoretical Investigations of Atomic and Molecular Properties with Emphasis on Many-Body Effects* (P017597); 24 months; $54,000
———; *Electronic Structure and Electron-Nuclear Multipole Interactions in the Solid State* (P018583); 24 months; $53,000
Keuffel, J. W.; *Cosmic Ray Muons and Neutrinos* (P024452); 12 months; $195,300
———; (P015946); 12 months; $195,300

VIRGINIA

UNIVERSITY OF VIRGINIA
Deaver, Bascom S. Jr.; *Macroscpoic Quantum Phenomena* (P016704); 24 months; $78,500
Stewart, John W.; *Development and Use of a Magnetic Densitometer for Cryogenic Fluids* (P016659); 18 months; $18,000

WASHINGTON

UNIVERSITY OF WASHINGTON
Dash, J. G.; *Helium Monolayers* (P019014); 24 months; $121,800
Dehmelt, Hans G.; *Reactions with Anistropic Projectiles as a Means for Polarizing and Analyzing Stored Ions* (P021467); 24 months; $64,600
Fortson, E. Norval; *Precision Radio Frequency Spectroscopy with Atomic Ions and Electrons by the Ion Storage, Exchange Collision Method* (P023287); 24 months; $51,600
Neddermeyer, S. H. and Lord, J. J.; *Cosmic Ray Investigations* (P017636); 12 months; $134,800
Williams, Robert W. and Cook, Victor; *Elementary Particle Physics Using High Energy Accelerators* (P016390); 12 months; $431,400

WISCONSIN

UNIVERSITY OF WISCONSIN, Madison
Barschall, H. H.; *Third International Symposium on Polarization Phenomena in Nuclear Reactions* (P016139); 12 months; $11,500
Bincer, Adam M. and McVoy, Kirk W.; *Research in Fundamental Particle Theory and in Nuclear Theory* (P015198); 12 months; $24,600

CHEMISTRY

Yen, William M.; *Spectroscopic Studies of Magnetically Ordered Materials* (P015426); 24 months; $79,000

UNIVERSITY OF WISCONSIN, Milwaukee

Greenler, Robert; *An Infrared and Raman Study of Molecular Adsorption and Surface Reactions* (P023292); 12 months; $13,500

WYOMING

UNIVERSITY OF WYOMING

Grandy, Walter T. Jr.; *Theoretical Research in Quantum Statistical Mechanics* (P014805); 24 months; $26,000

Schick, Lee H.; *Some Three-Baryon Systems* (P021289); 14 months; $15,400

Chemistry

ALABAMA

UNIVERSITY OF ALABAMA, Tuscaloosa

Abramovitch, Rudolph A.; *The Chemistry of Nitrenes and Related Intermediates* (P018557); 24 months; $47,600

ARIZONA

ARIZONA STATE UNIVERSITY

Lin, Sheng H.; *Luminescence, Anisotropic Excitations, Non-Stoichiometric Compounds* (P015044); 24 months; $40,700

O'Keeffe, Michael; *Chemical Kinetics of Solids* (P017530); 24 months; $34,800

UNIVERSITY OF ARIZONA

Barfield, Michael; *Theoretical Studies of Spin-Dependent Properties of Organic Molecules* (P018119); 24 months; $26,100

Bates, Robert B.; *Resonance-Stabilized Anions and Radicals* (P021115); 12 months; $37,800

Burke, Michael F.; *Gas-Solid Chromatography* (P017332); 24 months; $24,000

Hall, Henry K.; *Polymerization of Bicyclic Compounds by Opening of Strained C-C Single Bonds* (P018914); 24 months; $21,700

Steelink, Cornelius; *Stable Radical Intermediates in the Enzymic Oxidation of Phenols* (P016492); 24 months; $24,700

ARKANSAS

UNIVERSITY OF ARKANSAS

Johnson, D. A. and Carmichael, J. W. Jr.; *Photochemical and Spectroscopic Studies of Square Planar Complexes* (P017520); 24 months; $40,700

Schafer, Lothar; *Electron Diffraction Studies of Organometallic and Organometallic Derived Molecules in the Vapor Phase* (P018683); 24 months; $29,300

Siegel, Samuel; *Stereochemistry of Catalytic Hydrogenation and Exchange Reactions of Alkenes* (P015719); 24 months; $28,000

CALIFORNIA

CALIFORNIA INSTITUTE OF TECHNOLOGY

Bergman, Robert G.; *Bicycloheptatrienyl Systems* (P017578); 24 months; $29,500

Davidson, Norman and Samson, Sten; *Structure of Complex Intermetallic Compounds* (P017504); 24 months; $47,800

Goddard, William A. III; *Theoretical Calculations on Molecules, Atoms and Solids* (P015423); 24 months; $40,200

Gray, Harry B.; *Coordination Chemistry* (P14015X001); 12 months; $101,400

Hammond, George S.; *Photoreactions in Solution* (GP5992X003); 12 months; $45,800

Ireland, Robert E.; *Total Synthesis of Triterpenes* (P012281); 24 months; $69,300

McKoy, Vincent; *Theoretical Calculations on Atoms and Molecules* (P014602); 12 months; $17,100

Roberts, John D.; *Nuclear Magnetic Resonance Spectroscopy, Structures and Reaction Mechanism of Organic Compounds* (P022804); 12 months; $94,200

CALIFORNIA STATE COLLEGE AT LOS ANGELES

Onak, Thomas P.; *Pentaborane Chemistry* (P017704); 24 months; $20,500

Pine, Stanley H.; *Chemistry of Quaternary Ammonium Salts* (P019174); 24 months; $26,200

OCCIDENTAL COLLEGE

Amey, Ralph L.; *Local Liquid Structure in Aprotic Solvents* (P017467); 24 months; $18,400

SACRAMENTO STATE COLLEGE

Nelson, Robert F.; *Anodic Oxidation of Aromatic Amines* (P020606); 24 months; $25,000

STANFORD UNIVERSITY

Andersen, Hans C.; *The Collective Variables Method in Classical Statistical Mechanics* (P020058); 24 months; $27,200

Baldeschwieler, John D.; *Application of Magnetic Resonance Techniques to Problems in Structural Chemistry* (P023406); 12 months; $56,000

Johnson, William S.; *Synthetic Studies Related to Natural Products* (P07238X002); 12 months; $50,000

van Tamelen, Eugene E.; *Biogenesis and Biogenetically-Patterned Syntheses of Natural Products and Related Substances* (P23019X); 12 months; $42,200

Weinhold, Frank; *Error Bounds for Quantum—Mechanical Expectation Values* (P016112); 24 months; $30,600

UNIVERSITY OF CALIFORNIA, Berkeley

Giauque, William F.; *Thermodynamic and Magnetic Properties Particularly at Low Temperatures* (P023465); 12 months; $120,400

Heathcock, Clayton H.; *Polycyclodecane Chemistry* (P012224); 24 months; $22,000

McClain, William M.; *Two-Photon Molecular Spectroscopy* (P015077); 24 months; $41,000

Moore, C. Bradley; *Spectroscopic Studies of Energy Transfer* (P022735); 24 months; $84,000

Streitwieser, Andrew Jr.; *Rare Earth Sandwich Complexes* (P013369); 24 months; $97,200

UNIVERSITY OF CALIFORNIA, Davis

Musker, W. Kenneth; *Influence of Ligand Structure on the Coordination Properties of Various Elements* (P020523); 24 months; $33,600

Volman, David; *Photolysis of Chloropropenes; and Reactions of Hydroxyl Radicals* (P013974); 24 months; $24,700

Zweifel, George; *Synthesis and Properties of Organoalanes and Organoboranes Derived from Alkynes* (GP9398 001); $7,200

UNIVERSITY OF CALIFORNIA, Irvine

Doedens, Robert J.; *Structural Studies of Metal Complexes* (P020603); 24 months; $30,500

Lee, Edward K. C.; *Tracer and Luminescence Studies of Photochemically Generated Excited Molecules* (GP11390001); $11,000

Moore, Harold W.; *Chemistry of Azidoquinones and Related Compounds* (P019268); 24 months; $36,100

UNIVERSITY OF CALIFORNIA, Los Angeles

Cram, Donald J.; *Paracyclophane Chemistry* (GP7193X002); 12 months; $29,000

Hawthorne, M. Frederick; *Mechanism Studies Related to the Chemistry of Polyhedral Boranes* (P14373X); 24 months; $67,900

Kaesz, Herbert D.; *Transition Metal Carbonyl Hydrides* (P023267X); 12 months; $26,800

Kivelson, Daniel; *Electronic Paramagnetic Resonance Studies of Free Radicals* (P016346); 24 months; $74,700

Reiss, Howard; *Analysis and Development of the Theory of Nucleation* (P020884); 24 months; $23,900

UNIVERSITY OF CALIFORNIA, Riverside

Helmkamp, George K.; *Thiirenium Salts—Aromatic Heterocyclic Analogs of Cyclopropenyl Cations* (P017878); 24 months; $48,300

Okamura, William H.; *Synthesis, Structure, and Properties of Unsaturated Nitrogen Heterocycles* (P017576); 24 months; $29,600

Sawyer, Donald T.; *Nonaqueous Electrochemistry of Flavin Model Compounds* (P016114); 24 months; $38,500

Wing, Richard M.; *Inorganic Structural Chemistry* (P019254); 24 months $40,600

UNIVERSITY OF CALIFORNIA, San Diego

Faulkner, D. John; *Stereoselective Synthesis of Trans-Trisubstituted Olefinic Bonds* (P016355); 24 months; $22,500

Linck, Robert G.; *Inner-Sphere and the Outer-Sphere Electron Transfer Mechanisms* (P018040); 12 months; $15,000

Wheeler, John C.; *Statistical Mechanical and Thermodynamic Investigations of Phase Transitions and Critical Phenomena in Multi-component Systems* (P015429); 24 months; $34,900

Wilson, Kent; *Photofragment Spectroscopy of Diatomic Dissociative States* (P022524); 24 months; $70,000

UNIVERSITY OF CALIFORNIA, Santa Barbara

Bowers, Michael T.; *Ion Cyclotron Resonance* (P015628); 24 months; $27,000

Harris, David O.; *Microwave Spectroscopy and Molecular Structure* (P023473); 24 months; $35,000

Martin, Richard M.; *Electronic Energy Transfer in Crossed Molecular Beams* (P019581); 24 months; $30,000

Miller, Glenn H.; *Oxidation of Polymers* (P019663); 24 months; $26,000

UNIVERSITY OF SOUTHERN CALIFORNIA

Beaudet, Robert A.; *EPR Spectra of Triplet States* (P016136); 24 months; $37,500

Burg, Anton B.; *Chemical Consequences of Fluorocarbon Phosphines* (P017472); 24 months; $48,000

Dows, David A.; *Electric Field-Induced Spectra* (P019884); 24 months; $60,100

Schnepp, Otto; *Lattice Vibrations of Molecular Solids and Intermolecular Potentials* (P019662); 24 months; $68,500

COLORADO

COLORADO STATE UNIVERSITY

Daugherty, Ned A.; *Kinetics of Some Oxidation—Reduction Reactions of Tin (II) in Aqueous Perchloric Acid* (GP13900001); $1,495

——— *Kinetics of Some Oxidation—Reduction Reactions of Tin (II) in Aqueous Perchloric Acid* (P013900); 24 months; $35,000

Miller, Larry L.; *Vinyl Cation Energetics* (P018883); 24 months; $21,200

Skogerboe, R. K.; *Spectrochemical Applications of Low Power Microwave Excitation* (P021306); 24 months; $36,600

UNIVERSITY OF COLORADO

Damrauer, Robert; *Strained Organosilicon Rings—Synthesis and Mechanism* (P014311); 24 months; $24,300

DePuy, Charles H.; *Cyclopropane and Organometallic Chemistry* (P13783X); 12 months; $35,900

Hassner, Alfred; *Chemistry and Stereochemistry of Nitrogen Functions and Related Heterocycles* (P019253); 24 months; $41,800

King, Edward L.; *Complex Ions in Solution* (GP7185X002); 12 months; $26,800

Norman, Arlan D.; *Germanium—Silicon-Boron Hydrides and Phosphorus Cage Molecules* (P023575); 24 months; $46,400

CONNECTICUT

UNIVERSITY OF CONNECTICUT

Krause, Ronald A.; *Transition Metals-Polysulfide Chelates* (P020457); 24 months; $34,400

Masterton, William L.; *Ion Association and Ion-Molecular Interactions in Solutions of Complex Ion Electroyltes* (P012742); 12 months; $20,000

MATHEMATICAL AND PHYSICAL SCIENCES RESEARCH PROJECTS

YALE UNIVERSITY
McBride, J. Michael; *Crystal Chemistry of Radical Pairs* (P014607); 24 months; $26,500
Scott, A. I.; *Biosynthesis and Biogenetic-Type Synthesis Involving Oxidation of Aromatic Substrates* (P012642); 24 months; $65,500
Sinanoglu, Oktay; *Theoretical Chemistry* (P05899X003); 12 months; $60,300
Smith, Allan L.; *Molecular Emission Continua, Collisions of Excited States, and Molecular Ionization* (P014310); 24 months; $53,300
Suplinskas, Raymond J.; *Extensions of the Kinematic Model for Chemical Reaction* (P023275); 24 months; $30,200
Wiberg, Kenneth B.; *Mechanisms of Oxidative Processes* (GP8836X003); 12 months; $42,700

DELAWARE

UNIVERSITY OF DELAWARE
Burmeister, John L.; *Coordination Chemistry of Ambidentate Ligands* (P020607); 24 months; $25,400
Munson, Burnaby; *Chemical Ionization Mass Spectrometry* (P020231); 24 months; $35,000

DISTRICT OF COLUMBIA

CATHOLIC UNIVERSITY OF AMERICA
Sanders, William A.; *Calculation of Interatomic and Intermolecular Forces by Perturbation Theory* (P021298); 24 months; $30,600

GEORGETOWN UNIVERSITY
Horak, Vaclav; *Structure and Reactivity of Sulfur Heteroaromatics* (P019419); 24 months; $50,000

HOWARD UNIVERSITY
Eisner, Ulli; *Seven and Eight Membered Unsaturated Heterocycles* (P020008); 12 months; $15,200

NATIONAL ACADEMY OF SCIENCES, NATIONAL RESEARCH COUNCIL
Paul, Martin A.; *Conference on Computational Support for Research in Quantum Chemistry* (C310176); 12 months; $7,600
Wood, George W.; *Task Order for Support of the U. S. National Committee for Crystallography* (C310123001); 24 months; $10,300

FLORIDA

FLORIDA STATE UNIVERSITY
Clark, Ronald J.; *Metal Carbonyl-Phosphorus Trifluoride Chemistry* (P020276); 24 months; $28,000
Herz, Werner; *Terpene Chemistry* (P012582); 24 months; $47,300
Leffler, John E.; *Thermal and Photochemical Reactions of Free Radicals* (P015991); 24 months; $42,800
Quagliano, James V.; *Coordination Compounds Containing Positively Charged Ligands* (P022740); 24 months; $44,400
Walborsky, H. M.; *Dissolving Metal and Electrolytic Reductions* (P018118); 24 months; $39,800

UNIVERSITY OF FLORIDA
Bates, Roger G.; *Salt and Solvent Effects on Analytical Processes* (P014538); 24 months; $46,100
Butler, George B.; *Polymers Possessing a Thermally Initiated Flexibilizing Mechanism Resulting from Dissociation of Dative Bonds* (P017926); 24 months; $31,400
Cram, Stuart P.; *Gas Chromatographic Band Broadening Processes* (P014754); 24 months; $28,000
Deyrup, James A.; *Effects of Unshared Electrons on Adjacent Reacting Groups* (P017642); 24 months; $30,600
Dolbier, William R.; *Simultaneity of Cycloaddition Reactions* (P020598); 12 months; $14,500
Löwdin, Per-Olov and Öhrn, N. Yngve; *Quantum Chemistry of Atoms and Molecules* (P016666); 24 months; $56,000
Muschlitz, E. E. Jr.; *Reactions of Metastable Atoms and Molecules* (P021105); 12 months; $64,000
Person, Willis B.; *Infrared Intensity Studies* (P017818); 24 months; $52,100
Stoufer, R. Carl; *Spin-Pairing in First Row Transition Metal Complexes* (P018521); 12 months; $20,000
Vala, Martin; *Model Aromatic Carbonyl Compounds* (P012740); 24 months; $33,300

UNIVERSITY OF MIAMI
Criss, Cecil M.; *Thermodynamic Properties of Nonaqueous Solutions of Electrolytes* (P014537); 24 months; $38,000
Doepker, Richard D.; *Vacuum-Ultraviolet Photolysis of the Unsaturated Acyclic Hydrocarbons* (P020878); 24 months; $33,200

GEORGIA

EMORY UNIVERSITY
Clever, H. Lawrence; *Binary Nonelectrolyte Solutions* (P020604); 12 months; $11,900
Menger, Fred M.; *Mechanism of Chymotrypsin Action* (P011275); 24 months; $24,000

GEORGIA INSTITUTE OF TECHNOLOGY
Ashby, Eugene C.; *Organometallic Compounds—Composition in Solution, Mechanisms and Stereochemistry of Reaction* (P014795); 24 months; $40,000
Bertrand, J. Aaron; *Oxygen-Bridged Complexes of Transition Metal Ions* (P020885); 24 months; $45,000
Felton, Ronald H.; *Anions and Cations of Metalloporphyrines and Some Derivatives* (P017061); 24 months; $28,000
Grovenstein, Erling Jr., *Chemistry of Carbanions* (P019251); 24 months; $40,000
Liotta, Charles L.; *Mechanism of Transmission of Non-Conjugative Substituent Effects* (P014437); 24 months; $23,000

UNIVERSITY OF GEORGIA
King, R. Bruce; *Organometalic Derivatives of the Platinum Metals* (GP9662); 24 months; $38,500
Leyden, Donald E., Blount, Charles W. and Noakes, John E.; *Trace Element Analysis by Direct Determination on Ion Exchange Resins* (P024311); 24 months; $80,000
Story, Paul R.; *Peroxide Fragmentation Reactions* (P018911); 24 months; $31,100

HAWAII

UNIVERSITY OF HAWAII
Andermann, George; *Infrared Optical Properties of Crystals* (P013087); 24 months; $22,000
Hubbard, A. T.; *Thin Layer Electrochemistry in Molten Salt Solutions* (P019747); 24 months; $42,600
Liu, Robert S. H.; *Non-Radiative Processes in Electronically Excited Organic Molecules* (P014248); 24 months; $44,300
Schaleger, Larry L.; *Mechanisms of Hydrolysis* (P014693); 24 months; $26,900

IDAHO

UNIVERSITY OF IDAHO
Shreeve, Jean'ne M.; *Nonmetallic Fluorine Compounds* (P012647); 24 months; $51,900

ILLINOIS

ILLINOIS INSTITUTE OF TECHNOLOGY
Gutman, David; *Shock Tube and Mass Spectrometry Study of the Acetylene-Oxygen Reaction* (P023467); 24 months; $50,500
Larsen, Russell D.; *Statistical Spectral Analysis of Chemically-Related Time Series* (P016760); 24 months; $31,100

NORTHWESTERN UNIVERSITY
Frost, Arthur A.; *Electronic Energy of Small Molecules* (P019252); 24 months; $43,100
Ibers, James A.; *Structures and Properties of Transition Metal Complexes* (P020700); 24 months; $97,000
Marshall, James A.; *New Synthetic Approaches to Carbocyclic Compounds* (P016234); 24 months; $57,400
Pearson, Ralph G.; *Rates and Mechanisms of Very Rapid Inorganic Reactions* (GP6341X003); 12 months; $26,600
Smith, Donald E.; *Voltammetry with Varying Potentials* (P016281); 24 months; $65,000

QUINCY COLLEGE
Lang, Robert P.; *Spectrophotometric Study of Some Molecular Complexes of Iodine* (P017937); 24 months; $10,100

UNIVERSITY OF CHICAGO
Closs, Gerhard L.; *Small Ring Compounds* (P18719X); 12 months; $46,700
Hinze, Juergen; *Quantum Mechanical Studies of Molecular Structure* (P021108); 24 months; $44,000
Kleppa, Ole J.; *Thermodynamic Properties of Fused Salt Mixtures* (P014064); 24 months; $65,000
Levy, Donald H.; *Radio Frequency Spectroscopy of Excited States and Short-Lived Species* (P017891); 24 months; $63,800
Light, John C.; *Theoretical Chemistry—Scattering Processes' Density Matrices* (P021294); 24 months; $75,000
Meyer, Lothar; *Properties of Matter at Low Temperatures* (P019665); 24 months; $94,200
Yang, N. C.; *Photochemistry of Organic Compounds* (P016347); 24 months; $60,000

UNIVERSITY OF ILLINOIS, Chicago Circle
Baumgarten, Ronald J.; *New Approach to Deamination via a New Category of Nucleophilic Substitution Leaving Groups* (P017176); 24 months; $26,200
Carlin, Richard L.; *Specific Heats at Low Temperatures* (P012106); 24 months; $50,900
Matthews, Clifford N.; *Carbophosphoranes* (P016774); 24 months; $33,100
Moriarty, Robert M.; *Photochemical and Thermal Decomposition of Geminal Diazides and 1,5 Disubstituted Tetrazoles* (P018113); 24 months; $34,400

UNIVERSITY OF ILLINOIS, Urbana
Belford, R. Linn; *Unimolecular and Atom Abstraction Reactions* (P017875); 24 months; $67,200
Brown, Theodore L.; *Organometallic Chemistry* (GP6396X003); 12 months; $42,600
Flygare, W. H.; *High Resolution Microwave Spectroscopy* (P12382X001); 12 months; $67,000
Gutowsky, H. S.; *Nuclear Magnetic Resonance* (P016568X); 12 months; $48,500
Jonas, Jiri; *Nuclear Magnetic Resonance Study of Relaxation Phenomena* (GP12402001); $16,200
Laitinen, H. A.; *Surface Phenomena in Electroanalytical Chemistry* (P012831); 12 months; $29,200
Leonard, Nelson J.; *Structural Studies on Medium- and Small-Ring Compounds* (P08407X001); 12 months; $48,800
Malmstadt, H. V.; *Short-Time Phenomena in Sources of Spectrochemical Importance* (P018910); 24 months; $84,000
Marcus, Rudolph A.; *Theoretical Studies in Chemical Kinetics* (P020128); 24 months; $71,200
Martin, James C.; *Organic Chemistry* (P013331); 24 months; $51,700
Secrest, Don; *Computer Studies of Quantum Collisions* (P016458); 24 months; $25,800
Smith, Stanley G.; *Kinetics of Reactions of Grignard and Lithium Reagents* (P013329); 24 months; $35,000
Stucky, Galen D.; *Structural Properties of Group I, II and III Organometallic Compounds* (P015307); 24 months; $35,000

CHEMISTRY

Yardley, James T.; *Molecular Energy Transfer and Excited-State Dynamics* (P019068); 24 months; $45,000

INDIANA

INDIANA UNIVERSITY, Bloomington

Allerhand, Adam; *Magnetic Resonance and Relaxation Studies* (P017966); 24 months; $47,600

Kochi, Jay K.; *Oxidation and Reduction of Organic Compounds by Metal Species* (P16845X); 24 months; $132,100

Parmenter, Charles S.; *Energy Conversion Mechanisms in Polyatomic Molecules* (P022543); 24 months; $80,000

Shull, Harrison; *Atomic and Molecular Structure* (GP6650X003); 12 months; $54,300

Wehry, Earl L. Jr.; *Intermolecular Electronic Energy Transfer in Inorganic Coordination Compounds* (P08705001); 3 months; $4,500

PURDUE UNIVERSITY

Brewster, James H.; *Optically Active Styrenes and Stilbenes* (P017876); 24 months; $35,000

Brown, Herbert C.; *Structure and Reactivity* (GP6492X003); 12 months; $39,500

Cobble, James W.; *Thermodynamic and Kinetic Studies on Aqueous Solutions at Higher Temperatures* (P024609); 24 months; $40,000

Margerum, Dale W.; *Very Fast Substitution Reactions of Metal Complexes* (P14781X); 12 months; $38,800

Muller, Norbert; *Nuclear Magnetic Resonance Studies with Fluorine-Labelled Surfactants* (P019551); 24 months; $50,100

Perone, Sam P.; *Application of the On-Line Digital Computer to Diverse Problems in Chemical Research* (P021111); 24 months; $48,000

Robinson, William R.; *Organometallic Derivatives of Transition Metal Nitrosyls* (P017554); 24 months; $28,400

Rogers, L. B.; *On-Line Control for Chromatography* (P020727); 24 months; $100,000

Tobias, R. Stuart; *Sigma-Bonded Organometallics* (P023208); 24 months; $60,300

Truce, William E.; *Beta Sultones and Vinyl Sulfonates* (P012326); 12 months; $16,500

Walton, Richard A.; *Transition Metal Halides and Oxyhalides in Low Oxidation States* (P019422); 24 months; $26,300

UNIVERSITY OF NOTRE DAME

Hayes, Robert G.; *Effects of Pressure on Ligand Hyperfine Couplings* (P010063); 24 months; $24,000

IOWA

GRINNELL COLLEGE

Danforth, Joseph C.; *Kinetics and Chemistry of Solid Decompositions at Surfaces* (P009513001); $2,500

Erickson, Luther E.; *Conformational Analysis, Deprotonation and Racemization of Stable Diamagnetic Complexes* (P019229); 24 months; $16,500

IOWA STATE UNIVERSITY

Chapman, O. L.; *Meta-Stable Ground State Intermediates in Organic Photochemistry* (GP10164001); $8,200

Johnson, Dennis C.; *Electrochemical Study of Photochemical Reactions* (P018575); 24 months; $35,200

Trahanovsky, Walter S.; *Oxidation and Reduction of Organic Compounds* (P018031); 24 months; $35,000

Verkade, John G.; *Spectroscopic Studies of Phosphorus-Metal Bonding* (P023179); 24 months; $40,900

UNIVERSITY OF IOWA

Wawzonek, Stanley; *Electrochemical Preparation of 1,4-Dehydrobenzenes—Electrochemical Reduction of Organic Compounds in Propylene Carbonate* (P017317); 24 months; $53,000

KANSAS

KANSAS STATE UNIVERSITY

Cooks, R. G. and Setser, Donald W.; *Energy Partition in the Fragmentation of Organic Molecular Ions* (P016743); 24 months; $21,900

Lambert, Jack L.; *Insoluble Ion Association Compounds as Analytical Reagents* (P022734); 24 months; $35,700

McDonald, Richard N.; *Strained Ring Systems* (P010691); 24 months; $42,800

UNIVERSITY OF KANSAS

Everett, Grover W. Jr.; *Effects of Chiral Ligands on the Coordination Geometries of Some Transition Metal Complexes* (P019877); 24 months; $22,100

Harmony, Marlin D.; *Microwave Studies of N^{14} Quadrupole Coupling Constants* (P015127); 24 months; $30,000

Landgrebe, John A.; *Mechanistic Studies of Insertion Reactions by Divalent Carbon Intermediates* (P014112); 12 months; $16,000

KENTUCKY

UNIVERSITY OF KENTUCKY

Guthrie, Robert D.; *Carbanions, Electron Transfer vs. Proton Capture* (P017465); 24 months; $31,000

Niedenzu, Kurt; *The Formation of Borazines* (P014245); 24 months; $27,500

LOUISIANA

LOUISIANA STATE UNIVERSITY, Baton Rouge

Kestner, Neil R.; *Intermolecular Forces and Excess Electrons in Liquids* (P015279); 24 months; $34,600

Runnels, L. K.; *Dense Fluids and Phase Transitions* (P017026); 24 months; $41,700

Selbin, Joel; *Complexes of Transition Metal Oxocations* (P013275); 12 months; $19,200

West, Philip W.; *Selective and Sensitive Methods for the Determination of Sulfuric Acid Aerosols* (P018081); 24 months; $40,400

LOUISIANA STATE UNIVERSITY, New Orleans

Davis, Donald G.; *Electron Transfer Rates of Metalloporphyrins* (P019749); 24 months; $31,500

Kern, Ralph D. Jr.; *Exchange Reactions Using the Shock Tube Technique* (P023137); 24 months; $33,800

TULANE UNIVERSITY

Mague, Joel T.; *Oxidative Addition Reactions Using Ditertiary Phosphine and Arsine Complexes of Rhodium (I)* (P016380); 24 months; $27,400

Nugent, Maurice J.; *Steric and Electronic Effects in Paracyclophanes* (P016356); 24 months; $34,100

MARYLAND

JOHNS HOPKINS UNIVERSITY

Cowan, Dwaine O.; *Heavy Atom Effects in Organic Photochemistry* (P017751); 24 months; $29,000

Kokes, Richard J.; *Hydrogenation and Isomerization of Olefins Over Solid Catalysts* (P022830); 24 months; $34,000

UNIVERSITY OF MARYLAND

Grim, Samuel O.; *Coordination Compounds of Phosphorus* (P012539); 12 months; $18,500

Moore, John H. Jr.; *Inelastic, Non-Charge-Exchange Scattering of Ions from Small Molecules* (P019177); 24 months; $34,800

Staley, Stuart W.; *Hydrocarbon Anions* (P017899); 24 months; $50,400

MASSACHUSETTS

BOSTON UNIVERSITY

Gensler, Walter J.; *Bicyclo (2.1.0.) Pentane System* (P015631); 24 months; $31,000

Lowe, Marian A.; *Exciton-Magnon and Exciton-Exciton Transitions in Magnetically Ordered Insulators* (P017665); 24 months; $32,200

BRANDEIS UNIVERSITY

Grunwald, Ernest; *Fast Proton Transfer Reactions and Ion Pair Dissociation* (GP7381X002); 12 months; $32,800

Steel, Colin; *Primary Processes in the Photochemistry of Azo Compounds* (P018808); 24 months; $42,100

Tuttle, Thomas R. Jr.; *Spectroscopy of Metals and of Paramagnetic Anion Radical Salts in Polar Solvents* (P015517); 24 months; $39,000

CLARK UNIVERSITY

Allen, Harry C.; *Metal-Metal Interactions in Transition Metal Chelates* (P015516); 24 months; $34,800

HARVARD UNIVERSITY

Bartlett, Paul D.; *Mechanisms of Organic Reactions* (P08605X001); 12 months; $14,100

Corey, Elias J.; *Organic Chemistry—(A) Cyclic Structures Including Terpenes, (B) Organometallic Chemistry, (C) Photochemistry, (D) Synthetic Methodology* (P12692X001); 14 months; $100,000

Doering, Wm. von Eggers; *Mechanism of the Methylenecyclopropane and Related Rearrangements* (P18618X); 24 months; $62,000

Dolphin, David H.; *Pyrrole Chemistry* (P016761); 24 months; $20,900

Herschbach, Dudley R.; *Molecular Mechanics of Chemical Reactions* (GP5351X004); $25,000

Karplus, Martin; *Problems in Theoretical Chemistry* (P07907X002); 12 months; $46,900

Kistiakowsky, G. B.; *Gas Phase Reactions* (GP6424X003); 12 months; $41,500

Klemperer, William; *Molecular Structure and Molecular Collisions of Excited Electronic States of Molecules* (P018717); 24 months; $135,300

White, James D.; *Biogenetically Patterned Synthesis of Secondary Metabolites* (P015331); 24 months; $40,000

Wilson, E. Bright Jr.; *Molecular Spectroscopy and Pure Quantum Mechanical Theory* (P14012X001); 15 months; $90,500

MASSACHUSETTS INSTITUTE OF TECHNOLOGY

Amdur, I., Kinsey, J. L. and Oppenheim, I.; *Intermolecular Forces* (P019988); 12 months; $354,400

Baldwin, J. E.; *Bond Formation by Valence Rearrangement of Heterosystems* (P016781); 24 months; $35,000

Berchtold, Glenn A.; *Chemistry of Enamine N-Oxides* (P014880); 24 months; $50,100

Cotton, F. A.; *Inorganic and Organometallic Chemistry* (GP7034X002); 12 months; $63,800

Deutch, John M.; *Statistical Mechanical Investigations of Relaxation in Fluid Systems* (P018111); 24 months; $32,300

Dubrin, James W.; *Comparison of the Primary Reaction Modes of Internally and Translationally Excited Hydrogen Atoms* (P013277); 24 months; $31,000

Garland, Carl W.; *Ultrasonic Measurements Near Transition and Critical Points* (P013548); 24 months; $61,700

Holm, Richard H.; *Transition Metal Chemistry* (P018978); 12 months; $63,000

Kukolich, Stephen G.; *High Resolution Microwave Spectroscopy and Molecular Zeeman Effect Measurements* (P021110); 24 months; $35,100

Lord, Richard C.; *Methods and Applications of Far Infrared Spectroscopy* (P13473X001); 12 months; $50,000

Seyferth, Dietmar; *Unsaturated Organometallics* (GP6466X003); 12 months; $34,800

Silbey, Robert J.; *Theoretical Studies of the Excited States of Molecular Crystals* (P019874); 24 months; $37,600

Swain, C. Gardner; *Effect of Structural Changes in Reactants on the Structure of Transition States* (P014698); 24 months; $68,400

MATHEMATICAL AND PHYSICAL SCIENCES RESEARCH PROJECTS

NORTHEASTERN UNIVERSITY
DesMarteau, Darryl D.; *Chemistry of Xenon* (P023098); 24 months; $21,200

TUFTS UNIVERSITY
DeWald, Robert R.; *Rates of Electron-Reduction Reactions in Liquid Ammonia* (P020009); 24 months; $44,200
Gibb, Thomas R. P. Jr.; *Enrichment of Metals in Seawater by Foam Production* (P018623); 12 months; $14,100
Urry, Grant; *Perchloro-Polysilanes and Related Compounds* (P016577); 24 months; $35,700

UNIVERSITY OF MASSACHUSETTS
Archer, Ronald P.; *Isomerization in Coordination Compounds* (P020275); 12 months; $21,000
Cade, Paul E.; *Scattering Factors of Simple Molecules and Momentum-Space Quantum Chemistry* (P014606); 24 months; $31,800
Lillya, C. Peter; *Photoisomerization of Conjugated Enones and Dienones* (P020367); 24 months; $33,000
Macknight, William J.; *Role of Interchain Forces in Rubber Elasticity* (P022530); 24 months; $29,100
Rowell, Robert L.; *Colloid and Molecular Light Scattering* (P014110); 24 months; $35,000

WORCESTER POLYTECHNIC INSTITUTE
Bushweller, C. Hackett; *Multiheteroatomic Medium Rings and the Various Roles of the Lone Pair* (P018197); 24 months; $31,900

MICHIGAN

HOPE COLLEGE
Wettack, F. Sheldon; *Gas Phase Vibronic Energy Dissipation in Carbonyl Containing Compounds* (GP14308001); $1,055

MICHIGAN STATE UNIVERSITY
Brubaker, Carl H. Jr.; *Organometallic Compounds of Transition Elements* (P17422X); 12 months; $29,900
Crouch, Stanley R.; *Fast Kinetic Studies of Analytical Systems* (P018123); 24 months; $34,000
Legoff, Eugene; *Synthesis of Annellated Four-Membered Ring Hydrocarbons* (P017015); 24 months; $29,000
Nicholson, Richard S.; *Investigations of Organic Electrode Processes* (P010671001); $16,300
Schwendeman, Richard H.; *Molecular Structure and Rotational Relaxation by Microwave Spectroscopy* (P020883); 24 months; $48,700
——— *Internal Torsion by Microwave Spectroscopy* (GP8296 001); $32,900
Wagner, Peter J.; *Structure-Reactivity Relationships in Ketone Photochemistry* (P019580); 24 months; $50,400

OAKLAND UNIVERSITY
Harmon, Kenneth M.; *Conjugated Organoboron Hydride Ions and Polymers* (P016988); 12 months; $9,300

UNIVERSITY OF MICHIGAN
Bartell, Lawrence S.; *Electron Diffraction Study of Molecular Structure and Bonding* (P016030); 24 months; $93,500
Brintzinger, Hans H.; *Hydride Complexes of Group IV-VI Transition Metals* (P019421); 24 months; $81,500
Current, Jerry H.; *Spectroscopy Study of Small Radicals* (P015126); 12 months; $15,200
Dunn, Thomas M.; *Electronic Spectra of Inorganic Compounds* (P016089); 24 months; $61,300
Kopelman, Raoul; *Excitons in Mixed Molecular Crystals* (P018718); 24 months; $37,600
Mark, James E.; *Electrical Properties of Chain Molecules* (P016028); 24 months; $34,000
Morris, Michael D.; *Organometallic Electroanalytical Chemistry* (P014408); 12 months; $16,200
Overberger, Charles G.; *Cyclic Azo Compounds—Synthesis, Stereochemistry, and Decomposition Studies of Novel Ring Systems* (P012325); 24 months; $53,800
Smith, Peter A. S.; *Intramolecular Nitrenoid Reactions of Aryl Azides* (P015830); 24 months; $55,900

WAYNE STATE UNIVERSITY
Endicott, John F.; *Photochemistry of Inorganic Complex Ions* (P017082); 12 months; $17,200
Glick, Milton D.; *Structural Studies of Rare Earth Complexes* (P015070); 24 months; $27,000
Johnson, Carl R.; *Synthetic Sterochemical and Mechanistic Aspects of Sulfoxides, Sulfoxines and Sulfilimine* (P019623); 24 months; $55,200
Meyers, Albert I.; *Synthetic Utility of Heterocyclic Compounds. A Versatile Aldehyde Synthesis from Dihydro-1,3-Oxazines* (P022541); 12 months; $15,900
Oliver, John P.; *Reactivity and Bond Character of Organometallic Compounds* (P019299); 24 months; $39,600
Raban, Morton; *Torsional Stereoisomerism in Sulfenamides* (P017092); 24 months; $31,000

MINNESOTA

UNIVERSITY OF MINNESOTA
Bolton, James R.; *Electron Spin Resonance Study of Spin Relaxation and Chemical Kinetics* (GP8416 001); 6 months; $5,000
Britton, Doyle; *Structural Studies of Inorganic Cyanides and Related Compounds* (P016570); 24 months; $53,000
Hexter, Robert M.; *Phonon Interactions in Molecular Crystals* (P017553); 24 months; $51,800
Kreevoy, Maurice M.; *Solvent and Structure in Proton Transfer to Carbon and Hydrogen* (P013172); 24 months; $35,300
Meehan, E. J.; *Light Scattering Studies* (P014604); 24 months; $35,000
Prager, Stephen and Mead, C. Alden; *Theoretical Studies of Quantum Mechanical Systems Near the High-Mass Limit* (P017195); 24 months; $60,000

MISSOURI

UNIVERSITY OF MISSOURI, Columbia
Aue, Walter A.; *Chemically Bonded Chromatographic Supports* (P018616); 24 months; $25,600

WASHINGTON UNIVERSITY
Gutsche, C. David; *Intramolecular and Pseudo-Intramolecular Reactions* (P11087X001); 13 months; $34,200
Wahl, Arthur C.; *Oxidation-Reduction Reactions* (GP5939X003); 12 months; $41,200

MONTANA

MONTANA STATE UNIVERSITY
Anacker, Edward W.; *Surfactant Structure and Charge* (P018756); 24 months; $25,000

UNIVERSITY OF MONTANA
Shafizadeh, Fred; *Pyrolytic Reactions of Cellulosic Materials* (P015556); 24 months; $44,000
Waters, William L.; *Syntheic Utility of Organomercurials* (P018317); 24 months; $27,400
Woodbury, George W. Jr.; *Surface Tension Calculations for Lattice Models of Liquids* (P018913); 24 months; $21,800

NEW HAMPSHIRE

DARTMOUTH COLLEGE
Lemal, David M.; *Synthesis and Chemistry of* $(Cy)_6$ *and* $(Cy)_4$ *Valence Isomers* (P021296); 24 months; $59,200
Stockmayer, Walter H.; *Physical Chemistry of High Polymers* (P016232); 24 months; $66,000

UNIVERSITY OF NEW HAMPSHIRE
Owens, Charles W.; *Hot-Atom Chemistry of Solids* (P020061); 12 months; $15,200

NEW JERSEY

NEWARK COLLEGE OF ENGINEERING
Suchow, Lawrence; *Complex Rare Earth Chalcogenides and Pnictides* (P011527); 24 months; $25,100

PRINCETON UNIVERSITY
Jones, Maitland, Jr.; *Reactive Intermediates* (P012759); 24 months; $40,000
McClure, Donald S.; *Electronic Spectroscopy* (P016031); 24 months; $87,200

RUTGERS UNIVERSITY
Goodman, Lionel; *Studies of Electronic Promotion* (GP6301X003); 12 months; $39,300
Moss, Robert A.; *Diazotate and Deamination Chemistry* (P012645); 24 months; $33,000

NEW MEXICO

NEW MEXICO STATE UNIVERSITY
Ames, Lynford L.; *High-Temperature Thermodynamic Studies on the Rare-Earth Gaseous Monosulfides, Gaseous Dicarbides and Solid Hexaborides* (P012991); 24 months; $23,000

NEW YORK

ADELPHI UNIVERSITY
Landesberg, Joseph M.; *Transition Metals in Organic Chemistry* (P014741); 24 months; $22,000

CLARKSON COLLEGE OF TECHNOLOGY
Brunauer, Stephen; *Pore Structures of Adsorbents, Catalysts and Other Solids* (P014312); 12 months; $23,000
Matijevic, Egon; *Interactions of Complex Solute Species with Charged Interfaces* (P012220); 24 months; $30,700

COLUMBIA UNIVERSITY
Berne, Bruce J.; *Dynamics of Molecular Motion in Condensed Media* (P022881); 24 months; $42,000
Breslow, Ronald; *Intramolecular Solvation* (P017400); 24 months; $39,900
Pechukas, Philip; *Multidimensional WKB Methods in Molecular Quantum Mechanics* (P019097); 24 months; $28,500
Stern, Richard C.; *Molecular Beam Study of Inelastic Electron-Molecule Collisions* (P011631); 24 months; $44,300
Stork, Gilbert; *Synthetic Methods and Total Synthesis in Organic Chemistry* (P13785X001); 12 months; $59,700
Zare, Richard N.; *Molecular Fluorescence with Emphasis on Laser Excitation* (P016029); 24 months; $118,700

CORNELL UNIVERSITY
Bauer, Simon H. and Wilcox, C. F. Jr.; *Structure Analysis by Electron Diffraction* (P013471); 12 months; $32,100
Burlitch, James M.; *Metal—Metal Bonds* (P023513); 24 months; $36,400
Caldwell, Richard A.; *Photochemical Mechanisms* (P014796); 24 months; $28,100
Fay, Robert C.; *Stereochemistry of Some Metal Diketonates* (P016280); 24 months; $43,200
Goldstein, Melvin J.; *Isotopes and Concerted Reaction Mechanisms* (P015863); 12 months; $26,500
Miller, William T. Jr.; *Chemistry of Unsaturated Carbon-Fluorine Compounds* (P06501X003); 12 months; $34,000
Morrison, George H.; *Quantitative Mass Spectroscopy of Solids* (P06471X003); 12 months; $34,500
Scholer, Frederick R.; *Synthesis and Characterization of Metalloborane Polyhedra* (P017053); 24 months; $24,600
Sienko, M. J.; *Metal-Ammonia Solutions* (P017706); 24 months; $55,000

CHEMISTRY

CUNY, HUNTER COLLEGE
Wijnen, M. H. J.; *Hydrogen Atom Reactions with Chlorinated Olefins and the Photolysis of Chlorinated Compounds* (P014115); 24 months; $31,200

CUNY, QUEENS COLLEGE
Locke, David C.; *Selectivity in Liquid-Liquid Chromatography* (P017551); 24 months; $24,000

NEW YORK UNIVERSITY
Sundheim, Benson R.; *Photochemical Processes in Fused Salts* (P014615); 24 months; $37,200

POLYTECHNIC INSTITUTE OF BROOKLYN
Morawetz, Herbert; *Molecular Conformation by X-Ray Scattering from Solutions* (P019523); 12 months; $20,500

RENSSELAER POLYTECHNIC INSTITUTE
Bailey, Ronald A.; *Metal Ions in Fused Salts—Structure and Complexing* (P017528); 24 months; $34,600
Ferris, James P.; *Structure of the HCN Polymer and its Role in Chemical Evolution* (P019255); 24 months; $39,700
Harteck, Paul and Dondes, Seymour; *Photochemical Separation of Isotopes* (P014518); 24 months; $42,000
Wunderlich, Bernhard; *Heat Capacities of Solid Polymers* (P017206); 24 months; $46,000

STATE UNIVERSITY AT ALBANY
Bank, Shelton; *Reduction and Addition Reactions of Aromatic Radical Anions* (P019227); 24 months; $33,100
Frisch, Harry L.; *Chemical Instabilities and Their Study by Light Scattering* (P019881); 24 months; $59,200
Zuckerman, J. J.; *Organotin Compounds* (P016544); 24 months; $30,400

STATE UNIVERSITY AT BUFFALO
Nancollas, George H.; *Metal Complexes and Ion-Pairs* (P020405); 24 months; $33,000
Rechnitz, Garry A.; *Transient Phenomena at Ion-Selective Membrane Electrodes* (P011727); 24 months; $42,800
Wilkins, Ralph G.; *Rapid Inorganic Reactions in Solution* (P016580); 24 months; $56,000

STATE UNIVERSITY AT STONY BROOK
Fowler, Frank W.; *The Synthesis and Study of Non-Aromatic Unsaturated Heterocycles* (P020099); 24 months; $28,000
Haim, Albert; *Redox Reactions of Coordination Compounds in Aqueous Solution* (P009669); 24 months; $61,900
Hirota, Noboru; *Excited Molecules and Free Radicals* (P015365); 24 months; $40,000
Johnson, Philip M.; *Electronic Structure and Photochemical Decomposition of Organic Molecules* (P014439); 24 months; $43,300
Whitten, J. L.; *Theoretical Studies of Ground and Excited States of Polyatomic Molecules* (P018121); 24 months; $32,000

SYRACUSE UNIVERSITY
Dittmer, Donald C.; *Small-Ring Sulfur Compounds* (P020729); 24 months; $45,400
Goodisman, Jerry; *Statistical Calculations of Molecular Electronic Densities* (P020718); 24 months; $36,300

UNIVERSITY OF ROCHESTER
Muenter, John S.; *Molecular Beam Electric Resonance Spectroscopic Studies* (P018805); 24 months; $36,400
Schlessinger, R. H.; *Thiepins* (P013089); 24 months; $28,700

YESHIVA UNIVERSITY
Snyder, James P.; *Molecular and Electronic Structure of Ylides by NMR Spectroscopy* (P020403); 12 months; $14,100

NORTH CAROLINA

DUKE UNIVERSITY
Krigbaum, William R.; *X-Ray Diffraction Studies of Polymers* (P017873); 24 months; $42,200
Smith, Peter; *Reaction Kinetics by E.P.R. Spectroscopy* (P017579); 12 months; $20,300

EAST CAROLINA UNIVERSITY
Lamb, Robert C.; *Kinetics-Efficiency Experiments on Some Mixed Aliphatic Diacyl, Substituted-Benzoyl Peroxides* (P013845); 24 months; $20,100

NORTH CAROLINA STATE UNIVERSITY AT RALEIGH
Bumgardner, Carl L.; *Photochemical Reactions of Tetrafluorohydrazine* (P017316); 24 months; $47,800
Long, G. Gilbert and Bowen, Lawrence H.; *Mossbauer Spectroscopy of Antimony Compounds* (P020719); 24 months; $68,000

UNIVERSITY OF NORTH CAROLINA AT CHAPEL HILL
Buck, Richard P.; *Ion Selective Electrodes* (P020524); 24 months; $32,000
Hatfield, William E.; *Spin-Spin Coupling in Magnetically Condensed Complexes* (P022887); 24 months; $40,000
Isenhour, Thomas L.; *Computerized Learning Machines Applied to Chemical Problems* (P017674); 24 months; $47,000
Meyer, Thomas J.; *Electron Transfer Studies on Metal Complexes* (P017083); 24 months; $22,400
Rieke, Reuben D.; *Electrochemical Studies of 1,4-Diphosphoniacyclohexadiene-2,5 Salts* (P018112); 24 months; $22,500

UNIVERSITY OF NORTH CAROLINA AT GREENSBORO
Herman, Harvey B.; *Electrochemical Measurements of the Rate of Chemical Reactions* (P015374); 24 months; $30,600

OHIO

CASE WESTERN RESERVE UNIVERSITY
Dannley, Ralph L.; *Peroxides of Elements Other than Carbon* (P019018); 24 months; $30,600
Fackler, John P. Jr.; *Chemistry of Coordination Compounds* (P011701); 24 months; $56,600
Klopman, Gilles; *Self Consistent Field Calculations of Chemical Reactivity* (P019985); 24 months; $32,400
Lando, Jerome B.; *Crystal Structures of Fluorine-Containing Polyolefins* (P021301); 24 months; $33,600
Norlander, J. Eric; *Solvolysis in Trifluoroacetic Acid* (P020732); 12 months; $27,000
Olah, George A.; *Stable, Long-Lived Organic Ions* (P18098X); 24 months; $79,500
Yeager, Ernest; *Dynamic Properties of Molten Salts* (P014874); 24 months; $48,300

CLEVELAND STATE UNIVERSITY
Binkley, Roger W.; *Photochemical Studies of Azines and Anils* (P016664); 24 months; $27,800

KENT STATE UNIVERSITY
Gould, Edwin S.; *Metal Ion Catalysis of Oxygen Transfer Reactions* (P023136); 24 months; $43,600

OHIO STATE UNIVERSITY
Fraenkel, Gideon; *Dynamic and Chemical Behavior of Carbanions* (P016402); 24 months; $60,400
Hine, Jack; *Effect of Structure on Rates and Equilibria* (P014697); 24 months; $45,000
Meek, Devon W.; *Stereochemistry of Unusual Coordination Numbers in Transition Metal Complexes* (P023204); 24 months; $48,300
Newman, Melvin S.; *New Synthetic Reactions by the 3,2,1-Bicyclic Mechanism* (P12445X); 12 months; $35,400
Nielsen, Harald H. and Rao, K. N.; *Annual Symposia on Molecular Structure and Spectroscopy* (P016813); 36 months; $8,700
Wojcicki, Andrew; *Chemistry of Metal Carbonyls* (P022544); 24 months; $60,800

OHIO UNIVERSITY
Jewet, John G.; *Carbonium Ion Reactivity* (P021307); 12 months; $18,200
Kline, Robert J.; *Bridged Cyanide Complexes* (P021309); 12 months; $19,900

UNIVERSITY OF AKRON
Fetters, L. J. and McIntyre, Donald; *Synthesis and Structure of Block Polymers* (P018587); 24 months; $34,100
Garn, Paul D.; *Thermoanalytical Studies of Liquid Crystals and Thermal Decomposition Reactions* (P018059); 24 months; $52,800

UNIVERSITY OF CINCINNATI
Jaffe, Hans H.; *Semi-Empirical SCF Calculation of Electronic Spectra* (P015944); 24 months; $32,000

YOUNGSTOWN STATE UNIVERSITY
Rand, Leon; *Reaction Paths in the Kolbe Synthesis* (GP9202 001); $6,900

OKLAHOMA

UNIVERSITY OF OKLAHOMA
Dryhurst, Glenn; *Electrochemistry of Biologically Important Molecules* (P018570); 24 months; $47,000
Hagen, Arnulf P.; *High Pressure Synthesis of Inorganic and Organometallic Compounds* (P019873); 24 months; $25,700

OREGON

OREGON STATE UNIVERSITY
Hawkes, Stephen J.; *Optimization in Chromatography* (P015430); 12 months; $14,800
Marvell, Elliot N.; *Mechanisms of Thermal Reactions of Polyenes-Conformation and Reactivity in Bridged Medium Rings* (P015522); 24 months; $44,600
Scott, Allen B.; *Electron Traps in Polar Crystals* (P021303); 24 months; $34,100

UNIVERSITY OF OREGON
Baldwin, John E.; *Mechanisms of Cycloaddition Reactions* (P023021); 24 months; $39,800
Boekelheide, Virgil; *Aromatic Molecules Bearing Substituents within the Cavity of the Pi-Electron Cloud* (GP7062X002); 12 months; $40,300
Griffith, O. Hayes; *Electron Spin Resonance of Organic Inclusion Compounds* (P016341); 24 months; $40,000
Koenig, Thomas W.; *Reactive Radical Pairs* (GP11019001); $5,900
Noyes, Richard M.; *Reactions of Diatomic Molecules* (P022737); 24 months; $51,200
Philpott, Michael R.; *Theory of Molecular Excitons and Primary Photochemical Processes* (P020599); 24 months; $27,000

PENNSYLVANIA

BUCKNELL UNIVERSITY
Veening, Hans; *Liquid-Liquid Chromatography of Metal-Organic Compounds* (P018755); 24 months; $19,700

CARNEGIE MELLON UNIVERSITY
Bothner-By, Aksel A. and Fateley, William G.; *Rotation about Single Bonds* (P022943); 24 months; $65,300
Keisch, Bernard; *Isotope Mass Spectrometry to Trace the Source of Artists' Materials* (P013595); 24 months; $29,800
Mock, William L.; *Preparative and Mechanistic Sulfur Chemistry* (P019301); 16 months; $27,800
Stewart, Robert F.; *Valence Structure from Coherent X-Ray Scattering* (P022729); 24 months; $36,800
Van Dyke, Charles H.; *Mixed Hydrides of the Group IV Elements* (P012833); 24 months; $23,400

LEHIGH UNIVERSITY
Fowkes, Frederick M.; *Polymerization of Water* (P019176); 12 months; $16,992

MATHEMATICAL AND PHYSICAL SCIENCES RESEARCH PROJECTS

Ohnesorge, William E.; *Luminescence of Transition Metal Ion Complexes* (P022520); 24 months; $28,110

PENNSYLVANIA STATE UNIVERSITY
Fritz, J. J., Bernheim, Robert A. and Steele, William A.; *Low Temperature Research in Physical Chemistry* (P016088); 24 months; $150,000
Lampe, F. W.; *Collision Reactions of Electronically Excited Atoms and Molecules* (P017198); 24 months; $26,900
Richey, Herman G. Jr.; *Addition of Grignard and Lithium Reagents to Carbon-Carbon Multiple Bonds* (P015804); 24 months; $57,800

TEMPLE UNIVERSITY
Dori, Zvi; *The Photochemistry of Coordinated Azides* (P014875); 24 months; $30,500

UNIVERSITY OF PENNSYLVANIA
Donohue, Jerry; *Crystal and Molecular Structures of Substances of Unusual or Unknown Chemical Structure* (P015781); 12 months; $26,700
Hochstrasser, Robin M.; *Fifth Molecular Crystal Symposium* (P016235); 12 months; $5,000
Price, Charles C.; *Investigation of Chemical Properties of Thiabenzene Analogs* (P016236); 12 months; $13,700
Silvers, Stuart J.; *Molecular Level-Crossing and Optical Double Resonance Spectroscopy* (P013788); 24 months; $45,700
Thornton, Edward R.; *Origin and Interpretation of Isotope Effects; Solvolysis Mechanisms* (P022803); 24 months; $54,300

UNIVERSITY OF PITTSBURGH
Arnett, Edward M.; *Medium and Structural Effects in Organic Chemistry* (P06550X003); 7 months; $34,700
Coetzee, Johannes F.; *Reactions in Dipolar Aprotic Solvents* (P016342); 24 months; $70,000
Cohen, Theodore; *Organic Chemistry of Copper* (P022955); 24 months; $57,000
Dowd, Paul, *Trimethylenemethane and Tetramethylene-Ethane* (P023881); 12 months; $9,000

PUERTO RICO

UNIVERSITY OF PUERTO RICO, Piedras
Adam, Waldemar; *Novel Peroxide Heterocycles* (P017755); 24 months; $35,700

RHODE ISLAND

BROWN UNIVERSITY
Baird, James C. Jr.; *Experimental Atomic and Chemical Physics* (P014104); 24 months; $50,000
Eisenberg, Richard; *Organo-Transition Metal Chemistry* (P023139); 24 months; $45,600
Lawler, Ronald G. and Ward, Harold R.; *Chemically Induced Dynamic Nuclear Polarization (CIDNP) and Nuclear Spin Sorting (NSS)* (P020098); 24 months; $59,400
Rieger, Philip H.; *Magnetic Resonance Studies of Kinetics and Equilibria in Solution* (P016294); 24 months; $32,100

UNIVERSITY OF RHODE ISLAND
Nelson, Wilfred H.; *Rayleigh-Scattering—Depolarization Study of Molecular Structure in Solutions* (P019300); 12 months; $11,400

SOUTH CAROLINA

UNIVERSITY OF SOUTH CAROLINA
Bly, Robert S.; *Solvolytic Reactivity of Pi-Complexed Compounds* (P011920); 24 months; $35,200
Durig, James R.; *Far Infrared and Raman Spectra of Some Metallic Molecules of Group IVA Elements and Several Molecular Crystals* (P020723); 24 months; $46,300

SOUTH DAKOTA

UNIVERSITY OF SOUTH DAKOTA
Miller, Norman E.; *Borane Chemistry* (P014902); 24 months; $30,600

TENNESSEE

UNIVERSITY OF TENNESSEE
Larsen, John W.; *Salt Effects on the Products of a Unimolecular Solvolysis Reaction* (P010714); 24 months; $28,000
Schweitzer, George K. and Bull, William E.; *Photoelectron Spectroscopy of Coordination Compounds* (P015431); 24 months; $48,500

VANDERBILT UNIVERSITY
Tarbell, D. S.; *Chemistry of Phosphoric Carbonic Anhydrides, Dicarbonates, Tricarbonates and Related Compounds* (P015795); 24 months; $25,000
Wilson, David J.; *Energy Transfer and Chemical Reactions in Gases* (P013452); 24 months; $55,000

TEXAS

NORTH TEXAS STATE UNIVERSITY
Brady, William T.; *Ketene Chemistry* (P014016); 12 months; $9,200

RICE UNIVERSITY
Curl, Robert F.;*Microwave Studies of Reactive Simple Molecules and Conformational Isomerism* (GP6305X003); 12 months; $16,400
Hayes, Edward F.; *Quantum Mechanical Studies of Proton and Electron Transfer Reactions* (P018871); 24 months; $30,600

TEXAS A & M UNIVERSITY
Lunsford, Jack H.; *Gas-Solid Interactions at Semiconductor Surfaces* (P019875); 24 months; $24,300
——— (GP19875001); $300
Rose, Timothy L.; *Characterization of Primary Products from Photodissociation of Polyatomic Molecules* (P022523); 24 months; $39,800
Rowe, Marvin W.; *Isotopic Composition and Abundances of Noble Gases* (P018716); 24 months; $37,800
Schweikert, E. A.; *Charged Particle Activation Analysis* (P011630); 12 months; $12,000

UNIVERSITY OF HOUSTON
Kouri, Donald J.; *Quantum Mechanical Studies of Proton and Electron Transfer Reactions* (P018872); 24 months; $30,900
Veillon, Claude; *Atomic Fluorescence and High Resolution Atomic Absorption Spectrometry* (P017318); 24 months; $22,500

UNIVERSITY OF TEXAS AT AUSTIN
Bard, Allen J.; *Mechanisms of Organic Electrode Reactions* (GP6688X003); 12 months; $21,600
Bauld, Nathan L.; *Anion Radicals of "Unknown" Substrates* (P017596); 24 months; $21,800
Boggs, James E.; *Third Austin Symposium on Gas Phase Molecular Structure* (P017594); 12 months; $5,000
Matsen, F. A.; *The Aggregate Theory of Polyelectronic Systems* (P017593); 24 months; $45,600
Pettit, Rowland; *Some Aspects of Organometallic Chemistry* (P13373X001); 12 months; $32,100
White, John M.; *Photo-Induced Effects at Gas-Solid Interfaces* (P020370); 12 months; $17,000

UTAH

BRIGHAM YOUNG UNIVERSITY
Barnett, J. Dean, Hall, H. Tracy and Dalley, N. Kent; *High Pressure, High Temperature X-Ray Diffraction* (P015784); 24 months; $49,000

UNIVERSITY OF UTAH
Bentrude, Wesley G.; *Phosphoranyl-Radical Intermediates* (P022885); 24 months; $37,500
Eyring, Henry; *Liquids, Rate Processes, Optical Activity and High Pressure* (GP6496X003); 12 months; $30,400
Hadley, Steven G.; *Triplet States of Aromatic Heterocycles* (P019750); 24 months; $40,600
Walling, Cheves; *Organic Reaction Mechanisms* (P024300); 24 months; $136,000

UTAH STATE UNIVERSITY
Hansen, Wilford N.; *The Electrochemical Interface of Optically Transparent Electrodes* (P013767); 24 months; $37,000

VERMONT

MIDDLEBURY COLLEGE
Nelson, Ralph D. Jr.; *Far Infrared Refraction of Liquid Mixtures* (P022829); 12 months; $14,000

UNIVERSITY OF VERMONT
Allen, Christopher W.; *Exocyclic II-Interactions in Silylaminophosphonitriles* (P019943); 24 months; $27,000

VIRGINIA

UNIVERSITY OF VIRGINIA
Andrews, W. Lester S.; *Matrix Spectroscopic Studies of Chemical Intermediates* (P021304); 12 months; $18,800

VIRGINIA POLYTECHNIC INSTITUTE
Dessy, Raymond E.; *Inorganic Summer Symposium on Bio-Inorganic Chemistry* (P017527); 12 months; $3,000
Graybeal, J. D.; *Structural Investigations of Metal and Organometal Halides* (P013973); 24 months; $32,400
Wightman, James P.; *Thermodynamics of Adsorption from Nonelectrolyte Solutions onto Solids* (P021299); 12 months; $11,200

WASHINGTON

UNIVERSITY OF WASHINGTON
Eichinger, Bruce E.; *Macromolecular Solutions* (P020119); 24 months; $32,300
Gregory, Norman W.; *Vaporization Reactions* (GP6608X003); 12 months; $24,500
Rose, Norman J.; *Novel Metal Chelates* (P023209); 24 months; $42,900

WESTERN WASHINGTON STATE COLLEGE
Gerhold, G. A.; *Influence of Intermolecular Interactions on Electronic Spectra in Condensed Phases* (P022828); 24 months; $24,200

WEST VIRGINIA

MARSHALL UNIVERSITY
Douglass, James E.; *(Diamine) Boronium Ions—Study Concerning Their Preparation and Properties* (P014753); 12 months; $16,300

WISCONSIN

MARQUETTE UNIVERSITY
Schrader, David M.; *Positron Annihilation in Molecular Materials* (P012283); 24 months; $35,500

UNIVERSITY OF WISCONSIN, Madison
Bernstein, Richard B.; *Molecular Beam Scattering* (P016665); 12 months; $110,000
Crosley, David R.; *Level-Crossing Spectroscopy and Optical Radio-Frequency Double Resonance in Molecules* (P017790); 24 months; $45,600
Dahl, Lawrence F.; *Structural Systematics in Organometallic Complexes* (P019175); 12 months; $61,100
Evans, Dennis H.; *Coupling and Anodic Reactions of Phenols* (P019579); 24 months; $61,000
Ferry, John D.; *Molecular Motions in Polymers* (P010892); 24 months; $86,500
Gaines, Donald F.; *Studies in Borane Chemistry* (P012646); 24 months; $37,200
Nelsen, Stephen F.; *Electron Spin Resonance of Organic Radicals* (P017164); 12 months; $13,400

ANTHROPOLOGY

Noggle, Joseph H.; *Experimental Study of Nuclear Magnetic Relaxation* (P015068); 24 months; $27,000

Treichel, Paul M.; *Studies in Organometallic Chemistry* (P017207); 24 months; $43,300

Trost, Barry M.; *New Approaches to Small Ring Compounds* (P014609); 24 months; $31,900

Vedejs, Edwin; *New Methods for Hydrocarbon Synthesis* (P017018); 24 months; $29,500

Walters, John P.; *Short-Time Reactions in Spark Discharges* (P013975); 24 months; $138,900

Whitesides, Thomas H.; *Stabilization and Electrocyclic Reactivity of Complexed Ligands* (P016358); 24 months; $24,300

Whitlock, Howard W. Jr.; *Transition Metal Organometallies* (GP7443 001); 12 months; $8,900

Zimmerman, Howard E.; *Mechanistic and Exploratory Organic Photochemistry* (GP6797X002); 12 months; $30,500

WYOMING

UNIVERSITY OF WYOMING

Edmiston, Clyde; *Accurate Estimation of Properties of Large Molecules Using Optimumly Localized and Pseudonatural Molecular Orbitals* (P022940); 24 months; $32,700

Holt, Smith L.; *Electronic Structure of Transition Metal Ions* (P015432); 24 months; $36,100

Social Sciences Research Projects

Anthropology

ALASKA

UNIVERSITY OF ALASKA

Hippler, Arthur E.; *Urban Acculturation in Alaska* (GS3026); 12 months; $31,200

ARIZONA

NORTHERN ARIZONA UNIVERSITY

Ambler, J. Richard; *Archaeological Excavation of Dust Devil Cave* (GS3115); 12 months; $13,700

UNIVERSITY OF ARIZONA

Jelinek, Arthur J.; *The Prehistoric Cave Site of Tabun* (GS2696); 12 months; $7,000

Longacre, William A.; *Doctoral Dissertation Research in Anthropology* (GS2694); 12 months; $550

Vivian, Richard G.; *Prehistoric Water Conservation in Chaco Canyon* (GS3100); 12 months; $30,300

ARKANSAS

ARKANSAS ARCHEOLOGICAL SURVEY

Schambach, Frank; *Excavation of the Crenshaw Site, Arkansas* (GS2684); 12 months; $13,500

———; (GS2684 001); $6,025

CALIFORNIA

STANFORD UNIVERSITY

Befu, Harumi; *Institutional and Motivational Factors in Interpersonal Relations* (GS2370 001); $1,700

Wolf, Arthur P.; *Family Structure and Population Trends in Taiwan* (GS3041); 24 months; $92,300

UNIVERSITY OF CALIFORNIA, Berkeley

Anderson, James N.; *Research in Social Demography in Central Luzon, Philippines* (GS2995); 26 months; $17,500

Clark, J. Desmond; *Doctoral Dissertation Research in Anthropology* (GS2944); 12 months; $6,700

Emeneau, M. B.; *Doctoral Dissertation Research in Linguistics* (GS2946); 3 months; $2,650

Heizer, Robert F.; *Prehistoric Petroglyphs and Pictographs of California* (GS2708); 12 months; $6,100

Kay, Paul; *Doctoral Dissertation Research in Linguistics* (GS2930); 12 months; $850

Rowe, John H.; *Doctoral Dissertation Research in Anthropology* (GS3085); 12 months; $3,600

UNIVERSITY OF CALIFORNIA, Irvine

Colby, Benjamin N.; *The Structural Analysis of a Narrative Culture* (GS2984); 24 months; $61,800

UNIVERSITY OF CALIFORNIA, Los Angeles

Ladefoged, Peter N. and Fromkin, Victoria; *Linquistic Phonetics* (GS2859); 24 months; $60,500

UNIVERSITY OF CALIFORNIA, Riverside

Anderson, Eugene N.; *Ecology of a Fishing Community* (GS3111); 24 months; $26,000

UNIVERSITY OF CALIFORNIA, San Diego

Klima, Edward S.; *La Jolla Conference on Linguistic Theory and the Structure of English* (GS2845); 12 months; $9,000

Langdon, Margaret H.; *Conference on Hokan Languages* (GS2766); 12 months; $6,900

UNIVERSITY OF CALIFORNIA, Santa Cruz

Davenport, William H.; *Conference on Anthropology of the Bismarck Archipelago* (GS2938); 12 months; $10,100

Keesing, Roger M.; *Ethnography of Kwaio* (GS2990); 18 months; $32,000

UNIVERSITY OF CALIFORNIA, Santa Barbara

Fagan, Brian M.; *The Beginnings of Food Production in Sub-Saharan Africa* (GS3015); 18 months; $17,400

Spaulding, Albert C.; *Doctoral Dissertation Research in Anthropology* (GS2837); 18 months; $2,350

——— (GS2725); 12 months; $2,700

COLORADO

COLORADO COLLEGE

Nowak, Michael; *Archeology of Nunivak Island, Alaska* (GS3023); 18 months; $13,000

UNIVERSITY OF COLORADO

Benedict, James B. and Wheat, Joe Ben; *Prehistoric Man and Environment in the Colorado Rocky Mountains* (GS3052); 12 months; $25,000

Hester, James J.; *Northwest Coast Prehistory* (GS2448 002); $7,400

Wheat, Joe B.; *Archaeology of the Jurgens Site, Colorado* (GS3039); 12 months; $13,000

CONNECTICUT

HUMAN RELATIONS AREA FILES

Lebar, Frank; *Ethnographic Handbook of Insular Southeast Asiatic Cultures* (GS1763 001); 9 months; $9,400

UNIVERSITY OF CONNECTICUT

Bailit, Howard L.; *Collaborative Research on the Genetics of Human Populations in New Guinea* (GS2993); 24 months; $24,200

Laughlin, William S.; *Prehistory and Ecology of Anangula and Chaluka, Aleutian Islands* (GS2823); 24 months; $2,400

YALE UNIVERSITY

Dyen, Isidore; *Doctoral Dissertation Research in Linguistics* (GS2961); 12 months; $3,400

Huffman, Franklin E.; *Study of Mon-Khmer Dialects* (GS3068); 24 months; $7,200

Osgood, Cornelius; *Publication of 'The Chinese of Hong Kong—A Study of an Island Community'* (GN849); 12 months; $15,000

Rouse, Irving; *Doctoral Dissertation Research in Anthropology* (GS2924); 17 months; $5,650

DISTRICT OF COLUMBIA

CATHOLIC UNIVERSITY OF AMERICA

Gardner, William M.; *Archeology of the Potomac River Drainage System* (GS3020); 12 months; $26,100

CENTER FOR APPLIED LINGUISTICS

Sebeok, Thomas A.; *Preparation of 'Oceanic Linguistics'* (GS1522 002); $2,027

GEORGE WASHINGTON UNIVERSITY

Simons, Suzanne L.; *Cognatic Kinship at Sandia Pueblo* (GS3035); 24 months; $10,000

FLORIDA

UNIVERSITY OF FLORIDA

Fairbanks, Charles H.; *Doctoral Dissertation Research in Anthropology* (GS3105); 12 months; $3,600

Wing, Elizabeth S.; *Prehistoric Man-Animal Relationships in the Central Peruvian Andes* (GS3021); 12 months; $11,900

GEORGIA

UNIVERSITY OF GEORGIA

Caldwell, Joseph R.; *Doctoral Dissertation Research in Anthropology* (GS3024); 16 months; $4,200

Crawford, James M.; *Southeastern Indian Language Project* (GS3056); 12 months; $9,300

HAWAII

BERNICE P. BISHOP MUSEUM

Green, Roger C. and Yen, Douglas; *The Prehistory of the Southeast Solomon Islands* (GS2977); 24 months; $75,500

UNIVERSITY OF HAWAII

McKaughan, Howard P.; *Grammatical Analysis of the Jeh Language of South Vietnam* (GS2685); 12 months; $20,100

——— *Maranao Linguistic Studies* (GS3017); 18 months; $28,200

Parker, Gary J.; *Linguistic Study of the Huaylas-Conchucos Dialect of Quechua* (GS3034); 12 months; $6,900

Thompson, Laurence C.; *Linguistic Relationships* (GS2988); 24 months; $128,900

——— *Doctoral Dissertation Research in Linguistics* (GS2881); 18 months; $2,200

ILLINOIS

NORTHERN ILLINOIS UNIVERSITY

Gunnerson, James H.; *Apachean Archaeology* (GS3007); 24 months; $30,000

NORTHWESTERN UNIVERSITY

Sade, Donald S.; *Primate Social Patterns* (GS3114); 24 months; $51,800

——— (GS2377 001); $4,000

Struever, Stuart; *Doctoral Dissertation Research in Anthropology* (GS2769); 12 months; $9,200

——— *Excavation of the Koster Site* (GS3016); 12 months; $19,700

SOUTHERN ILLINOIS UNIVERSITY

Altschuler, Milton; *Ethnographic Survey of the Cayapa Indians, Ecuador* (GS3037); 3 months; $10,100

Dark, Philip J. C.; *Doctoral Dissertation Research in Anthropology* (GS2929); 12 months; $3,000

Guemple, D. Lee; *Ecological Adaptations and Social Patterns of the Belcher Islands Eskimo* (GS2686); 12 months; $11,800

Rands, Robert L.; *Mayan Ecology and Trade* (GS1455X001); 12 months; $8,000

Taylor, Walter W.; *Doctoral Dissertation Research in Anthropology* (GS2829); 17 months; $2,650

UNIVERSITY OF CHICAGO

Hamp, Eric P.; *Conference on Eskimo Linguistics* (GS2987); 12 months; $12,000

Howell, F. Clark; *Early Phases of Hominid Evolution* (GS2905); 24 months; $195,800

Merbs, Charles F.; *Doctoral Dissertation Research in Anthropology* (GS2831); 12 months; $1,000

Singer, Ronald; *Excavations at Clacton, a Middle Pleistocene Site* (GS2907); 24 months; $51,100

SOCIAL SCIENCES RESEARCH PROJECTS

Van Loon, Maurits; *Prehistory of the Euphrates Dam Reservoir* (GS2681); 12 months; $20,400

Zide, Norman H.; *Proto-Munda Phonology and Morphology* (GS2914); 20 months; $79,700

UNIVERSITY OF ILLINOIS, Urbana

Casagrande, Joseph B.; *The Indian in Colonial Ecuador and After Independence* (GS3049); 24 months; $43,600

Lathrap, Donald W.; *Doctoral Dissertation Research in Anthropology* (GS2922); 9 months; $4,300

——— (GS3072); 8 months; $6,100

——— (GS2777); 11 months; $2,200

Whitten, Norman E.; *Ethnography of the Quechua Indians of Eastern Ecuador* (GS2999); 24 months; $29,400

INDIANA

INDIANA UNIVERSITY, Bloomington

Driver, Harold E.; *Collaborative Research on North American Indian Societies* (GS2951); 24 months; $42,500

IOWA

UNIVERSITY OF IOWA

Helm, June and Howren, Robert; *Linguistic and Cultural Variation Among Athapaskan Indians* (GS3057); 24 months; $50,100

KANSAS

UNIVERSITY OF KANSAS

Bass, William M.; *Prehistoric Human Skeletal Materials from Plains Indian Sites* (GS2717); 16 months; $40,500

Kerley, Ellis R.; *Doctoral Dissertation Research in Physical Anthropology* (GS2731); 9 months; $500

Simonett, David S.; *Archaeology of San Miguel Island* (GS2820); 12 months; $2,800

Squier, Robert J.; *Olmec Development in the Los Tuxtlas Region, Mexico* (GS2895); 28 months; $135,000

MASSACHUSETTS

BOSTON UNIVERSITY

Hunt, Eva; *The Ethnohistory of Coxcatlan* (GS3000); 12 months; $18,800

HARVARD UNIVERSITY

Friedlaender, Jonathan S.; *Collaborative Research on the Genetics of Human Populations in New Guinea* (GS3088); 24 months; $15,900

Howells, William W.; *Doctoral Dissertation Research in Physical Anthropology* (GS3048); 6 months; $1,300

Lamberg-Karlovsky, C. C.; *Excavations at Tal-I-Yahya* (GS3087); 24 months; $48,800

Maybury-Lewis, David; *Doctoral Dissertation Research in Anthropology* (GS2842); 24 months; $6,200

Moseley, Michael E.; *Doctoral Dissertation Research in Anthropology* (GS2733); 12 months; $4,000

Williams, Stephen; *Archaeological Reconnaissance in the Boston Area* (GS2706); 24 months; $13,600

ROBERT S. PEABODY FOUNDATION FOR ARCHEOLOGY

MacNeish, Richard S.; *The Origins of Agriculture and Civilization in Highland Peru* (GS2927); 24 months; $154,500

MICHIGAN

UNIVERSITY OF MICHIGAN

Jorgensen, Joseph G.; *Collaborative Research on North American Indian Societies* (GS2952); 24 months; $31,500

———; *Doctoral Dissertation Research in Anthropology on Problems of Poverty* (GS2724); 12 months; $5,650

Parsons, Jeffrey R.; *Anthropological Investigation of An Andean Agricultural System* (GS2693); 20 months; $28,900

Schramm, Gene M.; *Doctoral Dissertation Research in Linguistics* (GS2841); 12 months; $2,450

Whallon, Robert Jr.; *Archaeological Investigations in East-Central Turkey* (GS3025); 12 months; $7,300

MINNESOTA

UNIVERSITY OF MINNESOTA

Sarles, Harvey B.; *Doctoral Dissertation Research in Anthropology* (GS2790); 22 months; $4,900

MISSOURI

UNIVERSITY OF MISSOURI, Columbia

Diehl, Richard A.; *Archaeological Investigation of Tula, Mexico* (GS2814); 15 months; $25,500

NEBRASKA

DANA COLLEGE

Lehmer, Donald J.; *Cultural Ecology of the Hidatsa Indians* (GS3003); 12 months; $26,800

NEVADA

DESERT RESEARCH INSTITUTE

Fowler, Don D.; *Archeological Excavations in Meadow Valley* (GS3029); 12 months; $18,100

NEW HAMPSHIRE

DARTMOUTH COLLEGE

Harp, Elmer Jr.; *Arctic Archaeology* (GS2915); 18 months; $28,900

NEW MEXICO

EASTERN NEW MEXICO UNIVERSITY

Irwin-Williams, Cynthia; *Origins of the Anasazi* (GS2416 001); $1,000

NEW YORK

BROOKDALE HOSPITAL CENTER

Valentine, Charles A.; *Ethnography of Low-Income Urban Afro-Americans* (GS2755); 6 months; $25,200

COLUMBIA UNIVERSITY

Fried, Morton H.; *Doctoral Dissertation Research in Anthropology* (GS2788); 24 months; $5,550

Harris, Marvin; *Authority and Superordination in Urban Families* (GS2916); 12 months; $35,000

Labov, William; *Doctoral Dissertation Research in Linguistics* (GS2887); 17 months; $4,000

Lanning, Edward P.; *Doctoral Dissertation Research in Anthropology* (GS2888); 16 months; $6,000

——— (GS2784); 12 months; $2,000

——— *Early Man in the Andean Area* (GS2560 001); $9,300

Murphy, Robert F.; *Doctoral Dissertation Research in Anthropology* (GS2950); 18 months; $6,450

Pitkin, Harvey; *Doctoral Dissertation Research in Anthropology* (GS2890); 12 months; $5,150

Solecki, Ralph S. and Perkins, Dexter Jr.; *Paleozoological Investigations from the Middle Paleolithic to the Neolithic in the Near East* (GS2828); 24 months; $36,500

CORNELL UNIVERSITY

Kennedy, Kenneth A. R.; *Ecology and Evolution of Prehistoric Man in Ceylon* (GS3109); 6 months; $3,800

CUNY, CENTRAL SYSTEM OFFICE

McLendon, Sally; *Doctoral Dissertation Research in Linguistics* (GS3118); 12 months; $6,000

Waterbury, Ronald; *Editing and Preparation for Publication of Malinowski's Oaxaca Field Notes* (GS3058); 6 months; $15,000

CUNY, HUNTER COLLEGE

Pasternak, Burton; *Social Consequences of Variation in Agricultural Irrigation* (GS2704); 24 months; $14,700

HOFSTRA UNIVERSITY

Turnbull, Colin M.; *Symbiosis and Opposition in Inter-Group Relations* (GS2983); 24 months; $66,900

STATE UNIVERSITY AT BINGHAMTON

Van Der Merwe, Nikolaas J.; *Archaeology of the Iron Age in the Palabora Area* (S002744); 24 months; $59,900

UNIVERSITY OF ROCHESTER

Hoben, Allan; *Formal Semantic Analysis of Deferential Behavior* (GS2296 001); $4,200

NORTH CAROLINA

UNIVERSITY OF NORTH CAROLINA AT CHAPEL HILL

Brockington, Donald L.; *Archaeology of the Oaxaca Coast, Mexico* (S002866); 24 months; $39,400

Coe, Joffre L.; *Doctoral Dissertation Research in Anthropology* (S003067); 4 months; $3,850

Goethals, Peter R.; *Doctoral Dissertation Research in Anthropology* (S002786); 24 months; $6,400

Gulick, John; *Cultural Influences on Human Fertility* (S003108); 14 months; $36,100

OHIO

CASE WESTERN RESERVE UNIVERSITY

Brose, David S.; *The Archaeology of Whittlesey Focus Sites, Ohio* (S003062); 12 months; $29,300

OKLAHOMA

GREAT PLAINS HISTORICAL ASSOCIATION

Hammatt, Hallett H.; *Archaeology of Domebo Canyon* (S002861); 12 months; $5,600

UNIVERSITY OF OKLAHOMA

Bell, Robert E.; *Preceramic Occupations of Highland Ecuador* (S003036); 24 months; $18,400

OREGON

UNIVERSITY OF OREGON

Aikens, C. Melvin; *Willamette Valley Prehistory* (S003009); 12 months; $30,400.

——— *Doctoral Dissertation Research in Anthropology* (S002948); 12 months; $8,250

Dorjahn, Vernon R.; *Doctoral Dissertation Research in Anthropology* (S002687); 18 months; $9,600

Johnson, Leroy Jr.; *Archaeology of the Klamath Basin* (S002997); 16 months; $44,500

PENNSYLVANIA

BRYN MAWR COLLEGE

De Laguna, Frederica; *Doctoral Dissertation Research in Anthropology* (S002712); 12 months; $2,500

UNIVERSITY OF PENNSYLVANIA

David, Nicholas C.; *Doctoral Dissertation Research in Anthropology* (S003122); 12 months; $7,500

Dyson, Robert H. Jr.; *Doctoral Dissertation Research in Anhtropology* (S003129); 12 $8,900

Fought, John G.; *Cholan Linguistics* (S002867); 24 months; $42,700

Goodenough, Ward; *Doctoral Dissertation Research in Linguistics* (S003031); 24 months; $5,500

Goodenough, Ward H.; *Doctoral Dissertation Research in Anthropology* (S002796); 18 months; $6,400

Hymes, Dell H.; *Doctoral Dissertation Research in Linguistics* (S002779); 12 months; $1,600

——— (S002909); 12 months; $1,850

Linares de Sapir, Olga; *Archaeological In-*

vestigations in Western Panama (S002846); 24 months; $30,700

Rainey, Froelich; *Development of Thermoluminescence Dating for Archaeology* (S002716); 24 months; $61,200

Southworth, Franklin C.; *South Asian Semantic Structures* (S002838); 33 months; $56,500

UNIVERSITY OF PITTSBURGH

Brown, L. Keith; *Urban Aspects of Family and Kinship in Japan* (S003002); 24 months; $51,700

Landy, David; *Doctoral Dissertation Research in Anthropology* (S002830); 24 months; $2,700

——— (S002690); 12 months; $4,050

Tuden, Arthur; *Doctoral Dissertation Research in Anthropology on Problems of Poverty* (GS2691); 15 months; $2,400

RHODE ISLAND

BROWN UNIVERSITY

Deetz, James; *Doctoral Dissertation Research in Anthropology* (S003043); 12 months; $9,000

Hicks, George L.; *Doctoral Dissertation Research in Anthropology* (S003110); 12 months; $750

TENNESSEE

VANDERBILT UNIVERSITY

Spores, Ronald M.; *The Mixtec Community and Intercommunity Relations in the Nochixtlan Valley* (S002849); 12 months; $42,500

TEXAS

RICE UNIVERSITY

Gamst, Frederick C.; *Social Consequences of Technological Change* (S003040); 12 months; $14,200

SOUTHERN METHODIST UNIVERSITY

Marks, Anthony E.; *Prehistory of Central Negev, Israel* (S003019); 12 months; $39,800

Shiner, Joel L.; *Doctoral Dissertation Research in Anthropology* (S002780); 10 months; $3,550

Trager, George L.; *Doctoral Dissertation Research in Linguistics* (S002722); 12 months; $2,600

UNIVERSITY OF TEXAS AT AUSTIN

Johnston, Francis E.; *Evolutionary Dynamics of Human Populations* (S003038); 6 months; $5,600

UTAH

UNIVERSITY OF UTAH

Jennings, Jesse D.; *Doctoral Dissertation Research in Anthropology* (S002711); 12 months; $1,700

VIRGINIA

COLLEGE OF WILLIAM AND MARY

Barka, Norman F. and McCary, Ben C.; *Archaeology of Eastern Virginia* (S002715); 24 months; $31,200

WASHINGTON

UNIVERSITY OF WASHINGTON

Atkins, John R.; *Doctoral Dissertation Research in Anthropology* (S003042); 19 montss; $2,800

Klein, Richard G.; *Excavation of Middle Stone Age Sites in South Africa* (S003013); 24 months; $90,300

Owen, Michael G.; *Syntactic and Semantic Studies of Yucatec* (S003022); 24 months; $38,100

WASHINGTON STATE UNIVERSITY

Bernard, H. Russell; *Doctoral Dissertation Research in Anthropology* (S002754); 15 months; $3,250

Irwin, Henry T.; *Archaeology of a Paleolithic Site in Spain* (S003005); 12 months; $50,000

Sibley, Willis E.; *Doctoral Dissertation Research in Anthropology* (S003093); 4 months; $1,800

WISCONSIN

UNIVERSITY OF WISCONSIN, Madison

Chard, Chester S.; *Doctoral Dissertation Research in Anthropology* (S002753); 5 months; $2,300

——— (S002734); 12 months; $8,000

Elmendorf, William W.; *Dynamics of Coast Salish Community Networks* (S003092); 12 months; $6,100

Thompson, Donald E.; *Late Prehistoric Occupation of the Eastern Slopes of the Andes* (S002836); 12 months; $37,100

UNIVERSITY OF WISCONSIN, Milwaukee

Fowler, Melvin L.; *Cahokia Site Archaeology* (S003006); 24 months; $51,700

WYOMING

UNIVERSITY OF WYOMING

Frison, George C.; *Archaeological Investigations at the Wardell Buffalo Kill Site, Wyoming* (GS003014); 12 months; $13,600

KENYA

NATIONAL MUSEUM CENTER FOR PREHISTORY AND PALEONTOLOGY

Leakey, L. S. B.; *Paleoanthropology in East Africa* (GS2774); 24 months; $57,700

ECONOMICS

Economics

CALIFORNIA

SAN DIEGO STATE COLLEGE

Venieris, John P.; *An Investigation of the Price Mechanism* (GS2870); 24 months; $37,300

STANFORD UNIVERSITY

Johnston, Bruce F.; *Doctoral Dissertation Research in Economics* (GS2826); 12 months; $6,800

Jones, William O.; *Doctoral Dissertation Research in Economics* (S002923); 12 months; $4,650

Kurz, Mordecai; *Efficient and Optimal Economic Growth* (GS02530001); $5,500

Lau, Lawrence; *Behavioral Equations and the Neoclassical Theory of Production and Consumption* (GS2874); 20 months; $42,200

Lind, Robert C. and Massell, Benton F.; *Neighborhood Dynamics* (GS2942); 22 months; $147,400

Massell, Benton F.; *Doctoral Dissertation Research in Economics on Problems of Poverty* (GS2931); 12 months; $700

UNIVERSITY OF CALIFORNIA, Berkeley

Akerlof, George; *Inflation and Investment* (GS2739); 24 months; $28,100

Break, George F.; *Doctoral Dissertation Research in Economics on Problems of Poverty* (GS2920); 12 months; $1,250

Douglas, Aaron; *Measuring the Capital Input for an Industry Production Function* (GS2729); 15 months; $21,500

Goldman, Steven M.; *Inflation and Individual Welfare* (GS2897); 15 months; $17,300

Harsanyi, John C.; *International Research Workshop in Game Theory* (GS2985); 12 months; $39,900

Letiche, John M.; *Sufficient Conditions for Trading Gains* (GS3018); 12 months; $19,600

Levy, Frank; *Doctoral Dissertation Research in Economics* (GS2921); 12 months; $1,100

———; (GS3069); 12 months; $1,850

Myers, John; *Doctoral Dissertation Research in Economics* (GS2783); 8 months; $900

Zarembka, Paul; *Econometric Modeling of Underdeveloped Economies* (GS2822); 24 months; $36,800

UNIVERSITY OF CALIFORNIA, Irvine

McCall, John; *Stochastic Economic Models of Poverty and Job Search* (GS2697); 24 months; $31,600

UNIVERSITY OF CALIFORNIA, San Diego

Bear, Donald V. T.; *Economic Theory and Distribution Lags* (GS2864); 24 months; $45,200

CONNECTICUT

WESLEYAN UNIVERSITY

Lovell, Michael C.; *Firm and Market Behavior* (GS2903); 24 months; $49,200

YALE UNIVERSITY

Brainard, William, Scarf, Herbert E. and Shubik, Martin; *Economic Theory and Econometrics* (S02000X001); 12 months; $242,200

Fellner, William J.; *The Theory of Technological Advance* (GS3112); 24 months; $40,000

Grether, David M. and Mieszkowski, Peter; *The Determinants of Real Estate Values* (GS3077); 12 months; $71,000

Ranis, Gustav; *Comparative Analysis of Economic Development* (GS2804); 24 months; $193,000

Shubik, Martin and Wolf, Gerrit; *Experimental Economic and Psychological Modeling in an Automated Laboratory* (GS2840); 12 months; $62,300

Truman, Edwin M. and Resnick, Stephen A.; *Collaborative Research on European Economic Integration* (GS2957); 16 months; $55,300

Wagner, Harvey M.; *Mathematical Programming and Mathematical Economics* (GS3032); 24 months; $60,500

DISTRICT OF COLUMBIA

AMERICAN UNIVERSITY

Bartfeld, Charles I.; *Doctoral Dissertation Research in Economics* (GS2919); 12 months; $6,100

BROOKINGS INSTITUTION

Fromm, Gary, Kuh, Edwin and Duesenberry, James S.; *An Econometric Model of the United States Economy* (GS3090); 12 months; $127,600

GEORGE WASHINGTON UNIVERSITY

Black, Guy; *Study of Private-Federal Research and Development Relationships* (GS2906); 9 months; $3,100

GEORGETOWN UNIVERSITY

Tella, Alfred J.; *Work Incentives and the Negative Income Tax* (GS2883); 12 months; $81,600

URBAN INSTITUTE

Hochman, Harold M.; *Collaborative Research in Utility Interdependence, Income Redistribution, and Fiscal Structure* (GS2805); 24 months; $45,000

GEORGIA

UNIVERSITY OF GEORGIA

Bonin, Joseph M. and Cohen, Harold A.; *An Economic Analysis of Seasonal Fluctuations in the U.S. Economy* (GS2835); 24 months; $79,800

ILLINOIS

NORTHWESTERN UNIVERSITY

Kihlstrom, Richard E. and Ledyard, John O.; *Consumer Information* (GS3046); 15 months; $29,800

SOUTHERN ILLINOIS UNIVERSITY, Edwardsville

Kohn, Robert E.; *A Model for Air Pollution Control* (GS2892); 24 months; $42,000

UNIVERSITY OF CHICAGO

Johnson, D. Gale; *Analysis of U. S. Agricultural Output and Resource Demand* (GS3071); 24 months; $75,600

SOCIAL SCIENCES RESEARCH PROJECTS

Weil, Roman L.; *Economic Programming* (GS2703); 24 months; $35,200

UNIVERSITY OF ILLINOIS

Nourse, Hugh O.; *Doctoral Dissertation Research in Economics on Problems of Poverty* (GS2954); 12 months; $650

INDIANA

PURDUE UNIVERSITY

Hendershott, Patric H. and Horwich, George; *Financial Markets in the Aggregate Economy* (GS2908); 24 months; $105,300

——— ; *Collaborative Research on a Model of the Money and Capital Markets* (GS2029 001); $9,300

IOWA

IOWA STATE UNIVERSITY

Thorbecke, Erik; *Collaborative Research on European Economic Integration* (GS2956); 16 months; $26,700

KANSAS

UNIVERSITY OF KANSAS

Richardson, David H.; *Collaborative Research on Distribution Theory of GCL Estimators in Simultaneous Equations* (GS2806); 24 months; $37,400

MARYLAND

JOHNS HOPKINS UNIVERSITY

Balassa, Bela; *Collaborative Research on European Economic Integration* (GS2959); 24 months; $47,400

MASSACHUSETTS

HARVARD UNIVERSITY

Arrow, Kenneth J. and Starrett, David A.; *Capital Theory and Intertemporal Resource Allocation* (GS2797); 24 months; $55,400

Dunlop, John T.; *Doctoral Dissertation Research in Economics* (GS2910); 12 months; $1,200

Griliches, Zvi; *Econometric Investigations of Technological Change* (S02762X); 24 months; $128,600

Jorgenson, Dale W.; *The Econometrics of Investment and Financial Valuation* (GS2802); 24 months; $99,400

Leontief, Wassily; *Basic Research on Input-Output Analysis* (GS1456X002); 12 months; $125,000

Pratt, John, Raiffa, Howard and Schlaifer, Robert; *Statistical Decision Theory* (GS2994); 24 months; $109,100

MASSACHUSETTS INSTITUTE OF TECHNOLOGY

Bhagwati, Jagdish N.; *Problems in the Theory of Trade and Welfare* (GS2978); 24 months; $20,900

Diamond, Peter A. and Foley, Duncan K.; *Theories of Public Finance, Uncertainty and Information* (GS2966); 24 months; $42,000

Thurow, Lester C.; *Determinants of Income Distribution* (GS2811); 24 months; $58,900

MICHIGAN

MICHIGAN STATE UNIVERSITY

Kreinin, M. E.; *Collaborative Research on European Economic Integration* (GS2958); 17 months; $27,300

UNIVERSITY OF MICHIGAN

Scherer, Frederic M.; *The Economics of Multi-Plant Operation* (GS2809); 24 months; $45,300

Sonquist, John A. and Morgan, James N.; *Advanced Multivariate Analysis Program Development* (GS2869); 12 months; $70,500

Stafford, Frank P. and Johnson, George E.; *Determinants of Accumulation of Human Capital* (GS3010); 17 months; $16,900

Stern, Robert M.; *International Economics* (GS3073); 24 months; $70,000

MINNESOTA

UNIVERSITY OF MINNESOTA

Chipman, John S.; *The Ordering of Portfolios and Investment Projects* (GS2973); 24 months; $44,300

MISSOURI

UNIVERSITY OF MISSOURI, Columbia

McQuigg, James D. and Pheric Sciences; *Determination of the Potential Socio-Economic Impact of Weather and Climate Modification* (A015413*); 24 months; $74,100

WASHINGTON UNIVERSITY

Bergstrom, Theodore; *Consumer Preference, Externality and General Equilibrium Mechanisms* (GS3070); 24 months; $21,800

NEW JERSEY

PRINCETON UNIVERSITY

Morgenstern, Oskar and Goldfeld, Stephen M.; *Stochastic Processes in Economics* (GS2799); 24 months; $146,300

NEW YORK

CORNELL UNIVERSITY

Davis, Tom E.; *Doctoral Dissertation Research in Economics* (GS2925); 12 months; $2,200

Kahn, Alfred E.; *Doctoral Dissertation Research in Economics* (GS2824); 12 months; $650

Liu, Ta-Chung; *A Monthly Econometric Model for the United States* (GS2872); 24 months; $75,700

Morse, Chandler; *Doctoral Dissertation Research in Economics* (GS2767); 12 months; $600

Robinson, Kenneth L.; *Doctoral Dissertation Research in Economics* (GS2918); 12 months; $2,200

NATIONAL BUREAU OF ECONOMIC RESEARCH

Lipsey, Robert E. and Weiss, Merle Y.; *The Relation of U. S. Manufacturing Abroad to U. S. Exports* (GS2935); 24 months; $74,900

Nadiri, Mohammed I.; *Inter-University Workshop on Applied Econometrics* (GS3028); 12 months; $39,400

STATE UNIVERSITY AT BUFFALO

Jen, Frank C.; *Innovation Pricing Strategies and Industry Structure* (GS2812); 7 months; $25,700

UNIVERSITY OF ROCHESTER

Jensen, Michael and Southwick, Lawrence; *Collaborative Research on Consumption, Production, and the Determination of Prices under Uncertainty* (GS2964); 12 months; $20,400

Maddala, G. S.; *Likelihood Methods in Pooling and Related Problems* (GS2894); 24 months; $54,100

Rose, Hugh; *Long-Run Macro-Economic Theory* (GS2756); 24 months; $44,900

NORTH CAROLINA

DUKE UNIVERSITY

Burdick, Donald S. and Naylor, Thomas H.; *Computer Simulation Experiments for Economic Systems* (GS1926 001); $11,900

Clark, Peter B.; *Flexible Exchange Rates and the Level of International Trade* (GS2996); 12 months; $7,000

Naylor, Thomas H., Burdick, Donald S. and Boughton, James M.; *Simulation Problems with Econometric Models* (GS2981); 24 months; $115,600

UNIVERSITY OF NORTH CAROLINA, Chapel Hill

Afriat, Sydney N.; *Analytical Theory of Consumption* (GS2195001); $4,100

Appleyard, Dennis R.; *Doctoral Dissertation Research in Economics* (GS2789); 12 months; $2,600

Gallman, Robert E.; *Efficiency in an Agricultural Export Region* (GS2730); 24 months; $51,100

PENNSYLVANIA

CARNEGIE-MELLON UNIVERSITY

Bronfenbrenner, Martin; *Income Distribution Theory* (GS2933); 12 months; $24,100

Lucas, Robert E. Jr., McGuire, Timothy W. and Rapping, Leonard A.; *Theory of the Aggregate Supply of Labor and Output* (GS2751); 24 months; $124,000

Roll, Richard W.; *Collaborative Research on Consumption, Production, and the Determination of Prices Under Uncertainty* (GS2962); 12 months; $20,000

PENNSYLVANIA STATE UNIVERSITY

Rodgers, James D.; *Collaborative Research in Utility Interdependence, Income Redistribution, and Fiscal Structure* (GS2974); 15 months; $14,900

Smith, James D.; *Estate Multiplier Estimation of the Distribution of Wealth* (GS2943); 12 months; $15,500

UNIVERSITY OF PENNSYLVANIA

Durand, John D. and Miller, Ann R.; *Comparative International Study of Labor Force Growth and Structure* (GS2787); 12 months; $57,800

Easterlin, Richard A.; *Doctoral Dissertation Research in Economics* (GS3116); 12 months; $1,700

Mansfield, Edwin; *Econometric Studies of Industrial Research and Technological Change* (GS2743); 36 months; $54,800

RHODE ISLAND

BROWN UNIVERSITY

Borts, George H. and Stein, Jerome L.; *Money and Economic Growth* (GS3083); 24 months; $100,000

Rohr, Robert J.; *Collaborative Research on Distribution Theory of GCL Estimators in Simultaneous Equations* (GS2807); 24 months; $40,400

TENNESSEE

UNIVERSITY OF TENNESSEE

Feiwel, George R.; *Industrialization and Central Planning in Bulgaria* (GS2970); 24 months; $34,100

TEXAS

SOUTHERN METHODIST UNIVERSITY

Huang, David S.; *Econometric Studies of the Residential Mortgage Market* (GS2834); 24 months; $54,100

UTAH

UTAH STATE UNIVERSITY

Gardner, B. Delworth; *Factors Influencing Residence Location of Farm Families* (GS2793); 24 months; $47,000

VIRGINIA

VIRGINIA POLYTECHNIC INSTITUTE

Tullock, Gordon; *Economic Analysis of Crime* (GS3079); 12 months; $43,500

WISCONSIN

UNIVERSITY OF WISCONSIN, Madison

Aigner, D. J.; *An Econometric Investigation of Short-Run Bank Behavior* (GS2878); 24 months; $62,500

David, Martin H. and Miller, Roger F.; *Variation in Disposable Incomes Under Alternative Tax Regimes* (GS2911); 24 months; $101,100

Nichols, Donald A.; *Equilibrium and Disequilibrium in Asset Markets* (GS2763); 24 months; $37,900

SOCIOLOGY AND SOCIAL PSYCHOLOGY

Sociology and Social Psychology

CALIFORNIA

CENTER FOR ADVANCED STUDY IN BEHAVIORAL SCIENCES
Davidson, Donald; *Foundations of Linguistics* (S002757Y; 10 months; $21,800

STANFORD UNIVERSITY
Alexander, C. Norman; *Symbolic Interaction Theory and Experimental Social Psychology* (S002759); 24 months; $48,000
Cohen, Elizabeth G.; *Generalization of Status Characteristics* (S002699); 24 months; $72,600
Ray, Michael L.; *Development of 'Unobtrusive' Measures of Attitude and Behavior* (S002683); 12 months; $7,500

U.S. INTERNATIONAL UNIVERSITY
Tedeschi, James T.; *Studies of Coercion and Inducement* (S003065); 24 months; $38,300

UNIVERSITY OF CALIFORNIA, Davis
Harrison, Albert A.; *The Exposure—Affect Relationship* (S002791); 24 months; $31,800

UNIVERSITY OF CALIFORNIA, Los Angeles
Gerard, Harold B.; *Speech Parapraxes and the Development of Language* (S002917); 24 months; $166,300
Kelley, Harold H.; *Studies of Social Relationships* (GS1121X003); 18 months; $43,000
Kuper, Leo; *Doctoral Dissertation Research in Sociology* (S002695); 18 months; $4,600

UNIVERSITY OF CALIFORNIA, Santa Barbara
McClintock, Charles G. and Messick, David M.: *The Study of Cooperative and Competitive Behavior* (S003061); 24 months; $50,800

UNIVERSITY OF SOUTHERN CALIFORNIA
Van Arsdol, Maurice D. Jr.; *Intrametropolitan Residential Mobility* (S002847); 24 months; $47,000

COLORADO

UNIVERSITY OF COLORADO
Elliott, Delbert S.; *Methodological Research in Sociometry* (S002876); 18 months; $44,200
McPhee, William N.; *Bibliographic Study of Formal Theory in Social Science* (S003051); 12 months; $14,600

CONNECTICUT

WESLEYAN UNIVERSITY
Hyman, Herbert H.; *Principles of Secondary Analysis of Survey Data* (S002772); 12 months; $3,000

YALE UNIVERSITY
Argyris, Chris; *Doctoral Dissertation Research in Administrative Sciences* (S003054); 12 months; $1,650
Day, Lincoln H.; *Social Determinants of Low Natality in Industrialized Countries* (S002692); 12 months; $20,600
Mettee, David R.; *Mediation of Interpersonal Attraction* (S002749); 24 months; $35,500
Oberschall, Anthony R.; *Social Structure and Innovation* (S003076); 24 months; $16,400

DISTRICT OF COLUMBIA

AMERICAN PSYCHOLOGICAL ASSOCIATION
Cook, Stuart W.; *Ethics of Research with Human Subjects* (B020066*); 24 months; $5,000

HOWARD UNIVERSITY
Gollin, Albert E.; *Conceptual and Historical Foundations of Public Opinion Research* (S002913); 11 months; $34,200

FLORIDA

UNIVERSITY OF FLORIDA
Silverman, Irwin; *Role Related Behavior in the Psychological Experiment* (S002718); 9 months; $6,100

HAWAII

UNIVERSITY OF HAWAII
Gallimore, Ronald; *Doctoral Dissertation Research in Psychology on Problems of Poverty* (GS2800); 12 months; $6,500

ILLINOIS

UNIVERSITY OF CHICAGO
Duncan, Starkey Jr.; *Nonverbal Communication* (S003033); 24 months; $49,300
McNeill, G. David; *Speech Rate in Child Language* (S002860); 24 months; $48,100
Meier, Paul; *Statistical Methodology in the Social Sciences* (S002818); 24 months; $135,000
Wiley, David E. and Bock, R. Darrell; *Multivariate Analysis of Qualitative Data* (S002900); 24 months; $104,700

UNIVERSITY OF ILLINOIS, Urbana
Davis, James H.; *Individual-Group Problem Solving* (S002747); 24 months; $49,600
Derber, Milton; *Doctoral Dissertation Research in Industrial Relations* (S003004); 12 months; $1,900

INDIANA

INDIANA UNIVERSITY, Bloomington
Burke, Peter J.; *Role Differentiation in Group Interaction* (S002742); 12 months; $15,000

PURDUE UNIVERSITY
Byrne, Donn E.; *A Reinforcement Model of Interpersonal Attraction* (S002752); 24 months; $56,000
Heslin, Richard; *Simulation Study of Perception and Attitude Change* (S002893); 24 months; $38,600

IOWA

UNIVERSITY OF IOWA
Pope, Hallowell; *Doctoral Dissertation Research in Sociology* (S003047); 12 months; $1,700

LOUISIANA

TULANE UNIVERSITY
Reissman, Leonard; (S002898); 8 months; $2,150

MASSACHUSETTS

HARVARD UNIVERSITY
Mishler, Elliot G.; *Social Context Effects on Language and Communication* (S003001); 24 months; $86,600
White, Harrison C.; *Mathematical Models of Structure and Process* (S002689); 24 months; $22,700

UNIVERSITY OF MASSACHUSETTS
Berger, Seymour M.; *Social Learning Through Observation* (S002746); 24 months; $40,700

MICHIGAN

HOPE COLLEGE
Myers, David G.; *Group Discussion Effects on Responses* (S002891); 24 months; $16,100

MICHIGAN STATE UNIVERSITY
Phillips, James L.; *Doctoral Dissertation Research in Psychology* (S002825); 18 months; $4,200
——— *Doctoral Dissertation Research in Social Psychology* (S002726); 12 months; $1,100
Rokeach, Milton; *Organization and Change in Values, Attitudes, and Behavior* (S003045); 24 months; $117,800

UNIVERSITY OF MICHIGAN
Blumenthal, Monica D., Kahn, Robert L. and Andrews, Frank M.; *Attitudes and Values Regarding Violence* (S002424001); $25,500
Duncan, Otis D.; *Causal Models in Social Research* (S002707); 24 months; $57,100
Gurin, Patricia; *Student Motivational Characteristics and College Experience* (S002877); 20 months; $36,900
Katz, Daniel; *Doctoral Dissertation Research in Psychology* (S002904); 17 months; $3,050
Lingoes, James C.; *Nonmetric Analysis in Social Science* (S002850); 24 months; $75,900
Pelz, Donald C.; *Detecting Causal Connections in Panel Data* (S002710); 22 months; $23,400
Reiss, Albert J. Jr.; *Discretionary Decisions in Legal Transactions* (S02771X); 24 months; $196,700
Zajonc, Robert B.; *Psychological Factors Associated with Frequently Encountered Stimuli* (S003119); 24 months; $94,800
Zander, Alvin F.; *Motivational Processes in Groups* (S003012); 24 months; $64,200

WAYNE STATE UNIVERSITY
Firestone, Ira J. and Kaplan, Kalman J.; *Attitude Gradient Structures* (S002852); 24 months; $73,900

MINNESOTA

UNIVERSITY OF MINNESOTA, Morris Campus
Klinger, Eric; *Structure and Determinants of Fantasy* (S002735); 24 months; $57,900

MISSOURI

UNIVERSITY OF MISSOURI, Columbia
Geen, Russell G.; *Stimulus and Arousal Determinants of Aggression* (S002748); 24 months; $55,000

NEBRASKA

UNIVERSITY OF NEBRASKA, Lincoln
Babchuk, Nicholas; *Doctoral Dissertation Research in Social Psychology* (S002848); 12 months; $8,200

NEW JERSEY

PRINCETON UNIVERSITY
Stone, Lawrence; *Changes in English Social Class Structure—1550–1750* (GS1559X002); 12 months; $32,600

RUTGERS UNIVERSITY
Toby, Jackson; *Delinquency in Industrial Societies* (GS1544X002); 12 months; $26,700

NEW YORK

COLUMBIA UNIVERSITY
Barton, Allen H. and Glaser, William A.; *The Brain Drain—An International Comparative Study* (S002889); 12 months; $80,400
Merton, Robert K.; *Sociology of Science* (S002736); 24 months; $145,000
Wallerstein, Immanuel; *Doctoral Dissertation Research in Sociology* (S002801); 15 months; $2,200

CORNELL UNIVERSITY
McGinnis, Robert; *Doctoral Dissertation Research in Sociology* (S003128); 12 months; $1,700
Meltzer, Leo, and Hayes, Donald P.; *Paralinguistic Dimensions of Group Interaction* (S003066); 24 months; $50,000
Whyte, William F. and Williams, Lawrence K.; *Social Structure and Change in Developing Areas* (S002884); 24 months; $178,400

CUNY, CITY COLLEGE
Dohrenwend, Barbara S.; *Interviewer-Respondent Interaction in Research Interviews* (S002896); 12 months; $16,100

NEW YORK UNIVERSITY
Smigel, Erwin O.; *Professionals in Large-Scale Organizations* (S002740); 12 months; $31,700

SOCIAL SCIENCE RESEARCH COUNCIL
Lehmann, Stanley; *Working Conference on Processes of Minority Influence in Groups* (S002819); 12 months; $9,940

STATE UNIVERSITY AT BINGHAMTON
Rehberg, Richard A.; *A Longitudinal Study of Adolescent Educational Expectations* (S002972); 24 months; $48,900

SOCIAL SCIENCES RESEARCH PROJECTS

STATE UNIVERSITY AT BUFFALO
Powell, Elwin H.; *Doctoral Dissertation Research in Sociology* (S003091); 12 months; $2,350
Pruitt, Dean G.; *Psychological Dynamics of Negotiation* (S002750); 24 months; $69,200
Raynor, Joel O.; *Effects of Distant Future Goals on Achievement Motivation* (S002863); 24 months; $39,700

SYRACUSE UNIVERSITY
Willie, Charles V.; *Doctoral Dissertation Research in Sociology* (S003121); 12 months; $5,150

TEACHERS COLLEGE
Hornstein, Harvey A.; *Experiments in the Social Psychology of Prosocial Behavior* (S002773); 24 months; $48,200

NORTH CAROLINA

DUKE UNIVERSITY
Jones, Edward E.; *Person Perception and Self-Presentation* (GS1114X003); 12 months; $27,400

NORTH CAROLINA STATE UNIVERSITY AT RALEIGH
Leventhal, Gerald S.; *Modes of Response to Inequity* (S002700); 24 months; $38,300

OHIO

CASE WESTERN RESERVE UNIVERSITY
Sussman, Marvin B.; *Doctoral Dissertation Research in Sociology* (S002705); 12 months; $1,450

OHIO STATE UNIVERSITY
Greenwald, Anthony G.; *Initial Opinion and Response to Persuasion* (S003050); 24 months; $45,400

PENNSYLVANIA

AMERICAN INSTITUTES FOR RESEARCH
Paydarfar, Ali A.; *Modern and Traditional Iran* (S002937); 12 months; $44,900

UNIVERSITY OF PITTSBURGH
Fararo, Thomas J.; *Doctoral Dissertation Research in Sociology* (S003095); 12 months; $800

SWARTHMORE COLLEGE
Gergen, Kenneth J.; *Effects of Aid on Recipient Attitudes* (S002803); 24 months; $74,000

RHODE ISLAND

BROWN UNIVERSITY
Weller, Robert H.; *In-Migration and Urban Growth* (S003078); 24 months; $14,500

TEXAS

UNIVERSITY OF TEXAS AT AUSTIN
Glenn, Norval D.; *Dortoral Dissertation Research in Sociology* (S002901); 12 months; $1,500

UTAH

UNIVERSITY OF UTAH
Anderson, Charles; *Doctoral Dissertation Research in Sociology on Problems of Poverty* (GS2934); 12 months; $850

WASHINGTON

UNIVERSITY OF WASHINGTON
Edelstein, Alex S.; *Communication Behavior and Perceptions of Conflict* (S002795); 18 months; $26,200

WESTERN WASHINGTON STATE COLLEGE
Mazur, D. Peter; *Demographic Analysis of USSR Ethnographic Data* (S003113); 26 months; $25,700

WISCONSIN

UNIVERSITY OF WISCONSIN, Madison
Berkowitz, Leonard; *Responsible Behavior in Dependency Relations* (GS1890X001); 12 months; $42,100
Hamilton, Richard F.; *Family, Class and Politics in West Germany* (S002792); 24 months; $59,300
Stone, Vernon A.; *Source-Content Orientation and Attitude Change* (GS1764 001); $2,100
Walster, Elaine C.; *Interpersonal Attraction* (S002932); 24 months; $83,000

History and Philosophy of Science

CALIFORNIA

CLAREMONT GRADUATE SCHOOL AND UNIVERSITY
Vickers, John M.; *Logical Analysis of Belief Change Criteria* (S002965); 14 months; $9,500

UNIVERSITY OF CALIFORNIA, Davis
Schwab, Richard N.; *Inventory of the Text and Plates of Diderot's 'Encyclopedie'* (S003055); 12 months; $13,500

UNIVERSITY OF CALIFORNIA, Los Angeles
Kaplan, David B.; *Non-Extensional Logic* (GS2721); 24 months; $24,000
Montague, Richard; *Studies in Metamathematics, Linguistics, and Philosophy of Science* (S002785); 24 months; $19,400
White, Lynn Jr.; *Doctoral Dissertation Research in the History of Technology* (S002991); 6 months; $1,000

CONNECTICUT

YALE UNIVERSITY
Aaboe, Asger; *Origins of Mathematical Astronomy* (S003053); 24 months; $21,000
Goldstein, Bernard R.; *History of Astronomy* (GS2025X001); 12 months; $10,200
Holmes, Frederic L.; *Development of Organic Analysis and Physiological Chemistry* (S003086); 24 months; $15,700

DISTRICT OF COLUMBIA

SMITHSONIAN INSTITUTION
Reingold, Nathan; *The Papers of Joseph Henry* (S01523X002); 12 months; $34,400

ILLINOIS

NORTHWESTERN UNIVERSITY
Daniels, George H.; *American Scientific Societies* (S003063); 12 months; $20,600

UNIVERSITY OF CHICAGO
Mehlberg, Henryk; *Measurement in Quantum Theory* (S002873); 9 months; $8,200
Schaffner, Kenneth F.; *The Logic and Epistemology of Molecular Biology* (S003099); 24 months; $21,100
Shapere, Dudley; *The Rationale of Scientific Development* (S003094); 24 months; $28,700

UNIVERSITY OF ILLINOIS, Urbana
Will, Frederick L.; *Philosophical Study of Inductive Claims and Justification* (S002886); 12 months; $22,100

INDIANA

INDIANA UNIVERSITY, Bloomington
Thoren, Victor E.; *Tycho Brahe's Work on Planetary Theory* (S002928); 24 months; $14,200

MASSACHUSETTS

HARVARD UNIVERSITY
Cohen, I. Bernard; *The Scientific Thought of Isaac Newton* (GS2063X001); 12 months; $19,100
Edsall, John T.; *Development of Biochemistry, 1905–1955* (S02723X); 24 months; $26,100
Holton, Gerald; *Genesis of Modern Physical Theory* (S02857); 24 months; $63,000
Murdoch, John E.; *Early History of the Measurement of Motion* (GS2815X); 24 months; $32,800

MICHIGAN

MICHIGAN STATE UNIVERSITY
Toulmin, Stephen E.; *Implications of the Neuro Sciences for the Theory of Knowledge* (S002782); 12 months; $33,600

MINNESOTA

UNIVERSITY OF MINNESOTA
Stuewer, Roger H.; *History of Optics* (S003098); 12 months; $4,900

MISSOURI

WASHINGTON UNIVERSITY
Allen, Garland E.; *The Embryological, Evolutionary, and Genetic Work of T. H. Morgan* (S002745); 24 months; $22,800

NEVADA

UNIVERSITY OF NEVADA, Reno Campus
Scott, William T.; *Scientific Methods in the Study of Precipitation Processes* (S002827); 24 months; $23,300

NEW JERSEY

PRINCETON UNIVERSITY
Kuhn, Thomas S.; *Doctoral Dissertation Research in History of Science* (S002713); 25 months; $1,000

NEW YORK

CORNELL UNIVERSITY
Black, Max; *Foundations of Theoretical Linguistics* (S002979); 14 months; $37,700

NEW YORK UNIVERSITY
Martin, Richard M.; *Formulation of Event-Logic* (S003096); 24 months; $15,500

POLYTECHNIC INSTITUTE OF BROOKLYN
Bromberg, Joan; *Studies in the History of Nuclear Theory—1928–1938* (S002992); 14 months; $9,400

UNIVERSITY OF ROCHESTER
Kyburg, Henry E. Jr.; *The Logical Foundations of Statistical Inference* (S002960); 24 months; $47,000
Lehrer, Keith E.; *The Application of Inductive Rules in Scientific Explanation and Rational Acceptance* (S003059); 15 months; $16,400

OHIO

ANTIOCH COLLEGE
Goldberg, Stanley; *Aspects of Response to Einstein's Special Theory of Relativity* (S002868); 24 months; $30,700

CASE WESTERN RESERVE UNIVERSITY
Schofield, Robert; *Doctoral Dissertation Research in the History of Science* (S002764); 12 months; $1,100
Stein, Howard; *Foundations of Physics* (S003101); 12 months; $15,300

OHIO STATE UNIVERSITY
Belkin, Johanna S. and Caley, Earle R.; *Early German Medical and Chemical Works* (S003082); 12 months; $16,800

PENNSYLVANIA

UNIVERSITY OF PENNSYLVANIA
Domotor, Zoltan; *Ordered Boolean Algebras and Their Applications* (S002936); 24 months; $21,700
Jeffrey, Richard C.; *Objective and Intersubjective Considerations in Decision Theory* (S003097); 24 months; $27,000
Thackray, Arnold W.; *Historical Relationships of Science, Technology and Society* (S003103); 24 months; $26,800

UNIVERSITY OF PITTSBURGH
Grunbaum, Adolf; *Philosophical Issues in Geometrodynamics and Cosmology* (S003107); 24 months; $25,300

RHODE ISLAND

BROWN UNIVERSITY
Chisholm, Roderick M.; *The Logic of Epistemic Preferability* (S002953); 24 months; $31,600

WASHINGTON

UNIVERSITY OF WASHINGTON
Hankins, Thomas L.; *Sir William Rowan Hamilton's Contributions to Science* (S002701); 17 months; $13,900

WISCONSIN

UNIVERSITY OF WISCONSIN, Madison
Lindberg, David C.; *Medieval and Renaissance Optics* (S002871); 17 months; $20,200

UNIVERSITY OF WISCONSIN, Milwaukee
Hull, David L.; *A Conceptual History of Taxonomy* (S003102); 12 months; $6,100

Geography

CALIFORNIA

UNIVERSITY OF CALIFORNIA, Irvine
Werner, Christian; *Formal Analysis and Optimization of Geographic Networks* (S002989); 14 months; $22,300

COLORADO

UNIVERSITY OF COLORADO
White, Gilbert F.; *Collaborative Research on Natural Hazards* (S02882X); 24 months; $96,200

FLORIDA

UNIVERSITY OF FLORIDA
McCune, Shannon; *Geographical Aspects of Changes in the Ryukyu Islands* (S002963); 12 months; $24,100

GEORGIA

UNIVERSITY OF GEORGIA
Hoy, Don R.; *Doctoral Dissertation Research in Geography* (S003008); 12 months; $1,800

ILLINOIS

NORTHWESTERN UNIVERSITY
Dacey, Michael F.; *Models of Urban Spatial Process* (S002967); 24 months; $76,700

UNIVERSITY OF CHICAGO
Berry, Brian J. L.; *Doctoral Dissertation Research in Geography* (S002813); 12 months; $4,000

INDIANA

BALL STATE UNIVERSITY
Carmin, Robert L.; *Symposium on the Future of Geographic Research in Latin America* (S002865); 12 months; $10,000

MASSACHUSETTS

CLARK UNIVERSITY
Cohen, Saul B.; *Doctoral Dissertation Research in Geography* (S002778); 12 months; $1,900

HARVARD UNIVERSITY
Warntz, William; *Geographical Patterns of Incomes for the United States* (S002833); 24 months; $99,900

MASSACHUSETTS INSTITUTE OF TECHNOLOGY
Carr, Stephen M.; *Doctoral Dissertation Research in Urban Studies and Planning* (S002821); 12 months; $2,650

OHIO

OHIO STATE UNIVERSITY
Rayner, John N. and Golledge, Reginald G.; *Spectral Analysis of Settlement Patterns* (S002781); 12 months; $37,300

POLITICAL SCIENCE

PENNSYLVANIA

PENNSYLVANIA STATE UNIVERSITY
Wernstedt, Frederick L.; *Doctoral Dissertation Research in Geography* (S002975); 12 months; $2,000

UNIVERSITY OF PENNSYLVANIA
Wolpert, Julian; *Interdependence in Locational Decisions* (S002758); 24 months; $60,600
———; *Doctoral Dissertation Research in Regional Science* (S002817); 14 months; $2,350

WASHINGTON

UNIVERSITY OF WASHINGTON
Morrill, Richard L.; *Experimental Derivation of Theoretical Surfaces* (S002853); 20 months; $49,700

WISCONSIN

UNIVERSITY OF WISCONSIN, Madison
Hudson, John C.; *Diffusion Processes in Urban Systems* (S002976); 12 months; $10,800

UNIVERSITY OF WISCONSIN, Milwaukee
Eidt, Robert C.; *Land Settlement Systems in West Central Europe* (S002839); 8 months; $8,100

Political Science

CALIFORNIA

STANFORD UNIVERSITY
Brody, Richard A.; *Issues in the Electoral Process* (S002855); 20 months; $87,200
Eulau, Heinz; *Doctoral Dissertation Research in Political Science* (S002765); 24 months; $2,800
———; *Decision Making in Small Groups* (S002698); 12 months; $44,600
———; *Doctoral Dissertation Research in Political Science* (S002926); 12 months; $1,800
Fagen, Richard R.; *Doctoral Dissertation Research in Political Science* (S002738); 12 months; $4,000

CONNECTICUT

UNIVERSITY OF CONNECTICUT
Ladd, Everett C. Jr.; *Doctoral Dissertation Research in Political Science* (S002851); 18 months; $3,200

FLORIDA

UNIVERSITY OF FLORIDA
Legg, Keith R.; *Doctoral Dissertation Research in Political Science* (S002912); 15 months; $2,750

HAWAII

UNIVERSITY OF HAWAII
Stauffer, Robert B.; *Doctoral Dissertation Research in Political Science* (S002856); 10 months; $7,000

ILLINOIS

NORTHWESTERN UNIVERSITY
Jacob, Herbert; *Citizen Orientations and Contact with Government* (S002702); 12 months; $19,750

UNIVERSITY OF CHICAGO
Easton, David; *Doctoral Dissertation Research in Political Science* (S002810); 12 months; $1,150
Lowi, Theodore J.; *Doctoral Dissertation Research in Political Science* (S002732); 12 months; $600
———; (S002688); 17 months; $700

UNIVERSITY OF ILLINOIS, Urbana
Monypenny, Phillip; *Doctoral Dissertation Research in Political Science* (S002727); 12 months; $2,700
Nagel, Stuart S.; *Effects of Alternative Legal Policies* (S002875); 36 months; $6,000

IOWA

UNIVERSITY OF IOWA
Schmidhauser, John; *Longitudinal Analyses of Judicial-Legislative Interaction* (S002862); 18 months; $52,500
Wahlke, John C.; *Doctoral Dissertation Research in Political Science* (S002761); 12 months; $6,900

KENTUCKY

UNIVERSITY OF KENTUCKY
Ulmer, S. Sidney; *Court Behavior Patterns* (S002682); 13 months; $41,700

MASSACHUSETTS

BOSTON UNIVERSITY
Zisk, Betty H.; *Simulation of Urban Bargaining Behavior* (S002794); 24 months; $45,000

MASSACHUSETTS INSTITUTE OF TECHNOLOGY
Rathjens, George W.; *Doctoral Dissertation Research in Political Science* (S003123); 12 months; $3,800

MICHIGAN

MICHIGAN STATE UNIVERSITY
Downes, Bryan T.; *Doctoral Dissertation Research in Political Science* (S003084); 15 months; $5,300
Ferguson, Leroy C.; *Doctoral Dissertation Research in Political Science* (S003120); 15 months; $1,700

UNIVERSITY OF MICHIGAN
Eldersveld, S. J.; *Comparative Study of Public Administrators* (GS2902X); 24 months; $200,900
Miller, Warren E.; *Seminars on Quantitative Political Science* (GZ1495*); 11 months; $70,000
———; *Research Conference on Small Natural-State Groups* (S002971); 12 months; $15,100
Ward, Robert E.; *Analysis of Behavioral Data on Japan* (S002776); 20 months; $35,800

MINNESOTA

UNIVERSITY OF MINNESOTA
Flanigan, William H. and Scott, Thomas M.; *Quantitative Analysis of Aggregate Data for Minnesota* (S002899); 12 months; $25,800
Holt, Robert T.; *Automata and Control Theory as Models for Organizational Decision Making* (S002955); 15 months; $55,700

MISSOURI

WASHINGTON UNIVERSITY
Kautsky, John H.; *Empirical Theory of Traditional Societies* (S002980); 14 months; $41,400

NORTH CAROLINA

UNIVERSITY OF NORTH CAROLINA
Beyle, Thad L.; *Doctoral Dissertation Research in Political Science on Problems of Poverty* (GS3011); 12 months; $3,600

NEW JERSEY

PRINCETON UNIVERSITY
Eckstein, Harry; *Comparative Study of Levels of Governmental Performance* (S003074); 24 months; $104,700
Lockwood, William W.; *Economic Development and Political Change in Asia* (S002714); 24 months; $25,000

NEW YORK

COLUMBIA UNIVERSITY
Chalmers, Douglas A.; *Doctoral Dissertation Research in Political Science* (S002844); 12 months; $750
Kesselman, Mark; *Regional Bases of French Political Parties* (GS2537 001); $6,400

SOCIAL SCIENCES RESEARCH PROJECTS

NEW YORK UNIVERSITY
Brams, Steven J.; *Collaborative Research on Comparative Voting Bodies* (S002798); 12 months; $14,600
Burrowes, Robert; *Multivariate Longitudinal Analysis of Conflict and Cooperation* (S002775); 24 months; $56,600

STATE UNIVERSITY AT BINGHAMTON
Banks, Arthur S.; *Multivariate Analysis of Cross-National Time-Series Data* (GS2420 001); $11,822

STATE UNIVERSITY AT STONY BROOK
Tanenhaus, Joseph; *Collaborative Research on Constitutional Courts* (S003132); 12 months; $1,100

SYRACUSE UNIVERSITY
O'Leary, Michael K.; *Collaborative Research on Comparative Voting Bodies* (GS2770); 12 months; $22,900

UNIVERSITY OF ROCHESTER
Goldberg, Arthur S.; *Testing of a Theoretical Model of Stability* (S002986); 12 months; $22,100
Riker, William H. and Ordeshook, Peter C.; *Systematic Political Theory* (S002720); 12 months; $17,700

OHIO

OHIO STATE UNIVERSITY
Hermann, Charles F. and Rosenau, James N.; *The Adaptation of National Societies* (S003117); 24 months; $92,300

PENNSYLVANIA

UNIVERSITY OF PENNSYLVANIA
Keim, Willard D.; *Doctoral Dissertation Research in Political Science* (S003060); 12 months; $1,100
Schwartz, David; *Research in Political Process* (S003089); 12 months; $1,000

TEXAS

UNIVERSITY OF TEXAS AT AUSTIN
Graham, Lawrence S.; *Doctoral Dissertation Research in Political Science* (S002760); 12 months; $3,350

WISCONSIN

UNIVERSITY OF WISCONSIN, Madison
Clausen, Aage R.; *Statistical Analysis of Influences on Voting Data* (GS2737); 24 months; $29,200

Special Projects in the Social Sciences

CALIFORNIA

STANFORD UNIVERSITY
Ferguson, Charles A. and Greenberg, Joseph H.; *Archival Research on Language Universals* (S002941); 24 months; $195,100

UNIVERSITY OF CALIFORNIA, Berkeley
Glock, Charles Y. and Nicholls, William L. II; *An Archive of Survey Data Collected in the Developing Countries* (S003104); 12 months; $47,900

UNIVERSITY OF CALIFORNIA, Los Angeles
Ladefoged, Peter; *Computerized Research on Speech* (S002741); 24 months; $80,200
———; (S002741001); $14,400

UNIVERSITY OF CALIFORNIA, San Diego
Stroll, Avrum and Klima, Edward; *Theoretical Investigation of Translation, Descriptions and Deep Structures* (S002982); 24 months; $58,200

CONNECTICUT

HUMAN RELATIONS AREA FILES
Ford, Clellan S.; *A Probability Sample of Files for Cross-Cultural Research* (S003075); 24 months; $78,200
———; *Second Hraf Conference on Cross-Cultural Research* (S002816); 12 months; $7,400

YALE UNIVERSITY
Ruggles, Richard; *Development of Machine-Readable Documentation Systems for Economic and Social Data* (GS2319 001); $50,700

DISTRICT OF COLUMBIA

NATIONAL ACADEMY OF ENGINEERING
Binns, Kerstin B.; *Symposium on 'The Quality of Life—Barriers and Constraints to the Use of Existing Technology in Balancing World Population and World Nutritional Resources'* (C310184*); 6 months; $3,000

INDIANA

INDIANA UNIVERSITY, Bloomington
Getman, Julius G. and Goldberg, Stephen B.; *Law-Related Study of Voting Behavior in National Labor Relations Board Elections* (S003030); 24 months; $203,400

MASSACHUSETTS

HARVARD UNIVERSITY
Kuno, Susumu; *Research in Mathematical Linguistics* (S002858); 24 months; $75,200
Watkins, Calvert; *Research on Indo-European Noun Morphology* (S002885); 24 months; $27,900

MASSACHUSETTS INSTITUTE OF TECHNOLOGY
Kuh, Edwin; *A Time-Shared Computer System for the Estimation, Testing and Simulation of Econometric Models* (GS2310 001); 4 months; $67,900
Weiner, Myron; *Archival Research Utilizing Processed Social Science Data on India* (S003080); 24 months; $135,800

MICHIGAN

UNIVERSITY OF MICHIGAN
Lakoff, George; *Generative Semantics* (S002939); 24 months; $58,800

NEW YORK

CORNELL UNIVERSITY
Silbey, Joel H.; *Collaborative Research on Archival Studies of Voting Behavior* (S002880); 24 months; $61,300

OHIO

UNIVERSITY OF CINCINNATI
Miller, Zane L.; *A Study of Urbanization in a River Valley* (S003106); 12 months; $16,500

WRIGHT STATE UNIVERSITY
Dorn Jacob H.; *Inventory of Historical Materials for Urban Research* (S002709); 12 months; $24,300

PENNSYLVANIA

AMERICAN INSTITUTE OF INDIAN STUDIES
Brown, W. Norman; *Regional Social Science Centers in India* (GS2655 001); 24 months; $20,000

CARNEGIE MELLON UNIVERSITY
Rosenthal, Howard L.; *Empirical Political Theory* (S002945); 24 months; $89,100

UNIVERSITY OF PENNSYLVANIA
Benson, Lee; *Collaborative Research on Archival Studies of Voting Behavior* (S002879); 24 months; $13,500

UNIVERSITY OF PITTSBURGH
Murdock, George P.; Barry, Herbert III and Tuden, Arthur; *Research in Cross-Cultural Coding and Sampling* (S002998); 24 months; $122,800

TEXAS

UNIVERSITY OF TEXAS AT AUSTIN
Cairns, Charles E.; *Theory of Phonological and Phonetic Universals* (S003044); 24 months; $25,800
Lehmann, W. P.; *Theoretical Investigation of Diachronic Syntax* (S003081); 12 months; $24,400

WASHINGTON

UNIVERSITY OF WASHINGTON
Horwood, Edgar M.; *Urban Analysis System Development* (S002832); 24 months; $151,000

WISCONSIN

UNIVERSITY OF WISCONSIN, Madison
Dennis, Jack S.; *A National Program Library and Central Program Inventory Service for the Social Sciences* (S002768); 24 months; $106,200

Engineering Research Projects

Engineering Chemistry

ALABAMA

UNIVERSITY OF ALABAMA, Tuscaloosa
Hatcher, William J. Jr.; *Research Initiation—Diffusivities in Zeolites by Chromatography* (K005424); 18 months; $15,000

ARIZONA

UNIVERSITY OF ARIZONA
Randolph, Alan D.; *Experimental and Theoretical Studies of Particle-Size Distributions in Crystallization Processes* (K016407); 24 months; $49,900

CALIFORNIA

STANFORD UNIVERSITY
Boudart, Michel J.; *Heterogeneous Catalysis by Transition Metals* (K017451); 24 months; $186,300

UNIVERSITY OF CALIFORNIA, Berkeley
Prausnitz, John M.; *Thermodynamic Properties of Fluid Mixtures for Chemical Engineering Design* (K014705); 24 months; $54,600

FLORIDA

UNIVERSITY OF FLORIDA
Gubbins, Keith E.; *Physical Properties of Gases Dissolved in Ionic Solutions* (K011997); 24 months; $36,100

GEORGIA

GEORGIA INSTITUTE OF TECHNOLOGY
Matteson, Michael J.; *Research Initiation—Application of High Voltage Electric Fields to the Removal of SO_2 and NO_2 from Combustion Effluent* (K005531); 18 months; $15,000

ILLINOIS

NORTHWESTERN UNIVERSITY
Bankoff, S. George and Graessley, W. W.; *Molecular Weight Distribution in Free Radical Systems* (K011290); 24 months; $57,700
Graessley, William W.; *Rheology of Polymer Systems* (K011290); 24 months; $56,700
Stein, Gilbert D.; *Research Initiation—Investigation of Small Molecular Clusters Formed by Homogeneous Nucleation* (K005892), 18 months; $13,300

UNIVERSITY OF ILLINOIS, Chicago Circle
Saxena, Satish C.; *Experimental Measurement of the Thermal Conductivity of Gases* (K012519); 24 months; $28,700

UNIVERSITY OF ILLINOIS, Urbana
Alkire, Richard Collin; *Research Initiation—Current Distribution During Electrophoretic Deposition of Insulating Films* (K005589); 18 months; $15,000
Eckert, Charles A.; *Molecular Thermodynamics of Kinetic Solvent Effects* (K014301); 24 months; $61,400

ENGINEERING CHEMISTRY

Hanratty, Thomas J.; *Two-Phase Flow* (K013748); 24 months; $44,800
———; *Wall Turbulence* (GK2813X001); 12 months; $36,200
Leland, Harry V.; *Research Initiation—Fate of Selected Organophosphate and Carbamate Insecticides in Surface Waters* (K005587); 18 months; $15,000

INDIANA

University of Notre Dame
Luks, Kraemer D.; *Molecular Theory of Transport Phenomena in Polyatomic Liquids* (K014212); 24 months; $39,000

IOWA

Iowa State University
Larson, Maurice A.; *Secondary Nucleation in Mixed Suspension Crystallization* (K014499); 24 months; $51,500

MARYLAND

University of Maryland
Gentry, James W.; *Research Initiation—Investigation of Oxidation Reactions in Solid Aerosol-Gas Systems in Near-Continuum Regimes* (K005449); 18 months; $15,000

MASSACHUSETTS

Massachusetts Institute of Technology
Baddour, Raymond F. and Modell, Michael; *Heterogeneous Catalysis* (GK1699X001); 12 months; $44,900
Brian, P. L. T. and Vivian, J. E.; *Gas Absorption with Simultaneous Chemical Reaction* (K011959); 24 months; $77,800
Reid, Robert C.; *Reactions of Irradiated, Dilute Fluorine-Oxygen Mixtures in Liquid Nitrogen with Organic Compounds* (K014551); 24 months; $33,600
Satterfield, Charles N.; *Chemical Processing in Trickle-Bed Reactors* (K012498); 24 months; $64,100
Yip, Sidney; *Frequency and Wavelength Dependent Fluctuations in Liquids and Gases* (K014915); 24 months; $56,100

Northeastern University
Blanc, Frederic C.; *Research Initiation—to Establish Efficient Electrolytic Control Systems for the Anaerobic Digestion Process* (K005555); 18 months; $14,900

University of Massachusetts
Lenz, Robert W.; *Reaction-Induced Crystallization of Polymers* (K016622); 24 months; $41,000

Worcester Polytechnic Institute
Ma, Y. H.; *Research Initiation—Rates of Sorption of Molecular Sieves in Multicomponent Sorbate Systems* (K005556); 18 months; $14,400

MICHIGAN

University of Michigan
Springer, George S.; *Measurements of Thermal Conductivities of Gases at High Temperatures* (K014006); 24 months; $47,600

MINNESOTA

University of Minnesota
Aris, Rutherford; *Adaptive Control of Chemical Reactors* (GK1367X002); 12 months; $40,500
Madden, Arthur J. Jr.; *Chemical Reactions in Two-Phase, Liquid-Liquid Systems* (K015253); 24 months; $41,300

MISSOURI

University of Missouri, Columbia
Luecke, R. H.; *Digital Process Control* (K013617); 24 months; $55,700
Novak, John T.; *Research Initiation—Lake Pollution by Internally Produced Organics* (K005377); 18 months; $15,000

Washington University
Spaeth, Edmund E.; *Research Initiation—an Investigation of Centrifugal Liquid-Liquid Extraction as a Novel Artificial Lung Design Concept* (K005563); 18 months; $15,000

NEW JERSEY

Princeton University
Johnson, Ernest F. and Mark, Peter; *Electronic Processes in Heterogeneous Catalysis* (K011101); 24 months; $54,700
Ollis, David F.; *Research Initiation—Field Ion Microscope Investigation of Multicomponent Catalyst Surfaces* (K005219); 18 months; $15,000
Wilkes, Garth L.; *Research Initiation—Molecular Interaction of Heparin with Polymeric Materials* (K005225); 18 months; $15,000

Rutgers University
Vieth, Wolf R.; *Structure—Property Relationships in Barrier Transport* (K014075); 12 months; $45,000

NEW YORK

Clarkson College of Technology
Estrin, Joseph and Youngquist, Gordon; *Experimental Investigation into the Fundamental Nature of Secondary Nucleation* (K013215); 24 months; $52,400

Cornell University
Watt, David M. Jr.; *Research Initiation—The Application of Electric Fields to the Study of Catalysis by Solids* (K005503); 18 months; $15,000

CUNY, City College
Williams, David J.; *Kinetic Studies of Emulsion Polymerization* (K017582); 24 months; $51,500

New York University
Treybal, Robert E.; *Mass Transfer in Continuously Operated Agitated Vessels* (K014007); 24 months; $27,700

Rensselaer Polytechnic Institute
Wotzak, Gregory P.; *Research Initiation—Cyclic Operation of Pollution-Control Reactors* (K005259); 18 months; $15,000

OHIO

Case Western Reserve University
Brosilow, Coleman B.; *Simulation and Control of Counter Current Separation Processes* (K013751); 12 months; $27,100
Walton, Alan G. and Baer, Eric; *Epitaxial Crystallization of Macromolecules* (K013296); 24 months; $52,400

OKLAHOMA

Oklahoma State University
Crynes, Billy L.; *Research Initiation—Photooxidation in Hydrocarbon-Nitrogen Oxide Systems—A Characterization of Reactor Wall Effects* (K005673); 18 months; $15,000

University of Oklahoma
Aldag, Arthur W. Jr.; *Research Initiation—Temperature-Programmed Desorption from Supported Metal Catalysts* (K005311); 18 months; $15,000
Robertson, James M.; *Research Initiation—Physiological and Structural Studies of Sphaerotilus Natans in Relation to the Bulking of Activated Sludge* (K005245); 18 months; $15,000

PENNSYLVANIA

Carnegie Mellon University
Brenner, Howard; *Transport Processes in Multiparticle Laminar Flow Systems* (K012583); 12 months; $35,900
Condiff, Duane W.; *Application of the Statistical Theory of Transport Processes* (K013941); 24 months; $46,000

Pennsylvania State University
Austin, Leonard G.; *Electrochemical Engineering of Porous Electrodes* (K011689); 24 months; $41,000
Becker, Philip M. and Heinsohn, Robert J.; *Effect of an Electric Field on the Chemical Kinetics of an Opposed Jet Diffusion Flame* (K019077); 24 months; $71,100

University of Pennsylvania
Myers, Alan L.; *Adsorption at the Liquid-Solid Interface* (K016758); 24 months. $58,700
Perlmutter, Daniel D.; *Stability of Tubular Recycle Reactors with Significant Axial or Transverse Transport Affects* (K010863); 24 months; $50,700

RHODE ISLAND

Brown University
Kestin, Joseph; *Transport Properties of Gases* (GK2133 001); 12 months; $57,700

SOUTH DAKOTA

South Dakota School of Mines and Technology
Fuerstenau, Maurice C.; *Adsorption Mechanisms of Anionic Surfactants on Insoluble Oxides and Silicates* (K024671); 24 months; $46,000

TENNESSEE

University of Tennessee
Hsu, Hsien-Wen; *Transport Phenomena and Thermodynamics in Zonal Centrifugation* (K011378); 24 months; $49,700
White, James L.; *Mechanism and Dispersion Processes in Chromatographic Fractionation of Polymers* (K011035); 24 months; $49,600

TEXAS

Rice University
Armeniades, Constantine D.; *Research Initiation—Structure-Property Relations of Ordered Polymers at Cryogenic Temperatures* (K005846); 18 months; $14,900
Jackson, Roy and Dyson, Derek C.; *Optimum Design Control and Operation of Chemical Reactors for Heterogeneously Catalyzed Chemical Reactions* (K012522); 24 months; $91,200
Kobayashi, Riki and Leland, Thomas W. Jr.; *Thermodynamic and Transport Properties of Non-Polar Compounds and Their Mixtures* (K018176); 24 months; $125,500
Leland, Thomas W. Jr. and Hightower, Joe W.; *Studies of Electronic Factors in Metal Oxide Catalysts* (K012858); 24 months; $63,400

University of Texas at Austin
Schechter, R. S.; *Hydrodynamic Stability* (K011562); 24 months; $28,000

UTAH

University of Utah
De Nevers, Noel; *A Study of Bubble Formation at Vibrated Orifices* (K014143); 24 months; $35,400
Miller Jan D.; *Research Initiation—an Electrokinetic Study of Calcium Fluoride* (K005239); 18 months; $14,100

VERMONT

University of Vermont
Jewell, William J.; *Estimating in Situ Nitrogen Fixation by Blue-Green Algae* (K025062); 24 months; $15,000

WASHINGTON

University of Washington
Finlayson, Bruce A.; *Nonsymmetric Stress Tensors and Thermal Mechanical Interactions* (K012517); 24 months; $47,300
Heidegger, William J.; *Dispersed Phase Mass Transfer* (K017145); 24 months; $55,300

ENGINEERING RESEARCH PROJECTS

WEST VIRGINIA

WEST VIRGINIA UNIVERSITY
Wen, C. Y.; *Dispersion of Non-Newtonian Liquids Flowing Through Fixed and Fluidized Beds* (K010977); 24 months; $53,000

WISCONSIN

MARQUETTE UNIVERSITY
Lewis, James L.; *Research Initiation—Mixed Intermolecular Constants from Acoustical Measurements* (K005416); 18 months; $14,900

UNIVERSITY OF WISCONSIN, Madison
Rudd, D. F.; *The Design of Processing Systems* (K019423); 24 months; $41,500

Engineering Energetics

ALABAMA

UNIVERSITY OF ALABAMA IN BIRMINGHAM
McCutcheon, Martin J.; *Research Initiation—Operating Characteristics of the Finite-Length Annular Induction MHD Generator* (K005329); 18 months; $15,000

ARIZONA

UNIVERSITY OF ARIZONA
Carlile, Robert N. and Johnson, V. R.; *Microwave Investigation of a Cylindrical Plasma Column* (K010918); 24 months; $46,000

CALIFORNIA

STANFORD UNIVERSITY
Crawford, F. W.; *Cyclotron Echoes and Wave/Wave Interaction Effects in Plasmas* (K016717); 24 months; $77,900
Sher, Rudolph; *Experiments on the Doppler Effect in Uranium Oxide and Thorium Capture and in Uranium-235* (K015979); 24 months; $48,400

UNIVERSITY OF CALIFORNIA, Berkeley
Ruby, Lawrence; *Studies in Reactor Dynamics* (K018980); 24 months; $51,200
Talobt, L. and Hurlbut, F. C.; *Study of Electron Temperature and Metastable Atoms in a Recombining Helium Plasma* (K015980); 12 months; $25,000

UNIVERSITY OF CALIFORNIA, Los Angeles
Knuth, Eldon L.; *Molecule-Surface Collisions with Translational Energies up to 20 Electron Volts* (KO15242); 12 months; $34,100

COLORADO

UNIVERSITY OF COLORADO
Gamow, Rustem Igor; *Research Initiation—Infrared Heat Receptor—Analysis of its Mode of Operation* (K005682); 18 months; $14,900
Timmerhaus, Klaus D. and Kropschot, Richard H.; *Thermal Transport Studies in Porous Media* (K002966); 24 months; $32,400

DISTRICT OF COLUMBIA

CATHOLIC UNIVERSITY OF AMERICA
Lee, Kai Fong; *Dispersion Relations of Inhomogeneous Plasmas and their Consequences* (K004592); 24 months; $38,400

FLORIDA

UNIVERSITY OF FLORIDA
Mockel, A. J.; *Application of Invariant Imbedding* (K010134); 24 months; $47,000

UNIVERSITY OF MIAMI
Adt,Robert R. Jr.; *Research Initiation—Irreversible Thermodynamic Analysis of the Heat Pipe* (K005823); 18 months; $15,000

GEORGIA

GEORGIA INSTITUTE OF TECHNOLOGY
Zuber, Novak; *Thermally Induced Flow Oscillations in Two Phase Flow Systems in Thermal Equilibrium* (K016023); 24 months; $72,000

HAWAII

UNIVERSITY OF HAWAII
Fand, Richard M.; *Simultaneous Boiling and Convection Heat Transfer from Bodies with Boundary Layer Separation* (K014546); 24 months; $50,000

ILLINOIS

NORTHWESTERN UNIVERSITY, Chicago Campus
Yu, E. Y.; *Rarefied Gas Dynamics* (K002831); 24 months; $28,000

UNIVERSITY OF ILLINOIS, Urbana
Chao, B. T.; *Thermal Response Behavior of Boundary Layer Flows* (K016270); 24 months; $67,900
Coleman, Paul D.; *Chemical Excitation of Molecules by Atomic Nitrogen* (K004331); 24 months; $55,000
Hudson, John L.; *Heat Transfer in Rotating Fluids* (K002505); 24 months; $31,500
Soo, Shao L.; *Carbon-Oxygen Fuel Cell* (K012698); 24 months; $62,500
Turnbull, Robert J.; *Research on Heat Transfer in Electric Fields* (K014624); 24 months; $35,300

IOWA

UNIVERSITY OF IOWA
Sayre, William W.; *Research Initiation—Mixing and Transfer Processes for Heated Effluents in Open Channel Flow* (K005918); 18 months; $15,000

KANSAS

KANSAS STATE UNIVERSITY
Kipp, John E.; *Heat Transfer and Pressure Drop for Liquid Hydrocarbons Flowing in an Annulus* (K011552); 24 months; $25,000
Shultis, John K.; *Research Initiation—Neutron Transport Theory without Azimuthal Symmetry* (K005557); 18 months; $14,500

UNIVERSITY OF KANSAS
Mesler, Russell B.; *Study of Bubble Shapes in Nucleate Boiling* (K010951); 24 months; $42,000

KENTUCKY

UNIVERSITY OF KENTUCKY
Birkebak, R. C. and Cremers, C. J.; *Thermophysical Properties of Frost Deposits* (K012989); 24 months; $82,600

LOUISIANA

TULANE UNIVERSITY
Seto, Y. J.; *Electromagnetic Wave Interaction with Shock Generated Plasma* (K002805); 24 months; $34,900

MASSACHUSETTS

MASSACHUSETTS INSTITUTE OF TECHNOLOGY
Olson, N. Thomas; *Ion Bombardment Induced Photon Emission* (K010001); 24 months; $49,200
Smullin, Louis D., Rose, David J. and Bekefi, George; *Plasma Dynamics* (K018185); 12 months; $312,000

UNIVERSITY OF MASSACHUSETTS
Rha, Chokyun; *Research Initiation—Investigation of Thermal and Rheological Properties of Proteinaceous Material* (K005702); 18 months; $15,000

MICHIGAN

MICHIGAN STATE UNIVERSITY
Asmussen, Jes; *Research Initiation—'Instabilities and Resonances in High Frequency Resonance Sustained Discharges'* (K005617); 18 months; $15,000
Chen, Kun-Mu; *Investigation of Electroacoustic Waves in Compressive Plasmas* (K002952); 24 months; $45,000

UNIVERSITY OF MICHIGAN
Arpaci, Vedat S.; *Elastoplastic Effects on Thermal Stability of Viscous Fluids* (K016094); 12 months; $19,700
Getty, W. D. and Rowe, Joseph E.; *Beam-Plasma Systems and Plasma Heating* (K015689); 24 months; $67,000

MINNESOTA

UNIVERSITY OF MINNESOTA
Eckert, Ernst R. G. and Pfender, E.; *Experimental Studies of Boundary Layers of a High Temperature Gas in the Presence of an Electric Field* (K015924); 24 months; $51,600
Goldstein, Robert J.; *Thermal Convection in Horizontal Fluid Layers* (K015252); 24 months; $50,000
Peterson, Edward W.; *Research Initiation—Perturbation of a Flowing Plasma by an Electrostatic Probe* (K005395); 18 months; $14,200

MISSOURI

UNIVERSITY OF MISSOURI, Columbia
Wollersheim, David E.; *Research Initiation—Free Convection from a Horizontal Cylinder to a Non-Newtonian Fluid* (K005454); 18 months; $15,000

UNIVERSITY OF MISSOURI, Rolla
Look, Dwight C. Jr.; *Research Initiation—Actual Surface Character and Reflectance* (K005238); 18 months; $15,000

MONTANA

MONTANA STATE UNIVERSITY
Genetti, William E.; *Research Initiation—Transpirational Momentum and Heat Transfer from a Rotating Cylinder* (K005321); 18 months; $15,000

NEW JERSEY

PRINCETON UNIVERSITY
Andres, R. P.; *Gas Phase Collision Processes by Means of Nozzle-Source Molecular Beams* (K004160); 24 months; $80,000

NEW YORK

COLUMBIA UNIVERSITY
Gross, R. A., Sen, A. K. and Schlesinger, S. P.; *Research on Waves, Instabilities and Fluctuations in Plasma* (GK1391X002); 12 months; $107,400
Teich, Melvin C.; *Linear and Nonlinear Detection of Infrared Laser Radiation* (K016649); 24 months; $47,700

COOPER UNION
Tan, Chor-Weng; *Heat and Mass Transfer in Magnetohydrodynamics* (K017000); 12 months; $21,000

CUNY, CITY COLLEGE
Cataldo, Joseph C.; *Research Initiation—'Spatial Distribution of a Streaming Plasma Column and the Accompanying Transverse Diffusion'* (K005230); 18 months; $15,000

RENSSELAER POLYTECHNIC INSTITUTE
Noon, Jack H.; *Electron Energy Loss Processes in Molecular Gas Plasmas* (K003834); 24 months; $50,000

STATE UNIVERSITY AT BUFFALO
Shaw, David T.; *Theoretical and Experimental Studies of Particle and Energy Transport in a Low-Voltage Cesium Discharge* (K016372); 24 months; $50,000

STATE UNIVERSITY AT STONY BROOK
Cess, Robert D.; *Study of Radiative Energy Transfer in Gases* (K016755); 24 months; $41,200

UNIVERSITY OF ROCHESTER
Lubin, Moshe J.; *A Theoretical and Experimental Study of Laser Plasmas* (K018578); 24 months; $91,300
Stroud, Carlos R.; *Research Initiation—Applications of Improved Semiclassical Radiation Theory* (K005459); 18 months; $14,800

ENGINEERING MECHANICS

NORTH CAROLINA

NORTH CAROLINA STATE UNIVERSITY AT RALEIGH

Mulligan, James C; *Research Initiation—Transient Freezing of Liquids in Internal Flow with Freeze Blockage* (K005665); 18 months; $15,000

Ozisik, M. N. and Siewert, C. E.; *The Application of Case s Normal Mode Expansion Technique to Radiative Heat Transfer Problems in Participating Media* (K011935); 24 months; $53,200

NORTH DAKOTA

NORTH DAKOTA STATE UNIVERSITY

Hugelman, Rodney D.; *Research Initiation—Induced Circulation through MHD Boundary Layer Control* (K005750); 18 months; $14,700

OHIO

OHIO STATE UNIVERSITY

Petrie, Stuart L.; *Mechanisms of Electronic Energy Transfer Between Molecules* (K002618); 24 months; $30,700

UNIVERSITY OF AKRON

Greene, Howard L.; *The Influence of Natural Convection Phenomena on the Flow of Newtonian and Non-Newtonian Fluids* (K015744); 15 months; $12,000

PENNSYLVANIA

UNIVERSITY OF PENNSYLVANIA

Altman, Manfred; *Energy Conversion* (K013566); 12 months; $68,000

Gitomer, Steven J.; *Research Initiation—'Instability Mechanisms in an Electron Cyclotron Resonance Plasma'* (K005475); 18 months; $14,700

SOUTH DAKOTA

SOUTH DAKOTA STATE UNIVERSITY

Hellickson, Mylo A.; *Research Initiation—Heat and Moisture Production in a Beef Confinement Unit* (K005891); 18 months; $15,000

TENNESSEE

TENNESSEE TECHNOLOGICAL UNIVERSITY

Pitts, Donald R.; *Research Initiation—Laminar Free Convective Heat Transfer from Cascaded Horizontal Cylinders* (K005640); 18 months; $15,000

TEXAS

LAMAR STATE COLLEGE OF TECHNOLOGY

Young, Fred; *Effects of Nucleation Site Geometry by a Simulation Technique* (K018048); 20 months; $17,900

TEXAS A & M UNIVERSITY

Carlson, Leland A.; *Research Initiation—Electron Temperature and End-Wall Pressure Effects on the Flow Behind Radiation—Coupled Reflected Shock Waves* (K005268); 18 months; $15,000

UNIVERSITY OF TEXAS, Arlington

Smith, Charles V. Jr.; *Research Initiation—High Voltage Electric Field Generators for Energy Conversion* (K005576); 18 months; $15,000

UNIVERSITY OF TEXAS, Austin

Brockmeier, Norman F.; *Microwave Plasma Chemistry—Measurements and Theoretical Predictions of Electrical Parameters and Conversion in a Methane-Ammonia Plasma* (K017445); 24 months; $23,400

Clark, William M. Jr.; *Research Initiation—Amplification of Ultrashort Pulses with* ND^{+3}: $POCl_3$ *Liquid Lasers* (K005342); 18 months; $14,200

Draper, E. Linn; *Research Initiation—Fission Product Yields from Twelve Heavy Nuclides* (K005477); 18 months; $14,900

Koen, Billy V.; *Research Initiation—Optimal Direct Digital Control of Distributed Parameter Nuclear Reactor Systems* (K005390); 18 months; $14,300

Wissler, Eugene H.; *Study of the Physical Properties of Submicroscopic Aerosol Particles* (K010099); 24 months; $30,000

UTAH

BRIGHAM YOUNG UNIVERSITY

Rogers, Vern C.; *Research Initiation—Element Identification with High Energy Neutrons* (K005889); 18 months; $13,700

WASHINGTON

UNIVERSITY OF WASHINGTON

Ahlstrom, Harlow G.; *Research in Magneto-Fluid Dynamics* (K020708); 24 months; $72,600

Babb, Albert L.; *Neutron Multiplying and Moderating Systems* (K019171); 24 months; $147,000

Decher, Reiner; *Research Initiation—End Loop Shorting Reduction in Nonequilibrium MHD Generators* (K005393); 18 months; $15,000

WASHINGTON STATE UNIVERSITY

Mandell, David A.; *Research Initiation—Radiation Heat Transfer within High-Temperature Gases* (K005276); 18 months; $15,000

WEST VIRGINIA

WEST VIRGINIA UNIVERSITY

Pappano, Alfred W.; *Research Initiation—Semi-Empirical Modelling of Thermal Neutron Cross Sections and Diffusion Parameters in Liquid Moderators* (K005678); 18 months; $14,900

Engineering Mechanics

ARKANSAS

UNIVERSITY OF ARKANSAS

Jong, Ing-Chang; *Vibrations and Nonconservative Stability Problems of Elastic Systems with Hysteresis* (K013533); 24 months; $21,800

CALIFORNIA

CALIFORNIA INSTITUTE OF TECHNOLOGY

Housner, George W. and Hudson, Donald E.; *A Comprehensive Research Program in Earthquake Engineering* (GK1197X); 12 months; $160,000

SACRAMENTO STATE COLLEGE

Niyogi, Pradyot K.; *Research Initiation—Statistical Evaluation of Load Factors for Structural Design* (K005657); 18 months; $14,900

SAN DIEGO STATE COLLEGE

Agarwal, Sohan L.; *Research Initiation—Soil-Structure Interaction of Pile Foundations for Offshore Structures* (K005340); 18 months; $15,000

STANFORD UNIVERSITY

Benjamin, Jack R. and Shah, Haresh; *Evaluation of Load Factors for Design of Structural Systems* (K004262); 24 months; $60,800

Johnston, James P.; *Structure and Stability of Turbulent Boundary Layers on Rotating Surfaces* (K016450); 24 months; $63,800

Linsley, Ray K.; *Application of Digital Simulation of the Hydrologic Cycle* (K017187); 24 months; $97,500

Roth, Bernard; *Kinematic Synthesis of Spatial Linkages* (K013930); 24 months; $65,700

UNIVERSITY OF CALIFONIA, Berkeley

Shephard, R. W. and Todd, David K.; *Optimal Determination of Stratified Groundwater Basin Characteristics* (K019836); 12 months; $15,500

UNIVERSITY OF CALIFORNIA, Los Angeles

Hart, Gary C.; *Research Initiation—Free Vibration Analysis of Framed Structures with Random Material and Geometric Properties* (K005513); 18 months; $15,000

Lin, T. H. and Ito, Y. Marvin; *Microstresses in Metals Under Monotonic and Reversed Loadings* (K015348); 24 months; $66,600

Singh, Awtar and Lee, Kenneth L.; *Reinforced Earth* (K013153); 24 months; $50,800

UNIVERSITY OF SOUTHERN CALIFORNIA

Browand, Frederick K.; *A Method for the Production of a Steady, Uniform Flow with a Density Gradient* (K004609); 24 months; $47,200

Masri, S. F.; *Impact Vibration Dampers in Aseismic Design* (K011553); 24 months; $37,200

Seide, Paul; *Stability of Thin Rings Under Nonuniform Loads* (K011957); 24 months; $33,400

COLORADO

COLORADO SCHOOL OF MINES

Kesic, Dragoljub M.; *Research Initiation—Structure of Scalar Fields Mixed by Turbulence for Arbitrary Schmidt Numbers* (K005873); 18 months; $15,000

COLORADO STATE UNIVERSITY

Bell, James M.; *Stability and Displacement of Statically Loaded Homogeneous Soil Structures* (K015793); 24 months; $75,000

Bodig, Jozsef and Goodman, James R.; *Predictions of Elastic Parameters for Wood* (K015785); 24 months; $90,100

Hardee, Addison G. Jr.; *Research Initiation—Diffusion of Spermatozoa in Cervical Mucus* (K005895); 18 months; $15,000

Histand, Michael B.; *Research Initiation—'Secondary Flows in the Aortic Arch'* (K005572); 18 months; $15,000

Shen, Hsieh W. and Todorovic, Petar; *Transport and Dispersion of Bed Materials in Open Channels* (K011499); 24 months; $87,200

Thomas, Charles W. and Albertson, Maurice L.; *Completion of Study of Civil Engineering Research Needs* (K016855); 12 months; $40,700

Tullis, J. Paul; *Cavitation Scale Effects for Generalized Valve Shapes* (K012513); 12 months; $43,500

Yevjevich, V.; *Large Continental Droughts* (K011564); 24 months; $185,500

UNIVERSITY OF COLORADO

Carley, James F.; *Biaxial Stretching of Viscoelastic Sheets* (K017661); 24 months; $26,500

Chanaud, Robert C.; *Research Initiation—'Sound Radiation from Flow Excited Ducting'* (K005242); 18 months; $14,700

Datta, S. K. and Jahsman, W. E.; *Wave Propagation in Solid Bodies* (K010872); 24 months; $65,500

Ko, Hon-Yim and Gerstle, Kurt H.; *Constitutive Relations for Rocks* (K014630); 24 months; $76,200

Snyder, Howard; *Hydrodynamics of Superfluid Helium* (K016592); 24 months; $70,000

UNIVERSITY OF DENVER

Kaplan, Michael A.; *Research Initiation—The Fracture of Ductile Metal Rods under Combined Axial Load and Hydrostatic Pressure* (K005710); 18 months; $15,000

CONNECTICUT

UNIVERSITY OF CONNECTICUT

Jeffers, Robert G.; *Research Initiation—'The Relation of Nonlinear to Linear Thin Elastic Shell Theory'* (K005623); 18 months; $15,000

DELAWARE

UNIVERSITY OF DELAWARE

Nowinski, J. L.; *Theory of Brittle Rupture and Biomechanics of Bones* (K013159); 24 months; $58,300

ENGINEERING RESEARCH PROJECTS

DISTRICT OF COLUMBIA

CATHOLIC UNIVERSITY OF AMERICA

Durelli, A. J.; *Moire and Holographic Techniques for Experimental Stress Analysis* (K016907); 24 months; $156,300

GEORGE WASHINGTON UNIVERSITY

Toriois, Theodore G.; *Local Buckling or Plasticity and the Ultimate Strength of Structures* (K016142); 24 months; $56,800

HOWARD UNIVERSITY

James, Clarence H. C.; *Research Initiation—Physical Effects of a Water Softener on a Natural Kaolinitic Clay* (K005425); 18 months; $15,000

NATIONAL ACADEMY OF SCIENCES

Coleman, John S.; *Task Order for Partial Support of the U.S. National Committee for the International Council for Building Research Studies and Documentation* (C310186); 12 months; $10,000

FLORIDA

FLORIDA STATE UNIVERSITY

Kranc, Stanley C.; *Research Initiation—Shock Waves in a Gas-Particle Suspension* (K005343); 18 months; $15,000

UNIVERSITY OF FLORIDA

Lindgren, E. Rune; *Structure of Turbulent Liquid Flows* (K004377); 12 months; $20,000

UNIVERSITY OF MIAMI

Freeman, Neil J.; *Load Transfer and Stress Concentrations in Composite Materials* (K016656); 24 months; $46,500

GEORGIA

GEORGIA INSTITUTE OF TECHNOLOGY

Anderson, Jerry M.; *Research Initiation—'Fracture in Rate Sensitive Materials'* (K005529); 18 months; $15,000

Williams, Kent C.; *Research Initiation—Turbulent-Acoustic Interactions* (K005524); 18 months; $15,000

Wulff, Wolfgang; *Research Initiation—'Turbulent Boundary Layer in a Rotating Fluid'* (K005525); 18 months; $15,000

HAWAII

UNIVERSITY OF HAWAII

Chiu, Arthur N. L.; *Dynamic Response of Free-Standing Structures to Wind Forces* (K013076); 24 months; $94,600

Seidl, Ludwig; *Prediction of the Motions of Moored Vessels in Irregular Sea* (K010528); 24 months; $40,800

ILLINOIS

ILLINOIS INSTITUTE OF TECHNOLOGY

Lavan, Zalman; *Effect of Rotation on the Stability of an Axial Flow Between Concentric Cylinders* (K016980); 24 months; $47,000

Moran, Thomas J.; *Research Initiation—Transient and Steady State Motions in Partially Dissipative Dynamical Systems* (K005865); 18 months; $15,000

UNIVERSITY OF ILLINOIS, Chicago Circle

Belytschko, Ted B.; *Research Initiation—Structural Sensitivity to Imperfections—A Finite Element Approach* (K005834); 18 months; $15,000

Lemke, Donald G.; *Research Initiation—Tandem and Axial Coupling of Turborotor Systems* (K005540); 18 months; $14,400

Uherka, Kenneth L.; *Research Initiation—'The Structure and Dynamics of Tornadoes'* (K005704); 18 months; $15,000

UNIVERSITY OF ILLINOIS, Urbana

Ang, Alfredo H. S.; *Probabilistic Aspects of Structural Mechanics Relevent to Safety and Design* (GK1812X001); 17 months; $45,000

Gallagher, Joseph P.; *Research Initiation—Corrosion Fatigue Crack Growth Rates Below KISCC* (K005584); 18 months; $15,000

Karara, H. M.; *Photogrammetric Potentials of Non-Metric Cameras* (K011655); 24 months; $60,000

McLaughlin, Philip V. Jr.; *Research Initiation—Plastic Behavior of Fibrous Composites* (K005582); 18 months; $15,000

Nielsen, N. N. and Sozen, Mete A.; *Effects of Earthquake Motions on Reinforced Concrete Buildings* (K015692); 12 months; $58,700

Schnobrich, W. C.; *Behavior of Reinforced Concrete Shells by the Use of Discrete Element Method* (K011190); 24 months; $47,200

Sidebottom, Omar M.; *Inelasticity Theories for Time Dependent and Time Independent Deformations* (K013613); 24 months; $45,800

INDIANIA

PURDUE UNIVERSITY

Bogdanoff, John L. and Sweet, A. L.; *Engineering Applications of Stochastics* (K014411); 24 months; $160,100

Brenchley, David L.; *Research Initiation—Dispersion of Pollutants from Tall Chimneys* (K005875); 18 months; $15,000

Goldschmidt, V. W.; *Turbulent Transport in Free Jets* (K019317); 24 months; $58,700

Schiff, Anshel J.; *Design, Development, and Testing of a Strong-Motion Seismograph* (K021033); 12 months; $25,500

Sun, Chin-Teh; *Research Initiation—'Mechanics of Composite Media'* (K005436); 18 months; $15,000

UNIVERSITY OF NOTRE DAME

Ariman, Teoman; *Thermal Stresses in Cylimdrical Shells with Cutouts* (K011194); 24 months; $30,500

Marley, Jerry J.; *Research Initiation—Activation Energy of Soil Creep* (K005763); 18 months; $15,000

IOWA

IOWA STATE UNIVERSITY

Demirel, Turgut and Handy, R. L.; *Adsorption Energies and Swelling Pressures of Montmorillonite* (K010191); 24 months; $43,200

Schmerr, Lester W. Jr.; *Research Initiation—Diffraction of Transient Elastic Waves by Wedges and Notches* (K005403); 18 months; $15,000

UNIVERSITY OF IOWA

Kane, Harrison; *Behavior of Loess in Confined Compression* (K015112); 24 months; $45,400

Kennedy, John F.; *Dynamics of Ice Covered Streams* (K017698); 24 months; $51,200

KANSAS

UNIVERSITY OF KANSAS

Reese, Charles D.; *Research Initiation—Combined Axial Compression and Torsional Buckling of Orthotropic Sandwich Cylinders and Cones with Clamped End Supports* (K005565); 18 months; $15,000

KENTUCKY

UNIVERSITY OF KENTUCKY

Beatty, Millard F. Jr.; *Special Topics in the General Theory of Elastic Stability* (K010172); 24 months; $42,500

Eichhorn, Roger; *Three Dimensional Turbulent Boundary Layers with Mass Transfer* (K015251); 24 months; $40,600

Kao, David T. Y.; *Research Initiation—'Solid Particles Moving in a Viscous Fluid'* (K005614); 18 months; $15,000

LOUISIANA

LOUISIANA POLYTECHNIC INSTITUTE

Lowther, James D.; *Research Initiation—'Entrainment and Droplet Diffusion in Two-Phase Flow'* (K005266); 18 months; $15,000

MARYLAND

JOHNS HOPKINS UNIVERSITY

Ericksen, J. L.; *Nonlinear Continuum Theories* (K013306); 24 months; $82,600

MASSACHUSETTS

MASSACHUSETTS INSTITUTE OF TECHNOLOGY

Argon, Ali S., Berg, Charles A. and McClintock, Frank A.; *The Mechanics of Ductile Fracture* (K01875X001); 18 months; $82,000

Kaufman, Roger E.; *Research Initiation—'Analysis and Synthesis of Mechanisms'* (K005661); 18 months; $14,600

Kerwin, Justin E.; *Unsteady Hydrodynamic Measurements* (K010524); 24 months; $81,600

Reinschmidt, Kenneth F.; *Optimization Methods for the Design of Urban Housing* (K016204); 24 months; $62,800

Wilson, D. G.; *Flow in Radial Diffusers* (K010322); 24 months; $30,800

UNIVERSITY OF MASSACHUSETTS

Goss, William P.; *Aerodynamic Performance of Gravity-Vacuum Tube Vehicles* (K024514); 18 months; $15,000

Kirchhoff, Robert H.; *Research Initiation—Dielectrophoretic Fluid Mechanics* (K005569); 18 months; $14,900

Mani, Ramani; *Research Initiation—Some Real Fluid Effects in Compressor Noise* (K005215); 18 months; $15,000

MICHIGAN

MICHIGAN STATE UNIVERSITY

Wen, Robert K.; *Dynamics of Three-Dimensional Structures* (K015698); 24 months; $53,700

MICHIGAN TECHNOLOGICAL UNIVERSITY

Krueger, Gordon P.; *Ultimate Strength Design of Timber Structures* (K010988); 24 months; $58,300

UNIVERSITY OF DETROIT

Kordyban, Eugene S.; *Research Initiation—'Aerodynamic Pressure Over Water Waves'* (K005693); 18 months; $15,000

UNIVERSITY OF MICHIGAN

Hammitt, Frederick G.; *Asymmethic Bubble Collapse* (K013081); 24 months; $57,300

Ogilvie, T. Francis; *Wave Resistance at Low Speed* (K014375); 12 months; $21,300

Sikarskie, David L.; *The Penetration of Anisotropic Brittle Rocks* (K017101); 24 months; $30,200

Streeter, Victor L. and Wylie, E. B.; *Transient Flow Through Open and Closed Conduits* (K014213); 24 months; $73,600

Woods, Richard D.; *Isolation of Structures from Earthquake Waves* (K016012); 24 months; $61,100

WAYNE STATE UNIVERSITY

Allen, Stuart J.; *Fluids with Microstructure* (K012990); 24 months; $27,200

Kline, Kenneth A.; *Theory of Blood Flow* (K012653); 24 months; $27,200

MINNESOTA

UNIVERSITY OF MINNESOTA

Beavers, Gordon S. and Sparrow, E. M.; *Coupled Flows in Ducts and Porous Media* (K013303); 24 months; $49,600

Joseph, Daniel D.; *Studies in Hydrodynamic Stability* (K012500); 24 months; $53,300

Pfannkuch, Hans-Olaf; *Electrical Conductivity in Saturated Porous Systems* (K011140); 24 months; $41,100

Ranz, William E.; *Basic Studies of Gas-Particulate Systems* (K010004); 24 months; $58,700

MISSOURI

ST. LOUIS UNIVERSITY

Pennell, Douglas G.; *Research Initiation—Residual Shear Strength of Stiff Clays* (K005325); 18 months; $15,000

Yu, Ching K.; *Research Initiation—The Behavior of Inelastic Beam-Columns Subjected to Cyclic Loads* (K005353); 18 months; months; $15,000

ENGINEERING MECHANICS

UNIVERSITY OF MISSOURI, Columbia

McBean, Robert P.; *Research Initiation—Analysis of Stiffened Plate Systems* (K005437); 18 months; $15,000

UNIVERSITY OF MISSOURI, Rolla

Kovacs, William D.; *Research Initiation—The Liquefaction Behavior of Saturated Sands in Cyclic Simple Shear* (K005434); 18 months; $15,000

WASHINGTON UNIVERSITY

Galambos, Theodore V.; *Fully Yielded Biaxially Loaded Steel Beams* (K010604); 24 months; $47,400

Szabo, Barna A.; *Research Initiation—The Local Solution Approach in the Finite Element Method* (K005258); 18 months; $15,000

NEW JERSEY

RUTGERS UNIVERSITY

Chen, Chuan F.; *Stability of Time-Dependent Rotational Flows* (K014275); 24 months; $48,000

NEW MEXICO

UNIVERSITY OF NEW MEXICO

Schreyer, Howard L.; *Elastic Shells with Large Strains* (K011254); 24 months; $32,500

NEW YORK

CLARKSON COLLEGE OF TECHNOLOGY

Thornton, William A.; *Synthesis of Structural Frameworks Under Time Dependent Loading* (K010007); 48 months; $43,200

COLUMBIA UNIVERSITY

Deresiewicz, Herbert; *Dispersion and Dissipation of Stress Waves in Solids* (K016361); 24 months; $40,600

Spillers, William R. and Al-Banna, S.; *Interactive Space Allocation System for Building Layout* (K018773); 24 months; $30,300

COOPER UNION

Grossman, Perry L.; *Research Initiation—'Nonlinear Vibration Analysis of Conical Shells'* (K005766); 18 months; $11,700

CORNELL UNIVERSITY

Bartel, Donald L.; *Research Initiation—Optimum Design of Mechanical Systems* (K005768); 18 months; $15,000

Dafermos, Constantine M.; *Research Initiation—'Asymptotic Stability of Processes'* (K005854); 18 months; $15,000

Nilson, Arthur H.; *Bond Stress-Slip Relations in Reinforced Concrete* (K017188); 12 months; $20,400

Rand, Richard H.; *Research Initiation—The Nonlinear Vibrations of Two Degree of Freedom Systems* (K005650); 18 months; $15,000

White, Richard N.; *Model Analysis of Reinforced Concrete Structures* (K013992); 24 months; $64,200

CUNY, CITY COLLEGE

Miller, C. A.; *Creep Effects in Indeterminate Concrete Structures* (K015458); 24 months; $45,000

Parnes, Raymond; *Wave Propagation Due to Moving Loads in Elastic Media* (K014514); 24 months; $36,400

NEW YORK UNIVERSITY

Bennett, Leon; *Creeping Flow Aerodynamics* (K010980); 24 months; $44,700

POLYTECHNIC INSTITUTE OF BROOKLYN

Drenick, Rudolph F.; *Prediction of Earthquake Resistance of Structures* (K014550); 24 months; $63,800

Kozin, F.; *Stability of Stochastic Systems* (K011549); 24 months; $79,200

RENSSELAER POLYTECHNIC INSTITUTE

Oakberg, Robert G.; *Research Initiation—Analysis of Building Frames with Shear Wall Assemblies* (K005441); 18 months; $15,000

Rubin, David; *Research Initiation—'The Thermodynamics of Inelastic Deformation'* (K005446); 18 months; $15,000

Sandor, George N.; *Kinematic Synthesis and Analysis of Mechanisms* (K012489); 24 months; $51,000

STATE UNIVERSITY AT BUFFALO

Bell, Adam C.; *Research Initiation—'Liquid Jet Interactions'* (K005908); 18 months; $15,000

Cheng, Ralph T.; *Variational Approach to Fluid Dynamics* (K011687); 24 months; $38,000

SYRACUSE UNIVERSITY

Evan-Iwanowski, R. M.; *Local Destabilizing Mechanisms in Shells* (K004565); 24 months; $61,400

UNION COLLEGE

Shanebrook, J. R.; *Three-Dimensional Turbulent Compressible Boundary Layers* (K012697); 24 months; $35,400

WEBB INSTITUTE OF NAVAL ARCHITECTURE

MacLean, Walter M.; *Research Initiation—Permanent Set Prediction for Long Plates Under Uniform Lateral Load* (K005836); 18 months; $15,000

Ward, Lawrence W.; *Ship Wave Resistance in Model and Full Scale* (GK2546 001); $12,300

NORTH CAROLINA

DUKE UNIVERSITY

Clough, Gerald W.; *Research Initiation—Failure Modes in Soils* (K005771); 18 months; $15,000

NORTH CAROLINA STATE UNIVERSITY AT RALEIGH

Rohrbach, Roger P.; *Research Initiation—The Use of Fluidics Phenomena in Agricultural Seed Metering* (K005813); 18 months; $15,000

OHIO

CASE WESTERN RESERVE UNIVERSITY

Janowitz, Gerald S.; *Research Initiation—The Coastal Boundary Layers in a Large Lake* (K005262); 18 months; $15,000

Prahl, Joseph M.; *Research Initiation—Thermal Discharges* (K005607); 18 months; $15,000

OHIO STATE UNIVERSITY

Fu, Whai Sang; *Asymmetric Mixed Boundary Value Problems in Thermoelasticity* (K010400); 24 months; $30,900

Ojalvo, Morris; *Shear Center and Warping Torsion Hypotheses as Applied to Curved Members* (K014633); 24 months; $32,900

Popelar, Carl H.; *Dynamic Stability of Freely Vibrating Thin-Walled Columns* (K004541); 24 months; $30,900

Yu, Lawrence K.; *Research Initiation—'The Oscillation of Fluid-Filled, Finitely-Strained Elastic Membranes* (K005691); 18 months; $15,000

OKLAHOMA

OKLAHOMA STATE UNIVERSITY

Cook, Echol E.; *Research Initiation—Removal of Organics by Trickling Filter* (K005413); 18 months; $15,000

PENNSYLVANIA

CARNEGIE MELLON UNIVERSITY

Krokosky, Edward M.; *Temperature Dependence of the Static Fatigue of Concrete* (K017859); 12 months; $19,900

Miller, Clarence A.; *Research Initiation—Interfacial Phenomena in Low-Tension Systems* (K005383); 18 months; $15,000

DREXEL UNIVERSITY

Koerner, Robert M.; *Research Initiation—Negative Skin Friction on Deep Foundations* (K005821); 18 months; $15,000

Wang, Albert S. D.; *Research Initiation—Large Dynamic Deformations of Elastic Incompressible Laminated Thick-Walled Shells* (K005788); 18 months; $15,000

LEHIGH UNIVERSITY

Beedle, Lynn S. and Lu, Le-Wu; *Research on the Design Problems of Tall Steel Buildings* (K019723); 24 months; $73,500

Chen, Wai-Fah; *Limit Analysis in Selected Concrete Problems* (K014274); 24 months; $37,500

Erdogan, Fazil; *Micromechanics and Fracture of Composite Materials* (K011977); 24 months; $62,500

Smith, Gerald F.; *Invariant-Theoretic Problems of Continuum Mechanics* (K011976); 24 months; $34,800

PENNSYLVANIA STATE UNIVERSITY

Morrow, Charles T.; *Research Initiation—Structural Mechanics of Selected Food Materials* (K005372); 18 months; $15,000

Phillips, Winfred M.; *Research Initiation—Hypersonic Transition Regime Flows* (K005559); 18 months; $15,000

UNIVERSITY OF PENNSYLVANIA

Soler, Alan I.; *Analysis of Thick Shells* (K014185); 24 months; $40,800

RHODE ISLAND

BROWN UNIVERSITY

Caswell, Bruce; *Asymptotic Analysis of Viscoelastic Flows* (K014758); 24 months; $45,000

Clifton, Rodney J.; *Dynamic Plastic Deformation Under Combined Stress States* (K010584); 24 months; $60,400

SOUTH CAROLINA

CLEMSON UNIVERSITY

Brandon, Craig A.; *Research Initiation—'The Effect of Suspended Particles on Turbulent Flow in a Pipe'* (K005839); 18 months; $15,000

UNIVERSITY OF SOUTH CAROLINA

Haines, Daniel W.; *Research Initiation—Wave Propagation in a Helical Coil* (K005408); 18 months; $14,200

SOUTH DAKOTA

SOUTH DAKOTA STATE UNIVERSITY

Lee, Peter Y.; *Research Initiation—A Laboratory Investigation of Clays with Double-Peak Compaction Curve* (K005714); 18 months; $15,000

TEXAS

TEXAS A & M UNIVERSITY

Hix, Charles Jr.; *A Study of Industrialized Housing Research* (K023698); 12 months; $34,000

UNIVERSITY OF TEXAS AT AUSTIN

Reese, Lymon C.; *Constitutive Relationships for Clay Soils* (K011927); 24 months; $35,700

UTAH

BRIGHAM YOUNG UNIVERSITY

Hanks, Richard W. and Cannon, John N.; *Transitional and Turbulent Flow of Non-Newtonian Fluids in Pipes and Parallel Plate Ducts* (K015893); $42,200

UNIVERSITY OF UTAH

Baer, Alva D.; *Study of the Regular Penetration of the Visous Sublayer by Utrbulent Fluctuations* (K010114); 24 months; $32,000

Chang, Po-Cheng; *Research Initiation—The Structure of Nonstationary Turbulent Flows* (K005502); 18 months; $15,000

Gascoigne, Harold E.; *Research Initiation—'Fringe Formation and Interpretation in Holographic Interferometry as Applied to the Deformation of Solids'* (K005423); 18 months; $15,000

Vyas, R. K.; *Membrane Analysis of Scalloped and Spherical Shells* (K016505); 24 months; $27,000

ENGINEERING RESEARCH PROJECTS

UTAH STATE UNIVERSITY
Jeppson, Roland W.; *Three Dimensional Free Surface Potential Flows* (K014707); 18 months; $23,800

VERMONT

UNIVERSITY OF VERMONT
Olson, James P.; *Research Initiation—Volume Change-Pore Pressure Relations for Saturated Cohesive Soils* (K005849); 18 months; $15,000

VIRGINIA

UNIVERSITY OF VIRGINIA
Dawson, Thomas H.; *Research Initiation—'The Role of Viscosity in Metal Deformations'* (K005773); 18 months; $15,000
Hahn, Eric J.; *Research Initiation—Dynamic Analysis of Flexible Damped Bearing Support Systems* (K005387); 18 months; $15,000
Harris, Wesley L. Sr.; *Research Initiation—The Near-Field Structure of Sonic Booms* (K005772); 18 months; $15,000

VIRGINIA POLYTECHNIC INSTITUTE
Brown, Eugene F.; *Research Initiation—The Transonic Flow Region in Truncated Conical Nozzles* (K005223); 18 months; $15,000
Werle, Michael J.; *Research Initiation—Second-Order Effects on Boundary Layer Stability* (K005224); 18 months; $15,000

WASHINGTON

UNIVERSITY OF WASHINGTON
Brien, Frederick B.; *Immersed Bodies Moving Through Liquid-Solid Suspensions* (K022895); 12 months; $8,900
Hawkins, Neil M.; *Shear and Moment Transfer Between Concrete Flat Plates and Columns* (K016373); 24 months; $52,200
Sherif, Mehmet A. and Bostrom, Robert C.; *Earthquake Engineering—Dynamic Soil Behavior and Microseismicity in the State of Washington* (K018177); 12 months; $59,900

WASHINGTON STATE UNIVERSITY
Roberson, John A.; *Effect of Turbulence on the Drag of Angular Blunt Bodies* (K010194); 24 months; $52,000

WEST VIRGINIA

WEST VIRGINIA UNIVERSITY
Hopkins, Gordon R.; *Research Initiation—'Stability of Fluid Conveying Tubes Subject to Periodic Nonconservative Forces'* (K005327); 18 months; $15,000

WISCONSIN

UNIVERSITY OF WISCONSIN, Madison
Christensen, Richard W.; *Mechanics of Granular Media* (K015085); 24 months; $40,100
Lodge, A. S.; *Polymer Solution Rheology* (K015611); 24 months; $61,800
Otis, David R.; *Thermal Damping in Gas-Filled Composite Materials During Impact Loading* (K014788); 24 months; $36,400
Sandor, Bela I.; *Research Initiation—Fatigue Damage Alleviation by Electrochemical Methods* (K005600); 18 months; $15,000
Vicker, John J. Jr. and Livermore, D. F.; *Analysis of Mechanisms by Matrix Methods* (K004552); 24 months; $48,300

UNIVERSITY OF WISCONSIN, Milwaukee
Balmer, Robert T.; *Research Initiation—Rotating Enclosures Containing the Immiscible Liquids* (K005231); 18 months; $14,700

Engineering Materials

CALIFORNIA

CALIFORNIA INSTITUTE OF TECHNOLOGY
Humphrey, F. B.; *Magnetic Flux Reversal in Thin Magnetic Films* (K015818); 24 months; $43,600

SAN DIEGO STATE COLLEGE
Ohnysty, Basil; *Effect of Large Temperature Gradients on the Movement of Pores and Inclusions in Metal Oxides* (K016756); 12 months; $14,200

STANFORD UNIVERSITY
Bube, Richard H.; *Photothermoelectric Effects in Semiconductors* (K017865); 24 months; $53,300
Parlee, N. A. D.; *Diffusion of Gases in Liquid Metals* (K017983); 12 months; $29,300
Pound, Guy M.; *Homogeneous Nucleation and Cluster Motions in Bulk Liquids* (K018895); 24 months; $55,300
Richards, Cedric W.; *Investigation of the Low-Temperature Dynamic-Mechanical Response of Hardened Cement Paste* (K011228); 24 months; $52,300
Spicer, William E.; *Optical and Transport Studies in Semiconductors* (GK2595X001); 12 months; $43,800

UNIVERSITY OF CALIFORNIA, Berkeley
Kobayashi, Shiro; *Variational Approach to the Solution of Plastic Deformation Processes* (KO14946); 24 months; $42,000
Sacks, Barry H.; *Research Initiation—Transferred Electron Effect in the Presence of a Strong Magnetic Field* (K005481); 18 months; $15,000
Wang, Shyh; *Optical Studies of Semiconductors and Dielectrics* (K013197); 24 months; $55,900

UNIVERSITY OF CALIFORNIA, Los Angeles
De Fontaine, D.; *Clustering and Ordering Instabilities in Multicomponent Systems* (K017146); 24 months; $54,700
Holm-Kennedy, James W.; *Research Initiation—Free Carrier Repopulation and Hot Carrier Transport in Semiconductors* (K005274); 18 months; $15,000
Sines, George; *Effect of Anisotropy on Defect Interactions* (K019905); 24 months; $50,000

UNIVERSITY OF SOUTHERN CALIFORNIA
Gershenzon, Murray; *Radiative Recombination in Semiconductors* (K012796); 24 months; $48,000
Wilcox, William R.; *Research on Urate Crystal Nucleation and Growth* (K017042); 24 months; $40,900

COLORADO

COLORADO SCHOOL OF MINES
Hager, John P.; *Thermodynamic Study of Vapor Complex Formation by Means of a New Transpiration-Mass Spectrometric Technique* (K020392); 24 months; $53,900
Moyzis, Joseph A.; *Research Initiation—Mossbauer Studies of Permalloy Alloys* (K005755); 18 months; $15,000

UNIVERSITY OF DENVER
Hepworth, Malcolm T.; *Determine and Correlate Thermodynamic Properties with Magnetic Properties in the Heusler Alloys* (KO12527); 12 months; $26,000
Newkirk, John B.; *Precipitation Control in Multicomponent Alloys* (K011304); 24 months; $98,300

CONNECTICUT

UNIVERSITY OF CONNECTICUT
Devereux, Owen F.; *Role of Environment in the Mechanical Behavior of Passive Films* (K020017); 24 months; $49,000
Strutt, Peter R.; *High Temperature Creep in Ordered Ternary B.C.C. Alloys* (K020412); 24 months; $53,900

YALE UNIVERSITY
Cargill, George S. III; *Research Initiation—Electronic Properties and Atomic Arrangements in Non-Crystalline Materials* (K005318); 18 months; $15,000

DELAWARE

UNIVERSITY OF DELAWARE
Schultz, Jerold M.; *Ultramolecular Structure and Behavior of Polymers* (K015240); 24 months; $60,000

FLORIDA

UNIVERSITY OF FLORIDA
Dehoff, R. T.; *Dynamics of Microstructural Change* (K016371); 12 months; $20,000
Guy, Albert G.; *Electrotransport in Alloys* (K015700); 24 months; $55,200

HAWAII

UNIVERSITY OF HAWAII
Larsen-Badse, Jorgen; *Abrasion of Metals* (K013685); 24 months; $59,000

ILLINOIS

NORTHWESTERN UNIVERSITY
Davidson, Theodore; *Research Initiation—Supramolecular Structure and Microplasticity of Polytetrafluoroethylene* (K005901); 18 months; $14,700
Owen, Walter S.; *Nucleation in Iron-Nickel and Related Alloys* (K023985); 24 months; $47,600
Wagner, J. Bruce Jr.; *Diffusion of Sulfur 35 in Single Crystals of NiO and CoO* (K023308); 24 months; $59,800

UNIVERSITY OF ILLINOIS, Urbana
Ehrlich, Gert; *Atomic Exploration of Crystal Surfaces* (K016593); 24 months; $60,000
Holonyak, Nick Jr.; *Luminescence and Laser Studies in III-V and II-VI Semiconductors* (K018960); 24 months; $60,000

MARYLAND

UNIVERSITY OF MARYLAND
Marcinkowski, M. J.; *Study of Grain Boundary Behavior During Plastic Deformation* (K013937); 24 months; $57,000

MASSACHUSETTS

MASSACHUSETTS INSTITUTE OF TECHNOLOGY
Suh, Nam P.; *Strain Rate Effects in Low Carbon B. C. C. Iron* (K019673); 12 months; $28,200
Wulff, John; *Corrosion-Resistant Titanium-Niobium Alloys* (GK1688X001); 12 months; $28,400
Yannas, Ioannis V.; *Engineering and Physical Chemistry of the Biological Solid State* (K019408); 24 months; $56,100

UNIVETSITY OF MASSACHUSETTS
Poli, Corrado and Boothroyd, Geoffrey; *Instability of the Cutting Process and Its Effect on Machine Tool Chatter* (K018894); 24 months; $50,100

MICHIGAN

MICHIGAN TECHNOLOGICAL UNIVERSITY
Shannette, Gary W.; *Research Initiation—Single Crystalline Elastic Constants of Nb-Mo and Nb-Ta Alloys* (K005888); 18 months; $15,000

UNIVERSITY OF MICHIGAN
Vincent, D. H.; *Fast Neutron Irradiation Effects in Cu-Fe Alloys* (K017073); 24 months; $51,800

WAYNE STATE UNIVERSITY
Yoon, Duk N.; *Research Initiation—Order-Disorder Critical Temperature in Ternary and Binary Alloys* (K005833); 18 months; $15,000

MINNESOTA

UNIVERSITY OF MINNESOTA
Iwasaki, I.; *Behavior of Platinum Electrodes as Redox Potential Indicators in Metallurgical Systems* (K019092); 18 months; $16,200
Toth, Louis E.; *Low Temperature Heat Capacity Measurements on Transition Metal Borides* (K011853); 24 months; $72,100

ENGINEERING SYSTEMS

MISSOURI

UNIVERSITY OF MISSOURI, Rolla
Wuttig, Manfred; *Interstitial Diffusion in Ferromagnetic Metals and Alloys* (K018460); 24 months; $59,400

WASHINGTON UNIVERSITY
Muller, Marcel W.; *Studies in Nonlinear Micromagnetics* (K015701); 24 months; $59,500

NEW JERSEY

STEVENS INSTITUTE OF TECHNOLOGY
Ohring, Milton; *Self Diffusion in Thin Films* (K016620); 24 months; $60,400

NEW YORK

POLYTECHNIC INSTITUTE OF BROOKLYN
Castleman, Louis S.; *Impurity-Dispersoid Interactions in Dispersed Phase Systems* (K014708); 24 months; $27,700

RENSSELAER POLYTECHNIC INSTITUTE
Hudson, John B.; *Vapor-Crystal Nucleation* (K013336); 24 months; $59,200

STATE UNIVERSITY AT BUFFALO
Ramalingam, Subbiah; *Research Initiation—Orthogonal Cutting of Dispersion Hardened Alloys* (K005917); 18 months; $15,000

STATE UNIVERSITY AT STONY BROOK
Carroll, T. Owen; *Research Initiation—Single—Domain Transient Effects in Liquid Crystals* (K005824); 18 months; $13,900

SYRACUSE UNIVERSITY
Vook, Richard W.; *X-Ray and Electron Microscope Studies of Thin Surface Layers* (K014627); 24 months; $57,300

UNIVERSITY OF ROCHESTER
Su, Goug-Jen; *Photochromic Mechanism in Glasses* (K019599); 24 months; $55,000

NORTH CAROLINA

DUKE UNIVERSITY
Hacker, Herbert Jr.; *Electric and Magnetic Properties of Rare Earth Compounds* (K014378); 24 months; $39,500

OHIO

CASE WESTERN RESERVE UNIVERSITY
Blackwell, John; *Research Initiation—a Study of the Fine Structure of Cellulose* (K005662); 18 months; $14,600
Geil, Phillip H.; *Polymer Morphology, Oriented and Amorphous Polymers* (K018636); 24 months; $92,200
Koenig, J. L.; *Structure of the Folded Polymer Chain* (K018603); 24 months; $52,700
Litt, Morton; *Crystalline Semi-Conducting and Conducting Polymers* (K013612); 24 months; $57,000

UNIVERSITY OF TOLEDO
Zrudsky, Donald R.; *Research Initiation—Low Temperature Thermal Conductivity and Thermoelectric Power Measurements of Cu-Ni Alloys* (K005592); 18 months; $15,000

YOUNGSTOWN STATE UNIVERSITY
Pejack, Edwin R.; *Research Initiation—the Behavior of Elastic Strip Material Under Tension and Moving Over a Set of Rolls* (K005319); 18 months; $14,700

PENNSYLVANIA

DREXEL UNIVERSITY
Herczfeld, Peter R.; *Transport Processes, Fluctuation Phenomena and Current Instabilities in Photoconductors* (K017861); 24 months; $53,000
Reynik, Robert J.; *Correlation of Liquid Metal Transport Properties Over Extensive Ranges of Temperature and Pressure* (K024366); 24 months; $60,000

LEHIGH UNIVERSITY
Chou, Y. T.; *Dislocation Theory and Application* (K016274); 24 months; $59,400
Notis, Michael R.; *Research Initiation—RF Sputtering of Nickel Alloy Films* (K005705); 18 months; $12,330

PENNSYLVANIA STATE UNIVERSITY
Fonash, Stephen J.; *Research Initiation—Elastic Strain Effects in Metal-Semiconductor Tunnel Contacts* (K005666); 18 months; $15,000
Roy, Della M.; *Theoretically Dense Cement and Pressure Strengthening* (K020408); 24 months; $55,400
Roy, Rustum, Barsch, Gerhard R. and Vedam, Kuppuswamy; *Engineering Materials at High Pressure* (GK1686X002); 12 months; $101,300
Simkovich, George; *Point Defects in Metal Sulfides* (K015360); 24 months; $44,000

UNIVERSITY OF PITTSBURGH
Magill, J. H.; *Crystallization and Morphology of a Siloxane Polymer* (K016999); 24 months; $30,000
Plazek, D. J.; *Linear Viscoelastic Response of Well-Defined Amorphous Polymeric Systems* (K016530); 24 months; $30,000

TENNESSEE

VANDERBILT UNIVERSITY
Lichter, Barry D.; *Electrical and Thermodynamic Properties of Liquid Semiconductors* (K018697); 24 months; $52,000

TEXAS

SOUTHERN METHODIST UNIVERSITY
Ashley, Kenneth L.; *Electron Recombination in Semiconductors Through Negatively Charged Recombination Centers* (K024145); 24 months; $51,700

UNIVERSITY OF TEXAS AT AUSTIN
Steinfink, Hugo; *Crystal Chemistry and Physical Properties of Rare Earth Systems* (K011376); 24 months; $51,200

UTAH

UNIVERSITY OF UTAH
Andrade, Joseph D.; *Research Initiation—Polymer Surface Engineering for Biological Compatibility* (K005464); 18 months; $15,000
Boyd, Richard H.; *Relaxation Processes in Polymers* (K019601); 24 months; $59,900
Chun, John S.; *Research Initiation—Mechanical Behavior of Tungsten at Low Temperatures* (K005420); 18 months; $15,000

VIRGINIA

UNIVERSITY OF VIRGINIA
Wilsdorf, Heinz G. F.; *Application of High Voltage Electron Microscopy to Metallurgical Problems* (K015694); 24 months; $45,000

VIRGINIA POLYTECHNIC INSTITUTE
Tenney, Darrel R.; *Research Initiation—Bimetallic Diffusion Zones in Cu-Pd System* (K005731); 18 months; $14,900

WASHINGTON

UNIVERSITY OF WASHINGTON
Polonis, Douglas H.; *Precipitation and Strengthening Processes in Alloys* (K012514); 24 months; $57,800
Schibli, Eugen G.; *Research Initiation—Determination of Impurity Profiles in Semiconductor Materials Containing Deep Levels* (K005293); 18 months; $15,000
Taggart, Raymond; *Grain Boundary Effects on the Nucleation of Fatigue Cracks* (K016858); 24 months; $40,000

WISCONSIN

MARQUETTE UNIVERSITY
Seitz, Martin A.; *Polarization Mechanisms and Mass Transport in Nonstoichiometric Zinc Oxide* (K016284); 24 months; $27,600

UNIVERSITY OF WISCONSIN, Madison
Clum, James A.; *Research Initiation—Study of Bainitic Microstructures in Some Commercial Steels* (K005629); 18 months; $15,000

UNIVERSITY OF WISCONSIN, Milwaukee
Chang, Y. Austin; *Thermodynamic Studies of Binary Intermetallic Phases with the F.C. Tetragonal Au-Cu I-Type (L10) Structure* (K017313); 24 months; $37,500
Pavelic, Vjekoslav; *Research Initiation—Thermal Process in Arc Welding with Metal Transfer* (K005484); 18 months; $15,000

Engineering Systems

ARIZONA

UNIVERSITY OF ARIZONA
Korn, Granino A.; *On-Line All Digital Simulation* (K015224); 24 months; $104,400

CALIFORNIA

SACRAMENTO STATE COLLEGE
Choma, John Jr.; *Research Initiation—Linear and Parametric Instabilities in Class C Transistor Amplifiers* (K005737); 18 months; $15,000

STANFORD UNIVERSITY
Eaves, B. Curtis; *Research Initiation—Complementary Pivot Theory of Fixed Points and Quadratic Programming* (K005695); 18 months; $15,000
Goodman, Joseph W.; *Optical Image Formation and Image Processing* (K018410); 12 months; $25,600
Gray, Robert M.; *Research Initiation—Information Rates of Autoregressive Sources* (K005452); 18 months; $15,000
Linvill, William K. and Luenberger, David G.; *Coherent System Research* (K016125); 12 months; $146,700
Veinott, Arthur F. Jr.; *Dynamic Inventory Models and Related Optimization Methods* (K018339); 24 months; $57,300

UNIVERSITY OF CALIFORNIA, Berkeley
Pederson, D. O. and Rohrer, R. A.; *Analog Integrated Circuit Design Automation* (K017931); 24 months; $50,000
Skoog, Ronald A.; *Research Initiation—Realization Theory and Time-Varying Network Synthesis* (K005786); 18 months; $15,000

UNIVERSITY OF CALIFORNIA, Davis
Mitra, Sanjit K.; *Active Filter Design for High Frequency Applications* (K014736); 24 months; $38,900

UNIVERSITY OF CALIFORNIA, Irvine
Barnes, Casper W.; *Spatial Signals and Imaging Systems* (K016005); 24 months; $69,100

UNIVERSITY OF CALIFORNIA, Los Angeles
Alexopoulos, Nicolaos G.; *Research Initiation—Electromagnetic Scattering from Radially Inhomogeneous Media* (K005433); 18 months; $15,000
Karplus, W. J., Vidal, J. J. and McNamee, L. P; *Hybrid-Computer Optimization of Distributed Parameter Systems* (K013156); 24 months; $86,200

UNIVERSITY OF CALIFORNIA, Santa Barbara
Kotzebue, Kenneth L.; *Characterization of Solid State Microwave Sources* (K015612); 24 months; $51,500

UNIVERSITY OF SOUTHERN CALIFORNIA
Davisson, Lee D.; *Nonparametric and Adaptive Techniques in Communications* (K014190); 24 months; $49,500
Lakin, Kenneth M.; *Research Initiation—Surface Wave Amplification* (K005751); 18 months; $15,000
Neustadt, Lucien W.; *Optimal Control Theory for Systems Utilizing Operator Equations* (K015787); 24 months; $52,100

ENGINEERING RESEARCH PROJECTS

COLORADO

COLORADO STATE UNIVERSITY

Childs, Dara W.; *Research Initiation—An Approximation Technique in Optimal Control* (K005560); 18 months; $15,000

UNIVERSITY OF DENVER

Parkins, George J. Jr.; *Research Initiation—Sensitivity Studies in Dynamic Systems* (K005460); 18 months; $15,000

CONNECTICUT

YALE UNIVERSITY

Denardo, Eric V. and Sobel, Matthew J.; *Structure and Computation of Optimal Policies in Sequential Decision Processes* (K013757); 24 months; $71,800

FLORIDA

UNIVERSITY OF FLORIDA

Chenette, Eugene R.; *Generation—Recombination Noise in Bipolar and Field Effect Transistors* (K014266); 24 months; $51,300

Roberts, Stephen D.; *Research Initiation—a Theoretical Basis for Facilities Location* (K005780); 18 months; $15,000

UNIVERSITY OF SOUTH FLORIDA

Garcia, Oscar N.; *Binary Codes for Improved Computer Reliability and Speed* (K015278); 14 months; $15,000

IDAHO

UNIVERSITY OF IDAHO

Maki, Gary K.; *Research Initiation—Fault Detection in Asychronous Sequential Circuits* (K005907); 18 months; $13,200

ILLINOIS

ILLINOIS INSTITUTE OF TECHNOLOGY

Costello, Daniel J.; *Research Initiation—Convolutional Coding in Reliable Communication* (K005265); 18 months; $15,000

NORTHWESTERN UNIVERSITY

Epstein, M.; *Transduction Techniques for Acoustic Surface-Wave Devices* (K010808); 24 months; $35,400

Pierskalla, William P.; *Ordering and Issuing Policies for Perishable Inventories* (K024792); 12 months; $16,600

UNIVERSITY OF ILLINOIS, Chicago Circle

Inada, Hitoshi; *Research Initiation—A Surface Wave Approach to High Frequency Scattering from Dielectric Bodies* (K005878); 18 months; $15,000

UNIVERSITY OF ILLINOIS, Urbana

Metze, Gernot; *Design of Diagnosable Digital systems* (K015459); 12 months; $40,000

Mittra, Raj; *Modified Wiener-Hopf Equation Applied to Boundary-Value Problems* (K015288); 24 months; $49,700

INDIANA

PURDUE UNIVERSITY

Fu, King-Sun, Kashyap, R. L. and Koivo, A. J.; *Control, Computation and Learning of Finite State Systems* (K018225); 24 months; $115,000

Gunshor, R. L. and Newhouse, V. L.; *Traveling-Wave Interaction Between Drifting Carriers in a Semiconductor and an External Superconducting Slow-Wave Circuit* (K011958); 24 months; $74,600

Vemuri, Venkateswararao; *Research Initiation—Hybrid Computer Methods for a Class of Functional Equations* (K005658); 18 months; $15,000

UNIVERSITY OF NOTRE DAME

Sain, Michael K. and Massey, James L.; *A Unified Study of Coding and Control Theories* (K013618) 24 months; $54,300

LOUISIANA

LOUISIANA STATE UNIVERSITY, Baton Rouge

Marshak, Alan H.; *Research Initiation—Synthesis of General Impurity Distributions Using a Two-Step Diffusion Process* (K005626); 18 months; $14,100

TULANE UNIVERSITY

Beck, Charles H.; *Investigation of Educational and Research Utilization of the Minuteman I Computer* (J000850*); 12 months; $15,000

MARYLAND

JOHNS HOPKINS UNIVERSITY

Davidson, Frederic M.; *Digital Camera* (K025063); 18 months; $15,000

Weiss, C. D.; *Topologically Constrained Switching Network Synthesis* (K012586); 24 months; $29,200

UNIVERSITY OF MARYLAND

Harger, Robert O.; *Optimum Methods of Optical Communication Through Atmospheric Turbulence* (K014920); 24 months; $66,200

Levine, William S.; *Research Initiation—Optimal Linear Output Feedback Controllers for Linear System* (K005375); 18 months; $14,800

Rao, T. R. N.; *Multi-Residue Codes for Self-Correcting Arithmetic Units* (K017308); 12 months; 12 months; $18,600

MASSACHUSETTS

MASSACHUSETTS INSTITUTE OF TECHNOLOGY

Drake, Alvin W.; *Operations Research for Public Systems* (K016471); 24 months; $121,100

Gould, L. A. and Evans, L. B.; *Control of Distributed Parameter Systems and Chemical Processes* (K014152); 12 months; $74,000

Hellman, Martin E.; *Research Initiation—Learning Under Finite Memory Constraints* (K005800); 18 months; $14,400

WORCESTER POLYTECHNIC INSTITUTE

Estes, Lee E.; *Research Initiation—Effect of Non-Linear Interactions on Photon Statistics* (K005517); 18 months; $14,400

MICHIGAN

UNIVERSITY OF MICHIGAN

McClamroch, Nathaniel H.; *Research Initiation—Control of Linear Hereditary Processes* (K005798); 18 months; $15,000

Phillips, Richard L.; *Non-Stationary Electric Arcs* (K014757); 24 months; $52,000

MISSOURI

UNIVERSITY OF MISSOURI, Rolla

Bourquin, Jack J.; *Research Initiation—Synthesis and Stability of a Class of Active Lumped-Distributed Systems* (K005335); 18 months; $15,000

Cunningham, David R.; *Research Initiation—Empirical Bayesian Estimation* (K005211); 18 months; $15,000

Fowler, Eddie R.; *Research Initiation—Transfer Function Synthesis* (K005466); 18 months; $15,000

Montgomery, David N.; *Research Initiation—Inductive Charging of Melting Hail* (K005236); 18 months; $15,000

WASHINGTON UNIVERSITY

Tarn, Tzyh Jong; *Research Initiation—Stochastic Optimal Control Using Mathematical Programming* (K005570); 18 months; $15,000

NEBRASKA

UNIVERSITY OF NEBRASKA, Lincoln

White, Alfred H.; *Research Initiation— Optimal Periodic Control of Distributed Systems* (K005267); 18 months; $15,000

NEW JERSEY

PRINCETON UNIVERSITY

Bruno, John L.; *Research Initiation—Criteria for Verification of the Correctness of Computer Programs* (K005535); 18 months; $15,000

Pavlidis, Theodosios; *Patterns Through the Use of Linguistic Techniques* (K013622); 24 months; $41,000

NEW MEXICO

NEW MEXICO STATE UNIVERSITY

Thomson, Wiley E.; *Research Initiation—Stability Analysis of Interconnected Systems* (K005652); 18 months; $15,000

NEW YORK

CLARKSON COLLEGE OF TECHNOLOGY

Farmer, Daniel E.; *Research Initiation—Fault Detection and Diagnosis for Digital Systems* (K005553); 18 months; $15,000

COLUMBIA UNIVERSITY

Newborn, Monroe M.; *Uniform Modular Automata* (K014342); 24 months; $45,300

Yang, Edward S.; *Semiconductor Field-Effect Devices* (K018772); 24 months; $46,600

CORNELL UNIVERSITY

Berger, Toby; *Rate Distortion Considering Sources with Memory* (K014449); 24 months; $42,600

Brown, Mark; *Research Initiation—Convergence of Stochastic Integrals* (K005758); 18 months; $15,000

Dunn, Joseph C.; *Research Initiation—Limit Cycling in Optimal Control Problems* (K005760); 18 months; $15,000

Szentirmai, G.; *Analysis and Design of Practical RC Active Networks* (K014414); 24 months; $47,200

CUNY, CITY COLLEGE

Oh, Se Jeung; *Computer-Aided Analysis of Dynamic Non-linear Network Response and Structural Stability in Design Automation* (K013620); 12 months; $21,000

NEW YORK UNIVERSITY

Ghausi, Mohammed S.; *Active Networks Compatible with Integrated Circuits* (K018465); 24 months; $64,800

Smith, Alvy R. III; *Research Initiation—LSI-Influenced Switching and Automata Theory* (K005406); 18 months; $15,000

POLYTECHNIC INSTITUTE OF BROOKLYN

Cassedy, E. S.; *Optical Parametric Phenomena* (K016018); 24 months; $36,700

Levis, Alexander H.; *Research Initiation—Optimal Control of Linear Systems with State-Dependent Sampling* (K005595); 18 months; $15,000

Maurer, Stewart J.; *Research Initiation—Radiation from a Dielectric Wedge by Ray-Optical Techniques* (K005787); 18 months; $15,000

STATE UNIVERSITY AT BUFFALO

Hiller, Lejaren; *Aanalysis and Synthesis of Audio Frequency Signals by Analog and Digital Techniques* (K014191); 24 months; $60,500

STATE UNIVERSITY AT STONY BROOK

Chang, Sheldon S. L.; *Multistage Decision Processes Using Fuzzy Dynamic Programming* (K016017); 24 months; $74,900

SYRACUSE UNIVERSITY

Cheng, David K.; *Optimization and Synthesis of Arrays Subject to Mutual Coupling* (K011245); 24 months; $42,500

Huang, Sheng-Cho; *Research Initiation—Solving Optimal Control Problems Using Epsilon-Technique* (K005613); 18 months; $15,000

NORTH CAROLINA

NORTH CAROLINA STATE UNIVERSITY AT RALEIGH

Gault, James W.; *Research Initiation—Fault Preprocessing and Self-Diagnosis of Digital Networks* (K05612); 18 months; $15,000

Huang, Sheng-Chao; *Research Initiation—Optimization of Multi-Server Queuing Systems* (K005611); 18 months; $15,000

SPECIAL ENGINEERING PROJECTS

OHIO

CASE WESTERN RESERVE UNIVERSITY
Windeknecht, Thomas G.; *Mathematical Theory of General Systems* (K013300); 24 months; $49,200

OHIO STATE UNIVERSITY
White, Lee J.; *Research Initiation—Optimum Data Allocation* (K005256); 18 months; $15,000

UNIVERSITY OF DAYTON
Crouch, Jack G.; *Research Initiation—Optimization of Approximating Functions for the Galerkin Method* (K005402); 18 months; $15,000
Fitz, Raymond; *Research Initiation—Digital Simulation of Nonlinear Estimation Problems* (K005397); 18 months; $15,000

UNIVERSITY OF TOLEDO
Solberg; James J.; *Research Initiation—Graphical Methods in the Analysis of Networks of Queues* (K005789); 18 months; $15,000

OKLAHOMA

OKLAHOMA STATE UNIVERSITY
Mulholland, Robert J.; *Research Initiation—Exponential Representation Theory for Linear Physical Systems* (K005283); 18 months; $15,000

UNIVERSITY OF OKLAHOMA
Mohler, Ronald R.; *Bilinear Control Processes* (K017866); 24 months; $59,500
Vargo, Paul M.; *Research Initiation—an Improved Graphically-Oriented Man-Computer Communication Device* (K005312); 18 months; $14,600

PENNSYLVANIA

CARNEGIE MELLON UNIVERSITY
Young, Tzay Y.; *Stochastic Approximation Techniques in Signal Detection and Pattern Recognition Studies* (K013510); 24 months; $49,600

LEHIGH UNIVERSITY
Dahlke, Walter E.; *Metal-Insulator-Semiconductor Structures* (K015864); 24 months; $57,200
———; (GK15864001); $25,200
Tzeng, Kenneth K.; *Research Initiation—Efficient Coding and Decoding Techniques* (K005647); 18 months; $11,546

PENNSYLVANIA STATE UNIVERSITY
Hulina, Paul T.; *Research Initiation—Synthesis of Asynchronous Sequential Circuits Using Transition Sensitive Flip-Flops* (K005713); 18 months; $15,000

PMC COLLEGES
Jefferis, Raymond P. III; *Research Initiation—Controllability and Observability of Model Distributed Systems* (K005507); 18 months; $15,000

UNIVERSITY OF PENNSYLVANIA
Vartanian, Michael M.; *Research Initiation—Fault Isolating Non-linear Electronic Networks* (K005551); 18 months; $15,000
Mihram, George A.; *Research Initiation—the Design and Analysis of Computer Simulations* (K005289); 18 months; $14,700

RHODE ISLAND

BROWN UNIVERSITY
Demeo, Edgar A.; *Research Initiation—Non-reciprocal Solid State Plasma Devices for Millimeter Wavelengths* (K005632); 18 months; $15,000
Savage, John E.; *Decoders for Error Correction* (K013162); 24 months; $34,500

UNIVERSITY OF RHODE ISLAND
Lindgren, Allen G.; *Stability of Steady Propagating Waveforms on Nonlinear Transmission Lines* (K012763); 12 months; $29,500

TEXAS

SOUTHERN METHODIST UNIVERSITY
Nardizzi, Louis R.; *Research Initiation—Engineering Systems with Optimal Computer Control* (K005608); 18 months; $15,000

TRINITY UNIVERSITY
Wakeland, William R.; *Research Initiation—Correlation of Performance Index Weighting Factors with System Response* (K005913); 18 months; $15,000

UNIVERSITY OF TEXAS AT AUSTIN
Coates, Clarence L.; *Threshold Gate Circuit and Logic Problems* (GK1146X002); 12 months; $66,200

UTAH

BRIGHAM YOUNG UNIVERSITY
Woodbury, Richard C.; *Trainable Recognition System for Optical Patterns* (K015221); 24 months; $25,900

WASHINGTON

UNIVERSITY OF WASHINGTON
Auth, David C.; *Research Initiation—Optical Generation of Intense SHF Acoustic Waves* (K005501); 18 months; $15,000
Lauritzen, Peter O.; *Avalanche Diode Noise and Failure Mode Studies* (K014545); 24 months; $29,300
Martin, Richard D.; *Research Initiation—Robust Detection and Estimation in Nearly Gaussian Noise Environments* (K005338); 18 months; $15,000

WISCONSIN

UNIVERSITY OF WISCONSIN, Madison
Seshadri, S. R. *Topics in Electromagnetic Fields* (K016203); 24 months; $38,100

UNIVERSITY OF WISCONSIN, Milwaukee
Vairavan, Kasivisvanathan; *Research Initiation—Finite Memory Sequential Machines with Application to Coding Theory* (K005651); 18 months; $14,200

Special Engineering Projects

ARIZONA

UNIVERSITY OF ARIZONA
Gensler, William G.; *Research Initiation—The Bioelectric Field, Ion Uptake and Ion Transport in Plants* (K005838); 18 months $15,000

CALIFORNIA

HEAT TRANSFER AND FLUID MECHANICS INSTITUTE
Sarpkaya, Turgut; *Heat Transfer and Fluid Mechanics Institute Conference* (K015111); 12 months; $1,500

STANFORD UNIVERSITY
Newcomb, Robert W.; *Special Session on Filters of the Third Asilomar Conference on Circuits and Systems* (K016240); 12 months; $900

UNIVERSITY OF CALIFORNIA, Berkeley
Oppenheim, A. K.; *Preparation of the Proceedings of the Second International Colloquium on Gas Dynamics of Explosions and Reactive Systems* (GN877); 6 months; $6,250

UNIVERSITY OF CALIFORNIA, Santa Barbara
Hickman, Roy S.; *Anomalous Rotational Excitation in Flames* (K015200); 24 months; $29,100

COLORADO

UNIVERSITY OF COLORADO
Kreith, Frank; *Increasing Water Supplies by Reducing Evapotranspiration from Living Plants* (K017184); 24 months; $48,500

UNIVERSITY OF COLORADO, Denver Center
Dare, Charles E.; *Research Initiation—Economic Evaluation of the Traffic-Actuated Speed Signal Funnel* (K005754); 18 months; $14,900

DISTRICT OF COLUMBIA

AMERICAN SOCIETY FOR CYBERNETICS
Fogel, Lawrence J.; *Conference on Cybernetics and Conflict Resolution* (K014267001); $700

NATIONAL ACADEMY OF ENGINEERING
Binns, Kerstin B.; *Partial Support of a Symposium on 'The Quality of Life—Barriers and Constraints to the Use of Existing Technology in Balancing World Population and World Nutritional Resources'* (C310184); 6 months; $3,000

NATIONAL ACADEMY OF ENGINEERING, COMMISSION ON ENGINEERING EDUCATION
Hall, Newman A.; *Task Order for Workshop on Social Directions for Technology* (C310182*); 12 months; $25,000

NATIONAL ACADEMY OF SCIENCES, NATIONAL RESEARCH COUNCIL
Cliffe, Robert A.; *Task Order for Partial Support of the Committee on Fire Research* (C310086007); 12 months; $13,000
———; *Partial Support of the U.S. National Committee for the International Institute of Refrigeration* (C310138003); 12 months; $5,000

FLORIDA

UNIVERSITY OF FLORIDA
Smith, Jack R.; *Automatic Analysis of the Electroencephalogram* (K015373); $79,800

GEORGIA

GEORGIA INSTITUTE OF TECHNOLOGY
Unger, Vernon E.; *Research Initiation—Mathematical Programming of Capital Budgeting Problems* (K005533); 18 months; $15,000

HAWAII

UNIVERSITY OF HAWAII
Watanabe, Michael S.; *International Conference on Pattern Recognition* (K019973); 18 months; $7,000

ILLINOIS

BRADLEY UNIVERSITY
Emanuel, Joseph T.; *Research Initiation—Visual and Auditory Sensory Inputs in Dichotic Decision Making* (K005864); 18 months; $12,300

NORTHWESTERN UNIVERSITY
Sorum, Marilyn J.; *Research Initiation—Estimating Probabilities of Misclassification* (K005339); 18 months; $14,300

UNIVERSITY OF ILLINOIS, Chicago Circle
Agarwal, Gyan C.; *Human Motor Coordination System* (K017581); 24 months; $41,300

UNIVERSITY OF ILLINOIS, Urbana
Carlson, Donald E.; *Controllable States of Elastic Heat Conductors* (K002502); 24 months; $26,000
Dunn, Floyd; *Interaction of Ultrasound and Biological Media* (K0019100); 24 months; $98,400
Kesler, Clyde E.; *Frontiers in Research and Practice in Plain Concrete* (K018461); 12 months; $8,500

INDIANA

PURDUE UNIVERSITY
Nachtigal, Chester L.; *Research Initiation—Active Control of Machine Tool Chatter* (K005699); 18 months; $15,000

IOWA

IOWA STATE UNIVERSITY
Brewer, Kenneth A.; *Research Initiation—Rate of Merging of Freeway Traffic Streams in Advance of a Through-Lane Closure* (K005451); 18 months; $15,000

ENVIRONMENTAL SCIENCES RESEARCH PROJECTS

MASSACHUSETTS

HARVARD UNIVERSITY

Drinker, Philip A.; *Collaborative Research in Gas Transfer and Cellular Injury in Extracorporeal Blood Oxygenation* (K013611); 24 months; $74,400

Emmons, Howard W.; *Sixth U.S. National Congress of Applied Mechanics* (K019358); 12 months; $3,000

MASSACHUSETTS INSTITUTE OF TECHNOLOGY

Crandall, Stephen H.; *Sound Reinforcement by Structural Interaction* (K010160); 24 months; $38,500

Moss, Richard A.; *Collaborative Research in Gas Transfer and Cellular Injury in Extracorporeal Blood Oxygenation* (K013610); 24 months; $64,800

Uhlmann, D. R.; *International Symposium —Behavior of Polymeric Solids Under Pressure* (K021450); 12 months; $5,950

MICHIGAN

UNIVERSITY OF DETROIT

Freeman, James J.; *Research Initiation—Ultrasonic Investigation of Stroke Volume* (K005496); 18 months; $15,000

MISSOURI

UNIVERSITY OF MISSOURI, Rolla

Gatley, William S.; *The Performance of Acoustic Filters* (K010844); 24 months; $26,000

NEBRASKA

UNIVERSITY OF NEBRASKA, Lincoln

Deshazer, James A.; *Research Initiation—Small Animal Calorimetery* (K005275); 18 months; $15,000

NEW JERSEY

NEWARK COLLEGE OF ENGINEERING

Marsh, Anthony H.; *Research Initiation—Determination of Human Visual Response to Threshold Line Signals* (K005697); 18 months; $13,800

NEW MEXICO

NEW MEXICO STATE UNIVERSITY

Freeburg, Robert S.; *Research Initiation—Modeling the Plant Micro-Environment* (K005861); 18 months; $14,800

NEW YORK

AMERICAN INSTITUTE OF CHEMICAL ENGINEERING

Van Antwerpen, F. J.; *Symposium on the Engineering Aspects of Electrical Discharges as Chemical Reactors* (K023890); 12 months; $2,100

AMERICAN POWDER METALLURGY INSTITUTE

Roll, Kempton H.; *International Powder Metallurgy Conference* (K020411); 12 months; $2,000

MANHATTAN COLLEGE

Stathis, Theodore C.; *Research Initiation—Nonlinear Theory Applied to Vascular Systems* (K005409); 18 months; $15,000

NEW YORK UNIVERSITY

Ehrenfeld, Sylvain; *Scheduling and Stopping Rules in Sequential Decision Problems* (K014073); 24 months; $63,800

SYRACUSE UNIVERSITY

Goel, Amrit L.; *Research Initiation—Optimal Cumulative Sum Procedures Based on Prior Distributions and Costs* (K005859); 18 months; $15,000

OHIO

AMERICAN SOCIETY FOR METALS

Paxton, H. W.; *Second International Conference on the Strength of Metals and Alloys* (K018047); 12 months; $1,600

PENNSYLVANIA

CARNEGIE MELLON UNIVERSITY

Longini, Richard L. and Zdrojkowski, Ronald; *Transcutaneous Measurement of Hemoglobin Oxygenation* (K013744); 24 months; $50,000

Shaw, M. C.; *Conference on Ultra Hard Tool Materials* (K024593); 12 months; $1,300

COMBUSTION INSTITUTE

Longwell, J. P.; *Thirteenth International Symposium on Combustion* (K016286); 12 months; $5,000

DREXEL UNIVERSITY

Moore, Thomas W.; *Research Initiation—The Reduction of Required Defibrillating Shock Energy by Accurately Timed, Staggered Shocks* (K005815); 18 months; $15,000

LEHIGH UNIVERSITY

Beedle, Lynn S.; *Conference on Stability Provisions of Design Specifications* (K021357); 12 months; $5,400

UNIVERSITY OF PENNSYLVANIA

Brockris, John O'M.; *Electrochemical Electricity Storers* (K016550); 24 months; $93,800

SOUTH CAROLINA

CLEMSON UNIVERSITY

Amoss, Donald C.; *Research Initiation—Detection and Estimation of Fetal Electrocardiograms* (K005863); 18 months; $15,000

TEXAS

SOUTHWEST RESEARCH INSTITUTE

Black, David L.; *Inter-American Conference on Materials Technology* (K016904); 12 months; $4,000

UNIVERSITY OF TEXAS AT AUSTIN

Allan, John J. III; *Research Initiation—Adaptive Mixture Control for Internal Combustion Engines* (K005538); 18 months; $14,900

VIRGINIA

AMERICAN SOCIETY OF PHOTOGRAMMETRY

Ghosh, Sanjib K.; *International Symposium on Photography and Navigation* (K016646); 12 months; $4,000

VIRGINIA POLYTECHNIC INSTITUTE

Jennelle, Ernest M.; *Research Initiation—An Evaluation of the BOD Test* (K005634); 18 months; $14,100

WEST VIRGINIA

WEST VIRGINIA UNIVERSITY

Diener, Robert G.; *Nondestructive Testing of Graft Unions in Trees* (K011757); 24 months; $31,000

Wegmann, Frederick J.; *Research Initiation —Use of Graph Theory Indices in Describing Transportation Network Structure* (K005330); 18 months; $15,000

WISCONSIN

MARQUETTE UNIVERSITY

Schlager, Kenneth J. and Sinha, Kumares C.; *Development of a Land Use Design Model* (K017166); 24 months; $43,000

UNIVERSITY OF WISCONSIN, Madison

Frank, Andrew A.; *Research Initiation—Stability and Control of Lower Limb Prosthetic and Orthodic Systems* (K005897); 18 months; $15,000

Environmental Sciences Research Projects

Atmospheric Sciences

ALASKA

UNIVERSITY OF ALASKA

Akasofu, Syun-Ichi; *Study of the Magnetospheric Substorm* (A017663); 12 months; $90,000

Davis, T. Neil; *Investigation of Auroral Zone Phenomena at Solar Maximum* (A011937001*); $4,600

Deehr, Charles S.; *Observations of the Predawn Enhancement of 6300 A (OI) Airglow* (A016394); 12 months; $13,100

Hook, Jerry L.; *Systematic Auroral Zone Measurements of Winds by Meteor Radar* (A016823); 18 months; $62,400

Mather, Keith B.; *Support for the Geophysical Institute* (A019475); 12 months; $80,600

Sheridan, Roger and Degen, Vladimir; *High Spectral Resolution Studies of Auroras Using an Image Intensifier Echelle Spectrograph* (A020564); 12 months; $14,300

Wescott, Eugene M. and Murcray, Wallace B.; *Field Line Tracing with Barium Ions* (A019265); 12 months; $84,300

Wilson, Charles R.; *Infrasonic Waves Within the Auroral Oval* (A016821); 12 months; $70,500

ARIZONA

UNIVERSITY OF ARIZONA

Gehrels, A. M. J.; *Project Polariscope* (P019108*); 12 months; $41,000

Reagan, John A. and Herman, Benjamin M.; *Lidar Probing of the Atmosphere* (A016764); 24 months; $132,200

CALIFORNIA

CALIFORNIA INSTITUTE OF TECHNOLOGY

Zirin, Harold; *Research in the Circulation of the Solar Atmosphere* (A024015); 12 months; $63,500

STANFORD RESEARCH INSTITUTE

Bates, Howard F.; *Radar Location of the Auroral Belt from the Northern Magnetic Pole* (A016269); 12 months; $57,700

STANFORD UNIVERSITY

Helliwell, R. A.; *VLF Measurements of Thermal Plasma and Geoelectric Fields in the Magnetosphere* (A018128); 24 months; $187,300

Villard, Oswald G. Jr.; *Geomagnetic PC 1 Micropulsation Propagation in the F2–Region Ionospheric Duct* (A017486); 12 months; $28,700

UNIVERSITY OF CALIFORNIA, Berkeley

Anderson, Kinsey A.; *Auroral Zone X-Rays —a Manifestation of the Magnetospheric Substorm* (A015225); 18 months; $27,800

Mozer, Forrest; *Measurements of Ionospheric Electric Fields on Rockets and Balloons* (A017328); 21 months; $114,100

Silver, Samuel; *Dynamic Observations of Solar Magnetic and Velocity Fields* (A016765); 24 months; $80,900

UNIVERSITY OF CALIFORNIA, Los Angeles

Bjerknes, Jacob; *Large Scale Atmosphere-Ocean Interaction* (A014354); 12 months; $19,200

Coleman, Paul J. Jr. and McPherron, Robert L.; *Study of Magnetic Field Measurements of Magnetospheric Fluctuations* (A017978); 12 months; $29,900

Holmboe, Jorgen; *Instability Mechanism of Large Scale Atmospheric Flow* (A017089); 12 months; $19,700

Kieffer, Hugh H.; *Spectra of Simulated Planetary Clouds and Frosts* (A021493); 24 months; $34,300

Pruppacher, Hans R.; *Wind Tunnel Investigation of Microphysical Processes in Clouds* (A018531); 24 months; $65,800

Sekera, Zdenek and Rao, C. R. Nagarhja; *Theoretical and Experimental Investigations of Atmospheric Radiative Transfer* (A016617); 24 months; $121,700

Siscoe, George; *Structure of Extended Solar Atmosphere* (A013554); 24 months; $61,100

Thorne, Richard M. and Venkateswaran, S. V.; *Ionospheric F–Region Effects of the*

ATMOSPHERIC SCIENCES

Total Solar Eclipse of March 7, 1970 (A020110); 12 months; $11,600

Venkateswaran, S. V.; *Movements and Morphology of Ionospheric Layers* (A018132); 24 months; $77,800

UNIVERSITY OF CALIFORNIA, San Diego

Booker, Henry G. and Rumsey, Victor H.; *Radiowave Diagnostic Studies of the Earth's Plasma Environment* (A020591); 24 months; $492,600

UNIVERSITY OF CALIFORNIA, San Diego
REVELLE COLLEGE

Suess, Hans E.; *Solar-Terrestrial Relationships Based Upon Secular Variations of the Atmospheric Carbon-14 Level* (A021161); 24 months; $30,300

COLORADO

COLORADO STATE UNIVERSITY

Reiter, Elmar R.; *Structures of Turbulence in the Free Atmosphere* (A023175); 12 months; $13,900

UNIVERSITY OF COLORADO

Barry, Roger G.; *Climatic Environment of the East Slope of the Colorado Front Range* (A015528); 21 months; $39,500

Benton, Edward R.; *Hydrodynamics of Rotating Fluids* (A016844); 24 months; $40,600

Rense, William A. and Rees, Manfred H.; *Theoretical Physics of the Upper Atmosphere* (A016290); 24 months; $90,800

UNIVERSITY OF DENVER

Roederer, Juan G.; *Particle and Field Asymmetries in the Outer Magnetosphere* (A017976); 18 months; $49,000

CONNECTICUT

CENTER FOR THE ENVIRONMENT AND MAN

Arnason, Geimundur; *Numerical Simulation of the Macrophysical and Microphysical Processes of Moist Convection* (A016408); 24 months; $134,900

YALE UNIVERSITY

Reifsnyder, William E.; *Quantity and Quality of Incident Light in Forest Sunflecks* (A015017); 12 months; $16,900

DISTRICT OF COLUMBIA

ARCTIC INSTITUTE OF NORTH AMERICA

Faylor, Robert C.; *Glacier-Climate Study of the Devon Island Ice Cap and its Relationship to the 'North Water'* (A019833); 12 months; $20,800

DEPARTMENT OF THE NAVY,
OFFICE OF NAVAL RESEARCH

Trumbull, Richard; *Churchill Skyhook Program* (AG18300001); 12 months; $59,100

NATIONAL ACADEMY OF SCIENCES,
NATIONAL RESEARCH COUNCIL

Odishaw, Hugh; *Task Order for Support of the Geophysics Research Board and its Affiliates* (C310049007); 12 months; $30,000

———; (C310049007*); 12 months; $60,400

Sievers, John R.; *Task Order for Support of the Committee on Atmospheric Sciences and the U.S. Committee for the Global Atmospheric Research Program* (C310009014); 12 months; $17,700

———; *Task Order for Partial Support of the Atmospheric Sciences Summer Study Review* (C310189); 7 months; $10,000

SMITHSONIAN INSTITUTION

Southworth, Richard B.; *Analysis of Photographic Observations of Comets* (GA753 001); $4,024

FLORIDA

FLORIDA ATLANTIC UNIVERSITY

Burnett, Clyde R.; *Measurement of the Abundance of Atomic Sodium in the Upper Atmosphere at Boca Raton, Florida* (A019361); 18 months; $14,300

FLORIDA STATE UNIVERSITY

Gille, John C.; *Temperature Wave Propagation in a Non-Gray Radiating Fluid and Related Studies* (A020213); 24 months; $77,800

UNIVERSITY OF MIAMI

Hirschberg, Joseph G.; *Interferometer Study of Coronal Emission Line Profiles at the Total Solar Eclipse of March 7, 1970* (A017977); 12 months; $12,600

Kraus, E. B.; *Air-Sea Interaction—the Planetary Boundary Layer* (A023169); 24 months; $176,300

Ostlund, H. Gote; *Tritium in Hurricanes* (A020622); 12 months; $37,000

HAWAII

UNIVERSITY OF HAWAII

Ramage, Colin S.; *Atmospheric Circulation Project for the International Indian Ocean Expedition* (GA386 005); $25,300

Steiger, Walter R.; *Dynamics and Maintenance of the Nighttime Ionosphere at Hawaii* (A016876); 12 months; $36,000

Yuen, Paul C. and Roelofs, Thomas H.; *Ionosphere Studies Using Total Electron Content* (A017330); 12 months; $70,400

IDAHO

UNIVERSITY OF IDAHO

Thomas Joe E.; *Collaborative Research in Infra-Sonic Waves in the Atmosphere-Pacific Northwest* (A023203); 24 months; $66,500

ILLINOIS

SOUTHERN ILLINOIS UNIVERSITY

Marshall, Lauriston C.; *Atmospheric Constituents of the Terrestrial Type Planets* (A016351); 12 months; $69,800

UNIVERSITY OF CHICAGO

Fultz, Dave; *Research in Meteorological Experimental Hydrodynamics* (A021160); 12 months; $97,400

Meyer, Peter; *Solar Modulation and Solar Production of High Energy Electrons* (A017210); 24 months; $155,700

Simpson, John A.; *Behavior of Cosmic Radiation in the Solar-Terrestrial Environment* (A020565); 5 months; $33,100

UNIVERSITY OF ILLINOIS, Urbana

Bowhill, Sidney A.; *Studies of Ionospheric Dynamics by Thompson Scatter* (A015526); 12 months; $43,000

Jones, Douglas M. A.; *Causes for Rainfall Enhancement Over Small-Scale Topographic Features* (A019494); 12 months; $19,400

Yeh, K. C. and Liu, C. H.; *Ionospheric Wave Propagation Problems* (A013723); 24 months; $73,000

INDIANA

BALL STATE UNIVERSITY

Hults, Malcom E.; *Shadow Bands During the March 7, 1970 Total Solar Eclipse* (A021168); 6 months; $1,600

PURDUE UNIVERSITY

Smith, Phillip J.; *Diagnostic Study of Synoptic Scale Atmospheric Processes* (A010933); 24 months; $56,000

IOWA

IOWA STATE UNIVERSITY

Yarger, Douglas N.; *Low Level Vertical Ozone Distribution Determined by Inversion of Backscattered Ultraviolet Radiation* (A013715); 24 months; $29,100

KANSAS

UNIVERSITY OF KANSAS

Beard, David B.; *Extraterrestrial Contributions to the Magnetic Environment* (A014029); 24 months; $63,200

MARYLAND

DEPARTMENT OF COMMERCE, ENVIRONMENTAL SCIENCE SERVICES ADMINISTRATION

Townsend, John W. Jr.; *Ionospheric Studies at Jicamarca Radio Observatory* (AG219); 12 months; $50,000

JOHNS HOPKINS UNIVERSITY

Phillips, Owen M.; *Geophysical Fluid Mechanics* (A016603); 24 months; $125,200

UNIVERSITY OF MARYLAND

Rosenberg, T. J.; *Distribution of Luminosity in Pulsating Auroral Forms* (A015807); 12 months; $14,500

Taylor, Leonard S.; *Electromagnetic Propagation in Atmospheric Turbulence* (A020776); 24 months; $30,800

MASSACHUSETTS

BOSTON COLLEGE

Carovillano, R. L. and Eather, R. H.; *Production Mechanism of Proton Auroras* (A018829); 20 months; $63,200

HARVARD UNIVERSITY

Dalgarno, A.; *Protoelectrons in the Atmosphere* (A021492); 12 months; $10,800

Goody, Richard M.; *Studies of Various Atmospheric Phenomena* (A013982); 12 months; $160,900

Menzel, Donald H.; *Spectrographic Study of the F and K Coronas at the Total Solar Eclipse of 7 March 1970* (A019477); 6 months; $13,700

MASSACHUSETTS INSTITUTE OF TECHNOLOGY

Malkus, Willem V. R.; *Geophysical Fluid Dynamics* (A022759); 12 months; $57,900

Starr, Victor P.; *Observational and Theoretical Studies of Planetary Atmospheres* (GA1310X001); 18 months; $182,700

OCEAN-ATMOSPHERE RESEARCH INSTITUTE

Schell, Irving I.; *Inertial Characteristics of the Circulation and Their Significance for Time-Averaged Long Range Forecasting* (A017802); 24 months; $45,900

UNIVERSITY OF MASSACHUSETTS

Goldenberg, H. Mark; *Measurement of the Solar Oblateness* (A016033); 12 months; $28,900

MISSOURI

UNIVERSITY OF MISSOURI, Columbia

Kung, Ernest C.; *Continued Diagnostic Study of the Atmospheric Energy Balance* (A015952); 24 months; $127,000

UNIVERSITY OF MISSOURI, Rolla

Levenson, Leonard L.; *Water Vapor-Surface Reactions, Dynamic and Steady State Studies with Molecular Beams and Quartz Crystal Microbalances* (A013948); 24 months; $65,800

NEW HAMPSHIRE

UNIVERSITY OF NEW HAMPSHIRE

Kaufmann, Richard L.; *Particle-Field Correlations Near the Magnetopause* (A014954); 24 months; $46,000

Lockwood, John A.; *Study of Intensity-Time Variations in the Cosmic Radiation* (A014900); 24 months; $51,000

NEW MEXICO

DEPARTMENT OF THE AIR FORCE,
AIR FORCE WEAPONS LABORATORY

Wiley, Robert; *Development of Satellite-Geomagnetic Field Line Intercept Computer Program* (AG222); 12 months; $10,000

NEW MEXICO INSTITUTE OF MINING AND TECHNOLOGY

Wilkening, Marvin H., Colgate, Stirling A. and Brook, Marx; *Support for Operation of Langmuir Laboratory for Atmospheric Research* (A016369); 12 months; $10,100

ENVIRONMENTAL SCIENCES RESEARCH PROJECTS

UNIVERSITY OF NEW MEXICO
Peterson, A. W.; *Infrared Scans of the Corona and Dust Emission Zones at the Eclipse of March 7, 1970* (A017291); 12 months; $14,200

NEW YORK

COLUMBIA UNIVERSITY
Donn, William L.; *Infrasonic, Gravity and Acoustic-Gravity Waves in the Lower Atmosphere and the Ionosphere* (A017454); 24 months; $150,100
Feely, Herbert W. and Broecker, Wallace S.; *The Use of Lead–210 to Trace Transport Processes in the Free Atmosphere* (A016776); 24 months; $32,500

CORNELL UNIVERSITY
Sudan, Ravindranath; *Instabilities in Plasmas* (A015981); 24 months; $85,700

DOWLING COLLEGE
Courten, Henry; *Cometary Investigation During the Total Solar Eclipse of March 7, 1970* (A016434); 12 months; $2,800
——— (GA16434001); $500

NEW YORK UNIVERSITY
Posmentier, Eric S.; *Study of Atmospheric Infrasound of 1–16 HZ* (A015118) 24 months; $46,000

STATE UNIVERSITY AT ALBANY
Blanchard, Duncan C.; *An Investigation of Water-to-Air Transfer of Nuclei, Organics and Iodine* (A023745); 12 months; $1,900
——— *Mechanism for the Water-to-Air Transfer and Concentration of Bacteria* (A023413); 12 months; $27,800
Jiusto, James E.; *Atmospheric Particulates* (A012735); 24 months; $78,600
Kamra, A. K.; *Electrification of Dust Storms* (A018667); 12 months; $25,000
Mohnen, Volker A.; *Ion Molecule Reactions of Atmospheric Importance* (A022760); 24 months; $94,500

STATE UNIVERSITY AT STONY BROOK
Simon, Michal; *Theoretical and Observational Study of the Solar Chromosphere* (A016222); 12 months; $14,900

NORTH CAROLINA

UNIVERSITY OF NORTH CAROLINA AT CHAPEL HILL
Ripperton, Lyman A.; *Chemical and Environmental Factors Affecting Ozone Concentration in the Lower Troposphere* (A014475); 24 months; $156,700

OHIO

CASE WESTERN RESERVE UNIVERSITY
Frye, Glenn M. Jr.; *Detection of High Energy Neutrons by a Spark Chamber, Time of Flight Method* (A019834); 12 months; $124,300

OHIO STATE UNIVERSITY
Seliga, Thomas A.; *Detection of Travelling Ionospheric Disturbances Resulting from the 1970 Solar Eclipse* (A023610); 12 months; $5,000

OKLAHOMA

UNIVERSITY OF OKLAHOMA
Fowler, Richard G.; *Investigation of Electron Hydrodynamic Waves* (A019474); 15 months; $20,500
Wilkins, Eugene M. and Sasaki, Yoshikazu; *Laboratory Investigations of Vortex Formation by Simulated Thermals* (A016350); 12 months; $17,300

OREGON

OREGON STATE UNIVERSITY
Fairchild, Clifford E.; *Dissociative Excitation of Oxyegn and Nitrogen* (A020327); 24 months; $33,300

PENNSYLVANIA

FRANKLIN INSTITUTE
Pomerantz, Martin A.; *Time Variations of the Primary Cosmic Radiation Near the North Geomagnetic Pole* (A020602); 12 months; $72,300

PENNSYLVANIA STATE UNIVERSITY
Ferraro, Anthony J. and Lee, H. S.; *Mid-Latitude D-Region Studies by Means of Cross-Modulation and Associated Programs* (A013885); 12 months; $57,500
Hagen, John P.; *Multi-Frequency Solar Observational Program* (A013714); 12 months; $38,700
——— *Reduction and Analysis of Data Taken at the March 7, 1970 Eclipse* (A023739); 12 months; $31,900
Lumley, John L. and Tennekes, Hendrik; *Investigation of Atmospheric Turbulence* (A018109); 24 months; $176,400

UNIVERSITY OF PITTSBURGH
Donahue, Thomas M.; *Airglow and Auroras in Planetary Atmospheres* (A018224); 12 months; $110,700

SOUTH CAROLINA

UNIVERSITY OF SOUTH CAROLINA
Safko, John L.; *Solar and Geomagnetic Measurements During the Total Eclipse of March 7, 1970* (A016771); 12 months; $3,500

TEXAS

RICE UNIVERSITY
Few, Arthur A. and Stebbings, R. F.; *Generation and Propagation of Acoustic Waves in the Earth's Lower Atmosphere* (A017979); 12 months; $48,200
Gordon, W. E.; *Atmospheric Processes in the Ionosphere and Lower Magnetosphere* (A017452); 24 months; $82,600

UNIVERSITY OF TEXAS AT AUSTIN
Jehn, Kenneth H. and Wagner, Norman K.; *Sea Breeze Investigation—Further Data Analysis* (A016167); 24 months; $69,200
Wagner, Norman K.; *Meteorological Studies of an Urban Atmosphere* (A016822); 24 months; $65,300

UNIVERSITY OF TEXAS AT DALLAS
Collins, Carl B.; *Neutral, Collision-Induced Recombination of Ions and Electrons* (A015434); 12 months; $52,700

UTAH

UNIVERSITY OF UTAH
Ketcham, Warren M.; *Observation of Microphysical Cloud Processes by Means of a Hologram System* (A014299); 12 months; $24,300

WASHINGTON

BATTELLE MEMORIAL INSTITUTE, BATTELLE-NORTHWEST
Smith, Leroi; *Analysis of Night Airglow Measurements from Argentina* (A017509); 12 months; $35,000

UNIVERSITY OF WASHINGTON
Businger, Joost A. and Fleagle, Robert G.; *Energy Transfer in the Planetary Boundary Layer* (A014680); 24 months; $349,400
Fritschen, Leo J.; *Measurement of Airflow in a Forest Clearing and Modeling Airflow in a Forest Canopy* (A022757); 24 months; $44,700

WASHINGTON STATE UNIVERSITY
Craine, Lloyd B.; *Collaborative Research in Infra-Sonic Waves in the Atmosphere—Pacific Northwest* (A023176); 24 months; $107,000

WISCONSIN

UNIVERSITY OF WISCONSIN, Madison
Birkemeier, William P., Sargeant, Douglas H. and Thomson, Dennis W.; *Microwave Tropo-scatter Probing of the Atmosphere* (A016658); 24 months; $317,600
Kutzbach, John E. and Bryson, Reid A.; *Interdisciplinary Research Program in Climatology* (A10651X001); 12 months; $220,700
Lettau, Heinz H. and Stearns, C. R.; *Studies of the Physical Structure of the Earth/Air Interface* (A010998); 24 months; $103,500
Roesler, F. L.; *High-Resolution Spectroscopic Studies in Aeronomy* (A015733); 24 months; $76,000

Earth Sciences

ALASKA

UNIVERSITY OF ALASKA
Mather, Keith B.; *Support for the Geophysical Institute* (A019475*); 12 months; $43,400

ARIZONA

ARIZONA STATE UNIVERSITY
Moore, Carleton B.; *Analysis of Selected Elements in Meteorites and Related Terrestrial Rocks* (A014389); 24 months; $50,400
Pewe, Troy L.; *Glacial Geology of the White Mountains Area, Eastern Arizona* (A013712); 26 months; $19,000

UNIVERSITY OF ARIZONA
Damon, Paul E.; *Carbon–14 Fluctuations in the Environment of Earth and Atmosphere* (A017485); 24 months; $50,000
Ferguson, Charles W.; *Dendrochronology of Bristlecone Pine* (A020618); 24 months; $45,000
Long, Austin and Damon, Paul E.; *Radiocarbon Chronology for Late Quaternary Time* (A016600); 24 months; $42,000

CALIFORNIA

CALIFORNIA INSTITUTE OF TECHNOLOGY
Albee, Arden L.; *Chemical Equilibrium in Coexisting Phases of Quartz-Muscovite Rock* (A012867); 24 months; $60,000
Silver, Leon T.; *Geologic History of the Southwestern Margin of North America* (A015989); 24 months; $70,000

HUMBOLDT STATE COLLEGE
Young, John C.; *Late Cenozoic Deformation in Coastal California North of Cape Mendocino* (A021479); 24 months; $14,600

SAN DIEGO STATE COLLEGE
Krummenacher, Daniel; *Geochronological and Geochemical Studies in Central Pacific* (A017823); 12 months; $9,500

STANFORD UNIVERSITY
Bartlett, Robert W.; *Mineral Thermodynamic Measurements Using Atomic Absorption* (A010899); 24 months; $52,700
Dickson, Frank W.; *Solubilities of Gold, Quartz, and Sulfides to 500 Degrees Centigrade—2000 Bars* (A015817); 24 months; $60,000
Johnson, Arvid M.; *Mechanical Analysis of Rock Deformation Associated with High-Level Intrusions* (A014899); 24 months; $55,000
Silberling, Norman J.; *Triassic Marine Invertebrate Faunas in Northwestern Nevada* (A019424); 24 months; $44,800

UNIVERSITY OF CALIFORNIA, Berkeley
Carmichael, I. S. E.; *Studies in Volcanic Petrology* (A011735); 24 months; $50,000
Gilbert, Charles M. and Reynolds, Mitchell W.; *Late Cenozoic Evolution of Structural Basins on the Western Margin of the Great Basin* (A019311); 24 months; $55,000
Johnson, Lane R.; *Seismic Body Waves* (A015717); 24 months; $31,400
Price, P. B.; *Geophysical Studies with Nuclear Tracks in Solids* (A023930); 12 months; $21,200

EARTH SCIENCES

UNIVERSITY OF CALIFORNIA, Davis
Moores, Eldridge M.; *Emplacement of Franciscan Type Rocks and Origin of Related Mafic and Ultramafic Rocks* (A014298); 24 months; $29,900

UNIVERSITY OF CALIFORNIA, Los Angeles
Ernst, Wallace G.; *Phase Equilibrium of Low-Grade Metamorphism—California Coast Ranges* (A011488); 24 months; $69,600
Knopoff, Leon; *Upper Mantle Structure in the Southern Andes* (A018673); 24 months; $117,600
——— *Surface-Wave Phase Velocities Across the United States* (A018672); 12 months; $28,900
Shreve, Ronald L.; *Statistical Geomorphology* (A018346); 24 months; $27,900
Slichter, Louis B.; *Regional Response to Earth Tides and Observations of Earth's Free Vibrations* (A014732); 12 months; $36,700
Wasson, John T.; *Origin and Composition of Meteorites* (A015731); 24 months; $60,200
Wetherill, George W.; *Precambrian Geochronology and Related Problems* (A012701); 24 months; $82,600

UNIVERSITY OF CALIFORNIA, San Diego
Brune, James N.; *Seismic Data Analysis* (A019473); 15 months; $51,400
Garrels, Robert M.; *Wet Synthesis of Silicates at Low Temperatures* (A020739); 12 months; $17,400
Gilbert, Freeman and Haubrich, Richard A.; *Observations of the Earth's Free Modes* (A015922); 24 months; $118,400
Hawkins, James W. Jr.; *Petrology and Geochemistry of Volcanic Rocks of Southwestern California and Northern Baja California* (A016120); 24 months; $26,300

UNIVERSITY OF CALIFORNIA, San Diego
REVELLE COLLEGE
Goodkind, John M.; *Gravimetric Research with a Superconducting Gravimeter* (A015985); 12 months; $33,200
Suess, Hans E.; *Determination of Tritium in Natural Waters* (A017689); 12 months; $40,800

UNIVERSITY OF CALIFORNIA, Santa Barbara
Tilton, George R.; *Isotopic Geochemistry of Lead* (A012478); 24 months; $55,500

UNIVERSITY OF SOUTHERN CALIFORNIA
Davis, Gregory A.; *Structural Development of the Southern Cordilleran Orogen, California* (A021401); 24 months; $9,700
Henyey, T. L. and Teng, Ta-Liang; *Thermal Diffusivity of Rocks and Minerals at Elevated Temperature and Pressure* (A018447); 24 months; $42,600

COLORADO

COLORADO SCHOOL OF MINES
Keller, George V.; *Electromagnetic Sounding of the Crust and Upper Mantle* (A013134); 24 months; $63,900
——— (GA13134001); 24 months; $12,003
Wildeman, Thomas R.; *Rare Earths in Rock Suites and Rock—Forming Mineral Systems* (A013572); 24 months; $31,900

UNIVERSITY OF COLORADO
Bradley, William C.; *Paleohydrology and Sedimentology of Lake Missoula Floods* (A021478); 12 months; $4,600
Walker, Theodore R.; *The Origin and Paleoclimatic Significance of Red Soils in Europe and North Africa* (A020738); 15 months; $17,900

CONNECTICUT

ST. JOSEPH COLLEGE
Murphy, Mary E.; *Identification of Branched, Cyclic and Aromatic Hydrocarbons in Geological Samples* (A024173); 12 months; $5,000

YALE UNIVERSITY
Carter, Neville L.; *Solid Flow in the Earth's Upper Mantle* (A015412); 24 months; $137,700

DISTRICT OF COLUMBIA

ARCTIC INSTITUTE OF NORTH AMERICA
Andersen, Bjorn G.; *Fennoscandian Moraine Project in Norway* (A025238); 24 months; $9,100

CARNEGIE INSTITUTION OF WASHINGTON
Aldrich, L. T.; *Studies in the Andes of the Conductivity Anomaly of the Earth's Crust and Mantle, as Measured by Magnetic Variations* (A015502); 24 months; $56,000

HOWARD UNIVERSITY
Thorpe, Arthur N.; *Natural Radiation Damage in Zircons* (A018612); 6 months; $4,800

NATIONAL ACADEMY OF SCIENCES,
NATIONAL RESEARCH COUNCIL
Berg, Joseph W. Jr.; *Partial Support of the Committee on the Geological Sciences* (C310188); 24 months; $33,100
——— ; *Partial Support for the Committee on Seismology* (C310142003); 33 months; $13,275
Bove, Albert; *Support of the U.S. National Committee for Geochemistry* (C310158001); 12 months; $13,900
Heindl, L. A.; *Support of the U.S. National Committee for the International Hydrological Decade* (C310153001); 12 months; $96,500
Odishaw, Hugh; *Support of the Geophysics Research Board and its Affiliates* (C310049007*); 12 months; $68,500

FLORIDA

FLORIDA STATE UNIVERSITY
Kennett, James P. and Watkins, Norman D.; *Paleomagnetic Polarity History and Biostratigraphy of Selected Upper Tertiary—Quaternary Marine Sequences in New Zealand* (A013093); 24 months; $52,900
Loper, David E.; *The Hydromagnetics of a Contained Rotating Fluid* (A018853); 24 months; $35,000

UNIVERSITY OF MIAMI
Harrison, Christopher G. A.; *Paleomagnetic Investigations of Puerto Rico* (A015948); 12 months; $22,100
Nagle, Frederick and Stipp, Jerry; *Andesite Genesis in Relation to Ocean-Floor Spreading and Island Arcs* (A013531); 12 months; $36,400

GEORGIA

UNIVERSITY OF GEORGIA
Frey, Robert W.; *Depositional Environments and Animal-Sediment Relationships in Holocene Salt Marshes* (A022710); 24 months; $26,700
Salotti, Charles A.; *A Chemical Study of a Footwall Core (Cherokee Mine, Ducktown, Tennessee)* (A015067); 12 months; $9,900

WEST GEORGIA COLLEGE
Crawford, Thomas J. and Medlin, Jack H.; *Stratigraphic, Structural, and Petrologic Relationships of Rock Units within and Adjacent to the Brevard Fault Zone* (A015917); 24 months; $27,800

HAWAII

UNIVERSITY OF HAWAII
Naughton, John J.; *Hawaiian Volcanic Gases and Volatiles* (A020316); 24 months; $35,200
Sutton, George H. and Walker, Daniel A.; *Seismological Investigation of the Northwestern and South Pacific Upper Mantle* (A012851); 24 months; $100,000

ILLINOIS

NORTHERN ILLINOIS UNIVERSITY
Goldich, Samuel S.; *Ages of Rocks Assigned to the Penokean Orogeny in Minnesota, Wisconsin and Michigan* (A012316); 24 months; $62,400

NORTHWESTERN UNIVERSITY
Dapples, Edward C.; *Petrology of Sandstones in the Franconia Formation of Western Wisconsin* (A016524); 24 months; $39,800

SOUTHERN ILLINOIS UNIVERSITY
Cohen, Arthur D.; *Origin, Description and Stratigraphy of the Peats of the Okefenokee Swamp* (A015984); 24 months; $36,800

UNIVERSITY OF CHICAGO
Hafner, Stefan S.; *Application of Nuclear Resonance Spectroscopy to Mineralogy* (A014811); 24 months; $75,000
Jamieson, John C.; *Physical Behavior of Solids Under Very High Pressure* (A016875); 24 months; $80,200
Wyllie, Peter J.; *Petrogenetic Links Between Carbonatites and Kimberlites* (A015718); 24 months; $65,000

UNIVERSITY OF ILLINOIS, Urbana
Klein, George Devries; *Hydraulic Factors Controlling the Migration of Dunes and Sand Waves in a Tide-Dominated Environment* (A021141); 18 months; $28,000
Stevenson, F. J.; *Biogeochemistry of Nitrogen* (A015707); 24 months; $30,000

INDIANA

INDIANA UNIVERSITY, Bloomington
Meinschein, Warren G.; *Biogeochemical Research* (B014867*); 24 months; $30,000
Vitaliano, Charles J.; *Geology of the Tobacco Root Mountains, Madison County, Montana* (A021369); 15 months; $13,300

IOWA

IOWA STATE UNIVERSITY
Vondra, Carl F.; *Stratigraphy of the Willwood Formation, and the Early Eocene Faunas and Faunal Succession, Bighorn Basin, Wyoming* (A015996); 12 months; $16,900

UNIVERSITY OF IOWA
Swett, Keene; *Comparative Petrology and Stratigraphy of Cambro-Ordovician Rocks of East Greenland, Scotland, and Newfoundland* (A016602); 12 months; $21,600

KANSAS

UNIVERSITY OF KANSAS
Van Schmus, Wm. Randall; *Geochronological and Geochemical Research* (A015951); 24 months; $35,500

LOUISIANA

LOUISIANA STATE UNIVERSITY, Baton Rouge
Van Den Bold, Willem A.; *Post Eocene Ostracoda of the Caribbean Region* (A016522); 36 months; $24,500

MAINE

UNIVERSITY OF MAINE, Orono
Denton, George H.; *Glacial Stratigraphy and Chronology, White River Valley, Alaska* (A021484); 14 months; $18,500
Hall, Bradford A.; *Dispersal Patterns and Paleoenvironments of Lower Devonian Clastic Rocks, Northern Maine* (A021119); 24 months; $31,800

MARYLAND

DEPARTMENT OF COMMERCE,
ENVIRONMENTAL SCIENCE SERVICES
ADMINISTRATION
Klaasse, James M.; *Support of the Foreign Standard Seismological Stations and Data Processing* (AG21600); 12 months; $283,500

JOHNS HOPKINS UNIVERSITY
Bricker, Owen P.; *Sedimentary and Diagenetic Record of Environmental Parameters in Recent Bahamian Tidal Flats* (A016143); 12 months; $44,700
Hunt, Charles B.; *Cenozoic History of Eastern Colorado* (A019172); 24 months; $27,500

ENVIRONMENTAL SCIENCES RESEARCH PROJECTS

MARYLAND ACADEMY OF SCIENCES

Gernant, Robert E. and Gibson, Thomas E.; *Environmental History of the Miocene of Southern Maryland* (A021483); 12 months; $12,500

UNIVERSITY OF MARYLAND

Elsasser, Walter M.; *New Mechanics of Orogeny* (A020740); 12 months; $28,000

MASSACHUSETTS

BOSTON UNIVERSITY

Brownlow, Arthur H.; *Sulfide-Silicate Relationships in the Boulder Batholith* (A018240); 24 months; $26,800

HARVARD UNIVERSITY

Barghoorn, Elso S.; *Precambrian Plant Fossils and the Organic Geochemistry of Precambrian Sediments* (A013821); 24 months; $54,000

Shankland, T. J.; *Vacuum Ultraviolet Spectroscopy of Minerals* (A012850); 24 months; $30,400

Siever, Raymond; *Petrology and Geochemistry of Diagenesis of Siliceous Sediments* (A012865); 24 months; $59,800

Thompson, James B., Jr. and Waldbaum, David R.; *Thermodynamic Properties of Rock-forming Minerals* (A012869); 24 months; $79,000

Thompson, James B., Jr. and Klein, Cornelius, Jr.; *Coexisting Minerals in Metamorphic Rocks* (A011435); 24 months; $65,900

MASSACHUSETTS INSTITUTE OF TECHNOLOGY

Aki, Keiiti; *Global Tectonic Synthesis from the Spectral Analysis of Earthquake Surface Waves* (A014812); 24 months; $74,900

Brace, William F.; *Creep and Attenuation in Partially Melted Rock* (A018342); 24 months; $143,800

Shapiro, Irwin I.; *Geophysical Applications of Atomic-Clock Radio Interferometry* (A015134); 24 months; $85,000

Wones, David R.; *Phase Equilibria of Biotites and Amphiboles* (A013092); 24 months; $60,000

Wuensch, Bernhardt J.; *Crystal Chemistry of Sulfosalts* (A022698); 24 months; $42,700

WOODS HOLE OCEANOGRAPHIC INSTITUTION

Von Herzen, Richard P. and Simmons, M. Gene; *Geothermal Investigations in Ocean Regions* (A016078); 24 months; $174,900

MICHIGAN

UNIVERSITY OF MICHIGAN

Peacor, Donald R.; *High Temperature Single-Crystal Investigations of Tektosilicate Structures* (A017402); 24 months; $39,500

MINNESOTA

UNIVERSITY OF MINNESOTA

Hooke, R. LeB.; *Flow Law of Polar Ice and Flow Field at Margin of the Barnes Ice Cap, Baffin Island, Canada* (A019310); 24 months; $22,800

Mooney, Harold M.; *Investigation of the Keweenawan Rocks of Southeastern Minnesota and Western Wisconsin* (A016521); 12 months; $3,100

Murthy, V. Rama and Pepin, Robert O.; *SR-Isotopes, Rare Gases, and Trace-Elements in the Upper Mantle* (A014950); 24 months; $40,800

UNIVERSITY OF MINNESOTA, Duluth Campus

Green, John C.; *Origin and Differentiation of Minnesota Keweenawan Lavas* (A013411); 24 months; $26,200

MISSOURI

ST. LOUIS UNIVERSITY

Nuttli, Otto W.; *Surface Wave Attenuation Studies Relevant to the Seismic Risk Problem in the Central and Eastern United States* (A020115); 24 months; $54,300

Stauder, William; *Research in Earth Strains and Focal Mechanisms* (A015816); 20 months; $56,000

UNIVERSITY OF MISSOURI, Columbia

Case, James E.; *Tectonophysics of Colombia and Panama* (A019308); 24 months; $43,100

Davis, Stanley N.; *Near-Surface Localization of Strain Along Joints in Bedrock* (A020737); 24 months; $24,200

UNIVERSITY OF MISSOURI, Rolla

Manuel, Oliver K.; *Noble Gases in the Earth and its Atmosphere* (A016618); 24 months; $56,900

WASHINGTON UNIVERSITY

Johns, William D.; *Reactions Between Clay Minerals and Organic Compounds at Low Temperatures (Below 200 Degrees C)* (A016163); 24 months; $55,000

MONTANA

MONTANA STATE UNIVERSITY

McMannis, William J.; *Precambrian History of the Pine Creek Metasedimentary Sequence, Park County, Montana* (A020735); 18 months; $15,200

UNIVERSITY OF MONTANA

Silverman, Arnold and Hyndman, Donald W.; *Petrochemical Study of Granitic Plutons of the Flint Creek Range, Montana* (A014401); 24 months; $50,900

NEW HAMPSHIRE

DARTMOUTH COLLEGE

Decker, Robert W.; *Remeasurement of Horizontal Ground Surface Deformation Lines in Iceland* (A026386); 12 months; $5,600

Drake, Charles L.; *The Meeting of the International Committee on Geodynamics* (A024966); 12 months; $11,600

NEW JERSEY

PRINCETON UNIVERSITY

Hollister, Lincoln S.; *Compositional Zoning in Garnet, Staurolite, Kyanite Andalusite, and Sillimanite, Kwoiek Area, British Columbia* (A020236); 12 months; $8,600

Phinney, Robert A.; *Magnetolluric Studies of Thermal Areas* (A018267); 15 months; $39,200

Vine, Frederick J.; *Acquire and Interpret an Aeromagnetic Survey of Cyprus* (A013255); 12 months; $14,200

RUTGERS UNIVERSITY, Camden Campus

Greenwood, Robert; *Synthesis of Crystalline Silica at Room Temperature* (A014553); 24 months; $14,600

NEW YORK

COLUMBIA UNIVERSITY

Anderson, Orson L.; *Investigation of the Effects of Magnetic Phase Transitions upon the Elastic Properties of Third Transition Metal Oxides* (A016082); 24 months; $49,400

Dalziel, Ian W. D. and Scholz, Christopher; *Dynamics of Fracture and Friction* (A021472); 24 months; $59,700

Drake, Charles L. and Sykes, Lynn R.; *Seismological Research on the East African Rift in Kenya* (A016520); 6 months; $14,600

Ewing, Maurice and Kuo, John T.; *Solid-Earth Tides Study* (A016478); 12 months; $40,000

Ewing, Maurice and Nowroozi, Ali A.; *Ocean Bottom Geophysical Studies* (A021473); 12 months; $62,400

Gast, Paul W. and Kay, Robert; *Lead and Strontium Isotope Compositions in Oceanic Volcanic Rocks, and Dispersed Elements* (A016457); 24 months; $64,500

Hass, Warren J.; *Operation of the Sub-Center of World Data Center 'A' for the International Upper Mantle Project* (A018275); 12 months; $17,300

Kay, Marshall; *Boulder-Bearing Mudstone—Central Volcanic Belt of Newfoundland and its Bearing on the Proto-Atlantic Problem* (A015350); 24 months; $10,300

STATE UNIVERSITY AT ALBANY

Means, Winthrop D.; *Experiments and Fabric Studies Relating to the Origin of Axial Plane Cleavages* (A015918); 24 months; $40,000

Putman, George W.; *Petrogenesis of the Superior Stock, Plumas County, California* (A012571); 24 months; $36,400

STATE UNIVERSITY AT BINGHAMTON

Bodine, Marc W. J.; *Geochemistry of Silicate Assemblages in Marine Evaporites* (A013090); 24 months; $27,000

Wu, Francis T.; *Source Mechanism of Earthquakes in Selected World-Wide Areas* (A018481); 24 months; $42,900

STATE UNIVERSITY AT STONY BROOK

Bence, A. E.; *Chemical and Petrological Investigation of Regional Metamorphism, Dutchess County, New York* (A015638); 24 months; $60,000

Palmer, Allison R.; *Physical Stratigraphy, Biostratigraphy and Paleoecology of the Carrara Formation and Correlative Units, Southern Great Basin, Nevada and California* (A020318); 24 months; $40,000

Papike, James J. and Prewitt, Charles T.; *Phase Transitions and Intracrystalline Equilibria in Pyroxenes and Amphiboles* (A012973); 24 months; $79,600

NORTH CAROLINA

UNIVERSITY OF NORTH CAROLINA AT CHAPEL HILL

Fullagar, Paul D.; *Geochronology Studies Related to the Geological Time-Scale* (A018448); 24 months; $32,300

Thomas Henry C.; *Diffusion and Conduction in Colloidal Suspensions* (A016569); 24 months; $43,800

OHIO

CASE WESTERN RESERVE UNIVERSITY

Aronson, James; *A Critical Test of the Sea Floor Spreading Hypothesis* (A024859); 18 months; $1,800

Green, H. W. II and Radcliffe, S. V.; *Transmission Electron Microscopy Study of the Deformation of Rock-Forming Minerals* (A013409); 24 months; $75,800

Nairn, Alan E. M.; *Paleomagnetic Investigation of Cenozoic Volcanics in the High Plateaus and Adjoining Great Basin of Southwestern Utah* (A013333); 24 months; $41,600

Stehli, F. G.; *Paleoclimatic Test of the Hypothesis of an Axial Magnetic Field in the Permian* (A020563); 24 months; $60,000

Stehli, Francis G. and Savin, Samuel M.; *Calibration of the Planetary Temperature Gradient, Upper Eocene and Maestrichtian* (A016827); 24 months; $80,000

OBERLIN COLLEGE

Skinner, William R.; *Polyphase Deformation and Metamorphism in the Precambrian Basement Complex of the Beartooth Mountains, Montana and Wyoming* (A019495); 24 months; $21,400

———; (GA19495001); $600

OKLAHOMA

SOUTHWESTERN STATE COLLEGE

Gunter, Bobby; *The Geochemistry of Hydrothermal Gases* (A016199); 24 months; $25,300

OREGON

OREGON STATE UNIVERSITY

Boucot, Arthur J.; *Stratigraphic and Paleontologic Study of the Siluro-Devonian of Bolivia* (A017764); 24 months; $26,700

———; *Taxonomic, Evolutionary, Animal Geographic Studies of Silurian-Lower Devonian Brachiopods and Gastropods* (A017455); 12 months; $40,100

EARTH SCIENCES

Couch, Richard; *A Seismic Refraction Study in the Alexander Archipelago, Southeast Alaska* (A026411); 12 months; $4,300

Dymond, Jack; *Potassium-Argon Ages of Oceanic Volcanic Rocks* (A017282); 12 months; $20,500

Johnson, J. Granville; *A Standard Biostratigraphic Sequence for the Lower and Middle Devonian in Western North America* (A017647;) 24 months; $35,100

UNIVERSITY OF OREGON

Goles, Gordon and McBirney, Alexander R.; *Geochemical Study of Volcanic Rocks from the Cascade Range* (A019382); 24 months; $50,000

PENNSYLVANIA

FRANKLIN AND MARSHALL COLLEGE

Moss, John H.; *Terrace Origin Studies, Shoshone River Basin, Wyoming* (A025173); 16 months; $2,950

Wiebe, R. A.; *Setting and Evolution of 'Granitic' Plutons in the Northern Appalachians, Cape Breton Island, Nova Scotia* (A021157); 24 months; $23,200

FRANKLIN INSTITUTE

Banerjee, Subir K.; *Study of Remanence in Fine Particles of Synthesized Iron Titanium Oxides* (A015950); 12 months; $63,200

LEHIGH UNIVERSITY

Goldstein, Joseph I.; *Iron-Nickel-Phosphorus Diffusion—Understanding Structure of Iron Meteorites* (A015349); 24 months; $53,400

PENNSYLVANIA STATE UNIVERSITY

Boettcher, Arthur L.; *Phase Transformations and the Generation of Magmas—Hydrothermal Pressures* (A012737); 24 months; $35,300

White, William B.; *Optical Spectra and Electron Transfer Processes in Iron-Containing Minerals* (A015923); 12 months; $11,300

Wright, Lauren A.; *Late Precambrian Paleoenvironments of the Southern Part of the Great Basin, California and Nevada* (A016119); 24 months; $31,500

UNIVERSITY OF PITTSBURGH

Bikerman, Michael; *K-AR Dating of Volcanic Rocks in Southwestern New Mexico* (A016002); 24 months; $30,100

Lidiak, Edward G.; *Experimental Study of the Role of Titanium in Basalts* (A012866); 24 months; $25,000

RHODE ISLAND

BROWN UNIVERSITY

Chinnery, Michael A.; *Investigations of Fault and Earthquake Mechanism* (A018870); 24 months; $34,300

Laporte, Leo F.; *Sedimentology and Stratigraphy of a Recent Carbonate Facies Mosaic, Cape Sable, Florida* (A014474); 12 months; $8,000

UNIVERSITY OF RHODE ISLAND

Kennett, J. P.; *Paleoclimatic and Biostratigraphic Studies of Sediments from The Gulf of Mexico* (A025175); 24 months; $20,000

SOUTH CAROLINA

CLEMSON UNIVERSITY

Griffin, Villard S. Jr.; *Geologic Relationships of the Inner Piedmont, Kings Mountain and Charlotte Belts, South Carolina* (A016164); 24 months; $24,000

Hatcher, Robert D. Jr.; *Structural, Stratigraphic and Metamorphic History of the Tallulah Falls Dome and Adjacent Area, Northeast Georgia* (A020321); 24 months; $30,000

UNIVERSITY OF SOUTH CAROLINA

Colquhoun, D. J.; *Age of the First Neogene Major Eustatic Drop in Sea Level* (A017941); 12 months; $10,700

Ferm, John C.; *Investigation of Carbonate Shelf, Clastic Beach-Barrier and Deltaic Facies in Carboniferous Rocks of Eastern Kentucky and Tennessee* (A020620); 24 months; $30,000

TEXAS

RICE UNIVERSITY

Burchfiel, B. C.; *Structural Development of the Southern Cordilleran Orogen* (A021375); 24 months; $24,100

Debremaecker, Jean-Claude; *Analysis of Earthquake Body Waves* (A016121); 24 months; $31,400

SOUTHERN METHODIST UNIVERSITY

Thorstenson, Donald C. and Mackenzie, Fred T.; *Geochemistry of Small Organic Molecules in Natural Waters* (A018083); 24 months; $30,000

TEXAS A & M UNIVERSITY

Sowers, George M.; *Criteria for Separating Bends from Buckles in Structural Geology* (A017326); 24 months; $30,000

UNIVERSITY OF TEXAS AT AUSTIN

Behrens, E. William and Land, Lynton S.; *Holocene Stratigraphy and Carbonate Sedimentation in South Texas* (A021390); 24 months; $30,800

Clabaugh, Stephen E., Long, Leon E. and McDowell, Fred W.; *K-AR Dating of Tertiary Volcanic Rocks of Western Texas* (A016080); 24 months; $38,300

Maxwell, John C.; *Emplacement of Franciscan-Type Rocks and Origin of Related Mafic and Ultramafic Rocks* (A024522); 24 months; $39,900

Parker, Patrick L.; *Stable Isotope Variations in Biogeochemical Systems* (A011414); 24 months; $40,700

Smith, . W. and Bostick, F. X. Jr.; *Magnetotelluric Tensor Method for Determining Subsurface Geological Features* (A017457); 12 months; $28,000

UNIVERSITY OF TEXAS AT DALLAS

Burek, Peter J.; *Paleomagnetism of the Middle East* (A013133); 12 months; $15,000

Dziewonski, Adam M.; *Detailed Study of Dispersion of the World Circling Surface Waves* (A015890); 24 months; $53,000

Helsley, Charles E.; *Paleomagnetic Studies of Continuous Stratigraphic Sequences of Paleozoic and Mesozoic Age* (A015999); 24 months; $80,000

Landisman, Mark; *Mantle Studies Related to the New Global Tectonics—Body Waves* (A022706); 12 months; $45,100

Pessagno, Emile A. Jr.; *Upper Cretaceous Radiolaria and Biostratigraphy of the California Coast Ranges* (A015998); 12 months; $20,000

Porath, H. and Dziewonski, Adam; *Geomagnetic Sounding of the Upper Mantle* (A018265); 24 months; $165,600

WEST TEXAS STATE UNIVERSITY

Schultz, Gerald A.; *Vertebrate Faunas Associated with Quarternary Volcanic Ash in the Texas Panhandle* (A020114); 24 months; $16,800

UTAH

BRIGHAM YOUNG UNIVERSITY

Best, Myron G. and Brimhall, W. H.; *Petrochemistry of Cenozoic Basalts, Colorado Plateaus, Arizona* (A015082); 24 months; $34,600

UNIVERSITY OF UTAH

Cook, Kenneth L.; *Operation of Uinta Basin Seismological Observatory* (A021120); 12 months; $48,300

Eardley, A. J.; *Coring the Pliocence—Pleistocene Transition in the Great Salt Lake Basin* (A015117); 12 months; $9,100

Shuey, Ralph T.; *Paleomagnetic Chronology and Correlation, Great Salt Lake Basin Sediments* (A016134); 12 months; $5,700

VIRGINIA

VIRGINIA POLYTECHNIC INSTITUTE

Gilbert, M. C.; *Stability of Calcic Amphiboles and Experimental Determination of Amphibole Miscibility Gaps* (A012479); 24 months; $40,600

Ribbe, P. H. and Gibbs, G. V.; *Nature and Variation of the SI-O Bond—Studies of Orthosilicates and Framework Structures* (A012702); 24 months; $57,800

Rich, Charles I.; *Potassium Selectivity as Related to Micromorphology of Mica-Vermiculites* (A018053); 24 months; $32,000

Robinson, Edwin S.; *Tidal Gravity Measurements in Southeastern United States* (A016118); 24 months; $25,700

WASHINGTON

UNIVERSITY OF WASHINGTON

Crosson, Robert S.; *Initial Study of Seismicity and Crustal Structure in Western Washington Using a Seismic Telemetry Network* (A012826); 24 months; $46,200

Evans, Bernard W.; *Microprobe Analysis of Selected Minerals in Rocks* (A016601); 12 months; $14,500

Lachapelle, Edward R.; *Glaciology, Micrometeorology and Related Climatology of the Blue Glacier, a Temperate, Alpine Glacier on Mt. Olympus, Washington* (A020319); 12 months; $20,000

Misch, Peter and Gresens, Randall L.; *Geochemical Investigation of Metamorphism in the North Cascades* (A011466); 24 months; $50,000

Porter, Stephen C.; *Pre-Wisconsin Glaciation, East-Central Cascade Range* (A013640); 12 months; $4,200

Porter, Stephen C. and Ugolini, Fiorenzo C.; *Quarternary Glaciation, Volcanism and Soil-Forming Processes on the Upper Slopes of Mauna Kea, Hawaii* (A020320); 24 months; $34,200

Stuiver, Minze; *Geochronometrical and Geophysical Applications of Isotopic Carbon* (A017910); 19 months; $21,100

Washburn, A. Lincoln; *Holocene Glacial Chronology and Isostatic Uplift Between Sondre Stromfjord and the Greenland Ice Sheet, West Greenland* (A020590); 12 months; $14,200

WEST VIRGINIA

MARSHALL UNIVERSITY

Bottino, Michael L.; *Geological Time-Scale and the Evaluation of Whole-Rock RB-SR Ages* (A016501); 24 months; $18,700

WISCONSIN

UNIVERSITY OF WISCONSIN, Madison

Bentley, Charles R. and Dowling, Forrest L.; *Magnetotelluric Measurements in Wisconsin and Vicinity* (A021372); 12 months; $33,900

Guidotti, Charles V.; *Petrology and Mineralogy of the Pelitic Schists of N.W. Maine* (A013415); 24 months; $25,000

Medaris, L. Gordon Jr.; *Element Partitioning Between Coexisting Minerals* (A014070); 24 months; $36,600

Smithson, Scott B.; *Three-Dimensional Studies of Granitic Plutons* (A012871); 24 months; $32,400

WYOMING

UNIVERSITY OF WYOMING

Bailey, Sturges W.; *Domain Structures in Plagioclase Feldspars* (A014071); 24 months; $30,000

Decker, Edward R.; *Heat Flow and Radioactivity Studies in Colorado and Wyoming* (A018450); 24 months; $60,500

Terry, Colin and Shive, Peter N.; *Mossbauer Studies of Magnetic Minerals* (A015732); 24 months; $64,800

SCOTLAND

UNIVERSITY OF EDINBURGH

Willmore, P. L.; *International Seismological Centre* (A018222); 12 months; $30,000

OCEANOGRAPHY RESEARCH PROJECTS

Oceanography Research Projects

Physical Oceanography

CALIFORNIA

U.S. NAVAL POSTGRADUTE SCHOOL
Preisendorfer, Rudolph W.; *An Atlas of Computed Oceanic Lightfields* (AG223); 18 months; $19,300
UNIVERSITY OF CALIFORNIA, Berkeley
Corcos, G. M.; *Dynamics of the Thermocline —Internal Waves and Turbulence* (A016490); 12 months; $22,400
UNIVERSITY OF CALIFORNIA, San Diego
Shepard, Francis P.; *Nature and Velocity of Currents in Submarine Canyons and Their Margins* (A019492); 12 months; $24,700
Smith, Raymond C.; *Measurement of Radiance Distributions in Natural Waters* (A019738); 12 months; $30,500
Tyler, John E.; *Optical Oceanography* (A019830); 12 months; $70,100

DISTRICT OF COLUMBIA

ARCTIC INSTITUTE OF NORTH AMERICA
Coachman, Lawrence K. and Dunbar, M. J.; *The Baffin Bay—North Water Project* (A016455); 12 months; $31,300

FLORIDA

NOVA UNIVERSITY
Niiler, Pearn P.; *Planetary and Internal Gravity Waves* (A014688); 24 months; $102,200
UNIVERSITY OF SOUTH FLORIDA
Carder, Kendall L.; *Loop Currents of the Eastern Gulf of Mexico* (A025991); 12 months; $5,000

HAWAII

UNIVERSITY OF HAWAII
Groves, Gordon W.; *Numerical Study of Barotropic Long Waves on a Rotating Ocean* (A017137); 24 months; $64,100
Stroup, Edward; *Time Variation in the Pacific Equatorial Undercurrent Along 145 Degrees W. Long* (A016038); 12 months; $26,400

ILLINOIS

UNIVERSITY OF CHICAGO
Platzman, George W.; *Global Oscillations of the World's Oceans* (A015995); 24 months; $67,000

MASSACHUSETTS

MASSACHUSETTS INSTITUTE OF TECHNOLOGY
Stommel, Henry M.; *Research in Oceanic Physics* (A021172); 12 months; $148,500
WOODS HOLE OCEANOGRAPHIC INSTITUTION
Bruce, John G.; *Study of the Somali Current and Associated Circulation During the Southwest Monsoon* (A017329); 24 months; $93,600
Metcalf, William G.; *Investigation of the Water Entering the Southeastern Caribbean* (A015730); 12 months; $92,000
Voorhis, Arthur D. and Schmitz, William J. Jr.; *Acoustic Dropsonde to Study the Oceanic Circulation* (A018533); 12 months; $55,400

OREGON

OREGON STATE UNIVERSITY
Bodvarsson, Gunnar; *Analytical Study of the Dynamics of Deep-Sea Currents in Coastal Regions in the Northeast Pacific* (A018129); 12 months; $13,500
Burt, Wayne V. and Pattullo, June G.; *Planning Session on Large-Scale, Long-Period, Ocean-Air Interaction Research in the Pacific Ocean* (A020872); 12 months; $4,400

RHODE ISLAND

UNIVERSITY OF RHODE ISLAND
Lambert, Richard B. and Stern, Melvin E.; *Studies of Vertical Mixing Processes in the Ocean* (A018766); 12 months; $49,700

TEXAS

GULF UNIVERSITIES RESEARCH CORP.
Sharp, James M.; *Development of a Comprehensive Plan for Implementing the Gulf Environmental Program* (A025776); 6 months; $100,000
TEXAS A & M UNIVERSITY
Ochiye, Takashi; *Studies on Circulation of the Gulf of Mexico and Caribbean Sea* (A013430); 12 months; $25,200
Reid, Robert O.; *Numerical Study of the Current Regime of the Western Cayman Sea* (A020569); 12 months; $16,500

WASHINGTON

UNIVERSITY OF WASHINGTON
Rattray, Maurice Jr. and Larsen, Lawrence H.; *Marine Hydrodynamics* (A022692); 12 months; $100,000

Submarine Geology

ALASKA

UNIVERSITY OF ALASKA
Reeburgh, William S.; *Gases in Interstitial Water as Indicators of Mixing in Sediments* (A019380); 24 months; $53,200

CALIFORNIA

UNIVERSITY OF CALIFORNIA, Los Angeles
Kaplan, I. R. and Presley, B. J.; *Ephemeral Properties of Interstitial Water From the Ocean Sediment Coring* (A020715); 12 months; $31,800
UNIVERSITY OF SOUTHERN CALIFORNIA
Gorsline, Donn S. and Drake, D. E.; *Suspended Sediment off Southern California* (A022842); 12 months; $7,800

CONNECTICUT

WESTERN CONNECTICUT STATE COLLEGE
Chen, Chin; *Quaternary Pteropods in Atlantic Deep-Sea Sediments* (A017483); 24 months; $10,131

DISTRICT OF COLUMBIA

GEORGE WASHINGTON UNIVERSITY
Siegel, Frederic R.; *Holocene Sedimentological and Geochemical Patterns in the Gulf of San Matias* (A016499); 24 months; $25,100

FLORIDA

FLORIDA STATE UNIVERSITY
Goodell, H. G.; *Wisconsin-Holocene Coastal Sedimentation Northeast Gulf of Mexico* (A015729); 24 months; $20,800
Scott, Robert B.; *Experimental Studies of the Alteration of Volcanic Glass in Submarine Environments* (A017691); 24 months; $30,700
UNIVERSITY OF MIAMI
Bader, Richard G. and Gerchakov, Sol M.; *Investigation of Organic Compounds and Minerals in Sediments and Natural Waters* (A014473); 24 months; $123,300
Bock, Wayne D.; *Roles of Benthonic Foraminifera in Ocean Sediment Cores* (A015228); 12 months; $34,000
Bostrom, Kurt G. V.; *Anomalous Sediments on Oceanic Ridges with Crustal Spreading* (A015248); 24 months; $140,300
Emiliani, Cesare; *Oxygen and Carbon Isotope Researches in Marine Geology* (A015226); 24 months; $163,500
——; *Curating of Marine Geological Collections and Thin-Sectioning of Oceanic Rocks* (A018854); 12 months; $18,400
Fisher, David E.; *Potassium/Argon and Rare Gas Studies on Deep-Sea Samples* (A015227); 24 months; $109,800
Hay, William W.; *High Resolution Probabilistic Biostratigraphy with Calcareous Nannofossils* (A015261); 24 months; $118,600
Neumann, A. C.; *Shallow Water Carbonate Environments and Quaternary Sea Levels, South Florida and Bahamas* (A015199); 24 months; $116,100

GEORGIA

UNIVERSITY OF GEORGIA
Sen Gupta, Barun K.; *Foraminiferal Distribution in the Quaternary Sediments of the Georgia Continental Shelf* (A016946); 24 months; $38,400

HAWAII

UNIVERSITY OF HAWAII
Moberly, Ralph Jr. and Resig, Johanna M.; *Chronology and Extent of Subsidence of the Hawaiian Ridge* (A010421); 24 months; $33,700

MASSACHUSETTS

WOODS HOLE OCEANOGRAPHIC INSTITUTION
Berggren, W. A. and Phillips, J. D.; *Biostratigraphic and Paleomagnetic Investigations on Deep Sea Cores* (A016098); 24 months; $93,000
Hunt, John M.; *Geochemistry of Interstitial Water in Cores Obtained from Deep Sea Drilling Project* (A014523); 24 months; $107,700
——; *Organic Geochemistry of Ocean Sediment Cores* (A019829); 12 months; $19,300
——; *Geophysical and Geochemical Study of Black Sea Mineral Deposits* (GA1659 001); 4 months; $10,400
Spencer, Derek W.; *Trace Element Distributions in the South Atlantic Ocean* (A013574); 24 months; $146,000
Thompson, Geoffrey; *Geochemical Study of Submarine Weathering of Rocks from the Mid-Atlantic Ridge* (A018731); 12 months; $24,600
Von Herzen, Richard P. and Degens, Egon T.; *Geochemical Study of Two African Rift Valley Lakes* (A019262); 12 months; $12,000

NEW YORK

COLUMBIA UNIVERSITY
Biscaye, Pierre E.; Lawrence, James R. and Broecker, Wallace S.; *Origins of Deep-Sea Sediments—Oxygen and Hydrogen Isotope Variations* (A017724); 24 months; $58,100
Broecker, Wallace S., Saito, Tsunemasa and Hunkins, Kenneth L.; *Integrated Study of Paleoclimates Using Oxygen Isotopes as a Tool* (A016769); 12 months; $40,800
Ericson, David; *Pleistocene Oceanography as Recorded in Deep-Sea Sediment Cores* (A015639); 12 months; $29,100
Ewing, Maurice, Ninkovich, Dragoslav and Burckle, Lloyd H.; *Late Cenozoic Biostratigraphy and Paleoecology of the Sea of Japan and Adjacent Areas of the Japanese Islands* (A020211); 24 months; $22,500
Ewing, Maurice, Hays, James D. and Saito, Tsunemasa; *Stratigraphic Studies of Deep-Sea Sediments* (A019690); 12 months; $174,800
Ewing, Maurice, Hays, James D. and Jacobs, Marian B.; *Orogenic and Climatic Events Indicated by Mineralogical Changes in Deep Sea Sediments* (A017197); 12 months; $34,600
Hays, James D. and Goll, Robert M.; *Biostratigraphic and Taxonomic Studies of Lower Tertiary Radiolaria* (A017122); 24 months; $43,200

BIOLOGICAL OCEANOGRAPHY

McIntyre, Andrew, Be, Allan W. H. and Ruddiman, William F.; *Electron Microscopy of Calcareous Plankton* (A014177); 24 months; $99,800

CUNY, QUEENS COLLEGE

Habib, Daniel; *Cretaceous Palynostratigraphy of National Ocean Sediment Coring Program Cores from the Horizon A Area* (A019941); 24 months; $34,700

McIntyre, Andrew; *Coccolith Carbonate and Ice Rafted Distributions in North Atlantic* (A014730); 24 months; $21,100

NORTH CAROLINA

DUKE UNIVERSITY

Pilkey, Orrin H.; *Sedimentation on the Hatteras Outer Ridge* (A012872); 24 months; $44,200

OREGON

OREGON STATE UNIVERSITY

Dymond, Jack R. and Heath, G. Ross; *Provenance of North Pacific Sediments* (A017643); 12 months; $56,500

Kulm, Laverne D. and Fowler, Gerald; *Sedimentation and Stratigraphy of Deep-Sea Environments in the Northeast Pacific* (A015926); 24 months; $94,200

Van Andel, Tjeerd H.; *Geologic Studies of Selected Parts of the Mid-Atlantic and Walvis Ridges in the South Atlantic* (A015316); 24 months; $117,000

RHODE ISLAND

BROWN UNIVERSITY

Imbrie, John; *Quantitative Paleoclimatic Analysis of Deep Sea Sediments* (A014853); 24 months; $89,300

WASHINGTON

UNIVERSITY OF WASHINGTON

Creager, Joe S. and McManus, Dean A.; *Stratigraphy of the Continental Shelves of the Laptev, East Siberian and Bering Sea* (A011126001); 12 months; $46,500

Creager, Joe S. and Ling, Hsin-Yi; *Curatorial Assistance for Marine Sediment Library* (A020201); 12 months; $11,500

Smith, J. Dungan; *A Field Investigation of Tidal and River Flow Over Non-Uniform Boundaries* (A014178); 24 months; $72,400

WASHINGTON STATE UNIVERSITY

Rosenberg, Yvonne H.; *Quaternary Oceanography and Geology of the Mediterranean Sea Basins* (A016500); 12 months; $18,700

Sorem, Ronald K.; *Investigation of Structure, Geochemistry, Mineralogy and Chemical Composition of Marine Manganese Nodules* (A014231); 24 months; $62,500

Submarine Geophysics

CALIFORNIA

UNIVERSITY OF CALIFORNIA, San Diego

Raitt, Russell W. and Shor, George G. Jr.; *Anisotropy Studies* (A011349001); 6 months; $40,000

CONNECTICUT

DEPARTMENT OF THE NAVY,
NAVY UNDERWATER SOUND LABORATORY

Mellen, Robert H. and Jones, Everett N.; *Deep Water Temperature and Sub-Bottom Research on Lake Tanganyika* (AG209); 12 months; $10,075

FLORIDA

UNIVERSITY OF MIAMI

Ball, Mahlon M.; *Geophysics of Eastern Bahama Crustal Transition Zone* (A019471); 12 months; $64,000

HAWAII

UNIVERSITY OF HAWAII

Malahoff, Alexander; *University of Hawaii Participation in Cruise of R/V Oceanographer, January–April 1970* (A020571); 12 months; $4,300

Woollard, George P. and Sutton, George H.; *Geophysical and Geological Study of the Darwin Rise* (A017879); 12 months; $160,100

MARYLAND

JOHNS HOPKINS UNIVERSITY

Pritchard, Donald W. and Schubel, J. R.; *Study to Assess the Importance of Gas Bubbles in the Marked Attenuation of Seismic Energy by Some Fine-Grained Sediments* (A021169); 5 months; $6,700

MASSACHUSETTS

WOODS HOLE OCEANOGRAPHIC INSTITUTION

Bowin, Carl O.; *Gravity Field at Sea and on Bordering Land Areas* (A012204); 24 months; $81,000

NEW JERSEY

PRINCETON UNIVERSITY

Morgan, W. J.; *Marine Geophysics at Princeton* (A016733); 12 months; $46,800

NEW YORK

COLUMBIA UNIVERSITY

Langseth, Marcus G., Ewing, Maurice and Epp, David E.; *Collection, Reduction and Interpretation of Ocean-Bottom Heat Flow Data* (A018765); 6 months; $19,500

Pitman, Walter and Worzel, J. Lamar; *Acquisition, Reduction and Interpretation of Marine Magnetic Data* (A019030); 6 months; $30,400

Talwani, Manik; *Acquisition of Continuous Gravity Data at Sea—Its Reduction and Interpretation* (A017761); 6 months; $59,900

WASHINGTON

UNIVERSITY OF WASHINGTON

Lister, C. R. B.; *Heat-Flow West of the Juan De Fuca Ridge* (GA1640 001); $900

Marine Chemistry

ALASKA

UNIVERSITY OF ALASKA

Burrell, David C. and Kinney, P. J.; *Logistic Support for Arctic Oceanography Program on U.S.C.G.C. Southwind Cruise to the North Barents Sea* (A024984); 5 months; $3,000

CALIFORNIA

UNIVERSITY OF CALIFORNIA, San Diego

Fisher, Frederick H.; *Electrical Conductance and Chemical Properties of Sea Water* (A018763); 12 months; $59,500

FLORIDA

UNIVERSITY OF MIAMI

Millero, Frank J.; *Study of Ion-Ion Interactions in Seawater* (A017386); 24 months; $48,300

———; (A023701); 6 months; $2,700

Ostlund, H. Gote; *Large Scale Exchange Processes in Sargasso Sea by Tritium Studies* (A018241); 24 months; $80,700

MASSACHUSETTS

WOODS HOLE OCEANOGRAPHIC INSTITUTION

Blumer, Max; *Organic Compounds in the Sea and in Marine Sediments* (A019472); 24 months; $99,100

Hunt, John M.; *Geochemical Ocean Section Study* (A015614); 12 months; $34,700

Mangelsdorf, Paul C. Jr. and Wilson, T. Roger S.; *Study of the Regional Variations of the Major Ions in Sea Water* (A018347); 12 months; $24,400

OHIO

UNIVERSITY OF CINCINNATI

Mark, Harry B. Jr.; *The In Situ Analysis of Trace Metal Ions in Sea Water Systems Employing an Electrochemical Preconcentration* (A025563); 12 months; $12,800

OREGON

OREGON STATE UNIVERSITY

Pytkowicz, Ricardo M.; *Chemical Equilibria in Sea Water* (A017011); 24 months; $95,700

RHODE ISLAND

UNIVERSITY OF RHODE ISLAND

Duce, Robert A.; *Trace Element Enrichment in Sea Surface Films* (A020000); 24 months; $45,100

Kester, Dana R.; *Research on the Colligative Properties of Seawater* (A019940); 24 months; $23,700

Biological Oceanography

ALASKA

UNIVERSITY OF ALASKA

Button, D. K.; *Microbial Growth Kinetics in Natural Waters* (B016540); 24 months; $88,200

Hood, Donald W.; *Oceanographic Research Vessel Operations* (B017160); 12 months; $49,650

BERMUDA

BERMUDA BIOLOGICAL STATION FOR RESEARCH

Deevey, Georgiana B. and Brooks, Albert L.; *Deep-Water Zooplankton of the Sargasso Sea* (B015575); 21 months; $32,100

CALIFORNIA

STANFORD UNIVERSITY

Wheeler, Ellsworth H. Jr.; *Support of the Stanford University Research Vessel* (GB8374 003); $9,000

———; *Support of the Stanford University Research Vessel* (GB8374X002); 12 months; $200,000

Wheeler, Ellsworth S.; *Stanford Biological Oceanography* (GB8408 001); 3 months; $10,000

UNIVERSITY OF CALIFORNIA, Davis

Hand, Cadet and Hamner, William M.; *Evolutionary Ecology of Medusae* (B022851); 24 months; $50,000

Lipps, Jere H.; *Ecologic and Ontogenetic Significance of Minor Element Concentrations in Modern Foraminiferal Tests* (B022852); 12 months; $18,400

UNIVERSITY OF CALIFORNIA, Irvine

Stephens, Grover S.; *Nutrition of Marine Phytoplankton* (B017263); 24 months; $58,800

UNIVERSITY OF CALIFORNIA, San Diego

Benson, Andrew A.; *Support for Physiological Research Ship, R/V Alpha Helix* (B022983); 7 months; $235,100

Benson, Andrew A. and Garey, Walter F.; *Support for Research Program for Physiological Laboratory Ship, R/V Alpha Helix* (B024816); 24 months; $290,000

———; *Equipment for Shipboard Field Work* (B024416); 12 months; $35,000

Hessler, Robert R.; *Deep-Sea Benthic Communities in the Pacific Ocean* (B014488); 24 months; $91,700

OCEANOGRAPHY RESEARCH PROJECTS

Nierenberg, William A. Benson, A. A. and Fager, E. W.; *Support of Graduate Research in Marine Biology and Oceanography* (B014921); 12 months; $67,500

Nierenberg, William A.; *Partial Support of Ship Operations for Research at Sea in Oceanography* (GA1300 003*); 12 months; $275,000

Parker, Frances L. and Phleger, Fred B.; *Applications of Foraminifera to Marine Processes and Marine History* (B021259); 24 months; $96,700

Scholander, Per F.; *Support for Research Program for Physiological Laboratory Ship, R/V Alpha Helix* (GB8400 002); $30,000

Strickland, John D. H. and Beers, John R.; *Micro-Zooplankters in the Marine Food Chain* (B019799); 12 months; $46,800

Tyler, John E.; *Primary Productivity and Radiant Energy* (B023434); 12 months; $5,528

Zobell, Claude E.; *Microbial Metabolism at Deep-Sea Pressures and Temperatures* (B016605); 12 months; $25,000

UNIVERSITY OF CALIFORNIA, Santa Barbara

Childress, James J.; *Studies on the Ecological Physiology of Midwater Crustaceans* (B018929); 24 months; $59,800

Davenport, Demorest; *Behavior of Motile Plankters* (B019645); 12 months; $15,500

UNIVERSITY OF SOUTHERN CALIFORNIA

Savage, Jay M.; *Support of Research Vessel Operation* (GB8206 002*); 12 months; $130,000

UNIVERSITY OF THE PACIFIC

Smith, Edmund H.; *Research Training in Marine Biology, Paleontology and Systematic Zoology* (B019344); 12 months; $9,900

CONNECTICUT

YALE UNIVERSITY

Rhoads, Donald C.; *Factors Controlling the Distribution of Benthic Trophic Groups in Shallow Marine Basins* (B020487); 24 months; $38,700

DISTRICT OF COLUMBIA

DEPARTMENT OF THE NAVY,
NATIONAL OCEANOGRAPHIC DATA CENTER

Slattery, F. L.; *Support of the National Oceanographic Data Center* (AG11500002*); 12 months; $27,500

DEPARTMENT OF THE NAVY,
OFFICE OF NAVAL RESEARCH

Trumbull, Richard; *Support of the Committee on Oceanography of the National Academy of Sciences* (AG221*); 9 months; $10,000

FLORIDA

FLORIDA PRESBYTERIAN COLLEGE

Ferguson, John C.; *Starfish Nutrition* (B014649); 12 months; $11,200

FLORIDA STATE UNIVERSITY

Menzies, Robert J.; *Similarities and Differences Between Atlantic and Pacific Decapod Faunas* (B019384); 12 months; $8,000

Oppenheimer, Carl H.; *Support of R/V Tursiops* (B020954*); 12 months; $33,800

MOTE MARINE LABORATORY

Gilbert, Perry W.; *Operation of a 33–Foot Oceanographic Vessel for Use in Marine Biological Research* (B017114); 24 months; $37,100

NOVA UNIVERSITY

Richardson, William S.; *Support for Operation of R/V Gulfstream* (A011222001*); 14 months; $7,900

Yentsch, Charles S.; *Factors Affecting the Production and Distribution of Chlorophyll Derivatives in the Marine Environment* (B016059); 24 months; $48,900

UNIVERSITY OF MIAMI

Bader, Richard G.; *Support of Research Vessel Operations* (GA4569 002*); 12 months; $325,400

Bunt, John S.; *Relationship of Photosynthesis to Respiration of Ocean Microalgae* (GB15896001); $2,400

———; (B015896); 12 months; $31,400

———; (B15896002); $5,000

Bunt, John S., Cooksey, Keith and Taylor, Barrie; *Physiology of Tropical Marine Blue-Green Algae* (B019800); 24 months; $94,600

Corcoran, Eugene F.; *Effluent Influences of the Mississippi River on the Chemistry and Biology of Adjacent Ocean Systems* (B020168); 12 months; $42,000

Zillioux, Edward J.; *Components of the Food Chain Supporting Larval Tuna* (B019646); 12 months; $37,100

GEORGIA

UNIVERSITY OF GEORGIA

Greene, Albert G. Jr.; *University of Georgia Marine Institute Research Vessels* (GA4497 002*); 12 months; $47,500

Johannes, Robert E.; *Nutrient Relations in Corals* (B014316); 24 months; $65,800

HAWAII

UNIVERSITY OF HAWAII

Caperon, John; *Nutrient Limited Growth and Population Dynamics of Marine Phytoplankton* (B015600); 24 months; $90,700

Chave, Keith E.; *Oceanographic-Ecologic Survey of Fanning Island, Equatorial Pacific* (B015581); 12 months; $32,800

Fournier, Robert O.; *Ecology and Systematics of Phytoflagellates Through an Annual Cycle* (B022628); 24 months; $50,600

Wollard, George P.; *Operational Support of Oceanographic Research Vessels* (GA453 002*); 12 months; $116,000

———; (GA1263 002); $22,600

Young, Richard; *Ecology of Mid-Water Cephalopods* (B020993); 24 months; $58,800

ILLINOIS

FIELD MUSEUM OF NATURAL HISTORY

McCammon, Helen M.; *Nutrient Utilization in Articulate Brachiopods* (B020067); 12 months; $19,000

MARYLAND

JOHNS HOPKINS UNIVERSITY

Pritchard, Donald W.; *Operation of the Research Vessels Ridgely Warfield, Maury and Lydia Louise II* (GA4506 001*); 12 months; $40,000

UNIVERSITY OF MARYLAND

Clark, Eugenie; *General Ecology and Behavior of Garden Eels, 'Gorgasia Sillneri', of the Red Sea* (B022910); 12 months; $9,000

MASSACHUSETTS

CLARK UNIVERSITY

Johansen, H. W.; *Effects of Environment on 'Corallina Officinalis' an Articulated Coralline Alga* (B015663); 24 months; $15,600

MARINE BIOLOGICAL LABORATORY

Carriker, Melbourne R.; *Systematic-Ecologic Studies of the Marine Biota of the Cape Cod Region* (B013250*); 12 months; $40,400

Steinbach, H. Burr; *Training Project in Experimental Marine Botany* (GZ1787*); 7 months; $14,570

WOODS HOLE OCEANOGRAPHIC INSTITUTION

Backus, Richard H.; *Horizontal Distribution of Mesopelagic Fishes in the North Atlantic Ocean and its Determining Factors* (B015764); 24 months; $189,900

Fye, Paul M.; *Support for the Operation of Oceanographic Research Vessels* (GA1298 005*); $10,100

———; (GA1298 004*); 12 months; $408,200

Guillard, Robert R. L.; *Comparative Environmental Physiology of Marine Phytoplankton* (B020488); 24 months; $83,400

Jannasch, Holger W.; *Microbial Transformations in Seawater* (B020956); 24 months; $157,400

Ryther, John H.; *Purchase of Biological Oceanographic Equipment* (B015487); 24 months; $37,300

Sanders, Howard L.; *Biology of the Deep Sea Benthos* (GB6027X003); 12 months; $100,000

Vaccaro, Ralph F.; *Heterotrophic Activity and Distribution of Dissolved Organic Carbon in the Sea* (B016539); 24 months; $75,100

Teal, John M.; *Energy Requirements of Marine Organisms* (B016161); 12 months; $44,200

Wall, David; *Biology and Paleontology of Marine Dinoflagellates and Hystrichospheres* (B020499); 24 months; $99,500

———; (GB7695 001); $12,400

MICHIGAN

UNIVERSITY OF MICHIGAN

Chandler, David C.; *Support of Two Research Vessels* (GA4507 002*); 12 months; $123,500

NEW HAMPSHIRE

UNIVERSITY OF NEW HAMPSHIRE

Croker, Robert A.; *Population Studies of Marine Amphipod Crustaceans* (B018590); 24 months; $21,900

Wood, Langley; *Conference on Pollution of Estuaries* (B023953); 12 months; $1,500

NEW JERSEY

RUTGERS UNIVERSITY

Gardiner, Lion F.; *Ecology and Systematics of the Deep-Sea Tanaidacea* (B020690); 24 months; $33,000

NEW YORK

COLUMBIA UNIVERSITY

Be, Allan W. H.; *Studies of Petropoda and Zooplankton Standing Stock in the South Atlantic* (B019237); 12 months; $26,100

——— *Ecology of Planktonic Foraminifera* (B013237); 24 months; $69,400

Worzel, J. Lamar; *Support of Research Vessels at Lamont-Doherty Geological Observatory* (GA1299 003*); 12 months; $100,500

CUNY, CITY COLLEGE

Lee, John J. and Tietjen, John H.; *Physiological Ecology of Marine Nematodes from Salt Marsh Epiphytic Communities* (B019245); 24 months; $73,800

HASKINS LABORATORIES INC.

Provasoli, Luigi; *Nutritional Relationships among Marine Organisms* (B019143); 24 months; $78,600

LONG ISLAND UNIVERSITY,
C. W. POST CENTER

Cahn, Phyllis H.; *Acoustico-Lateralis Function in Fish Orientation and Communication* (GB7420X001); 2 months; $6,100

RENSSELAER POLYTECHNIC INSTITUTION

Landau, Joseph V.; *Hydrostatic Pressure Effects on the Biosynthesis of Proteins and Nucleic Acids* (B018567*); 12 months; $10,000

NORTH CAROLINA

DUKE UNIVERSITY

Barber, Richard T.; *Scientific Research Training Program* (B017545); 12 months; $182,000

———; *Predoctoral and Postdoctoral Oceanographic Trainee Awards for Research Aboard R/V Eastward* (B017266); 18 months; $41,400

MARINE BIOLOGY

Barber, Richard T. and Newton, John G.; *Vessel Operation, R/V Eastward* (B017889); 12 months; $420,000
Costlow, John D. Jr.; *Research and Training in Marine Biology* (B018841); 12 months; $12,100

WRIGHTSVILLE MARINE BIOMEDICAL LABORATORY
Brauer, Ralph W.; *Conference on 'High Pressure Aquarium' Systems* (B023433); 12 months; $3,800

OREGON

OREGON STATE UNIVERSITY
Burt, Wayne V.; *Research Vessel Operations* (GA934 004*); 12 months; $211,400
Small, Lawrence; *Feeding niches for Neritic Microcrustacea* (B022984); 24 months; $63,500
Curl, Herbert Jr.; *Physiological Ecology of Dinoflagellate Bioluminescence* (B021430); 24 months; $40,100
Hedgepeth, Joel W.; *Research Training at the Marine Science Center* (B020237); 12 months; $8,500
McIntire, C. David; *Littoral Diatom Communities of the Yaquina River Estuary* (B018591); 24 months; $42,500

RHODE ISLAND

UNIVERSITY OF RHODE ISLAND
Knauss, John A.; *Partial Support R/V Trident* (A010030002*); 12 months; $113,600
Sieburth, John McN.; *Algal-Microbial Interactions in the Marine Food Web* (B018000); 24 months; $46,300

TEXAS

TEXAS A & M UNIVERSITY
Geyer, Richard A.; *Operations of Research Vessel Alaminos* (GA4544 003*); 12 months; $63,900
Wilson, William B.; *Growth of Marine Phytoplankton Cultures in a Chemostat* (B014668); 24 months; $47,200

VIRGINIA

COLLEGE OF WILLIAM AND MARY
Mangum, Charlotte P.; *Adaptations of Marine Interidal Invertebrates to Low Oxygen Conditions* (B020335); 24 months; $24,100

OLD DOMINION UNIVERSITY
Marshall, Harold G.; *Ecology of Coccolithophoridaceae in Atlantic Coastal Waters of the United States* (B013906); 24 months; $15,800

WASHINGTON

UNIVERSITY OF WASHINGTON
Dugdale, Richard C.; *Phytoplankton and Herbivorous Zooplankton Production Processes in the Sea* (B020182); 12 months; $210,500
Fernald, Robert L.; *Training Program for Graduate Students in the Marine Sciences at the Friday Harbor Laboratories* (B015353); 12 months; $27,200
Richards, Francis A.; *Oceanographic Vessel Operations* (GA1297 003*); 12 months; $224,200

Marine Biology

CALIFORNIA

CALIFORNIA ACADEMY OF SCIENCES
Eschmeyer, William N.; *Study of the Systematics of Indo-Pacific Scorpionfishes (Family Scorpaenidae)* (B015811); 24 months; $22,400

STANFORD RESEARCH INSTITUTE
Schusterman, R. J.; *Perceptual, Vocal, and Echo-Ranging Behavior of Seals and Sea Lions* (B017988); 6 months; $17,600

STANFORD UNIVERSITY
Epel, David; *Nature of Transformation of Hyaline Layer Substances in Sea Urchin Eggs* (B016155); 12 months; $18,200

UNIVERSITY OF CALIFORNIA, Berkeley
Alfert, Max; *Structure and Function of Cell Organelles During Maturation of Germ Cells of 'Urechis caupo'* (B012970); 18 months; $39,800

UNIVERSITY OF CALIFORNIA, San Diego
Benson, Andrew A.; *Radiochemical Investigations of Sulfocarbohydrate Metabolism* (B015500); 24 months; $46,000
Enns, Theodore; *Facilitated and Passive Gas Transport in Blood and Tissues* (B017494); 24 months; $39,000
Freeman, Gary L.; *Growth and Regeneration at a Cellular and Multicellular Level* (B020073); 12 months; $18,000
Humphreys, Tom D. II; *Biochemical Basis of Specific Cell Association* (B018008); 24 months; $40,000

UNIVERSITY OF CALIFORNIA, San Diego, REVELLE COLLEGE
Kamen, Martin D.; *Molecular Basis of Bacterial Photosynthesis* (GB7033X001); 12 months; $50,000

UNIVERSITY OF SOUTHERN CALIFORNIA
Garth, John S. and Haig, Janet; *Systematic Study of the Porcellanidae* (B016386); 12 months; $6,900

CONNECTICUT

YALE UNIVERSITY
Ramus, J. S.; *Developmental Studies of a Marine Red Alga* (B018144); 24 months; $30,000

FLORIDA

FLORIDA STATE UNIVERSITY
Menzies, Robert J.; *Systematics of Marine Isopod Crustacea* (B015761); 24 months; $38,000

UNIVERSITY OF FLORIDA
Brookbank, J. W.; *DNA Synthesis During the Development of Interordinal Hybrids* (B018006); 24 months; $31,000

UNIVERSITY OF MIAMI
Evans, David H.; *Ion and Water Balance of Euryhaline Teleost Fish* (B016839); 24 months; $50,000
Provenzano, Anthony J. Jr.; *Systematic Studies on Hermit Crabs and Other Decapod Crustaceans* (B7075X 001); 12 months; $44,200
Robins, C. Richard; *Oceanic Fishes of the Tropical Atlantic* (GB7015X001); 12 months; $32,400
Thomas, Lowell P.; *A Monographic Study of the Brittlestars of the Tropical Atlantic* (B016556); 24 months; $29,400

GEORGIA

UNIVERSITY OF GEORGIA
Cormier, Milton J.; *Chemistry and Enzymology of Bioluminescence* (GB7400X001); 12 months; $37,100
Travis, James; *Evolution of Digestive Enzymes* (B017956); 24 months; $32,000

ILLINOIS

UNIVERSITY OF ILLINOIS, Urbana
Whitt, Gregory S.; *Developmental Analyses of Teleost Lactate Dehydrogenase Isozymes* (B016425); 24 months; $45,000

KANSAS

UNIVERSITY OF KANSAS
Moore, Raymond C. and Teichert, Curt; *Treatise on Invertebrate Paleontology* (B018390); 12 months; $66,600

LOUISIANA

TULANE UNIVERSITY
Fingerman, Milton; *Regulation of Chromatophores in Crustaceans* (GB7595X001); 12 months; $25,000

MAINE

UNIVERSITY OF MAINE, Orono
Haynes, Julian F.; *Developmental and Functional Studies of Secretory Cells in Hydrozoan Polyps* (B016046); 24 months; $30,100
McCleave, James D.; *Perception of Weak Magnetic and Electric Fields and Measurement of Biological Clocks of Migratory Fishes* (B015685); 24 months; $35,800

MASSACHUSETTS

AMHERST COLLEGE
Ellis, Charles H. Jr.; *Genetic Control of Morphogenesis in Echinoderms* (B015108); 24 months; $18,800
Plough, Harold H.; *Ascidian Species on the Atlantic Continental Shelf* (GB6902X001); 12 months; $11,100

HARVARD UNIVERSITY
Hastings, J. W.; *Molecular Mechanisms in Biological Clocks* (B016512); 24 months; $96,000

MARINE BIOLOGICAL LABORATORY
Carriker, Melbourne R.; *Systematic-Ecologic Studies of the Marine Biota of the Cape Cod Region* (B013250); 12 months; $40,400
——— *Systematic-Ecologic Studies of the Marine Biota of the Cape Cod Region* (GB13250001*); $9,600

UNIVERSITY OF MASSACHUSETTS
Potswald, Herbert E.; *Regeneration in Annelids* (B018429); 24 months; $28,000

WOODS HOLE OCEANOGRAPHIC INSTITUTION
Carey, Francis G.; *Ecology of Warm-Bodied Fishes* (B014282); 24 months; $81,700
Williams, Timothy C.; *Marine Birds—Navigation and Flight Physiology* (B013246); 24 months; $55,000

MICHIGAN

UNIVERSITY OF MICHIGAN
Miller, Robert R.; *Systematics and Evolution of Poeciliid and Related Fishes* (GB6272X001); 10 months; $16,000

MINNESOTA

UNIVERSITY OF MINNESOTA
Herman, William S.; *Structure and Function of the Neuroendocrine System of Limulus Polyphemus* (B016607); 24 months; $39,000

MISSISSIPPI

GULF COAST RESEARCH LABORATORIES
Dawson, C. E.; *Distribution and Ecology of the Microdesmidae* (B015295); 24 months; $14,800

NEW HAMPSHIRE

DARTMOUTH COLLEGE
Ballard, William W.; *Comparative Studies of Gastrulation in Fish Embryos* (B015293); 24 months; $24,800

NEW JERSEY

PRINCETON UNIVERSITY
Johnson, Frank H. and Shimomura, Osamu; *Biochemistry of Luminescence Systems* (B015092); 24 months; $100,000

NEW YORK

CUNY, BROOKLYN COLLEGE
Collier, J. R.; *Gene Transcription in the Embryo of 'Ilyanassa Obsoleta'* (B015290); 24 months; $35,000

CUNY, HUNTER COLLEGE
Haschemeyer, Audrey E. V.; *Control of Protein Synthesis in Temperature Acclimation of Marine Organisms* (B014570); 20 months; $29,000

OCEANOGRAPHY RESEARCH PROJECTS

CUNY, MT. SINAI SCHOOL OF MEDICINE
Berger, E. R.; *Quantitative Electron Microscopic Analysis of Sea Urchin Gametes and Early Embryos* (B015659); 24 months; $30,000

STATE UNIVERSITY AT STONY BROOK
Edmunds, Leland N. Jr. *Periodic Enzyme Synthesis and the Control of the Ciradian Rhythm of Cell Division* (B012474); 24 months; $43,000

UNION COLLEGE
Rappaport, Raymond Jr.; *Mechanism of Cytokinesis in Animal Cells* (B014743); 24 months; $15,500

OREGON

OREGON STATE UNIVERSITY
Quatrano, Ralph S.; *Role of Gene Products in Cell Growth and Morphogenesis* (B014835); 24 months; $40,100

UNIVERSITY OF OREGON
Hoyle, Graham; *Neural Mechanisms Underlying Behavior* (B016962); 24 months; $83,300

PENNSYLVANIA

ACADEMY OF NATURAL SCIENCES OF PHILADELPHIA
Bohlke, James E.; *The Systematics of the Anguilliformes* (B017736); 24 months; $65,000
Foster, Neal R.; *Comparative Biosystematic Studies on Killifishes (Pisces, Cyprinodontidae)* (B015159); 24 months; $35,000
Hart, C. W. Jr. and Hart, Dabney G.; *A Systematic Study of Entocytherid Ostracods* (GB6943X001); 12 months; $12,600
Tyler, James C.; *Monograph on the Fishes of the Order Plectognathi* (B016190); 24 months; $36,000

PENNSYLVANIA STATE UNIVERSITY
Dunson, William A.; *Extrarenal Ionic Regulation* (B016653); 24 months; $23,000

RHODE ISLAND

BROWN UNIVERSITY
Green, Jonathan P.; *Morphological Color Changes in Crustacea* (B017994); 12 months; $10,000

VIRGINIA

WASHINGTON & LEE UNIVERSITY
Hickman, Cleveland P.; *Glomerular-Tubular Interactions in the Fish Kidney* (B017918); 24 months; $34,000

WASHINGTON

UNIVERSITY OF WASHINGTON
Kohn, Alan J.; *Indo-West Pacific Marine Mollusks of the Family Conidae* (B017735); 24 months; $32,000
Martin, Arthur W.; *Comparative Physiology of Cephalopod Reproduction and Excretion, and of Arachnid Circulation* (B017539); 24 months; $47,500

WISCONSIN

UNIVERSITY OF WISCONSIN, Madison
Hasler, Arthur D.; *Spatial Orientation of Fishes and its Sensory Bases* (GB7616X001); 12 months; $44,500

Oceanography, General

ALASKA

UNIVERSITY OF ALASKA
Hood, Donald W.; *Oceanographic Research Vessel Operations* (B017160*); 12 months; $49,650

CALIFORNIA

UNIVERSITY OF CALIFORNIA, San Diego
Nierenberg, William A.; *Partial Support of Ship Operations for Research at Sea in Oceanography* (GA1300 003*); 12 months; $825,000

UNIVERSITY OF SOUTHERN CALIFORNIA
Savage, Jay M., *Support of Research Vessel Operations* (GB8206 002*); 12 months; $70,000

DISTRICT OF COLUMBIA

DEPARTMENT OF THE NAVY, NATIONAL OCEANOGRAPHIC DATA CENTER
Austin, Thomas S.; *Support to World Data Center-A, Oceanography* (AG108 002); 12 months; $85,000
Slattery, F. L.; *Support of the National Oceanographic Data Center* (AG115 002); 12 months; $247,500

DEPARTMENT OF THE NAVY, OFFICE OF NAVAL RESEARCH
Trumbull, Richard; *Support of the Committee on Oceanography of the National Academy of Sciences* (AG221); 9 months; $10,000

NATIONAL ACADEMY OF SCIENCES
Vetter, Richard C.; *Task Order for Support of the Scientific Committee on Oceanic Research* (C310068007); 12 months; $3,000

FLORIDA

FLORIDA STATE UNIVERSITY
Oppenheimer, Carl H.; *Support of R/V Tursiops* (B020954*); 12 months; $15,000

NOVA UNIVERSITY
Richardson, William S.; *Support for Operation of R/V Gulfstream* (A011222001); 14 months; $11,900

UNIVERSITY OF MIAMI
Bader, Richard G.; *Support of Research Vessel Operations* (GA4569 002*); 12 months; $444,800

GEORGIA

UNIVERSITY OF GEORGIA
Greene, Albert G. Jr.; *University of Georgia Marine Institute Research Vessels* (GA4497 002); 12 months; $47,400

HAWAII

UNIVERSITY OF HAWAII
Woollard, George P.; *Operational Support of Oceanographic Research Vessels* (GA4530 002*); 12 months; $484,000

MARYLAND

JOHNS HOPKINS UNIVERSITY
Pritchard, Donald W.; *Operation of the Research Vessels Ridgely Warfield, Maury and Lydia Louise II* (GA4506 001); 12 months; $53,100

MASSACHUSETTS

WOODS HOLE OCEANOGRAPHIC INSTITUTION
Fye, Paul M.; *Support for the Operation of Oceanographic Research Vessels* (GA1298 004); 12 months; $691,800

MICHIGAN

UNIVERSITY OF MICHIGAN
Chandler, David C.; *Support of Two Research Vessels* (GA4507 002*); 12 months; $125,000

NEW YORK

COLUMBIA UNIVERSITY
Worzel, J. Lamar; *Support of Research Vessels at Lamont-Doherty Geological Observatory* (GA1299 003*); 12 months; $569,500

OREGON

OREGON STATE UNIVERSITY
Burt, Wayne V.; *Research Vessel Operations* (GA934 004*); 12 months; $208,600

RHODE ISLAND

UNIVERSITY OF RHODE ISLAND
Knauss, John A.; *Partial Support R/V Trident* (A010030002); 12 months; $170,400

TEXAS

TEXAS A & M UNIVERSITY
Geyer, Richard A.; *Operations of Research Vessel Alaminos* (GA4544 003); 12 months; $173,600

WASHINGTON

UNIVERSITY OF WASHINGTON
Richards, Francis A.; *Oceanographic Vessel Operations* (GA1297 003*); 12 months; $309,700

DEVELOPMENT AND IMPROVEMENT OF INSTITUTIONAL SCIENCE PROGRAMS

Institutional Grants for Science

ALABAMA

ALABAMA AGRICULTURAL AND MECHANICAL COLLEGE; (GU3198); $10,180
AUBURN UNIVERSITY; (GU3199); $55,825
SPRING HILL COLLEGE; (GU3200); $2,000
TROY STATE UNIVERSITY; (GU3201); $10,495
TUSKEGEE INSTITUTE; (GU3202); $12,865
UNIVERSITY OF ALABAMA, Birmingham; (GU3203); $28,465
UNIVERSITY OF ALABAMA, Huntsville; (GU3204); $18,845
UNIVERSITY OF ALABAMA, Tuscaloosa; (GU3205); $20,445

ALASKA

ALASKA METHODIST UNIVERSITY; (GU3206); $5,000
UNIVERSITY OF ALASKA; (GU3207); $80,440

ARIZONA

ARIZONA STATE UNIVERSITY; (GU3208); $38,985
NORTHERN ARIZONA UNIVERSITY; (GU3209); $10,540
PRESCOTT COLLEGE; (GU3210); $6,000
UNIVERSITY OF ARIZONA; (GU3211); $78,180

ARKANSAS

ARKANSAS STATE UNIVERSITY; (GU3212); $2,000
OUACHITA BAPTIST UNIVERSITY, (GU3213); $10,338
UNIVERSITY OF ARKANSAS; (GU3214); $49,825

CALIFORNIA

CABRILLO COLLEGE; (GU3215); $2,000
CALIFORNIA INSTITUTE OF TECHNOLOGY; (GU3216); $102,767
CALIFORNIA STATE COLLEGE AT FULLERTON; (GU3217); $11,058
CALIFORNIA STATE COLLEGE AT HAYWARD; (GU3218); $11,598
CALIFORNIA STATE COLLEGE AT LONG BEACH; (GU3219); $13,740
CALIFORNIA STATE COLLEGE AT LOS ANGELES; (GU3220); $16,545
CALIFORNIA STATE COLLEGE, San Bernardino; (GU3221); $10,495
CALIFORNIA STATE POLYTECHNIC COLLEGE; (GU3222); $10,180
——— ; (GU3223); $2,000
CERRITOS COLLEGE; (GU3224); $2,000
CHICO STATE COLLEGE; (GU3225); $10,315
CLAREMONT GRADUATE SCHOOL AND UNIVERSITY; (GU3226); $16,465
CLAREMONT MEN'S COLLEGE; (GU3227); $6,000
COLLEGE OF NOTRE DAME; (GU3228); $7,000
DIABLO VALLEY COLLEGE; (GU3229); $6,000
FRESNO STATE COLLEGE; (GU3230); $17,045
FULLERTON JUNIOR COLLEGE; (GU3231);$2,000
HARVEY MUDD COLLEGE; (GU3232); $13,924
HUMBOLDT STATE COLLEGE; (GU3233); $10,810
IMMACULATE HEART COLLEGE; (GU3234); $4,000
LOMA LINDA UNIVERSITY; (GU3235); $12,825
LOS ANGELES VALLEY COLLEGE; (GU3236); $2,000
MILLS COLLEGE; (GU3237); $10,518
OCCIDENTAL COLLEGE; (GU3238); $6,000
PACIFIC OAKS COLLEGE; (GU3239); $10,315
POMONA COLLEGE; (GU3240); $11,260
SACRAMENTO STATE COLLEGE; (GU3241); $11,938
SAN DIEGO STATE COLLEGE; (GU3242); $25,709
SAN FERNANDO VALLEY STATE COLLEGE; (GU3243); $10,518
SAN FRANCISCO STATE COLLEGE; (GU3244); $17,588
SAN JOSE STATE COLLEGE; (GU3245); $23,565
SANTA BARBARA CITY COLLEGE; (GU3246); $13,005
SONOMA STATE COLLEGE; (GU3247); $2,000
ST. MARY'S COLLEGE OF CALIFORNIA; (GU3248); $17,265
STANFORD UNIVERSITY; (GU3249); $116,953
UNIVERSITY OF CALIFORNIA, Berkeley; (GU3251); $114,783
UNIVERSITY OF CALIFORNIA, Davis; (GU3252); $74,540
UNIVERSITY OF CALIFORNIA, Irvine; (GU3253); $44,518
UNIVERSITY OF CALIFORNIA, Los Angeles; (GU3254); $113,084
UNIVERSITY OF CALIFORNIA, Riverside; (GU3255); $64,027
UNIVERSITY OF CALIFORNIA, San Diego; (GU3256); $112,005
UNIVERSITY OF CALIFORNIA, Santa Barbara; (GU3257); $76,802
UNIVERSITY OF CALIFORNIA, Santa Cruz; (GU3258); $31,961
UNIVERSITY OF CALIFORNIA SYSTEMS OFFICE; (GU3250); $32,983
UNIVERSITY OF REDLANDS; (GU3260); $11,890
UNIVERSITY OF SAN FRANCISCO; (GU3261); $12,225
UNIVERSITY OF SANTA CLARA; (GU3262); $11,553
UNIVERSITY OF SOUTHERN CALIFORNIA; (GU3263); $82,250
UNIVERSITY OF THE PACIFIC; (GU3259); $7,000

COLORADO

COLORADO COLLEGE; (GU3264); $10,428
COLORADO SCHOOL OF MINES; (GU3265); $16,106
COLORADO STATE UNIVERSITY; (GU3267); $84,830
LORETTO HEIGHTS COLLEGE; (GU3268); $11,485
TEMPLE BUELL COLLEGE; (GU3269); $10,405
UNIVERSITY OF COLORADO; (GU3270); $89,160
UNIVERSITY OF DENVER; (GU3271); $73,415
UNIVERSITY OF NORTHERN COLORADO; (GU3266); $11,148

CONNECTICUT

CENTRAL CONNECTICUT STATE COLLEGE; (GU3272); $2,000
CONNECTICUT COLLEGE; (GU3273); $10,630
FAIRFIELD UNIVERSITY; (GU3274); $10,630
NEW ENGLAND INSTITUTE; (GU3275); $12,445
ST. JOSEPH COLLEGE; (GU3276); $10,135
TRINITY COLLEGE; (GU3277); $15,924
UNIVERSITY OF CONNECTICUT; (GU3278); $47,728
UNIVERSITY OF HARTFORD; (GU3279); $11,643
WESLEYAN UNIVERSITY; (GU3280); $22,936
WESTERN CONNECTICUT STATE COLLEGE; (GU3281); $2,000
YALE UNIVERSITY; (GU3282); $96,300

DELAWARE

UNIVERSITY OF DELAWARE; (GU3283); $52,355

DISTRICT OF COLUMBIA

AMERICAN UNIVERSITY; (GU3284); $41,978
CATHOLIC UNIVERSITY OF AMERICA; (GU3285); $48,905
GALLAUDET COLLEGE; (GU3286); $14,085
GEORGE WASHINGTON UNIVERSITY; (GU3287); $68,425
GEORGETOWN UNIVERSITY; (GU3288); $35,485
HOWARD UNIVERSITY; (GU3289); $19,905
WASHINGTON TECHNICAL INSTITUTE; (GU3290); $11,710

FLORIDA

EMBRY-RIDDLE AERONAUTICAL INSTITUTE; (GU3291); $12,565
FLORIDA AGRICULTURAL & MECHANICAL UNIVERSITY; (GU3292); $10,284
FLORIDA ATLANTIC UNIVERSITY; (GU3293); $22,393
FLORIDA INSTITUTE OF TECHNOLOGY; (GU3294); $10,923
FLORIDA PRESBYTERIAN COLLEGE; (GU3295); $11,350
FLORIDA STATE UNIVERSITY; (GU3296); $99,445
FLORIDA TECHNOLOGICAL UNIVERSITY; (GU3297); $10,068
NEW COLLEGE; (GU3298); $10,113
NOVA UNIVERSITY; (GU3299); $17,885
STETSON UNIVERSITY; (GU3300); $8,000
UNIVERSITY OF FLORIDA; (GU3301); $85,425
UNIVERSITY OF MIAMI; (GU3302); $87,345
UNIVERSITY OF SOUTH FLORIDA; (GU3303); $17,745
UNIVERSITY OF WEST FLORIDA; (GU3304); $10,315

GEORGIA

ATLANTA UNIVERSITY; (GU3305); $11,800
AUGUSTA COLLEGE; (GU3306); $8,000
EMORY UNIVERSITY; (GU3307); $41,925
FORT VALLEY STATE COLLEGE; (GU3308); $10,338
GEORGIA INSTITUTE OF TECHNOLOGY; (GU3309); $68,765
GEORGIA SOUTHERN COLLEGE; (GU3310); $11,058
GEORGIA STATE UNIVERSITY; (GU3311); $12,065
MEDICAL COLLEGE OF GEORGIA; (GU3312); $12,385
MOREHOUSE COLLEGE; (GU3313); $10,405

DEVELOPMENT AND IMPROVEMENT OF INSTITUTIONAL SCIENCE PROGRAMS

UNIVERSITY OF GEORGIA; (GU3314); $82,305

WEST GEORGIA COLLEGE; (GU3315); $10,248

HAWAII

UNIVERSITY OF HAWAII; (GU3316); $83,840

IDAHO

IDAHO STATE UNIVERSITY; (GU3317); $11,899

UNIVERSITY OF IDAHO; (GU3318); $29,825

ILLINOIS

AUGUSTANA COLLEGE; (GU3319); $10,630

BRADLEY UNIVERSITY; (GU3320); $6,000

CITY COLLEGES OF CHICAGO; (GU3321); $13,405

DEPAUL UNIVERSITY; (GU3322); $13,565

EASTERN ILLINOIS UNIVERSITY; (GU3323); $10,248

ELMHURST COLLEGE; (GU3324); $10,270

EUREKA COLLEGE; (GU3325); $2,000

ILLINOIS COLLEGE; (GU3326); $2,000

ILLINOIS INSTITUTE OF TECHNOLOGY; (GU3327); $53,995

ILLINOIS STATE UNIVERSITY; (GU3328); $11,125

KENDALL COLLEGE; (GU3329); $5,000

KNOX COLLEGE; (GU3330); $10,743

LAKE FOREST COLLEGE; (GU3331); $4,000

LEWIS COLLEGE; (GU3332); $5,000

LOYOLA UNIVERSITY; (GU3333); $14,285

MACMURRAY COLLEGE; (GU3334); $6,000

MONMOUTH COLLEGE; (GU3335); $10,248

MUNDELEIN COLLEGE; (GU3336); $10,135

NORTH CENTRAL COLLEGE; (GU3337); $4,000

NORTHERN ILLINOIS UNIVERSITY; (GU3338); $24,933

NORTHWESTERN UNIVERSITY; (GU3339); $84,195

QUINCY COLLEGE; (GU3340); $10,248

ROOSEVELT UNIVERSITY; (GU3341); $10,990

ROSARY COLLEGE; (GU3342); $8,000

SOUTHERN ILLINOIS UNIVERSITY; (GU3343); $24,305

ST. PROCOPIUS COLLEGE; (GU3344); $11,643

ST. XAVIER COLLEGE; (GU3345); $10,068

UNIVERSITY OF CHICAGO; (GU3346); $109,715

UNIVERSITY OF ILLINOIS SYSTEM OFFICE; (GU3348); $24,636

UNIVERSITY OF ILLINOIS, Chicago Circle; (GU3347); $32,405

UNIVERSITY OF ILLINOIS, Urbana; (GU3349); $115,022

WESTERN ILLINOIS UNIVERSITY; (GU3350); $10,158

INDIANA

BALL STATE UNIVERSITY; (GU3351); $10,113

DEPAUW UNIVERSITY; (GU3352); $10,203

EARLHAM COLLEGE; (GU3353); $10,293

HANOVER COLLEGE; (GU3354); $4,000

INDIANA STATE UNIVERSITY; (GU3355); $11,463

INDIANA UNIVERSITY, Bloomington; (GU3356); $82,402

MANCHESTER COLLEGE; (GU3357); $10,090

PURDUE UNIVERSITY; (GU3358); $97,695

ROSE POLYTECHNIC INSTITUTE; (GU3359); $10,968

ST. JOSEPH'S COLLEGE; (GU3360); $10,158

TRI-STATE COLLEGE; (GU3361); $2,000

UNIVERSITY OF NOTRE DAME; (GU3362); $65,455

VALPARAISO UNIVERSITY; (GU3363); $8,000

WABASH COLLEGE; (GU3364); $10,023

IOWA

CENTRAL COLLEGE; (GU3366); $10,720

CLARKE COLLEGE; (GU3367); $10,248

COE COLLEGE; (GU3368); $10,765

CORNELL COLLEGE; (GU3369); $10,000

DORDT COLLEGE; (GU3370); $10,068

DRAKE UNIVERSITY; (GU3371); $11,215

GRINNELL COLLEGE; (GU3372); $11,463

IOWA STATE UNIVERSITY; (GU3373); $81,910

IOWA WESLEYAN COLLEGE; (GU3374); $10,720

KIRKWOOD COMMUNITY COLLEGE; (GU3365); $10,585

LORAS COLLEGE; (GU3375); $10,743

LUTHER COLLEGE; (GU3376); $5,000

MARYCREST COLLEGE; (GU3377); $10,585

UNIVERSITY OF IOWA; (GU3378); $81,465

UNIVERSITY OF NORTHERN IOWA; (GU3379); $10,203

WESTMAR COLLEGE; (GU3380); $2,000

KANSAS

BETHANY COLLEGE; (GU3381); $2,000

BETHEL COLLEGE; (GU3382); $10,090

FORT HAYS KANSAS STATE COLLEGE; (GU3383); $10,045

FRIENDS UNIVERSITY; (GU3384); $2,000

KANSAS STATE COLLEGE OF PITTSBURG; (GU3385); $7,000

KANSAS STATE TEACHERS COLLEGE; (GU3386); $10,315

KANSAS STATE UNIVERSITY; (GU3387); $59,645

ST. BENEDICT'S COLLEGE; (GU3388); $10,293

STERLING COLLEGE; (GU3389); $2,000

UNIVERSITY OF KANSAS; (GU3390); $65,652

WASHBURN UNIVERSITY OF TOPEKA; (GU3391); $7,000

WICHITA STATE UNIVERSITY; (GU3392); $10,608

KENTUCKY

BELLARMINE-URSULINE COLLEGE; (GU3393); $8,000

BEREA COLLEGE; (GU3394); $1,000

CAMPBELLSVILLE COLLEGE; (GU3395); $2,000

CENTRE COLLEGE OF KENTUCKY; (GU3396); $10,090

EASTERN KENTUCKY UNIVERSITY; (GU3397); $10,000

KENTUCKY STATE COLLEGE; (GU3398); $10,203

MOREHEAD STATE UNIVERSITY; (GU3399); $10,675

MURRAY STATE UNIVERSITY; (GU3400); $10,338

UNIVERSITY OF KENTUCKY; (GU3401); $78,920

UNIVERSITY OF LOUISVILLE; (GU3402); $27,605

WESTERN KENTUCKY UNIVERSITY; (GU3403); $2,000

LOUISIANA

CENTENARY COLLEGE OF LOUISIANA; (GU3404); $11,035

DELGADO COLLEGE; (GU3405); $13,185

FRANCIS T. NICHOLLS STATE COLLEGE; (GU3406); $12,534

GRAMBLING COLLEGE; (GU3407); $10,428

LOUISIANA STATE UNIVERSITY, Baton Rouge; (GU3410); $69,175

LOUISIANA STATE UNIVERSITY, New Orleans; (GU3411); $18,945

LOUISIANA POLYTECHNIC INSTITUTE; (GU3408); $13,222

LOYOLA UNIVERSITY; (GU3412); $12,186

NORTHEAST LOUISIANA STATE COLLEGE; (GU3413); $10,225

NORTHWESTERN STATE COLLEGE OF LOUISIANA; (GU3414); $10,113

SOUTHERN UNIVERSITY; (GU3415); $18,085

TULANE UNIVERSITY; (GU3416); $42,515

XAVIER UNIVERSITY OF LOUISIANA; (GU3417); $11,013

MAINE

BOWDOIN COLLEGE; (GU3418); $10,450

COLBY COLLEGE; (GU3419); $4,000

UNIVERSITY OF MAINE, Orono; (GU3420); $32,440

MARYLAND

CHARLES COUNTY COMMUNITY COLLEGE; (GU3421); $10,788

FROSTBURG STATE COLLEGE; (GU3422); $2,000

GOUCHER COLLEGE; (GU3423); $10,450

JOHNS HOPKINS UNIVERSITY; (GU3424); $90,215

LOYOLA COLLEGE; (GU3425); $10,495

MORGAN STATE COLLEGE; (GU3426); $10,090

UNIVERSITY OF MARYLAND; (GU3427); $100,845

WASHINGTON COLLEGE; (GU3428); $6,000

WOODSTOCK COLLEGE; (GU3409); $21,845

MASSACHUSETTS

AMHERST COLLEGE; (GU3430); $22,042

BOSTON COLLEGE; (GU3431); $27,025

BOSTON UNIVERSITY; (GU3432); $31,373

BRANDEIS UNIVERSITY; (GU3433); $65,141

CLARK UNIVERSITY; (GU3434); $30,572

COLLEGE OF THE HOLY CROSS; (GU3435); $10,810

EMMANUEL COLLEGE; (GU3436); $16,005

GORDON COLLEGE; (GU3437); $10,315

HARVARD UNIVERSITY; (GU3438); $111,121

LOWELL TECHNOLOGICAL INSTITUTE; (GU3439); $24,905

MASSACHUSETTS INSTITUTE OF TECHNOLOGY; (GU3440); $138,967

MERRIMACK COLLEGE; (GU3441); $10,203

MT. HOLYOKE COLLEGE; (GU3442); $10,450

NORTHEASTERN UNIVERSITY; (GU3443); $38,320

SMITH COLLEGE; (GU3444); $11,148

SOUTHEASTERN MASSACHUSETTS UNIVERSITY; (GU3445); $10,450

SPRINGFIELD COLLEGE; (GU3446); $2,000

TUFTS UNIVERSITY; (GU3447); $39,335

UNIVERSITY OF MASSACHUSETTS; (GU3448); $74,475

WELLESLEY COLLEGE; (GU3449); $10,495

WENTWORTH INSTITUTE; (GU3450); $15,685

WHEATON COLLEGE; (GU3451); $7,000

WILLIAMS COLLEGE; (GU3452); $10,000

WORCESTER POLYTECHNIC INSTITUTE; (GU3453); $17,340

MICHIGAN

ALBION COLLEGE; (GU3454); $10,135

ALMA COLLEGE; (GU3455); $6,000

CALVIN COLLEGE; (GU3456); $10,090

CENTRAL MICHIGAN UNIVERSITY; (GU3457); $11,238

CONCORDIA LUTHERAN JUNIOR COLLEGE; (GU3458); $2,000

DELTA COLLEGE; (GU3459); $2,000

EASTERN MICHIGAN UNIVERSITY; (GU3460); $10,225

GRAND VALLEY STATE COLLEGE; (GU3461); $2,000

HOPE COLLEGE; (GU3462); $10,135

INSTITUTIONAL GRANTS FOR SCIENCE

KALAMAZOO COLLEGE; (GU3463); $11,089
MERCY COLLEGE OF DETROIT; (GU3464); $10,720
MICHIGAN STATE UNIVERSITY; (GU3465); $102,811
MICHIGAN TECHNOLOGICAL UNIVERSITY; (GU3466;) $15,825
NORTHERN MICHIGAN UNIVERSITY; (GU3467); $2,000
OAKLAND UNIVERSITY; (GU3468); $19,985
UNIVERSITY OF DETROIT; (GU3469); $12,605
UNIVERSITY OF MICHIGAN; (GU3470); $114,132
WAYNE STATE UNIVERSITY; (GU3471); $55,499
WESTERN MICHIGAN UNIVERSITY; (GU3472); $15,505

MINNESOTA

AUGSBURG COLLEGE; (GU3473); $4,000
CARLETON COLLEGE; (GU3474); $12,263
COLLEGE OF ST. BENEDICT (GU3475); $2,000
COLLEGE OF ST. CATHERINE; (GU3476); $2,000
GUSTAVUS ADOLPHUS COLLEGE; (GU3477); $10,000
HAMLINE UNIVERSITY; (GU3478); $10,270
MACALESTER COLLEGE; (GU3479); $10,113
MOORHEAD STATE COLLEGE; (GU3480); $10,113
ST. MARY'S COLLEGE; (GU3481); $10,518
ST. OLAF COLLEGE; (GU3482); $10,810
UNIVERSITY OF MINNESOTA; (GU3483); $106,437
WINONA STATE COLLEGE; (GU3484); $10,000

MISSISSIPPI

ALCORN AGRICULTURAL & MECHANICAL COLLEGE; (GU3485); $12,105
MARY HOLMES COLLEGE; (GU3486); $16,785
MISSISSIPPI STATE UNIVERSITY; (GU3487); $52,995
TOUGALOO COLLEGE; (GU3488); $2,000
UNIVERSITY OF MISSISSIPPI; (GU3489); $20,725
UNIVERSITY OF SOUTHERN MISSISSIPPI; (GP3490); $10,698

MISSOURI

AVILA COLLEGE; (GU3491); $2,000
LINCOLN UNIVERSITY; (GU3492); $10,180
MINERAL AREA COLLEGE; (GU3493); $2,000
NORTHWEST MISSOURI STATE COLLEGE; (GU3494); $4,000
ROCKHURST COLLEGE; (GU3495); $2,000
SOUTHEAST MISSOURI STATE COLLEGE; (GU3496); $9,000
SOUTHWEST BAPTIST COLLEGE; (GU3497); $2,000
SOUTHWEST MISSOURI STATE COLLEGE; (GU3498); $10,158
ST. LOUIS UNIVERSITY; (GU3499); $26,034
UNIVERSITY OF MISSOURI, Columbia; (GU 3500); $66,575
UNIVERSITY OF MISSOURI, Kansas City; (GU3501); $13,865
UNIVERSITY OF MISSOURI, Rolla; (GU3502); $30,953
UNIVERSITY OF MISSOURI, St. Louis; (GU3503); $11,260
WASHINGTON UNIVERSITY; (GU3504); $70,675

MONTANA

MONTANA COLLEGE OF MINERAL SCIENCE; (GU3505); $10,158
MONTANA STATE UNIVERSITY; (GU3506); $43,465
UNIVERSITY OF MONTANA; (GU3507); $25,725

NEBRASKA

CHADRON STATE COLLEGE; (GU3508); $11,530
CONCORDIA TEACHERS COLLEGE; (GU3509); $9,000
CREIGHTON UNIVERSITY; (GU3510); $10,563
DANA COLLEGE; (GU3511); $7,000
NEBRASKA WESLEYAN UNIVERSITY; (GU3512); $2,000
UNIVERSITY OF NEBRASKA SYSTEM OFFICE; (GU3513); $16,245
UNIVERSITY OF NEBRASKA, Lincoln; (GU3514); $48,595
UNIVERSITY OF NEBRASKA, Omaha; (GU3515); $10,135

NEVADA

UNIVERSITY OF NEVADA, Reno Campus; (GU3517); $34,245
UNIVERSITY OF NEVADA SYSTEM OFFICE (GU3516); $38,740

NEW HAMPSHIRE

DARTMOUTH COLLEGE; (GU3518); $58,422
UNIVERSITY OF NEW HAMPSHIRE; (GU3519); $50,955

NEW JERSEY

COLLEGE OF ST. ELIZABETH; (GU3520); $3,000
DREW UNIVERSITY; (GU3521); $4,000
FAIRLEIGH DICKINSON UNIVERSITY; (GU3522); $10,315
MERCER COUNTY COMMUNITY COLLEGE; (GU3523); $9,000
MONMOUTH COLLEGE; (GU3524); $2,000
NEW JERSEY COLLEGE OF MEDICINE AND DENTISTRY; (GU3525); $11,440
NEWARK COLLEGE OF ENGINEERING; (GU3526); $12,758
PRINCETON UNIVERSITY; (GU3527); $98,910
RUTGERS UNIVERSITY; (GU3528); $78,310
SETON HALL UNIVERSITY; (GU3529); $10,720
STEVENS INSTITUTE OF TECHNOLOGY; (GU3530); $46,825

NEW MEXICO

EASTERN NEW MEXICO UNIVERSITY; (GU3531); $11,355
NEW MEXICO HIGHLANDS UNIVERSITY; (GU3532); $10,045
NEW MEXICO INSTITUTE OF MINING AND TECHNOLOGY; (GU3533); $41,365
NEW MEXICO STATE UNIVERSITY; (GU3534); $71,905
ST. JOHN'S COLLEGE; (GU3535); $2,000
UNIVERSITY OF ALBUQUERQUE; (GU3536); $11,170
UNIVERSITY OF NEW MEXICO; (GU3537); $67,655
WESTERN NEW MEXICO UNIVERSITY; (GU3538); $9,000

NEW YORK

ADELPHI UNIVERSITY; (GU3539); $17,436
AGRICULTURAL AND TECHNICAL COLLEGE, COBLESKILL; (GU3578); $10,225
ALBANY MEDICAL COLLEGE; (GU3540); $12,705
BARNARD COLLEGE; (GU3541); $11,688
CLARKSON COLLEGE OF TECHNOLOGY; (GU3542); $21,109
COLLEGE AT BROCKPORT; (GU3579); $10,203
COLLEGE AT BUFFALO; (GU3580); $12,565
COLLEGE AT CORTLAND; (GU3581); $10,068
COLLEGE AT FREDONIA; (GU3582); $11,800
COLLEGE AT NEW PALTZ; (GU3583); $3,000
COLLEGE AT ONEONTA; (GU3584); $2,000
COLLEGE AT PLATTSBURGH; (GU3585); $4,000
COLLEGE OF FORESTRY, SYRACUSE; (GU3586); $16,325
COLGATE UNIVERSITY; (GU3543); $10,473
COLUMBIA UNIVERSITY; (GU3544); $112,761
COOPER UNION; (GU3545); $10,405
CORNELL UNIVERSITY; (GU3546); $106,359
CUNY, BRONX COMMUNITY COLLEGE; (GU3548); $12,525
CUNY, BROOKLYN COLLEGE; (GU3549); $11,643
CUNY, CENTRAL SYSTEM OFFICE; (GU3547); $20,685
CUNY, CITY COLLEGE; (GU3550); $26,745
CUNY, HUNTER COLLEGE; (GU3552); $19,858
CUNY, QUEENS COLLEGE; (GU3553); $13,325
FORDHAM UNIVERSITY; (GU3554); $14,805
HAMILTON COLLEGE; (GU3555); $6,000
HOFSTRA UNIVERSITY; (GU3556); $9,000
ITHACA COLLEGE; (GU3557); $5,000
LEHMAN COLLEGE; (GU3551); $10,113
LEMOYNE COLLEGE; (GU3558); $10,000
LONG ISLAND UNIVERSITY; (GU3559); $13,845
MANHATTAN COLLEGE; (GU3560); $14,325
MANHATTANVILLE COLLEGE; (GU3561) $11,485
MOHAWK VALLEY COMMUNITY COLLEGE; (GU3562); $2,000
NAZARETH COLLEGE OF ROCHESTER; (GU3563); $2,000
NEW YORK INSTITUTE OF TECHNOLOGY; (GU3564); $23,845
NEW YORK MEDICAL COLLEGE; (GU3565); $18,505
NEW YORK UNIVERSITY; (GU3566); $104,308
NOTRE DAME COLLEGE, Staten Island; (GU3567); $2,000
POLYTECHNIC INSTITUTE OF BROOKLYN; (GU3568); $43,448
RENSSELAER POLYTECHNIC INSTITUTE; (GU3569); $65,475
ROCHESTER INSTITUTE OF TECHNOLOGY; (GU3570); $8,000
ROCKEFELLER UNIVERSITY; (GU3571); $33,710
SARAH LAWRENCE COLLEGE; (GU3572); $11,283
SKIDMORE COLLEGE; (GU3573); $10,068
ST. JOHN FISHER COLLEGE; (GU3574); $2,000
ST. JOHN'S UNIVERSITY; (GU3575); $11,768
ST. LAWRENCE UNIVERSITY; (GU3576); $7,000
STATE UNIVERSITY AT ALBANY; (GU3587); $42,364
STATE UNIVERSITY AT BINGHAMTON; (GU3588); $20,765
STATE UNIVERSITY AT BUFFALO; (GU3589); $53,705
STATE UNIVERSITY AT STONY BROOK; (GU3590); $55,992
STATE UNIVERSITY OF NEW YORK SYSTEM OFFICE; (GU3577); $14,605
SYRACUSE UNIVERSITY; (GU3591); $90,890
TEACHERS COLLEGE; (GU3592); $20,525
UNION COLLEGE; (GU3593); $10,450
UNIVERSITY OF ROCHESTER; (GU3594); $96,215
VASSAR COLLEGE; (GU3595); $10,518
WEBB INSTITUTE OF NAVAL ARCHITECTURE; (GU3596); $10,675
YESHIVA UNIVERSITY; (GU3597); $73,039

DEVELOPMENT AND IMPROVEMENT OF INSTITUTIONAL SCIENCE PROGRAMS

NORTH CAROLINA

APPALACHIAN STATE UNIVERSITY; (GU3598); $2,000

CAPE FEAR TECHNICAL INSTITUTE; (GU3599); $13,325

DUKE UNIVERSITY; (GU3600); $86,418

EAST CAROLINA UNIVERSITY; (GU3601); $12,805

ELON COLLEGE; (GU3602); $2,000

NORTH CAROLINA AGRICULTURAL AND TECHNICAL STATE UNIVERSITY; (GU3603); $10,698

NORTH CAROLINA STATE UNIVERSITY AT RALEIGH; (GU3610); $80,615

NORTH CAROLINA CENTRAL UNIVERSITY; (GU3604); $10,540

QUEENS COLLEGE; (GU3605); $10,068

ST. ANDREWS PRESBYTERIAN COLLEGE; (GU3606); $10,788

UNIVERSITY OF NORTH CAROLINA AT CHAPEL HILL; (GU3608); $54,544

UNIVERSITY OF NORTH CAROLINA AT GREENSBORO; (GU3609); $16,385

UNIVERSITY OF NORTH CAROLINA SYSTEM OFFICE; GU3607); $13,705

UNIVERSITY OF NORTH CAROLINA AT WILMINGTON; (GU3611); $2,000

WAKE FOREST UNIVERSITY; (GU3612); $15,025

NORTH DAKOTA

JAMESTOWN COLLEGE; (GU3613); $10,405

MINOT STATE COLLEGE; (GU3614); $7,000

NORTH DAKOTA STATE UNIVERSITY; (GU3615); $26,225

UNIVERSITY OF NORTH DAKOTA; (GU3616); $18,159

OHIO

ANTIOCH COLLEGE; (GU3617); $17,768

BLUFFTON COLLEGE; (GU3618); $10,518

BOWLING GREEN STATE UNIVERSITY; (GU3619); $12,965

CASE WESTERN RESERVE UNIVERSITY; (GU3620); $88,790

CENTRAL STATE UNIVERSITY; (GU3621); $10,023

CLEVELAND STATE UNIVERSITY; (GU3622); $4,000

COLLEGE OF WOOSTER; (GU3623); $10,203

DEFIANCE COLLEGE; (GU3624); $10,090

DENISON UNIVERSITY; (GU3625); $10,225

HEIDELBERG COLLEGE; (GU3626); $2,000

HIRAM COLLEGE; (GU3627); $2,000

JOHN CARROLL UNIVERSITY; (GU3628); $12,785

KENT STATE UNIVERSITY; (GU3629); $17,645

KENYON COLLEGE; (GU3630); $10,000

LAKE ERIE COLLEGE; (GU3631); $10,045

MIAMI UNIVERSITY; (GU3632); $14,769

MT. UNION COLLEGE; (GU3633) $4,000

MUSKINGUM COLLEGE; (GU3634); $4,000

OBERLIN COLLEGE; (GU3635); $12,982

OHIO NORTHERN UNIVERSITY; (GU3636); $2,000

OHIO STATE UNIVERSITY; (GU3637); $104,597

OHIO UNIVERSITY; (GU3638); $38,005

OHIO WESLEYAN UNIVERSITY; (GU3639); $12,305

OTTERBEIN COLLEGE; (GU3640); $10,045

UNIVERSITY OF AKRON; (GU3641); $19,328

UNIVERSITY OF CINCINNATI; (GU3642); $50,645

UNIVERSITY OF DAYTON; (GU3643); $66,585

UNIVERSITY OF TOLEDO; (GU3644; $16,665

WITTENBERG UNIVERSITY; ((GU3645); $7,000

WRIGHT STATE UNIVERSITY; (GU3646); $4,000

XAVIER UNIVERSITY; (GU3647); $10,113

YOUNGSTOWN STATE UNIVERSITY; (GU3648); $2,000

OKLAHOMA

BETHANY-NAZARENE COLLEGE; (GU3649); $2,000

CENTRAL STATE COLLEGE; (GU3650); $4,000

LANGSTON UNIVERSITY; (GU3651); $10,135

NORTHEASTERN STATE COLLEGE; (GU3652); $8,000

OKLAHOMA CITY UNIVERSITY; (GU3653); $11,418

OKLAHOMA STATE UNIVERSITY; (GU3654); $60,455

SOUTHWESTERN STATE COLLEGE; (GU3655); $10,383

UNIVERSITY OF OKLAHOMA; (GU3656); $53,765

UNIVERSITY OF TULSA; (GU3657); $13,880

OREGON

CENTRAL OREGON COMMUNITY COLLEGE; (GU3658); $2,000

EASTERN OREGON COLLEGE; (GU3659); $2,000

LINFIELD COLLEGE; (GU3660); $10,135

OREGON COLLEGE OF EDUCATION; (GU3661); $10,585

OREGON STATE UNIVERSITY; (GU3662); $90,550

PORTLAND STATE UNIVERSITY; (GU3663); $10,878

REED COLLEGE; (GU3664); $12,136

SOUTHERN OREGON COLLEGE; (GU3665); $10,180

UNIVERSITY OF OREGON; (GU3666); $79,421

UNIVERSITY OF PORTLAND; (GU3667); $14,825

WILLAMETTE UNIVERSITY; (GU3668); $4,000

PENNSYLVANIA

ALLEGHENY COLLEGE; (GU3669); $10,338

BRYN MAWR COLLEGE; (GU3670); $20,049

BUCKNELL UNIVERSITY; (GU3671); $11,619

CARNEGIE-MELLON UNIVERSITY; (GU3672); $86,520

CHATHAM COLLEGE; (GU3673); $3,000

CLARION STATE COLLEGE; (GU3674); $1,000

DICKINSON COLLEGE; (GU3675); $6,000

DREXEL UNIVERSITY; (GU3676); $25,385

DUQUESNE UNIVERSITY; (GU3677); $14,359

EAST STROUDSBURG STATE COLLEGE; (GU3678); $10,248

ELIZABETHTOWN COLLEGE; (GU3679); $5,000

FRANKLIN AND MARSHALL COLLEGE; (GU3680); $13,050

GENEVA COLLEGE; (GU3681); $9,000

GETTYSBURG COLLEGE; (GU3682); $3,000

HAHNEMANN MEDICAL COLLEGE AND HOSPITAL; (GU3683); $10,338

HAVERFORD COLLEGE; (GU3684); $13,230

JUNIATA COLLEGE; (GU3686); $10,000

LAFAYETTE COLLEGE; GU3687); $10,293

LEHIGH UNIVERSITY; (GU3688); $62,208

LINCOLN UNIVERSITY; (GU3689); $2,000

LYCOMING COLLEGE; (GU3690); $4,000

MERCYHURST COLLEGE; (GU3691); $2,000

MILLERSVILLE STATE COLLEGE; (GU3692); $10,045

MUHLENBERG COLLEGE; (GU3693); $6,000

PENNSYLVANIA STATE UNIVERSITY; (GU3695); $84,440

PHILADELPHIA COLLEGE OF PHARMACY AND SCIENCE; (GU3696); $4,000

PHILADELPHIA COLLEGE OF TEXTILES AND SCIENCE; (GU3697); $4,000

PMC COLLEGES; (GU3694); $10,113

ST. JOSEPH'S COLLEGE; (GU3698); $10,495

SWARTHMORE COLLEGE; (GU3699); $12,797

TEMPLE UNIVERSITY; (GU3700); $44,130

THIEL COLLEGE; (GU3701); $10,135

THOMAS JEFFERSON UNIVERSITY; (GU3685); $13,305

UNIVERSITY OF PENNSYLVANIA; (GU3702); $83,705

UNIVERSITY OF PITTSBURGH; (GU3703); $86,750

UNIVERSITY OF SCRANTON; (GU3704); $6,000

VILLANOVA UNIVERSITY; (GU3705); $15,365

WASHINGTON AND JEFFERSON COLLEGE; (GU3706); $10,045

WAYNESBURG COLLEGE; (GU3707); $10,045

WESTMINSTER COLLEGE; (GU3708); $2,000

WILSON COLLEGE; (GU3709); $10,180

WOMAN'S MEDICAL COLLEGE OF PENNSYLVANIA; (GU3710); $18,055

PUERTO RICO

UNIVERSITY OF PUERTO RICO, Mayaguez; (GU3829); $38,268

UNIVERSITY OF PUERTO RICO, Rio Piedras; (GU3830); $12,025

UNIVERSITY OF PUERTO RICO SYSTEM OFFICE; (GU3828); $27,945

RHODE ISLAND

BROWN UNIVERSITY; (GU3711); $69,227

PROVIDENCE COLLEGE; (GU3712); $7,000

RHODE ISLAND COLLEGE; (GU3713); $12,445

UNIVERSITY OF RHODE ISLAND; (GU3714); $54,305

SOUTH CAROLINA

CLEMSON UNIVERSITY; (GU3715); $46,265

COLLEGE OF CHARLESTON; (GU3716); $5,000

FURMAN UNIVERSITY; (GU3717); $7,000

MEDICAL COLLEGE OF SOUTH CAROLINA; (GU3718); $11,508

SOUTH CAROLINA STATE COLLEGE; (GU3719); $10,203

UNIVERSITY OF SOUTH CAROLINA; (GU3720); $21,985

WOFFORD COLLEGE; (GU3721); $10,248

SOUTH DAKOTA

AUGUSTANA COLLEGE; (GU3722); $10,090

BLACK HILLS STATE COLLEGE; (GU3723); $8,000

DAKOTA STATE COLLEGE; (GU3724); $10,068

SOUTH DAKOTA SCHOOL OF MINES AND TECHNOLOGY; (GU3725); $40,683

SOUTH DAKOTA STATE UNIVERSITY; (GU3726); $30,113

UNIVERSITY OF SOUTH DAKOTA; (GU3727); $12,428

TENNESSEE

CARSON-NEWMAN COLLEGE; (GU3728); $2,000

CHRISTIAN BROTHERS COLLEGE; (GU3729); $10,563

EAST TENNESSEE STATE UNIVERSITY; (GU3730); $10,203

GEORGE PEABODY COLLEGE FOR TEACHERS; (GU3731); $16,265

LE MOYNE-OWEN COLLEGE; (GU3732); $4,000

MARYVILLE COLLEGE; (GU3733); $10,248

MEMPHIS STATE UNIVERSITY; (GU3734); $14,365

MIDDLE TENNESSEE STATE UNIVERSITY; (GU3735); $5,000

GRADUATE SCIENCE FACILITIES

Southwestern at Memphis; (GU3736); $10,113

Tennessee State University; (GU3737); $10,203

Tennessee Technological University; (GU3738); $12,165

University of Tennessee at Martin; (GU3741); $12,425

University of Tennessee System Office; (GU3739); $17,725

University of Tennessee; (GU3740); $81,845

Vanderbilt University; (GU3742); $78,240

TEXAS

Austin College; (GU3743); $10,338

Baylor University; (GU3744); $38,740

Lamar State College of Technology; (GU3745); $10,090

Midwestern University; (GU3746); $5,000

North Texas State University; (GU3747); $12,605

Pan American College; (GU3748); $10,135

Prairie View A & M College; (GU3749); $10,315

Rice University; (GU3750); $83,475

Sam Houston State University; (GU3751); $7,000

Southern Methodist University; (GU3752); $39,965

Southwest Texas University; (GU3753); $3,000

Stephen F. Austin State University; (GU3754); $1,000

Sul Ross State University; (GU3755); $10,248

Texas A & M University; (GU3756); $92,665

Texas Christian University; (GU3757); $18,665

Texas Southern University; (GU3758); $10,248

Texas Tech University; (GU3759); $23,165

Texas Woman's University; (GU3760); $13,625

Trinity University; (GU3761); $6,000

University of Houston; (GU3762); $47,825

University of St. Thomas; (GU3763); $12,785

University of Texas, Arlington; (GU3765); $10,360

University of Texas at Austin; (GU3766); $98,415

University of Texas at El Paso; (GU3767); $21,925

University of Texas System Office; (GU3764); $36,115

UTAH

Brigham Young University; (GU3768); $28,386

University of Utah; (GU3769); $87,505

Utah State University; (GU3770); $55,065

Weber State College; (GU3771); $4,000

VERMONT

Bennington College; (GU3772); $2,000

Middlebury College; (GU3773); $12,054

University of Vermont; (GU3774); $31,775

Windham College; (GU3775); $2,000

VIRGIN ISLANDS

College of the Virgin Islands; (GU3831); $4,000

VIRGINIA

College of William and Mary; (GU3776); $18,752

Eastern Mennonite College; (GU3777); $10,675

Hampden-Sydney College; (GU3778); $8,000

Hollins College; (GU3779); $10,968

Madison College; (GU3780); $9,000

Old Dominion University; (GU3781); $14,705

Randolph-Macon College; (GU3782); $2,000

Roanoke College; (GU3783); $6,000

University of Virginia; (GU3784); $63,055

Virginia Commonwealth University; (GU3785); $20,625

Virginia Military Institute; (GU3786); $10,540

Virginia Polytechnic Institute; (GU3787); $54,765

Virginia State College; (GU3788); $10,180

Washington & Lee University; (GU3789); $8,000

WASHINGTON

Central Washington State College, (GU3790); $10,653

Gonzaga University; (GU3791); $10,360

Grays Harbor College; (GU3792); $2,000

Pacific Lutheran University; (GU3793); $9,000

Seattle Pacific College; (GU3794); $7,000

Seattle University; (GU3795); $6,000

University of Puget Sound; (GU3796); $2,000

University of Washington; (GU3797); $107,138

Washington State University; (GU3798); $55,805

Western Washington State College; (GU3799); $13,425

WEST VIRGINIA

Alderson-Broaddus College; (GU3800); $10,495

Davis & Elkins College; (GU3801); $10,068

Fairmont State College; (GU3802); $10,675

Marshall University; (GU3803); $10,450

West Liberty State College; (GU3804); $2,000

West Virginia Institute of Technology; (GU3805); $17,345

West Virginia University; (GU3806); $46,335

Wheeling College; (GU3807); $10,000

WISCONSIN

Alverno College; (GU3808); $2,000

Carroll College; (GU3809); $2,000

Dominican College; (GU3810); $10,135

Lawrence University; (GU3811); $10,000

Marquette University; (GU3812); $28,960

Ripon College; (GU3813); $10,270

St. Norbert College; (GU3814); $6,000

Stout State University; (GU3815); $12,665

University of Wisconsin, Madison; (GU3817); $109,134

University of Wisconsin, Milwaukee; (GU3818); $22,605

University of Wisconsin System Office; (GU3816); $10,293

Wisconsin State University, Stevens Point; (GU3823); $12,065

Wisconsin State University, Eau Claire; (GU3819); $2,000

Wisconsin State University, Oshkosh; (GU3820); $11,350

Wisconsin State University, Platteville; (GU3821); $4,000

Wisconsin State University, River Falls; (GU3822); $10,045

Wisconsin State University, Superior; (GU3824); $10,023

Wisconsin State University, Whitewater; (GU3825); $10,000

WYOMING

Casper College; (GU3826); $2,000

University of Wyoming; (GU3827); $45,810

Graduate Science Facilities

Life Science Facilities

CALIFORNIA

Claremont Graduate School and University;
Rice, Philip M.; *Construction of Botany Facilities* (GU3836); 12 months; $46,533

NEW YORK

American Museum of Natural History;
Nicholson, Thomas; *Renovation of Facilities for Ornithology* (GU3838); 12 months; $57,250

PENNSYLVANIA

University City Science Center
Peterson, Lysle H.; *Construction of the Monell Chemical Senses Center* (GU3197); 24 months; $300,000

Behavioral Sciences Facilities

MARYLAND

Johns Hopkins University
Gordon, Lincoln; *Renovation of Animal and Physiological Research Space and Shop Facilities for the Department of Psychology, the Johns Hopkins University* (GU3858); 12 months; $46,750

MASSACHUSETTS

Clark University
Jackson, Frederick H.; *Building Renovation (Old Library) for Geography, Economics, and History Center* (GU3194); 24 months; $297,000

Mathematics, Physics, and Astronomy Facilities

NEW YORK

State University at Albany
Gould, Samuel B.; *Construction of Graduate Extension to the Mathematics-Physics Building at Stony Brook* (GU3188); 36 months; $750,000

University of Rochester
Wallis, W. Allen; *Construction of a Mathematical Sciences Building* (GU3837); 36 months; $400,000

Chemistry, Atmospheric and Earth Science Facilities

MASSACHUSETTS

Harvard University
Pusey, Nathan M.; *Renovations for Organic Chemistry* (GU3189); 12 months; $96,110

OHIO

Case Western Reserve University
Morse, Robert W.; *Space Renovation for Geochronology and Palynology* (GU3834); 12 months; $40,174

DEVELOPMENT AND IMPROVEMENT OF INSTITUTIONAL SCIENCE PROGRAMS

PENNSYLVANIA

UNIVERSITY OF PENNSYLVANIA
Harnwell, G. P.; *Construction of a New Chemistry Building* (GU3196); 24 months; $750,000

WISCONSIN

UNIVERSITY OF WISCONSIN, Madison
Harrington, Fred H.; *Construction of a Geology-Geophysics Research Building* (GU3849); 36 months; $500,000

Engineering Facilities

CALIFORNIA

UNIVERSITY OF SOUTHERN CALIFORNIA
Topping, Norman; *Renovation of Biegler Hall of Engineering* (GU3839); 24 months; $157,300

PENNSYLVANIA

DREXEL UNIVERSITY
Gatlin, Carl; *Renovation of Applied Mechanics Research Laboratory* (GU3193); 24 months; $471,700

TENNESSEE

VANDERBILT UNIVERSITY
Heard, Alexander; *Renovation of Materials Science and Engineering Facilities* (GU3832); 12 months; $26,350

TEXAS

SOUTHERN METHODIST UNIVERSITY
Tate, Willis M.; *Renovation of Engineering Research Facilities* (GU3195); 12 months; $68,500

University Science Development Program

COLORADO

UNIVERSITY OF COLORADO
Manning, T. E.; *University Science Development Program for the University of Colorado* (GU1532 001); 24 months; $1,676,000

CONNECTICUT

UNIVERSITY OF CONNECTICUT
Babbidge, Homer D.; *Science Development Planning Grant* (GU3853); 12 months; $145,000

GEORGIA

UNIVERSITY OF GEORGIA
Davison, Fred C.; *University Science Development Program for the University of Georgia* (GU2590 001); 24 months; $2,276,000

NEW YORK

STATE UNIVERSITY AT STONY BROOK
Toll, John S.; *University Science Development Program for State University of New York at Stony Brook* (GU3850); 36 months; $2,000,000

Mathematics, Physics and Astronomy

ILLINOIS

NORTHWESTERN UNIVERSITY
Miller, J. Roscoe; *The Establishment of a University Science Development Program* (GU3851); 36 months; $525,000

MASSACHUSETTS

BRANDEIS UNIVERSITY
Schottland, C. I.; *University Science Development Program for Brandeis University* (GU3852); 36 months; $1,900,000

OREGON

UNIVERSITY OF OREGON
Clark, Robert; *University Science Development Program for the University of Oregon* (GU1146 001); 24 months; $2,748,000

Chemistry, Atmospheric and Earth Science

LOUISIANA

LOUISIANA STATE UNIVERSITY, Baton Rouge
Taylor, Cecil G.; *A University Science Development Supplementary Proposal* (GU1558 001); 24 months; $2,429,000

NEW YORK

UNIVERSITY OF ROCHESTER
Wallis, W. Allen; *University Science Development Program for the University of Rochester* (GU1154 002); 24 months; $1,205,000

Engineering

ILLINOIS

NORTHWESTERN UNIVERSITY
Miller, J. Roscoe; *The Establishment of a University Science Development Program* (GU3851*); 36 months; $975,000

Departmental Science Development Program

Life Sciences

MASSACHUSETTS

BOSTON UNIVERSITY
Fulton, George P.; *Departmental Science Development-Biological Sciences* (GU3846); 36 months; $650,000

Behavioral Sciences

DISTRICT OF COLUMBIA

GEORGETOWN UNIVERSITY
Fitzgerald, Thomas R.; *Science Development Program in the School of Languages and Linguistics* (GU3857); 36 months; $460,000

NEW HAMPSHIRE

UNIVERSITY OF NEW HAMPSHIRE
Erickson, Raymond L.; *Departmental Science Development—Psychology* (GU3845); 36 months; $480,000

NEW YORK

STATE UNIVERSITY AT BINGHAMTON
Latourette, John; *Departmental Science Development—Economics* (GU3844); 36 months; $390,000

OHIO

KENT STATE UNIVERSITY
Baron Seymour H.; *Departmental Science Development—Psychology* (GU3840); 36 months; $400,000

TEXAS

SOUTHERN METHODIST UNIVERSITY
Murphy, J. Carter; *Departmental Science Development—Economics* (GU3841); 36 months; $550,000

Mathematics, Physics and Astronomy

NEW JERSEY

STEVENS INSTITUTE OF TECHNOLOGY
Davis, J. H.; *Departmental Science Development—Physics* (GU3191); 36 months; $670,000

NEW YORK

YESHIVA UNIVERSITY
Lebowitz, Joel L.; *Departmental Science Development Program in Physics* (GU3835); 36 months; $900,000

VIRGINIA

COLLEGE OF WILLIAM AND MARY
Winter, R. G.; *Departmental Science Development—Physics* (GU3842); 36 months; $610,000

Chemistry, Atmospheric and Earth Science

ALASKA

UNIVERSITY OF ALASKA
Forbes, Robert B.; *Science Development Program in Geology* (GU3856); 36 months; $720,000

COLORADO

COLORADO SCHOOL OF MINES
Bisque, Ramon E.; *Science Development Program in Geoscience—Mineral Resources* (GU3854); 36 months; $700,000

HAWAII

UNIVERSITY OF HAWAII
Inskeep, Richard; *Science Development Program in Chemistry* (GU3855); 36 months; $606,000

OREGON

OREGON STATE UNIVERSITY
Shoemaker, David P.; *Departmental Science Development—Chemistry* (GU3848); 36 months; $600,000

VIRGINIA

VIRGINIA POLYTECHNIC INSTITUTE
Hahn, Marshall T. Jr.; *Departmental Science Development—Geological Sciences* (GU3192); 36 months; $500,000

Engineering

COLORADO

COLORADO STATE UNIVERSITY
Fead, F. W. N.; *Departmental Science Development—Civil Engineering* (GU3847); 36 months; $600,000

MISSISSIPPI

UNIVERSITY OF MISSISSIPPI
Butler, Chalmers N.; *Departmental Science Development Program in Electrical Engineering* (GU3833); 36 months; $400,000

PENNSYLVANIA

LEHIGH UNIVERSITY
Lewis, W. Deming; *Departmental Science Development—Mechanical Engineering and Mechanics* (GU3190); 48 months; $670,000

SOUTH CAROLINA

CLEMSON UNIVERSITY
Rich, L. G.; *Departmental Science Development—Engineering* (GU3843); 36 months; $650,000

COMPUTING ACTIVITIES IN EDUCATION AND RESEARCH

Education, Research and Training

ALABAMA

UNIVERSITY OF ALABAMA, Tuscaloosa
Seebeck, Charles L.; *Computer-Based Honors Program* (J000715); 24 months; $71,900

CALIFORNIA

PITZER COLLEGE
Rodman, John R. and Caporale, Rocco; *Faculty Postdoctoral Residency in Computer Science* (J000923); 24 months; $45,900

STANFORD UNIVERSITY
Massy, William F. and Nielsen, Norman R.; *Use of Computers in Management Science Education* (J000617); 24 months; $150,000
Miller, William F.; *Graphic Processing—Analysis, Languages, Data Structures* (J000687); 24 months; $79,000
Nielsen, Norman R.; *Development of a Simulation System for Use in Teaching Modeling and Analysis* (J000599); 19 months; $42,300

UNIVERSITY OF CALIFORNIA, Berkeley
Blum, Manuel; *Computability and Computational Complexity* (J000708); 24 months; $94,700
Burgoyne, Nicholas and Maurer, Ward D.; *Computer-Aided Mathematics* (J000821); 24 months; $90,400
Gwinn, William D., Streitwieser, A. Jr. and Pigford, Robert L.; *Use of Computers in Undergraduate Instruction in Chemistry* (J000615); 24 months; $174,500
Rosenfeld, A. H.; *Computer Support for a Research Center in Science Education* (J000765); 24 months; $124,600

UNIVERSITY OF CALIFORNIA, Los Angeles
Carlyle, J. W. and Greibach, Sheila A.; *Theory of Discrete Systems, Automata, and Formal Language* (J000803); 24 months; $126,300
Estrin, Gerald; *Modelling and Measurement of Computer Systems* (J000809); 24 months; $188,100

UNIVERSITY OF CALIFORNIA, Santa Cruz
Huffman, David A.; *Development of Graduate Program in Information and Computer Science* (J000483); 36 months; $333,000

UNIVERSITY OF SOUTHERN CALIFORNIA
Bellman, Richard; *Computer Training Program for High School Students from Disadvantaged Areas* (J000981); 12 months; $56,900

COLORADO

UNIVERSITY OF COLORADO
Bailey, Daniel E.; *Computer Laboratory for Instruction in Psychological Research* (J000453); 24 months; $300,000
Feng, Chuan C.; *Computing Science Curriculum Development Program* (J000146001); $19,100
Jordan, Harry F.; *Collaborative Research in Computational Methods for Functional Integrals* (P015507*); 12 months; $4,800
Waite, William M.; *Development of a Mobile Programming System* (J000625); 12 months; $19,100

CONNECTICUT

YALE UNIVERSITY
Peck, Merton J.; *Computer Use for Macro-Policy Simulation* (J000780); 24 months; $44,000

DISTRICT OF COLUMBIA

CONFERENCE BOARD OF THE MATHEMATICAL SCIENCES
Botts, Truman; *Contract for Supporting Services Relating to the Promoting and Conducting of a Series of Ten Regional Conferences in the Field of Mathematics* (C543000002); $9,000

D. C. PUBLIC SCHOOLS
Rice, William S.; *An Experimental Project on Para-Professional Training in Data Processing in Secondary Schools* (J000961); 6 months; $13,100

NATIONAL ACADEMY OF SCIENCES, NATIONAL RESEARCH COUNCIL
Cohen, Leon W.; *Task Order for Partial Support of Travel to the International Federation for Information Processing (IFIP) Congress in Ljubljana, Yugoslavia* (C310185*); 22 months; $4,955

NATIONAL ENDOWMENT FOR THE HUMANITIES
McArthur, Herbert; *A National Summer Training Institute for Humanistic Computation—American Council of Learned Societies* (AG224); 12 months; $25,000

HAWAII

UNIVERSITY OF HAWAII
Pager, Davis and Peterson, W. Wesley; *Theoretical and Practical Study of Programming Efficiency* (J000596); 24 months; $93,700

ILLINOIS

NORTHWESTERN UNIVERSITY
Lerner, Eugene M.; *Use of Computer Simulation in Management Science Education* (J000794); 24 months; $96,100

UNIVERSITY OF ILLINOIS, Urbana
Fosdick, Lloyd D.; *Collaborative Research in Computational Methods for Functional Integrals* (GP9665*); 12 months; $4,800
Gillies, D. B.; *Numerical Mathematics and Computer Science* (J000812); 24 months; $118,400
Muroga, Saburo; *Logical Design of Digital Networks by Large-Scale Integer Programming* (J000503); 24 months; $47,700
Robertson, J. E.; *Numerical Mathematics and Computer Sciences* (J000813); 24 months; $101,900

INDIANA

INDIANA UNIVERSITY, Bloomington
Bron, Walter E. and Lurie, Fred M.; *Experimental Study of the Application of Computers to Physics Laboratory Instruction* (J000854); 24 months; $75,600

PURDUE UNIVERSITY
Conte, S. D.; *Numerical Analysis Problem Solving System* (J000721); 21 months; $89,900
Schneider, Victor B.; *Use of a Formal Notation to Specify an Algol 68 Translator* (J000851); 26 months; $37,400

UNIVERSITY OF NOTRE DAME
Sayre, Kenneth M.; *Computer Recognition of Handwritten Sentences* (J000537); 24 months; $26,000

IOWA

UNIVERSITY OF IOWA
Mukhopadhyay, Amar; *Cellular Logic Array and Machines* (J000723); 24 months; $53,500

KANSAS

KANSAS STATE UNIVERSITY
Fisher, Paul S.; *Regional Conference on Translators and Translator Writing Systems* (P020046); 12 months; $9,400

UNIVERSITY OF KANSAS
Bavel, Zamir; *Structure, Connectivity and Homomorphisms of Automata* (J000639); 24 months; $65,000
Schweppe, Earl J.; *Conference on the Implementation of Computer Science Education* (J001063); 8 months; $11,700

LOUISIANA

TULANE UNIVERSITY
Beck, Charles H.; *Investigation of Educational and Research Utilization of the Minuteman I Computer* (J000850); 12 months; $15,300

MARYLAND

UNIVERSITY OF MARYLAND
Rosenfeld, Azriel and Pfaltz, John L.; *Picture Grammars, Web Grammars and Picture Description* (J000754); 24 months; $126,200

MASSACHUSETTS

BOLT BERANEK & NEWMAN INC.
Feurzeig, Wallace; *Contract for Programming-Languages as a Conceptual Framework for Teaching Mathematics* (C558000001); 2 months; $4,108
———; *Contract for the Continued Development of Programming Languages as a Conceptual Framework for Teaching Mathematics* (C615000); 12 months; $110,342

BOSTON UNIVERSITY
Esch, Robin E.; *Regional Conference on Numerical Analysis* (P019863); 12 months; $8,000

COLLEGE OF THE HOLY CROSS
Brooks, John E. and Perkins, Peter; *Residency and Computing Activities Development* (J000872); 24 months; $48,100

MASSACHUSETTS INSTITUTE OF TECHNOLOGY
Kuh, Edwin; *Feasibility Study on a National Computer Research Center for Economics and Management Science* (J000770); 12 months; $12,200
McIntosh, Stuart D.; *Data Structures and Data Management* (J000929); 18 months; $277,000
Papert, Seymour and Minsky, Marvin; *Computer-Based Curricular Research in Elementary Education* (J001049); 12 months; $49,700

MICHIGAN

MICHIGAN STATE UNIVERSITY
Beaman, John H.; *Computer Methods in Plant Systematics for Classroom Instruction*

COMPUTING ACTIVITIES IN EDUCATION AND RESEARCH

and Associated Research (J000573); 24 months; $84,400

UNIVERSITY OF MICHIGAN

Burks, A. W.; *A Computer Simulation System Based on Cellular Automation Arrays* (J000519); 24 months; $108,500

McFadden, James T.; *Development of a University Course in Management of Complex Biological Resources* (J000333); 24 months; $98,100

MINNESOTA

ST. MARY'S COLLEGE

Labelle, Joseph and Farrell, James; *To Develop Institutional Capability in Academic Computing and Self-Instruction* (J000920); 24 months; $46,600

MISSOURI

WEBSTER COLLEGE

Madden, Charles F.; *Computer Instruction for Junior and Senior High School Mathematics Teachers* (J000469); 12 months; $7,500

MONTANA

UNIVERSITY OF MONTANA

Banaugh, Robert P.; *Matrix Inversion and Boundary Value Problems* (J000506); 12 months; $11,300

NEW HAMPSHIRE

DARTMOUTH COLLEGE

Alverson, Hoyt S. and Fernandez, James W.; *Use of Computers in Introductory Anthropology* (J000719); 15 months; $24,200

Kreider, Donald L. and Luehrmann, Arthur W. Jr.; *Computer-Based Course Materials for Introductory University Mathematics, Physics and Engineering* (J000650); 24 months; $388,200

Meyers, Edmund D. Jr.; *Equipment for Project IMPRESS (Interdisciplinary Machine Processing for Research and Education in the Social Sciences)* (J000513); 24 months; $136,400

NEW JERSEY

MONMOUTH COLLEGE

Holt, Everett W. and Swartz, G. Boyd; *Computer Science Course Development and Faculty Training Project* (J000913); 24 months; $50,000

PRINCETON UNIVERSITY

Steiglitz, K. and Bruno, J.; *Problem Oriented Processors—Parallel Circulating Associative Memories* (J000965); 24 months; $60,000

NEW YORK

CORNELL UNIVERSITY

Dennis, J. E. Jr.; *Solutions of Nonlinear Equations by Newton-Like Methods* (J000844); 12 months; $7,150

NEW YORK UNIVERSITY

Freeman, Herbert; *Equipment for a Computer Science Research Facility* (J001009); 12 months; $69,400

ST. JOHN'S UNIVERSITY

Mills, Henry C. and Frechen, Joseph B.; *Using the Computer as a Research Tool* (J000882); 24 months; $44,700

STATE UNIVERSITY AT BUFFALO

Findler, Nicholas V.; *Extensions and Applications of an Associative Memory Language* (J000658); 24 months; $54,200

STATE UNIVERSITY AT STONY BROOK

Rohlf, F. James; *On-Line Computation for Statistical Analysis in Biology* (J000724); 12 months; $15,000

NORTH CAROLINA

NORTH CAROLINA STATE UNIVERSITY AT RALEIGH

Martin, Donald C.; *A Study of Interactive Analog Computer Terminals in Undergraduate Education* (J000473); 24 months; $100,400

NORTH DAKOTA

UNIVERSITY OF NORTH DAKOTA

Dietz, Conrad and Kemper, Gene A.; *Institute for Computer Supplemented Instruction* (J000836); 11 months; $27,600

OHIO

ANTIOCH COLLEGE

Greenlee, Howard S. and Carpenter, Richard N.; *Program for Academic Computer Usage Development* (J000876); 24 months; $43,700

BALDWIN-WALLACE COLLEGE

Miller, Richard I. and Gerhan, Richard C.; *Postdoctoral Faculty Residency in Computing and Computer Applications* (J000891); 24 months; $47,000

OREGON

OREGON STATE UNIVERSITY

Guthrie, Donald Jr.; *The Use of Computers in Teaching Statistics* (J000499); 27 months; $100,000

WILLAMETTE UNIVERSITY

Hafferkamp, Jack and Yungen, Walter A.; *The Innovative Expansion of the Use of the Computer in the Educational Program of a Liberal Arts College* (J000862); 24 months; $28,700

PENNSYLVANIA

CARNEGIE-MELLON UNIVERSITY

Loveland, Donald and Andrews, Peter B.; *Proof Procedures in Predicate Calculus and Type Theory* (J000580); 24 months; $47,800

PENNSYLVANIA STATE UNIVERSITY

Adams, William S.; *An Investigation of Digitally Controlled Analog Computer Patching Applicable to Hybrid Computation* (J000679); 24 months; $35,100

De Maine, Paul A. D.; *Decomposition, Classification and Indexing of Graphs and Pictures* (J000479); 24 months; $57,600

Gotterer, Malcolm H.; *Regional Conference on Phenomena that Need Basic Computational Theories* (P019991); 12 months; $8,000

Hammer, Preston C.; *Systems Theory and Applications* (J000797); 24 months; $52,400

Jones, Neil D.; *Theory of Computational Complexity* (J000508); 24 months; $24,400

UNIVERSITY OF PENNSYLVANIA

Carr, John W.; *Study of Multiprogramming System Design Principles* (J001141); 1 month; $10,100

RHODE ISLAND

AMERICAN MATHEMATICAL SOCIETY

Walker, Gordon L.; *Symposium on 'Computers in Algebra and Number Theory'* (P020418*); 18 months; $2,600

BROWN UNIVERSITY

Freiberger, Walter F.; *Studies in Computer Science* (J000807;) 12 months; $100,400

Grenander, Ulf; *Computational Probability and Statistics* (J000710); 24 months; $145,500

SOUTH CAROLINA

CLEMSON UNIVERSITY

Whitehurst, C. H.; *Computer Programming Instruction and Services for the Faculty and Students of Central Wesleyan College and Anderson College* (J000418); 12 months; $20,600

TEXAS

RICE UNIVERSITY

Jump, J. Robert; *A Study of Models of Cellular Computation Networks* (J000750); 24 months; $40,500

TRINITY UNIVERSITY

Thomas, Marion B. and Carter, Elmer B.; *To Maximize the Usefulness of the Computer Center to the Social Sciences and Urban Studies* (J000881); 24 months; $32,500

UNIVERSITY OF HOUSTON

Dawkins, G. S.; *Advanced Science Seminar Computer Simulation for System Analysis and Design* (GZ1497*); 11 months; $20,000

Donaghey, Charles E.; *List Processing Techniques in Scientific and Engineering Education* (J000189); 22 months; $23,400

UNIVERSITY OF TEXAS AT AUSTIN

Pratt, Terrence W.; *Formal Theory of Programming Language Definition* (J000778); 24 months; $23,100

Ramamoorthy, C. V. and Coates, C. L.; *Investigation of the User Microprogrammable Computing Systems* (J000492); 24 months; $92,500

Yeh, Raymond T.; *Graph Automata and their Decision Problems* (J000786); 12 months; $10,700

WASHINGTON

UNIVERSITY OF WASHINGTON

Horwood, Edgar M.; *Urban Analysis System Development* (S002832*); 24 months; $14,000

WISCONSIN

UNIVERSITY OF WISCONSIN, Madison

Dennis, Jack S.; *A National Program Library and Central Program Inventory Service* (S002768*); 24 months; $15,000

Uhr, Leonard; *Complex Information Processing and Models of Intelligence* (J000583); 21 months; $104,400

WYOMING

UNIVERSITY OF WYOMING

Meyer, E. G. and Rowland, J. H.; *Faculty Residency in Computing* (J000916); 24 months; $50,000

Institutional Computing Services

CALIFORNIA

UNIVERSITY OF CALIFORNIA, Los Angeles

Kehl, William B.; *Improvement of Computing Services* (J000543); 36 months; $1,000,000

COLORADO

COLORADO STATE UNIVERSITY

Marschner, B. W.; (J000651); 36 months; $400,000

FLORIDA

UNIVERSITY OF FLORIDA

Selfridge, R. G.; (J000681); 36 months; $450,000

ILLINOIS

UNIVERSITY OF ILLINOIS, Chicago Circle

Brown, Thomas; (J000757); 24 months; $100,000

KENTUCKY

THOMAS MORE COLLEGE

Graham, Richard; (J000382); 36 months; $57,400

MASSACHUSETTS

WORCESTER POLYTECHNIC INSTITUTE

Sondak, Norman E.; *Development of Time-Sharing Capability—Worcester Area Computation Center* (J000711); 36 months; $220,000

SPECIAL PROJECTS

MICHIGAN

Western Michigan University
Meagher, Jack; *Improvement of Computing Services* (J000394); 12 months; $5,000

MISSOURI

University of Missouri, Rolla
Lee, Ralph E.; (J000451); 36 months; $400,000

NEW YORK

University of Rochester
Van Atta, Bruce W.; (J000828); 24 months; $225,000

NORTH CAROLINA

North Carolina Agricultural and Techinacal State University
Beatty, George Jr.; (J000331); 36 months; $175,000

OHIO

Ohio University
Lilley, Robert W.; (J000352); 24 months; $80,000

PENNSYLVANIA

Albright College
Prine, Lewis E.; (J000401); 36 months; $35,000

Bucknell University
Staiano, Edward F.; (J000389); 36 months; $395,000

Susquehanna University
Growney, Wallace J.; *Remote-Terminal Computing Facility* (J000775); 24 months; $20,000

University of Pennsylvania
Freeman, David N.; *Improvement of Computing Services* (J000438); 36 months; $900,000

RHODE ISLAND

University of Rhode Island
Hemmerle, William J.; (J000419); 36 months; $200,000

TENNESSEE

University of Tennessee
Sherman, Gordon R.; (J000682); 36 months; $500,000

TEXAS

University of Texas at Austin
Young, David M.; (J000452); 24 months; $250,000

WEST VIRGINIA

Bethany College
Kurey, Joseph M.; (J000491); 24 months; $8,900

Bluefield State College
Hart, E. Franklin; *Participation in a Regional Computer Network* (J000487); 36 months; $30,000

West Virginia University
Barton, Jay; *Improvement of Computing Services* (J000500); 36 months; $930,000

WISCONSIN

Institute of Paper Chemistry
Strange, John G.; (J000369); 36 months; $150,000

Lawrence University
Hopkinson, Lee; (J000832); 24 months; $32,000

Special Projects

CALIFORNIA

California State College at Hayward
Southard, Thomas H.; *Pilot Regional Educational Computing Network* (J000800); 12 months; $55,700

College of Notre Dame
Chapin, June R.; *Interacting Patterns of Computer Support and Usage in U. S. Higher Education* (J000977); 12 months; $12,000

Mills College
Pillans, Helen; *Pilot Regional Educational Computing Network* (J000802); 12 months; $41,400

Rand Corporation
Levien, Roger E.; *Instructional Uses of Computers in Higher Education* (J000967); 12 months; $44,900

San Francisco State College
Westfall, John E.; *Pilot Regional Educational Computer Network* (J000799); 12 months; $120,100

Stanford University
Armer, Paul and Nielsen, Norman R.; *Pilot Regional Educational Computing Network* (J000798); 12 months; $114,700
Suppes, Patrick and Atkinson, Richard C.; *Research and Development in Computer-Assisted Instruction* (GJ443X 001); 12 months; $350,300

University of California, Irvine
Gelbaum, B. R.; *Experiment with Small Computers and Terminals in Math Education, Grades K—12* (J000717); 36 months; $217,000

University of California, Santa Barbara
Harris, David O.; *On-line Computer Network for Chemistry Education* (J000693); 12 months; $257,200

University of San Francisco
Haag, James N.; *Pilot Regional Educational Computing Network* (J000801); 12 months; $68,600

DISTRICT OF COLUMBIA

Conference Board of the Mathematical Sciences
Botts, Truman A.; *Study of High School Computer Education* (J000804); 24 months; $26,600

Office of Education
Molnar, Andrew R.; *Joint Support for the Educom Educational Information Network* (AG140 001); 15 months; $35,000

FLORIDA

Florida State University
Hansen, Duncan N.; *Systems Factors and Potentials of Computer-Assisted Instruction—Present and Future* (J000623); 12 months; $74,900
Schwarz, Guenter; *On-line Computer Network for Chemistry Education* (J000701); 12 months; $12,300

GEORGIA

Abraham Baldwin Agricultural College
Sherman, J. Dale; *Experiment in the Development of a Regional Computer Center* (J001031); 24 months; $14,000

Albany Junior College
Baxter, John L.; (J001032); 24 months; $14,000

Albany State College
Gilmore, Henry F.; (J001033); 24 months; $14,000

Armstrong State College
Anderson, Donald D.; (J001034); 24 months; $8,400

Berry College
Dickey, Ouida W.; (J000781); 24 months; $15,800

Brunswick Junior College
Monroe, D. M. Jr.; (J001035); 24 months; $3,200

Fort Valley State College
Hamza, Khidhir A. A.; (J001027); 24 months; $3,200

Georgia College at Milledgeville
Baarda, David G.; (J001025); 24 months; $14,000

Georgia Institute of Technology
Perlin, I. E.; (J001055); 24 months; $66,000
Sherry, Peter B.; *On-line Computer Network for Chemistry Education* (J000700); 12 months; $7,700

Georgia Southwestern College
Greene, J. Hubert; *Experiment in the Development of a Regional Computer Center* (J001036); 24 months; $14,000

Macon Junior College
Carlton, Jack K.; (J001038); 24 months; $3,200

Medical College of Georgia
Morse, Russell W.; (J001039); 24 months; $14,000

Middle Georgia College
Crawford, Harry D.; (J001040); 24 months; $3,200

Paine College
Dawson, Leonard E.; (J001149); 24 months; $1,500

Savannah State College
Wilson, Martha W.; (J001042); 24 months; $14,000

South Georgia College
Johnson, Robert R.; (J001043); 24 months; $3,600

Southern Regional Education Board
Hamblen, John W.; *Experiment of Ways of Supplying Computer Facilities to Small Colleges for Instructional Uses* (GJ269004); 18 months; $50,500
——— ; *Contract for Survey and Analysis of Computing Activities in United States Higher Education* (C604000); 24 months; $89,173

University of Georgia
Carmon, James L.; *Experiment in the Development of a Regional Computer Center* (J001054); 24 months; $66,000

University System of Georgia
Simpson, George L. Jr.; (J000608); 24 months; $233,200

Valdosta State College
Brooks, Sam W. III; (J001044); 24 months; $14,000

ILLINOIS

Illinois Institute of Technology
Von Weyssenhoff, Hanns and Bush, C. A.; *On-line Computer Network for Chemistry Education* (J000768); 12 months; $12,300

University of Illinois, Urbana
Bitzer, Donald L.; *Large-scale Computer Based Education Experiment* (J000974); 12 months; $430,000

IOWA

Iowa Wesleyan College
Mattson, Foster A.; *Development of a Regional Social Science Data Archive for Instruction* (J000636001); $4,275

Kirkwood Community College
Cunning, Charles J.; (GJ637 001); $3,700

Loras College
Schneider, Donald J.; *Development of a Social Science Data Archive for Instruction* (GJ635001); $3,500

Marycrest College
Miller, Robert F.; *Development of a Regional Social Science Data Archive for Instruction* (GJ634001); $2,900

St. Ambrose College
Svec, Fred J.; (GJ633001); $2,800

University of Dubuque
Cahill James Q. and Herum, Gary (J000638001); $3,500

COMPUTING ACTIVITIES IN EDUCATION AND RESEARCH

UNIVERSITY OF IOWA
Weeg, Gerard P.; *Conference on Computers in the College Curriculum* (J000927); 12 months; $64,300

LOUISIANA

LOUISIANA STATE UNIVERSITY, Baton Rouge
Runnels, L. K.; *On-line Computer Network for Chemistry Education* (J000699); 12 months; $11,500

MARYLAND

AMERICAN INSTITUTES FOR RESEARCH, Washington Office
Korotkin, Arthur L.; *Contract for an Investigation of Computing Activities in Secondary Schools* (C584000001); 4 months; $11,931

UNIVERSITY OF MARYLAND
Fowler, John M. and Blum, Ronald; *Conference on Computers in College Science Education—Physics and Mathematics* (J000985); 12 months; $19,300

MINNESOTA

UNIVERSITY OF MINNESOTA
Overend, John; *On-line Computer Network for Chemistry Education* (J000698); 12 months; $11,600

MISSOURI

UNIVERSITY OF MISSOURI, Columbia
Kim, Hyunyong; (J000697); 12 months; $6,400

NEW JERSEY

AMERICAN FEDERATION OF INFORMATION PROCESSING SOCIETIES
Gilchrist, Bruce; *Analytical Study of the Information Processing Field* (J000996); 12 months; $23,000

NEW YORK

SKIDMORE COLLEGE
Chu, Yu-Kuang; *Study and Evaluation of Student Response System in Undergraduate Education* (J000317); 24 months; $70,000

NORTH CAROLINA

APPALACHIAN STATE UNIVERSITY
Trivett, Boyd Clark; *Regional Development of Computer Uses in Higher Education* (J000727); 24 months; $16,900

BELMONT ABBEY COLLEGE
Mani, K. V.; (J000738); 24 months; $6,000

BENNETT COLLEGE
Watkins, Nellouise; (J000729); 24 months; $17,800

CAMPBELL COLLEGE
Jones, Vernon W.; (J000730); 24 months; $7,100

GUILFORD COLLEGE
Courtney, Fred I.; (J000731); 24 months; $6,500

NORTH CAROLINA AGRICULTURAL AND TECHNICAL STATE UNIVERSITY
Parker, Paul E.; (J000739); 24 months; $11,800

NORTH CAROLINA BOARD OF HIGHER EDUCATION
Parker, Louis T.; (J000726); 24 months; $344,000
Harrell, John D.; (J000732); 24 months; $6,200

NORTH CAROLINA CENTRAL UNIVERSITY

PFEIFFER COLLEGE
Riemann, J. Michael; (J000733); 24 months; $10,000

QUEENS COLLEGE
Fusaro, B. A.; (J000734); 24 months; $6,400

ST. ANDREWS PRESBYTERIAN COLLEGE
Rolland, William W.; (J000735); 24 months; $30,000

UNIVERSITY OF NORTH CAROLINA AT ASHEVILLE
Coyle, Francis J.; (J000728); 24 months; $9,400

UNIVERSITY OF NORTH CAROLINA AT CHAPEL HILL
Brooks, Frederick P. Jr. and Calingaert, Peter; *A Program of Experimental Explorations in Computer-Assisted Instruction* (J000755); 24 months; $197,900

WARREN WILSON COLLEGE
Cramer, George F.; *A Regional Development of Computer Uses in Higher Education* (J000736); 24 months; $3,400

PENNSYLVANIA

UNIVERSITY OF PITTSBURGH
Cooley, William W. and Glaser, Robert; *A Program of Research and Development in Computer-Assisted Instruction—Computer Requirements for Individualization in the Elementary School* (GJ540X 001); 12 months; $262,600
Dwyer, Thomas A.; *Regional Computing for Secondary School Systems—Phase II* (J001077); 12 months; $171,400
Johnson, K. Jeffrey; *On-line Computer Network for Chemistry Education* (J000696); 12 months; $12,600

TEXAS

UNIVERSITY OF TEXAS AT AUSTIN
Bunderson, C. Victor and Simmons, R. F.; *A Program of Research and Development in Computer-Assisted Instruction—A Study of Instructional Design and Response Systems for Computer-Based Tutorials* (GJ509X 001); 12 months; $235,200

VERMONT

UNIVERSITY OF VERMONT
Degrasse, Richard V.; *National Educational Teleprocessing Network Study* (J000947); 12 months; $35,900

VIRGINIA

HUMAN RESOURCES RESEARCH ORGANIZATION
Seidel, Robert J.; *Research on Instructional Decision Models* (J000774); 24 months; $303,900

WASHINGTON

WASHINGTON STATE UNIVERSITY
Stevens, Carl M.; *On-line Computer Network for Chemistry Education* (J000694); 12 months; $9,500

WISCONSIN

BELOIT COLLEGE
Spencer, Brock; (J000702); 12 months; $1,100

SPECIALIZED RESEARCH FACILITIES SUPPORT

Biological and Medical Sciences Facilities and Equipment

Research Facilities

MISSOURI

MISSOURI BOTANICAL GARDEN
Lewis, Walter H.; *Herbarium and Library Facilities at the Missouri Botanical Garden* (GB8792); 24 months; $600,000

Biological Equipment

AUSTRALIA

WALTER AND ELIZA HALL INSTITUTE OF MEDICAL RESEARCH
Miller, J. F. A. P.; *Equipment for Development of an Assay for the Immunosuppressive Capacity and Specificity of Antilymphocyte Serum* (B015016); 12 months; $4,500

CALIFORNIA

STANFORD UNIVERSITY
Feigen, George A.; *Equipment for Isolation, Characterization, and Mode of Action of Protein Venoms of Marine Animals* (B016364); 12 months; $12,000

UNIVERSITY OF CALIFORNIA, Davis
Brownson, Robert H.; *Equipment for Central Nervous System Feed-Back Mechanisms* (B024141); 18 months; $525

UNIVERSITY OF CALIFORNIA, Irvine
Thompson, Richard F.; *Equipment for a Computer Analysis of the Neural Bases of Behavior* (B014665001); $2,100

CONNECTICUT

YALE UNIVERSITY
Sibley, Charles G.; *Purchase of a Low Resolution Mass Spectrometer* (B024132); 12 months; $13,400

MARYLAND

UNIVERSITY OF MARYLAND
Aposhian, H. Vasken; *Equipment for Molecular and Cellular Biological Activities* (B015789); 12 months; $55,000

MICHIGAN

MICHIGAN STATE UNIVERSITY
Tolbert, N. E.; *Equipment—Preparative Zonal Ultracentrifuge* (B020809); 12 months; $21,700

UNIVERSITY OF MICHIGAN
Greeno, James G.; *Equipment for the Study of Storage and Retrieval in Verbal Learning and Cognitive Structures in Simple Mathematical Problem Solving* (B014824); 12 months; $39,100

MISSOURI

WASHINGTON UNIVERSITY
Banaszak, Leonard J.; *Equipment for X-Ray Structural Studies of Cytoplasmic Malate Dehydrogenase* (B024140); 12 months; $93,000

WISCONSIN

UNIVERSITY OF WISCONSIN, Madison
Borisy, Gary G.; *Equipment to Study Assembly and Function of Microtubules* (B019410); 12 months; $84,000

University Atmospheric Science Facilities and Equipment

ALASKA

UNIVERSITY OF ALASKA
Davis, T. Neil; *Ground Telemetry Station for Poker Flat Rocket Facility, Fairbanks, Alaska* (A017416); 12 months; $79,900

CALIFORNIA

STANFORD UNIVERSITY
Helliwell, R. A.; *Equipment for the Roberval VLF Observatory* (A017762); 12 months; $33,100

UNIVERSITY OF CALIFORNIA, Davis
Coulson, Kinsell L. and Myrup, Leonard O.; *Computer Control for Mobile Scientific Laboratory* (A013444); 12 months; $34,000

MISSOURI

UNIVERSITY OF MISSOURI, Rolla
Snow, William R.; *Crossed Molecular Beam Apparatus* (A013819); 12 months; $53,000

Social Science Research Facilities

NEW MEXICO

UNIVERSITY OF NEW MEXICO
Campbell, John M.; *Specialized Research Facility for the Anthropological and Archaeological Sciences* (S003064); 24 months; $300,000

Earth Sciences Research Facilities and Equipment

CALIFORNIA

CALIFORNIA INSTITUTE OF TECHNOLOGY
Silver, Leon T.; *Rehabilitation of Existing 12 inch Radius, Solid-source Mass Spectrometer* (A015988); 12 months; $29,600

MASSACHUSETTS

MASSACHUSETTS INSTITUTE OF TECHNOLOGY
Hurley, Patrick M.; *Data Reduction System Instrumentation* (A021381); 12 months; $13,800

MISSOURI

UNIVERSITY OF MISSOURI, Columbia
Himmelberg, Glen R.; *Request for Specialized Research Equipmet—Electron Microprobe* (A018445); 24 months; $60,000

Ocean Research Vessels and Facilities

Physical Oceanography

MASSACHUSETTS

WOODS HOLE OCEANOGRAPHIC INSTITUTION
Bowin, Carl O. and Rosenfeld, Melvin A.; *Acquisition of a Digitizing Table* (A018889); 12 months; $17,000

NEW YORK

COLUMBIA UNIVERSITY
Hayes, Dennis E. and Talwani, Manik; *Automatic Digitizing Equipment to Aid in the Processing of Echo-Sounding and Other Marine Geological and Geophysical Data* (A017644); 12 months; $8.500

OREGON

OREGON STATE UNIVERSITY
Burt, Wayne V.; *Shipboard Tethersonde Instrument System* (A022693); 12 months; $13,700

Neshyba, Steve; *Near-surface Ocean Current Meters for Use With Totem Buoys* (A023015); 12 months; $21,200

WASHINGTON

UNIVERSITY OF WASHINGTON
Smith, J. Dungan; *Purchase of Multichannel Tape System Recorder for Recording Analog Data* (A020416); 12 months; $21,500

Biological Oceanography

CALIFORNIA

UNIVERSITY OF CALIFORNIA, San Diego
Benson, Andrew A.; *Support for Physiological Research Research Ship, R/V Alpha Helix* (B022983*); 7 months; $15,000

UNIVERSITY OF SOUTHERN CALIFORNIA
Savage, Jay M.; *Support of Research Vessel Operations* (GB8206 003); $5,000

UNIVERSITY OF THE PACIFIC
Smith, Edmund H.; *Planning a Graduate tion Instrumentation* (B025868); 12 months; $10,000

FLORIDA

UNIVERSITY OF MIAMI
Bunt, John S.; *Purchase of Liquid Scintillation Instrumentation* (B024555); 12 months; $21,500

MASSACHUSETTS

WOODS HOLE OCEANOGRAPHIC INSTITUTION
Fye, Paul M.; *Support for the Operation of Oceanographic Research Vessels* (GA1298005); $14,200

NORTH CAROLINA

DUKE UNIVERSITY
Bookhout, C. G.; *Development of Duke University Marine Laboratory* (GB7812001); $27,500

——; *Scientific Research Training Program* (B017545001); $21,000

University Physics Research Facilities

High-Energy Facilities

CALIFORNIA

STANFORD UNIVERSITY
Fairbank, W. M. and Hofstadter, R.; *High Energy Physics and Cryogenic Research* (P022880*); 12 months; $751,000

MICHIGAN

UNIVERSITY OF MICHIGAN
Jones, Lawrence W.; *Nucleon-Nucleon Total Cross Sections* (P025196*); 12 months; $60,000

SPECIALIZED RESEARCH FACILITIES SUPPORT

NEW YORK

COLUMBIA UNIVERSITY
Devons, Samuel and Wu, Chien-Shung; *Research Program and Synchrocyclotron Operation* (P022786*); 12 months; $100,000
Lederman, Leon M.; *Synchrocyclotron Modernization* (GP7177 001); $100,000

UTAH

UNIVERSITY OF UTAH
Keuffel, J. W.; *Improvement of Cosmic Ray Facility* (P021087); 24 months; $68,200

Low-Energy Facilities

CALIFORNIA

STANFORD UNIVERSITY
Meyerhof, Walter E. and Hanna, Stanley S.; *Equipment for 18-MEV Tandem Van de Graaff Accelerator* (P023286); 12 months; $72,600

INDIANA

INDIANA UNIVERSITY, Bloomington
Miller, Daniel W. and Rickey, Martin E.; *Experimental Facilities for the Indiana Cyclotron* (P024846); 12 months; $1,102,000

NEW YORK

UNIVERSITY OF ROCHESTER
Gove, Harry E.; *Experimental Equipment for the Nuclear Structure Laboratory* (P025931); 12 months; $14,000

PENNSYLVANIA

UNIVERSITY OF PENNSYLVANIA
Middleton, Roy and Stephens, William; *Purchase of Van De Graaff Accelerator Tubes* (P018584); 12 months; $16,500

Atomic and Molecular Facilities

CALIFORNIA

STANFORD RESEARCH INSTITUTE
Smith, Felix T.; *Equipment for Atomic Collision Research* (P009676); 12 months; $20,000

CONNECTICUT

WESLEYEN UNIVERSITY
Hill, Henry A.; *Supporting Facilities at the Santa Catalina Laboratory for Experimental Relativity by Astrometry (SCLERA)* (P014565); 16 months; $30,700

NEW YORK

NEW YORK UNIVERSITY
Stroke, H. Henry; *Specialized Research Equipment for High-Resolution Spectroscopy* (P016676); 12 months; $24,200

WASHINGTON

UNIVERSITY OF WASHINGTON
Dehmelt, Hans G.; *Equipment for Research on Radio Frequency Spectroscopy of Stored Atoms and Ions* (P016669); 12 months; $50,000

Solid State Facilities

NEW JERSEY

PRINCETON UNIVERSITY
Schnatterly, Stephen E.; *Inelastic Electron Scattering Facility for Solid State Physics Research* (P015406); 12 months; $90,800

Chemistry Research Instruments

ALABAMA

UNIVERSITY OF ALABAMA, Tuscaloosa
Van Artsdalen, Ervin R.; *Purchase of an Automated X-Ray Diffractometer* (P018286); 12 months; $30,000

ARKANSAS

UNIVERSITY OF ARKANSAS
Meyer, Walter L.; *Purchase of a 100 MHz Nuclear Magnetic Resonance Spectrometer* (P018291); 12 months; $49,000

CALIFORNIA

CALIFORNIA INSTITUTE OF TECHNOLOGY
Hammond, George S.; *Purchase of an Ion Cyclotron Resonance Spectrometer* (P018398); 12 months; $32,800

UNIVERSITY OF CALIFORNIA, Riverside
Helmkamp, George K.; *Purchase of an Automated X-Ray Diffractometer* (P018035); 12 months; $40,000

UNIVERSITY OF CALIFORNIA, San Diego, REVELLE COLLEGE
Shuler, Kurt E.; *Purchase of a Mass Spectrometer* (P018245); 12 months; $15,100

UNIVERSITY OF CALIFORNIA, Santa Cruz
Bunnett, J. F.; *Purchase of a Spectropolarimeter* (P018115); 12 months; $20,000

UNIVERSITY OF CALIFORNIA, Santa Barbara
Bunton, Clifford A.; *Purchase of UV-VIS Spectrophotometer* (P018434); 12 months; $9,800

COLORADO

COLORADO STATE UNIVERSITY
Osteryoung, Robert A.; *Purchase of a Digital Data Acquisition System* (P018219); 12 months; $25,000

CONNECTICUT

UNIVERSITY OF CONNECTICUT
Vaughan, W. R.; *Purchase of Accessories to Up-grade a Mass Spectrometer* (P018332); 12 months; $40,000

WESLEYAN UNIVERSITY
Wharton, Peter S.; *Purchase of a Mass Spectrometer* (P018381); 12 months; $17,500

YALE UNIVERSITY
Wiberg, Kenneth B.; *Purchase of a Digital Data Acquisition System* (P018278); 12 months; $17,900

FLORIDA

FLORIDA STATE UNIVERSITY
Choppin, Gregory R.; *Purchase of an Electron Paramagnetic Resonance Spectrometer* (P017789); 12 months; $34,500

UNIVERSITY OF FLORIDA
Jones, W. M.; *Purchase of a Spectropolarimeter* (P018257); 12 months; $21,000

UNIVERSITY OF MIAMI
Stuckwisch, C. G.; *Purchase of a UV-VIS-NIR Spectrophotometer* (P018290); 12 months; $12,000

GEORGIA

GEORGIA INSTITUTE OF TECHNOLOGY
Spicer, William M.; *Purchase of an Electron Paramagnetic Resonance Spectrometer* (P018116); 12 months; $22,000

UNIVERSITY OF GEORGIA
Hill, R. K.; *Purchase of a Photoelectron Spectrometer* (P015861); 12 months; $40,000

HAWAII

UNIVERSITY OF HAWAII
Inskeep, Richard G.; *Purchase of an Automated X-Ray Diffractometer* (P018213); 12 months; $50,000

ILLINOIS

NORTHWESTERN UNIVERSITY
Basolo, Fred; *Purchase of a Signal Averaging Assembly* (P018384); 12 months; $25,000

UNIVERSITY OF CHICAGO
Nachtrieb, Norman H.; *Purchase of a Fourier Transform Accessory for an NMR Spectrometer* (P018307); 12 months; $30,000

UNIVERSITY OF ILLINOIS, Urbana
Gutowsky, H. S.; *A Laser Raman Spectrometer and Interferometer* (P018180); 12 months; $34,000

INDIANA

PURDUE UNIVERSITY
Foster, Joseph F.; *A Photoelectron Spectrometer* (P015335); 12 months; $50,000

UNIVERSITY OF NOTRE DAME
Magee, John L.; *An Instrument Modification for a 100 MHz NMR Spectrometer* (P018214); 12 months; $18,000

IOWA

UNIVERSITY OF IOWA
Duke, Frederick R.; *A Laser Accessory for a Raman Spectrometer* (P017684); 12 months; $10,500

KANSAS

KANSAS STATE UNIVERSITY
Daane, Adrian H.; *Purchase of a Nuclear Magnetic Resonance Spectrometer* (P018209); 12 months; $12,300

KENTUCKY

UNIVERSITY OF KENTUCKY
Kiser, Robert W.; *Purchase of an Electron Paramagnetic Resonance Spectrometer* (P018397); 12 months; $27,500

MARYLAND

JOHNS HOPKINS UNIVERSITY
Parr, Robert G.; *Purchase of a Recording Spectropolarimeter* (P018404); 12 months; $25,000

MASSACHUSETTS

HARVARD UNIVERSITY
Vanelli, Ronald E.; *Purchase of a Fourier Transform Accessory for an NMR Spectrometer* (P018403); 12 months; $49,700

MASSACHUSETTS INSTITUTE OF TECHNOLOGY
Ross, John; *Purchase of a 100 MHz Nuclear Magnetic Resonance Spectrometer with Fourier Transform Accessories* (P018380); 12 months; $59,000

NORTHEASTERN UNIVERSITY
Weiss, Karl: *Purchase of an Ultraviolet-Visible-Near Infrared Spectrophotometer* (P018296); 12 months; $9,800

TUFTS UNIVERSITY
Urry, Grant; *Purchase of a Mass Spectrometer* (P018255); 12 months; $23,000

UNIVERSITY OF MASSACHUSETTS
McEwen, William E.; *Purchase of a Medium Resolution Mass Spectrometer* (P018387); 12 months; $37,000

MICHIGAN

UNIVERSITY OF MICHIGAN
Overberger, Charles G.; *Purchase of an Automated X-Ray Diffractometer* (P018277); 12 months; $50,000

MINNESOTA

UNIVERSITY OF MINNESOTA
Hexter, R. M.; *Purchase of a Far Infrared Spectrophotometer* (P018325); 12 months; $17,200

ENGINEERING RESEARCH FACILITIES

MISSISSIPPI

UNIVERSITY OF MISSISSIPPI
Scott, Robert B. Jr.; *Purchase of an Electron Paramagnetic Resonance Spectrometer* (P018312); 12 months; $15,300

MISSOURI

UNIVERSITY OF MISSOURI, Columbia
Guyon, John; *Purchase of a Digital Data Acquisition System* (P018364); 12 months; $25,000

MONTANA

MONTANA STATE UNIVERSITY
Caughlan, Charles N.; *Purchase of an Amino Acid Analyzer* (P018093); 12 months; $13,000

NEBRASKA

UNIVERSITY OF NEBRASKA, Lincoln
Cromwell, Norman H.; *Purchase of a Digital Data Acquisition System* (P018383); 12 months; $25,000

NEVADA

UNIVERSITY OF NEVADA, Reno Campus
Guss, Cyrus O.; *Purchase of a Laser Raman Spectrometer* (P018299); 12 months; $15,000

NEW HAMPSHIRE

DARTMOUTH COLLEGE
Spencer, Thomas A.; *Purchase of Parts for a Multinuclear NMR Spectrometer* (P018303); 12 months; $29,100

NEW JERSEY

PRINCETON UNIVERSITY
Mislow, Kurt; *Purchase of a Vacuum Ultraviolet Spectrograph* (P018254); 12 months; $21,700

NEW MEXICO

NEW MEXICO STATE UNIVERSITY
Anex, Basil G.; *Purchase of a Recording Spectropolarimeter* (P018365), 12 months; $22,300

NEW YORK

COLUMBIA UNIVERSITY
Dailey, Benjamin P.; *Purchase of a Spectrofluorometer* (P018331); 12 months; $13,000

CUNY, HUNTER COLLEGE
Barrett, Edward J.; *Purchase of an Infrared Spectrophotometer* (P018327); 12 months; $19,800

SYRACUSE UNIVERSITY
Baker, W. A.; *Purchase of a Recording Spectropolarimeter* (P018452); 12 months; $22,500

NORTH CAROLINA

UNIVERSITY OF NORTH CAROLINA AT CHAPEL HILL
Little, William F.; *Purchase of a Laser Raman Spectrometer* (P018388); 12 months; $16,500

NORTH DAKOTA

NORTH DAKOTA STATE UNIVERSITY
Sugihara, James M.; *Purchase of Accessories to Upgrade a Mass Spectrometer* (P018369); 12 months; $22,000

OHIO

KENT STATE UNIVERSITY
Myers, Raymond R.; *Purchase of an Analytical Ultracentrifuge* (P018114); 12 months; $22,400

OKLAHOMA

OKLAHOMA STATE UNIVERSITY
Dermer, Otis C.; *Purchase of a 100 MHz Nuclear Magnetic Resonance Spectrometer* (P017641); 12 months; $35,000

UNIVERSITY OF OKLAHOMA
Ciereszko, Leon S.; *Purchase of an Automated X-Ray Diffractometer* (P018179); 12 months; $32,000

OREGON

UNIVERSITY OF OREGON
Boekelheide, Virgil; *Purchase of a Fourier Transform Infrared Spectrometer* (P018289); 12 months; $30,000

PENNSYLVANIA

UNIVERSITY OF PENNSYLVANIA
White, David; *Purchase of a Recording Spectropolarimeter* (P018167); 12 months; $25,000

VILLANOVA UNIVERSITY
Doyne, T. H.; *Purchase of an Infrared Spectrophotometer* (P017906); 12 months; $12,500

RHODE ISLAND

UNIVERSITY OF RHODE ISLAND
Rosie, Douglas M.; *Purchase of a Raman Spectrometer with Laser Source* (P018253); 12 months; $19,400

SOUTH CAROLINA

UNIVERSITY OF SOUTH CAROLINA
Bonner, O. D.; *Purchase of Components to Automate an Existing X-Ray Diffractometer* (P018507); 12 months; $30,500

TENNESSEE

UNIVERSITY OF TENNESSEE
Shirley, David A.; *Purchase of Components for a Photoelectron Spectrometer* (P018095); 12 months; $26,300

TEXAS

RICE UNIVERSITY
Margrave, John L.; *Purchase of a Laboratory Data Acquisition System* (P018372); 12 months; $40,000

UNIVERSITY OF HOUSTON
Wendlandt, W. W.; *Purchase of a Mass Spectrometer* (P018110); 12 months; $16,500

UTAH

UNIVERSITY OF UTAH
Grant, David M.; *Purchase of an Infrared Spectrophotometer* (P018435); 12 months; $13,000

VERMONT

UNIVERSITY OF VERMONT
Wulff, Claus A.; *Purchase of a Mass Spectrometer* (P017898); 12 months; $28,400

VIRGINIA

UNIVERSITY OF VIRGINIA
Martin, R. Bruce; *Purchase of a Laser Raman Spectrometer* (P018251); 12 months; $21,000

VIRGINIA POLYTECHNIC INSTITUTE
Clifford, Alan F.; *Purchase of an Electron Paramagnetic Resonance Spectrometer* (P018117); 12 months; $35,900

WASHINGTON

UNIVERSITY OF WASHINGTON
Schomaker, Verner; *Components for a Digital Data Acquisition System for a Mass Spectrometer* (P018433); 12 months; $42,300

WISCONSIN

UNIVERSITY OF WISCONSIN, Madison
Shain, Irving; *A Photoelectron Spectrometer* (P018396); 12 months; $55,000

Engineering Research Facilities

CALIFORNIA

STANFORD UNIVERSITY
Eustis, Robert H.; *Specialized Engineering Research Equipment* (K017019); 12 months; $41,000

UNIVERSITY OF CALIFORNIA, Berkeley
Angelakos, D. J.; *Specialized Engineering Research Equipment—Acquisition of Semiconductor Laboratory Research Equipment* (K019700); 12 months; $25,000

UNIVERSITY OF SOUTHERN CALIFORNIA
Pratt, William K.; *Specialized Research Equipment—Color Image Display Unit* (K019283); 12 months; $4,600

COLORADO

COLORADO SCHOOL OF MINES
Herold, Paul G.; *Purchase of Specialized Engineering Research Equipment* (K018080); 12 months; $9,500

COLORADO STATE UNIVERSITY
Cermak, J. E.; *Purchase of Data Acquisition Equipment for an Atmospheric-Surface-Layer Wind Tunnel* (K023679); 12 months; $15,900

CONNECTICUT

YALE UNIVERSITY
Schulz, George J.; *Data Handling System and Signal Averager* (K018239); 12 months; $27,400

FLORIDA

UNIVERSITY OF FLORIDA
Schaub, James H.; *Field Exploration Equipment for Soil Engineering* (K021365); 12 months; $11,500

UNIVERSITY OF SOUTH FLORIDA
Griffith, John E.; *Purchase of Photoelastic Transmission Polariscope System* (K014079); 12 months; $5,600

ILLINOIS

ILLINOIS INSTITUTE OF TECHNOLOGY
Dally, James W. and Broutman, Lawrence J.; *Purchase of High-Rate Servo-Control Loading System* (K015361); 12 months; $20,000

UNIVERSITY OF ILLINOIS, Urbana
Newmark, Nathan M.; *Purchase of Data Acquisition and Data Transmission Equipment* (K016707); 12 months; $62,100
Westwater, James W.; *Specialized Engineering Research Equipment* (K017692); 12 months; $25,500

INDIANA

PURDUE UNIVERSITY
Greenkorn, Robert A.; *Purchase of a Rheogoniometer* (K023322); 12 months; $27,500

UNIVERSITY OF NOTRE DAME
Szewczyk, Albin A.; *Purchase of a Tape Recorder for Turbulence Studies* (K012505); $13,700

IOWA

UNIVERSITY OF IOWA
Glover, John R.; *Purchase of Data Handling Facilities* (K016285); 12 months; $56,200

SPECIALIZED RESEARCH FACILITIES SUPPORT

MISSOURI

UNIVERSITY OF MISSOURI, Columbia
Dwyer, Samuel J. III; *Specialized Research Equipment to Improve Image Display Facility Utilizing Scan Conversion Devices* (K020401); 12 months; $26,400

UNIVERSITY OF MISSOURI, Rolla
Tracey, James H.; *Specialized Engineering Research Equipment-Computer Graphics Terminal* (K018187); 12 months; $26,600

NEW JERSEY

RUTGERS UNIVERSITY
Newman, B. A. and Sauer, J. A.; *Specialized Engineering Research Equipment* (K014442); 12 months; $8,000

NEW YORK

RENSSELAER POLYTECHNIC INSTITUTE
Corelli, John C.; *Cathode Ray Accelerator Facility* (K023084); 12 months; $4,600

OREGON

OREGON STATE UNIVERSITY
Wang, C. H.; *Purchase of Reactor Heat-Exchange Components to Enhance Materials Research Projects* (K024639); 12 months; $23,000

PENNSYLVANIA

DREXEL UNIVERSITY
Heckel, R. W.; *Purchase of an X-Ray Diffractometer* (K023748); 12 months; $6,100
Sun, Hun H.; *Specialized Research Equipment—Biomedical Engineering Computer Facility* (K023617); 12 months; $19,900

UNIVERSITY OF PENNSYLVANIA
Forsman, William C.; *Purchase of a Rheogoniometer* (K024070); 12 months; $27,500

TENNESSEE

UNIVERSITY OF TENNESSEE
Brooks, C. R.; *Purchase of a Scanning Electron Microscope* (K024912); 12 months; $20,000

TEXAS

UNIVERSITY OF HOUSTON
Finch, R. D. and Muster, D.; *Specialized Engineering Research Equipment—Acoustic Noise Studies* (K017798); 12 months; $26,900

UNIVERSITY OF TEXAS AT AUSTIN
Wilkov, M. A. and Ellison, S. P.; *Assistance in the Purchase of a Scanning Electron Microscope* (K012856); 12 months; $15,000

UTAH

UNIVERSITY OF UTAH
Sosin, A.; *Purchase of a Scanning Electron Microscope* (K018133); 12 months; $20,000

VIRGINIA

UNIVERSITY OF VIRGINIA
Meem, J. L.; *Increasing the Research Capabilities of the University of Virginia Nuclear Reactor Facility* (K018555); 12 months; $17,000

WEST VIRGINIA

WEST VIRGINIA UNIVERSITY
Kemp, Emory L.; *Specialized Engineering Research Equipment* (K023244); 12 months; $13,500

University Astronomy Research Facilities

DISTRICT OF COLUMBIA

CARNEGIE INSTITUTION OF WASHINGTON
Howard, Robert F.; *Modernization of the Optics of the 150-Foot Solar Tower Telescope at Mount Wilson* (P019487); 12 months; $31,400

MASSACHUSETTS

MASSACHUSETTS INSTITUTE OF TECHNOLOGY
Burke, Bernard F.; *Microwave Aperture Synthesis Facility* (P014589); 12 months; $76,000

Special Astronomy Equipment

MASSACHUSETTS

MASSACHUSETTS INSTITUTE OF TECHNOLOGY
Clark, George W.; *Construction of a Digital TV Photon Spectrum Scanner for Optical Observation of X-Ray Sources* (P015932); 18 months; $48,700

NEW MEXICO

NEW MEXICO STATE UNIVERSITY
Seeger, Charles L.; *Spectrograph for the New Mexico University Observatory* (P015437); 12 months; $12,500

OREGON

UNIVERSITY OF OREGON
Ebbighausen, E. G. and Donnelly, R. J.; *Purchase of Low Dispersion Spectrograph with Image Tube for the Pine Mountain Observatory* (P014621); 12 months; $21,200

PENNSYLVANIA

SWARTHMORE COLLEGE
Van De Kamp, Peter, *Construction of a Plate Storage Vault* (P017649); 12 months; $10,200

NATIONAL RESEARCH CENTERS

National Radio Astronomy Observatory

WEST VIRGINIA

NATIONAL RADIO ASTRONOMY OBSERVATORY
Heeschen, D. S.; *Management, Operation, and Maintenance of the National Radio Astronomy Observatory* (C450000009); 12 months; $5,800,000

National Center for Atmospheric Research

Site Development and Facilities

COLORADO

JEFFERSON COUNTY AIRPORT AUTHORITY
Huntsbarger, William; *Contract for Lease of Land for Use by NCAR Aircraft Facility, at Jefferson County Airport in Colorado* (CL1); $41,000

DISTRICT OF COLUMBIA

GENERAL SERVICES ADMINISTRATION
Schmidt, William A.; *Contract for Purchase of the Marshall Site for the National Center for Atmospheric Research* (CA11001); $21,000

General Operations

COLORADO

NATIONAL CENTER FOR ATMOSPHERIC RESEARCH
Firor, John; *Contract for Semi-Annual Payments on Land Use Agreement (Contract No 14-06-420-861) with the Bureau of Reclamation* (CA18); 6 months; $200
———; *Contract for the Management, Operation, and Maintenance of the National Center for Atmospheric Research, Boulder, Colorado* (C460000029); $75,000
———; (C460000031); $61,800
———; (C460000024); $7,200,000
———; (C460000023); $900,000
———; (C460000020); $3,000,000
———; (C460000028); $113,100

DISTRICT OF COLUMBIA

DEPARTMENT OF THE NAVY,
AIR SYSTEMS COMMAND
Appling, Carl; *Contract for Overhaul G.E. T64-10 Turboprop Engine for Buffalo Aircraft* (CA16000); 12 months; $18,000

NEW MEXICO

DEPARTMENT OF THE AIR FORCE,
HOLLOMON AIR FORCE BASE
Kanavy, C. C.; *Services Necessary to Measure Densities of Radar Images on Some 7500 Frames of 35-mm Film (C460)* (CA26000); $6,500

Kitt Peak National Observatory

ARIZONA

KITT PEAK NATIONAL OBSERVATORY
Mayall, Nicholas U.; *Contract for the Management, Operation and Maintenance of the Kitt Peak National Observatory, Tucson, Arizona* (C400000006); 12 months; $1,620,000
———; (C400000008); 12 months; $4,805,000

Cerro Tololo Inter-American Observatory

CERRO TOLOLO INTER-AMERICAN OBSERVATORY
Associated Universities for Research in Astronomy; *Contract for the Management, Operation, and Maintenance of the Cerro Tololo Inter-American Observatory, La Serena, Chile, South America (NSF C-525)* (C525000008); 12 months; $1,420,000
———; (C525000007); 12 months; $480,000

Arecibo Ionospheric Observatory

Site Development and Facilities

PUERTO RICO

ARECIBO OBSERVATORY
Pettengill, Gordon H.; *Contract for Operation and Maintenance of the Arecibo Observatory in Puerto Rico* (C600000003*); $75,000
———; (C600000002); $75,000

General Operation

NEW YORK

CORNELL UNIVERSITY
——— ; *Contract for Management, Operation and Maintenace of the Arecibo Ionospheric Observatory, Puerto Rico* (C600000); 12 months; $1,400,000

NATIONAL RESEARCH PROGRAMS

U.S. Antarctic Research Program

Scientific Research Programs

ALASKA

UNIVERSITY OF ALASKA

Akasofu, Syun-ichi; *Study of the Midday Auroras at the South Pole Station* (V024104); 12 months; $12,000

Hessler, V. P. and Heacock, R. R.; *Extra Low Frequency Studies at the Geomagnetic Poles* (A016496); 18 months; $37,500

Mather, Keith B. and Forbes, Robert B.; *Inaugural Symposium—the Geophysics and Geology of the Bering Sea Region* (A018785); 6 months; $12,500

Wescott, Eugene M. and Nielsen, Hans C. S.; *Seasonal Asymmetry of Aurora Conjugacy* (A020212); 12 months; $69,700

CALIFORNIA

DEPARTMENT OF THE NAVY, NAVAL UNDERSEAS R & D CENTER

Cummings, William C.; *Bioacoustics and Related Behavior of Subantarctic Cetaceans* (AG217); 12 months; $11,200

McDONNELL DOUGLAS ASTRONAUTICS

Masley, A. J.; *Contract for Conducting a Conjugate Point Riometer Program* (C393000005); 12 months; $50,069

SAN DIEGO NATURAL HISTORY MUSEUM

Gilmore, Raymond M.; *Population, Distribution and Behavior of Whales in Antarctic Waters* (A019306); 12 months; $42,300

STANFORD RESEARCH INSTITUTE

Bates, Howard F.; *Establishment of a Polar Auroral Radar System* (A010569); 18 months; $54,000

STANFORD UNIVERSITY

Helliwell, Robert A.; *Magnetospheric Research in Antarctica* (A019608); 12 months; $175,800

UNIVERSITY OF CALIFORNIA, Berkeley

Mozer, Forrest; *Measurement of Electric Fields on Balloons Flown from Eights, Antarctica* (A016489); 12 months; $16,500

Olcott, Harold S.; *Chlorinated Hydrocarbons—Patterns and Effects Upon the Reproductive Capacity of Antarctic Pelagic Sea Birds* (A014202); 12 months; $15,200

———; (V023898); 12 months; $20,700

UNIVERSITY OF CALIFORNIA, Davis

Feeney, Robert E.; *Comparative Biochemistry of Proteins (Antarctica)* (A022763); 12 months; $55,800

UNIVERSITY OF CALIFORNIA, Los Angeles

Norris, Kenneth S.; *Cetacea of Tierra del Fuego* (GA1420 002); 6 months; $2,300

Slichter, Louis B.; *Geophysical Measurements at the South Pole* (A017012); 12 months; $47,900

UNIVERSITY OF CALIFORNIA, Riverside

Rex, Robert W. and Margolis, Stanley V.; *Surface Microtextures of Sand Grains from the Southern Oceans* (A017084); 12 months; $8,800

UNIVERSITY OF CALIFORNIA, San Diego

Elsner, Robert W.; *Temperature Regulation in the Newborn Weddell Seal, 'Leptonychotes Weddelli'* (V023899); 12 months; $33,000

Hammel, Harold T.; *Reversible Freezing in Subantarctic Plant Tissue* (A019604); 12 months; $10,100

Hessler, Robert R.; *The Importance of Deep-Sea Isopod Taxa in Shallow Antarctic Waters* (A018946); 5 months; $2,700

COLORADO

DEPARTMENT OF COMMERCE, ENVIRONMENTAL SCIENCE SERVICES ADMINISTRATION

Campbell, Wallace H.; *Ultra Low Frequency (ULF) Studies in Antarctica* (AG97003); $5,000

UNIVERSITY OF COLORADO

Lemasurier, Wesley E.; *Petrology of Volcanic Rocks from Marie Byrd Land, Antarctica* (A021488); 12 months; $5,600

UNIVERSITY OF DENVER

Barcus, James R.; *Investigation of Variations in Low-Energy Cosmic Ray Cutoffs* (V024825); 12 months; $150,300

DISTRICT OF COLUMBIA

ARCTIC INSTITUTE OF NORTH AMERICA

Dalrymple, Paul C. and Kuhn, Michael; *Reduction and Analysis of Radiation Data, Plateau, Antarctica* (V023493); 12 months; $14,400

Millman, Peter M.; *Investigation of Auroral Conjugacy During Maximum Solar Activity* (A021487); 12 months; $15,000

BUREAU OF COMMERCIAL FISHERIES

Alton, Miles S.; *Antarctic Marine Resources Survey* (AG163001); 12 months; $98,200

DEPARTMENT OF COMMERCE, ENVIRONMENTAL SCIENCE SERVICES ADMINISTRATION

Nelson, James H.; *USARP Magnetic Observatories* (AG13700002); 12 months; $50,000

GEOLOGICAL SURVEY

James, Harold L.; *Geologic Mapping of the Lassiter Coast Area, Antarctica* (AG187001); 12 months; $86,500

Pecora, William T.; *Preparation, Printing, and Distribution of a Uniform Series of Antarctic Geologic Maps* (CA02000); 9 months; $4,800

SMITHSONIAN INSTITUTION

Wallen, I. E.; *Participation in USARP Expeditions* (GA4105001); 36 months; $1,846

———; *Cooperative Systematics Studies in Antarctic Biology* (GA1374001); $10,036

———; (GA15272001); $11,232

———; (A015272); 12 months; $41,600

FLORIDA

FLORIDA STATE UNIVERSITY

Goodell, H. Grant; *Marine Geology of the Southern Ocean* (A015703); 12 months; $78,900

———; *Contract for—Curatorship of 'Eltanin' Core Collection* (C564000001); 12 months; $17,100

Kennett, James P.; *Micropaleontology and Paleoenvironment of Southern Marine Sediments* (A015230); 12 months; $14,300

Watkins, N. D.; *Magnetic Properties of Antarctic Marine Sediments and Rocks* (A013132); 11 months; $33,700

UNIVERSITY OF MIAMI

Fell, Jack W.; *Distribution of Antarctic Marine Fungi* (A013675); 26 months; $26,700

IOWA

IOWA STATE UNIVERSITY

Baker, John R.; *Embryology and Incubation Behavior of the Adelie and Related Penguin Species* (V023744); 36 months; $54,300

KANSAS

UNIVERSITY OF KANSAS

Dort, Wakefield Jr.; *Glacial Geology of the McMurdo Area, Antarctica* (A016037); 12 months; $9,200

MARYLAND

DEPARTMENT OF COMMERCE, ENVIRONMENTAL SCIENCE SERVICES ADMINISTRATION

Campbell, W. H.; *Ultra Low Frequency (ULF) Studies in Antarctica* (AG97000004); 12 months; $65,900

Crary, J. H.; *VLF Studies in Antarctica* (AG98000003); 12 months; $2,600

Reid, George C.; *ERL Antarctic Riometer Program* (AG13900002); 12 months; $67,600

Rockney, Vaughn D.; *ESSA Antarctic Meteorology Research Program, 1970* (AG19800001); 8 months; $86,700

Weickman, H. K.; *ESSA Antarctic Meteorology Research Programs* (AG198002); 12 months; $152,300

Weyant, William S.; *A Detailed Study of the First Three Kilometers of Atmosphere Above the High Antarctic Plateau* (AG102001); $109

JOHNS HOPKINS UNIVERSITY

Sladen, William J. L.; *Antarctic Avian Population Studies* (A014622); 12 months; $38,100

UNIVERSITY OF MARYLAND

Rosenberg, T. J. and Matthews, D. L.; *Energetic Electron Precipitation Near the Plasmapause* (A019786); 15 months; $70,000

MASSACHUSETTS

CLARK UNIVERSITY

Gannutz, Theodore P.; *Photosynthesis and Respiration of Antaractic Lichens* (A013295); 12 months; $34,600

WOODS HOLE OCEANOGRAPHIC INSTITUTION

Warren, Bruce A.; *Physical Oceanography of Deep Antarctic Current, South Pacific Ocean* (A017772); 12 months; $58,000

———; (GA10824001); $1,100

MICHIGAN

UNIVERSITY OF MICHIGAN

Sparrow, Frederick K. Jr.; *Occurrence and Distribution of Zoospori Fungi in Antarctic Waters* (A016097); 24 months; $35,200

MINNESOTA

UNIVERSITY OF MINNESOTA

Erickson, Albert W. and Siniff, Donald B.; *Status and Population Dynamics of Antarctic Seals* (A016445); 6 months; $65,800

U.S. ANTARTIC RESEARCH PROGRAM

MONTANA

UNIVERSITY OF MONTANA
Honkala, Rudolf A.; *Melt as a Factor in Ice Loss at Anvers Island, Antarctica* (A016287); 12 months; $23,900

NEW HAMPSHIRE

DEPARTMENT OF THE ARMY,
TERRESTRIAL SCIENCE CENTER
Gow, Anthony J.; *Antarctic Ice Core Analysis and Related Glaciological Studies* (AG10500002); 12 months; $23,400
Hansen, B. Lyle; *Deep Core Drilling in Ice Project, Byrd Station, Antarctica* (AG106002); 12 months; $38,000
Langway, C. C. Jr.; *The Analysis of the Deep Ice Cores and Sub-ice Material* (AG21200); 12 months; $44,200
Langway, Chester C. Jr.; *Contract for Operation of the Central Ice Core Storage Facility* (CA02300); 12 months; $18,400

NEW JERSEY

PRINCETON UNIVERSITY
Judson, Sheldon; *Testing of Antarctic Ice Surges* (A014623); 7 months; $10,600

NEW MEXICO

UNIVERSITY OF NEW MEXICO
Frakes, Lawrence A.; *Marine Geology of the Weddell Sea* (A016497); 12 months; $42,000

NEW YORK

COLUMBIA UNIVERSITY
Dalziel, Ian W. D.; *Structural Studies in the Northwest Straits of Magellan, Chile* (A012301); 18 months; $23,600
Ewing, Maurice and Hayes, Dennis E.; *Geophysical Surveys and Studies from 'Eltanin'* (GV23334); 5 months; $79,700
Gordon, Arnold L.; *Collection, Processing and Analysis of Physical Oceanographic Data in Antarctic Waters* (GV19032); 5 months; $44,600
Hays, James D.; *Biostratigraphy and Mineralogy of Late Cenozoic Antarctic Deep-Sea Cores* (V021174); 9 months; $22,100

OHIO

CASE WESTERN RESERVE UNIVERSITY
Nairn, A. E. M.; *Paleomagnetism of Rocks from the Queen Alexandra Range, Antarctica* (A016598); 12 months; $18,600

OHIO STATE UNIVERSITY
Dalrymple, Paul C.; *Analysis of Plateau Micrometeorological Tower Data* (V024303); 12 months; $41,300
Dewart, Gilbert; *Geophysical and Glaciological Studies Along the Byrd Station Strain Net, Antarctica* (A014425); 13 months; $39,300
Klay, Jean-Roland and Orheim, Olav; *Effects of Recent Volcanic Eruptions on the Glaciers of Deception Island* (A014733); 16 months; $24,500
Rastorfer, James R.; *Physiological Studies of Antarctic Mosses* (A016000); 12 months; $18,600

OKLAHOMA

OKLAHOMA MEDICAL RESEARCH FOUNDATION
Shurley, Jay T.; *Quantitative Psychophysiological Sleep Patterns Study Under Polar Conditions* (A018698); 12 months; $54,100

OREGON

OREGON STATE UNIVERSITY
Curl, Herbert C. Jr.; *Physiological Ecology of Cryophilic Algae in Antarctica* (A014201); 24 months; $23,500

Hedgpeth, Joel W.; *Tidal Zone Ecology at Palmer Station, Anvers Island, Antarctica* (A018348); 30 months; $62,700

PENNSYLVANIA

FRANKLIN INSTITUTE
Pomerantz, Martin A.; *Investigations of Cosmic Ray Intensity Variations in Antarctica* (A021175); 12 months; $99,900

LEHIGH UNIVERSITY
Macnamara, E. Everett; *Pedology of Enderby Land, Antarctica* (A017208); 12 months; $8,703

RHODE ISLAND

UNIVERSITY OF RHODE ISLAND
Duce, Robert A.; *Trace Metals and Halogens in the Antarctic Atmosphere* (A020010); 12 months; $3,900

SOUTH CAROLINA

UNIVERSITY OF SOUTH CAROLINA
Conolly, John R.; *Sedimentation Patterns in the Southern Ocean Between Australia and Antarctica* (A019722); 8 months; $9,900

SOUTH DAKOTA

UNIVERSITY OF SOUTH DAKOTA
Rutford, Robert H.; *Geology of the Jones and Ellsworth Mountains, Antarctica* (V024156); 7 months; $4,600

SWITZERLAND

UNIVERSITY OF BERN
Oeschger, Hans; *Down Deep Bore Hole Carbon Dating Project at Byrd Station, Antarctica* (A016802); 21 months; $40,500

TEXAS

TEXAS A & M UNIVERSITY
Capurro, Luis R. A.; *Physical Oceanography of the Weddell Sea, Antarctica* (A016869); 12 months; $22,300
El-Sayed, Sayed Z. and Darnell, Rezneat M.; *Dynamics of Trophic Relations in the Southern Ocean* (A013836); 12 months; $71,000
Park, Tai Soo; *Systematic Studies of Antarctic Copepods* (A017085); 12 months; $25,800

UNIVERSITY OF TEXAS AT AUSTIN
Tucker, Arnold J.; *Contract for Geodetic and Upper Atmospheric Studies in Antarctica Using Artificial Earth Satellites* (C560000001); 9 months; $24,200

UNIVERSITY OF TEXAS AT DALLAS
Halpern, Martin; *Geochronologic-Geologic Studies in the Northwest Straits of Magellan, Chile* (A010529); 18 months; $57,100

UTAH

UTAH STATE UNIVERSITY
Muller-Schwarze, Dietland; *Anti-Predator and Social Behavior in Adelie Penguins* (V023494); 12 months; $30,500

VIRGINIA

VIRGINIA POLYTECHNIC INSTITUTE
Benoit, Robert E.; *Isolation and Physiology of Antarctic Microorganisms* (V025478); 12 months; $22,500
Parker, Bruce C.; *Limnological Investigations of Algal Communities* (A016768); 24 months; $49,200
Paterson, Robert A.; *Fresh Water and Marine Antarctic Fungi (Plankton Parasites and Aquatic Saprophytes)* (A016767); 24 months; $20,200

WASHINGTON

UNIVERSITY OF WASHINGTON
Swarm, H. Myron, Reynolds, Donald K. and Peden, Irene C.; *Radioscience Research in Antarctica* (A021176); 12 months; $60,300

WISCONSIN

UNIVERSITY OF WISCONSIN, Madison
Bentley, Charles R.; *Seismic Logging in Deep Drill Holes, Antarctica* (A014598); 22 months; $42,600
Bentley, Charles R. and Clough, John W.; *Electromagnetic Measurements at Byrd Station, Antarctica* (A014597); 12 months; $20,900
Craddock, Campbell; *Analysis of Antarctic Geologic Materials* (A015080); 15 months; $25,100
Schwerdtfeger, Werner; *Research in Antarctic Meteorology* (A016239); 12 months; $32,000

WISCONSIN STATE UNIVERSITY, Oshkosh
Laudon, Thomas S.; *Paleontology and Geochemistry of Ellsworth Land* (A014355); 17 months; $6,300

WYOMING

UNIVERSITY OF WYOMING
Houston, Robert S. and Smithson, Scott B.; *Geological and Geophysical Studies in Southern Victoria Land, Antarctica* (A019938); 16 months; $31,200

Geodesy and Cartography

DISTRICT OF COLUMBIA

DEPARTMENT OF THE ARMY,
CORPS OF ENGINEERS
Thornton, George E.; *Standard Geographic Nomenclature in the Antarctic for United States Use* (CA14000); 12 months; $24,500

GEOLOGICAL SURVEY
Baker, Arthur A.; *Antarctic Mapping Operations* (AG177001); 8 months; $349,600

NEW YORK

AMERICAN GEOGRAPHICAL SOCIETY
Bushnell, Vivian C.; *Contract for Antarctic Map Folio Series* (C293000008); 12 months; $95,500

Field Support

CALIFORNIA

HOLMES & NARVER INC.
Kelley, Charles W.; *Contract for Scientific Support Activities for the U.S. Antarctic Research Program* (C571000003); $11,031
———; *Scientific Support Activities for The U.S. Antarctic Research Program* (C571000002); 12 months; $419,563
———; *Contract for Scientific Support Activities for the U.S. Antarctic Research Program* (C571000001); $10,792

STANFORD UNIVERSITY
Lusignan, Bruce B.; *Contract for Design and Prototype Fabrication of an Automated Unmanned Antarctic Geophysical Laboratory* (C582000001); 10 months; $40,100
Peterson Allen M.; *Electronics Systems Engineering Studies* (A019603); 14 months; $45,800

DISTRICT OF COLUMBIA

AMERICAN EXPRESS COMPANY
Mummery, Desman L.; *Contract for Transportation Services Related to Non-Government Travel in the U.S. Antarctic Research Program* (C334000010); 12 months; $110,000

DEPARTMENT OF THE NAVY,
NAVY SUBSISTENCE OFFICE
Vogel, R. E.; *Rations Furnished USARP Personnel* (CA03000001); 12 months; $70,000

NATIONAL RESEARCH PROGRAMS

NEW JERSEY

Alpine Geophysical Association Inc.
Smolen, Walter; *Contract for the Furnishing of Services in Support of Scientific Program Conducted Aboard the USNS Eltanin* (C448000006); 2 months; $25,160
———; (C598000); 36 months; $392,498
———; (C598000002); 2 months; $30,000

Field Support—Vessel Program

DISTRICT OF COLUMBIA

Department of the Navy,
Military Sea Transport Service
Ramage, Lawson P.; *Contract for USNS Eltanin Operations and Maintenance* (CA013000); 3 months; $45,000
———; (CA01300002); 6 months; $600,000
———; (CA1300001); 3 months; $300,000

Department of the Navy,
Naval Ship Engineer Center
Diehl, William F.; *Contract for Services in Support of USNS 'Eltanin' Satellite Navigation System* (CA05000001); 12 months; $20,000

FLORIDA

Marine Acoustical Services
Ploegert, John; *Contract for Operation of the Antarctic Research Ship R/V Hero, and Palmer Station Laboratory* (C552000002); 12 months; $525,415

Field Support—Exchange Program

ALASKA

University of Alaska
Mather, Keith B.; *Support of Auroral Studies, 1970* (A019075); 12 months; $7,100

COLORADO

Department of Commerce,
Environmental Science Services Administration
Reid, George C. and Campbell, Wallace H.; *ESSA Research Laboratories Observing Program at Vostok* (AG206); 12 months; $6,700
Shapley, A. H.; *U.S. Exchange Scientist for Vostok Winter 1969* (AG166002); 3 months; $3,300

DISTRICT OF COLUMBIA

Arctic Institute of North America
Faylor, Robert C.; *Travel for Visiting Arctic Specialists* (V024798); 12 months; $4,800

General Services Administration,
National Archives and Records Service
Rhoads, James B.; *Regional Geography of the Lutzow-Holm Bay Area, Antarctica* (AG21100); 12 months; $3,060

MARYLAND

Johns Hopkins University
Sladen, William J. L.; *Graduate Studies in Ecology and Comparative Behavior* (A015613); 24 months; $12,600

MASSACHUSETTS

Massachusetts Institute of Technology
Simmons, Gene; *Support of Graduate Studies in Seismology* (A015956); 12 months; $5,700

MISSOURI

Washington University
Scharon, Leroy; *Rock Magnetic Studies in the Antarctic* (A016116); 8 months; $10,700

VIRGINIA

Roanoke College
Thompson, Jesse C. Jr.; *Ecology and Systematics of Antarctic Ciliated Protozoa* (A016495); 12 months; $18,991
———; (GA16495001); 18 months; $3,600

Field Support—Antarctic Information and Records Program

DISTRICT OF COLUMBIA

Arctic Institute of North America
Faylor, Robert C.; *Arctic Bibliography Project* (A020026); 3 months; $20,000

Library of Congress
Thuronyi, Geza T.; *Abstracting and Indexing Service for Current Antarctic Literature* (AG133 002); 12 months; $73,600

Smithsonian Institution
Fehlmann, H. A.; *Contract for Sorting of Collections from the U.S. Antarctic Research Program (USARP)* (C495000003); 12 months; $92,978
Landrum, B. J.; *Contract for Recording of Data for Specimens Collected During the U.S. Antarctic Research Program* (C494000004); 12 months; $49,319
Simkin, Thomas E.; *Contract for Processing of USARP Rock Samples* (C559000003); 12 months; $25,212

Field Support—Advisory Program

National Academy of Sciences,
National Research Council
Degoes, Louis; *Task Order for Committee on Polar Research* (C310036009); 12 months; $80,700

Arctic Ocean Research Program

Department of Transportation,
U. S. Coast Guard
Williams, James W.; *The Icebreaker Eastwind and the North Water Project* (CA27); 2 months; $30,000

WASHINGTON

University of Washington
Murphy, Stanley R. and Fletcher, J. O.; *Contract for the Arctic Ice Dynamics Joint Experiment (AIDJEX)* (C625000); 6 months; $100,000

Weather Modification

ARIZONA

Northern Arizona University
Layton, Richard G.; *Influence of the Surface Properties of Silver Iodide on Ice Nucleation* (A019074); 24 months; $31,700

CALIFORNIA

University of California, Los Angeles
Neiburger, Morris; *Growth of Precipitation Particles* (A019315); 24 months; $100,000

COLORADO

Colorado State University
Corrin, Myron; *Surface Properties of Heterogeneous Nuclei* (A019123); 12 months; $48,700
Grant, Lewis O.; *Heterogeneous Nucleation Research* (A020341); 12 months; $80,000
Sinclair, Peter C.; *Design of Descriptive and Theoretical Cloud Models* (A020330); 12 months; $149,400
———; *Joint Hail Suppression Research Program in Northeastern Colorado* (A020623); 12 months; $50,000

Department of Commerce,
Environmental Science Services Administration
Weickmann, Helmut K.; *Hail Modification Research in Northeastern Colorado* (AG19000001); 12 months; $37,000

National Center for Atmospheric Research
Firor, John; *Contract for the Management, Operation, and Maintenance of the National Center for Atmospheric Research, Boulder, Colorado* (C460000022); 12 months; $359,000
———; (C460000027); 12 months; $324,000

University of Colorado
Haas, J. Eugene; *Social Consequences of Planned Weather Modification* (A018724); 17 months; $52,300

DISTRICT OF COLUMBIA

National Academy of Sciences,
National Research Council
Sievers, John R.; *Task Order for Support of the Committee on Atmospheric Sciences and the U.S. Committee for the Global Atmospheric Research Program* (C310009014*); 12 months; $10,000

ILLINOIS

University of Chicago
Atlas, David; *Dual Wavelength Radar Hail Detector* (A019031); 12 months; $79,600
Braham, Roscoe R. Jr.; *Mass-Phase Partitioning in Natural Clouds* (A020470); 15 months; $245,800
Ginsburg, Norton S.; *Human Dimensions of Weather Modification* (A020971); 12 months; $700

University of Illinois, Urbana
Changnon, Stanley A. Jr.; *Study of Hailfall Data from a Hail-Rain Network* (A016917); 24 months; $115,000
Huff, Floyd A. and Changnon, Stanley A. Jr.; *Climatological Assessment of Urban Effects on Precipitation* (A018781); 24 months; $66,500
Mueller, Eugene A.; *Dual Wave Length Radar System Development for Hail and Related Atmospheric Conditions* (A018909); 18 months; $152,800
Schickedanz, Paul T.; *Theoretical Frequency Distributions for Rainfall Data* (A016168); 18 months; $37,500

MASSACHUSETTS

American Meteorological Society
Spengler, Kenneth C.; *National Conference on Weather Modification* (GA1323001); 25 months; $4,700

MISSOURI

University of Missouri, Columbia
McQuigg, James D.; *Determination of the Potential Socio-Economic Impact of Weather and Climate Modification* (A015413); 24 months; $74,100

MONTANA

Montana State University
Collins, Don D., Wright, John and Weaver, Theodore; *Ecological Effects of Weather Modification* (B020960); 12 months; $40,000

NEW MEXICO

New Mexico Institute of Mining and Technology
Brook, Marx; *Thunderstorm Electrification* (A018864); 24 months; $158,000
Moore, Charles B., Colgate, Stirling A. and Holmes, Charles; *Origin and Role of Elec-*

tricity in Clouds (A019037); 12 months; $163,700

Wilkening, Marvin H., and Brook, Marx; *Support for Operation of Langmuir Laboratory for Atmospheric Research* (A016369*); 12 months; $10,000

PENNSYLVANIA

LEHIGH UNIVERSITY

Zettlemoyer, A. C.; *Surface Chemistry of Ice Nucleation* (GA560X004); 12 months; $14,796

SOUTH DAKOTA

SOUTH DAKOTA SCHOOL OF MINES AND TECHNOLOGY

Davis, Briant L.; *Chemical Complexing of Silver Iodine-Alkali Iodide Aerosols Prepared for Cloud Seeding Purposes* (A018243); 24 months; $67,400

Schleusener, Richard A.; *Development of Research Aircraft for Hailstorm Penetrations* (A024651); 12 months; $82,000

WASHINGTON

UNIVERSITY OF WASHINGTON

Hobbs, Peter V.; *Cloud Physics and Weather Modification* (A017381); 12 months; $134,000

WYOMING

UNIVERSITY OF WYOMING

Veal, Donald L.; *Hailstorm Research* (A019105); 12 months; $59,400

Ocean Sediment Coring Program

PURCHASE ORDERS

Purchase Order; (P069573001); $10,000
———; (70)PO87; $4,081
———; (70)PO51; $36,265
———; (70PO108); $198
———; (70)PO83; $560
———; (70)PO67; $183
———; (70)PO22; $596
———; (70PO110); $73
———; (70)PO50; $8,379
———; (70)PO46; $15,907
———; (70)PO45; $780
———; (70PO104); $5,000
———; (70)PO69; $23
———; (70PO123); $2,500
———; (70PO106); $211

CALIFORNIA

UNIVERSITY OF CALIFORNIA, San Diego

Fisher, Robert L.; *Investigations of the North and Western Indian Ocean and Northwest Pacific—NOSC Site Surveys* (L026475); 18 months; $106,400

Nierenberg, William A.; *Contract for Drilling of Sediments and Shallow Basement Rocks in the Pacific and Atlantic Oceans and Adjacent Seas and the Necessary Curatorial Care, Examination and Distribution of the Resulting Cores* (C482000013); 12 months; $600,000
———; (C482000012); 12 months; $3,250,000
———; (C482000010); $1,500,000
———; (C482000009); $450,000
———; (C482000007); 36 months; $500,000

MARYLAND

DEPARTMENT OF COMMERCE, ENVIRONMENTAL SCIENCE SERVICES ADMINISTRATION

Rockney, Vaughn D.; *Meteorological Support of Deep-Sea Drilling Operations* (CA00700002); 3 months; $13,575
———; (CA07000003); 12 months; $47,500
———; (CA07000001); $9,700

International Biological Program

ARIZONA

ARIZONA STATE UNIVERSITY

Fouquette, M. J. Jr.; *Systematics of South American Amphibia* (GB12560001); 6 months; $7,300

UNIVERSITY OF ARIZONA

McGinnies, William G.; *Collection, Analysis and Synthesis of Plant Ecological Information Pertaining to Arid and Semiarid Portions of North America* (B016971); 12 months; $18,000

CALIFORNIA

UNIVERSITY OF CALIFORNIA, Berkeley

Huffaker, Carl B.; *Natural Enemies in Suppressing Spider Mites in Important Food Crop Ecosystems* (B020961); 24 months; $60,000
———; *Management of an Integrated, Inter-Institutional Program in Biological Control* (B019519); 12 months; $28,100

UNIVERSITY OF CALIFORNIA, Davis

Goldman, Charles R.; *Plankton Ecology and Land-Water Interaction in the Coniferous Forest Biome of Northern California* (B019052); 12 months; $44,700

UNIVERSITY OF CALIFORNIA, Riverside

DeBach, Paul; *Role of Entomophagous Insects in Population Regulation of Armored Scale Insects* (B014489); 24 months; $55,000

COLORADO

COLORADO STATE UNIVERSITY

Van Dyne, George M.; *Analysis of Structure and Function of Grassland Ecosystems* (B013096); 16 months; $1,800,000

CONNECTICUT

UNIVERSITY OF CONNECTICUT

Laughlin, William S.; *Aleut Adaptation to the Bering Land Bridge Coastal Configuration* (B018741); 12 months; $90,000

DISTRICT OF COLUMBIA

AMERICAN INSTITUTE OF BIOLOGICAL SCIENCE

Sprugel, George Jr.; *Conservation of Ecosystems* (B022981); 12 months; $52,600

NATIONAL ACADEMY OF SCIENCES, NATIONAL RESEARCH COUNCIL

Decarlo, Michael R.; *Task Order for Support of the International Biological Program* (C310098007); 12 months; $150,000

HAWAII

BERNICE P. BISHOP MUSEUM

Gressitt, J. Linsley; *Collaborative Research on Hawaii Terrestrial Biology Subprogram* (B023075); 12 months; $89,500

UNIVERSITY OF HAWAII

Berger, Andrew J. and Mueller-Dombois, Dieter; (B023230); 12 months; $140,200

INDIANA

INDIANA UNIVERSITY, Bloomington

Nelson, Craig E.; *Evolutionary Processes Contributing to Ecological and Species Diversity (Microhylidae)* (B018742); 24 months; $45,000

MARYLAND

ATOMIC ENERGY COMMISSION

Auerbach, Stanley I.; *Analysis of the Structure and Function of Ecosystems in the Deciduous Forest Biome* (AG19900001); 12 months; $143,000

Totter, John R.; *Area Program in Population Genetics* (AG220); 7 months; $62,500

NEW HAMPSHIRE

UNIVERSITY OF NEW HAMPSHIRE

Wood, Langley; *Dispersal, Speciation, and Acclimatization in Predatory Marine Gastropods* (B017482); 12 months; $81,000

NEW YORK

ROCKEFELLER UNIVERSITY

Dobzhansky, Theodosius and Ayala, Francisco J.; *Race and Species Formation in South American Drosophila* (B020694); 12 months; $39,000

NORTH CAROLINA

UNIVERSITY OF NORTH CAROLINA AT CHAPEL HILL

Lieth, Helmut; *Compilation of Data from Tropical Biomes Suitable for Use in Ecosystems Modeling* (B020068); 12 months; $3,600

TENNESSEE

ECOLOGICAL SOCIETY OF AMERICA

Huffaker, Carl B.; *Symposium on Theory and Practices of Biological Control* (B019554); 3 months; $3,000

TEXAS

UNIVERSITY OF TEXAS AT AUSTIN

Blair, W. F.; *Program on Convergent and Divergent Evolution* (B015768); 12 months; $115,000

Mabry, Tom J.; *Biochemical Systematic Investigations Emphasizing Studies of Plant Taxa Common to Both North and South America* (B016411); 24 months; $144,200

Wardlaf, Frank H.; *Publication of Brazilian Hylidae,' Volume I* (GN882*); 12 months; $8,000

UTAH

UTAH STATE UNIVERSITY

Wardlaw, Frank H.; *Publication of 'Brazilian and Function of Desert Ecosystems* (B015886); 12 months; $654,300

WASHINGTON

UNIVERSITY OF WASHINGTON

Dugdale, Richard C.; *Dynamics of Biological Production in Upwelling Ecosystems* (B018568); 12 months; $161,500

Gessel, Stanley P.; *Structure and Function of Coniferous Forest Ecosystems* (GB12075001); 6 months; $3,200

Global Atmospheric Research Program

COLORADO

COLORADO STATE UNIVERSITY

Cox, Stephen K. and Vonder Haar, Thomas H.; *Reduction and Analysis of the Wisconsin Radiation Experiment Data Collected During Bomex* (A018783); 24 months; $107,700

Gray, William M.; *Significance of Cumulonimbus Momentum Transport for Atmospheric Processes* (A019937); 24 months; $70,100

CONNECTICUT

CENTER FOR THE ENVIRONMENT AND MAN

Pandolfo, Joseph P.; *Numerical Model Studies of Air/Sea Interactions Utilizing Data from Project Bomex* (A018669); 12 months; $60,200

Robinson, G. D.; *Validity of Meteorological Prediction Equations and the Spectrum of Atmospheric Motion* (A017090); 12 months; $19,400

NATIONAL RESEARCH PROGRAMS

DISTRICT OF COLUMBIA

INDIVIDUAL AWARD
Kraichnan, Robert H.; *Turblence Dynamics and Predictability* (A016186); 12 months; $15,000

NATIONAL ACADEMY OF SCIENCES, NATIONAL RESEARCH COUNCIL
Sievers, John R.; *Task Order for Support of the Committee on Atmospheric Sciences and the U.S. Committee for the Global Atmospheric Research Program* (C310009014*); 12 months; $53,900

FLORIDA

FLORIDA STATE UNIVERSITY
Garstang, Michael, Gille, John and Warsh, Kenneth L.; *The Tropical Oceanic and Atmospheric Boundary Layers* (A019542); 24 months; $154,800
Krishnamurti, T. N.; *Global Tropical Weather Systems* (A017822); 24 months; $95,800

HAWAII

UNIVERSITY OF HAWAII
Murakami, Takio; *Experiment in Numerical Forecasting in the Tropics* (A018915); 24 months; $71,000

ILLINOIS

UNIVERSITY OF ILLINOIS, Urbana
Ogura, Yoshimitsu; *Numercial Study of Meso-Scale Convection* (A020328); 12 months; $94,800

MASSACHUSETTS

MASSACHUSETTS INSTITUTE OF TECHNOLOGY
Charney, Jule G.; *Theory of Large-Scale Atmospheric and Oceanic Processes* (GA402X 003); 12 months; $155,000
Mollo-Christensen, Erik; *Dynamics of the Processes in Air Sea Interaction* (A023339); 12 months; $79,000

MICHIGAN

UNIVERSITY OF MICHIGAN
Epstein, Edward S.; *Stochastic Dynamic Prediction* (A019248); 24 months; $134,400

NEVADA

DESERT RESEARCH INSTITUTE
Telford, James W.; *Jafna-Buffalo Thermal Project* (A011565); 12 months; $31,700

NEW JERSEY

PRINCETON UNIVERSITY
Mellor, George L.; *Experimental Study of Turbulent Flow Fields* (A014955); 24 months; $70,100
———; *Research in Geophysical Fluid Dynamics* (A014552); 24 months; $64,500

OKLAHOMA

UNIVERSITY OF OKLAHOMA
Eddy, Amos; *Investigation of Some Statistical Aspects of Meteorological Data Archiving* (A016034); 12 months; $17,800

WASHINGTON

UNIVERSITY OF WASHINGTON
Wallace, John M.; *Global Meteorology* (GA629X002); 12 months; $59,100

WISCONSIN

UNIVERSITY OF WISCONSIN, Madison
Hanson, Kirby and Vonder Haar, Thomas H.; *Reduction of Radiation Data from Bomex* (A019235); 12 months; $39,200
Hastenrath, S. L.; *Wind Structure and Atmospheric Energetics Over the Tropical Oceans* (A016370); 24 months; $101,100

Interdisciplinary Research Relevant to Problems of Our Society

CALIFORNIA

RAND CORPORATION
Morris, Deane N.; *The Growing Demand for Energy* (GI00044); 12 months; $180,000

UNIVERSITY OF CALIFORNIA, Davis
Musolf, Lloyd D. and Loomis, Robert S.; *Environmental Decision Making in the Lake Tahoe Basin* (GI00022); 12 months; $97,800
Watt, Kenneth E. F.; *Land Use and Energy Flow Component of a Model of Society* (GI00027); 24 months; $448,000

COLORADO

COLORADO STATE UNIVERSITY
Edwards, Harry W.; *Impact on Man of Environmental Contamination Caused by Lead* (GI00004); 24 months; $418,300

UNIVERSITY OF COLORADO
Brittin, Wesley E.; *Workshop on Research Problems in Air and Water Pollution* (GI00032); 12 months; $14,300

UNIVERSITY OF DENVER
Gilmore, John S.; *Public Policy Intervention in Inter-Industry Flows of Goods and Services to Reduce Pollution* (GI00011); 12 months; $81,100

DISTRICT OF COLUMBIA

GEORGE WASHINGTON UNIVERSITY
Mayo, Louis H.; *Pilot Research in Technology Assessment* (GI00041); 12 months; $110,000

ILLINOIS

UNIVERSITY OF ILLINOIS, Urbana
Metcalf, Robert L. and Ewing, Ben B.; *Interdisciplinary Study of Environmental Pollution by Lead* (GI00026); 15 months; $211,800

KANSAS

KANSAS STATE UNIVERSITY
Leachman, Robert B. and Williams, T. Alden; *Political and Scientific Effectiveness in Nuclear Materials Control* (GI00009); 24 months; $231,000

MARYLAND

APPLIED PHYSICS LABORATORY
Fristrom, R. M.; *Fire Problems Research and Synthesis* (GI00012); 24 months; $370,796

ATOMIC ENERGY COMMISSION
Liverman, James L.; *The Environment and Technological Assessment* (AG226); 12 months; $1,496,000

JOHNS HOPKINS UNIVERSITY
Roy, Robert H.; *Chesapeake Bay Research Planning* (GI00047); 4 months; $42,353

UNIVERSITY OF MARYLAND
Cronin, L. Eugene and Green, R. Lamar; *Chesapeake Bay Research Planning* (GI00048); 4 months; $55,000

MASSACHUSETTS

HARVARD UNIVERSITY
Thomas, Harold A.; *Environmental Systems Program* (GI00024); 24 months; $589,800

MASSACHUSETTS INSTITUTE OF TECHNOLOGY
Wilson, Carroll L.; *1970 Summer Study on Critical Environmental Problems* (GI00001); 12 months; $30,400

MICHIGAN

MICHIGAN STATE UNIVERSITY
Koenig, Herman E. and Cooper, William E.; *Design and Management of Environmental Systems* (GI00020); 24 months; $647,900

MONTANA

MONTANA STATE UNIVERSITY
Bradley, Charles C.; *Impact of Regional Development on a Semi-Primitive Environment* (I00038); 12 months; $110,000

NEW YORK

CLARKSON COLLEGE OF TECHNOLOGY
Leppert, George; *A Quantitative Model of Agropolis* (GI00008); 12 months; $70,000

CORNELL UNIVERSITY
Comar, Cyril L. and Auer, Peter L.; *National Energy Needs and Environmental Quality* (GI00018); 12 months; $189,250

STATE UNIVERSITY AT STONY BBOOK
Nathans, Robert, Ames, Edward and Beltrami, Edward J.; *Urban Science and Engineering* (GI00005); 24 months; $503,100

SYRACUSE UNIVERSITY RESEARCH CORP.
Teich, Albert H.; *Guidelines for a National Program of Environmental Research Laboratories* (GI00035); 12 months; $87,200

NATIONAL SEA GRANT PROGRAM

Sea Grant Institutional Support

CALIFORNIA

UNIVERSITY OF SOUTHERN CALIFORNIA
Tibby, Richard B.; *Sea Grant Institutional Support Program* (H000089); 12 months; $385,600

FLORIDA

UNIVERSITY OF MIAMI
Bader, Richard G.; (H000100); 12 months; $750,000

HAWAII

UNIVERSITY OF HAWAII
Gorter, Wytze; *Support of Institutional Sea Grant Program* (GH62); 12 months; $474,900
Murphy, Garth I.; *Development of Precious Coral Fishery in Hawaii* (H000080); 8 months; $76,300

MICHIGAN

UNIVERSITY OF MICHIGAN
McFadden, James T.; *Sea Grant Institutional Support* (GH98); 12 months; $719,400

NEW YORK

STATE UNIVERSITY AT STONY BROOK
Squires, Donald F.; *Sea Grant Institutional Planning* (H000069); 12 months; $18,700

OREGON

OREGON STATE UNIVERSITY
Frolander, Herbert F.; *Sea Grant Institutional Support* (H000097); 12 months; $1,399,700
———; (GH00045001); $8,400

RHODE ISLAND

UNIVERSITY OF RHODE ISLAND
Knauss, John A.; *Sea Grant Institutional Support* (H00099); 12 months; $900,000

WASHINGTON

UNIVERSITY OF WASHINGTON
Murphy, Stanley R.; *Sea Grant Institutional Support* (H000066); 12 months; $907,000
———; (GH66001); 7 months; $35,400

Sea Grant Project Support

ALASKA

UNIVERSITY OF ALASKA
Hickok, David; *Development of Alaska's Under-Utilized Arctic Marine Resources—A Coherent Area Project* (H000083); 12 months; $309,200

CALIFORNIA

CALIFORNIA COORDINATING COUNCIL ON HIGHER EDUCATION
Riese, Russell L.; *Report on Marine Sciences in California Institutions of Higher Education* (H000085); 12 months; $3,870

CALIFORNIA INSTITUTE OF TECHNOLOGY
North, Wheeler J.; *Restoration, Propagation and Management of Marine Algae* (GH0092); 24 months; $122,600

SAN DIEGO STATE COLLEGE
Farris, David A.; *Population Biology and Fishery of the California Spiny Lobster* (GH36001); 12 months; $5,000

SAN JOSE STATE COLLEGE
Harville, John P.; *A Coordinated Program of Education, Public Service and Pilot Research Activities for Monterey Bay and the Central California Coastal Region* (GH94); 12 months; $106,100

UNIVERSITY OF CALIFORNIA, Santa Barbara
Holmes, Robert W.; *Multiple Uses of Santa Barbara Channel Marine Resources* (H000095); 12 months; $287,700

DISTRICT OF COLUMBIA

AMERICAN ASSOCIATION OF JUNIOR COLLEGES
Pratt, Arden L.; *American Junior College Involvement in the Training of Marine Technicians* (GH2002); 5 months; $5,900

DEPARTMENT OF THE NAVY, OFFICE OF NAVAL RESEARCH
Trumbull, Richard; *Support of the Committee on Ocean Engineering of the National Academy of Engineerig* (AG185 001); 12 months; $15,000

GEORGETOWN UNIVERSITY
Colwell, R. R.; *Vibrio Parahaemolyticus in Chesapeake Bay—Isolation, Incidence and Pathogenicity* (H91); 24 months; $47,400

SMITHSONIAN INSTITUTION
Fehlmann, H. Adair and Wallen, I. E.; *Training Program for Marine Biological and Geological Technicians* (GH41 001); $6,750

SPORT FISHING INSTITUTE
Douglas, Philip A.; *Symposium on the Biological Significance of Estuaries* (H000070); 12 months; $8,150

WASHINGTON TECHNICAL INSTITUTE
Dennard, Cleveland L.; *Planning a Marine Science and Technology Program* (H000042001); $1,700

FLORIDA

FLORIDA ATLANTIC UNIVERSITY
Stephan, Charles R.; *Extension of a Cooperative Ocean Engineering Education Project* (H000084); 24 months; $180,000

UNIVERSITY OF SOUTH FLORIDA
Humm, Harold J.; *Experimental Cultivation of Red Algae of Economic Value in Florida Marine Waters* (H000068); 24 months; $32,400

GEORGIA

SKIDAWAY INSTITUTE OF OCEANOGRAPHY
Andrews, James W.; *A Study of the Nutritional, Environmental and Economical Requirements for the Intensive Aquaculture of Penaeid Shrimp* (H000073); 24 months; $139,000

UNIVERSITY OF GEORGIA
Noakes, John E.; *Geology* (H000071); 12 months; $60,700

HAWAII

OCEANIC FOUNDATION
Norris, Kenneth S.; *Food Resource Studies of Mullet and Milkfish* (H000076); 12 months; $64,712

Shehadeh, Ziad H. and Heath, Wallace G.; *Feasibility Pilot Project for a Method of Open Water Fish Farming* (H000078); 24 months; $63,747

LOUISIANA

FRANCIS T. NICHOLLS STATE COLLEGE
Harris, Alva H. and Rose, Curt D.; *Shrimp Production in Louisiana Salt-Marsh Impoundments Under Existing and Managed Conditions* (GH3001); $17,400
Rose, Curt D. and Harris, Alva H.; *Effects of Water Exchange and Blue Crab Control on Shrimp Production in Louisiana Salt-Marsh Impoundments* (H000075); 24 months; $96,300

MAINE

MAINE DEPARTMENT OF SEA AND SHORE FISHERIES
Harriman, Donald; *Marine Fisheries Extension Service* (H000072); 24 months; $94,800

SOUTHERN MAINE VOCATIONAL TECHNICAL INSTITUTE
Banerjee, Tapan; *Expansion of Applied Marine Biology and Oceanography Program in Technician Training* (H000035001); 4 months; $15,700

MASSACHUSETTS

BOSTON UNIVERSITY
Fulton, George P.; *Experimental Program for Counseling Toward Marine Careers* (H000074); 24 months; $13,575

MASSACHUSETTS INSTITUTE OF TECHNOLOGY
Keil, Alfred H.; *Ocean Utilization and Coastal Zone Development* (H000088); 12 months; $217,900

NEW ENGLAND COUNCIL INC.
Healey, Warren R.; *New England Regional Coastal Zone Management Conference, April 28 and 29, 1970 at New England Center for Continuing Education, Durham, New Hampshire* (GH81); 12 months; $4,300

WOODS HOLE OCEANOGRAPHIC INSTITUTION
Mather, Frank J. III; *Studies of Migrations and Populations of Certain Large Pelagic Fishes* (H000082); 24 months; $89,300

NEW HAMPSHIRE

UNIVERSITY OF NEW HAMPSHIRE
Corell, Robert W.; *The Science and Technology of Utilizing the Bottom Resources of the Continental Shelf* (H000077); 12 months; $90,000
Savage, Godfrey H.; *The Economic Aspects of the Feasibility, Design and Analysis of a Submerged Pipeline for Transporting Natural Gas Through a Deep Ocean Body* (H000064); 4 months; $6,900

NEW YORK

COLUMBIA UNIVERSITY
Brukholder, Paul R. and Roels, Oswald A.; *Research and Graduate Training in Food and Drugs from the Sea and Marine Pollution* (GH16001); $2,100
Roels, Oswald A.; *Artificial Upwelling* (H000087); 12 months; $242,100

COUNTIES OF NASSAU-SUFFOLK REGIONAL PLANNING BOARD
Williams, Clarke; *Development of Method-*

NATIONAL SEA GRANT PROGRAM

ologies for Planning for the Optimum Use of the Marine Resources of the Coastal Zone (H000063); 12 months; $129,800

NORTH CAROLINA

CAPE FEAR TECHNICAL INSTITUTE
Jordan, Arthur W.; *Improvement and Expansion of Marine Technology Curricula* (H000019001); 5 months; $126,400
———; *Contract for Operation of the R/V Undaunted During Tektite II* (C617000); 7 months; $70,667

UNIVERSITY OF NORTH CAROLINA AT CHAPEL HILL
Odum, H. T.. and Chustnut, A. F.; *Optimum Ecological Designs for Estuarine Systems of North Carolina* (GH18001); $50,900

PENNSYLVANIA

LEHIGH UNIVERSITY
Parks, James M. and Richards, Andrian F.; *Investigation of Geotechnical Properties in Two Sea Floor Demonstration Areas* (H000065); 12 months; $175,200

TEXAS

TEXAS A & M UNIVERSITY
Miloy, Leatha; *Preparation and Publication of Information Relating to NSF Sea Grant Program* (GN883*); 12 months; $35,200

UNIVERSITY OF TEXAS, Medical Branch, Galveston
Beckman, Edward L.; *Medical Aspects of Sustained Deepsea Operations* (H000079); 12 months; $124,500

VIRGIN ISLANDS

COLLEGE OF THE VIRGIN ISLANDS
Koblick, Ian G.; *Ecological Study for the Development of Lobster Management Techniques* (GH86); 12 months; $38,687

VIRGINIA

VIRGINIA INSTITUTE OF MARINE SCIENCE
Wood, John L. and Hargis, William J. Jr.; *Improved Management and Utilization of Estuarine Resources* (GH67); 12 months; $200,500

INTERNATIONAL COOPERATIVE SCIENTIFIC ACTIVITIES

Exchange of Scholars

INDIVIDUAL AWARD

Mathur, P. B.: *India–United States Exchange of Scientists and Engineers Programme* (GF384); 6 months; $2,710

Raizada, A. S.; *United States—India Exchange of Scientists and Engineers Programme* (GF403); 10 months; $3,810

ARIZONA

TRAVEL AWARD

UNIVERSITY OF ARIZONA

Kremp, Gerhard O. W.; *Discuss the Establishment of Palynological Data Archives and International Exchange of Palynological Data in Europe and Asia, March 7, 1970 to September 6, 1970* (GF379); 7 months; $1,700

CALIFORNIA

STANFORD UNIVERSITY

Noton, Bryan R.; *India-United States Exchange of Scientists and Engineers Programme* (GF378); 10 months; $1,301

UNIVERSITY OF CALIFORNIA, Berkeley

Hsiang, Wu-Yi; *Topology and Transformation Groups* (GF397); 6 months; $12,400

Snyder, William C.; *Ecology of 'Fusarium' Species in Taiwan with Reference to the 'Gibberella' Stage of 'F. Moniliforme' on Rice* (GF398); 24 months; $10,600

Kuh, Ernest S.; *Research in Nonlinear and Distributed Networks* (GF357); 11 months; $17,100

UNIVERSITY OF CALIFORNIA, San Diego, REVELLE COLLEGE

Hudson, Hugh S.; *Comparison of Solar X-Ray and Radio Emission* (GF374); 19 months; $7,900

UNIVERSITY OF CALIFORNIA, San Diego

McBeth, James W.; *A Study on the Viability and Growth of the California Red Abalone in Japan Using Japanese Aquaculture Techniques* (GF388); 12 months; $8,350

UNIVERSITY OF CALIFORNIA, Santa Barbara

Sherwin, John E.; *The Role of Auxin in Phytochrome Medicted Photomorphogenesis in Rice Coleoptiles* (GF360); 13 months; $8,400

DISTRICT OF COLUMBIA

NATIONAL ACADEMY OF SCIENCES

Rowan, E. C.; *International Organizations and Programs Project* (C310040009); $19,575

HOWARD UNIVERSITY

Robbins, Norman; *Study of Axonal Flow and the Trophic Influence of Nerve* (GF315 001); 3 months; $3,700

KANSAS

UNIVERSITY OF KANSAS

Friauf, Robert J.; *U. S.-India Exchange of Scientists and Engineers Programme* (GF375); 6 months; $1,421

MARYLAND

DEPARTMENT OF COMMERCE, ENVIRONMENTAL SCIENCE SERVICES ADMINISTRATION

Wait, James R.; *Research on the Theory of Propagation of Extremely Low Frequency (ELF) Electromagnetic Waves* (AG41001); $765

NATIONAL BUREAU OF STANDARDS

Lide, David R. Jr.; *U.S. Visit of Officers of the Soviet Commission on Tables of the Thermodynamic Properties of Gases* (AG225); 12 months; $1,650

JOHNS HOPKINS UNIVERSITY

Kaye, Kenneth S.; *Experimental Study of Turbulent Flows with 'Partly-Deterministic' Structure* (GF316001); 3 months; $2,400

UNIVERSITY OF MARYLAND

De Silva, Alan W.; *Research in Plasma Physics* (GF394); 12 months; $14,600

MICHIGAN

WAYNE STATE UNIVERSITY

Mayeda, Kazutoshi; *Genetic Analyses of Mongolian Gerbil* (GF392); 12 months; $15,600

MASSACHUSETTS

AMERICAN ACADEMY OF ARTS AND SCIENCES

Voss, John; *Support of U.S. Joint Committee for the International Center for Insect Physiology and Ecology in Nairobi, Kenya (ICIPE)* (GF407); 6 months; $16,870

MASSACHUSETTS INSTITUTE OF TECHNOLOGY

Chu, Lan Jen; *Engineering Education at National Chiao Tung University in Taiwan* (GF410); 36 months; $36,200

NEW YORK

ALBANY MEDICAL COLLEGE

Coulston, Frederick; *United States-India Exchange of Scientists and Engineers Programme* (GF381); 6 months; $1,321

COLUMBIA UNIVERSITY

Ewing, Maurice and Chen, Chin; *Marine Geology* (GF408); 10 months; $15,240

Greenbaum, Lowell M.; *Chemical Mediators in Inflammation* (GF389); 6 months; $2,200

Kim, Wan H.; *Development of Basic Building Block Design Techniques* (GF391); 12 months; $17,700

Sen, Amiya K.; *Shock Waves in Collisionless Plasma and Nonlinear Stability in Plasma* (GF361); 20 months; $13,800

CORNELL UNIVERSITY

Freed, Jack H.; *Theory of Spin Relaxation of Molecules and Related Phenomena* (GF363); 6 months; $6,400

STATE UNIVERSITY AT BUFFALO

Miles, Philip G.; *Morphogenesis of Higher Fungi* (GF396); 12 months; $15,700

PENNSYLVANIA

AIR POLLUTION CONTROL ASSOCIATION

Arch, Arnold; *Partial Support of Second International Clean Air Congress, International Union of Air Pollution Prevention Associations, Washington, D.C., December 6-11, 1970* (GF406); 12 months; $31,680

UNIVERSITY OF PENNSYLVANIA

Chance, Britton; *Interaction of Light-Induced Electron Transfer and Energy Conservation* (GF395); 24 months; $19,700

UNIVERSITY OF PITTSBURGH

Leighton, Joseph; *Organized Interaction Among Carcinoma Cells in Vitro* (GF393); 6 months; $2,100

RHODE ISLAND

INDIVIDUAL AWARD

Kushner, Harold J.; *United States-India Exchange of Scientists and Engineers Programme* (GF380); 8 months; $1,301

TENNESSEE

UNIVERSITY OF TENNESSEE

Goethert, B. H.; *U.S.-India Exchange of Scientists and Engineers Programme* (GF382); 6 months; $1,450

WASHINGTON

UNIVERSITY OF WASHINGTON

Richards, Francis A.; *Studies in Chemical Oceanography* (GF387); 15 months; $20,300

INDIA

INDIVIDUAL AWARD

Murthy, A. R. Vasudeva; *India-United States Exchange of Scientists and Engineers Programme* (GF386); 6 months; $2,010

Pillay, K. P. Ramakrishna; (GF383); 6 months; $1,810

Swaminathan, C. G.; (GF385); 6 months; $1,665

Basic Research Support

CALIFORNIA

UNIVERSITY OF CALIFORNIA, Berkeley

Bern, Howard A.; *Endocrine Control of Osmoregulation in Fishes* (GF372); 24 months; $15,000

UNIVERSITY OF CALIFORNIA, Los Angeles

Mintz, Yale; *Numerical Simulation of Interactions Between the Atmosphere and the Oceans* (GF369); 24 months; $5,500

HAWAII

UNIVERSITY OF HAWAII

Woollard, George P.; *Geomagnetic Variation Studies on Islands* (GF402); 12 months; $15,200

MASSACHUSETTS

MASSACHUSETTS INSTITUTE OF TECHNOLOGY

Stommel, Henry M.; *Treatise on the Kuroshio* (GF260001); 3 months; $6,300

NEW JERSEY

PRINCETON UNIVERSITY

Taylor, Edward C.; *The Chemistry of Furazanopyrimidines* (GF390); 24 months; $46,700

NEW YORK

POLYTECHNIC INSTITUTE OF BROOKLYN

Kozin, Frank; *Noise Stabilization of Non-Linear Systems* (GF373); 24 months; $2,400

OREGON

OREGON STATE UNIVERSITY

Conte, Frank P.; *Gill Development and its Role in Osmoregulation in Fishes* (GF371); 24 months; $16,000

INTERNATIONAL COOPERATIVE SCIENTIFIC ACTIVITIES

Near East and South Asia Exchange Program

INDIVIDUAL AWARDS
Badami, G. N.; *India-United States Exchange of Scientists and Engineers Programme* (GF365); 6 months; $2,122
Marathe, B. R.; (GF362); 6 months; $2,375
Mehta, D. J.; (GF366); 6 months; $2,675

ILLINOIS

UNIVERSITY OF ILLINOIS, Urbana
Maurer, Robert J.; (GF368); 6 months; $1,403

NEW YORK

NEW YORK UNIVERSITY
Shakun, Melvin F.; (GF370); 6 months; $1,287

LONG ISLAND JEWISH HOSPITAL
Meyer, Leo M.; (GF364); 6 months; $1,290

INDIA

TATA INSTITUTE OF FUNDAMENTAL RESEARCH
Toshar, B. V.; (GF359); 6 months; $1,288

East European Exchange Program

DISTRICT OF COLUMBIA

NATIONAL ACADEMY OF SCIENCES
Mitchell, Lawrence C.; *Exchange of Scientists Between the National Academy of Sciences, U.S.A. and the Academies of the Union of Soviet Socialist Republics, Poland, Yugoslavia, Romania, Hungary, Czechoslovakia, and Bulgaria* (C310039010); 12 months; $732,165

Research Development Program

DISTRICT OF COLUMBIA

NATIONAL ACADEMY OF SCIENCES
Rowan, E. C.; *International Organizations and Programs Project* (C310040008); 12 months; $130,115

MASSACHUSETTS

AMERICAN ACADEMY OF ARTS AND SCIENCES
Voss, John; *Conference on a Proposal to Establish an International Research Center on Insect Physiology and Endocrinology* (GF367); 12 months; $6,300
———; (GF367001); $1,845

Program Liaison and Support

DISTRICT OF COLUMBIA

COURTESY TRAVEL SERVICE
Marilley, Jane E.; *Contract for Administrative Services in Support of International Science Activities Sponsored by the National Science Foundation* (C374000009); 12 months; $150,000

SCIENCE INFORMATION SERVICE

Information Systems

DISTRICT OF COLUMBIA

AMERICAN INSTITUTE OF BIOLOGICAL SCIENCES
Olive, John R.; *An Information System for Flora North America (FNA)—Planning Stage* (GN812 001); $4,264

AMERICAN PSYCHOLOGICAL ASSOCIATION
Little, Kenneth B.; *Development of a National Information System for Psychology—Definition Phase* (GN772 003); 13 months; $294,150
———; *National Information System for Psychology-Planning and Staff Maintenance* (GN901); 5 months; $180,000

CONFERENCE BOARD OF THE MATHEMATICAL SCIENCES
Botts, Tuman A.; *Initial Planning Toward an National Information System in the Mathematical Sciences* (GN887); 12 months; $49,900

CENTER FOR APPLIED LINGUISTICS
Lotz, John; *An Information-System Program for the Language Sciences, Stage Three —Advanced System Design* (GN771 003); $295,900

MARYLAND

ENTOMOLOGICAL SOCIETY OF AMERICA
Foote, Richard H.; *A System Designed Entomological Data Center Feasibility Study (Phase II)* (GN866); 12 months; $87,150

MICHIGAN

ASSOCIATION OF ASIAN STUDIES
Steinberg, David Joel and Koh, Hesung C.; *Toward an Integrated Automated Information System for Asian Studies* (GN892); 29 months; $169,550

NEW YORK

AMERICAN INSTITUTE OF PHYSICS
Alt, Franz L.; *Maintenance of a Computer Store of Physics Information* (GN713004); 3 months; $53,600
Koch, H. William; *National Information Systems for Physics-Stage II Development* (GN864); 13 months; $640,200
———; *National Information System for Physics—Advance Planning* (GN863); 10 months; $98,300

ENGINEERING INDEX, INC.
Woods, Bill M.; *Development of Indexing Vocabularies in Electrical Engineering in Cooperation with the Institute of Electrical and Electronics Engineers (IEEE)* (GN850); 10 months; $17,000
———; (GN850 001); $4,800

INSTITUTE OF ELECTRICAL ENGINEERS
Fink, Donald G.; *Strengthening Information Services in Electrical and Electronics Engineering* (GN832002); 2 months; $18,666
———; (GN832001); $18,838

Chemical Information

DISTRICT OF COLUMBIA

AMERICAN CHEMICAL SOCIETY
Kuney, Joseph H.; *Implementation of the Role of the Computer in Scientific Publication* (GN891); 7 months; $59,600

ILLINOIS

ILLINOIS INSTITUTE OF TECHNOLOGY
Williams, Martha E.; *Expansion of Computer Search Center Capabilities and Services* (C554003); 5 months; $29,996

OHIO

AMERICAN CHEMICAL SOCIETY, CHEMICAL ABSTRACTS SERVICE
Tate, Fred A.; *Contract for Development of Computer-Managed Data Bases for the Fields of Chemistry and Chemical Engineering* (C583000003); $341,100
———; (C583000001); $1,196,400
———; *Contract for the Development of Computer-Based Information Handling Systems for Chemical and Chemical Engineering Information* (C586002); $333,200
———; (C586001); $159,500
———; *Contract for Experimental Development of a Mechanized Registry System for Chemical Compounds and for Research and Development in S lected Information Handling Problems* (C414000017); $13,695
———; *Experimental Research and Development Related to the National Chemical Information Program* (C521 010); 11 months; $326,200
———; (C521 009); $70,743

PENNSYLVANIA

UNIVERSITY OF PITTSBURGH
Arnett, Edward M.; *A Chemical Information Center Experimental Station* (GN738 002); 12 months; $235,000

Information Services

COLORADO

GEOLOGICAL SOCIETY OF AMERICA
Eckel, Edwin B.; *Publication of 'Bibliography and Index of Geology'* (GN888); 12 months; $75,000

DISTRICT OF COLUMBIA

AMERICAN GEOLOGICAL INSTITUTE
Hoover, Linn; *Geologic Reference File-Operation of a Computer-Based Bibliographic Data Bank* (GN667 001); 12 months; $311,000

AMERICAN GEOPHYSICAL UNION
Smith, Waldo E.; *Preparation of the U.S. Quadrennial Report to I.U.G.G. General Assembly* (GN899); 12 months; $25,000

AMERICAN PSYCHOLOGICAL ASSOCIATION
Little, Kenneth B.; *Preparation and Publication of 'Psychological Abstracts,' for Calendar Year 1969* (GN853); 12 months; $157,800
———; *National Information System for Psychology—Operations and Services* (GN900); 12 months; $119,000

INTERNATIONAL ASSOCIATION FOR PLANT TAXONOMY
Cowan, Richard S.; *Conversion of 'Index Nominum Genericorum' Data into Machine Readable Form* (GN886); 12 months; $45,300
———; *Preparation and Publication of Index Nominum Genericorum* (GN861); 12 months; $81,100

INDIANA

UNIVERSITY OF NOTRE DAME
Crovello, Theodore J.; *Automated Data File for Greene Herbarium Botanical Collection* (GN878); 12 months; $25,100

MINNESOTA

UNIVERSITY OF MINNESOTA
Eckert, E. R. G.; *Preparation of Annual Reviews and Current Bibliographies on Heat Transfer Research* (GN456 002); 24 months; $10,000

NEW JERSEY

HISTORY OF SCIENCE SOCIETY
Cohen, I. Bernard; *Preparation of a Cumulative Critical Bibliography of the History of Science* (GN814 001); $1,218

RUTGERS UNIVERSITY
Horowitz, Irving Louis; *Publication of Studies in Comparative International Development* (GN884); 18 months; $44,600

NEW YORK

AMERICAN INSTITUTE OF PHYSICS
Koch, H. William; *National Information System for Physics-Operations and Services* (GN865); 10 months; $265,300

AMERICAN MUSEUM OF NATURAL HISTORY
Atz, James W.; *Bibliographic Service in Ichthyology* (GN658 002); 12 months; $69,400
Dowling, Herndon G.; *Herpetological Information Search Systems* (GN707 001); 12 months; $114,500

CORNELL UNIVERSITY
Peakall, David B.; *Ecological Data Base for Ornithological Research* (GN859); 12 months; $24,000

ENGINEERING INDEX, INCORPORATED
Woods, Bill M.; *Operational Support to the Engineering Index Information System* (GN834 001); 12 months; $88,500

PENNSYLVANIA

CARNEGIE-MELLON UNIVERSITY
Lawrence, George H. M.; *Preparation of 'Bibliographic Huntiana'* (GN880); 12 months; $127,400

PENNSYLVANIA STATE UNIVERSITY
Traverse, Alfred; *Recording of Taxonomic, Stratigraphic, Morphological and Bibliographical Data for Palynology* (GN782 001); $30,000

RHODE ISLAND

AMERICAN MATHEMATICAL SOCIETY
Walker, Gordon L.; *Mathematical Offprint Service* (GN857); 12 months; $227,000

Foreign Science Information

Translations—Domestic

DISTRICT OF COLUMBIA

AMERICAN FISHERIES SOCIETY
Hutton, Robert F.; *Translation Journal, 'Hydrobiological Journal' (1970 Vol.)* (GN854); 18 months; $12,500

SCIENCE INFORMATION SERVICE

———; (GN854 001); $1,138

———; *Translation Journal, 'Problems of Ichthyology' (1970 Vol.)* (GN855); 18 months; $25,000

———; (GN855 001); $4,053

AMERICAN GEOLOGICAL INSTITUTE
Hoover, Linn; *Translation Journals—'International Geology Review', 'Doklady-Earth Science Sections', 'Paleontology', and 'Geochemistry International', 1970 Volumes* (GN876); 18 months; $123,800

AMERICAN GEOPHYSICAL UNION
Spilhaus, A. F. Jr.; *Translation Journals—'Atmospheric and Oceanic Physics,' 'Physics of the Solid Earth,' 'Geomagnetism and Aeronomy,' 'Geodesy and Aerophotography,' 'Oceanology,' 'Geotectonics,' and 'Soviet Hydrology-Selected Papers,' 1970 Issues* (GN860); 24 months; $101,000

OPTICAL SOCIETY OF AMERICA
Warga, Mary E.; *Translation Journal, Soviet Journal of Optical Technology, Volume 1969* (GN872); 17 months; $10,600

MARYLAND

ENTOMOLOGICAL SOCIETY OF AMERICA
Murdoch, Wallace P.; *Translation Journal, Entomological Review, Vol. 49 (1970)* (GN858); 18 months; $11,000

———; *Translation and Publication of the Chinese Journal, Acta Entomologica Sinica Vol. XIV* (GN645 001); $2,811

NEW YORK

AMERICAN GEOGRAPHICAL SOCIETY
Lakovitch, Joseph B. Jr.; *Selected Translation Journal, Soviet Geography-Review and Translation, Vol. XI (1970)* (GN867); 18 months; $21,000

AMERICAN SOCIETY OF CIVIL ENGINEERING
Wisely, William H.; *Translation Journal, Hydrotechnical Construction Vol. 1970* (GN866); 24 months; $11,800

PENNSYLVANIA

ARCTIC INSTITUTE OF NORTH AMERICA
Michael, Henry N.; *Translation and Editing of 'Anthropology of the North—Translation from Russian Sources, Vol. 10* (GN537 001); 18 months; $13,600

SOCIETY FOR INDUSTRIAL & APPLIED MATHEMATICS
Block, I. Edward; *Translation and Publication of Selected Articles on Control Theory in the Siam Journal of Control (SICON)'* (GN870); 24 months; $16,600

Translations—Foreign

INDIVIDUAL AWARD

Saad, Ismail; *Contract for Translation into English of Educational Literature, Compilation of Abstracts on Education in Pakistan, and Other Related Tasks as Mutually Agreed Upon by the Parties* (C505000007); $40,000

INDIA

INDIAN NATIONAL SCIENCE DOCUMENTARY CENTER
Parthasarathy, S.; *Contract for Translating into English & Printing in English Scientific & Technical Journals, Series, Books, Monographs, etc., Originally Published in Languages Other Than English—Preparation of Bibliographies on Indian Education, & Other Related Tasks as Mutually Agreed Upon by the Parties* (C466000013); $250,000

———; (C466000012); $300,000

ISRAEL

ISRAEL PROGRAM FOR SCIENTIFIC TRANSLATIONS
Rischin, Yitzhak; *Contract for Translating into English & Printing in English Scientific & Tech Journals, Articles, Books, Monographs, & Abstracts from Russian & Other Languages, Compilation of Bibliographies & Preparation of Abstracts from Russian & Other Languages, & Performance of Other Related Tasks as Mutually Agreed Upon by the Parties* (C503000010); 6 months; $400,000

———; *Contract for Translating into English & Printing in English Sci & Tech Journals, Articles, Books, Monographs, & Compilation of Bibliographies & Preparation of Abstracts from Russian & Other Languages & Performance of Other Related Tasks as Mutually Agreed Upon by the Parties* (C503000011*); $615,000

POLAND

CENTRAL INSTITUTE FOR SCIENCE TECHNOLOGY
Breda, Mieczyslaw; *Contract for Translating into English & Printing in English Scientific & Technical Journals, Articles, Monographs & Patents from Polish, Compilation of Bibliographies & Preparation of Abstracts from Polish & Performance of Other Related Mutually Agreed Upon Tasks* (C501000011); $250,000

———; (C501000010); $350,000

TUNISIA

AGENCE TUNISIENNE PUBLIC RELATIONS
Azzouz, Assedine; *Contract for Translation into English and Printing in English Scientific and Technical Journals, Articles and Monographs from Arabic, French and Other Languages, Compilation of Abstracts of Education in Tunisia and Other North African Countries and Other Related Tasks as Mutually Agreed Upon* (C504000008); $130,000

YUGOSLAVIA

NOLIT PUBLISHING HOUSE
Lasarevic, Sava D.; *Contract for Translation into English & Printing in English Scientific & Technical Journals, Articles & Mongraphs from Serbo-Croatian, Slovenian & Macedonian or Other Languages, Compilation of Bibliographies & Preparation of Abstracts from Original Languages & Performance of Other Mutually Agreed Upon Tasks* (C502000011); $22,500

———; (C502000010); $350,000

Services and Publications

BELGIUM

INTERNATIONAL COUNCIL OF SCIENTIFIC UNIONS
Poyen, Jeanne; *Continued Partial Support of the ICSU Abstracting Board* (GN898); 12 months; $10,000

DISTRICT OF COLUMBIA

CENTER FOR APPLIED LINGUISTICS
Roberts, A. Hood; *Support of the Linguistics Committee of the International Federation for Documentation* (N868); 12 months; $11,118

NATIONAL ACADEMY OF SCIENCES
Botts, Truman A.; *Preparation of a Revised Listing of U.S. Mathematicians for the World Directory of Mathematicians* (C310174); 6 months; $3,200

Coleman, John S.; *Contract for 'Support of the U.S. National Committee for the International Federation for Documentation (FID)'* (C310041010); 12 months; $50,100

Paul, Martin A.; *U.S. Support of the International Committee on Data for Science and Technology* (C310124002); 12 months; $20,000

GEORGIA

GEORGIA INSTITUTE OF TECHNOLOGY
Slamecka, Vladimir; *Updating and Completion of the Manuscript 'National Science Information Systems in Eastern Europe'* (GN885); 6 months; $2,008

ILLINOIS

JOHN CRERAR LIBRARY
Nowak, Ildiko D.; *Operation of the National Translations Center* (GN897); 12 months; $104,600

MICHIGAN

AMERICAN CONCRETE INSTITUTE
Shideler, Joseph J.; *Furthering Cement and Concrete Research Through International Exchange and Cooperative Efforts* (GN874); 6 months; $2,524

VIRGINIA

CLEARINGHOUSE FOR SCIENTIFIC AND TECHNICAL INFORMATION
Sauter, Hubert E.; *Operational Functions for NSF's Special Foreign Currency Science Information Program* (AG189001); 12 months; $55,700

Domestic Science Information

INDIVIDUAL AWARD
Casey, Florence B.; *Contract for a Compilation of Information Science Terms* (C605000); 5 months; $2,500

DISTRICT OF COLUMBIA

AMERICAN SOCIETY FOR INFORMATION SCIENCE
Bourne, Charles; *Innovative Presentations and Information Dissemination at a Technical Meeting* (GN848); 12 months; $9,670

LIBRARY OF CONGRESS
Mumford, L. Quincy; *One-Time Revision of Four General Directories of the National Referral Center* (AG203); 12 months; $78,700

SCIENCE COMMUNICATION INC.
Myatt, Dewitt O.; *Recommendations for National Science Information System Development* (C603000); 3 months; $2,500

SMITHSONIAN INSTITUTION
Translation Journal, 'Hydrobiological Journal' (1970 Vol.) (GN854); 18 months; $12,500

Hutton, Robert F.; *Translation Journal, 'Hydrobiological Journal,' (1970 Vol.)* (GN854 001); $1,138

Ripley, J. Dillon; *Contract for the Operation of the Science Information Exchange* (C437 012); 3 months; $400,000

———; (C437000011); 3 months; $400,000

———; (C437000013); 6 months; $737,367

NEW JERSEY

MATHEMATICA
Baumel, William J.; *A Cost Benefit Approach to Evaluation of Alternate Information Provision Procedures* (C606000); 8 months; $49,753

NORTH CAROLINA

UNIVERSITY OF NORTH CAROLINA AT CHAPEL HILL
Orne, Jerrold; *The American Standards Association, Section Subcommittee Z 39* (GN473004); 9 months; $12,400

Research and Studies

CALIFORNIA

STANFORD RESEARCH INSTITUTE
Coles, L. Stephen; *Techniques for Information Retrieval Using an Inferential Question-Answering System with Natural-Language Input* (GN895); 12 months; $32,000

STANFORD UNIVERSITY
Kincheloe, Wm. R. Jr.; *Research on Full-Text Time-Shared Scientific Information Retrieval* (GN894); 12 months; $22,000

SPECIAL PROJECTS

CONNECTICUT

YALE UNIVERSITY
Price, Derek J. DeSolla; *Statistical Studies of Citation and Other Characteristics of the World's Scientific Literature* (GN871); 12 months; $9,000

DISTRICT OF COLUMBIA

AMERICAN SOCIETY FOR INFORMATION SCIENCE
Koller, Herbert R.; *Workshop in Computer Controlled Photocomposition and Automatic Typesetting* (GN893); 12 months; $8,780

GEORGIA

GEORGIA INSTITUTE OF TECHNOLOGY
Slamecka, Vladimir; *Georgia Institute of Technology Science Information Research Center* (GN655002); 12 months; $290,000

INDIANA

PURDUE UNIVERSITY
Leimkuhler, Ferdinand F. and Reed, Phillip A.; *Operational Analysis of Information Systems* (GN759001); 12 months; $12,300

MARYLAND

JOHNS HOPKINS UNIVERSITY
Chapanis, Alphonse; *Human Communication in a Computer Environment* (GN890); 12 months; $35,000
Garvey, William D.; *A Behavioral Study of Scientific Communication* (GN514002); 16 months; $120,500

MASSACHUSETTS

MASSACHUSETTS INSTITUTE OF TECHNOLOGY
Allen, Thomas J.; *The Role of Information in Parallel R & D Projects* (GN597002); 12 months; $50,600
Overhage, Carl F. J.; *The Continuation of the Design, Development, and Evaluation of an Unconventional Library Catalog* (GN774002); 12 months; $100,000

MICHIGAN

UNIVERSITY OF MICHIGAN
Kochen, Manfred; *Integrative Mechanisms in Literature Growth* (GN879); 7 months; $35,000

NEW YORK

CORNELL UNIVERSITY
Salton, Gerald; *Text Analysis and Search Experiments in Automatic Document Retrieval (Using the SMART System)* (GN750001); 12 months; $16,000

OHIO

OHIO STATE UNIVERSITY
Yovits, Marshall C.; *Computer and Information Science Research Center* (GN534003); 12 months; $376,000

Special Projects

CALIFORNIA

STANFORD UNIVERSITY
Parker, Edwin B.; *Development of a Physics Information Retrieval System* (GN830001); 9 months; $370,100

DISTRICT OF COLUMBIA

ASSOCIATION OF RESEARCH LIBRARIES
McCarthy, Stephen A.; *A Study of the Characteristics, Costs, and Magnitude of Interlibrary Loans* (GN889); $53,000

GEORGIA

UNIVERSITY OF GEORGIA
Carmon, James L.; *To Expand the University of Georgia Information Center* (GN851001); $3,000
———; (GN851); 12 months; $250,000

ILLINOIS

UNIVERSITY OF CHICAGO
Fussler, Herman H.; *Development of an Integrated Computer-Based, Bibliographical Data System for a Large University Library* (GN556003); 12 months; $150,000

MINNESOTA

HAMLINE UNIVERSITY
Johnson, Herbert F. and King, Jack B.; *Information System Development Program for Hamline University Library* (GN873); 12 months; $99,300

NEW YORK

COLUMBIA UNIVERSITY
Fasana, Paul J.; *Library Systems Development for a Large Research Library* (GN694001); 18 months; $173,500

PENNSYLVANIA

LEHIGH UNIVERSITY
Hillman, Donald J.; *Campus-Based Information Retrieval System* (GN845); 12 months; $249,300

WASHINGTON

UNIVERSITY OF WASHINGTON
Rohn, Peter H.; *Computerized Treaty Information System* (GN783001); 12 months; $113,300

INTERNATIONAL SCIENTIFIC INFORMATON EXCHANGE (TRAVEL)

International Scientific Information Exchange (Travel)

INDIVIDUAL AWARDS

Amsterdam, Daniel; *NATO Advanced Study Institute on Microbeam Irradiation and Cellular Biology, Stresa, Italy, June 15-26, 1970* (GZ1806); 6 months; $498

Balch, Michael S.; *NATO Advanced Study Institute on Mathematical Models of Action and Reaction, Varenna, Italy, June 15-27, 1970* (GZ1812); 5 months; $458

Bishop, Albert B.; *International Symposium on Education and Training for Operational Research, Istanbul, Turkey, August 31-September 4, 1970* (GZ1809); 6 months; $828

Clough, John W.; *International Meeting on Radioglaciology, Lyngby, Denmark, May 11-13, 1970* (VO24129); 6 months; $594

Cofer, Rufus H. Jr.; *NATO Advanced Study Institute on Data Structures and Computer Systems, Marktoberdorf, Germany, August 17-28, 1970* (GZ1829); 6 months; $442

Cornell, Allin C.; *Conference on Dynamic Waves in Civil Engineering—Swansea, Wales, July 6-9, 1970* (KO24420); 6 months; $425

Damon, Richard W.; *Individual Travel to—Visit and Lecture at Various Institutes and Universities in the Soviet Union, April 20-May 6, 1970* (P024513); 6 months; $274

Davis, Richard G.; *NATO Advanced Study Institute on Odor Perception—Multi-Disciplinary Research Methods, Utrecht, the Netherlands, August 23-September 5, 1970* (GZ1837); 6 months; $490

De Rooy, Jacob; *NATO Advanced Study Institute on Recent Developments in Regional Science, Karlsruhe, West Germany, July 20-August 14, 1970* (GZ1823); 6 months; $338

Dimock, Jonathan; *NATO Advanced Study Institute on Theoretical Physics, Les Houches, France, July 5-August 29, 1970* (GZ1813); 6 months; $594

Duffie, John A.; *1970 International Solar Energy Society Conference, Melbourne, Australia—March 1-6, 1970* (K020876); 6 months; $700

Finley, James D.; *NATO Advanced Study Institute on Mathematical Physics, Istanbul, Turkey, August 10-21, 1970* (GZ1821); 6 months; $792

Friedman, David; *NATO Advanced Study Institute on Strong Interaction Physics, Heidelberg, Germany, July 20-31, 1970* (GZ1827); 6 months; $500

Frisch, Henry J.; *NATO Advanced Study Institute on Elementary Processes at High Energy, Erice (Trapani), Sicily, July 1-19, 1970* (GZ1825); 6 months; $683

Grammer, Garland Jr.; *NATO Advanced Study Institute on Hadronic Interactions of Electrons and Photons, Edinburgh, Scotland, July 26-August 15, 1970* (GZ1841); 6 months; $473

Griffiths, David J.; *NATO Advanced Study Institute on Strong Interaction Physics, Heidelberg, West Germany, July 19-31, 1970* (GZ1822); 6 months; $403

Herbich, John B.; *The World Dredging Conference, Singapore, July 6-10, 1970 & to visit Kyoto University, Ujigawa Hydraulic Laboratory & the Port & Harbor Research Institute in Kyoto & Yokosuka, Japan During the Period June 29–July 3, 1970* (H000090); 6 months; $1,179

Jaffe, Robert L.; *NATO Advanced Study Institute on Hadronic Interactions of Electrons and Photons, July 26-August 15, 1970* (GZ1835); 6 months; $520

Jenni, Donald A.; *NATO Advanced Study Institute on Odor Perception—Multidisciplinary Research Methods, Utrecht, the Netherlands, August 23-September 5, 1970* (GZ1832); 6 months; $563

Johnson, Wilbur V.; *International Congress on the Education of Teachers of Physics in Secondary Schools, Eger, Hungary, September 11-17, 1970* (GZ1798); 6 months; $620

Kenny, Patricia L.; *NATO Advanced Study Institute on Data Structures and Computer Systems, Munich, Germany, August 17-28, 1970* (GZ1826); 6 months; $501

Khatri, A. P.; *India-United States Exchange of Scientists and Engineers Programme* (GF377); 6 months; $3,285

Long, Wesley H.; *NATO Advanced Study Institute on Recent Developments in Regional Science, Karlsruhe, Germany, July 20-August 14, 1970* (GZ1819); 6 months; $384

Marshak, Marvin L.; *NATO Advanced Study Institute on Elementary Processes at High Energy, Erice (Trapani), Sicily, July 1-19, 1970* (GZ1824); 6 months; $412

Matz, David B.; *Symposium on Antarctic Geology, Oslo, Norway, August 6-15, 1970* (V025395); 6 months; $578

Mellema, D. Joel; *NATO Advanced Study Institute on Elementary Processes at High Energy Erice (Trapani), Sicily, July 1–19, 1970* (GZ1834); 6 months; $683

Miller, Neal E.; *Meeting of Central Council, International Brain Research Organization, Paris, France, December 1970* (GF409); 8 months; $464

Oines, Ole P. J.; *The International Congress on the Education of Teachers of Physics in Secondary Schools, Eger, Hungary, September 11-17, 1970* (GZ1817); 6 months; $864

Petraske, Eric W.; *NATO Advanced Study Institute on Elementary Processes at High Energy, Erice, Sicily, July 1-19, 1970* (GZ1840); 6 months; $683

Pickar, Arnold D.; *International Congress on the Education of Teachers of Physics in Secondary Schools, Eger, Hungary, September 11-17, 1970* (GZ1797); 6 months; $837

Pirie, Walter R.; *NATO Advanced Study Institute on Formulation and Assessment of Statistical Models for Experimental Psychology, The Hague, the Netherlands, August 5–21, 1970* (GZ1830); 6 months; $448

Richards, David R.; *NATO Advanced Study Institute on Hadronic Interactions of Electrons and Photons, Edinburgh, Scotland, July 26-August 15, 1970* (GZ1838); 6 months; $360

Rodkey, Leo Scott; *NATO Advanced Study Institute on Molecular and Cellular Aspects of Immunology, Thiverval-Grignon, France, August 23-31, 1970* (GZ1831); 6 months; $489

Roozen, Kenneth J.; *NATO Advanced Study Institute on the Uptake of Informative Molecules by Living Cells, Mol, Belgium, August 16-September 2, 1970* (GZ1828); 6 months; $435

Theil, Edward H.; *NATO Advanced Study Institute on Mathematical Physics, Istanbul, Turkey, August 10-21, 1970* (GZ1836); 6 months; $599

Tilton, Robert S.; *The International Congress on the Education of Teachers of Physics in Secondary Schools, Eger, Hungary, September 11-17, 1970* (GZ1818); 6 months; $817

Wiedemann, Carl F.; *NATO Advanced Study Institute on Formulation and Assessment of Statistical Models in Experimental Psychology, Amsterdam, the Netherlands, August 2–19, 1970* (GZ1820); 6 months; $381

Wilson, Walter S.; *NATO Advanced Study Institute on Algebraic Topology, Aarhus, Denmark, August 10-23, 1970* (GZ1796); 6 months; $403

Wortman, William R.; *NATO Advanced Study Institute on Theoretical Physics, Cargese, Corsica, France, June 28–July 25, 1970* (GZ1815); 6 months; $596

ALASKA

UNIVERSITY OF ALASKA

Wendler, Gerd; *Symposium on the Hydrology of Glaciers, Cambridge, England, September 7-13, 1969* (A017298); 6 months; $858

ARIZONA

UNIVERSITY OF ARIZONA

Bashkin, Stanley; *The Second International Conference on Atomic Physics, Oxford, England, July 21-24, 1970* (P025464); 6 months; $548

MUSEUM OF NORTHERN ARIZONA

Colbert, Edwin H.; *Second Symposium on Gondwana Stratigraphy and Paleontology being Held in Capetown and Johannesburg, South Africa During the Period July 3-24, 1970 and the Symposium on Antarctic Geology, Oslo, Norway, August 6-15, 1970* (V024272); 6 months; $1,631

LOWELL OBSERVATORY

Franz, Otto G.; *I.A.U. Meeting on Observational Methods of Visual Double Stars, Nice, France September 8-10, 1969* (P017261); 6 months; $638

UNIVERSITY OF ARIZONA

Kessler, John O.; *Third International Liquid Crystal Conference, Berlin, Germany, August 24-28, 1970* (P025611); 6 months; $128

ARIZONA STATE UNIVERSITY

Metzger, Darryl E.; *International Gas Turbine Conference—Brussels, Belgium, May 24-28, 1970* (K023757); 6 months; $510

UNIVERSITY OF ARIZONA

Steelink, Cornelius; *Seventh International Symposium on the Chemistry of Natural Products, Riga, Latvia, U.S.S.R., June 22-27, 1970* (P023498); 6 months; $730

ARKANSAS

UNIVERSITY OF ARKANSAS

Wolf, Helmut; *To Visit the Institute for Reactor Development in Karlsruhe, Germany, August 25, 1970-September 1, 1971* (K025066); 17 months; $720

BELGIUM

VONKARMAN INSTITUTE

Dietz, Robert O.; *Annual Meeting of the International Astronautics Federation—Mar Del Plata, Argentina— October 5-12, 1969* (K017777); 6 months; $600

CALIFORNIA

UNIVERSITY OF SOUTHERN CALIFORNIA

Bandy, Orville L.; *Symposium on Antarctic Geology, Oslo, Norway, August 6-15, 1970* (V024374); 6 months; $595

PASADENA FOUNDATION FOR MEDICAL RESEARCH

Berns, Michael W.; *International Travel Program-NATO Institute NATO Adv St Inst on Microbeam Irradiation & Cellular Biology June 6–27, 1970* (GZ1766); 6 months; $702

UNIVERSITY OF CALIFORNIA, Irvine

Bershad, Neil J.; *International Symposium on Information Theory, Noordwijk, the Netherlands, June 15-19, 1970* (K024274); 6 months; $200

STANFORD UNIVERSITY

Bienenstock, Arthur; *Symposium on Quantum Theory of the Disordered State, Trieste, Italy, April 11-13, 1970* (K024131); 6 months; $852

UNIVERSITY OF CALIFORNIA, San Diego

Bond, F. Thomas; *Seventh International Symposium on the Chemistry of Natural Products, Riga, Latvia, U.S.S.R., June 22-27, 1970* (P023499); 6 months; $938

UNIVERSITY OF CALIFORNIA, Los Angeles

Borko, Harold; *The Second Cranfield Conference on Mechanized Information Storage*

INTERNATIONAL SCIENTIFIC INFORMATION EXCHANGE (TRAVEL)

and Retrieval, Cranfield, England September 2-5, 1969 and Conference on Automated Publishing Systems, Newcastle Upon Tyne, England, September 7-13, 1969 (GN846); 6 months; $683

CALIFORNIA INSTITUTE OF TECHNOLOGY
Cameron, Roy E.; *Symposium on Taxonomy and Biology of Blue-Green Algae, Madras, India, January 8-13, 1970* (A018936); 6 months; $1,197

CALIFORNIA STATE COLLEGE AT LONG BEACH
Cebeci,, Tuncer; *Fourth International Heat Transfer Conference—Paris, France, August 31-September 5, 1970* (K025379); 6 months; $230

STANFORD UNIVERSITY
Chamberlin, Donald D.; *NATO Advanced Study Institute on the Architecture and Design of Digital Computers, Hyeres, France, August 25-September 6, 1969* (GZ1470); 6 months; $690

LOS ANGELES COUNTY MUSEUM OF NATURAL HISTORY
Collias, Elsie C.; *Fifteenth International Ornithological Congress, The Hague, the Netherlands, August 30-September 5, 1970* (B025177); 6 months; $595

UNIVERSITY OF CALIFORNIA, Los Angeles
Crocker, Stephen D.; *NATO Advanced Study Institute on Fundamental Aspects and Current Developments in Computing Programming, Copenhagen, Denmark, August 11-22, 1969* (GZ1453); 6 months; $598

STANFORD RESEARCH INSTITUTE
Cubicciotti, D.; *Meeting of the International Union of Pure and Applied Chemistry Commission on High Temperatures and Refractories, Karlsruhe, Germany, April 24 to May 1, 1970* (P023726); 6 months; $482

UNIVERSITY OF CALIFORNIA, Davis
Devries, Arthur Leland; *Symposium on Antarctic Oceanography, Tokyo, Japan, September 14-25, 1970* (V024630); 6 months; $722

UNIVERSITY OF CALIFORNIA, Los Angeles
Edwards, Donald K.; *Solar Energy Society International Conference—Melbourne, Australia March 2-6, 1970* (K020877); 6 months; $810

UNIVERSITY OF CALIFORNIA, San Diego
Finkel, Robert C.; *NATO Institute NATO Adv St Inst on the Moon & Planets Newcastle Upon Tyne Eng April 9-16, 1970* (GZ1530); 6 months; $527

GEOLOGICAL SURVEY
Ford, Arthur B.; *Symposium on Antarctic Geology, Oslo, Norway, August 6-15, 1970* (V024247); 6 months; $595

UNIVERSITY OF CALIFORNIA, Berkeley
Fuerstenau, Douglas W.; *IX International Mineral Processing Congress—Prague, Czechoslovakia June 1-6, 1970* (K024082); 6 months; $410

STANFORD UNIVERSITY
Golub, Gene Howard; *To Conduct Research During Sabbatical Leave in London, England and to Visit the Computing Center in Novosibirsk, U.S.S.R., January-August, 1971* (J001053); 17 months; $1,111

CALIFORNIA INSTITUTE OF TECHNOLOGY
Hammond, George S.; *International Symposium on University Chemical Education, Rome, Italy, October 16-19, 1969* (GZ1481); 6 months; $870

STANFORD UNIVERSITY
Harrison, Walter A.; *The Second Soviet Conference on Solid State Theory, Moscow, U.S.S.R., December 14-21, 1969* (P019196); 6 months; $305
Hughes, E. Barrie, *The Fourth International Symposium on Electron and Photon Interactions at High Energies at Liverpool, England, September 14-20, 1969* (P017228); 6 months; $664

RAND CORPORATION
Ikle, Fred C.; *Meeting of Methodology of Social Prognosis, United Nations Research Institute for Social Development, Geneva, Switzerland, February 16-20, 1970* (GF376); 6 months; $511

UNIVERSITY OF CALIFORNIA, Davis
Jensen, Gordon; *The Third International Congress of Primatology, Zurich, Switzerland, August 2-5, 1970* (B025108); 6 months; $884

STANFORD UNIVERSITY
Kays, William M.; *Fourth International Heat Transfer Conference—Paris, France, August 31-September 5, 1970* (K025254); 6 months; $800

UNIVERSITY OF SOUTHERN CALIFORNIA
Kim, Young B.; *The Twelfth International Conference on Low Temperature Physics to be Held in Kyoto, Japan, September 4-10, 1970* (P025708); 6 months; $722

UNIVERSITY OF CALIFORNIA, Los Angeles
Klinger, Allen; *The Symposium on Systems Engineering Approach to Computer Control, Kyoto, Japan, August 11-14, 1970* (K025635); 6 months; $750
Kolin, Alexander; *Symposium on Application of Magnetism in Bioengineering, Rehovot, Israel—December 9-11, 1969* (K018893); 6 months; $662

UNIVERSITY OF CALIFORNIA, San Diego
Kooyman, Gerald L.; *Symposium on the Functional Anatomy of Marine Mammals, Cambridge, England, July 16-18, 1970* (V024621); 6 months; $710

STANFORD UNIVERSITY
Lee, Erastus H.; *Twelfth British Theoretical Mechanics Colloquium—Norwich, England, March 23-26, 1970* (K021382); 6 months; $420
Loew, Gilda M. Harris; *NATO Advanced Study Institute on Fourth International Conference on Magnetic Resonance in Biological Systems, August 26-September 2, 1970* (GZ1791); 6 months; $514

UNIVERSITY OF CALIFORNIA, Berkeley
Mahan, Bruce H.; *International Symposium on University Chemical Education, Rome, Italy. October 16-19, 1969* (GZ1483); 6 months; $863

UNIVERSITY OF CALIFORNIA, Los Angeles
McNamee, Lawrence P.; *International Symposium on Computer Graphics, Middlesex, England, April 14-16, 1970* (J001021); 6 months; $710

STANFORD UNIVERSITY
Mistry, Nariman B.; *The Fourth International Symposium on Electron and Photon Interactions at High Energies in Liverpool, England, September 14-20, 1969* (P017229); 6 months; $468

UNIVERSITY OF CALIFORNIA, Davis
Mitra, Sanjit K.; *Fourth Colloquium on Microwave Communication—Budapest, Hungary, April 21-24, 1970* (K022918); 6 months; $350

UNIVERSITY OF CALIFORNIA, Los Angeles
Morrey, Charles B. Jr.; *Sixth General Assembly International Mathematical Union, Menton, France, August 28-30, 1970* (GF401); 6 months; $352

SYSTEM DEVELOPMENT CORP.
Mossmond, Marvin D.; *British Computer Society Symposium at Datafair '69, Manchester, England, August 25-29, 1969* (J000777); 6 months; $517

UNIVERSITY OF CALIFORNIA, Berkeley
Neyman, Jerzy; *The Thirty-Seventh Session of the International Statistical Institute, London, England, September 3-11, 1969* (P017066); 6 months; $590
Oswald, William J.; *The Third International Conference on Global Impacts of Applied Microbiology, Bombay, India, December 7-12, 1969* (B019511); 6 months; $1,224

STANFORD UNIVERSITY
Pettit, Joseph M.; *Popov Society Meeting, Moscow, U.S.S.R., May 26-June 12, 1970* (K024770); 6 months; $890

UNIVERSITY OF CALIFORNIA, Los Angeles
Popper, Daniel M.; *I.A.U. Meeting on Mass Loss and Evolution in Close Binaries, Elsinore, Denmark, September 15-19, 1969* (P017274); 6 months; $730

UNIVERSITY OF SOUTHERN CALIFORNIA
Porto, Sergio P. S.; *Seminar on the Theory of Impurity Centers in Solids, Tallinn, U.S.S.R., September 21-26, 1970* (P025634); 6 months; $800

STANFORD UNIVERSITY
Royden, Halsey L.; *The Romanian-Finnish Seminar on Teichmuller Space and Quasi-Conformal Mappings, Brasov, Romania, August 25-31, 1969* (P017065); 6 months; $748
Saal, Harry J.; *NATO Advanced Study Institute on Fundamental Aspects and Current Developments in Computer Programming, Lyngby, Copenhagen, Denmark, August 11-22, 1969* (GZ1463); 6 months; $598

CALIFORNIA INSTITUTE OF TECHNOLOGY
Sabersky, Rolf H.; *Fourth International Heat Transfer Conference—Paris, France, August 31-September 5, 1970* (K025258); 6 months; $750

UNIVERSITY OF CALIFORNIA, Riverside
Sawyer, Donald T.; *To Accept a Visiting Research Fellowship at Merton College, Oxford, England, and to Exchange Information with Scholars at Oxford, Gothenburg, and Oslo Universities, April 1-June 16, 1970* (P019979); 8 months; $755

UNIVERSITY OF CALIFORNIA. Santa Cruz
Scargle, Jeffrey D.; *Meeting on Pulsars and High Energy Activity in Supernovae Remnants, Rome, Italy, December 18-20, 1969* (P018966); 6 months; $432

UNIVERSITY OF CALIFORNIA. Irvine
Schinzinger, Roland; *The Symposium on Systems Engineering Approach to Computer Control, Kyoto, Japan, August 11-14, 1970* (K025636); 6 months; $400

CALIFORNIA INSTITUTE OF TECHNOLOGY
Schmidt, Maarten; *I.A.U. Symposium—'The Spiral Structure of our Galaxy', Basel, Switzerland, August 29-September 4, 1969* (P016990); 6 months; $577

UNIVERSITY OF SOUTHERN CALIFORNIA
Schnepp, O.; *Faraday Society Discussions on Motions in Molecular Crystals, Oxford, England, September 16-18, 1969* (P015340); 6 months; $725

CALIFORNIA INSTITUTE OF TECHNOLOGY
Schramm, David N.; *NATO Advanced Study Institute on the Moon and Planets, Newcastle Upon Tyne, England, April 9-16, 1970* (GZ1757); 6 months; $504

UNIVERSITY OF CALIFORNIA, Berkeley
Scordelis, Alexander C.; *International Colloquium on the Progress of Shell Structures in the Last Ten Years and Its Future Development, Madrid, Spain, September 30-October 3, 1969* (K017479); 6 months; $620

UNIVERSITY OF CALIFORNIA, Los Angeles
Scott, Robert L.; *Polymer Solutions, Manchester, England, April 15-17, 1970* (P021046); 6 months; $342

STANFORD UNIVERSITY
Smith, Alvy R. III; *NATO Advanced Study Institute on the Architecture and Design of Digital Computers, Hyeres, France, August 25-September 6, 1969* (GZ1472); 6 months; $592

INTERNATIONAL SCIENTIFIC INFORMATION EXCHANGE (TRAVEL)

UNIVERSITY OF CALIFORNIA, Los Angeles
Smith, Emil L.; *Thirteenth General Assembly, International Council of Scientific Unions, Madrid, Spain, September 24–29, 1970* (GF400); 6 months; $748

STANFORD UNIVERSITY
Smith, Grant Gill; *The Annual Joint Meeting of the Chemical Society and Royal Institute of Chemistry, Edinburgh, Scotland, April 7–10, 1970* (P021165); 6 months; $547

UNIVERSITY OF CALIFORNIA, Santa Barbara
Smith, Roulette W.; *'Congress of the Society for Programmed Instruction', Basel, Switzerland, May 26–30, 1970* (J001161); 6 months; $574

UNIVERSITY OF CALIFONIA, Berkeley
Spira, Philip M.; *NATO Advanced Study Institute on the Architecture and Design of Digital Computers, Hyeres, France, August 25–September 6, 1969* (GZ1460); 6 months; $363

UNIVERSITY OF CALIFORNIA, Davis
Starr, Mortimer P.; *The Third International Conference on Global Impacts of Applied Microbiology, Bombay, India, December 7–12, 1969* (B019509); 6 months; $1,225

STANFORD UNIVERSITY
Stocker, Bruce A. D.; *The Oholo Symposium on Microbial Toxins, Israel, March 1970* (B020942); 6 months; $954

UNIVERSITY OF CALIFORNIA, Berkeley
Thomas, Gareth; *The Seventh International Congress on Electron Microscopy, Grenoble, France, August 30–September 5, 1970* (K025250); 6 months; $520

SACRAMENTO STATE COLLEGE
Udvardy, Miklos D. F.; *Fifteenth International Ornithological Congress, The Hague, the Netherlands, August 30–September 5, 1970* (B025185); 6 months; $754

UNIVERSITY OF CALIFORNIA, Berkeley
Whinnery, John R.; *Popov Society Meeting, Moscow, U.S.S.R., May 26–June 12, 1970* (K024774); 6 months; $710

UNIVERSITY OF SOUTHERN CALIFORNIA
Wilmarth, Wayne K.; *International Conference on Mechanisms of Reactions in Solution, Canterbury, England, July 20–24, 1970* (P024673); 6 months; $564

UNIVERSITY OF CALIFORNIA, Los Angeles
Wittry, David B.; *The Seventh International Congress on Electron Microscopy, Grenoble, France, August 30–September 5, 1970* (K025260); 6 months; $440

CANADA

UNIVERSITY OF BRITISH COLUMBIA
Somero, George N.; *Symposium on Antarctic Oceanography, Tokyo, Japan, September 14–25, 1970* (V025127); 6 months; $722

COLORADO

UNIVERSITY OF COLORADO
Davey, William R.; *Stellar Evolution and Variable Stars, August 31–September 13, 1970* (GZ1792); 6 months; $482
Hassner, Alfred; *Lecture at Eizmann Institute and the Technion University, Rehovoth and Haifa, Israel, December 22, 1969–January 20, 1970* (P018675); 6 months; $500

UNIVERSITY OF DENVER
Johnson, Shirley A. Jr.; *To Collaborate and Study the Process of Technology Transfer with Two Recognized Authorities in Western Europe—Lund, Sweden and Birmingham, England, March 30, 1970–July 1, 1970* (K022812); 6 months; $560

UNIVERSITY OF COLORADO
Kridelbaugh, Stephen J.; *NATO Advanced Study Institute on the Moon and Planets, Newcastle Upon Tyne, England, April 9–16, 1970* (GZ1532); 6 months; $610
Lemasurier, Wesley E.; *Symposium on Antarctic Geology, Oslo, Norway, August 6–15, 1970* (V024249); 6 months; $533

COLORADO STATE UNIVERSITY
Miller, L. L.; *Symposium on Electrochemistry in Nonaqueous Media, Paris, France, July 6–11, 1970* (P025444); 6 months; $708

UNIVERSITY OF DENVER
Recht, Rodney F.; *Third International Conference on High Pressure—Aviemore, Scotland, May 11–15, 1970* (K023595); 6 months; $300

GEOLOGICAL SURVEY
Williams, Paul L.; *Symposium on Antarctic Geology, Oslo, Norway, August 6–15, 1970* (V024399); 6 months; $608

CONNECTICUT

ECONOMETRIC SOCIETY
Jorgenson, Dale W.; *The Second World Congress of the Econometric Society, Cambridge, England, September 8–14, 1970* (S002949); 12 months; $13,500

UNIVERSITY OF CONNECTICUT
Brush, Alan H.; *Fifteenth International Ornithological Congress, The Hague, the Netherlands, August 30–September 5, 1970* (B025176); 6 months; $375

YALE UNIVERSITY
Davidovits, Paul; *Israel Laser Symposium and Visit Science Laboratories in Israel and France, June 15–July 30, 1970* (K024120); 6 months; $300
Hirshfield, J. L.; *International Conference on the Physics of Quiescent Plasmas, Paris, France, September 8–12, 1969* (K016900); 6 months; $505
Scott, A. Ian; *Third Jamaican Symposium on Natural Products, Mona, Jamaica, January 5–9, 1970* (P019395); 6 months; $171

DELAWARE

UNIVERSITY OF DELAWARE
Sharnoff, Mark; *Symposium on Magneto-Optical Effects, London, England, December 11–12, 1969* (P018496); 6 months; $218

DISTRICT OF COLUMBIA

AMERICAN ANTHROPOLOGICAL ASSOCIATION
Reining, Conrad C.; *The Thirty-Ninth International Congress of Americanists, Lima, Peru, August 1970* (S002940); 12 months; $2,000

AMERICAN GEOPHYSICAL UNION
Smith, Waldo E.; *Partial Support of the Union General Assemblies for the International Union of Geodesy and Geophysics (IUGG)* (A013947001); $5,000

AMERICAN POLITICAL SCIENCE ASSOCIATION
Kirkpatrick, Evron M.; *Travel of Americans to Attend the Eighth World Congress of the International Political Science Association, Munich, Germany, August 30–September 5, 1970* (S002843); 12 months; $10,000

AMERICAN SOCIOLOGICAL ASSOCIATION
Volkart, E. H.; *Travel of American Sociologists to the Seventh World Congress of the International Sociological Association, Varna, Bulgaria, September 1–30, 1970* (S002854); 12 months; $15,000

NATIONAL ACADEMY OF SCIENCES, NATIONAL RESEARCH COUNCIL
Bartley, W. C.; *Task Order for Partial Support of U.S. Participation in the International Symposium on Solar-Terrestrial Physics* (C310179); 12 months; $15,100
Cohen, Leon W.; *Task Order for Partial Support of Travel of U.S. Participants to the International Congress of Mathematicians in Nice, France* (C310180); 10 months; $60,000
———; *Task Order for Partial Support of Travel to the International Federation for Information Processing (IFIP) Congress in Ljubljana, Yugoslavia* (C310185); 22 months; $5,000
Dyer, E. R.; *Task Order for Partial Support of U.S. Participation in the Fourteenth General Assembly of the International Astronomical Union, in Brighton, England, August 18–27, 1970* (C310181); 10 months; $24,350
Paul, Martin A.; *Partial Support of U.S. Participation in the International Symposium on the Autonomy and Biogenesis of Mitochondria and Chloroplasts, Canberra, Australia, December 8–12, 1969* (C310175); 7 months; $6,000

GEOLOGICAL SURVEY
Breger, Irving A.; *To Attend—'Third International Conference on the Origin of Life', Pont-A-Mousson, France, April 20–25, 1970* (A022973); 6 months; $365

GEORGE WASHINGTON UNIVERSITY
Heller, Robert B.; *Fourth International Ship Structures Congress—Tokyo, Japan, August 31–September 4, 1970* (K025067); 6 months; $660

NATIONAL COUNCIL OF TEACHERS OF MATHEMATICS
Hlavaty, Julius H.; *First International Congress on Mathematical Education, Lyons, France, August 24–30, 1969* (GZ1446); 6 months; $235

GEOLOGICAL SURVEY
Johnston, William D. Jr.; *'Commission on the Geological Map of the World', Paris, France, March 31–April 9, 1970* (A021100); 6 months; $502

NATIONAL FEDERATION OF SCIENCE ABSTRACTING AND INDEXING SERVICES
Keenan, Stella; *The Annual Meeting of the Abstracting Board of the International Council of Scientific Unions, Meeting to be Held in Rome, Italy, September 17–21, 1969* (GN847); 6 months; $646
Parme, Alfred L.; *International Colloquium on the Progress of Shell Structures in the Last Ten Years and its Future Development, Madrid, Spain, September 30–October 3, 1969* (K017481); 6 months; $430

AMERICAN GEOLOGICAL INSTITUTE
Rafter, Thomas F. Jr.; *The First General Assembly of the European Association of Earth Science Editors (EDITERRA), Ghent, Belgium, December 18–19, 1969* (GN862); 6 months; $496

GEORGE WASHINGTON UNIVERSITY
Weeks, David C.; *The International Conference on General Principles of Thesaurus Building, Warsaw, Poland, March 23–27, 1970* (GN875); 6 months; $470

FLORIDA

AMERICAN SOCIETY FOR HORTICULTURAL SCIENCE
Beattie, J. M.; *The Eighteenth International Horticultural Congress, Tel Aviv, Israel, March 18–25, 1970* (B020203); 12 months; $10,000

UNIVERSITY OF FLORIDA
Adams, E. Dwight; *The Twelfth International Conference on Low Temperature Physics to be Held in Kyoto, Japan, September 4–10, 1970* (P025536); 6 months; $641
Bates, Roger G.; *Symposium on Electrochemistry in Nonaqueous Media, Paris, France, July 6–11, 1970* (P024494); 6 months; $512
Battiste, Merle A.; *International Symposium on Aromaticity, Pseudoaromaticity, and Antiaromaticity, Jerusalem, Israel, March 31–April 3, 1970* (P021049); 6 months; $715

UNIVERSITY OF MIAMI
Bostrom, Kurt; *Eighth Congress of the International Union for Quaternary Research,*

INTERNATIONAL SCIENTIFIC INFORMATION EXCHANGE (TRAVEL)

Paris, France, August 30–September 5, 1969 (A016101); 6 months; $359

Fisher, David E.; *Symposium on the Chemistry and Mineralogy of Meteorites and Extraterrestrial Matter, London, England, April 6–8, 1970 and to Visit Scientific Institutions in Sheffield, Oxford, and London, England, Heidelberg, Germany, and Paris, France, April 9–19, 1970* (A022700); 6 months; $630

Neumann, A. Conrad; *Eighth Congress of the International Union for Quaternary Research, Paris, France, August 30–September 5, 1969* (A016474); 6 months; $565

O'Connor, Dennis M.; *The International Conference, 'Pacem in Maribus', in Malta, June 28 to July 3, 1970* (GH96); 6 months; $800

UNIVERSITY OF FLORIDA

Oliver, Calvin C.; *Twentieth Congress of the International Astronautical Federation, Mar Del Plata, Argentina, October 5–12, 1969* (K017503); 6 months; $600

Scott, Thomas A.; *The Sixteenth Colloquium Ampere on Magnetic Resonance, Bucharest, Romania, September 1–5, 1970* (P025561); 6 months; $599

Shands, Joseph W. Jr.; *The Oholo Symposium on Microbial Toxins, Israel, March 1970* (B020941); 6 months; $772

FLORIDA STATE UNIVERSITY

Skofronick, James G.; *The Second International Conference on Atomic Physics, Oxford, England, July 21–24, 1970* (P025463); 6 months; $275

UNIVERSITY OF MIAMI

Thomas, Terence M.; *Eighth Congress of the International Union for Quaternary Research, Paris, France, August 30–September 5, 1969* (A016475); 6 months; $440

FLORIDA STATE UNIVERSITY

Watkins, Norman D.; *Symposium on Antarctic Geology, Oslo, Norway, August 6–15, 1970* (V024499); 6 months; $408

UNIVERSITY OF FLORIDA

West, Frederick R.; *I.A.U. Meeting on Observational Methods of Visual Double Stars, Nice, France, September 8–10, 1969* (P017257); 6 months; $600

UNIVERSITY OF SOUTH FLORIDA

Wilson, Robert E.; *I.A.U. Meeting on Mass Loss and Evolution in Close Binaries, Elsinore, Denmark, September 15–19, 1969* (P017275); 6 months; $610

UNIVERSITY OF FLORIDA

Wood, Frank Bradshaw; (P017273); 6 months; $679

GEORGIA

UNIVERSITY OF GEORGIA

Allinger, Norman L.; *Colloquium on 'Conformational Analysis in Organic Molecules, Polymers and Crystals', Paris, France, June 24–July 3, 1970* (P024348); 6 months; $552

Frey, Robert W.; *The International Conference on Trace Fossils, Liverpool, England, January 6–8, 1970 and to Visit Scientific Personnel at the Senckenberg Institute, in Wilhelmshaven, Germany, January 11–13, 1970* (A020235); 6 months; $650

Hercules, David M.; *Fourth Heyrovsky Discussions, Bratislava, Czechoslovakia, May 18–22, 1970* (P018875); 6 months; $680

Howard, James D.; *International Conference on Trace Fossils, Liverpool, England, January 6–8, 1970* (A018216); 6 months; $496

Hoyt, John H.; *Eighth Congress of the International Association for Quaternary Research, Paris, France, August 30–September 5, 1969* (A016655); 6 months; $406

GEORGIA INSTITUTE OF TECHNOLOGY

Kezios, Stothe P.; *Fourth International Heat Transfer Conference, Paris, France, August 31–September 5, 1970* (K025255); 6 months; $330

EMORY UNIVERSITY

McClure, Harold M.; *The Third International Congress of Primatology, Zurich, Switzerland, August 2–5, 1970* (B025104); 6 months; $700

Perachio, Adrian A.; (B025105); 6 months; $700

HAWAII

UNIVERSITY OF HAWAII

Slepian, David; *Popov Society Meeting, Moscow, U.S.S.R., May 26–June 12, 1970* (K024769); 6 months; $710

ILLINOIS

UNIVERSITY OF ILLINOIS, Urbana

Ang, Alfredo H. S.; *Symposium on Limit State Design, London, England, September 11–12, 1969* (K017280); 6 months; $580

UNIVERSITY OF CHICAGO

Aschenbrenner, Richard A.; *Conference on 'Digital Computers of Very Great Power, Application, Construction and Technology' to be Held in Paris, France, September 8 and 9, 1969* (J000793); 6 months; $591

NORTHWESTERN UNIVERSITY

Bordwell, Frederick G.; *Fourth Symposium on Organic Sulphur, Venice, Italy, June 15–20, 1970* (P024566); 6 months; $485

UNIVERSITY OF ILLINOIS, Urbana

Chien, Robert T.; *International Symposium on Information Theory, Noordwijk, the Netherlands, June 15–19, 1970* (K024276); 6 months; $330

NORTHWESTERN UNIVERSITY

Friedman, Avner; *The Conference on Math Theory of Optimal Control and Minimal Surfaces, Tbilissi, U.S.S.R., September 15–18, 1969* (P017064); 6 months; $644

ILLINOIS STATE GEOLOGICAL SURVEY

Gluskoter, Harold J.; *The Meetings of the International Nomenclature and Analyses Commissions of the International Committee for Coal Petrology, Varna, Bulgaria, September 15–19, 1969* (A017297); 6 months; $758

NORTHWESTERN UNIVERSITY

Grau, Albert A.; *Meeting of the Working Group of the International Federation for Information Processing, Munich, Germany, July 14–18, 1970* (J000975); 6 months; $684

UNIVERSITY OF ILLINOIS, Urbana

Hanson, Alfred O.; *The Seventh International Conference on High Energy Accelerators in Yerevan, U.S.S.R., August 27 to September 2, 1969* (P017236); 6 months; $1,019

UNIVERSITY OF CHICAGO

Johnson, David L.; *NATO Advanced Study Institute on Amorphous Semi-Conductors, Ghent, Belgium, August 25–September 5, 1969* (GZ1464); 6 months; $491

Kaiser, E. T.; *International Conference on Mechanisms of Reactions in Solution, Canterbury, England, July 20–24, 1970* (P025598); 6 months; $592

UNIVERSITY OF ILLINOIS, Urbana

Kendeigh, S. Charles; *Fifteenth International Ornithological Congress, The Hague, the Netherlands, August 30–September 5, 1970* (B025180); 6 months; $625

NORTHWESTERN UNIVERSITY

Letsinger, Robert L.; *Third International Symposium on Photochemistry, St. Moritz, Switzerland, July 12–18, 1970* (P025447); 6 months; $676

ILLINOIS INSTITUTE OF TECHNOLOGY

Libertiny, George Z.; *Third International Conference on High Pressure—Aviemore, Scotland, May 11–15, 1970* (K023593); 6 months; $270

UNIVERSITY OF ILLINOIS, Urbana

Lo, Y. T.; *Fourth Colloquium on Microwave Communication—Budapest, Hungary, April 20–24, 1970* (K022919); 6 months; $710

NORTHERN ILLINOIS UNIVERSITY

McGinnis, Lyle D.; *Symposium on Antarctic Geology, Oslo, Norway, August 6–15, 1970* (V025394); 6 months; $497

SOUTHERN ILLINOIS UNIVERSITY

Meyers, Cal Y.; *Fourth Symposium on Organic Sulphur, Venice, Italy, June 15–20, 1970* (P024569); 6 months; $516

UNIVERSITY OF ILLINOIS, Urbana

Munse, W. H.; *Annual Assembly of the International Institute of Welding—Lausanne, Switzerland—July 12–18, 1970* (K024768); 6 months; $620

Paxton, Jack D.; *NATO Advanced Study Institute on Phytotoxins in Plant Diseases, Pugnochiuso, Italy, June 7–21, 1970* (GZ1771); 6 months; $604

NORTHWESTERN UNIVERSITY

Radnor, Michael; *International Meeting of the Institute of Management Sciences—London, England—June 29–July 3, 1970* (K024648); 6 months; $550

UNIVERSITY OF ILLINOIS, Urbana

Ravenhall, David Geoffrey; *The Fourth International Symposium on Electron and Photon Interactions at High Energies in Liverpool, England, September 14–20, 1969* (P017234); 6 months; $521

UNIVERSITY OF CHICAGO

Sachs, Robert G.; (P017232); 6 months; $463

Sally, Paul J. Jr.; *NATO Advanced Study Institute on Representations of LIE Groups, Namur, Belgium, September 1–12, 1969* (GZ1471); 6 months; $446

Thompson, Robert W.; *The Eleventh International Conference on Cosmic Rays, Budapest, Hungary, August 25–September 4, 1969* (P017235); 6 months; $576

Vandervoort, Peter Oliver; *I.A.U.—'The Spiral Structure of Our Galaxy', Basel, Switzerland, August 29–September 4, 1969* (P016991); 6 months; $435

UNIVERSITY OF ILLINOIS, Urbana

Westwater, J. W.; *Fourth International Heat Transfer Conference—Paris, France, August 31-September 5, 1970* (K025259); 6 months; $580

NORTHWESTERN UNIVERSITY

Woo, Chia-Wei; *The Twelfth International Conference on Low Temperature Physics to be Held in Kyoto, Japan, September 4-10, 1970* (P025540); 6 months; $265

UNIVERSITY OF CHICAGO

Zahler, Raphael; *NATO Advanced Study Institute in Algebraic Topology, Aarhus, Denmark, August 10-23, 1970* (GZ1794); 6 months; $497

NORTHWESTERN UNIVERSITY, Chicago Campus

Machol, Robert E.; *International Symposium on Education and Training for Operations Research, Istanbul, Turkey, August 31–September 4, 1970* (GZ1811); 6 months; $800

INDIANA

INDIANA UNIVERSITY, Bloomington

Bardin, Bryce M.; *The International Cyclotron Conference in Oxford, England, September 17-20, 1969* (P017568); 6 months; $542

Brock, Thomas D.; *The International Symposium on the Taxonomy and Biology of Blue-Green Algae, Madras, India, January 8-13, 1970* (B019817); 6 months; $1,368

Bron, Walter E.; *Tenth European Congress*

INTERNATIONAL SCIENTIFIC INFORMATION EXCHANGE (TRAVEL)

on Molecular Spectroscopy, Liege, Belgium, September 29-October 3, 1969 (P017256); 6 months; $480

UNIVERSITY OF NOTRE DAME

Gordon, Robert E.; *Representative Participation in First General Assembly, European Association of Editors of Biological Periodicals, London, England, April 1970* (GN881); 6 months; $530

INDIANA UNIVERSITY, Indianapolis

Hearne, Thomas M. Jr.; *NATO Instructional Conference on Finite Simple Groups, Oxford, England, September 2-20, 1969* (GZ1459); 6 months; $386

UNIVERSITY OF NOTRE DAME

Massey, James L.; *International Symposium on Information Theory, Noordwijk, the Netherlands, June 15-19, 1970* (K024279); 6 months; $300

PURDUE UNIVERSITY

Morrison, Harry; *Third International Symposium on Photochemistry, St. Moritz, Switzerland, July 12-18, 1970* (P023409); 6 months; $724

INDIANA UNIVERSITY, Bloomington

Parmenter, C. S.; *EUCHEM Conference on Molecular Spectroscopy, Cirencester, England, July 13-17, 1970* (P025445); 6 months; $606

PURDUE UNIVERSITY

Powell, Reginald W.; *Second European Conference on Thermophysical Properties of Solids at High Temperatures—Warrington, England—April 7-10, 1970* (K022917); 6 months; $430

INDIANA UNIVERSITY, Bloomington

Rickey, Martin E.; *The International Cyclotron Conference in Oxford, England, September 17-20, 1969* (P017567); 6 months; $533

Swihart, James C.; *The Twelfth International Conference on Low Temperature Physics to be held in Kyoto, Japan, September 4-10, 1970* (P025544); 6 months; $570

PURDUE UNIVERSITY

White, Joe Lloyd; *Second Conference of the International Committee on Studies of Bauxites and Aluminum-Oxide-Hydroxides, Budapest, Hungary, October 6-10, 1969* (A017294); 6 months; $672

IOWA

IOWA STATE UNIVERSITY

Baker, John R.; *Ninth International Embryology Conference, Moscow, U.S.S.R., August 25-29, 1969* (A016256); 6 months; $722

UNIVERSITY OF IOWA

Debault, L. E.; *NATO Institute NATO Adv St Institute on Microbeam Irradiation and Cellular Biology* (GZ1767); 6 months; $468

IOWA STATE UNIVERSITY

Nasser, Essam; *The Ninth International Conference on Phenomena in Ionized Gases, Bucharest, Romania, September 1-6, 1969* (K016898); 6 months; $845

Serovy, George K.; *International Gas Turbine Conference—Brussels, Belgium, May 24-29, 1970* (K023756); 6 months; $460

UNIVERSITY OF IOWA

Zweng, Marilyn J.; *First International Congress on Mathematical Education, Lyons, France, August 24-30, 1969* (GZ1451); 6 months; $471

KANSAS

UNIVERSITY OF KANSAS

Adams, Ralph N.; *Fourth Heyrovsky Discussions, Bratislava, Czechoslovakia, May 18-22, 1970* (P022915); 6 months; $688

WICHITA STATE UNIVERSITY

Tasch, Paul; *Second Symposium on Gondwana Stratigraphy & Paleontology Being Held in Capetown & Johannesburg, South Africa, July 3-24, 1970 & the Symposium on Antarctic Geology Being Held in Oslo, Norway, August 6-15, 1970* (V025391); 6 months; $1,000

UNIVERSITY OF KANSAS

Teichert, Curt; *Second Symposium on Gondwana Stratigraphy and Paleontology, Capetown, Johannesburg, South Africa, July 3-24, 1970* (A024524); 6 months; $940

KENTUCKY

EASTERN KENTUCKY UNIVERSITY

Nichols, Robert L.; *Symposium on Antarctic Geology, Oslo, Norway, August 6-15, 1970* (V024250); 6 months; $500

Wigley, Perry B.; *NATO Advanced Study Institute on the Paleogeography of the Permian, Mainz and Karlsruhe, Germany, August 3-22 1970* (GZ1805); 6 months; $470

LOUISIANA

LOUISIANA STATE UNIVERSITY, Medical Center

Heneghan, James B.; *NATO Advanced Study Institute on the Germ-Free Animal as a Tool in Research, Louvain, Belgium, September 14-20, 1969* (GZ1467); 6 months; $655

TULANE UNIVERSITY

Hofer, Helmut O.; *The Third International Congress of Primatology, Zurich, Switzerland, August 2-5, 1970* (B025101); 6 months; $535

LOUISIANA STATE UNIVERSITY, Baton Rouge

Huggett, R. W.; *The Eleventh International Conference on Cosmic Rays, Budapest, Hungary, August 25-September 4, 1969* (P017096); 6 months; $376

Kestner, Neil R.; *NATO Advanced Study Institute on New Directions in Atomic Physics, Ankara, Turkey, September 8-20, 1969* (GZ1445); 6 months; $685

TULANE UNIVERSITY

Lorenz, Rainer; *The Third International Congress of Primatology, Zurich, Switzerland, August 2-5, 1970* (B025103); 6 months; $487

LOUISIANA STATE UNIVERSITY, Baton Rouge

Runnels, L. K.; *Third International Liquid Crystal Conference, Berlin, West Germany, August 24-28, 1970* (P025495); 6 months; $725

Srinivasan, V. R.; *The Third International Conference on Global Impacts of Applied Microbiology, Bombay, India, December 7-12, 1969* (B019508); 6 months; $994

MAINE

UNIVERSITY OF MAINE, Orono

Borns, Harold W. Jr.; *Symposium on Antarctic Geology, Oslo, Norway, August 6-15, 1970* (V024245); 6 months; $433

MARYLAND

AMERICAN SOCIETY OF BIOLOGICAL CHEMISTRY

Harte, Robert A.; *Eighth International Congress of Biochemistry, Switzerland, September 1970* (B013866); 12 months; $10,000

UNIVERSITY OF MARYLAND

Glasspool, Richard; *NATO Advanced Study Institute on the Architecture and Design of Digital Computers, Hyeres, France, August 25-September 6, 1969* (GZ1461); 6 months; $390

Greenberg, Oscar W.; *Fifteenth International Conference on High Energy Physics, Kiev, U.S.S.R.; August 26-September 4, 1970* (P025628); 6 months; $534

Kim, Hogil; *The Seventh International Conference on High Energy Accelerators in Yerevan, U.S.S.R. August 27-September 2, 1969* (P017098); 6 months; $882

Mesztenyi, Charles K.; *NATO Advanced Study Institute on the Architecture and Design of Digital Computers, Hyeres, France, August 25-September 6, 1969* (GZ1474); 6 months; $406

JOHNS HOPKINS UNIVERSITY

Robinson, Dean W.; *Tenth European Congress on Molecular Spectroscopy, Liege, Belgium, September 29-October 3, 1969* (P017254); 6 months; $504

UNIVERSITY OF MARYLAND

Schmidt, Richard A.; *NATO Advanced Study Institute on Skilled Performance, Trieste, Italy, July 18, 1970 to August 1, 1970* (GZ1769); 6 months; $537

Sullivan, Woodruff T.; *NATO Advanced Study Institute on the Structure and Evolution of the Galaxy, Athens, Greece, September 8-19, 1969* (GZ1473); 6 months; $538

Westerhout, Gart; *I.A.U. Symposium—'The Spiral Structure of Our Galaxy,' Basel, Switzerland, August 29-September 4, 1969* (P016992); 6 months; $411

JOHNS HOPKINS UNIVERSITY

Wood, Robert C.; *The Fifteenth International Ornithological Congress, The Hague, the Netherlands, August 30-September 5, 1970* (V025060); 6 months; $537

UNIVERSITY OF MARYLAND

Zwanzig, Robert; *International Conference on Thermodynamics to be held at Cardiff, Wales, Great Britain, April 1-4, 1970* (P020027); 17 months; $419

MASSACHUSETTS

AMERICAN ACADEMY OF ARTS AND SCIENCES

Voss, John; *Conference on a Proposal to Establish an International Research Center on Insect Physiology and Endocrinology* (GF367*); 12 months; $6,000

MASSACHUSETTS INSTITUTE OF TECHNOLOGY

Amdur, Isadore; *The Second International Conference on Atomic Physics, Oxford, England, July 21-24, 1970* (P025461); 6 months; $343

BRANDEIS UNIVERSITY

Amit, Daniel J.; *The Twelfth International Conference on Low Temperature Physics to be Held in Kyoto, Japan, September 4-10, 1970* (P025542); 6 months; $490

HARVARD UNIVERSITY

Baum, Howard R.; *Seventh International Symposium on Rarefied Gas Dynamics, Pisa, Italy, June 29-July 3, 1970* (K024121); 6 months; $180

Bloch, Konrad E.; *Seventh General Assembly, International Union of Biochemistry, Interlaken, Switzerland, September 3-9, 1970* (GF404); 6 months; $390

MASSACHUSETTS INSTITUTE OF TECHNOLOGY

Cannon, John T.; *NATO Advanced Study Institute on Theoretical Physics, Les Houches, France, July 5-August 29, 1970* (GZ1782); 6 months; $584

BRANDEIS UNIVERSITY

Carlin, Stephen C.; *NATO Advanced Study Institute on Pyridine Nucleotide-Dependent Dehydrogenases, Konstanz, Germany, September 14-20, 1969* (GZ1466); 6 months; $494

HARVARD UNIVERSITY

Dalgarno, A.; *The Second International Conference on Atomic Physics, Oxford, England, July 21-24, 1970* (P025458); 6 months; $343

NORTHEASTERN UNIVERSITY

Dolansky, Ladislav; *International Congress on the Education of the Deaf—Stockholm, Sweden, August 12-21, 1970* (K025059); 6 months; $300

UNIVERSITY OF MASSACHUSETTS

Gluckstern, Robert; *The Seventh International Conference on High Energy Accelerators in Yerevan, U.S.S.R., August 27-September 2, 1969* (P017100); 6 months; $666

HARVARD UNIVERSITY

Goldberg, Leo; *To Lecture at Leningrad University and to Visit Pulkovo Observatory,*

INTERNATIONAL SCIENTIFIC INFORMATION EXCHANGE (TRAVEL)

Leningrad, U.S.S.R., September 30–October 14, 1969 (P017675); 6 months; $541

MASSACHUSETTS GENERAL HOSPITAL

Kalckar, Herman M.; *The Oholo Symposium on Microbial Toxins, Israel, March 1970* (B020936); 6 months; $592

UNIVERSITY OF MASSACHUSETTS

Karasz, Frank E.; *The Faraday Society General Discussion on Polymer Solutions, Manchester, England, April 15-17, 1970* (P021501); 6 months; $426

MASSACHUSETTS INSTITUTE OF TECHNOLOGY

King, John G.; *The Second International Conference on Atomic Physics, Oxford, England, July 21-24, 1970* (P025457); 6 months; $343

Kleppner, Daniel; (P025456); 6 months; $270

Lin, C. C.; *I.A.U. Symposium on the Spiral Structure of Our Galaxy, Basel, Switzerland, August 29-September 4, 1969* (P017364); 6 months; $564

HARVARD UNIVERSITY

Noyes, Robert W.; *NATO Advanced Study Institute on Physics of the Solar Corona, Lagonissi, Athens, Greece, September 6-17, 1970* (GZ1793); 6 months; $483

UNIVERSITY OF MASSACHUSETTS

Olson, Diane E.; *NATO Advanced Study Institute on Human Factors/Ergonomics-Recent Advances and their Application to Modern Technology, Palermo, Sicily, September 25-October 8, 1969* (GZ1457); 6 months; $419

HARVARD UNIVERSITY

Pappenheimer, Alwin M. Jr.; *The Oholo Symposium on Microbial Toxins, Israel, March 1970* (B020938); 6 months; $788

UNIVERSITY OF MASSACHUSETTS

Pichanick, Francis M. J.; *The Second International Conference on Atomic Physics, Oxford, England, July 21-24, 1970* (P025453); 6 months; $343

HARVARD UNIVERSITY

Revelle, Roger; *Thirteenth General Assembly, International Council of Scientific Unions, Madrid, Spain, September 24-29, 1970* (GF405); 6 months; $464

MASSACHUSETTS INSTITUTE OF TECHNOLOGY

Rose, Robert M.; *The Twelfth International Conference on Low Temperature Physics to be Held in Kyoto, Japan, September 4-10, 1970* (P025662); 6 months; $969

Ryder, John D.; *Popov Society Meeting, Moscow, U.S.S.R., May 26-June 12, 1970* (K024771); 6 months; $710

Seligman, Lawrence; *NATO Institute on the Arch & Design of Digital Computers, Hyeres, France, August 25–September 6, 1969* (GZ1469); 6 months; $431

UNIVERSITY OF MASSACHUSETTS

Stein, R. S.; *Third International Liquid Crystal Conference, Berlin, Germany, August 24-28, 1970, International Conference on Macromolecular Chemistry, Leiden, the Netherlands, August 31-September 4, 1970* (P025491); 6 months; $376

Taylor, Joseph H. Jr.; *Meeting on Pulsars and High Energy Activity in Supernovae Remnants, Rome, Italy, December 18-20, 1969* (P018967); 6 months; $563

HARVARD UNIVERSITY

Tinkham, Michael; *The Twelfth International Conference on Low Temperature Physics to be Held in Kyoto, Japan, September 4-10, 1970* (P025541); 6 months; $969

TUFTS UNIVERSITY

Tzannes, N. S.; *International Symposium on Information Theory—Noordwijk, the Netherlands, June 15-19, 1970* (K024278); 6 months; $420

MASSACHUSETTS INSTITUTE OF TECHNOLOGY

Uhlig, Herbert H.; *Third International Symposium on Passivity—Cambridge, England, July 6-10, 1970* (K024766); 6 months; $410

AMERICAN ACADEMY OF ARTS AND SCIENCES

Voss, John; *First and Second Meetings of the Organizing Group, International Center of Insect Physiology, and Ecology, the Netherlands and Sweden January 16-23, 1970* (B020447); 6 months; $454

UNIVERSITY OF MASSACHUSETTS

Vuilleumier, Francois; *Travel Support to Attend—Fifteenth International Ornithological Congress, The Hague, the Netherlands, August 30-September 5, 1970* (B025186); 6 months; $282

HARVARD UNIVERSITY

Williams, Carroll M.; *First and Second Meetings of the Organizing Group, International Center of Insect Physiology and Ecology, the Netherlands and Sweden, January 16-23, 1970* (B020448); 6 months; $464

MICHIGAN

UNIVERSITY OF MICHIGAN

Bartell, Lawrence S.; *Colloquium on 'Conformational Analysis in Organic Molecules, Polymers and Crystals,' Paris, France, June 24-July 3, 1970* (P024347); 6 months; $520

Clark, John A.; *Fourth International Heat Transfer Conference—Paris, France, August 31-September 5, 1970* (K025252); 6 months; $250

Disney, Ralph L.; *International Symposium on Education and Training for Operations Research, Istanbul, Turkey, August 31–September 4, 1970* (GZ1810); 6 months; $810

MICHIGAN STATE UNIVERSITY

Farnum, Donald G.; *Individual Travel to Present a Series of Invited Research Talks at Several Universities in Germany, November 9-22, 1969* (P018874); 10 months; $297

WAYNE STATE UNIVERSITY

Goodman, Morris; *The Third International Congress of Primatology, Zurich, Switzerland, August 2-5, 1970* (B025102); 6 months; $590

MICHIGAN STATE UNIVERSITY

Horne, Frederick H.; *Faraday Society General Discussion on Polymer Solutions, Manchester, England, April 15-17, 1970* (P023717); 6 months; $396

UNIVERSITY OF MICHIGAN

Jones, Lawrence W.; *The Eleventh International Conference on Cosmic Rays, Budapest, Hungary, August 25–September 4, 1969* (P017097); 6 months; $554

MICHIGAN STATE UNIVERSITY

Katz, Leo; *Colloquium on Combinatorial Mathematics, Balatonfured, Hungary, August 25–29, 1969 and the Thirty-Seventh Session of the International Statistical Institute, London, England, September 3-11, 1969* (P017068); 6 months; $700

Leroi, George E.; *Tenth European Congress on Molecular Spectroscopy, Liege, Belgium, September 29-October 3, 1969* (P017255); 6 months; $544

Linnell, Albert P.; *I.A.U. Meeting on Mass Loss and Evolution in Close Binaries, Elsinore, Denmark, September 15-19, 1969* (P017272); 6 months; $631

UNIVERSITY OF MICHIGAN

Longo, Michael J.; *Fifteenth International Conference on High Energy Physics, Kiev, U.S.S.R., August 26-September 4, 1970* (P025629); 6 months; $566

Merte, Herman Jr.; *Fourth International Heat Transfer Conference—Paris, France, August 31-September 5, 1970* (K025257); 6 months; $570

Rowe, Joseph E.; *Popov Society Meeting, Moscow, U.S.S.R., May 26-June 12, 1970* (K024773); 6 months; $710

Springer, George S.; *Seventh International Symposium on Rarefied Gas Dynamics, Pisa, Italy, June 29- July 3, 1970* (K024123); 6 months; $500

MICHIGAN STATE UNIVERSITY

Vance, Irvin E.; *First International Congress on Mathematical Education, Lyons, France, August 24–30, 1969* (GZ1447); 6 months; $579

Wagner, Peter J.; *Third International Symposium on Photochemistry, St. Moritz, Switzerland, July 12-18, 1970* (P023410); 10 months; $510

Yoder, Olen C.; *NATO Advanced Study Institute on Phytotoxins in Plant Diseases, Pugnochiuso, Italy, June 7-21, 1970* (GZ1768); 6 months; $614

MINNESOTA

UNIVERSITY OF MINNESOTA

Aris, Rutherford; *Chemical Engineering Conference—Melbourne and Sydney, Australia, August 18-26, 1970* (K024372); 6 months; $810

Erickson, Albert W.; *Symposium on Antarctic Oceanography, Tokyo, Japan, September 14-25, 1970* (V024631); 6 months; $900

Lawver, James E.; *IX International Mineral Processing Congress— Prague, Czechoslovakia June 1-6, 1970* (K024083); 6 months; $370

Luyten, Willem J.; *I.A.U. Meeting on Observational Methods of Visual Double Stars, Nice, France, September 8-10, 1969* (P017258); 6 months; $625

Siniff, Donald B.; *The Symposium on Antarctic Oceanography—Joint Oceanographic Assembly, Tokyo, Japan, September 14-25, 1970* (V025296), 6 months; $900

MACALESTER COLLEGE

Webers, Gerald F.; *Symposium on Antarctic Geology, Oslo, Norway, August 6-15, 1970* (V024253); 6 months; $519

MISSOURI

UNIVERSITY OF MISSOURI, Columbia

Feather, M. S.; *Fifth International Symposium on Carbohydrate Chemistry, Paris, France, August 17-22, 1970* (P025573); 6 months; $660

Goldberg, Herbert S.; *Second Meeting of the International Subcommittee on Gram-Negative Anaerobic Rod Bacilli, London, England, October 20-21, 1969* (B018099); 6 months; $550

WASHINGTON UNIVERSITY

Hoyt, Howard P. Jr.; *NATO Institute NATO Adv St. Inst. on the Moon & Planets Newcastle-upon-Tyne, England, April 9–16, 1970* (GZ1531); 6 months; $396

Jenkins, James A.; *The Conference on Functions of a Complex Variable, Oberwohlbach, Germany, February 20–28, 1970* (P022950); 6 months; $612

Lancaster, Glenn A.; *NATO Advanced Study Institute on Representations of LIE Groups, Namur, Belgium, September 1-12, 1969* (GZ1455); 6 months; $548

ST. LOUIS UNIVERSITY

Lee, Sook; *The Sixteenth Colloquium Ampere on Magnetic Resonance, Bucharest, Romania, September 1-5, 1970* (P025560); 6 months; $830

UNIVERSITY OF MISSOURI, Columbia

Peterson, Mark Richard; *Seventh International Congress of Agricultural Engineering, Baden-Baden, Germany, October 6-11, 1969* (K017403); 6 months; $128

ST. LOUIS UNIVERSITY

Ramsey, Robert B.; *NATO Advanced Study Institute on the Chemistry of Brain Development, Milan, Italy, September 9-19, 1970* (GZ1774); 6 months; $638

INTERNATIONAL SCIENTIFIC INFORMATION EXCHANGE (TRAVEL)

WASHINGTON UNIVERSITY
Scharnberger, Charles K.; *Symposium on Antarctic Geology, Oslo, Norway, August 6-15, 1970* (V024251); 6 months; $530
Sonnenwirth, Alexander C.; *Second Meeting of the International Subcommittee on Gram-Negative Anaerobic Rod Bacilli, London, England, October 20-21, 1969* (B018100); 6 months; $502
Zaborszky, John; *The Special Seminar on Learning Control, Nagoya, Japan, August 18-20, 1970* (K025058); 6 months; $445

NEBRASKA

UNIVERSITY OF NEBRASKA, Lincoln
Jaecks, Duane H.; *The Second International Conference on Atomic Physics, Oxford, England, July 21-24, 1970* (P025455); 6 months; $498

NEW HAMPSHIRE

UNIVERSITY OF NEW HAMPSHIRE
Andersen, Kenneth K.; *Fourth Symposium on Organic Sulphur, Venice, Italy, June 15-20, 1970* (P024567); 6 months; $634

DARTMOUTH COLLEGE
Garland, Stephen J.; *NATO Advanced Study Institute on Fundamental Aspects and Current Developments in Computer Programming, Copenhagen, Denmark, August 11-22, 1969* (GZ1462); 6 months; $443

DEPARTMENT OF THE ARMY, TERRESTRIAL SCIENCE CENTER
Gow, Anthony J.; *The Symposium on the Hydrology of Glaciers, Cambridge, England, September 7-12, 1969* (A015710); 6 months; $260

DARTMOUTH COLLEGE
Gribble, Gordon W.; *Seventh International Symposium on the Chemistry of Natural Products, Riga, Latvia, U.S.S.R., June 22-27, 1970* (P023497); 6 months; $591
Stockmayer, Walter H.; *International Symposium on Macromolecules, Leiden, the Netherlands, August 31-September 4, 1970* (P024345); 6 months; $345

NEW JERSEY

AMERICAN FEDERATION OF INFORMATION PROCESSING SOCIETIES
Gilchrist, Bruce; *Travel to Meetings of the International Federation of Information Processing Societies* (J000769); 12 months; $2,500

EDUCATIONAL TESTING SERVICE
Landis, Daniel; *International Conference of Production Research, Birmingham, England April 5-10, 1970* (GZ1528); 6 months; $444

PRINCETON UNIVERSITY
Barry, Simon; *NATO Advanced Study Institute on Theoretical Physics, Les Houches, France, July 5-August 29, 1970* (GZ1780); 6 months; $587

RUTGERS UNIVERSITY
Becker, Jerry P.; *First International Congress on Mathematical Education, Lyons, France, August 24-30, 1969* (GZ1448); 6 months; $548

PRINCETON UNIVERSITY
Bienkowski, George K.; *Seventh International Symposium on Rarefied Gas Dynamics, Pisa, Italy, June 29-July 3, 1970* (K024122); 6 months; $650

INSTITUTE FOR ADVANCED STUDY
Crater, Horace W.; *NATO Advanced Study Institute on Quantum Field Theory, Ajaccio, Corsica, June 27-July 25, 1970* (GZ1807); 6 months; $365

STEVENS INSTITUTE OF TECHNOLOGY
Gilman, Robert H.; *NATO Advanced Study Institute on Instructional Conference on Finite Simple Groups, Oxford, England, September 2-20, 1969* (GZ1479); 6 months; $330
Meissner, Hans; *The Twelfth International Conference on Low Temperature Physics to be Held in Kyoto, Japan, September 4-10, 1970* (P025533); 6 months; $500

RUTGERS UNIVERSITY
Stephen, Michael J.; (P025537); 6 months; $545

NEW MEXICO

UNIVERSITY OF NEW MEXICO
Blum, Julius R.; *The Thirty-Seventh Session of the International Statistical Institute, London, England, September 3-11, 1969* (P017067); 6 months; $683
Frakes, Lawrence A.; *'International Symposium on Stratigraphy and Paleontology of the Gondwana,' Capetown and Johannesburg, South Africa, July, 1970* (A020322); 8 months; $1,519

NEW MEXICO INSTITUTE OF MINING AND TECHNOLOGY
Wilkening, Marvin H.; *The International Symposium on Atmospheric Chemistry and Radioactivity, Heidelberg, Germany, September 8-13, 1969* (A016975); 6 months; $697

NEW YORK

NEW YORK ZOOLOGICAL SOCIETY
Fox, Paul J.; *Symposium on the Structure of the Crust and Mantle Beneath Inland and Marginal Seas, Madrid, Spain, September 4-6, 1969* (A017717); 6 months; $331

UNIVERSITY OF ROCHESTER
Agate, A. D.; *The Third International Conference on Global Impacts of Applied Microbiology, Bombay, India, December 7-12, 1969* (B019507); 6 months; $1,187

CORNELL UNIVERSITY
Ashcroft, Neil W.; *The International Conference on Amorphous and Liquid Semiconductors to be Held in Cambridge, England, September 24-27, 1969* (P017181); 6 months; $330

RENSSELAER POLYTECHNIC INSTITUTE
Bailey, Ronald A.; *EUCHEM Conference on Molten Salts, Colmar, France, May 10-15 1970* (P023123); 6 months; $556

CORNELL UNIVERSITY
Berger, Toby; *International Symposium on Information Theory, Noordwijk, The Netherlands, June 15-19, 1970* (K024273); 6 months; $200
Berkelman, Karl; *The Fifteenth International Conference on High Energy Physics, Kiev, U.S.S.R., August 26-September 4, 1970* (P025627); 6 months; $811

NEW YORK UNIVERSITY
Bernheimer, Alan W.; *The Oholo Symposium on Microbial Toxins, Israel, March 1970* (B020931); 6 months; $618

CORNELL UNIVERSITY
Brodsky, Stanley J.; *The Second International Conference on Atomic Physics, Oxford, England, July 21-24, 1970* (P025462); 6 months; $500

STATE UNIVERSITY AT BUFFALO
Buergel, Wolfgang; *NATO Advanced Study Institute on Theoretical Principles of Heterogeneous Catalysis on Semi-Conductors, Konstanz, Germany, August 31-September 12, 1970* (GZ1775); 6 months; $407

COLUMBIA UNIVERSITY
Burton, James J.; *NATO Advanced Study Institute on Electron Microscopy, Erice, Sicily, Italy, April 4-18, 1970* (GZ1762); 6 months; $439

STATE UNIVERSITY AT BUFFALO
Calkin, Parker E.; *Symposium on Antarctic Geology, Oslo, Norway, August 6-15, 1970* (V024363); 6 months; $100

DOWNSTATE MEDICAL CENTER
Craig, John P.; *The Oholo Symposium on Microbial Toxins, Israel, March 1970* (B020933); 6 months; $840

COLUMBIA UNIVERSITY
Dalziel, Ian W. D.; *Symposium on Antarctic Geology, Oslo, Norway, August 6-15, 1970* (V024373); 6 months; $410

UNIVERSITY OF ROCHESTER
Douglass, David H. Jr.; *The Twelfth International Conference on Low Temperature Physics to be Held in Kyoto, Japan, September 4-10, 1970* (P025545); 6 months; $1,008

NEW YORK UNIVERSITY
Ehrenfeld, Sylvain; *International Meeting of the Institute of Management Sciences, London, England, June 29-July 3, 1970* (K024649); 6 months; $420

CORNELL UNIVERSITY
Emlen, Stephen T.; *Fifteenth International Ornithological Congress, The Hague, the Netherlands, August 30-September 5, 1970* (B025179); 6 months; $514

ROCKEFELLER UNIVERSITY
Feinberg, Robert S.; *NATO Advanced Study Institute on Protein Structure and Function, Venice, Italy, August 20-September 1, 1970* (GZ1779); 6 months; $395

CORNELL UNIVERSITY
Gebhart, Benjamin; *Fourth International Heat Transfer Conference—Paris, France, August 31-September 5, 1970* (K025253); 6 months; $510

UNIVERSITY OF ROCHESTER
Gillman, Leonard; *First International Congress on Mathematical Education, Lyons, France, August 24-30, 1969* (GZ1449); 6 months; $233

CORNELL UNIVERSITY
Hand, Louis N.; *The Fourth International Symposium on Electron and Photon Interactions at High Energies, Liverpool, England, September 14-20, 1969* (P017226); 6 months; $468

COLUMBIA UNIVERSITY
Hayes, Dennis E.; *Symposium on Antarctic Geology, Oslo, Norway, August 6-15, 1970* (V024248); 6 months; $410
Hays, James D.; (V024396); 6 months; $410
———; *INQUA International Congress, Paris, France, August 30-September 5, 1969 and International Scientific Congress on the Volcano of Thera, Athens, Greece, September 15-22, 1969* (A017376); 6 months; $772

CORNELL UNIVERSITY
Ho, Paul S.; *International Conference on Atomic Transport in Solids and Liquids, Marstrand, Sweden, June 15-19, 1970* (K024163); 6 months; $400

RENSSELAER POLYTECHNIC INSTITUTE
Janz, George J.; *Twentieth Meeting of CITCE, Strasbourg, France, September 14-20, 1969* (P017001); 6 months; $410

AMERICAN SOCIETY OF MECHANICAL ENGINEERS
Juhasz, Stephen; *The Fourth International Conference of Heat Transfer, Versailles, France, August 31-September 5, 1970* (GN896); 6 months; $469

COLUMBIA UNIVERSITY
Kuo, John T.; *Sixth International Symposium on Earth Tides, Strasbourg, France, September 15-20, 1969* (A017295); 6 months; $440

NEW YORK UNIVERSITY
Landis, Fred; *Fourth International Heat Transfer Conference, Paris, France, August 31-September 5, 1970* (K025256); 6 months; $240

STATE UNIVERSITY AT BINGHAMTON
Lazoroff, Norman; *The International Symposium on the Taxonomy and Biology of Blue-*

INTERNATIONAL SCIENTIFIC INFORMATION EXCHANGE (TRAVEL)

Green Algae, Madras, India, January 8–13, 1970 (B019819); 6 months; $1,282

CORNELL UNIVERSITY
Loh, Eugene C.; *The Fourth International Symposium on Electron and Photon Interactions in High Energies in Liverpool, England from September 14–20, 1969* (P017227); 6 months; $468

COLUMBIA UNIVERSITY
Marshall, Thomas C.; *International Conference on the Physics of Quiescent Plasmas, Paris, France, September 8–12, 1969* (K016896); 6 months; $440

CORNELL UNIVERSITY
McDaniel, Boyce D.; *The Seventh International Conference on High Energy Accelerators in Yerevan, U.S.S.R., August 27–September 2, 1969* (P017099); 6 months; $926
Meinwald, Jerrold; *First and Second Meetings of the Organizing Group, International Center of Insect Physiology and Ecology, Netherlands and Sweden, January 16–23, 1970* (B020446); 6 months; $550

STATE UNIVERSITY AT STONY BROOK
Metcalf, Harold; *The Second International Conference on Atomic Physics, Oxford, England, July 21–24, 1970* (P025454); 6 months; $350

COLUMBIA UNIVERSITY
Ninkovich, Charles D.; *International-Science Congress on the Volcano of Thera/Athens, Greece, September 15–22, 1969* (A016726); 6 months; $480
Prendergast, Kevin H.; *I.A.U. Symposium—'The Spiral Structure of our Galaxy', Basel, Switzerland, August 29–September 4, 1969* (P016997); 6 months; $550

STATE UNIVERSITY AT ALBANY
Rabinowitch, Victor; *First and Second Meetings of the Organizing Group, International Center of Insect Physiology and Ecology, the Netherlands and Sweden, January 16–23, 1970* (B020717); 6 months; $578

COLUMBIA UNIVERSITY SYSTEM OFFICE
Rainwater, James, *The International Cyclotron Conference, Oxford, England, September 17–19, 1969* (P017429); 6 months; $465

CORNELL UNIVERSITY
Reppy, John D.; *The Twelfth International Conference on Low Temperature Physics to be Held in Kyoto, Japan, September 4–10, 1970* (P025543); 6 months; $1,010

COLUMBIA UNIVERSITY
Saito, Tsunemasa; *Symposium on Deep-Sea Sediments, Paris, France, August 30–September 5, 1969* (A017377); 6 months; $509

SYRACUSE UNIVERSITY
Schuerch, Conrad; *Fifth International Symposium on Carbohydrate Chemistry, Paris, France, August 17–22, 1970* (P025574); 6 months; $537

NEW YORK UNIVERSITY
Schuster, David I.; *Third International Symposium on Photochemistry, St. Moritz, Switzerland, July 12–18, 1970* (P023411); 6 months; $470

CORNELL UNIVERSITY
Stein, Peter C.; *The Fourth International Symposium on Electron and Photon Interactions at High Energies in Liverpool, England from September 14-20, 1969* (P017231); 6 months; $468
Steinkraus, Keith H.; *The Third International Conference on Global Impacts of Applied Microbiology, Bombay, India, December 7–12, 1969* (B019512); 6 months; $1,049

NEW YORK UNIVERSITY
Stoker, James J.; *The International Meeting of Floods, Their Prediction and the Defense of the Soil, Rome, Italy, November 23–30, 1969* (P017069); 6 months; $409

RENSSELAER POLYTECHNIC INSTITUTE
Strong, Robert L.; *International Conference on Mechanisms of Reactions in Solution, Canterbury, England, July 20–24, 1970* (P024674); 6 months; $439

NEW YORK UNIVERSITY
Sundheim, Benson R.; *EUCHEM Conference on Molten Salts, Colmar, France, May 10–15, 1970* (P021272); 6 months; $501
Thumim, Alfred I.; *Symposuim on Application of Magnetism in Bioengineering, Rehovot, Israel, December 9–11, 1970* (K018892,; 6 months; $240

CORNELL UNIVERSITY
Tigner, Maury; *The Seventh International Conference on High Energy Accelerators in Yerevan, U.S.S.R., August 27–September 2, 1969* (P017095); 6 months; $714

NEW YORK UNIVERSITY
Travis, Dennis; *NATO Instructional Conference on Finite Simple Groups, Oxford, England, September 2–20, 1969* (GZ1475); 6 months; $320

COLUMBIA UNIVERSITY
Vogeli, Bruce R.; *First International Congress on Mathematical Education, Lyon, France, August 24–30, 1969* (GZ1450); 6 months; $548

NEW YORK UNIVERSITY
Yagoda, Irwin; *International Symposium on Information Theory, Noordwijk, The Netherlands, June 15–19, 1970* (K024275); 6 months; $300

CORNELL UNIVERSITY
Yennie, Donald R.; *The Fourth International Symposium on Electron and Photon Interactions at High Energies, Liverpool England, September 14–20, 1969* (P017230); 6 months; $468

COLUMBIA UNIVERSITY SYSTEM OFFICE
Ziegler, Klaus; *The International Cyclotron Conference, Oxford, England, September 17–19, 1969* (P017430); 6 months; $403

NORTH CAROLINA

DUKE UNIVERSITY
Castlow, John D. Jr.; *The International Symposium on the Fertility of the Sea, Sao Paulo, Brazil, December 1–6, 1969* (B017442); 12 months; $5,000

NORTH CAROLINA STATE UNIVERSITY AT RALEIGH
Backman, Paul A.; *NATO Advanced Study Institute on Phytotoxins in Plant Diseases, Pugnochiuso, Italy, June 7–21, 1970* (GZ1770); 6 months; $592
Beatty, Kenneth Orion Jr.; *Fourth International Heat Transfer Conference, Paris, France, August 31,–September 5, 1970 and the International Conference on Heat and Mass Transfer, Dubrovnik, Yugoslavia, September 8–12, 1970* (K025251); 6 months; $460

ST. AUGUSTINE'S COLLEGE
Dillard, Margaret; *NATO Advanced Study Institute on Theoretical Physics, Les Houches, France, July 5–August 29, 1970* (GZ1781); 6 months; $664

NORTH CAROLINA FOUNDATION FOR MENTAL HEALTH RESEARCH
Oppenheim, Ronald W.; *Fifteenth International Ornithological Congress, The Hague, the Netherlands, August 30–September 5, 1970* (B025183); 6 months; $400

UNIVERSITY OF NORTH CAROLINA AT CHAPEL HILL
Stern, Arthur C.; *Workshop on Computer Control Systems for Preventing Air Pollution—Osaka, Japan, August 15, 1970* (K025057); 6 months; $380

NORTH DAKOTA

UNIVERSITY OF NORTH DAKOTA
Johnson, A. William; *International Symposium on Ylides, Leicester, England, July 14–16, 1970* (P024346); 6 months; $700

OHIO

CASE WESTERN RESERVE UNIVERSITY
Bodanszky, Miklos; *Natural Products Symposium, Gregynog, Wales, May 29–June 1, 1970* (P023829); 6 months; $519

KENT STATE UNIVERSITY
Brown, Glenn H.; *Third International Liquid Crystal Conference, Berlin, Germany, August 24–28, 1970* (P025494); 6 months; $437

OHIO STATE UNIVERSITY
Bull, Colin; *Symposium on Antarctic Geology, Oslo, Norway, August 6–15, 1970* (V024375); 6 months; $469
Busch, Daryle H.; *International Conference on Mechanisms of Reactions in Solution, Canterbury, England, July 20–24, 1970* (P024672); 6 months; $569
Coates, Donald A.; *Symposium on Antarctic Geology, Oslo, Norway, August 6–15, 1970* (V025396); 6 months; $484
Elliott, David; *Second International Symposium on Gondwana Stratigraphy and Paleontology, Capetown and Johannesburg, South Africa, July 3–24, 1970 and the Symposium on Antarctic Geology, Oslo, Norway, August 6–15, 1970* (V024371); 6 months; $1,403
Faure, Gunter; *Symposium on Antarctic Geology, Oslo, Norway, August 6–15, 1970* (V017538); 6 months; $484

KENT STATE UNIVERSITY
Franklin, W. M.; *Third International Liquid Crystal Conference, Berlin, Germany, August 24–28, 1970* (P025492); 6 months; $430

UNIVERSITY OF AKRON
Garn, Paul D.; *Committee on Standardization of the International Confederation for Thermal Analysis, London, England, October 1–3, 1969* (P017116); 6 months; $470

UNIVERSITY OF TOLEDO
Griffin, Claibourne E.; *Colloquium on Organophosphorus Compounds, Lodz, Poland, May 20–26, 1970* (P025140); 6 months; $249

OHIO STATE UNIVERSITY
Horton, Derek; *Fifth International Symposium on Carbohydrate Chemistry, Paris, France, August 17–22, 1970 and to Visit the Institute of Chemistry, Bratislava, Czechoslovakia, August 24–27, 1970* (P025575); 6 months; $430

CASE WESTERN RESERVE UNIVERSITY
Kuwana, Theodore; *Fourth Heyrovsky, Discussions, Bratislava, Czechoslovakia, May 18–22, 1970* (P021047); 6 months; $350

BOWLING GREEN STATE UNIVERSITY
Martin, Elden W.; *Fifteenth International Ornithological Congress, The Hague, the Netherlands, August 30–September 5, 1970* (B025181); 6 months; $388

CASE WESTERN RESERVE UNIVERSITY
McCuskey, S. W.; *I.A.U. Symposium—'The Spiral Structure of Our Galaxy,' Basel, Switzerland, August 29–September 4, 1969* (P016995); 6 months; $546

OHIO STATE UNIVERSITY
Mercer, John H.; *Symposium on Antarctic Geology, Oslo, Norway, August 6–15, 1970* (V024393); 6 months; $484
Orheim, Olav; *Symposium on Antarctic Geology, Oslo, Norway, August 6–15, 1970* (V025392); 6 months; $484

CASE WESTERN RESERVE UNIVERSITY
Ostrach, Simon; *Advisory Group for Aerospace Research and Development Meeting on Fluid Dynamics of Blood Circulation and*

INTERNATIONAL SCIENTIFIC INFORMATION EXCHANGE (TRAVEL)

Respiratory Flow, Naples, Italy, May 4–6, 1970 (K023614); 6 months; $570

OHIO STATE UNIVERSITY

Splettstoesser, John; *Symposium on Antarctic Geology, Oslo, Norway, August 6–15, 1970* (V025448); 6 months; $428

Taiganides, E. Paul; *Seventh International Congress of Agricultural Engineering, Baden-Baden, Germany, October 6–11, 1969* (K017404); 6 months; $480

OKLAHOMA

UNIVERSITY OF OKLAHOMA

Babb, Stanley E. Jr.; *Third International Conference on High Pressure—Aviemore, Scotland, May 11–15, 1970* (K023594); 6 months; $300

Hinshaw, Lerner B.; *The Oholo Symposium on Microbial Toxins, Israel, March 1970* (B020934); 6 months; $575

Payne, Robert B.; *Fifteenth International Ornithological Congress, The Hague, the Netherlands, August 30–September 5, 1970* (B025184); 6 months; $500

Weinheimer, Alfred J.; *Seventh International Symposium on the Chemistry of Natural Products, Riga, Latvia, U.S.S.R., June 22–27, 1970* (P023496); 6 months; $966

OREGON

UNIVERSITY OF OREGON

Castenholz, Richard W.; *The International Symposium on the Taxonomy and Biology of Blue-Green Algae, Madras, India, January 8–13, 1970* (B019818); 6 months; $1,202

OREGON STATE UNIVERSITY

Daniels, Malcolm; *Third International Symposium on Photochemistry, St. Moritz, Switzerland, July 12–18, 1970* (P025446); 6 months; $594

UNIVERSITY OF OREGON

Higgins, Richard J.; *The Twelfth International Conference on Low Temperature Physics to be Held in Kyoto, Japan, September 4–10, 1970* (P025532); 6 months; $442

PENNSYLVANIA

LEHIGH UNIVERSITY

Atano, Victor M.; *NATO Advanced Study Institute on Human Factors-Ergonomics—Recent Advances and Their Application to Modern Technology, Palermo, Sicily, September 25–October 8, 1969* (GZ1465); 6 months; $437

AMERICAN FORAGE AND GRASSLAND COUNCIL

Baylor, John E.; *Eleventh International Grassland Congress, Queensland, Australia, April 1970* (B021404); 12 months; $7,000

UNIVERSITY OF PENNSYLVANIA

Bockris, John O'M.; *Third International Symposium on Passivity—Cambridge, England, July 6–10, 1970 and to Visit Various Laboratories and Institutes in Germany, Poland, Yugoslavia, and Switzerland During the Period July 12–25, 1970* (K024767); 6 months; $540

ALBERT EINSTEIN MEDICAL CENTER

Boroff, Daniel A.; *The Second International Conference on Plant and Animal Toxins and the Oholo Symposium on Microbial Toxins, Israel, February–March 1970* (B020932); 6 months; $599

CARNEGIE MELLON UNIVERSITY

Coleman, Bernard D.; *International Conference on Thermodynamics to be Held at Cardiff, Wales, Great Britain, April 1–4, 1970* (P020028); 6 months; $458

UNIVERSITY OF PITTSBURGH

Garfunkel, Myron P.; *The Twelfth International Conference on Low Temperature Physics to be Held in Kyoto, Japan, September 4–10, 1970* (P025538); 6 months; $948

Gerjuoy, Edward; *The Second International Conference on Atomic Physics, Oxford, England, July 21–24, 1970* (P025460); 6 months; $548

LEHIGH UNIVERSITY

Hillman, Donald J.; *The Triennial Conference of the International Association of Technological University Libraries, Loughborough, Leicestershire, England, March 31–April 4, 1970* (N869); 6 months; $456

UNIVERSITY OF PITTSBURGH

Jeffrey, George Alan; *EUCHEM Conference on Carbohydrate Chemistry, Birmingham, England, September 15–19, 1969* (P016517); 6 months; $465

PENNSYLVANIA STATE UNIVERSITY

Jordan, Joseph; (P020970); 6 months; $542

ALBERT EINSTEIN MEDICAL CENTER

Kadis, Solomon; *The Oholo Symposium on Microbial Toxins, Israel, March 1970* (B020935); 6 months; $405

UNIVERSITY OF PENNSYLVANIA

Langenberg, D. N.; *The Twelfth International Conference on Low Temperature Physics to be Held in Kyoto, Japan, September 4–10, 1970* (P025539); 6 months; $900

LEHIGH UNIVERSITY

Luyben, William L.; *International Symposium on Distillation, Brighton, England, September 8–10, 1969* (K016899); 6 months; $300

UNIVERSITY OF PENNSYLVANIA

MacDiarmid, Alan G.; *To Visit and Lecture at Various Institutes and Universities in Moscow, Leningrad, and Tbilisi, U.S.S.R., East Berlin, Germany, and Prague, Czechoslovakia, June 8–26, 1970* (P025155); 6 months; $611

Middleton, Roy; *International Conference on Nuclear Reactions Induced by Heavy Ions, Heidelberg, Germany, July 15–18, 1969* (P016830); 6 months; $390

PENNSYLVANIA STATE UNIVERSITY

Pena, Jorge Augusto; *Seventh International Conference—Condensation and Ice Nuclei, Prague, Czechoslovakia, September, 17–20, 1969 ad Vienna, Austria, September 22–24, 1969* (A017622); 6 months; $453

CARNEGIE MELLON UNIVERSITY

Saibel, Edward; *Meeting of the International Organization of Economic and Cooperative Development Group on Wear, Delft, Holland, April 2–3, 1970* (K023363); 6 months; $300

LEHIGH UNIVERSITY

Salathe, Eric P.; *NATO Advanced Study Institute on Nonlinear Continuum Theories in Mechanics and Physics and Their Applications, (Bolzano), Italy, September 3–10, 1969* (GZ1454); 6 months; $609

Smith, Gerald F.; *Twelfth British Theoretical Mechanics Colloquium—Norwich, England, March 23–26, 1970* (K021383); 6 months; $310

PENNSYLVANIA STATE UNIVERSITY

Spackman, William; *'International Nomenclature and Analyses Commission', Varna, Bulgaria, September 15–20, 1969* (A017575); 6 months; $773

SWARTHMORE COLLEGE

Van De Kamp, Peter; *I.A.U. Meeting on Observational Methods of Visual Double Stars, Nice, France, September 8–10, 1969* (P017260); 6 months; $500

PENNSYLVANIA STATE UNIVERSITY

Voight, Barry; *Symposium of the International Society of Rock Mechanics, Oslo, Norway, September 23–25, 1969* (A017296); 6 months; $580

RHODE ISLAND

BROWN UNIVERSITY

Cooper, David B.; *International Symposium on Information Theory—Noordwijk, the Netherlands, June 15–19, 1970* (K024277); 6 months; $300

Dadanoff, Lec P.; *Second Soviet Conference on Solid State Theory, Moscow, U.S.S.R., December 14–21, 1969* (P019197); 8 months; $548

Savage, John E.; *International Symposium on Information Theory—Noordwijk, the Netherlands, June 15–19, 1970* (K024594); 6 months; $200

Schrier, Allan M.; *The Third International Congress of Primatology, Zurich, Switzerland, August 2–5, 1970* (B025106); 6 months; $594

SOUTH CAROLINA

CLEMSON UNIVERSITY

Andrews, John F.; *To Study and Do Research at the Water Pollution Research Laboratory, and Present a Seminar at the Water Research Assn, in Stevenage, England, December 29, 1969–June 15, 1970* (K019440); 12 months; $520

UNIVERSITY OF SOUTH CAROLINA

Conolly, John R.; *International Conference on Geophysics of the Earth and the Oceans, Sydney, Australia, January 19–23, 1970* (A020324); 3 months; $300

Edge, Ronald D.; *The International Conference on Atomic Collision Phenomena in Solids, Brighton, England, September 7–12, 1969* (P017776); 6 months; $385

Giles, Jr. Frederick H.; *Seminar on the Teaching of Physics in Schools, Copenhagen, Denmark, July 30–August 5, 1969* (GZ1458); 6 months; $594

SOUTH DAKOTA

SOUTH DAKOTA SCHOOL OF MINES AND TECHNOLOGY

Davis, Briant L.; *Seventh International Conference—Condensation and Ice Nuclei, Prague, Czechoslovakia, September 17–20, 1969 and Vienna, Austria, September 22–24, 1969* (A017621); 6 months; $725

UNIVERSITY OF SOUTH DAKOTA

Rutford, Robert H.; *Symposium on Antarctic Geology, Oslo, Norway, August 6–15, 1970* (V024394); 6 months; $487

TENNESSEE

UNIVERSITY OF TENNESSEE

Aiken, Robert M.; *World Conference on Computer Education, Amsterdam, the Netherlands, August 24–28, 1970* (J001105); 6 months; $558

MEMPHIS STATE UNIVERSITY

Bertelli, Domenick J.; *International Symposium on Aromaticity, Pseudoaromaticity, Antiaromaticity, Jerusalem, Israel, March 31–April 3, 1970* (P021048); 6 months; $733

AUSTIN PEAY STATE UNIVERSITY

Mayfield, Melburn R.; *International Congress on the Education of Teachers of Physics in Secondary Schools, Eger, Hungary, September 11–17, 1970* (GZ1814); 6 months; $682

TEXAS

RICE UNIVERSITY

Linsley, Jerald N.; *NATO Advanced Study Institute on Fundamental Aspects and Current Developments in Computer Programming, Copenhagen, Denmark, August 11–22, 1969* (GZ1456); 6 months; $545

UNIVERSITY OF TEXAS AT AUSTIN

Atkins, Patrick Riley; *Seventh International Conference on Condensation and Ice Nuclei, Prague, Czechoslovakia, September 17–20, 1969 ad Vienna, Austria, September 22–24, 1969* (A017787); 6 months; $600

De Vaucouleurs, Gerard H.; *I.A.U. Symposium—'The Spiral Structure of Our Galaxy', Basel, Switzerland, August 29–September 4, 1969* (P016993); 6 months; $564

TEXAS A & M UNIVERSITY

El-Sayed, Sayed Z.; *Symposium on Ant-*

INTERNATIONAL SCIENTIFIC INFORMATION EXCHANGE (TRAVEL)

arctic Oceanography, Tokyo, Japan, September 14–25, 1970 (V024629); 6 months; $909

UNIVERSITY OF TEXAS AT DALLAS
Halpern, Martin; *Symposium on Antarctic Geology, Oslo, Norway, August 6–15, 1970* (VO24397); 6 months; $584

UNIVERSITY OF TEXAS AT AUSTIN
Lagowski, J. J.; *Symposium on Electrochemistry in Nonaqueous Media, Paris, France, July 6–11, 1970* (P024493); 6 months; $821
Powers, Edward J. Jr.; *Ninth International Conference on Phenomena in Ionized Gases, Bucharest, Romania, September 1–6, 1969* (K017279); 6 months; $480

TEXAS INSTRUMENTS INC.
Schearer, Laird D.; *The Second International Conference on Atomic Physics, Oxford, England, July 21–24, 1970* (P025459); 6 months; $410

UNIVERSITY OF TEXAS AT AUSTIN
Selander, Robert K.; *Symposium on Variation in Mammalian Populations, London, England, November 14–15, 1969* (B017849); 6 months; $583

TEXAS TECH UNIVERSITY
Shine, Henry J.; *Fourth Symposium on Organic Sulphur, Venice, Italy, June 15–20, 1970* (P024568); 6 months; $450
Wade, F. Alton; *Symposium on Antarctic Geology, Oslo, Norway, August 6–15, 1970* (V024252); 6 months; $627

RICE UNIVERSITY
Warme, John E.; *International Trace Fossil Conference, Liverpool, England, January 6–8, 1970* (A016435); 8 months; $600

TEXAS TECH UNIVERSITY
Wilbanks, J. R.; *Symposium on Antarctic Geology, Oslo, Norway, August 6–15, 1970* (V025867); 6 months; $433

UTAH

UTAH STATE UNIVERSITY
Dixon, Keith L.; *Fifteenth International Ornithological Congress, The Hague, the Netherlands, August 30–September 5, 1970* (B025178); 6 months; $300

UNIVERSITY OF UTAH
Keuffel, Jack W.; *The Eleventh International Conference on Cosmic Rays, Budapest Hungary, August 25–September 4, 1969* (P016257); 6 months; $835
Parry, Robert W.; *International Symposium on Univ Chemical Education, Rome, Italy October 16–19, 1969* (GZ1482); 6 months; $460

BRIGHAM YOUNG UNIVERSITY
Smoot, L. Douglas; *Meeting of the D–5 Radar Interference Working Group of the Technical Tripartite Cooperation Program, Wescott, England, August 10–14, 1970* (K025056); 6 months; $610
Wilson, Arnold; *International Colloquium on the Progress of Shell Structures in the Last Ten Years and Its Future Development, Madrid, Spain, September 30–October 3, 1969* (K017480); 6 months; $510

VIRGINIA

UNIVERSITY OF VIRGINIA
Ianna, Philip A.; *NATO Advanced Study Institute on the Structure and Evolution of the Galaxy, Athens, Greece, September 8–19, 1969* (GZ1478); 12 months; $602

VIRGINIA POLYTECHNIC INSTITUTE
Pritsker, A. Alan B.; *Second International Congress on Project Planning by Network Analysis, Amsterdam, the Netherlands, October 6-10, 1969* (K017669); 6 months; $460

UNIVERSITY OF VIRGINIA
Roberts, William W. Jr.; *I.A.U. Symposium —'The Spiral Structure of Our Galaxy,' Basel, Switzerland, August 29-September 4, 1969* (P016994); 6 months; $450
Schatz, Paul N.; *Faraday Society Symposium on Magneto-Optical Effects, London, England, December 11-12, 1969* (P017935); 6 months; $262
Uhl, Vincent William; *Third International Congress of Chemical Engineering, Chemical Equipment and Automation, Marianske, Lazne, Czechoslovakia, September 15–22, 1969* (K016897); 6 months; $430

WASHINGTON

UNIVERSITY OF WASHINGTON
Bostrom, Robert C.; *International Symposium on Recent Crustal Movements and Associated Seismicity, Wellington, New Zealand, February 10-18, 1970* (A021099); 6 months; $740
Larsen, Lawrence H.; *Symposium on Physical Variability in the North Atlantic, Dublin, Ireland, September 25-27, 1969* (A015618); 6 months; $518
Lord, Jere J.; *The Eleventh International Conference on Cosmic Rays, Budapest, Hungary, August 25–September 4, 1969* (P016258); 6 months; $591

WEST VIRGINIA

WEST VIRGINIA UNIVERSITY
Moore, Desmond F.; *The Thirteenth International Congress of the International Federation of Engineering Societies of Automobile Technology, Brussels, Belgium, June 8-11, 1970 & the General Assembly of the International School of Production Research Being Held in Pisa & Turin, Italy, August 28-September 6, 1970* (K025055); 6 months; $550

WISCONSIN

UNIVERSITY OF WISCONSIN, Madison
Bock, Robert M.; *Seventh General Assembly, International Union of Biochemistry, Interlaken, Switzerland, September 3–9, 1970* (GF399); 6 months; $681
Harvey, John G.; *First International Congress on Mathematical Education, Lyons, France, August 24-30, 1969* (GZ1452); 6 months; $335
Hu, T. C.; *Seventh International Symposium on Mathematical Programming, The Hague, the Netherlands, September 14–18, 1970* (J001013); 6 months; $641

WISCONSIN STATE UNIVERSITY, Oshkosh
Laudon, Thomas S.; *Symposium on Antarctic Geology, Oslo, Norway, August 6-15, 1970* (V024395); 6 months; $546

UNIVERSITY OF WISCONSIN, Madison
Lenahan, John J.; *Fourth Australian Computer Conference, Adelaide, Australia, August 11-15, 1969* (J000772); 6 months; $1,350
McCormick, Garth P.; *Seventh International Symposium on Mathematical Programming, The Hague, the Netherlands, September 14–18, 1970* (J001015); 6 months; $331
Shain, Irving; *Fourth Heyrovsky Discussions, Bratislava, Czechoslovakia, May 18–22, 1970* (P018876); 6 months; $650
West, Robert C. Jr.; *International Symposium on Aromaticity, Pseudoaromaticity, and Antiaromaticity Jerusalem, Israel, March 31-April 3, 1970* (P021050); 6 months; $941
Zimmerman, Howard E.; *Third International Symposium on Photochemistry, St. Moritz, Switzerland, July 12-18, 1970* (P023412); 6 months; $618

WYOMING

UNIVERSITY OF WYOMING
Smithson, Scott B.; *Symposium on Antarctic Geology, Oslo, Norway, August 6-15, 1970* (V024398); 6 months; $646

SCIENCE EDUCATION PROGRAMS

Graduate Fellowships (Predoctoral)

NEW JERSEY

EDUCATIONAL TESTING SERVICE
Wiltsey, Robert G.; *Testing Program for FY–1971 Graduate Fellowship Applicants* (C62000); 12 months; $39,300

Graduate Traineeships

ALABAMA

AUBURN UNIVERSITY; Parker, W. V.; (GZ1533001); $10,120
———; (GZ1533); 18 months; $73,368

UNIVERSITY OF ALABAMA, Tuscaloosa; Rodgers, Eric; (GZ1534); 18 months; $63,088
———; (GZ780001); $948

ALASKA

UNIVERSITY OF ALASKA; Rae, K. M.; (GZ1535); 18 months; $27,422

ARIZONA

ARIZONA STATE UNIVERSITY; Burke, W. J.; (GZ1536001); $10,120
———; (GZ1536); 18 months; $144,275
———; (GZ782001); $1,178

UNIVERSITY OF ARIZONA; Bonneville, Frank A.; (GZ1537001); $10,120
———; (GZ1537); 18 months; $258,080

ARKANSAS

UNIVERSITY OF ARKANSAS; Harvey, Aubrey E.; (GZ1538); 18 months; $62,552
———; (GZ784001); $526

CALIFORNIA

CALIFORNIA INSTITUTE OF TECHNOLOGY; Bohnenblust, F.; (GZ1539); 18 months; $362,174

CLAREMONT GRADUATE SCHOOL AND UNIVERSITY; Rice, Philip M.; (GZ1540); 18 months; $45,326

LOMA LINDA UNIVERSITY; Stauffer, J. Paul; (GZ1541); 18 months; $10,460

STANFORD UNIVERSITY; Moses, Lincoln E.; (GZ1542); 18 months; $689,606

U.S. INTERNATIONAL UNIVERSITY; Rucker, W. Ray; (GZ1543); 18 months; $5,020

UNIVERSITY OF CALIFORNIA, Berkeley; Elberg, Sanford S.; (GZ1544); 18 months; $768,208

UNIVERSITY OF CALIFORNIA, Davis; Marr, Allen G.; (GZ1545001); $10,120
———; (GZ1545); 18 months; $203,338

UNIVERSITY OF CALIFORNIA, Irvine; Justice, K. E.; (GZ1546); 18 months; $59,270
———; (GZ1546001); $10,120
———; (GZ791001); $2,878

UNIVERSITY OF CALIFORNIA, Los Angeles; Carlisle, Donald; (GZ1547); 18 months; $434,183
———; (GZ1547001); $10,120

UNIVERSITY OF CALIFORNIA, Riverside; Hewitt, Robert R.; (GZ1548001); $10,120
———; (GZ1548); 18 months; $136,048

UNIVERSITY OF CALIFORNIA, San Diego; York, Herbert F.; (GZ1549); 18 months; $161,832

UNIVERSITY OF CALIFORNIA, Santa Cruz; Williamson, Stanley; (GZ1552); 18 months; $11,486

UNIVERSITY OF CALIFORNIA, San Francisco; Harper, Harold A.; (GZ1550); 18 months; $27,449

UNIVERSITY OF CALIFORNIA, Santa Barbara; Collins, Robert O.; (GZ1551001); $10,120
———; (GZ1551); 18 months; $116,982

UNIVERSITY OF SANTA CLARA; Parden, Robert J.; (GZ1554); 18 months; $22,162

UNIVERSITY OF SOUTHERN CALIFORNIA, Mayo, Charles G.; (GZ1555001); $10,120
———; (GZ1555); 18 months; $203,920

UNIVERSITY OF THE PACIFIC; McCrone, John D.; (GZ1553); 18 months; $17,133

COLORADO

COLORADO SCHOOL OF MINES; Jordan, A. Raymond; (GZ1556); 18 months; $38,293

COLORADO STATE UNIVERSITY; Bragonier, Wendell H.; (GZ1557); 18 months; $149,430
———; (GZ1557001); $10,120

UNIVERSITY OF COLORADO; Crowe, C. Lawson; (GZ1558001); $10,120
Love, William F.; (GZ1558); 18 months; $272,810

UNIVERSITY OF DENVER; Evers, Nathaniel H.; (GZ1559); 18 months; $39,643

CONNECTICUT

UNIVERSITY OF CONNECTICUT; Malone, Thomas F.; (GZ1560); 18 months; $154,420
Whetten, Nathan L.; (GZ1560001); $10,120
———; (GZ805001); $2,247

WESLEYAN UNIVERSITY; Kerr, William; (GZ1561); 18 months; $44,456

YALE UNIVERSITY; Justusson, J. William; (GZ1562); 18 months; $322,180

DELAWARE

UNIVERSITY OF DELAWARE; Kilpatrick, F. P.; (GZ1563); 18 months; $136,792
———; (GZ1563001); $10,120

DISTRICT OF COLUMBIA

AMERICAN UNIVERSITY; Boynton, Robert P.; (GZ1564); 18 months; $42,388

CATHOLIC UNIVERSITY OF AMERICA; O'Connor, James P.; (GZ1565001); $10,120
———; (GZ1565); 18 months; $119,101

GEORGE WASHINGTON UNIVERSITY; Burns, Arthur E.; (GZ1566); 18 months; $83,472
———; (GZ1566001); $10,120
———; (GZ811001); $347

GEORGETOWN UNIVERSITY; Porreco, Rocco E.; (GZ812001); $772
———; (GZ1567); 18 months; $88,030
———; (GZ1567001); $10,120

HOWARD UNIVERSITY; Miller, Carroll L.; (GZ1568); 18 months; $43,742

FLORIDA

FLORIDA STATE UNIVERSITY; Lewis, Thomas R.; (GZ1569); 18 months; $191,645
———; (GZ1569001); $10,120

NOVA UNIVERSITY; Wishart, Arthur W.; (GZ1570); 18 months; $10,460
———; (GZ815001); $686

UNIVERSITY OF FLORIDA; Hanson, Harold P.; (GZ816001); $844
———; (GZ1571001); $10,120
———; (GZ1571); 18 months; $243,368

UNIVERSITY OF MIAMI; Harrison, John A.; (GZ1572); 18 months; $95,648
———; (GZ1572001); $10,120

UNIVERSITY OF SOUTH FLORIDA; Lawton, Alfred H.; (GZ1573); 18 months; $12,602

GEORGIA

ATLANTA UNIVERSITY; Jarrett, Thomas D.; (GZ1574); 18 months; $17,070
———; (GZ818001); $778

EMORY UNIVERSITY; Bain, Carl E.; (GZ1575); 18 months; $81,645

GEORGIA INSTITUTE OF TECHNOLOGY; Webb, Sam C.; (GZ1576); 18 months; $228,345
———; (GZ1576001); $10,120

GEORGIA STATE UNIVERSITY; Rollow, J. W.; (GZ1577); 18 months; $16,949

MEDICAL COLLEGE OF GEORGIA; Behal, F. J.; (GZ1578); 18 months; $10,460

UNIVERSITY OF GEORGIA; Whitehead, T. H.; (GZ1579); 18 months; $138,037
———; (GZ1579001); $10,120

HAWAII

UNIVERSITY OF HAWAII; Rosenberg, Morton M.; (GZ1580); 18 months; $117,036
———; (GZ1580001); $10,120

IDAHO

UNIVERSITY OF IDAHO; Jackson, M. L.; (GZ1581); 18 months; $54,196
———; (GZ825001); $5,593

ILLINOIS

DEPAUL UNIVERSITY; Cortelyou, W. T.; (GZ1582); 18 months; $17,025

ILLINOIS INSTITUTE OF TECHNOLOGY; Grad, Arthur; (GZ1583001); $10,120
———; (GZ1583); 18 months; $175,304

ILLINOIS STATE UNIVERSITY; Helgeson, Arlan C.; (GZ1584); 18 months; $38,815

LOYOLA UNIVERSITY; Mariella, Raymond P.; (GZ1585); 18 months; $39,274

NORTHERN ILLINOIS UNIVERSITY; McIlrath, Wayne J.; (GZ1586); 18 months; $39,157

NORTHWESTERN UNIVERSITY; Baker, Robert H.; (GZ1587); 18 months; $399,053

SOUTHERN ILLINOIS UNIVERSITY; Rosenthal, Herbert H.; (GZ1588001); $10,120
———; (GZ1588); 18 months; $80,000
———; (GZ832001); $1,688

UNIVERSITY OF CHICAGO; O'Connell, Charles D.; (GZ1589); 18 months; $331,279

UNIVERSITY OF ILLINOIS, Chicago Circle; Otting, W. J.; (GZ1590); 18 months; $12,413

UNIVERSITY OF ILLINOIS, Medical Center; Binkley, S. B.; (GZ1591); 18 months; $5,020

UNIVERSITY OF ILLINOIS, Urbana; West, Vincent I.; (GZ1592001); $10,120
———; (GZ1592); 18 months; $689,218

INDIANA

INDIANA UNIVERSITY, Bloomington; Merritt Jr., Lynne L.; (GZ1593); 18 months; $283,518
———; (GZ1593001); $10,120

PURDUE UNIVERSITY; Gibbens, Victor E.; (GZ1594001); $10,120

GRADUATE TRAINEESHIPS

———; (GZ1594); 18 months; $608,346

UNIVERSITY OF NOTRE DAME; Beichner, Paul E.; (GZ1595); 18 months; $189,728

IOWA

IOWA STATE UNIVERSITY; Page, J. B.; (GZ1596); 18 months; $255,871
———; (GZ1596001); $10,120

UNIVERSITY OF IOWA; Spriestersbach, D. C.; (GZ1597001); $10,120
———; (GZ1597); 18 months; $211,330

KANSAS

KANSAS STATE UNIVERSITY; Noonan, John P.; (GZ1598001); $10,120
———; (GZ1598); 18 months; $147,720
———; (GZ840001); $2,673

UNIVERSITY OF KANSAS; Ellermeier, Robert D.; (GZ1599); 18 months; $172,228
———; (GZ1599001); $10,120

KENTUCKY

UNIVERSITY OF KENTUCKY; Cochran, Lewis W.; (GZ1600); 18 months; $109,964
———; (GZ1600001); $10,120

UNIVERSITY OF LOUISVILLE; Akers, Dee Ashley; (GZ1601); 18 months; $44,510

LOUISIANA

LOUISIANA STATE UNIVERSITY, Baton Rouge; Goodrich, Max; (GZ1603001); $10,120
———; (GZ1603); 18 months; $148,728

LOUISIANA STATE UNIVERSITY, New Orleans; Bobo, James R.; (GZ1605); 18 months; $27,359
———; (GZ846001); $488

LOUISIANA STATE UNIVERSITY, Medical Center; Coulson, Roland A.; (GZ1604); 18 months; $11,477

LOUISIANA POLYTECHNIC INSTITUTE; Hester, James L.; (GZ1602); 18 months; $12,458

LOYOLA UNIVERSITY; DiMaggio III, Anthony; (GZ1606); 18 months; $5,902

TULANE UNIVERSITY; Deener, David R.; (GZ1607001); $10,120
———; (GZ1607); 18 months; $108,329

MAINE

UNIVERSITY OF MAINE, Orono; Eggert, Franklin P.; (GZ1608); 18 months; $39,562
———; (GZ848001); $1,555

MARYLAND

JOHNS HOPKINS UNIVERSITY; Strider, Robert L.; (GZ1609001); $10,120
———; (GZ1609); 18 months; $221,555

UNIVERSITY OF MARYLAND; Pelczar Jr., Michael J.; (GZ1610); 18 months; $199,616
———; (GZ1610001); $10,120

MASSACHUSETTS

BOSTON COLLEGE; Aronoff, S.; (GZ1611); 18 months; $49,386

BOSTON UNIVERSITY; Kubzansky, Philip E.; (GZ1612); 18 months; $74,875
———; (GZ1612001); $10,120

BRANDEIS UNIVERSITY; Maher, Brendan A.; (GZ1613); 18 months; $104,524

CLARK UNIVERSITY; Cohen, Saul B.; (GZ1614); 18 months; $53,638

HARVARD UNIVERSITY; Sisson, Thomas K.; (GZ1615); 18 months; $287,617

LOWELL TECHNOLOGICAL INSTITUTE; Alexander, Edward L.; (GZ1616); 18 months; $16,728

MASSACHUSETTS INSTITUTE OF TECHNOLOGY; Baram, Michael S.; (GZ1617); 18 months; $763,452

NORTHEASTERN UNIVERSITY; King, William F.; (GZ1618); 18 months; $115,470
———; (GZ1618001); $10,120

TUFTS UNIVERSITY; McCarthy, Kathryn A.; (GZ1619); 18 months; $65,574
———; (GZ1619001); $10,120

UNIVERSITY OF MASSACHUSETTS; Gentile, Arthur C.; (GZ1620); 18 months; $148,539
———; (GZ1620001); $10,120

WORCESTER POLYTECHNIC INSTITUTE; Morton, Richard F.; (GZ1621); 18 months; $29,429

MICHIGAN

MICHIGAN STATE UNIVERSITY; Minkel, C. W.; (GZ1622); 18 months; $381,240
———; (GZ1622001); $10,120

MICHIGAN TECHNOLOGICAL UNIVERSITY; Yerg, Donald G.; (GZ1623); 18 months; $55,042
———; (GZ864001); $3,821

UNIVERSITY OF DETROIT; Mehlenbacher, Lyle E.; (GZ1624); 18 months; $38,455

UNIVERSITY OF MICHIGAN; Hay, G. E.; (GZ1625); 18 months; $722,371

WAYNE STATE UNIVERSITY; Burnham, Frederic B.; (GZ1626); 18 months; $156,364
Rumble, Thomas C.; (GZ1626001); $10,120

WESTERN MICHIGAN UNIVERSITY; Mallinson, George G.; (GZ1627); 18 months; $28,124
———; (GZ868001); $2,457

MINNESOTA

UNIVERSITY OF MINNESOTA; Crawford, Bryce Jr.; (GZ1628); 18 months; $435,071
———; (GZ1628001); $10,120

MISSISSIPPI

MISSISSIPPI STATE UNIVERSITY; Loftin, Marion T.; (GZ1629); 18 months; $56,842

UNIVERSITY OF MISSISSIPPI; Sam, Joseph; (GZ1630); 18 months; $44,492

UNIVERSITY OF SOUTHERN MISSISSIPPI; Lucas, Aubrey K.; (GZ1631); 18 months; $16,809

MISSOURI

ST. LOUIS UNIVERSITY; Eigel Jr., Edwin G.; (GZ1632); 18 months; $51,606

UNIVERSITY OF MISSOURI, Columbia; Bauman, John E. Jr.; (GZ1633); 18 months; $223,058
———; (GZ1633001); $10,120
———; (GZ874001); $3,263

UNIVERSITY OF MISSOURI, Kansas City; Dale, Wesley J.; (GZ875001); $1,124
———; (GZ1634); 18 months; $23,392

UNIVERSITY OF MISSOURI, Rolla; McFarland, Robert H.; (GZ1635001); $10,120
———; (GZ1635); 18 months; $99,630
———; (GZ876001); $566

WASHINGTON UNIVERSITY; Morrow, Ralph E.; (GZ1636); 18 months; $223,490
———; (GZ1636001); $10,120

MONTANA

MONTANA STATE UNIVERSITY; Goering, Kenneth J.; (GZ1637001); $10,120
———; (GZ1637); 18 months; $70,087
———; (GZ878001); $7,536

UNIVERSITY OF MONTANA; Stewart, John M.; (GZ879001); $2,268
———; (GZ1638); 18 months; $28,106

NEBRASKA

UNIVERSITY OF NEBRASKA, Lincoln; Drew, James V.; (GZ1639); 18 months; $108,747
McCashland, Benjamin W.; (GZ1639001); $10,120

NEVADA

UNIVERSITY OF NEVADA, Reno Campus; O'Brien, T. D.; (GZ1640); 18 months; $39,508

NEW HAMPSHIRE

DARTMOUTH COLLEGE; Hornig, James F.; (GZ1641); 18 months; $53,827
———; (GZ882001); $3,537

UNIVERSITY OF NEW HAMPSHIRE; Drew, William H.; (GZ883001); $1,441
———; (GZ1642); 18 months; $83,832
———; (GZ1642001); $10,120

NEW JERSEY

NEWARK COLLEGE OF ENGINEERING; Bedrosian, Alex; (GZ1643); 18 months; $27,152

PRINCETON UNIVERSITY; Edenfield, Edward J.; (GZ1644); 18 months; $304,932

RUTGERS UNIVERSITY; Torrey, Henry C.; (GZ1645); 18 months; $205,768
———; (GZ1645001); $10,120

SETON HALL UNIVERSITY; Connor, Joseph G.; (GZ1646); 18 months; $11,981

STEVENS INSTITUTE OF TECHNOLOGY; Booth, Eugene T.; (GZ1647001); $10,120
———; (GZ1647); 18 months; $100,872
———; (GZ888001); $129

NEW MEXICO

NEW MEXICO INSTITUTE OF MINING AND TECHNOLOGY; Holmes, Charles H.; (GZ1648); 18 months; $16,701
———; (GZ889001); $2,132

NEW MEXICO STATE UNIVERSITY; Thompson, Merrell E.; (GZ890001); $530
———; (GZ1649); 18 months; $72,049
———; (GZ1649001); $10,120

UNIVERSITY OF NEW MEXICO; Walker, Harold L.; (GZ1650001); $10,120
———; (GZ1650); 18 months; $105,145

NEW YORK

ADELPHI UNIVERSITY; Jennings, Manson Van B.; (GZ1651); 18 months; $22,978

ALFRED UNIVERSITY; Lawrence, W. G.; (GZ1652); 18 months; $15,900
———; (GZ893001); $1,210

CLARKSON COLLEGE OF TECHNOLOGY; Donaruma, L. Guy; (GZ1654); 18 months; $70,456

COLLEGE OF FORESTRY AT SYRACUSE; Webb, William L.; (GZ1670); 18 months, $53,638

COLUMBIA UNIVERSITY; Robey, Richard C.; (GZ1655); 18 months; $451,689

COOPER UNION; Wallace, F. A.; (GZ1656); 18 months; $10,460

CORNELL UNIVERSITY; Leurgans, Paul J.; (GZ1657); 18 months; $480,245

CUNY, CENTRAL SYSTEM OFFICE; Zupnick, Elliot; (GZ1653001); $10,120
———; (GZ1653); 18 months; $173,092

DOWNSTATE MEDICAL CENTER; Brooks, Chandler McC.; (GZ1671); 18 months; $10,460

FORDHAM UNIVERSITY; Finlay, James C.; (GZ1658); 18 months; $55,501

NEW SCHOOL FOR SOCIAL RESEARCH; Greenbaum, Joseph J.; (GZ1659); 18 months; $40,723

NEW YORK MEDICAL COLLEGE; Tabachnick, Milton; (GZ1660); 18 months; $22,018

NEW YORK UNIVERSITY; Chusid, Martin; (GZ1661); 18 months; $315,844
McCrone, Alistair W.; (GZ1661001); $10,120

POLYTECHNIC INSTITUTE OF BROOKLYN; Giordano, Anthony B.; (GZ1662); 18 months; $217,870

RENSSELAER POLYTECHNIC INSTITUTE; Wiberley, Stephen E.; (GZ1663); 18 months; $191,744
———; (GZ1663001); $10,120

ROCKEFELLER UNIVERSITY; Connelly, Clarence M.; (GZ1664); 18 months; $10,460

ST. BONAVENTURE UNIVERSITY; Anderson, Kenneth E.; (GZ1665); 18 months; $12,053
———; (GZ906001); $35

ST. JOHN'S UNIVERSITY; Medici, Paul T.; (GZ1666); 18 months; $39,472

SCIENCE EDUCATION PROGRAMS

STATE UNIVERSITY AT ALBANY; Flinton, Edgar W.; (GZ1667); 18 months; $28,124
———; (GZ909001); $1,665

STATE UNIVERSITY AT BINGHAMTON; Moore, Edward C.; (GZ1668); 18 months; $18,222

STATE UNIVERSITY AT BUFFALO; Holt, Andrew W.; (GZ1669); 18 months; $182,544
———; (GZ1669001); $10,120

STATE UNIVERSITY AT STONY BROOK; Weisinger, Herbert; (GZ1672001); $10,120
———; (GZ1672); 18 months; $72,112
———; (GZ913001); $949

SYRACUSE UNIVERSITY; Hough, W. Howard; (GZ1674); 18 months; $235,584
Wheadon, William C.; (GZ1674001); $10,120

UNION COLLEGE; Palmer, J. D.; (GZ1675); 18 months; $6,298

UNIVERSITY OF ROCHESTER; Spragg, S. D. S.; (GZ1676); 18 months; $223,445
———; (GZ1676001); $10,120

UPSTATE MEDICAL CENTER; Jacobs, Ross D.; (GZ1673); 18 months; $11,072

YESHIVA UNIVERSITY; Gittler, Joseph B.; (GZ1677); 18 months; $42,260

NORTH CAROLINA

DUKE UNIVERSITY; Harman, Charles M.; (GZ1678); 18 months; $154,240
Martin, David V.; (GZ1678001); $10,120

NORTH CAROLINA STATE UNIVERSITY AT RALEIGH; Peterson, Walter J.; (GZ1679001); $10,120
———; (GZ1679); 18 months; $261,687

UNIVERSITY OF NORTH CAROLINA AT CHAPEL HILL; Flora, Joseph M.; (GZ1680); 18 months; $198,339
———; (GZ1680001); $10,120

WAKE FOREST UNIVERSITY; Stroupe, Henry S.; (GZ1681); 18 months; $43,318

NORTH DAKOTA

NORTH DAKOTA STATE UNIVERSITY; Smith, Glenn S.; (GZ1682); 18 months; $38,383
———; (GZ923001); $2,929

UNIVERSITY OF NORTH DAKOTA; Johnson, A. William; (GZ924001); $2,499
———; (GZ1683); 18 months; $54,880

OHIO

BOWLING GREEN STATE UNIVERSITY; Jackson, William B.; (GZ1684); 18 months; $40,818
———; (GZ925001); $1,281

CASE WESTERN RESERVE UNIVERSITY; Hurley, Frank H.; (GZ1685); 18 months; $331,348

KENT STATE UNIVERSITY; White, John D.; (GZ1686); 18 months; $56,860

MIAMI UNIVERSITY; Wolverton, R. E.; (GZ1687); 18 months; $12,521

OHIO STATE UNIVERSITY; Protheroe, W. M.; (GZ1688001); $10,120
Roaden, Arliss L.; (GZ1688); 18 months; $424,171

OHIO UNIVERSITY; Cohn, Norman S.; (GZ1689); 18 months; $65,072

UNIVERSITY OF AKRON; Carson, R. C.; (GZ1690); 18 months; $23,842

UNIVERSITY OF CINCINNATI; Wessel, Robert H.; (GZ1691001); $10,120
———; (GZ1691); 18 months; $112,610

UNIVERSITY OF DAYTON; Stander, Joseph W.; (GZ1692); 18 months; $6,181

UNIVERSITY OF TOLEDO; Turin, John J.; (GZ1693); 18 months; $28,673

OKLAHOMA

OKLAHOMA STATE UNIVERSITY; Durham, Norman N.; (GZ1694001); $10,120
———; (GZ1694); 18 months; $173,819
———; (GZ933001); $4,990

UNIVERSITY OF OKLAHOMA; Riggs, Carl D.; (GZ934001); $3,751
———; (GZ1695001); $10,120
———; (GZ1695); 18 months; $143,708

OREGON

OREGON GRADUATE CENTER FOR STUDY AND RESEARCH; Scott, Arthur F.; (GZ1696); 18 months; $5,020

OREGON STATE UNIVERSITY; Bond, Carl E.; (GZ1697); 18 months; $160,459
———; (GZ1697001); $10,120
———; (GZ935001); $3,222

PORTLAND STATE UNIVERSITY; Harris, J. Kenneth; (GZ1698); 18 months; $12,278

UNIVERSITY OF OREGON; Fisk, Calvin W.; (GZ1699); 18 months; $143,612
———; (GZ1699001); $10,120
———; (GZ936001); $1,712

UNIVERSITY OF PORTLAND; Kehoe, J. A.; (GZ1700); 18 months; $10,460
———; (GZ937001); $610

PENNSYLVANIA

BRYN MAWR COLLEGE; Foster, Elizabeth R.; (GZ1701); 18 months; $27,728

CARNEGIE MELLON UNIVERSITY; Schatz, Edward R.; (GZ1702); 18 months; $272,765

DREXEL UNIVERSITY; Witzell, O. W.; (GZ1703); 18 months; $57,229

DUQUESNE UNIVERSITY; Gillis, Bernard T.; (GZ1704); 18 months; $11,378

HAHNEMANN MEDICAL COLLEGE AND HOSPITAL; Bondi, A.; (GZ1705); 18 months; $11,828
LEHIGH UNIVERSITY; Stout, Robert D.; (GZ1707); 18 months; $142,712

PENNSYLVANIA STATE UNIVERSITY; Howell Jr., B. F.; (GZ1708); 18 months; $431,218
———; (GZ1708001); $10,120

PHILADELPHIA COLLEGE OF PHARMACY AND SCIENCE; Reber, Louis A.; (GZ1709); 18 months; $11,693
———; (GZ945001); $410

TEMPLE UNIVERSITY; Harrington, George W.; (GZ1710); 18 months; $83,220
———; (GZ1710001); $10,120

THOMAS JEFFERSON UNIVERSITY; Baldridge, Robert C.; (GZ1706); 18 months; $6,208

UNIVERSITY OF PENNSYLVANIA; O'Kane, D. J.; (GZ1711); 18 months; $375,960

UNIVERSITY OF PITTSBURGH; McCoy, R. H.; (GZ1712); 18 months; $208,720
———; (GZ1712001); $10,120

VILLANOVA UNIVERSITY; Buford, Albert H.; (GZ1713); 18 months; $22,681
———; (GZ949001); $65

WOMAN'S MEDICAL COLLEGE OF PENNSYLVANIA; Goldman, David E.; (GZ1714); 18 months; $10,460

RHODE ISLAND

BROWN UNIVERSITY; Brennan, Michael J.; (GZ1715); 18 months; $201,592
———; (GZ1715001); $10,120

PROVIDENCE COLLEGE; Galkowski, Theodore T.; (GZ1716); 18 months; $11,378

UNIVERSITY OF RHODE ISLAND; Ferrante, William R.; (GZ1717); 18 months; $61,940

SOUTH CAROLINA

CLEMSON UNIVERSITY; Schwartz, A. E.; (GZ1718); 18 months; $108,734
———; (GZ1718001); $10,120

MEDICAL COLLEGE OF SOUTH CAROLINA; Kinard, F. W.; (GZ1719); 18 months; $5,983

UNIVERSITY OF SOUTH CAROLINA; Davis, H. W.; (GZ1720); 18 months; $63,830

SOUTH DAKOTA

SOUTH DAKOTA SCHOOL OF MINES; Gries, John Paul; (GZ1721); 18 months; $16,773
———; (GZ957001); $142

SOUTH DAKOTA STATE UNIVERSITY; Bailey, H. S.; (GZ958001); $624
———; (GZ1722); 18 months; $43,268

UNIVERSITY OF SOUTH DAKOTA; Cobb, Henry V.; (GZ1723); 18 months; $39,202
———; (GZ959001); $1,420

TENNESSEE

GEORGE PEABODY COLLEGE FOR TEACHERS; Tomlinson, Gus; (GZ960001); $2,865
———; (GZ1724); 18 months; $15,900
MEMPHIS STATE UNIVERSITY; Richardson, John W.; (GZ1725); 18 months; $24,040

UNIVERSITY OF TENNESSEE, Memphis; Alden, Roland H.; (GZ1727); 18 months; $6,033

UNIVERSITY OF TENNESSEE; Smith, Hilton A.; (GZ1726001); $10,120
———; (GZ1726); 18 months; $204,328

VANDERBILT UNIVERSITY; Thune, Leland E.; (GZ1728); 18 months; $148,170
———; (GZ1728001); $10,120

TEXAS

BAYLOR UNIVERSITY; Melnick, Joseph L.; (GZ1730); 18 months; $15,900
Toland, William G.; (GZ964001); $796
———; (GZ1729); 18 months; $27,395
NORTH TEXAS STATE UNIVERSITY; Toulouse, Robert B.; (GZ1731); 18 months; $39,544
———; (GZ966001); $79

RICE UNIVERSITY; Richter, G. H.; (GZ1732); 18 months; $177,234

SOUTHERN METHODIST UNIVERSITY; Albritton Jr., Claude C.; (GZ1733); 18 months; $74,524
———; (GZ1733001); $10,120
———; (GZ968001); $1,122

TEXAS A & M UNIVERSITY; Kunze, George W.; (GZ969001); $10,217
———; (GZ1734); 18 months; $150,597
———; (GZ1734001); $10,120

TEXAS CHRISTIAN UNIVERSITY; Wall, S. A.; (GZ1735); 18 months; $27,728
———; (GZ970001); $4,409

TEXAS TECH UNIVERSITY; Graves, Lawrence; (GZ1736); 18 months; $64,154

TEXAS WOMAN'S UNIVERSITY; Smith, A. A.; (GZ1737); 18 months; $11,441

UNIVERSITY OF HOUSTON; Bunn, Ronald F.; (GZ1738); 18 months; $83,670
———; (GZ1738001); $10,120
———; (GZ973001); $7,807

UNIVERSITY OF TEXAS, Arlington; Wyatt, J. L.; (GZ1739); 18 months; $6,028
UNIVERSITY OF TEXAS, Austin; Rogers, Lorene; (GZ1740001); $10,120
———; (GZ1740); 18 months; $402,110

UTAH

BRIGHAM YOUNG UNIVERSITY; Riddle, Chauncey C.; (GZ1741); 18 months; $64,532
———; (GZ975001); $4,865
UNIVERSITY OF UTAH; McMurrin, Sterling M.; (GZ1742001); $10,120
———; (GZ1742); 18 months; $186,292
UTAH STATE UNIVERSITY; Gardner, Eldon J.; (GZ1743); 18 months; $93,524
———; (GZ1743001); $10,120
———; (GZ977001); $12,621

VERMONT

UNIVERSITY OF VERMONT; Johnstone, Donald B.; (GZ1744); 18 months; $49,590

VIRGINIA

COLLEGE OF WILLIAM AND MARY; Selby, John E.; (GZ1745); 18 months; $43,268

UNIVERSITY OF VIRGINIA; Whitehead, W. D.; (GZ1747001); $10,120
———; (GZ1747); 18 months; $147,653

VIRGINIA COMMONWEALTH UNIVERSITY; Watts, Daniel T.; (GZ1746); 18 months; $16,962

VIRGINIA POLYTECHNIC INSTITUTE; Bull, Fred W.; (GZ1748); 18 months; $137,530
———; (GZ1748001); $10,120

WASHINGTON

UNIVERSITY OF WASHINGTON; McCarthy, Joseph L.; (GZ1749); 18 months; $428,545
———; (GZ1749001); $10,120

WASHINGTON STATE UNIVERSITY; Nyman, C. J.; (GZ1750); 18 months; $106,405
———; (GZ1750001); $10,120
———; (GZ984001); $1,483

WEST VIRGINIA

WEST VIRGINIA UNIVERSITY; Ludlum, John C.; (GZ1751); 18 months; $91,000
———; (GZ1751001); $10,120

WISCONSIN

INSTITUTE OF PAPER CHEMISTRY; Miller, Arild J.; (GZ1752); 18 months; $15,900

MARQUETTE UNIVERSITY; Quade, Quentin L.; (GZ1753); 18 months; $60,590

UNIVERSITY OF WISCONSIN, Madison; Bock, Robert M.; (GZ1754); 18 months; $550,883
———; (GZ1754001); $10,120

UNIVERSITY OF WISCONSIN, Milwaukee; Krill, Karl E.; (GZ1755001); $10,120
———; (GZ1755); 18 months; $66,492
———; (GZ989001); $1,093

WYOMING

UNIVERSITY OF WYOMING; Bruce, Robert H.; (GZ1756); 18 months; $61,292

Institutes and Research Participation for Science, Mathematics, and Engineering Teachers

College Teachers

ALABAMA

TUSKEGEE INSTITUTE
Gayed, Yacoub Kirollos; *Research Participation for College Teachers-Academic Year Phase* (GY7168); 24 months; $2,000

UNIVERSITY OF SOUTH ALABAMA
Miller, Nathan C.; (GY7117); 24 months; $2,000
Noelle, Gerald Lee; (GY7124); 24 months; $2,000

ARIZONA

ARIZONA STATE UNIVERSITY
Bender, Gordon L.; *Summer Institute in Desert Biology for College Teachers* (GY7040); 12 months; $41,930

MESA COMMUNITY COLLEGE
Bristol, Robert F.; *Research Participation for College Teachers-Academic Year Phase* (GY7084); 24 months; $2,000

NORTHERN ARIZONA UNIVERSITY
Bean, Kenneth Eugene; (GY7103); 24 months; $2,000
Troxler, Joseph R.; (GY7169); 24 months; $2,000

UNIVERSITY OF ARIZONA
Kisiel, Chester C.; *Summer Institute in Systems Hydrology for College Teachers* (GY7039); 12 months; $60,080
Sloane, Richard L.; *Summer Institute for College Teachers* (GY7046); 12 months; $34,200

ARKANSAS

ARKANSAS STATE UNIVERSITY
Greenwell, Donald Lee; *Research Participation for College Teachers-Academic Year Phase* (GY7126); 24 months; $2,000

UNIVERSITY OF ARKANSAS
Cordes, A. W.; *Research Participation for College Teachers-Summer Phase* (GY6923); 12 months; $18,900
Mix, Dwight F.; *Summer Institute in Electrical Engineering for College Teachers* (GY6998); 12 months; $24,970

CALIFORNIA

ASIA FOUNDATION
Coliver, Norman; *Foreign Participants Travel of Asian Participants in NSF Supplementary Training Program* (GW3090 002*); 3 months; $4,660

CALIFORNIA STATE COLLEGE AT LONG BEACH
Munsee, Jack H.; *Research Participation for College Teachers-Academic Year Phase* (GY7181); 24 months; $2,000

CALIFORNIA STATE COLLEGE AT LOS ANGELES
Cromwell, Leslie; *Summer Institute in Circuits and Devices for College Teachers* (GY6991); 12 months; $38,190

CALIFORNIA STATE POLYTECHNIC COLLEGE
Vollmar, Arnulf; *Research Participation for College Teachers-Academic Year Phase* (GY7067); 24 months; $2,000
Wills, Max T.; (GY7100); 24 months; $2,000

COLLEGE OF MARIN
Bezirjian, Onnig H.; (GY7106); 24 months; $2,000

CUESTA COLLEGE
Fisher, Richard Lee; (GY7194); 24 months; $2,000

FRESNO STATE COLLEGE
Ziegler, Stanley Martin; (GY7121); 24 months; $2,000

HUMBOLDT STATE COLLEGE
Norris, Daniel H.; (GY7109); 24 months; $2,000

LOMA LINDA UNIVERSITY
Botimer, Laurence Wallace; (GY7139); 24 months; $2,000

PACIFIC COLLEGE
Isaak, Daniel; (GY7129); 24 months; $2,000

SACRAMENTO STATE COLLEGE
Hurley, C. Robert; (GY7136); 24 months; $2,000

SAN DIEGO STATE COLLEGE
Carlson, John R.; (GY7125); 24 months; $2,000
Snodgrass, Herschel R.; *Summer Institute in Macroscopic Quantum Physics for College Teachers* (GY6992); 12 months; $48,620

SAN JOSE STATE COLLEGE
Lange, L. H.; *Summer Institute in Mathematics for College Teachers* (GY6997); 12 months; $41,870

STANFORD UNIVERSITY
Benjamin, Jack R.; *Short Course in the Use of Probability and Statistics in Civil Engineering for College Teachers* (GY6979); 12 months; $24,650
Epel, David; *Summer Institute in Marine Biology for College Teachers* (GY7027); 12 months; $47,540
Levin, Henry; *Summer Institute in Economics for College Teachers* (GY6995); 12 months; $44,170
Rambo, William R.; *Research Participation for College Teachers-Summer Phase* (GY6910); 12 months; $31,600

UNIVERSITY OF CALIFORNIA, Berkeley
Brink, David L.; (GY6906); 12 months; $20,560
Schrock, Virgil E.; (GY6927); 12 months; $20,570

UNIVERSITY OF CALIFORNIA, Riverside
Peplies, Robert W.; *Short Course on the Geographic Applications of Remote Sensing for College Teachers* (GY6977); 12 months; $41,050

UNIVERSITY OF CALIFORNIA, Santa Barbara
Meyer, William H.; *Short Course in Mathematics for College Teachers* (GY6973); 24 months; $62,270

UNIVERSITY OF SAN FRANCISCO
Gruhn, Thomas Albin; *Research Participation for College Teachers-Academic Year Phase* (GY7184); 24 months; $2,000

UNIVERSITY OF SANTA CLARA
Siljak, Dragoslav; *Short Course in Engineering for College Teachers* (GY6974); 12 months; $13,700

COLORADO

COLORADO COLLEGE
Sterling, Daniel J.; *Summer Institute in Mathematics for College Teachers* (GY7029); 12 months; $36,560

COLORADO STATE UNIVERSITY
McClellan, J. Forbes; *Summer Institute in Field Biology for College Teachers* (GY7044); 12 months; $40,020

LORETTO HEIGHTS COLLEGE
Salzman, John D.; *Research Participation for College Teachers-Academic Year Phase* (GY7118); 24 months; $2,000

MESA COLLEGE
Johnson, James B.; (GY7131); 24 months; $2,000

UNIVERSITY OF COLORADO
Bailey, Daniel E.; *Summer Institute on Computer Science in Social and Behavioral Science Research for College Teachers* (GY7001); 12 months; $49,920
Bailey, Daniel E. and McMillan, Claude; *Summer Institute in Computer Science in Social and Behavioral Science Research for College Teachers* (GY3994 001); $300
Erdelyi, Edward A.; *Summer Institute on Electromagnetic Fields in Devices for College Teachers* (GY6985); 12 months; $43,780
Ives, Jack D.; *Research Participation for College Teachers-Summer Phase* (GY6913); 12 months; $15,600
Kelso, Alec J.; *Summer Institute in Anthropology for College Teachers* (GY7032); 12 months; $65,640
Lauer, B. E.; *Summer Institute in Chemical and Mechanical Engineering for College Teachers* (GY7037); 12 months; $39,650
McMillan, C. and Plane, D. R.; *Summer Institute in Management Science and Operations Research for College Teachers* (GY7016); 12 months; $44,200
Meek, John S.; *Research Participation for College Teachers-Summer Phase* (GY6932); 12 months; $27,250

CONNECTICUT

YALE UNIVERSITY
Galston, Arthur W.; *Short Course for College Teachers in Botany* (GY5471001); $300

DELAWARE

UNIVERSITY OF DELAWARE
Mather, John R.; *Short Course in Climatology for College Teachers* (GY6961); 12 months; $34,150

DISTRICT OF COLUMBIA

ARCTIC INSTITUTE OF NORTH AMERICA
Hills, T. L.; *Summer Institute in Geography for College Teachers* (GY7045); 12 months; $13,930
Ragle, Richard H.; *Research Participation for College Teachers-Summer Phase* (GY6948); 12 months; $18,040

SCIENCE EDUCATION PROGRAMS

SMITHSONIAN INSTITUTION
Mayr, Otto; *Summer Institute on Methods and Problems in the History of Technology for College Teachers* (GY7006); 12 months; $14,670

FLORIDA

FLORIDA STATE UNIVERSITY
Miles Jr., E. P.; *Summer Institute on Calculus and the Computer for College Teachers* (GY7019); 12 months; $50,580
Scarborough, B. B.; *Research Participation for College Teachers-Summer Phase* (GY6922); 12 months; $19,850

MIAMI-DADE JUNIOR COLLEGE
Christensen, Robert F.; *Research Participation for College Teachers-Academic Year Phase* (GY7130); 24 months; $2,000

ST. LEO COLLEGE
Adisesh, Setty R.; (GY7203); 24 months; $2,000

UNIVERSITY OF FLORIDA
Ballard, Stanley S.; *Research Participation for College Teachers-Summer Phase* (GY6938); 12 months; $15,350
Brey Jr., Wallace S.; (GY6940); 12 months; $21,800

UNIVERSITY OF MIAMI
Jones, Albert C. and Iversen, Edwin S.; (GY6954); 12 months; $30,340

UNIVERSITY OF WEST FLORIDA
Birdwhistell, Ralph K.; *Short Course in Spectroscopy and the Structure of Molecules for College Teachers* (GY6969); 12 months; $9,260

GEORGIA

GEORGIA SOUTHERN COLLEGE
Bennett, Sara Neville; *Research Participation for College Teachers-Academic Year Phase* (GY7076); 24 months; $2,000

MORRIS BROWN COLLEGE
Anderson, Gloria L.; (GY7107); 24 months; $2,000

UNIVERSITY OF GEORGIA
Hinton, Don B.; *Summer Institute in Mathematics for College Teachers* (GY6987); 12 months; $39,950
Michaels, Gene E.; *Research Participation for College Teachers-Summer Phase* (GY6936); 12 months; $15,350

ILLINOIS

ATOMIC ENERGY COMMISSION,
Chicago Operations Office
Miller, Shelby A.; *Short Course in Symmetry and Group Theory in Chemistry for College Teachers* (AG207); 12 months; $8,100
———; *Short Course in Nuclear Radiation Detection for College Teachers* (AG208); 12 months; $10,800

BLACKBURN COLLEGE
Singh, Dilbagh; *Research Participation for College Teachers-Academic Year Phase* (GY7143); 24 months; $2,000

ELMHURST COLLEGE
Granchoff, John C.; (GY7178); 24 months; $2,000

ILLINOIS INSTITUTE OF TECHNOLOGY
Darsow, William F.; *Research Participation for College Teachers-Summer Phase* (GY6944); 12 months; $20,000
Neubert, Theodore J.; (GY6950); 12 months; $28,780
Roush, Allan H.; (GY6939); 12 months; $15,600
Weinstock, Harold; *Summer Institute in Computer Use in Undergraduate Instruction in Physics for College Teachers* (GY7036); 12 months; $47,030

ILLINOIS STATE UNIVERSITY
Morris, Charles E.; *Summer Institute in Mathematics for College Teachers* (GY7028); 12 months; $61,460

ILLINOIS WESLEYAN UNIVERSITY
Frank, Forrest J.; *Research Participation for College Teachers-Academic Year Phase* (GY7186); 24 months; $2,000

MACMURRAY COLLEGE
Snyder, Richard D.; (GY7179); 24 months; $2,000

NORTHWESTERN UNIVERSITY
Krizek, Raymond J.; *Summer Institute in Physical Properties and Behavior of Soils for College Teachers* (GY7000); 12 months; $56,780

QUINCY COLLEGE
Cardillo, Frances M.; *Research Participation for College Teachers-Academic Year Phase* (GY7137); 24 months; $2,000

UNIVERSITY OF ILLINOIS, Urbana
Ang, Alfredo H-S.; *Summer Institute in Probabilistic Structural Mechanics and Design for College Teachers* (GY7012); 12 months; $56,040
Dobrovolny, Jerry S.; *Summer Institute in Electronics and Machine Design for College Teachers* (GY7010); 12 months; $55,600
Malmstadt, Howard V.; *Short Course in Electronics for College Teachers* (GY6962); 12 months; $38,390
Steggerda, F. R.; *Research Participation for College Teachers-Summer Phase* (GY6941); 12 months; $19,140
Zaring, Wilson M.; *Summer Institute in Mathematics for College Teachers* (GY7008); 12 months; $72,610

WHEATON COLLEGE
Devries, David Andrew; *Research Participation for College Teachers-Academic Year Phase* (GY7054); 24 months; $2,000
Sanderson, William Ashman; (GY7149); 24 months; $2,000

INDIANA

EARLHAM COLLEGE
Posnick, Gregory Meredith; (GY7153); 24 months; $2,000
Whitcomb, Stuart E.; *Summer Institute in Physical Science for College Teachers* (GY7022); 12 months; $49,030

MANCHESTER COLLEGE
Miller, Edward George; *Research Participation for College Teachers-Academic Year Phase* (GY7095); 24 months; $2,000

PURDUE UNIVERSITY
Perone, S. P.; *Short Course in Digital Computers in Chemical Instrumentation for College Teachers* (GY6967); 12 months; $30,550
Wiley, Jay W.; *Summer Institute in Economic Education for College Teachers* (GY6994); 12 months; $41,320

PURDUE UNIVERSITY, Calumet Center
Mitchell, George R.; *Research Participation for College Teachers—Academic Year Phase* (GY7081); 24 months; $2,000

PURDUE UNIVERSITY, Fort Wayne Center
Longroy, Allan L.; (GY7105); 24 months; $2,000

ST. MARY'S COLLEGE
Feigl, Dorothy M.; (GY7115); 24 months; $2,000

IOWA

BUENA VISTA COLLEGE
Staudinger, W. Leroy; (GY7175); 24 months; $2,000

CENTRAL COLLEGE
Wilson, David E.; (GY7170); 24 months; $2,000

COE COLLEGE
Nee, Timothy Ling-Sun; (GY7083); 24 months; $2,000

SIMPSON COLLEGE
Head, James Crawford; (GY7164); 24 months; $2,000

UNIVERSITY OF IOWA
Porter, J. R.; *Research Participation for College Teachers-Summer Phase* (GY6951); 12 months; $17,200

WARTBURG COLLEGE
Eiben, Galen James; *Research Participation for College Teachers-Academic Year Phase* (GY7145); 24 months; $2,000

KANSAS

KANSAS STATE UNIVERSITY
Lambert, Jack L.; *Research Participation for College Teachers-Summer Phase* (GY6909); 12 months; $21,260

OTTAWA UNIVERSITY
Lewis, Tom Blackwell; *Research Participation for College Teachers-Academic Year Phase* (GY7098); 24 months; $2,000

UNIVERSITY OF KANSAS
Wiseman, Gordon G.; *Summer Institute in Physics for College Teachers* (GY7015); 12 months; $46,070

WASHBURN UNIVERSITY OF TOPEKA
Shores, Richard; *Research Participation for College Teachers-Academic Year Phase* (GY7069); 24 months; $2,000

KENTUCKY

BELLARMINE-URSULINE COLLEGE
Kupchella, Charles E.; (GY7123); 24 months; $2,000
Rauckhorst, William H.; (GY7060); 24 months; $2,000

THOMAS MORE COLLEGE
Groskreutz, Harvey E.; (GY7090); 24 months; $2,000

WESTERN KENTUCKY UNIVERSITY
Wilkerson, Claude Manuel, (GY7108); 24 months; $2,000

LOUISIANA

LOUISIANA STATE UNIVERSITY, Baton Rouge
Grenchik, Raymond T.; *Summer Institute in General Astronomy for College Teachers* (GY7023); 12 months; $33,570
Nauman, Robert V.; *Summer Institute in Chemistry for College Teachers* (GY7004); 12 months; $45,900
———; *Research Participation for College Teachers-Summer Phase* (GY6921); 12 months; $25,250
Ralph, Dorr C.; (GY6911); 12 months; $18,400

NORTHEAST LOUISIANA STATE COLLEGE
Vanlandingham, Samuel L.; *Research Participation for College Teachers-Academic Year Phase* (GY7142); 24 months; $2,000

NORTHWESTERN STATE COLLEGE OF LOUISIANA
Browder, James Steve; (GY7078); 24 months; $2,000

XAVIER UNIVERSITY OF LOUISIANA
Malmstrom, Mary Carl; (GY 7127); 24 months; $2,000

MARYLAND

FROSTBURG STATE COLLEGE
Tate, Robert L.; (GY7114); 24 months; $2,000

HARFORD JUNIOR COLLEGE
Grimm III, Floyd H.; (GY7077); 24 months; $2,000

MORGAN STATE COLLEGE
Dixon, Robert M.; (GY7155); 24 months; $2,000

INSTITUTES AND RESEARCH PARTICIPATION FOR SCIENCE, MATHEMATICS, AND ENGINEERING TEACHERS

UNIVERSITY OF MARYLAND
Connors, Philip I.; *Summer Institute in Physics for College Teachers* (GY7038); 12 months; $39,240
Laster, Howard; *Research Participation for College Teachers in Physics and Astronomy* (GY5389001); $2,400
———; *Research Participation for College Teachers-Summer Phase* (GY6945); 12 months; $26,500

MASSACHUSETTS

BOSTON UNIVERSITY
Coulter, Lowell V.; (GY6908); 12 months; $26,500

BRANDEIS UNIVERSITY
Steel, Colin; (GY6952); 12 months; $27,100

EMMANUEL COLLEGE
Heidt, Lawrence J.; *Short Course in Photochemistry and Solar Energy Conversion for College Teachers* (GY6983); 12 months; $13,580

STATE COLLEGE AT LOWELL
Lee, Siu-Lam; *Research Participation for College Teachers-Academic Year Phase* (GY7085); 24 months; $2,000

TUFTS UNIVERSITY
Evans, Gordon G.; *Research Participation for College Teachers-Summer Phase* (GY6933); 12 months; $15,900
Illinger, Karl H.; *Summer Institute on Modern Aspects of Physical Chemistry for College Teachers* (GY7030); 12 months; $33,680

WELLESLEY COLLEGE
Webster, Eleanor R.; *Institute for Chemistry Teacher Retrainees* (GY6989); 24 months; $62,080

WENTWORTH INSTITUTE
Thomson, Charles M.; *Summer Institute in Electronic Technology for College Teachers* (GY6993); 12 months; $40,950

WORCESTER POLYTECHNIC INSTITUTE
Richardson, Glen A.; *Summer Institute in Electrical Engineering for College Teachers* (GY7048); 12 months; $45,460

MICHIGAN

ADRIAN COLLEGE
Pickering, Ed R.; *Research Participation for College Teachers-Academic Year Phase* (GY7066); 24 months; $2,000
Xavier, K. S.; (GY7092); 24 months; $2,000

ALBION COLLEGE
Aiuto, Russell; *Short Course in Genetics for College Teachers* (GY6965); 12 months; $16,290
Woodward Jr., Addison Ely; *Research Participation for College Teachers-Academic Year Phase* (GY7156); 24 months; $2,000

ANDREWS UNIVERSITY
Rowland, Sattley Clark; (GY7132); 24 months; $2,000

CALVIN COLLEGE
Kromminga, Albion J.; (GY7071); 24 months; $2,000
Van Till, Howard J.; (GY7072); 24 months; $2,000

GRAND RAPIDS JUNIOR COLLEGE
Bruder, Karl Fritz; (GY7205); 24 months; $2,000

HOPE COLLEGE
Hoepfinger, Lynn M.; (GY7183); 24 months; $2,000

KALAMAZOO COLLEGE
Buskirk, Allen V.; (GY7088); 24 months; $2,000

LAKE MICHIGAN COLLEGE
Bair, Clark Larry; (GY7055); 24 months; $2,000

MICHIGAN TECHNOLOGICAL UNIVERSITY, Lake Superior State College
Chandra, Purna; (GY7154); 24 months; $2,000

MICHIGAN STATE UNIVERSITY
Haynes, Sherwood K.; *Research Participation for College Teachers—Summer Phase* (GY6912); 12 months; $19,380

NAZARETH COLLEGE
Zeleznik, Pauline; *Research Participation for College Teachers—Academic Year Phase* (GY7096); 24 months; $2,000

OAKLAND UNIVERSITY
Feeman, George F.; *Summer Institute in Mathematics for College Teachers* (GY6996); 12 months; $34,220

OLIVET COLLEGE
Wellfare, Robert Wayne; *Research Participation for College Teachers—Academic Year Phase* (GY7150); 24 months; $2,000

UNIVERSITY OF MICHIGAN
Comstock, Craig; *Summer Institute in Mathematics for College Teachers* (GY7026); 12 months; $41,350
Coombs, Clyde H.; *Summer Institute in Mathematical Psychology for College Teachers* (GY7049); 12 months; $34,320
———; (GY5441001); $1,400
Lowry, Robert J.; *Research Participation for College Teachers—Summer Phase* (GY6931); 12 months; $17,400
Papsdorf, James D.; (GY6929); 12 months; $27,450

UNIVERSITY OF MICHIGAN, Flint College
Kren, Robert M.; *Research Participation for College Teachers—Academic Year Phase* (GY7102); 24 months; $2,000

WAYNE STATE UNIVERSITY
Parsons, Willard H.; *Short Course in Geology for College Teachers* (GY6984); 12 months; $24,280

MINNESOTA

BETHEL COLLEGE AND SEMINARY
Rodgers, James Edward; *Research Participation for College Teachers—Academic Year Phase* (GY7161); 24 months; $2,000

CONCORDIA COLLEGE
Heuer, Charles Vernon; (GY7141); 24 months; $2,000

FERGUS FALLS STATE JUNIOR COLLEGE
McKeag, Roderick James; (GY7193); 24 months; $2,000

GUSTAVUS ADOLPHUS COLLEGE
Splittgerber, Allan Gene; (GY7099); 24 months; $2,000

MANKATO STATE COLLEGE
Inman, Fred W.; *Summer Institute in Physics for College Teachers* (GY7033); 12 months; $29,500

UNIVERSITY OF MINNESOTA
Lindgren, Bernard W.; *Summer Institute in Statistics and Probability for College Teachers* (GY7042); 12 months; $46,660

MISSISSIPPI

HOLMES JUNIOR COLLEGE
Shaffer Jr., C. Edward; *Research Participation for College Teachers—Academic Year Phase* (GY7068); 24 months; $2,000

MISSOURI

MISSOURI BAPTIST COLLEGE
Pope, Kenneth Harden; (GY7195); 24 months; $2,000

MISSOURI WESTERN COLLEGE
Lambing, Larry L.; (GY7204); 24 months; $2,000

UNIVERSITY OF MISSOURI, Rolla
Hill, Otto H.; *Research Participation for College Teachers—Summer Phase* (GY6919); 12 months; $15,650
Lee, Ralph E.; *Summer Institute in Numerical and Statistical Methods of Digital Computing and Programming Languages for College Teachers* (GY6986); 12 months; $58,380

MONTANA

UNIVERSITY OF MONTANA
Behan, Mark J.; *Summer Institute in Botany for College Teachers* (GY6999); 12 months; $47,390

NEW JERSEY

FAIRLEIGH DICKINSON UNIVERSITY, Florham-Madison Campus
Multer, H. G. and Weiss, M. P.; *Short Course in Modern Carbonate Sediments and Related Carbonate Rocks for College Teachers* (GY6970); 12 months; $17,340

PATERSON STATE COLLEGE
Callahan, Robert F.; *Research Participation for College Teachers—Academic Year Phase* (GY7133); 24 months; $2,000
Merijanian, Ashot; (GY7056); 24 months; $2,000

RUTGERS UNIVERSITY
Singley, Mark E.; *Short Course in Systems Analysis for Agriculture and Land Use Planning for College Teachers* (GY6975); 12 months; $19,790

TRENTON STATE COLLEGE
Star, Aura Estelle; *Research Participation for College Teachers—Academic Year Phase* (GY7173); 24 months; $2,000

UPSALA COLLEGE
Gordon, Roger P.; (GY7091); 24 months; $2,000

NEW MEXICO

EASTERN NEW MEXICO UNIVERSITY
Russell, Thomas W.; (GY7101); 24 months; $2,000

NEW MEXICO STATE UNIVERSITY
Thomas, John D.; *Summer Institute in Mathematics for College Teachers* (GY7018); 12 months; $43,290

NEW YORK

AGRICULTURAL & TECHNICAL COLLEGE AT CANTON
Selleck, Erwin A.; *Research Participation for College Teachers—Academic Year Phase* (GY7163); 24 months; $2,000

CLARKSON COLLEGE OF TECHNOLOGY
Misiaszek, Edward T.; *Summer Institute in Basic Engineering for College Teachers* (GY7009); 12 months; $63,650

COLLEGE AT FREDONIA
Neveu, Maurice Cleon; *Research Participation for College Teachers—Academic Year Phase* (GY7144); 24 months; $2,000

COLLEGE AT ONEONTA
Chiang, Joseph F.; (GY7160); 24 months; $2,000
Dixon, William B.; (GY7159); 24 months; $2,000
Wang, Charles T. P.; (GY7082); 24 months; $2,000

CUNY, BROOKLYN COLLEGE
Rosen, Milton J.; *Research Participation for College Teachers—Summer Phase* (GY6907); 12 months; $17,100

CUNY, CITY COLLEGE
Erlbach, Erich; *Research Participation for College Teachers—Academic Year Phase* (GY7158); 24 months; $2,000

CUNY, QUEENS COLLEGE
Lanson, Robert Nathan; (GY7148); 24 months; $2,000

SCIENCE EDUCATION PROGRAMS

CUNY, QUEENSBORO COMMUNITY COLLEGE
Feit, Julius; (GY7075); 24 months; $2,000

HAMILTON COLLEGE
Kinnel, Robin B.; (GY7063); 24 months; $2,000

HOBART AND WM. SMITH COLLEGE
Campbell, Larry E.; (GY7180); 24 months; $2,000

MARIST COLLEGE
Turley, Hugh Patrick; (GY7177); 24 months; $2,000

NEW YORK UNIVERSITY
Novikoff, Albert; *Summer Institute in Mathematics for College Teachers* (GY7043); 12 months; $35,410

POLYTECHNIC INSTITUTE OF BROOKLYN
Jolls, Kenneth R.; *Short Course in Science and Engineering for College Teachers* (GY6982); 12 months; $19,010

RENSSELAER POLYTECHNIC INSTITUTE
Richtol, Herbert H.; *Summer Institute in Instrumental Methods of Analysis in Chemistry for College Teachers* (GY7020); 12 months; $31,000

ROCHESTER INSTITUTE OF TECHNOLOGY
Henderson, John T.; *Summer Institute in Technical Engineering for College Teachers* (GY7047); 12 months; $48,770

SKIDMORE COLLEGE
Spears, William C.; *Summer Institute in Child Psychology for College Teachers* (GY7031); 12 months; $34,490

ST. BONAVENTURE UNIVERSITY
Budzinski, Walter V.; *Research Participation for College Teachers—Academic Year Phase* (GY7079); 24 months; $2,000

SYRACUSE UNIVERSITY
Evan-Iwanowski, R. M.; *Research Participation for College Teachers—Summer Phase* (GY6946); 12 months; $14,400
Vook, Richard W.; *Short Course in Electron Microscopy and Diffraction for College Teachers* (GY6972); 12 months; $5,760
———; *Research Participation for College Teachers—Summer Phase* (GY6943); 12 months; $12,700

UNIVERSITY OF ROCHESTER
Thompson, Brian J.; *Short Course in Optics for College Teachers* (GY6963); 12 months; $13,780

YORK COLLEGE
Gerardi, Gary J.; *Research Participation for College Teachers—Academic Year Phase* (GY7104); 24 months; $2,000

NORTH CAROLINA

CHOWAN COLLEGE
Faucette, Deaton Francis; (GY7197); 24 months; $2,000
Hazelton, George L.; (GY7196); 24 months; $2,000

DUKE UNIVERSITY
Strobel, Howard A.; *Summer Institute on Molecular Structure Elucidation by Physical Methods for College Teachers* (GY7035); 12 months; $33,540

EAST CAROLINA UNIVERSITY
Heckel, Edgar; *Research Participation for College Teachers—Academic Year Phase* (GY7087); 24 months; $2,000

NORTH CAROLINA STATE UNIVERSITY AT RALEIGH
Craig, H. Bradford; *Research Participation for College Teachers—Summer Phase* (GY6918); 12 months; $21,550

NORTH CAROLINA WESLEYAN COLLEGE
Baxter Jr., John E.; *Research Participation for College Teachers—Academic Year Phase* (GY7187); 24 months; $2,000

PFEIFFER COLLEGE
Riemann, James M.; (GY7057); 24 months; $2,000

UNIVERSITY OF NORTH CAROLINA AT CHAPEL HILL
McCormick, J. Frank; *Short Course in Ecology for College Teachers* (GY6978); 12 months; $26,250
Mewborn, Ancel C.; *Summer Institute in Mathematics for College Teachers* (GY6990); 12 months; $36,300
Rowan, Lawrence G.; *Short Course in Electronics for College Teachers* (GY6976); 12 months; $29,980

NORTH DAKOTA

UNIVERSITY OF NORTH DAKOTA
Stewart, James A.; *Research Participation for College Teachers—Summer Phase* (GY6930); 12 months; $6,080

OHIO

BOWLING GREEN STATE UNIVERSITY
Harmon, Shirley Ann; *Research Participation for College Teachers—Academic Year Phase* (GY7146); 24 months, $2,000

CASE WESTERN RESERVE UNIVERSITY
Kuwana, Theodore; *Research Participation for College Teachers—Summer Phase* (GY6937); 12 months; $21,800
McGervey, John D.; (GY6925); 12 months; $19,500

CEDARVILLE COLLEGE
Frye, Laurel Bert; *Research Participation for College Teachers—Academic Year Phase* (GY7192); 24 months; $2,000
Helmick, Larry S.; (GY7191); 24 months; $2,000

MUSKINGUM COLLEGE
Crandell, Merrell E.; (GY7167); 24 months; $2,000

OHIO DOMINICAN COLLEGE
Matesich, Mary Andrew; (GY7162); 24 months; $2,000

OHIO STATE UNIVERSITY
King, Leslie J.; *Short Course on Models of Urban Spatial Structure and Ecology for College Teachers* (GY6960); 12 months; $18,580

OHIO UNIVERSITY
Dorf, Richard C.; *Short Course in Engineering and the Social Sciences for College Teachers* (GY6968); 12 months; $15,500
Harlan, William H.; *Summer Institute in the Social Sciences for College Teachers* (GY7013); 12 months; $38,960

OKLAHOMA

CAMERON STATE AGRICULTURAL COLLEGE
Hawk, Ira Lee; *Research Participation for College Teachers—Academic Year Phase* (GY7201); 24 months; $2,000

OKLAHOMA STATE UNIVERSITY
Jewett, John; *Summer Institute in Mathematics for College Teachers* (GY7025); 12 months; $42,430
Todd, Glenn W.; *Research Participation for College Teachers—Summer Phase* (GY6916); 12 months; $20,300

UNIVERSITY OF OKLAHOMA
Andree, Richard V.; (GY6924); 12 months; $26,250
Loren, G. Hill; (GY6935); 12 months; $10,150

OREGON

OREGON STATE UNIVERSITY
Krueger, James H.; (GY6928); 12 months; $27,250
Logan, A. V.; *Summer Institute for College Teachers* (GY4025001); $390
MacDonald, Donald L.; *Research Participation for College Teachers—Summer Phase* (GY6920); 12 months; $21,800
Stephen, W. P.; (GY6942); 12 months; $23,080

PORTLAND COMMUNITY COLLEGE
Powell, Charlie Roy; *Research Participation for College Teachers—Academic Year Phase* (GY7073); 24 months; $2,000

UNIVERSITY OF OREGON
Moursund, A. F.; *Summer Institute in Mathematics for College Teachers* (GY7011); 12 months; $40,640

PENNSYLVANIA

BUCKS COUNTY COMMUNITY COLLEGE
Greenhalgh, Samuel; *Research Participation for College Teachers—Academic Year Phase* (GY7200); 24 months; $2,000

CARNEGIE MELLON UNIVERSITY
Moore, Richard A.; *Summer Institute in Mathematics for Junior College Teachers* (GY7041); 12 months; $53,360

DICKINSON COLLEGE
Wolf, Neil S.; *Research Participation for College Teachers—Academic Year Phase* (GY7174); 24 months; $2,000

GROVE CITY COLLEGE
Lim, Benjamin S.; (GY7165); 24 months; $2,000

HAVERFORD COLLEGE
Rowe Jr., Preston Brainard; (GY7147); 24 months; $2,000

LAFAYETTE COLLEGE
Majumdar, Shyamal K.; (GY7188); 24 months; $2,000

LOCK HAVEN STATE COLLEGE
Newcomer, Charles Alfred; (GY7166); 24 months; $2,000

LYCOMING COLLEGE
Getchell, Charles L.; (GY7061); 24 months; $2,000

PENNSYLVANIA STATE UNIVERSITY
Thornton, Charles P.; *Summer Institute in Mineralogy, Geochemistry and Petrology for College Teachers* (GY7034); 12 months; $32,860

SHIPPENSBURG STATE COLLEGE
Beres, James A.; *Research Participation for College Teachers—Academic Year Phase* (GY7189); 24 months; $2,000
Rill, Morris Eugene; (GY7059); 24 months; $2,000

SUSQUEHANNA UNIVERSITY
McGrath, Thomas F.; (GY7171); 24 months; $2,000

UNIVERSITY OF SCRANTON
MacEntee, Francis Joseph; (GY7172); 24 months; $2,000

WESTMINSTER COLLEGE
Gray, David Bryce; (GY7151); 24 months; $2,000
Harms, Clarence E.; (GY7111); 24 months; $2,000

WILSON COLLEGE
Gondal, Surinder K.; (GY7152); 24 months; $2,000

PUERTO RICO

UNIVERSITY OF PUERTO RICO, Rio Piedras
Lugo, Herminio Lugo; *Summer Institute in Marine Biology and Tropical Ecology for College Teachers* (GY7005); 12 months; $31,450

RHODE ISLAND

BROWN UNIVERSITY
Quinn, John J.; *Research Participation for College Teachers—Summer Phase* (GY6905); 12 months; $14,460

INSTITUTES AND RESEARCH PARTICIPATION FOR SCIENCE, MATHEMATICS, AND ENGINEERING TEACHERS

SOUTH CAROLINA

CLAFLIN COLLEGE
Singh, Dharmdeo Narayan; *Research Participation for College Teachers—Academic Year Phase* (GY7122); 24 months; $2,000

FURMAN UNIVERSITY
Brewer, Charles Lee; (GY7094); 24 months; $2,000

UNIVERSITY OF SOUTH CAROLINA
Schuette, Oswald F.; *Research Participation for College Teachers—Summer Phase* (GY6955); 12 months; $25,750
Smith Jr., Alfred G.; *Summer Institute in Economics for College Teachers* (GY7024); 12 months; $43,400

SOUTH DAKOTA

SOUTH DAKOTA STATE UNIVERSITY
Bruce, James D.; *Research Participation for College Teachers—Academic Year Phase* (GY7058); 24 months; $2,000

TENNESSEE

CLEVELAND STATE COMMUNITY COLLEGE
Speight, David Larry; (GY7190); 24 months; $2,000

TENNESSEE TECHNOLOGICAL UNIVERSITY
Farrar, David T.; (GY7182); 24 months; $2,000

VANDERBILT UNIVERSITY
Bogitsh, Burton J.; *Short Course in Zoology for College Teachers* (GY6971); 12 months; $18,600
Hamilton, Joseph H.; *Research Participation for College Teachers—Summer Phase* (GY6917); 12 months; $10,290
Shahan, Ewing P.; *Summer Institute in Economics for College Teachers* (GY7002); 12 months; $29,920
Wesson, J. R. and Bryant, B. F.; *Summer Institute in Mathematics for College Teachers* (GY7021); 12 months; $34,100

TEXAS

DALLAS COUNTY JUNIOR COLLEGE DISTRICT, El Centro College
Schaar, Peter Leslie; *Research Participation for College Teachers—Academic Year Phase* (GY7202); 24 months; $2,000

MCLENNAN COMMUNITY COLLEGE
Schwarz, James Robert; (GY7199); 24 months; $2,000

MCMURRAY COLLEGE
Sharp Jr., A. C.; (GY7089); 24 months; $2,000

SAM HOUSTON STATE UNIVERSITY
Souchek, Julianne; (GY7062); 24 months; $2,000

SAN ANTONIO COLLEGE
Lincoln, James P.; (GY7198); 24 months; $2,000

SOUTHWEST TEXAS UNIVERSITY
Cude, Willis A.; (GY7064); 24 months; $2,000

SOUTHWESTERN UNIVERSITY
Hilgeman, Frederick R.; (GY7119); 24 months; $2,000

TEXAS A & M UNIVERSITY
Ham, Joe S.; *Research Participation for College Teachers—Summer Phase* (GY6914); 12 months; $15,400
Weekes, Donald F.; *Summer Institute in Physics for College Teachers* (GY6988); 12 months; $65,950

TEXAS WOMAN'S UNIVERSITY
Foster, Norman G.; *Summer Institute in Analytical Instrumentation for College Teachers* (GY7003); 12 months; $41,680

UNIVERSITY OF TEXAS AT AUSTIN
Bold, Harold C.; *Research Participation for College Teachers—Summer Phase* (GY6904); 12 months; $24,500
Little, R. N.; (GY6949); 12 months; $24,700

UTAH

WEBER STATE COLLEGE
Hobbs, Earnest Lagrande; *Research Participation for College Teachers—Academic Year Phase* (GY7176); 24 months; $2,000

VIRGINIA

HAMPDEN-SYDNEY COLLEGE
Mayo IV, Thomas Tabb; (GY7080); 24 months; $2,000

MADISON COLLEGE
Garrison, Norman Eugene; (GY7074); 24 months; $2,000

OLD DOMINION UNIVERSITY
Young Jr., Eddie H.; (GY7157); 24 months; $2,000

UNIVERSITY OF RICHMOND
Mateer, Richard A.; (GY7116); 24 months; $2,000

UNIVERSITY OF VIRGINIA, George Mason College
Papaconstantopoulos, D.; (GY7086); 24 months; $2,000

VIRGINIA INSTITUTE OF MARINE SCIENCE
Bailey, Robert S.; *Research Participation for College Teachers—Summer Phase* (GY6915); 12 months; $15,560

VIRGINIA MILITARY INSTITUTE
Gregorek Jr., Edward S.; *Research Participation for College Teachers—Academic Year Phase* (GY7128); 24 months; $2,000
Settle Jr., Frank A.; (GY7070); 24 months; $2,000

VIRGINIA POLYTECHNIC INSTITUTE
Frederick, Daniel; *Short Course on Recent Developments in Continuum Mechanics for College Teachers* (GY6966); 12 months; $24,060
———; *Short Course on Recent Developments in Continuum Mechanics for College Teachers in Engineering, Mathematics and Physics* (GY5480001); $400
Herndon, James F.; *Summer Institute in Mathematical Applications in Political Science for College Teachers* (GY7017); 12 months; $46,990

WASHINGTON

PACIFIC NORTHWEST ASSOCIATION FOR COLLEGE PHYSICS
Johnson, Wilbur V.; *Summer Institute for College Teachers in Physics* (GY5503 001); $3,090

UNIVERSITY OF WASHINGTON
Dubisch, Roy; *Summer Institute in Mathematics for College Teachers* (GY7014); 12 months; $37,690

WALLA WALLA COLLEGE
Forss, Carl A.; *Research Participation for College Teachers—Academic Year Phase* (GY7065); 24 months; $2,000

WASHINGTON STATE UNIVERSITY
Stacy, Gardner W.; *Research Participation for College Teachers—Summer Phase* (GY6926); 12 months; $10,900

WHITWORTH COLLEGE
Johnson, William Allen; *Research Participation for College Teachers—Academic Year Phase* (GY7110); 24 months; $2,000

WEST VIRGINIA

WEST VIRGINIA INSTITUTE OF TECHNOLOGY
Yu, Juin Sheng; (GY7134); 24 months; $2,000

WISCONSIN

CARTHAGE COLLEGE
Souter, Walter Richard; (GY7112); 24 months; $2,000

HOLY FAMILY COLLEGE
Shimondle, Loretta; (GY7113); 24 months; $2,000

UNIVERSITY OF WISCONSIN, Green Bay, Fox Valley Branch
Herrick, Daniel Lance; (GY7138); 24 months; $2,000

UNIVERSITY OF WISCONSIN CENTER, Marshfield-Wood County Campus
Halloran, Donal W.; (GY7135); 24 months; $2,000

UNIVERSITY OF WISCONSIN, Madison
Smart, J. R.; *Summer Institute in Mathematics for College Teachers* (GY7007); 12 months; $51,130
Wang, Chu-Kia; *Short Course in Computer Matrix Methods in Structural Mechanics for College Teachers* (GY6980); 12 months; $21,250

UNIVERSITY OF WISCONSIN, Milwaukee
Jaggard, R. A.; *Short Course in Physics for College Teachers* (GY6981); 12 months; $14,720

VITERBO COLLEGE
Bahr, Sr. Janice; *Research Participation for College Teachers—Academic Year Phase* (GY7093); 24 months; $2,000

WISCONSIN STATE UNIVERSITY, Stevens Point
Harris, Joseph Belknap; (GY7140); 24 months; $2,000
Radtke, Douglas D.; (GY7097); 24 months; $2,000

WISCONSIN STATE UNIVERSITY, Platteville
Likes, James Gary; (GY7185); 24 months; $2,000

WYOMING

UNIVERSITY OF WYOMING
Howatson, John; *Research Participation for College Teachers—Summer Phase* (GY6947); 12 months; $9,900
Prowse, Derek J.; (GY6953); 12 months; $12,880

Secondary School Teachers

ALABAMA

ALABAMA AGRICULTURAL AND MECHANICAL COLLEGE
Smoot, Henrene Ellington; *Summer Institute for Secondary School Teachers* (GW4542); 11 months; $44,207
———; *In-Service Institute in Earth Science and Mathematics for Secondary School Teachers* (GW5494); 17 months; $6,866

ALABAMA STATE UNIVERSITY
Pratt, Charles; *In-Service Institute in Chemistry, for Secondary School Teachers* (GW5515); 17 months; $10,844

AUBURN UNIVERSITY
Butz, R. K.; *Summer Institute in Mathematics for Secondary School Teachers* (GW4637); 11 months; $40,084
Carr, Howard E.; *Summer Institute in Physics for Secondary School Teachers* (GW4695); 11 months; $44,190

MOBILE COLLEGE
Smith, George A.; *In-Service Institute in Mathematics for Secondary School Teachers* (GW5486); 17 months; $7,228

TUSKEGEE INSTITUTE
Koons, Lawrence F.; *Summer Institute in Chemistry for Secondary School Teachers* (GW4646); 11 months; $30,949
Taylor, Grady W.; *Summer Institute in Agricultural Science for Secondary School Teachers* (GW4890); 11 months; $42,324

UNIVERSITY OF ALABAMA IN BIRMINGHAM
Settine, Robert L.; *In-Service Institute in Chemistry for Secondary School Teachers* (GW5548); 17 months; $12,264

SCIENCE EDUCATION PROGRAMS

UNIVERSITY OF ALABAMA, Tuscaloosa
Baker, Truman D.; *Summer Institute in the Use of Computers in Mathematics and Science Instruction for Secondary School Teachers* (GW4893); 11 months; $44,727

ALASKA

ALASKA METHODIST UNIVERSITY
Lewis, William A.; *In-Service Institute in Physical Science for Secondary School Teachers* (GW5477); 13 months; $4,441

UNIVERSITY OF ALASKA, Anchorage Regional Center
Stickney, Roland F.; *In-Service Institute in Biological Sciences/Special Emphasis to Arctic Ecology for Secondary School Teachers* (GW5677); 17 months; $5,933

ARIZONA

ARIZONA STATE UNIVERSITY
Bedient, Jack D.; *In-Service Institute in Mathematics for Secondary School Teachers* (GW5416); 17 months; $16,459
———; (GW5843); 26 months; $16,459
Leathers, Chester R.; *Summer Institutes in Biochemistry and Cellular Biology for Secondary School Teachers* (GW4773); 11 months; $55,723
Miller Paul T.; *Summer Institute in Geology for Secondary School Teachers* (GW4842); 11 months; $46,515
Rawls, William S.; *In-Service Institute in Experimental Optics for Secondary School Teachers* (GW5661); 17 months; $8,591
Smith, Lehi T.; *Academic Year Institute for Secondary School Teachers* (GW4821); 24 months; $125,706

NORTHERN ARIZONA UNIVERSITY
Beus, Stanley S.; *Summer Institute for Secondary School Teachers* (GW4578); 11 months; $44,043
Willis, W. R.; *Summer Institute in Harvard Project Physics for Secondary School Teachers* (GW4682); 11 months; $49,246

UNIVERSITY OF ARIZONA
Hoshaw, Robert W.; *In-Service Institute in Science for Secondary School Teachers* (GW5626); 17 months; $17,166
———; (GW5812); 26 months; $17,166
Seeley, Millard G.; *Summer Institute in Science for Secondary School Teachers* (GW4589); 11 months; $81,980
Steinbrenner, Arthur H.; *Summer Institute for Secondary School Teachers* (GW4543); 11 months; $71,536
———; *In-Service Institute in Mathematics for Secondary School Teachers* (GW5749); 26 months; $6,385
———; (GW5422); 17 months; $7,050
———; (GW5390); 17 months; $6,196

ARKANSAS

UNIVERSITY OF ARKANSAS
Orton, W. R.; *Academic Year Institute for Secondary School Teachers* (GW4828); 22 months; $130,962
Scroggs, James E.; *Summer Institute for Secondary School Teachers* (GW4544); 11 months; $51,457
Sharrah, Paul C.; *Summer Institute in Basic Atomic and Nuclear Physics for Secondary School Teachers* (GW4970); 11 months; $26,345
Venus, Charles E.; *In-Service Institute in Applied Principles of Economic Development for Secondary School Teachers* (GW5672); 17 months; $6,464

CALIFORNIA

CALIFORNIA STATE COLLEGE AT FULLERTON
Benson, Russell V.; *Summer Institute in Mathematics for Secondary School Teachers* (GW4867); 11 months; $30,237
Lafky, John D.; *In-Service Institute in Economics for Secondary School Teachers* (GW5700); 17 months; $7,785
Lepper, Robert E.; *In-Service Institute in Earth Science for Secondary School Teachers* (GW5587); 17 months; $13,481
Turner, George C.; *In-Service Institute in Human Ecology for Secondary School Teachers* (GW5731); 26 months; $9,607
———; (GW5447); 17 months; $9,607
———; *Academic Year Institute in Biology for Secondary School Teachers* (GW5851); 26 months; $32,013

CALIFORNIA STATE COLLEGE AT HAYWARD
Purvis, C. T.; *In-Service Institute in Mathematics for Secondary School Teachers* (GW5411); 17 months; $12,660
———; (GW5732); 26 months; $12,998

CALIFORNIA STATE COLLEGE AT LONG BEACH
Bauer, Roger D.; *Summer Institute in Biochemistry for Secondary School Teachers* (GW4843); 11 months; $35,621

CALIFORNIA STATE COLLEGE AT SAN BERNARDINO
Stanton, Gordon E.; *In-Service Institute in Sociology for Secondary School Teachers* (GW5586); 17 months; $12,021

DOMINICAN COLLEGE AT SAN RAFAEL
Aquinas, M.; *In-Service Institute in Man and His Marine Environmental Developments for Secondary School Teachers* (GW5678); 17 months; $6,248

FRESNO STATE COLLEGE
Labarre, Anthony E. Jr.; *Summer Institute in Mathematics for Secondary School Teachers* (GW5003); 11 months; $90,373
Lyles Jr., Samuel P.; *In-Service Institute in Physics for Secondary School Teachers* (GW5625); 17 months; $8,846

HARVEY MUDD COLLEGE
Alford, Jack L.; *Summer Institute in Engineering Concepts Curriculum Project for Secondary School Teachers* (GW4722); 11 months; $35,139

HUMBOLDT STATE COLLEGE
Beilfuss, Erwin R.; *Summer Institute in Marine Biology and Oceanography for Secondary School Teachers* (GW5037); 11 months; $60,545

SACRAMENTO STATE COLLEGE
Baum, Milton S.; *In-Service Institute in Economics for Secondary School Teachers* (GW5671); 17 months; $8,230

SAN DIEGO STATE COLLEGE
Becker, Gerald A.; *Academic Year Institute for Secondary School Teachers* (GW4795); 22 months; $147,929
———; *Summer Institute in Mathematics for Secondary School Teachers* (GW4901); 11 months; $56,924
Branstetter, R. Dean; *In-Service Institute in Mathematics for Secondary School Teachers* (GW5670); 17 months; $22,602
Cox, George W.; *Summer Institute in Field Ecology for Secondary School Teachers* (GW4940); 11 months; $46,319
Daub, C. T.; *Summer Institute in Astronomy for Secondary School Teachers* (GW4683); 11 months; $56,500
Dessel, Norman F.; *Summer Institute in Harvard Project Physics for Secondary School Teachers* (GW4892); 11 months; $52,174
Eidemiller, D. I.; *In-Service Institute in Physical Geography for Secondary School Teachers* (GW5669); 17 months; $9,884
———; (GW5831); 26 months; $9,884
Finch Jr., William A.; *In-Service Institute in Earth Science for Secondary School Teachers* (GW5458); 17 months; $11,611
Gifford, Adam; *In-Service Institute in Economics for Secondary School Teachers* (GW5537); 17 months; $8,006
Ingmanson, Dale E.; *Summer Institute for Secondary School Teachers* (GW5016); 11 months; $36,851
McClintic, Joseph O.; *In-Service Institute in Economics for Secondary School Teachers* (GW5699); 17 months; $8,844
Smith Jr., Louis E.; *Summer Institute in Physics for Secondary School Teachers* (GW4774); 11 months; $56,240

SAN FRANCISCO STATE COLLEGE
Pestrong, Raymond; *In-Service Institute in Earth Science for Secondary School Teachers* (GW5547); 17 months; $12,168

SAN JOSE STATE COLLEGE
Fowler, Kenneth A.; *Summer Institute in Mathematics for Secondary School Teachers* (GW5038); 11 months; $36,445
Kramer, Max; *Academic Year Institute for Secondary School Teachers, Pre-Service Institute for Secondary School Teachers* (GW4793); 22 months; $145,446
Mings, Turley; *Summer Institute in Economics for Secondary School Teachers* (GW4924); 11 months; $45,030
———; *In-Service Institute in Economics Content and Teaching Strategies for Secondary School Teachers* (GW5698); 17 months; $7,472
Pickering, Marjorie A.; *In-Service Institute in Mathematics for Secondary School Teachers* (GW5398); 17 months; $23,728
Smart, James R.; *Summer Institute in Mathematics for Secondary School Teachers* (GW5051); 11 months; $58,165

SONOMA STATE COLLEGE
Brumbaugh, Joe H.; *Summer Institute in Field Biology and Ecology for Secondary School Teachers* (GW4704); 11 months; $54,820
Duncan, Donald G.; *In-Service Institute in Mathematics for Secondary School Teachers* (GW5514); 17 months; $6,604

STANFORD UNIVERSITY
Bacon, Harold M.; *Summer Institute in Mathematics for Secondary School Teachers* (GW4684); 11 months; $58,740
Baxter, Charles; *Summer Institute in Research in Space Biology for Secondary School Teachers* (GW5008); 11 months; $41,277
Thomas, John H.; *Summer Conference in Science for the Disadvantaged for Science Supervisors* (GW5340); 18 months; $15,189

UNIVERSITY OF CALIFORNIA, Berkeley
Fry, Wayne L.; *Summer and In-Service Institute in Earth Science for Secondary School Teachers* (GW5034); 19 months; $64,110
———; (GW5034*); 19 months; $9,420
———; *In-Service Institute in Earth Science for Secondary School Teachers* (GW5752); 26 months; $9,420
Helmholz, A. C.; *Summer Institute in Basic Physics for Secondary School Teachers* (GW4664); 11 months; $54,749
Kelly, Lola S.; *Summer Institute in Radiation Biology for Secondary School Teachers* (GW4992); 11 months; $29,083
Knight, C. Arthur; *Summer Institute for Secondary School Teachers* (GW4580); 11 months; $66,724
Pimentel, George C.; (GW4579); 11 months; $52,152
Portis, Alan M.; *Summer Institute in Physics, Chemistry and Biology for Secondary School Teachers* (GW4623); 11 months; $56,256
Stone, Edward C.; *In-Service Institute in Biology for Secondary School Teachers* (GW5862); 26 months; $7,513
———; (GW5555); 17 months; $7,525

UNIVERSITY OF REDLANDS
Sanderson Jr., Judson; *Summer Institute in Mathematics for Secondary School Teachers* (GW4868); 11 months; $30,548

INSTITUTES AND RESEARCH PARTICIPATION FOR SCIENCE, MATHEMATICS, AND ENGINEERING TEACHERS

Trolan, J. Kenneth; *In-Service Institute in Physics for Secondary School Teachers* (GW5779); 26 months; $9,450
——— ; (GW5413); 17 months; $9,450

UNIVERSITY OF SAN FRANCISCO
Farrell, Edward J.; *Summer Conference in Geometry for Secondary School Teachers and Supervisors* (GW5323); 10 months; $12,312
——— ; *In-Service Institute in Mathematics for Secondary School Teachers* (GW5391); 17 months; $6,766
——— ; (GW5751); 26 months; $6,766
——— ; *Summer Institute in Mathematics for Secondary School Teachers* (GW4904); 11 months; $38,266

UNIVERSITY OF SANTA CLARA
Iwamoto, Kichiro K.; *Summer Institute in Principles of Sociology for Secondary School Teachers* (GW4923); 11 months; $41,951
Klosinski, Leonard F.; *In-Service Institute in Mathematics for Secondary School Teachers* (GW5842); 26 months; $20,991
——— ; *Summer and In-Service Institute in Mathematics for Secondary School Teachers* (GW4642*); 21 months; $22,727
Klosinski, Leonard F. and Alexanderson, G. L.; (GW4642); 21 months; $52,308

UNIVERSITY OF SOUTHERN CALIFORNIA
Gorsline, Donn S.; *Summer Institute for Secondary School Teachers* (GW4844); 11 months; $49,442
White, Paul A.; *In-Service Institute in Mathematics for Secondary School Teachers* (GW5780); 26 months; $24,693
——— ; (GW5421); 17 months; $24,693
——— ; *Summer Institute in Mathematics for Secondary School Teachers* (GW5006); 11 months; $54,340

UNIVERSITY OF THE PACIFIC
Arvey, M. Dale; *In-Service Institute in Biology for Secondary School Teachers* (GW5459); 17 months; $6,275
——— ; (GW5713); 26 months; $6,275

COLORADO

COLORADO SCHOOL OF MINES
Burnett, Jerrold J.; *In-Service Institute in Physics (General, Intermediate and Harvard Project) for Secondary School Teachers* (GW5493); 17 months; $16,360
——— ; *In-Service Institute in Physics for Secondary School Teachers* (GW5821); 26 months; $17,050
Lucas, George B.; *In-Service Institute in Earth Science, Physical Science and Computer Science for Secondary School Teachers* (GW5810); 26 months; $19,428
———; (GW5595); 17 months; $19,428
Trexler, David W.; *Summer Institute in Chemistry and Geology for Secondary School Teachers* (GW4769); 11 months; $42,315

COLORADO STATE UNIVERSITY
Painter, Richard J.; *Summer Institute for Secondary School Teachers* (GW4545); 11 months; $64,992
Swanson, V. B.; *Summer Institute in Genetics for Secondary School Teachers* (GW4775); 11 months; $33,089

FORT LEWIS COLLEGE
Bushnell, Donald D.; *In-Service Institute in Mathematics for Secondary School Teachers* (GW5585); 17 months; $8,093
——— ; (GW5760); 26 months; $8,092

SOCIAL SCIENCE EDUCATION CONSORTIUM
Stevens Jr., W. Williams; *Summer Conference in New Social Science Curriculum Material for Social Science Supervisors* (GW5335); 10 months; $15,202

SOUTHERN COLORADO STATE COLLEGE
Fisher, John; *In-Service Institute in Mathematics for Secondary School Teachers* (GW5402); 17 months; $7,308

TEMPLE BUELL COLLEGE
Polt, James M.; *Summer Institute in Psychological Research and Methods for Secondary School Teachers* (GW5007); 11 months; $50,188

UNIVERSITY OF COLORADO
Bradley, William C.; *In-Service Institute in Earth Science for Secondary School Teachers* (GW5420); 17 months; $16,740
Mayer, William V.; *Summer Conference on Pre-Service and In-Service Preparation of Biology Teachers for Science Supervisors and College Faculty* (GW5328); 10 months; $8,785
——— ; *Summer Conference in Biological Science and Society for Secondary School Teachers* (GW5329); 10 months; $14,419
——— ; *Summer Institute in Biology for Secondary School Teachers* (GW4776); 11 months; $47,314
Wailes, James R.; *Summer Conference on National Curricular Studies for Science Supervisors and College Teacher Education Personnel* (GW5330); 10 months; $15,620

UNIVERSITY OF COLORADO,
Colorado Springs Center
Sherman, Jack E.; *In-Service Institute in Physical Science for Secondary School Teachers* (GW5524); 17 months; $7,174

UNIVERSITY OF NORTHERN COLORADO
Beel, John A.; *Summer Institute in Chemistry for Secondary School Teachers* (GW4845); 11 months; $58,790
James, M. Lynn; *In-Service Institute in Computer Science for Secondary School Teachers* (GW5482); 17 months; $9,361
Shropshire, Kenneth L.; *In-Service Institute in Earth Science for Secondary School Teachers* (GW5584); 17 months; $20,899

CONNECTICUT

FAIRFIELD UNIVERSITY
Bolger, Robert E.; *In-Service Institute in Mathematics for Secondary School Teachers* (GW5405); 17 months; $8,734
——— ; (GW5800); 26 months; $8,930

SOUTHERN CONNECTICUT STATE COLLEGE
Leskowitz, Irving; *Summer Institute in Biology for Secondary School Teachers* (GW4990); 11 months; $40,896

ST. JOSEPH COLLEGE
Markham, M. Clare; *Summer and In-Service Institute in Environmental Studies for Secondary School Teachers* (GW5057*); 11 months; $3,014
——— ; (GW5057); 11 months; $6,446
——— ; *In-Service Institute in Physical Science for Secondary School Teachers* (GW5801); 26 months; $12,300
——— ; (GW5462); 17 months; $12,310

UNIVERSITY OF CONNECTICUT
Wolfe, Martin S.; *Summer Institute in Mathematics for Secondary School Teachers* (GW4971); 11 months; $51,570

WESLEYAN UNIVERSITY
Cronin, James E.; *Academic Year Institute for Secondary School Teachers* (GW4840); 24 months; $84,824
——— ; *Summer Institute in Mathematics and Science for Secondary School Teachers* (GW4648); 11 months; $84,643
——— ; *Summer Institute for Secondary School Teachers* (GW4647); 11 months; $21,094

DELAWARE

UNIVERSITY OF DELAWARE
Brown, John A.; *In-Service Institute in Mathematics for Secondary School Teachers* (GW5875); 25 months; $6,718
——— ; (GW5523); 17 months; $6,740

Uffelman, Robert L.; *In-Service Project in Junior High School Science* (GW5711); 15 months; $24,500
Yolles, Seymour; *In-Service Institute in Chemistry for Secondary School Teachers* (GW5583); 17 months; $13,450
——— ; *Summer Institute in Chemistry for Secondary School Teachers* (GW4888); 11 months; $32,343

DISTRICT OF COLUMBIA

AMERICAN UNIVERSITY
Motz, Annabelle B.; *In-Service Institute in Sociology for Secondary School Teachers* (GW5554); 17 months; $18,053
——— ; (GW5876); 26 months; $18,179
——— ; (GW4276 001); $7,430
——— ; *Summer Conference in Sociology for Secondary School Teachers* (GW5342); 10 months; $17,252
Schubert, Leo; *In-Service Institute in Science and Mathematics for Secondary School Teachers* (GW5406); 17 months; $23,023
——— ; *Summer Institute for Secondary School Teachers* (GW4666); 11 months; $57,376
——— ; *In-Service Institute in Science and Mathematics for Secondary School Teachers* (GW5798); 26 months; $23,624

CATHOLIC UNIVERSITY OF AMERICA
Moller, Raymond W.; *Summer Institute in Mathematics for Secondary School Teachers* (GW4663); 11 months; $55,825

GEORGE WASHINGTON UNIVERSITY
Bari, Ruth A.; *In-Service Institute in Mathematics for Secondary School Teachers* (GW4277 001); 2 months; $6,040
——— ; (GW5668); 17 months; $13,531
——— ; *Summer Institute in Mathematics for Secondary School Teachers* (GW5024); 11 months; $34,033

HOWARD UNIVERSITY
Taylor, Marie C.; *Summer Institute for Secondary School Teachers* (GW4546); 11 months; $34,704

JOINT BOARD ON SCIENCE EDUCATION
Boek, Jean K.; *In-Service Conferences on Science and Mathematics Education* (GW4899); 10 months; $3,422

SURVEYS AND RESEARCH CORPORATION
Ehrman, Libert; *Contract for the Processing of Information Related to the Proposal Receipt, Evaluation, and Award Procedures of the National Science Foundation's Science Education Programs* (C433011); 15 months; $26,411

FLORIDA

FLORIDA ATLANTIC UNIVERSITY
Clark, Samuel F.; *Summer Institute in Science for Secondary School Teachers* (GW4922); 11 months; $88,649
——— ; *In-Service Institute in Physical and Environmental Sciences for Secondary School Teachers* (GW5437); 17 months; $17,244

FLORIDA INSTITUTE OF TECHNOLOGY
Potter, James G.; *Summer Institute in Physics for Secondary School Teachers* (GW4671); 11 months; $49,210

FLORIDA STATE UNIVERSITY
Kalin, Robert; *Academic Year Institute for Secondary School Teachers* (GW4835); 24 months; $111,297
——— ; *Academic Year Institute in Mathematics for Secondary School Teachers* (GW5852); 26 months; $89,702
Westmeyer, Paul; *In-Service Institute in BSCS-Special Materials for Secondary School Teachers* (GW5792); 26 months; $24,426
——— ; *Summer Institute in Harvard Project Physics for Secondary School Teachers* (GW4926); 11 months; $47,558

SCIENCE EDUCATION PROGRAMS

———; *In-Service Institute in BSCS-Special Materials for Secondary School Teachers* (GW5460); 17 months; $24,073

———; *Summer Institute in New Curriculum Developments for Science Supervisors* (GW4878); 11 months; $36,165

Wills, Herbert; *Summer Institute in Computers and Computing for Secondary School Teachers* (GW4696); 11 months; $52,938

FLORIDA TECHNOLOGICAL UNIVERSITY

Bolte, John R.; *Summer Institute in Harvard Project Physics for Secondary School Teachers* (GW4993); 11 months; $51,064

UNIVERSITY OF FLORIDA

Rappenecker, Caspar; *Summer Institute in Earth Science for Secondary School Teachers* (GW4939); 11 months; $54,309

———; *In-Service Institute in Mathematics for Secondary School Teachers* (GW5786); 26 months; $42,981

———; *In-Service Institute in Introductory Physical Science for Secondary School Teachers* (GW5777); 26 months; $15,373

———; *In-Service Institute in Earth Science and General Sciences for Secondary School Teachers* (GW5715); 26 months; $46,608

———; *In-Service Institute in Introductory Physical Science for Secondary School Teachers* (GW5513); 17 months; $11,982

———; *In-Service Institute in Earth Science for Secondary School Teachers* (GW5512); 17 months; $45,270

———; *In-Service Institute in Mathematics for Secondary School Teachers* (GW5536); 17 months; $41,759

GEORGIA

ATLANTA UNIVERSITY

Frederick, Lafayette; *Academic Year Institute for Secondary School Teachers* (GW4831); 24 months; $133,450

CLARK COLLEGE

Puri, O. P.; *Summer Institute in Physics for Secondary School Teachers* (GW4723); 11 months; $31,300

———; *In-Service Institute in Physical Science for Secondary School Teachers* (GW5457); 17 months; $5,800

EMORY UNIVERSITY

Downes, John P.; *Summer Institute in Mathematics for Secondary School Teachers* (GW4846); 11 months; $49,915

Lester, Charles T.; *In-Service Institute in Mathematics for Secondary School Teachers* (GW5667); 17 months; $8,651

GEORGIA STATE UNIVERSITY

Hassard, John R.; *In-Service Institute in Earth Science for Secondary School Teachers* (GW5666); 17 months; $13,898

MORRIS BROWN COLLEGE

Penn, James H.; *Summer Institute in Science and Mathematics for Secondary School Teachers* (GW4943); 11 months; $54,680

SHORTER COLLEGE

Greear, Philip F. C.; *Summer Conference in Natural Resource Use Education for Secondary School Teachers* (GW5336); 10 months; $6,875

UNIVERSITY OF GEORGIA

Kelly, Paul E.; *Summer Institute in Sociology for Secondary School Teachers* (GW4914); 11 months; $35,826

Koelsche, Charles L.; *Summer Institute in Science for Secondary School Teachers* (GW4724); 11 months; $37,028

———; *In-Service Institute in Biological, Physical, and Earth Sciences for Secondary School Teachers* (GW5594); 17 months; $34,801

———; (GW5811); 26 months; $34,802

Waggoner, W. H.; *Summer Institute in Mathematics and Science for Secondary School Teachers* (GW4955); 11 months; $58,745

Westfall, Jonathan J.; *Academic Year Institute in Science and Mathematics* (GW3504 001); $1,538

———; (GW4803 001); $1,538

———; *Academic Year Institute for Secondary School Teachers* (GW4803); 24 months; $187,881

HAWAII

UNIVERSITY OF HAWAII

Koehler, Dorothy I.; *Summer Institute in Mathematics for Secondary School Teachers* (GW4988); 11 months; $32,681

Lamoureux, Charles H.; *Summer Institute in Science and Mathematics for Secondary and Elementary Teachers* (GW5045); 11 months; $69,734

IDAHO

COLLEGE OF IDAHO

Henry, Boyd; *In-Service Institute in Mathematics for Secondary School Teachers* (GW5702); 17 months; $14,256

———; (GW5737); 26 months; $14,256

IDAHO STATE UNIVERSITY

Hilzman, John; *Summer Institute in Physical Science for Secondary School Teachers* (GW4962); 11 months; $46,010

UNIVERSITY OF IDAHO

Anderegg, Doyle E.; *Academic Year Institute for Secondary School Teachers* (GW4815); 24 months; $94,840

Tylutki, Edmund E.; *Summer Institute in Mathematics and Science for Secondary School Teachers* (GW4711); 11 months; $83,117

ILLINOIS

DEPAUL UNIVERSITY

Buccino, Alphonse; *In-Service Institute in Mathematics for Secondary School Teachers* (GW5813); 26 months; $13,105

———; (GW5466); 17 months; $11,921

Semrad, Joseph E.; *In-Service Institute in Biological Sciences for Secondary School Teachers* (GW5396); 17 months; $10,527

———; (GW5808); 22 months; $10,527

Stinchcomb, Thomas G.; *Summer and In-Service Institute in Physics for Secondary School Teachers* (GW5019*); 22 months; $3,533

———; (GW5019); 22 months; $8,568

———; *1969 Summer and In-Service Institute in Physics for Secondary School Teachers* (GW3651 001); $1,027

EASTERN ILLINOIS UNIVERSITY

Baker, Weldon N.; *Summer Institute in Botany, Chemistry, Physics, and Zoology for Secondary School Teachers* (GW4960); 11 months; $80,400

Waddell, Robert C.; *Summer and In-Service Institute in Physics for Secondary School Teachers* (GW4994*); 22 months; $3,070

———; (GW4994); 22 months; $10,707

GREENVILLE COLLEGE

Miller, Ralph J.; *Summer and In-Service Institute in Physics and Physical Science for Secondary School Teachers* (GW5035*); 22 months; $6,475

———; (GW5035); 22 months; $85,148

ILLINOIS INSTITUTE OF TECHNOLOGY

Reingold, Haim; *Summer Institute in Mathematics for Secondary School Teachers* (GW4590); 11 months; $108,550

———; *In-Service Institute in Mathematics, Physics, Chemistry, and Computers for Secondary School Teachers* (GW5784); 26 months; $94,500

———; (GW5522); 17 months; $90,622

Tagliacozzo, Daisy M.; *Academic Year Institute for Secondary School Teachers* (GW4786); 22 months; $144,626

Torda, Florence; *Academic Year Institute in Sociology for Secondary School Teachers* (GW5861); 25 months; $103,338

Wilcox, L. R.; *Academic Year Institute for Secondary School Teachers* (GW4784); 24 months; $102,801

Zwicker, Earl; *Summer and In-Service Institute in Physics for Secondary School Teachers* (GW4972*); 22 months; $2,886

———; (GW4972); 22 months; $12,374

ILLINOIS STATE UNIVERSITY

Poe, Douglas; *In-Service Institute in Basic Economics for Secondary School Teachers* (GW5623); 17 months; $5,083

KNOX COLLEGE

Bryan, Robert E.; *Summer Institute in Mathematics for Secondary School Teachers* (GW4624); 11 months; $50,131

Priestley, Herbert; *Summer Institute for Secondary School Teachers* (GW4547); 11 months; $71,200

LAKE FOREST COLLEGE

Coutts, John W.; *Summer Institute in Chemistry for Secondary School Teachers* (GW4753); 11 months; $42,904

Shively, Ralph L.; *Summer Institute in Mathematics for Secondary School Teachers* (GW4697); 11 months; $52,979

———; *In-Service Institute in Mathematics for Secondary School Teachers* (GW5881); 25 months; $6,364

———; *In-Service Institute in Mathematics for Secondary School Teachers* (GW5401); 17 months; $6,750

MONMOUTH COLLEGE

Johnson, A. Franklin; *In-Service Institute in Physics for Secondary School Teachers* (GW5694); 17 months; $7,079

NORTHEASTERN ILLINOIS STATE COLLEGE

Qutub, Musa Y.; *In-Service Institute in Earth Science Using ESCP Approach for Secondary School Teachers* (GW5660); 17 months; $11,360

NORTHERN ILLINOIS UNIVERSITY

Behr, Merlyn J.; *Summer Institute in Mathematics for Secondary School Teachers* (GW4766); 11 months; $61,993

Rolf, Frederick W.; *Summer Institute in Physical Science for Secondary School Teachers* (GW4670); 11 months; $64,510

NORTHWESTERN UNIVERSITY

Hildebrandt, E. H. C.; *Summer Institute for Secondary School Teachers* (GW4548); 11 months; $64,280

ROOSEVELT UNIVERSITY

Estin, Robert W.; *Summer Institute in Physics and Mathematics for Secondary School Teachers* (GW4879); 11 months; $42,857

———; *In-Service Institute in Physics, Chemistry and Mathematics for Secondary School Teachers* (GW5444); 17 months; $28,270

———; (GW5846); 26 months; $28,258

SOUTHERN ILLINOIS UNIVERSITY

Harris Jr., Stanley E.; *In-Servcie Institute in Earth Science for Secondary School Teachers* (GW5456); 17 months; $16,910

Jones, David L.; *Summer Institute for Secondary School Teachers* (GW4930); 11 months; $53,649

McDaniel, Wilbur C.; *Summer Institute in Mathematics for Secondary School Teachers* (GW4880); 11 months; $52,859

SOUTHERN ILLINOIS UNIVERSITY, Edwardsville Campus

Pendergrass, R. N.; (GW5017); 11 months; $51,775

Phillips, Paul H.; *In-Service Institute in Mathematics for Secondary School Teachers* (GW5863); 26 months; $12,587

———; (GW5546); 17 months; $12,587

Zurheide, Frederick W.; *Summer and In-*

INSTITUTES AND RESEARCH PARTICIPATION FOR SCIENCE, MATHEMATICS, AND ENGINEERING TEACHERS

Service Institute in Physics for Secondary School Teachers (GW5048*); 22 months; $6,334
——— ; (GW5048); 22 months; $13,104

UNIVERSITY OF CHICAGO
Putnam, Alfred L.; *Academic Year Institute for Secondary School Teachers* (GW4813); 24 months; $90,207

UNIVERSITY OF ILLINOIS, Chicago Circle
Hart, Alice; *In-Service Institute in Mathematics for Secondary School Teachers* (GW5790); 26 months; $18,000
——— ; (GW5455); 17 months; $18,000
Setton, Henry A.; *In-Service Institute in Preparation for the Course, 'The Man-Made World' for Secondary School Teachers* (GW5624); 17 months; $9,140

UNIVERSITY OF ILLINOIS, Urbana
Beberman, Max; *Summer Institute in Vector Geometry for Secondary School Teachers* (GW4952); 19 months; $33,860
——— ; *Summer Institute in Mathematics for Secondary School Teachers* (GW4987); 11 months; $125,000
Bitzer, Donald L.; *Summer Institute in Engineering Concepts Curriculum Project for Secondary School Teachers* (GW5054); 11 months; $37,633
Moyer, M. Eugene; *In-Service Project in Economics for Secondary School Teachers* (GW4536); 13 months; $16,709
Paden, Donald W.; *Summer Institute in Economics for Secondary School Teachers* (GW4755); 11 months; $61,050
Zaring, Wilson; *In-Service Institute in Computer Science for Secondary School Teachers* (GW5665); 19 months; $13,850
——— ; *Academic Year Institute for Secondary School Teachers* (GW4841); 22 months; $203,842
——— ; *Academic Year Institute in Mathematics for Secondary School Teachers* (GW5858); 25 months; $133,962

WESTERN ILLINOIS UNIVERSITY
Niss, James F.; *Academic Year Institute for Secondary School Teachers* (GW4796); 22 months; $123,677
Wendt, Arnold; *Summer Institute for Secondary School Teachers* (GW4598); 11 months; $44,755
Wylie, Douglas W.; *Summer and In-Service Institute in Physics for Secondary School Teachers* (GW5055); 22 months; $9,175
——— ; (GW5055*); 22 months; $2,709

INDIANA

BALL STATE UNIVERSITY
Avila, Ramon L.; *In-Service Institute in Mathematics for Secondary School Teachers* (GW5664); 17 months; $5,594
McCormick, Roy L.; *Summer Institute in Mathematics for Secondary School Teachers* (GW4672); 11 months; $42,159
Mertens, Thomas R.; *Summer Institute in Biology for Secondary School Teachers* (GW4698); 11 months; $74,008
Nisbet, Jerry J.; *Academic Year Institute for Secondary School Teachers* (GW4820); 22 months; $125,107

BUTLER UNIVERSITY
Swartz, Howard A.; *Summer Institute in Radioisotope Techniques for Secondary School Teachers* (GW5031); 11 months; $35,539

DEPAUW UNIVERSITY
Cook, Donald J.; *Summer Institute for Secondary School Teachers* (GW5027); 11 months; $49,095

EARLHAM COLLEGE
Fishback, W. T.; *Summer Institute in Mathematics for Secondary School Teachers* (GW4710); 11 months; $30,448

FRANKLIN COLLEGE
Park, Richard M.; (GW4764); 11 months; $45,090

INDIANA STATE UNIVERSITY
Poorman, Lawrence E.; *In-Service Institute in Physics for Secondary School Teachers* (GW5591); 17 months; $8,370
——— ; *In-Service Institute in Physics and Biophysics for Secondary School Teachers* (GW5832); 26 months; $8,370

INDIANA UNIVERSITY, Bloomington
Droste, John B.; *Summer Institute in Earth Science for Secondary School Teachers* (GW4777); 11 months; $64,733
McClung, L. S.; *Summer Institute in Microbiology for Secondary School Teachers* (GW4673); 11 months; $34,583
Reshkin, Mark; *Summer Institute for Secondary School Teachers* (GW4591); 11 months; $53,704
Wentworth, Rupert A. D.; *Summer Institute in Chemistry for Secondary School Teachers* (GW4869); 11 months; $33,713
Wilcox, Marie S.; *Summer Institute in Mathematics for Secondary School Teachers* (GW4725); 11 months; $49,917

INDIANA UNIVERSITY, Northwest
Mizrahi, Abshalom; *In-Servcie Institute in Mathematics for Secondary School Teachers* (GW5553); 17 months; $7,652

NATIONAL ASSOCIATION OF BIOLOGY TEACHERS
Koffler, Henry; *In-Service Institute in NABT Regional Seminars in Biology for Secondary School Teachers* (GW5701); 17 months; $30,564

PURDUE UNIVERSITY
Herron, J. D.; *Summer Institute in Chemistry for Secondary School Teachers* (GW4881); 11 months; $69,477
Keller, M. Wiles; *Summer Institute for Secondary School Teachers* (GW4551); 11 months; $71,716
——— ; *In-Service Institute in Mathematics for Secondary School Teachers* (GW5827); 26 months; $90,154
——— ; (GW5424); 17 months; $90,162
——— ; (GW5423); 7 months; $22,008
Lefler, Ralph W.; *Summer Institute for Secondary School Teachers* (GW4549); 11 months; $84,695
Olson, J. Bennet; (GW4550); 11 months; $68,136

UNIVERSITY OF NOTRE DAME
Anthony, Robert L.; (GW4552); 11 months; $51,724
Goetz, Abraham; *Academic Year Institute for Secondray School Teachers* (GW4802); 22 months; $168,179
——— ; *Summer Institute for Secondary School Teachers* (GW4902); 11 months; $100,516
Hofman, Emil T.; (GW4912); 11 months; $85,124
Murphy, Michael J.; *Summer Institute in Earth Sciences for Secondary School Teachers* (GW4632); 11 months; $56,029
——— ; *In-Service Institute in Earth Science for Secondary School Teachers* (GW5759); 26 months; $10,296
Murphy, Rev. Michael J.; (GW5446); 17 months; $10,294

VALPARAISO UNIVERSITY
Hallerberg, Arthur E.; *Summer Institute for Secondary School Teachers* (GW4554); 11 months; $44,142
Krekeler, Carl H.; *Summer Institute in Biology for Secondary School Teachers* (GW4768); 11 months; $40,540

IOWA

CLARKE COLLEGE
Keller, Sr. Mary K.; *Summer Institute for Secondary School Teachers* (GW4674); 11 months; $29,389

DRAKE UNIVERSITY
Canfield, Earle L.; *In-Service Institute in Science and Mathematics for Secondary School Teachers* (GW5395); 17 months; $7,805
——— ; *In-Service Institute in Biology for Secondary School Teachers* (GW5765); 26 months; $7,805
Gillam, B. E.; *Summer Institute in Mathematics for Secondary School Teachers* (GW4740); 11 months; $63,361
Rogers, Rodney A.; *Summer Institute for Secondary School Teachers* (GW4553); 11 months; $64,999

IOWA STATE UNIVERSITY
Hotchkiss, Donald.; *Summer Institute in Probability, Statistics and Computer Science for Secondary School Teachers* (GW4778); 11 months; $39,714
Hussey, Keith M.; *Summer Institute in Earth Science for Secondary School Teachers* (GW4870); 11 months; $64,478

LUTHER COLLEGE
Pilgrim, Donald H.; *Summer Institute in Mathematics for Secondary School Teachers* (GW4779); 11 months; $30,546

UNIVERSITY OF IOWA
Wilmeth, J. Richard; *Summer Institute in Sociology for Secondary School Teachers* (GW4931); 11 months; $43,965
Yager, Robert E.; *Academic Year Institute for Secondary School Teachers* (GW4781); 24 months; $66,085
——— ; *Summer Institute in Science for Secondary School Teachers* (GW4758); 11 months; $60,249
——— ; *Summer Institute in Molecular Biology for Secondary School Teachers* (GW4950); 20 months; $68,115
——— ; *In-Service Institute in Earth Science for Secondary School Teachers* (GW5754); 26 months; $11,806
——— ; *Academic Year Institute in Mathematics and Science for Secondary School Teachers* (GW5849); 26 months; $55,715
——— ; *In-Service Institute in Basic Sciences for Secondary School Teachers* (GW5535); 17 months; $13,299
——— ; *In-Service Institute in Earth Science for Secondary School Teachers* (GW5506); 17 months; $11,806
——— ; *In-Service Institute in Basic Sciences for Secondary School Teachers* (GW5741); 26 months; $13,299
Zweng, Marilyn J.; *Summer Conference in Computer Science for Secondary School Teachers* (GW5324); 10 months; $13,996

UNIVERSITY OF NORTHERN IOWA
Anderson, Wayne I.; *In-Service Institute in Earth Science for Secondary School Teachers* (GW5592); 17 months; $9,652
Hanson, Robert W.; *Academic Year Institute in Physical Science and Earth Science for Secondary School Teachers* (GW4791); 24 months; $199,390
Tepaske, E. Russell; *Summer Institute in Life Science for Secondary School Teachers* (GW4726); 11 months; $49,210

KANSAS

FORT HAYS KANSAS STATE COLLEGE
Toalson, W.; *Summer Institute in Mathematics for Secondary School Teachers* (GW4633); 11 months; $62,742

KANSAS STATE COLLEGE OF PITTSBURG
Smith, R. G.; *Summer Institute in Biology, Chemistry and Mathematics for Secondary School Teachers* (GW4599); 11 months; $87,000
Thomas, Harold L.; *In-Service Institute in Mathematics for Secondary School Teachers* (GW5582); 17 months; $6,181

KANSAS STATE TEACHERS COLLEGE
Bridge, Thomas E.; *In-Service Institute in Earth and Environmental Sciences for Sec-*

SCIENCE EDUCATION PROGRAMS

———; *In-Service Institute in Earth and Environmental Science for Secondary School Teachers* (GW5419); 17 months; $10,557

Cram, S. Winston; *Summer Institute in Physical Science for Secondary School Teachers* (GW4592); 11 months; $65,473

Durst, Harold; *In-Service Institute in Biology, Mathematics, and Physical Science for Secondary School Teachers* (GW5454); 17 months; $19,022

———; *In-Service Institute in Science and Mathematics for Secondary School Teachers* (GW5727); 26 months; $18,133

———; *Summer Institute in Environmental Biology and Human Ecology for Secondary School Teachers* (GW4903); 11 months; $54,102

Spencer, Dwight L.; *Summer Institute in Biology, Mathematics, and Earth Science for Secondary School Teachers* (GW4649); 11 months; $22,216

KANSAS STATE UNIVERSITY

Chelikowsky, J. R.; *Summer Institute in Earth Science for Secondary School Teachers* (GW4956); 11 months; $55,450

Dixon, Lyle J.; *In-Service Institute in Mathematics for Secondary School Teachers* (GW5593); 17 months; $21,732

Fuller, Leonard E.; *Summer Institute in Mathematics for Secondary School Teachers* (GW4727); 11 months; $64,000

Moore, Arnold J.; *Summer Institute in Physics for Secondary School Teachers* (GW4650); 11 months; $56,690

Walters, Charles P.; *In-Service Institute in Earth Science for Secondary School Teachers* (GW5697); 17 months; $8,636

NATIONAL ASSOCIATION OF BIOLOGY TEACHERS

Koffler, Henry; *In-Service Conference in Biology for Secondary School Teachers* (GW4537); 13 months; $32,347

WASHBURN UNIVERSITY OF TOPEKA

Greene, Laura Z.; *Summer Institute in Mathematics for Secondary School Teachers* (GW4667); 11 months; $61,287

KENTUCKY

GEORGETOWN COLLEGE

Dobyns, Roy A.; *In-Service Institute in Mathematics for Secondary School Teachers* (GW5481); 17 months; $5,680

MURRAY STATE UNIVERSITY

Blackburn, Walter E.; *Summer Institute for Secondary School Teachers* (GW4555); 11 months; $51,776

———; *Summer Institute in Natural Sciences for Secondary School Teachers* (GW4630); 11 months; $49,430

Robertson, Harold G.; *Summer Institute in Mathematics for Secondary School Teachers* (GW5058); 11 months; $35,097

UNIVERSITY OF LOUISVILLE

Clay, William M.; *Summer Institute in Science of Inland Waters for Secondary School Teachers* (GW4871); 11 months; $44,448

LOUISIANA

DILLARD UNIVERSITY

Bryan, Clifford R.; *Summer Institute in Biology for Secondary School Teachers* (GW4728); 11 months; $47,170

GRAMBLING COLLEGE

Douglas, Samuel H.; *Summer Institute in Mathematics for Secondary School Teachers* (GW4765); 11 months; $44,187

———; *In-Service Institute in Mathematics for Secondary School Teachers* (GW5521); 17 months; $10,237

———; (GW5877); 26 months; $10,230

LOUISIANA STATE UNIVERSITY, Baton Rouge

Bertrand, Alvin L.; *Summer Institute in Sociology for Secondary School Teachers* (GW4967); 11 months; $36,152

Gosselink, James G.; *Summer Institute for Secondary School Teachers* (GW4556); 11 months; $94,395

Karnes, Houston T.; *Academic Year Institute for Secondary School Teachers Pre-Service Institute for Secondary School Teachers* (GW4805); 24 months; $191,333

———; *Academic Year Institute in Mathematics for Secondary School Teachers* (GW5857); 25 months; $92,452

Mitchell, Benjamin E.; *Summer Institute for Secondary School Teachers* (GW4557); 11 months; $65,969

Roberts, J. Harvey; *Summer Institute in Life Sciences for Secondary School Teachers* (GW4745); 11 months; $54,748

LOUISIANA POLYTECHNIC INSTITUTE

Davis, Billy J.; *Summer Institute in Environmental Biology for Secondary School Teachers* (GW4937); 11 months; $42,321

LOYOLA UNIVERSITY

Benedetto, F. A.; *In-Service Institute for Secondary School Teachers* (GW4315 001); $2,927

———; *In-Service Institute in Introductory Physical Science (Televised) for Secondary School Teachers* (GW5533); 17 months; $22,843

———; *In-Service Institute in Physics for Secondary School Teachers* (GW5534); 17 months; $11,453

———; (GW5734); 26 months; $11,453

DiMaggio III, Anthony; *In-Service Institute in Chemistry for Secondary School Teachers* (GW5410); 17 months; $11,261

———; (GW5733); 26 months; $11,576

———; *Summer Institute in Physical Science for Secondary School Teachers* (GW5018); 11 months; $44,727

Todd, Lewis; *In-Service Institute in Mathematics for Secondary School Teachers* (GW5511); 17 months; $18,873

———; (GW5778); 26 months; $18,873

MCNEESE STATE COLLEGE

Spencer, S. M.; *In-Service Institute in Biology and Mathematics for Secondary School Teachers* (GW5773); 26 months; $15,080

———; (GW5461); 17 months; $15,350

NORTHEAST LOUISIANA STATE COLLEGE

Byrd, David S.; *In-Service Institute in Chemistry for Secondary School Teachers* (GW5435); 17 months; $6,208

Harmon, Frank L.; *In-Service Institute in Mathematics for Secondary School Teachers* (GW5590); 17 months; $7,606

———; *Summer Institute in Mathematics for Secondary School Teachers* (GW5043); 11 months; $38,934

NORTHWESTERN STATE COLLEGE OF LOUISIANA

Whittington Jr, Russell; *In-Service Institute in Mathematics for Secondary School Teachers* (GW5404); 17 months; $11,531

SOUTHEASTERN LOUISIANA COLLEGE

Andrews, Newton S.; (GW5415); 17 months; $8,998

———; (GW5878); 26 months; $8,990

SOUTHERN UNIVERSITY

Bowen, Princess M.; *In-Service Institute in Geography for Secondary School Teachers* (GW5589); 17 months; $7,903

Roddy, Leon; *Summer Institute for Secondary School Teachers* (GW4558); 11 months; $60,878

Ruffin, Spaulding M.; *Summer Institute in Chemistry for Secondary School Teachers* (GW4712); 11 months; $27,425

White, Vandon E.; *In-Service Institute in Earth Science for Secondary School Teachers* (GW5693); 17 months; $8,933

TULANE UNIVERSITY

Young, Gail S.; *Summer Institute in Mathematics for Secondary School Teachers* (GW4882); 11 months; $53,778

MAINE

BOWDOIN COLLEGE

Chittim, Richard L.; (GW4685); 11 months; $58,419

Grobe, Charles A.; *Academic Year Institute for Secondary School Teachers* (GW4865); 22 months; $81,597

Gustafson, Alton H.; *Summer Institute for Secondary School Teachers* (GW4581); 11 months; $39,648

Sheats, John E.; (GW4625); 11 months; $36,765

COLBY COLLEGE

Reid, Evans B.; *Summer Institute in Science and Mathematics for Secondary School Teachers* (GW4699); 11 months; $84,435

UNIVERSITY OF MAINE, Orono

Mairhuber, John C.; *In-Service Institute in Mathematics for Secondary School Teachers* (GW5835); 26 months; $10,650

———; *Summer and In-Service Institute in Mathematics for Secondary School Teachers* (GW5044*); 22 months; $10,651

———; (GW5044); 22 months; $45,142

UNIVERSITY OF MAINE, Portland

Rogers, Paul C.; *In-Service Institute in Mathematics for Secondary School Teachers* (GW5520); 17 months; $8,038

MARYLAND

MORGAN STATE COLLEGE

Fraser, Thomas P.; *Summer and In-Service Institute in Science and Mathematics for Secondary School Teachers* (GW4688*); 21 months; $30,627

———; (GW4688); 21 months; $70,363

Jerkins, Kenneth F.; *In-Service Institute in Science for Secondary School Teachers* (GW5803); 26 months; $31,105

Ritter, Fredric A.; *In-Service Institute in Geography for Secondary School Teachers* (GW5663); 17 months; $7,897

Talbot, Walter R.; *Academic Year Institute for Secondary School Teachers* (GW4837); 24 months; $113,870

ST. JOSEPH COLLEGE

Scully, Marcella; *Summer Institute in Biology for Secondary School Teachers* (GW4872); 20 months; $32,325

UNIVERSITY OF MARYLAND

DeSilva, Alan W.; *In-Service Institute in Physics for Secondary School Teachers* (GW5622); 17 months; $9,690

Dilavore, Philip; *In-Service Institute in Harvard Project Physics for Secondary School Teachers* (GW5644); 17 months; $10,941

Fey, James; *Summer Institute in Mathematics for Secondary School Teachers* (GW4913); 11 months; $41,346

Gardner, Marjorie H.; *In-Service Institute in Chemistry for Secondary School Teachers* (GW5545); 17 months; $11,491

Good, Richard A.; *In-Service Institute in Mathematics for Secondary School Teachers* (GW5662); 17 months; $14,020

———; (GW5755); 26 months; $14,016

Lockard, J. David; *Academic Year Institute for Science Supervisors* (GW5855); 26 months; $99,645

———; *Summer Institute in Biology for Secondary School Teachers* (GW4945); 11 months; $61,580

———; *Academic Year Institute for Secondary School Teachers* (GW4811); 22 months; $95,740

Maccini, John A.; *In-Service Institute in Earth Science for Secondary School Teachers* (GW5802); 26 months; $18,702

———; *In-Service Institute in Astronomy and Oceanography for Secondary School Teachers* (GW5795); 26 months; $9,888

INSTITUTES AND RESEARCH PARTICIPATION FOR SCIENCE, MATHEMATICS, AND ENGINEERING TEACHERS

———; (GW5510); 17 months; $9,888
———; *In-Service Institute in Earth Science for Secondary School Teachers* (GW5588); 17 months; $18,415
———; *In-Service Institute in Meteorology/Ecology for Secondary School Teachers* (GW5495); 17 months; $9,741

UNIVERSITY OF MARYLAND, Baltimore County Campus
Horelick, Brindell; *Summer Institute in Mathematics for Secondary School Teachers* (GW5015); 11 months; $41,485

WESTERN MARYLAND COLLEGE
Sturdivant, Harwell P.; *Summer Institute for Secondary School Teachers* (GW4559); 11 months; $42,181

MASSACHUSETTS

AMERICAN INTERNATIONAL COLLEGE
Benjamin, Henry A.; *In-Service Institute in Chemistry for Secondary School Teachers* (GW5652); 17 months; $5,665
———; *In-Service Institute in Physical Science for Secondary School Teachers* (GW5775); 26 months; $5,700

ASSUMPTION COLLEGE
Doyle, George A.; *Summer Institute in Economics for Secondary School Teachers* (GW4944); 11 months; $38,013

BOSTON COLLEGE
Bezuszka, Rev. Stanley J.; *In-Service Institute in Mathematics for Secondary School Teachers* (GW5425); 17 months; $16,248
———; *Academic Year Institute for Secondary School Teachers* (GW4833); 24 months; $151,840
———; *In-Service Institute in Mathematics for Secondary School Teachers* (GW5739); 26 months; $16,248
———; *Summer Institute for Secondary School Teachers* (GW4582); 11 months; $55,323
Fimian Jr., Walter J.; *Summer Institute in Radiation Biology for Secondary School Teachers* (GW4954); 11 months; $24,656
Skehan, James W.; *Summer Institute in Earth Science for Secondary School Teachers* (GW4729); 11 months; $35,352
———; *In-Service Institute in Earth Science for Secondary School Teachers* (GW5491); 17 months; $13,273

BOSTON UNIVERSITY
Fitzgerald, J. Franklin; *In-Service Institute in Computer Mathematics for Secondary School Teachers* (GW5653); 17 months; $10,709
———; *In-Service Institute in Mathematics for Secondary School Teachers* (GW5509); 13 months; $3,763
Gheith, Mohamed A.; *Summer Institute for Secondary School Teachers* (GW4583); 11 months; $43,425
Hein, George E.; *In-Service Institute in Earth and General Science for Secondary School Teachers* (GW5544); 17 months; $12,432

COLLEGE OF THE HOLY CROSS
Dewey, Daniel G.; *In-Service Institute in Mathematics for Secondary School Teachers* (GW5561); 17 months; $5,265
———; *Summer Institute in Mathematics for Secondary School Teachers* (GW4759); 11 months; $49,895
———; *In-Service Institute in Mathematics for Secondary School Teachers* (GW5838); 26 months; $5,265
Flavin, John W.; *In-Service Institute in Biology for Secondary School Teachers* (GW5872); 25 months; $6,213
———; (GW5417); 17 months; $6,283
Gunter, Roy C.; *Summer Institute in Physics for Secondary School Teachers* (GW4651); 11 months; $45,330
MacDonnell, Robert B.; *In-Service Institute in Earth Science for Secondary School Teachers* (GW5451); 17 months; $6,875
———; *Summer Institute in Biology, Chemistry and Physics for Secondary School Teachers* (GW4847); 11 months; $59,739
———; *In-Service Institute in Physical Science for Secondary School Teachers* (GW5748); 26 months; $6,875

EASTERN NAZARENE COLLEGE
Lashley, Gerald E.; *In-Service Institute in Science and Mathematics for Secondary School Teachers* (GW5746); 16 months; $18,453
———; *In-Service Institute in Mathematics for Secondary School Teachers* (GW5487); 17 months; $18,358

HARVARD UNIVERSITY
Holton, Gerald; *Summer Institute in Harvard Project Physics for Secondary School Teachers* (GW4770); 11 months; $49,404
Watson, Fletcher G.; *Academic Year Institute for Secondary School Teachers* (GW4785); 22 months; $72,238
———; *Academic Year Institute in Physical Sciences for Secondary School Teachers* (GW5859); 25 months; $41,723

SIMMONS COLLEGE
Brauner, Phyllis A.; *Summer Institute in Chemistry for Secondary School Teachers* (GW4908); 11 months; $25,400
Desua, Frank C.; *Summer Institute in Mathematics for Secondary School Teachers* (GW4848); 11 months; $26,888

STATE COLLEGE AT BOSTON
Fairbanks III, George A.; *In-Service Institute in Physics and Physical Science for Secondary School Teachers* (GW5603); 17 months; $10,146

STATE COLLEGE AT BRIDGEWATER
Boutilier, Robert F.; *Summer Institute in Field Work in Earth Science for Secondary School Teachers* (GW4749); 11 months $23,605
Chiccarelli, Joseph B.; *In-Service Institute in Biology, Chemistry, Earth Science, Mathematics and Physics for Secondary School Teachers* (GW5730); 26 months; $33,725
———; (GW5692); 17 months; $33,762
Maier, Emanuel; *Summer Institute in Geography for Secondary School Teachers* (GW4980); 11 months; $34,547

STATE COLLEGE AT FRAMINGHAM
Murray, John T.; *In-Service Institute in Descriptive Astronomy for Secondary School Teachers* (GW5480); 17 months; $6,906

WORCESTER POLYTECHNIC INSTITUTE
Morton, Richard F.; *Summer and In-Service Institute in Science and Mathematics for Secondary School Teachers* (GW4979); 21 months; $94,469
———; (GW4979*); 21 months; $13,658
———; *In-Service Institute in Physics and Biology for Secondary School Teachers* (GW5745); 26 months; $13,614

MICHIGAN

ANDREWS UNIVERSITY
Jones, Harold T.; *Summer Institute for Secondary School Teachers* (GW4686); 11 months; $32,410
Jorgensen, Roy A.; *In-Service Institute in Mathematics for Secondary School Teachers* (GW5581); 17 months; $5,678

CALVIN COLLEGE
Sinke, Carl J.; *Summer Institute in Mathematics and Biology for Secondary School Teachers* (GW5056); 11 months; $33,115

CENTRAL MICHIGAN UNIVERSITY
Serier, Lester H.; *Summer Institute in Mathematics for Secondary School Teachers* (GW4947); 11 months; $38,841

HOPE COLLEGE
Folkert, Jay E.; (GW4639); 11 months; $50,554
Jekel, Eugene C.; *Summer Institute in Chemistry for Secondary School Teachers* (GW4657); 11 months; $60,170

MICHIGAN TECHNOLOGICAL UNIVERSITY
Brown, Robert T.; *Summer Institute in General Ecology Principles and Techniques for Secondary School Teachers* (GW4652); 11 months; $39,597
Gale, Calvin W.; *In-Service Institute in Physics, Philosophy of Science and Political Science for Secondary School Teachers* (GW5651); 17 months; $6,607
Wyble, D. O.; *Summer Institute in Earth Science for Secondary School Teachers* (GW4662); 11 months; $55,903

MICHIGAN STATE UNIVERSITY
Chamberlain, Von Del; *Summer Institute in Planetarium Instruction for Secondary School Teachers* (GW4849); 11 months; $32,350
Peabody, Frank R.; *Summer Institute in Biology for Secondary School Teachers* (GW4953); 11 months; $109,419
Smith, Jane E.; *Summer Institute in Earth Science for Secondary School Teachers* (GW5042); 11 months; $43,440
Stonehouse, Harold; *In-Service Institute in Earth Science for Secondary School Teachers* (GW5601); 17 months; $14,781
Wagner, John; *Summer Institute in Mathematics for Secondary School Teachers* (GW4705); 11 months; $39,714
———; *Summer Conference in Mathematics for Leaders of In-Service Programs in Elementary School Mathematics* (GW5337); 10 months; $14,874
Woodby, Lauren G.; *In-Service Institute in Mathematics for Secondary School Teachers* (GW5602); 17 months; $10,147
———; *Summer Institute in Mathematics for Secondary School Teachers* (GW4949); 11 months; $31,397
———; *In-Service Institute in Mathematics for Secondary School Teachers* (GW5794); 26 months; $10,147

UNIVERSITY OF DETROIT
Mehlenbacher, Lyle E.; (GW5785); 26 months; $12,635
———; *Academic Year Institute for Secondary School Teachers* (GW4782); 22 months; $102,027
———; *In-Service Institute in Mathematics for Secondary School Teachers* (GW5403); 17 months; $7,908
———; *Summer and In-Service Institute in Mathematics for Secondary School Teachers* (GW4643*); 21 months; $12,635
———; (GW4643); 21 months; $44,258

UNIVERSITY OF MICHIGAN
Brumfiel, Charles; *In-Service Project in Mathematics for Junior High School Teachers* (GW5867); 13 months; $9,390
Krause, Eugene F.; *Academic Year Institute for Secondary School Teachers* (GW4866); 24 months; $80,065

WAYNE STATE UNIVERSITY
Hooper, Henry O.; *In-Service Institute in Physics for Secondary School Teachers* (GW5814); 26 months; $10,704
———; (GW5393); 17 months; $10,704
———; *In-Service Institute in Physical Science for Secondary School Teachers* (GW5479); 7 months; $7,392
Mayeda, K.; *In-Service Institute in Biology for Secondary School Teachers* (GW5742); 26 months; $10,260
Rossmoore, Harold W.; (GW5490); 17 months; $10,260
Slaby, Harold T.; *In-Service Institute in Mathematics for Secondary School Teachers* (GW5815); 26 months; $9,897
———; *Summer and In-Service Institute in Mathematics for Secondary School Teachers* (GW4989*); 22 months; $9,897

SCIENCE EDUCATION PROGRAMS

———; (GW4989); 22 months; $64,144

Stevens, Calvin L.; *Academic Year Institute for Secondary School Teachers* (GW4799); 24 months; $73,865

WESTERN MICHIGAN UNIVERSITY

Brawer, Milton J.; *Summer Institute in Sociology for Secondary School Teachers* (GW4957); 11 months; $49,013

Clarke, A. Bruce; *Academic Year Institute for Secondary School Teachers* (GW4823); 22 months; $97,614

———; *Summer Institute in Mathematics for Secondary School Teachers* (GW4668); 11 months; $68,864

Fisk, Franklin G.; *Summer Institute in History and Philosophy of Science for Secondary School Teachers* (GW4850); 11 months; $44,129

Goodnight, Clarence J.; *Summer Institute in Biology for Secondary School Teachers* (GW4959); 11 months; $53,715

———; *In-Service Institute in Biology for Secondary School Teachers* (GW5771); 26 months; $13,503

———; (GW5439); 17 months; $13,503

Helburn, Nicholas; *Summer Institute for Secondary School Teachers* (GW5004); 11 months; $52,426

Mallinson, George G.; *In-Service Institute in Science and Mathematics for Secondary School Teachers* (GW5445); 17 months; $16,601

———; *Summer Institute in Physical Sciences and Related Mathematics for Secondary School Teachers* (GW4593); 11 months; $56,538

———; *In-Service Institute in Science and Mathematics for Secondary School Teachers* (GW5738); 26 months; $16,601

MINNESOTA

BEMIDJI STATE COLLEGE

Smith, Gerald J.; *In-Service Institute in Mathematics, Biology, Chemistry, and Physical Science for Secondary School Teachers* (GW5736); 26 months; $19,681

———; *In-Service Institute in Mathematics, Biology, Chemistry, and Earth Science for Secondary School Teachers* (GW5691); 17 months; $19,681

Winter, Wesley W.; *Summer Institute in Harvard Project Physics for Secondary School Teachers* (GW4713); 11 months; $46,115

CARLETON COLLEGE

Mathews, Robert T.; *Summer Institute for Secondary School Teachers* (GW4560); 11 months; $49,402

CENTRAL MINNESOTA EDUCATIONAL R&D COUNCIL

Laselle, Clarence A.; *In-Service Institute in FORTRAN Computer Language for Secondary School Teachers* (GW5695); 17 months; $6,890

COLLEGE OF ST. TERESA

Clarus, Mary; *Summer Conference in Transition Metal Chemistry for Secondary School Teachers* (GW5333); 10 months; $11,885

COLLEGE OF ST. THOMAS

Allen, Martin; *Summer Institute for Secondary School Teachers* (GW4675); 11 months; $53,707

MACALESTER COLLEGE

Hastings, Russell B.; *Summer Institute in General Science for Secondary School Teachers* (GW5032); 11 months; $50,839

MANKATO STATE COLLEGE

Henderson, Donald L.; *In-Service Institute in Computer Curriculum Development Project for Secondary School Teachers* (GW5560); 17 months; $3,611

Kuhn, David J.; *In-Service Institute in Biological Science for Secondary School Teachers* (GW5559); 17 months; $8,480

———; *Summer Institute in Biological Science for Secondary School Teachers* (GW5021); 11 months; $33,785

ST. CLOUD STATE COLLEGE

Hopkins, Harold; *In-Service Institute in Biology for Secondary School Teachers* (GW5621); 17 months; $6,657

———; (GW5804); 26 months; $6,657

———; *Summer Institute in Biology, Chemistry, and Physics for Secondary School Teachers* (GW4976); 11 months; $62,809

Stennes, Florence; *Summer Conference in Computer Programming for Teachers of Secondary School Mathematics* (GW5332); 10 months; $12,760

ST. MARY'S COLLEGE

Severin, Charles; *Summer Institute in Biology for Secondary School Teachers* (GW4897); 11 months; $80,000

UNIVERSITY OF MINNESOTA

Hobbie, Russell K.; *Summer Institute in Physics for Secondary School Teachers* (GW4973); 11 months; $74,844

Lewis, Darrell R.; *In-Service Institute in Intermediate Economics and Economic Education for Secondary School Teachers* (GW5501); 17 months; $8,724

———; (GW5834); 26 months; $8,685

Phinney, William C.; *In-Service Institute in Environmental Studies for Secondary School Teachers* (GW5650); 17 months; $16,093

WINONA STATE COLLEGE

Bayer, Thomas N.; *Summer Institute in the Earth Sciences for Secondary School Teachers* (GW4905); 11 months; $38,394

MISSISSIPPI

ALCORN AGRICULTURAL & MECHANICAL COLLEGE

Parker, Henry L.; *In-Service Institute in Biology and Mathematics for Secondary School Teachers* (GW5580); 17 months; $15,097

———; *Summer Institute in Biology and Chemistry for Secondary School Teachers* (GW4700); 11 months; $78,796

JACKSON STATE COLLEGE

Uzodinma, John E.; *Summer Institute in Biology for Secondary School Teachers* (GW4948); 11 months; $47,760

MISSISSIPPI STATE UNIVERSITY

Sheely, Clyde Q.; *Summer Institute in Science and Mathematics for Secondary School Teachers* (GW4873); 11 months; $89,811

RUST COLLEGE

Cook, John D.; *In-Service Institute in the Adaptation of Innovative Social Science Curriculum Materials for Use by Secondary School Teachers* (GW5600); 17 months; $10,358

UNIVERSITY OF MISSISSIPPI

Hein, Harold C.; *In-Service Institute in Earth Science for Secondary School Teachers* (GW5467); 17 months; $8,788

Longest, William D.; *Summer Institute in Science and Mathematics for Secondary School Teachers* (GW4730); 11 months; $103,530

Norman, William H.; *Academic Year Institute for Secondary School Teachers* (GW4819); 24 months; $192,731

UNIVERSITY OF SOUTHERN MISSISSIPPI

Sonnier, Isadore L.; *In-Service Institute in Intermediate School Science Programs for Secondary School Teachers* (GW5551); 17 months; $15,373

———; *In-Service Institute in Earth Science for Secondary School Teachers* (GW5870); 25 months; $15,373

Webster, Porter G.; *Summer Institute in Mathematics for Secondary School Teachers* (GW4874); 11 months; $56,005

MISSOURI

CENTRAL MISSOURI STATE COLLEGE

Emerson, John W.; *In-Service Institute in ESCP Approach to Earth Science for Secondary School Teachers* (GW5598); 17 months; $5,957

DRURY COLLEGE

Johnson, Elbert B.; *In-Service Institute in Basic Economics for Secondary School Teachers* (GW5647); 17 months; $4,501

NORTHEAST MISSOURI STATE COLLEGE

Rosebery, Dean A.; *In-Service Institute in ESCP Approach to Earth Science for Secondary School Teachers* (GW5577); 17 months; $6,072

———; *Summer Institute for Secondary School Teachers* (GW4594); 11 months; $43,672

NORTHWEST MISSOURI STATE COLLEGE

Mallory, Bob F.; *In-Service Institute in ESCP Approach to Earth Science for Secondary School Teachers* (GW5604); 17 months; $5,957

ROCKHURST COLLEGE

Doyle, William C.; *Summer Institute for Secondary School Teachers* (GW4608); 11 months; $31,367

SOUTHEAST MISSOURI STATE COLLEGE

Knox, B. Ray; *In-Service Institute in ESCP Approach to Earth Science for Secondary School Teachers* (GW5659); 17 months; $5,957

SOUTHWEST MISSOURI STATE COLLEGE

Kurtz, Vincent E.; (GW5619); 17 months; $5,957

ST. LOUIS UNIVERSITY

Andrews, John J.; *In-Service Institute in Mathematics for Secondary School Teachers* (GW5828); 26 months; $11,270

———; (GW5452); 17 months; $11,270

Brill Jr., Kenneth G.; *In-Service Institute in ESCP Approach to Earth Science for Secondary School Teachers* (GW5596); 17 months; $5,986

Regan, Francis; *Summer Institute in Mathematics for Secondary School Teachers* (GW4635); 11 months; $60,513

UNIVERSITY OF MISSOURI, Columbia

Guyon, John C.; *Summer Institute in Chemistry and Physics for Secondary School Teachers* (GW4731); 11 months; $49,138

Johnson, Clayton H.; *In-Service Institute in ESCP Approach to Earth Science for Secondary School Teachers* (GW5543); 17 months; $39,670

Kuhlman, John M.; *Academic Year Institute for Secondary School Teachers* (GW4824); 24 months; $137,268

———; *Academic Year Institute in Advanced Economics for Secondary School Teachers* (GW5854); 26 months; $118,341

Leake, John B.; *In-Service Project in Physics for Secondary School Teachers* (GW4540); 25 months; $16,203

O'Connor, William C.; *In-Service Institute in Basic Economics for Secondary School Teachers* (GW5620); 17 months; $20,628

UNIVERSITY OF MISSOURI, Kansas City

Anderson, Robert G.; *In-Service Institute in Biology and Mathematics for Secondary School Teachers* (GW5578); 17 months; $23,966

———; (GW5874); 25 months; $23,966

Darby, Orville L.; *In-Service Institute in Basic Economics for Secondary School Teachers* (GW5649); 17 months; $4,372

Parizek, Eldon J.; *In-Service Institute in ESCP Approach to Earth Science for Secondary School Teachers* (GW5597); 17 months; $6,286

UNIVERSITY OF MISSOURI, Rolla

Beveridge, Thomas; *In-Service Institute in ESCP Approach to Earth Science for Sec-*

INSTITUTES AND RESEARCH PARTICIPATION FOR SCIENCE, MATHEMATICS, AND ENGINEERING TEACHERS

ondary School Teachers (GW5579); 17 months; $6,286

Fuller, Harold Q.; *Summer Institute in Science and Mathematics for Secondary School Teachers* (GW4626); 11 months; $98,083

UNIVERSITY OF MISSOURI, St. Louis

McKenna, Joseph P.; *In-Service Institute in Basic Economics for Secondary School Teachers* (GW5648); 17 months; $5,258

Rigden, John S.; *In-Service Institute in the Physical Sciences for Secondary School Teachers* (GW5690); 17 months; $11,414

———; *In-Service Project in Science for Secondary School Teachers* (GW4534); 12 months; $11,027

WASHINGTON UNIVERSITY

Gouldner, Helen P.; *In-Service Institute in Sociology for Secondary School Teachers* (GW5599); 17 months; $7,287

MONTANA

EASTERN MONTANA COLLEGE

Peterson, Oliver; *In-Service Institute in Abstract Algebra and Probability—Statistics for Secondary School Teachers* (GW5400); 17 months; $6,365

MONTANA STATE UNIVERSITY

Amend, John R.; *Summer Institute in Nuclear and Instrumental Chemistry for Secondary School Teachers* (GW4911); 11 months; $31,423

UNIVERSITY OF MONTANA

Fields, Robert W.; *Summer Conference in Field Geology for Secondary School Teachers* (GW5326); 10 months; $18,860

Myers, William M.; *Academic Year Institute for Secondary School Teachers* (GW4794); 24 months; $91,384

———; *Summer Institute for Secondary School Teachers* (GW4595); 11 months; $85,497

Preece, Sherman J. Jr.; *Summer Institute in Biological Science for Secondary School Teachers* (GW4883); 11 months; $50,528

Solberg, Richard A.; *Summer Institute in Field Botany for Secondary School Teachers* (GW4687); 11 months; $21,180

NEBRASKA

CHADRON STATE COLLEGE

Hughes, Eugene M.; *In-Service Institute in Science for Secondary School Teachers* (GW5542); 17 months; $24,428

———; *In-Service Institute in Physical Science for Secondary School Teachers* (GW5766); 28 months; $24,428

NEBRASKA WESLEYAN UNIVERSITY

French, Walter R.; *Summer Institute in Introductory Physical Science for Secondary School Teachers* (GW4734); 11 months; $64,876

Moore, Carroll; *Summer Institute in Physical Science for Seconday School Teachers* (GW4748); 11 months; $18,162

UNIVERSITY OF NEBRASKA, Lincoln

Davidson, John F.; *Summer Institute in Botany for Secondary School Teachers* (GW4851); 11 months; $39,709

Mientka, Walter E.; *Summer Institute in Mathematics for Secondary School Teachers* (GW4714); 11 months; $59,534

UNIVERSITY OF NEBRASKA, Omaha

Brooks, Merle E.; *In-Service Institute in Science and Mathematics for Secondary School Teachers* (GW5817); 26 months; $21,424

———; (GW5465); 17 months; $21,424

NEVADA

UNIVERSITY OF NEVADA, Reno Campus

Tompson, R. N.; *Summer Institute for Secondary School Teachers* (GW4561); 11 months; $49,027

NEW HAMPSHIRE

DARTMOUTH COLLEGE

Ballard, William W.; *In-Service Institute in Biology for Secondary School Teachers* (GW5767); 26 months; $8,379

———; (GW5657); 17 months; $8,324

NEW ENGLAND COLLEGE

O'Donnell, Thomas M.; *In-Service Institute in Computer Programming for Secondary School Teachers* (GW5485); 17 months; $8,026

PLYMOUTH STATE COLLEGE

Davis, Alan Hale; *Summer Conference of the New England Association of Chemistry Teachers* (GW4541); 3 months; $4,059

Wixson, Eldwin A.; *In-Service Institute in Mathematics for Secondary School Teachers* (GW5532); 17 months; $16,036

SUFFOLK UNIVERSITY

West II, Arthur J.; *Summer Institute in Biology of Marine Organisms for Secondary School Teachers* (GW4859); 11 months; $29,740

UNIVERSITY OF NEW HAMPSHIRE

Amell, Alexander R.; *Summer Institute in Chemistry for Secondary School Teachers* (GW4638); 11 months; $80,879

Balomenos, Richard H.; *In-Service Institute in Mathematics for Secondary School Teachers* (GW5839); 26 months; $10,004

———; *Summer and In-Service Institute in Mathematics for Secondary School Teachers* (GW5041*); 22 months; $10,004

———; (GW5041); 22 months; $84,016

NEW JERSEY

DREW UNIVERSITY

Greenspan, Bernard; *Summer Institute in Mathematics for Secondary School Teachers* (GW4951); 11 months; $54,446

———; *In-Service Institute in Mathematics for Secondary School Teachers* (GW5440); 17 months; $7,303

FAIRLEIGH DICKINSON UNIVERSITY

Rappaport, Max; *In-Service Institute in Modern Electronic Instrumentation for Secondary School Teachers* (GW5426); 17 months; $4,873

MONMOUTH COLLEGE

Wulf, Leo M.; *In-Service Institute in Mathematics for Secondary School Teachers* (GW5879); 26 months; $7,498

———; (GW5689); 17 months; $7,498

MONTCLAIR STATE COLLEGE

Clifford, Paul C.; (GW5676); 17 months; $7,200

Maletsky, Evan M.; *Summer Institutes in Mathematics for Secondary School Teachers* (GW4653); 11 months; $74,210

Zabady, Albert; *In-Service Institute in Modern Chemistry for Secondary School Teachers* (GW5658); 17 months; $10,107

NEWARK COLLEGE OF ENGINEERING

Konove, Carl; *In-Service Institute in Mathematics for Secondary School Teachers* (GW5558); 17 months; $4,588

Landsman, Leon D.; *In-Service Institute in Physics for Secondary School Teachers* (GW5675); 17 months; $6,682

PRINCETON UNIVERSITY

Dorf, Erling; *Summer Conference in Physical and Historical Geology for Secondary School Teachers* (GW5334); 10 months; $25,427

RUTGERS UNIVERSITY

Barlaz, Joshua; *In-Service Institute in Mathematics for Secondary School Teachers* (GW5646); 17 months; $22,557

———; *Summer Institute in Mathematics for Secondary School Teachers* (GW4690); 11 months; $85,935

Smith, Bennett L.; *In-Service Institute in Earth Science for Secondary School Teachers* (GW5557); 17 months; $12,790

Wolfson, Kenneth G.; *Academic Year Institute for Secondary School Teachers* (GW4800); 24 months; $154,241

RUTGERS, THE STATE UNIVERSITY, Newark Campus

Wiles, William W.; *Summer Institute in Earth Sciences for Secondary School Teachers* (GW4689); 11 months; $40,917

SETON HALL UNIVERSITY

Andrushkiw, Joseph W.; *Summer Institute in Mathematics for Secondary School Teachers* (GW5020); 11 months; $55,592

ST. PETER'S COLLEGE

Grant, James J.; *In-Service Institute in Physics for Secondary School Teachers* (GW5484); 17 months; $6,771

Varrichio, Francis A.; *In-Service Institute in Mathematics for Secondary School Teachers* (GW5464); 17 months; $11,555

STEVENS INSTITUTE OF TECHNOLOGY

Seavy, Robert H.; *In-Service Institute in Chemistry, Calculus and Computer Science for Secondary School Teachers* (GW5407); 17 months; $13,108

———; *Summer Institute in Chemistry, Mathematics, and Physics for Secondary School Teachers* (GW4910); 11 months; $25,790

VITRO LABORATORIES

Finkel, S. I.; *Contract for a Survey to Ascertain the Educational and Professional Backgrounds, and to Obtain Information on the Operating Environment of Teachers of Secondary School Science in Public and Private Schools of the United States During the School Year 1968-69* (C565000002); 4 months; $19,333

NEW MEXICO

NEW MEXICO INSTITUTE OF MINING AND TECHNOLOGY

Smith, Clay T.; *Summer Conference in Earth Science for Secondary School Teachers* (GW5341); 10 months; $18,618

———; *In-Service Institute in Earth Science for Secondary School Teachers* (GW5696); 17 months; $14,341

NEW MEXICO HIGHLANDS UNIVERSITY

Amai, Robert L. S.; *Summer Institute in Science and Mathematics for Secondary School Teachers* (GW5014); 11 months; $45,321

UNIVERSITY OF NEW MEXICO

Mitchell, Merle; *In-Service Institute in Mathematics for Secondary School Teachers* (GW5757); 26 months; $7,218

———; (GW5427); 17 months; $7,522

———; *Summer Institute for Secondary School Teachers* (GW4562); 11 months; $51,769

NEW YORK

ADELPHI UNIVERSITY

Bettelheim, Frederick A.; *Summer and In-Service Institute in Chemistry for Secondary School Teachers* (GW5012*); 22 months; $10,358

———; (GW5012); 22 months; $36,930

———; *In-Service Institute in Chemistry for Secondary School Teachers* (GW5830); 26 months; $10,357

Pohle, Frederick V.; *Summer and In-Service Institute in Mathematics for Secondary School Teachers* (GW5047*); 22 months; $32,826

———; (GW5047); 22 months; $45,395

Sirkin, Leslie A.; *In-Service Institute in the Earth Sciences for Secondary School Teachers* (GW5531); 17 months; $8,011

AMERICAN MUSEUM OF NATURAL HISTORY, Hayden Planetarium

SCIENCE EDUCATION PROGRAMS

Branley, Franklyn M.; *In-Service Institute in Astronomy for Secondary School Teachers* (GW4383 001); $1,007
———; (GW5471 001); $1,007
———; (GW5471); 17 months; $5,450

BROOKHAVEN LABORATORY
Hudis, Jerome; *In-Service Institute in Chemistry for Secondary School Teachers* (GW5566); 17 months; $7,350

CLARKSON COLLEGE OF TECHNOLOGY
Jones Jr., George L.; *Summer Institute for Secondary School Teachers* (GW4563); 11 months; $99,975

COLLEGE AT BUFFALO
Laug, George M.; *Summer Institute in Environmental Biology for Secondary School Teachers* (GW4708); 11 months; $30,314
Orgren, James; *Summer Institute in Earth Science for Secondary School Teachers* (GW4852); 11 months; $38,828
Schefler, William C.; *In-Service Institute in Selected Topics from New York State Junior High Science Curriculum for Secondary School Teachers* (GW5567); 17 months; $7,711
———; *In-Service Institute in Problems in Population and Environmental Pollution for Secondary School Teachers* (GW5673); 17 months; $4,021
Zingaro, Joseph S.; *In-Service Institute for Secondary School Teachers* (GW5688); 17 months; $7,171
———; *In-Service Institute in Interdisciplinary Science for Secondary School Teachers* (GW5743); 26 months; $7,171

COLLEGE AT FREDONIA
Connelly Jr., John J.; *Summer Institute in Introductory Physical Science for Secondary School Teachers* (GW5040); 11 months; $36,042

COLLEGE AT GENESEO
Reid, Archibald; *In-Service Institute in Biology and Chemistry for Secondary School Teachers* (GW5576); 17 months; $15,518
———; *In-Service Institute in Biology for Secondary School Teachers* (GW5845); 26 months; $8,982

COLLEGE AT ONEONTA
Egan, Frank; *Summer Institute in Mathematics for Secondary School Teachers* (GW4691); 11 months; $32,311

COLLEGE AT OSWEGO
Cox, Donald D.; *Summer Institute in Modern Biology for Secondary School Teachers* (GW4885); 11 months; $42,882
Swift, J. Nathan; *In-Service Institute in Earth Science for Secondary School Teachers* (GW5478); 17 months; $14,275

COLLEGE AT POTSDAM
Haskins, Elmer E.; *Summer Institute in Mathematics for Secondary School Teachers* (GW4742); 11 months; $40,104
Major, Alexander G.; *Summer Institute in Biology for Secondary School Teachers* (GW4701); 11 months; $67,006
Wolff, Peter C.; *Summer Institute in Principles of Animal Behavior for Secondary School Teachers* (GW4709); 11 months; $40,679

COLGATE UNIVERSITY
Downie, Edwin J.; *Summer Institute in Mathematics for Secondary School Teachers* (GW5000); 11 months; $46,974

CORNELL UNIVERSITY
Cooke, W. Donald; *Academic Year Institute for Secondary School Teachers* (GW4792); 22 months; $223,681
Shaw, R. William; *Summer Institute in Astronomy and Atmospheric Science for Secondary School Teachers* (GW4676); 11 months; $30,198

CUNY, BROOKLYN COLLEGE
Halpern, Alvin; *Summer and In-Service Institute in Physics for Secondary School Teachers* (GW5022*); 20 months; $12,385
———; (GW5022); 20 months; $38,552
Jordan, Meyer; *In-Service Institute in Mathematics for Secondary School Teachers* (GW5807); 26 months; $59,717
———; (GW5556); 17 months; $58,137
———; (GW5519); 8 months; $21,364

CUNY, CENTRAL SYSTEM OFFICE
Gabai, Hyman; *Summer Institute in Mathematics for Secondary School Teachers* (GW4692); 11 months; $29,478

CUNY, CITY COLLEGE
Kremer, Chester B.; *Summer Institute in Science and Mathematics for Secondary School Teachers* (GW4757); 11 months; $58,755

FORDHAM UNIVERSITY
Brande, Edward W.; *In-Service Institute in Mathematics for Secondary School Teachers* (GW5816); 26 months; $23,783
———; *Summer and In-Service Institute in Mathematics for Secondary School Teachers* (GW4996*); 21 months; $23,783
———; (GW4996); 21 months; $64,284

HAMILTON COLLEGE
Gere, Brewster H.; *Summer Institute in Mathematics for Secondary School Teachers* (GW4743); 11 months; $55,560

HOFSTRA UNIVERSITY
Davis, Jerry B.; *In-Service Institute in Biology for Secondary School Teachers* (GW5819); 26 months; $7,348
———; (GW5483); 17 months; $7,348

IONA COLLEGE
Murphy, James J.; *In-Service Institute in Computer Science for Secondary School Teachers* (GW5614); 17 months; $5,635

MANHATTAN COLLEGE
Crowe, George J.; *In-Service Institute in Physics for Secondary School Teachers* (GW5841); 26 months; $8,202
———; *Summer and In-Service Institute in Physics for Secondary School Teachers* (GW4644); 21 months; $29,610
———; (GW4644*); 21 months; $8,202
McNamara, John B.; *In-Service Institute in Mathematics for Secondary School Teachers* (GW5687); 17 months; $14,291
———; (GW5806); 26 months; $13,423
Titone, Luke V.; *In-Service Institute in Physics for Secondary School Teachers* (GW5770); 26 months; $7,516
———; (GW5408); 17 months; $7,515

MARITIME COLLEGE
Degani, Meir H.; *In-Service Institute in Computer Mathematics for Secondary School Teachers* (GW5613); 17 months; $7,446

NEW YORK UNIVERSITY
Hausner, Melvin; *In-Service Institute in Mathematics for Secondary School Teachers* (GW5499); 17 months; $72,295
———; (GW5782); 26 months; $72,293

PACE COLLEGE
Hallenbeck, Luke J.; *In-Service Institute in Biology for Secondary School Teachers* (GW5656); 17 months; $10,989
Konde, Anthony; *In-Service Institute in Chemistry, Astronomy and Geology for Secondary School Teachers* (GW5824); 26 months; $21,663
———; (GW5508); 17 months; $21,663
Trebatoski, Alice M.; *Summer Conference in Ecology for Secondary School Teachers* (GW5325); 10 months; $11,876

POLYTECHNIC INSTITUTE OF BROOKLYN
Fischer, George J.; *Summer Institute in Metallic and Polymeric Materials Science for Secondary School Teachers* (GW4875); 11 months; $30,868
Liao, Thomas; *Summer Institute in Engineering Concepts Curriculum Project for Secondary School Teachers* (GW4706); 11 months; $30,353
Piel, E. J.; *In-Service Institute in Engineering Concepts Curriculum Project for Secondary School Teachers* (GW5530); 17 months; $11,396

RENSSELAER POLYTECHNIC INSTITUTE
Bunce, Stanley C.; *Summer Institute in the Natural Sciences for Secondary School Teachers* (GW4998); 11 months; $128,679
Eppenstein, Walter; *Summer Institute in Harvard Project Physics for Secondary School Teachers* (GW4707); 11 months; $54,598

ST. LAWRENCE UNIVERSITY
Bowers, Landon E.; *Summer Institute in Microbiology and Biochemistry for Secondary School Teachers* (GW4732); 11 months; $30,866

STATE UNIVERSITY AT ALBANY
Nurnberger, Robert G.; *In-Service Institute in Problems of Air Pollution for Secondary School Teachers* (GW5674); 17 months; $9,420

STATE UNIVERSITY AT BUFFALO
Montague, Harriet F.; *Summer Institute for Secondary School Teachers* (GW4564); 11 months; $48,754

SYRACUSE UNIVERSITY
Collette, Alfred T.; *Summer Institute in Biology for Secondary School Teachers* (GW4909); 11 months; $46,630
———; *Academic Year Institute for Secondary School Teachers* (GW4825); 24 months; $205,850
Exner, Robert M.; *Summer Institute for Secondary School Teachers* (GW4565); 11 months, $49,873
Ginsburg, Nathan; *Summer Institute in Physics and Chemistry for Secondary School Teachers* (GW4600); 11 months; $61,335
Manwaring, James R.; *In-Service Institute in Environmental Sciences for Secondary School Teachers* (GW5507); 17 months; $12,090
Schaff, John; *Summer Institute for Secondary School Teachers* (GW4733); 11 months; $43,425

TEACHERS COLLEGE
Vogeli, Bruce R.; *Summer Institute in SSMCIS Mathematics for Secondary School Teachers* (GW4946); 11 months; $56,845
———; *In-Service Institute in Mathematics for Secondary School Teachers* (GW5679); 17 months; $13,790

UNION COLLEGE
Gillette, Edwin F.; *In-Service Institute in Chemistry, Earth Science and Mathematics for Secondary School Teachers* (GW5645); 17 months; $15,425
———; *Summer Institute for Secondary School Teachers* (GW4601); 11 months; $143,425
———; *In-Service Institute in Science and Mathematics for Secondary School Teachers* (GW5764); 26 months; $15,425

UNIVERSITY OF ROCHESTER
Montean, John J.; *In-Service Institute in Earth Science for Secondary School Teachers* (GW5549); 17 months; $10,169
———; *Academic Year Institute for Secondary School Teachers* (GW4816); 24 months; $94,779

VASSAR COLLEGE
Johnsen, John H.; *Summer Institute in Geological Evolution of Eastern North America—for Secondary School Teachers* (GW4659); 11 months; $39,595

YESHIVA UNIVERSITY
Gelbart, Abe; *In-Service Institute in Mathematics and Physics for Secondary School Teachers* (GW5822); 26 months; $138,583
———; (GW5450); 17 months; $166,301

INSTITUTES AND RESEARCH PARTICIPATION FOR SCIENCE, MATHEMATICS, AND ENGINEERING TEACHERS

NORTH CAROLINA

BENNETT COLLEGE
Sayles, J. Henry; *In-Service Institute in Computer Mathematics for Secondary School Teachers* (GW5428); 17 months; $8,604

DUKE UNIVERSITY
Reynolds, Thomas D.; *Summer Institute in Science and Mathematics for High School Teachers of Science and Secondary School Teachers* (GW4760); 11 months; $109,172

EAST CAROLINA UNIVERSITY
Dough, Robert L.; *In-Service Institute in Physics for Secondary School Teachers* (GW5643); 17 months; $13,041
——— ; *Summer and In-Service Institute in Harvard Project Physics for Secondary School Teachers* (GW5005); 22 months; $38,649
——— ; (GW5005*); 22 months; $6,259
Sowell, Katye O.; *In--Service Institute in Mathematics for Secondary School Teachers* (GW5469); 17 months; $11,236

NORTH CAROLINA ACADEMY OF SCIENCE
Reardon, Anna Joyce; *In-Service Institute in Physics for Secondary School Teachers* (GW5618); 12 months; $5,260
——— ; *In-Service Conference on Recent Advances in Physics for Secondary School Teachers* (GW4539); 11 months; $5,030

NORTH CAROLINA AGRICULTURAL & TECHNICAL STATE UNIVERSITY
Graves, Artis P.; *Summer Institute in Biology for Secondary School Teachers* (GW4754); 11 months; $68,724

NORTH CAROLINA STATE UNIVERSITY AT RALEIGH
Blakeway, Edward G.; *In-Service Institute in ECCP for Secondary School Teachers* (GW5575); 17 months; $9,419
Shannon, Henry A.; *Summer Institute in Earth Sciences for Secondary School Teachers* (GW4938); 20 months; $44,106

NORTH CAROLINA CENTRAL UNIVERSITY
Townes, Mary M.; *Summer Institute in Biology and Mathematics for Secondary School Teachers* (GW4999); 11 months; $50,711

ST. AUGUSTINE'S COLLEGE
Jones Jr., Joseph; *In-Service Institute in Biochemistry and Cellular Biology for Secondary School Teachers* (GW5686); 17 months; $7,228

UNIVERSITY OF NORTH CAROLINA AT ASHEVILLE
Squibb, S. Dexter; *In-Service Institute in Chemistry for Secondary School Teachers* (GW5518); 17 months; $6,392
——— ; (GW5823); 26 months; $6,392

UNIVERSITY OF NORTH CAROLINA AT CHAPEL HILL
Hounshell, Paul B.; *In-Service Conferences for Leadership Personnel in Science for Secondary School Teachers* (GW5565); 17 months; $13,688
——— ; *Summer Institute in Biology and Physical Science for Secondary School Teachers* (GW5001); 11 months; $64,439
——— ; *Academic Year Institute for Secondary School Teachers* (GW4818); 22 months; $162,263
——— ; *In-Service Conferences in Biology, Chemistry, and Science Supervision for Secondary School Teachers* (GW4535); 12 months; $13,859
Ingram, Roy L.; *Summer Institute in Geology for Secondary School Teachers* (GW4602); 11 months; $42,427
White, William A.; *In-Service Institute in Physical Geology for Secondary School Teachers* (GW5781); 26 months; $8,832
——— ; (GW5443); 17 months; $8,852

UNIVERSITY OF NORTH CAROLINA AT GREENSBORO
Vanselow, Clarence H.; *Summer Institute in Chemistry, CHEM Study for Secondary School Teachers* (GW4677); 11 months; $45,246

WAKE FOREST UNIVERSITY
Gentry, Ivey C.; *Summer Institute for Secondary School Teachers* (GW4584); 11 months; $64,867
Seelbinder, Ben M.; *In-Service Institute in Mathematics for Secondary School Teachers* (GW5473); 17 months; $8,674
——— ; (GW5797); 26 months; $9,077
Williams, George P.; *Summer Institute for Secondary School Teachers* (GW4566); 11 months; $26,003

WESTERN CAROLINA UNIVERSITY
Chapman, John J.; *Summer Institute in Earth Science for Secondary School Teachers* (GW4966); 11 months; $48,567
Marshall, J. Ronald; *In-Service Institute in Mathematics for Secondary School Teachers* (GW5449); 17 month; $6,676

NORTH DAKOTA

NORTH DAKOTA STATE UNIVERSITY
Broberg, Joel W.; *Summer Institute in Science for Secondary School Teachers* (GW4929); 11 months; $72,995
Koob, Robert D.; *In-Service Institute in Chemistry for Secondary School Teachers* (GW5772); 26 months; $6,658
——— ; (GW5612); 17 months; $6,658

UNIVERSITY OF NORTH DAKOTA
Jacob, Arthur F.; *In-Service Institute in Earth Science (Oceanography) for Secondary School Teachers* (GW5611); 17 months; $11,504
Wardner, C. A.; *Summer Institute in a Broad Study of Science for Secondary School Teachers* (GW4995); 11 months; $80,819
——— ; *Academic Year Institute for Secondary School Teachers* (GW4801); 24 months; $186,963

OHIO

AMERICAN SOCIETY FOR METALS
Mueller, William M.; *Summer Conference on the Solid State Physics of Metals for Secondary School Teachers* (GW5343); 10 months; $16,471

ANTIOCH COLLEGE
Corwin, James F.; *Summer Institute in General Science for Secondary School Teachers* (GW4669); 11 months; $96,032

BALDWIN-WALLACE COLLEGE
Miller, John W.; *Summer Conference on the Laboratory Culture and Use of Marine Organisms for Secondary School Teachers* (GW5331); 10 months; $6,591

BOWLING GREEN STATE UNIVERSITY
Hall, W. H.; *Summer Institute in Physical Chemistry and Other Advanced Chemistry Courses for Secondary School Teachers* (GW4603); 11 months; $75,283
Kirby, William A.; *Summer Institute in Mathematics for Secondary School Teachers* (GW5030); 11 months; $67,777
Leetch, J. Frederick; *Academic Year Institute for Secondary School Teachers* (GW4814); 24 months; $186,563
Snyder, Eldon E.; *In-Service Institute in Sociology for Secondary School Teachers* (GW5609); 17 months; $7,377

CAPITAL UNIVERSITY
Sievert, Carl F.; *Summer Institute for Secondary School Teachers* (GW4567); 11 months; $43,060

CASE WESTERN RESERVE UNIVERSITY
Guenther, Paul E.; *Summer Institute in Mathematics for Secondary School Teachers* (GW4761); 11 months; $38,558

CENTRAL STATE UNIVERSITY
Holmes, Defield T.; *Summer Institute in Laboratory Training Program for Secondary School Teachers* (GW4876); 11 months; $37,868

CLEVELAND STATE UNIVERSITY
VanVoorhis, W. R.; *Summer and In-Service Institute in Mathematics for Secondary School Teachers* (GW4978*); 22 months; $13,473
——— ; (GW4978); 22 months; $34,225

DENISON UNIVERSITY
King, Paul G.; *Summer Institute in Economics and Urban Problems for Secondary School Teachers* (GW4884); 11 months; $45,852

JOHN CARROLL UNIVERSITY
Nash, Harry C.; *Summer and In-Service Institute in Physics for Secondary School Teachers* (GW4991*); 21 months; $14,394
——— ; (GW4991); 21 months; $34,613
——— ; *In-Service Institute in Physics for Secondary School Teachers* (GW5642); 17 months; $14,340
——— ; *In-Service Institute in Modern Physics for Secondary School Teachers* (GW5564); 8 months; $5,645
——— ; *In-Service Institute in Physics for Secondary School Teachers* (GW5763); 26 months; $14,931
——— ; (GW5799); 26 months; $12,481

KENT STATE UNIVERSITY
Cummins, Kenneth B.; *In-Service Institute in Mathematics for Secondary School Teachers* (GW5761); 26 months; $10,230
——— ; (GW5617); 17 months; $10,230
——— ; *Summer Institute in Mathematics for Secondary School Teachers* (GW4660); 11 months; $63,743
——— ; *In-Service Institute in Mathematics for Secondary School Teachers* (GW4413 001); $1,200

MIAMI UNIVERSITY
Vaughn, Charles M.; *In-Service Institute in Environmental Science for Secondary School Teachers* (GW5684); 17 months; $9,970
——— ; *In-Service Institute in Comparative Animal Physiology and Modern Microbiology for Secondary School Teachers* (GW5756); 26 months; $9,784
——— ; *In-Service Institute in Developmental Biology and Modern Genetics for Secondary School Teachers* (GW5434); 17 months; $9,738
Weidner, Bruce V.; *Summer Institute for ondary School Teachers* (GW4604); 11 months; $86,238

OHIO DOMINICAN COLLEGE
Matesich, Mary Andrew; *In-Service Institute in Chemistry for Secondary School Teachers* (GW5517); 17 months; $5,110

OHIO STATE UNIVERSITY
Kerr, Douglas S.; *Summer Institute in Computer and Information Science for Secondary School Teachers* (GW4715); 11 months; $52,680
Mayer, Victor J.; *In-Service Institute in Earth Science for Secondary School Teachers* (GW5412); 17 months; $10,253
——— ; *Summer Institute for Secondary School Teachers* (GW4894); 11 months; $41,946
Riley, William R.; *In-Service Institute in Physics for Secondary School Teachers* (GW5528); 17 months; $10,482
——— ; (GW5787); 26 months; $10,461
Riner, John; *Summer Institute for Secondary School Teachers* (GW4585); 11 months; $104,339
Schlessinger, Fred R.; *Academic Year Institute for Secondary School Teachers* (GW4808); 22 months; $137,126
Spieker, Edmund M.; *Summer Institute for Secondary School Teachers* (GW4568); 11 months; $48,670

OHIO UNIVERSITY
Malcom, Paul S.; (GW4678); 11 months; $78,509
Skinner, Jr., Ray; *In-Service Institute in*

SCIENCE EDUCATION PROGRAMS

Earth Science for Secondary School Teachers (GW5453); 17 months; $20,259

Warmke, Roman F.; *In-Service Institute in Economic Education and Related Social Sciences for Secondary School Teachers* (GW5873); 25 months; $15,238

———; (GW5616); 17 months; $15,238

OHIO WESLEYAN UNIVERSITY

Russell, Leonard N.; *Summer Institute in Physical Science for Secondary School Teachers* (GW4918); 11 months; $65,948

Staley, David H.; *In-Service Institute in Mathematics for Secondary School Teachers* (GW5529); 17 months; $13,469

UNIVERSITY OF AKRON

Griffin, C. Frank; *In-Service Institute in Harvard Project Physics for Secondary School Teachers* (GW5496); 17 months; $8,504

UNIVERSITY OF CINCINNATI

Rolwing, Raymond H.; *In-Service Institute in Mathematics for Secondary School Teachers* (GW5468); 17 months; $18,446

———; *Summer Institute in Mathematics for Secondary School Teachers* (GW4744); 11 months; $17,607

———; *Academic Year Institute for Secondary School Teachers* (GW4834); 24 months; $100,555

———; *In-Service Institute in Mathematics for Secondary School Teachers* (GW5774); 26 months; $19,525

UNIVERSITY OF DAYTON

Mushenheim, H. G.; *Summer Institute in Mathematics for Secondary School Teachers* (GW4762); 11 months; $48,004

UNIVERSITY OF TOLEDO

Foster, Edward S.; *In-Service Institute in Physical Science for Secondary School Teachers* (GW5610); 17 months; $10,137

OKLAHOMA

NORTHWESTERN STATE COLLEGE

Rogers, Stearns W.; *In-Service Institute in Physics for Secondary School Teachers* (GW5436); 17 months; $6,110

———; *In-Service Institute in Mathematical and Laboratory Techniques in Physical Science for Secondary School Teachers* (GW5833); 26 months; $6,110

OKLAHOMA STATE UNIVERSITY

Bruneau, L. Herbert; *Academic Year Institute for Secondary School Teachers* (GW4807); 24 months; $94,155

Crockett, Jerry J.; *In-Service Institute in Sociology for Secondary School Teachers* (GW5488); 17 months; $11,779

Millington, Clayton; *Summer Institute in Economics for Secondary School Teachers* (GW4968); 11 months; $56,423

Stone, John E.; *Summer Institute in Earth Science for Secondary School Teachers* (GW5049); 21 months; $55,817

SOUTHEASTERN STATE COLLEGE

Dwight, Leslie A.; *Summer Institute in Mathematics for Secondary School Teachers* (GW4925); 11 months; $70,423

SOUTHWESTERN STATE COLLEGE

McKellips, Raymond L.; *Summer Institute for Secondary School Teachers* (GW4569); 11 months; $56,509

Reynolds, Earl A.; *Summer Institute in Chemistry for Secondary School Teachers* (GW4605); 11 months; $57,662

UNIVERSITY OF OKLAHOMA

Andree, Richard V.; *Summer Institute in Computer Programming and Related Mathematics for Secondary School Teachers* (GW5050); 11 months; $73,143

———; *In-Service Institute in Computer Programming and Related Mathematics for Secondary School Teachers* (GW5768); 26 months; $30,539

———; (GW5574); 17 months; $30,539

Larsh, Howard W.; *Summer Institute in Biology for Secondary School Teachers* (GW5039); 11 months; $78,373

———; *Academic Year Institute for Secondary School Teachers* (GW4830); 24 months; $99,471

Levy, Gene; (GW4806); 22 months; $102,735

———; *Summer Institute in Mathematics for Secondary School Teachers* (GW4665); 11 months; $56,166

Mankin, Charles J.; *Academic Year Institute for Secondary School Teachers* (GW4788); 24 months; $117,010

———; *Summer Institute in Earth Science for Secondary School Teachers* (GW4928); 11 months; $64,509

———; (GW5052); 11 months; $72,058

———; *In-Service Institute in Earth Science for Secondary School Teachers* (GW5685); 17 months; $15,060

———; *Academic Year Institute in Earth Science for Secondary School Teachers* (GW5850); 26 months; $100,228

———; *In-Service Institute in Earth Science for Secondary School Teachers* (GW5820); 26 months; $16,745

OREGON

MARYLHURST COLLEGE

Colbert, Loretta Ann; *In-Service Institute in Mathematics for Secondary School Teachers* (GW5442); 17 months; $11,575

———; (GW5825); 26 months; $12,857

OREGON COLLEGE OF EDUCATION

Chatham, Ronald L.; *In-Service Institute in World Regional Geography for Secondary School Teachers* (GW5641); 17 months; $9,008

Gallagher, James W.; *Summer Institute in Geography for Secondary School Teachers* (GW5026); 11 months; $50,546

Redbird, Helen M.; *In-Service Institute in Sociology of Race Relations for Secondary School Teachers* (GW5573); 17 months; $9,755

———; (GW5868); 26 months; $9,752

OREGON STATE UNIVERSITY

Fox, Fred W.; *Summer Institute for Secondary School Teachers* (GW4606); 11 months; $69,621

Simons, William H.; *Summer Institute in Mathematics for Secondary School Teachers* (GW4716); 11 months; $71,580

Williamson, Stanley E.; *Academic Year Institute for Secondary School Teachers* (GW4826); 24 months; $202,250

PORTLAND STATE UNIVERSITY

Fiasca, Michael; *In-Service Conference in Environmental Science for State Supervisors of Science* (GW5377); 8 months; $21,824

Harris, J. Kenneth; *Summer Institute in Mathematics of Computers for Secondary School Teachers* (GW4927); 11 months; $46,794

———; *In-Service Institute in Mathematics for Secondary School Teachers* (GW5864); 25 months; $7,617

———; (GW5683); 17 months; $7,617

Vanatta, Robert O.; *In-Service Institute in Earth Science for Secondary School Teachers* (GW5655); 17 months; $9,335

REED COLLEGE

Davis, Kenneth E.; *In-Service Institute in Physics, Mathematics and Chemistry for Secondary School Teachers* (GW5826); 26 months; $15,744

———; (GW5500); 17 months; $15,745

SOUTHERN OREGON COLLEGE

McCoy, Robert A.; *Summer Institute in Geometry for Secondary School Teachers* (GW4631); 11 months; $57,390

Purdom, William B.; *Summer Institute in Biological and Earth Sciences for Secondary School Teachers* (GW4941); 11 months; $50,740

UNIVERSITY OF OREGON

Moursund, A. F.; *Academic Year Institute for Secondary School Teachers* (GW4798); 24 months; $115,590

———; *Summer and In-Service Institute in Mathematics for Secondary School Teachers* (GW4586*); 21 months; $8,103

———; (GW4586); 21 months; $71,613

———; *Academic Year Institute in Mathematics for Secondary School Teachers* (GW5860); 25 months; $103,335

———; *In-Service Institute in Mathematics for Secondary School Teachers* (GW5750); 26 months; $8,476

Moursund, David G.; *Summer Institute in Computer Science for Secondary School Teachers* (GW4717); 11 months; $54,541

Tepfer, Sanford S.; *Academic Year Institute for Secondary School Teachers* (GW4804); 24 months; $102,211

UNIVERSITY OF PORTLAND

Wack, Paul E.; *Summer Institute in Integrated Physics and Chemistry for Secondary School Teachers* (GW4596); 11 months; $71,439

PENNSYLVANIA

BEAVER COLLEGE

Breyer, Arthur C.; *Summer Institute in Chemistry for Secondary School Teachers* (GW4763); 11 months; $59,187

BUCKNELL UNIVERSITY

Kieft, Lester, *Summer Institute for Secondary School Teachers* (GW4587); 11 months; $67,962

CARLOW COLLEGE

Uricchio, William A.; *In-Service Institute in Biology for Secondary School Teachers* (GW5865); 25 months; $9,323

———; (GW5438); 17 months; $9,782

DREXEL UNIVERSITY

Fields, Ewaugh F.; *In-Service Institute in Mathematics for Secondary School Teachers* (GW5681); 17 months; $8,790

Hamman, Robert L.; *Summer Institute in Economics for Secondary School Teachers* (GW4780); 11 months; $42,444

EDINBORO STATE COLLEGE

Schneider, Michael C.; *Summer Institute in Earth Science for Secondary School Teachers* (GW4997); 11 months; $43,798

———; *In-Service Institute in Earth Science for Secondary School Teachers* (GW5429); 17 months; $13,055

FRANKLIN AND MARSHALL COLLEGE

Hood, Richard F.; *Summer Institute in Physics for Secondary School Teachers* (GW4702); 11 months; $27,815

Kauffman, Marvin E.; *Summer Institute for Secondary School Teachers* (GW4607); 11 months; $61,140

GANNON COLLEGE

Zagorski, Stanley J.; *In-Service Institute in Environmental Science for Secondary School Teachers* (GW5527); 17 months; $13,350

GETTYSBURG COLLEGE

Holder, Leonard I.; *Summer Institute in Mathematics for Secondary School Teachers* (GW4746); 11 months; $51,561

INDIANA UNIVERSITY OF PENNSYLVANIA

Hoyt, John P.; *Summer Institute in Mathematics for Secondary School Teachers* (GW4895); 11 months; $40,050

LAFAYETTE COLLEGE

Mugridge, Larry R.; *In-Service Institute in Mathematics for Secondary School Teachers* (GW5504); 17 months; $8,384

———; (GW5840); 26 months; $8,384

Shaughnessy, Edward P.; *Summer Institute in Mathematics for Secondary School Teachers* (GW4969); 11 months; $53,160

INSTITUTES AND RESEARCH PARTICIPATION FOR SCIENCE, MATHEMATICS, AND ENGINEERING TEACHERS

LEHIGH UNIVERSITY
Taylor, Douglas H.; (GW4636); 11 months; $38,310

LINCOLN UNIVERSITY
Harrison, Joseph L.; *Summer Institute in the General Sciences for Secondary School Teachers* (GW4886); 11 months; $41,856

MILLERSVILLE STATE COLLEGE
McIlwaine, William B.; *Summer Institute in Earth Science for Secondary School Teachers* (GW4860); 11 months; $34,279
———; *In-Service Institute in Chemistry, Biology, and Earth Science for Secondary School Teachers* (GW5805); 26 months; $22,400
———; (GW5608); 17 months; $22,400

PENNSYLVANIA STATE UNIVERSITY
Kieren, Thomas E.; *Summer Institute in Computer Science for Key Teachers and Supervisors in Secondary Schools* (GW4942); 11 months; $39,667
Lancaster, Otis E.; *Summer Institute in Engineering Concepts for Secondary School Teachers* (GW4887); 20 months; $36,857
Mack, Sidney F.; *Summer Institute in Mathematics for Secondary School Teachers* (GW4735); 11 months; $44,807
———; *Academic Year Institute for Secondary School Teachers* (GW4810); 22 months; $95,383
Miller, E. Willard; (GW4790); 24 months; $105,614
Shigley, James W.; *Summer Institute in the Biological Sciences for Secondary School Teachers* (GW4853); 11 months; $38,031

PENNSYLVANIA STATE UNIVERSITY, Behrend Campus
Balmer, Louis W.; *In-Service Institute in Mathematics for Secondary School Teachers* (GW5682); 10 months; $6,541
———; *In-Service Conference in Science and Mathematics for Secondary School Teachers* (GW4538); 6 months; $6,517

PHILADELPHIA COLLEGE OF PHARMACY AND SCIENCE
Chase, Grafton D.; *Summer Institute in Radioisotope Technology for Secondary School Teachers* (GW4975); 11 months; $34,670

SHIPPENSBURG STATE COLLEGE
Sieber, James L.; *Summer Institute in Computer Science for Secondary School Teachers* (GW4736); 11 months; $47,000
———; *In-Service Institute in Mathematics for Secondary School Teachers* (GW5498); 17 months; $9,360

TEMPLE UNIVERSITY
Schmuckler, Joseph S.; *In-Service Institute for Secondary School Teachers* (GW5654); 17 months; $7,072
Sutman, Frank X.; *Academic Year Institute for Secondary School Teachers* (GW4832); 22 months; $119,987
———; *Academic Year Institute for Training Science Supervisors for Inner City Schools* (GW5856); 26 months; $71,485
Wurster, Marie A.; *In-Service Institute in Mathematics for Secondary School Teachers* (GW5793); 26 months; $18,810
———; (GW5563); 17 months; $18,507

UNIVERSITY OF PENNSYLVANIA
Girault, Emily S.; *In-Service Institute in Sociology for Secondary School Teachers* (GW5639); 17 months; $10,173
———; (GW5869); 25 months; $10,242

UNIVERSITY OF PITTSBURGH
Knipp, John C.; *In-Service Institute in Mathematics for Secondary School Teachers* (GW5414); 17 months; $7,842
———; *Summer Institute for Secondary School Teachers* (GW4570); 11 months; $27,606

UNIVERSITY OF SCRANTON
Strickland, Harry B.; *In-Service Institute in Biology, Chemistry, Mathematics and Physics for Secondary School Teachers* (GW5433); 17 months; $20,892
———; (GW5740); 26 months; $21,180

VILLANOVA UNIVERSITY
Markham, James J.; *In-Service Institute in Science and Mathematics for Secondary School Teachers* (GW5550); 17 months; $27,463

WEST CHESTER STATE COLLEGE
Branton, Richard G.; *In-Service Institute in Mathematics for Secondary School Teachers* (GW5735); 26 months; $25,148
Filano, Albert E.; (GW5409); 17 months; $24,418
———; *Summer Institute in Mathematics for Secondary School Teachers* (GW4896); 11 months; $33,378
Rickert, Russell K.; *In-Service Institute in Physical Science for Secondary School Teachers* (GW5640); 17 months; $7,021
———; (GW5638); 17 months; $8,668

WILKES COLLEGE
Bruch, Alvan; *In-Service Institute in Earth Science for Secondary School Teachers* (GW5540); 17 months; $6,790
Michelini, Francis J.; *Summer Institute in Biology for Secondary School Teachers* (GW4737); 11 months; $46,596

PUERTO RICO

CATHOLIC UNIVERSITY OF PUERTO RICO
Escabi, Luis A.; *In-Service Institute in Radiobiology and Topics in Inorganic Chemistry for Secondary School Teachers* (GW5562); 17 months; $6,045

INTER-AMERICAN UNIVERSITY OF PUERTO RICO
Velez, Ismael; *Summer Institute in Biology for Secondary School Teachers* (GW4906); 11 months; $46,736

UNIVERSITY OF PUERTO RICO, Rio Piedras
Bobonis, Augusto; *Summer and In-Service Institute in Mathematics and Physics for Secondary School Teachers* (GW5010*); 22 months; $15,685
———; (GW5010); 22 months; $94,691

UNIVERSITY OF PUERTO RICO SYSTEM OFFICE, Cayey Regional College
Garcia, Mariano; *In-Service Institute in Mathematics for Secondary School Teachers* (GW5463); 17 months; $9,200
———; *Summer Institute in Mathematics for Secondary School Teachers* (GW4693); 11 months; $61,200
———; *In-Service Institute in Mathematics for Secondary School Teachers* (GW5796); 26 months; $9,200

RHODE ISLAND

BROWN UNIVERSITY
Clapp, Leallyn B.; *Summer Institute for Secondary School Teachers* (GW4609); 11 months; $44,720

NEW ENGLAND ASSOCIATION OF CHEMISTRY TEACHERS
Meinhold, Russell; *Summer Conference for Chemistry Teachers* (GW5327); 10 months; $5,370

RHODE ISLAND COLLEGE
Bierden, James E.; *In-Service Institute in Mathematics for Secondary School Teachers* (GW5497); 17 months; $7,689
———; (GW5758); 26 months; $7,689

SOUTH CAROLINA

THE CITADEL, THE MILITARY COLLEGE OF SOUTH CAROLINA
Ingraham, J. R.; (GW5637); 17 months; $7,885
———; (GW5744); 26 months; $8,367

CLAFLIN COLLEGE
Smith Sr., Hampton D.; *Summer Institute in Science and Mathematics for Secondary School Teachers* (GW4877); 11 months $38,859

CLEMSON UNIVERSITY
Flatt, James L.; *In-Service Institute in Mathematics for Secondary School Teachers* (GW5472); 17 months; $11,219
Graben, H. W.; *Summer Institute in Physics for Secondary School Teachers* (GW4965); 11 months; $56,741

SOUTH CAROLINA STATE COLLEGE
Roache, Lewie C.; *Summer Institute in Science for Secondary School Teachers* (GW4752); 11 months; $74,859
———; *In-Service Institute in Science for Secondary School Teachers* (GW5394); 17 months; $20,390

UNIVERSITY OF SOUTH CAROLINA
Davies, Tudor T.; *In-Service Institute in Earth Science for Secondary School Teachers* (GW5431); 17 months; $30,276
Silvernail, Richard G.; *Summer Institute for Secondary School Teachers* (GW4985); 11 months; $38,172
Williams, W. L.; *Summer Institute in Mathematics for Secondary School Teachers* (GW4861); 11 months; $88,323
———; *Academic Year Institute for Secondary School Teachers* (GW4817); 24 months; $184,530

SOUTH DAKOTA

SOUTH DAKOTA SCHOOL OF MINES AND TECHNOLOGY
Lowe, Clifford; *Summer Institute in Physics for Secondary School Teachers* (GW4771); 11 months; $31,330
Mickelson, John C.; *Summer Institute in Geology for Secondary School Teachers* (GW4898); 11 months; $43,829

UNIVERSITY OF SOUTH DAKOTA
Coker, E. Howard; *Summer Institute in Chemistry, Mathematics and Physics for Secondary School Teachers* (GW4610); 11 months; $37,681
Connors, Howard W.; *Summer Institute in Earth Science, Physical Science, and Mathematics for Secondary School Teachers of General Science* (GW4634); 11 months; $71,196
Ekman, William E.; *Summer Institute in Mathematics for Secondary School Teachers* (GW4661); 11 months; $40,202
Estee, Charles R.; *Summer Institute in Chemistry (CHEMS) for Secondary School Teachers* (GW4655); 11 months; $41,711
Schmulbach, James C.; *Academic Year Institute for Secondary School Teachers* (GW4822); 24 months; $189,454
Stevenson, Robert E.; *In-Service Institute in Biology, Chemistry, Earth Science, Mathematics and Physics for Secondary School Teachers* (GW5526); 17 months; $26,048
Van Bruggen, Theodore; *Summer Institute in Biology for Secondary School Teachers* (GW4679); 11 months; $44,042

TENNESSEE

CHRISTIAN BROTHERS COLLEGE
Doody, Edward; *Summer Institute in Introductory Physical Science for Secondary School Teachers* (GW4751); 11 months; $56,774

EAST TENNESSEE STATE UNIVERSITY
Hartsell, Lester C.; *Summer Institute in Mathematics for Secondary School Teachers* (GW4738); 11 months; $44,567
———; *In-Service Institute in Mathematics for Secondary School Teachers* (GW5525); 17 months; $5,049

FISK UNIVERSITY
Deshpande, Krishnanath B.; *Academic Year Institute for Secondary School Teachers* (GW4838); 22 months; $97,388

SCIENCE EDUCATION PROGRAMS

Hull Jr., George; *Summer Institute in Biology, Chemistry, and Physics for Secondary School Teachers* (GW4627); 11 months; $82,005

GEORGE PEABODY COLLEGE FOR TEACHERS

Tomlinson, Gus; *Summer Institute in Physical Sciences for Secondary School Teachers* (GW4588); 11 months; $55,525

——— ; *Summer Institute in Science for Secondary School Teachers* (GW4935); 11 months; $78,218

KNOXVILLE COLLEGE

Harvey, Robert H.; *In-Service Institute in Chemistry and Physical Science for Secondary School Teachers* (GW5636); 17 months; $8,402

——— ; *In-Service Institute in Discovery Approach in Modern Mathematics for Secondary School Teachers* (GW5572); 17 months; $8,180

——— ; *In-Service Institute in Chemistry and Physical Science for Secondary School Teachers* (GW5871); 25 months; $8,402

LANE COLLEGE

Douglass Jr., John; *Summer Institute in Mathematics for Secondary School Teachers* (GW4891); 11 months; $48,205

MEMPHIS STATE UNIVERSITY

Houk, Larry W.; *In-Service Institute in Principles of Chemistry and Organic Chemistry for Secondary School Teachers* (GW5503); 17 months; $10,181

Jermann, William H.; *In-Service Institute in Engineering Science (ECCP) for Secondary School Teachers* (GW5474); 17 months; $6,122

MIDDLE TENNESSEE STATE UNIVERSITY

Wiser, J. Eldred; *Summer Institute in Mathematics and Science for Secondary School Teachers* (GW4862); 11 months; $99,912

UNIVERSITY OF TENNESSEE

Eaves, Edgar D.; *Summer Institute in Mathematics for Secondary School Teachers* (GW4934); 11 months; $62,100

UNIVERSITY OF THE SOUTH

Ellis, Eric H.; *Summer Institute in Science and Mathematics for Secondary School Teachers* (GW4772); 11 months; $69,803

VANDERBILT UNIVERSITY

Ray, Oakley S.; *Summer Conference in Psychobiology for Secondary School Teachers* (GW5338); 10 months; $17,579

——— ; *Special Project in Drug Abuse Education and Behavior for Secondary School Science Teachers and Supervisors* (GW5211); 18 months; $22,377

TEXAS

BAYLOR UNIVERSITY

Tweedie, Virgil L.; *Summer Institute in Biology and Chemistry for Secondary School Teachers* (GW4864); 11 months $52,476

EAST TEXAS STATE UNIVERSITY

Rohrer, Charles S.; *In-Service Institute in Chemistry, Earth Science, and Physics for Secondary School Teachers* (GW5430); 17 months; $15,661

——— ; *Summer Institute in Physical Science for Secondary School Teachers* (GW4640); 11 months; $54,497

HUSTON-TILLOTSON COLLEGE

Morton, J. H.; *Summer Institute in Science and Mathematics for Secondary School Teachers* (GW4718); 11 months; $43,027

INCARNATE WORD COLLEGE

Armer, Joseph Marie; *In-Service Institute in Biology and Chemistry for Secondary School Teachers* (GW5441); 17 months; $9,691

——— ; (GW5783); 26 months; $9,489

Faust, Claude M.; *In-Service Institute in Mathematics for Secondary School Teachers* (GW5753); 26 months; $4,796

——— ; (GW5392); 17 months; $4,997

LAMAR STATE COLLEGE OF TECHNOLOGY

Matthews, William H.; *Summer Institute in ESCP Earth Science for Secondary School Teachers* (GW4658); 11 months; $42,415

Rigney, Carl J.; *Summer Institute in Physics and Astronomy for Secondary School Teachers* (GW4611); 11 months; $41,758

NORTH TEXAS STATE UNIVERSITY

Cochran, Kendall P.; *Summer Institute for Secondary School Teachers* (GW4571); 11 months; $33,763

Cowan, Paul J.; *In-Service Institute in Earth Science for Secondary School Teachers* (GW5635); 17 months; $10,695

Escue, R. B. Jr.; *Summer Institute in General Science for Secondary School Teachers* (GW4703); 11 months; $40,750

Luker, William A.; *In-Service Institute in Economics for Secondary School Teachers* (GW5615); 17 months; $8,205

Schlueter, Edgar A.; *In-Service Institute in Biology for Secondary School Teachers* (GW5502); 17 months; $7,915

Vest, Floyd R.; *In-Service Institute in Mathematics for Secondary School Teachers* (GW5634); 17 months; $8,122

PRAIRIE VIEW A & M COLLEGE

O'Banion, E. E.; *Summer Institute in the Sciences for Secondary School Teachers* (GW5023); 11 months; $39,695

Stewart, A. D.; *Summer Institute in Mathematics for Secondary School Teachers* (GW4747); 11 months; $24,506

SAM HOUSTON STATE UNIVERSITY

Manka, Charles K.; *Summer Institute in Astronomy for Secondary School Teachers* (GW4612); 11 months; $31,553

SOUTHERN METHODIST UNIVERSITY

Stallcup, William B.; *Summer Institute for Secondary School Teachers* (GW4597); 11 months; $36,987

TEXAS A & M UNIVERSITY

Schroeder, Melvin C.; *Summer Institute in the Earth Sciences for Secondary School Teachers* (GW4936); 11 months; $73,115

——— ; *Academic Year Institute for Secondary School Teachers* (GW4797); 24 months; $190,775

Sicilio, Fred; *Summer Institute in Chemistry (CHEM Study) for Secondary School Teachers* (GW4680); 11 months; $50,755

TEXAS ARTS & INDUSTRY UNIVERSITY

Kruse, Olan E.; *Summer Institute for Secondary School Teachers* (GW4572); 11 months; $35,128

TEXAS CHRISTIAN UNIVERSITY

Colquitt, Landon A.; *In-Service Institute in Mathematics for Secondary School Teachers* (GW5541); 17 months; $5,876

——— ; (GW5809); 26 months; $5,876

TEXAS SOUTHERN UNIVERSITY

Clarkson, Llayron L.; *Summer Institute in Mathematics for Secondary School Teachers* (GW4983); 11 months; $52,115

Terry, Robert J.; *In-Service Institute in Biology for Secondary School Teachers* (GW5747); 26 months; $7,639

——— ; *Summer Institute in Biology for Secondary School Teachers* (GW4628); 11 months; $72,400

——— ; *In-Service Institute in Biology for Secondary School Teachers* (GW5516); 17 months; $7,639

TEXAS TECH UNIVERSITY

Prior, Paul V.; *Summer Institute in Biology for Secondary School Teachers* (GW4958); 11 months; $42,966

Riggs, Charles L.; *In-Service Institute in Mathematics for Secondary School Teachers* (GW5538); 17 months; $8,191

TEXAS WOMAN'S UNIVERSITY

Christy, John H.; *Summer Institute in Mathematics for Secondary School Teachers* (GW4964); 11 months; $26,361

Davis, Ethelyn; *Summer Institute in Sociology for Secondary School Teachers* (GW5033); 11 months; $30,989

TRINITY UNIVERSITY

Gaedke, Rudolph M.; *Summer Institute in Harvard Project Physics for Secondary School Teachers* (GW5002); 11 months; $42,073

UNIVERSITY OF HOUSTON

Bishop, Margaret S.; *Academic Year Institute for Secondary School Teachers* (GW4836); 24 months; $126,981

Creswell, John L.; *In-Service Institute in UICSM Mathematics for Secondary School Teachers* (GW5607); 8 months; $8,420

——— ; (GW5470); 17 months; $9,097

——— ; (GW5844); 26 months; $9,097

Myrick, H. Nugent; *Summer Conference in Environmental Pollution Control Technology for Secondary School Science Teachers* (GW5339); 10 months; $16,316

Rogers, Curtis A.; *Summer Institute for Secondary School Teachers* (GW4573); 11 months; $42,757

Solliday, James R.; *In-Service Institute in Earth Science for Secondary School Teachers* (GW5418); 17 months; $17,176

UNIVERSITY OF TEXAS AT AUSTIN

Guy Jr., William T.; *Summer Institute in Mathematics for Secondary School Teachers* (GW4981); 11 months; $56,161

Lee, Addison E.; *Academic Year Institute for Secondary Teachers* (GW4827); 24 months; $183,679

——— ; *Summer Institute in Science for Secondary School Teachers* (GW4750); 11 months; $74,918

UNIVERSITY OF TEXAS AT EL PASO

Nymann, James E.; *Summer Institute in Mathematics for Secondary School Teachers* (GW4681); 11 months; $35,190

Strain, William S.; *Summer Institute in Earth Science for Secondary School Teachers* (GW5025); 11 months; $28,729

UTAH

SOUTHERN UTAH STATE COLLEGE

Cooper, Laurence C.; *In-Service Institute in Earth Science for Secondary School Teachers* (GW5605); 17 months; $13,141

UNIVERSITY OF UTAH

Derrick, William R.; *In-Service Institute in Mathematics for Secondary School Teachers* (GW5492); 17 months; $9,219

——— ; (GW5818); 26 months; $9,219

Parmley, Thomas J.; *Academic Year Institute for Secondary School Teachers* (GW4812); 22 months; $140,601

Pendleton, Robert C.; *Summer Institute in Radiation Biology for Secondary School Teachers* (GW4977); 11 months; $28,261

UTAH STATE UNIVERSITY

Broadbent, Marden; *In-Service Institute in Mathematics for Secondary School Teachers* (GW5837); 26 months; $8,492

——— ; *In-Service Institute in Modern Biology for Secondary School Teachers* (GW5836); 26 months; $9,651

——— ; *In-Service Institute in Mathematics for Secondary School Teachers* (GW5571); 17 months; $10,153

——— ; *In-Service Institute in Modern Biology for Secondary School Teachers* (GW5633); 17 months; $8,042

Elich, Joe; *Summer Institute for Secondary School Teachers* (GW4574); 11 months; $73,269

Hunsaker, Neville C.; *In-Service Institute in Mathematics for Secondary School Teachers* (GW5880); 25 months; $6,703

——— ; *Summer Institute in Mathematics*

for Secondary School Teachers (GW4974); 21 months; $69,757

———; *In-Service Institute in Mathematics for Secondary School Teachers* (GW5476); 17 months; $6,703

Kim, Yun; *Summer Institute in Population Studies for Secondary School Teachers* (GW5053); 11 months; $44,200

VERMONT

UNIVERSITY OF VERMONT

Schoonmaker, N. James; *Summer Institute in Mathematics for Secondary School Teachers* (GW4961); 11 months; $75,294

VIRGINIA

COLLEGE OF WILLIAM AND MARY

Cato, Benjamin R. Jr.; *Summer Institute in the Sciences and Mathematics for Secondary School Teachers* (GW4854); 11 months; $91,315

Reynolds, Thomas L.; *In-Service Institute in Mathematics for Secondary School Teachers* (GW5448); 17 months; $7,482

———; (GW5776); 26 months; $7,723

HAMPTON INSTITUTE

Bonner, Robert D.; *Summer Institute in Biology for Secondary School Teachers* (GW4986); 11 months; $42,466

Fields, Victor H.; *In-Service Institute in Chemistry and Mathematics for Secondary School Teachers* (GW5762); 26 months; $16,203

———; (GW5397); 17 months; $16,203

———; *Summer Institute in Chemistry, Mathematics and Physics for Secondary School Teachers* (GW4645); 11 months; $81,054

MADISON COLLEGE

Sanders, William M.; *Summer Institute in Mathematics for Secondary School Teachers* (GW4916); 11 months; $51,890

OLD DOMINION UNIVERSITY

Clark, Allen K.; *Summer Institute in Chemistry for Secondary School Teachers* (GW4756); 11 months; $32,768

RANDOLPH-MACON WOMAN'S COLLEGE

Whidden, Helen L.; *Summer Institute in the Sciences and Mathematics for Secondary School Teachers* (GW4767); 11 months; $74,022

UNIVERSITY OF VIRGINIA

Lowry, William C.; *In-Service Conference in Mathematics for State Supervisors of Mathematics* (GW4921); 6 months; $28,797

Thompson, Ertle; *Academic Year Institute for Secondary School Teachers* (GW4783); 24 months; $145,209

———; *In-Service Institute in Biology, Chemistry, Earth Science, Physical Science, Physics, and Mathematics for Secondary School Teachers* (GW5769); 26 months; $73,927

———; (GW5632); 17 months; $72,431

VIRGINIA POLYTECHNIC INSTITUTE

Shockley, James E.; *Summer Institute in Mathematics for Secondary School Teachers* (GW4933); 11 months; $55,466

VIRGINIA STATE COLLEGE

Dunn, Richard H.; *Summer Institute in Biology for Secondary School Teachers* (GW4920); 11 months; $81,869

Gipson, Mack; *Summer Institute for Secondary School Teachers* (GW4575); 11 months; $57,393

Hamlett, Hunter D.; *In-Service Institute in Biology, Earth Science and Mathematics for Secondary School Teachers* (GW5432); 17 months; $20,440

Woodson Jr., Bernard R.; *Academic Year Institute for Secondary School Teachers* (GW4839); 24 months; $130,772

WASHINGTON

CENTRAL WASHINGTON STATE COLLEGE

Shrader, John S.; *In-Service Institute in Earth Science for Secondary School Teachers* (GW5631); 17 months; $11,950

FORT WRIGHT COLLEGE OF THE HOLY NAME

Gautereaux, Ione; *In-Service Institute in Environmental Biology, Chemistry, Earth Science and Mathematics for Secondary School Teachers* (GW5789); 26 months; $14,000

Gautereaux, M. Eugene; (GW5399); 17 months; $13,642

PACIFIC LUTHERAN UNIVERSITY

Herzog, John O.; *In-Service Institute in Mathematics for Secondary School Teachers* (GW5475); 17 months; $11,014

———; (GW5829); 26 months; $11,013

SEATTLE PACIFIC COLLEGE

Crichton, James H.; *In-Service Institute in Physics for Secondary School Teachers* (GW5539); 17 months; $4,629

Shaw, Ross F.; *Summer Institute in Marine Biology for Secondary School Teachers* (GW4863); 11 months; $42,256

SEATTLE UNIVERSITY

Cowgill, James J.; *Summer Institute in Chemistry, Physics and Mathematics for Secondary School Teachers* (GW4963); 11 months; $81,803

UNIVERSITY OF WASHINGTON

Arons, Arnold; *Summer Institute in Physics for Secondary School Teachers* (GW5036); 11 months; $63,762

Fleming, Richard H.; *Summer Institute in 'World of Water' for Secondary School Teachers* (GW4889); 11 months; $67,658

Johnson, David L.; *Summer Institute in Engineering Concepts (ECCP) for Secondary School Teachers* (GW4855); 11 months; $36,129

Stoebe, Thomas G.; *In-Service Institute in Materials Science for Secondary School Teachers* (GW5569); 17 months; $9,321

WASHINGTON STATE UNIVERSITY

Batey Jr., Harry H.; *Summer Institute in Science for Secondary School Teachers* (GW4719); 11 months; $107,707

Robertson, Jack M.; *Summer Institute for Secondary School Teachers* (GW4576); 11 months; $70,486

WESTERN WASHINGTON STATE COLLEGE

Chaney, Robin; *Summer Institute in Mathematics for Secondary School Teachers* (GW4613); 11 months; $59,575

Christman, Robert A.; *Summer Institute in Earth Science for Secondary School Teachers* (GW4984); 11 months; $54,328

Craswell, Keith J.; *Academic Year Institute for Secondary School Teachers* (GW4787); 24 months; $95,186

Gelder, Harvey M.; *Summer Institute in Mathematics for Secondary School Teachers* (GW4856); 11 months; $54,686

WEST VIRGINIA

BETHANY COLLEGE

Spence, John A.; *Summer Institute in Physical Science for Secondary School Teachers* (GW4739); 11 months; $50,283

MARSHALL UNIVERSITY

Ward, Harold E.; *Summer Institute in Biology for Secondary School Teachers* (GW4982); 11 months; $48,150

WEST VIRGINIA UNIVERSITY

Peters, I. D.; *In-Service Institute in Mathematics and Physics for Secondary School Teachers* (GW5630); 17 months; $30,239

———; (GW5791); 26 months; $30,239

Peters, I. Dee; *Summer and Academic Year Institute in Mathematics for Secondary School Teachers* (GW5009*); 22 months; $89,689

———; (GW5009); 22 months; $59,440

———; *In-Service Institute in Mathematics for Secondary School Teachers* (GW4482 001); $1,550

WHEELING COLLEGE

Loner, Charles J.; *In-Service Institute in Chemistry for Secondary School Teachers* (GW5505); 17 months; $4,027

WISCONSIN

MARQUETTE UNIVERSITY

Bournique, Raymond A.; *Summer Institute in Chemistry for Secondary School Teachers* (GW5028); 11 months; $18,955

Connellan, Miriam E.; *Summer Institute in Mathematics for Secondary School Teachers* (GW4641); 11 months; $38,792

RIPON COLLEGE

Carley, David W.; *Summer Institute in Chemistry for Secondary School Teachers* (GW4656); 11 months; $37,742

UNIVERSITY OF WISCONSIN, Madison

Harvey, John G.; *Summer Institute in Mathematics for Secondary School Teachers* (GW5013); 11 months; $43,810

———; *Academic Year Institute for Secondary School Teachers* (GW4809); 24 months; $142,532

———; *Academic Year Institute in Mathematics and Science for Secondary School Teachers* (GW5853); 29 months; $120,160

Wuerger, William W.; *Summer Institute in Science (Engineering Concepts Curriculum Project) for Secondary School Teachers* (GW4857); 11 months; $49,916

UNIVERSITY OF WISCONSIN, Milwaukee

Jaggard, R. A.; *Summer Institute in Physics for Secondary School Teachers* (GW4858); 11 months; $60,673

———; *In-Service Institute in Harvard Project Physics & Astronomy for Secondary School Teachers* (GW5489); 17 months; $8,918

WISCONSIN STATE UNIVERSITY SYSTEM OFFICE

Fay, Marcus J.; *Summer Institute in Field Biology and Ecology for Secondary School Teachers* (GW4919); 11 months; $52,504

WISCONSIN STATE UNIVERSITY, Eau Claire

Goranson, Leonard D.; *Summer Institute in Geography for Secondary School Teachers* (GW5011); 11 months; $64,475

Wick, Marshall E.; *In-Service Institute in Mathematics for Secondary School Teachers* (GW5606); 8 months; $10,772

WISCONSIN STATE UNIVERSITY, Oshkosh

Crimmins, Timothy F.; *Summer Institute in Chemistry for Secondary School Teachers* (GW4915); 11 months; $44,246

WISCONSIN STATE UNIVERSITY, River Falls

Dollahon, James C.; *Summer Institute in Basic and Agricultural Sciences for Secondary School Teachers* (GW5029); 11 months; $49,650

Hall, Lyle C.; *In-Service Institute in Basic Quantum Mechanics for Secondary School Teachers* (GW5628); 29 months; $5,533

WISCONSIN STATE UNIVERSITY, Superior

Brieske, Phillip R.; *In-Service Institute in Harvard Project Physics for Secondary School Teachers* (GW5627); 17 months; $57,357

———; *Summer Institute in Harvard Project Physics for Secondary School Teachers* (GW4907); 11 months; $65,664

———; *Academic Year Institute for Secondary School Teachers* (GW4829); 22 months; $123,508

———; *In-Service Institute in Introductory Physical Science for Secondary School Teachers* (GW5788); 26 months; $57,357

Cowie, James B.; *In-Service Institute in Narcotics Education for Secondary School Teachers* (GW5680); 17 months; $5,723

SCIENCE EDUCATION PROGRAMS

Dailey, Donald M.; *In-Service Institute in AAAS Materials for Secondary School Teachers* (GW5629); 8 months; $6,414
———; *In-Service Science Education Implementation Project for Teachers* (GW5866); 13 months; $3,032
Tychsen, Paul C.; *Summer Institute in Earth Science for Secondary School Teachers* (GW4917); 11 months; $43,709

WYOMING

UNIVERSITY OF WYOMING
Guenther, William C.; *Summer Institute in Probability and Statistics for Secondary School Teachers* (GW4694); 11 months; $57,485
Harding, Samuel W.; *Summer Institute for Secondary School Teachers* (GW4720); 11 months; $83,625
Jenkins, Terry L.; (GW4654); 11 months; $68,666
Pancoe, William L.; *Summer Institute in Radiation Biology for Secondary School Teachers* (GW4932); 11 months; $33,254
Prowse, Derek J.; *Academic Year Institute for Secondary School Teachers* (GW4789); 24 months; $135,306
Tillery, Bill W.; *In-Service Institute in Physical Science (ISCS), for Secondary School Teachers* (GW5568); 17 months; $18,285

Special Programs

CALIFORNIA

ASIA FOUNDATION
Coliver, Norman; *Foreign Participants, Travel of Asian Participants in NSF Supplementary Training Program* (GW3090002); 3 months; $7,976

COLORADO

UNIVERSITY OF COLORADO
Hill, A. David; *A Briefing Conference for Geography Staff on NSF Education Projects* (GW5721); 5 months; $5,481

DISTRICT OF COLUMBIA

AMERICAN ASSOCIATION FOR THE ADVANCEMENT OF SCIENCE
Kabisch, William T.; *Visiting Foreign Scientists Project* (GW2267002); 12 months; $38,764

AMERICAN FRIENDS OF THE MIDDLE EAST
Parker, Orin D.; *Foreign Participants, Travel/Middle East Participants to NSF Supplementary Training Programs* (GW5380); 13 months; $6,500

AMERICAN GEOLOGICAL INSTITUTE
Thompson, John F.; *A Briefing Conference for Earth Science Staff of NSF Education Projects* (GW5722); 5 months; $7,660

ORGANIZATION OF AMERICAN STATES—PAN AMERICAN UNION
Perkinson, Jesse D.; *Foreign Participants, Travel of Latin American Participants to NSF Supplementary Training Programs* (GW3089002); 3 months; $5,000

MASSACHUSETTS

HARVARD UNIVERSITY
Maybury, Robert H.; *A Briefing Conference for Physics Staff on NSF Education Projects* (GW5723); 5 months; $12,149

MICHIGAN

MICHIGAN TECHNOLOGICAL UNIVERSITY
Brown, Robert, T.; *1969 Summer Institute in Ecology for Secondary School Teachers* (GW3695001); $372

MINNESOTA

MACALESTER COLLEGE
Hastings, Russell B.; *1969 Summer Institute in General Science for Secondary School Teachers* (GW3706001); $600

NEBRASKA

NEBRASKA WESLEYAN UNIVERSITY
French, Walter R.; *1969 Summer Institute in Introductory Physical Science for Secondary School Teachers* (GW3728001); $373

PENNSYLVANIA

U.S.-SOUTH AFRICA LEADERSHIP EXCHANGE PROGRAM
Brewer, James C.; *Foreign Participants, Travel/South African Participants in NSF Supplementary Training Programs* (GW5381); 13 months; $5,500

Pre-Service Teacher Education

DISTRICT OF COLUMBIA

AMERICAN GEOLOGICAL INSTITUTE
Romey, William D.; *Earth Science Teacher Preparation Project* (GY7688); 39 months; $419,670

ILLINOIS

KNOX COLLEGE
Priestley, Herbert; *A Coordinated Plan for Secondary School Science Teacher Training* (GY7689); 27 months; $160,155

MICHIGAN

EASTERN MICHIGAN UNIVERSITY
Breedlove, Charles B.; *Program for the Preparation of Prospective Teachers of the Physical Sciences* (GY8424); 39 months; $73,600

MISSOURI

TARKIO COLLEGE
Shinpoch, John R.; *Increased Classroom Experience for Pre-Service Elementary Teachers* (GY7696); 26 months; $24,300

TENNESSEE

AUSTIN PEAY STATE UNIVERSITY
Mayfield, M. R.; *Pre-Service Teacher Education, Physics the Program for Teachers* (GY5356002); 6 months; $33,050
———; (GY7694); 11 months; $139,843

WASHINGTON

UNIVERSITY OF WASHINGTON
Arons, Arnold B.; *A Coordinated Program of Instruction in Physical Science and Science Teaching for Pre-Service Elementary Teachers* (GY7687); 30 months; $143,200

Science Education for Undergraduate Students

Undergraduate Research Participation

ALABAMA

UNIVERSITY OF ALABAMA, Tuscaloosa; Gambrell Jr., S. C.; (GY7604); 11 months; $8,800
Garner, Robert H.; (GY7249); 20 months; $10,000

ARIZONA

ARIZONA STATE UNIVERSITY; Reiser, Castle O.; (GY7409); 11 months; $8,520

NORTHERN ARIZONA UNIVERSITY; Wettaw, John F.; (GY7382); 11 months; $7,010

PRESCOTT COLLEGE; Euler, Robert C.; (GY7520); 11 months; $8,000

UNIVERSITY OF ARIZONA; Evans, Daniel D.; (GY7258); 11 months; $7,200

ARKANSAS

UNIVERSITY OF ARKANSAS; Sims, Leslie B.; (GY7327); 11 months; $12,980
Yeargan, J. R.; (GY7406); 11 months; $7,720

CALIFORNIA

CALIFORNIA INSTITUTE OF TECHNOLOGY; Clark Donald S.; (GY7360); 11 months; $9,120
Davidson, Norman; (GY7351); 11 months; $18,000
Epstein, Samuel; (GY7328); 11 months; $6,130
Sinsheimer, Robert L.; (GY7352); 11 months; $15,120

CALIFORNIA STATE COLLEGE AT FULLERTON; Langworthy, William B.; (GY7596); 11 months; $8,660

CALIFORNIA STATE COLLEGE AT LOS ANGELES; Moye, Anthony J.; (GY7218); 11 months; $7,320

FRESNO STATE COLLEGE; Russell, Kenneth H.; (GY7590); 11 months; $11,750

HARVEY MUDD COLLEGE; Greever, John; (GY7542); 11 months; $10,780
Kubota, Mitsuru; (GY7269); 11 months; $9,120

IMMACULATE HEART COLLEGE; Green Sr. Agnes Ann; (GY7489); 11 months; $3,200

OCCIDENTAL COLLEGE; Amey, Ralph L.; (GY7584); 11 months; $6,560

POMONA COLLEGE; Beilby, Alvin L.; (GY7347); 11 months; $9,720
Ogier, Walter T.; (GY7504); 11 months; $9,120

SACRAMENTO STATE COLLEGE; Fish, Richard W.; (GY7401); 11 months; $7,900

SAN DIEGO STATE COLLEGE; Collier, Boyd D.; (GY7487); 11 months; $10,360
Walba, Harold; (GY7391); 11 months; $21,560

SAN JOSE STATE COLLEGE; Albert, Norman E.; (GY7559); 11 months; $9,220

STANFORD UNIVERSITY; Lee, Welton L.; (GY7288); 11 months; $15,400
Shepard, O. Cutler; (GY7495); 11 months; $12,990

UNIVERSITY OF CALIFORNIA, Berkeley; Kenyon, George L.; (GY7410); 11 months; $16,820

UNIVERSITY OF CALIFORNIA, Davis; Ragland, Thomas E.; (GY7490); 11 months; $7,770

UNIVERSITY OF CALIFORNIA, Los Angeles; Bayes, Kyle D.; (GY7585); 11 months; $25,450
Carlisle, Donald; (GY7287); 11 months; $5,600
Pickett, M. J.; (GY7272); 11 months; $10,160

UNIVERSITY OF CALIFORNIA, Riverside; Fung, Sun-Yiu; (GY7491); 11 months; $9,960
Rettig, Michael F.; (GY7519); 11 months; $16,230
Rhine, Ramon J.; (GY7361); 11 months; $10,400

UNIVERSITY OF CALIFORNIA, Santa Barbara; Kaska, William C.; (GY7257); 11 months; $9,940

UNIVERSITY OF REDLANDS; Stafford, Robert C.; (GY7324); 11 months; $2,810

UNIVERSITY OF SANTA CLARA; De Bouvere, Karel L.; (GY7268); 11 months; $6,300

UNIVERSITY OF SOUTHERN CALIFORNIA; Wolf, Walter; (GY7420); 11 months; $14,590

SCIENCE EDUCATION FOR UNDERGRADUATE STUDENTS

COLORADO

COLORADO COLLEGE; Roberts, Carl L.; (GY7325); 11 months; $6,080
Taber, Richard L.; (GY7432); 11 months; $4,760

COLORADO STATE UNIVERSITY; Hill, Duane W.; (GY7568); 11 months; $17,970
McCallum, Malcolm E.; (GY7572); 11 months; $5,900

ROCKY MOUNTAIN BIOLOGICAL LABORATORY; Enders, Robert K.; (GY7304); 11 months; $9,580

TEMPLE BUELL COLLEGE; Greenspoon, Joel; (GY7236); 16 months; $4,680

UNIVERSITY OF COLORADO; Ashby, Neil; (GY7265); 20 months; $11,500
Hassner, Alfred; (GY7227); 11 months; $11,200
Ives, Jack D.; (GY7465); 11 months; $10,320

CONNECTICUT

CONNECTICUT COLLEGE; MacKinnon, John R.; (GY7593); 11 months; $7,740

FAIRFIELD UNIVERSITY; Boggio, Joseph E.; (GY7242); 11 months; $3,400

ST. JOSEPH COLLEGE; Markham, Sr. Maria Clare; (GY7241); 11 months; $4,200

UNIVERSITY OF CONNECTICUT; Brush, Alan H.; (GY7533); 11 months; $10,780
Chance, Norman A.; (GY7470); 16 months; $6,480

WESLEYAN UNIVERSITY; Curott, David R.; (GY7630); 11 months; $6,340
Leermakers, Peter A.; (GY7477); 11 months; $4,930

YALE UNIVERSITY; Graham, William R.; (GY7359); 20 months; $7,320
Hay, George A.; (GY7259); 11 months; $8,400
Lane, Robert E.; (GY7626); 11 months; $12,000

DISTRICT OF COLUMBIA

HOWARD UNIVERSITY; Hambright, Peter; (GY7511); 11 months; $8,540

SMITHSONIAN INSTITUTION; Ritterbush, Philip C.; (GY7622); 11 months; $24,910
———; (GY6056001); $564
———; (GY6040001); $319

FLORIDA

FLORIDA STATE UNIVERSITY; Grosslight, Joseph H.; (GY7260); 11 months; $11,200

UNIVERSITY OF FLORIDA; Dolbier Jr., William R.; (GY7547); 11 months; $17,280

Eisenberg, Martin A.; (GY7317); 10 months; $22,510
Elliott, Paul R.; (GY7538); 15 months; $13,170

UNIVERSITY OF SOUTH FLORIDA; Worrell, Jay H.; (GY7621); 11 months; $15,200

GEORGIA

EMORY UNIVERSITY; Day Jr., R. A.; (GY7321); 11 months; $8,480

GEORGIA COLLEGE AT MILLEDGEVILLE; Lipscomb, Harriett; (GY7320); 11 months; $6,500

UNIVERSITY OF GEORGIA; Cadwallader, D. E.; (GY7536); 11 months; $10,490
Coward, S. J.; (GY7350); 11 months; $8,400
Gibbons, J. Whitfield; (GY7372); 11 months; $9,300
Scott, Alan; (GY7240); 20 months; $6,740

HAWAII

UNIVERSITY OF HAWAII; Gilje, John W.; (GY7517); 11 months; $14,560
Griffin, P. Bion; (GY7548); 11 months; $15,600

IDAHO

IDAHO STATE UNIVERSITY; Swanson Jr., Earl H.; (GY7443); 11 months; $5,150

UNIVERSITY OF IDAHO; Bopp, Gordon R.; (GY7349); 11 months; $6,080

ILLINOIS

BRADLEY UNIVERSITY; Bjorklund, Richard G.; (GY7215); 11 months; $6,080

DEPAUL UNIVERSITY; Georgakis, Constantine, (GY7348); 11 months; $5,000
Stinchcomb, Thomas G.; (GY7512); 11 months; $7,600

FIELD MUSEUM OF NATURAL HISTORY; Martin, Paul S.; (GY7225); 11 months; $18,250

ILLINOIS INSTITUTE OF TECHNOLOGY; Neubert, Theodore J.; (GY7482); 11 months; $10,200
Stueben, Edwin F.; (GY7291); 20 months; $11,360

KNOX COLLEGE; Schramm, Peter; (GY7471); 11 months; $3,410

LEWIS COLLEGE; Hogan, H. Philip; (GY7214); 11 months; $4,560

NORTH CENTRAL COLLEGE; Doty, Barbara; (GY7528); 11 months; $3,300

NORTHERN ILLINOIS UNIVERSITY; Kevill, Dennis N.; (GY7244); 11 months; $10,800
Mittler, Sidney; (GY7580); 20 months; $12,970

NORTHWESTERN UNIVERSITY; Goldberg, Erwin; (GY7602); 11 months; $12,260
Lambert, Joseph B.; (GY7346); 11 months; $12,670

SOUTHERN ILLINOIS UNIVERSITY; Caskey, Albert L.; (GY7637); 35 months; $7,980
McClary, Dan O.; (GY7478); 11 months; $5,900

UNIVERSITY OF CHICAGO; Harberger, Arnold C.; (GY7609); 11 months; $18,240
Yang, N. C.; (GY7232); 20 months; $15,200

UNIVERSITY OF ILLINOIS, Chicago Circle; Hadley, Elmer B.; (GY7234), 11 months; $18,240

UNIVERSITY OF ILLINOIS, Urbana; Eades, David; (GY7283); 11 months; $7,800
Gallagher, J. P.; (GY7550); 21 months; $5,100
Malpass, Roy S.; (GY7537); 11 months; $17,280

INDIANA

EARLHAM COLLEGE; Gifford, Cameron E.; (GY7354); 11 months; $10,560
Martin, Charles W.; (GY7518); 11 months; $6,180
Smith, H. Warren; (GY7451); 11 months; $6,130

INDIANA UNIVERSITY, Bloomington; Konetzka, Walter A.; (GY7421); 11 months; $13,920
Shiner Jr., V. J.; (GY7629); 11 months; $17,740
Springer, George; (GY7610); 11 months; $16,450

PURDUE UNIVERSITY; Cwalina, Gustav E.; (GY7323); 11 months; $5,900
Goldschmidt, Victor W.; (GY7342); 16 months; $7,360

UNIVERSITY OF NOTRE DAME; Fairley, William M.; (GY7496); 11 months; $4,660
Lauer, Kenneth R.; (GY7587); 11 months; $3,570
Mihelich, John W.; (GY7353); 11 months; $4,500
Peretti, E. A.; (GY7344); 11 months; $4,550
Thorson, Ralph E.; (GY7216); 11 months; $6,600
Yang, K. T.; (GY7441); 20 months; $12,200

IOWA

COE COLLEGE; Watkins, S. R.; (GY7248); 11 months; $4,200

CORNELL COLLEGE; Lyon, David L.; (GY7284); 11 months; $3,780

DORDT COLLEGE; Geels, Edwin J.; (GY7217); 11 months; $3,240

GRINNELL COLLEGE; Erickson, Luther E.; (GY7431); 11 months; $4,000

IOWA STATE UNIVERSITY; Barton, Thomas J.; (GY7380); 11 months; $19,890
Hartman, Paul A.; (GY7250); 11 months; $6,600
Rougvie, Malcolm A.; (GY7226); 11 months; $7,200

LUTHER COLLEGE; Lecander, Ronald; (GY7316); 11 months; $2,640

UNIVERSITY OF IOWA; Guillory, J. Keith; (GY7544); 11 months; $9,520
Schulz, Rudolph W.; (GY7251); 11 months; $12,000

KANSAS

KANSAS STATE UNIVERSITY; Schrenk, William G.; (GY7469); 11 months; $12,100

ST. BENEDICT'S COLLEGE; Senecal, Gerard; (GY7247); 20 months; $3,400

UNIVERSITY OF KANSAS; Bass, William M.; (GY7416); 11 months; $2,800
Friesen, Benjamin S.; (GY7411); 11 months; $4,040
Price, G. Baley; (GY7296); 11 months; $5,370
Wiley, Robert A.; (GY7238); 11 months; $7,600
Willhite, G. Paul; (GY7640); 11 months; $6,300

WICHITA STATE UNIVERSITY; Carper, W. R.; (GY7466); 11 months; $6,570

KENTUCKY

THOMAS MORE COLLEGE; Miner, George K.; (GY7618); 11 months; $4,860

UNIVERSITY OF KENTUCKY; Kostenbauder, H. B.; (GY7437); 11 months; $9,500
Weaver, Ralph H.; (GY7237); 11 months; $9,940
Wood, Don J.; (GY7488); 44 months; $18,860

UNIVERSITY OF LOUISVILLE; Lang, Calvin A.; (GY7418); 11 months; $13,870

LOUISIANA

LOUISIANA STATE UNIVERSITY, Baton Rouge; Kent Jr., George C.; (GY7255); 11 months; $7,560

LOUISIANA POLYTECHNIC INSTITUTE; Galli, Anthony J.; (GY7403); 11 months; $8,300

LOYOLA UNIVERSITY; Benedetto, F. A.; (GY7385); 11 months; $5,600
Gary Jr., Lee P.; (GY7571); 11 months; $5,340

TULANE UNIVERSITY; Bamforth, Stuart S.; (GY7554); 11 months; $8,080

MAINE

BOWDOIN COLLEGE; Huntington, Charles E.; (GY7254); 11 months; $11,200

JACKSON LABORATORY; Waymouth, Charity; (GY7614); 15 months; $21,390
———; (GY6053001); $1,848

MARYLAND

UNIVERSITY OF MARYLAND; Brown, Joshua R. C.; (GY7371); 11 months; $10,000
Jarvis, Bruce B.; (GY7335); 11 months; $15,260
Karp, Carol; (GY7623); 11 months; $13,820
Koo, Ted S. Y.; (GY7340); 11 months; $8,570

SCIENCE EDUCATION PROGRAMS

MASSACHUSETTS

AMHERST COLLEGE; Snellgrove, Richard A.; (GY7338); 11 months; $7,350

BRANDEIS UNIVERSITY; Heller, Peter; (GY7315); 11 months; $9,310

CHILD'S CANCER RESEARCH FOUNDATION; Yerganian, George; (GY7330); 11 months; $8,000

CLARK UNIVERSITY; Curtis, Joseph C.; (GY7370); 11 months; $7,200
Gerber, Stanford N.; (GY7392); 11 months; $5,390
Kaplan, Bernard; (GY7599); 11 months; $8,490

HARVARD UNIVERSITY; Vogt, Evon Z.; (GY7231); 11 months; $22,800
Watkins, Calvert; (GY7583); 11 months; $6,880

MASSACHUSETTS INSTITUTE OF TECHNOLOGY; Lodish, Harvey F.; (GY7462); 11 months; $9,310
Ross, John; (GY7497); 11 months; $12,260

MT. HOLYOKE COLLEGE; Kopp, Richard W.; (GY7298); 11 months; $7,180

NORTHEASTERN UNIVERSITY; Moyer, Samuel E.; (GY7509); 12 months; $9,720

REGIS COLLEGE; McCarthy, Sr. Viterbo; (GY7563); 45 months; $3,680

SIMMONS COLLEGE; Milburn, Josephine F.; (GY7639); 45 months; $12,900

SMITH COLLEGE; Fleck, George; (GY7378); 11 months; $5,520

STATE COLLEGE AT BRIDGEWATER; Boutilier, Robert F.; (GY7430); 16 months; $5,420

TUFTS UNIVERSITY; Cormack, Allan M.; (GY7428); 11 months; $3,640
Nickerson, Norton H.; (GY7600); 11 months; $11,880

UNIVERSITY OF MASSACHUSETTS; Bartlett, L. M.; (GY7297); 11 months; $17,080
Hoffman, Allan R.; (GY7245); 11 months; $7,920

WELLESLEY COLLEGE; Friedman, Lawrence B.; (GY7429); 11 months; $4,700
Padvkula, Helen A.; (GY7253); 11 months; $8,400

WHEATON COLLEGE; Pearson, Myrna S.; (GY7252); 11 months; $8,400

WILLIAMS COLLEGE; Brown, Fielding; (GY7369); 11 months; $2,000
Crider, Andrew B.; (GY7499); 11 months; $3,400
Labine, Patricia A.; (GY7493); 11 months; $5,530

MICHIGAN

ADRIAN COLLEGE; Husband, Robert W.; (GY7395); 11 months; $1,300
Miller, Robert C.; (GY7433); 11 months; $2,540

ALMA COLLEGE; Kapp, Ronald O.; (GY7368); 11 months; $6,080

EASTERN MICHIGAN UNIVERSITY; Giles, Richard A.; (GY7404); 45 months; $14,890

HOPE COLLEGE; Brink, Irwin J.; (GY7314); 11 months; $7,970

KALAMAZOO COLLEGE; Palmer, David W.; (GY7256); 20 months; $2,400

MICHIGAN TECHNOLOGICAL UNIVERSITY; Erbisch, Frederic H.; (GY7552); 11 months; $4,020

MICHIGAN STATE UNIVERSITY; Barch, A. M.; (GY7527); 11 months; $15,000
Brown, James A.; (GY7526); 11 months; $9,360
Kelly, W. H.; (GY7521); 11 months; $13,990
Kindel, Paul K.; (GY7367); 11 months; $14,250
Tsai, Chester; (GY7450); 11 months; $21,000

NORTHERN MICHIGAN UNIVERSITY; Barry, Roger D.; (GY7389); 11 months; $7,750
Gill, Gordon; (GY7366); 11 months; $1,800

OAKLAND UNIVERSITY; Butterworth, F. M.; (GY7480); 11 months; $11,540
Tomboulian, Paul; (GY7438); 11 months; $7,510

UNIVERSITY OF MICHIGAN; Burckhalter, J. H.; (GY7365); 11 months; $6,350
Ford, Richard I.; (GY7576); 20 months; $18,240
Larimore, Ann E.; (GY7384); 11 months; $4,460

WAYNE STATE UNIVERSITY; Fraser, Winifred D.; (GY7292); 10 months; $12,670
Hooper, Henry O.; (GY7500); 11 months; $7,460

WESTERN MICHIGAN UNIVERSITY; Kent, Neil D.; (GY7507); 11 months; $10,920

MINNESOTA

CARLETON COLLEGE; Finholt, James E.; (GY7570); 11 months; $8,760
Thomas, Thurlo B.; (GY7364); 11 months; $7,200

MACALESTER COLLEGE; Slowinski, Emil J.; (GY7417); 11 months; $5,270

ST. OLAF COLLEGE; Larson, Loren; (GY7363); 11 months; $12,160
Petersen, Arnold J.; (GY7425); 11 months; $7,600
Rossing, Thomas D.; (GY7362); 11 months; $7,600
Tarr, Donald A.; (GY7476); 11 months; $6,960

UNIVERSITY OF MINNESOTA; Henderson, Lavell M.; (GY7331); 15 months; $8,200
Keller, Kenneth H.; (GY7468); 11 months; $7,140

UNIVERSITY OF MINNESOTA, Duluth Campus; Caple, Ronald; (GY7562); 11 months; $9,490

MISSISSIPPI

UNIVERSITY OF MISSISSIPPI; Borne, Ronald F.; (GY7597); 23 months; $12,000

MISSOURI

SOUTHEAST MISSOURI STATE COLLEGE; Froemsdorf, Donald H.; (GY7472); 11 months; $7,700

UNIVERSITY OF MISSOURI, Columbia; Creighton, Donald L.; (GY7394); 13 months; $10,780
Kaiser, Edwin M.; (GY7281); 11 months; $13,200

UNIVERSITY OF MISSOURI, Kansas City; Olson, Philip G.; (GY7591); 45 months; $9,600

UNIVERSITY OF MISSOURI, Rolla; Bertnolli, E. C.; (GY7445); 11 months; $4,830
Fuller, Harold Q.; (GY7455); 11 months; $9,150
Senne Jr., Joseph H.; (GY7310); 11 months; $11,830
Wellek, Robert M.; (GY7642); 11 months; $6,140

UNIVERSITY OF MISSOURI, St. Louis; Block, Eric; (GY7594); 11 months; $11,050

WASHINGTON UNIVERSITY; Kardos, John L.; (GY7501); 11 months; $6,090

MONTANA

MONTANA STATE UNIVERSITY; Jennings, Paul W.; (GY7358); 11 months; $10,400

UNIVERSITY OF MONTANA; Eddleman, L. E.; (GY7399); 11 months; $7,050

NEBRASKA

UNIVERSITY OF NEBRASKA, Lincoln; Drew, James V.; (GY7243); 11 months; $4,880
Ullman, Frank E.; (GY7503); 11 months; $6,330

NEVADA

UNIVERSITY OF NEVADA, Reno Campus; Burkhart, Richard D.; (GY7588); 11 months; $5,300

NEW HAMPSHIRE

DARTMOUTH COLLEGE; Cleland, Robert L.; (GY7524); 11 months; $9,300
Leaton, R. N.; (GY7532); 11 months; $5,460

UNIVERSITY OF NEW HAMPSHIRE; Haaland, Gordon A.; (GY7319); 11 months; $4,760
Munroe, M. Evans; (GY7239); 20 months; $13,000

NEW JERSEY

PRINCETON UNIVERSITY; Bonini, William E.; (GY7280); 11 months; $3,040
Bruce, Victor; (GY7565); 11 months; $6,830
Lockard, W. Duane; (GY7620); 19 months; $10,430
Tucker, Albert W.; (GY7481); 11 months; $7,760

RUTGERS UNIVERSITY; Page, Robert H.; (GY7306); 11 months; $11,340
Wohl, Ronald A.; (GY7556); 11 months; $12,2000

RUTGERS THE STATE UNIVERSITY, Newark Campus; Kluiber, Rudolph W.; (GY7412); 11 months; $8,230

SETON HALL UNIVERSITY; Ewing, Galen W.; (GY7516); 11 months; $11,700

UPSALA COLLEGE; Borowitz, Grace; (GY7460); 11 months; $5,790

NEW MEXICO

NEW MEXICO STATE UNIVERSITY; Ames, Lynford L.; (GY7329); 11 months; $8,520
Ford, C. Quentin; (GY7598); 23 months; $10,640

UNIVERSITY OF NEW MEXICO; Ellis, Henry C.; (GY7408); 11 months; $8,130

NEW YORK

ADELPHI UNIVERSITY; Lacey, Richard J.; (GY7474); 11 months; $11,700

AMERICAN MUSEUM OF NATURAL HISTORY; Shaw, Evelyn; (GY7219); 20 months; $2,830

CANISIUS COLLEGE; Bieron, Joseph F.; (GY7575); 11 months; $6,100

CLARKSON COLLEGE OF TECHNOLOGY; Nunge, Richard J.; (GY7318); 11 months; $9,630
Partch, Richard E.; (GY7464); 11 months; $7,500

COLLEGE OF FORESTRY AT SYRACUSE; Simeone, John B.; (GY7286); 11 months; $17,600

COLLEGE OF PHARMACEUTICAL SCIENCE; Patel, Dhun B.; (GY7387); 11 months; $7,600

COLD SPRING HARBOR LABORATORY; Werner, Rudolf; (GY7603); 11 months; $16,030

CORNELL UNIVERSITY; Kostiner, Edward S.; (GY7634); 11 months; $16,770
Stycos, J. M.; (GY7615); 20 months; $12,800

CUNY, BROOKLYN COLLEGE; Rosen, Milton J.; (GY7540); 11 months; $17,090

CUNY, CITY COLLEGE; Menkes, Sherwood B.; (GY7285); 11 months; $11,200
Wilen, Samuel H.; (GY7643); 11 months; $28,000

CUNY, HUNTER COLLEGE; Fuchs, Helmuth; (GY7534); 11 months; $8,330

FORDHAM UNIVERSITY; Weber, Alfons; (GY7635); 11 months; $9,380

HAMILTON COLLEGE; Denney, Donald J.; (GY7506); 11 months; $8,500

HEALTH RESEARCH INC., Roswell Park Division; Mirand, Edwin A.; (GY7625); 11 months; $34,800

ITHACA COLLEGE; Koch, Heinz F.; (GY7413); 11 months; $6,100

SCIENCE EDUCATION FOR UNDERGRADUATE STUDENTS

MANHATTAN COLLEGE; Batt, Bro. C. William; (GY7607); 20 months; $9,120

NEW YORK UNIVERSITY; Landis, Fred; (GY7502); 11 months; $16,610
O'Malley, Robert E. Jr.; (GY7628); 11 months; $15,520
Underwood, Graham R.; (GY7295); 11 months; $10,590
Yarmus, L.; (GY7579); 11 months; $5,990

POLYTECHNIC INSTITUTE OF BROOKLYN; Boorstyn, Robert R.; (GY7434); 11 months; $18,300
Goldman, James A.; (GY7578); 11 months; $9,870
Koplik, B.; (GY7531); 11 months; $9,380
Krieger, J. B.; (GY7582); 11 months; $11,240
Nardo, Sebastian V.; (GY7439); 11 months; $8,480

RENSSELAER POLYTECHNIC INSTITUTE; Richtol, Herbert H.; (GY7484); 11 months; $10,230
Yergin, Paul F.; (GY7581); 11 months; $14,400

STATE UNIVERSITY AT ALBANY; Andrews, C. L.; (GY7376); 11 months; $4,610
Zenner, Walter P.; (GY7617); 11 months; $5,100

STATE UNIVERSITY AT BINGHAMTON; Sterling Jr., Nicholas J.; (GY7444); 11 months; $7,110

STATE UNIVERSITY AT STONY BROOK; Bonner, F. T.; (GY7235); 20 months; $15,000

SYRACUSE UNIVERSITY; Dosanjh, D.; (GY7449); 11 months; $9,190
Schroder, Klaus; (GY7605); 11 months; $8,830

UNION COLLEGE; Lanese, John G.; (GY7448); 11 months; $11,670

UNIVERSITY OF ROCHESTER; Efran, Jay S.; (GY7308); 11 months; $19,280

NORTH CAROLINA

NORTH CAROLINA STATE UNIVERSITY AT RALEIGH; Doolittle, Jesse S.; (GY7632), 11 months; $13,480
Freedman, Leon D.; (GY7551); 11 months; $11,700
Maki, T. E.; (GY7279); 11 months; $7,000

UNIVERSITY OF NORTH CAROLINA AT CHAPEL HILL; Bell, C. Ritchie; (GY7396); 11 months; $9,440
Carter, Lewis F.; (GY7535); 11 months; $12,600
Hatfield, William E.; (GY7311); 11 months; $16,360

UNIVERSITY OF NORTH CAROLINA AT GREENSBORO; Puterbaugh, Walter H.; (GY7555); 11 months; $10,630

NORTH DAKOTA

MINOT STATE COLLEGE; Leiby, Paul D.; (GY7386); 11 months; $5,850

NORTH DAKOTA STATE UNIVERSITY; Maricich, Tom J.; (GY7454); 11 months; $8,240

UNIVERSITY OF NORTH DAKOTA; Karner, Frank R.; (GY7619); 11 months; $5,410
Stenberg, Virgil I.; (GY7415); 11 months; $7,810

OHIO

ANTIOCH COLLEGE; Corwin, James F.; (GY7541); 19 months; $7,510
Taylor, Charles E.; (GY7638); 20 months; $6,140

BOWLING GREEN STATE UNIVERSITY; Badia, Pietro; (GY7606); 11 months; $8,600

CASE WESTERN RESERVE UNIVERSITY; Davis, Robert P.; (GY7379); 11 months; $20,420
McGervey, John D.; (GY7543); 11 months; $14,070
Wallace, John F.; (GY7345); 11 months; $8,850

COLLEGE OF WOOSTER; Elwell, David L.; (GY7505); 11 months; $1,940
Leach, David A.; (GY7545); 11 months; $5,700
Moke, Charles B.; (GY7278); 11 months; $2,800
Powell, David L.; (GY7494); 11 months; $8,360

DENISON UNIVERSITY; Alrutz, Robert W.; (GY7457); 11 months; $7,400
Bork, Kennard B.; (GY7560); 11 months; $2,950
Grant Jr., Roderick M.; (GY7483); 11 months; $2,960

HIRAM COLLEGE; Barrow Jr., James H.; (GY7313); 11 months; $6,940

JOHN CARROLL UNIVERSITY; Carome, Edward F.; (GY7275); 11 months; $7,620

OBERLIN COLLEGE; High, Lee R.; (GY7405); 11 months; $3,480
Schoonmaker, R. C.; (GY7300); 11 months; $8,400
Sherman, Thomas F.; (GY7549); 11 months; $7,210
Warner, Robert E.; (GY7357); 11 months; $7,000

OHIO STATE UNIVERSITY; Fleisher, Belton M.; (GY7290); 20 months; $8,500
Malspeis, Louis; (GY7446); 11 months; $22,680
Ouellette, Robert J.; (GY7213); 11 months; $20,550
Wigen, Philip E.; (GY7326); 11 months; $19,720

OHIO WESLEYAN UNIVERSITY; Wilson, Lauren R.; (GY7601); 11 months; $5,110

UNIVERSITY OF DAYTON; Boehman, Louis; (GY7561); 11 months; $5,590
Chantell, Charles J.; (GY7525); 11 months; $9,840

WRIGHT STATE UNIVERSITY; Karl, David J.; (GY7586); 11 months; $4,870

YOUNGSTOWN STATE UNIVERSITY; Kelley Jr., George W.; (GY7608); 44 months; $14,400

OKLAHOMA

OKLAHOMA CITY UNIVERSITY; Branch, John C.; (GY7508); 11 months; $9,630

OKLAHOMA STATE UNIVERSITY; Martin, Joel J.; (GY7627); 11 months; $6,000

SOUTHWESTERN STATE COLLEGE; White, Harold M.; (GY7220); 20 months; $10,200

UNIVERSITY OF OKLAHOMA; Jerner, R. Craig; (GY7458); 11 months; $11,680

OREGON

LINFIELD COLLEGE; Boling, John L.; (GY7277); 11 months; $1,400

OREGON STATE UNIVERSITY; Denison, William C.; (GY7641); 11 months; $4,800
Parsons, Theran D.; (GY7301); 11 months; $11,000

PORTLAND STATE UNIVERSITY; Levinson, A. S.; (GY7523); 11 months; $7,740

REED COLLEGE; Brehm, Bert G.; (GY7381); 11 months; $9,520
Hoffman, Dennis G.; (GY7456); 11 months; $8,660
Hunt, Burrowes; (GY7426); 11 months; $4,530
Squier, Leslie H.; (GY7427); 11 months; $10,330

UNIVERSITY OF OREGON; Carroll, George C.; (GY7302); 11 months; $26,200
——— ; (GY7302001); $350
King, Gerald A.; (GY7624); 13 months; $10,150
Koenig, Thomas W.; (GY7513); 11 months; $10,100
Moursund, A. F.; (GY7305); 11 months; $12,090

PENNSYLVANIA

ALBERT EINSTEIN MEDICAL CENTER; Ajl, Samuel J.; (GY5859001); $1,737

ALLEGHENY COLLEGE; Parsons, William H.; (GY7273); 11 months; $3,360

BRYN MAWR COLLEGE; Berliner, Ernst; (GY7261); 11 months; $5,200
Conner, Robert L.; (GY7453); 11 months; $8,920
Gonzalez, R. C.; (GY7337); 11 months; $4,730

BUCKNELL UNIVERSITY; Candland, Douglas K.; (GY7514); 11 months; $9,630
Harclerode, Jack E.; (GY7262); 11 months; $8,400

CARNEGIE MELLON UNIVERSITY; Paxton, Harold W.; (GY7530); 11 months; $4,000

DICKINSON COLLEGE; Long, Howard C.; (GY7276); 20 months; $3,640
Meyer, Marvin W.; (GY7423); 11 months; $8,600

DREXEL UNIVERSITY; Keitel, Glenn H.; (GY7452); 14 months; $13,990
Winters, Lawrence J.; (GY7612); 18 months; $13,640

DUQUESNE UNIVERSITY; Frank, Sylvan G.; (GY7334); 11 months; $1,600

ELIZABETHTOWN COLLEGE; Spangler, Martin O. L.; (GY7299); 11 months; $4,560

FRANKLIN AND MARSHALL COLLEGE; Brookshire, Kenneth H.; (GY7564); 11 months; $7,170
Ritter, Dale F.; (GY7475); 11 months; $8,280

FRANKLIN INSTITUTE; Danforth, William E.; (GY7293); 11 months; $7,710

GENEVA COLLEGE; Adams, Roy M.; (GY7375); 11 months; $7,200

HAVERFORD COLLEGE; Chesick, John P.; (GY7263); 11 months; $4,200
Green, Louis C.; (GY7463); 11 months; $3,450

HOLY FAMILY COLLEGE; Lontz, John F.; (GY7467); 11 months; $2,860

JUNIATA COLLEGE; Rockwell, Kenneth; (GY7422); 11 months; $7,370

LAFAYETTE COLLEGE; Grant, Raymond W.; (GY7307); 11 months; $4,330
Miller, Thomas G.; (GY7631); 11 months; $4,300

LEHIGH UNIVERSITY; Kalnins, Arturs; (GY7356); 11 months; $2,640
Tall, Lambert; (GY7459); 11 months; $7,420
Wood, John D.; (GY7282); 11 months; $7,200

LYCOMING COLLEGE; Kim, Moo Ung; (GY7289); 17 months; $3,220

MUHLENBERG COLLEGE; Smart, G. N. Russell; (GY7374); 11 months; $6,570

PENNSYLVANIA STATE UNIVERSITY; Schipper, Lowell; (GY7447); 11 months; $9,040

SWARTHMORE COLLEGE; Rawson, Kenneth S.; (GY7339); 11 months; $8,430

TEMPLE UNIVERSITY; Dubeck, Leroy W.; (GY7486); 44 months; $14,640

UNIVERSITY OF PENNSYLVANIA; Myers, Alan L.; (GY7333); 11 months; $7,740

RHODE ISLAND

BROWN UNIVERSITY; Ciabattoni, J.; (GY7546); 11 months; $17,420
Engen, Trygg; (GY7567); 11 months; $14,100
Seidel, G. M.; (GY7312); 11 months; $7,550

PROVIDENCE COLLEGE; Galkowski, Theodore T.; (GY7229); 11 months; $6,600

SCIENCE EDUCATION PROGRAMS

UNIVERSITY OF RHODE ISLAND; Lal, Harbans; (GY7461); 11 months; $15,280
Polk, Charles; (GY7397); 11 months; $9,500
Test, Frederick L.; (GY7592); 43 months; $11,620

SOUTH CAROLINA

CLEMSON UNIVERSITY; Hare Jr., William R.; (GY7473); 11 months; $6,580

FURMAN UNIVERSITY; Pyron, R. Scott; (GY7436); 11 months; $7,080
Stratton, Lewis P.; (GY7419); 11 months; $5,680

UNIVERSITY OF SOUTH CAROLINA; Woodward Jr., E. C.; (GY7529); 11 months; $3,980

SOUTH DAKOTA

AUGUSTANA COLLEGE; Gildseth, Wayne M.; (GY7230); 11 months; $5,600
Proscott, Lansing M.; (GY7577); 11 months; $8,980

TENNESSEE

ATOMIC ENERGY COMMISSION, Oak Ridge Operations Office; Rayburn, Louis A.; (AG215); 11 months, $10,900

SOUTHWESTERN AT MEMPHIS; Amy, Robert L.; (GY7522); 11 months; $7,570

TENNESSEE TECHNOLOGICAL UNIVERSITY; Sissom, Leighton E.; (GY7440); 11 months; $13,370

UNIVERSITY OF TENNESSEE; Jones, Arthur W.; (GY7355); 11 months; $11,000
Kleinfelter, D. C.; (GY7573); 11 months; $13,420

VANDERBILT UNIVERSITY; Heiser, A. M.; (GY7341); 11 months; $6,370
Lichter, Barry D.; (GY7336); 11 months; $10,790
Peterson, Richard A.; (GY7390); 17 months; $11,700
Thune, Leland E.; (GY7377); 11 months; $10,670

TEXAS

RICE UNIVERSITY; Bourne Jr., Henry C.; (GY7309); 11 months; $9,420
Glass, Graham P.; (GY7558); 11 months; $8,400

SAM HOUSTON STATE UNIVERSITY; Johnson, James E.; (GY7492); 11 months; $5,870

SOUTHERN METHODIST UNIVERSITY; Boyd, James R.; (GY7595); 11 months; $2,640
Turlington, B. L.; (GY7383); 44 months; $18,830

TEXAS TECH UNIVERSITY; Kristiansen, Magne; (GY7485); 11 months; $10,830

TEXAS WOMAN'S UNIVERSITY; Cockerline, Alan W.; (GY7267); 11 months; $7,600

TRINITY UNIVERSITY; Gaedke, Rudolph M.; (GY7613); 11 months; $6,740

UNIVERSITY OF TEXAS AT AUSTIN; Dougal, Arwin A.; (GY7616); 11 months; $5,160
Manosevitz, Martin; (GY7407); 11 months; $12,670

UTAH

BRIGHAM YOUNG UNIVERSITY; Gardner, John H.; (GY7393); 11 months; $15,270

UNIVERSITY OF UTAH; Dodd, David H.; (GY7510); 11 months; $16,820

UTAH STATE UNIVERSITY; Helm, William T.; (GY7271); 12 months; $9,760

WEBER STATE COLLEGE; Jackson, L. E.; (GY7515); 11 months; $13,580

VERMONT

MIDDLEBURY COLLEGE; Scaife, Charles W. J.; (GY7553); 11 months; $7,370

UNIVERSITY OF VERMONT; Kuehne, Martin E.; (GY7398); 11 months; $11,080

WINDHAM COLLEGE; Shaw Jr., Charles E.; (GY7402); 11 months; $2,960

VIRGINIA

COLLEGE OF WILLIAM AND MARY; Byrd, Mitchell A.; (GY7228); 11 months; $6,000
Goodwin, Bruce K.; (GY7388); 11 months; $4,790
Tyree Jr., S. Young; (GY7343); 20 months; $8,400

HAMPDEN-SYDNEY COLLEGE; Smith Jr., Homer A.; (GY7539); 11 months; $6,180

HOLLINS COLLEGE; Webster, Ronald L.; (GY7589); 11 months; $5,290

OLD DOMINION UNIVERSITY; Clark, Allen K.; (GY7424); 11 months; $6,300
Levy, Gerald F.; (GY7498); 11 months; $7,920
Zaneveld, J. S.; (GY7223); 11 months; $7,600

UNIVERSITY OF VIRGINIA; Lovelace, Eugene A.; (GY7414); 10 months; $8,360
Morton Jr., Harold S.; (GY7479); 11 months; $7,710

VIRGINIA INSTITUTE OF MARINE SCIENCE; Bailey, Robert S.; (GY7294); 11 months; $13,610

VIRGINIA POLYTECHNIC INSTITUTE; Brice Jr., Luther K.; (GY7270); 11 months; $6,000
Comparin, Robert A.; (GY7569); 11 months; $9,430

WASHINGTON & LEE UNIVERSITY; Shillington, James K.; (GY7221); 11 months; $5,600

WASHINGTON

PACIFIC LUTHERN UNIVERSITY; Giddings, William P.; (GY7633); 11 months; $7,400

SEATTLE UNIVERSITY; Steckler, Bernard M.; (GY7322); 11 months; $5,180

UNIVERSITY OF WASHINGTON; Groman, Neal B.; (GY7442); 11 months; $4,930
Nelson, Wendel L.; (GY7400); 11 months; $9,460
Schomaker, Verner; (GY7224); 11 months; $9,600

WASHINGTON STATE UNIVERSITY; Bhatia, Vishnu N.; (GY7266); 11 months; $3,660
Stevens, Carl M.; (GY7246); 11 months; $7,620

WEST VIRGINIA

WEST VIRGINIA UNIVERSITY; Muth, Chester W.; (GY7566); 11 months; $9,070

WHEELING COLLEGE; Loner, Charles J.; (GY7222); 16 months; $6,800

WISCONSIN

BELOIT COLLEGE; Spencer, Brock; (GY7373); 13 months; $6,920
Stenstrom, Richard C.; (GY7332); 20 months; $3,990

LAWRENCE UNIVERSITY; Brandenberger, J. R.; (GY7435); 11 months; $3,720

UNIVERSITY OF WISCONSIN, Madison; Connors, Kenneth A.; (GY7574); 11 months; $9,640
Shain, Irving; (GY7611); 11 months; $21,870
Shea, Daniel F.; (GY7303); 11 months; $13,380

UNIVERSITY OF WISCONSIN, Milwaukee; Chang, Y. Austin; (GY7636); 11 months; $4,230
Dittman, Richard H.; (GY7274); 11 months; $6,600
Madison, Harry L.; (GY7557); 11 months; $9,960

WISCONSIN STATE UNIVERSITY, Stevens Point; Chander, Jagdish; (GY7233); 16 months; $1,200

WYOMING

UNIVERSITY OF WYOMING; Howatson, John; (GY7264); 11 months; $3,600

Science Education for Secondary School Students

Student Science Training Program

ARIZONA

NORTHERN ARIZONA UNIVERSITY; Orosz, Robert A.; (GW5290); 11 months; $11,818

UNIVERSITY OF ARIZONA; Younggren, Newell A.; (GW5246); 10 months; $10,869

CALIFORNIA

CALIFORNIA STATE COLLEGE AT LOS ANGELES; Killgrove, R. B.; (GW5262); 11 months; $19,887

COMMISSION FOR ADVANCED SCIENCE TRAINING; McKee, Ralph W.; (GW5231); 11 months; $8,240
Sobel, Harry; (GW4132001); $1,887

HARVEY MUDD COLLEGE; Bell, Graydon D.; (GW5232); 11 months; $12,736

HUMBOLDT STATE COLLEGE; Butler, John E.; (GW5306); 11 months; $26,489

SALK INSTITUTE FOR BIOLOGICAL STUDIES; Hyman, Robert; (GW5234); 11 months; $2,200

SAN DIEGO STATE COLLEGE; Deaton, Edmund I.; (GW5216); 11 months; $17,871
Spangler, John A.; (GW5286); 11 months; $17,325

SAN JOSE STATE COLLEGE; Watanabe, Ronald S.; (GW5215); 10 months; $16,410

UNIVERSITY OF CALIFORNIA, Berkeley; Wolf, Frantisek; (GW5250); 10 months; $16,165

UNIVERSITY OF SOUTHERN CALIFORNIA; Merz, Robert C.; (GW5295); 11 months; $15,525

COLORADO

COLORADO COLLEGE; Hilt, Richard L.; (GW5222); 11 months; $15,198

COLORADO STATE UNIVERSITY; Walter, Richard; (GW4616); 12 months; $2,000

UNIVERSITY OF COLORADO; Ives, Jack D.; (GW5281); 11 months; $7,097

CONNECTICUT

UNIVERSITY OF BRIDGEPORT; Tucci, James V.; (GW5282); 11 months; $27,402

WESLEYAN UNIVERSITY; Leadbetter, Edward R.; (GW5301); 19 months; $16,330

DISTRICT OF COLUMBIA

AMERICAN UNIVERSITY; Schubert, Leo; (GW5227); 11 months; $8,428

FLORIDA

FLORIDA INSTITUTE OF TECHNOLOGY; Woodbridge, David D.; (GW5292); 11 months; $12,520

FLORIDA STATE UNIVERSITY; Green, George F. Jr.; (GW5310); 11 months; $19,106
Schlitt, Dorothy; (GW5279); 11 months; $15,493

UNIVERSITY OF FLORIDA; Arnold, Luther A.; (GW5271); 11 months; $12,984

GEORGIA

EMORY UNIVERSITY; Neff, Mary F.; (GW5275); 11 months; $10,064

MORRIS BROWN COLLEGE; Payne, W. F.; (GW5318); 11 months; $13,990

UNIVERSITY OF GEORGIA; Heric, E. L.; (GW5319); 10 months; $15,063

HAWAII

UNIVERSITY OF HAWAII; Hylin, John W.; (GW5278); 11 months; $20,031

SCIENCE EDUCATION FOR SECONDARY SCHOOL STUDENTS

IDAHO

UNIVERSITY OF IDAHO; Browne, Michael E.; (GW5229); 11 months; $18,061

ILLINOIS

CHICAGO PARK DISTRICT, Adler Planetarium; Burns, Jay; (GW5317); 21 months; $15,345
———; (GW4004001); $3,213

FIELD MUSEUM OF NATURAL HISTORY; Smith, Harriet; (GW5294); 11 months; $8,291

ILLINOIS INSTITUTE OF TECHNOLOGY; Caton, Willis B.; (GW5247); 11 months; $22,040
Reingold, Haim; (GW5225); 21 months; $17,900

KNOX COLLEGE; Green, D. Wayne; (GW5316); 11 months; $17,489

MACMURRAY COLLEGE; Kohlbecker, Eugene E.; (GW5237); 11 months; $22,015

NORTHERN ILLINOIS UNIVERSITY; Spangler, Charles W.; (GW5221); 11 months; $17,664

NORTHWESTERN UNIVERSITY; Rutherford, David; (GW5280); 11 months; $10,236

UNIVERSITY OF ILLINOIS, Urbana; Dobrovolny, Jerry S.; (GW5241); 11 months; $14,921
Parr, James T.; (GW5251); 11 months; $17,362

INDIANA

INDIANA UNIVERSITY, Bloomington; Klinge, Paul; (GW5312); 11 months; $25,384

PURDUE UNIVERSITY; Eggert, Dean A.; (GW5226); 11 months; $21,125

IOWA

COE COLLEGE; Watkins, Stanley R.; (GW5263); 11 months; $15,538

LUTHER COLLEGE; Tjostem, John L.; (GW5244); 11 months; $9,675

KANSAS

KANSAS STATE TEACHERS COLLEGE, Oram, S. Winston; (GW5269); 11 months; $10,715

UNIVERSITY OF KANSAS; Middaugh, Richard L.; (GW5265); 11 months; $28,726

KENTUCKY

ASBURY COLLEGE; Ray, J. Paul; (GW5283); 10 months; $25,948

BELLARMINE-URSULINE COLLEGE; Daly, John M.; (GW5239); 11 months; $8,845

LOUISIANA

LOUISIANA STATE UNIVERSITY, Baton Rouge; Day, M, Clyde; (GW5254); 11 months; $34,120

LOYOLA UNIVERSITY; Christman, John F.; (GW5293); 11 months; $32,359

MAINE

JACKSON LABORATORY; Waymouth, Charity; (GW5235); 11 months; $18,445

MARYLAND

MORGAN STATE COLLEGE; King, John W.; (GW5285); 11 months; $13,733

MASSACHUSETTS

FOUNDATION FOR RESEARCH ON THE NERVOUS SYSTEM; Bogoch, Samuel; (GW4018002); 12 months; $7,200
———; (GW4018001); $8,490

STONEHILL COLLEGE; Litchfield, Marshall B.; (GW5249); 11 months; $19,095

MICHIGAN

MICHIGAN STATE UNIVERSITY; Carroll, Tom W.; (GW5259); 11 months; $14,987
Peebles, Charles R.; (GW5274); 11 months; $37,504

NORTHERN MICHIGAN UNIVERSITY; Farrell, John P.; (GW5308); 11 months; $20,388

OAKLAND UNIVERSITY; McKay, James H.; (GW5264); 11 months; $10,984

WAYNE STATE UNIVERSITY; Roellig, Leonard O.; (GW5220); 11 months; $18,920

WESTERN MICHIGAN UNIVERSITY; Kent, Neil D.; (GW5256); 11 months; $32,833
Mallinson, George G.; (GW5267); 11 months; $19,096

MINNESOTA

ST. OLAF COLLEGE; Enger, Thomas P.; (GW5245); 11 months; $12,407

MISSISSIPPI

MISSISSIPPI STATE COLLEGE FOR WOMEN; Sherman, Harry L.; (GW5300); 11 months; $8,722

NEVADA

FORESTA INSTITUTE OF OCEAN STUDIES; Miller, Richard Gordon; (GW4025001); $3,219

NEW HAMPSHIRE

UNIVERSITY OF NEW HAMPSHIRE; Bennett Jr., Albert B.; (GW5277); 11 months; $23,293

NEW JERSEY

RUTGERS UNIVERSITY; Greitzer, Samuel L.; (GW5309); 11 months; $19,717

STEVENS INSTITUTE OF TECHNOLOGY; White, Myron E.; (GW5296); 11 months; $24,674
———; (GW5261); 20 months; $9,817

NEW MEXICO

NEW MEXICO STATE UNIVERSITY; Cleveland, E. L.; (GW5284); 11 months; $18,682

NEW YORK

AMERICAN MUSEUM OF NATURAL HISTORY; Branley, Franklyn M.; (GW5314); 18 months; $12,650
———; (GW4027001); $4,035

BOYCE THOMPSON INSTITUTE FOR PLANT RESEARCH; App, Alva A.; (GW5260); 11 months; $8,076

BROOKLYN CHILDREN'S MUSEUM; Laruffa, Anthony; (GW5238); 11 months; $12,460

COLLEGE AT FREDONIA; Keller, Roy A.; (GW5268); 11 months; $7,220
Vreeland, Howard W.; (GW 5228); 11 months; $10,353

CORNELL UNIVERSITY; Richardson, Robert C.; (GW5217001); $275
———; (GW5217); 11 months; $12,215

CUNY, CENTRAL SYSTEM OFFICE; Jordan, Meyer; (GW5304); 21 months; $53,507

HEALTH RESEARCH INC., Roswell Park Division; Mirand, Edwin A.; (GW5302); 11 months; $19,620

MANHATTAN COLLEGE; Henderson, A. Peter; (GW5224); 11 months; $16,792

NEW YORK UNIVERSITY; Novikoff; Albert B.; (GW5299); 19 months; $14,564

POLYTECHNIC INSTITUTE OF BROOKLYN; Hauser, Norbert; (GW5272); 11 months; $13,159

SYRACUSE UNIVERSITY; Trischka, John W.; (GW5218); 11 months; $28,894

WALDEMAR MEDICAL RESEARCH FOUNDATION; Gross, Leo; (GW5230); 11 months; $18,510

YESHIVA UNIVERSITY; Tendler, Moses D.; (GW5320); 11 months; $12,267

NORTH CAROLINA

BENNETT COLLEGE; Sayles, J. Henry; (GW5236); 10 months; $22,893

EAST CAROLINA UNIVERSITY; Daugherty, Patricia A.; (GW5322); 18 months; $8,736

UNIVERSITY OF NORTH CAROLINA AT CHAPEL HILL; Jicha, Donald C.; (GW5252); 10 months; $22,925

NORTH DAKOTA

NORTH DAKOTA STATE UNIVERSITY; Hill, Loren W.; (GW5219); 11 months; $7,193

OHIO

OHIO STATE UNIVERSITY; Brown, Harold D.; (GW5276); 11 months; $32,044
Yarrington, Paul T.; (GW5287001); $250
———; (GW5287); 11 months; $13,874

OHIO UNIVERSITY; Tong, James Y.; (GW5240); 11 months; $15,786

UNIVERSITY OF AKRON; Burrowbridge, Donald R.; (GW5255); 10 months; $17,217

OKLAHOMA

UNIVERSITY OF OKLAHOMA; Iverson, Lloyd A.; (GW5266); 11 months; $18,000

OREGON

OREGON STATE UNIVERSITY; Enlows, Harold E.; (GW5257); 11 months; $12,074

PENNSYLVANIA

BUCKNELL UNIVERSITY; Kieft, Lester; (GW5297); 11 months; $12,141

CARNEGIE MELLON UNIVERSITY; McKinney, David S.; (GW5291); 11 months; $9,548

CLARION STATE COLLEGE; Konitzky, Gustav A.; (GW5233); 11 months; $8,672

DREXEL UNIVERSITY; Bednar, J. Bee; (GW5298); 11 months; $12,259

HAHNEMANN MEDICAL COLLEGE AND HOSPITAL; Satinsky, Victor P.; (GW5315); 20 months; $26,201

LEHIGH UNIVERSITY; Wilansky, Albert; (GW5213); 11 months; $10,459

UNIVERSITY OF PENNSYLVANIA; Faul, Henry; (GW5214); 19 months; $6,044

RHODE ISLAND

UNIVERSITY OF RHODE ISLAND; Knauss, John A.; (GW5288); 11 months; $15,271

SOUTH CAROLINA

CLEMSON UNIVERSITY; Camper, N. Dwight; (GW5248); 11 months; $12,442

TENNESSEE

AUSTIN PEAY STATE UNIVERSITY; Woodward, Ernest L.; (GW5289); 11 months; $17,094

UNIVERSITY OF TENNESSEE; McLean, Robert A.; (GW5242); 11 months; $17,366

TEXAS

BAYLOR UNIVERSITY; Schwetman, Herbert D.; (GW5258); 11 months; $12,058

HARDIN-SIMMONS UNIVERSITY; Robinson, Charles D.; (GW5321); 10 months; $9,152

PRAIRIE VIEW A & M COLLEGE; O'Banion, E. E.; (GW5313); 11 months; $9,756

RICE UNIVERSITY; Adams, John A. S.; (GW5273); 11 months; $17,264

SOUTHERN METHODIST UNIVERSITY; Palas, Frank J.; (GW5307); 11 months; $12,812

TEXAS A & M UNIVERSITY; Smith, Fred E.; (GW5253); 11 months; $13,304
Thompson, J. George H.; (GW5223); 11 months; $12,992

UNIVERSITY OF TEXAS AT AUSTIN; Spear, Irwin; (GW5311); 11 months; $15,750

SCIENCE EDUCATION PROGRAMS

UTAH

UTAH STATE UNIVERSITY; Cannon, Lawrence O.; (GW5243); 18 months; $15,520

VIRGINIA

UNIVERSITY OF VIRGINIA; Stevenson, Edward C.; (GW5270); 11 months; $15,763

VIRGINIA POLYTECHNIC INSTITUTE; Potter, L. M.; (GW5303); 11 months; $11,650

WASHINGTON

UNIVERSITY OF WASHINGTON; Jayne, Benjamin A.; (GW5305); 11 months; $14,214

WHITMAN COLLEGE; Keiser Jr., Victor H.; (GW5212); 11 months; $13,623

Supplementary Science Projects for Students

ALABAMA

STILLMAN COLLEGE
Weaver, Patricia L.; *A Mathematics Project for Secondary School Students* (GW5710); 5 months; $2,094

UNIVERSITY OF ALABAMA, Tuscaloosa
Plunkett, Robert L.; *Alabama Mathematics Talent Search* (GW5060); 9 months; $9,723

ARIZONA

ARIZONA ACADEMY OF SCIENCE
Smith, David T.; *Supplementary Science Projects for Students* (GW4622); 12 months; $2,070

DISTRICT OF COLUMBIA

AMERICAN ASSOCIATION FOR THE ADVANCEMENT OF SCIENCE
Kabisch, William T.; *Holiday Science Lecture Series* (GE5915004); 14 months; $16,995

AMERICAN COUNCIL ON EDUCATION
Astin, Alexander W.; *A Program of Longitudinal Research on the Higher Educational System* (GR89*); 12 months; $25,000

OTHER REIMBURSEMENTS
Project Australia 1970—Participant Stipends and Domestic Travel for May for the President's Australian Science Scholars (MOR7013); $4,300
Expenses Relating to President's Australian Science Scholars, August 19, 1970 (70PO943); 3 months; $450
Travel and Stipend for 1969 Project Australia Scholars (MOR001); 6 months; $6,305

FLORIDA

UNIVERSITY OF MIAMI
Meyer, Herman; *Accelerated Mathematics Project for Talented Secondary School Students* (GW5717); 19 months; $16,175

HAWAII

HAWAIIAN ACADEMY OF SCIENCE
Carr, Albert B.; *Student Science Seminar Project* (GW4523001); $200

IOWA

IOWA ACADEMY OF SCIENCE
Starr, Frank W.; *Junior Academy of Science Project* (GW4614); 12 months; $2,000

KANSAS

KANSAS ACADEMY OF SCIENCE
Welsch, Marlin; *Junior Academy of Science Project* (GW4615); 12 months; $2,014

LOUISIANA

LOUISIANA ACADEMY OF SCIENCE
Cousins, Genevieve; *Junior Academy of Science Project* (GW4619); 12 months; $2,050

MARYLAND

MARYLAND ACADEMY OF SCIENCES
Whiteford, Edith B.; *Student Science Seminar Project* (GW4577); 13 months; $9,105

MISSOURI

ACADEMY OF SCIENCE OF ST. LOUIS
Brazier, Donn P.; *Science Enrichment Projects for Students* (GW4506001); $1,493

UNIVERSITY OF MISSOURI, Kansas City
Nahrstedt, Gary W.; *Supplementary Science Projects for Students, Junior Academy of Science* (GW5367); 9 months; $1,977

NEW JERSEY

NEW JERSEY ACADEMY OF SCIENCE
Mitchell, Denis; *Junior Academy of Science Project* (GW4618); 12 months; $2,000

NEW YORK

COLUMBIA UNIVERSITY
Sachs, Allan M.; *Science Honors Program* (GW5703); 16 months; $35,000

OHIO

OHIO ACADEMY OF SCIENCE
Acker, Gerald; *Junior Academy of Science Project* (GW4617); 12 months; $2,000

OKLAHOMA

OKLAHOMA ACADEMY OF SCIENCE
Crockett, Jerry J.; *Junior Academy of Science Project* (GW4620001); $122
———; (GW4620); 12 months; $1,785

PENNSYLVANIA

HAHNEMANN MEDICAL COLLEGE AND HOSPITAL
Satinsky, Victor P.; *Academic Year Science-Oriented Course for Twelfth Grade Students* (GW5724); 19 months; $35,452

TENNESSEE

TENNESSEE ACADEMY OF SCIENCE
Bailey, John H.; *Junior Academy of Science Project* (GW4621); 12 months; $2,000

Cooperative College—School Science Program

ALABAMA

SPRING HILL COLLEGE; Furman, Walter L.; (GW5114); 19 months; $31,964

TROY STATE UNIVERSITY; Wilkes, James C.; (GW5115); 19 months; $33,834

UNIVERSITY OF ALABAMA, Birmingham; Burford, Ernest; (GW5119); 19 months; $20,217

UNIVERSITY OF ALABAMA, Tuscaloosa; Garner, Robert H.; (GW5149); 19 months; $25,806

ARIZONA

ARIZONA STATE UNIVERSITY; Smith, Lehi T.; (GW4065002*); $3,329
———; (GW4065002); $20,629

UNIVERSITY OF ARIZONA; Chilcott, John H.; (GW5090); 19 months; $40,030
McCullough Jr., Edgar J.; (GW5153); 19 months; $29,618
Yang, Tien Wei; (GW5123); 15 months; $29,938

ARKANSAS

UNIVERSITY OF ARKANSAS; Hines, Sallylee; (GW5136); 19 months; $27,256

CALIFORNIA

CALIFORNIA STATE COLLEGE AT FULLERTON; Lay, L. Clark; (GW5204); 19 months; $40,411

CALIFORNIA STATE COLLEGE AT HAYWARD; Scudder, Harvey I.; (GW5155); 19 months; $24,368

CALIFORNIA STATE COLLEGE; Coash, John R.; (GW5206); 19 months; $40,962

CHICO STATE COLLEGE; Cook, Lloyd M.; (GW5124); 19 months; $39,592

COLLEGE OF NOTRE DAME; Chapin, June R.; (GW5105); 19 months; $25,602

SAN DIEGO STATE COLLEGE; Burton, Charles R.; (GW5093); 19 months; $43,439

SAN FRANCISCO STATE COLLEGE; Kingsley, Rembert B.; (GW5166); 19 months; $20,658

UNIVERSITY OF CALIFORNIA, Berkeley; Karplus, Robert; (GW5084); 19 months; $43,488
———; (GW5086); 19 months; $41,723
Laetsch, Watson M.; (GW5085); 19 months; $58,054
Wolf, Frantisek; (GW5704*); 15 months; $23,741

UNIVERSITY OF CALIFORNIA, Los Angeles; Green, John W.; (GW5188); 19 months; $85,172
———; (GW5188 001); $1,595

COLORADO

UNIVERSITY OF COLORADO; Anderson, Ronald D.; (GW5072); 14 months; $39,595
Feng, Chuan C.; (GW5106); 16 months; $11,206
Maler, G. J.; (GW5177); 19 months; $53,528
———; (GW5177001); $6,268

UNIVERSITY OF DENVER; Hoffman, Ruth Irene; (GW5134); 14 months; $28,829

UNIVERSITY OF NORTHERN COLORADO; Fry, Richard K.; (GW5191); 19 months; $41,759

CONNECTICUT

UNIVERSITY OF BRIDGEPORT; Menzel, E. Wesley; (GW5088); 19 months; $27,193

DISTRICT OF COLUMBIA

AMERICAN UNIVERSITY; Schubert, Leo; (GW5068); 19 months; $32,326

HOWARD UNIVERSITY; Eagleson, Halson V.; (GW5172); 19 months; $32,567

FLORIDA

FLORIDA ATLANTIC UNIVERSITY; Clark, Samuel F.; (GW5133); 19 months; $33,347

FLORIDA INSTITUTE OF TECHNOLOGY; Woodbridge, David D.; (GW5148); 19 months; $60,123

UNIVERSITY OF FLORIDA; Bingham, N. E.; (GW5151); 19 months; $74,823

UNIVERSITY OF SOUTH FLORIDA; Kopp, E. W.; (GW5183); 19 months; $39,553

GEORGIA

GEORGIA INSTITUTE OF TECHNOLOGY; Jensen, Alton P.; (GW5065); 19 months; $29,602
Larson, Ronal W.; (GW5113); 19 months; $31,747

OGLETHORPE COLLEGE; Wheeler, George F.; (GW5127); 19 months; $46,624

HAWAII

UNIVERSITY OF HAWAII; Campbell, Robert L.; (GW5159); 19 months; $41,325

IDAHO

NORTHWEST NAZARENE COLLEGE; Marks, Darrell L.; (GW5110); 19 months; $33,220

UNIVERSITY OF IDAHO; Browne, M. E.; (GW5102); 19 months; $54,670

ILLINOIS

ILLINOIS INSTITUTE OF TECHNOLOGY; Greeno, C. L.; (GW5144); 19 months; $54,220

COOPERATIVE COLLEGE—SCHOOL SCIENCE PROGRAM

Machtinger, Lawrence A.; (GW5082); 19 months; $16,419

NORTHWESTERN UNIVERSITY; Cember, Herman; (GW5179); 19 months; $62,164

ROOSEVELT UNIVERSITY; Estin, Robert W.; (GW5121); 19 months; $31,805

UNIVERSITY OF ILLINOIS, Urbana; Beberman, Max; (GW5157); 19 months; $50,803
——; (GW5181); 19 months; $21,571
——; (GW5180); 19 months; $24,480

INDIANA

INDIANA INSTITUTE OF TECHNOLOGY; Williams, Walter J. Jr.; (GW5141); 19 months; $33,652

PURDUE UNIVERSITY Indianapolis Center; Alton, Elaine V.; (GW5140); 19 months; $32,584
——; (GW5140 001); $90
——; (GW5140*); 19 months; $13,962

VALPARAISO UNIVERSITY; Knudten, Richard D.; (GW5184); 19 months; $61,928

IOWA

UNIVERSITY OF IOWA; Yager, Robert E.; (GW5098); 19 months; $40,133

UNIVERSITY OF NORTHERN IOWA; Unruh, Roy D.; (GW5063); 19 months; $30,135

KANSAS

KANSAS STATE TEACHERS COLLEGE; Menhusen, Bernadette; (GW5069); 14 months; $16,996

KANSAS STATE UNIVERSITY; James, Robert K.; (GW5190); 19 months; $16,571

KENTUCKY

EASTERN KENTUCKY UNIVERSITY; George, Ted; (GW5104); 19 months; $44,184

LOUISIANA

FRANCIS T. NICHOLLS STATE COLLEGE; Ohmer, Merlin M.; (GW5199); 19 months; $46,876

LOUISIANA STATE UNIVERSITY, Baton Rouge; Wells, Darthon V.; (GW5129); 19 months; $39,919

NORTHEAST LOUISIANA STATE COLLEGE; Mapp, Marcus; (GW5071); 19 months; $27,898

NORTHWESTERN STATE COLLEGE OF LOUISIANA; Waskon, John D.; (GW5147); 19 months; $24,999

MARYLAND

UNIVERSITY OF MARYLAND; Gardner, Marjorie H.; (GW5132); 19 months; $53,504

MASSACHUSETTS

EASTERN NAZARENE COLLEGE; Jablonski, John R.; (GW5142); 19 months; $53,515

STATE COLLEGE AT BRIDGEWATER; Weygand, George A.; (GW5062); 19 months; $43,655

STATE COLLEGE AT LOWELL; Klee, Lucille H.; (GW5200); 19 months; $39,420

UNIVERSITY OF MASSACHUSETTS; Konicek, Richard D.; (GW5135); 19 months; $29,389

MICHIGAN

CALVIN COLLEGE; Ehlers, Vernon J.; (GW5094); 19 months; $32,767

MICHIGAN STATE UNIVERSITY; Brehm, Shirley; (GW5158); 19 months; $27,081
Vance, Irvin E.; (GW5388); 19 months; $43,671

UNIVERSITY OF DETROIT; Lytle III, Archie K.; (GW5126); 19 months; $68,572

MINNESOTA

BETHEL COLLEGE AND SEMINARY; Carlson, Philip R.; (GW5161); 19 months; $40,000

ST. CLOUD STATE COLLEGE; Highsmith, Robert J.; (GW5097); 18 months; $23,296
——; (GW5097*); 18 months; $5,824

UNIVERSITY OF MINNESOTA; Phinney, William C.; (GW5128); 19 months; $23,720

WINONA STATE COLLEGE; Hamerski, David E.; (GW5108*); 19 months; $5,468
——; (GW5108); 19 months; $38,279

MISSISSIPPI

MISSISSIPPI STATE COLLEGE FOR WOMEN; Sherman, Harry L.; (GW5139); 19 months; $10,425

UNIVERSITY OF SOUTHERN MISSISSIPPI; Craven, Bobby E.; (GW5145); 19 months; $29,451
Sonnier, Isadore L.; (GW5066); 19 months; $48,237

MISSOURI

CENTRAL MISSOURI STATE COLLEGE; Hopping, Joe M.; (GW5100); 20 months; $33,581

SOUTHEAST MISSOURI STATE COLLEGE; Bahn Jr., E. Lawrence; (GW5070); 19 months; $33,341

UNIVERSITY OF MISSOURI, Columbia; Leake, John Benjamin; (GW5103); 8 months; $5,293

NEW JERSEY

TRENTON STATE COLLEGE; Pregger, Fred T.; (GW5096); 19 months; $36,132

NEW YORK

COLLEGE AT GENESEO; Oakes, Russell C.; (GW5109*); 19 months; $1,723
——; (GW5109); 19 months; $15,514

COLLEGE CENTER OF FINGER LAKES; Potter, Louise F.; (GW5130); 19 months; $33,739

COLLEGE OF ST. ROSE; McGrath, John F.; (GW5080); 19 months; $32,550

CUNY, BROOKLYN COLLEGE; Gavurin, Lester L.; (GW5197); 19 months; $47,858

CUNY, HUNTER COLLEGE; Bulkin, Bernard J.; (GW5092); 19 months; $39,203

HOFSTRA UNIVERSITY; Sparberg, Esther B.; (GW5077); 19 months; $16,551

STATE UNIVERSITY AT ALBANY; Boehm, Thomas; (GW5075); 19 months; $13,705
——; (GW5208); $822

STATE UNIVERSITY AT STONY BROOK; Paldy, Lester G.; (GW5150); 19 months; $49,874

SYRACUSE UNIVERSITY; Davis, Robert B.; (GW5164); 19 months; $18,564
——; (GW5165); 19 months; $45,490
——; (GW5164*); 19 months; $9,282
——; (GW5194); 19 months; $28,603
——; (GW5173*); 19 months; $4,304
——; (GW5194*); 19 months; $14,301
——; (GW5173); 19 months; $8,609
——; (GW5165*); 19 months; $22,745
——; (GW4900); 11 months; $11,845
——; (GW4900*); 11 months; $3,948

WALDEMAR MEDICAL RESEARCH FOUNDATION; Gross, Leo; (GW5193); 19 months; $25,703

YORK COLLEGE; Pomilla, Frank R.; (GW5143); 19 months; $47,178

NORTH CAROLINA

APPALACHIAN STATE UNIVERSITY; Derrick, F. Ray; (GW5101); 19 months; $17,493

EAST CAROLINA UNIVERSITY; Mattheis, Floyd E.; (GW5081); 19 months; $75,941

NORTH DAKOTA

MINOT STATE COLLEGE; Clausen, Eric; (GW5154); 19 months; $36,282

OHIO

KENT STATE UNIVERSITY; Duffy, Norman V.; (GW5074); 19 months; $19,193

OHIO STATE UNIVERSITY; Mayer, Victor J.; (GW5117); 19 months; $44,547

OHIO UNIVERSITY; Light, Kenneth H.; (GW5138); 19 months; $45,658
Skinner Jr., Ray; (GW5170); 19 months; $31,973
——; (GW5170001); $5,400
Witters, Weldon L.; (GW5073); 19 months; $35,386

UNIVERSITY OF AKRON; Jackson, Jim L.; (GW5064); 19 months; $61,273

OKLAHOMA

EAST CENTRAL STATE COLLEGE; Stafford, Don G.; (GW5116); 19 months; $31,925

OKLAHOMA STATE UNIVERSITY; Crockett, Jerry; (GW5083); 19 months; $46,426

UNIVERSITY OF OKLAHOMA; Renner, John W.; (GW5178); 19 months; $33,771

OREGON

EASTERN OREGON COLLEGE; Bolen, Virgil A.; (GW5107); 19 months; $14,417

PORTLAND STATE UNIVERSITY; Fiasca, Michael; (GW5198); 19 months; $32,659

PENNSYLVANIA

CARLOW COLLEGE; Uricchio, William A.; (GW5168); 19 months; $56,124

HAHNEMANN MEDICAL COLLEGE AND HOSPITAL; Satinsky, Victor P.; (GW5187); 19 months; $59,308

PENNSYLVANIA STATE UNIVERSITY; Alfke, Dorothy; (GW5169); 19 months; $38,567
Jester, William A.; (GW5205); 19 months; $33,385

PMC COLLEGES; Smyth, M. P.; (GW5171); 19 months; $32,949

UNIVERSITY OF PITTSBURGH; Flint, Norman K.; (GW5091); 21 months; $37,921

WEST CHESTER STATE COLLEGE; Fasnacht, Wesley E.; (GW5079); 19 months; $35,620

WILKES COLLEGE; Holden, Stanley J.; (GW5112); 19 months; $29,641

RHODE ISLAND

UNIVERSITY OF RHODE ISLAND; Hemmerle, William J.; (GW5111); 19 months; $32,271

SOUTH CAROLINA

CLEMSON UNIVERSITY; Flatt, James L.; (GW5152); 19 months; $31,157

SOUTH DAKOTA

AUGUSTANA COLLEGE; Lindell, Verlyn L.; (GW5067); 19 months; $28,894

TENNESSEE

AUSTIN PEAY STATE UNIVERSITY; Stokes, William G.; (GW5146); 14 months; $24,229

CHRISTIAN BROTHERS COLLEGE; Staub, Robert; (GW5078); 19 months; $38,438

EAST TENNESSEE STATE UNIVERSITY; Hartsell, Lester C.; (GW5137); 19 months; $19,642

MEMPHIS STATE UNIVERSITY; Kirksey, H. Graden; (GW5125); 19 months; $40,262
Sobol, John A.; (GW5131); 19 months; $47,911

UNIVERSITY OF TENNESSEE AT CHATTANOOGA; Wilson, Robert Lake; (GW5089); 19 months; $30,101

TEXAS

ANGELO STATE UNIVERSITY; Young, Bernard T.; (GW5160); 19 months; $35,052

NORTH TEXAS STATE UNIVERSITY; Nunley, B. G.; (GW5185); 19 months; $30,730

STEPHEN F. AUSTIN STATE UNIVERSITY; Layton, W. I.; (GW5122); 19 months; $28,814

SCIENCE EDUCATION PROGRAMS

UNIVERSITY OF HOUSTON; Schirner, Silas W.; (GW5156); 19 months; $48,774

UNIVERSITY OF TEXAS AT AUSTIN; Little, R. N.; (GW5192); 19 months; $39,976

UNIVERSITY OF TEXAS AT EL PASO; Bolen, Max C.; (GW5201); 19 months; $34,282

UTAH

BRIGHAM YOUNG UNIVERSITY; Wight, Theodore A.; (GW5076); 19 months; $69,930

VIRGINIA

OLD DOMINION UNIVERSITY; Pittman, Melvin A.; (GW5087); 10 months; $33,112

WEST VIRGINIA

MARSHALL UNIVERSITY; Martin, Donald C.; (GW5095); 19 months; $27,598

WEST VIRGINIA STATE COLLEGE; Kagen, Herbert P.; (GW5167); 19 months; $37,217

WEST VIRGINIA UNIVERSITY; Eaves, J. C.; (GW5118); 19 months; $57,845

WISCONSIN

UNIVERSITY OF WISCONSIN, Madison; Harvey, John G.; (GW5099); 10 months; $36,039

WISCONSIN STATE UNIVERSITY, Oshkosh; Beck, Eugene J.; (GW5120); 19 months; $29,415

Specialized Advanced Science Education Projects

Advanced Science Education Projects

CALIFORNIA

GREATER LOS ANGELES CONSORTIUM
Dewey, Richard E., *Urban Studies Examining the Parameters of an Emerging Discipline* (GZ1529); 12 months; $11,160

MARINE BIOLOGICAL LABORATORY
Case, James F.; *Training Project in Experimental Invertebrate Zoology* (GZ1788); 7 months; $20,160

STANFORD UNIVERSITY
Rivers, William L.; *Interdisciplinary Course Development in Mass Communication* (GZ1488); 15 months; $31,500
Rubin, David M., Sachs, David P. and Krupp, Jan R.; *Study on Mass Media Coverage of Environmental Problems* (GZ1777); 16 months; $53,590
Sturrock, Peter A.; *A Graduate Program in Astrophysics* (GZ1485001); $4,250
———; (GZ1485); 21 months; $33,600

UNIVERSITY OF CALIFORNIA, Berkeley
Reif, Frederick; *A New Graduate Program in Science and Mathematics Education* (GZ1816); 16 months; $29,700

UNIVERSITY OF CALIFORNIA, San Diego
Saltman, Paul D.; *Design of a Series of Films on the Relation of Science and Technology to the Problems of Society* (GZ1808); 14 months; $22,930

UNIVERSITY OF CALIFORNIA, Santa Barbara
Harris, Richard L.; *Summer Field Training Seminar for Political Science Graduate Students Specializing in Foreign Studies* (GZ1804); 6 months; $9,760
Holmes, Robert W.; *Symposium on Oil Pollution of the Sea* (GZ1786); 15 months; $6,130

COLORADO

COLORADO STATE UNIVERSITY
Frayer, Warren E.; *Symposium on the Development and Implementation of Courses and Curricula in Natural-Resource Biometry* (GZ1477); 17 months; $6,020

Miller, Lee D.; *Special Project in Graduate Education, the Technology & Application of Remote Sensing of Natural Resources* (GZ1760); 30 months; $112,300

FEDERATION OF ROCKY MOUNTAIN STATES INC.
Partridge, William S.; *Publication of Rocky Mountain Science Council Science Newsletter* (GT11*); 12 months; $1,500

UNIVERSITY OF DENVER
Mersky, Roy M.; *Law Librarianship Institute* (GZ1764); 6 months; $12,930

DISTRICT OF COLUMBIA

AMERICAN ASSOCIATION FOR THE ADVANCEMENT OF SCIENCE
Berl, Walter G.; *Public Understanding of Science Activities/The 1969 AAAS Meeting Related to Public Understanding of Science* (GZ1486); 4 months; $10,200

AMERICAN INSTITUTE OF BIOLOGICAL SCIENCES
Olive, John R.; *Workshop Conference on Environmental Education* (GZ1480); 15 months; $6,320
———; *Status of Biological Field Stations in Teaching and Research* (B023195*); 12 months; $6,250

CIVIL SERVICE COMMISSION
Hampton, Robert E.; *Exhibit on the Work of Federal Scientists and Engineers* (AG204001); 12 months; $2,000
———; (AG204); 12 months; $4,000

NATIONAL ENDOWMENT FOR THE HUMANITIES
Keeney, Barnaby; *Public Understanding of Science, a Dialogue on the Identity and Dignity of Man* (AG214); 6 months; $7,640

FLORIDA

AMERICAN COLLEGE PUBLIC RELATIONS ASSOCIATION
Lynch, Robert R.; *Conference on Interpreting Science for the Media* (GZ1375001); $503

GEORGIA

GEORGIA STATE UNIVERSITY
Hadley, Joseph H.; *Strengthening of Graduate Instructional Program in Physics* (GZ1800); 27 months; $73,970

HAWAII

UNIVERSITY OF HAWAII
Armstrong, R. Warwick; *Special Project in Graduate Education Interdisciplinary Program in Human Ecology* (GZ1487); 21 months; $54,920

IDAHO

IDAHO STATE UNIVERSITY
Hartman, Alan M.; *Psychology Graduate Education Project* (GZ685001); 15 months; $12,410

ILLINOIS

NORTHWESTERN UNIVERSITY
Brown, Laurie M.; *Advanced Science Seminar, Symposium on Elementary Particles* (GZ1489); 6 months; $4,025

INDIANA

INDIANA UNIVERSITY, Bloomington
Caldwell, Lynton K.; *Curriculum Development in the Study of the Interaction of Science and Society* (GZ1772); 18 months; $29,440

PURDUE UNIVERSITY
Ferris, Virginia R.; *Nematode Reference Collection* (GZ416001); 24 months; $10,000

KANSAS

WICHITA STATE UNIVERSITY
Loper, Gerald D.; *Expansion of The Master's Degree Program in Physics* (GZ702001); 12 months; $36,200

MASSACHUSETTS

AMERICAN METEOROLOGICAL SOCIETY
Spengler, Kenneth C.; *Science Writers Seminars on Atmospheric Sciences* (GZ394001); $2,075

HARVARD UNIVERSITY
Price, Don K. and Zeckhauser, Richard; *Science and Public Policy Program, Curriculum Development for Public Policy Program, Faculty Research Seminar in Analytic Methods and Public Policy* (GR88*); 36 months; $99,478

UNIVERSITY OF MASSACHUSETTS
Nash, William A.; *Development of New Courses in Ocean Engineering* (GZ1759); 15 months; $16,840

MICHIGAN

AMERICAN SOCIETY OF AGRICULTURAL ENGINEERS
Curry, R. Bruce; *Symposium on Graduate Education in Agricultural Engineering* (GZ1468); 18 months; $18,250

MICHIGAN STATE UNIVERSITY
Krupka, Lawrence R. and McClary, Andrew; *Summer Courses on Science, Technology, and Human Values* (GZ1763); 13 months; $35,050
Sparrow, Frederick K.; *Advanced Course in the Physiological Ecology of Plants* (GZ1758); 9 months; $4,830

MISSOURI

WASHINGTON UNIVERSITY
Hoelscher, Erwin C.; *Graduate Education Program in Building Environmental Systems Program* (GZ684001); 27 months; $45,740

MONTANA

WESTERN MONTANA COMMISSION FOR PUBLIC INFORMATION
Gordon, Clarence C.; *A Project to Inform the Citizens of Montana of the Scientific Aspects of Environmental Pollution* (GZ1773); 24 months; $49,550

NEBRASKA

CREIGHTON UNIVERSITY
Scheerer, Anne E.; *Joint Educational-Industrial-Governmental Science Symposium* (GZ1527); 6 months; $2,260

NEW JERSEY

RUTGERS UNIVERSITY
Hamilton, Leonard; *Advanced Laboratory Methods in Experimental and Physiological Psychology* (GZ1484); 15 months; $24,265

NEW MEXICO

ROCKY MOUNTAIN MATHEMATICS CONSORTIUM
Epstein, Bernard; *Advanced Science Seminar, Summer Seminar on Reproducing Kernels in Analysis and Probability* (GZ1490); 11 months; $54,200

UNIVERSITY OF NEW MEXICO
Tomasson, Richard F.; *Student Research Allocations Fund* (GZ1761); 15 months $2,700

NEW YORK

CORNELL UNIVERSITY
Pimentel, David; *Pest Population Ecology—An Inter-University Training Program* (GZ1371001); $6,400

INSTITUTE ON MAN AND SCIENCE
Clinchy, Everett R.; *Summer Session on the Quality of Life—Science, Technology and Values* (GZ1789); 7 months; $25,250

NEW SCHOOL FOR SOCIAL RESEARCH
Williams, Curtis A.; *Design and Evaluation of an Experimental Curriculum for Public Understanding of Science* (GZ634 002); 14 months; $9,150

SPECIALIZED ADVANCED SCIENCE EDUCATION PROJECTS

OHIO

CASE WESTERN RESERVE UNIVERSITY
Douglas, Robert G.; *Symposium on Deep Sea Drilling Project—Results from the Atlantic and Pacific* (GZ1765); 3 months; $1,800

OREGON

UNIVERSITY OF OREGON
Sampson, Garth C.; *A Teaching Collection for Old World Prehistory* (GZ1833); 6 months; $9,250

PENNSYLVANIA

CARNEGIE MELLON UNIVERSITY
Artman, Joseph O.; *Graduate Instructional Laboratory in Quantum and Optical Electronics* (GZ1526); 14 months; $25,000

PENNSYLVANIA STATE UNIVERSITY
Roy, Rustum; *Pilot Science-Society Presentations at Two National Assemblies* (GZ1799); 32 months; $10,640

TENNESSEE

AMERICAN SOCIETY OF PLANT TAXONOMY
Lewis, F. Harlan; *Advanced Science Seminar, A Summer Institute in Systematic Biology* (GZ1516); 11 months; $18,105

ATOMIC ENERGY COMMISSION,
Oak Ridge Operations Office
Grigorieff, Vladimir W.; *Public Understanding of Science—Tutorials on the Impact of Science and Technology on Society* (AG218); 15 months; $21,050

UNIVERSITY OF TENNESSEE
Tanner, James T.; *Ecology for Planners and Engineers* (GZ1778); 30 months; $18,270

WASHINGTON

UNIVERSITY OF WASHINGTON
Jayne, Benjamin A.; *Development of Instructional Materials for Graduate Study in the Physics of Wood and Fiber Composite Materials* (GZ1803); 38 months; $72,555

WISCONSIN

UNIVERSITY OF WISCONSIN, Madison
Harvey, John G.; *A Study and Analysis of the R. L. Moore Instructional Technique* (GZ1790); 17 months; $6,700

Advanced Training Projects

ARIZONA

UNIVERSITY OF ARIZONA
Thompson, R. H.; *Advanced Field Training in Archaeology* (GZ1493); 11 months; $21,625

CALIFORNIA

UNIVERSITY O6 CALIFORNIA, Irvine
Bunker, D. L.; *Winter Course in Gas Kinetics* (GZ1499); 5 months; $11,190

UNIVERSITY OF CALIFORNIA, San Diego
Stern, H.; *Summer Workshop on Molecular Techniques in Developmental Biology* (GZ1505); 11 months; $26,200

COLORADO

COLORADO STATE UNIVERSITY
Williams, J. S.; *Summer Seminar for Workers in Statistics* (GZ1522); 23 months; $9,985

ROCKY MOUNTAIN MATHEMATICS CONSORTIUM, INC.
Epstein, B.; *Summer Seminar on Reproducing Kernels in Analysis* (GZ1490); 11 months; $54,200

UNIVERSITY OF COLORADO
Thomas, G. E.; *Summer Institute in Planetary Atmospheres* (GZ1503); 11 months; $16,340

CONNECTICUT

YALE UNIVERSITY
Rhoads, D. C.; *Field Course in Marine Ecology for Paleontologists* (GZ1506); 11 months; $9,975

DISTRICT OF COLUMBIA

AMERICAN GEOLOGICAL INSTITUTE
Donnelly, T. W.; *International Field Institute—Eastern Caribbean Island Arc* (GZ1511;) 11 months; $75,250

ASSOCIATION OF AMERICAN GEOGRAPHERS
Berry, B. J. L.; *Seminar on the Geographical Analysis of Major U.S. Metropolitan Areas* (GZ1500); 11 months; $26,675

ASSOCIATION OF AMERICAN LAW SCHOOLS
Yegge, R. B.; *Institute on Social Science Methods in Legal Education* (GZ1517); 11 months; $16,200

LINGUISTIC SOCIETY OF AMERICA
Lehiste, I.; *Seminar in Linguistics* (GZ1491); 11 months; $55,380

FLORIDA

AMERICAN SOCIETY OF PLANT TAXONOMISTS
Lewis, F. H.; *A Summer Institute in Systematic Biology* (GZ1516); 11 months; $18,105

ORGANIZATION FOR TROPICAL STUDIES, INC.
Spencer, J. T.; *A Continuing Program of Graduate Education in Tropical Studies in Central America* (GZ1802); 16 months; $361,055

UNIVERSITY OF FLORIDA
Hren, J. J.; *Short Course on Field Ion Microscopy and Related Fields* (GZ1502); 11 months; $9,715

UNIVERSITY OF MIAMI
Teas, H. J.; *Advanced Science Seminar in Tropical Botany* (GZ1523); 11 months; $19,450

ILLINOIS

NORTHWESTERN UNIVERSITY
Brown, L. M.; *Symposium on Elementary Particles* (GZ1489); 6 months; $4,025
Dacey, M. F.; *Seminar Series in Quantitative Geography* (GZ7682); 12 months; $3,720

UNIVERSITY OF ILLINOIS
Jakle, J. A.; *Field Research Training Project in Geography* (GZ1776); 6 months; $14,690

KANSAS

UNIVERSITY OF KANSAS
Jones, J. K., Jr.; *Field Methods for Vertebrate Biologists* (GZ1512); 11 months; $5,000

MAINE

BOWDOIN COLLEGE
Johnson, R. W.; *Summer 1970 Advanced Seminar in Finite Groups* (GZ1524); 11 months; $90,960

MARYLAND

UNIVERSITY OF MARYLAND
Lippincott, E. R.; *Laser Raman Institute and Workshop* (GZ1519); 11 months; $4,500

MASSACHUSETTS

BRANDEIS UNIVERSITY
Deser, S.; *1970 Brandeis University Summer Institute in Theoretical Physics* (GZ1521); 11 months; $58,770

MARINE BIOLOGICAL LABORATORY
Case, J. F.; *Training Project in Experimental Invertebrate Zoology* (GZ1788); 7 months; $20,160
Siegelman, H. W.; *Training Project in Experimental Marine Botany* (GZ1787); 7 months; $15,000

UNIVERSITY OF MASSACHUSETTS
Nash, W. A.; *Earthquake Engineering* (GZ1510); 11 months; $20,225

WOODS HOLE OCEANOGRAPHIC INSTITUTION
Malkus, W. V. R.; *Summer Programs in Geophysical Fluid Dynamics* (GZ1494); 11 months; $46,660
Maxwell, A. E.; *Postdoctoral Research Training Program in Oceanography* (GZ1508); 23 months; $62,670

MICHIGAN

MICHIGAN STATE UNIVERSITY
Miller, M.; *Summer Institute of Glaciological and Arctic Sciences* (GZ1498); 11 months; $35,955

UNIVERSITY OF MICHIGAN
Miller, W. E.; *Seminars on Quantitative Political Science* (GZ1495); 11 months; $23,000

MINNESOTA

UNIVERSITY OF MINNESOTA
Jenkins, J. J.; *Theories for Higher Mental Processes* (GZ1507); 11 months; $18,710

MISSOURI

ASSOCIATED UNIVERSITIES FOR INTERNATIONAL EDUCATION
Mulligan, J. E.; *Tropical Studies in British Honduras* (GZ1783); 7 months; $10,000

MONTANA

MONTANA STATE UNIVERSITY
Swenson, R. J.; *Summer Workshop in Statistical Physics* (GZ1504); 11 months; $19,545

NEVADA

UNIVERSITY OF NEVADA, Reno Campus
d'Azevedo, W. L.; *Field Training in Cultural Anthropology* (GZ1518); 11 months; $22,130

NORTH CAROLINA

NORTH CAROLINA STATE UNIVERSITY
Rabb, R. L.; *A Research and Training Conference on Concepts of Pest Management* (GZ1520); 11 months; $10,365

OREGON

UNIVERSITY OF OREGON
Young, P. D. *Anthropology Field School in Ixmiquilpan, Mexico* (GZ1514); 11 months; $19,700

PENNSYLVANIA

UNIVERSITY OF PITTSBURGH
Plotnicov, L.; *Field Training in Cultural Anthropology* (GZ1496); 11 months; $19,295

RHODE ISLAND

AMERICAN MATHEMATICAL SOCIETY
Walker, G. L.; *1970 Summer Seminar on Mathematical Problems in Geophysical Sciences* (GZ1509); 11 months; $26,910

TEXAS

UNIVERSITY OF HOUSTON
Dawkins, G. S.; *Computer Simulation for System Analysis and Design* (GZ1497); 11 months; $23,600

UNIVERSITY OF TEXAS AT AUSTIN
Shieve, W. C.; *Advanced School for Statistical Mechanics and Thermodynamics* (GZ1501); 11 months; $21,955

UTAH

UTAH STATE UNIVERSITY
Riley, J. P.; *Advanced Science Seminar for Hydrology Professors* (GZ1525); 11 months; $22,820

SCIENCE EDUCATION PROGRAMS

WASHINGTON

WASHINGTON STATE UNIVERSITY
Bernard, H. R.; *Graduate Field School in Anthropology in the Vicinity of Spokane, Washington* (GZ1513); 11 months; $25,040

WISCONSIN

UNIVERSITY OF WISCONSIN
Grossman, J. B.; *1970 Summer Institute in Behavioral Science and Law* (GZ1515); 11 months; $40,505

CANADA

CANADIAN MATHEMATICAL CONGRESS
Maranda, J.; *U.S. Participation in Seminar on Commutative Algebra and Algebraic Geometry* (GZ1492); 11 months; $14,700

BERMUDA

BERMUDA BIOLOGICAL STATION
Mackenzie, F. T.; *Graduate Training in the Interrelationships of Marine Organisms and Sediments* (GZ1784); 6 months; $25,545

Special Projects in Science Education

ALABAMA

STILLMAN COLLEGE
Weaver, Patricia L.; *A Mathematics Project for Secondary School Students* (GW5710*); 5 months; $20,496

CALIFORNIA

CALIFORNIA INSTITUTE OF TECHNOLOGY
Wood, David S.; *An Ecological Study of the Automobile in American Society* (GY8439); 15 months; $43,165

UNIVERSITY OF CALIFORNIA, Berkeley
Blackwell, David; *Berkeley Summer Statistics Program for Disadvantaged Students* (GY7644); 15 months; $36,245
Wolf, Frantisek; *Training in Mathematics of Secondary School Teachers of Talented Students* (GW5704); 15 months; $7,099

VENTURA COUNTY INDEPENDENT EDUCATION COUNCIL
Brisby, William L.; *A Marine Biology Institute for Secondary School Teachers and Students* (GW4525001); $250

COLORADO

COLORADO STATE UNIVERSITY
Frayer, Warren E.; *Symposium on the Development and Implementation of Courses and Curricula in Natural-Resource Biometry* (GZ1477*); 17 months; $5,000

UNIVERSITY OF COLORADO
Haas, John A.; *Seminar Program on Social Science Teaching* (GY7680); 16 months; $32,695
Mayer, William V.; *Summer Conference on Pre-Service and In-Service Preparation of Biology Teachers for Science Supervisors and College Faculty* (GW5328*); 10 months; $8,785
Wailes, James R.; *Summer Conference on National Curricular Studies for Science Supervisors and College Teacher Education Personnel* (GW5330*); 10 months; $2,650
——— ; *Conference for Secondary School Principals on New Science and Mathematics Curricula* (GW5372); 9 months; $15,209
———; *Conference for Elementary Education Administrators on New Elementary Science Curricula* (GW5373); 9 months; $15,324

DISTRICT OF COLUMBIA

AMERICAN ASSOCIATION FOR THE ADVANCEMENT OF SCIENCE
Mayor, John R.; *Special Projects in Undergraduate Science Education Feasibility Study of Short Courses for College Teachers* (C602000); 6 months; $14,540

AMERICAN INSTITUTE OF BIOLOGICAL SCIENCES
Busser, John H.; *Series of Tutorial Lectures and Coordinated Workshops on Computers for the Biologists* (GY7685); 12 months; $2,445

INSTITUTE FOR SERVICES TO EDUCATION
Humphries, Frederick S.; *A General Education Revision Project in Support of Institutional Change in Thirteen Predominantly Negro Colleges* (GY8432); 12 months; $77,148

JOINT BOARD ON SCIENCE EDUCATION
Boek, Jean K.; *In-Service Conferences on Science and Mathematics Education* (GW4899*); 10 months; $1,120

NATIONAL SCIENCE TEACHERS ASSOCIATION
Novak, Joseph D.; *Study of Exemplary Science Facilities for Schools* (GW5716001); $1,400

SCIENCE SERVICE INC.
Sherburne, E. G. Jr.; *National Conference on Science Youth Activities* (GW4629); 10 months; $14,000
——— ; (GW4629 001); $256

SMITHSONIAN INSTITUTION
Taylor, Frank A.; *A Study of Opportunities for Extending Museum Contributions to Precollege Science Education* (GW5046); 10 months; $17,800

FLORIDA

FLORIDA STATE UNIVERSITY
Westmeyer, Paul; *Conference for Secondary School Principals on New Science and Mathematics Curricula* (GW5382); 7 months; $18,487

HAWAII

UNIVERSITY OF HAWAII
Lamoureux, Charles H.; *Summer Institute in Science and Mathematics for Secondary and Elementary Teachers* (GW5045*); 11 months; $3,070

ILLINOIS

AMERICAN SOCIETY FOR ENGINEERING EDUCATION
Knepler, Henry; *Liberal Arts Education for Engineers* (GY7206); 13 months; $24,032

ATOMIC ENERGY COMMISSION, Chicago Operations Office
Miller, Shelby A.; *Undergraduate January Term Program* (AG210); 30 months; $28,275

NATIONAL ACADEMY OF ENGINEERING, Commission on Engineering Education
Hall, Newman A.; *Task Order for Workshop on Social Directions for Technology* (C310182); 12 months; $26,025

UNIVERSITY OF ILLINOIS, Chicago Circle
Walter, Robert L.; *Conference on Teaching Chemistry to Underprepared Students* (GY7684); 8 months; $2,365

INDIANA

INDIANA UNIVERSITY, Bloomington
Engle, Shirley H.; *Conference for Curriculum Directors and Supervisors on New Social Studies Curricula* (GW5370); 9 months; $18,869
Springer, George; *Meeting on Assistance to Developing Colleges* (GY7053); 6 months; $1,545

IOWA

UNIVERSITY OF IOWA
McAdam, John E.; *Conference for Secondary School Principals and Other School Administrators on New Science and Mathematics Curricula* (GW5371); 9 months; $27,534
——— ; (GW4233001); 12 months; $5,812

KENTUCKY

CENTRE COLLEGE OF KENTUCKY
Matheny, Larry R. and Wakefield, Jerry R.; *Student Originated Research on Kentucky's Wild Rivers* (GY8442); 3 months; $7,030

LOUISIANA

GRAMBLING COLLEGE
Kennedy, Amos P.; *Brookhaven Semester Program for Students and Faculty Members from Developing Colleges and Universities* (GY7668); 21 months; $50,985

MARYLAND

MORGAN STATE COLLEGE
Talbot, Walter R.; *Conference on Mathematics in Developing Colleges* (GY7672); 13 months; $28,650

MASSACHUSETTS

BOSTON UNIVERSITY
Delicata, Dino; *Student Originated Research on Eutrophication Stimulation* (GY8441); 6 months; $6,555

EDUCATION DEVELOPMENT CENTER
Haber-Schaim, Uri; *Supervisory Teacher Workshop Undergraduate Program* (GY7670); 22 months; $46,435

HAMPSHIRE COLLEGE
Birney, Robert C.; *The Creation and Evaluation of a Field Study Program for Social Sciences* (GY7212); 15 months; $22,680
Hafner, Everett M.; *AAAS Symposium on Undergraduate Environmental Science* (GY7207001); $820
——— ; *Stipends for Student Travel, AAAS Symposium* (GY7207); 3 months; $8,230

HARVARD UNIVERSITY
Speyer, James L.; *Student Originated Research in Human Ecology* (GY8440); 6 months; $5,980
Zinberg, P. Doty; *Trends in Undergraduate Concentration in Science & Their Origins* (GY7667); 15 months; $1,800

MICHIGAN

MICHIGAN STATE UNIVERSITY
Newton, David; *National Conference on Innovation in Teacher Education Programs in the Sciences* (GY7690); 14 months; $8,445

OAKLAND UNIVERSITY
Pino, L. N.; *Student Originated and Managed Study of Air Pollution and Vehicular Propulsion* (GY8443); 6 months; $8,370

MISSOURI

MIDCONTINENT REGIONAL EDUCATIONAL LABORATORY
Roberts, Richard A.; *Training of In-Service Biology Teachers to Work With the Academically Unsuccessful Student* (GW4721); 12 months; $10,634

NEBRASKA

KEARNEY STATE COLLEGE
Kuecker, John F.; *Nebraska Science Education Program—Curriculum Development in Physical Sciences* (GY7665); 13 months; $44,535

UNIVERSITY OF NEBRASKA, Lincoln
McCurdy, Donald W.; *Implementation of the Nebraska Physical Science Project* (GW4522001); 17 months; $40,192

NEW HAMPSHIRE

DARTMOUTH COLLEGE
King, Allen L.; *International Working Seminar on the Role of the History of Physics in Physics Education* (GY7691); 13 months; $12,620

COLLEGE SCIENCE IMPROVEMENT PROGRAMS

UNIVERSITY OF NEW HAMPSHIRE
Smith M. Daniel; *Conference for Secondary School Principals on New Science and Mathematics Curricula* (GW5376); 9 months; $25,817

NEW JERSEY

GLASSBORO STATE COLLEGE
Renlund, Robert N.; *Inter-Institutional Association New Jersey Consortium Development Program* (GY7660); 17 months; $32,345

NEW YORK

JOINT COUNCIL ON ECONOMIC EDUCATION
Saunders, Phillip; *Lasting Effectiveness of Introductory Economics Courses* (GY7208); 35 months; $60,460

NATIONAL SCIENCE TEACHERS ASSOCIATION
Novak, Joseph D.; *Study of Exemplary Science Facilities for Schools* (GW5716); 18 months; $58,750

RENSSELAER POLYTECHNIC INSTITUTE
Becker, Martin; *A Professional Program in Nuclear Engineering and Science for Disadvantaged Students* (GY6896); 24 months; $49,135

SYRACUSE UNIVERSITY
Davis, Robert B.; *Training in Madison Project Mathematics for Resource People in Several Large Cities* (GW5709); 6 months; $58,299

NORTH CAROLINA

EAST CAROLINA UNIVERSITY
Helms, R. M.; *Solar Eclipse Conference* (GY7210); 10 months; $15,405

NORTH CAROLINA STATE UNIVERSITY AT RALEIGH
Park, Hubert V.; *Academic Year Institute for Mathematics Teacher Retrainees* (GY8425); 16 months; $42,550

REGIONAL EDUCATION LABORATORY FOR THE CAROLINAS AND VIRGINIA
Straley, Joseph W.; *Revitalization of Freshman-Sophomore Physics at Twenty Colleges* (GY8435); 17 months; $147,800
———; (GY6165001); $23,174

OHIO

HEIDELBERG COLLEGE
Reno, Martin, Stanforth, Robert, and King, Gayle; *Student Originated Research on the Sandusky River* (GY7683); 6 months; $4,755

OKLAHOMA

MIDCONTINENT REGIONAL EDUCATION LABORATORY
Roberts, Richard A.; *Training of In-Service Biology Teachers to Work With the Academically Unsuccessful Student* (GW4721001); $1,412

UNIVERSITY OF OKLAHOMA
Mankin, Charles J.; *Development of Educational Guidebook Series on Geology of Oklahoma* (GW5726); 16 months; $15,821

OREGON

OREGON MUSEUM OF SCIENCE AND INDUSTRY
Whitney Jr., Hartwell H.; *A School Year Science Investigation Project* (GW5706); 16 months; $16,500
———; (GW4526001); $4,125

PACIFIC UNIVERSITY
Malcolm, David R.; *Acquisition and Operation of a Center for Ecological Studies* (GY8300); 15 months; $32,000

PORTLAND STATE UNIVERSITY
Allen, John Eliot; *Conference for Secondary School Principals on New Science and Mathematics Curricula* (GW5374); 9 months; $18,537

TENNESSEE

MEMPHIS STATE UNIVERSITY
Schwartz, Donald; (GW5383); 7 months; $21,178

TENNESSEE TECHNOLOGICAL UNIVERSITY
Martin, Robert E.; *Undergraduate Education in Science Special Projects Program* (GY7679); 39 months; $297,900

VIRGINIA

UNIVERSITY OF VIRGINIA
Thompson, Ertle; *Conference for Secondary School Principals on New Science and Mathematics Curricula* (GW5375); 9 months; $19,577

WISCONSIN

UNIVERSITY OF WISCONSIN, Milwaukee
Kovacic, Peter; *Undergraduate-Graduate Research Collaboration Program* (GY7663); 34 months; $55,615

College Science Improvement Programs

ALABAMA

UNIVERSITY OF ALABAMA IN BIRMINGHAM; Fattig, W. Donald and Shannon, John H.; (GY7652); 32 months; $89,000
Segal, Arthur C. and Schultz, Norman L.; (GY7651); 32 months; $81,500

CALIFORNIA

UNIVERSITY OF CALIFORNIA, Santa Barbara; Gilbert, William J.; (GY7658); 15 months; $25,400

DISTRICT OF COLUMBIA

AMERICAN COUNCIL ON EDUCATION; Astin, Alexander W.; (GR89*); 12 months; $25,000

FLORIDA

FLORIDA ATLANTIC UNIVERSITY; Michels, Kenneth M. and Banter, John C.; (GY6899); 36 months; $255,300

FLORIDA INSTITUTE OF TECHNOLOGY; Miller, John E.; (GY7681); 39 months; $197,600

ILLINOIS

ASSOCIATED COLLEGES OF THE MIDWEST; Hunter, Robert; (GY8426); 39 months; $107,400
Stewart, Blair; (GY8428); 39 months; $164,900

KNOX COLLEGE; Salter, Lewis S.; (GY7676); 39 months; $180,600

NORTHERN ILLINOIS UNIVERSITY; Redmore, Fred; (GY7647); 32 months; $116,200

INDIANA

BALL STATE UNIVERSITY; Carmin, Robert L.; (GY7674); 39 months; $211,900

IOWA

DRAKE UNIVERSITY; Vandenbranden, Robert J. and Parker, David; (GY7648); 23 months; $48,100

UNIVERSITY OF NORTHERN IOWA; Burham, Robert L.; (GY7653); 23 months; $58,600

LOUISIANA

NORTHEAST LOUISIANA STATE COLLEGE; Dupree, Daniel E.; (GY6898); 36 months; $201,400

MARYLAND

UNIVERSITY OF MARYLAND; Dixon, Peggy; (GY7649); 23 months; $18,900

MASSACHUSETTS

AMHERST COLLEGE; Brophy, Gerald P.; (GY7657); 39 months; $88,500
Irvine, William M.; (GY8427); 27 months; $75,900

SIMMONS COLLEGE; Piper, James U.; (GY8430); 39 months; $62,500

MICHIGAN

UNIVERSITY OF MICHIGAN, Dearborn Campus; Hertzler, Emanuel C. and Sutherland, Roger A.; (GY7645); 23 months; $33,900

MINNESOTA

CONCORDIA COLLEGE; Homann, H. Robert; (GY7675); 39 months; $152,900

GUSTAVUS ADOLPHUS COLLEGE; Kendall, John S.; (GY8436); 39 months; $245,200

MANKATO STATE COLLEGE; Sala, Rexford Q.; (GY7654); 23 months; $50,300

NEW HAMPSHIRE

NEW HAMPSHIRE COLLEGE AND UNIVERSITY COUNCIL; West II, Arthur J.; (GY8429); 39 months; $75,700

NEW YORK

ST. JOHN FISHER COLLEGE; Heininger Jr., Clarence G.; (GY7655); 15 months; $31,700

ST. LAWRENCE UNIVERSITY; Baker, D. K.; (GY7211); 24 months; $184,600

NORTH CAROLINA

ST. AUGUSTINE'S COLLEGE; Gipson, Jeffery; (GY7656); 39 months; $70,400

NORTH DAKOTA

MINOT STATE COLLEGE; Leiby, Paul D.; (GY8422); 39 months; $150,200

OHIO

OHIO NORTHERN UNIVERSITY; Glass, Robert J.; (GY7669); 38 months; $246,600

OHIO WESLEYAN UNIVERSITY; Lisensky, Robert P.; (GY7673); 39 months; $287,400

OKLAHOMA

UNIVERSITY OF TULSA; McKay, Edward S.; (GY8434); 39 months; $225,500

RHODE ISLAND

PROVIDENCE COLLEGE; Galkowski, Theodore T.; (GY7678); 39 months; $266,600

SOUTH CAROLINA

FURMAN UNIVERSITY; Patterson, C. Stuart; (GY7120); 36 months; $275,100

WOFFORD COLLEGE; Stephens, B. G.; (GY7666); 36 months; $295,500

TENNESSEE

CHRISTIAN BROTHERS COLLEGE; Althaus, Bro. H. Louis; (GY7686); 39 months; $114,400

KING COLLEGE; Burke, Edward W. Jr.; (GY6897); 36 months; $120,500

VANDERBILT UNIVERSITY; Hanson, Harold N.; (GY8431); 7 months; $12,600

TEXAS

SOUTHWEST TEXAS UNIVERSITY; Norris Jr., W. E.; (GY7695); 39 months; $244,800

STEPHEN F. AUSTIN STATE UNIVERSITY; Layton, W. I.; (GY7692); 39 months; $239,200

TEXAS ASSOCIATION FOR GRADUATE EDUCATION AND RESEARCH; Edwards, Frank C.; (GY7659); 39 months; $172,500

SCIENCE EDUCATION PROGRAMS

TRINITY UNIVERSITY; Andrews, Robert V.; (GY7697); 39 months; $152,800

UNIVERSITY OF HOUSTON; Welch, Jay N.; (GY7650); 23 months; $134,300

VERMONT

UNIVERSITY OF VERMONT; Essler, Warren O.; (GY7209); 36 months; $206,300

VIRGINIA

HOLLINS COLLEGE; Stewart, Roberta A.; (GY6900); 36 months; $143,600

OLD DOMINION UNIVERSITY; Pittman, Melvin A.; (GY8437); 39 months; $201,300

SWEET BRIAR COLLEGE; Belcher, Jane C.; (GY7661); 39 months; $203,900

VIRGINIA MILITARY INSTITUTE; Minnix, Richard B.; (GY7677); 39 months; $232,700

VIRGINIA POLYTECHNIC INSTITUTE; Tucker, James F. and Brandon, William R.; (GY7646); 23 months; $50,000

Instructional Scientific Equipment for Undergraduate Education

ALABAMA

ALEXANDER CITY STATE JUNIOR COLLEGE; Widder, Arlon Arnold; (GY8272); 25 months; $4,000

AUBURN UNIVERSITY; Phillips, Charles L.; (GY8112); 25 months; $3,900
Vachon, R. I.; (GY8126); 25 months; $4,100

BIRMINGHAM SOUTHERN COLLEGE; Bailey, Paul C.; (GY7840); 25 months; $7,900

MILES COLLEGE; Arrington Jr., Richard; (GY7915); 25 months; $3,100

SPRING HILL COLLEGE; Kane, John W.; (GY7728); 25 months; $3,100

UNIVERSITY OF ALABAMA, Birmingham; McCutcheon, Martin J.; (GY8107); 25 months; $1,700
Talbot, Thomas F.; (GY8110); 25 months; $6,500

UNIVERSITY OF ALABAMA, Tuscaloosa; Barfield, B. F.; (GY8117); 25 months; $5,700
Cox, Robert M.; (GY8113); 25 months; $5,400
Gambrell Jr., Samuel C.; (GY7786); 25 months; $4,800
Green, Margaret; (GY7838); 25 months; $12,500
Keith, Warren G.; (GY8111); 25 months; $6,300

ARIZONA

ARIZONA STATE UNIVERSITY; Berman, Neil S.; (GY7787); 25 months; $5,000
Craig Jr., Samuel E.; (GY7785); 25 months; $7,200
Sheridan, Michael F.; (GY7727); 25 months; $1,800

NORTHERN ARIZONA UNIVERSITY; Barnes, Charles W.; (GY7726); 25 months; $6,000
Griffith, Charles R.; (GY8057); 25 months; $3,100

ARKANSAS

ARKANSAS COLLEGE; Seibert, Daniel E.; (GY8229); 25 months; $1,300

ARKANSAS POLYTECHNIC COLLEGE; Trigg, William Walker; (GY8236); 25 months; $3,800

UNIVERSITY OF ARKANSAS; Bower, Raymond K.; (GY7837); 25 months; $3,300
Couper, James R.; (GY7784); 25 months; $8,000
Gilbrech, Donald A.; (GY7968); 25 months; $3,800
Phillips, J. R.; (GY7835); 25 months; $4,900
Piper, E. L.; (GY8172); 25 months; $4,800
Sabbe, Wayne E.; (GY833); 25 months; $2,100
Thoma, John A.; (GY8019); 25 months; $6,900

CALIFORNIA

AMERICAN RIVER COLLEGE; Kong, Ronald A.; (GY8192); 25 months; $1,200

AZUSA PACIFIC COLLEGE; Lamar, Daniel S.; (GY8233); 25 months; $1,900

CALIFORNIA STATE COLLEGE AT LONG BEACH; Osborne, Douglas; (GY7941); 25 months; $1,400

CALIFORNIA STATE COLLEGE AT LOS ANGELES; Davis, Terry E.; (GY7932); 25 months; $8,800

CERRITOS COLLEGE; Peter, James R.; (GY8027); 25 months; $1,700

CHICO STATE COLLEGE; Hauser, Rolland K.; (GY8206); 25 months; $1,100
Nazzaro, J. R.; (GY8224); 25 months; $4,400

CONTRA COSTA COLLEGE; Seemann, D. J.; (GY7949); 25 months; $4,900

DIABLO VALLEY COLLEGE; Wheeler, Charles W.; (GY8269); 25 months; $10,600

FRESNO CITY COLLEGE; Bengel, Earl C.; (GY8212); 25 months; $20,900
Robinson, Elroy B.; (GY8200); 25 months; $2,600

FRESNO STATE COLLEGE; Weitzman, Raymond S.; (GY7799); 25 months; $1,900

HUMBOLDT STATE COLLEGE; Allen, William V.; (GY8215); 25 months; $8,700

LOYOLA UNIVERSITY OF LOS ANGELES; Schwartz, J. R.; (GY8289); 25 months; $6,000

MODESTO JUNIOR COLLEGE; Osner, Henry (GY8298); 25 months; $4,900

MOUNT SAN ANTONIO COLLEGE; Allen, George; (GY7942); 25 months; $9,600

PASADENA CITY COLLEGE; Eaton Jr., Clyde B.; (GY7939); 25 months; $21,300

SACRAMENTO STATE COLLEGE; Hu, Chien Y.; (GY7870); 25 months; $2,400
Newcomb, Charles P.; (GY8068); 25 months; $4,400

SAN DIEGO STATE COLLEGE; Awbrey, Frank T.; (GY7850); 25 months; $9,300
Catlett, Robert H.; (GY7817); 25 months; $5,000
Leach, Larry L.; (GY8064); 25 months; $500
Lillegraven, Jason A.; (GY8221); 25 months; $600
Neel, James W.; (GY8175); 25 months; $1,200
Templin, Jacques; (GY7963); 25 months; $5,700

SONOMA STATE COLLEGE; Dunning Jr., John R.; (GY8209); 25 months; $25,000
Hermans, Colin O.; (GY8176); 25 months; $1,800

STANFORD UNIVERSITY; Schawlow, A. L.; (GY8245); 25 months; $6,700

UNIVERSITY OF CALIFORNIA, Berkeley; Smith, O. J. M.; (GY7770); 25 months; $8,500

UNIVERSITY OF CALIFORNIA, San Diego; Raskin, Jeffrey F.; (GY7852); 25 months; $25,000

UNIVERSITY OF CALIFORNIA, Santa Barbara; Fenech, H.; (GY7977); 25 months; $10,600
MacDonald, Keith B.; (GY7899); 25 months; $6,500
Mellichamp, D. A.; (GY7783); 25 months; $7,800

UNIVERSITY OF SOUTHERN CALIFORNIA; Browand, Frederick; (GY7961); 25 months; $7,700
Cohen, Jordan L.; (GY8031); 25 months; $6,800

UNIVERSITY OF THE PACIFIC; Beauchamp, Kenneth L.; (GY8222); 25 months; $2,900

VICTOR VALLEY COLLEGE; Baartz, Alice Louise; (GY8032); 25 months; $2,800

WHITTIER COLLEGE; Bender, David; (GY8247); 25 months; $4,400
Graham, Laurence; (GY8279); 25 months; $6,500

COLORADO

COLORADO STATE UNIVERSITY; Avery, David D.; (GY8034); 25 months; $12,500
Culver, Roger B.; (GY8070); 25 months; $2,600
Maxwell, Lee M.; (GY8197); 25 months; $25,000
McAllister, G. L.; (GY7973); 25 months; $5,900
Winder, Dale R.; (GY8075); 25 months; $22,000

OTERO JUNIOR COLLEGE; Andersen, William C.; (GY8290); 25 months; $1,600

TEMPLE BUELL COLLEGE; Connally, Roy E.; (GY7717); 25 months; $2,500

UNIVERSITY OF COLORADO; Chanaud, Robert; (GY8142); 25 months; $2,800
Krug, Richard F.; (GY7800); 25 months; $1,100
Porter, Keith R.; (GY7834); 25 months; $25,000
Sweetman, Richard H.; (GY7923); 25 months; $600
———; (GY8063); 25 months; $4,400
Uberoi, Mahinder S.; (GY7974); 25 months; $4,200

UNIVERSITY OF COLORADO, Colorado Springs Center; Sherman, Robert J.; (GY7892); 25 months; $3,500

UNIVERSITY OF COLORADO, Denver Center; Fahrion, Nell G.; (GY8035); 25 months; $3,800
Gabe, Robert A.; (GY7782); 25 months; $7,700

UNIVERSITY OF DENVER; Calvert, J. B.; (GY7867); 25 months; $6,700
Moe, Maynard L.; (GY8132); 25 months; $8,600

UNIVERSITY OF NORTHERN COLORADO; Cobb, L. Glen; (GY7725); 25 months; $3,500
Dietz, Richard D.; (GY7866); 25 months; $5,400

CONNECTICUT

FAIRFIELD UNIVERSITY; Schurdak, John J.; (GY7921); 25 months; $18,200

WESLEYAN UNIVERSITY; Page, Thornton; (GY7920); 25 months; $15,000

DELAWARE

UNIVERSITY OF DELAWARE; Robinson, David M.; (GY8108); 25 months; $6,300

DISTRICT OF COLUMBIA

GEORGE WASHINGTON UNIVERSITY; Coates, A. G.; (GY8293); 25 months; $7,100

HOWARD UNIVERSITY; Talbert, Preston T.; (GY8058); 25 months; $24,300

FLORIDA

FLORIDA AGRICULTURAL AND MECHANICAL UNIVERSITY; Beech, J. Alan; (GY8170); 25 months; $10,500
Day, James L.; (GY7738); 25 months; $4,200

FLORIDA ATLANTIC UNIVERSITY; Hartt, W. H.; (GY8250); 25 months; $4,600

FLORIDA INSTITUTE OF TECHNOLOGY; Revay Jr., Andrew W.; (GY8210); 25 months; $5,900

FLORIDA STATE UNIVERSITY; Leysieffer, Frederick W.; (GY8213); 25 months; $10,800

INSTRUCTIONAL SCIENTIFIC EQUIPMENT FOR UNDERGRADUATE EDUCATION

FLORIDA TECHNOLOGICAL UNIVERSITY; Cunningham, Glenn N.; (GY8292); 25 months; $5,400
Erickson, Ernest E.; (GY8116); 25 months; $7,200
Rexroad, Harvey N.; (GY7868); 25 months; $3,800

NEW COLLEGE; Gorfein, David S.; (GY8046); 25 months; $3,300

PALM BEACH JUNIOR COLLEGE; Truchelut, George B.; (GY8231); 25 months; $2,500

UNIVERSITY OF FLORIDA; Gaither, Robert B.; (GY7771); 25 months; $15,600
Leslie, Gerald R.; (GY8203); 25 months; $5,800
Winefordner, James D.; (GY8003); 25 months; $22,000

UNIVERSITY OF SOUTH FLORIDA; Anderson, Melvin W.; (GY8115); 25 months; $5,600
Dawes, Clinton J.; (GY8169); 25 months; $3,500

GEORGIA

CLARK COLLEGE; Hubert, Charles E.; (GY8167); 25 months; $8,300

COLUMBUS COLLEGE; Smith, Earle C.; (GY7737); 25 months; $2,300

GEORGIA COLLEGE AT MILLEDGEVILLE; Cotter, David J.; (GY8166); 25 months; $9,500

GEOGRIA INSTITUTE OF TECHNOLOGY; Braden, C. H.; (GY8138); 25 months; $15,000
Bush, Aubrey M.; (GY8152); 25 months; $23,800
Sturrock, Peter E.; (GY7743); 25 months; $18,500
Thomas, Edward; (GY7869); 25 months; $9,900
Young, R. A.; (GY8096); 25 months; $12,400

GEORGIA STATE UNIVERSITY; Fairchild, Donald L.; (GY8258); 25 months; $4,700
Hopkins Jr., Harry P.; (GY8066); 25 months; $6,000

MIDDLE GEORGIA COLLEGE; Husa Jr., William J.; (GY8291); 25 months; $4,000

MORRIS BROWN COLLEGE; Geer, Radford M.; (GY7720); 25 months; $3,000

GUAM

UNIVERSITY OF GUAM; Eldredge, L. G.; (GY8158); 25 months; $8,200

HAWAII

UNIVERSITY OF HAWAII; Htun, K. M.; (GY8121); 25 months; $24,500

UNIVERSITY OF HAWAII, Honolulu Community College; Mikasa, Henry Y.; (GY8253); 25 months; $2,300

IDAHO

COLLEGE OF SOUTHERN IDAHO; Campbell, Chester L.; (GY8026); 25 months; $1,300
Strope, Marvin B.; (GY7798); 25 months; $2,700

UNIVERSITY OF IDAHO; Davis, Lawrence W.; (GY8077); 25 months; $7,400
Ingerson, T. E.; (GY7891); 25 months; $5,800

ILLINOIS

BLACKBURN COLLEGE; Werner Jr., William E.; (GY7831); 25 months; $3,800

CITY COLLEGES OF CHICAGO, Loop College; Deyoung, Edwin L.; (GY8232); 25 months; $3,700

ILLINOIS INSTITUTE OF TECHNOLOGY; Filler, Robert; (GY7740); 25 months; $10,600
Hayashi, Teru; (GY8159); 25 months; $16,000
Langdon, W. M.; (GY8062); 25 months; $6,200

ILLINOIS STATE UNIVERSITY; Buffington, John D.; (GY7859); 25 months; $5,700
Reiter, Richard C.; (GY8020); 25 months; $7,000

KNOX COLLEGE; Boyd, John W.; (GY8080); 25 months; $2,100
Demott, Lawrence I.; (GY7900); 25 months; $5,600
Neumiller Jr., Harry J.; (GY8056); 25 months; $15,400

LOYOLA UNIVERSITY; Ijomah, B. I. C.; (GY7944); 25 months; $6,100

MCHENRY COUNTY COLLEGE; Konitzer, John; (GY7764); 25 months; $1,800

MUNDELEIN COLLEGE; Murphy, Sr., Mary N.; (GY7830); 25 months; $2,000

NORTHERN ILLINOIS UNIVERSITY; Banovetz, James M.; (GY8051); 25 months; $10,500

OLNEY CENTRAL COLLEGE; Stencel Jr., John E.; (GY8278); 25 months; $3,600

SOUTHERN ILLINOIS UNIVERSITY; Elkins, Donald M.; (GY8160); 25 months; $1,900

SOUTHERN ILLINOIS UNIVERSITY, Edwardsville Campus; Rockman, Charles M.; (GY8090); 25 months; $3,400
Sanders, Steven G.; (GY8081); 25 months; $4,500
Wilbraham, Antony C.; (GY8234); 25 months; $11,400
Wittig, Gertrude C.; (GY8154); 25 months; $8,300

SPOON RIVER COLLEGE; Kauffman, Harry F.; (GY8139); 25 months; $600

UNIVERSITY OF ILLINOIS, Urbana; Bateman, Paul T.; (GY8135); 25 months; $4,900
Boileau, Richard A.; (GY8262); 25 months; $9,100
Clark, J. M. Jr.; (GY8010); 25 months; $22,600
Kaplan, Samuel; (GY8155); 25 months; $25,000

WESTERN ILLINOIS UNIVERSITY; Wylie, Douglas W.; (GY8086); 25 months; $14,000

INDIANA

BALL STATE UNIVERSITY; Smith Jr., Charles Edward; (GY8244); 25 months; $4,300

DEPAUW UNIVERSITY; Kuempel, John R.; (GY7746); 25 months; $3,900

INDIANA UNIVERSITY, Bloomington; Pataki, Louis; (GY7855); 25 months; $1,200
Vitaliano, Charles J.; (GY8042); 25 months; $1,500
Weber, Ronald E.; (GY8201); 25 months; $22,900

PURDUE UNIVERSITY, Calumet Center; Battenburg, J. R.; (GY7954); 25 months; $21,900

ROSE POLYTECHNIC INSTITUTE; Llewellyn, Ralph A.; (GY8218); 25 months; $11,600

TAYLOR UNIVERSITY; Nussbaum, Elmer; (GY7885); 25 months; $2,200

TRI-STATE COLLEGE; Hill, William W.; (GY7769); 25 months; $5,100

UNIVERSITY OF EVANSVILLE; Weller, Lowell E.; (GY7735); 25 months; $12,000

UNIVERSITY OF NOTRE DAME; Crosson, Frederick J.; (GY7801); 25 months; $24,000

VALPARAISO UNIVERSITY; Lehmann, Gilbert; (GY7793); 25 months; $4,300
Mortimer, Kenneth; (GY8078); 25 months; $2,000

IOWA

COE COLLEGE; Keiser, Jeffrey E.; (GY7797); 25 months; $3,600

CORNELL COLLEGE; Ault, Addison; (GY7936); 25 months; $2,600
Deskin, William A.; (GY7736); 25 months; $5,000

IOWA STATE UNIVERSITY; Hardy, Rolland L.; (GY8141); 25 months; $9,100
Mathews, Jerold C.; (GY7962); 25 months; $25,000
Stephenson, David T.; (GY8129); 25 months; $12,300

SIMPSON COLLEGE; Delisle, Donald; (GY8165); 25 months; $1,500
Dittmer, Donald R.; (GY7719); 25 months; $2,100

UNIVERSITY OF IOWA; Eyman, Earl D.; (GY7792); 25 months; $18,500
Gonzalez, Nancie L.; (GY8054); 25 months; $4,900
Pflaum, Ronald T.; (GY7739); 25 months; $3,000
Shannon, Lyle W.; (GY7802); 25 months; $5,500

KANSAS

BETHEL COLLEGE; Adler, Robert G.; (GY8008); 25 months; $1,900

INDEPENDENCE COMMUNITY JUNIOR COLLEGE; Osborn, Russell S.; (GY7945); 25 months; $2,600

KANSAS STATE COLLEGE OF PITTSBURG; Walker, Joe M.; (GY8265); 25 months; $5,700

ST. BENEDICT'S COLLEGE; Sauerwein, Werner; (GY8194); 25 months; $2,100

TABOR COLLEGE; Johnson, William J. (GY8263); 25 months; $2,500

UNIVERSITY OF KANSAS; Wiseman, Gordon G.; (GY7903); 25 months; $15,300
Yochim, J. M.; (GY8273); 25 months; $17,000

WICHITA STATE UNIVERSITY; Carper, W. R.; (GY7748); 25 months; $4,400
Loper, Gerald D.; (GY8208); 25 months; $3,900

KENTUCKY

ALICE LLOYD COLLEGE; Osborn, Gerald; (GY8190); 25 months; $1,700

BRESCIA COLLEGE; Armendarez, Peter X.; (GY7996); 25 months; $1,800

THOMAS MORE COLLEGE; Miner, George K.; (GY7814); 25 months; $9,000

UNIVERSITY OF LOUISVILLE; Taylor, K. Grant; (GY8285); 25 months; $13,800

WESTERN KENTUCKY UNIVERSITY; Davis, Chester L.; (GY7956); 25 months; $9,100

LOUISIANA

LOUISIANA STATE UNIVERSITY, Baton Rouge; Carver, Dale R.; (GY7768); 25 months; $10,400
Greenberg, David B.; (GY7790); 25 months; $9,600
Hart, George F.; (GY8040); 25 months; $1,900
Kamel, Adel M.; (GY7791); 25 months; $5,000

LOUISIANA STATE UNIVERSITY, New Orleans; Siddall III, T. H.; (GY8005); 25 months; $14,800
St. Cyr II, William W.; (GY7964); 25 months; $8,900

LOUISIANA POLYTECHNIC INSTITUTE; Flournoy, Robert W.; (GY7836); 25 months; $5,700

MCNEESE STATE COLLEGE; Bogle, Tommy E.; (GY8091); 25 months; $4,100

NORTHEAST LOUISIANA STATE COLLEGE; Depoe, C. E.; (GY7829); 25 months; $6,600

NORTHWESTERN STATE COLLEGE OF LOUISIANA; Noble, Robert E.; (GY7894); 25 months; $4,100

SOUTHEASTERN LOUISIANA COLLEGE; Wallace, W. R.; (GY7818); 25 months; $5,600

UNIVERSITY OF SOUTHWESTERN LOUISIANA; Bernard, Davy L.; (GY8092); 25 months; $3,500
Cosper, Sammie W.; (GY8104); 25 months; $6,000
Fang, Cheng-Shen; (GY7789); 25 months; $2,200

SCIENCE EDUCATION PROGRAMS

MAINE

BATES COLLEGE; Morrison, R. F.; (GY8043); 25 months; $2,200

BOWDOIN COLLEGE; Gustafson, Alton H.; (GY7896); 25 months; $7,000

COLBY COLLEGE; Allen, Donald B.; (GY7724); 25 months; $5,900
Metz, Roger N.; (GY7931); 25 months; $1,100

UNIVERSITY OF MAINE, Orono; McIntyre, Gary A.; (GY7828); 25 months; $6,900
Wolfhagen, James L.; (GY7898); 25 months; $4,500

MARYLAND

ALLEGANY COMMUNITY COLLEGE; Pawlowski, Anthony T.; (GY8255); 25 months; $2,700

UNIVERSITY OF MARYLAND; Ohaver, T. C.; (GY8235); 25 months; $7,500

VILLA JULIE COLLEGE; Tauber, Daniel R.; (GY8198); 25 months; $500

MASSACHUSETTS

BOSTON UNIVERSITY; Speisman, Joseph C.; (GY8055); 25 months; $22,300

LOWELL TECHNOLOGICAL INSTITUTE; Coleman, Robert M.; (GY8286); 25 months; $10,000

MASSACHUSETTS INSTITUTE OF TECHNOLOGY; Ross, John; (GY7995); 25 months; $12,000
Walton, William U.; (GY7904); 25 months; $21,900

MT. HOLYOKE COLLEGE; Williamson, Kenneth L.; (GY7851); 25 months; $12,500

SMITH COLLEGE; Josephs, Jess J.; (GY7948); 25 months; $7,500

STATE COLLEGE AT BRIDGEWATER; Chipman, Wilmon B.; (GY7847); 25 months; $6,500

TUFTS UNIVERSITY; Johnson, Roger N.; (GY7860); 25 months; $6,900

UNIVERSITY OF MASSACHUSETTS; Barnes, Ramon M.; (GY8059); 25 months; $25,000
Glorioso, R. M.; (GY7788); 25 months; $25,000
Russell, George A.; (GY7765); 25 months; $16,500

WELLESLEY COLLEGE; Brown, Judith C.; (GY7813); 25 months; $2,400
Hicks, Sonja E.; (GY7827); 25 months; $6,300

WESTERN NEW ENGLAND COLLEGE; Azar, Robert C.; (GY7780); 25 months; $1,600
Sundberg Jr., Henry L.; (GY7781); 25 months; $1,100

WHEATON COLLEGE; Landis, Harry M.; (GY7930); 25 months; $600

MICHIGAN

ADRIAN COLLEGE; Madole, R. F.; (GY7902); 25 months; $1,800

ALBION COLLEGE; Woodward Jr., Addison E.; (GY8259); 25 months; $4,900

ALMA COLLEGE; Toller, Louis; (GY7927); 25 months; $4,900

CALVIN COLLEGE; Griffioen, Roger D.; (GY7875); 25 months; $10,500

DELTA COLLEGE; Devinney, Robert H.; (GY7912); 25 months; $2,500
Northrup, Richard; (GY7897); 25 months; $5,700

EASTERN MICHIGAN UNIVERSITY; Scott, Ronald M.; (GY7751); 25 months; $6,300

GRAND VALLEY STATE COLLEGE; Richmond, Gary D.; (GY7757); 25 months; $1,800

MICHIGAN TECHNOLOGICAL UNIVERSITY; Clark, John R.; (GY7779); 25 months; $14,100
D'Angelo, Henry; (GY8134); 25 months; $18,500
Oswald, James A.; (GY7772); 25 months; $12,600
Timm, Robert F.; (GY7906); 25 months; $11,500

MICHIGAN TECHNOLOGICAL UNIVERSITY, Lake Superior State College; Weatherby, Gerald D.; (GY8017); 25 months; $7,100

MICHIGAN STATE UNIVERSITY; Fisher, P. David; (GY7909); 25 months; $8,000
Hart, Harold; (GY7758); 25 months; $11,400

OAKLAND UNIVERSITY; Weng, Tung H.; (GY7946); 25 months; $19,700
Witt, Howard R.; (GY7947); 25 months; $5,900

UNIVERSITY OF MICHIGAN; Caddell, Robert; (GY8120); 25 months; $7,100
Evans, Lary L.; (GY7846); 25 months; $7,600
Westervelt, F. H.; (GY8145); 25 months; $14,700

WAYNE STATE UNIVERSITY; Furlong, Robert B.; (GY8039); 25 months; $16,100
Johnston, Ray E.; (GY7808); 25 months; $3,100
Stagner, Ross; (GY7937); 25 months; $9,400
Zak, Bennie; (GY8184); 25 months; $12,400

WESTERN MICHIGAN UNIVERSITY; Dickason, David; (GY7803); 25 months; $3,100
Kent, Neil D.; (GY8033); 25 months; $6,000

MINNESOTA

AUGSBURG COLLEGE; Sulerud, Ralph L.; (GY7895); 25 months; $3,600

BEMIDJI STATE COLLEGE; Feistner, Sam; (GY7955); 25 months; $2,800

GUSTAVUS ADOLPHUS COLLEGE; Kendall, John; (GY8049); 25 months; $4,000

MACALESTER COLLEGE; Welch, Claude A.; (GY8164); 25 months; $12,000

MANKATO STATE COLLEGE; Gordon, Donald; (GY7826); 25 months; $1,700
Krabbenhoft, Kenneth L.; (GY8180); 25 months; $5,000
Rucker, William B.; (GY8037); 25 months; $24,300

MESABI STATE JUNIOR COLLEGE; Norton, Gary; (GY8028); 25 months; $1,500

SOUTHWEST MINNESOTA STATE COLLEGE; Mitchell, C. Dean; (GY8061); 25 months; $16,200

ST. CLOUD STATE COLLEGE; Goodrich, Herbert; (GY7878); 25 months; $5,500

UNIVERSITY OF MINNESOTA; Goodman, Lawrence E.; (GY8195); 25 months; $5,900
Gordon, Joan; (GY7890); 25 months; $6,200
McTavish, D. G.; (GY7796); 25 months; $3,900
Roll, Peter G.; (GY8082); 25 months; $13,600
Smith, L. H.; (GY8179); 25 months; $6,300

UNIVERSITY OF MINNESOTA, Duluth Campus; Marsden, Ralph W.; (GY7721); 25 months; $1,700

MISSISSIPPI

RUST COLLEGE; Cook, John; (GY8053); 25 months; $4,400

TOUGALOO COLLEGE; Morse, Claire; (GY8228); 25 months; $4,900

UNIVERSITY OF MISSISSIPPI; Bostian, Harry; (GY8125); 25 months; $2,200
Kelly, Robert E.; (GY8100); 25 months; $4,600
Little, John L.; (GY7994); 25 months; $1,100
McLaughlin, Kenneth P.; (GY8095); 25 months; $6,800

MISSOURI

CENTRAL MISSOURI STATE COLLEGE; Bell, M. D.; (GY7924); 25 months; $5,000

NORTHEAST MISSOURI STATE COLLEGE; Walker, Donald E.; (GY8211); 25 months; $6,200

PARK COLLEGE; Jirgal Jr., George H.; (GY7926); 25 months; $1,800

ROCKHURST COLLEGE; Hancox, Robert R.; (GY8131); 25 months; $1,100
Hill, J. J.; (GY7812); 25 months; $900

SOUTHWEST MISSOURI STATE COLLEGE; Robinson, Orin R.; (GY7778) 25 months; $2,100
Stevenson, Robert T.; (GY8264); 25 months; $16,000

STATE FAIR COMMUNITY COLLEGE; McDowell, Mathew E.; (GY8296); 25 months; $4,000

TARKIO COLLEGE; Blewitt, Harry Lyon; (GY8001); 25 months; $2,500

UNIVERSITY OF MISSOURI, Columbia; Harris, Franklin D.; (GY7975); 25 months; $12,800

UNIVERSITY OF MISSOURI, Kansas City; Niu, H. P.; (GY7773); 25 months; $5,400
Westermann, Edwin J.; (GY7928); 25 months $21,400

UNIVERSITY OF MISSOURI, Rolla; Tranter, William H.; (GY7766); 25 months; $10,600

UNIVERSITY OF MISSOURI, St. Louis; Corey, Joyce Y.; (GY8012); 25 months; $6,100

MONTANA

CARROLL COLLEGE; Bowman, Noel E.; (GY8072); 25 months; $1,800

MONTANA COLLEGE OF MINERAL SCIENCE; Griffiths, Vernon; (GY7983); 25 months; $1,000
———; (GY7985); 25 months; $1,300
Twidwell, L. G.; (GY7877); 25 months; $1,100

MONTANA STATE UNIVERSITY; Pagenkopf, Gordon K.; (GY7747); 25 months; $1,500
Videon, Fred F.; (GY7982); 25 months; $22,600

UNIVERSITY OF MONTANA; Margrave Jr., Thomas E.; (GY7811); 25 months; $12,700
Nakamura, Mitsuru J.; (GY7889); 25 months; $7,000

NEBRASKA

CHADRON STATE COLLEGE; Struempler, Arthur W.; (GY7744); 25 months; $3,000
———; (GY8103); 25 months; $1,600

CREIGHTON UNIVERSITY; Cipolla, Sam J.; (GY7810); 25 months; $6,200

NEBRASKA WESLEYAN UNIVERSITY; Warwick, Robert J.; (GY8223); 25 months; $3,500

UNIVERSITY OF NEBRASKA, Lincoln; Booth, Alan; (GY8202); 25 months; $25,000
Caldwell, Warren W.; (GY8189); 25 months; $2,400
Pao, Yen-Ching; (GY7862); 25 months; $10,400
Sonderegger, Theo; (GY7713); 25 months; $5,500
Wilkins, Charles L.; (GY8018); 25 months; $21,800

UNIVERSITY OF NEBRASKA, Omaha; Hendricks, Shelton E.; (GY7716); 25 months; $1,800
Shroder, John F.; (GY8287); 25 months; $2,300

NEVADA

UNIVERSITY OF NEVADA, Reno Campus; Shifley Jr., L. H.; (GY8252); 25 months; $2,900

UNIVERSITY OF NEVADA, Las Vegas; Goldman, Aaron; (GY7718); 25 months; $21,600
Skaggs, Robert L.; (GY8199); 25 months; $15,000

INSTRUCTIONAL SCIENTIFIC EQUIPMENT FOR UNDERGRADUATE EDUCATION

NEW HAMPSHIRE

DARTMOUTH COLLEGE; Lemal, David M.; (GY8266); 25 months; $4,700

UNIVERSITY OF NEW HAMPSHIRE; Goodrich, Robert W.; (GY8248); 25 months; $21,400
Wochholz, Harold F.; (GY7734); 25 months; $10,500

NEW JERSEY

DREW UNIVERSITY; Baker, E. G. Stanley; (GY8275); 25 months; $4,500
Rohrs, Harold C.; (GY8151); 25 months; $5,000

GEORGIAN COURT COLLEGE; Coakley, Sr. Mary P.; (GY8254); 25 months; $4,200

PRINCETON UNIVERSITY; Lo, Arthur W.; (GY8249); 25 months; $24,800
Mark, Peter; (GY7775); 25 months; $7,700

RIDER COLLEGE; Carlson, James H.; (GY7879); 25 months; $1,900
Mayer, Thomas C.; (GY7880); 25 months; $1,200

RUTGERS UNIVERSITY; Flaherty, Charles F.; (GY8204); 25 months; $10,100
Lalancette, Roger A.; (GY7992); 25 months; $5,400
Polymeropoulos, C. E.; (GY7706); 25 months; $10,000
Psuty, Norbert P.; (GY7935); 25 months; $5,100
Rosenberg, David; (GY8147); 25 months; $13,100

RUTGERS, THE STATE UNIVERSITY, Douglass College; McDonald, Neil A.; (GY8052); 25 months; $1,600

RUTGERS, THE STATE UNIVERSITY, Newark Campus; Hall, James C.; (GY8294); 25 months; $14,600

SOMERSET COUNTY COLLEGE; Whitaker, Jo Anne; (GY7886); 25 months; $8,200

STEVENS INSTITUTE OF TECHNOLOGY; Meissner, Hans; (GY8094); 25 months; $14,600

NEW MEXICO

NEW MEXICO INSTITUTE OF MINING AND TECHNOLOGY; Budding, A. J.; (GY7795); 25 months; $8,800

NEW MEXICO HIGHLANDS UNIVERSITY; Shields, Lora M.; (GY7815); 25 months; $2,700

NEW MEXICO STATE UNIVERSITY; Cuffey, James; (GY7730); 25 months; $1,800
Davis, Dennis D.; (GY7997); 25 months; $2,900
Folster, Harry G.; (GY7774); 25 months; $5,900
Wilson, Donald B.; (GY8122); 25 months; $5,400

UNIVERSITY OF NEW MEXICO; Gurbaxani, Shyam H.; (GY8136); 25 months; $2,600

NEW YORK

CLARKSON COLLEGE OF TECHNOLOGY; Cotellessa, R. F.; (GY7981); 25 months; $20,000
Erian, Fadel F.; (GY7881); 25 months; $8,700

COLLEGE AT ONEONTA; Harman, Willard N.; (GY8156); 25 months; $3,100

COLLEGE AT OSWEGO; Weber, Peter G.; (GY7807); 25 months; $6,300

COLLEGE OF FORESTRY AT SYRACUSE; Berglund, John V.; (GY8153); 25 months; $12,100

COLGATE UNIVERSITY; Cochran, John C.; (GY7845); 25 months; $19,400

COLUMBIA UNIVERSITY; Nevin, John A.; (GY8230); 25 months; $25,000

CORNELL UNIVERSITY; Cotts, Robert M.; (GY8099); 25 months; $4,600

CUNY, CITY COLLEGE; Sank, Diane; (GY8270); 25 months; $6,500

CUNY, HUNTER COLLEGE; Beveridge, David L.; (GY8162); 25 months; $8,800
Dolciani, Mary P.; (GY7958); 25 months; $6,100
Mandriota, Frank J.; (GY7712); 25 months; $8,300

HOFSTRA UNIVERSITY; Goldstein, Stanley P.; (GY7776); 25 months; $1,700
Rosman, Howard; (GY8000); 25 months; $15,100
Weissman, David E.; (GY7882); 25 months; $2,200

LEHMAN COLLEGE; Bowman II, James F.; (GY7919); 25 months; $15,500
Engelke, Charles E.; (GY7731); 25 months; $9,800
Knehr, C. A.; (GY7715); 25 months; $2,800

LEMOYNE COLLEGE, Murray, Thomas E.; (GY7809); 25 months; $2,400

LONG ISLAND UNIVERSITY, Brooklyn Center; Howard, Brian; (GY7990); 25 months; $6,800

LONG ISLAND UNIVERSITY, C. W. Post Center; Meiselman, Newton; (GY8277); 25 months; $10,100

LONG ISLAND UNIVERSITY, Southampton Center; Berkebile, C. Alan; (GY7916); 25 months; $4,300
Haresign, Thomas; (GY8150); 25 months; $3,600
Siegel, Alvin; (GY7950); 25 months; $5,100
Welker, J. R.; (GY7951); 25 months; $16,000

MOHAWK VALLEY COMMUNITY COLLEGE; Kosiewicz, Raymond; (GY8246); 25 months; $4,300

NEW YORK UNIVERSITY; Davis, Thomas W.; (GY7998); 25 months; $7,600

POLYTECHNIC INSTITUTE OF BROOKLYN; Jolls, Kenneth R.; (GY8118); 25 months; $15,900
Mikochik, Stephen T.; (GY7883); 25 months; $2,700
Weiss, Gerald; (GY7980); 25 months; $19,400

RENSSELAER POLYTECHNIC INSTITUTE; Altwicker, Elmar R.; (GY8128); 25 months; $11,500
Janz, G. J.; (GY7999); 25 months; $9,300

ROCHESTER INSTITUTE OF TECHNOLOGY; Goldblatt, Norman R.; (GY8069); 25 months; $5,900

SKIDMORE COLLEGE; Johnson, Kenneth G.; (GY8227); 25 months; $7,200

ST. LAWRENCE UNIVERSITY; Beiswenger, Hugo A.; (GY7714); 25 months; $2,400
Parker, Francis D.; (GY7959); 25 months; $3,900
Stradling, Samuel S.; (GY7750); 25 months; $2,400

STATE UNIVERSITY AT BUFFALO; Steegmann Jr., A. T.; (GY7853); 25 months; $3,600

STATE UNIVERSITY AT STONY BROOK; Strom, Stephen E.; (GY8089); 25 months; $12,500

SYRACUSE UNIVERSITY; Bickart, T. A.; (GY7767); 25 months; $22,900

NORTH CAROLINA

CATAWBA COLLEGE; Privett, Donald R.; (GY7913); 25 months; $7,400

LOUISBURG COLLEGE; Pruette, C. Ray; (GY7873); 25 months; $200

NORTH CAROLINA STATE UNIVERSITY AT RALEIGH; Benson Jr., Ray Braman; (GY7777); 25 months; $25,000
Cavaroc Jr., Victor V.; (GY7914); 25 months; $9,600
Wertz, Dennis W.; (GY7857); 25 months; $19,000

PFEIFFER COLLEGE; Horne, Edward E.; (GY7943); 25 months; $2,500

UNIVERSITY OF NORTH CAROLINA AT CHARLOTTE; Burson, S. L.; (GY7842); 25 months; $6,400

NORTH DAKOTA

JAMESTOWN COLLEGE; Mason, Harry; (GY7804); 25 months; $4,300

MINOT STATE COLLEGE; Sheldon, Richard W.; (GY8038); 25 months; $1,100

NORTH DAKOTA STATE UNIVERSITY; Bromel, Mary C.; (GY7819); 25 months; $2,600
Campbell, Edward C.; (GY7874); 25 months; $2,200
Knoeck, John; (GY7893); 25 months; $7,100
Pfister, Philip C.; (GY7971); 25 months; $4,600

UNIVERSITY OF NORTH DAKOTA; Oring, Lewis W.; (GY8178); 25 months; $3,500

OHIO

BALDWIN-WALLACE COLLEGE; Proctor, David G.; (GY8280); 25 months; $900

CASE WESTERN RESERVE UNIVERSITY; Kuwana, J. Fackler; (GY7762); 25 months; $11,600

CLEVELAND STATE UNIVERSITY; Goradia, Chandra P.; (GY7707); 25 months; $14,600

DENISON UNIVERSITY; Graham, Charles E.; (GY8226); 25 months; $1,700

KENYON COLLEGE; Jegla, Thomas C.; (GY8281); 25 months; $2,500

MARY MANSE COLLEGE; Francis, Sr. John; (GY8074); 25 months; $1,000
Godar, Edith M.; (GY7876); 25 months; $2,100

MUSKINGUM COLLEGE; Landolt, Robert George; (GY7905); 25 months; $4,300

OBERLIN COLLEGE; Skinner, William R.; (GY8025); 25 months; $4,100

OHIO STATE UNIVERSITY; Pettyjohn, W. A.; (GY7722); 25 months; $3,200
Roark, Terry P.; (GY7863); 25 months; $3,200

OHIO UNIVERSITY; Ingham, Robert K.; (GY8282); 25 months; $7,200
Lawrence, Roy A.; (GY7708); 25 months; $2,000
Lewis, Paul; (GY7952); 25 months; $4,700

OHIO WESLEYAN UNIVERSITY; Freed, James M.; (GY7823); 25 months; $7,200
Holm, Roger D.; (GY8014); 25 months; $4,700

UNIVERSITY OF AKRON; Pinnick, Harry T.; (GY8101); 25 months; $8,000
Sarikelle, Simsek; (GY7884); 25 months; $4,000

UNIVERSITY OF CINCINNATI; Joiner, William C. H.; (GY8093); 25 months; $13,300

UNIVERSITY OF DAYTON; Bajpai, Praphulla K.; (GY8157); 25 months; $11,700

UNIVERSITY OF TOLEDO; Bennett, Gary F.; (GY8109); 25 months; $3,100

WILMINGTON COLLEGE; Wood, Thomas K.; (GY8182); 25 months; $7,800

WITTENBERG UNIVERSITY; Westneat, David F.; (GY8002); 25 months; $4,700
——— ; (GY8021); 25 months; $3,500

WRIGHT STATE UNIVERSITY; Kemp, Edward H.; (GY7711); 25 months; $10,200

YOUNGSTOWN STATE UNIVERSITY; Kelley, George W.; (GY8183); 25 months; $6,200

OKLAHOMA

OKLAHOMA STATE UNIVERSITY; Fisher, Donald D.; (GY8193); 25 months; $25,000
Gee, Lynn L.; (GY7806); 25 months; $9,700

SCIENCE EDUCATION PROGRAMS

Hecock, Richard D.; (GY8271); 25 months; $5,400
Hodnett, Ernest M.; (GY7755); 25 months; $3,500
Thompson, T. B.; (GY7910); 25 months; $1,800

SEMINOLE JUNIOR COLLEGE; Trammell, Bob L.; (GY8274); 25 months; $3,400

SOUTHEASTERN STATE COLLEGE; Robinson, Jack L.; (GY8237); 25 months; $7,000

UNIVERSITY OF OKLAHOMA; Christensen, James H.; (GY8149); 25 months; $9,300

UNIVERSITY OF TULSA; Hartman, Roger D.; (GY8098); 25 months; $4,700
Levengood, C. A.; (GY7839); 25 months; $6,800
Philoon, Wallace C.; (GY7705); 25 months; $3,600

OREGON

MT. HOOD COMMUNITY COLLEGE; Davis, Don; (GY7940); 25 months; $5,800

OREGON STATE UNIVERSITY; Amort, D.; (GY7704); 25 months; $4,500

OREGON TECHNICAL INSTITUTE; Jackson, Harry J.; (GY7701); 25 months; $2,700
Whitwer, Donald H.; (GY7957); 25 months; $6,100

PACIFIC UNIVERSITY; Meyer-Arendt, Jurgen R.; (GY8260); 25 months; $5,000

REED COLLEGE; Squier, Leslie; (GY7710); 25 months; $21,700

UNIVERSITY OF OREGON; Boekelheide, Virgil; (GY8023); 25 months; $16,000

WILLAMETTE UNIVERSITY; Duell, Paul M.; (GY7749); 25 months; $2,300

PENNSYLVANIA

ALBRIGHT COLLEGE; Hall, John S.; (GY8048); 25 months; $3,600

BRYN MAWR COLLEGE; Gonzalez, R. C.; (GY7732); 25 months; $11,000

BUCKNELL UNIVERSITY; Harclerode, Jack E.; (GY7825); 25 months; $6,700
Leshner, Alan I., (GY7856); 25 months; $6,900
Marchand, Denis E.; (GY7723); 25 months; $4,500
Schwensfeir Jr., Robert J.; (GY7901); 25 months; $4,100

CARNEGIE MELLON UNIVERSITY; Shaw, M. C.; (GY8119); 25 months; $21,700

DREXEL UNIVERSITY; Matula, Richard A.; (GY7709); 25 months; $20,300

DUQUESNE UNIVERSITY; Raizen, Eileen C.; (GY8276); 25 months; $10,700
Wang, Jin Tsai; (GY7759); 25 months; $6,500

FRANKLIN AND MARSHALL COLLEGE; Senko, Monte G.; (GY8047); 25 months; $8,700

JUNIATA COLLEGE; Norris, Wilfred G.; (GY7925); 25 months; $3,100

LAFAYETTE COLLEGE; Grant, Raymond S.; (GY8225); 25 months; $6,600

LEHIGH UNIVERSITY; Bryen, Stephen D; (GY8256); 25 months; $4,300
Aquinas, Sr., M.; (GY8177); 25 months; $2,300
Diefenderfer, A. James; (GY7756); 25 months; $8,200
Ondria, John G.; (GY7970); 25 months; $24,800

MORAVIAN COLLEGE; Ridge, Jack R.; (GY8087); 25 months; $5,000

MUHLENBERG COLLEGE; Boyer, Robert A.; (GY8085); 25 months; $7,900
Shive, Donald W.; (GY8013); 25 months; $4,800

PMC COLLEGES; Jefferis III, Raymond P.; (GY8106); 25 months; $23,400
Thornton, Elizabeth K.; (GY8060); 25 months; $4,800

SLIPPERY ROCK STATE COLLEGE; Bushnell, Kent O.; (GY8044); 25 months; $2,400

ST. FRANCIS COLLEGE; Duryea, William R.; (GY7843); 25 months; $1,700

SWARTHMORE COLLEGE; Hammons, James H.; (GY8022); 25 months; $15,100

TEMPLE UNIVERSITY; Feldman, Stuart; (GY8015); 25 months; $6,400

THIEL COLLEGE; Bennett, R. B.; (GY8004); 25 months; $11,900

VILLANOVA UNIVERSITY; Rice, William J.; (GY7702); 25 months; $3,000

WESTMINSTER COLLEGE; Dewitt, H. D.; (GY7848); 25 months; $6,100

WILKES COLLEGE; Wong, Bing K.; (GY7934); 25 months; $500

PUERTO RICO

INTER-AMERICAN UNIVERSITY OF PUERTO RICO; Verter, Herbert S.; (GY8242); 25 months; $6,400

RHODE ISLAND

ROGER WILLIAMS COLLEGE; Callahan, J. D.; (GY8030); 25 months; $2,500
Hetzler, Charles; (GY8207); 25 months; $500

UNIVERSITY OF RHODE ISLAND; Hill, Robert B.; (GY8261); 25 months; $11,600

SOUTH CAROLINA

COLLEGE OF CHARLESTON; Gibson, Gerald W.; (GY8243); 25 months; $4,900
Kirkland, F. Ronald; (GY7871); 25 months; $4,700

FURMAN UNIVERSITY; Brantley, William H.; (GY7864); 25 months; $4,200
Brewer, Charles L.; (GY7858); 25 months; $3,600
Price Jr., Van; (GY7911); 25 months; $4,300

WINTHROP COLLEGE; Tutwiler, Frank B.; (GY8011); 25 months; $12,000

SOUTH DAKOTA

AUGUSTANA COLLEGE; Kintner, Robert Roy; (GY8006); 25 months; $10,200
Nelson, V. R.; (GY7865); 25 months; $1,400

SOUTH DAKOTA SCHOOL OF MINES AND TECHNOLOGY; Hamel, James V.; (GY7703); 25 months; $9,600
Riemenschneider, Albert L.; (GY7965); 25 months; $10,600
Thorson, Donald A.; (GY7989); 25 months; $11,700

SOUTH DAKOTA STATE UNIVERSITY; Johnson, Emory E.; (GY7854); 25 months; $4,000
Sandfort, John F.; (GY7966); 25 months; $10,800

UNIVERSITY OF SOUTH DAKOTA; Carraher, Charles E., (GY7741); 25 months; $5,000
French, Gilbert M.; (GY7729); 25 months; $5,500
Heisinger, James F.; (GY7824); 25 months; $5,000

TENNESSEE

CHRISTIAN BROTHERS COLLEGE; Brown, Ray W. Jr.; (GY8127); 25 months; $2,900
Ventura, H. John; (GY8114); 25 months; $3,100

JACKSON STATE COMMUNITY COLLEGE; Palmer, Ray A.; (GY8295); 25 months; $4,000

MEMPHIS STATE UNIVERSITY; Doyle, Jack E.; (GY7960); 25 months; $17,000
Lounsbury, R. W.; (GY8196); 25 months: $3,100
Rushton, Priscilla S.; (GY8171); 25 months; $18,600
Slater, Carl D.; (GY7742); 25 months; $2,100

MIDDLE TENNESSEE STATE UNIVERSITY; Moody, Thomas L.; (GY8102); 25 months; $8,000

TENNESSEE TECHNOLOGICAL UNIVERSITY; Ballal, S. K.; (GY7887); 25 months; $5,900
Hunter, Gordon E.; (GY7794); 25 months; $5,700
Mitchell, William S.; (GY7978); 25 months; $7,000

UNIVERSITY OF TENNESSEE AT MARTIN; Wakim, Jubran M.; (GY8016); 25 months; $3,500

UNIVERSITY OF TENNESSEE; Hoffman, Graham W.; (GY8088); 25 months; $22,800
Meyer, Bernadine H.; (GY8067); 25 months; $6,100
Smith Jr., Buford; (GY7993); 25 months; $4,600
Speckhart, M. Milligan; (GY7933); 25 months; $5,300

VANDERBILT UNIVERSITY; Baker Jr., William Roy; (GY7938); 25 months; $9,100
Clement Jr., William M.; (GY7861); 25 months; $3,000

TEXAS

BISHOP COLLEGE; Tunstall, Lucille H.; (GY8065); 25 months; $5,400

COOKE COUNTY JUNIOR COLLEGE; Stanley, William; (GY8268); 25 months; $2,000

DALLAS COUNTY JUNIOR COLLEGE DISTRICT, El Centro College; Gonzalez, Carlos; (GY8140); 25 months; $4,200

HOUSTON BAPTIST COLLEGE; Modisette, Jerry L.; (GY8214); 25 months; $3,700

MCMURRY COLLEGE; Jones Jr., W. Norton; (GY8239); 25 months; $4,100

NORTH TEXAS STATE UNIVERSITY; Marshall, James L.; (GY7816); 25 months; $11,000

SAN ANTONIO COLLEGE; Howard, Charles; (GY8238); 25 months; $6,700

SOUTHERN METHODIST UNIVERSITY; Nardizzi, Louis R.; (GY8251); 25 months; $12,500

TEXAS A & M UNIVERSITY; Ables, Ernest D.; (GY7822); 25 months; $3,700
Davies, David K.; (GY7908); 25 months; $3,200
Dunlap, Wayne A.; (GY8123); 25 months; $7,700
Fahlquist, Davis A.; (GY8219); 25 months; $4,900
Herbich, John B.; (GY7987); 25 months; $7,500
Howes, J. R.; (GY8173); 25 months; $10,700
Martell, A. E.; (GY7754); 25 months; $14,300
Vinson, S. Bradleigh; (GY7841); 25 months; $5,000

TEXAS TECH UNIVERSITY; Reis, L. A.; (GY7700); 25 months; $1,400

TEXAS TECHNOLOGICAL COLLEGE; Kristiansen, M.; (GY7979); 25 months; $7,000

TRINITY UNIVERSITY; Lovell, F. M.; (GY8267); 25 months; $14,500

UNIVERSITY OF TEXAS AT ARLINGTON; Anderson, Jay E. Jr.; (GY8216); 25 months; $3,200

UNIVERSITY OF TEXAS AT AUSTIN; Hougen, Joel O.; (GY8130); 25 months; $22,000

UTAH

UNIVERSITY OF UTAH; Gibbs, Peter; (GY7918); 25 months; $8,900
Ridd, Merrill K.; (GY8217); 25 months; $5,000

COURSE CONTENT IMPROVEMENT

UTAH STATE UNIVERSITY; Fletcher, William I.; (GY7988); 25 months; $9,000
Spendlove, Rex S.; (GY8174); 25 months; $7,700

WEBER STATE COLLEGE; Neff, Thomas R.; (GY8220); 25 months; $3,800

VERMONT

BENNINGTON COLLEGE; Cornwell, Robert G.; (GY7922); 25 months; $5,500

UNIVERSITY OF VERMONT; Ellis, David M.; (GY8144); 25 months; $1,300
Juenker, D. W.; (GY7872); 25 months; $9,500
Taylor, Charles F.; (GY8133); 25 months; $8,200

WINDHAM COLLEGE; Barcik, J. D.; (GY8041); 25 months; $6,100
Lawson, Fay A.; (GY8146); 25 months; $4,000
Rice, Michael; (GY7929); 25 months; $2,100
Shaw Jr., Charles E.; (GY8045); 25 months; $2,800
Westing, Arthur H.; (GY8148); 25 months; $4,900

VIRGINIA

COLLEGE OF WILLIAM AND MARY; Tyree Jr., S. Y.; (GY7753); 25 months; $17,400

OLD DOMINION UNIVERSITY; Cupschalk, Stephen G.; (GY7991); 25 months; $15,800
Torres, Arnold L.; (GY7849); 25 months; $7,800

SOUTHWEST VIRGINIA COMMUNITY COLLEGE; Cox, Curtis; (GY8283); 25 months; $1,300
Estep, Gary; (GY8163); 25 months; $200

UNIVERSITY OF RICHMOND; Taylor, J. J.; (GY8083); 25 months; $21,300

UNIVERSITY OF VIRGINIA, George Mason College; Mielczarek, Eugenie V.; (GY8097); 25 months; $6,100

VIRGINIA COMMONWEALTH UNIVERSITY; Reed, James R.; (GY8181); 25 months; $3,500

VIRGINIA POLYTECHNIC INSTITUTE; Pavlik, W. B.; (GY8036); 25 months; $4,700
Phillips, Jean A.; (GY8168); 25 months; $3,000

VIRGINIA STATE COLLEGE; Davenport, James C.; (GY8079); 25 months; $9,000

WASHINGTON

CLARK COLLEGE; Nelson, William D.; (GY8240); 25 months; $1,600

EASTERN WASHINGTON STATE COLLEGE; Parker, Ormond J.; (GY8007); 25 months; $10,900
Schwalm, Dennis E.; (GY7986); 25 months; $5,800

PACIFIC LUTHERAN UNIVERSITY; Jensen, Joann S.; (GY7888); 25 months; $3,000

SEATTLE PACIFIC COLLEGE; Anderson, Roger H.; (GY7805); 25 months; $1,600

SEATTLE UNIVERSITY; Schroeder, David W.; (GY7976); 25 months; $5,700

UNIVERSITY OF PUGET SOUND; Nigh, Wesley G.; (GY8241); 25 months; $6,800

UNIVERSITY OF WASHINGTON; Colcord, J. E.; (GY7984); 25 months; $5,600
Larsen, L. H.; (GY8205); 25 months; $11,000

WHITWORTH COLLEGE; Johnson, William A.; (GY8009); 25 months; $2,600

WEST VIRGINIA

WEST VIRGINIA UNIVERSITY; Kowal, Norman E.; (GY7820); 25 months; $4,900
Norman, Charles; (GY7832); 25 months; $5,000
Pavlovic, Arthur S.; (GY8105); 25 months; $12,500

Traynelis, Vincent J.; (GY7752); 25 months; $10,400
Walters, Richard E.; (GY7917); 25 months; $3,500

WISCONSIN

BELOIT COLLEGE; Dobson, D. A.; (GY8076); 25 months; $10,000
Woodard, H. H.; (GY7907); 25 months; $6,800

CARROLL COLLEGE; Bayer, Richard; (GY8024); 25 months; $5,100
Christoph, Roy J.; (GY8186); 25 months; $7,100

RIPON COLLEGE; Alexander Jr., William A.; (GY8050); 25 months; $800

STOUT STATE UNIVERSITY; Faris, John J.; (GY8071); 25 months; $1,400
Lowry, Edward M.; (GY8188); 25 months; $300
Nitz, Otto W.; (GY8161); 25 months; $1,600

UNIVERSITY OF WISCONSIN, Green Bay; Starkey, Ronald; (GY8288); 25 months; $10,600

UNIVERSITY OF WISCONSIN, Waukesha County Campus; Grotz, Leonard C.; (GY7745); 25 months; $3,600

UNIVERSITY OF WISCONSIN, Madison; Allen, Paul J.; (GY8187); 25 months; $3,000
Parmentier, R. D.; (GY7967); 25 months; $5,100
Perrin, John H.; (GY8284); 25 months; $3,900
Rongstad, Orrin James; (GY7821); 25 months; $1,000
Shain, Irving; (GY7763); 25 months; $18,300
Webster, John G.; (GY8191); 25 months; $4,200

UNIVERSITY OF WISCONSIN, Milwaukee; Roderick, Gilbert L.; (GY8124); 25 months; $7,000
Sherman, D. R.; (GY7060); 25 months; $13,700
Sorensen Jr., Arthur; (GY8143); 25 months; $16,300

WISCONSIN STATE UNIVERSITY, Eau Claire; Mertz, Marvin C.; (GY8073); 25 months; $8,200

WISCONSIN STATE UNIVERSITY, Eau Claire; Bakken, Arnold; (GY8185); 25 months; $8,000

WISCONSIN STATE UNIVERSITY, La Crosse; Fletcher, Richard; (GY7844); 25 months; $10,600

WISCONSIN STATE UNIVERSITY, Oshkosh; Gueths, James E.; (GY8084); 25 months; $6,500

WISCONSIN STATE UNIVERSITY, Platteville; Al-Khafaji, A. Amir; (GY8137); 25 months; $6,500
McCullough, Earl S.; (GY7972); 25 months; $12,100

WISCONSIN STATE UNIVERSITY, Superior; Bahnick, Donald A.; (GY7761); 25 months; $3,300
Brieske, Phillip R.; (GY8257); 25 months; $1,100
Dickas, Albert B.; (GY7733); 25 months; $4,400

WISCONSIN STATE UNIVERSITY, Whitewater; Joern, William A.; (GY7760); 25 months; $4,400

WYOMING

NORTHWEST COMMUNITY COLLEGE; Vining, Wesley; (GY8297); 25 months; $4,400

UNIVERSITY OF WYOMING; Prowse, Derek J.; (GY7953); 25 months; $17,800

Course Content Improvement

College and University Studies

CALIFORNIA

CALIFORNIA INSTITUTE OF TECHNOLOGY
Zirin, Harold; *Solar Atmosphere Movie* (GY8423); 12 months; $11,265

CONNECTICUT

YALE UNIVERSITY
Peck, Merton J.; *Computer Use for Macro-Policy Simulation* (J000780*); 24 months; $43,900

DISTRICT OF COLUMBIA

AMERICAN ASSOCIATION FOR THE ADVANCEMENT OF SCIENCE
Mayor, John R.; *Development of New Guidelines for Preparation Programs for Teachers of Secondary School Science* (GY7050); 16 months; $93,370

AMERICAN CHEMICAL SOCIETY, Division of Chemical Education
Wall, Frederick T.; *Course Content Improvement, College-Chemistry Conference on Education in Chemistry* (GY7671); 6 months; $19,500

AMERICAN INSTITUTE OF BIOLOGICAL SCIENCES
Kormondy, Edward J. and Kollros, Jerry; *The Commission on Undergraduate Education in the Biological Sciences* (GY6787 002); $32,633
———; *Commission on Undergraduate Education in the Biological Sciences* (GY6787 003); 12 months; $320,000
Olive, John R.; *Office of Biological Education* (GY6872001); 24 months; $90,000

ASSOCIATION OF AMERICAN GEOGRAPHERS
Lounsbury, John F.; *Commission on College Geography* (GE8216006); 18 months; $192,500

ENVIRONMETRICS
Pickett, Robert A.; *Urban Operational Simulation Curriculum Development Project* (GY8433); 25 months; $49,713

MATHEMATICAL ASSOCIATION OF AMERICA
Boas, Ralph P.; *Committee on the Undergraduate Program in Mathematics* (G23827-006); 36 months; $817,625

FLORIDA

FLORIDA STATE UNIVERSITY
Schwarz, Guenter; *Development of a Computer Oriented Intermediate Level Course in Differential Equations* (GY8438); 12 months; $23,890
———; *Development of a Computer Related Calculus Sequence* (GY3696002); 14 months; $39,520

IDAHO

IDAHO STATE UNIVERSITY
Swanson Jr., Earl H.; *Course Content Improvement, College-Social Sciences Educational Films of Experiments in Flintworking* (GY7051); 12 months; $49,880

INDIANA

PURDUE UNIVERSITY
Postlethwait, S. N.; *Biological Sciences* (GY7664); 37 months; $837,091

MARYLAND

UNIVERSITY OF MARYLAND
Fowler, John M.; *Commission on College Physics* (GY3793004); 14 months; $313,600

MASSACHUSETTS

EDUCATION DEVELOPMENT CENTER
Max, Nelson, L.; *Topology Film Project* (GY7699); 33 months; $265,150

SCIENCE EDUCATION PROGRAMS

MICHIGAN

UNIVERSITY OF MICHIGAN, Flint College
DeGraaf, Donald E.; *Flint Introductory Physics Sequence* (GY3792004); 17 months; $69,040

NEW YORK

ASSOCIATION FOR COMPUTING MACHINERY
Viavant, William; *Computer Science and Computer Facilities Consulting Service for Small Schools* (GY7052); 20 months; $69,135

POLYTECHNIC INSTITUTE OF BROOKLYN
Bregman, Judith; *Course Content Improvement, College-Physics A Computer Animated Film on the Quantum Mechanical Harmonic Oscillator* (GY7682); 18 months; $23,732

RENSSELAER POLYTECHNIC INSTITUTE
Leitner, Alfred; *Films Demonstrating Atomic Structure Via Atomic Spectra* (GY6934); 24 months; $87,080

OHIO

UNIVERSITY OF AKRON
Grunberg, Emile; *Teaching Introductory Economics Through Laboratory Work* (GY7662); 32 months; $53,385

PENNSYLVANIA

PENNSYLVANIA STATE UNIVERSITY
McKinstry, Herbert A.; *Single Concept Films on Materials Science/Engineering* (GY7698); 34 months; $250,900

SOUTH DAKOTA

SOUTH DAKOTA STATE UNIVERSITY
Nelson, Gorman R.; *Evaluation and Continued Development of Geometric and Graphic Aids for College Calculus* (GY6901); 12 months; $23,370

TEXAS

TEXAS TECH UNIVERSITY
Hagler, M. Kristiansen; *Course Content Improvement, College-Engineering Development of Laser Experiments for Undergraduate Electrical Engineering Studies* (GY4761001); 11 months; $9,200

Secondary School Studies

CALIFORNIA

HARVEY MUDD COLLEGE
Alford, Jack L.; *Interdisciplinary Conference for School Decision Makers on Man-Made World* (GW5385); 6 months; $2,490

STANFORD UNIVERSITY
Gross, Richard and Girault, Emily; *Social Sciences Resource Personnel Workshop in Social Resources* (GW5347); 17 months; $56,700

UNIVERSITY OF CALIFORNIA, Berkeley
Rosenfeld, Arthur H.; *Computer-Based Self-Instructional Course for Supplementary Training for Secondary School Teachers* (GW5061); 12 months; $124,200
Whinnery, John R.; *Interdisciplinary Conference for School Decision Makers on Man-Made World* (GW5720); 10 months; $10,110

COLORADO

UNIVERSITY OF NORTHERN COLORADO
Halvorson, Peter L.; *Social Sciences High School Geography Project Leadership Conference* (GW5351); 17 months; $31,000

DISTRICT OF COLUMBIA

AMERICAN GEOLOGICAL INSTITUTE
Samples, Wm. D.; *Earth Science and Astronomy Environmental Studies for Urban Youth* (GW5387); 13 months; $170,600

FLORIDA

FLORIDA STATE UNIVERSITY
Burkman, Ernest; *Intermediate Science Curriculum Study* (GW4235002); 15 months; $447,400
Burkman, P. Westmeyer and Burkman, Ernest; *Two Leadership Workshops in the Intermediate Science Curriculum Study* (GW5355); 17 months; $28,390
Rollins, James H.; *Implementation Program for the University of Illinois Junior High School Math Project* (GW5345*); 17 months; $28,564
——— ; (GW5345); 17 months; $6,336

ILLINOIS

UNIVERSITY OF ILLINOIS, Chicago Circle
Setton, Henry A.; *Interdisciplinary Conference for School Decision Makers on Man-made World* (GW5386); 10 months; $1,910

UNIVERSITY OF ILLINOIS, Urbana
Beberman, Max; *Implementation Program for Junior High School Mathematics* (GW4502001); $18,000

INDIANA

INDIANA UNIVERSITY, Bloomington
Hanvey, Robert G.; *An Anthropology Case Materials Project* (GW5059); 20 months; $111,400

PURDUE UNIVERSITY
Herron, J. Dudley; *Leadership Training in the Intermediate Science Curriculum Study* (GW5356); 17 months; $20,790

KENTUCKY

MOREHEAD STATE UNIVERSITY
Nail, Billy R.; *Implementation Program for the University of Illinois Junior High School Math Project* (GW5360); 17 months; $39,740

MARYLAND

UNIVERSITY OF MARYLAND
Lockard, J. David; *Leadership Workshop in the Intermediate Science Study Materials* (GW5344); 17 months; $19,970

MASSACHUSETTS

AMERICAN METEOROLOGICAL SOCIETY
Spengler, Kenneth C.; *Development of Education Films in Meteorology* (G13696003); $18,000

EDUCATION DEVELOPMENT CENTER
Haber-Schaim, Uri; *An Implementation Project in the Second Year Physical Science Course* (GW5357); 17 months; $68,700
——— ; *Completion of a Second-Year Course in Physical Science to Follow the Introductory Physical Science Course* (GW2187004); 24 months; $104,400

HARVARD UNIVERSITY
Holton, Gerald, Maybury, Robert H. and Watson, Fletcher G.; *Course Content Improvement (Non-Limitation)* (GW5210); 17 months; $32,700

MICHIGAN

AMERICAN SOCIOLOGICAL ASSOCIATION
Angell, Robert C.; *Sociological Resources for the Social Studies* (GE5186007); $14,400
——— ; *Sociological Resources for the Social Studies* (GE5186008); 20 months; $520,200

WESTERN MICHIGAN UNIVERSITY
Vuicich, George; *A Leadership Workshop in High School Geography Project Materials* (GW5354); 17 months; $49,600

NEW HAMPSHIRE

DARTMOUTH COLLEGE
Kurtz, Thomas E.; *Demonstration and Experimentation in Computer Training and Use in Secondary Schools* (GW2246002); $19,200

NEW YORK

POLYTECHNIC INSTITUTE OF BROOKLYN
Piel, Emil J.; *Conference for School Decision Makers on Man-made World* (GW5719); 10 months; $3,590

STATE UNIVERSITY AT ALBANY
Nurnberger, Robert G.; *Workshop to Prepare a Study Guide on Problems of Air Pollution* (GW5705); 10 months; $33,800

TEACHERS COLLEGE
Fehr, Howard F.; *Secondary School Mathematics Curriculum Improvement Study* (GW4533); 24 months; $331,000

NORTH CAROLINA

UNIVERSITY OF NORTH CAROLINA AT CHAPEL HILL
Schlechty, Phillip C.; *Resource Personnel Workshop in Social Resources* (GW5350); 17 months; $53,400

OHIO

OHIO STATE UNIVERSITY
Riley, William R.; *Conference for School Decision Makers on Man-made World* (GW5389); 6 months; $3,410

OREGON

OREGON STATE UNIVERSITY
Wilson, Howard; *Implementation Program for the University of Illinois Junior High School Math Project* (GW5358); 17 months; $44,600

PORTLAND STATE UNIVERSITY
Dittmer, Karl; *Portland Interdisciplinary Science Project* (GW4216001); 18 months; $52,900

RHODE ISLAND

BROWN UNIVERSITY
Clapp, Leallyn B.; *Resource Personnel Workshop for the Chemical Bond Approach* (GW5348); 17 months; $79,000

TENNESSEE

MEMPHIS STATE UNIVERSITY
Jermann, William H.; *Conference for School Decision Makers on Man-made World* (GW5718); 8 months; $3,660

WISCONSIN

UNIVERSITY OF WISCONSIN, Madison
Marshall Jr., W. R.; (GW5384); 6 months; $4,440

Elementary and Junior High School Studies

ALABAMA

AUBURN UNIVERSITY
Ellisor, Mildred; *Summer Institute for Teacher Training and Curriculum Dissemination Using Man, A Course of Study* (GW5379); 17 months; $26,100

ARIZONA

ARIZONA STATE UNIVERSITY
Podlich, William F.; *Summer Institute for Teacher Training and Curriculum Dissemination Using Man, A Course of Study* (GW5365); 17 months; $25,800

COURSE CONTENT IMPROVEMENT

CALIFORNIA

SAN FRANCISCO STATE COLLEGE
Haan, Aubrey and Webb, Sylvester L.; *A Leadership Workshop in Diffusion of the 'Science Curriculum Improvement Study' Materials in Urban Areas* (GW5349); 17 months; $47,600

STANFORD UNIVERSITY
Suppes, Patrick; *Experimental Teaching of Mathematics in the Elementary School* (G18709009); 14 months; $170,000

COLORADO

TEMPLE BUELL COLLEGE
Fitzgerald, Thomas A.; *Leadership Preparation for the Implementation 'Man—A Course of Study'* (GW5708); 12 months; $52,300

UNIVERSITY OF COLORADO, Denver Center
McGlathery, Glenn E.; *Leadership Workshop for Science Curriculum Improvement* (GW5363); 17 months; $39,900

CONNECTICUT

CENTRAL CONNECTICUT STATE COLLEGE
Reilley, Dennen; *Leadership Preparation for the Implementation of 'Man—A Course of Study'* (GW5848); 17 months; $50,100

DISTRICT OF COLUMBIA

AMERICAN ASSOCIATION FOR THE ADVANCEMENT OF SCIENCE
Mayor, John R.; *Commission on Science Education* (G22286012); 12 months; $110,500

WASHINGTON SCHOOL OF PSYCHIATRY
Cort, H. Russell, Jr.; *A Feasibility Study to Identify Approaches to Evaluation of 'Man—A Course of Study'* (GW5707); 11 months; $14,000

FLORIDA

FLORIDA STATE UNIVERSITY
Harrison, Robert S.; *Leadership Preparation for the Implementation 'Man—A Course of Study'* (GW5847); 14 months; $39,169

ILLINOIS

UNIVERSITY OF ILLINOIS, Urbana
Beberman, Max; *Elementary School Mathematics and Science* (GW4532); 13 months; $112,000

MASSACHUSETTS

EASTERN NAZARENE COLLEGE
Tracy, Thomas R.; *Workshop in Implementation of Elementary School Science and Mathematics Curricula* (GW5352); 17 months; $25,000

EDUCATION DEVELOPMENT CENTER
Devore, Irven; *A Secondary School Course, 'Exploring Human Nature'* (GW5209); 12 months; $300,000
Griffith, Joe; *Leadership Preparation for the Implementation of Elementary Science Study* (GW5361); 17 months; $64,800
Lomon, Earle L.; *Unified Science and Mathematics Materials for Elementary Schools* (GW5207); 18 months; $145,000
Watson, Frank J.; *Elementary Science Study* (G21815014); 12 months; $249,300

MINNESOTA

UNIVERSITY OF MINNESOTA
Werntz, James H. Jr.; *Minnesota Mathematics and Science Teaching Project* (GE3013); 12 months; $160,500

MISSOURI

NORTHEAST MISSOURI STATE COLLEGE
Evans, Denmen C.; *Workshop in Implementation of Elementary School Science and Mathematics Curricula* (GW5346); 17 months; $15,900

NEW YORK

EASTERN REGIONAL INSTITUTE FOR EDUCATION
Herlihy, John G.; *Leadership Preparation for the Implementation of 'Man—A Course of Study'* (GW5378); 17 months; $69,500

NEW YORK UNIVERSITY
Randolph Lynne; *Summer Institute for Teacher Training and Curriculum Dissemination Using 'Man—A Course of Study'* (GW5364); 17 months; $22,900

OREGON

UNIVERSITY OF OREGON
Harris, William H.; *Leadership Preparation for the Implementation of 'Man—A Course of Study'* (GW5368); 17 months; $75,400

TENNESSEE

GEORGE PEABODY COLLEGE FOR TEACHERS
Farnen Jr., Russell F.; *Leadership Preparation for the Implementation of 'Man—A Course of Study'* (GW5366); 17 months; $56,700

TEXAS

OUR LADY OF THE LAKE COLLEGE
Pena, Karen; *Teacher Training in and Dissemination of 'Man—A Course of Study'* (GW5362); 17 months; $23,600

VERMONT

UNIVERSITY OF VERMONT
Agne, Russell M.; *Leadership Development in 'Science—A Process Approach'* (GW5369); 17 months; $38,100

WASHINGTON

SEATTLE PACIFIC COLLEGE
Harris, Richard; *Workshop in Implementation of Elementary School Science and Mathematics Curricula* (GW5359); 17 months; $21,000

WISCONSIN

EDGEWOOD COLLEGE
McSweeney, Jean; (GW5353); 17 months; $20,500

Supplementary Teaching Aids

CALIFORNIA

COLLEGE OF THE HOLY NAME
Gaffney, Sr. Alma Rose; *Cooperative College School Science* (GW5189); 19 months; $39,367

INSTITUTE OF MATHEMATICAL STATISTICS
Olkin, Ingram; *Visiting Scientists (Colleges)* (GY8416); 15 months; $8,350

UNIVERSITY OF CALIFORNIA, Davis
Perkes, Victor Aston; *Cooperative College School Science* (GW5182); 19 months; $25,303

CONNECTICUT

UNIVERSITY OF CONNECTICUT
Dyrli, Odvard Egil; (GW5162); 19 months; $36,599

DISTRICT OF COLUMBIA

AMERICAN GEOLOGICAL INSTITUTE
Shipman, Ross L.; *Visiting Geological Scientist Program* (GY8405); 15 months; $15,000

AMERICAN GEOPHYSICAL UNION
Spilhaus Jr., A. F.; *Visiting Scientist Program in Geophysics* (GY8406); 15 months; $23,500

AMERICAN INSTITUTE OF BIOLOGICAL SCIENCES
Olive, John R.; *Visiting Scientists Program* (GY8407); 15 months; $14,000

AMERICAN PSYCHOLOGICAL ASSOCIATION
Simmons, William; *Visiting Scientists (Colleges) Program* (GY8410); 15 months; $15,000

AMERICAN SOCIETY FOR ENGINEERING EDUCATION
Bradley Jr., Francis X.; *Visiting Scientists (Colleges)* (GY8412); 15 months; $7,350

AMERICAN SOCIOLOGICAL ASSOCIATION
Volkart, Edmund H.; *Visiting Scientists (Colleges)* (GY8414); 15 months; $10,000

ASSOCIATION OF AMERICAN GEOGRAPHERS
Natoli, Salvatore J.; *Visiting Scientists (Colleges)* (GY8415); 15 months; $5,000

CATHOLIC UNIVERSITY OF AMERICA
Goddu, Roland J.; *Cooperative College School Science* (GW5195); 19 months; $38,387

FEDERAL CITY COLLEGE
Bolten, M.; *Cooperative College School Science* (GW5203); 19 months; $30,005

SOCIETY OF AMERICAN FORESTERS
Theoe, Donald R.; *Visiting Scientists Program* (GY8419); 15 months; $6,000

INDIANA

PURDUE UNIVERSITY
Devito, Alfred; *Cooperative College School Science* (GW5176); 19 months; $32,647

KENTUCKY

AMERICAN ANTHROPOLOGICAL ASSOCIATION
Richards, Cara E.; *Visiting Scientists (Colleges)* (GY8401); 15 months; $15,000

MARYLAND

OPERATIONS RESEARCH SOCIETY OF AMERICA
Schrady, David A.; *Visiting Scientists (Colleges)* (GY8418); 15 months; $6,700

MASSACHUSETTS

AMERICAN METEOROLOGICAL SOCIETY
Spengler, Kenneth C.; *Visiting Scientists Program in Meteorology* (GY8409); 15 months; $2,465

NEW YORK

AMERICAN INSTITUTE OF PHYSICS
Strassenburg, Arnold A.; *Visiting Scientists Program in Physics* (GY8408); 15 months; $31,400

COLLEGE AT OSWEGO
O'Donnell, Raymond T.; *Cooperative College School Science* (GW5186); 19 months; $34,640

CUNY, BROOKLYN COLLEGE
Sharefkin, Belle D.; (GW5174); 19 months; $35,943

MATHEMATICAL ASSOCIATION OF AMERICA
Pownall, Malcolm W.; *Visiting Scientists (Colleges)* (GY8417); 15 months; $28,250

NORTH CAROLINA

SOCIETY OF WOOD SCIENCE AND TECHNOLOGY
Barefoot, A. C.; *Visiting Scientists (Colleges)* (GY8421); 15 months; $3,400

UNIVERSITY OF NORTH CAROLINA AT CHARLOTTE
Clay, James W.; *Cooperative College School Science* (GW5202); 19 months; $48,531

OHIO

AMERICAN CHEMICAL SOCIETY
Glasoe, Paul K.; *Visiting Scientists (Colleges)* (GY8403); 15 months; $20,000

AMERICAN SOCIETY OF PHOTOGRAMMETRY
Mintzer, Olin W.; (GY8413); 15 months; $2,490

SCIENCE EDUCATION PROGRAMS

PENNSYLVANIA

AMERICAN ECONOMIC ASSOCIATION
Saunders, Phillip; *Visiting Scientist Program in Economics* (GY8404); 15 months; $7,025

SOCIETY FOR INDUSTRIAL AND APPLIED MATHEMATICS
Diprima, Richard C.; *Visiting Scientist Program* (GY8420); 15 months; $15,075

WILSON COLLEGE
Giles Jr., Lester A.; *Cooperative College School Science* (GW5163); 19 months; $30,815

SOUTH DAKOTA

BLACK HILLS STATE COLLEGE
Follette, Everett L.; (GW5196); 19 months; $20,978

UTAH

SOUTHERN UTAH STATE COLLEGE
Lebaron, George; (GW5175); 19 months; $33,316

VIRGINIA

AMERICAN ASTRONOMICAL SOCIETY
Fredrick, Laurence W.; *Visiting Scientists (Colleges) Program* (GY8402); 15 months; $21,200

WISCONSIN

AMERICAN SOCIETY OF AGRONOMY
Stelly, Matthias; *Visiting Scientists (Colleges)* (GY8411); 15 months; $10,000

Combined Elementary Through College Level Studies

CALIFORNIA

STANFORD UNIVERSITY
Begle, E. G.; *School Mathematics Study Group* (G18758016); 12 months; $641,000

COLORADO

SOCIAL SCIENCE EDUCATIONAL CONSORTIUM
Morrissett, Irving; *Improve Analysis and Use of Social Science Curriculum Materials, K–12* (GW2277002); 24 months; $297,800

MASSACHUSETTS

EDUCATION DEVELOPMENT CENTER
Dow, Peter B.; *A Curriculum for Teacher Education in the Social Studies* (GE3430010); 12 months; $150,600

MINNESOTA

CARLETON COLLEGE
Berwald, Helen D.; *Video Tape Project* (GW5725); 25 months; $102,400

NEW YORK

CUNY, CITY COLLEGE
Sacks, Martin and Lee, John J.; *Ecological Film Project* (GW5712); 20 months; $51,400

TEACHERS COLLEGE
Rosskopf, Myron F.; *Conference on Piaget-type Research* (GW5729); 6 months; $20,000

STUDIES OF NATIONAL RESOURCES FOR SCIENCE AND TECHNOLOGY

Studies of Scientific and Technical Manpower Resources

DISTRICT OF COLUMBIA

IMMIGRATION AND NATURALIZATION SERVICE
Loughran, E. A.; *Contract for Reproduction of Statistical Punch Cards Relating to Immigrant Professional, Technical, and Kindred Workers and Nonimmigrant Tabulations for Fiscal Year 1969* (CA17); 3 months; $650

NATIONAL ACADEMY OF SCIENCES, NATIONAL RESEARCH COUNCIL
Boercker, Fred D.; *Task Order for Coding, Processing, and Tabulation of Data Resulting from the Survey of Doctoral Employment* (C310178); 6 months; $7,110

MARYLAND

BUREAU OF THE CENSUS
Brown, George H.; *Interagency Agreement for Support of a Study 'Planning for the 1970 Postcensal Study of Scientific, Engineering and Technical Manpower'* (CA08000001); 12 months; $100,000
———; *Contract for Study of Immigrant Scientists and Engineers in the United States* (CA19000); 6 months; $37,600

MASSACHUSETTS

MASSACHUSETTS INSTITUTE OF TECHNOLOGY
Brown, Sanborn C.; *A Symposium on the Manpower Supply, Need, and Utilization of Graduate Scientists and Engineers* (GR90); 6 months; $3,400

UTAH

INSTITUTE FOR BEHAVIORAL RESEARCH
Taylor, Calvin W.; *Eighth National Creativity Conference* (GR92); 12 months; $6,450

Studies of Educational and Research Institutions

COLORADO

WESTERN INTERSTATE COMMISSION ON HIGHER EDUCATION
Lawrence, Ben; *A National Invitational Research Training Seminar on the Outputs of Higher Education—Their Proxies, Measurements and Evaluation* (GR82); 12 months; $6,450

DISTRICT OF COLUMBIA

AMERICAN COUNCIL ON EDUCATION
Astin, Alexander W.; *A Program of Longitudinal Research on the Higher Educational System* (GR89); 12 months; $50,000

CONFERENCE BOARD OF THE MATHEMATICAL SCIENCES
Botts, Truman A.; *Study of Undergraduate Training in the Mathematical Sciences* (GR91); 19 months; $29,000

COUNCIL OF GRADUATE SCHOOLS IN THE UNITED STATES
McCarthy, Joseph L.; *A Study of the Unit Costs of Graduate Education* (GR80); 17 months; $70,800

NATIONAL ACADEMY OF SCIENCES, NATIONAL RESEARCH COUNCIL
Boercker, F. D.; *Task Order for Doctorate Survey Program* (310082006); 12 months; $68,200

OFFICE OF EDUCATION
Metz, Stafford; *Secondary School Course Enrollment Study* (CA24000); 5 months; $35,000

Studies of Public and Private Funding of Science and Technology

MARYLAND

BUREAU OF THE CENSUS
Gretton, Owen C.; *Contract for Annual Survey of Research and Development in American Industry Covering the Period April 1, 1970, Through January 31, 1971* (CA04000001); 10 months; $85,000
McNelis, David P.; *Survey of State Government R & D Activities, FY's 1967 and 1968* (AG165001); $15,000
Tolli, Donna; *Survey of Research and Development Activities of Local Governments, Fiscal Years 1968 and 1969* (CA15); 11 months; $48,000

Studies of Interactions of Science and Technology With Society

CALIFORNIA

CALIFORNIA STATE OFFICE OF RESEARCH
Monagan, Bob; *Role of State Legislature in Formulating National Science Policies* (GT6); 14 months; $10,000

COLORADO

FEDERATION OF ROCKY MOUNTAIN STATES INC.
Partridge, William S.; *Publication of Rocky Mountain Science Council Science Newsletter* (GT11); 12 months; $1,500

CONNECTICUT

INSTITUTE FOR THE FUTURE
Enzer, Selwyn; *Contract for Preparing and Conducting a Pilot Test Simulation Conference* (C624000); 6 months; $9,655
Gordon, Theeodore J. and Kramish, Arnold; *A Study of Potential Impacts of Science* (GR70001); $58,000

DISTRICT OF COLUMBIA

URBAN INSTITUTE
Hatry, Harry P.; *Seminar on Application of Science and Technology to Local Governments* (GT9); 6 months; $1,400

GEORGIA

GEORGIA SCIENCE AND TECHNOLOGY COMMISSION
Mock, John E.; *National Conference on Goals, Policies, and Programs of Federal, State, and Local Science Agencies* (GR79); 12 months; $19,500

SOUTHERN INTERSTATE NUCLEAR BOARD
Rogers, Wyatt M. Jr.; *Role of Government in Promoting Regional Development Through Science & Technology and Recommendations for Interstate Science & Technology Policies and Programs* (GR68); 15 months; $70,000

ILLINOIS

COMPUTER HORIZONS INC.
Narin, Francis; *Contract for Exploration of the Possibility of Generating Importance and Utilization Measures by Citation Indexing of Approximately 250 Journals in the Physical Sciences* (C627000); 14 months; $93,024

ILLINOIS BOARD OF HIGHER EDUCATION
Holderman, James B.; *Midwest Regional Conference on Science, Technology and State Government—Achieving Environmental Quality in a Developing Economy* (GT8); 12 months; $15,000

KENTUCKY

COUNCIL OF STATE GOVERNMENTS
Belsley, Lyle; *Science and Technology and State Government* (GR77); 12 months; $212,000

LOUISIANA

LOUISIANA BOARD ON NUCLEAR ENERGY
Whittinghill, Donald J.; *Opportunity of the 70's—Public Policy and Environment* (GT4); 12 months; $15,000

MONTANA

GOVERNORS COMMITTEE ON SCIENCE AND TECHNOLOGY
Huffman, Roy E.; *Seminar Series on Science, Technology and State Government in Montana* (GT1); 12 months; $10,000

NEW YORK

NEW YORK STATE EDUCATION DEPARTMENT
Kille, Frank R.; *Survey of Scientific and Technological Advice and Application in Local Government* (GT10); 12 months; $12,000

OKLAHOMA

FRONTIERS OF SCIENCE FOUNDATION OF OKLAHOMA
Hadley, Garland R.; *Oklahoma Project to Develop a Policy Structure for Science and Technology* (GT5); 12 months; $15,000

PENNSYLVANIA

PENNSYLVANIA STATE UNIVERSITY
Engel, Alfred J.; *Intergovernment Relations in the Determination of Air Pollution Research* (GT3); 24 months; $41,800
Roy, Rustum; *Motion Picture Film on Pennsylvania's Science and Technology Advisory Program* (GT7); 12 months; $5,000

VIRGINIA

STATE COUNCIL ON HIGHER EDUCATION FOR VIRGINIA
Barnes, Dennis W.; *Science, Technology and State Policy—Environmental Quality in Virginia* (GT2); 12 months; $25,000

Studies of Science Policies and Programs

CALIFORNIA

STANFORD UNIVERSITY
Dunn, D. A., Parker, E. B. and Rosse, J. N.; *Communications Technology and Public Policy* (GR86); 24 months; $191,100

STUDIES OF NATIONAL RESOURCES FOR SCIENCE AND TECHNOLOGY

UNIVERSITY OF CALIFORNIA, Berkeley
La Porte, T. R.; *Planning for a Program in Science, Technology and Public Affairs* (GR78); 12 months; $9,600

DISTRICT OF COLUMBIA

NATIONAL ACADEMY OF SCIENCES
Green, Robert E.; *Task Order for Support of the Committee on Science and Public Policy* (C310027007); 16 months; $110,000
Odishaw, Hugh; *Task Order for Partial Support for a Study of Nuclear Physics and the Planning of a Prospective Physics Survey* (C310172001); 12 months; $40,000
Starr, Chauncey, *Task Order for a Continuing Review by the Committee on Public Engineering Policy of the Engineering Needs of the U.S., as Well as the Application of Engineering to Critical Public Problems* (C310134003); 12 months; $80,000

NATIONAL ACADEMY OF SCIENCES, NATIONAL RESEARCH COUNCIL
Dunham, Charles L. and Marshall, Louise H.; *Task Order for Partial Support for a Survey of Graduate Academic Courses in Scientific Disciplines Related to Brain and Behavior* (C310183); 6 months; $1,137

SURVEYS AND RESEARCH CORP.
Watkins, Ralph; *Contract for the Study of the Organization and Support of Scientific Research and Development in Communist China* (C317000004); $4,842
———; *Contract for Completion of the Study of the Organization and Support of Scientific Research and Development in Communist China* (C399000004); $2,234

MASSACHUSETTS

HARVARD UNIVERSITY
Price, Don K., Bator, Francis, and Zeckhauser, Richard; *Science and Public Policy Program, Curriculum Development for Public Policy Program, Faculty Research Seminar in Analytic Methods and Public Policy* (GR88); 36 months; $99,822

National Register of Scientific and Technical Personnel

DISTRICT OF COLUMBIA

AMERICAN ANTHROPOLOGICAL ASSOCIATION
Lehman, Edward J.; *Contract for Establishment, Operation, and Maintenance of the Anthropology Section of the National Register of Scientific and Technical Personnel* (C459000005); 12 months; $3,842

AMERICAN CHEMICAL SOCIETY
Stanerson, B. R.; *Contract for Operation and Maintenance of the Chemical Sciences Section of the National Register of Scientific and Technical Personnel* (C592); 12 months; $135,700

AMERICAN GEOLOGICAL INSTITUTE
Hoover, Linn; *Contract for Operation and Maintenance of the Geological Sciences Section of the National Register of Scientific and Technical Personnel* (C591); 12 months; $61,525

AMERICAN INSTITUTE OF BIOLOGICAL SCIENCES
Olive, John R.; *Contract for Operation and Maintenance of the Biological Sciences Section of the National Register of Scientific and Technical Personnel* (C588); 12 months; $80,311

AMERICAN PSYCHOLOGICAL ASSOCIATION
Little, Kenneth B.; *Contract for the Operation and Maintenance of the Psychological Sciences Section of the National Register of Scientific and Technical Personnel* (C597); 12 months; $62,500

AMERICAN SOCIOLOGICAL ASSOCIATION
Volkart, E. H.; *Contract for Operation and Maintenance of the Sociological Sciences Section of the National Register of Scientific and Technical Personnel* (C594); 12 months; $29,337
———; (C386000007); $1,350

CLEMENTS PRINTING CO., INC.
Thompson, William F.; *Contract for Printing Services for the 1968 National Register of Scientific and Technical Personnel and Specialties List* (C544000002); 4 months; $30,278

CENTER FOR APPLIED LINGUISTICS
Lotz, John; *Contract for Operation and Maintenance of the Linguistics Section of the National Register of Scientific and Technical Personnel* (C590); 12 months; $13,697

POST OFFICE DEPARTMENT
Postmaster; *Interagency Agreement for Health Unit Services for the National Register Records Center at Raleigh, North Carolina* (CA25); 12 months; $460

MARYLAND

FEDERATION OF AMERICAN SOCIETIES FOR EXPERIMENTAL BIOLOGY
McManus, J. F. A.; *Contract for the Operation and Maintenance of the Experimental Biology Section of the National Register of Scientific and Technical Personnel* (C589); 12 months; $42,937

WESTAT RESEARCH ANALYSTS INC.
Bryant, Edward C.; *Contract for a Study of the Use of Sampling in Surveys of Scientific and Technical Personnel* (C587000001); 5 months; $10,695

NEW YORK

ENGINEERS JOINT COUNCIL
Frey, Carl; *Contract for the Operation and Maintenance of the Engineers Section of the National Register of Scientific and Technical Personnel* (C407000005); 12 months; $25,000

NORTH CAROLINA

NORTH CAROLINA STATE UNIVERSITY AT RALEIGH
Martin, Leroy B. Jr.; *Contract for Maintenance and Operation of a Records Center for the National Register of Scientific and Technical Personnel* (C361000007); 12 months; $375,976

RHODE ISLAND

AMERICAN MATHEMATICAL SOCIETY
Walker, Gordon L.; *Contract for Operation and Maintenance of the Mathematical, Statistical and Computer Sciences Section of the National Register of Scientific and Technical Personnel* (C574000001); 12 months; $92,432

FELLOWSHIP AND TRAINEESHIP AWARDS OFFERED

Name and Field of Study of Individuals Offered National Science Foundation Fellowships by State of Permanent Residence and Program, Fiscal Year 1970

ALABAMA

Graduate

ALPHIN, HENRIETTA B., Chemistry
CLARK, DAVID B., Biology, General
CROWSON, LAWRENCE D., Mathematics
DECKER, STEPHEN K., Physics
DOBBS, GARY H. III, Biology, General
EDDY, LYNNE J., Physiology
EVANS, RICHARD W., Biology, General
LYRENE, PAUL M., Life Sciences
SMITH, NANTELLE E., Chemistry
SUMMERVILLE, RICHARD H., Chemistry
WADE LEROY G. Jr., Chemistry
WHEELER, WILLIAM H., Mathematics
WIESCHAUS, ERIC F., Genetics
WILLIAMS, EDWIN S., Linguistics

Science Faculty

GUNTER, SHIRLEY E., Microbiology
HERRING, BRUCE E., Engineering
JETT, ARTHUR V., Jr., Mathematics
MARYLAND, WALLACE, Jr., Mathematics

ALASKA

Graduate

ERICSON, CAROL A., Zoology
HEFLINGER, BRUCE L., Engineering

ARIZONA

Graduate

BETHUNE, DONALD S., Physics
BRIGGS, NANCY D., Earth Sciences
BUCK, MARSHALL W., Mathematics
DOIDGE, JAMES A., Biology, General
ELLIS, DAVID H., Zoology
FLOWER, RICHARD A., Engineering
FOSTER, LESLIE V., Mathematics
FRYE, CHARLES L., Chemistry
GOTTESFELD, ALLEN S., Earth Sciences
GROTH, EDWARD J., III, Physics
KRONENFELD, RICHARD L., Physics
LIBERTINI, LOUIS J., Chemistry
MASTERS, WILLIAM M., Psychology
PACE, WILSON D., Engineering
PORTER, JOHN R., Botany
SCHRENK, LOREN C., Engineering
SMITH, ROBERT L., Mathematics
THOMSON, ROSS D., Economics
WINOGRAD, ISAAC J., Earth Sciences
WOODS, CHARLES L., Physics

Postdoctoral

DELTON, MARY H., Chemistry

Science Faculty

BEUS, STANLEY S., Earth Sciences
BUSECK, PETER R., Earth Sciences
TRENT, DEE D., Earth Sciences

ARKANSAS

Graduate

ANGLIN, MELVIN D., Psychology
ASHCRAFT, KEITH E., Physiology
ENGLES, CHARLES R., Engineering
PFEIFER, JAMES B., Physics
RICE, GARY W., Chemistry
VINSON, JOHN W., Chemistry

Science Faculty

COGBURN, CECIL O., Engineering
PEARSON, JAMES V., Engineering

CALIFORNIA

Graduate

ALEXANDER, ROGER K., Mathematics
ALPHA, SIGMA R., Chemistry
ANDERSON, CARL L., Engineering
ANGELO, RAYMOND L., Physics
ANTOGNINI, JAMES A., Psychology
ARNOLDUSSEN, THOMAS C., Engineering
BAECHER, GREGORY B., Engineering
BAGRASH, FRANK M., Psychology
BAILIS, ELLIOT I., Mathematics
BAIN, MICHAEL L., Mathematics
BARON, AILEEN G., Anthropology and Archeology
BASS, GAIL W., Urban and Regional Planning
BEESON, MICHAEL J., Mathematics
BENNETT, ALBERT F., Life Sciences
BERRY, LEE A., Physics
BEWLEY, TRUMAN F., Mathematics
BIENIEK, RONALD J., Physics
BINDER, IRWIN, Chemistry
BLEAKNEY, THOMAS H., Physics
BLESSUM, MARGARET F., History and Philosophy of Science
BLUMENTHAL, GEORGE R., Physics
BRACKENBURY, ROBERT W., Biology, General
BREZINSKI, DONALD P., Biophysics
BRODERSEN, ROBERT W., Physics
BROGAN, PATRICK A., Linguistics
BROKER, THOMAS R., Biochemistry
BROWN, KENNETH S., Biochemistry
BROWN, MARIAN V., Economics
BRUCE, KIM B., Mathematics
BRUTLAG, DOUGLAS L., Biochemistry
BUCKHOLTZ, THOMAS J., Physics
BUDD, SUSAN M., Physiology
BUETTNER, HARLEY M., Engineering
BURKE, SHEILA A., Economics
CABRERA, BLAS, Physics
CAHN, ROBERT N., Physics
CALMAN, JOHN R., Zoology
CAMERON, ROBERT A., Biology, General
CAMMACK, DAVID A., Engineering
CAMPBELL, ROBERT D., Chemistry
CANTOR, MURRAY R., Mathematics
CAPEN, RONALD L., Zoology
CARLSON, CHARLES W., Earth Sciences
CARNE, ROGER S., Mathematics
CARSON, DONALD S., Engineering
CASBERG, RONALD V., Engineering
CASE, TED J., Biology, General
CEDERQUIST, JOHN N., Earth Sciences
CHAPLIN, STEPHEN J., Biology, General
CHARTOCK, MICHAEL A., Biology, General
CHODOROW, NANCY J., Sociology
CHONG, DARRYL G., Engineering
CHRISTENSEN, WAYNE L., Biology, General
CHU, STEVEN, Physics
CHUCK, ALLAN, Mathematics
CISNE, JOHN L., Earth Sciences
CLARK, MARYCLAIRE K., Genetics
CLARKE, STEVEN G., Biochemistry
CLOUGH, GENE A., Physics
COHEN, BRUCE I., Physics
COOK, PHILIP J., Economics
CORNISH, JAN A., Mathematics
COUGHLIN, MICHAEL D., Biology, General
COX, JAMES L., Biology, General
CRANDALL, RICHARD E., Physics
CRANE, MICHAEL A., Engineering
CZAPLICKI, JON S., Anthropology and Archeology
DAVE, NIKHIL, Biology, General
DAVEY, ROBERT F., Engineering
DAVIS, DONALD M., Mathematics
DECHENE, BRENT E., Linguistics
DEMARTINI, EDWARD E., Biology, General
DIAMOND, SHIRLEY G., Psychology
DILLION, JERRY L., Engineering
DOBERNE, LEONARD, Medical Sciences
DRISCOLL, CHARLES F., Physics
DUNHAM, DOUGLAS J., Mathematics
DUTCH, STEVEN I., Earth Sciences
EATON, TIMOTHY R., Mathematics
ECKART, MARK J., Physics
ELFVING, DONALD C., Botany
ELGIN, ROBERT L., Physics
ELGIN, SARAH C., Biochemistry
ELLIS, RAYMOND W., Engineering
ENDICOTT, KIRK M., Anthropology and Archeology
ENDLER, JOHN A., Genetics
ENGMANN, DOUGLAS J., Economics
ESHERICK, PETER, Chemistry
ESSIG, FREDERICK B., Botany
ESTER, MICHAEL R., Anthropology and Archeology
EVANS, NEAL J., Physics
FAIRBANK, WILLIAM M. JR., Physics
FARBER, MICHAEL B., Biochemistry
FARNBACH, JOHN S., Engineering
FARRINGTON, GEORGE L., Chemistry
FELLMAN, LEONARD A., Mathematics
FISCHER, CLAUDE S., Sociology
FISH, MARTIN P., Mathematics
FISK, STEPHEN T., Mathematics
FLYNN, THOMAS E., Mathematics
FOGEL, BARRY S., Mathematics
FOLLANSBEE, STEPHEN E., Biochemistry
FOOR, FORREST Jr., Biochemistry
FOSTER, RACHEL I., Botany
FRALEY, ROBERT A., Mathematics
FREEBORN, W. PHELPS, Earth Sciences
FREEDMAN, MICHAEL H., Mathematics
FREEMAN, JAY R., Physics
FREESE, RALPH S., Mathematics
FRIEDMAN, DANIEL, Mathematics
FROEHLICH, JEFFERY W., Anthropology and Archeology
FRUCHTER, JONATHAN S., Chemistry
FULLER, SAMUEL H., Engineering
GANTT, RICHARD F., Zoology
GECHTER, JERRY, Mathematics
GEE, KATHLEEN T., Physics
GELINAS, RICHARD E., Biology, General
GELLER, MARGARET J., Physics
GERSTEIN, DEAN R., Sociology
GILL, JAMES B., Earth Sciences
GISH, WALTER C., Physics
GONZALES, LINDA W., Physiology
GOODMAN, ROBERT L., Physiology
GOULD, JEFFREY L., Economics
GREEN, MARK L., Mathematics
GREISEN, ERIC W., Astronomy
GRIFFITH, OWEN W., Biochemistry

FELLOWSHIP AND TRAINEESHIP AWARDS

GROBSTEIN, PAUL, Biology, General
GUCKENHEIMER, JEAN E., Mathematics
GULLAHORN, GORDON E., Astronomy
HAAS, DAVID F., Sociology
HAFFERTY, WILLIAM M., Engineering
HAGER, WILLIAM W., Mathematics
HANNAH, ERIC C., Physics
HANSEN, CHRISTOPHER B., Microbiology
HANSMA, PAUL K., Physics
HANSON, ROBERT B., Astronomy
HARRIS, RONALD M., Biology, General
HAVENS, ALAN D., Biology, General
HAY, PHILIP J., Chemistry
HAYNOR, DAVID R., Mathematics
HAYS, ROBERT L., Biology, General
HEDGECOCK, DENNIS, Biology, General
HILLIS, ALAN P., Engineering
HINTON, WILLIAM F., Botany
HITZL, LINDA C., Engineering
HITZL, THOMAS N., Anthropology and Archeology
HJELM, REX P. JR., Biochemistry
HOAGLAND, KAREN E., Physiology
HOCKSTRA, DALE J., Mathematics
HOEL, ELVIN L., Chemistry
HOLTZ, JAMES Z., Physics
HOLWERDA, ROBERT A., Chemistry
HOWARD, FRED P., Physics
HUBBARD, ANN L., Biology, General
HUDGIN, RICHARD H., Physics
HUESTIS, DAVID L., Chemistry
HUESTIS, WRAY H., Biophysics
IBANEZ, PAUL, Engineering
ISAMAN, DAVID L., Engineering
JAFFE, WALTER J., Astronomy
JARDIN, STEPHEN C., Engineering
JOESTEN, RAYMOND L., Earth Sciences
JOHNSON, MICHAEL A., Physics
JOURIS, WILLIAM, Engineering
JUSTICE, JEFFREY D., Physics
KAHN, NANCY, Mathematics
KAPLAN, ROBERT A., Psychology
KARAIAN, CHARLES H., Engineering
KAUFFMAN, GEORGE E., Mathematics
KELLMAN, SANFORD A., Astronomy
KENNEDY, HELEN A., Botany
KEY, SYDNEY J., Economics
KIBBY, SALLY S., Microbiology
KIMBALL, RALPH B., Engineering
KING, DAVID L., Chemistry
KIRCHANSKI, STEFAN J., Botany
KIRSCHNER, MARC W., Biochemistry
KISKIS, JOSEPH E. JR., Physics
KISKIS, RONALD C., Chemistry
KLEIN, FREDERICK W., Earth Sciences
KLEIN, WILLIAM L., Biology, General
KLIMEK, PETER E., Mathematics
KOSSLYN, STEPHEN M., Psychology
KRAVIF, DIANE, Linguistics
KREMER, MARY G., Psychology
KRONSTADT, MICHAEL, Chemistry
KRUGER, VALERIE P., Engineering
KURTAK, DANIEL C., Zoology
KYTE, JACK E., Biochemistry
LABINGER, JAY A., Chemistry
LAMBERT, STEVEN J., Earth Sciences
LANNON, JOSEPH E., Physics
LAROCK, RICHARD C., Chemistry
LEIN, LAURA, Anthropology and Archeology
LEON, JEFFREY S., Mathematics
LEONARD, JACK E., Chemistry
LEONG, DAVID C., Engineering
LEONG, JOHN K., Mathematics
LESGOLD, ALAN M., Psychology
LEVENSON, MARC D., Physics
LEVIN, RICHARD C., Economics
LEWIS, STEVEN M., Biophysics
LEY, RICHARD W., Mathematics
LIEBMAN, JEFFREY M., Psychology
LIEF, ROBERT E., Engineering
LIPPE, WILLIAM R., Psychology
LIVAK, RONALD J., Metallurgy
LIVINGSTON, DENNIS M., Biochemistry
LO, THOMAS K., Physics
LOBBAN, PETER E., Biochemistry
LOGAN, JOHN R., Sociology
LOTT, DEBORAH A., Psychology
LUNDY, ROBERT T., Anthropology and Archeology
MACDONALD, KENNETH C., Earth Sciences
MACQUEEN, DAVID B., Mathematics
MALCOLMSON, PETER, Mathematics
MARYNICK, DENNIS S., Chemistry
MATHER, JOHN C., Physics
MCCAMMON, JAMES A., Chemistry
MCCREERY, RICHARD L., Chemistry
MCFARLAND, BENTSON H., Biophysics
MCGUIRE, ROBERT E., Physics
MENAS, TELIS K., Mathematics
MEREDITH, RUSSELL D., Mathematics
METS, LAURENS J., Biochemistry
MEYER, STEPHEN F., Physics
MILLAR, DARYL B., Psychology
MILLER, JOHN H., Biology, General
MILLER, VICTOR S., Mathematics
MINTZ, LEE G., Economics
MOORE, TERRENCE N., Mathematics
MOSHER, JAMES M., Physics
MUNRO, ALLEN, Psychology
MUNRO, PAMELA L., Linguistics
MURPHY, WILLIAM I., Biochemistry
MYERS, DALE W., Mathematics
NEBEKER, HENRY G., Engineering
NELSON, JAMES A., Engineering
NELSON, MICHAEL A., Psychology
NEVILLE, RICHARD C., Engineering
NEWHALL, NICHOLAS S., Mathematics
NEWMAN, LEON S. JR., Mathematics
OKRAND, MARC, Linguistics
O'NEILL, VIRGINIA J., Mathematics
PALKOVIC, LAWRENCE A., Biology, General
PARKER, GEORGE D., Mathematics
PAYNE, MICHAEL T., Biochemistry
PEARSON, CAROL A., Biology, General
PEARSON, PETER K., Chemistry
PENALOSA, JAVIER, Botany
PERCIVAL, FRANK W., Botany
PETERSEN, GLENN T., Anthropology and Archeology
PICCOLO, MARYLYNN M., Anthropology and Archeology
PILLSBURY, MICHAEL P., Political Science
PITELKA, LOUIS F., Physiology
POOL, JEREMY D., Anthropology and Archeology
POWERS, DANA A., Chemistry
POWERS, THOMAS W., Chemistry
PRAGER, ELLEN M., Biochemistry
PRIMAKOFF, PAUL, Biochemistry
RAFFERTY, JAMES M., Psychology
RAICHART, DENNIS W., Chemistry
REIGER, LLOYD A., Earth Sciences
REICH, CARY J., Chemistry
REKERS, GEORGE A., Psychology
RICE, ROBERT H., Biology, General
RICE, THOMAS R., Engineering
RICHARDSON, JEFFERY H., Chemistry
ROSSUM, DAVID P., Biology, General
ROTHBART, GEORGE B., Physics
RUSBULT, CRAIG F., Chemistry
RUSSELL, DAVID L., Mathematics
SAMET, HANAN, Mathematics
SCANDELLA, CARL J., Biochemistry
SCHADE, CRISTY M., Engineering
SCHOR, ROBERT H., Physiology
SCHWEICKERT, RICHARD A., Earth Sciences
SEIDEN, ABRAHAM, Physics
SELVIG, SUSAN E., Biology, General
SHAFER, GLENN R., Mathematics
SHAFFER, DAVID B., Astronomy
SHALLON, CAROLINE R., Mathematics
SHELTON, JAMES C. III, Physics
SHERROD, SHELBY A., Chemistry
SHIELDS, GREGORY A., Astronomy
SIERK, ARNOLD J., Physics
SILVER, BRUCE R., Physics
SIMONTON, DEAN K., Psychology
SIRBU, MARVIN A., JR., Engineering
SMITH, DAVID C., Biology, General
SMITH, DAVID H., Physics
SMITH, ELEANOR M., Biology, General
SMITH, GARY R., Physics
SMITH, LEVERETT R., Chemistry
SMITH, ROGER A., Physics
SOHOLT, LARS F., Biology, General
SPAFFORD, DAVID C., Physiology
SPRINGSTON, FREDERICK J., Psychology
STEIN, SAMUEL R., Physics
STEVENS, JOHN C., Physics
STEWART, HUGH B., Mathematics
STEWART, JOAN G., Botany
STONE, MILBURN J., Political Science
STORMER, JOHN C. JR., Earth Sciences
STRATTON, ALAN J., Earth Sciences
STUDEBAKER, JOEL F., Chemistry
SWANSON, RICHARD M., Engineering
SWITZER, MARILYN F., Biology, General
SZOLOVITS, PETER, Mathematics
TABACHNICK, BARBARA J., Psychology
TAYLOR, LAURENCE R., Mathematics
TAYLOR, MICHAEL E., Mathematics
TAYLOR, ROBERT J., Biology, General
THORN, MARTHAJO R., Biology, General
THORNE, RODERICK E., Engineering
THYKEN, ROBERT J. JR., Political Science
TIEMAN, SUZANNAH B., Psychology
TILLEY, JEFFERSON W., Chemistry
TOBER, JAMES A., Economics
TOPIC, JOHN R., Anthropology and Archeology
TORIGOE, ERNEST W., Physics
TORNHEIM, KEITH, Biochemistry
TRAUT, THOMAS W., Biology, General
TREISMAN, PHILIP M., Mathematics
TURITZIN, STEPHEN N., Biology, General
UNZELMAN, JAYNE M., Biology, General
VILLANI, DANIEL D., Engineering
WADE, BRADFORD W., Mathematics
WAGNER, LAWRENCE F. JR., Engineering
WAITES, ROBERT F., Physics
WALBOT, VIRGINIA E., Biology, General
WALKER, ROBERT C., Physics
WALLACE, JON M., Physics
WAPLES, DOUGLAS W., Chemistry
WARD, MICHAEL P., Economics
WEBSTER, STEVEN K., Physiology
WEBSTER, THOMAS P. III, Zoology
WEISS, BRIAN E., Anthropology and Archeology
WEISSMAN, KAREN G., Biology, General
WENDSCHUH, PETER H., Chemistry
WERNAU, WILLIAM C., Engineering
WEST, RICHARD W., Economics
WESTER, GENE W., Engineering
WHITE, FLORENCE V., Zoology
WHITMORE, CHARLES S., Political Science
WICKSTROM, ERIC, Biochemistry
WILD, JAMES R., Biology, General
WILHELM, NEIL C., Engineering
WILLCUTT, GORDON J. JR,. Engineering
WILLIAMS, JOSEPH T., Economics
WILLSON, STEPHEN J., Mathematics
WILSNACK, RICHARD W., Sociology
WILSON, CHRISTOPHER P., Astronomy
WITMAN, GEORGE B. III, Biology, General
WOLCOTT, THOMAS G., Biology, General
WRIGHT, WILLIAM A., Earth Sciences
YANG, PAUL C., Mathematics
YOUNT, JAMES C., Earth Sciences
ZADEH, NORMAN, Mathematics
ZUCCARELLI, ANTHONY J., Biophysics
ZUCKER, ROBERT S., Zoology
ZUCKERMAN, GREGG J., Mathematics
ZVESPER, JOHN R., Political Science

Postdoctoral

ALKAITIS, ALGIS A., Biology, General
BELL, ROBERT M., Biochemistry
BRENNER, MICHAEL, Biology, General
CALDWELL, CHARLES W., Engineering
CARLITZ, ROBERT D., Physics
CHANEY, STEPHEN G., Microbiology
CHASE, CLEMENT G., Earth Sciences
COLLIER, JANE F., Anthropology and Archeology
CORONITI, FERDINAND V., Physics

FELLOWSHIP AND TRAINEESHIP AWARDS

DIRLAM, JOHN P., Chemistry
ELLIS, JAMES E., Psychology
GARDNER, DAVID D., Linguistics
HAHN, ALEXANDER J., Mathematics
HINKLE, DAVID C., Biochemistry
JAEGER, DAVID A., Chemistry
KERSHAW, DAVID S., Physics
LEE, KENNETH L., Engineering
LEVY, DONALD J., Physics
LILLYWHITE, HARVEY B., Biology, General
MACDONALD, JOHN R., Engineering
MATTHEWS, JUNE L., Physics
MOLZ, FRED J., Earth Sciences
SCANDELLA, CARL J., Biology, General
STRATHMANN, RICHARD R., Zoology
TANSEY, MICHAEL R., Microbiology
TIMMER, LAVERN W., Botany
WARD, SAMUEL, Biology, General
WARNER, PHILIP M., Chemistry
WESELOH, RONALD M., Zoology
WIGHTMAN, FREDERIC L., Psychology
WIMBUSH, ARNOLD M., Earth Sciences

Senior Postdoctoral

CICOUREL, AARON V., Social Science
ERNST, WALLACE G., Earth Sciences
FAGEN, RICHARD R., Economics
HAMMEL, EUGENE A., Anthropology and Archeology
HELLWARTH, ROBERT W., Physics
HOLLAND, JOHN J., Biology, General
HSIAO, THEODORE C., Biology, General
JACKSON, EVERETT D., Earth Sciences
OLDHAM, WILLIAM G., Engineering
PHILLIPS, NORMAN E., Physics
RABINOWITZ, JESSE C., Biology, General
SCHAWLOW, ARTHUR L., Physics
SHIRLEY, DAVID A., Physics
STALLINGS, JOHN R., Mathematics
STOCKING, C. R., Botany
WELCH, WILLIAM J., Astronomy
WHITE, ROBERT M., Physics

Science Faculty

BALDWIN, EVELYN J., Earth Sciences
BELL, GRAYDON D., Physics
BENTLEY, DONALD L., Mathematics
BERGAUIST, LOIS M., Microbiology
BISSEY, JACK E., Chemistry
BORRELLI, ROBERT L., Mathematics
CURREN, TERENCE B., Anthropology and Archeology
EVANS, RONALD L., Biology, General
GALLIN, DANIEL, Mathematics
GARRETT, ROGER E., Engineering
HAMERNIK, ROBERT E., Engineering
HOSHIZAKI, BARBARA J., Botany
LEACH, DONALD P., Engineering
LUK, KING S., Engineering
MULVIHILL, MICHAEL E., Engineering
PORTER, CHARLES W., Zoology
RIVERA, DOROTHY M., Mathematics
SMITH, STUART B., Sociology
WASSEL, GUSTAV N., Engineering

COLORADO

Graduate

ADLER, NANCY E., Psychology
ANDERSON, MICHAEL P., Mathematics
APPELT, JUDY P., Geography
AVERY, MICHAEL V., Mathematics
CORRUCCINI, LINTON R., Physics
DENDY, LESLIE A., Biochemistry
DISESSA, ANDREA A., Physics
EDGAR, GERALD A., Mathematics
ELDRIDGE, ROGER L., International Relations
FEINSINGER, PETER, Zoology
FINZER, WILLIAM F., Physics
GRANTHAM, GARY D., Chemistry
HAGEDORN, ALFRED A. III, Chemistry
HATTAN, DAVID E., Engineering
HUBBARD, MONT JR., Engineering
LOVELL, KATHRYN L., Biophysics
MARCUS, FREDERICK B., Physics
MENNEMEYER, PAUL F., Engineering
MURAHATA, RICHARD I., Medical Sciences
PARKER, TIMOTHY W., Psychology
PHELPS, ROBERT W., Biophysics
REVIER, CHARLES F., Economics
ROSENBERG, JAMES E., Mathematics
SAMPLE, JUDITH A., Zoology
SETCHELL, ROBERT E., Engineering
SHAKLEE, JAMES B., Biology, General
SILVER, JOYCE M., Biology, General
WILLIAMS, ROBERT H., Chemistry

Postdoctoral

BULL, THOMAS E., Chemistry
FLETCHER, VERNON R., Chemistry

Science Faculty

ARONHIME, PETER B., Engineering
BAKER, JAMES R., Engineering
CAMPBELL, JAMES B., Biology, General
MCINTIRE, DEAN P., Mathematics
TOLAR, ROBERT A., Mathematics
WILLIAMS, NEVILLE A. JR., Mathematics

CONNECTICUT

Graduate

ALDINS, JANIS, Physics
ASCHER, WILLIAM L., Political Science
AVERILL, BRUCE A., Chemistry
BAKER, MYRON C., Biology, General
BARSTOW, DAVID R., Mathematics
BEARDSLEY, GEORGE P., Chemistry
BONVILLIAN, JOHN D., Psychology
BRILMAYER, ROBERTA M., Mathematics
BURNAP, CHARLES A., Physics
CHOUINARD, LEO G. II, Mathematics
CLELAND, STORY, Biology, General
DAVIS, RICHARD S., Physics
DONNEE, LOUISE H., Psychology
DORION, FRANCIS A., Economics
DUTTON, RICHARD L., Mathematics
FUTIA, CARL A., Economics
GENSEL, PATRICIA A., Botany
GERSICK, KELIN E., Psychology
GIERASCH, LILA M., Biophysics
GOODEARL, KENNETH R., Mathematics
GREIF, JEFFREY M., Physics
GROVES, SAMUEL T., Botany
HOLBROOK, BRUCE E., Anthropology and Archeology
HOLDER, RICHARD W., Chemistry
HUBELBANK, MARK, Engineering
HUFFMAN, MARCIA R., Medical Sciences
HUTCHISON, KEITH W., Microbiology
JAFFE, ROBERT L., Physics
KATZ, RICHARD S., Political Science
KEIRNS, JAMES J., Biophysics
KELLER, THOMAS M., Physics
KENNEDY, MARIAN L., Psychology
LEBETKIN, LEWIS M., Biochemistry
LIEBER, MICHAEL, Anthropology and Archeology
MCNAUGHT, WILLIAM, Economics
MORTON, MARILYN J., Political Science
MOSTOW, MARK A., Biology, General
NAGY, WILLIAM E., Linguistics
OVERHOLT, WILLIAM H., Political Science
PHILLIPS, RICHARD D., Biology, General
PIZER, ARNOLD K., Mathematics
POLLACK, EDWARD E., Psychology
POZEN, ROBERT C., Social Science
PROBER, DANIEL E., Physics
PUKKILA, PATRICIA J., Biology, General
RACKOVSKY, SHALOM R., Chemistry
ROSEN, KENNETH T., Economics
ROTHE, PAUL H., Engineering
SALOP, STEVEN C., Economics
SAVITZKY, STEPHEN R., Mathematics
SCHWABEL, RONALD C., Engineering
SHAPIRO, STUART L., Astronomy
SHECTMAN, STEPHEN A., Astronomy
SHIPMAN, HENRY L., Astronomy
SHMOOKLER, ROBERT J. Biology, General
SOLIN, ROBERT N., Mathematics
TURNROSE, BARRY E., Astronomy
TYLER, LINDA P., Biology, General
VARNI, JOHN G., Psychology
VERDERY, KATHERINE M., Anthropology and Archeology
WALKER, HENRY M., Mathematics
WALKER, RODERICK S., Engineering
WILBUR, HENRY M., Zoology
WILLIAMS, JUNE H., Biology, General

Postdoctoral

BYLEBYL, JEROME J., History and Philosophy of Science
LANDOWNE, DAVID, Physiology
MALOZEMOFF, ALEXIS P., Physics
NORDHAUS, WILLIAM D., Economics
SPEAR, PETER D., Biology, General
ZARET, THOMAS M., Biology, General

Senior Postdoctoral

MURDOCK, BENNET B. JR., Psychology
WAKSMAN, BYRON H., Medical Sciences

Science Faculty

BERLINGHOFF, WILLIAM P., Mathematics
DEVILLIERS, RAOUL A., Mathematics
FOX, JOHN N., Physics
LAMBRAKIS, CONSTANTINE, Engineering
SMAIL, J. KENNETH, Biology, General

DELAWARE

Graduate

COURSEN, MARK A., Mathematics
DENEEF, CHARLES P., Physics
FENTON, WAYNE A., Biochemistry
JUMARS, PETER A., Biology, General
MORGAN, MICHAEL S., Engineering
NIEDZIELSKI, MARIE S., History and Philosophy of Science
PODGORSKI, WILLIAM A., Engineering
SHIMSHICK, EDWARD J., Chemistry
WELLS, JOHN V., Economics

Postdoctoral

PARKINSON, JOHN S. JR., Biology, General
WOOD, DAVID D., Biology, General

DISTRICT OF COLUMBIA

Graduate

CHAMBERS, FRANK W., Physics
DIRKS, JEAN A., Psychology
GREENWALD, BRUCE C., Economics
HOLLICK, ANN L., International Relations
JONES, MARIE T., International Relations
KOCHER, ELIZABETH C., Psychology
MARCUSS, ROSEMARY D., Economics
MARGOLIES, DAVID S., Mathematics
MCCLELLAND, JAMES L., Psychology
MCCLENON, ROBERT C., History and Philosophy of Science
MONTGOMERY, HARRY M., Social Science
NONOSHITA, RUTH V., Psychology

FLORIDA

Graduate

BARFIELD, MARY A., Physiology
BATCHELDER, JOHN B. JR., Chemistry
BERMAN, MARILYN I., Psychology
BLOOM, STEPHEN A., Biology, General
FAIRBANKS, CHARLES H. JR., Political Science
FISCHER, ROBERT A., Mathematics
FOX, THOMAS O., Biochemistry
GANCARZ, ALEXANDER J. JR., Earth Sciences
GRIFFIN, JERRY H., Engineering
GUTHRIE, STEWART E., Anthropology and Archeology
HANNA, CHARLES C., Mathematics
HARWELL, MARK A., Biology, General
HILLMAN, JEFFREY D., Microbiology
JAMISON, ROBERT E., Mathematics
LAYTON, RALPH P., Physics
MARTINO, EDWARD R., Psychology
MASON, DOUGLAS C., Chemistry
MAY, JAMES P., Earth Sciences
MAY, MICHAEL L., Zoology
MCCORY, ROBERT T., Mathematics
MCDIARMID, MERCEDES F., Zoology

FELLOWSHIP AND TRAINEESHIP AWARDS

McKinney, William L., Mathematics
Neill, Darryl B., Psychology
Norman, Joe G. Jr., Chemistry
Osborn, Lynne M., Psychology
Penney, Richard C., Mathematics
Pollack, Frederick J., Mathematics
Rackow, David L., Psychology
Radomski, Mark S., Physics
Ramirez, Ronald M., Engineering
Schneider, William M., Anthropology and Archeology
Settle, Russell F., Economics
Shalloway, David I., Physics
Stark, Eugene E. Jr., Engineering
Staros, James V., Biophysics
Tracy, David H., Physics
Tresemer, David W., Sociology
White, Susan C., Zoology

Senior Postdoctoral

Savage, I. R., Sociology

Science Faculty

Clapp, David E., Engineering
Hartman, John P., Engineering
McCord, William M., Astronomy
Renne, Roger L., Mathematics

GEORGIA

Graduate

Beale, James T., Mathematics
Brown, Evelyn W., Zoology
Child, George I., Biology, General
Collins, Nicholas C., Biology, General
Cook, Della C., Anthropology and Archeology
Day, Michael A., Mathematics
Dickerson, Frank M., Economics
Donkar, Eli N., Mathematics
Gettinger, Joshua S., Mathematics
Gill, John T., Mathematics
Goldstein, Sheldon, Physics
Harkins, Catherine W., Sociology
Holland, Charles J., Mathematics
Juricek, Diane K., Genetics
Leavell, Kenneth H., Chemistry
Pace, Jack R., Mathematics
Rhodes, Allan D., Mathematics
Shavell, Steven, M., Economics
Speciner, Michael, Mathematics
Tennant, Julia E., Mathematics

Postdoctoral

Pirages, Dennis C., Political Science

Science Faculty

Allison, James M., Engineering
Astin, Janice T., Mathematics
Crowell, James B., Jr., Mathematics
Hatcher, Martha T., Microbiology

HAWAII

Graduate

Chance, Kelly V., Chemistry
Chang, Rita Y., Mathematics
Hawkins, Emily A., Linguistics
Ho, Thomas I., Mathematics
Kau, Randall K., Economics
Little, Georgiandra, Zoology
Minami, Sharon A., Biology, General
Shen, Gilbert, Physics
Stoecker, Diane B., Microbiology
Yamanaka, Kathleen T., Psychology
Yano, Kenneth T., Engineering

Science Faculty

Stoutemyer, David R., Mathematics

IDAHO

Graduate

Carlson, James A., Mathematics
Curtin, Daniel J., Engineering
Evans, Paul D., Chemistry
Ford, Lawrence H., Physics
Simpson, Jack W., Physics
Sneed, Paul G., Anthropology and Archeology
Vetter, Arthur M., Jr., Physics

Postdoctoral

McCalley, Roderick C., Biochemistry
McDonald, Dennis L., Biology, General
Ramshaw, John D., Chemistry

Science Faculty

Hathaway, Cecil W., Engineering
Kratz, Lawrence J., Mathematics

ILLINOIS

Graduate

Anderson, Bryon D., Physics
Ash, John F., Biology, General
Assink, Roger A., Chemistry
Austin, Jared A., Chemistry
Baltaxe, James B., Anthropology and Archeology
Beard, Richard V., Engineering
Bedford, Eric D., Mathematics
Berbaum, Michael L., Psychology
Berman, Stephen A., Physics
Berntsen, Jon H., Chemistry
Bonney, Charles A., Sociology
Bowen, Marshall A., Physics
Brown, Leo D., Biochemistry
Burzotta, Linda L., Sociology
Bush, Bruce L., Physics
Byrn, Stephen R., Chemistry
Cachel, Susan M., Anthropology and Archeology
Cannell, Lynell E., Physics
Carhart, Raymond E., Chemistry
Cates, Marshall L., Mathematics
Cecchi, Joseph L., Physics
Charlson, Gary S., Engineering
Cheney, Carol, Physiology
Chevalier, Roger A., Astronomy
Chodzko, Carol A., Biology, General
Clusin, William T., Psychology
Collins, Charles P., Economics
Cook, Charles L., Chemistry
Cox, Paul H., Physics
Davis, Phillip H., Chemistry
DeBolt, Lawrence C., Chemistry
Dellutri, Dale A., Mathematics
Diemente, Damon L., Chemistry
Dieterich, David A., Chemistry
Dodgson, Jerry B., Biochemistry
Dorschel, Craig A., Chemistry
Dudgeon, Dan E., Engineering
Dudzik, Michael A., Physics
Duvivier, Joseph A., Physics
Dyer, Eric L., Biology, General
Erb, Sara J., Chemistry
Erxleben, Edward J., Jr., Physics
Estes, Gerald M., Engineering
Evans, Ronald J., Mathematics
Evenson, Paul A., Physics
Farkas, Daniel R., Mathematics
Fehling, Michael R., Psychology
Field, Robert C., Mathematics
Flasar, F. Michael, Physics
Fleissner, William G., Mathematics
Forbis, Richard M., Chemistry
Gaede, Bruce J., Chemistry
Gannon, Thomas F., Engineering
Garvey, Patrick M., Chemistry
Gelbart, Nina R., History and Philosophy of Science
Giles, Roscoe C., III, Physics
Gilkey, Peter B., Mathematics
Goldman, Barbara M., Biology, General
Grossman, George L., Mathematics
Hall, Richard B., Agriculture
Hanson, Donald A., Engineering
Harrison, Virginia, Life Sciences
Hartman, Raymond S., Economics
Havel, James J., Chemistry
Hayes, Alice B., Botany
Heitsch, James L., Mathematics
Henkin, Bruce M., Chemistry
Herring, Susan W., Life Sciences
Hoffman, Alan W., Physics
Hoffman, Philip T., Mathematics
Hollander, Clifford R., Mathematics
Holt, Robert W., Psychology
Holze, Gordon H., Engineering
Horvitz, Howard R., Genetics
Ingle, James D., Jr., Chemistry
Jach, Terrence J., Physics
Jaffey, Stephen M., Mathematics
Johnson, Margaret J., Sociology
Juhlin, Kenton D., Mathematics
Kamm, Kenneth S., Chemistry
Kammler, David W., Mathematics
Kanofsky, Jeffrey R., Chemistry
Karpus, James T., Zoology
Katz, Jack S., Sociology
Kazmer, Daniel R., Economics
King, Paul L., Mathematics
Kirchenberg, Ralph J., Zoology
Klee, Howard, Jr., Engineering
Klun, Joseph R., Psychology
Kobliska, Gary K., Mathematics
Kohn, Peter M., Mathematics
Korth, Bruce A., Psychology
Koziol, James A., Mathematics
Kutner, David H., Jr., Psychology
Lager, George A., Earth Sciences
Landahl, Carol A., Zoology
Langacker, Paul G., Physics
Larson, Rachel P., Economics
Leben, William R., Linguistics
Lentz, Bernard F., Economics
Leone, Stephen R., Chemistry
Levi, Judith N., Linguistics
Levin, George B., Chemistry
Lindquist, Jeffrey C., Physics
Lundeen, Stephen R., Physics
Lyons, Terry G., Mathematics
Maestro, Vittorio T., Anthropology and Archeology
Magnuski, Henry S., Mathematics
Maier, Thomas O., Chemistry
Malmborg, Cheryl L., Anthropology and Archeology
Manelis, Leon M., Psychology
Mariella, Raymond P., Jr., Chemistry
Markunas, Albert L., Engineering
Martin, Nancy C., Genetics
Masover, Gerald K., Biology, General
Mathews, James M., Mathematics
Matthews, Howard W., Political Science
McAdam, Peter L., Engineering
Meldgin, Mark J., Biochemistry
Meltzer, Paul S., Life Sciences
Mercer, Robert L., Mathematics
Merkel, Robert R., Chemistry
Metler, Thomas J., Chemistry
Mocella, Michael T., Chemistry
Molnar, David A., Economics
Muirhead, Katharine A., Chemistry
Muller, Kenneth J., Biophysics
Mungall, William S., Chemistry
Mutti, John H., Economics
Nachman, Paul M., Astronomy
Newman, Paul D., Chemistry
Noren, Richard W., Chemistry
Norris, Charles H., Earth Sciences
Pannatoni, Ronald F., Mathematics
Patterson, George S., Chemistry
Pazdra, Ronald R., Physics
Pigage, Lee C., Earth Sciences
Pivorunas, August, Earth Sciences
Porter, Thomas L., Anthropology and Archeology
Putnam, Daniel E., Mathematics
Rehmer, Judith A., Earth Sciences
Reisinger, Lee W., Engineering
Renner, Terrence A., Chemistry
Rice, David P., Chemistry
Rocke, David M., Mathematics
Rosen, Harvey S., Economics
Sachs, Edward S., Chemistry
Sackett, Philip B., Chemistry
Sato, Vicki L., Biology, General
Schermer, Timothy N., Mathematics

FELLOWSHIP AND TRAINEESHIP AWARDS

SCHMITT, PEGGY H., Genetics
SIEGEL, JEREMY J., Economics
SINGER, RICHARD A., Biology, General
SMITH, DAVID J., Mathematics
SMITH, IRL W., Jr., Physics
SMITH, STEPHEN A., Earth Sciences
SORCE, JOHN W., Engineering
SPRIETSMA, SUZANNE R., Psychology
SAUNDERS, KIM D., Earth Sciences
SPRUGEL, DOUGLAS G., Biology, General
STARACE, ANTHONY F., Physics
STROMQUIST, WALTER R., Mathematics
STULL, WILLIAM J., Economics
THORNBURG, KATIE A., Chemistry
TREES, SCOTT C., Genetics
TUCKER, JOHN R., Physics
TROYER, KATHRYN, Psychology
TYLKE, JOHN A., Engineering
WAKERLY, JOHN F., Engineering
WARNOCK, WILLIAM W., Astronomy
WEINSTEIN, NEIL D., Chemistry
WERES, OLEH, Chemistry
WHITMORE, DOUGLAS M., Physics
WILCOX, ROBERT B., Jr., Political Science
WILLEMSEN, JORGE F., Physics
WILSON, JOHN W., III, Zoology
WODARCZYK, FRANCIS J., Chemistry
YESKE, RONALD A., Engineering
YOUNG, BAMBI M., Biochemistry
ZIMMER, BARBARA J., Zoology
ZIMMERMAN, ROGER P., Biology, General

Postdoctoral

BOYLE, WILLIAM J., Jr., Chemistry
ELLIOTT, STEVEN P., Chemistry
HOLCOMB, ALLEN G., Chemistry
KOCHMAN, STANLEY O., Mathematics
KOUBA, JOSEPH E., Chemistry
NELSON, THOMAS O., Psychology
PATTERSON, PAUL H., Physiology
PODOLSKY, WILLIAM J., Physics
REVZIN, ARNOLD, Chemistry
RITSCHER, JAMES S., Chemistry
ROCHESTER, LEON S., Physics
SCOTT, GARY W., Chemistry
SMITH, GERALD R., Biology, General
WHELAND, ROBERT C., Chemistry

Senior Postdoctoral

ADELMAN, IRMA, Economics
KRUSKAL, WILLIAM H., Mathematics
LASHOF, RICHARD K., Mathematics
SYPHERD, PAUL S., Biology, General
TURKEVICH, ANTHONY L., Chemistry

Science Faculty

ALICH, Sr., AGNES A., Chemistry
GRIER, JAMES B., Mathematics
JONES, CHARLES V., Mathematics
MYERS, LEONARD D., Mathematics
PORTER, SAMUEL E., Chemistry
SMITH, ALBERT J., Biology, General
TAITT, HENRY A., Astronomy
WOODS, SARAH H., Chemistry

INDIANA

Graduate

ADDINGTON, DAVID V., Engineering
BALTZ, BERNARD L., Mathematics
BARTELT, MARK L., Mathematics
BUHLER, JERRY D., Chemistry
BUIKSTRA, JANE E., Anthropology and Archeology
BUNES, LEONARD A., Chemistry
BUSH, ROBERT L., Physics
CHISCON, MARTHA O., Microbiology
DURLAND, LESLIE L., Mathematics
EGAN, DENNIS E., Psychology
FOWLER, MICHAEL L., Earth Sciences
FRANSON, JAMES D., Physics
GAUGER, MICHAEL A., Mathematics
GOLDMAN, NEIL M., Mathematics
GOLOMB, DEBORAH, Mathematics
GUNTHER, GARY R., Biochemistry
GUNTHER, MICHAEL D., Mathematics
HAGEN, DAVID C., Biochemistry
HARROLD, FRANK B., Anthropology and Archeology
HOOKER, ROBERT P., Chemistry
KENNEDY, ALBERT J., Chemistry
KLEIER, DANIEL A., Chemistry
LANCASTER, RONALD L., Mathematics
MAGUIRE, MICHAEL E., Biochemistry
MAJDA, ANDREW J., Mathematics
MCMANAMA, CAROL S., Psychology
NARCOWICH, FRANCIS J., Physics
NARDI, JAMES B., Biology, General
NEUHAUSER, BARBARA J., Physics
PERSHING, DAVID W., Engineering
RHODE, RICHARD D., Mathematics
RIEGER, CHARLES J., III, Mathematics
ROBERTS, THOMAS M., Biochemistry
ROSEVEAR, WILLIAM H., Physics
RYAN, ELLEN B., Psychology
SACKS, RICHARD A., Physics
SANDERS, LISA A., Chemistry
STOCKMEYER, LARRY J., Mathematics
TOMKOVICH, RUSSELL, Chemistry
WAITE, EMILY M., Sociology
WARREN, STEPHEN G., Chemistry
WHITE, DAVID H., Chemistry
WIEN, RICHARD W., Jr., Chemistry

Postdoctoral

MORSS, LESTER R., Chemistry
WALTZ, RONALD E. Physics
YOST, WILLIAM A., Psychology

Senior Postdoctoral

AXELROD, BERNARD, Biochemistry
PASTO, DANIEL J., Chemistry

Science Faculty

BEAUREGARD, LAURENT A., History and Philosophy of Science
JOBE, EVAN K., History and Philosophy of Science
MILLER, RANDAL P., Mathematics
ROGERS, JOSEPH E., Jr., Chemistry
WORTHLEY, WARREN W., Engineering

IOWA

Graduate

ANDERSON, DANIEL D., Mathematics
BAINBRIDGE CRAIG W., Mathematics
BAINBRIDGE, DUDLEY J., Mathematics
BOUSKA, AMY SUSAN, Sociology
CARPENTER, PATRICIA A., Psychology
CASJENS, SHERWOOD R., Biochemistry
CECH, THOMAS R., Chemistry
COLLINS, DAVID W. Physics
COOK, CHERYL A., Economics
FEE, EVERETT J., Zoology
FLEMING, MARK W., Physics
FORD, PENELOPE B., Anthropology and Archeology
JACOBSON, GARY R. Biochemistry
JENKS, SUSAN E., Earth Sciences
KEIDERLING, TIMOTHY A., Chemistry
KING, HARRISS T., Physics
KNAACK, LORETTA J., Biology, General
LANE RICHARD H., Anthropology and Archeology
MAGID, LINDA L., Chemistry
MALONE, WILLIAM J., Chemistry
MARKS, JAMES P., Mathematics
MATTHIAS, ROBERT F., Earth Sciences
MIDLAND, MICHAEL M., Chemistry
MILLER RORERT L., Physics
OSTREM, DENNIS L., Biochemistry
PETERSON, MARK A., Physics
POWELL, JEFFREY R., Genetics
RIEDER, MARY D., Economics
SCHLEGEL, ROBERT A., Biology General
SCHLIE, THEODORE W., Social Science
STRAND, TIMOTHY C., Physics
TOGEAS, JAMES B., Chemistry
WOODWARD, MICHAEL D., Botany

Postdoctoral

HINRICHSEN JOHN J., Mathematics
IRWIN, GALEN A., Political Science
MOSES, RONALD W., Jr., Physics

Science Faculty

MCDOUGALL, DAVID W., Engineering
SCHAEFER, JOSEPH A., Physics
WETMORE, CLIFFORD M., Biology, General

KANSAS

Graduate

BRUNE, DANIEL C., Botany
CARTER, JOSEPH G., Earth Sciences
CHELIKOWSKY, JAMES R., Physics
COOLEY, RALPH E., Linguistics
HAMMONS, GLENN T., Economics
HOMER, WILLIAM D., II, Mathematics
HOSEIN, BARBARA H., Genetics
JOHNSON, LINDA M., Biology, General
KORNFELD, JUDITH R., Linguistics
LIEBERT, JAMES W., Astronomy
LUSK, MARY D., Psychology
MITCHELL JERRY K., Physics
MUELLER, PAUL A., Earth Sciences
NYE, WILLIAM W., Economics
RICHARDSON, LOU B., Genetics
SHAPLEY, JOHN R., Chemistry
STUMP, DANIEL R., Physics

Postdoctoral

KERFOOT, WILLIAM B., Biology, General
NORTON, JOHN L., Physics

Science Faculty

PICKERILL, MAX E., Chemistry
ROSS, RICHARD, Engineering
SENECAL, REV. GERARD J., Physics

KENTUCKY

Graduate

BATTLE, GUY A., III, Mathematics
BECHER, JOHN H., Engineering
BEYER, CARL F., Jr., Biochemistry
BRYSON, JULIETTE A., Chemistry
CAMPBELL, GERALD A., Physics
CONNOR, DAN E., Chemistry
COOK, THOMAS J., Chemistry
ELLIOTT, ELLEN J., Biochemistry
GOAD, LARRY E., Astronomy
GOTT, JOHN R., III, Astronomy
HILGEFORD, ERIC J., Zoology
HILL, WILLIAM H., Biology, General
JOHNSON, NANCY E., Psychology
KESSELMAN, JONATHAN R., Economics
LYNCH, DENIS A. Jr., Chemistry
POWERS, SUE G., Biochemistry
SCOTT, ROBERT A., Engineering
SMITH, FRANK G., III, Engineering
THOMAS, JAMES W., Mathematics
UNGER, EARL M., Economics
ZELESKY, BEVERLY S., Biology, General

Science Faculty

MINER, GEORGE K., Physics
PAYNE, MARVIN G., Physics
PIERCE, RONALD E., Mathematics

LOUISIANA

Graduate

ARMSTRONG, ROBERT C., Engineering
BELL, THOMAS L., Physics
DURAND, CLARENCE O., Mathematics
FIASCONARO, JAMES G., Engineering
GROVES, ROBERT M., Sociology
JACKS, KENNETH A., Medical Sciences
JOHN, MIRIAM E., Engineering
JUNG, MICHAEL E., Chemistry
KELLEY, WALTER G., Mathematics
MCGINNIS, EILEEN, B., Mathematics
ROUX, STANLEY J., Botany
RUSSO, RAY L., Mathematics
SAUSSY, GORDON A., Economics
SNODGRASS, MARY E., Psychology
ST. AMANT, JOSEPH L., Biology, General
VANVOORHIS, DAVID C., Mathematics

FELLOWSHIP AND TRAINEESHIP AWARDS

Postdoctoral

Kennedy, Frank S., Biochemistry

Science Faculty

Diem, John E., Mathematics
Malmstrom, Sr., Mary C., Chemistry
Newman, Rogers J., Mathematics
Smith, Charles R., Mathematics
Webb, George R., Engineering

MAINE

Graduate

Jensen, Harbo P., Chemistry
Marston, Richard C., Economics
Moberg, William K., Chemistry
Palmer, Chester I. Jr., Mathematics
Siegel, Joseph, Engineering

Science Faculty

McLeod, Roger D., Physics

MARYLAND

Graduate

Abramson, Jean E., Mathematics
Anderson, Wendell L., Mathematics
Asquith, Kenneth P., Economics
Berkus, Michael D., Mathematics
Blachly, Alice C., Biology, General
Blank, Esther P., Psychology
Bonwit, Karen M., Mathematics
Braly, Kenneth A., Astronomy
Buhrman, Robert A., Physics
Cooper, Martin D., Physics
Damon, James N., Mathematics
Dowsett, Frederick R., Jr., Earth Sciences
Dunning, John O., Biology, General
Edgar, Thomas F., Engineering
Eisenberg, Akiva M., Chemistry
Ellison, Anthony M., Biology, General
Faber, Sandra M., Astronomy
Fain, Margery J., Biology, General
Feinsilver, Philip J., Mathematics
Ferguson, John B., Biology, General
Fischer, Eric A., Biology, General
Frank, Elise J., Psychology
Frey, Mary A., Physiology
Fullen, Robert E., Economics
Gotts, Harvey S., Physics
Green, Sheldon J., Chemistry
Haberman, Carol B., Social Science
Hart, Edwin F., Engineering
Henry, Susan A., Genetics
Hill, Marianne T., Economics
Irvine, Judith T., Anthropology and Archeology
Jackel, Janet P., Physics
Jacobson, Richard M., Chemistry
Karo, Douglas P., Physics
Kidwell, Mark E., Mathematics
Krafsur, Elliot S., Zoology
Lagarias, Jeffrey C., Mathematics
Lange, Louis G., III, Chemistry
Lazarus, Stephen M., Physics
Leonberger, Frederick J., Engineering
Levin, David A., Social Science
Levinson, Lorraine E., Urban and Regional Planning
Maiorana, Virginia C., Zoology
Maling, Joan M., Linguistics
Millner, Alan R., Engineering
Moss, Philip I., Economics
Myers, Philip, Zoology
Ogus, Arthur E., Mathematics
Overbey, Charles M., Psychology
Parman, Susan M., Anthropology and Archeology
Posey, Karen W., Mathematics
Quinn, Joseph F., Economics
Ramsay, Jean L., Chemistry
Roedder, Spencer W., Physics
Salwin, Arthur E., Chemistry
Schaps, Mary E., Mathematics
Schmidt, Brian K., Mathematics
Shanks, William H., Chemistry
Simons, Gary, Chemistry
Sollner, Barbara D, Microbiology
Stahler, James H., Engineering
Taylor, Russell H., Mathematics
Teitelbaum, Phyllis M., Sociology
Valentine, Page C., Jr., Earth Sciences
Vogelstein, Bert, Mathematics
Walpole, Peter H., Physics
Ward, William W., Biochemistry
Wilson, Catharine, A., Political Science
Wolman, Anne S., Economics

Postdoctoral

Kinnersley, William M., Physics

Science Faculty

Fowler, Charles A., III, Engineering
Soliman, Ibrahim M., Engineering

MASSACHUSETTS

Graduate

Audet, James P., Chemistry
Baker, Brenda S., Mathematics
Bartlett, Paul A., Chemistry
Beals, Jacquelyn W., Psychology
Berkowitz, Harriet F., Sociology
Bivins, Joseph T., Sociology
Blumberg, Peter M., Biochemistry
Bogrow, Paul A., Psychology
Borken, Richard J., Physics
Brenner, Norman M., Astronomy
Brooks, Alison S., Anthropology and Archeology
Brown, Kenneth S., Mathematics
Butman, Bradford, Earth Sciences
Byrne, Hugh M., Earth Sciences
Cane, David E., Chemistry
Chamberlain, Gary E., Economics
Cheney, Maynard C., Chemistry
Clapp, Charles H., Chemistry
Cohen, Jonathan B., Chemistry
Conrad, Geoffrey W., Anthropology and Archeology
Cushing, Steven, Linguistics
Davidson, Gordon K., Engineering
Deaton, Paul F., Engineering
Dinerman, Ina R., Anthropology and Archeology
Doherty, Paul M., Physics
Earle, Timothy K., Anthropology and Archeology
Eaton, Sandra S., Chemistry
Eilbert, Richard F., Physics
Ekman, Kenneth E., Mathematics
Entin, Stephen J., Economics
Feldman, Theodore S., Mathematics
Field, Alexander J., Economics
Foote, Stephen L., Psychology
Fowler, Elizabeth, Biochemistry
Frank, Robert E., Chemistry
Friedman, Sheldon I., Economics
Gagne, Robert R., Chemistry
Gilman, Antonio, Anthropology and Archeology
Godfrey, Laurie A., Anthropology and Archeology
Goff, Christopher G., Biochemistry
Gonick, Lawrence R., Mathematics
Gorin, Ralph E., Engineering
Greep, Nancy C., Physiology
Gurwitz, Sharon B., Psychology
Hakim, Frances S., Zoology
Haseltine, William A., Biophysics
Henry, Charles S., Biology, General
Holbrook, Sally J., Biology, General
Howell, John A., Physics
Huie, Ben T., Chemistry
Jarvis, Leo M., Linguistics
Jayne, Lance W., Mathematics
Jewett, Susan, Biology, General
Johnson, Jerry W., Engineering
Johnson, Louis G., Physics
Jordan, Kenneth D., Chemistry
Kates, James M., Engineering
Kelley, James A., Chemistry
Kemler, Deborah G., Psychology
Kiester, Alan R., Biology, General
Kirshner, Robert P., Astronomy
Kleinberg, Evelyn G., Biology, General
Kra, Ethan E., Mathematics
Kroch, Anthony S., Linguistics
Labelle, Edward F. Jr., Biochemistry
Lasnik, Howard B., Linguistics
Leavitt, Richard P., Physics
Loomis, Jack M., Psychology
Love, Evelyn S., Earth Sciences
Luft, Harold S., Economics
Lumbert, Robert B., Mathematics
Lynch, Nancy A., Mathematics
MacWilliams, Harry K., Biology, General
Mains, Elizabeth E., Biophysics
Mains, Richard E., Biology, General
Manski, Charles F., Economics
Markiewicz, Robert S., Physics
Martin, Alice N., Physiology
Maslak, Samuel H., Engineering
Masterman, Hugh C., Engineering
Matteo, Charles C., Microbiology
McGaffigan, Edward Jr., Physics
Miller, Harold M., Engineering
Minkley, Edwin G. Jr., Chemistry
Moissides, Lydia E., Chemistry
Moulton, Peter F., Engineering
Murphy, Gerald J., Chemistry
Murphy, Michael R., Psychology
Myers, Jerome F., Mathematics
Nicoli, David F., Physics
Nutter, Deborah W., Political Science
Ouderkirk, Andrew E., Engineering
Overcamp, Thomas J., Engineering
Paine, John B. III, Chemistry
Paul, Gerald, Physics
Picard, Raymond L., Engineering
Pierson, Gary O., Chemistry
Prival, Michael J., Microbiology
Reaume, David M., Economics
Reddy, Richard D., Sociology
Reich, Michael, Economics
Reid, Thomas S., Biology, General
Reintjes, John F., Physics
Renner, Carl A., Chemistry
Renshaw, Bruce B., Mathematics
Rierdan, Jill E., Psychology
Rockland, Charles, Mathematics
Rosenfield, Donald B., Engineering
Rosenstein, Marvin, Atmospheric Sciences
Rosenthal, Alan I., Chemistry
Rowan, Robert III, Chemistry
Ruth, Gregory R., Engineering
Schnell, Charles E., Biology, General
Shalen, Peter B., Mathematics
Shea, Susan M., Urb. and Reg. Planning
Shillman, Robert J., Engineering
Shore, Douglas L., Chemistry
Siggia, Eric D., Physics
Skerry, Malcolm A., Physics
Smith-Johannsen, Robert, Biophysics
Snider, Barry B., Chemistry
Spencer, David A., Mathematics
Stark, Rexford A., Chemistry
Steele, Robert S., Psychology
Steinbach, Joseph H., Biology, General
Strickland, William J., Psychology
Sullivan, Bradley J., Physiology
Thresher, Ronald E., Biology, General
Toll, Charles H., Mathematics
Tossell, John A., Chemistry
Trivers, Robert L., Biology, General
Udin, Susan B., Physiology
Vetterling, William T., Physics
Wacks, Kenneth P., Engineering
Walker, William W. Jr., Engineering
Warren, Ira D., Chemistry
Weidner, Donald J., Earth Sciences
Weinberg, Marc S., Engineering
Weinstein Milton C., Mathematics
Weiss, Jeffrey M., Physics
Wentworth, Thomas R., Biology, General
Wicke, Brian G., Chemistry

FELLOWSHIP AND TRAINEESHIP AWARDS

WIESEL, WILLIAM E. JR., Astronomy
WILLEMAIN, THOMAS R., Engineering
WING, RENA R., Psychology
WOFSY, STEVEN C., Chemistry
WOLNIK, WALTER J., Engineering
WUONOLA, MARK A., Chemistry
ZAGIER, DON B., Mathematics
ZUERCHER, JOSEPH C., Engineering

Postdoctoral

ANDERSON, STUART D., Biophysics
BONNER, WILLIAM M., Biology, General
BOYD, RICHARD N., History and Philosophy of Science
BROWER, RICHARD C., Physics
COOPER, ROBERT L., Linguistics
ESSENBERG, MARGARET K., Microbiology
GOODENOUGH, DANIEL A., Biology, General
GRIECO, PAUL A., Chemistry
OSEROFF, ALLAN R., Biophysics
REBEK, JULIUS JR., Chemistry
SOLOMON, SEAN C., Earth Sciences
STOFFOLANO, JOHN G. JR., Zoology

Science Faculty

FABOS, JULIUS G., Urb. and Reg. Planning
GORE, RICHARD Z., Earth Sciences
HARRIS, DENTON B., Engineering
NIEMI, EUGENE E. JR., Engineering
PRYOR, MARILYN Z., Physiology
SHEPARD, JUNE S., Biology, General
ZIMMERMAN, CLAIRE, Psychology

MICHIGAN

Graduate

ADDICOTT, JOHN F., Biology, General
ANDERSON, CHARLES F., Chemistry
ARNOLD, STEVAN J., Biology, General
ASCH, DAVID L., Anthropology and Archeology
AYOUB, MARY M., Mathematics
BECK, KAREN L., Sociology
BEGNOCHE, CATHERINE A., Sociology
BEISWENGER, RONALD E., Zoology
BEKOFSKE, KEITH L., Engineering
BENNINGER, LARRY K., Earth Sciences
BISHOP, CHRISTINE E., Economics
BLASIUS, KARL R., Astronomy
BLUMHAGEN, DAN W., Genetics
BOOKSTEIN, FRED L., Sociology
BRUBAKER, LINDA B., Biology, General
BURKE, JONATHAN A., Chemistry
CARROLL, ALLEN M., Physics
CENZER, DOUGLAS A., Mathematics
COLESTOCK, PATRICK L., Engineering
COOPER, LYNN A., Psychology
COOPER, MICHAEL R., Earth Sciences
COWAN, DAVID P., Earth Sciences
DUNKER, ALAN M., Chemistry
EAMES, CONSTANCE D., Zoology
EVANS, BENNY D., Mathematics
FROHARDT, DANIEL E., Mathematics
FROHLICH, MICHAEL W., Botany
GOULD, THOMAS H., Engineering
GREENLEE, THOMAS R., Physics
GRIFFITH, WALTER L., Mathematics
GROST, MICHAEL E., Mathematics
HAILE, DARRELL E., Mathematics
HALL, MARK W., Urb. and Reg. Planning
HANTLER, SIDNEY L., Mathematics
HARWOOD, DANE L., Psychology
HASSELBACH, CLARENCE G., Physics
HEPPENHEIMER, THOMAS A., Engineering
HOLDERNESS, JAMES H., Engineering
HUDSON, BRUCE S., Chemistry
HUDSON, SUZANNE S., Chemistry
HUNT, DONALD E., Psychology
JACKSON, SYDNEY V., Chemistry
JONAH, CAROLINE A., Engineering
KENNEDY, MICHAEL P., Economics
KERFOOT, WILSON C., Zoology
KLATZKY, ROBERTA L., Psychology
KNIRK, DWAYNE L., Chemistry
KRIEGER, GARY A., Political Science
KUIPERS, BENJAMIN J., Mathematics
KUTKUS, MINDAUGAS J., Engineering
LIEDER, CHARLES A., Chemistry
LITVEN, JOSEPH A., Psychology
MAGLOTT, DONNA R., Biology, General
MASTERSON, LAWRENCE J., Geography
MATTERN, DANIELL L., Chemistry
MCTAVISH, JEANNE E., Psychology
MONTAGUE, MICHAEL J., Botany
NEERING, MICHAEL J., Engineering
NEMVALTS, KALLE, Mathematics
OEGEMA, THEODORE R., Biochemistry
OTTO, NORMAN C., Engineering
PLATT, TERRY, Biochemistry
POWERS, GARY J., Engineering
PUGH, EDWARD N., Psychology
QUIRK, JEFFERY A., Engineering
REINOEHL, JOHN H., Mathematics
REITMAN, LARRY N., Chemistry
REWOLDT, GREGORY M., Physics
RIVERS, LYNN J., Biology, General
RODMAN, JAMES E., Botany
SAEGERT, SUSAN C., Psychology
SANDERS, PATRICIA D., Mathematics
SHERMAN, ROLAND W., Psychology
SHILLER, ROBERT J., Economics
SIPE, JOHN E., Physics
SKOCPOL, THEDA R., Sociology
STEWART, JOHN A., Sociology
STRICKLAND, SIDNEY, Biochemistry
SUGGS, JOHN W., Chemistry
SWEANY, RAY L., Chemistry
TABER, ROBERT C., Physics
TANNER, JANE E., Psychology
TURK, JOHN S., Physics
ULLMAN, VICTOR P., Biochemistry
VANBREE, KENNETH A., Engineering
VERNIER, PAUL T., Microbiology
WEPFER, RICHARD W., Mathematics
WESTON, RAFAEL R., Economics
WINTER, KENNETH A., Psychology
WORK, DALE E., Chemistry

Postdoctoral

JACO, WILLIAM H., Mathematics
JENSEN, DOUGLAS A., Physics
MATTHEWS, ROWENA G., Biochemistry
PASSMORE, HOWARD C., Genetics
SCHUMAN, MARILYN, Chemistry
VITT, DALE H., Botany

Senior Postdoctoral

BLINDER, SEYMOUR M., Chemistry
HECHT, KARL T., Physics
LOCKWOOD, JOHN L., Botany
ROOT, WILLIAM L., Engineering
ZAJONC, ROBERT B., Psychology

Science Faculty

BALEK, RICHARD W., Biology, General
CRUMP, JOHN W., Chemistry
DELESPINASSE, PAUL F., Social Science
GRIFFIOEN, ROGER D., Physics
HEIMS, STEVE P., History and Philosophy of Science
KLEIN, ROY S., History and Philosophy of Science
MCDONALD, JAMES R., Geography

MINNESOTA

Graduate

ANDERSON, PAUL A., Economics
ANDERSON, RICHARD J., Chemistry
BERNTSON, GARY G., Psychology
BIEGING, JOHN H., Astronomy
BJORKMAN, JAMES W., Political Science
BOWMAN, DANIEL C., Anthropology and Archeology
BROWN, DENNIS P., Mathematics
COHEN, FAY G., Anthropology and Archeology
CONNOR, JAMES L., Psychology
COOK, THOMAS J., Mathematics
DAIGNEAU, PATRICIA J., Chemistry
DEVRIES, CYRIL P., Physics
DRESSEL, THOMAS D., Engineering
EICHTEN, ESTIA J., Physics
FAREL, RICHARD A., Engineering
FIRESTONE, ROGER M., Mathematics
GUENTHNER, THOMAS M., Medical Sciences
HASTIE, REID K., Psychology
HELD, ISAAC M., Physics
HELLER, ERIC J., Chemistry
HELQUIST, PAUL M., Chemistry
ILENDA, CASMIR S., Chemistry
JOHNSON, JERRY L., Mathematics
KELSEY, CYNTHIA C., Anthropology and Archeology
KIRST, HERBERT A., Chemistry
LAMBERTON, CHARLES E., Economics
LAMMERS, ROBERTA K., Biology, General
LAURIE, CECELIA A., Mathematics
LETOURNEAU, PAUL C., Biology, General
LIPPA, ERIK A., Mathematics
MAHLE, SUSAN K., Genetics
MCCASLIN, STANLEY J., Physics
MCCREE, ANN V., Biology, General
MIKKELSON, JAMES M., Engineering
MOTTL, MICHAEL J., Earth Sciences
NESS, LINDA A., Mathematics
NESS, STEPHEN A., Mathematics
NEWBOLD, FRED R., Engineering
OMVEDT, NEIL J., Mathematics
OOTHOUDT, MICHAEL A., Physics
PETERSON, ROLF O., Biology, General
POSNANSKY, CARLA J., Psychology
RAUCHER, STANLEY, Chemistry
RUHA, JEFFREY N., Mathematics
SANDELL, NILS R. JR., Engineering
STONE, DOUGLAS R., Engineering
SUNDAY, DANIEL M. JR., Mathematics
THOMPSON, FRANK J., Political Science
THURBER, CHARYL B., Psychology
VERANTH, JOSEPH L., Engineering
VOSS, RICHARD F., Physics
WALCZAK, MICHAEL R., Chemistry

Postdoctoral

BECCHETTI, FREDERICK D., Physics
HUTCHINSON, CHARLES R., Chemistry

Science Faculty

OLSON, DUANE N., Physics
SCHUSTER, SEYMOUR, Mathematics
STEEN, LYNN, A., Mathematics
WILL, ROBERT E., Economics

MISSISSIPPI

Graduate

ALLEN, THOMAS L., Economics
ARNETT, JOHN F., Chemistry
DENHAM, WOODROW W. Jr., Anthropology and Archeology
LUCAS, ROBERT M., Biophysics

Science Faculty

GRUCHY, DAVID F., Genetics
MILLER, ELDON L., Mathematics
WIRTH, THOMAS H., Chemistry

MISSOURI

Graduate

ABRAMSON, FRED G., Engineering
BAUER, GERALD L., Engineering
BECKNER, WILLIAM E., Mathematics
BIRNBAUM, HOWARD G., Economics
BROWN, RANDY L., Physics
CURTRIGHT, THOMAS L., Physics
DAY, HENRY P., Engineering
DEAN, WALTER K. JR., Chemistry
DOLL, JIMMIE D., Chemistry
DRAZEN, MARK, Mathematics
EHRHARDT, GARY J., Chemistry
FARRAR, JAMES M., Chemistry
FEE, DARRELL C., Chemistry
FIENUP, JAMES R., Physics
FULLING, STEPHEN A., Physics
HAMNER, PHILLIP G., Engineering

FELLOWSHIP AND TRAINEESHIP AWARDS

HONEA, ELMONT G., Earth Sciences
IMMELE, JOHN D., Chemistry
KELSEY, DONALD R., Chemistry
LYTTLE, TERRENCE W., Genetics
MANNING, FRANK B., Engineering
MILLER, KENNETH G., Mathematics
MUNCH, CHARLOTTE S., Biology, General
PARKS, DAVID R., Physics
POLINSKY, ALAN M., Economics
ROBBINS, MEDFORD D., Chemistry
RYLAND, STEPHEN L., Earth Sciences
SCHARLEMANN, ERNST T., Astronomy
SCHARLEMANN, MARTIN G., Mathematics
SCHRAMM, DAVID N., Physics
SCHROEDER, GEORGE W., Engineering
SMARR, LARRY L., Physics
TEMPLETON, ALAN R., Genetics
WEAVER, KEITH A., Physics
WEFEL, JOHN P., Physics
WHITE, JAMES E. Mathematics
WOODRUFF, ROBERT A., Chemistry

Postdoctoral

EARLY, THOMAS O., Earth Sciences

Science Faculty

ANDALAFTE, EDWARD Z., Mathematics
CORNWELL, LARRY W., Mathematics
GERBER, LAURA L., Mathematics
GIBBS, DAVID A., Physics
KIRK, GERALD R., Medical Sciences
LITTLE, BILLY F., Chemistry
MCGHEE, FLIN C., Engineering
STOCKTON, RAYMOND E., Mathematics
THURMAN, ROBERT E., Physics

MONTANA

Graduate

BETTS, BURR J., Zoology
BOND, ALAN B., Zoology
CLARK, JAMES W., Physics
HARTUNG, JAMES R., Earth Sciences
HILL, JESSE K., Astronomy
KLOVSTAD, JOHN W., Engineering
POPE, EMILY N., Linguistics
SHIMEK, RONALD L., Biology, General
SPARHAWK, FRANKLIN J., Sociology

NEBRASKA

Graduate

BULLOCK, ARTHUR M., Mathematics
COY, STEPHEN L., Chemistry
GIBLIN, DARYL E., Biochemistry
GRUENHAGE, GARY F., Mathematics
KEIFER, STEPHEN C., Engineering
RAPP, VICKI S., Microbiology
RIPS, LANCE J., Psychology
RIZLEY, ROSS C., Psychology
ROGGE, ALLEN E., Anthropology and Archeology
SEAVER, DAVID A., Psychology
SHARP, MARGARET A., Botany
TOMAS, PETER A., Mathematics
WEHRBEIN, WILLIAM M., Physics

Science Faculty

HEPWORTH, J. L., Biology, General

NEVADA

Graduate

CARTER, MARK E., Psychology
WILLSON, LEE M., Astronomy
WILSON, WARREN G., Physics

NEW HAMPSHIRE

Graduate

BICKEL, POLLY M., Anthropology and Archeology
GRUBER, PETER J., Botany
LIMBERT, DOUGLAS A., Physics
PARADOWSKI, ROBERT J., History and Philosophy of Science
TUCKER, THOMAS W., Mathematics

Postdoctoral

PRITCHARD, JOHN B., Physiology

Science Faculty

WALKE, RAYMOND H., Chemistry

NEW JERSEY

Graduate

ANDERSEN, F., ALAN, Biophysics
APGAR, WILLIAM C. Jr., Economics
ARNOLD, HAMILTON W., Engineering
AUDESIRK, GERALD J., Physiology
BARABAS, ARTHUR H., Earth Sciences
BELASCO, ELLIOT P., Physics
BERGER, ALAN E., Mathematics
BILANIN, ALAN J., Engineering
BIRCHARD, BRUCE A., Anthropology and Archeology
BLOCH, SUSAN L., Mathematics
BOGAN, ELIZABETH C., Economics
BOLES, JOHN F., Engineering
BRADSHAW, ROBERT W., Engineering
BREIGER, RONALD L., Sociology
BRENNER, LARRY B., Mathematics
BRISSON, LAWRENCE J., Mathematics
CARAZZONE, JAMES J., Physics
CHERLIN, GREGORY L., Mathematics
COOK, JOHN S., Mathematics
CORWIN, THOMAS L., Mathematics
DEVANEY, MICHAEL J., Mathematics
DOLDE, WALTER C., Jr., Economics
DRIEDGER, PAUL E., Chemistry
DUFF, DONALD G., Engineering
DURANA, STEPHEN C., Chemistry
DWYER, WILLIAM G., Mathematics
EHRENKRANZ, JEAN F., Linguistics
ELLIS, RICHARD S., Mathematics
EPSTEIN, RICHARD G., Mathematics
FARRAR, GLENNYS R., Physics
FELDER, HARRY III, Mathematics
FISCHLER, IRA S., Psychology
FLOODY, OWEN R., Psychology
FRAM, DAVID M., Physics
FRANKLIN, HOWARD M., Mathematics
GELLER, SUSAN C., Mathematics
GILMARTIN, KEVIN J., Psychology
GOLDRICH, NANCY R., Zoology
GOLDSTEIN, ANDREW C., Engineering
GORDON, WILLIAM J., Mathematics
GREENBERG, SANFORD N., Political Science
GREENBERG, WILLIAM J., Political Science
GREGOR, JOHN M., Physics
GROSS, KENNETH W., Biology, General
GURIN, DOUGLAS, B., Urban and Regional Planning
HARRIS, DANIEL C., Chemistry
HARTMAN, WILLIAM H., Physics
HARTWICK, ROBERT F., Biology, General
HAUCK, WALTER W., Jr., Mathematics
HAZEN, ROBERT M., Earth Sciences
HENDRICK MICHAEL E., Chemistry
HERRICK, GLENN A., Biology, General
HISIGER, ROBERT S., Engineering
HLADKY, STEPHEN B., Biophysics
HOWITT, ARNOLD M., Political Science
JACOBS, STEPHEN A., Chemistry
JAGACINSKI, RICHARD J., Psychology
KAUFMANN, SIDNEY J., Urban and Regional Planning
KENDALL, DOUGLAS S., Chemistry
KESSLER, Sr. MARIE I., Zoology
KNAUER, SCOTT C., Engineering
KRAMER, STEVEN D., Physics
LANDAU, IRA T., Psychology
LAWN, RICHARD M., Astronomy
LEWIS, JAMES B., Chemistry
LIEBERMAN, JUDY, Physics
LINDEN, LAWERENCE H., Engineering
LOMBARDERO, DAVID A., Mathematics
LOOMIS, TIMOTHY P., Earth Sciences
LUKOFF, DAVID G., Anthropology and Archeology
LYONS, HAROLD W., Biology, General
MAHER, MARY ANNE T., Mathematics
MALBROCK, JANE C., Mathematics
MARTIN, STEPHEN, Economics
MARTIN, THOMAS F., Biophysics
MELOSH, HENRY J., IV, Physics
MERRIAM, GEORGE R., III, Earth Sciences
METZGER, LOUIS S., Engineering
MODRESKI, PETER J., Earth Sciences
MOLDENKE, ALISON F., Biology, General
MORTOLA, ALBERT P., Chemistry
MOYER, WAYNE A., Biology, General
MRAW, STEPHEN C., Chemistry
MURRAY, JAY D., Earth Sciences
NEAREY, TERRANCE M., Linguistics
OETTING, JOHN D., Engineering
OWENS, AARON J., Physics
PATT, STEVEN L., Chemistry
PENNOTTI, RAYMOND J., Engineering
PERLMAN, DAVID A., Mathematics
PETERSON, CHARLES H., Biology, General
PETRILLO, EDWARD W., Jr., Chemistry
RAM, JEFFREY, L., Biochemistry
ROBOCK, ALAN D., Atmospheric Sciences
SCHAFFHAUSEN, BRIAN S., Biochemistry
SCHLENKER, BARRY R., Psychology
SEMEL, VICKI G., Political Science
SEPKOSKI, JOSEPH J., Jr., Earth Sciences
SERWER, DANIEL P., History and Philosophy of Science
SHULMAN, HERBERT B., Mathematics
SOMMESE, ANDREW J., Mathematics
SPARROW, DAVID A., Physics
SWANSON, ROGER C., Mathematics
SZPER, DOUGLAS A., Mathematics
THOMPSON, ANNE M., Chemistry
ULLRICH, ROBERT C., Biology, General
UMBREIT, JAY N., Biochemistry
VANDEWATER, PAUL N., Economics
VANDUYNE, LINDA S., Biochemistry
VANWIE, DONALD G., Biology, General
VEROSUB, KENNETH L., Physics
VOLLBRECHT, JUDITH A., Anthropology and Archeology
WACHTEL, ALAN L., Physics
WACHTER, KENNETH W., Mathematics
WADT, WILLARD R., Chemistry
WALD, ROBERT M., Physics
WALLACE, BRUCE G., Medical Sciences
WALTER, HOWARD A., Jr., Physics
WARNER, SHARON A., Anthropology and Archeology
WHITE, CARL R., Earth Sciences
WILSON, AMY A., Sociology
WINARSKY, NORMAN D., Mathematics

Postdoctoral

BATTERMAN, STEVEN C., Engineering
GRANDY, RICHARD E., History and Philosophy of Science
HARDESTY, DONALD R., Engineering
MADEY, JOHN M., Physics
PAIGE, FRANK E., Jr., Physics
RHOADS, ROBERT E., Biochemistry
RIDGWAY, ROBERT W., Chemistry
SINGER, DAVID A., Mathematics
SPITZER, NICHOLAS C., Biology, General

Senior Postdoctoral

GOLDFELD, STEPHEN M., Economics

Science Faculty

DAMBROSA, MICHAEL J., Mathematics
DEMETROPOULOS, ANDREW, Mathematics
DEVOS, ROBERT M., Mathemntics
FRIEDRICHS, ROBERT W., History and Philosophy of Science
GUILFOYLE, RICHARD H., Mathematics

NEW MEXICO

Graduate

ADLER, ROBERT E., Psychology
BENNETT, ALLEN, Botany
CHILDRESS, CHARLES L., Mathematics
DARDEN, THOMAS A., Mathematics
FURCHNER, CAROL S., Psychology

FELLOWSHIP AND TRAINEESHIP AWARDS

GARRETT, HENRY B., Physics
HANNAH, MARSHA J., Mathematics
HUFFMAN, WILLIAM C., Mathematics
LEGANT, PATRICIA, Psychology
LEVINE, MARCELLA S., Economics
MACQUIGG, DAVID R., Engineering
MODRICH, PAUL L., Biochemistry
O'ROURKE, PETER J., Mathematics
WOLLMAN, ERIC R., Physics

Science Faculty

CALTON, WILLIAM G., Mathematics

NEW YORK

Graduate

ADAMS, GEORGE G., Engineering
AISSEN, JUDITH L., Linguistics
AJAMI, ALFRED M., Jr., Biology, General
AKULA, JOHN L., Sociology
ALOFF, SIMON I., Mathematics
ALTMAN, LESLIE V., Psychology
ANDERSON, JOHN G., Earth Sciences
ASIMOV, DANIEL A., Mathematics
AUGENLICHT, LEONARD H., Biology, General
AUSUBEL, FREDERICK M., Microbiology
BAKER, ALAN P., Mathematics
BARKER, WILLIAM H., Mathematics
BARNARD, WILLIAM D., Earth Sciences
BARSA, EDWARD A., Chemistry
BASSEIN, RICHARD, Mathematics
BAUMEL, ROBERT T., Physics
BAXTER, MILTON, Linguistics
BEGLEY, RICHARD F., Physics
BEKENSTEIN, JACOB D., Physics
BERLINER, JEFFREY E., Engineering
BERMAN, STEPHEN J., Mathematics
BIRKEN, STEVEN, Microbiology
BIRNBAUM, ABRAHAM A., Physics
BISHOP, PETER B., Mathematics
BIXBY, ROBERT E., Mathematics
BLAIR, PHILIP J., Jr., Anthropology and Archeology
BLAND, ROBERT G., Mathematics
BLOCH, ROBERT J., Biochemistry
BLUMENFELD, BARRY J., Physics
BONO, JAMES J., History and Philosophy of Science
BOOTHE, JOAN N., Economics
BORIE, ELLEN D., Mathematics
BOUCHARD, ROBERT A., Biology, General
BOYLE, BRIAN E., Engineering
BRAND, RONALD K., Mathematics
BROWN, MICHAEL L., Mathematics
BRUMER, PAUL W., Chemistry
BURKE, WYLIE G., Genetics
BURNS, ROBERT P., History and Philosophy of Science
CALABRESE, RONALD L., Biology, General
CALDAROLA, PATRICIA C., Biology, General
CALLAHAN, JAMES J., Mathematics
CANDIOTTI, ALAN, Mathematics
CARROLL, JOHN S., Psychology
CARROLL, JOSEPH E., Mathematics
CARY, PAUL R., Biochemistry
CASAMASSIMA, SALVATORE, Engineering
CASCIONE, ROSA, Psychology
CHALMER, PAUL D., Chemistry
CHANG, RICHARD, Physics
CHANG, THEODORE C., Mathematics
CHENG, JULIAN, Physics
COCHRAN, JAMES R., Earth Sciences
COHEN, ELI, Mathematics
COLLIER, MARJORIE M., Biology, General
COOK, LESLIE P., Mathematics
COOPERSTEIN, BRUCE N., Mathematics
CORCORAN, RICHARD J., Chemistry
CRANE, GODFREY A., Biochemistry
CRAVENS, THOMAS E., Astronomy
CRESWELL, KAREN M., Physiology
CUCITI, PEGGY L., Political Science
DAVIDS, WINSTON G., Mathematics
DAVIS, MARK F., Engineering
DAVIS, RANDALL, Physics
DEAN, ROBERT J., Biology, General
DEINHARDT, CAROL L., Social Science
DEVOE, RALPH G., Physics
DUBOSE, CLARENCE R., Urban and Regional Planning
DUNN, JOSEPH L., Mathematics
DURISEN, RICHARD H., Astronomy
DUSHANE, THEODORE E., Mathematics
EDELSTEIN, MARTIN, Mathematics
EDENBERG, HOWARD J., Biology, General
EELLS, RICHARD C., Mathematics
EICKWORT, KATHLEEN R., Biology, General
ELLIS, MARTIN H., Mathematics
EPHRAIM, ROBERT M., Mathematics
EPSTEIN, IRVING R., Chemistry
ERENRICH, EVELYN S., Chemistry
EWELL, PETER T., Political Science
FAIER, MARGO B., Economics
FARACI, ROBERT A., Linguistics
FARRELL, EILEEN R., Anthropology and Archeology
FELSINGER, NEAL, Mathematics
FERRANTE, JEANNE, Mathematics
FINK, BARBARA R., Mathematics
FISANICK, GEORGIA J., Chemistry
FISKE, GARY S., Mathematics
FONTANA, BENEDETTO, Political Science
FOX, JOHN D., Sociology
FOX, LAUREL R., Biology, General
FRANCESCHETTI, DONALD R., Chemistry
FRANK, RICHARD A., Psychology
FRIEDMAN, JEFFREY F., Physics
FRIEDMAN, JULIETTE, Chemistry
FRIMER, ARYEH A., Chemistry
FROIMOWITZ, MARK, Chemistry
GABOW, HAROLD N., Mathematics
GAGANIDZE, NINA F., Biology, General
GAGEN, PAUL F., Engineering
GALLANT, ROBERT P., Engineering
GARBADE, KENNETH D., Economics
GARBINCIUS, PETER H., Physics
GELLER, DENNIS P., Mathematics
GERVER, JOSEPH L., Mathematics
GERVER,, MICHAEL J., Physics
GLENBOCKI, ALEXANDER J., Engineering
GLENN, PAUL G., Mathematics
GODFREY, DONALD A., Physiology
GOLDBERG, JONATHAN, Economics
GOLDSTEIN, SETH M., Economics
GORDON, SHELLEY M., Biology, General
GOTTESMAN, SUSAN, Genetics
GOULD, ROY R., Chemistry
GRAESER, LAIRD, Chemistry
GREENSPAN, BEVERLY N., Physiology
GRIMALDI, JOHN J., Chemistry
GRISSING, EDWARD L., Jr., Physics
GROSSBERG, KENNETH A., Political Science
GURRIA, GEORGE M., Chemistry
GUTMAN, GEORGE A., Biology, General
HAHN, ROBERT A., Anthropology and Archeology
HAKIM, RAZIEL S., Biology, General
HALFANT, MATTHEW, Mathematics
HALPER, CARYL M., Psychology
HAMBURGER, HOWARD R., Mathematics
HARBUS, FREDRIC I., Physics
HARPAZ, NOAM, Biophysics
HARPER, JAMES D., Mathematics
HARRINGTON, JONATHAN E., Biology, General
HARROW, KEITH, Mathematics
HART, MICHAEL H., Astronomy
HARTMAN, NEIL, Genetics
HARWOOD, JONATHAN H., Microbiology
HASSETT, JAMES M., Psychology
HAYNES, JOHN B., Jr., Physics
HEITNER, HOWARD I., Chemistry
HELLER, PETER S., Economics
HERBST, ERIC, Chemistry
HERTZ, VICTOR J., Political Science
HEUMANN, MILTON, Political Science
HINGERTY, BRIAN E., Physics
HIRSCHLAND, EDWARD C., Linguistics
HODES, HAROLD T., History and Philosophy of Science
HOEGERMAN, STANTON F., Genetics
HOFFMAN, JEFFREY A., Astronomy
HORN, RICHARD J., Physiology
HU, EVELYN L., Physics
HULL, WILLIAM E., Chemistry
HYMOWECH, MARVIN, Mathematics
INSELMAN, BARBARA R., Psychology
IVES, SHIRLEY M., Chemistry
JACKSON, ROBERT C., Biochemistry
JASKOW, PAUL L., Economics
JAYNE, JOHN E., Mathematics
JEFFORDS, DONALD J., Mathematics
JEKOWSKY, ELIOT, Chemistry
JOHNSON, DAVID H., Physics
JOHNSON, SUZANNE B., Psychology
JOHNSTON, RALPH A., Mathematics
JOSS, PAUL C., Astronomy
KAMINSKY, MARK E., Physics
KANAREK, KATHRYN B., Engineering
KAPPELMAN, ALLAN H., Chemistry
KATZ, JEROME M., Mathematics
KATZ, JONATHAN I., Astronomy
KELLEY, DARCY B., Psychology
KENDLER, KAREN S., Sociology
KENNEDY, GEORGE G., Zoology
KENNEDY, KENNETH W., JR., Mathematics
KERSTEIN, ALAN R., Physics
KING, RONALD F., Political Science
KLAVAN, ARTHUR R., Engineering
KLINGER, JOHN S., Physics
KOBLITZ, NEAL I., Mathematics
KOCH, STEPHEN A., Chemistry
KOGUT, JOHN B., Physics
KOPLIK, JOEL I., Physics
KOVAC, MARK P., Physiology
KRAMER, KENNETH B., Mathematics
KRAUS, RICHARD C., Political Science
KRIGER, AMY R., Economics
KULLMAN, BRIAN C., Mathematics
KUTNER, MARC L., Physics
LAKE, ALVIN E., III, Psychology
LANDES, ELISABETH M., Economics
LANDESMAN, PETER, Mathematics
LANDISMAN, LAURIE, Microbiology
LARKIN, RONALD P., Psychology
LASKER, ROZ D., Biochemistry
LASLEY, ERIC L., Physics
LATTA, CYNTHIA M., Economics
LAUNAY, ROBERT G., Anthropology and Archeology
LAWRENCE, CHARLES B., Biology, General
LAX, ROBERT F., Mathematics
LEIBOVICH, LEWIS, Mathematics
LENTZ, PEGGY A., Geography
LERCH, PETER F., Mathematics
LEUNG, ELEANOR H., Psychology
LEVINE, MICHAEL S., Psychology
LEVINE, SUSAN P., Anthropology and Archeology
LEWIS, DAVID B., Urban and Reginnal Planning
LI, JADE, Biophysics
LIEBERMAN, RICHARD B., Engineering
LINKER, ARTHUR, Physics
LIU, HSIAO-PING, Anthropology and Archeology
LIZARDI, PAUL M., Biology, General
LORING, DEN/S W., Mathematics
LUDEL, JACQUELINE, Psychology
LUDUEN, RICHARD F., Biology, General
MANDELL, GORDON K., Engineering
MANEVITZ, LARRY M., Mathematics
MANFIELD, PHILIP E., Mathematics
MANNIX, JUDITH A., Psychology
MARCHASE, RICHARD B., Biophysics
MARCHESE, JOSEPH C., History and Philosophy of Science
MARMORINO, GEORGE O., Earth Sciences
MASLEY, JOHN M., Mathematics
MAYER, GARRY F., Biology, General
MAYNES, GORDON G., Chemistry
MAZAIKA, PAUL K., Mathematics
MCALLISTER, MARY C., Biology, General
MCBREARTY, JOHN B., Sociology
MCCREERY, JOHN L., Anthropology and Archeology
MCDONOUGH, JUDITH J., Botany
MCLAUGHLIN, JOHN B., Physics

FELLOWSHIP AND TRAINEESHIP AWARDS

McSweeney, Frances K., Psychology
Mead, Lawrence M., Political Science
Meldon, Jerry H., Engineering
Merel, Gail, Political Science
Michaels, Ira A., Chemistry
Miesowicz, Frederick M., Chemistry
Miller, Haynes R., Mathematics
Milmoe, Susan E., Psychology
Mindes, Paula, Psychology
Mineka, Susan, Psychology
Minsker, Steven, Mathematics
Mokrzycki, Kenneth L., Political Science
Moreno, Carlos J., Mathematics
Morrissey, Richard F., Sociology
Mosco, Vincent C., Sociology
Murray, Marc M., Earth Sciences
Nation, James B., Mathematics
Neuffer, David V., Physics
Nicotri, Mary E., Biology, General
Novick, Stewart E., Chemistry
Oberst, Daniel J., Linguistics
Odell, Edward W., Mathematics
Odessey, Richard, Biochemistry
Olson, John S., Biochemistry
O'Neill, Lawrence H., Physics
Otis, John N., Physics
Paisner, Jeffrey A., Physics
Palm, Gregory K., Economics
Pangle, Thomas L., Political Science
Parrish, James D., Biology, General
Parshall, Brian J., Mathematics
Paul, James A., Political Science
Perlo, Stanley, Mathematics
Perry, Charles S., Sociology
Perry, Mary J., Biology, General
Phillies, George D., Biophysics
Pigott, Charles A., Economics
Pike, Carl S., Botany
Pilkonis, Paul A., Psychology
Pinkham, George S., Engineering
Plaisted, David A., Mathematics
Platzblatt, Irene M., Chemistry
Ploss, Harry L., Physics
Politzer, Hugh D., Physics
Poucher, John S., Physics
Price, Herbert, Anthropology and Archeology
Prior, Ronald L., Biology, General
Prosnitz, Donald, Engineering
Rabinowitz, Stanley, Mathematics
Rackoff, Charles W., Mathematics
Raiken, Eliot M., Physics
Rand, James B., Genetics
Rauscher, M. Kathryn, Chemistry
Reed, Adam V., Psychology
Rescigno, Thomas N., Chemistry
Ribet, Kenneth A., Mathematics
Riesbeck, Christopher K., Mathematics
Rivest, Ronald L., Mathematics
Roberts, Jerry B., Biophysics
Rose, Harvey A., Physics
Rosen, David M., Anthropology and Archeology
Rosen, Jay S., Mathematics
Rosen, Richard D., Atmospheric Sciences
Rosenberg, Alan H., Biophysics
Rosenberg, Robert C., Chemistry
Rosenblum, Harvey S., Atmospheric Sciences
Rosenkranz, Philip W., Engineering
Rossen, Robert H., Engineering
Rothenberg, Allan F., Physics
Rothman, Golda R., Psychology
Rowen, Louis H., Mathematics
Ruben, David J., Chemistry
Sandel, Frederick L., Engineering
Saracino, Daniel H., Mathematics
Scharfsteln, Joel, Mathematics
Schechter, Diane, Linguistics
Scheim, David E., Mathematics
Scheinberg, Norman R., Engineering
Scherr, David E., Mathematics
Schoenblum, Jeffrey A., Political Science
Schoenfeld, Susan L., Biology, General
Schuessler, Robert W., Economics
Schulman, Lloyd L., Atmospheric Sciences
Schwartz, Jeffrey, Chemistry
Schwartz, Joan D., Biology, General
Schwartzkroin, Philip A., Life Sciences
Serkes, Ira M., Engineering
Shorenstein, Rosalind G., Biochemistry
Shriver, Bruce D., Mathematics
Shustek, Leonard J., Mathematics
Siegelbaum, Lewis H., Political Science
Sieradzki, Terrie C., History and Philosophy of Science
Sigwalt, Elinor S., Anthropology and Archeology
Silkworth, Mary L., Genetics
Silverstein, Alvin J., Chemistry
Simpson, Raymond W., Engineering
Singer, Michael F., Mathematics
Smith, Mark S., Mathematics
Snapp, Barbara D., Zoology
Sokoloff, David R., Physics
Sola, Kenneth E., Psychology
Solomon, Ronald M., Mathematics
Sonn, Miriam R., Sociology
Sontz, Ann L., Anthropology and Archeology
Stamper, Martha R., Microbiology
Stehn, Robert A., Zoology
Steinberg, Robert A., Biology, General
Stickler, Albert C., Engineering
Stoeckly, Beth B., Physics
Straley, Susan C., Botany
Straus, David M., Physics
Strauss, Monty J., Mathematics
Stronger, Leo, Linguistics
Suter, Kathryn A., Psychology
Suter, Laurance J., Physics
Sverdlove, Ronald, Mathematics
Szatrowski, Ted H., Mathematics
Tanenbaum, James R., Social Science
Tenenbaum, Aaron M., Mathematics
Thiel, David W., Mathematics
Thompson, Mark S., Economics
Traube, Elizabeth G., Anthropology and Archeology
Trauber, Philip, Mathematics
Trentacosta, Joseph D., Engineering
Trowbridge, Lee D., Chemistry
Tsang, James C., Physics
Valeo, Ernest J., Physics
Viavattene, Ronald L., Chemistry
Vigdor, Steven E., Physics
Vinick, Fredric J., Chemistry
Walker, David, Earth Sciences
Wand, Mitchell, Mathematics
Warren, Michael E., Engineering
Waser, Peter M., Physiology
Wasserman, Neil H., Physics
Weinberg, Erick J., Physics
Weinberger, David B., Mathematics
Weiner, Alan M., Biochemistry
Weinreich, Michael, Engineering
Weintraub, Jeff A., Sociology
Weiss, Harvey W., Anthropology and Archeology
Weiss, Joan M., Economics
Weiss, Richard S., Mathematics
Weitzner, Eric R., Mathematics
Wellander, Charles R., Physics
Werner, Paul D., Psychology
Wilczek, Frank A., Mathematics
Wilemski, Gerald R., Chemistry
Wilson, Karl A., Biochemistry
Winter, Ruth H., Biology, General
Wiseman, Andrew, Biology, General
Woodbury, Hanni J., Anthropology and Archeology
Wurzburger, Benjamin W., Economics
Yanof, Arnold W., Physics
Yee, David, Chemistry
Yellen, Janet L., Economics
Young, Curtis J., Engineering
Zakheim, Dov S., International Relations
Zauderer, Maurice, Biophysics
Zimmer, Robert J., Mathematics
Zlotnick, Fred M., Mathematics
Zucker, Steven M., Mathematics

Postdoctoral

Drew, Donald A., Mathematics
Dyer, Robert S., Physiology
Foit, Franklin F., Jr., Earth Sciences
Freedman, Michael P., Economics
Halpern, Jerald A., Biology, General
Jakobsson, Eric G., Physiology
Kirby, Edward P., Biochemistry
Kleinman, Arthur M., History and Philosophy of Science
Lees, Jack A., Mathematics
Levitt, Morris R., Physics
Maglio, Vincent J., Earth Sciences
Mosser, Jerry L., Biology, General
Quirk, William J., Astronomy
Rosenthal, Hanan, Physics
Schaich, William L., Physics
Schwartz, Jeffrey, Chemistry
Silver, Allan A., Sociology
Sweedler, Moss E., Mathematics
Varmus, Harold E., Microbiology
Walsh, Christopher T., Biochemistry
Weitzman, Martin L., Economics
Wiesenfeld, John R., Chemistry
Wolfowitz, Paul D., International Relations

Senior Postdoctoral

Ballantyne, Joseph M., Engineering
Goldhaber, Alfred S., Physics
Kuo, John T., Earth Sciences
Ling, Frederick S., Engineering
Saunders, William H., Chemistry
Sievers, Albert J., Physics
Tang, Chung-Liang, Physics
Thorndike, Edward H., Physics
White, David H., Physics

Science Faculty

Albrecht, Andreas C., Chemistry
Beeson, Roberta J., Biolngy, General
Cannon, James A., Physics
Carlton, Allan M., Mathematics
Chamberlain, Harry D., Biology, General
Dorrance, Robert W., Jr., Biology, General
Dunn, Brian F., Mathematics
Gilman, Robert E., Chemistry
Mahoney, Robert P., Microbiology
McFee, Richard, Engineering
McLelland, James M., Earth Sciences
Monaghan, William, Physics
Panzeca, Philip J., Engineering
Pitt, Joel H., Mathematics
Ricardo, Henry J., Mathematics
Rosenfeld, Robert, Mathematics
Stack, William J., Engineering
Supple, Jerome H., Chemistry
Wasiutynski, C. M., Mathematics
Weiss, David M., Sociology
Wilson, Stephen O., Geography

NORTH CAROLINA

Graduate

Ackermann, Arthur F., Jr., Mathematics
Broughton, Harold S., Chemistry
Carroll, Felix A., Jr., Chemistry
Cline, Alice A., Engineering
Cromartie, William J., Biology, General
Crouse, Gray F., Chemistry
Culberson, Charles H., Earth Sciences
Dawson, Daniel J., Chemistry
Dellinger, Charles E., Engineering
Dillard, Sandra A., Physics
Dobbs, Betty J., History and Philosophy of Science
Harper, Harry W., Psychology
Hecky, Robert E., Biology, General
Hosley, Rebecca A., Biology, General
Kinsey, Bill H., Jr., Agriculture
Kushman, John E., Economics
Mann, Margaret E., Mathematics
Matthews, Roger H., Chemistry
McGinty, David J., Chemistry
Meares, Claude F., Chemistry
Merritt, Frank S., Physics
Parrott, George C., Physiology
Plow, Eric P., Mathematics
Powers, Phyllis M., Microbiology

FELLOWSHIP AND TRAINEESHIP AWARDS

ROSENSON, LEON M., Zoology
SAWYER, JOHN W., Mathematics
SCANDELLA, DOROTHEA H., Biology, General
SETZER, CHARLES B., Mathematics
SMITH, DONALD J., JR., Political Science
STINSON, THOMAS W., III, Physics
TURNER, ROY S., History and Philosophy of Science
WESTMORELAND, DAVID G., Chemistry
WILCOX, CLARK R., Mathematics
WRIGHT, WILLIAM V., Mathematics

Postdoctoral

BELL, WALTER H. Jr., Linguistics
CIALDINI, ROBERT B., Psychology
PULLIAM, HOWARD R., Genetics
RHODES, CARL D. Jr., Biochemistry
SMOUSE, PETER E., Biology, General

Senior Postdoctoral

HUFSCHMIDT, MAYNARD M., Urban and Regional Planning

Science Faculty

COOPER, PETER P., II Anthropology and Archeology
PASOUR, ERNEST C., JR., Economics
PRESTON, DOROTHY K., Mathematics

NORTH DAKOTA

Graduate

COMITA, JEAN J., Biochemistry
KNAPP, ROBERT E., Mathematics
MILLER, DOUGLAS K., Biochemistry
RAMSEY, LAURENCE T., Mathematics
SCHLIPF, JOHN S., Mathematics
THUNBERG, ALLEN L., Biochemistry
WATLAND, ROSS T., Engineering
WILLIAMSON, PATRICK L., Biochemistry
ZACREP, DOUGLAS P., Physics

Science Faculty

EBELING, KENNETH A., Engineering

OHIO

Graduate

ADAMS, ELEANOR M., Social Science, General
BECKER, PAUL K., Psychology
BENKART, GEORGIA M., Mathematics
BERBERICH, JOEL J., Physiology
BOYS, WILLIAM E., Linguistics
BRADY, JOHN B., Earth Sciences
BREITENBACH, JOHN L., Medical Sciences
BROWN, CANDICE E., Microbiology
BUMILLER, CARL L., Mathematics
CALL, FREDERICK W., Mathematics
CAMPBELL, PAUL J., Mathematics
CASE, DAVID A., Chemistry
CHALKER, BRUCE E., Biology, General
CHANAN, GARY A., Physics
CLARK, BRIAN O., Physics
COE, RICHARD D., Economics
COLE, CHARLES N., JR., Biology, General
COREY, BRIAN E., Physics
CORN, MARY L., Genetics
CUMMINS, PHILIP A., Economics
DALY, THOMAS P., Physics
DASCH, GREGORY A., Biology, General
DEPIERRE, JOSEPH W., Biochemistry
DESMARAIS, THOMAS A., Engineering
DICKINSON, BRADLEY W., Engineering
DYKE, THOMAS R., Chemistry
EASTMAN, JOHN A., Engineering
EINSTEIN, GLORIA C., Sociology
ELLERBROCK, LEROY A., Botany
FARR, ROBERT A., Chemistry
FIELD, ROBERT A., Physics
FIRMENT, M/CHAEL J., Social Science
FLIERL, GLENN R., Physics
FONTANA, ROBERT E. JR., Engineering
FOUCHE, DANIEL G., Physics
FRIEDBERG, CHARLES B., Physics
FRYDL, EDWARD J., Economics
FULLER, WILLIAM C., Chemistry
GALEHOUSE, DANIEL C., Physics
GARVER, CRAIG M., Earth Sciences
GLANTZ, STANTON A., Engineering
GOETZ, DAVID W., Chemistry
HAAS, ROBERT, Mathematics
HAMMOND, NANCY J., Political Science
HAUEISEN, DONALD C., Physics
HAVACH, GEORGE A., Chemistry
HEDRICK, CHARLES L., Physics
HEITHAUS, EARL R., Biology, General
HERTZ, RAYMOND K., Chemistry
HEYWOOD, JANET L., Biochemistry
HOFFMAN, MICHAEL J., Mathematics
HYNDMAN, JANET R., Biology, General
IGLAUER, CAROL, Psychology
ISAACSON, ERIC J., Mathematics
JOHNSON, JAMES D., Physics
KALT, MARVIN R., Zoology
KAMINSKI, PAUL G., Engineering
KAST, STEVEN J., Engineering
KLOSTERMAN, PETER S., Biochemistry
KNOKE, DAVID H., Sociology
KONOPKA, RONALD J., Biochemistry
KOSTANSEK, EDWARD C., Chemistry
KRAUS, MARVIN C., Economics
LABAVITCH, JOHN M., Botany
LAPENAS, GEORGE N., Physiology
LEAMAN, KEVIN D., Earth Sciences
LEVINTHAL, CHARLES F., Psychology
LICHSTEIN, MICHAEL L., Economics
LINDER, THOMAS M., Physiology
LOCKHART, JAMES M., Physics
LOWDERMILK, W. H., Physics
MACKIE, MARTHA A., Economics
MARTIN, DAVID U., Physics
MARTIN, TERRENCE W., Physics
MEDVICK, PATRICIA A., Biology, General
MEUTH, MARK L., Biology, General
MILBRODT, THOMAS O., Physics
MILLER, MARILYN, Botany
MINNS, RICHARD A., Chemistry
MOLLENKOPF, JOHN H., Political Science
MOONEY, JOSEPH F., Mathematics
MORSE, LARRY E., Botany
MOTL, MARY L., Genetics
MOYE, PATRICIA A., Anthropology and Archeology
MUHLEMAN, LINDA J., Anthropology and Archeology
MYERS, LORA G., Zoology
NEWELL, STEVEN Y., Biology, General
OLSON, DONALD W., Physics
O'REILLY, JAMES E., Chemistry
PALMER, LAWRENCE G., Biology, General
PANTLE, CURTIS R., Physiology
PEGG, WILLIAM J., Chemistry
PETTI, RICHARD J., Mathematics
PINKEL, DANIEL, Physics
PRERO, AARON J., Economics
RAWSKI, THOMAS G., Economics
REID, JOSEPH D. JR., Economics
RUDOLPH, LEE N., Mathematics
RYCHENER, MICHAEL D., Mathematics
SALSBURY, DAVID L., Mathematics
SANTNER, THOMAS J., Mathematics
SEIFERAS, JOEL I., Mathematics
SEJNOWSKI, TERRENCE J., Physics
SHEPHERD, REX E., Chemistry
SHERMAN, SHERRILL A., Psychology
STEEGE, DEBORAH A., Biology, General
STORRIE, BRIAN, Biology, General
TEICHMOELLER, JOHN G., Economics
THORN, CHARLES B. III, Physics
TIEFERT, MARJORIE A., Biology, General
TIEMEIER, DAVID C., Biochemistry
TORONTO, ELLEN K., Psychology
UMANS, STEPHEN D., Engineering
VANVECHTEN, DEBORAH, Physics
VARIAN, HAL R., Economics
VIROST, RICHARD A., Engineering
VOLCK, ERIC W., Botany
WEHRLY, THOMAS E., Mathematics
WHITNEY, RONALD F., Mathematics
WILSON, WILLIAM F., Mathematics
WINSTEL, ROBERT A., Biology, General
WORMAN, WALTER E., Physics
WRIGHT, NEIL R., Economics
WRIGHT, PHILEMON K. III, Metallurgy
WUEST, JAMES D., Chemistry
YUN, DAVID Y., Mathematics

Postdoctoral

PENWELL, RICHARD C., Engineering
SCOGIN, RONNIE L., Botany
SELISKAR, CARL J., Biophysics
SIMONS, JOHN P., Chemistry
WINOGRAD, NICHOLAS, Chemistry

Science Faculty

COMFORT, JOHN C., Mathematics
CRAIG, NORMAN C., Chemistry
FARRINGTON, FRANKLIN D., Engineering
HILL, WILLIAM M. II, Mathematics
KOCH, ARTHUR T., Chemistry
MAISEL, JAMES E., Engineering
NOLTE, BYRON H., Engineering
TEPE, FRANK R. JR., Engineering
WETZEL, DANIEL E., Physics

OKLAHOMA

Graduate

ATON, THOMAS J., Physics
BOATRIGHT, JULIE A., Sociology
BRATTON, TIMOTHY L., Engineering
COSGROVE, ANN M., Botany
EDMONDS, ALLAN L., Mathematics
EDMONDS, DON E., Psychology
FAGIN, RONALD, Mathematics
GOULD, JAMES L., Zoology
HIVELY, RAY M., Physics
HUMPHREY, GILBERT E., Mathematics
KEELER, THEODORE E., Economics
MARPLE, STANLEY L. JR., Engineering
MCGALLIARD, RUSSELL L., Engineering
MCWILLIAMS, JAMES C., Mathematics
MURPHY, JOHN B., Agriculture
OSPOVAT, DOV, History and Philosophy of Science
ROSS, MARK J., Anthropology and Archeology
SPALL, RICHARD D., Zoology
WOODARD, JAMES B. Jr., Engineering

Science Faculty

BOATRIGHT, KIRK E., Engineering
DUNCAN, WILLIAM P., Chemistry
MOORE, JESSE C., Chemistry
NOBLE, DONALD J., Physiology
PAINE, MYRON D., Engineering
REDDEN, CARL R., Sociology

OREGON

Graduate

ALWARD, EUGENE R., Mathematics
AYRES, WILLIAM A., Engineering
BEARD, ELIZABETH C., Zoology
BRANSTATOR, GRANT W., Mathematics
BUFTON, LINDA L., Genetics
CHRISTIANSEN, DANIEL S., Economics
CRAVEN, THOMAS C., Mathematics
DODDS, STANLEY A., Physics
EDMONSTON, BARRY J., Social Science
EERKES, GARY L., Mathematics
FAUGHT, WILLIAM S. III, Mathematics
HALL, JOHN H., Earth Sciences
HARCOMBE, PAUL A., Biology, General
HAWKINS, JOHN P., Anthropology and Archeology
HERMENS, KENNETH A., Engineering
KLEIN, NORMAN, Anthropology and Archeology
KLEMM, RICHARD A., Chemistry
KURJAN, PHILIP M., Physics
LAING, KATHERINE M., Earth Sciences
LEE, EDMUND G., Mathematics
LINDSAY, BRUCE G., Mathematics
LINK, REGINA G., Earth Sciences
MATE, BRUCE R., Biology, General
MCKAY, GORDON A., Earth Sciences
NEUBAUER, PAUL R., Linguistics
PETERSON, JAMES L., Mathematics

FELLOWSHIP AND TRAINEESHIP AWARDS

PHILIPS, SUSAN U., Anthropology and Archeology
PIERSON, BEVERLY K., Biology, General
PRIBNOW, DAVID G., Biology, General
REITER, CAROL R., Anthropology and Archeology
SCHWARZ, GERALD W., Mathematics
SENDERS, CARLA J., Psychology
SHORT, ROBERT W., Physics
SPEEDIE, MARILYN W., Biochemistry
STRAND, LARRY L., Botany
WANNIER, PETER G., Biophysics
WATSON, BRUCE A., Biochemistry
WEBB, THOMAS R., Chemistry
WEST, LLOYD A., Chemistry
WHITELEY, NORMAN M., Biochemistry
WILSON, MARK L., History and Philosophy of Science
WOLD, RONALD O., Earth Sciences

Postdoctoral

RAYMOND, JONATHAN C., Microbiology
SCOTT, WALTER A., Biology, General

Senior Postdoctoral

KOENIG, THOMAS W., Chemistry
PYTKOWICZ, RICARDO M., Earth Sciences

Science Faculty

HARTVIGSON, ZENAS R., Mathematics
MCINTIRE, CHARLES D., Mathematics

PENNSYLVANIA

Graduate

AHLSTROM, JAMES C., Physics
ARMSTRONG, JOHN III, Linguistics
BALL, WILLIAM A., Psychology
BALTUSCHECK, CLIFFORD M., Physics
BATES, STEPHEN C., Engineering
BELL, DAVID M., Biology, General
BERGMANN, JACOB, Mathematics
BERKOWITZ, SUSAN G., Anthropology and Archeology
BERRIER, JOHN V., Chemistry
BETZ, JOAN L., Biochemistry
BEZMAN, RICHARD D., Chemistry
BEZMAN, SUSAN A., Chemistry
BIERBAUM, VERONICA M., Chemistry
BLACKADAR, BRUCE E., Mathematics
BORGER, JUDITH A., Biology, General
BOWKER, JEFFREY C., Engineering
BRENNER, MALCOLM W., Psychology
BROWN, JAMES W., Physics
BROWN, KENNETH L., Biochemistry
BROWN, WILLIAM S., Engineering
BUCK, BARBARA R., History and Philosophy of Science
BURGGRAF, BRUCE F., Engineering
BYRNE, SHEILA, Zoology
CAFFENTZIS, CONSTANTINE, History and Philosophy of Science
CARTER, JOHN L., Mathematics
CATLIN, PAUL A., Mathematics
CHAPMAN, SALLY, Chemistry
CHENCINSKI, KURT, Chemistry
CHUBB, JUDITH A., International Relations
CLARK, GEOFFREY A., Anthropology and Archeology
COLELLA, FRANK J., Political Science
CONFALONE, PASQUALE N., Chemistry
COOK, ROBERT C., Chemistry
CORSARO, FELICIA A., Chemistry
D'ANGELO, LOUIS P. JR., Mathematics
DAVIS, DUANE H., Chemistry
DEBARTOLO, GILBERT F., Economics
DETWILER, JOHN S., Engineering
DITZLER, WILLIAM R., Physics
DODSON, PETER J., Earth Sciences
DOLINAR, SAMUEL J., Engineering
EISELE, GEORGE II, Chemistry
EVANS, BEN E., Chemistry
FIELDS, JOHN R., Physics
FORD, MICHAEL E., Chemistry
FRANKEL, DONALD S. JR., Chemistry
FREIMAN, MARC P., Economics
GEER, KEVIN A., Physics
GEIST, RICHARD F., Engineering
GERHART, DAVID Z., Biology, General
GERRA, RALPH A. JR., Mathematics
GILBERT, JAMES L., Physics
GOLDBERG, EUGENE L., Mathematics
GOLDBERG, KENNETH, Mathematics
GORDON, LOUIS, Mathematics
GRAETZ, SARAH J., Biology, General
GRIESS, ROBERT L. JR., Mathematics
GULDIN, RICHARD W., Agriculture
HAAS, CAROL K., Chemistry
HALL, HORACE T., Chemistry
HAWKINS, ROBERT D., Psychology
HEFFNER, LINDA JEAN, Physiology
HEINTZBERGER, EDWARD H., Mathematics
HENRETTA, JOHN C., Sociology
HILBORN, ROBERT C., Physics
HOBURG, JAMES F., Engineering
HOFFER, CAROL P., Anthropology and Archeology
HOFFMAN, PAULA L., Physiology
HOSTETLER, RALPH B., Chemistry
INGRAM, GREGORY K., Economics
KAPLAN, JEFFREY A., Engineering
KARNOFSKY, JOEL R., Mathematics
KERLICK, GEORGE D., Physics
KETTNER, JAMES E., Mathematics
KIMBLE, ROBERT J. JR., Mathematics
KOCHMAN, FRED, Mathematics
KRECKER, DONALD K., Mathematics
KROUSE, CHERYL H., Psychology
KUBERT, DANIEL S., Mathematics
KUNESH, CHARLES J., Chemistry
LAKOWICZ, JOSEPH R. JR., Chemistry
LAMPERT, RICHARD H., Engineering
LAWSON, REBECCA F., Zoology
LEWIS, DALE M., Earth Sciences
LEWIS, JAMES A., Biochemistry
LO, BERNARD, History and Philosophy of Science
MAERKER, JOHN M., Engineering
MAHER, KATHLEEN V., Biochemistry
MALINOSKI, SR. B., Zoology
MANDELL, STEWART, Mathematics
MARCUS, RITA R., Biochemistry
MASTRO, ANDREA M., Physiology
MASTROCOLA, ANTONIETTA, Chemistry
MAURER, DAPHINE, M., Psychology
MCCARTHY, KAY L., Medical Sciences
MELVIN, JONATHAN D., Physics
MENDELSSOHN, MARVIN, Mathematics
MEYERS, JAMES H., Earth Sciences
MILLER, EDWARD Y., Mathematics
MIRKIN, MITCHELL I., Engineering
MIZIORKO, HENRY M., Biochemistry
MURTAUGH, TERENCE S., Chemistry
NEILSON, BRUCE J., Engineering
NOVOSELLER, DANIEL E., Physics
NUNEMACHER, JEFFREY L., Mathematics
OTTON, JAMES K., Earth Sciences
PARSONS, WILLIAM G., Mathematics
PASTIN, MARK J., History and Philosophy of Science
PERRI, ALBERT J., Physics
PHILLIPS, WILLIAM D., Physics
PRICKETT, MARTHA E., Anthropology and Archeology
PYERITZ, REED E., Biology, General
QUAILE, JAMES P., Engineering
RAFSKY, LAWRENCE C., Mathematics
REHR, JOHN J. JR., Physics
REUSS, ROBERT H., Chemistry
ROHRBAUGH, DENNIS K., Chemistry
ROSEN, ALBERT E., Physics
RUDNICK, LAWRENCE, Physics
SAMUELS, SAM L., Engineering
SANDEL, BONNIE B., Chemistry
SANDER, JOAN M., Mathematics
SANTAVICCA, DOMENIC A., Engineering
SCHAUFFLER, WILLIAM M., Anthropology and Archeology
SCHMIDT, RICHARD A., Engineering
SCHONBACH, DAVE I., Engineering
SCHROEDER, FRIEDHELM, Biochemistry
SCHWARTZ, BARRY J., Psychology
SEKELA, ALBERT M., Engineering
SHAFFER, GARY R., Earth Sciences
SHOGAN, ANDREW W., Mathematics
SICILIANO, THOMAS V., History and Philosophy of Science
SIMON, RICHARD M., Engineering
SKVORETZ, JOHN V. JR., Sociology
SMOLIAR, STEPHEN W., Mathematics
SQUIRES, STEPHEN L., Engineering
SULLIVAN, CHARLES L., Chemistry
TAYLOR, CHRISTOPHER A., Economics
TELSCH, RICHARD W., Engineering
TENNITY, KENNETH J., Linguistics
TINSMAN, JAMES H., Anthropology and Archeology
TOWNSEND, JOHN S., Physics
TURNBAUGH, WILLIAM A., Anthropology and Archeology
TYSON, JOHN J., Chemistry
UMIKER, DONNA J., Anthropology and Archeology
VANCE, BILLIE J., Psychology
VERBEEK, EARL R., Earth Sciences
VERMEYCHUK, J. GREGORY, Engineering
WASSON, ROBERT A., Engineering
WEINSTEIN, FRANK C., Physics
WEINSTEIN, MICHAEL M., Economics
WEIR, RICHARD A., Chemistry
WEISS, RANDALL D., Economics
WEISS, RICHARD M., Mathematics
WELSH, JAMES B., Anthropology and Archeology
WERNER, MATTHEW L., Earth Sciences
WHALEN, ROBERT G., Biochemistry
WHIPPLE, WILLIAM R., History and Philosophy of Science
WHITE, WILLIAM I., Chemistry
WHITMER, CANDACE D., Psychology
WOOD, MICHAEL L., Mathematics
WYVRATT, MATTHEW J., Chemistry
ZAKIAN, VIRGINIA A., Biology, General

Postdoctoral

ARONS, JONATHAN, Astronomy
JENKINS, JERRY A., Chemistry
KRISTAN, WILLIAM B. JR., Physiology
MACKENZIE, DAVID R., Botany
MYERS, JUDITH H., Biology, General
RODEWALD, RICHARD D., Biology, General
SNYDER, ALLAN W., Biophysics
TOTON, EDWARD T., Physics

Senior Postdoctoral

GREEN, PAUL B., Botany
MALLORY, FRANK B., Chemistry
RASMUSSEN, HOWARD, Biochemistry

Science Faculty

EISENSTEIN, BRUCE A., Engineering
ENGELSON, IRVING, Engineering
FAGOT, WILFRED C., Mathematics
HARTLEY, HAROLD M., Physics
MILLER, THOMAS G., Chemistry
MOSER, JOSEPH G., Mathematics
OSTLING, EDWARD G., Mathematics
PARKER, GARY E., Biology, General
RANCK, JOHN P., Chemistry
ROY, NINA M., Mathematics
SOYSTER, ALLEN L., Engineering
TREXLER, JOHN P., Earth Sciences
WESCHLER, SR. MARY C., Chemistry

PUERTO RICO

Graduate

D'EUSTACHIO, PETER G., Biochemistry
TOLEDO, DOMINGO, Mathematics

Science Faculty

ACOSTA, CANDIDA R., Microbiology
VELEZDESANTIAGO, MARCEL, Chemistry

RHODE ISLAND

Graduate

AIKIN, JOHN O., Psychology
BARRY, ROBERTA G., Linguistics

FELLOWSHIP AND TRAINEESHIP AWARDS

Basener, Richard F., Mathematics
Benedek, Roy, Physics
Cooper, Robert A., Chemistry
Magee, Francis R. Jr., Engineering
McNamee, Paul E., Chemistry
Wing, Thomas G., Psychology

Postdoctoral

Hollins, Mark, Psychology

Science Faculty

Camp, David S., Psychology

SOUTH CAROLINA

Graduate

Coss, Harold T., Zoology
Gaver, Kenneth M., Economics
Jeffcoat, Robert L., Engineering
Killingsworth, Robert B., Economics
Maynard, James B., Earth Sciences
McCollum, William N., Chemistry
O'Bryan, Nelson B. Jr., Chemistry
Rice, Richard W., Engineering
Zucker, Charles L., Mathematics

Postdoctoral

Leff, Herbert L., Psychology

Science Faculty

Ellison, William L., Biology, General
Harris, Jerry L., Physics

SOUTH DAKOTA

Graduate

Elder, Robert T., Biochemistry
Elder, Vincent A., Chemistry
Kuhlman, Myron I., Engineering
Powers, Dale R., Chemistry
Schrader, Elaine C., Chemistry
Sivertson, Larry M., Engineering
Snyder, Lee R., Medical Sciences

Science Faculty

Johnson, Leland G., Physiology
Yocom, Kenneth L., Mathematics

TENNESSEE

Graduate

Arabie, Clay P., Psychology
Beavers, Alex N. Jr., Engineering
Bennett, Betsy D., Pathology
Bray, James C., Engineering
Garrison, Sidney C., Mathematics
Gottlieb, Dorma J., Genetics
Jones, Richard R., Chemistry
Lightman, Alan P., Physics
Rainey, Petrie M., Chemistry
Robinson, Frank P., IV, Binlogy, General
Thoennes, David J., Chemistry

Postdoctoral

Hammond, Allen L., Atmospheric Sciences

Science Faculty

Callis, Charles P., Physics
Herrell, Astor Y., Chemistry
Kinloch, John, Mathematics
Turner, Barbara H., Biology, General
Young, David P., History and Philosophy of Science

TEXAS

Graduate

Baldwin, Otha D., Earth Sciences
Barnes, Robert H., Anthropology and Archeology
Baumgardner, John R., Engineering
Berscheidt, Tim R., Political Science
Clarke, Mildred A., Chemistry
Clarke, Thomas C., Chemistry
Cohen, James S., Physics
Cozzens, William A., Social Science
Creech, David L., Agriculture
Curry, Stephen M., Physics
Dahlberg, Kenneth E., Mathematics
Danna, Kathleen J., Biology, General
Danziger, Sheldon H., Economics
Dipboye, Robert L., Psychology
Doyle, Lynn C., Biology, General
Drinnan, Carol L., Biochemistry
Dunne, George C., Earth Sciences
Eanes, Robert S., Engineering
Follstaedt, David M., Physics
Gierisch, Bobby M., Sociology
Hamilton, Karen K., Mathematics
Hammerstrom, Thomas M., Mathematics
Hankamer, Jorge E., Linguistics
Hollifield, Patrick H., Linguistics
Hughes, Henry G. III, Physics
James, John A., Economics
Jordan, George S., Mathematics
Jordon, Lawrence M., Chemistry
Kazen, Phyllis M., Sociology
Ladner, Robert C., Chemistry
Leeman, William P., Earth Sciences
Loewenstein, Patricia L., Chemistry
Logue, John A., Political Science
Luke, Ronald T., Economics
Lynch, Paul M., Engineering
Mayer, James M., Chemistry
McClure, Suzanne, Genetics
McDonald, James D., Chemistry
McManigal, James G., Agriculture
Melton, Lynn A., Chemistry
Melton, William C., Economics
Messenger, Thomas J., Chemistry
Moore, Linda R., Biology, General
Norris, H. L., III, Engineering
Norton, Jack R., Chemistry
Oiesen, James F., Economics
Parrish, David K., Earth Sciences
Pitts, Jon T., Mathematics
Price, Thomas G., Jr., Engineering
Raschke, Erin C., Biochemistry
Raschke, William C., Biochemistry
Reeve, Scott C., Earth Sciences
Rylander, Henry G., Engineering
Smith, Jane A., Microbiology
Smith, Stephen D., Mathematics
Stern, Lawrence D., Economics
Streit, Gerald E., Chemistry
Tankersley, Lawrence L., Physics
Thorpe, James C., Mathematics
Udovic, Joseph D., Biology, General
Vaughn, William J., Mathematics
Vos, Robert G., Engineering
Wadsworth, Adrian R., Mathematics
Waide, Jack B., Biology, General
Warshaw, Stephen J., Biology, General
White, Gary Lynn, Biochemistry
Wilson, John F., Psychology
Winchester, Paul D., Earth Sciences
Wise, James D., Jr., Engineering
Ziegler, Regina G., Biochemistry

Postdoctoral

Ellison, John R., Genetics
Forristall, George Z., Engineering
Fuchs, James A., Biology, General
Greenberg, Harvey J., Engineering
Lagow, Richard J., Chemistry
McGill, Thomas C., Jr., Physics
Voelker, Robert A., Genetics
Weisheit, Jon C., Physics

Senior Postdoctoral

Cox, David J., Biochemistry

Science Faculty

Adams, Jasper E., Mathematics
Coleman, Max W., Mathematics
Dvoracek, Marvin J., Earth Sciences
Frye, Bernard L., Microbiology
Kesler, Oren B., Mathematics
Ramirez, Samuel A., Life Sciences
Read, David R., Mathematics

UTAH

Graduate

Ashby, Ned T., Political Science
Bargerton, Cecil B., Physics
Barrett, Wayne W., Mathematics
Christensen, Clark G., Astronomy
Crowley, Joan E., Psychology
Fitz, Franklin K., Biology, General
Folland, Gerald B., Mathematics
Gilchrist, Glen J., Economics
Israelsen, Lyle D., Economics
James, Laurence P., Earth Sciences
Linford, Rulon K., Engineering
Merrill, Hyde M., Engineering
Petersen, Harold C., Economics
Rawlings, Jill, Chemistry
Siirola, Jeffrey J., Engineering
Smith, Craig C., Engineering
Stacey, Dennis W., Biochemistry
Streeper, Richard D., Chemistry

Postdoctoral

MacKnight, Martha P., Biology, General
West, Neil E., Biology, General

Senior Postdoctoral

Williams, Max L., Jr., Engineering

Science Faculty

Johnson, Morris R., Engineering
Kimball, Richard N., Engineering
Kopp, Richard S., Earth Sciences

VERMONT

Graduate

Hanson, Susan E., Geography
Lubin, Adam, Mathematics
Shepardson, Wilfred B., Mathematics
Whitney, James A., Earth Sciences

Postdoctoral

Juhasz, Joseph B., Psychology

VIRGIN ISLANDS

Graduate

Kean, Orville E., Mathematics

VIRGINIA

Graduate

Agresta, Steven J., Economics
Bausum, David R., Mathematics
Blevins, Robert D., Engineering
Buchwalter, Stephen L., Chemistry
Carlson, William E., Mathematics
Collins, Francis S., Chemistry
Cromartie, Thomas H., Chemistry
Dieffenbach, Bruce C., Economics
Downey, Peter J., Mathematics
Gallant, Stephen I., Mathematics
Greenbaum, Hyman A., Mathematics
Grosser, Deborah A., Biology, General
Grossman, Jerrold W., Mathematics
Guess, Harry A., Mathematics
Hawley, Danny L., Physics
Heins, Marjorie A., History and Philosophy of Science
Henle, James M., Mathematics
Johnston, James C., III, Psychology
Joyner, William H., Jr., Mathematics
Kingston, George C., Engineering
Kurtz, David W., Physics
Larsen, Richard A., Chemistry
Lind, Douglas A., Mathematics
Lysy, Dusan G., Physics
Monk, Leonard G., Mathematics
Neu, Carl R., Economics
Petsko, Gregory A., Biology, General
Quick, Roy F., Jr., Engineering
Reedy, Christopher L., Mathematics
Sachs, Steven G., Social Science
Sassaman, Clay A., Zoology
Seccombe, Stephen D., Engineering
Shostak, Robert E., Mathematics
Silverman, Mark P., Chemistry
Suter, Glenn, W., II, Biology, General
Swanson, Claude V., Jr., Physics
Tait, Richard H., Physics

FELLOWSHIP AND TRAINEESHIP AWARDS

VANHOVER, KATHLEEN I., Psychology
WHEELER, GEORGE M., Engineering
WIKSWO, JOHN P., JR., Physics
WILLETT, JOHN C., Atmospheric Sciences
WILSON, WILLIAM K., Mathematics
WINN, JOHN S., Chemistry
WOBUS, RICHARD L., Atmospheric Sciences
ZEUL, CAROLYN R., Sociology

Postdoctoral

GAMMON, RICHARD H., Astronomy
HAMRICK, GARY C., Mathematics
JOHNSON, MIKKEL B., Physics

Science Faculty

BARNES, ANNIE S., Anthropology and Archeology
BARTLETT, CHARLES S., JR., Earth Sciences
EISS, NORMAN S., JR., Engineering
WETMORE, STANLEY I., JR., Chemistry

WASHINGTON

Graduate

ACHEN, CHRISTOPHER H., Political Science
ADOLPHSON, ALAN C., Mathematics
BALDWIN, KEITH G., Earth Sciences
BETHGE, PAUL H., Chemistry
BOOTH, DOUGLAS E., Economics
BOWERS, SUSAN K., Biology, General
BURTON, DORIS E., Economics
CAMPBELL, GREGORY L., Engineering
COHEN, EDWARD H., Biology, General
COKELET, EDWARD D., Earth Sciences
CUNNINGHAM, CHRISTOPHER, Physics
DAVIS, BRIAN C., Chemistry
DAVIS, DENNY C., Engineering
DICKSON, LAWRENCE J., Mathematics
DONALDSON, GREGORY J., Physics
DUFFY, CLARENCE J., Earth Sciences
EPTON, MICHAEL A., Mathematics
FEINBERG, JERRY M., Mathematics
FOWLER, BRADLEY C., Mathematics
GUMERMAN, RAYMOND J., Engineering
HARDWIDGE, EDWARD A., Chemistry
HARDY, JOHN T., Botany
HECHT, JULIA A., Anthropology and Archeology
HUDSON, MARGARET L., Botany
KAMMEYER, PETER C., Mathematics
KAUTZ, RICHARD L., Engineering
KING, ROBERT E., Anthropology and Archeology
KLEE, WENDY P., Linguistics
KOLBE, ANSFRID L., Economics
LAMB, MARY M., Zoology
LERWICK, STUART N., Mathematics
LINN, JOHN C., Engineering
LONG, PHILIP E., Earth Sciences
MACOMBER, WALTER G., Engineering
MAIZELS, NANCY M., Biophysics
MAJORS, JOHN E., Physics
MALCOLM, JOHN M., Mathematics
MARSTON, PHILIP L., Physics
MAXWELL, JEAN A., Anthropology and Archeology
MCCOLLOM, ROBERT L., Earth Sciences
MCCOY, JOHN W., Biology, General
MILES, FRANK B., Mathematics
MILTON, KIMBALL A., Physics
NELSON, RICHARD W., Economics
OSBORNE, MASON S., Mathematics
OSHEROFF, DOUGLAS D., Physics
OTTO, CHARLOTTE F., Chemistry
OTTO, JAMES R., JR., Mathematics
REECK, GERALD R., Biochemistry
SARI, SEPPO O., Physics
SHAW, GEORGE H., Earth Sciences
SHELTON, WENDY C., Psychology
SYRDAL, DANIEL D., Chemistry
TABER, DOUGLASS F., Chemistry
THOMPSON, NORMAN L., Urb. and Reg. Planning
THORN, JAMES V., Physics
VANCE, RICHARD R., Biology, General
WILLIAMSON, LORNA J., Chemistry
WOIROL, GREGORY R., Economics
WOODIN, SARAH A., Biology, General

Postdoctoral

GETHMANN, RICHARD C., Genetics
RADKE, CLAYTON J., Engineering
THOMPSON, WILLIAM F., Botany

Senior Postdoctoral

STERN, EDWARD A., Physics
VINCOW, GERSHON, Chemistry

Science Faculty

BLOW, RICHARD T. III, Engineering
MARTIN, GORDON W., Biochemistry
MONTZINGO, LLOYD J. Jr., Mathematics

WEST VIRGINIA

Graduate

ANDERSON, WILLIAM T., Engineering
PROCTOR, EUGENIA C., Psychology

Postdoctnral

DUBA, ALFRED G., Earth Sciences
FREDERICK, SUE E., Botany

WISCONSIN

Graduate

AIST, JAMES R., Botany
BERNDT, ERNST R., Economics
BLAEDEL, KENNETH L., Engineering
BOETTCHER, ROBERT J., Chemistry
BROWN, TERRENCE L., Physics
CAMPEN, JAMES T., Economics
CULLERS, ROBERT L., Earth Sciences
DANNER, FREDERICK W., Psychology
DEVEREAUX, WILLIAM R., Chemistry
DOSCH, DAVID L., Engineering
DUNCAN, DANA L., Anthropology and Archeology
EVERS, MARK, Sociology
FEHL, DAVID L., Physics
FOAT, LAWRENCE M., Mathematics
GILBERTSON, ROGER L., Earth Sciences
GLORVIGEN, BRADLEY W., Chemistry
GODSIL, JAMES J., Political Science
GREGORY, DAVID A., Anthropology and Archeology
GROVE, KENNETH W., Earth Sciences
HAHN, BRIAN K., Chemistry
HANSEN, EVERETT M., Botany
HERB, REBECCA A., Mathematics
HIBBARD, WILLIAM L., Mathematics
JACOBS, LOIS J., Genetics
KEATING, EDWARD L., Urban and Regional Planning
KEPPEL, ROBERT A., Chemistry
KINSINGER, JAMES A., Chemistry
KNOBLAUCH, TOM E., Mathematics
KRACHT, WILLIAM R., Chemistry
LINDERMAN, JOHN P., Mathematics
LISIAK, KENNETH P., Engineering
LORENZ, ROBERT D., Engineering
MANTEI, NED A., Biology, General
MARKOWSKI, GREGORY R., Engineering
MARSHALL DELMAR, Physics
MARTENS, SUSAN K., Mathematics
MARTINIAK, LEONARD J., Mathematics
MATZ, DAVID J., Engineering
MAYER, JEROME F., Engineering
MCCARTNEY, NANCY G., Biology, General
MEWALDT, RICHARD A., Physics
MILLER, JOANNE, Sociology
MORRIS, RICHARD V., Chemistry
NAPS, THOMAS L., Mathematics
NEWTON, JAMES D., Mathematics
NIRSCHL, JOSEPH P., Engineering
OTTENSMANN, JOHN R., Urban and Regional Planning
PEDERSON, MARY L., Zoology
PEET, ROBERT K., Biology, General
PILARSKI, LINDA M., Biochemistry
POTTS, MARY A., Biochemistry
PRAG, RAYMOND L., Agriculture
PROCTOR, ALAN E., Chemistry
ROEHRDANZ, RICHARD L., Biology, General
ROSSMAN, GEORGE R., Chemistry
SCANLON, WILLIAM J., Chemistry
SCHICK, THOMAS E., Economics
SCHLATTER, JAMES C., Engineering
SCHMIDT, PETER J., Economics
SHAPIRO, DANIEL B., Mathematics
SIME, GARY E., Mathematics
SIMS, JOHN E., Biology, General
SLEJKO, FRANK L., Chemistry
ST. AMANT FRANCIS C. Jr., Physics
STEPHENSON, JACK E., Engineering
STIEFEL, LEANNA, Economics
SUNQUIST, MARK L., Engineering
SUTTON, JAMES E., Biology, General
TYLER, JAMES V., Physics
WANG, HELEN P., Mathematics
WARNER, HAROLD R. Jr., Engineering
BRAY, JAMES C., Engineering
WEGENKE, ROLF W., Political Science
WHEELER, ROBERT L., Mathematics
WIERZBA, DENNIS P., Mathematics
WILHELMS, JANE P., Biology, General
WILSON, RICHARD H., Biology, General
WOLF, SUSAN W., Biology, General
WORLD, JAMES P., Genetics

Postdoctoral

KUELBS, JAMES D., Mathematics
MASSA, DENNIS J., Biochemistry
WEBB, THOMPSON III, Earth Sciences

Senior Postdoctoral

LEVIN, JACOB J., Mathematics
RANNEY, JOSEPH A., Political Science
SEQUEIRA, LUIS, Botany
SESHADRI, SENGADU R., Engineering
WHITLOCK, HOWARD W. Jr., Biology, General

Science Faculty

BUSBY, EDWARD O., Engineering
DEBOTH, GENE A., Mathematics
HOOVER, KENNETH R., Political Science
KUNDERT, KENNETH R., Mathematics
NELSON, DEAN P., Chemistry
SENGENBERGER, DAVID L., Earth Sciences
SUKOW, WAYNE W., Physics
TRENN, THADDEUS J., History and Philosophy of Science
WILSON, ANITA K., Agriculture

WYOMING

Graduate

AHLBRANDT, THOMAS S., Earth Sciences
ARBIC, BERNARD J., Mathematics
BOUGHN, STEPHEN P., Physics
CARPENTER, GAIL A., Mathematics
WHITNEY, ALAN R., Engineering
WILLIAMS, ERNEST H. Jr., Biology, General

Senior Postdoctoral

EDMISTON, CLYDE K., Chemistry

Science Faculty

RHODINE, CHARLES N., Engineering

FELLOWSHIP AND TRAINEESHIP AWARDS

Fellowship Awards by Program and Field, Fiscal Year 1970

	Total	Graduate	Postdoctoral	Senior Postdoctoral	Science Faculty	Senior Foreign Scientists
Engineering Sciences	327	262	9	6	44	6
Aeronautical	28	23	1	0	3	1
Agricultural	5	0	—	0	4	—
Ceramic	0	1	—	—	—	—
Chemical	50	49	1	—	0	—
Civil	15	7	1	0	7	—
Electrical	110	100	0	0	9	1
Electronics	29	24	0	3	2	—
Engineering Mechanics	13	12	1	0	0	—
Engineering Science	3	2	0	0	1	—
Hydraulic	2	—	—	0	1	1
Industrial	5	1	—	—	4	—
Materials	6	3	1	1	1	—
Mechanical	26	20	2	0	3	1
Metallurgical	2	2	0	0	0	—
Nuclear	11	9	0	0	1	1
Petroleum	0	—	—	0	—	—
Sanitary	0	0	—	—	0	0
Engineering, Other	22	9	2	2	8	1
Mathematical Sciences	552	470	10	4	60	8
Algebra or Number Theory	112	93	2	0	13	4
Analysis	106	89	1	1	13	2
Applications of Mathematics	46	40	1	0	5	0
Computer Science	76	70	0	0	6	—
Geometry	21	18	0	—	3	—
Logic or Foundations	33	31	0	—	2	—
Probability and Statistics	30	20	0	1	8	1
Topology	64	50	6	2	5	1
Mathematics, Other	64	59	0	0	5	0
Physical Sciences	965	791	64	25	56	29
Physical Sciences, General	—	0	—	—	0	—
Astronomy	43	36	3	1	2	1
Atmospheric Sciences, General	1	1	0	—	0	—
Meteorology	7	6	1	0	—	—
Chemistry	408	338	27	8	25	10
Earth Sciences	108	82	8	3	11	4
Physical Oceanography	15	13	1	1	0	—
Metallurgy	3	2	0	0	0	1
Physics	380	313	24	12	18	13
Subtotal, EMP Sciences	(1844)	(1523)	(83)	(35)	(160)	(43)
Life and Medical Sciences	627	508	63	13	30	13
Agriculture	9	7	0	0	1	1
Biochemistry	110	96	10	3	1	—
Biology, General	130	108	16	5	10	0
Biological Oceanography	21	20	1	0	0	—
Biophysics	34	27	4	0	0	3
Botany	48	36	6	4	1	1
Ecology	78	67	7	—	4	—
Genetics	36	29	5	0	1	1
Microbiology	30	19	5	0	6	0
Pathology	1	1	0	0	0	—
Physiology	41	32	6	0	3	—
Zoology	59	51	3	0	1	4
Life Sciences, Other	8	6	0	0	1	1
Medical Sciences	13	9	0	1	1	2
Social Sciences & Psychology	611	551	23	10	22	5
Anthropology	68	63	1	1	2	1
Archeology	22	21	0	0	1	—
Economics	144	135	3	3	2	1
Geography	6	4	0	0	2	—
History & Philosophy of Science	33	22	4	0	7	—
Linguistics	38	35	3	0	0	—
Political Science	51	46	2	1	1	1
Sociology	52	46	1	1	3	1
Social Sciences, Other	34	28	1	2	2	1
Psychology	163	151	8	2	2	0
Subtotal, Life & Social Sci.	(1238)	(1059)	(86)	(23)	(52)	(18)
Totals, Fellowship Programs	3082	2582	169	58	212	61

FELLOWSHIP AND TRAINEESHIP AWARDS

Graduate Traineeship Awards, Field Unspecified Fiscal Year 1970

	Total	Graduate	Postdoctoral	Senior Postdoctoral	Science Faculty	Senior Foreign Scientists
9- or 12-months Graduate Traineeships (for above fields)	5301					
Fiscal Year 1970	2075					
Continuations of:						
Fiscal Year 1969	889					
Fiscal Year 1968	1162					
Fiscal Year 1967	1175					
Summer Traineeships for Graduate Teaching Assistants, Fiscal Year 1970						
Fiscal Year 1970	938					
Subtotals	(6239)					
Grand Totals, Fellowship and Traineeship Programs	9321	2582	169	58	212	61

Name, Country, Field of Study, and U.S. Institution of Affiliation of Individuals Offered National Science Foundation Fellowships for Senior Foreign Scientists,* Fiscal Year 1970

ARGENTINA

FAVRET, EWALD A., *Genetics*, Washington State University

AUSTRALIA

DIESENDORF, WALTER, *Engineering*, Rensselaer Polytechnic Institute
OATES, WILLIAM A., *Metallurgy*, University of Vermont
SPICER, BRIAN, *Physics*, Catholic University of America
WATSON-MUNRO, CHARLES N., *Physics*, Texas Tech University
WILLIAMS, ARTHUR, *Engineering*, University of Miami

CZECHOSLOVAKIA

KUKLA, GEORGE, *Earth Sciences*, Columbia University
MASEK, JIRI, *Chemistry*, Stanford University
PRIBIL, RUDOLF, *Chemistry*, Oklahoma State University
ZAHRADNIK, RUDOLF, *Chemistry*, Louisiana State University in New Orleans

FINLAND

LEHTO, OLLI E., *Mathematics*, Case Western Reserve University

FRANCE

DELANGE, HUBERT, *Mathematics*, University of Illinois, Urbana
VERDIER, JEAN-LOUIS, *Mathematics*, Brandeis University

HUNGARY

BOKONYI, SANDOR, *Zoology*, University of California, Los Angeles

INDIA

KRISHNAN, RAPPAL S., *Physics*, North Texas State University
MURTY, C. R. K., *Physics*, University of Maine
PANDALAI, K. A. V., *Engineering*, George Washington University
SODHA, MAHENDRA S., *Physics*, Drexel University

ISRAEL

REICHAW, MEIR, *Mathematics*, Bowling Green State University
SHAMIR, JACOB, *Chemistry*, Memphis State University
WEINREB, ARYE, *Physics*, University of Notre Dame

ITALY

BARSOTTI, IACOPO, *Mathematics*, Yale University
MONTANARI, FERNANO, *Chemistry*, Southern Illinois University

JAPAN

AOKI, KEN-ICHIRO, *Earth Sciences*, University of New Mexico
IZAWA, KEISUKE**, *Engineering*, Stevens Institute of Technology
MORITA, TOHRU, *Physics*, Ohio University
NAKAJIMA, TAKESHI, *Chemistry*, North Dakota
NAKAMURA, TERUTARO, *Physics*, University of Idaho
SAKAI, AKIRA, *Botany*, University of Minnesota
TABATA, YONEHO, *Engineering*, University of Maryland

NETHERLANDS

BRUSSAARD, PIETER J., *Physics*, Duke University
DEJONG, FRITS J., *Economics*, Florida State University
KUPERUS, MAX, *Astronomy*, University of Wyoming
VOS, JOHANNES J., *Biophysics*, Indiana University

NORWAY

HAUGE, EIVING H., *Physics*, The Rockefeller University

POLAND

RACZKA, RICHARD, *Physics*, University of Colorado

SWEDEN

ALGVERE, KARL V., *Agriculture*, State University College of Forestry at Syracuse University
DEDIJER, STEVAN, *Sociology*, Dartmouth College
HELANDER, D. HERBERT, *Medical Science*, University of Alabama in Birmingham

SWITZERLAND

HUBER, PETER, *Mathematics*, Princeton University

UNITED KINGDOM

BARLTOP, JOHN A., *Chemistry*, University of Nevada
BLACK, DUNCAN, *Political Science*, Virginia Polytechnic Institute
EMELEUS, HARRY J., *Chemistry*, Marquette University
FYFE, WILLIAM S., *Earth Sciences*, The Pennsylvania State University
JOSEPHSON, BRIAN D., *Physics*, Cornell University
NEEDHAM, RODNEY, *Anthropology*, University of California, Riverside
OLLIS, W. DAVID, *Chemistry*, Wesleyan University
SWINNERTON-DYER, HENRY P. F., *Mathematics*, Harvard University
YOUNGSON, ALEXANDER J., *Social Sciences, other*, University of South Carolina

WEST GERMANY

BONHOEFFER, FRIEDERICH, *Biophysiis*, University of Utah
DINGHAS, ALEXANDER, *Mathematics*, Fordham University
HAHN, FRITZ P. K., *Medical Science*, Tulane University
HELLWINKEL, DIETER, *Chemistry*, Yeshiva University
JANDER, RUDOLF, *Zoology*, University of Kansas
KINNE, OTTO, *Sife Sciences, other*, Arizona State University
KOMNICK, HANS, *Zoology*, Colorado State University
MOSONYI, EMIL F., *Engineering*, University of Wisconsin at Milwaukee
MUULSCHLEGEL, BERNHARD, *Physics*, University of California, Santa Barbara
MULLER, ADOLF, *Biophysics*, Georgia State University
SCHERER, GERHARD, *Zoology*, South Dakota State University

U. S. S. R.

GORSHKOV, GEORGE, *Earth Sciences*, University of Oregon

*Fellowships are for tenure at U.S. Institutions.
**Deceased, prior to entering on tenure.

FELLOWSHIP AND TRAINEESHIP AWARDS

Institutions Granted Traineeships and Chosen by Fellowship Awardees, Fiscal Year 1970

UNITED STATES INSTITUTIONS	TOTALS	TRAINEESHIPS: Graduate (9- or 12-mos.) New	TRAINEESHIPS: Graduate (9- or 12-mos.) Continuation	TRAINEESHIPS: Graduate Teaching Assts. (Summer)	FELLOWSHIPS: Graduate	FELLOWSHIPS: Postdoctoral	FELLOWSHIPS: Senior Postdoctoral	FELLOWSHIPS: Science Faculty
ALABAMA								
Auburn University	20	7	8	5	—	—	—	—
University of Alabama	19	4	7	5	1	—	—	2
ALASKA								
University of Alaska	7	2	3	1	1	—	—	—
ARIZONA								
Arizona State University	36	11	17	7	1	—	—	—
University of Arizona	72	19	30	10	9	—	—	4
ARKANSAS								
University of Arkansas	20	4	7	4	2	—	—	3
CALIFORNIA								
California Institute of Technology	177	25	43	3	102	3	1	—
Center for Advanced Study	4	—	—	—	—	—	4	—
Claremont Graduate School	12	2	6	3	1	—	—	—
Loma Linda University	3	1	1	—	—	—	—	1
Stanford University	456	48	80	9	294	14	6	5
University of California, Berkeley	371	52	90	16	197	10	1	5
University of California, Davis	54	16	23	7	8	—	—	—
University of California, Irvine	21	7	6	2	5	—	—	1
University of California, Los Angeles	123	32	50	11	25	—	1	4
University of California, Riverside	40	12	15	4	8	—	—	1
University of California, San Diego	67	12	18	4	26	6	—	1
University of California, San Francisco	11	2	3	1	1	4	—	—
University of California, Santa Barbara	36	10	13	6	6	—	—	1
University of California, Santa Cruz	11	1	1	1	5	1	1	1
University of the Pacific	4	1	2	1	—	—	—	—
University of Santa Clara	6	2	2	1	—	—	—	1
University of Southern California	53	15	24	6	6	—	—	2
U.S. International University	2	1	—	—	1	—	—	—
COLORADO								
Colorado School of Mines	8	2	5	1	—	—	—	—
Colorado State University	38	12	17	6	1	—	—	2
National Center for Atmospheric Research	1	—	—	—	—	1	—	—
University of Colorado	77	20	32	10	10	1	—	4
University of Denver	10	2	5	2	1	—	—	—
CONNECTICUT								
University of Connecticut	39	12	18	7	2	—	—	—
Wesleyan University	13	3	5	2	1	—	—	2
Yale University	191	23	38	—	120	6	—	4
DELAWARE								
University of Delaware	32	11	16	4	1	—	—	—
DISTRICT OF COLUMBIA								
American University	12	2	5	4	1	—	—	—
Catholic University	27	10	14	3	—	—	—	—
George Washington University	24	8	9	4	2	—	—	—
Georgetown University	21	8	10	3	—	—	—	—
Howard University	9	2	6	1	—	—	—	—
Smithsonian Institution	1	—	—	—	—	1	—	—
FLORIDA								
Florida State University	46	15	22	5	3	—	—	1
Nova University	2	1	1	—	—	—	—	—
University of Florida	60	18	28	8	2	—	1	3
University of Miami	32	8	11	4	9	—	—	—
University of South Florida	6	1	1	2	2	—	—	—
GEORGIA								
Atlanta University	4	1	2	1	—	—	—	—
Emory University	22	6	9	2	4	—	—	1
Georgia Institute of Technology	50	17	27	5	1	—	—	—
Georgia State College	4	1	2	1	—	—	—	—
Medical College of Georgia	2	1	1	—	—	—	—	—
University of Georgia	38	12	15	7	2	—	—	2
HAWAII								
University of Hawaii	35	10	13	6	5	1	—	—
IDAHO								
University of Idaho	13	4	6	2	—	—	—	1
ILLINOIS								
Argonne National Laboratory	1	—	—	—	—	—	—	1
DePaul University	5	1	2	1	1	—	—	—
Eastern Illinois University	1	—	—	—	—	—	—	1
Illinois Institute of Technology	41	14	20	4	2	—	—	1
Illinois State University	9	2	5	2	—	—	—	—
Loyola University	9	2	5	2	—	—	—	—
Northern Illinois University	10	2	5	3	—	—	—	—
Northwestern University	95	28	46	7	11	1	—	2
Southern Illinois University	22	—	9	6	—	—	—	—
University of Chicago	157	23	38	6	85	2	1	2
University of Illinois	201	81	48	18	52	—	—	2
University of Illinois/Chicago Circle	9	1	1	2	4	—	—	1
University of Illinois/Medical Center	1	1	—	—	—	—	—	—
INDIANA								
Ball State University	1	—		—	—	—	—	1
Indiana University	79	21	32	11	11	1	—	3
Purdue University	151	43	70	17	15	1	—	5
University of Notre Dame	41	13	22	4	2	—	—	—
IOWA								
Iowa State University	69	19	29	11	9	—	—	1
University of Iowa	53	16	24	9	3	—	—	1
KANSAS								
Kansas State University	38	12	17	5	1	—	—	3
University of Kansas	48	13	20	8	5	—	—	2

FELLOWSHIP AND TRAINEESHIP AWARDS

Institutions Granted Traineeships and Chosen by Fellowship Awardees, Fiscal Year 1970—Continued

	TOTALS	TRAINEESHIPS: Graduate (9– or New	TRAINEESHIPS: Graduate (9– or Continuatinn	TRAINEESHIPS: Graduate Teaching Assistants (Sum-	FELLOWSHIPS: Graduate	FELLOWSHIPS: Post-uoctoral	FELLOWSHIPS: Senior Post-doctoral	FELLOWSHIPS: Science Faculty
KENTUCKY								
University of Kentucky	30	9	13	6	—	—	—	2
University of Louisville	10	3	5	2	—	—	—	—
LOUISIANA								
Lousiana Polytechnic Institute	4	1	1	2	—	—	—	—
Louisiana State University	39	12	17	7	—	1	—	2
Louisiana State University/Medical Center	3	1	1	1	—	—	—	—
Louisiana State University in New Orleans	6	2	3	1	—	—	—	—
Loyola University	2	1	—	1	—	—	—	—
Tulane University	29	10	12	3	3	—	—	1
MAINE								
University of Maine	10	2	5	2	1	—	—	—
MARYLAND								
Johns Hopkins University	74	17	26	5	24	1	—	1
University of Maryland	56	14	23	12	6	1	—	—
MASSACHUSETTS								
Boston College	12	4	5	2	1	—	—	—
Boston University	23	7	8	5	1	—	—	2
Brandeis University	36	7	12	3	13	1	—	—
Children's Medical Center	1	—	—	—	—	1	—	—
Clark University	15	4	6	1	2	—	—	2
Harvard University	429	19	33	9	349	14	3	2
Lowell Technical Institute	4	1	2	1	—	—	—	—
Massachusetts Institute of Technology	449	52	89	12	288	6	—	2
Northeastern University	30	10	13	5	2	—	—	—
Tufts University	18		7	2	1	1	—	—
University of Massachusetts	37	12	17	7	—	—	—	1
Woods Hole Oceanographic Institution	1	—	—	—	—	—	—	1
Worcester Polytechnic Institute	9	2	3	2	—	—	—	2
MICHIGAN								
Michigan State University	103	27	44	14	17	1	—	—
Michigan Technological University	12	4	6	2	—	—	—	—
University of Detroit	9	2	5	1	—	—	—	1
University of Michigan	234	49	84	17	76	4	—	4
Wayne State University	42	12	18	8	3	—	—	1
Western Michigan University	7	2	3	2	—	—	—	—
MINNESOTA								
University of Minnesota	112	31	51	13	17	—	—	—
University of Minnesota/St. Paul	1	—	—	—	1	—	—	—
MISSISSIPPI								
Mississippi State University	14	4	6	4	—	—	—	—
University of Mississippi	10	3	5	2	—	—	—	—
University of Southern Mississippi	4	1	2	1	—	—	—	—
MISSOURI								
St. Louis University	14	3	6	4	1	—	—	—
University of Missouri, Columbia	57	17	26	9	3	—	—	2
University of Missouri, Kansas City	6	1	3	2	—	—	—	—
University of Missouri, Rolla	29	8	12	5	1	—	—	3
Washington University	58	17	26	5	6	2	—	2
MONTANA								
Montana State University	17	7	8	2	—	—	—	—
University of Montana	9	2	3	2	2	—	—	—
NEBRASKA								
Creighton University	1	—	—	—	1	—	—	—
University of Nebraska	31	8	13	7	3	—	—	—
NEVADA								
University of Nevada	10	2	5	2	1	—	—	—
NEW HAMPSHIRE								
Dartmouth College	15	4	6	1	4	—	—	—
University of New Hampshire	22	8	9	4	—	—	—	1
NEW JERSEY								
Institute for Advanced Study	5	—	—	—	—	5	—	—
Newark Colleges	1	—	—	—	1	—	—	—
Newark College of Engineering	6	2	3	1	—	—	—	—
Princeton University	204	21	36	4	136	5	—	—
Rutgers, The State University	51	16	23	9	2	1	—	2
Seton Hall University	3	1	1	1	—	—	—	—
Stevens Institute of Technology	25	8	12	4	—	—	—	—
NEW MEXICO								
New Mexico Institute of M&T	4	1	2	1	—	—	—	1
New Mexico State University	20	7	8	3	—	—	—	—
University of New Mexico	27	9	12	5	1	—	1	2
NEW YORK								
Adelphi University	5	1	3	1	—	—	—	—
Alfred University	3	1	2	—	—	—	—	—
Brooklyn College	1	—	—	—	1	—	—	—
City University of New York	43	13	20	8	2	—	—	—
Clarkson College of Technology	17	5	8	2	2	—	—	—
Columbia University	141	31	53	7	47	3	—	—
Cooper Union	2	1	1	—	—	—	—	—
Cornell University	212	3	56	13	103	1	—	—
Fordham University	17	4	6	3	2	—	—	5
New School for Social Research	10	2	5	3	—	—	—	2
New York Medical College	5	2	2	1	—	—	—	—
New York University	88	23	37	8	16	—	—	—
Polytechnic Institute of Brooklyn	49	15	25	5	2	—	—	4
Rensselaer Polytechnic Institute	46	15	22	6	2	—	—	2
Rockefeller University	26	1	1	—	24	—	—	1
St. Bonaventure University	3	1	1	1	—	—	—	—
St. John's University	10	2	5	2	1	—	—	—

FELLOWSHIP AND TRAINEESHIP AWARDS

Institutions Granted Traineeships and Chosen by Fellowship Awardees, Fiscal Year 1970—Continued

	TOTALS	TRAINEESHIPS			FELLOWSHIPS			
		Graduate (9- or 12-mos.)		Graduate Teaching Assts. (Summer)				
		New	Continuation		Graduate	Post-doctoral	Senior Post-doctoral	Science Faculty
State University of New York:								
SUNY at Albany	11	4	6	1	—	—	—	—
College of Forestry at Syracuse	3	1	1	—	—	—	—	1
Downstate Medical Center	8	2	3	2	—	—	—	1
SUNY at Binghamton	6	1	2	2	1	—	—	—
SUNY at Buffalo	46	14	21	7	3	—	—	1
SUNY at Stony Brook	25	7	8	4	5	1	—	—
Upstate Medical Center	4	1	1	1	—	1	—	—
Syracuse University	53	18	26	8	1	—	—	—
Union College and University	2	1	—	1	—	—	—	—
University of Rochester	52	17	26	5	3	—	—	1
Yeshiva University	10	3	5	—	—	—	—	2
NORTH CAROLINA								
Duke University	46	12	18	4	10	1	—	1
North Carolina State University at Raleigh	60	20	30	7	2	—	—	1
University of North Carolina	53	16	22	7	7	—	—	1
Wake Forest University	9	3	5	1	—	—	—	—
NORTH DAKOTA								
University of North Dakota	9	2	5	1	1	—	—	—
North Dakota State University	12	4	6	2	—	—	—	—
OHIO								
Bowling Green State University	10	2	5	3	—	—	—	—
Case Western Reserve University	82	22	39	6	11	1	—	3
Kent State University	14	4	6	4	—	—	—	—
Miami University	4	1	1	2	—	—	—	—
Ohio State University	103	30	48	17	5	—	—	3
Ohio University	17	4	7	6	—	—	—	—
University of Akron	6	1	3	2	—	—	—	—
University of Cincinnati	29	9	13	6	—	—	—	1
University of Dayton	3	1	—	1	—	—	—	1
University of Toledo	7	2	3	2	—	—	—	—
OKLAHOMA								
Oklahoma State University	46	14	20	6	1	—	—	5
University of Oklahoma	38	11	17	7	1	—	—	2
OREGON								
Oregon Graduate Center	1	1	—	—	—	—	—	—
Oregon State University	45	12	19	7	3	—	—	4
Portland State University	4	1	1	2	—	—	—	—
University of Oregon	46	12	16	6	10	1	—	1
University of Portland	2	1	1	—	—	—	—	—
PENNSYLVANIA								
Bryn Mawr College	9	2	3	1	1	—	—	2
Carnegie-Mellon University	67	18	32	5	11	—	—	1
Drexel Institute of Technology	14	4	6	3	1	—	—	—
Duquesne University	3	1	1	1	—	—	—	—
Hahnemann Medical College and Hospital	3	1	1	1	—	—	—	—
Lehigh University	35	10	16	4	3	—	—	2
Pennsylvania State University	107	32	49	11	14	—	—	1
Philadelphia College of Pharmacy and Science	3	1	1	1	—	—	—	—
Temple University	22	8	9	5	—	—	—	—
Thomas Jefferson University	2	1	—	1	—	—	—	—
University of Pennsylvania	110	26	43	8	1	1	—	1
University of Pittsburgh	55	16	23	8	1	—	—	1
Villanova University	5	1	3	1	—	—	—	—
Woman's Medical College	2	1	1	—	—	—	—	—
PUERTO RICO								
University of Puerto Rico	1	—	—	—	—	—	—	1
RHODE ISLAND								
Brown University	57	16	23	4	13	—	—	1
Providence College	3	1	1	1	—	—	—	—
University of Rhode Island	15	4	7	3	31	—	1	—
SOUTH CAROLINA								
Clemson University	26	10	12	3	7	—	—	—
Medical University of South Carolina	2	1	—	1	—	—	—	—
	16	4	7	5	—	—	—	—
SOUTH DAKOTA								
South Dàkota School of Mines	4	1	2	1	—	—	—	—
South Dakota State University	9	3	5	1	13	—	—	—
University of South Dakota	9	2	5	2	—	—	—	—
TENNESSEE								
George Peabody College	3	1	2	—	—	—	—	—
Memphis State University	7	1	3	3	—	—	—	—
Oak Ridge National Laboratory	1	—	—	—	—	—	—	—
University of Tennessee	50	16	23	8	2	—	—	1
University of Tennessee/Memphis	2	1	—	1	—	—	—	—
Vanderbilt University	37	12	17	4	1	—	—	3
TEXAS								
Baylor University	7	2	3	1	—	—	—	—
Baylor University/Medical College	3	1	2	—	1	—	—	—
Lamar State College of Technology	1	—	—	—	1	—	—	—
North Texas State University	9	2	5	2	—	—	—	—
Rice University	61	12	21	3	24	—	—	1
Southern Methodist University	19	7	8	4	—	—	—	—
Texas A and M University	40	11	18	7	2	—	—	2
Texas Christian University	6	2	3	1	—	—	—	—
Texas Tech University	18	4	7	6	—	—	—	1
Texas Woman's University	3	1	1	1	—	—	1	—
University of Houston	25	8	9	5	1	—	—	2
University of Texas/Arlington	2	1	—	1	—	—	—	—
University of Texas/Austin	94	29	47	11	6	1	—	—

FELLOWSHIP AND TRAINEESHIP AWARDS

Institutions Granted Traineeships and Chosen by Fellowship Awardees, Fiscal Year 1970—Continued

	TOTALS	TRAINEESHIPS: Graduate (9- or 12-mos.) New	TRAINEESHIPS: Graduate (9- or 12-mos.) Continuation	TRAINEESHIPS: Graduate Teaching Assts. (Summer)	FELLOWSHIPS: Graduate	FELLOWSHIPS: Post-doctoral	FELLOWSHIPS: Senior Post-doctoral	FELLOWSHIPS: Science Faculty
UTAH								
Brigham Young University	17	4	7	6	—	—	—	—
University of Utah	46	14	22	6	2	1	—	1
Utah State University	27	8	11	4	2	—	—	
VERMONT								
University of Vermont	11	3	6	2	—	—		—
VIRGINIA								
College of William and Mary	9	3	5	1	—	—		—
University of Virginia	36	12	17	5	—	—		1
Virginia Commonwealth University	4	1	2	1	—	—		—
Virginia Polytechnic Institute	33	11	16	5	—	1		—
WASHINGTON								
University of Washington	136	32	49	11	36	3	—	5
Washington State University	29	9	12		—	—		1
WEST VIRGINIA								
West Virginia University		8	10	6	—	—		1
WISCONSIN								
Institute of Paper Chemistry	3	1	2	—	—	—	—	—
Marquette University	14	4	7	2	1	—	—	—
University of Wisconsin, Madison	212	30	64	17	86	2	—	4
University of Wisconsin, Milwaukee	20	7	7	4	2	—	—	—
WYOMING								
University of Wyoming	17	4	7	3	2	—		1
TOTAL AT U.S. INSTITUTIONS	9,123	2,075	3,226	938	2,552	115	22	195
TOTAL CHOOSING FOREIGN INSTITUTIONS (See following pages)	137	—	—	—	30	54	36	17
GRAND TOTAL	9,260	2,075	3,226	938	2,582	169	58	212

FELLOWSHIP AND TRAINEESHIP AWARDS

Foreign Institutions Chosen by Fellowship Awardees, Fiscal Year 1970

COUNTRY AND INSTITUTION	TOTALS	Graduate	Post-doctoral	Senior Post-doctoral	Science Faculty
AFRICA					
Fourah Bay College	1	—	1	—	—
University of Khartoum	1	—	1	—	—
AUSTRALIA					
University of Adelaide	2	1	—	1	—
Australian National University	1	1	—	—	—
University of Sidney	1	—	1	—	—
AUSTRIA					
Vienna University	1	—	1	—	—
BELGIUM					
Catholic University of Louvain	1	—	—	1	—
Royal Observatory of Belgium	1	—	—	1	—
BERMUDA					
Bermuda Biological Station	1	—	—	1	—
CANADA					
National Research Council	1	—	—	—	1
University of British Columbia	2	2	—	—	—
University of Toronto	1	—	—	—	1
DENMARK					
Aarus University	1	—	1	—	—
University of Copenhagen	4	—	3	—	1
FINLAND					
Institute of Technology	1	—	—	1	—
FRANCE					
Institute of Physio-Chemical Biology	1	—	—	1	—
University of Paris	3	—	—	2	1
GERMANY					
Albert Ludwig University at Freiburg	1	—	1	—	—
Deutsches Elektronen-Synchroton	1	—	1	—	—
Free University of Berlin	1	—	—	1	—
Institute for Plasma Physics	1	—	1	—	—
Rhenish Westphalian Technical University	1	1	—	—	—
Technical Institute at Munich	1	—	—	1	—
University of Konstanz	1	—	1	—	—
GUATEMALA					
Institute of Nutrition of Central America and Panama	1	1	—	—	—
ISRAEL					
Hebrew University	2	—	2	—	—
Israel Institute of Technology	2	—	1	1	—
ITALY					
International Centre for Theoretical Physics	1	—	1	—	—
THE NETHERLANDS					
National Defense Organization	1	—	1	—	—
State University of Leyden	3	1	1	—	1
NEW ZEALAND					
University of Otago	1	—	1	—	—
Victoria University of Wellington	1	1	—	—	—
SWEDEN					
Gothenburg University	1	—	—	1	—
Mittag-Leflar Institute	1	—	—	—	1
Royal Caroline Institute of Medicine and Surgery	1	—	1	—	—
SWITZERLAND					
CERN	4	—	4	—	—
Swiss Federal Institute of Technology	2	1	1	—	—
University of Basel	2	—	—	2	—
University of Geneva	2	—	—	2	—
University of Zurich	1	—	1	—	—
World Health Organization	1	—	—	1	—
UNION OF SOVIET SOCIALIST REPUBLICS					
Computing Center, USSR Academy of Sciences	1	—	—	—	—
Institute for High Energy Physics	1	—	1	1	—
UNITED KINGDOM					
Medical Research Council	2	1	1	—	—
Mullard Observatory	1	—	—	1	—
National Institute for Medical Research	1	—	—	1	—
National Institute of Oceanography	1	—	1	—	—
Queen's University of Belfast	1	—	1	—	—
University of Bristol	3	—	3	—	—
University of Cambridge	21	6	7	6	2
University of Durham	1	—	1	—	—
University of East Anglia	1	—	—	—	1
University of Edinburgh	2	1	—	1	—
University of Glasgow	3	1	1	—	1
University of Leeds	1	—	—	1	—
University of Leicester	1	—	1	—	—
University of London	15	5	3	4	3
University of Lund	1	—	1	—	—
University of Oxford	12	4	4	2	2
University of Reading	2	—	—	2	—
University of Southhampton	1	—	1	—	—
University of Stirling	1	—	1	—	—
University of Sussex	3	2	—	—	1
University of Warwick	1	1	—	—	—
University College of Swansea	1	—	—	—	1
YUGOSLAVIA					
University of Ljubljana	1	—	1	—	—
TOTAL CHOOSING FOREIGN INSTITUTIONS	137	30	54	36	17

FELLOWSHIP AND TRAINEESHIP AWARDS

Present or Most Recent Institutional Affiliation of Individuals Offered Science Faculty, Senior Postdoctoral, and Postdoctoral Fellowships During Fiscal Year 1970

U.S. INSTITUTIONS	Science Faculty	Senior Post-doctoral	Post-doctoral
ALABAMA			
Alabama State University	1	—	—
Auburn University	1	—	—
Tuskegee Institute	1	—	—
University of Alabama	1	—	—
ARIZONA			
Arizona State University	1	—	—
Northern Arizona University	1	—	—
ARKANSAS			
John Brown University	1	—	—
University of Arkansas	1	—	—
CALIFORNIA			
California Institute of Technology	—	—	2
University of California, Los Angeles	1	—	—
California State Polytechnic Institute	1	—	—
Chapman College	1	—	—
Citrus Junior College	1	—	—
College of San Mateo	1	—	—
El Camino College	1	—	—
Foothill College	1	—	—
Fresno State Cnllege	1	—	—
Harvey Mudd College	2	—	—
Hughes Research Laboratories	—	1	—
Los Angeles City College	1	—	—
Los Angeles Valley College	1	—	—
Loyola University, Los Angeles	1	—	—
Pomona College	1	—	—
Rio Hondo Junior College	1	—	—
San Diego State College	1	—	—
San Jose State College	1	—	—
Stanford University	—	3	8
University of California, Berkeley	—	7	9
University of California, Davis	1	2	2
University of California, Los Angeles	—	1	8
University of California, Riverside	—	1	3
University of California, San Diego	—	1	—
University of California, Santa Barbara	—	1	—
University of Pacific	1	—	—
University of San Francisco	1	—	—
COLORADO			
Colorado College	1	—	—
University of Northern Colorado	1	—	—
Fort Lewis College	1	—	—
University of Colorado	1	—	1
CONNECTICUT			
Fairfield University	1	—	—
New Haven College	1	—	—
Sacred Heart University	1	—	—
Southern Connecticut State Colege	1	—	—
Yale University	—	1	6
DISTRICT OF COLUMBIA			
George Washington University	1	—	1
Howard University	—	—	—
FLORIDA			
Florida State University	—	1	—
Florida Tech University	2	—	—
New College	1	—	—
College of Orlando	1	—	—
University of Florida	—	—	1
University of Miami	—	—	1
GEORGIA			
Gainesville Junior College	1	—	—
Georgia State University	2	—	—
University of Georgia	1	—	1
West Georgia College	1	—	—
HAWAII			
University of Hawaii	1	—	—
IDAHO			
Idaho State University	1	—	—
University of Idaho	1	—	—
ILLINOIS			
Eastern Illinois University	1	—	—
Illinois State University, Normal	1	—	—
Northern Illinois University	1	—	—
Northwestern University	—	1	2
Principia College	2	—	—
Roosevelt University	1	—	—
Southeastern Illinois College	1	—	—
University of Chicago	—	3	8
University of Illinois, Urbana	—	1	3
Western Illinois	1	—	—
Wheaton College, Illinois	1	—	—
INDIANA			
Ball State University	1	—	—
Earlham College	1	—	—
Franklin College of Indiana	1	—	—
Indiana University	—	—	1
Purdue University	1	1	2
Tri-State College	1	—	—
University of Notre Dame	—	1	1
IOWA			
Loras College	1	—	—
University of Iowa	1	—	1
Wartburg College	1	—	—
KANSAS			
College of Emporia	1	—	—
Kansas State College, Pittsburg	1	—	—
St. Benedict's College of Kansas	1	—	—
University of Kansas	—	—	1
Wichita State University	1	—	—
KENTUCKY			
Berea College	1	—	—
Eastern Kentucky University	1	—	—
Thomas More College	1	—	—
University of Kentucky	1	—	—
LOUISIANA			
Centenary College, Louisiana	1	—	—
Northeast Louisiana State College	1	—	—
Southern University	1	—	—
Tulane University of Louisiana	2	—	1
Xavier University, Louisiana	1	—	—
MARYLAND			
Johns Hopkins University	—	—	3
U.S. Naval Academy	1	—	—
MASSACHUSETTS			
Brandeis University	—	—	1
Child's Medical Center, Boston	—	—	1
Harvard University	—	—	14
Lowell Technological Institute	2	—	—
Massachusetts Institute of Technology	—	—	6
Mt. Holyoke College	1	—	—
North Shore Community College	1	—	—
University of Massachusetts	3	—	2
Wellesley College	1	—	—
MICHIGAN			
Adrian College	1	—	—
Albion College	1	—	—
Calvin College	1	—	—
Eastern Michigan University	1	—	—
Michigan State University	—	1	—
University of Detroit	1	—	—
University of Michigan	—	4	6
Wayne State University	1	—	—
Western Michigan University	1	—	—
MINNESOTA			
Carleton College	2	—	—
College Station Scholastica	1	—	—
St. Olaf College	2	—	—
University of Minnesota/Minneapolis	—	—	1
MISSISSIPPI			
Mary Holmes Junior College	1	—	—
University of Mississippi	1	—	—
William Carey College	1	—	—
MISSOURI			
Culver-Stockton College	1	—	—
Metropolitan Junior College	1	—	—
Northwest Missouri State College	1	—	—
Southwest Baptist College	1	—	—
Stephens College	1	—	—
Southwest Missouri State College	1	—	—
University of Missouri, St. Louis	1	—	—
NEBRASKA			
Chadron State College	1	—	—
NEW HAMPSHIRE			
Nathaniel Hawthorne College	1	—	—
NEW JERSEY			
Drew University	1	—	—
Institute for Advanced Study	—	—	1
Mercer County Community College	1	—	—
Monmouth College of New Jersey	1	—	—
Montclair State College	1	—	—
Newark College of Engineering	1	—	—
Princeton University	—	1	3
Seton Hall University	1	—	—
NEW MEXICO			
Eastern New Mexico University	1	—	—
NEW YORK			
Colgate University	1	—	—
Columbia University	—	1	5
Cornell University	1	4	2
Herkimer County Community College	1	—	—
Hofstra University	1	—	—
Long Island University	1	—	—
Manhattan College	1	—	—
Mohawk Valley Community College	1	—	—
Nassau Community College	1	—	—
Pace College	1	—	—
Rensselaer Polytechnic Institute	—	1	1
Rochester Institute of Technology	2	—	—
Rockefeller University	—	—	3
Skidmore College	1	—	—

FELLOWSHIP AND TRAINEESHIP AWARDS

Present or Most Recent Institutional Affiliation of Individuals Offered Science Faculty, Senior Postdoctoral, and Postdoctoral Fellowships During Fiscal Year 1970—Continued

U.S. INSTITUTIONS	Science Faculty	Senior Post-doctoral	Post-doctoral
NEW YORK—(Continued)			
Staten Island Community College	1	—	—
SUNY at Albany	1	—	—
SUNY at Buffalo	—	—	1
SUNY at Stony Brook	1	1	—
SUNY, College at Fredonia	1	—	—
SUNY, College at New Paltz	1	—	—
Syracuse University	1	—	1
University of Rochester	—	2	—
Utica College	1	—	—
Yeshiva University	—	—	1
NORTH CAROLINA			
Catawba College	1	—	—
Duke University	—	—	2
Meredith College	1	—	—
North Carolina State University,	1	—	1
Raleigh	—	1	2
University of North Carolina,			
Chapel Hill	1	—	—
NORTH DAKOTA			
North Dakota State University	—	—	—
OHIO			
Case Western Reserve	1	—	3
Cedarville College	1	—	—
Ohio State University	1	—	—
Hiram College	1	—	—
Oberlin College	1	—	—
Ohio Dominican College	1	—	—
Ohio Northern University	1	—	—
Ohio State University	—	—	—
Ohio University	2	—	1
University of Cincinnati	—	—	—
OKLAHOMA			
East Central State College	1	—	—
Northeastern State College	1	—	—
Oklahoma Panhandle State College	1	—	—
Oklahoma State University	1	—	—
Phillips University	1	—	—
Southwestern State College of Oklahoma	1	—	—
OREGON			
Oregon College of Education	2	—	—
Oregon State University	1	1	1
University of Oregon	—	1	2
PENNSYLVANIA			
Beaver College	1	—	—
Bryn Mawr College	—	1	—
Carnegie-Mellon University	—	—	1
Drexel Institute of Technology	1	—	—
Eastern Baptist College of Penn	1	—	—
Elizabethtown College	1	—	—
Juniata College	2	—	—
Lafayette College	2	—	—
Mercyhurst College	1	—	—
Penn State University	1	—	1
University of Pennsylvania	—	2	4
University of Pittsburgh	—	—	1
Villanova University	1	—	—
Waynesburg College	1	—	—
West Chester State College	1	—	—
PUERTO RICO			
Catholic University of Puerto Rico	1	—	—
University of Puerto Rico	1	—	—
RHODE ISLAND			
Brown University	—	—	1
University of Rhode Island	1	—	—
SOUTH CAROLINA			
Erskine College	1	—	—
SOUTH DAKOTA			
Augustana College of South Dakota	1	—	—
South Dakota State University	1	—	—
TENNESSEE			
East Tennessee State University	1	—	—
Geo. Peabody College for Teachers	1	—	—
Maryville College of Tennessee	1	—	—
Knoxville College	1	—	—
University of Tennessee, Martin	1	—	—
TEXAS			
Lamar State College of Technology	1	—	—
Rice University	—	—	3
Sam Houston State University	1	—	—
Southern Methodist University	—	—	1
Stephen F. Austin State University	1	—	—
Texas A&M University	—	—	1
Texas Tech University	1	—	—
University of Texas, Arlington	1	—	—
University of Texas, Austin	1	1	2
University of Texas, El Paso	1	—	—
UTAH			
College of Southern Utah	1	—	—
University of Utah	—	1	1
Utah State University of Agriculture	—	—	—
and Applied Science	1	—	1
Westminster College of Utah	1	—	—
VERMONT			
Bennington College	—	—	1
VIRGINIA			
Emory and Henry College	1	—	—
Hampton Institute	1	—	—
University of Virginia	—	—	1
Virginia Millitary Institute	1	—	—
Virginia Polytechnic Institute	1	—	—
WASHINGTON			
Seattle Community College	1	—	—
Seattle Pacific College	2	—	—
University of Washington	—	2	2
WISCONSIN			
St. Norbert College	2	—	—
Stout State University	1	—	—
University of Wisconsin, Madison	—	5	8
Wisconsin State University, Eau Claire	1	—	—
Wisconsin State University, Plattville	2	—	—
Wisconsin State University, River Falls	1	—	—
Wisconsin State Univ., Stevens Point	1	—	—
Wisconsin State University, Whitewater	1	—	—
WYOMING			
University of Wyoming	1	1	—
U. S. GOVERNMENT			
Department of HEW	—	—	2
Natick Laboratories, Massachusetts	—	—	1
Naval Medical Research Institute	—	—	1
U. S. Geological Survey, Menlo Park	—	1	—
FOREIGN INSTITUTIONS			
CANADA			
University of Toronto, Canada	—	1	1
DENMARK			
Copenhagen, Denmark	—	—	1
ENGLAND			
University of Bristol	—	—	1
University of Cambridge	—	—	1
University of Durham	—	—	1
University of London	—	—	1
SCOTLAND			
University of Glasgow, Scotland	—	—	1
SWITZERLAND			
CERN	—	—	1
	212	58	169

SOURCE: NSF, "National Science Foundation Grants and Awards 1970," 1971.

Part 3

PERSONNEL, ADVISORY COMMITTEES AND PANELS

ORGANIZATION

NATIONAL SCIENCE FOUNDATION

NATIONAL SCIENCE BOARD

DIRECTOR

DEPUTY DIRECTOR

GENERAL COUNSEL

OFFICE OF GOVERNMENT & PUBLIC PROGRAMS

EXECUTIVE COUNCIL
Secretariat

- OFFICE OF ECONOMIC, MANPOWER, AND SPECIAL STUDIES
  - OFFICE OF ECONOMIC AND MANPOWER STUDIES
  - OFFICE OF SPECIAL STUDIES
- ASSISTANT DIRECTOR FOR INSTITUTIONAL PROGRAMS
  - DIVISION OF INSTITUTIONAL DEVELOPMENT
  - DIVISION OF INSTITUTIONAL RESOURCES
- ASSISTANT DIRECTOR FOR EDUCATION
  - DIVISION OF GRADUATE EDUCATION IN SCIENCE
  - DIVISION OF UNDERGRADUATE EDUCATION IN SCIENCE
  - DIVISION OF PRE-COLLEGE EDUCATION IN SCIENCE
- ASSISTANT DIRECTOR FOR NATIONAL AND INTERNATIONAL PROGRAMS
  - OFFICE OF SEA GRANT PROGRAMS
  - OFFICE OF POLAR PROGRAMS
  - OFFICE OF INTERNATIONAL PROGRAMS
  - OFFICE OF NATIONAL CENTERS & FACILITIES OPERATIONS
  - OFFICE OF COMPUTING ACTIVITIES
  - OFFICE OF SCIENCE INFORMATION SERVICE
  - OFFICE FOR INTNAT'L DECADE OF OCEAN EXPLORATION
  - OFFICE OF INTERGOVERNMENTAL SCIENCE PROGRAMS
- ASSISTANT DIRECTOR FOR RESEARCH
  - DIVISION OF BIOLOGICAL & MEDICAL SCIENCES
  - DIVISION OF ENGINEERING
  - DIVISION OF ENVIRONMENTAL SCIENCES
  - DIVISION OF MATHEMATICAL & PHYSICAL SCIENCES
  - DIVISION OF SOCIAL SCIENCES
  - OFFICE OF INTERDISCIPLINARY RESEARCH
- ASSISTANT DIRECTOR FOR ADMINISTRATION
  - OFFICE OF BUDGET, PROGRAMMING, AND ANALYSIS
  - GRANTS & CONTRACTS OFFICE
  - AUDIT OFFICE
  - MANAGEMENT ANALYSIS OFFICE
  - ADMINISTRATIVE SERVICES OFFICE
  - FINANCIAL MANAGEMENT OFFICE
  - PERSONNEL OFFICE
  - DATA MANAGEMENT SYSTEMS OFFICE
  - PROGRAM REVIEW OFFICE
  - HEALTH SERVICE

NATIONAL SCIENCE FOUNDATION

1800 G STREET, N.W.

WASHINGTON, D.C. 20550

DIRECT-IN-DIALING 63 AND EXTENSION

AREA CODE 202

INFORMATION: 655-4000

NSF Cable Address: NATSCIFOUN, Washington, D.C.

NSF RCA Telex Call Number: 24521-NASCF UR

NSF Western Union Call Letters: MBP

OFFICE OF THE DIRECTOR

Director, William D. McElroy Rm 520 ... 24001
Deputy Director, Raymond L. Bisplinghoff Rm 520 ... 24376
Special Assistant, Donald E. Cunningham Rm 518 ... 24384
Special Assistant, Richard A. Edwards Rm 518 ... 24368
Special Assistant, Lawton M. Hartman, III Rm 505 ... 25778
Special Assistant, David E. Ryer Rm 519 ... 24394
Administrative Assistant, Mrs. Lois J. Hamaty Rm 520 ... 24363
Personal Secretary, Miss Doris McCarn Rm 520 ... 24001
Secretary, Mrs. Mary E. Bundy Rm 520 ... 24376
Secretary, Mrs. Maydie O. Hughes Rm 520 ... 24001
Special Services Coordinator, Bradley Phillips Rm 523 ... 24047

Executive Secretary, Executive Council, Douglas L. Brooks Rm 504 ... 24010
Deputy Executive Secretary, Executive Council, Bodo Bartocha Rm 504 ... 25762
Secretary, National Science Board, Miss Vernice Anderson Rm 546 ... 25840

OFFICE OF THE GENERAL COUNSEL

General Counsel, William J. Hoff Rm 501 ... 24386
Deputy General Counsel, Charles F. Brown Rm 501 ... 24388
Deputy General Counsel Charles Maechling, Jr. Rm 501 ... 24398

Assistant Counsel:
John B. Farmakidies Rm 501 ... 24393
Arthur J. Kusinski Rm 501 ... 24396
Mrs. Maryann B. Lloyd Rm 501 ... 24397

OFFICE OF GOVERNMENT AND PUBLIC PROGRAMS

Director, Clarence C. Ohlke Rm 526 ... 27320
Deputy for Government Liaison, Theodore W. Wirths Rm 526 ... 27320
Deputy for Public Programs, Edward R. Trapnell Rm 526 ... 27320
Institution Relations Officer, George E. Arnstein Rm 526 ... 27320
Special Assistant, Alfred Rosenthal Rm 527 ... 27320
Administrative Assistant, Mrs. Telula E. Eddins Rm 526 ... 25780

CONGRESSIONAL LIAISON OFFICE

Head (Acting), Theodore W. Wirths Rm 526 ... 27320
Congressional Liaison Specialist, Richard E. Stephens Rm 526 ... 27320
Congressional Liaison Specialist, Miss Elizabeth S. Hunt Rm 526 ... 25780

PUBLIC AFFAIRS OFFICE

Head (Vacant) Rm 527 ... 24100
Public Information Officer, Roland D. Paine, Jr. Rm 527 ... 24100
Public Service Officer, Nathan Kassack Rm 527 ... 25722
Press Officer, Jack Renirie Rm 527 ... 25728

PUBLIC UNDERSTANDING OF SCIENCE OFFICE

Head, Albert H. Rosenthal Rm 526 ... 27320

PUBLICATIONS RESOURCE OFFICE

Head, Jack Kratchman Rm 543 ... 27390

INSTITUTIONAL PROGRAMS

Assistant Director, Louis Levin Rm 516 ... 27304
Special Assistant, Mrs. Mildred C. Allen Rm 516 ... 27304
Deputy Assistant Director, Howard E. Page Rm 420 ... 24342
Executive Assistant, J. Merton England Rm 420 ... 24364
Special Assistant for Evaluation, Denzel D. Smith Rm 420 ... 24032
Administrative Officer, Mrs. Lena R. Stratton Rm 420 ... 24370

DIVISION OF INSTITUTIONAL DEVELOPMENT

Division Director, William V. Consolazio Rm 419 ... 24345
Deputy Division Director, Joshua M. Leise Rm 418 ... 24380
Special Assistant, Joseph F. Carrabino Rm 419 ... 24060

Program Management:
George W. Baker Rm 417 ... 24345
Richard A. Carrigam Rm 414 ... 24355
Paul G. Cheatham Rm 414 ... 24355
Bernard Chern Rm 417 ... 24355
Jerome S. Daen Rm 414 ... 24355
Richard H. Hall Rm 414 ... 24355
John F. Lance Rm 415 ... 24355
Ralph H. Long, Jr. Rm 417 ... 24355
George E. Brosseau, Jr. Rm415 ... 24355

DIVISION OF INSTITUTIONAL RESOURCES

Acting Division Director, Harold Horowitz Rm 421 ... 24061
Deputy Division Director, Harold Horowitz Rm 421 ... 24061
Special Assistant, Elmer G. Havens Rm 421 ... 24061

Institutional Planning and Evaluation:
Harold A. Spuhler Rm 423 ... 27367
George A. Livingston Rm 423 ... 27367

Institutional Grants for Science:
Miss Patricia E. Nicely Rm 412 ... 24360
Joseph G. Danek Rm 412 ... 24360

Facilities Evaluation:
Lloyd O. Herwig Rm 411 ... 27364
Frederic A. Leonard Rm 411 ... 27364
S. A. Heider Rm 410 ... 27364
Albert P. Bregida Rm 410 ... 27364
Caldwell N. Dugan Rm 410 ... 27364

EDUCATION

Assistant Director, Lloyd G. Humphreys Rm 506 24174
- Deputy Assistant Director, Thomas D. Fontaine Rm 620 25888
- Executive Assistant to the Deputy, Keith R. Kelson Rm 620 25868
- Special Assistant, Albert T. Young Rm 620 25898
- Special Assistant, Mrs. Senta A. Raizen Rm 506 24176
- Administrative Officer, Mrs. Frances O. Watts Rm 620 25891

DIVISION OF GRADUATE EDUCATION IN SCIENCE

Division Director, Howard D. Kramer Rm 616 .... **25730**
Deputy Division Director, Francis G. O'Brien Rm 618 .... **25734**
Administrative Officer, Miss Honora F. Thompson Rm 614 .... **25743**

Advanced Science Education Program

Acting Program Director, Mrs. Alice P. Withrow Rm 610 25747
- **Associate Program Directors:**
  - **Miss M.** Joan Callanan Rm 609 .... **25748**
  - Terence L. Porter Rm 610 25738

Graduate Fellowships and Traineeships Program

Program Director, Douglas S. Chapin Rm 612 .... **25732**

Faculty and Postdoctoral Fellowships Program

Program Director, Hall Taylor Rm 616 .... **25700**

Senior Fellowships Program

Program Director, Mrs. Marjory R. Benedict Rm 616 .... **25745**

Program Management Unit

Head, Mrs. Margaret Covey Thompson Rm 614 .... **25757**

DIVISION OF UNDERGRADUATE EDUCATION IN SCIENCE

Division Director, Lyle W. Phillips Rm 623 .... **25928**
- **Deputy Division Director,** Alfred F. Borg Rm 623 .... **25929**
- **Professional Associate** (Program Coordinator) **Lafe R. Edmunds** Rm 623 .... **25924**
- **Staff Assistant, Mrs.** Margaret C. Thompson Rm 623 .... **25902**

Pre-Service Teacher Education Program

Program Director, Donald C. McGuire Rm 623 25987

College Teacher Program

Program Director, Reinhard L. Korgen Rm 627 .... **25904**
- **Associate Program Director,** Arthur C. Hoffman Rm 627 .... **25905**
- **Associate Program Director,** (Vacant) Rm 627 25903

Undergraduate Instructional Programs

Program Director, Alexander J. Barton Rm 625 25911
- **Associate Program Director (URP),** Leo A. Sciuchetti Rm 625 25915
- **Associate Program Director (UISE),** Harold Zallen Rm 625 25913
- **Associate Program Director,** Charles H. Dickens Rm 625 25915
- Associate Program Director, Gerald A. Edwards Rm 625 .... [illegible]

EDUCATION *(Continued)*

College Science Curriculum Improvement Program

Acting Program Director, Leo Baggerly Rm 627 .... **25908**
- **Associate Program Director,** John L. Snyder Rm 627 .... **25909**
- Associate Program Director, Robert J. Toft Rm 627 25909
- **Assistant Program Director,** Gregg Edwards Rm 627 .... **25914**

College Science Improvement Programs

Program Director, James C. Kellett, Jr. Rm 625 .... **25916**
- **Associate Program Director,** William H. Adams Rm 625 .... **25919**
- **Associate Program Director,** Robert F. Watson Rm 625 .... **25917**

DIVISION OF PRE-COLLEGE EDUCATION IN SCIENCE

Division Director, Charles A. Whitmer Rm 601 25726
- Senior Staff Associate, Howard J. Hausman Rm 601 25720
- **Staff Assistant, Mrs.** Phyllis L. Johnson Rm 601 .... **25957**
- **Professional Assistant,** Lowell G. Kraegel Rm 601 .... **25980**
- **Administrative Officer,** Mrs. Wilda G. Tehaan Rm 601 .... **25716**

Summer Study Program

Program Director, William E. Morrell Rm 649 .... **25978**
- **Associate** Program Director and Coordinator for Foreign Activities, **Lewis A. Gist** Rm 649 .... **25954**
- Associate Program Director, Samuel W. Harding Rm 649 25955

Academic Year Study Program

Program Director, Michael M. Frodyma Rm 607 25970
- Associate Program Directors:
  - Theodore L. Reid Rm 607 25958
  - Alphonse Buccino Rm 607 25959

Pre-College Course Content Improvement Program

Program Director, Laurence O. Binder Rm 605 .... **25878**
- Associate Program Director: Raymond J. Hannapel Rm 605 .... **25879**
- Professional Assistants: Mrs. Jean B. Intermaggio Rm 605 .... **25865**
  - Miss Mary M. Kohlerman Rm 605 .... **25864**

Student and Cooperative Program

Program Director (Acting), Walter L. Gillespie Rm 605 25867
- **Associate Program Directors:**
  - Robert B. Garrabrant Rm 607 25887
  - Glen H. Cannell Rm 605 25875
  - Robert H. Harvey Rm 605 25875

DIVISION OF SCIENCE RESOURCES AND POLICY STUDIES

Division Director, Charles E. Falk Rm 551 25770
- **Special Assistant,** Vera L. Klingsberg Rm 551 .... **27334**

Office of Economic and Manpower Studies

Head, Thomas J. Mills Rm 401 24164
Deputy Head, Leonard Lederman Rm 401 24350
Chief Statistician, Sidney A. Jaffe Rm 409 24170
Staff Associate, Mrs. Kathryn S. Arnow Rm 452 24166

Editorial and Inquiries Unit
Editor, Lionel C. Bischoff Rm 403 24143

STATISTICAL SURVEYS AND REPORTS SECTION

Head, Kenneth Sanow Rm 451 24172

Government Studies Group
Study Director, Benjamin L. Olsen Rm 641 24178

Universities and Nonprofit Institutions Studies Group
Study Director, Joseph H. Schuster Rm 448 24080

Industry Studies Group
Study Director, Thomas J. Hogan Rm 405-A 24058

Analytical Studies Group
Study Director, Lawrence Seymour Rm 447 24077

SPONSORED SURVEYS AND STUDIES SECTION

Head, Robert W. Cain Rm 407 24334
Staff Associate, Zola Bronson Rm 402 24024

Scientific Manpower Studies Group
Study Director, Norman Seltzer Rm 404 24390

Science Education Studies Group
Study Director, Justin C. Lewis Rm 408 24324

National Register Group
Study Director, Milton Levine Rm 406 24160

Special Analysis Group
Study Director, (Vacant) Rm 409 24352
Staff Associate, James J. Zwolenik Rm 641 24150

Office of Policy Studies

Acting Head, Charles E. Falk Rm 551 25770

NATIONAL AND INTERNATIONAL PROGRAMS

Assistant Director, Thomas B. Owen Rm 510 27300
Executive Assistant, Richard J. Green Rm 510 27302
Deputy Assistant Director, T. O. Jones Rm 703 24180
Acting Special Assistant, H. S. Francis, Jr. Rm 703 24316
Special Assistant, G. R. Toney Rm 703 24238
Administrative Assistant, Mrs. Charlotte Morgan Rm 702 27359

NATIONAL OCEANOGRAPHIC LABORATORY SYSTEM

Project Officer, Miss Mary K. Johrde Rm 339 24202

OFFICE FOR THE INTERNATIONAL DECADE OF OCEAN EXPLORATION

Head, Feenan D. Jennings Rm 701 27356
Special Assistant, John R. Twiss, Jr. Rm 701 27357

OFFICE OF POLAR PROGRAMS

Acting Head, Louis O. Quam Rm 340 24221
Deputy Head, Philip M. Smith Rm 340 24235
Arctic Interagency Coordinator, Price Lewis, Jr. Rm 340 24258
Administrative Assistant, Lorraine P. Curtin Rm 340 27362

Science Programs
Program Manager, Biological Sciences, George A. Llano Rm 311 24194
Program Manager, Atmospheric Sciences, Ray R. Heer, Jr. Rm 340 24246
Program Manager, Earth Sciences, Mort D. Turner Rm 340 24256
Polar Information Service
Director, Kurt G. Sandved Rm 345 24220
Technical Director (vacant), Rm 345 24239

Field Operations Program
Program Manager, Kendall N. Moulton Rm 340 24247
Associate Program Manager, William T. Austin Rm 340 24206
Associate Program Manager—International and Cartographic Affairs,
Walter R. Seelig Rm 340 24250
NSF Representative, New Zealand, Richard L. Penney Rm 340 24255
NSF Representative, Antarctica, D. Christopher Shepherd Rm 340 24255
Manager, Antarctic Vessel Operations, Merle R. Dawson Rm 340 24203
Special Projects Officer, Robert L. Dale Rm 340 24257
Polar Engineering Officer, Jerry W. Huffman Rm 340 24207

OFFICE OF COMPUTING ACTIVITIES

Head, John R. Pasta Rm 710 25960
Special Assistant, Christof N. Schubert Rm 710 25960

Computer Science and Engineering Section
Head, Kent Curtis Rm 707 27346

Theoretical Computer Science Program
Program Director (Acting), Kent Curtis Rm 707 27346

Software and Programming Systems Program
Program Director, Thomas A. Keenan Rm 707 27346

Computer Systems Design Program
Program Director, John R. Lehmann Rm 707 27346

Computer Innovation in Education Section
Head, Arthur S. Melmed Rm 707 25790

Computer Technology and Systems Program
Program Directer (Acting), Arthur S. Melmed Rm 707 25790

Computer-Oriented Curricular Activities Program
Program Director, Andrew R. Molnar Rm 707 25790

Regional Cooperative Computing Activities Program
Program Director, Lawrence H. Oliver Rm 707 25790

Computer Applications in Research Section
Head, D. Don Aufenkamp Rm 707 25985

Special Research Resources Program
Program Director (Vacant) Rm 707 25985

Techniques and Systems Program
Program Director (Vacant) Rm 707 25985

OFFICE OF INTERNATIONAL PROGRAMS

Head, Arthur Roe Rm 536 .... **25798**
Deputy Head, Ernest R. Sohns Rm 536 25814
Administrative Assositant, Miss Mildred Bosilevac Rm 536 25805

Cooperative Science Activities

U.S.-India Program
Program Director, Gordon Hiebert Rm 532 .... **25792**

U.S.-Japan Program and U.S.-Republic of China Program
Program Director, J. E. O'Connell Rm 532 .... **25782**
Staff Associate, Richard R. Ries Rm 532 25806

U.S.-Australia Program and U.S.-Italy Program
Staff Associate, R. R. Ronkin Rm 532 25807

U.S.-Romania Program, U.S.-France Program, and U.S.-USSR/East European Program
Professional Associate, Robert Hull Rm 532 25813

Scientific Liaison

Professional Associates:
Duncan Clement Rm 532 25811
Robert Hull Rm 532 25813
Warren Thompson Rm 532 25829

NSF/New Delhi
NSF Science Liaison Staff
USAID New Delhi
c/o Agency for International Development
Washington, D. C. 20523

Head, Max Hellman
Staff Associates:
E. A. Ashby
William Blanpied
John McGreal
Harold Huneke

NSF/Tokyo
Science Liaison Staff
National Science Foundation
c/o American Embassy
APO San Francisco 96503
Head, Henry Birnbaum
Staff Associate, Arthur F. Findeis

OFFICE OF NATIONAL CENTERS AND FACILITIES OPERATIONS

Head, Daniel Hunt Rm 706 25717
Administrative Assistant, Mrs. Opal A. Chidester Rm 706 25719

Ocean Sediment Coring Program
Project Officer, Joe D. Sides Rm 706 27339
Field Project Officer, A. R. McLerran (FTS) A.C. 714, 293-5493

National Centers for Research Astronomy
Project Officer, Ronald R. LaCount Rm 706 25712

National Center for Atmospheric Research
Project Officer, Giorgio Tesi Rm 706 27340

OFFICE OF INTERGOVERNMENTAL SCIENCE PROGRAMS

Head, M. Frank Hersman Rm 536 25768

Local Government Program
Acting Program Manager, Miss Delores A. Gregory Rm 536 25768

OFFICE OF SCIENCE INFORMATION SERVICE

Annex 1

(2430 Pennsylvania Avenue, N.W.)

Head, Burton W. Adkinson Rm 203 25824
Deputy Head, Henry J. Dubester Rm 203 25826
Administrative Officer, Mrs. Dorothy B. W. Blumenthal Rm 203 25835

Staff Coordination Group
Senior Staff Associate William S. Barker Rm 203 25826

Research and University Information Systems Program
Program Director Edward C. Weiss Rm 213 25818
Associate Program Director Thomas W. Quigley, Jr. Rm 212 25844

Special Foreign Currency Program
Program Director Eugene Pronko Rm 215 25706
Associate Program Directors Charles Zalar Rm 216 25830
Robert F. Kan Rm 215 25706

Information Services Program
Program Director Gordon B. Ward Rm 220 25850

Information Systems Program
Program Director Harold E. Bamford, Jr. Rm 201 25816
Associate Program Directors Paul D. Olejar Rm 201 25800
John A. Scopino Rm 234 25800

RESEARCH

Assistant Director, Edward C. Creutz Rm 512 27342
Deputy Assistant Director, Edward P. Todd Rm 320 24240
Executive Assistant to the Deputy, Jerome H. Fregeau Rm 320 24248
Senior Staff Associate, Eugene L. Hess Rm 320 25988
Senior Staff Associate (Planning), Wayne R. Gruner Rm 320 24270
Special Assistant, Leonard F. Gardner Rm 320 24278
Administrative Officer, Edward W. Hilton Rm 320 24243

DIVISION OF ENVIRONMENTAL SCIENCES

Division Director, A. P. Crary Rm 339 24295

Head, Fred D. White Rm 312 24198

Aeronomy Program
Program Director (Acting), Neil M. Brice Rm 312 24185

Meteorology Program
Program Director (Acting), H. Frank Eden Rm 312 24190
Associate Program Director, Richard E. Orville Rm 312 24190

Solar Terrestrial Research Program
Program Director, Neil M. Brice Rm 312 24184

Weather Modification Program
Program Director, Peter H. Wyckoff Rm 312 24147

Interdepartmental Committee for Atmospheric Sciences
Executive Secretary, Sherman W. Betts Rm 312 24189

National Center for Atmospheric Research
Scientific Coordinator, Glenn E. Stout Rm 312 24189

Earth Sciences Section

Head, William E. Benson Rm 310 24210

Geochemistry Program
Program Director, Alvin Van Valkenburg Rm 310 24213

Geology Program
Program Director, Richard G. Ray Rm 310 24218

Geophysics Program
Program Director, David B. Slemmons Rm 311 24219

Oceanography Section

Head, P. Kilho Park Rm 346 24227

Physical Oceanography Program
Program Director, W. Bruce McAlister Rm 346 24236

Biological Oceanography Program
Program Director, Dirk Frankenberg Rm 346 24236
Assistant Program Director, Miss Jean T. DeBell Rm 346 25858

Submarine Geology and Geophysics Program
Program Director, A. Conrad Neumann Rm 346 24215

DIVISION OF BIOLOGICAL AND MEDICAL SCIENCES

Division Director, Harve J. Carlson Rm 325 24338
Professional Assistant, Mrs. Wardella Doscher Rm 325 24337
Deputy Division Director, John W. Mehl Rm 325 24326
Planning Officer, William J. Riemer Rm 325 24332
Administrative Officer, Mrs. Dora J. Hruz Rm 325 24300

Cellular Biology Section

Head, Herman W. Lewis Rm 326 24200

Developmental Biology Program
Program Director, Richard W. Siegel Rm 326 24314
Assistant Program Director, Mrs. Anne H. Schauer Rm 326 24314

Program Director (Acting), Herman W. Lewis Rm 326 24200
Associate Program Director, Mrs. Mary Wolff Rm 326 24200

Ecology and Systematic Biology

Head, Walter H. Hodge Rm 330-332 27318

Ecosystem Analysis Program
Program Director, Charles F. Cooper Rm 330 25854
Associate Program Director, Mrs. Josephine K. Doherty Rm 330 25854

General Ecology Program
Program Director, John L. Brooks Rm 330 27324
Assistant Program Director, Mrs. Joan Jordan Rm 330 27324

Systematic Biology Program
Program Director (Vacant) Rm 332 25846
Associate Program Director, Loran Anderson Rm 332 25846
Associate Program Director, William E. Sievers Rm 332 25846
Assistant Program Director, Ray S. Birdsong Rm 332 25846

Molecular Biology Section

Head, Sigmund R. Suskind Rm 329 24260

Biochemistry Program
Program Director (Acting), Sigmund R. Suskind Rm 329 24260
Associate Program Director, Mrs. Estella K. Engel Rm 329 24260

Biophysics Program
Program Director, Eloise E. Clark Rm 329 24260
Assistant to the Program Director, Miss Brenda C. Flam Rm 329 24260

Physiological Processes Section

Head, David B. Tyler Rm 323 24298

Regulatory Biology Program
Program Director (Acting), David B. Tyler Rm 323 24298
Associate Program Director, Miss Jeanette Ruth Rm 323 24298

Metabolic Biology Program
Program Director, Elijah B. Romanoff Rm 323 24312
Associate Program Director, Miss Cecilia W. Spearing Rm 323 24312

Psychobiology Program

Program Director, Henry S. Odbert Rm 333 24264
Associate Program Director, James H. Brown Rm 333 24264
Assistant Program Director, Miss Mary Ann Sestili Rm 333 24264

DIVISION OF ENGINEERING

Division Director, John M. Ide Rm 336 ... 24280
Administrative Officer, Mrs. Beatrice Jenkins Rm 336 ... 24280

Engineering Chemistry Program
Program Director, Lewis G. Mayfield Rm 336 ... 24280
Associate Program Director (Vacant) Rm 336 ... 24280

Engineering Energetics Program
Program Director (Acting), R. E. Rostenbach Rm 336 ... 24280

Engineering Materials Program
Program Director, Israel Warshaw Rm 336 ... 24280
Associate Program Director, Robert J. Reynik, Rm 336 ... 24280

Engineering Mechanics Program
Program Director, Michael P. Gaus Rm 336 ... 24280
Associate Program Director, George K. Lea Rm 336 ... 24280
Associate Program Director, Charles A. Babendreier Rm 336 ... 24280
Assistant Program Director, Charles C. Thiel, Jr. Rm 336 ... 24280

Engineering Systems Program
Program Director, Gilbert B. Devey Rm 336 ... 24280
Associate Program Director, Elias Schutzman Rm 336 ... 24280

Special Programs
Program Director, Morris S. Ojalvo Rm 336 ... 24280

DIVISION OF SOCIAL SCIENCES

Division Director, Howard H. Hines Rm 316 ... 24286
Special Assistant, Mrs. Bertha W. Rubinstein Rm 316 ... 24289
Administrative Officer, Mrs. Dorothy S. Lang Rm 316 ... 24294

Anthropology Program
Program Director, Richard W. Lieban Rm 318 ... 24208
Associate Program Director, Mrs. Mary Greene Rm 317 ... 24208

Economics Program
Program Director, James H. Blackman Rm 317 ... 25968
Assistant Program Director, Miss Norma M. Katelvero Rm 317 ... 24104

Geography Program
Program Director, Howard H. Hines Rm 316 ... 24286
Assistant Program Director, Mrs. Bertha W. Rubinstein Rm 316 ... 24289

History and Philosophy of Science Program
Special Consultant, Dudley Shapere Rm 317 ... 24182
Assistant Program Director, Mrs. Mary Greene Rm 317 ... 24182

Political Science Program
Program Director, William A. Lucas Rm 319 ... 25714
Assistant Program Director, Mrs. Bertha W. Rubinstein Rm 316 ... 24289

Sociology and Social Psychology Program
Program Director, John C. Scott Rm 319 ... 24204
Assistant Program Director, Mrs. Kathleen L. Schwartzman Rm 319 ... 24204

DIVISION OF MATHEMATICAL AND PHYSICAL SCIENCES

Division Director, William E. Wright Rm 352 ... 24320
Deputy Division Director (vacant) Rm 352 ... 24320
Executive Assistant, Andrew W. Swago Rm 352 ... 24320
Administrative Officer, Charles W. Hines Rm 341 ... 24230

Astronomy Section

Head, Robert Fleischer Rm 351 ... 24196
Assistant Program Director, Miss Marjorie Williams Rm 351 ... 24186

National Astronomy Observatories
Scientific Coordinator, Gerald F. Anderson Rm 351 ... 27332

Solar System Astronomy Program
Program Director, Harold H. Lane Rm 301 ... 24186

Stars and Stella Evolution Program
Program Director (Acting), Robert Fleischer Rm 351 ... 24196

Stellar Systems and Motions Program
Program Director, Harold H. Lane Rm 301 ... 24186

Galactic and Extragalactic Astronomy Program
Program Director, James P. Wright Rm 351 ... 24192

Astronomical Instrumentation and Development Program
Program Director (Acting), Robert Fleischer Rm 351 ... 24196

Chemistry Section

Head, M. Kent Wilson Rm 348 ... 24262

Chemical Dynamics Program
Program Director, Donald A. Speer Rm 343 ... 24272

Chemical Instrumentation and Analysis Program
Program Director, Richard S. Nicholson Rm 348 ... 24276

Chemical Thermodynamics Program
Program Director (Vacant) Rm 343 ... 24272

Quantum Chemistry Program
Program Director, William H. Cramer Rm 348 ... 24266

Structural Chemistry Program
Program Director, O. William Adams Rm 348 ... 24266

Synthetic Chemistry Program
Program Director, Oren F. Williams Rm 348 ... 24276
Associate Program Director, John S. Showell Rm 348 ... 24276

Mathematical Sciences Section

Head, William H. Pell Rm 303 ... 27377

Program Directors:
Barnett R. Akins Rm 303 ... 27377
Ralph M. Krause Rm 303 ... 27377
William G. Rosen Rm 301 ... 27377

MATHEMATICAL AND PHYSICAL SCIENCES *(Continued)*

Physics Section

Head, Paul F. Donovan Rm 307 24310

Atomic, Molecular, and Plasma Physics Program
Program Director, Rolf M. Sinclair Rm 305 24317

Elementary Particle Physics Program
Program Director, J. Howard McMillen Rm 305 24317

Intermediate and High-Energy Physics Program
Program Director, Marcel Bardon Rm 305 24317

Nuclear Physics Program
Program Director, William S. Rodney Rm 305 24317

Solid State and Low Temperature Physics Program
Program Director, Howard W. Etzel Rm 307 24302

Theoretical Physics Program
Associate Program Director, Harold S. Zapolsky Rm 307 24302
Associate Program Director, Angelo Bardasis Rm 307 24300

OFFICE OF INTERDISCIPLINARY RESEARCH

Head, Joel A. Snow Rm 213 25863
Administrative Officer, Mrs. Marilyn J. Einhorn Rm 213 27350
Staff Associate, Joseph Coates Rm 213 27350
Staff Associate, James D. Cowhig Rm 213 27350
Staff Associate, Jesse C. Denton Rm 213 27350
Staff Associate, Mrs. Gladys G. Handy Rm 213 27350
Staff Associate, Richard C. Kolf Rm 213 27350
Staff Associate, Charles T. Owens Rm 213 27350
Staff Associate, Robert Rabin Rm 213 27350

ADMINISTRATION

Assistant Director, Bernard Sisco Rm 525 25710
Deputy Assistant Director, T. E. Jenkins Rm 525 25766
Special Assistant, Calvin C. Jones Rm 540 27315
Administrative Manager, F. C. Sheppard Rm 525 25760
Director of Equal Employment Opportunity, Howard S. Schilling Rm 220 25998

ADMINISTRATIVE SERVICES OFFICE

Administrative Services Officer, Howard Tihila Rm 220 24130
Deputy Administrative Services Officer, John T. Harrigan Rm 220 24130
Distribution Section, Joseph N. Newman Rm 233 24152
Library, Mrs. Frances M. Pentecost Rm 219 24070
Printing and Reproduction Section, John C. Holmes Rm 231 24116
Supply and Maintenance Section, T. E. Harrison Rm 237 24222
Travel Service Section, Miss Bonnie S. Hutchinson Rm 221 24140
Central Processing Section, Mrs. Carol Manuel (Acting) Rm 223 24120

AUDIT OFFICE

Audit Officer, Robert B. Boyden Rm 645 24072
Deputy, Warren J. Hynes Rm 645 24072
Senior Audit Manager, Francis E. Carlson (Inst. & Educ. Programs) Rm 645 24096
Senior Audit Manager, Louis Siegel (Liaison, Spec. Studies, & Admin. Activities) Rm 645 24148
Senior Audit Manager, James L. Stennett (Research, Nat'l. & Internat'l. Programs) Rm 645 24068
Audit Managers:
Ralph E. Casey Rm 645 24148
John G. Myers Rm 645 24088
John J. Patermaster Rm 645 24148

DATA MANAGEMENT SYSTEMS OFFICE

Data Management Systems Officer, E. W. Barrett Rm 437 25990
Deputy, R. W. H. Lee Rm 437 25990

Computer Services Section

Section Chief, David Staudt Rm 440 27380
Operations Group
Kenneth F. Bootes Rm 440 27380
I/O Control Group
(Vacant) Rm 440 27380
Programming & Systems Support Group
Head, Dennis Chin Rm 440 27380

Information Systems Section

Section Chief, Richard Robertson Rm 440 27371

FINANCIAL MANAGEMENT OFFICE

Financial Management Officer, Kenneth B. Foster Rm 432 24012
Deputy Financial Management Officer, Howard R. Copperman Rm 432 24366
Finncial Systems Staff, Thomas M. Ryan Rm 432 24374
Award Accounting Section, William Ward Rm 434 24006
General Accounting Section, Ralph Burke Rm 433-A 24038
Payroll and Voucher Examination Section, Mrs. Lorraine M. Gilbert Rm 436 24146

GRANTS AND CONTRACTS OFFICE

Grants and Contracts Officer, Wilbur W. Bolton, Jr. Rm 640 25772
Staff Assistant, William E. Fee, Jr. Rm 640 25920
Staff Assistant, William B. Cole, Jr. Rm 640 25920

Grants Branch

Grants Administrator, Gaylord L. Ellis Rm 640 25920
Special Assistant, Harry Hyman Rm 640 25920
Grants Administration Section
Head, George B. Bush, Jr. Rm 634 25950
Deputy Head, James L. Bostick Rm 634 25950

Staff Assistant, William K. Sprague, Jr. Rm 634 ... 25950
Grants Manager, Area #1, Charlotte R. Raymond Rm 638 ... 25900
Grants Manager, Area #2, Nicholas J. Dassoulas Rm 638 ... 25900
Grants Manager, Area #3, Eugene C. Stewart Rm 634 ... 25965
Grants Manager, Area #4, Francis Naughten Rm 634 ... 25965
Grants Manager, Area #5, Barbara V. Hyland Rm 634 ... 25938
Grants Manager, Area #6, Earl Anderson Rm 634 ... 25940

Records and Reports Section
Head, Richard E. Grabe Rm 633 ... 25926

Project Property Section
Head, Paul Ashby Rm 630 ... 24108

Contracts Branch

Contracts Administrator, Robert D. Newton Rm 630 ... 25872
Special Assistant, Robert A. Mitchelitch Rm 630 ... 25884
NCAR, KPNO, CTIO, Gerald A. Greenwood Rm 630 ... 25884
NRAD, AO, DSDP, OPP, Morris T. Phillips Rm 630 ... 25860
General Contracts, James L. Vitol Rm 630 ... 25892

HEALTH SERVICE

Director, James W. Long, M.D. Rm 439 ... 24041
Nurse-in-Charge, Mrs. Jean Westfall, R.N. Rm 439 ... 24040

MANAGEMENT ANALYSIS OFFICE

Management Analysis Officer, George Pilarinos Rm 549 ... 24110
Head, Management Surveys and Systems Section, Fred K. Murakami Rm 549 ... 24110
Head, Management Services Section, John E. Kirsch Rm 549 ... 24026
Forms Unit Rm 549 ... 24026

OFFICE OF BUDGET, PROGRAMMING, AND PLANNING ANALYSIS

Head, Arley T. Bever Rm 426 ... 25764

Budget Office

Head, Walton M. Hudson Rm 429 ... 24086
Deputy Head, Van A. Neely Rm 429 ... 25705
Budget Analysts:
George A. Barber Rm 429 ... 24084
James B. Geraus Rm 429 ... 25705
Barbara Saunders Rm 429 ... 24084
Special Projects Officer, Mrs. Nilda Gatbonton Rm 429 ... 24084

Programming Office

Head, Syl McNinch, Jr. Rm 428 ... 24098
Section Head, Research Programs, Lawrence Cohen Rm 428 ... 25876
Program Analysis Officer, Charles T. Henney Rm 428 ... 25876
Section Head, Nat'l & Int'l Programs, Oscar C. Vigen Rm 428 ... 24379
Program Analyst, G. Michael Montross Rm 428 ... 24379
Section Head, Educational and Institutional Programs,

Plans and Analysis Office

Head, Harry J. Piccariello Rm 425 ... 25774
Deputy Head (Vacant) Rm 425 ... 25774
Staff Associates:
William D. Cummins Rm 425 ... 25784
Donald E. Cunningham Rm 425 ... 25785
Emanuel Haynes Rm 425 ... 25777
Robert W. Lamson Rm 425 ... 25776
Mrs. Mary Parramore Rm 426 ... 24050
Mrs. Gloria S. Seeman Rm 425 ... 25793
Bernard R. Stein Rm 425 ... 25754

PERSONNEL OFFICE

Personnel Officer (Acting), George Pilarinos Rm 211 ... 24136
Employment and Position Management Section
E. Paul Broglio (OGC, AD/A) Rm 211 ... 24119
William M. Dorie (AD/E, AD/NI) Rm 212 ... 24124
Duncan M. McGregor (O/D, GPP, AD/R, AD/I) Rm 211 ... 24118
Personnel Management Assistance Section
Frederick Becker (O/D, GPP, AD/R, AD/I) Rm 212 ... 24090
Henry F. Blackner (AD/E, AD/NI) Rm 211 ... 24106
Mrs. Bessie B. White (OGC, AD/A) Rm 208 ... 24107

PROGRAM REVIEW OFFICE

Program Review Officer, Lewis P. Jones Rm 540 ... 27315
Chief, Visual Information, Emmett Lucas Rm 540 ... 27315
Visual Information Assistant, William E. Lewis Rm 540 ... 27315
Review Center Receptionist, Mrs. Marie M. McIntosh Rm 540 ... 27316

COMMERCE DEPARTMENT

NATIONAL OCEANIC AND ATMOSPHERE ADMINISTRATION

Office of Sea Grant Programs

Head, Robert B. Abel Rm 537 ... 25944
Deputy Head, Harold L. Goodwin Rm 537 ... 25948
Program Director, Arthur G. Alexious Rm 537 ... 25944
Program Director, Robert D. Wildman Rm 537 ... 25944

NATIONAL SCIENCE FOUNDATION ORGANIZATIONAL ABBREVIATIONS

Administrative Services Office	ASO
Assistant Director for Administration	AD/A
Assistant Director for Education	AD/E
Assistant Director for Institutional Programs	AD/I
Assistant Director for National & International Programs	AD/NI
Assistant Director for Research	AD/R
Audit Office	ADT
Biological & Medical Sciences, Division of	BMS
Budget, Programming, and Analysis, Office of	BPA
Computing Activities, Office of	OCA
Data Management Systems Office	DMS
Director, Office of the	O/D
Economic & Manpower Studies, Office of	EMS
Engineering, Division of	ENG
Environmental Sciences, Division of	DES
Financial Management Office	FMO
General Counsel, Office of	OGC
Government and Public Programs, Office of	GPP
Graduate Education in Science, Division of	GES
Grants & Contracts Office	GCO
Interdisciplinary Research, Office of	OIR
Intergovernmental Science Programs, Office of	ISP
International Decade of Ocean Exploration, Office for	IDOE
International Programs, Office of	OIP
Management Analysis Office	MAO
Mathematical & Physical Sciences, Division of	MPS
National Centers & Facilities Operation, Office of	NCF
Personnel Office	PER
Polar Programs, Office of	OPP
Policy Studies, Office of	OPS
Pre-College Education in Science, Division of	PES
Program Review Office	PRO
Science Information Service, Office of	SIS
Science Resources and Policy Studies, Division of	SRPS
Sea Grant Programs, Office of (NOAA)	OSG
Social Sciences, Division of	SOC
Undergraduate Education in Science, Division of	UES

Name	DIVISION	ROOM NO	EXT
A			
Abel, Robert B.	OSG	537	25944
Abell, Bruce R.	GPP	543	27390
Aborn, Murray	SOC	318	24216
Adams, Elizabeth B., Mrs.	GES	611	25756
Adams, O. William	MPS	348	24266
Adams, William H.	UES	625	25919
Adkinson, Burton W.	SIS	A1-203	25824
Agins, Barnett R.	MPS	303	27377
Alexander, Alisann, Miss	UES	625	25918
Alexander, James E.	ASO	237	24222
Alexiou, Arthur G.	OSG	537	25944
Allen, Martha J., Miss	OPP	340	24247
Allen, Mildred C., Mrs.	AD/I	516	27304
Alston, Barbara H., Mrs.	EMS	448	24081
Anderson, Betty L., Mrs.	AD/I	414	24355
Anderson, Earl K.	GCO	634	25940
Anderson, Gerald F.	MPS	351	27332
Andeson, Loran C.	BMS	332	25846
Anderson, R. Gail, Miss	MAO	548	24110
Anderson, Vernice, Miss	O/D	546	25840
Andrews, John H.	EMS	408	24331
Anholt, Gladys B., Mrs.	UES	627	25908
Archer, Thersa C., Mrs.	UES	625	25918
Arnow, Kathryn, Mrs.	EMS	452	24166
Arnstein, George E.	GPP	526	27320
Arthur, Donald G.	DMS	440	27371
Ashby, Paul	GCO	630	24108
Askins, Robert E.	DMS	440	27380
Asrael, Aaron R.	GCO	630	25892
Aughenbaugh, Charlotte M., Mrs.	AD/R	320	24278
Aufenkamp, Darrel D.	OCA	707	25985
Austin, William T.	OPP	340	24206
B			
Babendreier, Charles A.	ENG	336	24280
Baggerly, Leo L.	UES	627	25908
Bailey, Margaret R., Mrs.	AD/NI	510	27300
Baird, John B. P.	EMS	405-B	24336
Baisey, Evelyn, Mrs.	ASO	223	24120
Baker, George W.	AD/I	417	24355
Baker, Martena K., Mrs.	DES	311	24219
Balwanz, Karen R., Mrs.	PES	601	25713
Bamford, Harold E., Jr.	SIS	A1-201	25800
Banks, Elizabeth L., Mrs.	DES	312	24184
Barbour, Maria L., Mrs.	ISP	536	25768
Bardasis, Angelo	MPS	307	24302
Bardon, Marcel	MPS	305	24317
Barker, Gayle F., Miss	EMS	452	24166
Barker, William S.	SIS	A1-203	25836
Barrett, Catherine I., Mrs.	GES	612	25732
Barrett, Edgar W.	DMS	437	25990
Barries, Joel L.	EMS	404	24330
Bartocha, Bodo	O/D	504	25762
Bartlett, Alice L., Miss	AD/R	320	24278
Bartock, Ann K., Mrs.	PES	601	25720
Barton, Alexander, J.	UES	625	25911
Barton, Audrey J., Mrs.	GCO	638	25900
Bean, Margaret C., Mrs.	OIP	531	25812
Beaudry, Viola M., Miss	OPP	311	24194
Becker, Frederick, Jr.	PER	212	24090
Benedict, Marjory R., Mrs.	GES	616	25745
Bennett, Francis J.	DMS	440	27380
Bennett, Ruth E., Miss	PES	605	25875
Benson, William E.	DES	310	24210
Benton, Joyce E., Miss	PES	649	25982
Berghaus, Leona M., Miss	SIS	A1-212	25844
Berryman, Sadie V., Mrs.	ASO	232	24129
Betancourt, Bonnie N., Mrs.	AD/I	417	24355
Betts, Sherman W.	DES	312	24189
Bever, Arley T.	BPA	426	25764
Bierly, Eugene W.	DES	312	24190
Biggar, Ronald S., Jr.	EMS	448	24082
Biggin, Nancy K., Mrs.	AD/I	417	24355
Bigler, Calista J., Miss	SIS	A1-216	25830
Binder, Laurence O.	PES	605	25878
Birdas, Christina, Miss	PES	649	25979
Birdsong, Ray S.	BMS	332	25846
Bischoff, Lionel C.	EMS	403	24143
Bisplinghoff, Raymond L.	O/D	520	24376
Bittner, Joan G., Mrs.	AD/R	512	27342
Black, Hazel A., Mrs.	ENG	336	24280
Blackman, James H.	SOC	317	25968
Blackner, Henry F.	PER	211	24106
Blair, Barbee, Miss	BPA	428	25876
Blakey, Carl A.	ASO	233	24129
Blumenthal, Dorothy B. W., Mrs.	SIS	A1-203	25835
Bollinger, Constance M., Mrs.	MPS	303	27377
Bolton, Wilbur W., Jr.	GCO	640	25772
Bootes, Kenneth F.	DMS	440	27380

Name	DIVISION	ROOM NO	EXT
B			
Boring, Carol, Mrs.	MAO	547	24026
Bosilevac, M. Mildred, Miss	OIP	536	25805
Bostick, James L.	GCO	634	25950
Boutchyard, Linda L., Mrs.	GPP	526	27320
Bowman, Leo	FMO	431	24037
Boyden, Mary M., Miss	EMS	403	24168
Boyden, Robert B.	ADT	645	24072
Brandenberg, Vashti S., Mrs.	GPP	527	25728
Bregida, Albert P.	AD/I	410	27364
Brice, Neil M.	DES	312	24184
Brickey, Janet, Miss	SOC	319	24204
Briggs, Kay E., Mrs.	ASO	221	24140
Broglio, E. Paul	PER	211	24119
Bronson, Zola	EMS	402	24024
Brooks, Ann M., Miss	SOC	316	24286
Brooks, Douglas L.	O/D	504	24010
Brooks, Jean L., Mrs.	MAO	549	24110
Brooks, John L.	BMS	330	27324
Brosnan, Carol R., Miss	GES	613	25740
Brosseau, George E., Jr.	AD/I	415	24355
Brown, Charles F.	OGC	501	24388
Brown, Darlene C., Miss	FMO	436	24146
Brown, Georgianna M., Mrs.	AD/NI	339	24237
Brown, James H.	BMS	333	24264
Brown, Joan M., Mrs.	PER	211	24136
Brown, John James	EMS	406	24161
Brown, Judith A., Miss	PER	212	24124
Brown, Lavonia A., Miss	PER	212	24125
Brown, Violet C., Mrs.	OGC	501	24387
Bruce, Phyllis A., Miss	DMS	440	27380
Bruning, William C.	GCO	630	25892
Buccino, Alphonse	PES	607	25959
Buckus, Mary G., Mrs.	UES	623	25987
Buie, Aileen M., Miss	GCO	634	25938
Bundy, Mary E., Mrs.	O/D	520	24376
Burger, Alma R., Miss	GPP	526	27322
Burke, Ralph G., Jr.	FMO	433A	24038
Burnisky, Stephen G.	GCO	638	25900
Burns, Mary E., Miss	AD/I	414	24355
Burns, Robert W.	OIP	531	25828
Burton, S. Justine, Miss	PES	601	25727
Bush, George B., Jr.	GCO	634	25950
Butler, M. Veronica, Miss	AD/I	412	24360
Byrd, Maxine L., Mrs.	AD/I	420	24342

Name	DIVISION	ROOM NO	EXT
C			
Cahill, Golda M., Miss	ASO	237	24222
Cain, Robert W.	EMS	407	24334
Calfee, Mary F., Mrs.	ASO	223	24120
Callanan, Margaret J., Miss	GES	609	25748
Callaway, M., Patricia, Miss	BMS	333	24264
Campbell, Edith B., Mrs.	DES	312	24147
Canby, Patricia B., Mrs.	FMO	434	24019
Cannell, Glen H.	PES	605	25875
Cardwell, Charlene S., Mrs.	DES	339	24295
Carlson, Francis E.	ADT	645	24096
Carlson, Harve J.	BMS	325	24338
Carrabino, Joseph F.	AD/I	419	24060
Carrigan, Richard A.	AD/I	414	24355
Carroll, Margaret V., Mrs.	AD/R	320	24240
Carter, James B.	ASO	233	24152
Carter, Rosa N., Mrs.	GCO	634	25965
Casey, Ralph E.	ADT	645	24148
Castle, Isabel H., Mrs.	SIS	A1-213	25818
Cavagrotti, Frances R., Mrs.	O/D	518	24384
Cephas, Theodosia D., Mrs.	EMS	405A	24045
Certo, Catherine W., Mrs.	ENG	336	24280
Chapin, Douglas S.	GES	612	25732
Chapman, Lula E., Miss	OIP	531	25813
Charles, Mary Louise, Mrs.	GES	610	25736
Cheatham, Paul G.	AD/I	414	24355
Chern, Bernard	AD/I	417	24355
Chidester, Opal A., Mrs.	NCF	706	25719
Chin, Dennis E.	DMS	440	27380
Chirichiello, John R.	EMS	405	24045
Chisley, Ivy E., Mrs.	ASO	221	27361
Chittenden, Andrew D.	ADT	645	24148
Cima, Louis	SIS	A1-222	25850
Clark, Darla J., Miss	GPP	526	27320
Clark, Eloise E., Miss	BMS	329	24260
Clem, Georgia R., Miss	ASO	221	24140
Clement, Duncan	OIP	532	25811
Clemons, Grace, Miss	ADT	645	24097
Clemons, Lynne, Mrs.	BMS	325	24337
Coates, Joseph F.	OIR	213	27350
Cohen, Lawrence	BPA	428	25876
Cohen, Marlin A.	DMS	440	27371
Colbert, Lynne T., Miss	GCO	640	25927
Colbertsen, Cecilie E., Mrs.	BMS	329	24260
Cole, William B., Jr.	GCO	640	25920
Coleman, Lawrence D.	OPP	344	24242

Name	DIVISION	ROOM NO	EXT
C			
Commins, William D.	BPA	425	25784
Consolazio, William V.	AD/I	419	24345
Cook, Charles	ASO	237	24023
Cook, M. Kathryn, Miss	DMS	440	27380
Cooper, Charles F.	BMS	330	25854
Copperman, Howard R.	FMO	432	24366
Cooperman, Ida D., Mrs.	ADT	645	24088
Corbin, Mary Anne, Mrs.	FMO	434	24007
Cordle, Elisabeth B., Miss	O/D	546	25840
Covington, Katie L., Mrs.	OGC	501	24397
Cowhig, James D.	OIR	213	27350
Cox, Lillian T., Mrs.	MPS	351	24186
Cramer, William H.	MPS	348	24266
Crampton, Mary Jane, Miss	OCA	707	25985
Crary, Albert P.	DES	339	24295
Creutz, Edward C.	AD/R	512	27342
Cross, Willamette J., Miss	OPP	340	24234
Crowell, Judith Mrs.	UES	625	25916
Cuatrecasas, Martha N., Mrs.	EMS	448	24005
Cunningham, Donald E.	O/D	518	24384
Curland, Bertrand J.	EMS	641	24150
Curtin, Lorraine P., Mrs.	OPP	340	27362
Curtis, Judith M., Mrs.	O/D	505	25779
Curtis, Kent K.	OCA	707	27346
Curtis, Thomas	DMS	440	27371
Custer, Paul E.	DMS	440	27371
D			
Daen, Jerome S.	AD/I	414	24355
Dale, Robert L.	OPP	340	24257
Danek, Joseph G.	AD/I	412	24360
Daniel, Rebecca S., Mrs.	ASO	220	24130
Dante, Katherine J., Miss	DMS	440	27371
Dassoulas, Nicholas J.	GCO	638	25900
Dawson, M. Frances, Miss	SOC	318	24208
Dawson, Merle R.	OPP	340	24203
Day, Shirley Y., Mrs.	OPP	340	24203
DeBell, Jean T., Miss	DES	346	25858
DeHerrera, Lena A., Miss	GCO	630	25884
Delaney, Rita C., Mrs.	DES	312	24185
Dennis, Betty, Miss	OIR	213	27350
Dennis, Eleanor P., Mrs.	AD/R	512	27342
Dennis, Elizabeth A., Mrs.	BMS	323	24298
Denton, Jesse C.	OIR	213	27350
Deutsch, Alan J.	DMS	440	27380
Devey, Gilbert B.	ENG	336	24280
Devine, Lola C., Mrs.	AD/R	320	24248
Devlin, Lynda L., Miss	FMO	432	24366
Dickens, Charles H.	UES	625	25915
Dickey, Marlene B., Mrs.	AD/NI	703	24180
Dickson, Theresa C., Miss	DMS	440	27373
Diehl, Phelps A.	ASO	223	24120
Dixson, Bonnie J., Miss	OIP	536	25798
Dodd, Walter H.	GPP	527	25704
Doherty, Josephine K., Mrs.	BMS	330	25854
Dominelli, Joan A., Miss	BMS	332	25858
Donovan, Paul F.	MPS	307	24310
Dooling, Thomas A.	GPP	543	27390
Dorie, William M.	PER	212	24124
Doschek, Wardella W.	BMS	325	24337
Drury, Robynne A., Mrs.	GES	611	25756
Dryzer, Bella E., Mrs.	GES	612	27310
Dubester, Henry J.	SIS	A1-203	25826
Duckett, Hope W., Mrs.	ASO	223	24120
Dugan, Caldwell N.	AD/I	410	27364
Dulan, Charles J.	ADT	645	24068
Dunmore, Henry L.	ASO	237	24023
Duval, Suzanne M., Miss	EMS	406	24160
E			
Eden, H. Frank	DES	312	24190
Ebenfield, Helene, Miss	EMS	452	24166
Eddins, Telula E., Mrs.	GPP	526	25780
Edmonds, Janice A., Mrs.	ASO	223	24120
Edmunds, Lafe R.	UES	623	25924
Edwards, Bernice J., Mrs.	ASO	231	24116
Edwards, Gerald A.	UES	625	27330
Edwards, Gregg	UES	623	25935
Edwards, Lola V., Miss	EMS	408	24325
Edwards, Margaret S., Mrs.	UES	625	25919
Edwards, Martha E., Miss	FMO	436	24157
Edwards, Richard A.	O/D	518	24368
Edwards, Roberta, Miss	DES	312	24236
Einhorn, Marilyn J., Mrs.	OIR	213	27350
Eisman, Eleanor F., Mrs.	AD/I	516	27304
Elliott, Terry D.	UES	627	25896
Ellis, Gaylord L.	GCO	640	25920
Endres, Hazel L., Mrs.	DES	339	24274
Engel, Estella K., Mrs.	BMS	329	24260

Name	DIVISION	ROOM NO	EXT
E			
England, J. Merton	AD/I	420	24364
Essrick, Carol B., Miss	SOC	317	24182
Etzel, Howard W.	MPS	307	24302
Evans, Priscilla E.	AD/I	420	24370
Evans, Ruth A., Mrs.	GES	609	25747
F			
Falk, Charles E.	SRPS	551	25770
Farmakides, John B.	OGC	501	24393
Fatzinger, Harold R.	FMO	434	24015
Feather, Gerald E.	ASO	231	24116
Fee, William E., Jr.	GCO	640	25920
Field, William H.	ENG	336	24280
Fisk, Mary R., Mrs.	EMS	406	24160
Flack, Mary Ann, Miss	GPP	543	27390
Flam, Brenda C., Miss	BMS	329	24260
Fleischer, Robert	MPS	351	24196
Fleming, Herman G.	MAO	547	24026
Flurry, Laverne, Miss	DES	332	25858
Foggo, Helen T., Mrs.	BMS	325	24332
Fontaine, Thomas D.	AD/E	620	25888
Forbes, Mary J., Miss	MAO	224	24010
Ford, Albert C.	DMS	440	27380
Ford, Melvin D.	ASO	233	24152
Forrest, Harley M.	ASO	232	24128
Foster, Daniel L.	ADT	645	24088
Foster, Kenneth B.	FMO	432	24012
Foster, Penny D., Mrs.	EMS	448	24083
Foster, Raymond	O/D	523	24047
Fox, Gloria A., Miss	GES	612	27310
Francer, Maria H., Miss	ASO	223	24120
Francis, Henry S., Jr.	AD/NI	703	24316
Frankenberg, Dirk	DES	346	24236
Franklin, Leila M., Miss	PES	605	25878
Franklin, Micheal E.	DMS	440	27380
Franko, Stephen J.	GCO	630	25860
Freeman, Evelyn V., Mrs.	PES	651	25980
Fregeau, Jerome H.	AD/R	320	24248
Frentz, George H., Jr.	DMS	440	27371
Frick, Sondra, Miss	GCO	634	25950
Friedman, Lester	EMS	448	24083
Friedman, Norman W.	EMS	641	24151
Frodyma, Michael M.	PES	607	25970
Frost, Charles W., Jr.	GCO	630	24108
Frost, Marsha E., Mrs.	PER	212	24090

Name	DIVISION	ROOM NO	EXT
F			
Fuller, Corine B., Mrs.	EMS	403	24168
Fulwood, Ruth E., Mrs.	DES	310	24211
G			
Galinn, Myra B., Mrs.	GCO	634	25950
Gannon, Joseph P.	EMS	404	24392
Gardner, Byron I.	ADT	645	24096
Gardner, Leonard F.	AD/R	320	24278
Gardner, Patricia E., Miss	GCO	634	25927
Garland, Herman	ASO	237	24023
Garrabrant, Robert B.	PES	607	25887
Garver, Deborah E., Mrs.	MPS	348	24272
Gatbonton, Nilda S., Mrs.	BPA	429	24084
Gatton, Janet L., Mrs.	GES	618	25734
Gaudreau, Mary C., Mrs.	GES	616	25700
Gaus, Michael P.	ENG	336	24280
Gearhart, Ann E., Miss	PES	601	25713
Geary, Martin V.	GCO	634	25940
Gee, Margaret E., Mrs.	PER	208	24114
Gerasimou, Helen T., Mrs.	OPP	340	24228
Geraus, James B.	BPA	429	25705
Getz, Diana L., Miss	BPA	426	24050
Gibson, Daniella N., Miss	FMO	432	24012
Gilbert, Lorraine M., Mrs.	FMO	436	24146
Gillespie, Walter L.	PES	605	25867
Gist, Lewis A.	PES	649	25954
Goldblatt, Mildred F., Miss	AD/I	411	27364
Goodwin, Harold L.	OSG	537	25948
Gorham, Geraldine, Miss	MAO	549	24110
Gorski, William	DMS	440	27371
Gould, David E.	BPA	428	25759
Gould, Joan E., Mrs.	MPS	303	27377
Grabe, Richard E.	GCO	633	25926
Graham, Faithy A., Mrs.	AD/I	420	24370
Graham, Ora M., Mrs.	AD/E	620	25868
Graves, Frank	GCO	634	25965
Green, Jackye M., Mrs.	BMS	326	24200
Green, Mildred M., Miss	BMS	325	24338
Green, Richard J.	AD/NI	510	27302
Greene, Dorothy M., Mrs.	PES	605	25870
Greene, Mary W., Mrs.	SOC	318	24208
Greene, Robert J.	GCO	634	25926
Greenwald, Ernest	OSG	537	25944
Greenwood, Gerald A.	GCO	630	25884

Name	DIVISION	ROOM NO	EXT
G			
Gregory, Dolores A., Miss	ISP	536	25768
Griffin, Geraldine E., Mrs.	SOC	316	24294
Grimes, Vicki Lynn, Miss	SIS	A1-215	25706
Gruner, Wayne R.	AD/R	320	24270
Gunsher, Stanley D.	ADT	645	24144
H			
Hale, Shirley A., Miss	ASO	231	24116
Hall, Earlene J., Mrs.	PES	649	25982
Hall, Richard A.	AD/I	414	24355
Hamaty, Joyce, Mrs.	NSB	546	25840
Hamaty, Lois J., Mrs.	O/D	520	24363
Hamilton, Lela J., Miss	SIS	A1-203	25834
Handy, Gladys G., Mrs.	OIR	213	27350
Hanna, Lila M., Miss	SIS	A1-203	25824
Hannapel, Raymond J.	PES	605	25879
Hannington, Patricia R., Mrs.	MAO	549	24110
Harb, Victoria N., Mrs.	FMO	435	24035
Harding, Samuel W.	PES	649	25955
Hardy, Robert B.	GCO	634	25938
Harrigan, John T.	ASO	220	24130
Harrington, Laura, Mrs.	AD/A	525	25710
Harrison, Johnnie M., Mrs.	FMO	431	24038
Harrison, T. Eugene	ASO	237	24222
Hartman, Lawton M., III	O/D	505	25778
Hartman, Patricia L., Mrs.	OPP	340	24235
Hartmeyer, Charlotte R., Mrs.	AD/NI	510	27300
Harvey, Robert H.	PES	605	25875
Hausman, Howard J.	PES	601	25720
Havens, Elmer G.	AD/I	412	24360
Havens, Lizabeth G., Mrs.	MPS	348	24266
Hawkins, Pearline H., Mrs.	EMS	641	24178
Hayden, Lucille D., Mrs.	AD/A	220	25998
Haynes, Emanuel	BPA	425	25777
Hays, Veda C., Miss	AD/I	410	27364
Heckman, Florence, Miss	ASO	219	24070
Heenan, William F.	SIS	A1-233	25800
Heer, Ray R., Jr.	OPP	340	24246
Heider, S .A.	AD/I	410	27364
Heller, Patricia L., Mrs.	ASO	221	24140
Henney, Charles T.	BPA	428	25876
Henry, Edward H.	ASO	231	24116
Hersman, M. Frank	ISP	536	25768
Herwig, Lloyd O.	AD/I	411	27364

Name	DIVISION	ROOM NO	EXT
H			
Hess, Eugene L.	AD/R	320	25988
Hickman, Diane E., Miss	PES	607	25970
Hiebert, Gordon	OIP	532	25792
Hilgert, Cecelia, Mrs.	EMS	447	24077
Hill, Eunice U.	ADT	645	24144
Hill, Rene F., Mrs.	AD/E	506	24174
Hilliard, Bill R.	FMO	431	24038
Hilton, Edward W.	AD/R	320	24243
Hilton, Viola G., Mrs.	DMS	440	27380
Hines, Charles W.	MPS	341	24230
Hines, Howard H.	SOC	316	24286
Hodge, Priscilla A., Miss	AD/R	320	24243
Hodge, Walter H.	BMS	330	27318
Hoff, William J.	OGC	501	24386
Hoffman, Arthur C.	UES	627	25905
Hoffman, Joan M., Miss	FMO	432	24012
Hogan, Thomas J.	EMS	405	24058
Holmes, John C.	ASO	231	24116
Holmes, Linda G., Mrs.	SIS	A1-203	25836
Horowitz, Harold	AD/I	421	24061
Horsley, Presocia C., Mrs.	BMS	325	24326
Hruz, Dora J., Mrs.	BMS	325	24300
Huckenpahler, James G.	EMS	448	24083
Hudleson, Dorothy M., Mrs.	BPA	428	25759
Hudson, Renate G., Miss	ASO	231	24116
Hudson, Walton M.	BPA	429	24086
Huey, Bobby	ASO	237	24222
Huffman, James A.	DMS	440	27371
Huffman, Jerry W.	OPP	340	24207
Hughes, Maydie O., Mrs.	O/D	520	24001
Hull, Robert F.	OIP	532	25813
Humphreys, Lloyd G.	AD/E	506	24174
Hunt, Daniel, Jr.	NCF	706	25717
Hunt, Elizabeth S., Miss	GPP	526	25780
Hunter, Janice L., Mrs.	EMS	447	24077
Hunter, Margaret J., Mrs.	OIP	532	25782
Hurley, David H.	ASO	223	24120
Hurley, Phyllis J., Mrs.	MPS	348	24266
Hutchinson, Bonnie S., Miss	ASO	221	24140
Hutchinson, Florence J., Mrs.	FMO	431	24037
Hutchinson, Linda S., Mrs.	MPS	351	24188
Hyland, Barbara V., Mrs.	GCO	634	25938
Hyman, Harry	GCO	640	25920
Hynes, Warren J.	ADT	645	24072

Name	DIVISION	ROOM NO	EXT
I			
Ide, John M.	ENG	336	24280
Imrich, Phyllis L., Mrs.	GCO	630	25892
Inabinet, Mary L., Mrs.	GCO	634	25965
Intermaggio, Jean B., Mrs.	PES	605	25865
J			
Jackson, Arcelia G., Miss	AD/NI	702	27359
Jackson, Donald E.	FMO	431	24016
Jackson, Edith V. W., Mrs.	FMO	431	24017
Jackson, Martha A.	PES	607	25958
Jackson, Phyllis, Miss	BMS	330	25854
Jaffe, Florence K., Mrs.	MPS	305	24317
Jaffe, Sidney A.	EMS	409	24170
Jarrott, Ethel S., Miss	ENG	336	24280
Jenkins, Beatrice H., Mrs.	ENG	336	24280
Jenkins, Thomas E.	AD/A	525	25766
Jennings, Feenan D.	IDOE	701	27356
Jeter, Gail D., Miss	PES	649	25982
Jeter, Peggy J., Miss	NCF	706	25717
Johnson, Dianne S., Mrs.	UES	625	25911
Johnson, LaVerne P., Mrs.	GPP	527	24100
Johnson, Phyllis L., Mrs.	PES	601	25957
Johnson, Therra A., Miss	FMO	431	24038
Johnson, Yvonne C., Miss	BPA	425	25772
Johrde, Mary K., Miss	AD/NI	339	24202
Jones, Anna D., Miss	MPS	342	24290
Jones, Birdie P., Mrs.	MPS	342	24290
Jones, Calvin C.	AD/A	540	27315
Jones, Clarice D., Mrs.	ENG	336	24280
Jones, Dorothy P., Mrs.	DMS	440	27336
Jones, Gladys B., Mrs.	ASO	223	24120
Jones, Lewis P.	PRO	540	27315
Jones, Melvin F.	EMS	449	24076
Jones, Mina Y., Mrs.	BPA	428	24098
Jones, Myles E.	ASO	231	24116
Jones, T. O.	AD/NI	703	24180
Jordan, Joan M., Mrs.	BMS	330	27324
Joy, James H.	SIS	A1-233	25800
K			
Kan, Robert F.	SIS	A1-215	25706
Kane, Judith A., Mrs.	BMS	323	24298
Kapp, Ruth Y., Mrs.	ASO	223	24120
Kassack, Nathan	GPP	527	25722
Katelvero, Norma M.	SOC	317	24104
Keenan, Thomas A.	OCA	707	27346
Keller, Marlene Mrs.	MPS	307	24302
Kellett, James C., Jr.	UES	625	25916
Kelly, Eleanor, Miss	OPP	345	24220
Kelly, Patricia A., Mrs.	DES	310	24213
Kelly, Shirley L., Mrs.	UES	627	25914
Kelson, Keith R.	AD/E	620	25868
Kennedy, Winifred B., Mrs.	MPS	305	24317
Keola, Patricia A. L., Mrs.	PES	605	25864
Kerstiens, Donald E.	GCO	634	25965
Kiel, Esther C., Mrs.	FMO	434	24019
Killen, Benetta Joan, Mrs.	BPA	426	25764
King, Richard E.	FMO	431	24374
Kirby, Carolyn P., Mrs.	PES	607	25977
Kirk, Ladson E., Jr.	ASO	231	24116
Kirkpatrick, Lois D., Mrs.	ASO	223	24120
Kirsch, John E.	MAO	548	24026
Klingsberg Vera I., Mrs.	SRPS	551	27334
Knott, Dorothy A., Mrs.	EMS	641	24150
Kohlerman, Mary M., Miss	PES	605	25864
Kolf, Richard C.	OIR	213	27350
Korengold, Barbara L., Mrs.	GPP	526	27320
Korgen, Reinhard L.	UES	627	25904
Kozlowski, Joseph P.	EMS	405	24004
Kraegel, Lowell G.	PES	601	25980
Kraft, Donald E.	ASO	237	24222
Kramer, Hildegard B., Mrs.	OIP	532	25795
Kramer, Howard D.	GES	616	25730
Krasnow, David P.	MAO	549	24110
Kratchman, Jack	GPP	543	27390
Krause, Ralph M.	MPS	303	27377
Krug, Vera K., Mrs.	BMS	333	24264
Kruithoff, Idele E., Mrs.	GCO	634	25940
Krum, Herman	ADT	645	24068
Kuehnle, George	FMO	434	24014
Kusinski, Arthur J.	OGC	501	24396
L			
Labanics, Dagny J., Mrs.	PER	212	24125
Lagguth, Richard V.	DMS	440	27371
Lahti, Vellamo M., Miss	EMS	409	24170
LaCount, Ronald R.	NCF	706	25712
Lamson, Robert W.	BPA	425	25776
Lance, John F.	AD/I	415	24355
Lane, Harold H.	MPS	301	24186
Lang, Dorothy S., Mrs.	SOC	316	24294

Name	DIVISION	ROOM NO	EXT
L			
Langland, Ila M., Mrs.	GCO	634	25936
Lashley, Myrtle D., Miss	UES	625	25915
LaTorre, Mary B., Mrs.	PES	607	25970
Layton, June H., Mrs.	GCO	637	25933
LeCounte, Mitchell	ASO	237	24023
Lea, George K.	ENG	336	24280
Lederman, Leonard L.	EMS	401	24350
Lee, Ellen M., Mrs.	EMS	401	24165
Lee, Marlene R., Miss	FMO	431	24374
Lee, Patricia A., Mrs.	GPP	526	25781
Lee, Richard W. H.	DMS	437	25990
Lefebvre, Albert C.	SIS	A1-222	25850
Lehmann, John R.	OCA	707	27346
Leise, Joshua M.	AD/I	418	24380
Leonard, Frederic A.	AD/I	411	27364
Leonard, Minnie S., Mrs.	ASO	223	24120
Lerch, Jean L., Mrs.	DES	346	24227
Letson, Jerry W.	ASO	219	24070
Levin, Louis	AD/I	516	27304
Levine, Milton	EMS	406	24160
Lewis, Herman W.	BMS	326	24200
Lewis, Justin C.	EMS	408	24324
Lewis, Mary G., Miss	PES	605	25870
Lewis, Opal L., Miss	OIR	213	27350
Lewis, Price, Jr.	OPP	340	24258
Lewis, William E.	PRO	540	27315
Lewis, Zelma E., Mrs.	UES	627	25915
Liddell, Beverly E., Mrs.	SOC	317	25969
Lieban, Richard W.	SOC	318	24208
Lilly, Leon B.	ADT	645	24068
Lindsay, Felix H. I.	EMS	408	24328
Lindsey, Margaret B., Mrs.	ASO	223	24120
Lipscomb, Barbara Jean, Mrs.	ENG	336	24280
Livingston, Dorothy B., Mrs.	GES	616	25730
Livingston, George A.	AD/I	423	27367
Llano, George A.	OPP	311	24194
Lloyd, Maryann B., Mrs.	OGC	501	24397
Long, Aaron L.	DMS	440	27380
Long, Barbara A., Mrs.	DMS	440	27336
Long, James W.	HS	439	24041
Long, Ralph H., Jr.	AD/I	417	24355
Loycano, Robert J.	EMS	447	24079
Lucas, Emmett, W.	PRO	540	27315
Lucas, William A.	SOC	319	25714
Lyon, Hugh L.	GCO	634	25926
M			
MacGregor, Duncan M.	PER	211	24118
Mackin, Paul J.	EMS	409	24352
Maechling, Charles, Jr.	OGC	501	24398
Magat, Claude S.	ADT	645	24144
Maio, Frank L., Jr.	ASO	223	24120
Maloney, Betty W., Mrs.	O/D	546	25840
Manouelian, Denise R., Mrs.	UES	627	25904
Manuel, Carol A., Mrs.	ASO	223	24120
Markey, Mary E., Mrs.	FMO	431	24017
Marlow, Carole B., Mrs.	UES	623	25928
Maroney, Rebecca L., Miss	BMS	328•	24300
Martin, Doris M., Mrs.	OIP	536	25814
Martin, Geraldine Z., Mrs.	MPS	303	27377
Martin, M. Delores, Mrs.	DMS	440	27380
Mason, Alice H., Mrs.	BMS	330	27324
Maxfield, M. Patricia, Miss	BMS	326	24314
Mayes, Charles U.	ASO	223	24120
Mayfield, Lewis G.	ENG	336	24280
McAlister, W. Bruce	DES	346	24236
McAuliffe, Myra J., Miss	OIR	213	27350
McCarn, Doris, Miss	O/D	520	24001
McCarthy, Maureen E., Miss	OPP	340	24258
McClure, Mary B., Mrs.	UES	627	25914
McDougal, Leroy	ASO	223	24120
McElroy, William D.	O/D	520	24001
McGuire, Donald C.	UES	623	25987
McIntosh, Marie M., Mrs.	PRO	540	27315
McKinney, Alice M., Miss	MPS	342	24290
McKoy, Oliver	O/D	232	24128
McManus, Earl T.	ASO	231	24116
McMillen, J. Howard	MPS	305	24317
McNabb, Lyn O., Miss	SOC	318	24216
McNeil, Eugenia M., Miss	DES	346	24236
McNinch, Syl, Jr.	BPA	428	24098
McPherson, Grace M., Miss	FMO	436	24146
Mehl, John W.	BMS	325	24326
Melby, Hilda H., Mrs.	AD/I	419	24345
Melendez, Dorothy D., Mrs.	GES	616	25744
Melin, Mildred S., Mrs.	ASO	223	24120
Melmed, Arthur S.	OCA	707	25790
Mercer, Lucille J., Mrs.	DMS	440	27371
Merida, Doris E., Mrs.	GPP	526	27320
Merrell, Marie L., Mrs.	OCA	707	25790
Michelitch, Robert A.	GCO	630	25884

Name	DIVISION	ROOM NO	EXT
M			
Middleton, Norma J., Mrs.	BMS	329	24260
Mieremet, Marian P., Mrs.	EMS	405	24045
Mikuluk, Marya, Miss	OGC	501	24398
Miller, Charles L.	GCO	638	25900
Miller, Susan W., Mrs.	BMS	325	24332
Mills, Thomas J.	EMS	401	24164
Millspaugh, Elva A., Mrs.	PES	605	25867
Minnifield, Ramona L., Mrs.	OCA	707	25960
Missouri, Evelyn H., Mrs.	SIS	A1-203	25836
Molitor, Wilma W., Mrs.	EMS	404	24391
Molnar, Andrew R.	OCA	707	25960
Monroe, Mary S., Mrs.	SRPS	551	25770
Montross, G. Michael	BPA	428	24379
Mooney, Mary C., Mrs.	ADT	645	24148
Moore, Ruth C., Mrs.	MAO	544	24110
Morano, Darleen F., Miss	DMS	437	25990
Morgan, Charlotte, Mrs.	AD/NI	702	27359
Morgret, Helen T., Mrs.	GCO	630	25872
Morrell, William E.	PES	649	25978
Morton, Dorothy T., Mrs.	SIS	A1-215	25706
Mostakis, Michael J.	MAO	547	24110
Moulton, Kendall N.	OPP	340	24247
Mulqueen, Jane, Miss	DMS	440	27380
Mulyihill, Sheila A., Miss	OSG	537	25944
Murakami, Fred K.	MAO	549	24110
Myers, John G.	ADT	645	24096
N			
Napper, Louis Q.	DMS	440	27380
Natoli, Anthony J.	DMS	440	27380
Naughten, Francis G.	GCO	633	25965
Neal, Mary F., Mrs.	DES	312	24189
Neely, Van A.	BPA	429	25705
Neuman, A. Conrad	DES	346	24215
Newman, Joseph N.	ASO	233	24152
Newton, Robert D.	GCO	630	25872
Nicely, Patricia E., Miss	AD/I	412	24360
Nicholson, Richard S.	MPS	348	24276
Nieberding, Rose Mary L., Mrs.	OCA	710	25960
Norment, Susan E., Mrs.	FMO	434	24006

Name	DIVISION	ROOM NO	EXT
O			
Oberg, Gloria G., Miss	BPA	428	24379
O'Brien, Francis G.	GES	618	25734
O'Brien, Inez C., Miss	SIS	A1-217	25830
O'Clery, Thelma, Mrs.	GES	611	25756
O'Connell, J. E.	OIP	532	25782
O'Doherty, John K.	GPP	543	27390
Odbert, Henry S.	BMS	333	24264
Ohlke, Clarence C.	GPP	526	27320
Ojalvo, Morris S.	ENG	336	24280
Oldes, Dora L., Miss	PER	208	24114
Oliver, Lawrence H.	OCA	707	25790
Oliveras, Carmen Ana, Miss	OPP	340	24256
Olsen, Benjamin L.	EMS	641	24178
Olson, Pauline A., Miss	GPP	527	25722
Olsoni, Karl E.	SIS	A1-220	25850
O'Malley, Helen C., Mrs.	EMS	401	24350
O'Neill, Thomas H. R.	DES	312	24185
Orr, Jane, Mrs.	BMS	325	24338
Orville, Richard E.	DES	312	24190
Owen, Thomas B.	AD/NI	510	27300
Owens, Charles T.	OIR	213	27350
Owens, Rosemary E., Miss	SRPS	551	27334
P			
Page, Howard E.	AD/I	420	24342
Paine, Roland D., Jr.	GPP	527	24100
Palmer, Christine, Miss	DMS	437	25990
Park, P. Kilho	DES	346	24227
Parke, Ann, Miss	UES	627	25896
Parker, Mary D., Miss	SIS	A1-201	25803
Parramore, Mary L., Mrs.	BPA	426	24050
Parrott, Olga M., Mrs.	GES	612	27310
Pasta, John R.	OCA	710	25960
Patermaster, John J.	ADT	645	24144
Patterson, Rosa C., Mrs.	GCO	634	25938
Payne, Elizabeth G., Mrs.	MPS	352	24320
Pearce, Bill M.	MAO	548	24110
Pell, William H.	MPS	303	27377
Pendleton, A. Deloise, Miss	BMS	323	24312
Penney, Richard L.	OPP	340	24255
Pentecost, Frances M., Mrs.	ASO	219	24070
Petrocci, Mary S., Mrs.	OPP	345	24220

Name	DIVISION	ROOM NO	EXT
P			
Pfeiffer, Leonard M.	DMS	440	27336
Phillips, Bradley	O/D	523	24047
Phillips, Dixie M., Mrs.	FMO	434	24007
Phillips, Lyle W.	UES	623	25928
Phillips, Morris T.	GCO	630	25860
Piccariello, Harry J.	BPA	425	25774
Pike, Elizabeth A., Mrs.	SIS	A1-215	25706
Pilarinos, George	MAO	549	24110
Pittle, Ethel R., Mrs.	PES	649	25957
Plater, Barbara D., Mrs.	DES	339	24295
Plummer, Mary S., Mrs.	PER	211	24064
Pollock, Donald K.	SIS	A1-212	25844
Poloway, Irving	ADT	645	24092
Porter, Terence L.	GES	610	25738
Potter, Helen B., Miss	AD/R	320	25988
Powell, Charolette S., Mrs.	GCO	638	25900
Prentiss, Susan E., Miss	FMO	434	24006
Presnell, Margen, Miss	GCO	634	25950
Preston, Warren H.	ASO	232	24129
Prokop, Ivan, R.	ADT	645	24092
Pronko, Eugene	SIS	A1-215	25706
Pruett, Patricia A., Mrs.	OIP	532	25806
Pugh, Jane E., Miss	EMS	641	24150
Puma, Daniel R.	GES	618	25737
Purcell, Norma L., Mrs.	OIR	213	27350
Puslowski, Xavier J.	EMS	447	24077
Q			
Quam, Louis O.	OPP	340	24221
Quigley, Thomas W., Jr.	SIS	A1-212	25844
R			
Rabin, Robert	OIR	213	27350
Raizen, Senta A., Mrs.	AD/E	506	24176
Ramey, Nancy Z., Mrs.	ASO	219	24070
Ray, Richard G.	DES	310	24218
Raymond, Charlotte R., Mrs.	GCO	638	25900
Reed, Carolyn A., Miss	ASO	237	24222
Reese, Edward B.	ADT	645	24088
Reese, Fred E.	ADT	645	24088
Reid, J. Sylvia, Miss	ENG	336	24280
Reid, Theodore L.	PES	607	25958
Renaud, Janet M., Mrs.	OIR	213	27350
Renirie, Jack	GPP	527	25728
Rexroad, Helen H., Mrs.	HS	439	24040
Reynik, Robert J.	ENG	336	24280
Rhodes, Agnes D., Mrs.	DES	312	24198
Rhodes, Barbara H., Mrs.	ASO	221	24140
Rhodes, Sarah N., Mrs.	SIS	A1-234	25800
Rice, Virginia F., Miss	GES	613	25740
Rich, Jacquelyn S., Miss	ASO	231	24116
Richardson, Charlesetta J., Mrs.	OIP	532	25812
Richardson, Udel	ASO	231	24116
Richardson, Vivian D., Miss	UES	627	25896
Riemer, William J.	BMS	325	24332
Ries, Richard R.	OIP	532	25807
Rison, Kathryn R., Mrs.	BMS	332	25846
Rizzo, Debrah C., Miss	PES	607	25887
Roberts, Edith J., Mrs.	MPS	348	24262
Roberts, L. Diane, Mrs.	BMS	332	27324
Roberts, M. Adelaide, Miss	MPS	352	24320
Robertson, Diana R., Miss	DMS	440	27371
Robertson, Earl M.	ASO	232	24128
Robertson, Richard D.	DMS	440	24129
Robinson, Emma H., Miss	BPA	425	25776
Robinson, John W.	GPP	527	25703
Robinson, Queen M., Mrs.	GES	610	25739
Rodman, Pauline M., Mrs.	PER	208	24114
Rodney, William S.	MPS	305	24317
Roe, Arthur	OIP	536	25798
Romanoff, Elijah B.	BMS	323	24312
Romeo, Edna M., Mrs.	AD/NI	703	24316
Rones, Gwendolyn, Mrs.	DES	332	25858
Ronkin, Raphael R.	OIP	532	25807
Roseń, William G.	MPS	301	27377
Rosenthal, Albert H.	GPP	526	27320
Rosenthal, Alfred	GPP	527	27320
Ross, Luba A., Mrs.	GCO	638	25900
Rosseel, Anne M., Mrs.	GES	613	25740
Rostenbach, Royal E.	ENG	336	24280
Rosul, Joan M., Miss	PER	211	24118
Rothchild, Micheal	DMS	440	27380
Rothenberg, Nellie R., Mrs.	GES	609	25749
Routhenstein, Irene, Miss	GCO	630	25872
Rubinstein, Bertha W., Mrs.	SOC	316	24289

Name	DIVISION	ROOM NO	EXT
R			
Ruh, Marion B., Mrs.	EMS	449	24076
Ruth, H. Jeanette, Miss	BMS	323	24298
Ryan, Thomas M.	FMO	432	24374
Ryba, Theresa M., Miss	PES	605	25865
Ryer, David E.	O/D	518	24394
Ryff, John V.	MPS	303	27377
S			
Sale, Suzanne H., Mrs.	EMS	447	24077
Samson, Napoleon	O/D	232	24128
Sanders, Sharon Lynne, Mrs.	O/D	519	24394
Sandved, Kurt G.	OPP	345	24220
Sanow, Kenneth P.	EMS	451	24172
Sansbury, Belinda L., Miss	BMS	329	24260
Santos, Robert O.	EMS	405	24045
Saunders, Barbara A., Mrs.	BPA	429	24084
Savory, Carol A., Mrs.	MPS	341	24230
Schauer, Anne H., Mrs.	BMS	326	24314
Schiavone, Joseph J.	ADT	645	24092
Schilling, Howard S.	AD/A	220	25998
Schirnhofer, Edythe S., Mrs.	OPP	340	24221
Schlusser, Freda P., Miss	DMS	440	27380
Schubert, Christof N.	OCA	710	25960
Schuster, Joseph H.	EMS	448	24080
Schutt, Joseph J.	ADT	645	24088
Schutzman, Elias	ENG	336	24280
Schwartzman, Kathleen L., Mrs.	SOC	319	24204
Sciuchetti, Leo A.	UES	625	25915
Scopino, John A.	SIS	A1-234	25800
Scott, Barbara G., Miss	GCO	637	25933
Scott, John C.	SOC	318	24204
Scott, Patricia L.	OSG	537	25948
Scott, Richard P.	ASO	219	24070
Seelig, Walter R.	OPP	340	24250
Seeman, Gloria S., Mrs.	BPA	425	25793
Selcuk, Selim M.	SIS	A1-216	25706
Seltzer, Norman	EMS	404	24390
Sera, Mary E., Mrs.	BMS	323	24298
Sestili, Mary Ann, Miss	BMS	333	24264
Seymour, Lawrence A.	EMS	447	24077
Shannon, Mary C., Mrs.	BPA	429	25705
Share, Irene P., Mrs.	ENG	336	24280
Shaw, Mary L., Mrs.	GES	615	25741
Shedrick, Charlotte A.	PER	211	24106
Shelton, Robert W.	GCO	634	25926
Shepherd, Donald C.	OPP	346	24255
Sheppard, Frank C.	AD/A	525	25760
Sherman, Mildred C., Miss	AD/E	620	25888
Shifflett, Carol A., Mrs.	PER	212	24125
Showell, John S.	MPS	348	24276
Sias, Lillian D., Mrs.	UES	628	25896
Sickles, Patricia T., Mrs.	AD/I	417	24355
Sides, Joe D.	NCF	706	27339
Siegel, Louis	ADT	645	24148
Siegel, Richard W.	BMS	326	24314
Sievers, William E.	BMS	332	25846
Sills, Beatrice S., Mrs.	O/D	504	24010
Simmons, Jane U., Mrs.	ASO	223	24120
Simmons, Joan G., Mrs.	AD/A	525	25760
Sinclair, Rolf M.	MPS	305	24317
Sisco, Bernard	AD/A	525	25710
Sivchev, Mary, Miss	AD/I	421	24061
Slater, Whitney S.	BPA	428	25759
Slemmons, David B.	DES	311	24219
Sloper, Marleen I., Mrs.	PER	212	24124
Smith, Denzel, D.	AD/I	420	24032
Smith, Mulinda L., Mrs.	ADT	549	24148
Smith, Philip M.	OPP	340	24235
Smith, Susan D., Miss	OIP	532	25789
Snow, Joel A.	OIR	213	25863
Snuffer, Nancy M., Mrs.	FMO	436	24156
Snyder, R. Craig	MAO	547	24026
Snyder, John L.	UES	627	25909
Sohns, Ernest R.	OIP	536	25814
Solomon, Mary A., Miss	MPS	348	24276
Sordo, Martha J., Miss	GPP	543	27392
Spahr, Nancy R., Mrs.	DMS	440	27336
Spearing, Cecilia W., Miss	BMS	323	24312
Speck, S. Martin	GCO	630	24108
Speer, Donald A.	MPS	343	24272
Sprague, William K., Jr.	GCO	634	25950
Springer, Roland S.	GCO	638	25900
Spuhler, Harold A.	AD/I	423	27367
Stabnow, Barbara A., Mrs.	DES	310	24218
Staudt, David A.	DMS	440	25880
Staudte, Gladys M. J., Mrs.	AD/I	423	27367
Steadman, Eris M., Miss	GES	612	27310
Stein, Bernard R.	BPA	425	25754
Stennett, James L.	ADT	645	24068

Name	DIVISION	ROOM NO	EXT
S			
Stephens, Patricia A., Miss	GCO	634	25965
Stephens, Richard E.	GPP	526	27320
Stevens, Janell C., Mrs.	AD/I	420	24370
Stewart, Christine C., Mrs.	EMS	404	24392
Stewart, Eugene C.	GCO	634	25940
Stewart, William L.	EMS	447	24077
Stoddard, Eleanor H., Miss	EMS	641	24151
Stout, Glenn E.	DES	312	24189
Strait, Shirley R., Mrs.	AD/I	410	27364
Stratton, Lena R., Mrs.	AD/I	420	24370
Stutsman, Jane, Miss	PES	649	25982
Sulkin, Naomi A., Mrs.	SRPS	408	24328
Suskind, Sigmund R.	BMS	329	24260
Swago, Andrew W.	MPS	352	24320
Swartout, Bonnie M., Miss	GCO	634	25940
Szwast, Rose M., Miss	AD/A	525	25766
T			
Tabor, Douglas E.	GCO	634	25965
Taylor, Denise E., Miss	GCO	634	25926
Taylor, Hall	GES	616	25700
Taylor, Hannah E., Mrs.	DES	312	24190
Taylor, Maggie H., Mrs.	GCO	634	25965
Tehaan, Wilda G., Mrs.	PES	601	25716
Terry, Peggy J., Miss	GCO	640	25772
Tesi, Giorgio	NCF	706	27340
Thiel, Charles C., Jr.	ENG	336	24280
Thomas, Leon R.	FMO	435	24008
Thompson, Honora F., Miss	GES	614	25743
Thompson, Margaret Covey, Mrs.	GES	614	25757
Thompson, Margaret Cullen, Mrs.	UES	623	25902
Thompson, Marie M., Mrs.	OCA	710	25960
Thompson, Roselene C., Mrs.	AD/I	415	24355
Thompson, Warren E.	OIP	532	25829
Thomson, Rena C., Mrs.	GPP	527	25702
Tihila, Howard	ASO	220	24130
Todd, Edward P.	AD/R	320	24240
Toft, Robert J.	UES	627	25909
Tolton, Betty A., Mrs.	PES	605	25879
Tominosky, Audrey G., Mrs.	MPS	351	24196
Tompkins, Richard W.	ASO	231	24116
Toney, George R.	AD/NI	703	27360
Towler, Andrea M., Miss	AD/NI	703	27360
Trapnell, Edward R.	GPP	526	27320
Trent, Lorraine B., Mrs.	AD/I	532	25782

Name	DIVISION	ROOM NO	EXT
T			
Trout, L. Marie, Mrs.	ASO	237	24222
Troy, Alice S., Mrs.	FMO	434	24007
Trueheart, Russell, Jr.	ASO	231	24116
Trujillo, Marie P., Mrs.	GES	614	25757
Tune, Elizabeth S., Mrs.	O/D	504	24010
Turner, Mort D.	OPP	340	24256
Twiss, John R., Jr.	IDOE	701	27357
Tyler, David B.	BMS	323	24298
Tyler, Lottie J.	OGC	501	24388
Tyson, Faith D., Mrs.	GES	614	25740
U			
Utterback, Howard T., Jr.	MAO	544	24110
V			
Vaden, William	ASO	232	24129
Van Belleghem, Daniel J.	DMS	440	27371
Van Valkenburg, Alvin	DES	310	24213
Vance, Lovina, Miss	ASO	221	24140
Vanderberry, Anne J., Miss	OPP	340	24250
Verderber, Robert C.	SIS	A1-214	25818
Vermillion, Nancy E., Miss	MPS	305	24317
Vigen, Oscar C.	BPA	428	24379
Vitol, James L.	GCO	630	24892
Voit, Carol, Miss	OSG	537	25944
W			
Wagner, Leonore U., Mrs.	EMS	447	24079
Walker, Maurice M.	ASO	232	24129
Wallace, Deena M., Miss	IDOE	701	27356
Wallace, Marie M., Miss	ASO	223	24120
Waller, Pearl R., Mrs.	EMS	407	24334
Walston, Helen A., Mrs.	GCO	634	25938
Walters, Mary C., Miss	AD/A	525	25710
Ward, Gordon B.	SIS	A1-220	25850
Ward, Mary H., Mrs.	ASO	223	24120
Ward, William H.	FMO	434	24006
Warshaw, Israel	ENG	336	24280
Washington, Diane P., Mrs.	AD/A	525	25710
Watson, Robert F.	UES	625	25917
Watts, Frances O., Mrs.	AD/E	620	25891
Weal, Dolly J., Miss	DMS	440	27336
Weatherwax, Sarah J., Miss	GES	615	25741

Name	DIVISION	ROOM NO	EXT
W			
Wehn, M. Elise, Mrs.	AD/I	423	27367
Weiss, Edward C.	SIS	A1-213	25818
Wells, Betty Lou, Miss	BMS	323	24312
Wells, Mildred D., Mrs.	EMS	451	24172
Wells, Ruth O., Mrs.	DMS	440	27380
Wendell, Joseph A.	FMO	433	24016
Westfall, Jean, Mrs.	HS	439	24040
Wharton, Flora J., Mrs.	MPS	307	24302
Wheeler, Richard E.	ASO	232	24129
White, Bessie B.	PER	208	24107
White, Fred D.	DES	312	24198
White, Jearldine R., Mrs.	GES	615	25741
White, Lillian P., Mrs.	ENG	336	24280
White, Robert F.	ASO	232	24129
White, Sarah O., Mrs.	FMO	435	24008
Whitehead, Francis	ASO	223	24120
Whitmer, Charles A.	PES	601	25726
Wilcox, Peggy, Miss	GCO	633	25936
Wildman, Robert D.	OSG	631	25944
Wilkerson, Cora L.	AD/I	421	24061
Wilkins, Melinda H., Mrs.	DMS	440	27336
Williams, Ann M., Mrs.	NCF	706	25719
Williams, Charles W., Jr.	O/D	520	24363
Williams, Delores S., Mrs.	BPA	425	25774
Williams, Doris, M., Mrs.	FMO	431	24017
Williams, Elizabeth J., Miss	EMS	403	24168
Williams, Marjorie, Miss	MPS	351	24186
Williams, Oren F.	MPS	348	24276
Wilson, Azeal J., Mrs.	PER	439	24040
Wilson, Bernadette B., Mrs.	MAO	549	24110
Wilson, M. Kent	MPS	348	24262
Winslow, Ina M., Mrs.	BPA	425	25793
Wirths, Theodore W.	GPP	526	27320
Withrow, Alice P., Mrs.	GES	610	25747
Wolff, Mary L., Mrs.	BMS	326	24200
Woodall, Irene P., Mrs.	EMS	449	24076
Woods, Shirley, Miss	GCO	640	25920
Woodson, Robert A.	DMS	440	27380
Worthington, Randall	SIS	A1-220	25850
Worthington, Virginia A., Mrs.	UES	623	25929
Wright, Alfreda, Miss	PES	601	25957
Wright, James P.	MPS	351	24192
Wright, Valerie, Miss	BMS	328	24300
Wright, William E.	MPS	352	24320
Wyckoff, Peter H.	DES	312	24147
Y			
Young, Albert T.	AD/E	620	25898
Young, Beverly A., Miss	OSG	537	25948
Young, Lois E.	GCO	630	25860
Young, Simuel H.	FMO	435	24007
Z			
Zajac, Wayne D.	EMS	641	24178
Zalar, Charles	SIS	A1-216	25706
Zallen, Harold	UES	625	25913
Zapolsky, Harold S.	MPS	307	24302
Zito, Sandra L., Miss	UES	623	25924
Zwolenik, James J.	SRPS	641	24150

SOURCE: NSF, "National Science Foundation Telephone Directory."

National Science Board

Terms Expire May 10, 1972

CHARLES F. JONES, Vice Chairman of the Board, Humble Oil & Refining Co., Houston, Tex.

THOMAS F. JONES, Jr., President, University of South Carolina, Columbia, S.C.

*ROBERT S. MORISON, Professor of Science and Society, Program on Science, Technology, and Society, Cornell University, Ithaca, N.Y.

E. R. PIORE, Vice President and Chief Scientist, International Business Machines Corp., Armonk, N.Y.

JOSEPH M. REYNOLDS, Boyd Professor of Physics and Vice President for Instruction and Research, Louisiana State University, Baton Rouge, La.

ATHELSTAN F. SPILHAUS, Post Office Box 887, Palm Beach, Fla.

H. GUYFORD STEVER, President, Carnegie-Mellon University, Pittsburgh, Pa.

RICHARD H. SULLIVAN, Assistant to the President, Carnegie Corporation of New York, New York, N.Y.

Terms Expire May 10, 1974

R. H. BING, Rudolph E. Langer Professor of Mathematics, University of Wisconsin, Madison, Wis.

*HARVEY BROOKS, Gordon McKay Professor of Applied Physics and Dean of Engineering and Applied Physics, Harvard University, Cambridge, Mass.

WILLIAM A. FOWLER, Institute Professor of Physics, California Institute of Technology, Pasadena, Calif.

NORMAN HACKERMAN, President, William Marsh Rice University, Houston, Tex.

PHILIP HANDLER, President, National Academy of Sciences, Washington, D.C.

JAMES G. MARCH, David Jacks Professor of Higher Education, Political Science, and Sociology, School of Education, Stanford University, Stanford, Calif.

GROVER E. MURRAY, President, Texas Tech University, Lubbock, Tex.

FREDERICK E. SMITH, Professor of Advanced Environmental Studies in Resources and Ecology, Graduate School of Design, Harvard University, Cambridge, Mass.

Terms Expire May 10, 1976

*H. E. CARTER (Chairman, National Science Board), Vice Chancellor for Academic Affairs, University of Illinois, Champaign, Ill.

ROBERT A. CHARPIE, President, Cabot Corp., Boston, Mass.

LLOYD M. COOKE, Director of Urban Affairs, Union Carbide Corp., New York, N.Y.

ROBERT H. DICKE, Cyrus Fogg Brackett Professor of Physics, Department of Physics, Princeton University, Princeton, N.J.

DAVID M. GATES, Director, Missouri Botanical Garden, St. Louis, Mo.

*ROGER W. HEYNS (Vice Chairman, National Science Board), Chancellor, University of California at Berkeley, Berkeley, Calif.

FRANK PRESS, Chairman, Department of Earth and Planetary Sciences, Massachusetts Institute of Technology, Cambridge, Mass.

F. P. THIEME, President, University of Colorado, Boulder, Colo.

Member Ex Officio

*W. D. MCELROY, Director, National Science Foundation, Washington, D.C. (Chairman, Executive Committee)

* * *

VERNICE ANDERSON, Secretary, National Science Board, National Science Foundation, Washington, D.C.

*Member, Executive Committee.

Advisory Committees and Panels

ADVISORY COMMITTEE FOR BIOLOGICAL AND MEDICAL SCIENCES

David M. Gates
Director, Missouri Botanical Garden
St. Louis, Mo.

Nelson G. Hairston
Director, Museum of Zoology
University of Michigan
Ann Arbor, Mich.

J. Woodland Hastings (Chairman)
Biological Laboratories
Harvard University
Cambridge, Mass.

Anton Lang
Director, AEC Plant Research Laboratory
Michigan State University
East Lansing, Mich.

Lawrence R. Pomeroy
Department of Zoology
University of Georgia
Athens, Ga.

John W. Saunders (Vice Chairman)
Department of Biological Sciences
State University of New York
Albany, N.Y.

George Sayers
Department of Physiology
Case Western Reserve University
Cleveland, Ohio

Charles G. Sibley
Peabody Museum
Yale University
New Haven, Conn.

Richard L. Solomon
Department of Psychology
University of Pennsylvania
Philadelphia, Pa.

Norton D. Zinder
Rockefeller University
New York, N.Y.

ADVISORY COMMITTEE FOR COMPUTING ACTIVITIES

Gordon W. Blackwell
President
Furman University
Greenville, S.C.

Samuel D. Conte
Computer Sciences Department
Purdue University
Lafayette, Ind.

Sidney Fernbach
Head, Computation Division
Lawrence Radiation Laboratory
Livermore, Calif.

Wayne Holtzman
Dean, College of Education
University of Texas
Austin, Tex.

Thurston E. Manning
Vice President for Research and Planning
University of Colorado
Boulder, Colo.

Dwaine Marvick
Professor of Political Science
University of California, Los Angeles
Los Angeles, Calif.

Alan J. Perlis
Department of Computer Science
Carnegie-Mellon University
Pittsburgh, Pa.

Louis T. Rader
Department of Electrical Engineering
University of Virginia
Charlottesville, Va.

Ottis W. Rechard
Computer Science Department
Washington State University
Pullman, Wash.

J. T. Schwartz
Courant Institute of Mathematical Sciences
New York University
New York, N.Y.

Harrison Shull (Chairman)
Dean of the Graduate School
Indiana University
Bloomington, Ind.

Patrick C. Suppes
Institute for Mathematical Studies in the Social Sciences
Stanford University
Stanford, Calif.

ADVISORY COMMITTEE FOR ENGINEERING

Donald A. Dahlstrom
Vice President
EIMCO Corp.
Salt Lake City, Utah

Daniel C. Drucker (Deputy Chairman)
Dean, College of Engineering
University of Illinois
Urbana, Ill.

Arthur E. Humphrey
Department of Chemical Engineering
University of Pennsylvania
Philadelphia, Pa.

William K. Linvill
Department of Engineering Economic Systems
Stanford University
Palo Alto, Calif.

Charles L. Miller
Department of Civil Engineering
Massachusetts Institute of Technology
Cambridge, Mass.

Rustum Roy
Materials Research Laboratory
Pennsylvania State University
University Park, Pa.

Robert M. Saunders
School of Engineering
University of California, Irvine
Irvine, Calif.

Winfield W. Tyler (Chairman)
Vice President for Research
Xerox Corp.
Rochester, N.Y.

Robert E. Uhrig
Dean of Engineering
University of Florida
Gainesville, Fla.

Max L. Williams, Jr.
Dean, College of Engineering
University of Utah
Salt Lake City, Utah

ADVISORY COMMITTEE FOR ENVIRONMENTAL SCIENCES

Clarence R. Allen
Division of Geological Sciences
California Institute of Technology
Pasadena, Calif.

Richard M. Goody
Division of Engineering and Applied Physics
Harvard University
Cambridge, Mass.

Glenn R. Hilst
Traveler's Research Center, Inc.
Hartford, Conn.

John R. Hogness (Vice Chairman)
Executive Vice President
University of Washington
Seattle, Wash.

Richard H. Jahns (Chairman)
Dean, School of Earth Sciences
Stanford University
Stanford, Calif.

James Warren McKie
Department of Economics
Vanderbilt University
Nashville, Tenn.

Carl Oppenheimer
Department of Oceanography
Florida State University
Tallahassee, Fla.

Robert O. Reid
Texas A&M University
College Station, Tex.

Gilbert F. White
Department of Behavioral Science
University of Colorado
Boulder. Colo.

ADVISORY COMMITTEE FOR INSTITUTIONAL PROGRAMS

John A. D. Cooper
President
Association of American Medical Colleges
Washington, D.C.

William C. Friday
President
University of North Carolina at Chapel Hill
Chapel Hill, N.C.

James M. Hester
President
New York University
New York, N.Y.

Lyle H. Lanier
Executive Vice President
University of Illinois at Urbana
Urbana, Ill.

W. Deming Lewis
President
Lehigh University
Bethlehem, Pa.

Joseph McCarthy
Dean of Graduate School
University of Washington
Seattle, Wash.

Edgar F. Shannon, Jr. (Chairman)
President
University of Virginia
Charlottesville, Va.

George L. Simpson, Jr.
Chancellor
University of Georgia System
Atlanta, Ga.

Calvin A. Vanderwerf
President
Hope College
Holland, Mich.

ADVISORY COMMITTEE FOR MATHEMATICAL AND PHYSICAL SCIENCES

Bart J. Bok
Director, Steward Observatory
University of Arizona
Tucson, Ariz.

Gerald M. Clemence
Yale University Observatory
New Haven, Conn.

Paul J. Flory
Department of Chemistry
Stanford University
Palo Atlo, Calif.

Herman H. Goldstine (Chairman)
Research Division
IBM Corporation
Yorktown Heights, N.Y.

Irving Kaplansky
Department of Mathematics
University of Chicago
Chicago, Ill.

Leon M. Lederman
Department of Physics
Columbia University
New York, N.Y.

Howard Reiss
Department of Chemistry
University of California, Los Angeles
Los Angeles, Calif.

William P. Slichter
Bell Telephone Laboratories
Murray Hill, N.J.

George H. Vineyard (Vice Chairman)
Brookhaven National Laboratory
Upton, Long Island, N.Y.

ADVISORY COMMITTEE FOR PLANNING

Paul M. Doty (Vice Chairman)
Department of Chemistry
Harvard University
Cambridge, Mass.

Wayland C. Griffith (Chairman)
Vice President for Research and Technology
Lockheed Aircraft Corp.
Sunnyvale, Calif.

Milton Harris
Chairman of the Board of Directors
American Chemical Society
Washington, D.C.

Allyn W. Kimball
Dean of Arts and Sciences
Johns Hopkins University
Baltimore, Md.

Edwin Mansfield
Department of Economics
University of Pennsylvania
Philadelphia, Pa.

Emmanuel G. Mesthene
Program on Technology and Society
Harvard University
Cambridge, Mass.

David Z. Robinson
Vice President for Academic Affairs
New York University
New York, N.Y.

Robert L. Sproull
Provost
University of Rochester
Rochester, N.Y.

M. H. Trytten
Consultant to the President of the National Academy of Sciences
Washington, D.C.

Aaron B. Wildavsky
Dean, Graduate School of Public Affairs
University of California, Berkeley
Berkeley, Calif.

ADVISORY COMMITTEE FOR SCIENCE EDUCATION

George H. Baird
President and Executive Director
Educational Research Council
Cleveland, Ohio

Lionel V. Baldwin
Dean, College of Engineering
Colorado State University
Fort Collins, Colo.

H. Russell Beatty
President, Wentworth Institute
Boston, Mass.

Herbert J. Greenberg (Chairman)
Department of Mathematics
University of Denver
Denver, Colo.

John K. Hulm
Westinghouse Electric Corp.
Pittsburgh, Pa.

Allan A. Kuusisto
President
Hobart and William Smith Colleges
Geneva, N.Y.

J. Stanley Marshall
President, Florida State University
Tallahassee, Fla.

James W. Mayo
Department of Physics
Morehouse College
Atlanta, Ga.

James F. Nickerson
President, Mankato State College
Mankato, Minn.

Donald W. Stotler
Science Supervisor
Portland Public Schools
Portland, Oreg.

Allen F. Strehler
Dean of Graduate Studies
Carnegie-Mellon University
Pittsburgh, Pa.

Marion E. Marts
Dean of Summer Quarter
University of Washington
Seattle, Wash.

John C. McKinney
Department of Sociology and Anthropology
Duke University
Durham, N.C.

Advisory Committee for Social Sciences

Roger C. Buck
Department of History and Philosophy of Science
Indiana University
Bloomington, Ind.

Eugene A. Hammel
Department of Anthropology
University of California, Berkeley
Berkeley, Calif.

Albert H. Hastorf
Dean, School of Humanities and Science
Stanford University
Palo Alto, Calif.

David G. Hays
Chairman of Linguistics Program
State University of New York, Buffalo
Buffalo, N.Y.

Robert E. Lane
Department of Political Science
Yale University
New Haven, Conn.

Robert H. Strotz (Chairman)
Dean, College of Arts and Sciences
Northwestern University
Evanston, Ill.

Anthony F. C. Wallace (Vice Chairman)
Department of Anthropology
University of Pennsylvania
Philadelphia, Pa.

Robert B. Yegge
College of Law
University of Denver
Denver, Colo.

Science Information Council

Burton W. Adkinson
Head, Office of Science Information Service
National Science Foundation
Washington, D.C.

Herbert S. Bailey, Jr. (Chairman)
Director
Princeton University Press
Princeton, N.J.

Carey Croneis
Chancellor
Rice University
Houston, Tex.

Martin M. Cummings
Director
National Library of Medicine
Bethesda, Md.

Victor J. Danilov
Industrial Research, Inc.
Beverly Shores, Ind.

Bowen C. Dees
President
The Franklin Institute
Philadelphia, Pa.

Amitai W. Etzioni
Department of Sociology
Columbia University
New York, N Y.

Herman H. Fussler
University of Chicago Library
University of Chicago
Chicago, Ill.

Robert E. Gordon
Associate Dean
College of Science
University of Notre Dame
Notre Dame, Ind.

Clifford Grobstein
Vice Chancellor for Health Sciences and Dean of the School of Medicine
University of California, San Diego
La Jolla, Calif.

H. William Koch
Director
American Institute of Physics
New York, N.Y.

J. C. R. Licklider
Professor of Electrical Engineering
Massachusetts Institute of Technology
Cambridge, Mass.

L. Quincy Mumford
The Librarian of Congress
Washington, D.C.

John W. Murdock
Department of Economics and Information Research
Battelle Memorial Institute
Columbus, Ohio

Byron Riegel
Chemical Research and Development
Searle & Co.
Chicago, Ill.

John Sherrod
Director
National Agricultural Library
U.S. Department of Agriculture
Beltsville, Md.

John C. Weaver
President
University of Missouri
Columbia, Mo.

Leo Weins
President
The H. W. Wilson Co.
Bronx, N.Y.

F. Karl Willenbrock
Provost
Faculty of Engineering and Applied Sciences
State University of New York
Buffalo, N.Y.

Advisory Panel for Anthropology

Helen Codere
Department of Anthropology
Brandeis University
Waltham, Mass.

Harold C. Conklin
Department of Anthropology
Yale University
New Haven, Conn.

E. Mott Davis
Department of Anthropology
University of Texas at Austin
Austin, Tex.

Kent V. Flannery
Museum of Anthropology
University of Michigan
Ann Arbor, Mich.

John M. Roberts
Department of Anthropology
Cornell University
Ithaca, N.Y.

Gerald E. Williams
Department of Anthropology
University of Rochester
Rochester, N.Y.

Advisory Panel for Astronomy

Alan H. Barrett
Research Laboratory of Electronics
Massachusetts Institute of Technology
Cambridge, Mass.

Frank J. Kerr
Department of Physics and Astronomy
University of Maryland
College Park, Md.

Robert P. Kraft
Lick Observatory
University of California, Santa Cruz
Santa Cruz, Calif.

Charles R. O'Dell (Chairman)
Yerkes Observatory
Williams Bay, Wis.

Maarten Schmidt
California Institute of Technology
Pasadena, Calif.

Arne Slettebak
Perkins Observatory
Delaware, Ohio

Alexander G. Smith
Department of Physics
University of Florida
Gainesville, Fla.

Harlan J. Smith
Department of Astronomy
University of Texas
Austin, Tex.

James W. Warwick
Department of Astro-Geophysics
University of Colorado
Boulder, Colo.

Advisory Panel for Atmospheric Sciences

Alfred K. Blackadar
Department of Meteorology
Pennsylvania State University
University Park, Pa.

Neil M. Brice
Center for Radiophysics and Space Research
Cornell University
Ithaca, N.Y.

Robert L. Chasson
Department of Physics
University of Denver
Denver, Colo.

Thomas A. Donohue
Department of Physics
University of Pittsburgh
Pittsburgh, Pa.

Lewis O. Grant
Department of Atmospheric Sciences
Colorado State University
Fort Collins, Colo.

William E. Gordon
Vice President
Rice University
Houston, Tex.

James E. McDonald
Institute of Atmospheric Physics
University of Arizona
Tucson, Ariz.

Richard J. Reed
Department of Atmospheric Sciences
University of Washington
Seattle, Wash.

W. R. Derrick Sewell
Department of Geography
University of Victoria
Victoria, British Columbia
Canada

Advisory Panel for Biochemistry

Luigi C. Gorini
Department of Bacteriology
Harvard Medical School
Boston, Mass.

Elvin A. Kabat
Department of Microbiology
Columbia University
New York, N.Y.

Richard Y. Morita
Department of Microbiology
Oregon State University
Corvallis, Oreg.

Masayasu Nomura
Department of Genetics
University of Wisconsin
Madison, Wis.

Robert P. Perry
Institute for Cancer Research
Philadelphia, Pa.

John L. Westley
Department of Biochemistry
University of Chicago
Chicago, Ill.

Advisory Panel for Biological Oceanography

Edward Chin
Institute of Natural Resources
University of Georgia
Athens, Ga.

Rita R. Colwell
Department of Biology
Georgetown University
Washington, D.C.

Richard W. Eppley
Institute of Marine Resources
University of California, San Diego
La Jolla, Calif.

George D. Grice
Woods Hole Oceanographic Institution
Woods Hole, Mass.

Frederick A. Kalber
Aquatic Sciences, Inc.
Boca Raton, Fla.

Advisory Panel for Biophysics

Peter F. Curran
Department of Physiology
Yale University
New Haven, Conn.

Frederick L. Crane
Department of Biology
Purdue University
Lafayette, Ind.

David R. Davies
Laboratory of Molecular Biology
National Institutes of Health
Bethesda, Md.

R. Bruce Martin
Department of Chemistry
University of Virginia
Charlottesville, Va.

Carl Louis Schildkraut
Department of Cell Biology
Albert Einstein College of Medicine
Yeshiva University
New York, N.Y.

Robert F. Steiner
Division of Biochemistry
Naval Medical Research Institute
Bethesda, Md.

ADVISORY PANEL FOR CHEMISTRY

Theodore L. Brown
Department of Chemistry
University of Illinois
Urbana, Ill.

Robert L. Burwell, Jr.
Department of Chemistry
Northwestern University
Evanston, Ill.

James P. Collman
Department of Chemistry
Stanford University
Stanford, Calif.

John M. Deutch
Department of Chemistry
Massachusetts Institute of Technology
Cambridge, Mass.

Richard L. Hinman
Union Carbide Corp.
Tarrytown Technical Center
Tarrytown, N.Y.

Roald Hoffman
Department of Chemistry
Cornell University
Ithaca, N.Y.

Ronald E. Kagarise
Chemistry Division
U.S. Naval Research Laboratory
Washington, D.C.

William R. Krigbaum
Department of Chemistry
Duke University
Durham, N.C.

Larry L. Miller
Department of Chemistry
Colorado State University
Fort Collins, Colo.

Richard M. Noyes
Department of Chemistry
University of Oregon
Eugene, Oreg.

Robert W. Parry
Department of Chemistry
University of Utah
Salt Lake City, Utah

Irving Shain (Chairman)
Department of Chemistry
University of Wisconsin
Madison, Wis.

ADVISORY PANEL FOR COMPUTER SCIENCE

Bruce W. Arden
University of Michigan Computer Center
Ann Arbor, Mich.

William F. Atchison
Director, Computer Science Center
University of Maryland
College Park, Md.

Samuel D. Conte (Chairman)
Computer Science Department
Purdue University
Lafayette, Ind.

Juris Hartmanis
Computer Science Department
Cornell University
Ithaca, N.Y.

Harry D. Huskey
Computer Center
University of California, Santa Cruz
Santa Cruz, Calif.

Marvin Minsky
Computer Science Department
Massachusetts Institute of Technology
Cambridge, Mass.

ADVISORY PANEL FOR DEVELOPMENTAL BIOLOGY

Ursula K. Abbott
Department of Avian Sciences
University of California, Davis
Davis, Calif.

Fotis C. Kafatos
Biological Laboratories
Harvard University
Cambridge, Mass.

James W. Lash
Department of Anatomy
University of Pennsylvania
Philadelphia, Pa.

Peter M. Ray
Department of Biological Sciences
Stanford University
Palo Alto, Calif.

Lionel I. Rebhun
Department of Biology
University of Virginia
Charlottesville, Va.

Richard A. Rifkind
Department of Medicine
Columbia University
New York, N.Y.

Robert H. Rownd
Laboratory of Molecular Biology
University of Wisconsin
Madison, Wis.

Joseph E. Varner
MSU/AEC Plant Research Laboratory
Michigan State University
East Lansing, Mich.

ADVISORY PANEL FOR EARTH SCIENCES

Don L. Anderson
Division of Geological Sciences
California Institute of Technology
Pasadena, Calif.

Sydney P. Clark, Jr.
Department of Geophysics
Yale University
New Haven, Conn.

Allan V. Cox
Department of Geophysics
Stanford University
Stanford, Calif.

Hans P. Eugster
Department of Geology
Johns Hopkins University
Baltimore, Md.

Robert N. Ginsburg
School of Marine and Atmospheric Sciences
University of Miami
Miami, Fla.

Stanley R. Hart
Department of Terrestrial Magnetism
Carnegie Institution of Washington
Washington, D.C.

Dallas L. Peck
U.S. Geological Survey
Washington, D.C.

David M. Raup
Department of Geology
University of Rochester
Rochester, N.Y.

M. Gene Simmons
Department of Geology and Geophysics
Massachusetts Institute of Technology
Cambridge, Mass.

George H. Sutton
Hawaii Institute of Geophysics
University of Hawaii
Honolulu, Hawaii

Advisory Panel for Ecology

Thomas Brock
Department of Bacteriology
Indiana University
Bloomington, Ind.

Lincoln P. Brower
Department of Biology
Amherst College
Amherst, Mass.

George H. Lauff
Kellogg Biological Station
Michigan State University
Hickory Corners, Mich.

Paul S. Martin
Geochronology Laboratories
University of Arizona
Tucson, Ariz.

Harold A. Mooney
Department of Biological Sciences
Stanford University
Palo Alto, Calif.

Kenneth Norris
Department of Zoology
University of California, Los Angeles
Los Angeles, Calif.

Luigi Provasoli
Osborn Memorial Laboratories
Yale University
New Haven, Conn.

David E. Reichle
Oak Ridge National Laboratory
Oak Ridge, Tenn.

Lawrence B. Slobodkin
Evolution and Ecology Program
State University of New York
Stony Brook, N.Y.

Earl L. Stone
Department of Agronomy
Cornell University
Ithaca, N.Y.

Advisory Panel for Economics

Robert H. Haveman
Resources for the Future, Inc.
Washington, D.C.

Charles C. Holt
Urban Institute
Washington, D.C.

Daniel L. McFadden
Department of Economics
University of California, Berkeley
Berkeley, Calif.

Marc Nerlove
Department of Economics
University of Chicago
Chicago, Ill.

Vernon L. Smith
Department of Economics
University of Massachusetts
Amherst, Mass.

Jaroslav Vanek
Department of Economics
Cornell University
Ithaca, N.Y.

Advisory Panel for Genetic Biology

David Baltimore
Department of Biology
Massachusetts Institute of Technology
Cambridge, Mass.

Cedric I. Davern
Professor of Biology
University of California, Santa Cruz
Santa Cruz, Calif.

Rowland H. Davis
Department of Botany
University of Michigan
Ann Arbor, Mich.

Burke H. Judd
Department of Zoology
University of Texas
Austin, Tex.

Richard C. Lewontin
Department of Zoology
University of Chicago
Chicago, Ill.

Oliver E. Nelson
Laboratory of Genetics
University of Wisconsin
Madison, Wis.

Herschel L. Roman
Department of Genetics
University of Washington
Seattle, Wash.

Charles Yanofsky
Department of Biological Sciences
Stanford University
Stanford, Calif.

Advisory Panel for History and Philosophy of Science

Peter Achinstein
Department of Philosophy
Johns Hopkins University
Baltimore, Md.

Garland E. Allen
Department of Biology
Washington University
St. Louis, Mo.

Roderick Chisholm
Department of Philosophy
Brown University
Providence, R.I.

Robert S. Cohen
Department of Physics
Boston University
Boston, Mass.

Thomas S. Kuhn
Program in History and Philosophy of Science
Princeton University
Princeton, N.J.

Henry E. Kyburg
Department of Philosophy
University of Rochester
Rochester, N.Y.

Advisory Panel for Institutional Computing Services

Richard Andree
Department of Mathematics
University of Oklahoma
Norman, Okla.

Wilfrid J. Dixon
School of Medicine
University of California, Los Angeles
Los Angeles, Calif.

William S. Dorn
Department of Mathematics
University of Denver
Denver, Colo.

Earle C. Fowler
Department of Physics
Duke University
Durham, N.C.

Edwin Kuh
Department of Economics
Massachusetts Institute of Technology
Cambridge, Mass.

David E. Lamb
Department of Statistics and Computer Science
University of Delaware
Newark, Del.

Max Vernon Mathews
Bell Telephone Laboratories, Inc.
Murray Hill, N.J.

Gilbert D. McCann
Willis H. Booth Computing Center
California Institute of Technology
Pasadena, Calif.

Ernest P. Miles, Jr.
Computing Center and Department of Mathematics
Florida State University
Tallahassee, Fla.

Benjamin Mittman
Director Vogelback Computer Center
Northwestern University
Evanston, Ill.

Ottis W. Rechard (Chairman)
Computer Science Department
Washington State University
Pullman, Wash.

Zevi Salsburg
Department of Chemistry
Rice University
Houston, Tex.

Sally Y. Sedelow
Department of Computer and Information Science
University of North Carolina
Chapel Hill, N.C.

ADVISORY COUNCIL FOR MANPOWER AND EDUCATION STUDIES PROGRAMS

Harold Goldstein
Bureau of Labor Statistics
U.S. Department of Labor
Washington, D.C.

Raymond Jacobson
Civil Service Commission
Washington, D.C.

Albert Kay
The Pentagon
Washington, D.C.

Thomas J. Mills (Chairman)
Office of Economic and Manpower Studies
National Science Foundation
Washington, D.C.

William Mirengoff
U.S. Department of Labor
Washington, D.C.

Elliot S. Pierce
U.S. Atomic Energy Commission
Washington, D.C.

Robert H. Rankin
Selective Service Commission
Washington, D.C.

Conrad Taeuber
Bureau of the Census
U.S. Department of Commerce
Washington, D.C.

Margaret West
National Institutes of Health
Bethesda, Md.

ADVISORY PANEL FOR MATHEMATICAL SCIENCES

Richard D. Anderson
Department of Mathematics
Louisiana State University
Baton Rouge, La.

R. Creighton Buck
Department of Mathematics
University of Wisconsin
Madison, Wis.

Murray Gerstenhaber
Department of Mathematics
University of Pennsylvania
Philadelphia, Pa.

Robert C. James
Department of Mathematics
Claremont Graduate School and University Center
Claremont, Calif.

William J. LeVeque
Department of Mathematics
University of Michigan
Ann Arbor, Mich.

Gerald J. Lieberman
Department of Statistics and Operations Research
Stanford University
Stanford, Calif.

Richard S. Palais
Department of Mathematics
Brandeis University
Waltham, Mass.

Abraham Robinson
Department of Mathematics
Yale University
New Haven, Conn.

Hans F. Weinberger
Department of Mathematics
University of Minnesota
Minneapolis, Minn.

ADVISORY PANEL FOR METABOLIC BIOLOGY

Robert W. Bernlohr
Department of Microbiology
University of Minnesota
Minneapolis, Minn.

Oscar M. Hechter
Institute for Biomedical Research
American Medical Association
Chicago, Ill.

Rachmiel Levine
Department of Medicine
New York Medical College
New York, N.Y.

Richard L. Malvin
Department of Physiology
University of Michigan
Ann Arbor, Mich.

Harry Rudney
Department of Biological Chemistry
University of Cincinnati
Cincinnati, Ohio

Anthony San Pietro
Department of Botany
Indiana University
Bloomington, Ind.

Charles L. Wadkins
Department of Biochemistry
University of Arkansas
Little Rock, Ark.

Willis A. Wood
Department of Biochemistry
Michigan State University
East Lansing, Mich.

ADVISORY PANEL FOR OCEANOGRAPHY

Robert S. Arthur
Scripps Institution of Oceanography
University of California, San Diego
La Jolla, Calif.

John V. Byrne
Department of Oceanography
Oregon State University
Corvallis, Oreg.

John E. Nafe
Department of Geology
Lamont-Doherty Geological Observatory
Columbia University
Palisades, N.Y.

Worth D. Nowlin
Department of Oceanography
Texas A&M University
College Station, Tex.

H. Gote Ostlund
Dorothy H. and Lewis Rosenstiel School of Marine and Atmospheric Sciences
University of Miami
Miami, Fla.

Karl K. Turekian
Department of Geology
Yale University
New Haven, Conn.

ADVISORY PANEL FOR OCEANOGRAPHIC FACILITIES

Parke A. Dickey
Department of Geology
University of Tulsa
Tulsa, Okla.

Robert G. Paquette
AC Electronics Defense Research Laboratories
Goleta, Calif.

Max Silverman
Scripps Institution of Oceanography
University of California, San Diego
La Jolla, Calif.

Warren C. Thompson
Department of Meteorology and Oceanography
U.S. Naval Postgraduate School
Monterey, Calif.

William R. Walton
Pan American Petroleum Corp.
Tulsa Okla.

Advisory Panel for Physics

Manfred A. Biondi
Department of Physics
University of Pittsburgh
Pittsburgh, Pa.

Charles K. Bockelman
Nuclear Structure Laboratory
Yale University
New Haven, Conn.

Walter L. Brown
Bell Telephone Laboratories
Murray Hill, N.J.

Herbert B. Callen (Chairman)
Department of Physics
University of Pennsylvania
Philadelphia, Pa.

W. Dale Compton
Department of Physics
University of Illinois
Urbana, Ill.

Russell J. Donnelly
Department of Physics
University of Oregon
Eugene, Oreg.

David Feldman
Department of Physics
Brown University
Providence, R.I.

Ronald Geballe
Department of Physics
University of Washington
Seattle, Wash.

Harry E. Gove
Nuclear Structure Research Laboratory
University of Rochester
Rochester, N.Y.

Burton J. Moyer
Lawrence Radiation Laboratory
University of California, Berkeley
Berkeley, Calif.

Richard Wilson
Department of Physics
Harvard University
Cambridge, Mass.

Advisory Panel for Polar Programs

Laurence M. Gould (Chairman)
Professor of Geology
College of Mines
University of Arizona
Tucson, Ariz.

Paul C. Daniels
Lakeville, Conn.

Laurence Irving
Institute of Arctic Biology
University of Alaska
College, Alaska

Heinz H. Lettau
Department of Meteorology
University of Wisconsin
Madison, Wis.

Ernst Stuhlinger
George C. Marshall Space Flight Center
National Aeronautics and Space Administration
Huntsville, Ala.

Advisory Panel for Political Science

Frank Munger
Department of Political Science
University of Florida
Gainesville, Fla.

James Prothro
Institute for Research in Social Science
University of North Carolina
Chapel Hill, N.C.

Robert E. Scott
Department of Political Science
University of Illinois
Urbana, Ill.

S. Sidney Ulmer
Department of Political Science
University of Kentucky
Lexington, Ky.

Sidney Verba
Department of Political Science
University of Chicago
Chicago, Ill.

John C. Wahlke
Department of Political Science
University of Iowa
Iowa City, Iowa

Advisory Panel for Psychobiology

Jacob Beck
Department of Psychology
University of Oregon
Eugene, Oreg.

Russell M. Church
Department of Psychology
Brown University
Providence, R.I.

Bert F. Green
Department of Psychology
Johns Hopkins University
Baltimore, Md.

Frank McKinney
Museum of Natural History
University of Minnesota
Minneapolis, Minn.

Donald A. Riley
Department of Psychology
University of California, Berkeley
Berkeley, Calif.

Allen W. Stokes
Department of Wildlife Resources
Utah State University
Logan, Utah

Garth J. Thomas
Center for Brain Research
University of Rochester
Rochester, N.Y.

Richard F. Thompson
Department of Psychobiology
University of California, Irvine
Irvine, Calif.

Ad Hoc Review Committee for Radio Astronomy Facilities

Bart J. Bok
Steward Observatory
University of Arizona
Tucson, Ariz.

Stirling A. Colgate
New Mexico Institute of Mining and Technology
Socorro, N. Mex.

Robert H. Dicke (Chairman)
Department of Physics
Princeton University
Princeton, N.J.

Rudolph Kompfner
Electronics and Radio Research
Bell Telephone Laboratories, Inc.
Holmdel, N.J.

William W. Morgan
Yerkes Observatory
Williams Bay, Wis.

Eugene N. Parker
Enrico Fermi Institute for Nuclear Studies
University of Chicago
Chicago, Ill.

Merle A. Tuve
Department of Terrestrial Magnetism
Carnegie Institution of Washington
Washington, D.C.

Gart Westerhout
Astronomy Program
University of Maryland
College Park, Md.

ADVISORY PANEL FOR REGULATORY BIOLOGY

Samuel Aronoff
Boston College
Chestnut Hill, Mass.

John M. Brookhart
Department of Physiology
University of Oregon
Portland, Oreg.

Sam L. Clark
Department of Nutrition
Harvard School of Public Health
Boston, Mass.

J. M. Daly
Department of Biochemistry and Nutrition
University of Nebraska
Lincoln, Nebr.

William Etkin
Department of Anatomy
Albert Einstein College of Medicine
Yeshiva University
New York, N.Y.

Harold T. Hammel
Physiological Research Laboratory
Scripps Institution of Oceanography
University of California, San Diego
La Jolla, Calif.

Riley Housewright
Fort Detrick
Frederick, Md.

Herbert A. Roeller
Department of Biology
Texas A&M University
College Station, Tex.

Gerhard Werner
Department of Pharmacology
University of Pittsburgh
Pittsburgh, Pa.

ADVISORY PANEL FOR SCIENCE DEVELOPMENT

William Bevan
Vice President and Provost
Johns Hopkins University
Baltimore, Md.

Theodore H. Bullock
University of California, San Diego
San Diego, Calif.

Bryce L. Crawford, Jr.
Dean of the Graduate School
University of Minnesota
Minneapolis, Minn.

James D. Ebert
Department of Embryology
Carnegie Institution of Washington
Baltimore, Md.

Philip M. Hauser
Department of Sociology
University of Chicago
Chicago, Ill.

Howard R. Neville
President
Claremont Men's College
Claremont, Calif.

Joseph M. Pettit
Dean, School of Engineering
Stanford University
Palo Alto, Calif.

Ragnar Rollefson
Department of Physics
University of Wisconsin
Madison, Wis.

O. Meredith Wilson
Director, Center for Advanced Study in the Behavioral Sciences
Stanford, Calif.

ADVISORY PANEL FOR SEA GRANT INSTITUTIONAL SUPPORT

Sanford S. Atwood
President
Emory University
Atlanta, Ga.

Thomas Barrow
Humble Oil & Refining Co.
Houston, Tex.

Phillip Eisenberg
Hydronautics, Inc.
Laurel, Md.

Roy Gaul
Westinghouse Research Laboratories
Ocean Research Laboratory
San Diego, Calif.

LeVan Griffis
Vice Provost
Southern Methodist University
Dallas, Tex.

Joseph E. Henderson
Applied Physics Laboratory
University of Washington
Seattle, Wash.

David S. Potter
General Motors Corp.
Milwaukee, Wis.

Jack Ruina
Vice President for Special Laboratories
Massachusetts Institute of Technology
Cambridge, Mass.

H. Burr Steinbach
Woods Hole Oceanographic Institute
Woods Hole, Mass.

James H. Wakelin, Jr.
Ryan Aeronautical Co.
Washington, D.C.

ADVISORY PANEL FOR SEA GRANT PROJECT SUPPORT

Lewis M. Alexander
Department of Geography
University of Rhode Island
Kingston, R.I.

John H. Busser
American Institute of Biological Sciences
Washington, D.C.

Joseph Morton Caldwell
Coastal Engineering Research Center
U.S. Army Corps of Engineers
Washington, D.C.

Lincoln D. Cathers
Deep Submergence Systems Project Office
Chevy Chase, Md.

Francis T. Christy, Jr.
Resources for the Future, Inc.
Washington, D.C.

Lewis Eugene Cronin
Director, Natural Resources Institute
Chesapeake Biological Laboratory
Solomons, Md.

James A. Crutchfield
Department of Economics
University of Washington
Seattle, Wash.

Jacob J. Dykstra
Point Judith Fisherman's Cooperative Association
Point Judith, R.I.

Sylvia A. Earle
Farlow Herbarium
Harvard University
Cambridge, Mass.

Haven Emerson
Oceans General, Inc.
Miami, Fla

Herbert Frolander
Office of Dean of Research
Oregon State University
Corvallis, Oreg.

Roger W. Fulling
E. I. DuPont de Nemours & Co.
Wilmington, Del.

William Gaither
University of Delaware
Newark, Del.

Donald Hood
Institute of Marine Science
University of Alaska
College, Alaska

John Dove Isaacs
Scripps Institution of Oceanography
University of California
La Jolla, Calif.

Paul M. Jacobs
The Gorton Corp.
Gloucester, Mass.

Milton G. Johnson
Environmental Science Services Administration
Rockville, Md.

S. Russell Keim
Committee on Ocean Engineering
National Academy of Engineering
Washington, D.C.

William J. Hargis, Jr.
Virginia Institute of Marine Science
Gloucester Point, Va.

John Lyman
Department of Environmental Sciences and Engineering
University of North Carolina
Chapel Hill, N.C.

Arthur E. Maxwell
Woods Hole Oceanographic Institution
Woods Hole, Mass.

H. Crane Miller
U.S. Senate
Washington, D.C.

Johnes K. Moore
Marblehead, Mass.

Michael Neushul, Jr.
Department of Biology
University of California, Santa Barbara
Santa Barbara, Calif.

Ross F. Nigrelli
New York Zoological Society
Brooklyn, N.Y.

Arthur Francis Novak
Baton Rouge, La.

H. T. Odum
Department of Zoology
University of North Carolina
Chapel Hill, N.C.

John Padan
Bureau of Mines
U.S. Department of the Interior
Tiberon, Calif.

Gerard R. Pomerat
San Francisco, Calif.

Lawrence Pomeroy
Department of Zoology
University of Georgia
Athens, Ga.

Adrian Richards
Center for Marine Studies
Lehigh University
Bethlehem, Pa.

Russell Otto Sinhuber
Department of Food Science and Technology
Oregon State University
Corvallis, Oreg.

James Marion Snodgrass
Scripps Institution of Oceanography
University of California
La Jolla, Calif.

Harris B. Stewart, Jr.
Atlantic Oceanographic Laboratories
Miami, Fla.

Richard Timme
Interstate Electronics Corp.
Anaheim, Calif.

Merrill A. True
Department of Biology
Tulane University
New Orleans, La.

Allyn Collins Vine
Woods Hole Oceanographic Institution
Woods Hole, Mass.

David Wallace
New York State Conservation Department
Ronhomkoma, N.Y.

Elizabeth M. Wallace
Oyster Institute of North America
Sayville, Long Island, N.Y.

Advisory Panel for Sociology and Social Psychology

J. Stacy Adams
School of Business Administration
University of North Carolina
Chapel Hill, N.C.

Herbert L. Costner
Department of Sociology
University of Washington
Seattle, Wash.

Morton Deutsch
Department of Psychology
Columbia University
New York, N.Y.

Kenneth J. Gergen
Department of Psychology
Swarthmore College
Swarthmore, Pa.

Warren O. Hagstrom
Department of Sociology
University of Wisconsin
Madison, Wis.

Leonard Reissman
Department of Sociology
Tulane University
New Orleans, La.

Philip C. Sagi
Department of Sociology
University of Pennsylvania
Philadelphia, Pa.

Robert B. Zajonc
Research Center for Group Dynamics
Institute for Social Research
University of Michigan
Ann Arbor, Mich.

Advisory Panel for Systematic Biology

Harlan P. Banks
Division of Biological Sciences
Cornell University
Ithaca, N.Y.

Theodore J. Crovello
Department of Biology
University of Notre Dame
Notre Dame, Ind.

David E. Fairbrothers
Department of Botany
Rutgers, the State University
New Brunswick, N.J.

Morris Goodman
Department of Anatomy
Wayne State University
Detroit, Mich.

Paul D. Hurd, Jr.
Department of Entomology
University of California, Berkeley
Berkeley, Calif.

Richard F. Johnston
Museum of Natural History
University of Kansas
Lawrence, Kans.

Eugene N. Kozloff
Friday Harbor Laboratories
University of Washington
Friday Harbor, Wash.

Malcolm C. McKenna
Department of Geology and Paleontology
American Museum of Natural History
New York, N.Y.

SOURCE: NSF, "National Science Foundation Annual Report 1970," 1971.

Part 4

NSF SUPPORT PROGRAMS

I. SCIENTIFIC RESEARCH

The National Science Foundation provides comprehensive support to research in all the sciences. Mechanisms through which research is supported include: (1) project grants to scientists, primarily at universities and colleges; (2) five university-administered, Government-owned National Research Centers available to all qualified scientists; (3) federally administered cooperative National Research Programs of a specialized nature. In addition, the Foundation assists in the procurement of specialized research equipment and facilities. The Foundation is the only Federal agency that is expressly authorized by statute to initiate and support scientific research that is not mission-oriented.

The Foundation considers all proposals for the support of research, regardless of source. The majority of such requests are submitted by U.S. universities and colleges on behalf of individual scientists or groups of scientists on their faculties. Foundation policy is to emphasize research that contributes to graduate and postdoctoral education in the sciences. Support of research at foreign institutions is provided only when it is clearly in the interest of science in the United States.

Research proposals are considered primarily on the basis of scientific merit. Scientific merit is assessed according to the promise of significant scientific results, the possible scientific impact, the probable opening of a new field, the educational by-products, and potential applications.

Programs to support scientific research are described in this section, with the following exceptions. Research in scientific information is found in section VI under that heading. Research related to development of marine resources and to computing activities is listed separately under section IV, Combined Scientific Research and Education Programs, inasmuch as they combine research and education functions within a single program.

Scientific Research Projects

The National Science Foundation awards grants to support research in science, engineering, and mathematics. On rare occasions research support may take the form of a contract rather than a grant; proposals directed at grants or contracts are prepared in an identical manner.

A research project grant may support either a specific research project or general research in a coherent area of science.

Research support is given to the full spectrum of sciences, including:

Biological & Medical Sciences
(excludes clinical aspects)
cellular biology; envircnmental and systematic biology; molecular biology; physiological processes; psychobiology; biological oceanography

Engineering
engineering chemistry; engineering energetics; engineering materials; engineering mechanics; engineering systems

Mathematical & Physical Sciences
astronomy; chemistry; mathematics; physics

Social Sciences
anthropology; economic and social geography; economics; history and philosophy of science; linguistics; political science; social psychology; sociology

Environmental Sciences
atmospheric sciences; earth sciences; oceanography

Institutions are required to share in the cost of each research project supported by an NSF grant; this may be accomplished by a contribution to any cost element in the project, direct or indirect. Before submitting a proposal for research support the pamphlet **Grants for Scientific Research** should be consulted. The Foundation does not provide standard application forms for research proposals.

Grants normally provide support for periods up to 24 months, but under certain circumstances can be made for periods up to a maximum of 60 months. For projects of high scientific merit initial funding may be for two years with assurance of support for the full term of the project, contingent upon the availability of funds and the scientific progress of the research.

Eligibility

Proposals may be submitted by colleges and universities and by academically related nonprofit research organizations. The conditions under which support is occasionally provided to other types of organizations and to individuals is described in the NSF pamphlet **Grants for Scientific Research,** available from the address below. Inquiry may also be made directly to the Assistant Director for Research.

Deadlines

Proposals may be submitted at any time. Approximately three to six months are required for consideration of a proposal. Proposals requesting renewal support should be submitted at least six months in advance of the anticipated termination date of the existing grant in order to assure uninterrupted support.

Additional Information

Communications may be addressed to the appropriate division: Division of Biological and Medical Sciences, Division of Engineering, Division of Environmental Sciences, Division of Mathematical and Physical Sciences, or Division of Social Sciences; National Science Foundation, Washington, D. C. 20550.

Interdisciplinary Research Relevant to the Problems of Our Society (IRRPOS)

The National Science Foundation awards grants and contracts to support basic and applied interdisciplinary research on projects seeking to increase the fund of knowledge needed to resolve an important problem of society. The Foundation does not intend at present to specify problem areas for study. Prospective grantees should consider their particular capabilities as they apply to problems of national concern. The disciplines required for the interdisciplinary research effort will be largely determined by the nature of the societal problem under study.

The IRRPOS program complements other NSF programs for support of scientific research in specific disciplines and, therefore, the interdisciplinary character of the proposed research usually will be an important factor in considering a project for support under this program. Grants normally provide support for a period up to 24 months. Projects of specific interest to the missions of other Federal agencies normally will not be supported. Social action programs will not be supported.

Institutions are required to share in the cost of each research project supported by an NSF research grant. This may be accomplished by a contribution to any cost element in the project, direct or indirect.

Eligibility

Proposals for grants or contracts for research support may be submitted by colleges and universities and by academically related nonprofit research organizations. The conditions under which support is occasionally provided to other types of organizations and to individuals are described in the NSF pamphlet **Grants for Scientific Research,** available from the Foundation.

Deadlines

Proposals may be submitted at any time. However prior discussion with the Office of Interdisciplinary Research is recommended. Approximately three to six months are required for consideration of a proposal.

Additional Information

In addition to the NSF pamphlet **Grants for Scientific Research,** supplementary guidelines for the IRRPOS program are available. Communications may be addressed to: Office of Interdisciplinary Research, National Science Foundation, Washington, D. C. 20550.

International Decade of Ocean Exploration

In support of the International Decade of Ocean Exploration, the National Science Foundation awards grants and contracts to colleges and universities for broad, interdisciplinary, cooperative programs of ocean research and exploration with emphasis on environmental quality, environmental prediction, and seabed assessment. Priority is placed on rapid analysis and data sharing. The Decade is unique in that it recognizes that a major share of world oceanographic effort must be devoted to globally planned and coordinated study of the ocean as a system, for the benefit of mankind.

The long-range goals of the Decade are:

(1) to preserve the ocean environment by accelerating scientific observation of the natural state of the ocean and its interactions with the continental margins;

(2) to develop and improve an ocean forecasting and monitoring system, to facilitate prediction of oceanographic and atmospheric conditions, and to reduce hazards to life and property and permit more effective use of marine resources;

(3) to expand seabed assessment activities, to permit better management of ocean mineral exploration and exploitation;

(4) to improve worldwide oceanographic data exchange;

(5) to increase opportunities for international sharing of responsibilities and costs for ocean exploration and assure better use of limited exploration resources.

The United States national program will be coordinated closely with the Long-term and Expanded Program of Ocean Exploration and Research of the International Oceanographic Commission of UNESCO. The Comprehensive Outline of the Scope of this program was endorsed by the General Assembly of the United Nations in December 1969. The International Decade of Ocean Exploration has been approved as an important element of this program. The criteria of the Expanded Program which could be applied as appropriate in selecting cooperative projects, and which could also serve as criteria for the Decade, are:

"(1) Member States are willing to participate actively in the project;

(2) The project can be carried out most effectively through international cooperative action;

(3) The project has a sound scientific basis and is well designed to yield significant new information;

(4) The project will provide information and understanding that will contribute to the goal of enhanced utilization of the ocean and its resources;

(5) The project will help meet the needs of developing countries."

"A project that satisfied all those criteria would be an extremely strong candidate for inclusion in the Expanded Program. It will not be necessary in each case that all criteria be met, but the willingness of Member States to participate is clearly essential."

The Decade is a new program, funded for the first time in the fiscal year beginning July 1, 1970.

Additional Information

Communications may be addressed to: Office for the International Decade of Ocean Exploration, National Science Foundation, Washington, D. C. 20550.

U.S. Antarctic Research Program

The National Science Foundation awards grants to support research projects in all fields of science pertinent to the Antarctic, including both field work in Antarctica and study in the United States of specimens or data already gathered. On occasion research support may take the form of a contract rather than a grant.

The U.S. Antarctic Research Program supports research projects in the fields of: behavioral sciences; biology; cartography; geology; glaciology; meteorology; oceanography; terrestrial physics; upper atmosphere physics. Logistic support operations for scientific and other programs in Antarctica are carried out by the U.S. Navy.

Support is also given for Antarctic science information activities (translating, abstracting, indexing of literature, cataloguing, sorting, preservation of specimens) and for polar research centers.

Institutions are required to share in the cost of research projects supported by an NSF grant.

Grants are normally made for a period of 12 months, but under certain circumstances can be made for periods up to a maximum of 60 months. For projects of high scientific merit initial funding may be for two years with assurance of support for the full term of the project, contingent upon the availability of funds and the scientific progress of the research.

Eligibility

Proposals for grants or contracts for research project support may be submitted by colleges and universities and by academically related nonprofit research organizations. The conditions under which support is occasionally provided to other types of organizations are described in the NSF pamphlet **Grants for Scientific Research,** available from the Foundation.

Before submitting a proposal for research support, the pamphlet **Grants for Scientific Research** should be consulted. Scientists are encouraged to discuss their plans by letter or in person before submitting formal proposals. The Foundation does not provide standard application forms for proposals.

Deadlines

Proposals should be submitted by February 1 for work in Antarctica during the following austral summer (October to February).

Additional Information

Communications may be addressed to: Office of Polar Programs, National Science Foundation, Washington, D. C. 20550.

Arctic Research Program

The National Science Foundation has been assigned responsibility as lead agency for Arctic environmental research, with the advice of the Interagency Arctic Research Coordinating Committee (IARCC). The Foundation has accordingly established the Arctic Research Program to provide support for academic research and to coordinate the Foundation program with those of other Federal agencies through IARCC. This program will be initiated in fiscal year 1971.

The Foundation has in the past supported activities in the Arctic region through grants and contracts awarded by existing program elements of various offices and divisions. These programs will continue to support such activities. Proposals for research projects in a specific scientific discipline should be addressed to the appropriate division of the Assistant Director for Research. (Page 314) The Arctic Research Program will support projects of an interdisciplinary nature, including field investigations that require logistic arrangements and/or interagency or international cooperation, as well as the subsequent analysis of data.

The program of academic research will react to problems of the Arctic seas and pack ice, tundra ecosystems, geomagnetic phenomena, snow, ice and permafrost phenomena, and other scientific problems related to the physical and biological aspects of cold-dominated environment, and man's impact upon them. Support is also given for Arctic science information activities.

Eligibility

Proposals for grants or contracts for research project support may be submitted by colleges and universities and by academically related nonprofit research organizations. Grants are normally made for a period of 12 months, but under certain circumstances can be made for periods up to a maximum of 60 months. Institutions are required to share in the cost of research projects supported by an NSF grant.

Additional Information

Communications may be addressed to: Office of Polar Programs, National Science Foundation, Washington, D. C. 20550.

Ocean Sediment Coring Program

The National Science Foundation awards grants for studies of cores obtained by rotary drilling techniques from the oceans in order to increase our knowledge of the ocean basins and the earth as a whole, utilizing material gained by penetrations through the sedimentary layer.

This program is conducted as a national research program. The acquisition of the core material is funded by the Foundation. Grants are made to support the analyses of the cores. At the present time the Foundation has contracted with the University of California Scripps Institution of Oceanography to perform drilling operations in the Atlantic Ocean, Pacific Ocean, and adjacent seas; prepare preliminary core descriptions, and distribute the material to interested parties.

Drilling operations began in August 1968 and will be undertaken over a four-year period. It is planned to obtain cores through the sedimentary layer, with very short penetrations into the underlying crystalline basement at some sites.

Among the research problems subject to investigation with analyses of the core material are: the history of ocean basins; the hypotheses of sea-floor spreading and continental drift; the origin and history of such fundamental features as mid-ocean ridges, abyssal plains, continental margins; and the formation of submarine mineral deposits.

Eligibility

Proposals for grants for studies of the core material may be submitted by colleges and universities, nonacademic nonprofit organizations, individual scientists, and others.

Deadlines

Proposals may be submitted at any time; approximately six months are required to consider a proposal.

Additional Information

Communications may be addressed to: Office of National Centers and Facilities Operations, National Science Foundation, Washington, D. C. 20550, or University of California, San Diego, Scripps Institution of Oceanography, La Jolla, Calif. 92037.

Global Atmospheric Research Program (GARP)

The National Science Foundation awards grants to support research projects involving the general circulation of the atmosphere to improve the capability of long-range weather prediction, to explore the feasibility of large-scale weather and climate modification, and to promote the education and training of atmospheric scientists.

The Global Atmospheric Research Program (GARP) is a long-term commitment by many nations. Within the United States, by formal agreement among Federal agencies, the Foundation is the primary agency for the support of non-Federal research in the program, particularly at universities. The Department of Commerce is the primary agency for Federal activities.

Grants are normally made for periods up to 24 months, but under certain circumstances can be made for periods up to a maximum of 60 months. For projects of high scientific merit, initial funding may be for two years with assurance of support for the full term of the project, contingent upon the availability of funds and the scientific progress of the research.

Eligibility

Institutions eligible to submit proposals under GARP are colleges and universities; nonacademic, nonprofit organizations; and individual scientists. Occasionally NSF sponsors supporting efforts by other Government agencies, particularly for field programs. Institutions are required to share in the cost of their research projects supported by an NSF research grant; this may be accomplished by a contribution to any cost element in the project, direct or indirect.

Before submitting a research proposal, the NSF pamphlet **Grants for Scientific Research,** available from the Foundation, should be consulted. The Foundation does not provide standard application forms for research proposals.

Deadlines

Proposals may be submitted at any time; approximately three months are required for consideration of a proposal.

Additional Information

Communications may be addressed to: Division of Environmental Sciences, National Science Foundation, Washington, D. C. 20550.

International Biological Program

The National Science Foundation awards grants to support research projects that are part of the U.S. participation in the International Biological Program (IBP). The theme of IBP is the study of "the biological basis of productivity and human welfare," and the major portion of the program is in the area of ecosystem analysis.

The International Biological Program was proposed by the International Council of Scientific Unions in 1964; there are 55 nations now participating in the program. The U.S. National Committee for the International Biological Program, established by the National Academy of Sciences-National Research Council, assists in planning U.S. participation in IBP.

Eligibility

Proposals for grants for research projects under IBP may be submitted by colleges and universities and by academically related nonprofit research organizations. The conditions under which support may be provided to other types of organizations and to individuals are described in the NSF pamphlet **Grants for Scientific Research,** available from the Foundation.

Before submitting a research proposal the pamphlet **Grants for Scientific Research** should be consulted. The Foundation does not provide standard application forms for research proposals.

Institutions are required to share in the cost of each research project supported by an NSF research grant; this may be accomplished by a contribution to any cost element in the project, direct or indirect.

Deadlines

Proposals may be submitted at any time; approximately six months are required for consideration of a proposal. Grants are normally made for periods up to 24 months.

Additional Information

Communications may be addressed to: Division of Biological and Medical Sciences, National Science Foundation, Washington, D. C. 20550.

Weather Modification Research

The National Science Foundation awards grants and contracts to support studies, basic and applied research, and evaluation in the field of weather modification.

Weather modification research support constitutes a national research program, and the program has resulted in a broad effort extending across the entire spectrum of weather modification problems.

Eligibility

Institutions eligible to submit proposals under the weather modification program are colleges and universities, which have conducted much of the research; and nonacademic, nonprofit organizations. In addition NSF occasionally sponsors the efforts of other Government agencies in weather modification.

Before submitting a research proposal the NSF pamphlet **Grants for Scientific Research,** available from the Foundation, should be consulted. The Foundation does not provide standard application forms for research proposals.

Deadlines

Proposals may be submitted at any time; approximately three months are required for consideration of a proposal.

Additional Information

Communications may be addressed to: Weather Modification Program, Division of Environmental Sciences, National Science Foundation, Washington, D. C. 20550.

Earthquake Engineering

The National Science Foundation awards grants and contracts to support research in earthquake engineering.

Earthquake engineering research support constitutes a national research program which has as its objectives the development of engineering knowledge and the transfer of this knowledge into practice, in order to minimize deaths, injuries, social disruption, and economic losses resulting from earthquakes. Investigations may cover subject areas such as socioeconomic aspects of earthquakes; earthquake ground motion; soil mechanics, structural dynamics analysis and design related to earthquakes; effects of earthquakes on earth structures, coastal and inland waters, utilities and public service facilities; and post-earthquake inspection and study.

Support may be requested for research or for specialized facilities for earthquake engineering. There are no standard application forms for research proposals.

Eligibility

Guidelines on eligibility and proposal preparation are contained in the NSF pamphlet **Grants for Scientific Research** which may be obtained from the Foundation.

Deadlines

Proposals may be submitted at any time during the year.

Additional Information

Some background information on current earthquake engineering research is contained in the reports **Report on NSF-UCEER Conference on Earthquake Engineering Research,** accession number PB-186-181, and the NAE report **Earthquake Engineering Research,** accession unmber PB-188-636, which are available from the Federal Clearinghouse for Scientific and Technical Information, Springfield, Va. 22151. Communications with NSF may be addressed to: Earthquake Engineering Program, Division of Engineering, National Science Foundation, Washington, D. C. 20550.

Engineering Research Initiation Grants

The National Science Foundation awards grants to encourage the development of meritorious graduate research programs by engineering faculty members.

The usual duration of a grant will include the first summer, and the following academic year and summer. The grant amount will not normally exceed $15,000.

Eligibility

Proposals may be submitted by institutions of higher education that award graduate degrees in engineering on behalf of faculty members who:

(1) Are members of the teaching faculty;

(2) Have received the Ph.D. degree within the past three years (excluding active-duty time in the U.S. Armed Forces), or have completed all requirements for the Ph.D. degree;

(3) Have had no substantial research support.

Deadlines

Instructions for preparing engineering research initiation proposals are available in early October from the office listed below. Application deadline is early December. Awards are made in mid-March.

Additional Information

Pamphlet **Engineering Research Initiation Grants,** NSF 69-22. Communications may be addressed to: Division of Engineering, National Science Foundation, Washington, D. C. 20550.

Doctoral Dissertation Research in the Social Sciences, Systematic Biology, Ecology, Oceanography, Earth Sciences, and Atmospheric Sciences.

The National Science Foundation awards grants to improve the scientific quality of dissertations in the social sciences and certain sciences involving extensive field work and to make possible the use of larger quantities of better quality data. Grants are awarded for periods up to 18 months. Grant funds may not be used as a stipend for the doctoral candidate, although he may receive support from other sources.

In collaboration with the Office of Economic Opportunity, special grants are also awarded by the Foundation in support of doctoral thesis research centrally related to problems of poverty.

Eligibility

Proposals for the support of dissertation research in the social sciences, systematic biology, ecology, oceanography, earth sciences, and atmospheric sciences and dissertation research on poverty may be submitted by universities on behalf of doctoral candidates. The proposal should be initiated by the dissertation advisor, department chairman, or chairman of the departmental committee on doctoral degrees. The student should have made plans for his dissertation.

Deadlines

Proposals may be submitted at any time; one or more grant requests may be made in a single proposal if the budget for each request is set forth separately. Four months should be allowed for processing the grant application.

Additional Information

A leaflet that sets forth application procedures is available from the Foundation. Communications may be addressed to: Division of Biological and Medical Sciences, Division of Environmental Sciences, or Division of Social Sciences, National Science Foundation, Washington, D. C. 20550.

Specialized Research Facilities and Equipment Program

The National Science Foundation awards grants for specialized research facilities and major items of research equipment.

Facilities supported under this program are those required for highly specialized scientific purposes, as distinct from laboratory buildings used in normal academic research programs. Examples are: nuclear reactors, controlled-environment biological laboratories, oceanographic research vessels and marine research and supporting facilities, mobile laboratories, off-campus research facilities, and unique one-of-a-kind research facilities. Grants may provide for construction or modernization of facilities.

Equipment support may be provided where a research tool is needed by several investigators in a department. Examples are: electron microscopes, mass spectrometers, cryogenic equipment, and special-purpose computers.

The National Science Foundation encourages local contributions from non-Federal funds whenever possible; however, there is no fixed requirement as to the amount of funds that institutions must contribute.

Before submitting a proposal for specialized research facilities and equipment the NSF pamphlet **Grants for Scientific Research** should be consulted. The Foundation does not provide standard application forms for research facilities and equipment proposals.

Eligibility

Institutions eligible to submit proposals are colleges and universities offering graduate studies (though in exceptional circumstances colleges and universities without graduate programs may be eligible), associations of colleges and universities, and nonprofit research institutions such as research museums.

Deadlines

Proposals may be submitted at any time. Approximately four to six months are required for consideration of a proposal.

Additional Information

The NSF pamphlet **Grants for Scientific Research** is available from the Foundation. Communications may be addressed to the appropriate division: Division of Biological and Medical Sciences, Division of Engineering, Division of Environmental Sciences, Division of Mathematical and Physical Sciences, or Division of Social Sciences; National Science Foundation, Washington, D. C. 20550.

National Center for Atmospheric Research

The National Science Foundation supports the National Center for Atmospheric Research as an independent, interdisciplinary research center that serves as a focal point for an expanding national research effort in the atmospheric sciences. NCAR offers support services, fellowships, and research facilities to qualified scientists working in the field of atmospheric research.

Headquarters and major laboratories of NCAR are located in Boulder, Colo. NCAR's research and operations are worldwide; several permanent field stations are maintained. Support of NCAR is provided under the terms of a contract between the Foundation and the University Corporation for Atmospheric Research (UCAR), a nonprofit corporation.

Research and facilities programs of NCAR are carried out by four groups: The Laboratory of Atmospheric Sciences (LAS), the High Altitude Observatory (HAO), the Facilities Laboratory (FAL), and the Advanced Study Program (ASP). LAS is concerned primarily with the earth's atmosphere up to an altitude of about 60 miles. HOA is interested in the sun and the regions between the sun and the earth and operates a permanent observing station at Climax, Colo., equipped with a 16-inch corona scope. Other NCAR facilities available to assist visiting scientists include a Scientific Balloon Flight Station at Palestine, Tex., a Computing Facility, and an Aviation Facility.

In addition to conducting its own research programs, NCAR participates in a number of atmospheric research efforts conducted by Government agencies, university scientists, and research groups on a national or international scale. These major efforts include computer simulation of atmospheric motions, large-scale meteorological experiments, and broad research programs underlying the national efforts in weather and climate modification. More than 400 scientists, engineers, technicians, and support personnel comprise the NCAR staff.

Eligibility

Visiting scientists study and conduct research at NCAR under fellowships and research programs. NCAR facilities are available to qualified U.S. scientists, subject to scheduling feasibility.

Additional Information

Communications may be addressed to: Office of National Centers and Facilities Operations, National Science Foundation, Washington, D. C. 20550, or Director, National Center for Atmospheric Research, Boulder, Colo. 80302.

Kitt Peak National Observatory

The National Science Foundation supports the Kitt Peak National Observatory, an independent national research center whose optical telescopes and associated equipment are available to all qualified U.S. scientists.

Headquarters of KPNO is in Tucson, Ariz.; observing facilities are located atop Kitt Peak about 45 miles southwest of Tucson. KPNO is supported under the terms of a contract between the Foundation and the Association of Universities for Research in Astronomy, Inc. (AURA).

Major astronomical instruments at Kitt Peak include the world's largest solar telescope, a 50-inch remote-controlled reflecting telescope, and an 84-inch and two 36-inch reflecting telescopes. A 150-inch reflecting telescope is scheduled to be completed in the spring of 1972. KPNO has a small staff of resident scientists.

Eligibility

KPNO makes up to 60 percent of the observing time on each instrument available for the use of visiting scientists. All qualified U.S. scientists and on occasion foreign visitors may use the instruments, subject to priorities based on the scientific merit of the proposed research, the capability of the instruments to do the work, and available time.

Additional Information

Communications may be addressed to: Office of National Centers and Facilities Operations, National Science Foundation, Washington, D. C. 20550, or Director, Kitt Peak National Observatory, 950 North Cherry Avenue, Tucson, Ariz. 85717.

National Radio Astronomy Observatory

The National Science Foundation supports the National Radio Astronomy Observatory, an independent national research center through which Government-owned radio astronomy facilities are made available to all qualified U.S. scientists. NRAO provides scientists with the large radio antennas, receivers, and other equipment needed to detect, measure, and identify radio waves from outer space.

Headquarters for NRAO is in Charlottesville, Va.; observing facilities are located primarily in Green Bank, W. Va. NRAO is supported under the terms of a contract between the Foundation and Associated Universities, Inc., a nonprofit corporation.

Major research facilities at NRAO include a 140-foot highly precise, fully steerable radio telescope; an interferometer consisting of three fully steerable 85-foot telescopes with a baseline of 5,000 to 9,000 feet; and a 300-foot radio telescope rotatable in a north-south plane. A 36-foot radio telescope operating at millimeter wavelengths is located at the Kitt Peak National Observatory near Tucson, Ariz. NRAO has a small staff of astronomers.

Eligibility

NRAO makes up to 70 percent of the observing time on each instrument available for the use of visiting scientists. All qualified U.S. scientists and on occasion foreign visitors may use the instruments, subject to priorities based on the scientific merit of the proposed research, capability of the instruments to do the work proposed, and time available.

Additional Information

Communications may be addressed to: Office of National Centers and Facilities Operations, National Science Foundation, Washington, D. C. 20550, or Director, National Radio Astronomy Observatory, Charlottesville, Va. 22901.

Cerro Tololo Inter-American Observatory

The National Science Foundation supports the Cerro Tololo Inter-American Observatory, an independent research center whose optical telescopes and related facilities are available to all qualified scientists from the United States, Chile, and elsewhere in Latin America. CTIO provides astronomers with the opportunity to observe those parts of the Southern Hemisphere skies which are not visible or not adequately observable from the United States, using telescopes made available by the Federal Government and other organizations.

The Cerro Tololo Observatory is located on a 7,200-foot mountain in the foothills of the Andes Mountains about 300 miles north of Santiago, Chile. The administrative headquarters is in the coastal city of La Serena, about 60 miles away. CTIO is supported under the terms of a contract between the Foundation and the Association of Universities for Research in Astronomy, Inc., (AURA), which also operates Kitt Peak National Observatory. Close ties are maintained with the University of Chile.

Major astronomical instruments at Cerro Tololo include 60-inch and 36-inch telescopes, two 16-inch telescopes, and a 24-inch Schmidt camera on long-term loan from the University of Michigan. A 150-inch reflecting telescope is scheduled to be completed in 1973. Cerro Tololo has a small permanent staff of U.S. scientists.

Eligibility

Most of the observing time at Cerro Tololo is used by visiting astronomers. Qualified scientists may use the instruments subject to priorities based on the scientific merit of the proposed research, capability of the instruments to do the work proposed, and available time.

Additional Information

Communications may be addressed to: Office of National Centers and Facilities Operations, National Science Foundation, Washington, D. C. 20550, or Director, Kitt Peak National Observatory, 950 North Cherry Avenue, Tucson, Ariz. 85717.

National Arecibo Observatory

The National Science Foundation supports the National Arecibo Observatory, an independent national research center through which Government-owned facilities are made available to all qualified U.S. scientists for the conducting of radio astronomy, radar astronomy, and aeronomy research. The observatory is managed and operated by Cornell University under contract with the Foundation.

The world's largest reflector, a 1,000-foot diameter spherical fixed telescope, is located at Arecibo Observatory. The immense size of this research instrument has enabled it to make unique and significant contributions to our understanding of the earth's atmosphere, the solar system, and radio sources outside of the solar system.

The major objective of the Arecibo Observatory is to make available on a national basis radio and radar astronomy facilities that will enable it to be a major contributor to new discoveries in the fields of ionospheric studies, lunar and planetary radar, and radio astronomy.

Future planning for the observatory includes new and important installations. The most important is the upgrading of the present 1,000-foot reflector. A new surface will enable the telescope to be operated at 10-cm. wavelength in lieu of the present 70-cm. wavelength. Also, included in the planning are additional office and laboratory space, more visiting scientists' quarters, dining and recreational facilities, and improved facilities for visitors to Arecibo. All of the above facilities will enable the Arecibo Observatory to become a truly national research center.

Additional Information

Communications may be addressed to: Office of National Centers and Facilities Operations, National Science Foundation, Washington, D. C. 20550, or Director, Arecibo Observatory, Box 995, Arecibo, Puerto Rico 00612.

II. SCIENCE EDUCATION

The ability of the Nation to achieve in science is directly related to its ability to maintain an adequate supply of well trained scientific personnel. Hence it is vitally important that academic institutions provide science education that keeps pace with advances in science and technology and prepares individuals for their various professional responsibilities. A major task of the National Science Foundation is to strengthen education in science at all levels.

Over the years the Foundation has experimented with a number of approaches to improving science education and has aimed at providing the kind of assistance to individuals and institutions that best meets current needs. The Foundation's educational support efforts seek to:

> Further the training of highly able graduate students and established scientists;
>
> Improve the subject-matter competence of teachers of science, mathematics, or engineering at the various educational levels;
>
> Provide modern instructional materials and courses;
>
> Provide special training opportunities for increasing the scientific knowledge and experience of talented undergraduate and high school students;
>
> Improve science instruction at the undergraduate level by assisting institutions in acquiring modern instructional scientific equipment.

Programs to meet these objectives are described on the following pages, grouped by the level of education supported.

Graduate Fellowships

The National Science Foundation awards Graduate Fellowships for study or work leading to a master's or doctoral degree in the mathematical, physical, medical, biological, engineering, and social sciences and in the history and philosophy of science. Awards will not be made in clinical, education, or business fields, nor in history or social work, nor for work toward medical or law degrees.

Graduate Fellowships are awarded on the basis of the applicant's ability as evidenced by academic records, letters of recommendation, and scores obtained in examinations designed to measure scientific aptitude and achievement.

Fellowships are awarded for full-time study or research at appropriate nonprofit U.S. or foreign institutions of higher education.

Graduate Fellowships are awarded for one or two years, with a 9 or 12 months' tenure in the fellowship year. The basic 12-month stipend for graduate fellows is $2,400 for the first year level, $2,600 for the intermediate level, and $2,800 for the terminal level graduate student. A travel and dependency allowance may also be provided. A fellow may receive concurrently additional educational training remuneration from the Veterans Administration and may receive supplementation of his stipend from institutional funds according to his year of residence at the institution.

Eligibility

Graduate Fellowships are offered only to individuals who: (1) are citizens or nationals of the United States; (2) have demonstrated ability and special aptitude for advanced training in the sciences; and (3) have been or will be admitted to graduate status by the institution selected.

Deadlines

A brochure on the Graduate Fellowship Program is available each year in October from the Foundation. Applications must be submitted to the Fellowship Office, National Research Council, 2101 Constitution Ave., N.W., Washington, D. C. 20418, by late November. NSF announces the awards in March. Each application must include a complete transcript of college and university records, and a proposed plan for graduate study or research.

Additional Information

Communications may be addressed to: Division of Graduate Education in Science, National Science Foundation, Washington, D. C. 20550.

Graduate Traineeships

The National Science Foundation awards grants that enable universities to provide Graduate Traineeships in the mathematical, physical, medical, biological, engineering, and social sciences, and in the history and philosophy of science. Awards will not be made in clinical, education, or business fields, nor in history or social work, nor for work toward medical or law degrees.

Traineeships are awarded to individuals by the institution, not by the National Science Foundation. Trainees may be appointed for part-time or full-time 9- or 12-month tenure periods only. The basic 12-month stipends to be paid from NSF funds to trainees are: first year level, $2,400; intermediate level, $2,600; and terminal year level, $2,800. A dependency allowance may also be provided. An institution may supplement a trainee's stipend from institutional funds according to his year of residence at the traineeship institution.

No funds are available for new starts in fiscal year 1971; however, NSF expects to provide an estimated 3,500 continuing Graduate Traineeships in 1971-72 which would be made to institutions receiving traineeship grants in fiscal year 1970.

Eligibility

Institutions Institutions eligible to submit proposals for Graduate Traineeships are universities that confer doctoral degrees in science. Proposals are submitted on behalf of departments of science or engineering, and a separate proposal is required for each department or comparable unit.

Individuals To be eligible for tenure under a Graduate Traineeship an individual: (1) must be a citizen or national of the United States; (2) must be enrolled in a program leading to an advanced degree in science; and (3) must be affiliated with the institution at which he receives his appointment.

Deadlines

Institutions An announcement containing application materials is available in August from the Foundation. The proposal closing date is in October; grants are made in February for the following academic year.

Individuals A list of institutions in which Graduate Traineeships are available may be obtained in February from the Foundation. Individuals wishing to apply for a Graduate Traineeship should request application forms, brochures, or other information from the institution in which he is, or intends to be, enrolled. The deadline for receipt of applications is established by the institution. Traineeship appointments normally must be made before the opening of the fall term.

Additional Information

Communications may be addressed to: Division of Graduate Education in Science, National Science Foundation, Washington, D. C. 20550.

Summer Traineeships for Graduate Teaching Assistants

The National Science Foundation awards grants that enable universities to provide Summer Traineeships for Graduate Teaching Assistants. Traineeships are awarded to individual trainees by the grantee institution, not by the Foundation.

A trainee's stipend may not exceed $85 per week, and may not be less than $50 per week; the rate within these limits is set by the institution. The trainee may not teach or perform any assistantship services while on tenure.

Eligibility

Institutions Institutions eligible to submit proposals for Summer Traineeships for Graduate Teaching Assistants are universities that confer doctoral degrees in science. Proposals are submitted on behalf of their departments of mathematics, science, or engineering, and a separate proposal is required for each department or comparable unit.

Individuals To be eligible for tenure under a Summer Traineeship for Graduate Teaching Assistants an individual: (1) must be a citizen or national of the United States; (2) must be a graduate student enrolled in a program leading to an advanced degree in the sciences (physical, biological, medical, and social), engineering, mathematics, or interdisciplinary areas; (3) must be making satisfactory progress toward an advanced degree; (4) must have at least one academic year as a graduate teaching assistant; and (5) must not be enrolled in programs leading to the M.D., J.D., L.L.B., D.D.S., D.V.M. degrees nor for study leading to degrees in business administration, clinical medicine, clinical psychology, social work, teaching, or science education.

Deadlines

Institutions An announcement containing application materials for institutions is available in August from the Foundation. The proposal closing date is in October; grants are made in mid-February for the following summer.

Individuals A list of institutions in which summer traineeships are available may be obtained in February from the Foundation. Individuals wishing to apply for a Summer Traineeship for Graduate Teaching Assistants should request application forms, brochures or other information from the institution in which he is enrolled. The deadline for receipt of applications is established by the institution.

Additional Information

Communications may be addressed to: Division of Graduate Education in Science, National Science Foundation, Washington, D. C. 20550.

Postdoctoral Fellowships

The National Science Foundation awards Postdoctoral Fellowships for study or work in mathematics, the sciences (physical, biological, medical, and social), engineering, and interdisciplinary areas. Fellowships are not awarded in clinical, education, or business areas.

The evaluation of applicants is based on their ability as evidenced by academic records, letters of recommendation, and other indications of scientific competence.

Fellowships are awarded for full-time scientific study or research at appropriate nonprofit U.S. or foreign institutions.

The usual tenure of a postdoctoral fellow is 12 months. However, tenures from 6 to 24 months are available upon adequate justification. The normal stipend for a postdoctoral fellow is $6,500 per year. A limited travel and dependency allowance may also be provided. A person may not hold a postdoctoral fellowship for a total period of more than two years in any five consecutive years.

Eligibility

Postdoctoral Fellowships are offered only to individuals who: (1) are citizens or nationals of the United States; (2) hold a doctoral degree in one of the basic fields of science, mathematics, or engineering, **or** have scientific training or research experience equivalent to a doctorate, **or** hold a degree such as M.D., D.D.S., or D.V.M. and desire to obtain further training for a career in research, **or** hold a legal degree, J.D. or L.L.B., and desire to obtain further training for a career in research which employs the methodology of the social sciences or which interrelates with research in the natural or social sciences; and (3) present an acceptable plan of study or research at the postdoctoral academic level.

Deadlines

A brochure on the Postdoctoral Fellowship Program is available each year in October from the Foundation. Applications must be submitted to the Fellowship Office, National Research Council, 2101 Constitution Avenue, N.W., Washington, D. C. 20418, by early December of each year. Awards are announced by the National Science Foundation in March.

Additional Information

Communications may be addressed to: Division of Graduate Education in Science, National Science Foundation, Washington, D. C. 20550.

Senior Postdoctoral Fellowships

The National Science Foundation awards Senior Postdoctoral Fellowships for study or research in mathematics, the sciences (physical, biological, medical, and social), engineering, or interdisciplinary areas. Fellowships are not awarded in clinical, education, or business areas.

The evaluation of applicants is based on their ability as evidenced by letters of recommendation, previous scientific accomplishments, and other indications of scholarly activity.

Fellowships are awarded for full-time scientific study or research at appropriate nonprofit U.S. or foreign institutions.

The usual tenure of a senior postdoctoral fellow is 9 to 12 months. However, tenures from 3 to 24 months are available upon adequate justification. Stipends are based on the applicant's earned income; the Foundation's contribution does not exceed $1,250 per month. A limited travel allowance is provided. An individual who has held a Senior Postdoctoral Fellowship for a total period of two years is ineligible to hold another such fellowship for a period of five years beyond the termination of his most recent fellowship.

Eligibility

Senior Postdoctoral Fellowships are offered only to individuals who: (1) are citizens or nationals of the United States; (2) have held a doctoral degree in one of the basic fields of science, mathematics, or engineering for at least five years, **or** have scientific training or research experience equivalent to a doctorate of at least five years standing, **or** hold a degree such as M.D., D.D.S., or D.V.M. for a period of at least five years, and desire to obtain further training for a career in research, and present plans for study or research in science of the same level of merit as presented by applicants trained in the basic sciences, **or** hold a legal degree, J.D. or L.L.B., for a five-year period and desire to obtain further training for a career in research which employs the methodology of the social sciences or which interrelates with research in the natural or social sciences, and present plans for study or research of the same level of merit as presented by applicants trained in the basic sciences; and (3) present an acceptable plan of research at the postdoctoral academic level.

Deadlines

A brochure is available each year in August from the Foundation. Applications must be submitted to the Division of Graduate Education in Science, National Science Foundation, Washington, D. C. 20550, by early October of each year. Awards are announced by the Foundation in December.

Additional Information

Communications may be addressed to: Division of Graduate Education in Science, National Science Foundation, Washington, D. C. 20550.

Science Faculty Fellowships

The National Science Foundation awards Science Faculty Fellowships for study or work in mathematics, the sciences (physical, biological, medical, and social), engineering, and interdisciplinary areas. Applied and empirical studies in the field of law which employ the methodology of the social sciences or which interrelate with research in the natural or social sciences are acceptable. Fellowships are not awarded in clinical, education, or business areas. The primary purpose of Science Faculty Fellowships is to help college teachers enhance their teaching effectiveness, rather than to provide research support.

The evaluation of applicants is based on their ability (as evidenced by letters of recommendation, academic records, and professional activity) and on the proposed Activities Program. Fellows are selected from either of two groups: those holding a Ph.D. and those not holding such a degree.

Fellowships are awarded for full-time scientific study or research at appropriate nonprofit U.S. or foreign institutions. Arrangements for affiliation with the fellowship institution are the responsibility of the fellow.

The usual tenure of a Science Faculty Fellowship is 9 to 12 months; however, tenures from 3 to a maximum of 15 months are available. Teachers who are unable to apply for fellowships tenable during all or part of an academic year may wish to consider the provision in this program which allows awardees to undertake fellowship studies either in one summer, or in two or three consecutive summer periods. Stipends are based on the applicant's earned income; the Foundation's contribution does not exceed $1,250 per month. A limited travel allowance is also provided.

Eligibility

Science Faculty Fellowships are offered only to individuals who: (1) are citizens or nationals of the United States; (2) hold a baccalaureate degree or its equivalent; (3) have had three or more academic years' experience in teaching science, mathematics, or engineering as a full-time staff member at the collegiate (including junior college) level; and (4) intend to continue teaching.

Deadlines

A brochure on the program is available each year in August from the Foundation. Applications must be submitted to the Division of Graduate Education in Science, National Science Foundation, Washington, D. C. 20550, by early October of each year. Awards are announced in December. Each applicant must submit an "Activities Program"—an individual plan of graduate or postdoctoral level study and/or research.

Additional Information

Communications may be addressed to: Division of Graduate Education in Science, National Science Foundation, Washington, D. C. 20550.

North Atlantic Treaty Organization (NATO) Postdoctoral Fellowships in Science

In cooperation with the Department of State, the National Science Foundation awards NATO Postdoctoral Fellowships in Science for scientific study or work in mathematics, the sciences (physical, biological, medical, and social), engineering, or interdisciplinary areas. Fellowships are not awarded for support of work toward the M.D., D.V.M., or D.D.S. degrees, nor for support of residency training or similar work leading to qualification in a clinical field.

The NATO fellowship program is designed to assist in obtaining a closer collaboration among the scientists of the NATO nations. Fellowships are awarded for full-time scientific study or work at nonprofit scientific institutions located in foreign countries that are members of or are cooperating with NATO.

Evaluation of applicants will be based on their academic records, letters of recommendation, and ability to carry out the activities program. Consideration is also given to proposed fellowship activities that promote international science cooperation.

The tenure of a NATO Postdoctoral Fellowship in Science is normally 9 or 12 months; in no case may it be less than 6 or more than 12 months. Fellows may begin fellowship activities at any time within one year following announcement of the award. The stipend is $6,500 for a full year. A limited travel and dependency allowance may also be provided. During their tenures NATO fellows may not receive remuneration from another fellowship, scholarship, or similar award, or a Federal grant.

Eligibility

NATO Postdoctoral Fellowships in Science, awarded by the National Science Foundation, are offered only to individuals who: (1) are citizens or nationals of the United States; (2) have demonstrated ability and special aptitude for advanced training in the sciences; (3) have a doctoral degree in one of the qualifying fields of science; **or** have had scientific training and research experience equivalent to that represented by the science doctorate; **or** have a degree such as M.D., D.D.S., or D.V.M. and desire to obtain further training for a career in research.

This program is designed primarily for applicants who have received their doctorates within the past five years. Each applicant must submit an outline of his proposed study under the fellowship and complete transcripts of his college and university records.

Deadlines

A brochure is available each year in July from the Foundation. Applications must be submitted to the Fellowship Office, National Research Council, 2101 Constitution Avenue, N.W., Washington, D. C. 20418 by late September of each year. Awards are announced by the National Science Foundation in November.

Additional Information

Communications may be addressed to: Division of Graduate Education in Science, National Science Foundation, Washington, D. C. 20550.

North Atlantic Treaty Organization (NATO) Senior Fellowships in Science

In cooperation with the Department of State, the National Science Foundation awards NATO Senior Fellowships in Science for the study of new scientific techniques and developments at nonprofit research and educational institutions in other NATO nations, or in countries cooperating with NATO. Awards are made for study or work in mathematics, the sciences (physical, biological, medical, and social), engineering, or interdisciplinary areas. Fellowships are not awarded in clinical, education, or business areas.

Tenure for NATO Senior Fellowships in Science normally ranges from one to three months; in unusual circumstances a tenure of less than four weeks or a maximum of six months may be approved. A fellow will receive a subsistence allowance of $16 for each day of tenure, and is permitted to receive his regular salary and/or appropriate allowances provided by his nominating institution. A travel allowance is also provided.

Eligibility

Any U.S. educational institution that offers a postbaccalaureate degree in one of the sciences, or any nonprofit scientific research institution, may nominate for an award a staff member who: (1) is a citizen or national of the United States; (2) has a professional standing in the field with which his fellowship would be concerned; (3) has had at least five years' experience in research, teaching, or relevant professional work; and (4) has linguistic abilities necessary for profitable discussion with colleagues in the countries he proposes to visit. The institutional nomination form requires a statement, from the president or other appropriate official of the nominating institution, showing the expected benefits to the institution if the fellowship were awarded.

Deadlines

Applications may be submitted to the Division of Graduate Education in Science, National Science Foundation, Washington, D. C. 20550, between mid-July and January 31 of the following year. The Foundation notifies applicants within four months as to the outcome of their applications.

Additional Information

Communications may be addressed to: Division of Graduate Education in Science, National Science Foundation, Washington, D. C. 20550.

Advanced Training Projects

The National Science Foundation awards grants for Advanced Training Projects which provide opportunities for advanced and specialized education and research training in the sciences (physical, biological, medical, and social), mathematics, engineering, or interdisciplinary areas. The projects supplement graduate school curricula and enable participants to pursue science subjects in depth. They convene for periods of time ranging from one week to an academic year.

Most of the projects supported to date have consisted of colloquia or conferences, special lecture or laboratory courses, and field training, but other formats are eligible for consideration. Participants may receive stipends or per diem and travel costs.

Eligibility

Institutions Organizations eligible to apply for grants to support Advanced Training Projects are colleges, universities, and other appropriate nonprofit organizations.

Individuals Participants are selected by the grantee organization on a regional or national basis. The professional levels of the participants receiving grant support are limited to graduate faculty, young postdoctorals, and predoctoral students from colleges and universities. Scientists from Government and industry may attend, but no travel or subsistence support is available from Foundation funds.

Deadlines

Institutions An announcement containing application materials is available in March from the Foundation. The closing date for receipt of proposals is mid-June; announcement of awards is made in mid-November.

Individuals Lists of Advanced Training Projects are available in December from the Foundation after award announcements.

Additional Information

Communications may be addressed to: Division of Graduate Education in Science, National Science Foundation, Washington, D. C. 20550.

North Atlantic Treaty Organization (NATO) Advanced Study Institute Participant Grants

The National Science Foundation awards grants to enable U.S. scientists to attend certain NATO Advanced Study Institutes. These meetings, held usually during the summer and varying in length from one to eight weeks, permit exhaustive treatment of a given scientific topic by individuals whose reputations are worldwide.

An international travel grant normally covers the cost of round-trip air fare between the point of origin in the United States and the institute, based on jet-economy, if applicable, or excursion rates. U.S. flag carriers must be used for overseas travel. Per diem is not paid by the Foundation, but in some cases may be available from the NATO institute.

Eligibility

Institutes Each year the National Science Foundation selects certain institutes to receive support for participant-travel and invites the institute director to recommend U.S. participants for such awards. The Foundation then invites the recommended participants to apply for international travel grants.

Individuals To be eligible to receive an NSF international travel grant to attend a NATO Advanced Study Institute, an individual must be: (1) a citizen or national of the United States, and (2) an outstanding young scientist (graduate or recent postdoctoral student).

In addition, individual institutes have specific academic prerequisites for admission. Their announcements should be consulted for details.

Deadlines

Announcements of NATO Advanced Study Institutes are frequently posted by the departments of colleges and universities and also are printed in professional and academic journals. The Foundation makes available a list of these institutes annually around March. Individuals wishing to attend a NATO Advanced Study Institute should request information from the institute director. The deadline for receipt of applications for admission is established by the institute.

Additional Information

Communications may be addressed to: Division of Graduate Education in Science, National Science Foundation, Washington, D. C. 20550.

Senior Foreign Scientist Fellowships

The National Science Foundation awards Senior Foreign Scientist Fellowships to foreign scientists whose formal training, teaching, and research experiences are of sufficient distinction to enable them to make significant contributions to science education and scientific research at U.S. universities. Awards are made to scientists whose fields of specialization are in mathematics, the sciences (physical, biological, medical, and social), engineering, or interdisciplinary areas.

Tenures from 5 to 12 months may be requested. Stipends are commensurate with the level of salaries paid to U.S. faculty members of similar rank and status at the host institution, but the Foundation contribution will not exceed $1,250 per month. A limited travel allowance is paid for the fellow and accompanying dependents. While on tenure a fellow may not accept remuneration from another fellowship, assistantship, or similar award from within the United States.

Eligibility

Institutions Participating institutions are those which grant a Ph.D. in the sciences and indicate a desire to participate in the program.

Individuals Senior Foreign Scientist Fellowships are offered only to individuals who: (1) have already demonstrated outstanding ability in a field of science supported by this program, (2) either have held a doctoral degree for five or more years in one of the fields supported by the program **or** have had research or teaching experience over a 10-year period comparable to the training represented by such a degree, (3) are in good health and have a proficiency in the English language sufficient for carrying out the activities proposed, and (4) are nominated by a U.S. institution participating in the program for an award at that institution. Citizens or permanent residents of the United States are not eligible.

Deadlines

Applicants apply to a U.S. participating institution. A brochure listing the participating institutions for the next two years is available from the Foundation. Nominations by the participating U.S. institutions are accepted and evaluated by the Foundation periodically.

Additional Information

Communications may be addressed to: Division of Graduate Education in Science, National Science Foundation, Washington, D. C. 20550.

Special Projects in Graduate Education

The National Science Foundation awards grants to upgrade the quality of science instructional programs at the graduate level and to identify and support new approaches for improving graduate science education. No specific criteria are established; creative and novel approaches are encouraged.

Examples of appropriate projects are: strengthening a graduate degree program by the development of new or special course offerings, including the design and preparation of films and other educational aids; improvement of graduate-level training programs for prospective junior college and college science teachers; developing model courses for incorporation into graduate programs in other institutions; initiating or strengthening interinstitutional programs; the development of courses on the relationship between science and society; and supporting special conferences or studies on national problems in graduate education.

Eligibility

Institutions eligible to submit proposals for Special Projects in Graduate Education are universities and colleges, and other appropriate nonprofit organizations or professional scientific societies. Before preparing a formal proposal the project should be discussed informally with NSF project staff, and/or a preliminary proposal should be submitted. An announcement containing application materials is available from the Foundation.

Deadlines

Proposals may be submitted at any time. The period of time required for processing a proposal varies greatly; some proposals require six months or more before a decision can be reached.

Additional Information

Communications may be addressed to: Division of Graduate Education in Science, National Science Foundation, Washington, D. C. 20550.

Summer Institutes for College Teachers

The National Science Foundation awards grants to support summer institutes in advanced-level science, mathematics, and engineering courses for college teachers. Such courses permit exploration in depth of those areas that may have become particularly significant for the reorganization and strengthening of the college curriculum. The duration of the summer institutes varies considerably, but the average is seven weeks.

In the Summer Institutes for College Teachers program, a maximum stipend of $75 per week is paid to predoctoral participants; $100 per week to postdoctoral participants. Dependency and travel allowances are also provided.

Eligibility

Institutions Institutions eligible to apply for grants to support summer institutes are normally colleges and universities with graduate programs where staffing, laboratories, and libraries are adequate for the advanced nature of the work.

Individuals To be eligible to participate in Summer Institutes for College Teachers an individual must be a U.S. college teacher of one of the sciences (biological, medical, physical, or social), mathematics, or engineering. Teachers at junior or community colleges or technical schools are eligible. A limited number of college teachers who are foreign nationals may be accepted as participants in these institutes.

In addition, individual summer institutes have established specific academic prerequisites for admission; their brochures should be consulted for details.

Deadlines

Institutions An announcement containing application materials is available in March from the Foundation. The application deadline is June 1. Grants for Summer Institutes for College Teachers are made in October for the following summer.

Individuals A list of institutions offering Summer Institutes for College Teachers is given in the Directory of College Teacher Programs published annually in late November; the directory is available from the Foundation.

Individuals wishing to apply for admission should request brochures, application forms, and other information from the project director in charge of the institute. The deadline for receipt of applications is established by each local project director.

Participants are selected by the institutions involved, not by the Foundation.

Additional Information

Communications may be addressed to: Division of Undergraduate Education in Science, National Science Foundation, Washington, D. C. 20550.

Short Courses for College Teachers

The National Science Foundation awards grants for short courses in science, mathematics, and engineering for college teachers. The courses are under the direction of highly competent research scientists who provide specialized short-term instructional programs (less than four weeks' duration) covering recent advances in selected areas of their scientific fields.

Scheduling of courses is arranged for time periods that are convenient for college teachers, such as early or late summer, or during the academic year.

In the Short Courses for College Teachers program, a maximum stipend of $75 per week is paid to predoctoral participants; $100 per week to postdoctoral participants. A travel allowance is also paid, but a dependency allowance is not provided.

Eligibility

Institutions Institutions eligible to apply for grants to support short courses are normally colleges and universities with graduate programs where staffing, laboratories, and libraries are adequate for the advanced nature of the work.

Individuals To be eligible to participate in Short courses for College Teachers an individual must be a U.S. college teacher of one of the sciences (biological, medical, physical, or social), mathematics, or engineering. Teachers at junior or community colleges or technical schools are eligible.

In addition, individual short courses have established specific academic prerequisites for admission; their brochures should be consulted for details.

Deadlines

Institutions An announcement containing application materials is available in March from the Foundation. The application deadline is June 1. Grants for Short Courses for College Teachers are made in October for the following summer and academic year.

Individuals A list of institutions offering Short Courses for College Teachers is given in the Directory of College Teacher Programs published annually in late November; the directory is available from the Foundation.

Individuals wishing to apply for admission should request brochures, application forms, and other information from the project director in charge of the project. The deadline for receipt of applications is set by the project director. Participants are selected by the institutions involved, not the Foundation.

Additional Information

Communications may be addressed to: Division of Undergraduate Education in Science, National Science Foundation, Washington, D. C. 20550.

Short Courses for College Teachers

The National Science Foundation awards grants for short courses in science, mathematics, and engineering for college teachers. The courses are under the direction of highly competent research scientists who provide specialized short-term instructional programs (less than four weeks' duration) covering recent advances in selected areas of their scientific fields.

Scheduling of courses is arranged for time periods that are convenient for college teachers, such as early or late summer, or during the academic year.

In the Short Courses for College Teachers program, a maximum stipend of $75 per week is paid to predoctoral participants; $100 per week to postdoctoral participants. A travel allowance is also paid, but a dependency allowance is not provided.

Eligibility

Institutions Institutions eligible to apply for grants to support short courses are normally colleges and universities with graduate programs where staffing, laboratories, and libraries are adequate for the advanced nature of the work.

Individuals To be eligible to participate in Short courses for College Teachers an individual must be a U.S. college teacher of one of the sciences (biological, medical, physical, or social), mathematics, or engineering. Teachers at junior or community colleges or technical schools are eligible.

In addition, individual short courses have established specific academic prerequisites for admission; their brochures should be consulted for details.

Deadlines

Institutions An announcement containing application materials is available in March from the Foundation. The application deadline is June 1. Grants for Short Courses for College Teachers are made in October for the following summer and academic year.

Individuals A list of institutions offering Short Courses for College Teachers is given in the Directory of College Teacher Programs published annually in late November; the directory is available from the Foundation.

Individuals wishing to apply for admission should request brochures, application forms, and other information from the project director in charge of the project. The deadline for receipt of applications is set by the project director. Participants are selected by the institutions involved, not the Foundation.

Additional Information

Communications may be addressed to: Division of Undergraduate Education in Science, National Science Foundation, Washington, D. C. 20550.

In-Service Seminars for College Teachers
(Program Suspended During Fiscal Year 1971)

The National Science Foundation awards grants for in-service seminars designed to meet professional needs of college faculty who are located within commuting distance of the institution conducting the seminar. These seminars enable the teacher-participants, while teaching on a full-time basis, to obtain additional knowledge of subject-matter in their scientific disciplines and to become acquainted with new textbooks and laboratory equipment.

No stipends or dependency allowances are paid to participants. Travel allowances are ordinarily available.

Eligibility

Institutions Institutions eligible to apply for grants to support In-Service Seminars for College Teachers are colleges and universities that offer graduate programs in the sciences involved.

Individuals To be eligible to attend an In-Service Seminar for College Teachers an individual: (1) must be a U.S. college teacher of one of the sciences (biological, medical, physical, or social), mathematics, or engineering; and (2) must live within commuting distance of the in-service seminar. Teachers at junior or community colleges or technical schools are eligible.

In addition, individual in-service seminars have established specific academic prerequisites for admission; their directors should be consulted for details.

Deadlines

Institutions An announcement containing application materials is available in March from the Foundation. The application deadline is June 1. Grants for In-Service Seminars for College Teachers are made in October for the following academic year.

Reinstatement of the program will be announced to colleges and universities well in advance of proposal receipt deadline.

Individuals A list of institutions offering In-Service Seminars for College Teachers is given in the Directory of College Teacher Programs published annually in late November; the directory is available from the Foundation. Individuals wishing to apply for admission should request information and application forms from the seminar director. The deadline for receipt of applications is established by the seminar director. Participants are selected by the institutions involved, not the Foundation.

Additional Information

Communications may be addressed to: Division of Undergraduate Education in Science, National Science Foundation, Washington, D. C. 20550.

Research Participation for College Teachers

The National Science Foundation awards grants to provide summer research opportunities for college teachers who have adequate subject-matter knowledge in their scientific disciplines but have limited opportunity for research during the academic year.

Many of the grantee institutions are authorized to nominate participants to receive academic year extensions. Each of these awards is intended to encourage the summer participant to continue his research in the ensuing two academic years at his home institution with grants to the home institution for those selected.

The maximum stipend is $75 per week for predoctoral participants, and $100 per week for postdoctoral participants. Dependency and travel allowances are also provided.

Eligibility

Institutions Institutions eligible to apply for grants to support Research Participation for College Teachers are normally colleges and universities that offer the Ph.D. degree, or field stations and research laboratories affiliated with them.

Individuals To be eligible to attend research participation projects for college teachers an individual: (1) must have a master's degree in the scientific field of the intended research participation; and (2) must be a U.S. college teacher of one of the sciences (biological, medical, physical, or social), mathematics, or engineering.

In addition, individual research participation projects have established specific academic prerequisites for admission; their brochures should be consulted for details.

Teachers at junior or community colleges or technical schools are eligible.

Research participation projects are carried on at both the predoctoral and postdoctoral levels.

Deadlines

Institutions An announcement containing application materials is available in March from the Foundation. The application deadline is June 1. Grants in support of Research Participation for College Teachers are made in October for the following summer.

Individuals A list of institutions offering summer research participation projects and academic year extensions for college teachers is given in the Directory of College Teacher Programs published annually in late November; the directory is available from the Foundation.

Individuals wishing to apply for admission should request brochures, application forms, and other information from the director of the research project. The deadline for receipt of applications is established by the project director. Participants are selected by the institutions involved, not the Foundation.

Additional Information

Communications may be addressed to: Division of Undergraduate Education in Science, National Science Foundation, Washington, D. C. 20550.

Undergraduate Research Participation

The National Science Foundation awards grants that provide undergraduate students with research or independent study opportunities under the guidance of competent research directors.

Undergraduate Research Participation grants are awarded principally for full-time (usually summer) projects of at least eight weeks' duration. A small number of part-time academic year projects are also supported. (Academic year grants are available only to institutional departments that need Foundation support to test the effectiveness of student research as a device for training science majors, with a view toward incorporating such training into the regular curriculum. Academic year projects may be a component of a year-round grant, or be separate. They are supported for a maximum of three years, with no renewal. Institutions are limited to a maximum of six academic year projects.)

Full-time undergraduate research participants may receive stipends at a rate not to exceed $60 per week, up to a maximum of 12 weeks. No stipends for academic-year (part-time) participants are provided by the Foundation.

Eligibility

Institutions Organizations eligible to apply for an Undergraduate Research Participation grant are four-year colleges, universities, and nonprofit research institutions.

Individuals To be eligible to participate in an Undergraduate Research Participation project an individual must be a full-time undergraduate student and be well-grounded in science. A student may apply for full-time projects at institutions other than the one he attends, and for projects in disciplines other than his major field.

Since each research participation project establishes specific academic prerequisites for student admission, the appropriate project director should be consulted for details.

Deadlines

Institutions An Undergraduate Research Participation announcement containing application materials is available in late May from the Foundation. The date for receipt of applications is in early September; the grant award date is in late December.

Individuals A list of institutions conducting undergraduate research participation projects is available from the Foundation in February for the following summer and academic year.

Individuals wishing to apply for admission should request brochures, application forms, and other information from the project director. Application deadlines are set by the project director. Participants are selected by the institutions involved, not the Foundation.

Additional Information

Communications may be addressed to: Division of Undergraduate Education in Science, National Science Foundation, Washington, D. C. 20550.

Student-Originated Studies

The National Science Foundation awards grants (effective July 1, 1970) in a competitive program for the support of student-originated research in environmental problems. The program seeks to advance two basic objectives: (1) to encourage serious students of science to express in productive ways their growing concern for the environmental well-being of the Nation; and (2) to provide support for groups of college and university students who can demonstrate their readiness to assume increasing responsibility for their own educational development.

Projects must: (a) meet standards of intellectual rigor, replicability, and—to a reasonable degree—originality; (b) be organized around a single problem or group of logically related problems concerned with the quality of the environment (physical or social); (c) be interdisciplinary in nature; and (d) be student-originated and student directed.

Eligibility

Institutions Groups of science students in four-year colleges and universities are eligible to apply for Student-Originated Studies (SOS) grants. Guidelines are being kept as brief and straightforward as possible to permit maximum diversity and flexibility in the projects proposed. A **group** of students wishing to ally themselves for a summer's work of 10 to 12 weeks must submit a proposal describing the project they envision.

Each project must name a Student Project Director and a (faculty) Project Advisor. Physical facilities and fiscal services must be provided by a college or university that agrees to accept the grant on behalf of the student group and to serve as its host. Both undergraduate and graduate students may participate in SOS projects. Participants may be drawn exclusively from the student body of the host institution, but inclusion of students from other institutions is encouraged.

Individuals Students not affiliated with a group applying for SOS support may be accepted for one of the projects supported by the Foundation. A list of the projects that will operate each summer will be mailed to individual inquirers in February. Such individuals must then apply to the Student Project Director of the activity in which they are interested to ascertain what vacancies are available; learn what talents, qualities, or prerequisites are required by the project; secure applications materials, and the like. Individual participants will be selected by local project officials—not by the National Science Foundation.

Deadlines

The date for receipt of applications is in late October; the grant award date is in late January.

Additional Information

A Student-Originated Studies announcement containing applications materials is available from the Foundation. Communications may be addressed to: Student-Originated Studies Program, Division of Undergraduate Education in Science, National Science Foundation, Washington, D. C. 20550.

Undergraduate Science Curriculum Improvement

The National Science Foundation awards grants for projects to improve science education, course content and curricula in the biological, engineering, mathematical, physical, and social sciences, in the history and philosophy of science, and in interdisciplinary approaches to the above areas.

Projects supported include studies of problems and ways to initiate appropriate efforts to solve them, development and evaluation of innovative approaches to science teaching, and the development of aids for the presentation of new subject matter—particularly new science courses, course segments, instructional techniques and materials.

Three factors are given primary consideration in assessing the merit of a proposed project: (1) demonstration that the activity will be of high quality; (2) the proposal involves the time, effort, and leadership of scientists distinguished as investigators or teachers in their respective fields; and (3) gives promise of bringing about substantial improvement and wide impact on undergraduate science education throughout the nation, or making progress with an unusual new idea. Projects that are principally local in nature cannot be supported unless their novelty and implications as a model are execptional.

Eligibility

Institutions eligible to submit proposals for Undergraduate Science Curriculum Improvement are colleges, universities, and other nonprofit institutions and organizations. Prospective proposers are encouraged to describe their projects in a preliminary document in sufficient detail so that the Foundation can determine whether a formal proposal can be considered. This document should discuss the rationale, the personnel, the amount and nature of support requested, the expected outcome, as well as plans for the evaluation and for the dissemination of the ideas and materials produced.

Deadlines

Proposals may be submitted at any time; processing of a proposal requires approximately six months.

Additional Information

Communications may be addressed to: Division of Undergraduate Education in Science, National Science Foundation, Washington, D. C. 20550.

Pre-Service Teacher Education Program

The National Science Foundation awards grants to improve programs for the preparation of prospective pre-college science teachers, by emphasizing both increased knowledge of the subject matter and greater skill in organizing and presenting course materials. The objectives of the program are to develop the type of curricular change at colleges which will increase the scientific competence of graduates and at the same time provide the pedagogical preparation essential to their performance as teachers of science.

Projects under the Pre-Service Teacher Education Program (UPSTEP) may include any activity or combination of activities calculated to improve the preparation of undergraduate students for careers as elementary or secondary school science teachers. A proposal should show that both education and science departments will be jointly involved in producing graduates who are thoroughly prepared both substantively and pedagogically to become science teachers.

Experience has shown that the problems of science teacher education are multifaceted, and that all the facets are more likely to be considered if the design and execution of improvement projects includes representatives of all the groups that will be affected. Projects already underway include many of the following activities: recruitment that informs students and strengthens communication with their teachers; teaching science and education courses through the use of instructional methods that derive from the subject being taught; providing a diversity of teaching and other classroom experience before the required student teaching; developing strong collaboration of college faculty and supervising teachers; including in the undergraduate preparation a thorough grounding in the more modern elementary and secondary courses and curricula; maintaining close liaison with new graduates and their administrative superiors; supporting pre-service and in-service teachers by assembling materials and advisors for continuing self-renewal and professional growth.

Eligibility

Institutions eligible to submit proposals under UPSTEP are four-year colleges and universities that have, or are actively planning, elementary or secondary school teacher education programs in the sciences. Proposals may also be submitted by existing or ad hoc consortia of institutions. A brochure containing suggestions for submission of proposals is available from the Foundation. It is suggested that the proposed project first be described in a preliminary proposal with sufficient detail to permit the Foundation to determine whether a formal proposal should be submitted.

Deadlines

Proposals may be submitted at any time; processing requires six to nine months.

Additional Information

Communications may be addressed to: Division of Undergraduate Education in Science, National Science Foundation, Washington, D. C. 20550.

Visiting Scientists (Colleges)

The National Science Foundation awards grants to provide for visits of productive and creative scientists to colleges and small universities for two to three days to give lectures; hold seminars; confer with students, administrators, and instructors; and to aid in other ways in motivating students toward the pursuit of careers in science and teaching science. The program is directed primarily to those colleges and universities in which educational opportunities are more limited than in larger or more amply equipped institutions.

Eligibility

Organizations Organizations eligible to submit proposals for the Visiting Scientists (Colleges) program are national scientific and professional societies in the sciences (biological, medical, physical, and social), engineering, and mathematics.

Institutions Institutions eligible to obtain visits by visiting scientists are junior colleges, technical schools, four-year colleges, and universities.

A list of organizations that have received grants under the Visiting Scientists (Colleges) program is available from the Foundation. Institutions wishing to apply for visits by scientists should make their requests through the project director of the appropriate organization holding a Foundation grant under this program. Visits are arranged by the organization, not by the Foundation.

Deadlines

The deadline for submission of proposals by organizations is early October; grants are awarded in March for the following academic year.

Additional Information

Communications may be adressed to: Division of Undergraduate Education in Science, National Science Foundation, Washington, D. C. 20550.

Special Projects in Undergraduate Science Education

The National Science Foundation awards grants for experimental and developmental projects in undergraduate education that represent promising innovations or totally new approaches. These may possibly involve several elements of, or variants of, existing NSF programs.

Among the projects that have been supported have been special education projects for college teachers, special projects for college students, and various special activities in science teaching centers where facilities are provided for cooperative work by faculties from many different institutions. Science education programs of interinstitutional associations which enable groups of colleges and universities to pool their resources in undertaking the solution of mutual problems in science education, previously supported through the Special Projects Program, should now be directed to the College Science Improvement Program.

Eligibility

Institutions eligible to submit proposals for Special Projects in Undergraduate Science Education are colleges and universities and other appropriate nonprofit institutions and organizations. It is suggested that the proposed project first be described in sufficient detail in a preliminary document so that the Foundation can determine whether a formal proposal can be considered.

Deadlines

Proposals may be submitted at any time; processing of a proposal requires from four to six months.

Additional Information

An information sheet describing the program is available from the Foundation. Communications may be addressed to: Division of Undergraduate Education in Science, National Science Foundation, Washington, D. C. 20550.

Undergraduate Instructional Scientific Equipment
(Program Suspended During Fiscal Year 1971)

The National Science Foundation awards grants to assist institutions of higher education to significantly improve science curricula at the undergraduate level by providing funds to purchase instructional equipment needed to implement the improvement. Not more than 50 percent of the cost of the equipment will be funded by the Foundation, and the institution's matching funds must be derived from non-Federal sources.

Eligibility

Institutions eligible to submit proposals for Undergraduate Instructional Scientific Equipment are junior colleges, colleges, and universities.

Deadlines

An announcement containing application materials is available in October from the Foundation. The closing date for receipt of applications is in late January; the award date is mid-May.

Reinstatement of the program will be announced to colleges and universities well in advance of proposal receipt deadline.

Additional Information

Communications may be addressed to: Division of Undergraduate Education in Science, National Science Foundation, Washington, D. C. 20550.

Academic Year Institutes for Secondary School Teachers

The National Science Foundation awards grants for institutes conducted full time during the school year to provide specially designed science or mathematics programs for secondary school teachers or supervisors. Some institutes include an additional related summer program to enable selected participants to complete the requirements for an advanced degree.

Academic year institutes are available for teachers or supervisors interested in concentrating on a single discipline or studying several related disciplines, and teachers seeking specific educational objectives, such as science supervision.

For experienced teachers a maximum stipend of $3,000 is paid for the period September 1 to June 30. Dependency, book, and travel allowances are also provided.

Eligibility

Institutions Institutions eligible to apply for grants to conduct Academic Year Institutes for Secondary School Teachers are colleges and universities which offer appropriate graduate-level work.

Individuals To be eligible to attend an Academic Year Institute for Secondary School Teachers an individual: (1) must be presently employed as a teacher or supervisor of science or mathematics in grades 7-12 with at least three years of teaching experience; (2) must ordinarily have received a bachelor's degree; and (3) must not have attended a previous NSF-supported academic year instiitute, two or more summers of a sequential program in summer institutes leading to an advanced degree, or any three summer institutes during the five years preceding the academic year in question.

In addition, individual institutes establish specific academic prerequisites for admission; their brochures should be consulted for details.

Deadlines

Institutions An academic year institute announcement containing grant application materials is available in mid-March from the Foundation. The deadline for receipt of proposals is June 1. Grants are made in mid-September for the following academic year.

Individuals A directory of institutions offering academic year institutes is available in October for the following academic year from the Foundation.

Individuals wishing to apply for admission should request brochures, application forms, and other information from the appropriate institute director. Completed application forms must be mailed no later than January 20; successful applicants will be notified of awards by February 15. Participants are selected by the institutions involved, not the Foundation.

Additional Information

Communications may be addressed to: Division of Pre-College Education in Science, National Science Foundation, Washington, D. C. 20550.

Summer Institutes and Conferences for Secondary School Teachers

The National Science Foundation awards grants to support institutes and conferences which provide opportunities for the supplementary training of secondary school teachers and supervisors of science and mathematics during the summer months.

Summer institutes provide for a variety of needs, such as: subject-matter updating, in-depth advanced training, remedial study, knowledge of new curriculum materials and teaching methods, assistance to teachers in developing materials adapted to their own needs, experience in research, and the development of leadership and supervisory capability. Most of these institutes offer a single summer of study. About one-third are sequential institutes involving a planned program of study for the same participants for several summers.

Summer conferences are specialized in nature, and of short duration, usually one to four weeks.

Summer institute participants may receive a maximum stipend of $75 per week, plus dependency and travel allowances. Summer conference participants receive an allowance toward the costs of room, board, and travel. Participants in institutes or conferences are exempt from the payment of tuition or academic fees.

Eligibility

Institutions Institutions eligible to apply for summer institute or conference grants are colleges and universities which grant at least a baccalaureate-level degree, and appropriate nonprofit organizations.

Individuals To be eligible to attend a Summer Institute or Conference for Secondary School Teachers, an individual must be currently employed as a teacher or a science or mathematics supervisor at the secondary school level, grades 7-12. A teacher must be employed at least half-time and teach at least one full course in science or mathematics; however, eligibility is also extended to applicants who teach substantial amounts of social science in at least one secondary school course.

Priority among applicants for summer institutes is given to those who have not previously received stipends in summer institutes or academic year institutes, except for qualified returnees in sequential institutes. Preference is given to individuals who have taught for at least three years.

In addition, individual institutes establish specific academic prerequisites for admission; their brochures should be consulted for details.

Deadlines

Institutions A summer institute announcement containing grant application materials is available in mid-March from the Foundation. The deadline for receipt of proposals is July 1. Grants for summer institutes are awarded in November for the following summer. Proposals for summer conferences should be submitted by October 1, except in unusual circumstances.

Individuals A directory of institutions offering summer institutes and conferences is available in December for the following summer from the Foundation. Individuals wishing to apply for admission should request brochures, application forms, and other information from the appropriate institute director. Completed application forms must be mailed no later than February 15. Participants are selected by the institutions involved, not the Foundation.

Additional Information

Communications may be addressed to: Division of Pre-College Education in Science, National Science Foundation, Washington, D. C. 20550.

In-Service Institutes for Secondary School Teachers

The National Science Foundation awards grants that provide supplemental science or mathematics instruction for secondary school teachers or supervisors at times so chosen that teachers may participate in a program of study without interference with their classroom duties.

Although many in-service institutes meet once a week for periods of two to four hours for a full academic year, others are held in vacation periods or irregularly, as they are not restricted to a particular schedule format. These institutes enable teachers to obtain additional knowledge of subject matter and/or to become acquainted with important new textual and laboratory materials developed by a number of course content study groups.

There are no tuition and fees charged for participating teachers. A book purchase allowance and a travel allowance for commuting expenses are provided.

Eligibility

Institutions Organizations eligible to apply for grants to support In-Service Institutes for Secondary School Teachers are universities and colleges that grant at least a baccalaureate-level degree and other appropriate nonprofit organizations.

Individuals To be eligible to attend an In-Service Institute for Secondary School Teachers an individual must be a supervisor or teacher of science or mathematics in grades 7-12. In addition, individual institutes establish specific academic prerequisites for admission; their brochures should be consulted for details.

Deadlines

Institutions An in-service institute announcement containing grant application materials is available in early September from the Foundation. The deadline for receipt of proposals is in early November; grants are awarded in March for the following academic year.

Individuals A directory of institutions offering in-service institutes is available in April for the following academic year from the Foundation. Individuals wishing to apply for admission should request brochures, application forms, and other information from the appropriate institute director. The deadline for receipt of applications is established by each institute director and is given in the specific institute brochure. Participants are selected by the institutions involved, not the Foundation.

Additional Information

Communications may be addressed to: Division of Pre-College Education in Science, National Science Foundation, Washington, D. C. 20550.

Cooperative College-School Science Program

The National Science Foundation awards grants that enable school systems and nearby colleges or universities to work cooperatively to bring about significant improvements in the science or mathematics programs of the school systems. Projects may focus on elementary or secondary school programs.

Projects are expected to reflect the needs and plans of the cooperating school systems. They may be designed to strengthen current courses of study, to adapt new materials to local use, to prepare teachers in subject matter relevant to the school system's instructional needs, or to accomplish a combination of activities.

Projects may provide for the training of key staff members of the school systems in the summer and follow-up activities during the school year when the strengthened or new programs are implemented. Orientation activities may take place in the spring preceding the summer phase.

Eligibility

Institutions Institutions eligible to submit proposals to the Cooperative College-School Science Program are universities, colleges and other appropriate nonprofit organizations. Grants are not made directly to elementary or secondary school systems, but close collaboration between schools and the grantee institution in designing the proposal and carrying out the project is essential.

Individuals To be eligible for participation in a project, a teacher must be employed by the collaborating school system. Selection of participants is made jointly by the cooperating institution and the local school system.

Deadlines

Institutions An announcement containing grant application materials is available in April from the Foundation. The deadline for receipt of proposals is in mid-August; grants are awarded in December for the following spring, summer, or academic year.

Individuals Announcements of grants and qualifications for participants are made locally and disseminated to eligible teachers.

Additional Information

Communications may be addressed to: Division of Pre-College Education in Science, National Science Foundation, Washington, D. C. 20550.

Pre-College Course Content Improvement

The National Science Foundation awards grants for projects that encourage and assist scientists and engineers working with educators to create new approaches to science, engineering, and mathematics instruction.

Examples of projects that have received support are: (1) projects for the development of course segments dealing with new approaches to subject-matter presentations through written materials, film, television, laboratory experiments and equipment, or programmed approaches; (2) projects to develop complete model courses or course sequences, using many types of learning and teaching aids; (3) small-scale experimental projects, typically limited in subject-matter scope and academic level, whose primary purpose is the investigation of innovative approaches to science teaching; (4) committee and conference studies designed to identify problems in a given field and to formulate guidelines for the evolution of modern instructional programs; (5) planning and coordination projects designed to develop basic guidelines for course improvement, to stimulate the initiation of appropriate projects, to correlate independent developmental projects, and to facilitate wide dissemination of the results of such efforts; (6) projects to develop leadership qualities and strengthen the backgrounds in science curricula of resource people who, when well versed in the philosophy and materials of one or more curriculum study groups, are then able to aid local schools and school teachers in adopting—and adapting—new courses and materials of their own choice to achieve special goals; and (7) projects whose studies of the learning process can be expected to be useful in developing improved curriculum materials.

Eligibility

Institutions eligible to submit proposals for Pre-College Course Content Improvement projects are colleges and universities and other appropriate nonprofit organizations. Elementary and secondary schools, school systems, and State departments of education are normally excluded as grantees, although the involvement of schools and teachers in all phases of the development of materials is essential.

Deadlines

Proposals may be submitted at any time. Prospective proposers are encouraged to describe their projects in a preliminary document so that the Foundation can determine whether a formal proposal can be considered.

Additional Information

Communications may be addressed to: Division of Pre-College Education in Science, National Science Foundation, Washington, D. C. 20550.

Student Science Training Program (Pre-College)

The National Science Foundation awards grants that provide advanced educational opportunities for superior secondary school students. These activities, usually conducted at the grantee institution, encourage student participation in either scientific research or special course work.

Training is usually offered during the summer in sessions of at least five weeks' duration, although academic year projects may also be supported. Research participation projects afford the student the opportunity to work with experienced scientific investigators and to obtain firsthand knowledge of research methods and techniques. Course-oriented projects present subject matter at a level more advanced than can be expected in high school.

Costs of instruction are paid by the Foundation; the student is expected to pay his own expenses for room, board, and travel. Financial assistance is available for students who otherwise would be unable to attend.

Eligibility

Institutions Institutions eligible to apply for grants under the Student Science Training Program (Pre-College) are universities and colleges which grant at least a baccalaureate-level degree, and other appropriate nonprofit organizations.

Individuals To be eligible to participate in a student science training project an individual must be a high-ability secondary school student, as evidenced by school records. Summer projects are open only to students who will be completing their junior year (11th grade) at the time of application. Academic year projects are open to students from the 10th, 11th, and 12th grades.

In addition, individual projects establish specific academic and other prerequisites for admission; their brochures should be consulted for details.

Deadlines

Institutions An announcement containing grant application materials is available in April from the Foundation. The deadline for receipt of proposals is late August. Grants are awarded in December for the following summer and academic year.

Individuals A directory of institutions offering science training programs for high-ability secondary school students is available in January for the following summer from the Foundation.

Individuals wishing to apply for admission should request brochures, application forms, and other information from the appropriate project director. The deadline for receipt of application forms is established by the project director and usually falls between March 1 and April 1 for summer projects. Participants are selected by the institutions involved, not the Foundation.

Additional Information

Communications may be addressed to: Division of Pre-College Education in Science, National Science Foundation, Washington, D. C. 20550.

Special Projects in Pre-College Science Education

The National Science Foundation awards grants to encourage unusual experiments in pre-college science or mathematics education. A limited number of projects that promise quality improvement in these fields are supported each year. Activities that do not fit the guidelines of the more structured programs supported by the Foundation, or that combine elements of these programs in an unusual manner, are eligible under this program.

Projects may include support for: (a) improving the training and effectiveness of teachers, principals, administrators, and resource personnel; (b) motivational and enrichment activities for high-ability or for underprivileged students, such as the Holiday Science Lecture Series administered by the American Association for the Advancement of Science; (c) experimental activities to explore the use of new teaching techniques and educational materials; and (d) other activities having as their objective the enhancement of school science education. There are no specific guidelines for the special projects program as fresh ideas are not easily categorized in advance.

Eligibility

Institutions eligible to apply for grants to support Special Projects in Pre-College Science Education are universities and colleges which grant at least a baccalaureate-level degree, and other appropriate nonprofit organizations. Before a formal proposal is submitted, a detailed outline of the project containing its main features and an estimate of the cost should be sent to the Foundation.

Deadlines

There are no deadlines for the receipt of proposals; at least four months are needed to process a proposal.

Additional Information

Communications may be addressed to: Division of Pre-College Education in Science, National Science Foundation, Washington, D. C. 20550.

III. INSTITUTIONAL SCIENCE

The growth of science in the United States is dependent upon the strength of the Nation's colleges and universities. While grants for individual research projects may serve to strengthen specific segments of these institutions, they do not provide the broadly based and comprehensive support needed for the general improvement of science programs.

To help meet these broader needs, several Federal agencies in recent years have begun programs of institutional support. The Foundation shares in this activity through its programs of Science Development, College Science Improvement, and Institutional Grants for Science. The development of such programs has been among the noteworthy changes in the Foundation's program structure in the last few years. The Science Development Program is also responding to the need to find solutions to social problems by emphasizing the improvement of social science departments, the strengthening of interdisciplinary activities in graduate research and education, and the fostering of problem-oriented centers or institutes.

In all these institutional programs, the Foundation tries to avoid any encroachment upon the autonomy of universities and colleges. Its aim is to provide institutions with the means to carry out their own plans for development and to adapt to their own changing needs.

The institutional programs are described on the following pages.

Science Development Program

The National Science Foundation awards grants to aid in improving the quality of research and educational activity in individual areas of science and engineering in institutions that are already engaged in such activities at the graduate level.

The Science Development Program will emphasize the marked improvement and strengthening of individual departments, programs, and areas of science and engineering normally associated with academic institutions. It will also support unusual developmental efforts that couple research and education to the solution of problems associated with national needs. Under exceptional conditions, the Foundation will entertain proposals for the development of interdisciplinary or multidisciplinary groupings, programs, centers, or institutes oriented toward national problems.

The Science Development Program is separated into four categories: (1) mathematics, the natural sciences, and engineering; (2) the social sciences; (3) interdisciplinary or multidisciplinary activities formed by two or more subject-matter areas of science and engineering; and (4) activities organized as programs, centers, institutes, etc., that are directly related to solving national problems.

All four categories assume close association between research and science education at the appropriate levels. Although the program aims primarily at graduate-level education and research, improvement of undergraduate instructional activities as a concomitant aspect is expected.

In selecting a grantee, the Foundation will give major consideration to the following criteria:

(1) evidence of carefully developed plans for a major upgrading of the individual department or area of science to a significant level of quality within a three- to five-year period;

(2) the presence of sufficient scientific strength in the department, group, program, or area to serve as a realistic base for the proposed development; and

(3) evidence of adequate financial resources to give reasonable assurance that the goals stated in the proposal can be achieved and maintained.

Eligibility

Except for those universities already recognized as being outstanding, all institutions of higher education in the United States and its territories and possessions are eligible under category 1, if they have graduate programs in science or engineering. All institutions of higher education having approved graduate programs in science or engineering are eligible under categories 2 and 3. Any appropriate grouping of individuals or organizations is eligible to apply for support under category 4.

A proposal may request support for up to three years of the development plan. It may include any items deemed necessary to accomplish the proposed improvement during this period. However, support for facilities is limited to modest renovations in categories 1-3, when the facilities are part of the development plan. New construction is limited to category 4.

A brochure containing detailed suggestions for submission of proposals is available from the address listed below.

Deadlines

Proposals under the Science Development Program may be submitted at any time.

Additional Information

Communications may be addressed to: Division of Institutional Development, National Science Foundation, Washington, D. C. 20550.

College Science Improvement Programs

The National Science Foundation awards grants to accelerate development of the science, mathematics, and engineering capabilities of predominantly undergraduate institutions. Proposals should contain a coherent and realistic plan for accomplishing this accelerated development. The plan may include any combination of activities calculated to improve the preparation of college students for careers in science or science teaching.

The College Science Improvement Programs (COSIP) consist of three programs with separate eligibility and closing dates, as follows.

A. Individual Institutional Projects The maximum duration of a grant is three years; grants will not exceed an average of $100,000 per year. Only one proposal from an institution will be considered at any one time.

Eligibility

Colleges and universities that have strong baccalaureate programs in the sciences, and that did not grant more than 10 Ph.D. degrees in the sciences during the academic years 1961-62 to 1963-64 inclusive, are eligible for grants. Preference is given to institutions awarding 100 or more science baccalaureates in the most recent three-year period for which data are available.

Deadlines

Proposals may be submitted at any time; processing requires approximately six to nine months.

B. Interinstitutional Projects in Four-Year Colleges The maximum duration of a grant is three years; grants will not exceed an average of $100,000 per year.

Eligibility

Formal and ad hoc associations of four-year colleges and universities for cooperative science projects are eligible for grants. Proposals should request support for institutional development projects that are better accomplished by groups of institutions than by individual institutions.

Institutions eligible to participate are four-year colleges and universities with baccalaureate programs in the sciences that have not granted more than 10 Ph.D. degrees in the sciences during academic years 1961-62 to 1963-64 inclusive. However, a university that has exceeded that number may serve as advisor to a group of eligible institutions.

Deadlines

Deadlines for submission of proposals under Interinstitutional Projects are mid-February and mid-October; awards are made in early June or mid-January.

C. Cooperative Projects for Two-Year Colleges Regional groupings of two-year colleges may participate with a nearby college or university in cooperative projects to accelerate faculty development and related course content improvement. Each proposal should deal with a single science discipline. No two-year college will be permitted to participate concurrently in more than two cooperative projects. Grants are limited to a duration of three years.

Eligibility

The cooperative four-year institution preferably is one that grants the master's degree or Ph.D. in the appropriate science field. It serves as the grantee institution and contributes leadership to the project. Two-year colleges eligible to participate are those that offer college-parallel courses in science for transfer credit.

Deadlines

Deadline for submission of proposals under the Cooperative Projects for Two-Year Colleges is mid-October; awards are announced in mid-January.

Additional Information

An announcement containing guidelines for the submission of proposals and application materials is available from the Foundation. Communications may be addressed to: Division of Undergraduate Education in Science, National Science Foundation, Washington, D. C. 20550.

Institutional Grants for Science Program

The National Science Foundation awards grants for broad institutional use to colleges and universities, based on total Federal research awards. These are flexible funds for use at the discretion of the institution to strengthen and balance science programs of research and education. The funds may not be used for indirect costs.

The grants are computed by a formula that provides 100 percent of the first $10,000 in Federal research awards to an institution, plus a graduated percentage of the amount of awards in excess of $10,000. More than 600 institutions participate annually in the Institutional Grants for Science Program.

Eligibility

Institutions eligible to apply for grants under the Institutional Grants for Science Program are colleges and universities receiving Federal research awards, excluding those of the Public Health Service, during the previous year (July 1-June 30). Grants made by the Foundation through its programs of Undergraduate Research Participation and Research Participation for College Teachers also establish eligibility for Institutional Grants and are included in the base for their computation.

Deadlines

Announcements are available in April of each year from the address listed below. The application deadline is June 30. Grants are announced in October.

Additional Information

Communications may be addressed to: Division of Institutional Resources, National Science Foundation, Washington, D. C. 20550.

IV. COMBINED SCIENTIFIC RESEARCH AND EDUCATION PROGRAMS

Computing Activities in Education and Research

The National Science Foundation awards grants and contracts for a variety of computer-oriented projects in education and research. Support is provided for research in computer science, for projects involving computers in educational innovations, for faculty and student training, for cooperative efforts to make computing services more widely available, and for a few studies and conferences.

1. **Research and Education** Grants are awarded to encourage high-quality research in computer science; software, hardware, and theoretical projects are all eligible for support. Grants typically include provision for graduate students involved in the research project.

Projects concerned with applications of computers and computer science can be considered when this component is unusual and innovative, but not if a computer is used in a routine fashion. For example, support may be provided for an educational effort that involves a basic restructuring of a science course or sequence of courses because of a new approach that depends heavily on a computer for problem-solving simulation, or data acquisition. Another area of application is computer-assisted or computer-managed instruction: this program has focused attention on long-range research and development, and on efforts to reduce costs.

2. **Regional Cooperative Projects and Other Training Activities** For several years the Foundation has encouraged regional cooperation to make existing computer facilities accessible to nearby institutions. A typical project is centered about a university with a strong computer operation, and involves a number of nearby colleges. The participating institutions collaborate in faculty training programs and in exploration of computer use in a variety of courses.

Training projects in other formats can also be supported. These are usually designed for a special group of participants, and do not follow the traditional pattern of summer institutes for teachers.

3. **Special Computing Services for Research and Science Education** The traditional program that has assisted universities and colleges with the improvement of campus computing facilities is being reoriented. In the future, emphasis will be placed on facilities that furnish special services accessible to scientists on a number of campuses. Proposals will be considered if the projects are concerned with unique resources, innovative approaches to providing service, or novel approaches to reducing academic computing costs.

Eligibility

Institutions eligible to submit proposals under the computing activities program include colleges and universities, consortia of such institutions, and nonprofit organizations. Institutions are encouraged to describe their projects in a preliminary document so that the Foundation can determine whether a formal proposal can be considered. The NSF pamphlet **Grants for Computing Activities** should be consulted prior to submission of a formal proposal.

Deadlines

Proposals may be submitted at any time and generally require three to nine months for evaluation before notification.

Additional Information

Communications may be addressed to: Office of Computing Activities, National Science Foundation, Washington, D. C. 20550.

National Sea Grant Program

The National Science Foundation awards grants for research and development, education and training, and advisory services in all fields related to the development and utilization of marine resources and the marine environment. The term **marine environment** is defined to include the oceans, the continental shelf of the United States, the Great Lakes, and the seabed and subsoil of the submarine areas adjacent to the coasts of the United States and its island territories to the depth of 200 meters, or beyond that limit, to waters that permit the exploitation of natural resources.

This activity was authorized by Congress in the National Sea Grant Colleges and Program Act of 1966, Public Law 89-688, and extended by Public Law 90-477.

The research program is applied to the following:

(1) Development, conservation, and economic utilization of the resources of the marine environment (including definition and harvesting of natural stocks of living organisms and minerals, and aquaculture of commercially valuable species).

(2) Marine technology and engineering.

(3) Estuarine and coastal utilization (including recreation and multiple-use problems).

(4) Economic, legal, medical, and sociological considerations related to the management and development of natural resources of the marine environment.

The education program emphasizes engineering over basic science, and includes two-year technician training programs. The total educational objective is very broad, encompassing training in law, economics, and administration as they apply to ocean-related problems.

For operational purposes, the National Sea Grant Program is divided into two categories: Sea Grant Institutional Support and Sea Grant Project Support.

1. **Sea Grant Institutional Support** is focused on institutions engaged in comprehensive marine resources programs that include research, education, and advisory services. Such institutions are expected to provide leadership and scientific and technological resources for marine activities within their regions. Small grants are also available to provide support for planning institutional programs.

2. **Sea Grant Project Support** is directed to individual projects in marine resource development. In general, each project will be a well-defined research, study, education, advisory, or training activity expected to produce information, techniques, methods, or systems applicable to marine resource development. Also eligible for support are coherent area projects which provide multidisciplinary approaches to pressing problems or immediate opportunities in marine resources.

Eligibility

Institutions eligible to submit proposals under this program include universities, colleges, junior colleges, technical schools, institutes, laboratories, and public or private agencies. An announcement containing application materials is available from the Foundation.

Deadlines

Proposals may be submitted at any time.

Additional Information

Communications may be addressed to: Office of Sea Grant Programs, National Science Foundation, Washington, D. C. 20550.

V. INTERNATIONAL SCIENCE

United States-Australia Cooperative Science Program

The National Science Foundation coordinates the participation of U.S. scientists in the United States-Australia Agreement for Scientific and Technical Cooperation. The purpose of the agreement is to provide additional opportunities to exchange ideas, information, skills and techniques; to collaborate on problems of mutual interest, to work together in unique environments and to utilize special facilities. As the program develops, the agreement will be implemented through detailed arrangements in such specific areas as environmental sciences, agricultural sciences, biological and medical sciences, and innovative approaches to science education. Scientists may be in civil agencies of the government or in academic or other institutions.

Joint activities which further cooperation may include

(1) Exchange of scientists and technical experts.

(2) Pursuit of joint research activities.

(3) Convocation of joint seminars.

Funds for the support of the activities of American scientists may come from any U.S. source, which includes but is not confined to the regular research support programs of the Foundation. Proposals are submitted to the appropriate funding agency or institution in accordance with its normal procedures.

Each activity in the program requires approval by the Foundation and by the Commonwealth Department of Science and Education, the executive agencies responsible for carrying out the terms of the agreement in the United States and Australia. Nothing in the agreement is intended to prejudice other arrangements for scientific and technical cooperation between the two countries.

Additional Information

Communications may be addressed to: United States-Australia Program, Office of International Programs, National Science Foundation, Washington, D. C. 20550.

United States-Republic of China Cooperative Science Program

The National Science Foundation awards grants to support the participation of U.S. scientists in the United States-Republic of China Cooperative Science Program. Chinese funds support Chinese scientists participating in the program.

Three types of projects are supported in the program. They are:

(1) Cooperative research in all areas of the natural sciences;

(2) Visiting scientists:

- (a) short-term visitors to serve as consultants, to lecture, to participate in symposia or special workshops, and similar activities in all fields; but limited to U.S. scientists already in East Asia for other purposes;
- (b) long-term visitors from six months to one year to instruct at the graduate level and to conduct collaborative research in all areas of the natural sciences.

(3) Scientific seminars on any appropriate scientific subject including science education.

A brochure describing the United States-Republic of China Cooperative Science Program is available from the Foundation, together with instructions and guidelines for submitting proposals.

Eligibility

Those eligible to submit proposals are colleges and universities, individual scientists, or groups of scientists. The program is aimed primarily at the academic scientist; however, others may be considered. An informal inquiry to the Foundation should be made before a formal proposal is submitted. All projects involving both United States and Chinese scientists are jointly funded and must be approved by the Foundation and the National Science Council in Taipei.

Deadlines

Proposals may be submitted at any time, but with the exception of (2) (a) above; approximately six months are needed for consideration.

Additional Information

Communications may be addressed to: United States-Republic of China Cooperative Science Program, Office of International Programs, National Science Foundation, Washington, D. C. 20550.

United States-France Exchange of Scientists Program

The National Science Foundation and the Centre National de la Recherche Scientifique jointly sponsor exchange of scientists for study or research in the mathematical, physical, chemical, and engineering sciences and in the biological sciences exclusive of the medical sciences. Awards are not made in the social or medical sciences or in education or business fields.

Eligibility

Eligible individuals are citizens or nationals of the United States and France who will have earned a doctoral degree or its equivalent normally not more than five years prior to the commencement of the exchange visit. Eligible institutions are, for American candidates, any appropriate nonprofit French institution. Appropriate nonprofit institutions are institutions of higher education; government research institutes, laboratories, or centers; and privately sponsored nonprofit institutes. The period of the exchange visit is normally between five and fifteen months. French candidates may obtain information and application materials from the Centre National de la Recherche Scientifique. American candidates may obtain information and application materials from the address below.

Additional Information

Communications may be addressed to: Office of International Programs, National Science Foundation, Washington, D. C. 20550

United States-India Exchange of Scientists and Engineers

The National Science Foundation administers the participation of U.S. scientists and engineers in a program of short-term exchanges. In India the program is administered by the Council for Scientific and Industrial Research. These organizations are jointly responsible for approving each exchange visit. The National Science Foundation pays only travel costs of U.S. scientists to and from India. Within India, expenses are covered by the local hosts. A brochure describing this program is available from the Foundation.

Eligibility

Individual senior scientists and engineers are eligibile to submit proposals. The evaluation of requests is based on the applicant's professional qualifications and the merit of the proposed activity in India.

Deadlines

Proposals may be submitted at any time.

Additional Information

Communications may be addressed to: United States-India Exchange of Scientists and Engineers, Office of International Programs, National Science Foundation, Washington, D. C. 20550.

United States-Italy Cooperative Science Program

The National Science Foundation coordinates the participation of U.S. scientists and institutions in the United States-Italy Cooperative Science Program.

The objectives of the program are to promote cooperation between scientists of the two countries for peaceful purposes and to provide additional opportunities for them to exchange ideas, skills and techniques; to attack problems of particular mutual interest; to work together in unique environments and to utilize special facilities.

Types of projects included in this program are

(1) Joint science research projects.

(2) Exchange of scientists, in connection with approved projects.

(3) Seminars to exchange information and plan cooperative research.

Each activity in the program involves participation by scientists of both countries and requires approval by the Foundation and by the Consiglio Nazionale delle Ricerche, the executive agencies responsible for carrying out the terms of the agreement in the United States and Italy. Nothing in the agreement is intended to prejudice other arrangements for scientific cooperation between the two countries.

Funds for the support of the activities of American scientists may come from any U.S. source, which includes but is not confined to the regular research support programs of the Foundation. Proposals are submitted to the appropriate funding agency or institution in accordance with its normal procedures. At the same time, the U.S. investigator sends a copy of his proposal to the address below, together with a copy of the joint Application Form, signed by him and the Italian principal investigator.

Additional Information

Communications may be addressed to: United States-Italy Cooperative Science Program, Office of International Programs, National Science Foundation, Washington, D. C. 20550.

United States-Japan Cooperative Science Program

The National Science Foundation awards grants to support the participation of U.S. scientists in the United States-Japan Cooperative Science Program. Japanese funds support Japanese scientists participating in the program.

Three types of projects are included in the program. They are:

(1) Cooperative research in all areas of the natural sciences;

(2) Scientific seminars;

(3) Visiting scientists.

The joint United States-Japan Committee stresses the need for more American scientists working for extended periods in Japanese laboratories.

A brochure describing the United States-Japan Cooperative Science Program is available from the Foundation, together with instructions and guidelines for submitting proposals.

Eligibility

American scientists in any fundamental scientific discipline may apply to the Foundation for support of a research or training project to be carried out in a Japanese institution. Assurance must be given that the Japanese institution can accommodate the American participant.

Organizations eligible to submit proposals are colleges and universities, nonprofit research institutions, individual scientists, or groups of scientists. This program is aimed primarily at the academic scientist; however, others may be considered. All projects involving both United States and Japanese scientists are jointly funded and must be approved both by the Foundation and the Japan Society for the Promotion of Science.

Deadlines

Proposals may be submitted at any time; approximately six months are needed to consider a proposal.

Additional Information

Communications may be addressed to: United States-Japan Cooperative Science Program, Office of International Programs, National Science Foundation, Washington, D. C. 20550.

United States-Romania Science and Technology Exchange and Cooperation

Jointly with the National Council for Scientific Research of Romania, the National Science Foundation supports exchanges of scientists and cooperative scientific activities between scientists and institutions of the United States and Romania. Exchange and cooperation may be in those fields of science in which the Foundation supports scientific research projects.

Eligibility

Proposals for grants for cooperative scientific activities may be submitted by colleges and universities and by academically related nonprofit research organizations. Individual scientists employed by such institutions and organizations may apply for support of exchange visits to Romania.

Additional Information

Communications may be addressed to: Office of International Programs, National Science Foundation, Washington, D. C. 20550.

International Science Education Assistance Programs

The National Science Foundation provides, upon request, advisory services to the Agency for International Development and administers certain regional or country programs with funds transferred by AID.

These activities take place under the terms of an agreement concluded in 1965 between NSF and AID, which has the purpose of permitting AID to draw on the experience of NSF in devising methods of improving science education through teacher training programs and the development of new and improved science curricular materials.

Present activities under the agreement are:

Cooperative Program for the Improvement of Science Education in India. Under an agreement between NSF, AID, and the Government of India, the Foundation recruits and assigns technical and administrative personnel for long-term positions with the Science Liaison Staff in New Delhi and for short-term consultant assignments with specific projects in India. Some short-term consultants are employed to assist with projects in curriculum design, the production of texts and teaching aids, the design of teacher training programs, and college and university development. They serve at various places in India, at different times of the year, and usually for periods less than three months. In addition, some consultants assist with summer institutes, held usually during the period April to July, chiefly in connection with the projects named above.

Eligibility

U.S. scientists and science teachers with extensive experience in educational projects in the United States or abroad are eligible to apply. If selected they are appointed as consultants to the National Science Foundation and will be assigned to specific projects in India. These consultants will receive travel and living expenses and a consulting fee for the duration of their service of India.

Science Education Support-Worldwide. Under an agreement with the Office of Education and Human Resources of AID's Technical Assistance Bureau, the Foundation administers a small program that provides services or special studies performed by selected institutional grantees or consultants to help AID fulfill its responsibilities for assisting the developing countries in which it works to improve the quality of their scientific and technical education.

The limited funds available are already fully committed, but the Foundation will accept informal proposals which will be reviewed for possible inclusion in the future. These informal proposals should be limited to projects in science education improvement which provide for studies or services of general applicability in AID cooperating countries.

Additional Information

Communications may be addressed to: Office of International Programs, National Science Foundation, Washington, D. C. 20550.

International Travel Grants

The National Science Foundation awards international travel grants to assist scientists to go abroad for one of the following purposes:

(1) Attending international scientific congresses and meetings;

(2) Obtaining or exchanging information in the areas of basic research, science education, science information or information relating to international scientific programs and associated activities;

(3) Cooperating in international scientific activities.

International travel is defined as all travel outside the United States and its possessions, except Mexico and Canada.

NSF each year selects certain meetings, in areas of particular interest to the Foundation, for which participant support may be granted.

International travel grants made to individuals are based on, and normally limited to, the equivalent cost of jet-economy air transportation from the city where the traveler resides, or is employed, to his destination abroad and return. A per diem may be paid when an individual is traveling as a representative of the U.S. Government. Travel must be by U.S. flag carriers, except in special circumstances.

Eligibility

Requests for international travel grants may be submitted by individual U.S. scientists or by nonprofit organizations (usually professional societies). When a request is submitted by an individual U.S. scientist, NSF form 9-1, Application for International Travel Grant, available from the Foundation, should be used.

Deadlines

Approximately a month is required to process requests, but those for travel to meetings should be submitted well in advance because evaluation of requests normally occurs several months before the meeting date.

Additional Information

Communications may be addressed to the appropriate division or office: Division of Biological and Medical Sciences; Division of Engineering; Division of Environmental Sciences; Division of Mathematical and Physical Sciences; Division of Social Sciences; Office of Science Information Service; Office of International Programs; Office of Computing Activities; Office of Polar Programs; Advanced Science Education Program, Division of Graduate Education in Science; National Science Foundation, Washington, D. C. 20550.

VI. SCIENCE INFORMATION

Science Information Service

The National Science Foundation awards grants and contracts to improve the dissemination of scientific information. Foundation support may be provided for the following activities:

(1) Development and improvement of information systems.

(2) Operational support for information systems and services, and the publication of results of original research, including journals and monographs; production and publication of abstracts, indexes, and other bibliographic aids.

(3) Research in science information, including both theoretical and applied aspects.

(4) Translation and republication in English of significant foreign scientific literature, preparation and publication of reference aids dealing with important foreign and international research institutions and agencies—their activities, organization, and staff.

The Foundation's pamphlets **Improving the Dissemination of Scientific Information** and **Grants for Scientific Research** should be consulted for additional information on scientific information programs and instructions for submission of proposals.

Eligibility

Institutions eligible to submit proposals are professional scientific and technical societies, universities and colleges, and organizations both for profit and not for profit. Organizations that plan to submit proposals are encouraged to discuss their ideas informally with the appropriate staff members before preparing formal proposals.

Deadlines

Proposals may be submitted at any time; approximately three months are required to consider a proposal.

Additional Information

Communications may be addressed to: Office of Science Information Service, National Science Foundation, Washington, D. C. 20550.

VII. SPECIAL PROGRAMS

Scientific Conference Grants

The National Science Foundation awards grants to support conferences and symposia that bring together leading scientists who are pioneering in new or incompletely explored fields of science.

The Foundation does not provide support for regular meetings of scientific societies. Support for special conferences should be requested only if regular meetings of professional societies do not provide the necessary forum.

Eligibility

Proposals for support for scientific conferences may be submitted by colleges and universities, nonprofit research institutions, or scientific or professional societies. Concomitant support by several Federal agencies or private organizations is permissible.

Deadlines

Proposals for Scientific Conference Grants may be submitted at any time.

Additional Information

Communications may be addressed to the following divisions or offices as appropriate: Division of Biological and Medical Sciences; Division of Engineering; Division of Environmental Sciences; Division of Mathematical and Physical Sciences; Division of Social Sciences; Office of Computing Activities; or Office of Science Information Service; National Science Foundation, Washington, D. C. 20550.

Intergovernmental Science Programs

The National Science Foundation awards grants to enable State and local levels of government to develop improved programs and institutions for applying science and technology to governmental problems, and for implementing recommendations or utilizing information resulting from NSF programs.

Objectives of Intergovernmental Science Programs are:

(1) To advance the understanding of public issues and problems having scientific and technological content at the State and local levels of government, and to assess needs and opportunities for more effective application of science and technology;

(2) To demonstrate innovative science and technology planning and decision making processes related to State, local, and regional problems;

(3) To stimulate selected State and local governments' experimentation, on a pilot basis, with science and technology systems in the context of their own needs and resources;

(4) To encourage adoption of new systems which show promise for enhancing State and local ability to incorporate science and technology into public programs;

(5) To improve communication between persons and groups concerned with science and technology at the Federal, State, and local levels of government.

The proposal activity must involve a problem of general interest to State and local governments. Preference will be given to innovative approaches looking toward the development of models for governmental use of science and technology. Activities supported may include research projects, manpower and education programs (involving State and local government officials), technology assessment and forecasting studies, and planning studies to help develop innovative policies and programs for State and local governments. Institutional support will be provided to assist in establishment of centers for governmental science policy planning. Conferences and seminar projects will also be supported.

Beginning in fiscal years 1971 or 1972, it is expected that pilot experiments will be supported (a) to enable State and local governments to install systems for applying science and technology in public decisionmaking processes, or (b) to improve mechanisms for R&D transfer between units of government, colleges and universities, and the private sector.

Eligibility

Proposals may be submitted by units of State and local governments and their regional organizations, legislatures, law schools, State academies of science, colleges and universities that grant at least a baccalaureate-level degree in science, and nonprofit institutions. Proposals from academic institutions in association with a unit of government will be given preference; however, awards will be made under other organizational arrangements. There is no requirement for matching funds, but normally applicants are required to share in the cost of any proposed activity.

Proposals may be submitted to other Federal agencies for partial support and to NSF for those activities that fall outside the program scope of other Federal agencies.

Deadlines

Proposals may be submitted at any time; processing of a proposal requires approximately six months. Informal inquiry to the Foundation may be made to determine whether or not a potential project would qualify for support under NSF Intergovernmental Science Programs.

Additional Information

In addition to the NSF pamphlet **Grants for Scientific Research,** supplementary guidelines for preparation of proposals are available. Communications may be addressed to: Office of Intergovernmental Science Programs, National Science Foundation, Washington, D. C. 20550.

Public Understanding of Science Program

The National Science Foundation awards grants to support a greater understanding of science and the relationship of science and technology to the problems of society.

Proposed projects must reflect accepted standards of scientific objectivity, and a well defined communications mission involving the American public.

A limited number of projects are funded which focus on either increasing the scientific knowledge of news media personnel or promoting the exchange of ideas through seminars and conferences between scientists and laymen on science policy issues of national and regional import.

Program guidelines are flexible; proposals for consideration by other Foundation programs which have built-in public understanding of science components are encouraged. Projects should be described in sufficient detail in preliminary correspondence prior to the submission of a formal proposal.

Eligibility

Institutions eligible to submit proposals are colleges and universities and independent, non-profit organizations.

Deadlines

Proposals may be submitted at any time.

Additional Information

An announcement containing a more detailed description is available from the Foundation. Communications may be addressed to: Public Understanding of Science Office, Office of Government and Public Programs, National Science Foundation, Washington, D. C. 20550.

University Science Planning and Policy Program

The National Science Foundation awards grants for the purpose of developing the Nation's capabilities for research and training in the area of science planning and/or science policy.

The area of science policy is defined broadly in terms of the Foundation's statutory responsibilities: ". . . to appraise the impact of research upon industrial development and upon the general welfare"; and ". . . the Director shall recommend and encourage the pursuit of national policies for the promotion of basic research and education in the sciences." The areas of interest involve both the overall administration and support of scientific activities, and the manner in which these activities impinge on, and are affected by, the social, economic, and legal structure of the Nation.

Activities eligible for support under this program may take a variety of forms. Grants may be awarded for research projects concerning science planning, science policy issues, and the techniques and methodologies appropriate thereto. The research, which typically should be interdisciplinary in character, may be conducted by faculty members and graduate students working either individually or in groups.

At institutions where a sufficient potential already exists, interdisciplinary programs may be supported to conduct coherent efforts involving a variety of research projects, research seminars, and possibly the development of related curricula. Typically, junior and senior faculty members and graduate students would be involved together in these activities. Institutions are encouraged to examine their capabilities in this area realistically and to develop imaginative concepts and proposals for extending them.

Eligibility

Proposals for University Science Planning and Policy grants may be submitted by colleges and universities that grant at least a baccalaureate-level degree in science and mathematics.

Deadlines

A proposal may be submitted at any time; approximately six months are required to consider and process a proposal.

Additional Information

Communications may be addressed to: Office of Plans and Analysis, National Science Foundation, Washington, D. C. 20550.

SOURCE: NSF, "Guide to Programs," 1970.

Part 5

NSF LEGISLATION

TITLE 42.—THE PUBLIC HEALTH AND WELFARE

Chapter 16.—NATIONAL SCIENCE FOUNDATION

---

§ 1861. Establishment; composition.

There is established in the executive branch of the Government an independent agency to be known as the National Science Foundation (hereinafter referred to as the "Foundation"). The Foundation shall consist of a National Science Board (hereinafter referred to as the "Board") and a Director. (May 10, 1950, ch. 171, § 2, 64 Stat. 149.)

SHORT TITLE

Section 1 of act May 10, 1950, provided that act May 10, 1950, which is classified to this chapter, should be popularly known as the "National Science Foundation Act of 1950."

REORGANIZATION PLAN NO. 2 OF 1962

Eff. June 8, 1962, 27 F.R. 5419, 76 Stat. 1253, as amended Aug. 14, 1964, Pub. L. 88–426, title III, § 305(41), 78 Stat. 427

Prepared by the President and transmitted to the Senate and the House of Representatives in Congress assembled, March 29, 1962, pursuant to the provisions of the Reorganization Act of 1949, 63 Stat. 203, as amended [sections 133z to 133z–15 of Title 5].

CERTAIN SCIENCE AGENCIES AND FUNCTIONS

PART I. OFFICE OF SCIENCE AND TECHNOLOGY

Section 1. Office of Science and Technology. There is hereby established in the Executive Office of the President the Office of Science and Technology, hereafter in this Part referred to as the Office.

Sec. 2. Director and deputy. (a) There shall be at the head of the Office the Director of the Office of Science and Technology, hereafter in this Part referred to as the Director. The Director shall be appointed by the President by and with the advice and consent of the Senate.

(b) There shall be in the Office a Deputy Director of

the Office of Science and Technology, who shall be appointed by the President by and with the advice and consent of the Senate. The Deputy Director shall perform such functions as the Director may from time to time prescribe and shall act as Director during the absence or disability of the Director or in the event of vacancy in the office of Director.

(c) No person shall while holding office as Director or Deputy Director engage in any other business, vocation, or employment.

Sec. 3. Transfer and performance of functions. (a) There are hereby transferred from the National Science Foundation to the Director:

(1) So much of the functions conferred upon the Foundation by the provisions of section 3(a)(1) of the National Science Foundation Act of 1950 (42 U.S.C. 1862(a)(1) [section 1862(a)(1) of this title]) as will enable the Director to advise and assist the President in achieving coordinated Federal policies for the promotion of basic research and education in the sciences.

(2) The functions conferred upon the Foundation by that part of section 3(a)(6) of the National Science Foundation Act of 1950 (42 U.S.C. 1862(a)(6)) which reads as follows: "to evaluate scientific research programs undertaken by agencies of the Federal Government."

(b) In carrying out the functions transferred by the provisions of section 3(a) of this reorganization plan, the Director shall assist the President as he may request with respect to the coordination of Federal scientific and technological functions and agencies.

(c) The Director may from time to time make such provisions as he deems appropriate authorizing the performance of any of his functions by any other officer, or by any employee or agency, of the Office.

Sec. 4. Personnel. The Director may appoint employees necessary for the work of the Office under the classified civil service and fix their compensation in accordance with the classification laws.

Part II. National Science Foundation

Sec. 21. Executive Committee. (a) There is hereby established the Executive Committee of the National Science Board, hereafter in this Part referred to as the Executive Committee, which shall be composed of five voting members. Four of the members shall be elected as hereinafter provided. The Director provided for in section 22 of this reorganization plan, ex officio, shall be the fifth member and the chairman of the Executive Committee.

(b) At its annual meeting held in 1964 and at each of its succeeding annual meetings the National Science Board, hereafter in this Part referred to as the Board, shall elect two of its members as members of the Executive Committee, and the Executive Committee members so elected shall hold office for two years from the date of their election. Any person who has been a member of the Executive Committee (established by this reorganization plan) for six consecutive years shall thereafter be ineligible for service as a member thereof during the two-year period following the expiration of such sixth year. For the purposes of this subsection, the period between any two consecutive annual meetings of the Board shall be deemed to be one year.

(c) At is first meeting held after the effective date of this section the Board shall elect four of its members as members of the Executive Committee. As designated by the Board, two of the Executive Committee members so elected shall hold office as such members until the date of the annual meeting of the Board held in 1964 and the other two members so elected shall hold such office until the annual meeting of the Board held in 1965.

(d) Any person elected as a member of the Executive Committee to fill a vacancy occuring prior to the expiration of the term for which his predecessor was elected shall be elected for the remainder of such term.

(e) The functions conferred upon the Executive Committee now existing under the provisions of the National Science Foundation Act of 1950 [this chapter], by the provisions of section 6 of the National Science Foundation Act of 1950 (42 U.S.C. 1865) or otherwise, are hereby transferred to the Executive Committee established by the provisions of this Part; and the authority of the National Science Board to assign its powers and functions to the now-existing Executive Commitee, and statutory limitations upon such assignment, shall hereafter be applicable to the Executive Commitee established by the provisions of this Part.

Sec. 22. Director. (a) There is hereby established in the National Science Foundation a new office with the title of Director of the National Science Foundation. The Director of the National Science Foundation, hereafter in this Part referred to as the Director, shall be appointed by the President by and with the advice and consent of the Senate. Before any person is appointed as Director the President shall afford the Board an opportunity to make recommendations to him with respect to such appointment. The Director shall serve for a term of six years unless sooner removed by the President. The Director shall not engage in any business, vocation or employment other than that of serving as such Director, nor shall he, except with the approval of the Board, hold any office in, or act in any capacity for, any organization, agency, or institution with which the Foundation makes any contract or other arrangement under the National Science Foundation Act of 1950 [this chapter].

(b) Except to the extent inconsistent with the provisions of section 23(b)(2) of this reorganization plan, all functions of the office of Director of the National Science Foundation abolished by the provisions of 23(a)(2) hereof are hereby transferred to the office of Director established by the provisions of subsection (a) of this section.

(c) The Director, ex officio, shall be an additional member of the Board and, except in respect of compensation and tenure, shall be coordinate with other members of the Board. He shall be a voting member of the Board and shall be eligible for election by the Board as chairman or vice chairman of the Board.

Sec. 23. Abolitions. (a) The following agencies, now existing under the National Science Foundation Act of 1950 [this chapter], are hereby abolished:

(1) The Executive Committee of the National Science Board (section 6 of Act; 42 U.S.C. 1865 [section 1865 of this title]).

(2) The office of Director of the National Science Foundation (sections 2 and 5 of Act; 42 U.S.C. 1861, 1864).

(b) There are also hereby abolished:

(1) The functions conferred upon the National Science Board by that part of section 6(a) of the National Science Foundation Act of 1950 (42 U.S.C. 1865(a)) which reads "The Board is authorized to appoint from among its members an Executive Committee".

(2) The functions of the Director of the National Science Foundation provided for in sections 4(a) and 5(a) of the National Science Foundation Act of 1950 (42 U.S.C. 1863(a), 1864(a) with respect to serving as a nonvoting member of the Board and his functions with respect to serving as a nonvoting member of the Executive Committee provided for in section 6(b) of that Act (42 U.S.C. 1865(b)).

(3) So much of the functions conferred upon divisional committees by the provisions of section 8(d) of the National Science Foundation Act of 1950 (42 U.S.C. 1867(d)) as consists of making recommendations to, and advising and consulting with, the Board.

(c) The provisions of sections 23(a)(1) and 23(b)(1) hereof shall become effective on the date of the first meeting of the Board held after the effective date of the other provisions of this reorganization plan.

Part III. Transitional Provisions

Sec. 31. Incidental transfers. (a) So much of the personnel, property, records, and unexpended balances of appropriations, allocations, and other funds employed, held, used, available, or to be made available, in connection with the functions transferred by the provisions of section 3 of this reorganization plan as the Director of the Bureau of the Budget shall determine shall be transferred to the Office of Science and Technology at such time or times as the said Director shall direct.

(b) Such further measures and dispositions as the Director of the Bureau of the Budget shall deem to be necessary in order to effectuate the transfers provided for in subsection (a) of this section shall be carried out in such manner as he shall direct and by such agencies as he shall designate.

Sec. 32. Interim officers. (a) The President may authorize any person who immediately prior to the effective date of Part I of the reorganization plan holds a position in the Executive Office of the President to act as Director of the Office of Science and Technology until the office of Director is for the first time filled pursuant to the provisions of this reorganization plan or by recess appointment, as the case may be.

(b) The President may authorize any person who immediately prior to the effective date of section 22 of this reorganization plan holds any office existing under the provisions of the National Science Foundation Act of 1950 [this chapter] to act as Director of the National Science Foundation until the Office of Director is for the first time filled pursuant to the provisions of this reorganization plan or by recess appointment, as the case may be.

(c) The President may authorize any person who serves in an acting capacity under the foregoing provisions of this section to receive the compensation attached to the office in respect of which he so serves. Such compensation, if authorized, shall be in lieu of, but not in addition to, other compensation from the United States to which such person may be entitled.

§ 1862. Functions; reports.

(a) The Foundation is authorized and directed—

(1) to develop and encourage the pursuit of a national policy for the promotion of basic research and education in the sciences;

(2) to initiate and support basic scientific research and programs to strengthen scientific research potential in the mathematical, physical, medical, biological, engineering, and other sciences, by making contracts or other arrangements (including grants, loans, and other forms of assistance) to support such scientific activities and to appraise the impact of research upon industrial development and upon the general welfare;

(3) at the request of the Secretary of Defense, to initiate and support specific scientific research activities in connection with matters relating to the national defense by making contracts or other arrangements (including grants, loans, and other forms of assistance) for the conduct of such scientific research;

(4) to award, as provided in section 1869 of this title, scholarships and graduate fellowships in the mathematical, physical, medical, biological, engineering, and other sciences;

(5) to foster the interchange of scientific information among scientists in the United States and foreign countries;

(6) to evaluate scientific research programs undertaken by agencies of the Federal Government, and to correlate the Foundation's scientific research programs with those undertaken by individuals and by public and private research groups;

(7) to establish such special commissions as the Board may from time to time deem necessary for the purposes of this chapter;

(8) to maintain a register of scientific and technical personnel and in other ways provide a central clearinghouse for information covering all scientific and technical personnel in the United States, including its Territories and possessions;

(9) to initiate and support a program of study, research, and evaluation in the field of weather modification, giving particular attention to areas that have experienced floods, drought, hail, lightning, fog, tornadoes, hurricanes, or other weather phenomena, and to report annually to the President and the Congress thereon.

(b) In exercising the authority and discharging the functions referred to in subsection (a) of this section, it shall be one of the objectives of the Foundation to strengthen basic research and education in the sciences, including independent research by individuals, throughout the United States, including its Territories and possessions, and to avoid undue concentration of such research and education.

(c) The Foundation shall render an annual report to the President for submission on or before the 15th day of January of each year to the Congress, summarizing the activities of the Foundation and making such recommendations as it may deem appropriate. Such report shall include (1) minority views and recommendations if any, of members of the Board, and (2) information as to the acquisition and disposition by the Foundation of any patents and patent rights. (May 10, 1950, ch. 171, **§ 3, 64** Stat. 149; July 11, 1958, Pub. L. 85–510, § 1, 72 Stat. 353; Sept. 8, 1959, Pub. L. 86–232, § 1, 73 Stat. 467.)

AMENDMENTS

1959—Subsec. (a)(2). Pub. L. 86–232 clarified the Foundation's authority to support programs to strengthen scientific research potential.

1958—Subsec. (a)(9). Pub. L. 85–510 added subsec. (a)(9).

TRANSFER OF FUNCTIONS

Transfer of those functions under subsection (a)(1) of this section from the Foundation to the Director of the Office of Science and Technology as will enable him to advise and assist the President in achieving coordinated Federal policies for the promotion of basic research and education in the sciences, see section 3(a)(1) of 1962 Reorg. Plan No. 2, eff. June 8, 1962, 27 F.R. 5419, 76 Stat. 1253, set out as a note under section 1861 of this title.

Functions under subsection (a)(6) of this section relating to evaluation of scientific research programs undertaken by agencies of the Federal Government transferred to the Director of the Office of Science and Technology by section 3(a)(2) of 1962 Reorg. Plan No. 2, set out as a note under section 1861 of this title.

EMERGENCY PREPAREDNESS FUNCTIONS

Ex. Ord. No. 11095, Feb. 26, 1963, 28 F.R. 1859, directed the Director of the National Science Foundation to prepare national emergency plans and develop preparedness programs covering functions assigned to him by the Executive Order, designed to develop a state of readiness with respect to all conditions of national emergency, including attack upon the United States.

INVESTIGATION OF NEED FOR GEOPHYSICAL INSTITUTE IN TERRITORY OF HAWAII

Joint Res. Aug. 1, 1956, ch. 865, 70 Stat. 922, directed the National Science Foundation to conduct an investigation into the need for and the feasibility and usefulness of a geophysical institute located in the Territory of Hawaii. The Foundation was required to report the results of its investigations, together with its recommendations based thereon, to the Congress not later than 9 months after Aug. 1, 1956.

EX. ORD. NO. 10521. ADMINISTRATION OF SCIENTIFIC RESEARCH

Ex. Ord. No. 10521, Mar. 17, 1954, 19 F.R. 1499, as amended by Ex. Ord. No. 10807, § 6(b), Mar. 13, 1959, 24 F.R. 1899, provided:

SECTION 1. The National Science Foundation (hereinafter referred to as the Foundation) shall from time to time recommend to the President policies for the promotion and support of basic research and education in the sciences, including policies with respect to furnishing guidance toward defining the responsibilities of the Federal Government in the conduct and support of basic scientific research.

Sec. 2. The Foundation shall continue to make comprehensive studies and recommendations regarding the Nation's scientific research effort and its resources for scientific activities, including facilities and scientific personnel, and its foreseeable scientific needs, with particular attention to the extent of the Federal Government's activities and the resulting effects upon trained scientific personnel. In making such studies, the Foundation shall make full use of existing sources of information and research facilities within the Federal Government.

Sec. 3. The Foundation, in concert with each Federal agency concerned, shall review the basic scientific research programs and activities of the Federal Government in order, among other purposes, to formulate methods for strengthening the administration of such programs and activities by the responsible agencies, and to study areas of basic research where gaps or undesirable overlapping of support may exist, and shall recommend to the heads of agencies concerning the support given to basic research.

Sec. 4. As now or hereafter authorized or permitted by law, the Foundation shall be increasingly responsible for providing support by the Federal Government for general-purpose basic research through contracts and grants. The conduct and support by other Federal agencies of basic research in areas which are closely related to their missions is recognized as important and desirable, especially in response to current national needs, and shall continue.

Sec. 5. The Foundation, in consultation with educational institutions, the heads of Federal agencies, and the Commissioner of Education of the Department of Health, Education, and Welfare, shall study the effects upon educational institutions of Federal policies and administration of contracts and grants for scientific research and development, and shall recommend policies and procedures which will promote the attainment of general national research objectives and realization of the research needs of Federal agencies while safeguarding the strength and independence of the Nation's institutions of learning.

Sec. 6. The head of each Federal agency engaged in scientific research shall make certain that effective executive, organizational, and fiscal practices exist to ensure (a) that the Foundation is consulted on policies concerning the support of basic research, (b) that approved scientific research programs conducted by the agency are reviewed continuously in order to preserve priorities in research efforts and to adjust programs to meet changing conditions without imposing unnecessary added burdens on budgetary and other resources, (c) that applied research and development shall be undertaken with sufficient consideration of the underlying basic research and such other factors as relative urgency, project costs, and availability of manpower and facilities, and (d) that, subject to considerations of security and applicable law, adequate dissemination shall be made within the Federal Government of reports on the nature and progress of research projects as an aid to the efficiency and economy of the overall Federal scientific research program.

Sec. 7. Federal agencies supporting or engaging in scientific research shall, with the assistance of the Foundation, cooperate in an effort to improve the methods of classification and reporting of scientific research projects and activities, subject to the requirements of security of information.

Sec. 8. To facilitate the efficient use of scientific research equipment and facilities held by Federal agencies:

(a) the head of each such agency engaged in scientific research shall, to the extent practicable, encourage and facilitate the sharing with other Federal agencies of major equipment and facilities; and

(b) a Federal agency shall procure new major equipment or facilities for scientific research purposes only after taking suitable steps to ascertain that the need cannot be met adequately from existing inventories or facilities of its own or of other agencies; and

(c) the Interdepartmental Committee on Scientific Research and Development shall take necessary steps to ensure that each Federal agency engaged directly in scientific research is kept informed of selected major equipment and facilities which could serve the needs of more than one agency. Each Federal agency possessing such equipment and facilities shall maintain appropriate records to assist other agencies in arranging for their joint use or exchange.

Sec. 9. The heads of the respective Federal agencies shall make such reports concerning activities within the purview of this order as may be required by the President.

Sec. 10. The National Science Foundation shall provide leadership in the effective coordination of the scientific information activities of the Federal Government with a view to improving the availability and dissemination of scientific information. Federal agencies shall cooperate with and assist the National Science Foundation in the performance of this function, to the extent permitted by law.

Ex. Ord. No. 10807. Federal Council for Science and Technology

Ex. Ord. No. 10807, Mar. 13, 1959, 24 F.R. 1897, provided:

Section 1. *Establishment of Council.* (a) There is hereby established the Federal Council for Science and Technology (hereinafter referred to as the Council).

(b) The Council shall be composed of the following-designated members: (1) the Special Assistant to the President for Science and Technology, (2) one representative of each of the following-named departments, who shall be designated by the Secretary of the Department concerned and shall be an official of the Department of policy rank: the Departments of Defense, the Interior, Agriculture, Commerce, and Health, Education, and Welfare, (3) the Director of the National Science Foundation, (4) the Administrator of the National Aeronautics and Space Administration, and (5) a representative of the Atomic Energy Commission, who shall be the Chairman of the Commission or another member of the Commission designated by the Chairman. A representative of the Secretary of State designated by the Secretary and a representative of the Director of the Bureau of the Budget designated by the Director may attend meetings of the Council as observers.

(c) The Chairman of the Council (hereinafter referred to as the Chairman) shall be designated by the President from time to time from among the members thereof. The Chairman may make provision for another member of the Council, with the consent of such member, to act temporarily as Chairman.

(d) The Chairman (1) may request the head of any Federal agency not named in section 2(b) of this order to designate a representative to participate in meetings or parts of meetings of the Council concerned with matters of substantial interest to the agency, and (2) may invite other persons to attend meetings of the Council.

(e) The Council shall meet at the call of the Chairman.

Sec. 2. *Functions of Council.* (a) The Council shall consider problems and developments in the fields of science and technology and related activities affecting more than one Federal agency or concerning the over-all advancement of the Nation's science and technology, and shall recommend policies and other measures (1) to provide more effective planning and administration of Federal scientific and technological programs, (2) to identify research needs including areas of research requiring additional emphasis, (3) to achieve more effective utilization of the scientific and technological resources and facilities of Federal agencies, including the elimination of unnecessary duplication, and (4) to further international cooperation in science and technology. In developing such policies and measures the Council, after consulting, when considered appropriate by the Chairman, the National Academy of Sciences, the President's Science Advisory Committee, and other organizations, shall consider (i) the effects of Federal research and development policies and programs on non-Federal programs and institutions, (ii) long-range program plans designed to meet the scientific and technological needs of the Federal Government, including manpower and capital requirements, and (iii) the effects of non-Federal programs in science and technology upon Federal research and development policies and programs.

(b) The Council shall consider and recommend measures for the effective implementation of Federal policies concerning the administration and conduct of Federal programs in science and technology.

(c) The Council shall perform such other related duties as shall be assigned, consonant with law, by the President or by the Chairman.

(d) The Chairman shall, from time to time, submit to the President such of the Council's recommendations or reports as require the attention of the President by reason of their importance or character.

SEC. 3. *Agency assistance to Council.* (a) For the purpose of effectuating this order, each Federal agency represented on the Council shall furnish necessary assistance to the Council in consonance with section 214 of the act of May 3, 1945, 59 Stat. 134 (31 U.S.C. § 691). Such assistance may include (1) detailing employees to the Council to perform such functions, consistent with the purposes of this order, as the Chairman may assign to them, and (2) undertaking, upon request of the Chairman, such special studies for the Council as come within the functions herein assigned to the Council.

(b) Upon request of the Chairman, the heads of Federal agencies shall, so far as practicable, provide the Council with information and reports relating to the scientific and technological activities of the respective agencies.

SEC. 4. *Standing committees and panels.* For the purpose of conducting studies and making reports as directed by the Chairman, standing committees and panels of the Council may be established in consonance with the provisions of section 214 of the act of May 3, 1945, 59 Stat. 134 (31 U.S.C. § 691). At least one such standing committee shall be composed of scientist-administrators representing Federal agencies, shall provide a forum for consideration of common administrative policies and procedures relating to Federal research and development activities and for formulation of recommendations thereon, and shall perform such other related functions as may be assigned to it by the Chairman of the Council.

SEC. 5. *Security procedures.* The Chairman shall establish procedures to insure the security of classified information used by or in the custody of the Council or employees under its jurisdiction.

SEC. 6. *Other orders; construction of orders.* (a) Executive Order No. 9912 of December 24, 1947, entitled "Establishing the Interdepartmental Committee on Scientific Research and Development," is hereby revoked.

(b) Executive Order No. 10521 of March 17, 1954 [set out as a note under this section], entitled "Administration of Scientific Research by Agencies of the Federal Government," is hereby amended:

(1) By substituting for section 1 thereof the following: "SECTION 1. The National Science Foundation (hereinafter referred to as the Foundation) shall from time to time recommend to the President policies for the promotion and support of basic research and education in the sciences, including policies with respect to furnishing guidance toward defining the responsibilities of the Federal Government in the conduct and support of basic scientific research."

(2) By inserting before the words "scientific research programs and activities" in section 3 thereof the word "basic".

(3) (i) By adding the word "and" at the end of paragraph (a) of section 8 thereof, (ii) by deleting the semicolon and the word "and" at the end of paragraph (b) of section 8 and inserting in lieu thereof a period, and (iii) by revoking paragraph (c) of section 8.

(4) By adding at the end of the order a new section 10 reading as follows:

"SEC. 10. The National Science Foundation shall provide leadership in the effective coordination of the scientific information activities of the Federal Government with a view to improving the availability and dissemination of scientific information Federal agencies shall cooperate with and assist the National Science Foundation in the performance of this function, to the extent permitted by law."

(c) The provisions of Executive Order No. 10521, as hereby amended, shall not limit the functions of the Council under this order. The provisions of this order shall not limit the functions of any Federal agency or officer under Executive Order No. 10521, as hereby amended.

(d) The Council shall be advisory to the President and to the heads of Federal agencies represented on the Council; accordingly, this order shall not be construed as subjecting any agency, officer, or function to control by the Council.

DWIGHT D. EISENHOWER

FUNCTIONS AS NOT LIMITED

Section 6(c) of Ex. Ord. No. 10827, set out as a note under this section, provided that Ex. Ord. No. 10521 should not limit functions of Federal Council for Science and Technology under Ex. Ord. No. 10827, and that Ex. Ord. No. 10827 should not limit functions of any agency or officer under Ex. Ord. No. 10521.

§ 1863. National Science Board.

(a) Composition; appointment; qualifications.

The Board shall consist of twenty-four members to be appointed by the President, by and with the advice and consent of the Senate, and of the Director ex officio, and shall, except as otherwise provided in this chapter, exercise the authority granted to the Foundation by this chapter. The persons nominated for appointment as members (1) shall be eminent in the fields of the basic sciences, medical science, engineering, agriculture, education, or public affairs; (2) shall be selected solely on the basis of established records of distinguished service; and (3) shall be so selected as to provide representation of the views of scientific leaders in all areas of the Nation. The President is requested, in the making of nominations of persons for appointment as members, to give due consideration to any recommendations for nomination which may be submitted to him by the National Academy of Sciences, the Association of Land Grant Colleges and Universities, the National Association of State Universities, the Association of American Colleges, or by other scientific or educational organizations.

(b) Term of office.

The term of office of each voting member of the Board shall be six years, except that (1) any member appointed to fill a vacancy occurring prior to the expiration of the term for which his predecessor was appointed shall be appointed for the remainder of such term; and (2) the terms of office of the members first taking office after May 10, 1950, shall expire, as designated by the President at the time of appointment, eight at the end of two years, eight at the end of four years, and eight at the end of six years, after May 10, 1950. Any person who has been a member of the Board for twelve consecutive years shall thereafter be ineligible for appointment during the two-year period following the expiration of such twelfth year.

(c) Executed.

(d) Meetings.

The Board shall meet annually on the third Monday in May, unless, prior to May 10 in any year, the Chairman has set the annual meeting for a day in May, other than the third Monday, and at such other times as the Chairman may determine, but he shall also call a meeting whenever one-third of the members so request in writing. A majority of the voting members of the Board shall constitute a quorum. Each member shall be given notice, by registered mail or by certified mail, mailed to his last known address of record not less than fifteen days prior to any meeting, of the call of such meeting.

(e) Election of Chairman and Vice Chairman; term; vacancy.

An election of the Chairman and Vice Chairman of the Board shall take place at the first meeting of the National Science Board following enactment of this legislation. Thereafter such election shall take place at the second annual meeting occurring after each such election. The Vice Chairman shall perform the duties of the Chairman in his absence. In case a vacancy occurs in the chairmanship or vice chairmanship, the Board shall elect a member to fill such vacancy. (May 10, 1950, ch. 171, § 4, 64 Stat. 150; Sept. 8, 1959, Pub. L. 86–232, § 2, 73 Stat. 467; June 11, 1960, Pub. L. 86–507, § 1 (36), 74 Stat. 202.)

REFERENCES IN TEXT

"Enactment of this legislation", referred to in subsec. (e), means enactment of Pub. L. 86–232, which was approved on Sept. 8, 1959.

CODIFICATION

Subsec. (c) provided that "the President shall call the first meeting of the Board, at which the first order of business shall be the election of a chairman and a vice chairman" and is now covered by subsec. (e) of this section.

AMENDMENTS

1960—Subsec. (d). Pub. L. 86–507 inserted "or by certified mail" following "registered mail."

1959—Subsec. (d). Pub. L. 86–232 changed the annual meeting of the Board from the first Monday in December to the third Monday or other designated day in May.

Subsec. (e). Pub. L. 86–232 substituted provision for an election of a Chairman and Vice Chairman of the Board at the first meeting of the Board following enactment of Pub. L. 86–232 and at each second annual meeting thereafter in place of provision for election of the first Chairman and Vice Chairman to serve until the first Monday in December next succeeding the date of election and for election of subsequent officers for terms of two years thereafter.

§ 1864. Director of Foundation; appointment; tenure; powers and duties.

(a) There shall be a Director of the Foundation who shall be appointed by the President, by and with the advice and consent of the Senate. The Board may make recommendations to the President with respect to the appointment of the Director, and the Director shall not be appointed until the Board has had an opportunity to make such recommendations. He shall serve as a nonvoting ex officio member of the Board. In addition thereto he shall be the chief executive officer of the Foundation. The Director shall serve for a term of six years unless sooner removed by the President.

(b) In addition to the powers and duties specifically vested in him by this chapter, the Director shall, in accordance with the policies established by the Board, exercise the powers granted by sections 1869 and 1870 of this title, together with such other powers and duties as may be delegated to him by the Board; but no final action shall be taken by the Director in the exercise of any power granted by section 1869 or 1870(c) of this title unless in each instance the Board has reviewed and approved the action proposed to be taken, or such action is taken pursuant to the terms of a delegation of authority from the Board or the Executive Committee to the Director. (May 10, 1950, ch. 171, § 5, 64 Stat. 151; Sept. 8, 1959, Pub. L. 86–232, § 3, 73 Stat. 467.)

CODIFICATION

Provisions of this section which prescribed the annual compensation of the Director were omitted to conform to the provisions of the Federal Executive Salary Schedule. See section 2210 et seq. of Title 5, Executive Departments and Government Officers and Employees.

AMENDMENTS

1959—Subsec. (b). Pub. L. 86–232 provided for delegation of authority from the Board or the Executive Committee to the Director.

TRANSFER OF FUNCTIONS

Office of Director of National Science Foundation established under the provisions of this section abolished and functions transferred to Director of National Science Foundation appointed pursuant to 1962 Reorg. Plan No. 2, see section 22 (a), (b) of 1962 Reorg. Plan No. 2, eff. June 8, 1962, 27 F.R. 5419, 76 Stat. 1253, set out as a note under section 1861 of this title.

§ 1865. Power of Board to create committees.

(a) Executive Committee; assignment of powers and functions; exception.

The Board is authorized to appoint from among its members an Executive Committee, and to assign to the Executive Committee such of the powers and functions granted to the Board by this chapter as it deems appropriate; except that the Board may not assign to the Executive Committee the function of establishing policies.

(b) Executive Committee; composition; term of office; eligibility for renomination; representation of diverse interests; reports.

If an Executive Committee is established by the Board—

(1) Such Committee shall consist of the Director, as a nonvoting ex officio member, not less than five nor more than nine other members elected by the Board from among their number.

(2) The term of office of each voting member of such Committee shall be two years, except that (A) any member elected to fill a vacancy occurring prior to the expiration of the term for which his predecessor was elected shall be elected for the remainder of such term; and (B) the term of office of four of the members first elected after May 10, 1950, shall be one year.

(3) Any person who has been a member of such Committee for six consecutive years shall thereafter be ineligible for election during the two-year period following the expiration of such sixth year.

(4) The membership of such Committee shall, so far as practicable, be representative of diverse interests and shall be so chosen as to provide representation, so far as practicable, for all areas of the Nation.

(5) Such Committee shall render an annual report to the Board, and such other reports as it may deem necessary, summarizing its activities and making such recommendations as it may deem appropriate. Minority views and recommendations, if any, of members of the Executive Committee shall be included in such reports.

(c) Additional committees.

The Board is authorized to appoint from among its members or otherwise such committees as it deems necessary, and to assign to committees so appointed such survey and advisory functions as the Board deems appropriate for the purposes of this chapter. (May 10, 1950, ch. 171, § 6, 64 Stat. 151; Sept. 8, 1959, Pub. L. 86–232, § 4, 73 Stat. 467.)

AMENDMENTS

1959—Subsec. (a). Pub. L. 86–232 eliminated the prohibition against assignment to the Executive Committee of the function of review and approval.

Subsec. (b)(1). Pub. L. 86–232 authorized the Board to have an Executive Committee consisting of from five to nine members rather than the fixed number of nine.

TRANSFER OF FUNCTIONS

Executive Committee of National Science Board appointed under the provisions of this section abolished and functions conferred by this section transferred to Executive Committee of National Science Board established by 1962 Reorg. Plan No. 2, see sections 21(e) and 23(a)(1) of 1962 Reorg. Plan No. 2, eff. June 8, 1962, 27 F.R. 5419, 76 Stat. 1253, set out as a note under section 1861 of this title.

§ 1866. Divisions within Foundation.

(a) Until otherwise provided by the Board there shall be within the Foundation the following divisions:

(1) A Division of Medical Research;

(2) A Division of Mathematical, Physical, and Engineering Sciences;

(3) A Division of Biological Sciences; and

(4) A Division of Scientific Personnel and Education, which shall be concerned with programs of the Foundation relating to the granting of scholarships and graduate fellowships in the mathematical, physical, medical, biological, engineering, and other sciences.

(b) There shall also be within the Foundation such other divisions as the Board may, from time to time, deem necessary. (May 10, 1950, ch. 171, § 7, 64 Stat. 152.)

§ 1867. Divisional committees; composition; terms of office; chairmen; rules; duties; recommendations.

(a) There shall be a committee for each division of the Foundation.

(b) Each divisional committee shall be appointed by the Board and shall consist of not less than five persons who may be members or nonmembers of the Board.

(c) The terms of members of each divisional committee shall be two years. Each divisional committee shall annually elect its own chairman from among its own members and shall prescribe its own rules of procedure subject to such restrictions as may be prescribed by the Board.

(d) Each divisional committee shall make recommendations to, and advise and consult with, the Board and the Director with respect to matters relating to the program of its division. (May 10, 1950, ch. 171, § 8, 64 Stat. 152.)

§ 1868. Special commissions; composition; chairman and vice chairman; duties.

(a) Each special commission established pursuant to section 1862 (a) (7) of this title shall consist of eleven members appointed by the Board, six of whom shall be eminent scientists and five of whom shall be persons other than scientists. Each special commission shall choose its own chairman and vice chairman.

(b) It shall be the duty of each such special commission to make a comprehensive survey of research, both public and private, being carried on in its field, and to formulate and recommend to the Foundation at the earliest practicable date an over-all research program in its field. (May 10, 1950, ch. 171, § 9, 64 Stat. 152.)

§ 1869. Scholarships and graduate fellowships.

The Foundation is authorized to award, within the limits of funds made available specifically for such purpose pursuant to section 1875 of this title, scholarships and graduate fellowships for scientific study or scientific work in the mathematical, physical, medical, biological, engineering, and other sciences at appropriate nonprofit American or nonprofit foreign institutions selected by the recipient of such aid, for stated periods of time. Persons shall be selected for such scholarships and fellowships from among citizens or nationals of the United States, and such selections shall be made solely on the basis of ability; but in any case in which two or more applicants for scholarships or fellowships, as the case may be, are deemed by the Foundation to be possessed of substantially equal ability, and there are not sufficient scholarships or fellowships, as the case may be, available to grant one to each of such applicants, the available scholarship or scholarships or fellowship or fellowships shall be awarded to the applicants in such manner as will tend to result in a wide distribution of scholarships and fellowships among the States, Territories, possessions, and the District of Columbia. Nothing contained in this chapter shall prohibit the Foundation from refusing or revoking a scholarship or fellowship award, in whole or in part, in the case of any applicant or recipient, if the Board is of the opinion that such award is not in the best interests of the United States. (May 10, 1950, ch. 171, § 10, 64 Stat. 152; Sept. 8, 1959, Pub. L. 86–232, § 5, 73 Stat. 468; June 29, 1960, Pub. L. 86–550, 74 Stat. 256; Oct. 16, 1962, Pub. L. 87–835, § 2, 76 Stat. 1070.)

AMENDMENTS

1962—Pub. L. 87–835 authorized the Foundation to refuse or revoke a scholarship or fellowship award if they believe such award is not in the best interests of the United States.

1960—Pub. L. 86–550 authorized the selection of nationals for scholarships and fellowships.

1959—Pub. L. 86–232 substituted "appropriate" for "accredited" and deleted "of higher education" following "foreign institutions".

§ 1870. General authority of Foundation.

The Foundation shall have the authority, within the limits of available appropriations, to do all things necessary to carry out the provisions of this chapter, including, but without being limited thereto, the authority—

(a) to prescribe such rules and regulations as it deems necessary governing the manner of its operations and its organization and personnel;

(b) to make such expenditures as may be necessary for administering the provisions of this chapter;

(c) to enter into contracts or other arrangements, or modifications thereof, for the carrying on, by organizations or individuals in the United States and foreign countries, including other government agencies of the United States and of foreign countries, of such basic scientific research activities as the Foundation deems necessary to carry out the purposes of this chapter, and, at the request of the Secretary of Defense, specific scientific research activities in connection with matters relating to the national defense, and, when deemed appropriate by the Foundation, such contracts or other arrangements, or modifications thereof, may be entered into without legal consideration, without performance or other bonds, and without regard to section 5 of Title 41;

(d) to make advance, progress, and other payments which relate to scientific research without regard to the provisions of sectiton 529 of Title 31;

(e) to acquire by purchase, lease, loan, gift, or condemnation, and to hold and dispose of by grant, sale, lease, or loan, real and personal property of all kinds necessary for, or resulting from, the exercise of authority granted by this chapter;

(f) to receive and use funds donated by others, if such funds are donated without restriction other than that they be used in furtherance of one or more of the general purposes of the Foundation;

(g) to publish or arrange for the publication of scientific and technical information so as to further the full dissemination of information of scientific value consistent with the national interest, without regard to the provisions of section 111 of Title 44;

(h) to accept and utilize the services of voluntary and uncompensated personnel and to provide transportation and subsistence as authorized by section 73b–2 of Title 5 for persons serving without compensation; and

(i) to prescribe, with the approval of the Comptroller General of the United States, the extent to which vouchers for funds expended under contracts for scientific research shall be subject to itemization or substantiation prior to payment, without regard to the limitations of other laws relating to the expenditure of public funds and accounting therefor. (May 10, 1950, ch. 171, § 11, 64 Stat. 153; Sept. 8, 1959, Pub. L. 86–232, § 6, 73 Stat. 468.)

AMENDMENTS

1959—Subsec. (e). Pub. L. 86–232 included acquisition of property by condemnation.

§ 1871. Patent rights; protection of public interest or equities of individuals or organizations; employees barred.

(a) Each contract or other arrangement executed pursuant to this chapter which relates to scientific research shall contain provisions governing the disposition of inventions produced thereunder in a manner calculated to protect the public interest and the equities of the individual or organization with which the contract or other arrangement is executed: *Provided, however,* That nothing in this chapter shall be construed to authorize the Foundation to enter into any contractual or other arrangement inconsistent with any provision of law affecting the issuance or use of patents.

(b) No officer or employee of the Foundation shall acquire, retain, or transfer any rights, under the patent laws of the United States or otherwise, in any invention which he may make or produce in connection with performing his assigned activities and which is directly related to the subject matter thereof: *Provided, however,* That this subsection shall not be construed to prevent any officer or employee of the Foundation from executing any application for patent on any such invention for the purpose of assigning the same to the Government or its nominee in accordance with such rules and regulations as the Director may establish. (May 10, 1950, ch. 171, § 12, 64 Stat. 154.)

§ 1872. International cooperation and coordination with foreign policy.

(a) The Foundation is authorized to cooperate in any international scientific activities consistent with the purposes of this chapter and to expend for such international scientific activities such sums within the limit of appropriated funds as the Foundation may deem desirable. The Director, with the approval of the Board, may defray the expenses of representatives of Government agencies and other organizations and of individual scientists to accredited international scientific congresses and meetings whenever he deem [1] it necessary in the promotion of the objectives of this chapter. In this connection, with the approval of the Secretary of State, the Foundation may undertake programs granting fellowships to, or making other similar arrangements with, foreign nationals for scientific study or scientific work in the United States without regard to section 1869 of this title or the affidavit of allegiance to the United States required by section 1874(d)(2) of this title.

(b) (1) The authority to enter into contracts or other arrangements with organizations or individuals in foreign countries and with agencies of foreign countries, as provided in section 1870 (c) of this title, and the authority to cooperate in international scientific activities as provided in subsection (a) of this section, shall be exercised only with the approval of the Secretary of State, to the end that such authority shall be exercised in such manner as is consistent with the foreign policy objectives of the United States.

(2) If, in the exercise of the authority referred to in paragraph (1) of this subsection, negotiation with foreign countries or agencies thereof becomes necessary, such negotiation shall be carried on by the Secretary of State in consultation with the Director. (May 10, 1950, ch. 171, § 13, 64 Stat. 154; Sept. 8, 1959, Pub. L. 86–232, § 7, 73 Stat. 468.)

AMENDMENTS

1959—Subsec. (a). Pub. L. 86–232 authorized the Foundation, with approval of the Secretary of State, to cooperate in scientific activities rather than scientific research activities, and to grant fellowships or make other arrangements with foreign nationals for scientific study or scientific work in the United States.

Subsec. (b)(1). Pub. L. 86–232 deleted "research" from the phrase "scientific research activities."

[1] So in original.

§ 1872a. Weather modification.

(a) Consultations.

In carrying out the provisions of section 1862 (a) (9) of this title, the Foundation shall consult with meteorologists and scientists in private life and with agencies of Government interested in, or affected by, experimental research in the field of weather control.

(b) Research programs.

Research programs to carry out the purposes of section 1862 (a) (9) of this title, whether conducted by the Foundation or by other Government agencies or departments, may be accomplished through contracts with, or grants to, private or public institutions or agencies, including but not limited to cooperative programs with any State through such instrumentalities as may be designated by the governor of such State.

(c) Acceptance of gifts.

For the purposes of section 1862 (a) (9) of this title, the Foundation is authorized to accept as a gift, money, material, or services: *Provided,* That notwithstanding section 1870 (f) of this title, use of any such gift, if the donor so specifies, may be restricted or limited to certain projects or areas.

(d) Loan of property.

For the purposes of section 1862 (a) (9) of this title, other agencies of the Government are authorized to loan to the Foundation without reimbursement, and the Foundation is authorized to accept and make use of, such property and personnel as may be deemed useful, with the approval of the Director of the Bureau of the Budget.

(e) Hearings; oaths or affirmations.

The Director of the Foundation, or any employee of the Foundation designated by him, may for the purpose of carrying out the provisions of section 1862 (a) (9) of this title hold such hearings and sit and act at such times and places and take such testimony as he shall deem advisable. The Director or any employee of the Foundation designated by him may administer oaths or affirmations to witnesses appearing before the Director or such employee.

(f) Documentary evidence; contempt; enforcement of subpena; jurisdiction; witness fees; violations and penalties; public records.

(1) The Director of the Foundation may obtain by regulation, subpena, or otherwise such information in the form of testimony, books, records, or other writings, may require the keeping of and furnishing such reports and records, and may make such inspections of the books, records, and other writings and premises or property of any person or persons as may be deemed necessary or appropriate by him to carry out the provisions of section 1862 (a) (9) of this title, but this authority shall not be exercised if adequate and authoritative data are available from any Federal agency. In case of contumacy by, or refusal to obey a subpena served upon, any person referred to in this subsection, the district court of the United States for any district in which such person is found or resides or transacts business, upon application by the Director, shall have jurisdiction to issue an order requiring such person to appear and give testimony or to appear and produce documents, or both; and any failure to obey such order of the court may be punished by such court as a contempt thereof.

(2) The production of a person's books, records, or other documentary evidence shall not be required at any place other than the place where such person usually keeps them, if, prior to the return date specified in the regulations, subpena, or other document issued with respect thereto, such person furnishes the Foundation with a true copy of such books, records, or other documentary evidence (certified by such person under oath to be a true and correct copy) or enters into a stipulation with the Director as to the information contained in such books, records, or other documentary evidence. Witnesses shall be paid the same fees and mileage that are paid witnesses in the courts of the United States.

(3) Any person who willfully performs any act prohibited or willfully fails to perform any act required by the above provisions of this subsection, or any regulation issued thereunder, shall upon conviction be fined not more than $500.

(4) Information contained in any statement, report, record, or other document furnished pursuant to this subsection shall be available for public inspection, except (A) information authorized or required by statute to be withheld and (B) information classified in accordance with law to protect the national security. The foregoing sentence shall not be interpreted to authorize or require the publication, divulging, or disclosure of any information described in section 1905 of Title 18, except that the Director may disclose information described in such section 1905, furnished pursuant to this subsection, whenever he determines that the withholding thereof would be contrary to the purposes of this section and section 1862 (a) (9) of this title. (May 10, 1950, ch. 171, § 14, as added July 11, 1958, Pub. L. 85–510, § 2, 72 Stat. 353.)

§ 1873. Employment of personnel.

(a) Appointment and compensation.

The Director shall, in accordance with such policies as the Board shall from time to time prescribe, appoint and fix the compensation of such personnel as may be necessary to carry out the provisions of this chapter. Such appointments shall be made and such compensation shall be fixed in accordance with the provisions of the civil-service laws and regulations and the Classification Act of 1949: *Provided,* That the Director may, in accordance with such policies as the Board shall from time to time prescribe, employ such technical and professional personnel and fix their compensation, without regard to such laws, as he may deem necessary for the discharge of the responsibilities of the Foundation under this chapter. The Deputy Director hereinafter provided for, and the members of the divisional committees and special commissions, shall be appointed without regard to the civil-service laws or regulations. Neither the Director nor the Deputy Director shall engage in any other business, vocation, or employment than that of serving as such Director or Deputy Director, as the case may be;

nor shall the Director or Deputy Director, except with the approval of the Board, hold any office in, or act in any capacity for, any organization, agency, or institution with which the Foundation makes any contract or other arrangement under this chapter.

(b) Deputy Director; appointment; duties.

The Director may appoint, with the approval of the Board, a Deputy Director who shall perform such functions as the Director, with the approval of the Board, may prescribe and shall be Acting Director during the absence or disability of the Director or in the event of a vacancy in the Office of the Director.

(c) Operation of laboratories and pilot plants.

The Foundation shall not, itself, operate any laboratories or pilot plants.

(d) Compensation of members of Board and divisional committees.

The members of the Board, and the members of each divisional committee, or special commission, shall receive compensation at the rate of $50 for each day engaged in the business of the Foundation pursuant to authorization of the Foundation, and shall be allowed travel expenses as authorized by section 73b–2 of Title 5.

(e) Federal officers as members of divisional committees and special commissions.

Persons holding other offices in the executive branch of the Federal Government may serve as members of the divisional committees and special commissions, but they shall not receive remuneration for their services as such members during any period for which they receive compensation for their services in such other offices.

(f) Exemption from provisions of sections 281, 283, or 284 of Title 18 and section 99 of Title 5.

Service of an individual as a member of the Board, of a divisional committee, or of a special commission shall not be considered as service bringing him within the provisions of sections 281, 283, or 284 of Title 18 or section 99 of Title 5, unless the act of such individual, which by such section is made unlawful when performed by an individual referred to in such section, is with respect to any particular matter which directly involves the Foundation or in which the Foundation is directly interested.

(g) Utilization of appropriations in making contracts.

In making contracts or other arrangements for scientific research, the Foundation shall utilize appropriations available therefor in such manner as will in its discretion best realize the objectives of (1) having the work performed by organizations, agencies, and institutions, or individuals in the United States or foreign countries, including Government agencies of the United States and of foreign countries, qualified by training and experience to achieve the results desired, (2) strengthening the research staff of organizations, particularly nonprofit organizations, in the States, Territories, possessions, and the District of Columbia, (3) aiding institutions, agencies, or organizations which, if aided, will advance basic research, and (4) encouraging independent basic research by individuals.

(h) Transfer of research funds of other Government departments or agencies.

Funds available to any department or agency of the Government for scientific or technical research, or the provision of facilities therefor, shall be available for transfer, with the approval of the head of the department or agency involved, in whole or in part, to the Foundation for such use as is consistent with the purposes for which such funds were provided, and funds so transferred shall be expendable by the Foundation for the purposes for which the transfer was made, and, until such time as an appropriation is made available directly to the Foundation, for general administrative expenses of the Foundation without regard to limitations otherwise applicable to such funds.

(i) Transfer of National Roster of Scientific and Specialized Personnel.

The National Roster of Scientific and Specialized Personnel shall be transferred from the United States Employment Service to the Foundation, together with such records and property as have been utilized or are available for use in the administration of such roster as may be determined by the President. The transfer provided for in this subsection shall take effect at such time or times as the President shall direct. **(May 10, 1950, ch. 171, § 15, formerly § 14, 64 Stat. 154, renumbered July 11, 1958, Pub. L. 85–510, § 2, 72 Stat. 353, and amended Sept. 8, 1959, Pub. L. 86–232, § 8, 73 Stat. 469.)**

REFERENCES IN TEXT

The civil-service laws, referred to in the text, are classified generally to Title 5, Executive Departments and Government Officers and Employees.

The Classification Act of 1949, referred to in the text, is classified to chapter 21 of Title 5.

AMENDMENTS

1959—Subsec. (d). Pub L. 86–232 increased compensation for $25 to $50 per diem.

§ 1874. Security provisions.

(a) Nuclear energy research and development.

The Foundation shall not support any research or development activity in the field of nuclear energy, nor shall it exercise any authority pursuant to section 1870 (e) of this title in respect to that field, without first having obtained the concurrence of the Atomic Energy Commission that such activity will not adversely affect the common defense and security. To the extent that such activity involves restricted data as defined in the Atomic Energy Act of 1946 the provisions of that Act regarding the control of the dissemination of restricted data and the security clearance of those individuals to be given access to restricted data shall be applicable. Nothing in this chapter shall supersede or modify any provision of the Atomic Energy Act of 1946.

(b) Research relating to national defense.

(1) In the case of scientific or technical research activities under this chapter in connection with matters relating to the national defense, with respect to which funds have been transferred to the Foundation from the Department of Defense in accordance with the provisions of section 1873 (h) of this title, the Secretary of Defense shall establish such security requirements and safeguards, includ-

ing restrictions with respect to access to information and property, as he deems necessary.

(2) In the case of scientific research activities under this chapter in connection with matters relating to the national defense other than research activities referred to in paragraph (1) of this subsection, the Foundation shall establish such security requirements and safeguards, including restrictions with respect to access to information and property, as it deems necessary.

(3) Any agency of the Government exercising investigatory functions is authorized to make such investigations and reports as may be requested by the Foundation in connection with the enforcement of security requirements and safeguards, including restrictions with respect to access to information and property, established under paragraph (1) or (2) of this subsection.

(c) Clearance of personnel by Civil Service Commission.

No employee of the Foundation shall be permitted to have access to information or property with respect to which access restrictions have been established under subsection (b) (1) or (2) of this section until the Civil Service Commission shall have made an investigation into the character, associations, and loyalty of such individual and shall have reported the findings of said investigation to the Foundation, and the Foundation shall have determined that permitting such individual to have access to such information or property will not endanger the common defense and security.

(d) Oath and statement prerequisite to acceptance of scholarship or fellowship; ineligibility of Communist organization members; penalties.

(1) No part of any funds appropriated or otherwise made available for expenditure by the Foundation under authority of this chapter shall be used to make payments under any scholarship or fellowship awarded to any individual under section 1869 of this title, unless such individual—

(A) has taken and subscribed to an oath or affirmation in the following form: "I do solemnly swear (or affirm) that I bear true faith and allegiance to the United States of America and will support and defend the Constitution and laws of the United States against all its enemies, foreign and domestic"; and

(B) has provided the Foundation (in the case of applications made on or after October 1, 1962) with a full statement regarding any crimes of which he has ever been convicted (other than crimes committed before attaining sixteen years of age and minor traffic violations for which a fine of $25 or less was imposed) and regarding any criminal charges punishable by confinement of thirty days or more which may be pending against him at the time of his application for such scholarship or fellowship.

The provisions of section 1001 of Title 18, shall be applicable with respect to the oath or affirmation and statement herein required.

(2) (A) When any Communist organization, as defined in section 782(5) of Title 50, is registered or there is in effect a final order of the Subversive Activities Control Board requiring such organization to register, it shall be unlawful for any member of such organization with knowledge or notice that such organization is so registered or that such order has become final (i) to make application for any scholarship or fellowship which is to be awarded from funds part or all of which are appropriated or otherwise made available for expenditure under the authority of section 1869 of this title, or (ii) to use or attempt to use any such award.

(B) Whoever violates subparagraph (A) of this paragraph shall be fined not more than $10,000, or imprisoned not more than five years, or both. (May 10, 1950, ch. 171, § 16, formerly § 15, 64 Stat. 156; Apr. 5, 1952, ch. 159, § 1, 66 Stat. 43, renumbered July 11, 1958, Pub. L. 85–510, § 2, 72 Stat. 353, and amended Oct. 16, 1962, Pub. L. 87–835, § 1, 76 Stat. 1069.)

REFERENCES IN TEXT

The Atomic Energy Act of 1946, referred to in subsec. (a), was act Aug. 1, 1946, ch. 724, 60 Stat. 755. The act was amended generally by act Aug. 30, 1954, ch. 1073, 68 Stat. 919, to be known as the Atomic Energy Act of 1954, and is classified to chapter 23 of this title.

"That Act", referred to in said subsec. (a), refers to the Atomic Energy Act of 1946.

AMENDMENTS

1962—Subsec. (d). Pub. L. 87–835 designated existing provisions as par. (1), added the reference to section 1869 of this title, and substituted the requirement, for applications made on or after Oct. 1, 1962, of a full statement regarding convictions for crimes, other than any committed before age 16 or for minor traffic violations, and any criminal charges punishable by thirty days confinement, or more, pending at time of application for scholarship or fellowship, for the requirement of an affidavit stating the affiant did not believe in, and was not a member or supporter of any organization believing in, or teaching, the violent overthrow of the United States Government, or by any illegal means, in such par. (1), and added par. (2).

1952—Subsec. (c). Act Apr. 5, 1952, substituted the "Civil Service Commission" for the "Federal Bureau of Investigation".

§ 1875. Appropriations.

(a) To enable the Foundation to carry out its powers and duties, there is authorized to be appropriated to the Foundation, out of any money in the Treasury not otherwise appropriated, such sums as may be necessary to carry out the provisions of this chapter.

(b) Appropriations made pursuant to the authority provided in subsection (a) of this section shall remain available for obligation, for expenditure, or for obligation and expenditure, for such period or periods as may be specified in the Acts making such appropriations. (May 10, 1950, ch. 171, § 17, formerly § 16, 64 Stat. 157; Aug. 8, 1953, ch. 377, 67 Stat. 488, renumbered July 11, 1958, Pub. L. 85–510, § 2, 72 Stat. 353.)

AMENDMENTS

1953—Subsec. (a). Act Aug. 8, 1953, removed the $15 million limitation on the amount of the annual appropriations.

§ 1876. Science Information Service; functions.

The National Science Foundation shall establish a Science Information Service. The Foundation, through such Service, shall (1) provide, or arrange for the provision of, indexing, abstracting, translating, and other services leading to a more effective dissemination of scientific information, and (2) undertake programs to develop new or improved methods, including mechanized systems, for making scientific information available. (Pub. L. 85–864, title IX, § 901, Sept. 2, 1958, 72 Stat. 1601.)

CODIFICATION

Section was enacted as part of the National Defense Education Act of 1958, Pub. L. 85–864, and not as part of this chapter, which constitutes the National Science Foundation Act of 1950.

§ 1877. Science Information Council.

(a) Establishment; membership; elections and appointments; tenure; reappointment.

The National Science Foundation shall establish, in the Foundation, a Science Information Council (hereafter in sections 1876—1879 of this title referred to as the "Council") consisting of the Librarian of Congress, the director of the National Library of Medicine, the director of the Department of Agriculture library, and the head of the Science Information Service, each of whom shall be ex officio members, and fifteen members appointed by the Director of the National Science Foundation. The Council shall annually elect one of the appointed members to serve as chairman until the next election. Six of the appointed members shall be leaders in the fields of fundamental science, six shall be leaders in the fields of librarianship and scientific documentation, and three shall be outstanding representatives of the lay public who have demonstrated interest in the problems of communication. Each appointed member of such Council shall hold office for a term of four years, except that (1) any member appointed to fill a vacancy occurring prior to the expiration of the term for which his predecessor was appointed shall be appointed only for the remainder of such term, and (2) that of the members first appointed, four shall hold office for a term of three years, four shall hold office for a term of two years, and three shall hold office for a term of one year, as designated by the Director of the National Science Foundation at the time of appointment. No appointed member of the Council shall be eligible for reappointment until a year has elapsed since the end of his preceding term.

(b) Duties and meetings.

It shall be the duty of the Council to advise, to consult with, and to make recommendations to, the head of the Science Information Service. The Council shall meet at least twice each year, and at such other times as the majority thereof deems appropriate.

(c) Compensation and allowance for expenses.

Persons appointed to the Council shall, while serving on business of the Council, receive compensation at rates fixed by the National Science Foundation, but not to exceed $50 per day, and shall also be entitled to receive an allowance for actual and necessary travel and subsistence expenses while so serving away from their places of residence. (Pub. L. 85–864, title IX, § 902, Sept. 2, 1958, 72 Stat. 1601.)

CODIFICATION

Section was enacted as part of the National Defense Education Act of 1958, Pub. L. 85–864, and not as part of this chapter, which constitutes the National Science Foundation Act of 1950.

§ 1878. Functions relating to Science Information Service and Council.

In carrying out its functions under sections 1876—1879 of this title, the National Science Foundation shall have the same power and authority it has under this chapter to carry out its functions under this chapter. (Pub. L. 85–864, title IX, § 903, Sept. 2, 1958, 72 Stat. 1601.)

CODIFICATION

Section was enacted as part of the National Defense Education Act of 1958, Pub. L. 85–864, and not as part of this chapter, which constitutes the National Science Foundation Act of 1950.

§ 1879. Appropriations for Science Information Service and Council.

There are authorized to be appropriated for the fiscal year ending June 30, 1959, and for each succeeding fiscal year, such sums as may be necessary to carry out the provisions of sections 1876—1879 of this title. (Pub. L. 85–864, title IX, § 904, Sept. 2, 1958, 72 Stat. 1602.)

CODIFICATION

Section was enacted as part of the National Defense Education Act of 1958, Pub. L. 85–864, and not as part of this chapter, which constitutes the National Science Foundation Act of 1950.

§ 1880. National Medal of Science.

There is established a National Medal of Science (hereinafter referred to as the "medal"), which shall be of such design and materials and bear such inscriptions as the President, on the basis of recommendations submitted by the National Science Foundation, may prescribe, and shall be awarded as provided in section 1881 of this title. (Pub. L. 86–209, § 1, Aug. 25, 1959, 73 Stat. 431.)

CODIFICATION

Section was not enacted as a part of the National Science Foundation Act of 1950 which comprises this chapter.

§ 1881. Same; award; number; citizenship; ceremonies.

(a) The President shall from time to time award the medal, on the basis of recommendations received from the National Academy of Sciences or on the basis of such other information and evidence as he deems appropriate, to individuals who in his judgment are deserving of special recognition by reason of their outstanding contributions to knowledge in the physical, biological, mathematical, or engineering sciences.

(b) Not more than twenty individuals may be awarded the medal in any one calendar year.

(c) An individual may not be awarded the medal unless at the time such award is made he—

(1) is a citizen or other national of the United States; or

(2) is an alien lawfully admitted to the United States for permanent residence who (A) has filed an application for petition for naturalization in the manner prescribed by section 1445(b) of Title 8 and (B) is not permanently ineligible to become a citizen of the United States.

(d) The presentation of the award shall be made by the President with such ceremonies as he may deem proper, including attendance by appropriate

Members of Congress. (Pub. L. 86–209, § 2, Aug. 25, 1959, 73 Stat. 431.)

CODIFICATION

Section was not enacted as a part of the National Science Foundation Act of 1950 which comprises this chapter.

Chapter 16A.—GRANTS FOR SUPPORT OF SCIENTIFIC RESEARCH

§ 1891. Authorization to make grants.

The head of each agency of the Federal Government, authorized to enter into contracts for basic scientific research at nonprofit institutions of higher education, or at nonprofit organizations whose primary purpose is the conduct of scientific research, is authorized, where it is deemed to be in furtherance of the objectives of the agency, to make grants to such institutions or organizations for the support of such basic scientific research. (Pub. L. 85–934, § 1, Sept. 6, 1958, 72 Stat. 1793.)

§ 1892. Same; title to equipment.

Authority to make grants or contracts for the conduct of basic or applied scientific research at nonprofit institutions of higher education, or at nonprofit organizations whose primary purpose is the conduct of scientific research, shall include discretionary authority, where it is deemed to be in furtherance of the objectives of the agency, to vest in such institutions or organizations, without further obligation to the Government, or on such other terms and conditions as the agency deems appropriate, title to equipment purchased with such grant or contract funds. (Pub. L. 85–934, § 2, Sept. 6, 1958, 72 Stat. 1793.)

§ 1893. Annual report to Congress; contents.

Each agency or department of the Federal Government exercising authority granted by this chapter shall make an annual report on or before June 30th of each year to the appropriate committees of both Houses of Congress. Such report shall set forth therein, for the preceding year, the number of grants made pursuant to the authority provided in section 1891 of this title, the dollar amount of such grants, and the institutions in which title to equipment was vested pursuant to section 1892 of this title. (Pub. L. 85–934, § 3, Sept. 6, 1958, 72 Stat. 1793.)

SOURCE: United States Code, 1964 Edition. 1965.

TITLE 42.—THE PUBLIC HEALTH AND WELFARE

Chapter 16.—NATIONAL SCIENCE FOUNDATION (Supplement)

§ 1862. Functions.

(a) Initiation and support of studies and programs; scholarships; current register of scientific and technical personnel.

The Foundation is authorized and directed—

(1) to initiate and support basic scientific research and programs to strengthen scientific research potential in the mathematical, physical, medical, biological, engineering, social, and other sciences, by making contracts or other arrangements (including grants, loans, and other forms of assistance) to support such scientific activities and to appraise the impact of research upon industrial development and upon the general welfare;

(2) to award, as provided in section 1869 of this title, scholarships and graduate fellowships in the mathematical, physical, medical, biological, engineering, social, and other sciences;

(3) to foster the interchange of scientific information among scientists in the United States and foreign countries;

(4) to foster and support the development and use of computer and other scientific methods and technologies, primarily for research and education in the sciences;

(5) to evaluate the status and needs of the various sciences as evidenced by programs, projects, and studies undertaken by agencies of the Federal Government, by indviduals, and by public and private research groups, employing by grant or contract such consulting services as it may deem necessary for the purpose of such evaluations; and to take into consideration the results of such evaluations in correlating the research and educational programs undertaken or supported by the Foundation with programs, projects, and studies undertaken by agencies of the Federal Government, by individuals, and by public and private research groups;

(6) to maintain a current register of scientific and technical personnel, and in other ways to provide a central clearinghouse for the collection, interpretation, and analysis of data on the availability of, and the current and projected need for, scientific and technical resources in the United States, and to provide a source of information for policy formulation by other agencies of the Federal Government; and

(7) to initiate and maintain a program for the determination of the total amount of money for scientific research, including money allocated for the construction of the facilities wherein such research is conducted, received by each educational institution and appropriate nonprofit organization in the United States, by grant, contract, or other arrangement from agencies of the Federal Government, and to report annually thereon to the President and the Congress.

(b) Contracts, grants, loans, etc., for scientific activities; financing of programs.

The Foundation is authorized to initiate and support specific scientific acitivities in connection with matters relating to international cooperation or national security by making contracts or other arrangements (including grants, loans, and other forms of assistance) for the conduct of such scientific activities when initiated or supported pursuant to requests made by the Secretary of State or the Secretary of Defense shall be financed solely from funds transferred to the Foundation by the requesting Secretary as provided in section 1873(g) of this title, and any such activities shall be unclassified and shall be identified by the Foundation as being undertaken at the request of the appropriate Secretary.

(c) Scientific research programs at academic and other nonprofit institutions; applied scientific research programs by Presidential directive; employment of consulting services; coordination of activities.

In additional to the authority contained in subsections (a) and (b) of this section, the Foundation is authorized to initiate and support scientific research, including applied research, at academic and other nonprofit institutions. When so directed by the President, the Foundation is further authorized to support, through other appropriate organizations, applied scientific research relevant to national problems involving the public interest. In exercising the authority contained in this subsection, the Foundation may employ by grant or contract such consulting services as it deems necessary, and shall coordinate and correlate its activities with respect to any such problem with other agencies of the Federal Government undertaking similar programs in that field.

(d) Promotion of basic research and education in the sciences.

The Board and the Director shall recommend and encourage the pursuit of national policies for the promotion of basic research and education in the sciences.

(e) Balancing of research and educational activities in the sciences.

In exercising the authority and discharging the functions referred to in the foregoing subsections, it shall be one of the objectives of the Foundation to strengthen research and education in the sciences, including independent research by individuals, throughout the United States, and to avoid undue concentration of such research and education.

(f) Annual report to the President and Congress.

The Foundation shall render an annual report to the President for submission on or before the 15th day of January of each year to the Congress summarizing the activities of the Foundation and making such recommendations as it may deem appropriate. Such report shall include information as to the acquisition and disposition by the Foundation of any patents and patent rights. (As amended July 8, 1968, Pub. L. 90–407, § 1, 82 Stat. 360.)

AMENDMENTS

1968—Subsec. (a)(1). Pub. L. 90–407 redesignated former subsec. (a)(2) as (a)(1), and, as so redesignated. added social sciences to the enumerated list of sciences. Former subsec. (a)(1) was redesignated as (d).

Subsec. (a)(2). Pub. L. 90–407 redesignated former subsec. (a)(4) as (a)(2), and, as so redesignated, added social sciences to the enumerated list of sciences. Former subsec. (a)(2) was redesignated as (a)(1).

Subsec. (a)(3). Pub. L. 90–407 redesignated former subsec. (a)(5) as (a)(3). Former subsec. (a)(3) was redesignated as (b).

Subsec. (a)(4). Pub. L. 90–407 added subsec. (a)(4) Former subsec. (a)(4) was redesignated as (a)(2).

Subsec. (a)(5). Pub. L. 90–407 redesignated former subsec. (a)(6) as (a)(5), and, as so redesignated, provided for the employment of consulting services, by grant or contract, to assist in the evaluation of the status and needs of the various sciences as evidenced by the programs and studies undertaken by agencies of the government, by individuals, and by public and private research groups, and provided for the consideration of the results of such evaluations in the correlation of the Foundation's programs with those undertaken by agencies of the government, as well as those undertaken by individuals and by public and private research groups. Former subsec. (a)(5) was redesignated as (a)(3).

Subsec. (a)(6). Pub. L. 90–407 redesignated former subsec. (a)(8) as (a)(6), and, as so redesignated, provided that the register of scientific and technical personnel shall be current, and authorized the Foundation to analyze and interpret the collected data on the availability of, and the current and projected need for, scientific and technical resources in the United States and to make such information available to other agencies of the government for policy formulation. Former subsec. (a)(6) was redesignated as (a)(5).

Subsec. (a)(7). Pub. L. 90–407 added subsec. (a)(7). Former subsec. (a)(7), which provided for the establishment of such special commissions as the Board may from to time deem necessary for the purposes of this chapter, was eliminated.

Subsec. (a)(8). Pub. L. 90–407 redesignated former subsec. (a)(8) as (a)(6).

Subsec. (a)(9). Pub. L. 90–407 struck out subsec. (a)(9), which authorized the Foundation to initiate and support a program of study, research, and evaluation in the field of weather modification, with particular attention to areas experiencing floods, drought, etc., and to report annually to the President and the Congress thereon.

Subsec. (b). Pub. L. 90–407 redesignated former subsec. (a)(3) as (b), and, as so redesignated, substituted provisions authorizing the Foundation to initiate and support specific scientific activities in matters related to international cooperation or national security for provisions authorizing the Foundation to initiate and support only scientific research activities, only in matters related to national defense and only when requested to do so by the Secretary of Defense, and added provisions specifying the manner of financing such scientific activities. Former subsec. (b) redesignated (e).

Subsec. (c). Pub. L. 90–407 added subsec. (c). Former subsec. (c) redesignated (f).

Subsec. (d). Pub. L. 90–407 redesignated former subsec. (a)(1) as (d), and, as so redesignated, substituted provisions authorizing the Board and the Director to recommend and encourage national policies promoting basic research and education in the sciences for provisions authorizing and directing the Foundation to develop and encourage such policies.

Subsec. (e). Pub. L. 90–407 redesignated former subsec. (b) as (e), and, as so redesignated, substituted "the foregoing subsections" for "subsection (a) of this section", "strengthen research" for "strengthen basic research", and struck out the reference to the territories and possessions of the United States.

Subsec. (f). Pub. L. 90–407 redesignated former subsec. (c) as (f), and, as so redesignated, struck out the provision requiring the report to include the minority views and recommendations if any, of members of the Board.

CONTINUATION OF AUTHORIZATION FOR WEATHER MODIFICATION PROGRAMS; REPEAL

Section 11(1) of Pub. L. 90–407 provided in part that the authorization for the programs initiated under former subsec. (a)(9) of this section shall continue in effect until Sept. 1, 1968 for the purposes of section 1872a of this title.

CONTINUATION OF EXISTING OFFICES, PROCEDURES, AND ORGANIZATION OF THE NATIONAL SCIENCE FOUNDATION

Section 16 of Pub. L. 90–407 provided that: "Except as otherwise specifically provided therein, the amendments made by this Act [which enacted section 1864a of this title, amended this section, sections 1863—1866, 1868—1870, 1872—1875, and 1877 of this title, sections 5313,

5314, and 5316 of title 5, Government Organization and Employees, repealed sections 1867, and 1872a of this title, and enacted provisions set out as a note under section 5313 of Title 5] are intended to continue in effect under the National Science Foundation Act of 1950 [this chapter] the existing offices, procedures, and organization of the National Science Foundation as provided by such Act, [this chapter] part II of Reorganization Plan Numbered 2 of 1962, and Reorganization Plan Numbered 5 of 1965 [set out in Appendix to Title 5]. From and after the date of the enactment of this Act [July 18, 1968], part II of Reorganization Plan Numbered 2 of 1962, and Reorganization Plan Numbered 5 of 1965, shall be of no force or effect; but nothing in this Act shall alter or affect any transfers of functions made by part I of such Reorganization Plan Numbererd 2 of 1962."

EMERGENCY PREPAREDNESS FUNCTIONS

For assignment of certain emergency preparedness functions to the Director of the National Science Foundation, see Parts 1, 23, and 30 of Ex. Ord. No. 11490, Oct. 28, 1969, 34 F.R. 17567, set out as a note under section 2292 of Title 50, Appendix, War and National Defense.

EX. ORD. NO. 10807. FEDERAL COUNCIL FOR SCIENCE AND TECHNOLOGY

Ex. Ord. No. 10807, Mar. 13, 1959, 24 F.R. 1897, as amended by Ex. Ord. No. 11381, Nov. 8, 1967, 32 F.R. 15629, provided:

SECTION 1. *Establishment of Council.* (a) There is hereby established the Federal Council for Science and Technology (hereinafter referred to as the Council).

(b) The Council shall be composed of the Special Assistant to the President for Science and Technology and one representative of each of the following: Department of Agriculture, Department of Commerce, Department of Defense, Department of Health, Education, and Welfare, Department of Housing and Urban Development. Department of the Interior, Department of State, Department of Transportation, Atomic Energy Commission, National Aeronautics and Space Administration, and National Science Foundation. Each such representative shall be an official of policy rank designated by the head of the Federal agency concerned, and, in the case of the Atomic Energy Commission, shall be its Chairman or another member of the Commission designated by the Chairman of the Commission. A representative of the Director of the Bureau of the Budget designated by the Director may attend meetings of the Council as an observer.

* * * * *

§ 1863. National Science Board.

(a) Composition; appointment; establishment of policies of the Foundation.

The Board shall consist of twenty-four members to be appointed by the President, by and with the advice and consent of the Senate, and of the Director ex officio. In addition to any powers and functions otherwise granted to it by this chapter, the Board shall establish the policies of the Foundation.

(b) Executive Committee; delegation of powers and functions.

The Board shall have an Executive Committee as provided in section 1865 of this title, and may delegate to it or to the Director or both such of the powers and functions granted to the Board by this chapter as it deems appropriate.

(c) Qualifications for Board membership; recommendations.

The persons nominated for appointment as members of the Board (1) shall be eminent in the fields of the basic, medical, or social sciences, engineering, agriculture, education, research management, or public affairs; (2) shall be selected solely on the basis of established records of distinguished service; and (3) shall be so selected as to provide representation of the views of scientific leaders in all areas of the Nation. The President is requested, in the making of nominations of persons for appointment as members, to give due consideration to any recommendations for nomination which may be submitted to him by the National Academy of Sciences, the National Association of State Universities and Land Grant Colleges, the Association of American Universities, the Association of American Colleges, the Association of State Colleges and Universities, or by other scientific or educational organizations.

(d) Term of office; reappointment.

The term of office of each member of the Board shall be six years; except that any member appointed to fill a vacancy occurring prior to the expiration of the term for which his predecessor was appointed shall be appointed for the remainder of such term. Any person, other than the Director, who has been a member of the Board for twelve consecutive years shall thereafter be ineligible for appointment during the two-year period following the expiration of such twelfth year.

(e) Meetings; quorum; notice.

The Board shall meet annually on the third Monday in May unless, prior to May 10 in any year, the Chairman has set the annual meeting for a day in May other than the third Monday, and at such other times as the Chairman may determine, but he shall also call a meeting whenever one-third of the members so request in writing. A majority of the members of the Board shall constitute a quorum. Each member shall be given notice, by registered mail or certified mail mailed to his last known address of record not less than fifteen days prior to any meeting, of the call such meeting.

(f) Election of Chairman and Vice Chairman; vacancy.

The election of the Chairman and Vice Chairman of the Board shall take place at each annual meeting occurring in an even-numbered year. The Vice Chairman shall perform the duties of the Chairman in his absence. In case a vacancy occurs in the chairmanship or vice chairmanship, the Board shall elect a member to fill such vacancy.

(g) Annual report to the President and Congress; recommendations.

The Board shall render an annual report to the President, for submission on or before the 31st day of January of each year to the Congress, on the status and health of science and its various disciplines. Such report shall include an assessment of such matters as national scientific resources and trained manpower, progress in selected areas of basic scientific research, and an indication of those aspects of such progress which might be applied to the needs of American society. The report may include such recommendations as the Board may deem timely and appropriate.

(h) Appointment and assignment of staff; compensation; security requirements.

The Board may, with the concurrence of a majority of its members, permit the appointment of a staff consisting of not more than five professional staff members and such clerical staff members as may be necessary. Such staff shall be appointed by the Director and assigned at the direction of the

Board. The professional members of such staff may be appointed without regard to the provisions of Title 5 governing appointments in the competitive service, and the provisions of chapter 51 of Title 5 relating to classification, and compensated at a rate not exceeding the appropriate rate provided for individuals in grade GS-15 of the General Schedule under section 5332 of Title 5, as may be necessary to provide for the performance of such duties as may be prescribed by the Board in connection with the exercise of its powers and functions under this chapter. Each appointment under this subsection shall be subject to the same security requirements as those required for personnel of the Foundation appointed under section 1873(a) of this title.

(i) Special commissions.

The Board is authorized to establish such special commissions as it may from time to time deem necessary for the purposes of this chapter.

(j) Committees; survey and advisory functions.

The Board is also authorized to appoint from among its members such committees as it deems necessary, and to assign to committees so appointed such survey and advisory functions as the Board deems appropriate to assist it in exercising its powers and functions under this chapter. (As amended July 18, 1968, Pub. L. 90-407, § 2, 82 Stat. 361.)

AMENDMENTS

1968—Subsec. (a). Pub. L. 90-407 substituted provisions which authorized the Board to establish the policies of the Foundation in addition to any powers and functions otherwise granted to it by this chapter, for provisions which authorized the Board, except as otherwise provided by this chapter, to exercise the authority granted to the Foundation by this chapter. Provisions of this subsection, which enumerated the qualifications of persons nominated for appointment to the Board and provided for the specified organizations to make recommendations to the President of individuals qualified for nomination, were designated as subsec. (c).

Subsec. (b). Pub. L. 90-407 added subsec. (b). Former subsec. (b) was redesignated as (d).

Subsec. (c). Pub. 90-407 redesignated provisions of former subsec. (a) as (c), and, as so redesignated, added social science and research management to the enumerated fields of eminence, and substituted "the National Association of State Universities and Land Grant Colleges, the Association of American Universities, the Association of American Colleges, the Association of State Colleges and Universities" for "the Association of Land Grant Colleges and Universities, the National Association of State Universities, the Association of American Colleges". Former subsec. (c), which provided that "The President shall call the first meeting of the Board, at which the first order of business shall be the election of a chairman and a vice chairman", was eliminated as executed.

Subsec. (d). Pub. L. 90-407 redesignated former subsec. (b) as (d), and, as so redesignated, substituted "term of office of each member" for "term of office of each voting member", struck out "the terms of office of the members first taking office after May 10, 1950, shall expire, as designated by the President at the time of appointment, eight at the end of two years, eight at the end of four years, and eight at the end of six years, after May 10, 1950", and provided for the exemption of the Director from the prohibition against reappointment within two years following twelve consecutive years of Board membership. Former subsec. (d) was redesignated as (e).

Subsec. (e). Pub. L. 90-407 redesignated former subsec. (d) as (e), and, as so redesignated, substituted "A majority of the members of the Board shall constitute a quorum" for "A majority of the voting members of the Board shall constitute a quorum". Former subsec. (e) was redesignated as (f).

Subsec. (f). Pub. L. 90-407 redesignated former subsec. (e) as (f), and, as so designated, substituted provisions that the election of the Chairman and Vice Chairman take place at each annual meeting occurring in an even-numbered year for provisions that their election take place at the first meeting of the National Science Board following the enactment of Pub. L. 86-232, and that thereafter such election take place at the second annual meeting occurring after each such election.

Subsecs. (g)—(j). Pub. L. 90-407 added subsecs. (g)—(j).

CONTINUATION OF EXISTING OFFICES, PROCEDURES, AND ORGANIZATION OF THE NATIONAL SCIENCE FOUNDATION

Amendment by Pub. L. 90-407 intended to continue in effect the existing offices, procedures, and organization of the Foundation, see section 16 of Pub. L. 90-407, set out as a note under section 1862 of this tile.

SECTION REFERRED TO IN OTHER SECTIONS

This section is referred to in sections 1868, 1873 of this title.

§ 1864. Director of Foundation.

(a) Appointment; compensation; term of office.

The Director of the Foundation (referred to in this chapter as the "Director") shall be appointed by the President, by and with the advice and consent of the Senate. Before any person is appointed as Director, the President shall afford the Board an opportunity to make recommendations to him with respect to such appointment. The Director shall receive basic pay at the rate provided for level II of the Executive Schedule under section 5313 of Title 5, and shall serve for a term of six years unless sooner removed by the President.

(b) Exercise of authority of Foundation; actions as final and binding upon the Foundation.

Except as otherwise specifically provided in this chapter (1) the Director shall exercise all of the authority granted to the Foundation by this chapter (including any powers and functions which may be delegated to him by the Board), and (2) all actions taken by the Director pursuant to the provisions of this chapter (or pursuant to the terms of a delegation from the Board) shall be final and binding upon the Foundation.

(c) Delegation and redelegation of functions.

The Director may from time to time make such provisions as he deems appropriate authorizing the performance by any other officer, agency, or employee of the Foundation of any of his functions under this chapter, including functions delegated to him by the Board; except that the Director may not redelegate policymaking functions delegated to him by the Board.

(d) Formulation of programs.

The formulation of programs in conformance with the policies of the Foundation shall be carried out by the Director in consultation with the Board.

(e) Authority to contract, grant, etc.; limitations and conditions; waiver.

The Director shall not make any contract, grant, or other arrangement pursuant to section 1870(c) of this title without the prior approval of the Board, except that a grant, contract, or other arrangement involving a total commitment of less than $2,000,000, or less than $500,000 in any one year, or a commitment of such lesser amount or amounts and subject to such other conditions as the Board in its discretion may from time to time determine to be appropriate and publish in the Federal Register, may be made if such action is taken pursuant to the terms and conditions set forth by the Board, and if each such action is reported to the Board at the Board meeting next following such action.

(f) Status; power to vote and hold office.

The Director, in his capacity as ex officio member of the Board, shall, except with respect to compensation and tenure, be coordinate with the other members of the Board. He shall be a voting member of the Board and shall be eligible for election by the Board as Chairman or Vice Chairman of the Board. (As amended July 18, 1968, Pub. L. 90–407, § 3, 82 Stat. 362.)

AMENDMENTS

1968—Subsec. (a). Pub. L. 90–407 added the provision prescribing the annual rate of compensation of the Director, and struck out the provisions authorizing the Director to serve as a nonvoting ex officio member of the Board and as the chief executive officer of the Foundation.

Subsec. (b). Pub. L. 90–407 substituted provisions authorizing the Director, except as otherwise provided, to exercise all of the authority granted to the Foundation by this chapter and to take action final and binding upon the Foundation for provisions authorizing the Director, in addition to the powers and duties specifically vested in him by this chapter, to exercise the powers granted by sections 1869 or 1870 (c) of this title and such other powers and duties delegated by the Board to him, and the proviso that no action taken by the Director pursuant to section 1869 or 1870 (c) shall be final unless in each instance the Board has reviewed and approved the action proposed to be taken, or such action is taken pursuant to the terms of a delegation of authority from the Board or the Executive Committee to the Director.

Subsecs. (c)—(f). Pub. L. 90–407 added subsecs. (c)—(f).

EFFECTIVE DATE OF 1968 AMENDMENT

Amendment by Pub. L. 90–407, insofar as related to rates of basic pay, effective on the first day of the first calendar month which begins on or after July 18, 1968, see section 15(a)(4), set out as a note under section 5313 of Title 5, Government Organization and Employees.

TRANSFER OF FUNCTIONS

Authority of Director of the National Science Foundation, from time to time, to make appropriate provisions authorizing the performance by any other officer, or by any agency or employee, of the National Science Foundation of any of his functions (including functions delegated to him by the National Science Board), see Reorg. Plan No. 5 of 1965, eff. July 27, 1965, 30 F.R. 9355, 79 Stat. 1323, set out as a note under section 1867 of this title.

CONTINUATION OF EXISTING OFFICES, PROCEDURES, AND ORGANIZATION OF THE NATIONAL SCIENCE FOUNDATION

Amendment by Pub. L. 90–407 intended to continue in effect the existing offices, procedures, and organization of the Foundation, see section 16 of Pub. L. 90–407, set out as a note under section 1862 of this title.

§ 1864a. Deputy Director of the Foundation; Assistant Directors; appointment; compensation; powers and duties.

(a) There shall be a Deputy Director of the Foundation (referred to in this chapter as the "Deputy Director"), who shall be appointed by the President, by and with the advice and consent of the Senate. Before any person is appointed as Deputy Director, the President shall afford the Board and the Director an opportunity to make recommendations to him with respect to such appointment. The Deputy Director shall receive basic pay at the rate provided for level III of the Executive Schedule under section 5314 of Title 5, and shall perform such duties and exercise such powers as the Director may prescribe. The Deputy Director shall act for, and exercise the powers of, the Director during the absence or disability of the Director or in the event of a vacancy in the office of Director.

(b) There shall be four Assistant Directors of the Foundation (each referred to in this chapter as an "Assistant Director"), who shall be appointed by the President, by and with the advice and consent of the Senate. Before any person is appointed as an Assistant Director, the President shall afford the Board and the Director an opportunity to make recommendations to him with respect to such appointment. Each Assistant Director shall receive basic pay at the rate provided for level V of the Executive Schedule under section 5316 of Title 5, and shall perform such duties and exercise such powers as the Director may prescribe. (May 10, 1950, ch. 171, § 6, as added July 18, 1968 Pub. L. 90–407, § 4, 82 Stat. 363.)

EFFECTIVE DATE

Section, insofar as related to rates of basic pay, effective on the first day of the first calendar month which begins on or after July 18, 1968, see section 15(a)(4) of Pub. L. 90–407, set out as a note under section 5313 of Title 5, Government Organization and Employees.

CONTINUATION OF EXISTING OFFICES, PROCEDURES, AND ORGANIZATION OF THE NATIONAL SCIENCE FOUNDATION

Amendment by Pub. L. 90–407 intended to continue in effect the existing offices, procedures, and organization of the Foundation, see section 16 of Pub. L. 90–407, set out as a note under section 1862 of this title.

§ 1865. Executive Committee.

(a) Composition; powers and function; membership; chairman.

There shall be an Executive Committee of the Board (referred to in this chapter as the "Executive Committee"), which shall be composed of five members and shall exercise such powers and functions as may be delegated to it by the Board. Four of the members shall be elected as provided in subsection (b) of this section, and the Director ex officio shall be the fifth member and the chairman of the Executive Committee.

(b) Election to membership; term of office; eligibility for reelection.

At each of its annual meetings the Board shall elect two of its members as members of the Executive Committee, and the Executive Committee members so elected shall hold office for two years from the date of their election. Any person, other than the Director, who has been a member of the Executive Committee for six consecutive years shall thereafter be ineligible for service as a member thereof during the two-year period following the expiration of such sixth year. For the purposes of this subsection, the period between any two consecutive annual meetings of the Board shall be deemed to be one year.

(c) Term of vacancy appointment.

Any person elected as a member of the Executive Committee to fill a vacancy occuring prior to the expiration of the term for which his predecessor was elected shall be elected for the remainder of such term.

(d) Reports; minority views.

The Executive Committee shall render an annual report to the Board, and such other reports as it may deem necessary, summarizing its activities and making such recommendations as it may deem appropriate. Minority views and recommendations, if any, of members of the Executive Committee shall be included in such reports. (May 10, 1950, ch. 171, § 7, formerly § 6, 64 Stat. 151, amended Sept. 8, 1959, Pub. L. 86-232, § 4, 73 Stat. 467, renumbered and amended July 18, 1968, Pub. L. 90-407, §§ 4, 5, 82 Stat. 363, 364.)

AMENDMENTS

1968—Subsec. (a). Pub. L. 90-407, § 5, made mandatory the organization of the Executive Committee, struck out the prohibition that the Board may not assign to the Executive Committee the function of establishing policies, and added the provisions setting forth the number of members, their manner of election, and the status of the Director.

Subsec. (b). Pub. L. 90-407, § 5, substituted provisions that the Board elect two members as members of the Executive Committee at its annual meeting, with the period between any two consecutive annual meetings to be deemed one year, for provisions covering the composition of the Executive Committee, setting forth a special one year term of office for four members first elected after May 10, 1950, and directing that the membership of the Committee represent diverse interests and areas. Provisions of former subsecs. (b) (2) (A) and (b) (5) were redesignated as subsecs. (c) and (d), respectively.

Subsecs. (c). Pub. L. 90-407, § 5, redesignated former subsec. (b) (2) (A) as (c), and, as so redesignated substituted "Any person elected as a member of the Executive Committee" for "any member elected". Former subsec. (c) authorizing the Board to appoint such additional committees as it deems necessary, and to delegate to such committees survey and advisory functions as it deems appropriate, was eliminated.

Subsec. (d). Pub. L. 90-407, § 5, redesignated former subsec. (b) (5) as (d), and, as so redesignated, substituted "The Executive Committee' for "Such Committee".

CONTINUATION OF EXISTING OFFICES, PROCEDURES, AND ORGANIZATION OF THE NATIONAL SCIENCE FOUNDATION

Amendment by Pub. L. 90-407 intended to continue in effect the existing offices, procedures, and organization of the Foundation, see section 16 of Pub. L. 90-407, set out as a note under section 1862 of this title.

SECTION REFERRED TO IN OTHER SECTIONS

This section is referred to in section 1863 of this title.

§ 1866. Divisions within Foundation.

There shall be within the Foundation such Divisions as the Director, in consultation with the Board, may from time to time determine. (May 10, 1950, ch. 171, § 8, formerly § 7, 64 Stat. 152, renumbered and amended July 18, 1968, Pub. L. 90-407, §§ 4, 6, 82 Stat. 363, 364.)

AMENDMENTS

1968—Pub. L. 90-407, § 6, substituted provisions that there be within the Foundation such divisions as the Director, in consultation with the Board, may from time to time determine for provisions that, unless otherwise provided by the Board, there be within the Foundation a Division of Medical Research, a Division of Mathematical, Physical, and Engineering Sciences, a Division of Biological Sciences, a Division of Scientific Personnel and Education, and such other divisions as the Board deems necessary.

CONTINUATION OF EXISTING OFFICES, PROCEDURES, AND ORGANIZATION OF THE NATIONAL SCIENCE FOUNDATION

Amendment by Pub. L. 90-407 intended to continue in effect the existing offices, procedures, and organization of the Foundation, see section 16 of Pub. L. 90-407, set out as a note under section 1862 of this title.

§ 1867. Repealed. Pub. L. 90-407, § 4, July 18, 1968, 82 Stat. 363.

Section, act May 10, 1950, ch. 171, § 8, 64 Stat. 152, authorized a committee for each division of the Foundation, and provided for the composition, terms of office, chairmenship, rules of procedure, and powers and duties of each divisional committee.

CONTINUATION OF EXISTING OFFICES, PROCEDURES, AND ORGANIZATION OF THE NATIONAL SCIENCE FOUNDATION

Amendment by Pub. L. 90-407 intended to continue in effect the existing offices, procedures, and organization of the Foundation, see section 16 of Pub. L. 90-407, set out as a note under section 1862 of this title.

REORGANIZATION PLAN NO. 5 OF 1965

Eff. July 27, 1965, 30 F.R. 9355, 79 Stat. 1323.

Prepared by the President and transmitted to the Senate and the House of Representatives in Congress assembled, May 27, 1965, pursuant to the provisions of the Reorganization Act of 1949, 63 Stat. 203, as amended [see section 901 et seq. of Title 5, Government Organization and Employees].

NATIONAL SCIENCE FOUNDATION

SECTION 1. ABOLITION OF COMMITTEES

There are hereby abolished all functions of the (divisional) committees provided for in section 8 of the National Science Foundation Act of 1950 (64 Stat. 152; 42 U.S.C. 1867), all functions with respect to the appointment of committees under that section, and all committees now existing under that section. The Director of the National Science Foundation shall make such provisions as he shall deem necessary respecting the winding up of any outstanding affairs of the committees abolished by this section.

SEC. 2. AUTHORITY TO DELEGATE

The Director of the National Science Foundation may from time to time make such provisions as he shall deem appropriate authorizing the performance by any other officer, or by any agency or employee, of the National Science Foundation of any of his functions (including functions delegated to him by the National Science Board).

MESSAGE OF THE PRESIDENT

To the Congress of the United States:

I transmit herewith Reorganization Plan No. 5 of 1965, prepared in accordance with the provisions of the Reorganization Act of 1949, as amended, and providing for certain reorganizations relating to the National Science Foundation.

The plan contains two reorganization measures. First all committees provided for in section 8 of the National Science Foundation Act of 1950 would be abolished. That section provides that there shall be a committee for each division of the Foundation, having not less than five members who are appointed by the National Science Board for 2-year terms. Section 8, as affected by section 23(b) (3) of Reorganization Plan No. 2 of 1962 (76 Stat. 1255), directs each such committee to make recommendations to and advise and consult with the Director of the National Science Foundation with respect to matters relating to the program of its division. Originally the Foundation had three such committees, corresponding to its three divisions. With the growth of the Foundation, five additional divisions have been established; consequently the Foundation, in accordance with the requirements of section 8, now has eight divisional committees. This multiplication in the number of committees has proved cumbersome. For example, three committees are now concerned with scientific personnel and education matters instead of the original one committee, even though one committee is all that is required to meet the Foundation's needs in this area. The elimination of the various statutory divisional committees will simplify the structure of the Foundation and improve its administration.

The second reorganization measure contained in the accompanying reorganization plan would empower the Director of the National Science Foundation to delegate functions vested in him by law or delegated to him by the National Science Board. The expanding responsibilities of the Foundation and the Director indicate that it is necessary that the Director clearly have such authority.

Upon the taking effect of the reorganization plan, the National Science Foundation will institute such new arrangements, in lieu of the divisional committees now required by law, as it deems appropriate. Such new arrangements may include the establishment of committees under section 6 of the National Science Foundation Act of 1950 and such other devices for obtaining advice as may be available to the Foundation.

After investigation, I have found and hereby declare that each reorganization included in the reorganization plan transmittted herewith is necessary to accomplish one or more of the purposes set forth in section 2(a) of the Reorganization Act of 1949, as amended.

The reorganization plan will permit more effective management of the affairs of the National Science Foundation. It is, however, impracticable to specify or itemize at this time the reductions of expenditures which it is probable will be brought about by the taking effect of the reorganizations included in the reorganization plan.

The statutory authority for the exercise of certain functions which would be abolished by section 1 of the reorganization plan is contained in section 8 of the National Science Foundation Act of 1950, 64 Stat, 152.

I recommend that the Congress allow the reorganization plan to become effective.

LYNDON B. JOHNSON.

THE WHITE HOUSE, May 27, 1965.

§ 1868. Special commissions; composition; chairman and vice chairman; duties.

(a) Each special commission established pursuant to section 1863(i) of this title shall consist of eleven members appointed by the Board, six of whom shall be eminent scientists and five of whom shall be persons other than scientists. Each special commission shall choose its own chairman and vice chairman.

* * * * *

(As amended July 18, 1968, Pub. L. 90–407, § 7, 82 Stat. 364.)

AMENDMENTS

1968—Subsec. (a). Pub. L. 90–407 substituted "section 1863 (i) of this title" for "section 1862 (a)(7) of this title".

CONTINUATION OF EXISTING OFFICES, PROCEDURES, AND ORGANIZATION OF THE NATIONAL SCIENCE FOUNDATION

Amendment by Pub. L. 90–407 intended to continue in effect the existing offices, procedures, and organization of the Foundation, see section 16 of Pub. L. 90–407, set out as a note under section 1862 of this title.

§ 1869. Scholarships and graduate fellowships.

The Foundation is authorized to award, within the limits of funds made available specifically for such purpose pursuant to section 1875 of this title, scholarships and graduate fellowships for scientific study or scientific work in the mathematical, physical, medical, biological, engineering, social, and other sciences at appropriate nonprofit American or nonprofit foreign institutions selected by the recipient of such aid, for stated periods of time. Persons shall be selected for such scholarships and fellowships from among citizens or nationals of the United States, and such selections shall be made solely on the basis of ability; but in any case in which two or more applicants for scholarships or fellowships, as the case may be, are deemed by the Foundation to be possessed of substantially equal ability, and there are not sufficient scholarships or fellowships, as the case may be, available to grant one to each of such applicants, the available scholarship or scholarships or fellowship or fellowships shall be awarded to the applicants in such manner as will tend to result in a wide distribution of scholarships and fellowships throughout the United States. Nothing contained in this chapter shall prohibit the Foundation from refusing or revoking a scholarship or fellowship award, in whole or in part, in the case of any applicant or recipient, if the Board is of the opinion that such award is not in the best interests of the United States. (As amended July 18, 1968, Pub. L. 90–407, § 8, 82 Stat. 364.)

AMENDMENTS

1968—Pub. L. 90–407 added social sciences to the enumerated list of sciences, and substituted "throughout the United States" for "among the States, Territories, possessions, and the District of Columbia".

CONTINUATION OF EXISTING OFFICES, PROCEDURES, AND ORGANIZATION OF THE NATIONAL SCIENCE FOUNDATION

Amendment by Pub. L. 90–407 intended to continue in effect the existing offices, procedures, and organization of the Foundation, see section 16 of Pub. L. 90–407, set out as a note under section 1862 of this title.

SECTION REFERRED TO IN OTHER SECTIONS

This section is referred to in section 1862 of this title.

§ 1870. General authority of Foundation.

The Foundation shall have the authority, within the limits of available appropriations, to do all things necessary to carry out the provisions of this chapter, including, but without being limited thereto, the authority—

* * * * *

(c) to enter into contracts or other arrangements, or modifications thereof, for the carrying on, by organizations or individuals in the United States and foreign countries, including other government agencies of the United States and of foreign countries, of such scientific activities as the Foundation deems necessary to carry out the purposes of this chapter, and, at the request of the Secretary of State or Secretary of Defense, specific scientific activities in connection with matters relating to international cooperation or national security, and, when deemed appropriate by the Foundation, such contracts or other arrangements, or modifications thereof may be entered into without legal consideration, without performance or other bonds, and without regard to section 5 of Title 41;

(d) to make advance, progress, and other payments which relate to scientific activities without regard to the provisions of section 529 of Title 31;

* * * * *

(h) to accept and utilize the services of voluntary and uncompensated personnel and to provide transportation and subsistence as authorized by section 5703 of Title 5, for persons serving without compensation;

(i) to prescribe, with the approval of the Comptroller General of the United States, the extent to which vouchers for funds expended under contracts for scientific research shall be subject to itemization or substantation prior to payment, without regard to the limitations of other laws relating to the expenditure of public funds and accounting therefor; and

(j) to arrange with and reimburse the heads of other Federal agencies for the performance of any activity which the Foundation is authorized to conduct. (As amended July 18, 1968, Pub. L. 90–407, § 9, 82 Stat. 365.)

AMENDMENTS

1968—Subsec. (c). Pub. L. 90–407 substituted "scientific activities" for "basic scientific research activities" and "scientific research activities", "international cooperation or national security" for "national defense", and added "Secretary of State" following "at the request of the".

Subsec. (d). Pub. L. 90–407 substituted "activities" for

Subsec. (h). Pub. L. 90–407 substituted "section 5703 of Title 5" for section 73b–2 of Title 5".

Subsec. (j). Pub. L. 90–407 added subsec. (j).

CONTINUATION OF EXISTING OFFICES, PROCEDURES, AND ORGANIZATION OF THE NATIONAL SCIENCE FOUNDATION

Amendment by Pub. L. 90–407 intended to continue in effect the existing offices, procedures, and organization of the Foundation, see section 16 of Pub. L. 90–407, set out as a note under section 1862 of this title.

SECTION REFERRED TO IN OTHER SECTIONS

This section is referred to in sections 1864, 1874 of this title.

§ 1872. International cooperation and coordination with foreign policy.

(a) The Foundation is authorized to cooperate in any international scientific activities consistent with the purposes of this chapter and to expend for such international scientific activities such sums within the limit of appropriated funds as the Foundation may deem desirable. The Director may defray the expenses of representatives of Government agencies and other organizations and of individual scientists to accredited international scientific congresses and meetings whenever he deem [1] it necessary in the promotion of the objectives of this chapter. In this connection, with the approval of the Secretary of State, the Foundation may undertake programs granting fellowships to, or making other similar arrangements with, foreign nationals for scientific study or scientific work in the United States without regard to section 1869 of this title or the affidavit of allegiance to the United States required by section 1874(d)(2) of this title.

* * * * *

(As amended July 18, 1968, Pub. L. 90–407, § 10, 82 Stat. 365.)

AMENDMENTS

1968—Subsec. (a). Pub. L. 90–407 struck out ", with the approval of the Board," following "The Director", and substituted "section 15(d)(2) of this Act" for "section 16(d)(2) of this Act", which resulted in no substantive change in the text of the present section, since, for purposes of classification, provision was translated as "section 1874(d)(2) of this title" by prior amendment.

CONTINUATION OF EXISTING OFFICES, PROCEDURES, AND ORGANIZATION OF THE NATIONAL SCIENCE FOUNDATION

Amendment by Pub. L. 90–407 intended to continue in effect the existing offices, procedures, and organization of the Foundation, see section 16 of Pub. L. 90–407, set out as a note under section 1862 of this title.

§ 1872a. Repealed. Pub. L. 90–407, § 11(1), July 18, 1968, 82 Stat. 365.

Section, act May 10, 1950, ch. 171, § 14, as added July 11, 1958, Pub. L. 85–510, § 2, 72 Stat. 353, authorized the Foundation, in carrying out a program of study, research, and evaluation in the field of weather modification, to consult with meteorologists and scientists, make contracts and grants, accept gifts, loan property, conduct hearings, and subpoena books and records.

EFFECTIVE DATE OF REPEAL

Section 11 (1) of Pub. L. 90–407 provided in part that the repeal of section 14 of the National Science Foundation Act of 1950 [this section] was effective September 1, 1968, and that provisions authorizing the foundation to initiate and support programs in the field of weather modification should remain in effect until September 1, 1968 for purpose of this section.

CONTINUATION OF EXISTING OFFICES, PROCEDURES, AND ORGANIZATION OF THE NATIONAL SCIENCE FOUNDATION

Amendment by Pub. L. 90–407 intended to continue in effect the existing offices, procedures, and organization of the Foundation, see section 16 of Pub. L. 90–407, set out as a note under section 1862 of this title.

§ 1873. Employment of personnel.

(a) Appointment; compensation; application of civil service laws; technical and professional personnel; members of special commissions.

The Director shall, in accordance with such policies as the Board shall from time to time prescribe, appoint and fix the compensation of such personnel as may be necessary to carry out the provisions of this chapter. Except as provided in section 1863(h) of this title, such appointments shall be made and such compensation shall be fixed in accordance with the provisions of Title 5, governing appointments in the competitive service, and the provisions of chapter 51 and subchapter III of chapter 53 of Title 5 relating to classification and General Schedule pay rates: *Provided*, That the Director may, in accordance with such policies as the Board shall from time to time prescribe, employ such technical and professional personnel and fix their compensation, without regard to such provisions, as he may deem necessary for the discharge of the responsibilities of the Foundation under this chapter. The members of the special commissions shall be appointed without regard to the provisions of Title 5, governing appointments in the competitive service.

(b) Outside employment and activities.

Neither the Director, the Deputy Director, nor any Assistant Director shall engage in any other business, vocation, or employment while serving in such position; nor shall the Director, the Deputy Director, or any Assistant Director, except with the approval of the Board, hold any office in, or act in any capacity for, any organization, agency, or institution with which the Foundation makes any grant, contract, or other arrangement under this chapter.

(c) Operation of laboratories and pilot plants.

The Foundation shall not, itself, operate any laboratories or pilot plants.

(d) Compensation of members of Board and special commissions.

The members of the Board and the members of each special commission shall receive compensation at the rate of $100 for each day engaged in the business of the Foundation pursuant to authorization of the Foundation and shall be allowed travel expenses as authorized by section 5703 of Title 5.

(e) Federal officers as members of special commissions; compensation.

Persons holding other offices in the executive branch of the Federal Government may serve as

[1] So in original.

members of special commissions, but they shall not receive remuneration for their services as such members during any period for which they receive compensation for their services in such other offices.

(f) Utilization of appropriations in making contracts.

In making contracts or other arrangements for scientific research, the Foundation shall utilize appropriations available therefor in such manner as will in its discretion best realize the objectives of (1) having the work performed by organizations, agencies, and institutions, or individuals in the United States or foreign countries, including Government agencies of the United States and of foreign countries, qualified by training and experience to achieve the results desired, (2) strengthening the research staff of organizations, particularly nonprofit organizations, in the United States, (3) adding institutions, agencies, or organizations which, if aided, will advance scientific research, and (4) encouraging independent scientific research by individuals.

(g) Transfer of research funds of other Government departments or agencies.

Funds available to any department or agency of the Government for scientific or technical research, or the provision of facilities therefor, shall be available for transfer, with the approval of the head of the department or agency involved, in whole or in part, to the Foundation for such use as is consistent with the purposes for which such funds were provided, and funds so transferred shall be expendable by the Foundation for the purposes for which the transfer was made.

(h) Definition.

For purposes of this chapter, the term "United States" when used in a geographical sense means the States, the District of Columbia, the Commonwealth of Puerto Rico, and all territories and possessions of the United States.

(i) Expiration of authorization.

Notwithstanding any other provision of law, the authorization of any appropriation to the Foundation shall expire (unless an earlier expiration is specifically provided) at the close of the second fiscal year following the fiscal year for which the authorization was enacted, to the extent that such appropriation has not theretofore actually been made. (May 10, 1950, ch. 171, § 14, 64 Stat. 154, renumbered § 15, July 11, 1958, Pub. L. 85–510, § 2, 72 Stat. 353, and amended Sept. 8, 1959, Pub. L. 86–232, § 8, 73 Stat. 469, and renumbered § 14 and amended July 18, 1968, Pub. L. 90–407, §§ 11(2), 12, 82 Stat. 365, 366; Nov. 18, 1969, Pub. L. 91–120, § 3, 83 Stat. 203.)

AMENDMENTS

1969—Subsec. (i). Pub. L. 91–120 added subsec. (i).

1968—Subsec. (a). Pub. L. 90–407, § 12, substituted provisions making applicable chapter 51 and subchapter III of chapter 53 of Title 5, relating to classification and General Schedule pay rates, for provisions making applicable the civil-service laws and regulations and the Classification Act of 1949, and provisions that the members of special commissions be appointed without regard to the provisions of Title 5, governing appointments in the competitive service, for provisions that the Deputy Director, and members of divisional committees and special commissions be appointed without regard to the civil-service laws or regulations. Provisions this subsection, relating to outside employment and activities of certain specified officers of the Foundation, were designated as subsec. (b).

Subsec. (b). Pub. L. 90–407, § 12, redesignated provisions of former subsec. (a) as (b), and, as so redesignated, added Assistant Directors to the specified officers of the Foundation prohibited from engaging in outside employment and activities. Former subsec. (b), providing for the appointment of a Deputy Director, was eliminated.

Subsec. (d). Pub. L. 90–407, § 12, eliminated applicability to members of each divisional committee, and substituted "$100" for "$50" and "section 5703" for "section 73b–2".

Subsec. (e). Pub. L. 90–407, § 12, struck out "the divisional committees and" following "may serve as members of".

Subsec. (f). Pub. L. 90–407, § 12, redesignated former subsec. (g) as (f), and, as so redesignated, in cl. (2) substituted "United States" for "States, Territories, possessions, and the District of Columbia", in cl. (3) substituted "advance scientific research" for "advance basic research", and in cl. (4) substituted "independent scientific research" for "independent basic research". Former subsec. (f), exempting members of the Board, divisional committees, or special commissions form the provisions of former sections 281, 283, or 284 of Title 18 or former section 99 of Title 5, unless the act made unlawful by the aforementioned former sections directly involved or directly interested the Foundation, was eliminated.

Subsec. (g). Pub. L. 90–407 redesignated former subsec. (h) as (g), and, as so redesignated, struck out "and, until such time as an appropriation is made available directly to the Foundation, for general administrative expenses of the Foundation without regard to limitations otherwise applicable to such funds" following "the purposes for which the transfer was made". Former subsec. (g) was redesignated as (f).

Subsec. (h). Pub. L. 90–407 added subsec. (h). Former subsec. (h) was redesignated as (g).

Subsec. (i). Pub. L. 90–407 struck out subsec. (i), which provided for the transfer of the National Roster of Scientific and Specialized Personnel from the United States Employment Service to the Foundation.

TRANSFER OF FUNCTIONS

Authority of Director of the National Science Foundation, from time to time, to make appropriate provisions authorizing the performance by any other officer, or by any agency or employee, of the National Science Foundation of any of his functions (including functions delegated to him by the National Science Board), see Reorg. Plan No. 5 of 1965, eff. July 27, 1965, 30 F.R. 9355, 79 Stat. 1323, set out as a note under section 1867 of this title.

CONTINUATION OF EXISTING OFFICES, PROCEDURES, AND ORGANIZATION OF THE NATIONAL SCIENCE FOUNDATION

Amendment by Pub. L. 90–407 intended to continue in effect the existing offices, procedures, and organization of the Foundation, see section 16 of Pub. L. 90–407, set out as a note under section 1862 of this title.

SECTION REFERRED TO IN OTHER SECTIONS

This section is referred to in sections 1862, 1863, 1874 of this title.

§ 1874. Security provisions.

(a) Nuclear energy research and development.

The Foundation shall not support any research or development activity in the field of nuclear energy, nor shall it exercise any authority pursuant to section 1870 (e) of this title in respect to that field, without first having obtained the concurrence of the Atomic Energy Commission that such activity will not adversely affect the common defense and security. To the extent that such activity involves restricted data as defined in the Atomic Energy Act of 1954 the provisions of that Act regarding the control of the dissemination of restricted data and the security clearance of those individuals to be given access to restricted data shall be applicable. Nothing in this chapter shall supersede or modify any provision of the Atomic Energy Act of 1954.

(b) Research relating to national defense.

(1) In the case of scientific or technical research activities under this chapter in connection with matters relating to the national defense, with respect to which funds have been transferred to the Foundation from the Department of Defense in accordance with the provisions of section 1873 (g) of this title, the Secretary of Defense shall establish such security requirements and safeguards, including restrictions with respect to access to information and property, as he deems necessary.

* * * * *

(May 10, 1950, ch. 171, § 15, 64 Stat. 156; Apr. 5, 1952, ch. 159, § 1, 66 Stat. 43, renumbered § 16, July 11, 1958, Pub. L. 85–510, § 2, 72 Stat. 353, and amended Oct. 16, 1962, Pub. L. 87–835, § 1, 76 Stat. 1069, and renumbered § 15 and amended July 18, 1968, Pub. L. 90–407, §§ 11(2), 13, 82 Stat. 365, 366.)

AMENDMENTS

1968—Subsec. (a). Pub. L. 90–407, § 13, substituted "1954" for "1946".

Subsec. (b)(1). Pub. L. 90–407, § 13, substituted "section 1873 (g) of this title" for section 1873 (h) of this title".

CONTINUATION OF EXISTING OFFICES, PROCEDURES, AND ORGANIZATION OF THE NATIONAL SCIENCE FOUNDATION

Amendment by Pub. L. 90–407 intended to continue in effect the existing offices, procedures, and organization of the Foundation, see section 16 of Pub. L. 90–407, set out as a note under section 1862 of this title.

SECTION REFERRED TO IN OTHER SECTIONS

This section is referred to in section 1872 of this title and in title 5 section 1304.

§ 1875. Appropriations.

(a) To enable the Foundation to carry out its powers and duties, there is hereby authorized to be appropriated to the Foundation for the fiscal year ending June 30, 1969, the sum of $525,000,000; but for the fiscal year ending June 30, 1970, and each subsequent fiscal year, only such sums may be appropriated as the Congress may hereafter authorize by law. Sums authorized by this subsection shall be in addition to sums authorized by section 1122(b)(1) of Title 33.

* * * * *

(May 10, 1950, ch. 171, § 16, 64 Stat. 157; Aug. 8, 1953, ch. 377, 67 Stat. 488, renumbered § 17, July 11, 1958, Pub. L. 85–510, § 2, 72 Stat. 353, and renumbered § 16, and amended July 18, 1968, Pub. L. 90–407, §§ 11 (2), (14), 82 Stat. 365, 366.)

CODIFICATION

Section 1122(b)(1) was, in the original, section 201(b)(1) of the Marine Resources and Engineering Development Act of 1966. For purposes of classification, section 201(b)(1) of the Marine Resources and Engineering Development Act of 1966 was translated as section 1122(b)(1) of Title 33 as the probable intent of Congress.

AMENDMENTS

1968—Subsec. (a). Pub. L. 90–407, § 14, substituted provisions authorizing the appropriation of funds for the fiscal year ending June 30, 1969, June 30, 1970, and each subsequent fiscal year, such sums to be in addition to sums authorized by section 1122(b)(1) of Title 33, for provisions authorizing the appropriation of such sums as may be necessary to carry out the provisions of this chapter out of any money in the Treasury not otherwise appropriated.

CONTINUATION OF EXISTING OFFICES, PROCEDURES, AND ORGANIZATION OF THE NATIONAL SCIENCE FOUNDATION

Amendment by Pub. L. 90–407 intended to continue in effect the existing offices, procedures, and organization of the Foundation, see section 16 of Pub. L. 90–407, set out as a note under section 1862 of this title.

SECTION REFERRED TO IN OTHER SECTIONS

This section is referred to in section 1869 of this title.

§ 1877. Science Information Council.

* * * * *

(c) Compensation and allowance for expenses.

Persons appointed to the Council shall, while serving on business of the Council, receive compensation at rates fixed by the National Science Foundation, but not to exceed $100 per day, and shall also be entitled to receive an allowance for actual and necessary travel and subsistence expenses while so serving away from their places of residence. (As amended Pub. L. 90–407, § 15(b), July 18, 1968, 82 Stat. 367.)

AMENDMENTS

1968—Subsec. (c). Pub. L. 90–407 substituted "$100" for "$50".

CONTINUATION OF EXISTING OFFICES, PROCEDURES, AND ORGANIZATION OF THE NATIONAL SCIENCE FOUNDATION

Amendment by Pub. L. 90–407, intended to continue in effect the existing offices, procedures, and organization of the Foundation, see section 16 of Pub. L. 90–407, set out as a note under section 1862 of this title.

§ 1882. Information furnished to Congressional committees.

Notwithstanding any provision of this chapter, or any other provision of law, the Director of the National Science Foundation shall keep the Committee on Science and Astronautics of the House of Representatives and the Committee on Labor and Public Welfare of the Senate fully and currently informed with respect to all of the activities of the National Science Foundation. (Pub. L. 91–120, § 6, Nov. 18, 1969, 83 Stat. 203.)

CODIFICATION

Section was enacted as a part of the National Science Foundation Authorization Act, 1970, and not as a part of the National Science Foundation Act of 1950 which comprises this chapter.

Chapter 16B.—CONTRACTS FOR SCIENTIFIC AND TECHNOLOGICAL RESEARCH [New]

Sec.
1900. Interior Department programs.
(a) Authorization for research contracts.
(b) Capabilities of prospective contractors; advice and assistance, coordination of research, lines of inquiry, and cooperation.
(c) Research reports or publications.
(d) Limitation; submission to Congress.
1900a. Rules and regulations.
1900b. Amendment, modification, or repeal of authorizations for execution of contracts for research.

§ 1900. Interior Department programs.

(a) Authorization for research contracts.

The Secretary of the Interior is authorized to enter into contracts with educational institutions, public or private agencies or organizations, or persons for the conduct of scientific or technological research into any aspect of the problems related to the programs of the Department of the Interior which are authorized by statute.

(b) Capabilities of prospective contractors; advice and assistance, coordination of research, lines of inquiry, and cooperation.

The Secretary shall require a showing that the institutions, agencies, organizations, or persons with which he expects to enter into contracts pursuant to this section have the capability of doing effective work. He shall furnish such advice and assistance as he believes will best carry out the mission of the Department of the Interior, participate in coordinating all research initiated under this section, indicate the lines of inquiry which seem to him most important, and encourage and assist in the establishment and maintenance of cooperation by and between the institutions, agencies, organizations, or persons and between them and other research organizations, the United States Department of the Interior, and other Federal agencies.

(c) Research reports or publications.

The Secretary may from time to time disseminate in the form of reports or publications to public or private agencies or organizations, or individuals such information as he deems desirable on the research carried out pursuant to this section.

(d) Limitation; submission to Congress.

No contract involving more than $25,000 shall be executed under subsection (a) of this section prior to thirty calendar days from the date the same is submitted to the President of the Senate and the Speaker of the House of Representatives and said thirty calendar days shall not include days on which either the Senate or the House of Representatives is not in session because of an adjournment of more than three calendar days to a day certain or an adjournment sine die. (Pub. L. 89–672, § 1, Oct. 15, 1966, 80 Stat. 951.)

§ 1900a. Rules and regulations.

The Secretary shall prescribe such rules and regulations as he deems necessary to carry out the provisions of this chapter. (Pub. L. 89–672, § 2, Oct. 15, 1966, 80 Stat. 951.)

§ 1900b. Amendment, modification, or repeal of authorizations for execution of contracts for research.

Nothing contained in this chapter is intended to amend, modify, or repeal any provisions of law administered by the Secretary of the Interior which authorize the making of contracts for research. (Pub. L. 89–672, § 3, Oct. 15, 1966, 80 Stat. 951.)

SOURCE: United States Code, 1964 Edition, Supplement V. 1970.

Appendix A

NSF
Career Opportunities

Training Programs Within the Foundation

The extent to which the Foundation carries out its mission depends on the continuous acquisition of competent professional and administrative personnel, including members of the graduating classes of our colleges and universities. To replenish and maintain its staff of qualified personnel, the National Science Foundation has established the training programs described on the following pages. Not all of these programs are operational each year since they depend on need, funds, and position ceilings.

Aside from the initial training and orientation associated with trainee positions, certain other specialized training and employee development programs are available. Trainees may be enrolled in a variety of government and non-government training courses related to their permanent work assignments. Cost of all authorized training courses will be paid by the Foundation including those offered by local universities in the Washington, D. C. area.

Trainee positions will start at the GS-5 or GS-7 level, depending upon the candidate qualifications. Persons successfully completing the training programs are assigned to permanent positions and promoted to the next higher grade. Thereafter, they are provided with an environment in which they can assume as much responsibility as their abilities allow and where promotional opportunities are excellent. If you are seeking a challenging and rewarding future, you should consider a career with the National Science Foundation.

Accountant and Auditor Trainees

Individuals selected for this program receive assignments in the Financial Management Office or Audit Office. The modern accounting and auditing techniques as well as the professional character of the personnel operating these systems provide the trainee with an excellent environment for personal professional development.

The Financial Management Office trainee works on projects involving the processing and recording of expenditures, the analysis and reporting of transactions, allotment accounting, and grant accounting. Also, the trainee learns about budget operations, data processing, and the Foundation's financial arrangements with colleges and universities, non-profit organizations, and grantees as well as contractors who undertake major scientific projects.

The Audit Office trainee receives assignments involving comprehensive audits of:

- National Science Foundation cost-reimbursable contracts with commercial and industrial organizations;
- Grants to colleges and universities;
- Grants and contracts to non-profit organizations;
- The National Science Foundation's National Research Centers;
- And the Foundation's financial and managerial programs.

The primary purpose of an audit is to determine whether the grantees and contractors have properly discharged their financial responsibilities in the use of National Science Foundation funds and whether they have adhered to the Foundation's legal requirements and policies. Internal audits are designed, as a tool of sound management, to evaluate the effectiveness of the National Science Foundation's accounting, financial, and managerial controls.

Computer-Systems Trainees

Individuals selected for this program are assigned to the Data Management Systems Office. This office, under the Assistant Director for Administration, serves as a focal point for the development of automated data management systems. The Office possesses a staff of computer-systems analysts, computer specialists and programmers who are responsible for translating broadly based specifications for data and information systems into computer systems for use on the IBM 360 System under development at the Foundation.

The Foundation's on-site computer facilities offer trainees three significant advantages. First, all trainees receive instruction in programming. Second, the opportunity exists for trainees to familiarize themselves with the operation of the IBM 360 System, and third, the computer is available during prime time for testing and debugging of programs. A computer-systems trainee receives instruction in writing and testing computer programs and shares the responsibility for "systems analysis" that must be done in connection with every project. In practice, this means that the trainee works directly with systems analysts within the Data Management Systems Office, as well as representatives of other offices of the Foundation.

Economist Trainees

Individuals selected for this program go to the Office of Economic and Manpower Studies. This office collects and analyzes data on virtually all aspects of science and technology, including scientific manpower, expenditures for research and development, technological transfer, and science education. Increasing emphasis on the economic impact of research and development has focused attention on the many problems challenging the talents of economists. Economic and statistical studies are assuming increasing importance, as management tools, in the conduct of national affairs.

The Economist Trainee becomes directly involved with such analytical studies. The trainee also participates with senior staff members in planning and implementing surveys designed to compile data on the volume and distribution of research and development expenditures, and on the development, characteristics, and deployment of the scientific labor force. Compilation of such data precedes the analyses of the relationships between research and development, and other economic variables; as well as between the levels of scientific activity and the manpower needs of our country. The trainee has frequent contacts with officials from other Federal agencies, universities and colleges, industrial research laboratories, and nonprofit research organizations.

Science-Education Trainees

Individuals selected for this program are assigned to one of three Education Divisions; Pre-College, Undergraduate, or Graduate. Because the trainee receives successive assignments in each of the Division's programs or activities, he acquires familiarity with the mission and activities of the Foundation in general and the Division in particular.

As the trainees' overall knowledge of programs increases, opportunities to work in the reporting and planning activities of the Division are afforded. Examples of typical trainee assignments follow:

- Preparation of non-operational type correspondence.
- Participation in meetings with professional staff.
- Compilation and interpretation of data.
- Specialized studies relating to planning.
- Modification and review of programs.
- Assisting in the formation of budget estimates and in the preparation of detailed background data for budget hearings before Congressional Committees.

Throughout the entire training period, the trainee communicates with professional personnel—educators, scientists, consultants, and administrators.

Management Trainees

Individuals selected for this program receive rotational assignments in the offices of the Assistant Director for Administration. The trainee rotates, for example, in several of the following offices: the Grants and Contracts Office, the Management Analysis Office, the Personnel Office, and the Administrative Services Office. Training assignments proceed from the general to the particular specific assignments and scheduled conferences are given to reinforce the basic information which the trainee receives. At the completion of the rotational assignments, the trainee has developed not only a facility for comprehending functional and organizational relationships, but also a capacity to deal with problem assignments in each of the previously mentioned administrative offices. A short description of the work performed in these offices follows:

1. **Grants and Contracts Office:** Responsible for negotiating and monitoring administrative aspects of grants for the support of various scientific endeavors and for negotiating and administering contracts relating to Foundation activities.
2. **Management Analysis Office:** Responsible for a variety of management services, including organization and methods studies, manpower analysis, and administration of Foundation records.
3. **Personnel Office:** Responsible for the development and implementation of the Foundation's personnel program.
4. **Administrative Services:** Responsible for a variety of Administrative Services including distributions, printing and reproduction, supply and maintenance, and travel.

Grant Management Trainees

Individuals selected for this program are assigned to the Grants Branch of the Grants and Contracts Office. This branch represents the Foundation in liaison with the business offices of educational institutions and scientific organizations, and it negotiates and monitors administrative aspects of grants for the support of various scientific endeavors such as research projects, construction of science facilities, and education in the sciences.

During his development, the trainee receives work assignments of progressive complexity accompanied by on-the-job instruction and specialized reading material. The training assignments are designed to provide an understanding of each job and its relationship to other duties and responsibilities performed in both the Grants Branch and other offices throughout the Foundation. Subject matter covered includes analysis of grant proposals to determine allowability and reasonableness of cost and identification of problems with respect to income resulting from grants or contracts. In exploring these subject areas, the trainees come in contact with professional personnel working in other units of the Foundation, and other governmental and non-governmental agencies.

Benefits

To make the years ahead still more interesting and rewarding, the trainee has in addition to his salary, training, and challenging work assignments the following additional benefits:

Vacation: During the first three years of Federal employment you are allowed 13 workdays a year of annual leave each year. For the next 12 years you are allowed 20 workdays of annual leave each year. After 15 years of service you are allowed 26 workdays of annual leave each year. You may accumulate, on a year-end basis, leave up to a total of 30 days.

Holidays: There are eight holidays in each calendar year. The holidays are New Year's Day, Washington's Birthday, Memorial Day, Independence Day, Labor Day, Veterans' Day, Thanksgiving and Christmas. If a holiday falls on a weekend, the preceding Friday or following Monday is designated as a holiday.

Sick Leave: Sick leave with pay is credited at the rate of 13 workdays each year. Sick leave is permitted to accrue without a limitation on the amount accumulated.

Retirement: Employees with permanent appointments who have a regular scheduled tour of duty are covered by the Civil Service Retirement System. The employee's share of the cost of retirement is 7 percent of his gross earnings. The Government contributes the remainder of the amount necessary to provide liberal retirement benefits. This retirement plan compares very favorably with any plan in private industry.

Life & Hospitalization Insurance: Term life insurance is available in dollar amounts based upon the annual salary of the employee. (Annual salary rounded off to the next higher thousand plus two additional thousand. For example, an employee earning $11,233 per annum would be eligible for $14,000 in term life insurance.) The cost of such insurance is 27½ cents per thousand dollars of coverage per bi-weekly pay period. An additional $10,000 of term life insurance is optionally available at additional cost. There are up to 15 different hospital and medical insurance plans from which to choose. Employees are provided specific information concerning hospitalization and medical insurance when they report for duty.

Incentive Awards: The Foundation operates a modern incentive awards program to give recognition to employees demonstrating exceptional ability. The award system includes letters of commendation, suggestion awards, liberal cash bonuses for sustained superior performance and special achievement, extra within-grade pay increases for high quality performance, and the Foundation's highest honor, the Distinguished Service Award.

All of these awards are designed to show that the Foundation appreciates the work of its employees and gives special recognition to those who take the lead.

Life in Washington, D. C.

If one plans a career in Government, there is no better place to start than in Washington, D. C.

Greater Washington, consisting both of the District of Columbia and the surrounding Maryland and Virginia suburbs, comprises a community with a population in excess of two and a half million people. Almost any type of housing is available from downtown high-rise apartments to suburban communities. Washington is a city of libraries, galleries, museums, and memorials. Cultural assets include the Library of Congress, the Corcoran Art Gallery, and the Phillips Gallery. The Smithsonian Institution maintains the National Gallery of Art, the Freer Art Gallery, the Museum of Natural History, the Museum of History and Technology, and the National Zoological Park. Many ceremonial and business functions of Government are available to the public including welcoming heads of state, visiting embassies and legations, and attending sessions of the Congress and Congressional Committee meetings.

The Washington area abounds in recreational facilities. These facilities are varied and include picnic areas, tennis courts, golf courses, hockey fields, swimming pools, baseball, football, and extensive areas for boating. Washington is represented by major league professional teams in football, baseball, and soccer. Fishing enthusiasts have excellent facilities in the upper Potomac River, nearby trout streams, and the Chesapeake Bay. The State of Maryland maintains about 10,000 acres of public hunting lands and Virginia offers more than a million and a half acres where deer, grouse, quail, ducks and geese may be taken. The Delaware and Maryland Atlantic Ocean Beaches are approximately three hours away by automobile and the Catoctin Mountains in Maryland and Blue Ridge Mountains of Virginia are even closer.

In addition, the NSF Employees' Association sponsors many social, recreational, and educational activities. The NSF Federal Credit Union provides employees with a convenient method of saving money, at favorable interest rates, as well as readily available loan facilities.

How to Apply

1. An application for Federal Employment (Standard Form 171) as well as the Application for the Federal Service Entrance Examination may be obtained at College Placement Offices, Federal Post Offices, and offices of the U.S. Civil Service Commission.

2. Candidates who have passed the Federal Service Entrance Examination or its equivalent should submit, at their earliest convenience, a signed and completed Standard Form 171 either to the National Science Foundation representative when he visits the campus or to the Personnel Office, National Science Foundation, 1800 G Street, N.W., Washington, D. C. 20550.

For additional information, candidates may consult the **Federal Career Directory—A Guide for College Students.** This directory is available for reference in College Placement Offices and libraries.

SOURCE: NSF, "Career Opportunities with the National Science Foundation."

INDEX

A

B

C

D

E

F

T

U

W